PROGRESS IN CYBERNETICS AND SYSTEMS RESEARCH

Volume III

General Systems Methodology
Fuzzy Mathematics and Fuzzy Systems
Biocybernetics and Theoretical Neurobiology

PROGRESS IN CYBERNETICS AND SYSTEMS RESEARCH

Volume I *R. Trappl and F. R. Pichler,* eds.
General Systems • Engineering Systems • Biocybernetics and Neural Systems

Volume II *R. Trappl and F. de P. Hanika,* eds.
Socio-economic Systems • Cognition and Learning • Systems Education • Organization and Management

Volume III *R. Trappl, G. J. Klir, and L. Ricciardi,* eds.
General Systems Methodology • Fuzzy Mathematics and Fuzzy Systems • Biocybernetics and Theoretical Neurobiology

Volume IV *R. Trappl and G. Pask,* eds.
Cybernetics of Cognition and Learning • Structure and Dynamics of Socio-economic Systems • Health Care Systems • Engineering Systems Methodology

PROGRESS IN CYBERNETICS AND SYSTEMS RESEARCH

Volume III

General Systems Methodology
Fuzzy Mathematics and Fuzzy Systems
Biocybernetics and Theoretical Neurobiology

Edited by

ROBERT TRAPPL

Professor of Medical Cybernetics
University of Vienna Medical School, Austria

GEORGE J. KLIR

Professor and Chairman, Systems Science Department
State University of New York at Binghamton

LUIGI RICCIARDI

Professor of Cybernetics
Istituto di Scienze dell'Informazione
Universita di Salerno, Italy

WITH INTRODUCTION BY

P. B. CHECKLAND

HEMISPHERE PUBLISHING CORPORATION

Washington London

A HALSTED PRESS BOOK

JOHN WILEY & SONS

New York London Sydney Toronto

Copyright © 1978 by Hemisphere Publishing Corporation.
All rights reserved.
No part of this book may be reproduced in any form,
by photostat, microform, retrieval system, or any other means,
without the prior written permission of the publisher.

Hemisphere Publishing Corporation
1025 Vermont Ave., N.W., Washington, D.C. 20005

Distributed solely by Halsted Press, a Division of John Wiley & Sons, Inc., New York.

1 2 3 4 5 6 7 8 9 0 LILI 7 8 3 2 1 0 9 8

Library of Congress Cataloging in Publication Data

Main entry under title:

Progress in cybernetics and systems research.

Vol. 2 edited by R. Trappl and F. de P. Hanika; v. 3 by R. Trappl, G. J. Klir, and L. Ricciardi.
Papers presented at a symposium organized by the Austrian Society for Cybernetic Studies.
Sponsored by the Bundesministerium für Wissenschaft und Forschung, and others.
Includes bibliographical references and indexes.
1. Cybernetics–Congresses. 2. System theory–Congresses. I. Trappl, Robert. II. Pichler, Franz R. III. Hanika, Francis de Paula. IV. Osterreichische Studiengesellschaft für kybernetik. V. Austria. Bundesministerium für Wissenschaft und Forschung.
Q300.P75 001.53 75-6641
ISBN 0-470-88475-4 (v. I)
0-470-88476-2 (v. II)
0-470-26371-7 (v. III)
0-470-99380-4 (v. IV)

Printed in the United States of America

This Symposium was organized by the Austrian Society for Cybernetic Studies

PATRONS

Dr. Hertha Firnberg
Federal Minister of Science and Research

Prim. Dr. Ingrid Leodolter
Federal Minister of Health and Environmental Protection

Leopold Gratz
Mayor of the City of Vienna

Magn. Prof. Dr. Franz Seitelberger
Rektor of the University of Vienna

CHAIRMAN

Prof. Dr. Robert Trappl
President, Österreichische Studiengesellschaft für Kybernetik

SPONSORS

Bundesministerium für Wissenschaft und Forschung
Gemeinde Wien
IBM
Österreichische Mineralölverwaltung

PROGRAMME COMMITTEE

Prof. B. Banathy (USA)
Dr. G. Chroust (Austria)
Prof. W. Dressler (Austria)
Prof. H. Drischel (GDR)
Acad. V. Glushkov (USSR)
Prof. F. de P. Hanika (UK)
Prof. L. Heinrich (Austria)
Dr. J.P.C. Kleijnen (Nthld.)
Prof. G. Klir (USA)
Prof. M'Pherson (UK)
M. McGoff (USA)
Prof. J.G. Miller (USA)
Prof. J. Milsum (Canada)
Prof. G. Pask (UK)
Prof. F. Pichler (Austria)
Prof. L. Ricciardi (Italy)
Dr. N. Rozsenich (Austria)
Prof. R. Trappl (Austria)
Prof. T. Vámos (Hungary)
Prof. H.-J. Zimmermann (BRD)

ORGANIZING COMMITTEE

U. Angerstein Dr. G. Chroust S. Cornock Dr. M. Holzer P.E. Martin M. Martin
E. Engelhardt A. de P. Hanika Prof. F. de P. Hanika G. Nauer Prof. R. Trappl I. Wenko

Contents

List of authors

Andrew, Dr. A.M., University of Reading, Whiteknights, Reading, Berks, UK.

Araki, Tetsuo, Research Institute of Electrical Communication, Tohoku University, Sendai, Japan.

Aracil, J., Department of Automatic Control, E.T.S. Ing. Industriales, University of Seville, Avda. Raina Mercedes, Seville, Spain.

Argentesi, Dr. Flavio, 21020 Centro Euratom di Ispra, Varese, Italy.

Arigoni, Prof. Anio O., Via G. Boccaccio 9, 44100 Ferrara, Italy.

Ascoli, Dr. C., C.N.R. Laboratorio per lo Studio delle Proprieta' Fisiche di Biomolecole e Cellule, Via F. Buonarroti 9, 56100 Pisa, Italy.

Babcock, Anita K. PhD., State University of New York at Buffalo, Biophysics Dept., 4234 Ridge Lea Road, Buffalo, New York 14214, USA.

Basar, E., Institute of Biophysics, Hacettepe University, Ankara, Turkey.

Biondi, E., Istituto di Elettrotecnica ed Elettronica, Politecnico di Milano, Milano, Italy.

Bogdanski, Prof. Casimir, 43 rue de l'Abbé-Grégoire, Hotel Orient, 75006 Paris, France.

Bollmann, Dr. Peter, Technische Universität Berlin, Otto Suhr-Allee 18/20, D-1 Berlin 10, FRG.

Bona, B., Centro Elaborazione Numerale dei Segnali (CNR), Istituto di Elettrotecnico Nazionale Galileo Ferraris, Torino, Italy.

Braitenberg, Prof. Dr. Valentin, Max-Planck-Institut für Biologische Kybernetik der Universität Tübingen, Spemannstr. 33, D-74 Tübingen, FRG.

Broekstra, Dr. G., Graduate School of Management, Poortweg 6-8, Delft, The Netherlands.

Camacho, E.F., Department of Automatic Control, E.T.S. Ing. Industriales, University of Seville, Avda. Raina Mercedes, Seville, Spain.

Capocelli, Prof. Renato M., Laboratorio di Cibernetica del C.N.R., Via Toiano 2, 80072 Arco Felice (Napoli), Italy.

Carlsson, Christer, Åbo Swedish University, School of Economics, Henriksgatan 7, 20500 Åbo 50, Finland.

Carpenter, Dr. Gail A., Dept. of Mathematics, M.I.T., 2-336, Cambridge, Massachusetts 02139, USA.

Chillemi, S., Laboratorio di Cibernetica, 80072 Arco Felice (Napoli), Italy.

Cordella, L., Laboratorio di Cibernetica, 80072 Arco Felice (Napoli), Italy.

Curnow, R.C., Science Policy Research Unit, Mantel Building, University of Sussex, Sussex, UK.

Dacquino, G.F., Istituto di Elettrotecnica ed Elettronica, Politecnico di Milano, Piazza Leonardo da Vinci 32, 20133 Milano, Italy.

Della Riccia, Prof. Giacomo, Laboratorio di Cibernetica del C.N.R., Via Toiano 2, 80072 Arco Felice (Napoli), Italy.

Düchting, Prof. Dr. W., Lehrstuhl für Elektronik, Gesamthochschule Siegen, Fischbacherbergstr. 2, 59 Siegen, FRG.

Dufosse, Dr. Michel, Laboratoire de Physiologie, 91 Bvd. de l'Hopital, 75013 Paris, France.

Fevrier, Prof. Paulette, Université Pierre et Marie Curie, 63 rue de Cevennes, 75075 Paris, France.

Flohr, Prof. Dr. Hans W., Studienbereich 3 (Biologie) der Universität Bremen, Achterstr., NW2, D-2800 Bremen 33, FRG.

Fritschka, Dr. Emanuel, Freie Universität Berlin, Institut für Neuropathologie (WE15), Klinikum Steglitz (FB2), 1 Berlin 45, Hindenburgdamm 30, FRG.

Gaines, Prof. B.R., Dept. of Electrical Engineering Science, University of Essex, Colchester, Essex, UK.

Gal, Prof. Dr. Thomas, Kupferstraße 28, 51 Aachen, FRG.

Gasparski, Prof. Dr. W., Tot Winsmego 10-15, 01-711 Warsaw, Poland.

Gawroński, Ryszard, Politechnika Świętokrzyska, Kielce, Poland.

Goguen, Joseph A., University of Colorado, Department of Anatomy, School of Medicine, 4200 East Ninth Avenue, Denver, Colorado 80220, USA.

Golze, Dr. Ulrich, Institut für Praktische Mathematik, Technische Universität Hannover, 3 Hannover, FRG.

Gönder, A., Institute of Biophysics, Hacettepe University, Ankara, Turkey.

Grandori, Ferdinando, Istituto di Elettronica, Via Ponzio 34/5, 20133 Milano, Italy.

Greguss, Dr. Pal, Applied Biophysics Laboratory, Technical University Budapest, H-1111 Budapest, Kruspér u. 2-4, Hungary.

Grossberg, Prof. Stephen, Boston University, College of Liberal Arts, 264 Bay State Road, 4th Floor, Boston, Mass. 02215, USA.

Hajek, Prof. Dr. Otmar, Dept. of Mathematics, Case Western Reserve University, Cleveland, Ohio 44106, USA.

Hamacher, Dipl.-Math. H., Institut für Wirtschaftswissenschaften, Lehrstuhl für Unternehmensforschung, RWTH Aachen, Templergraben 64, 51 Aachen, FRG.

Hamalainen, Dipl. Ing. Raimo P., Helsinki University of Technology, Systems Theory Laboratory, 02150 Espoo 15, Finland.

Hanen, Albert, Laboratoire de Physiologie, 91 Bvd. de l'Hopital, 75013 Paris, France.

Harao, Masateru, Research Institute of Electrical Communication, Tohoku University, Sendai, Japan.

Harth, Prof. Erich, Physics Department, Syracuse University, Syracuse, New York 13210, USA.

Hirsch, Dr. Helmut, Bundesministerium für Handel, Gewerbe und Industrie, Abteilung VI/2, Schwarzenbergplatz 1, 1010 Vienna, Austria.

Hofferberth, Bernhard, Freie Universität Berlin, Institut für Neuropathologie (WE15), Klinikum Steglitz (FB2), Hindenburgdamm 30, 1 Berlin 45, FRG.

Hölscher, Prof. H., Physiologisches Institut der Universität Bonn, Nußallee 11, 53 Bonn, FRG.

Holden, Dr. A.V., University of Leeds, Dept. of Physiology, Leeds LS2 9JT, UK.

Inaudi, D., Gruppo Nazionale Informatica Matematica (CNR), Istituto Scienza dell'Informazione, Università di Torino, Italy.

Kitagawa, Prof. Dr. Tosio, Research Institute of Fundamental Information Science, Kyushu University, Fukuoka, Japan.

Kittler, Prof. Josef, Ecole Nationale Supérieure des Télécommunications, Pièce B-222, 46 Rue Barrault, 75634 Paris Cedex 13, France.

Klir, Prof. Dr. George J., School of Advanced Technology, State University of New York at Binghamton, Binghamton, N.Y. 13901, USA.

Konrad, E., Technische Universität Berlin, Otto Suhr-Allee 18/20, D-1 Berlin 10, FRG.

Lasker, Prof. George Eric, University of Windsor, Windsor 11, Ontario, Canada.

Leitsch, Dr. Alexander, Interfakultäres Rechenzentrum der Universität Wien, Universitätsstraße 7, 1010 Vienna, Austria.

Llinás, Rudolfo, Division of Neurobiology, Dept. of Physiology and Biophysics, University of Iowa, USA.

Macukow, Bohdan, Politechnika Warszawska, Warszawa, Poland.

Mamdani, E.H., Dept. of Electrical and Electronic Engineering, Queen Mary College (University of London), Mile End Road, London E1 4NS, UK.

Matakas, Dr. Frank, Freie Universität Berlin, Institut für Neuropathologie (WE15), Klinikum Steglitz (FB2), Hindenburgdamm 30, 1 Berlin 45, FRG.

McLean, J. Michael, Science Policy Research Unit, Mantel Building, University of Sussex, Sussex, UK.

Milanese, Prof., Istituto Elettrotecnico Nazionale Galileo Ferraris, Corso Massimo d'Azeglio 42, 10125 Torino, Italy.

Milgram, Dr. M., Département de Mathématiques Appliquées et d'Informatique, Université de Technologie, 25 rue Eugène-Jacquet, 60206 Compiègne, France.

Montes, C.G., Dept. of Automatic Control, E.T.S. Ing. Industriales, University of Seville, Avda. Raina Mercedes, Seville, Spain.

Müller, A., Istituto di Elettrotecnica ed Elettronica, Politecnico di Milano, Milano, Italy.

Noguchi, Shoichi, Research Institute of Electrical Communication, Tohoku University, Sendai, Japan.

Nurminen, Markku I., Turku School of Economics, Computer Science, Rehtorinpellontie 5, 20500 Turku 50, Finland.

Pacut, Andrzej, Ph.D., Warsaw Technical University, Institute of Mathematics, Dept. of Mathematical Methods in Informatics, Pl. Jednośći Robotniczej 1, 00-661 Warsaw, Poland.

Palm, Dr. G., Max-Planck-Institut für biologische Kybernetik, Spemannstraße 38, 74 Tübingen, FRG.

Pedotti, A., Istituto di Elettronica, Via Ponzio 34/5, 20133 Milano, Italy.

Pellionisz, Dr. András, First Dept. of Anatomy, Semmelweis University, Medical School, Tüzoltó utca 58, 1450 Budapest, Hungary.

Petracchi, D., C.N.R. Laboratorio per lo Studio delle Proprieta' Fisiche di Biomolecole e Cellule, Via F. Buonarroti 9, 56100 Pisa, Italy.

Pfaffelhuber, Prof. Dr. E., Institute of Information Sciences, University of Tübingen, 74 Tübingen, FRG.

Priese, Dr. Lutz, Fachgebiet Systemtheorie und Systemtechnik, Abteilung Raumplanung, Universität Dortmund, Postfach 500500, 46 Dortmund 50, FRG.

Procyk, T., University of London, Queen Mary College, Dept. of Electrical and Electronic Engineering, Mile End Road, London E1 4NS, UK.

Ransom, Dr. Robert, University of Sussex, Falmer, Brighton BN1 9RM, Sussex, UK.

Ricciardi, Prof. Dr. Luigi, Istituto di Scienze dell' Informazione, Universita di Salerno, Salerno, Italy.

Rödding, Prof. Dr. Dieter, and Prof. Dr. Walburga, Universität Münster, Roxelerstraße 64, 44 Münster, FRG.

Röder, U., Institut für Experimentalphysik der Technischen Hochschule Darmstadt, Schloßgartenstraße 7, 61 Darmstadt, FRG.

Rona, M., Dept. of Physics, Middle East Technical University, Ankara, Turkey.

Rosen, Prof. Dr. Robert, Dept. of Physiology and Biophysics, Dalhousie University, Halifax, Nova Scotia B3H 4H7, Canada.

Sagaama, Ing. Sadok, Ecole Nationale Supérieure des Mines de Saint-Etienne, 158bis cours Fauriel, 42023 St. Etienne, France.

Sato, Dr. Masako, Dept. of Mathematics, Faculty of Engineering Science, University of Osaka, Prefecture, Mozu-Ume Cho, Sakai, Osaka, Japan.

Sato, Shunsuke, Faculty of Engineering Science, Osaka University, Toyonaka, Osaka 500, Japan.

Sharp, Dr. J.A., System Dynamics Research Group, University of Bradford, Bradford, UK.

Shepherd, P., Science Policy Research Unit, Mantel Building, University of Sussex, Sussex, UK.

Sommer, Dipl.-Kfm. Goetz, Institut für Wirtschaftswissenschaften, Lehrstuhl für Unternehmensforschung, Templergraben 64, 51 Aachen, FRG.

Sugeno, Dr.-Ing. Michio, Dept. of Electrical and Electronic Engineering, Queen Mary College, Mile End Road, London E1 4NS, UK.

Taddei-Ferretti, Prof. C., Laboratorio di Cibernetica, 80072 Arco Felice (Napoli), Italy.

Terano, T., Dept. of Control Engineering, Tokyo Institute of Technology, Tokyo, Japan.

Trockel, C., Physiologisches Institut der Universität Bonn, Nußallee 11, 53 Bonn, FRG.

Ungan, P., Institute of Biophysics, Hacettepe University, Ankara, Turkey.

Uranishi, Michiyoshi, Faculty of Engineering Science, Osaka University, Toyonaka, Osaka 500, Japan.

Uyttenhove, Prof. H.J.J., State University of New York at Binghamton, Binghamton, New York 13901, USA.

Varela, Francisco, University of Colorado, Department of Anatomy, School of Medicine, 4200 East Ninth Avenue, Denver, Colorado 80220, USA.

Villa, A., Istituto Elettrotecnico Nazionale Galileo Ferraris, Corso Massimo d'Azeglio 42, 10125 Torino, Italy.

Walker, Prof. Dr. Crayton C., University of Connecticut, U-41, Storrs, Connecticut 06268, USA.

Waśniowski, Dr. Ryszard, Technical University of Wroclaw, Wybrzeze Wyspianskiego 27, 50-370 Wroclaw, Poland.

Wess, Dr. Othmar, Gesellschaft für Strahlen- und Umweltforschung, Abteilung Kohärente Optik, Ingolstädter Landstrasse 1, D-8042 Neuherberg, FRG.

Willigan, Rev. J. Dennis, Dept. of Sociology, University of North Carolina, Chapel Hill, N.C. 27514, USA.

Witten, I.H., Man-Machine Systems Laboratory, Dept. of Electrical Engineering Sciences, Colchester, Essex, UK.

Zeleny, Prof. Milan, Columbia University, Graduate School of Business, New York, N.Y. 10027, USA.

Zimmermann, Prof. Dr. H.J., Lehrstuhl für Unternehmensforschung, Institut für Wirtschaftswissenschaften der Technischen Hochschule Aachen, Templergraben 64, 51 Aachen, FRG.

KEYNOTE INTRODUCTION

The systems view and the systems method: must they remain separate?

PROFESSOR P.B. CHECKLAND
Department of Systems,
University of Lancaster, UK

I face some difficulties in making some remarks which are approximately relevant to all of you, whether you are bio-cyberneticians, general systems theorists, management problem-solvers, neurobiologists, control engineers, fuzzy mathematicians or computer enthusiasts. The danger is that anything so general will be devoid of content; and this brings an added difficulty because what I wish to say is that the systems movement—300 members of which are gathered here this morning—is too ready to generalize, not ready enough to test its generalizations. That will be the content of my sermon.

I use the word 'sermon' deliberately because I'm sure you will agree that from a pulpit of this magnificence in a hall of this splendour only sermons are possible! I shall therefore extract from the later stages of my talk a quotation which can serve as the text for my sermon. Here it is; it was written more than 300 years ago by Isaac Newton in a letter to Roger Cotes:

> If, instead of sending the observations of able seamen to able mathematicians on land, the land would send able mathematicians to sea, it would signify much more to the improvement of navigation and the safety of men's lives and estates on that element.

That is my text; here is my sermon...

I

American advertizing executives talk of 'running an idea up the flag pole' in order to see whether anyone salutes. It has been true for some years now that whenever a 'systems approach' is run up the flag pole, most people salute. No one ever is heard boasting that they do *not* use a systems approach, and cogent attacks on the idea of systems thinking are all too rare. I was delighted, therefore, recently to come across some words of Jacques Monod, Nobel Prize-winning molecular biologist:

> ...What I consider completely sterile is the attitude, for instance, of Bertalanffy (but he is not the only one), who is going around and jumping around for years saying that all the analytical science and molecular biology does not really get to interesting results; let's talk in terms of general systems theory. Now I was struck by this term and I talked to some systems theorists and informationists and so on, and they all agree that there is not and there cannot be anything such as a general systems theory; it's impossible. Or, if it existed, it would be meaningless. [1]

Alas, this criticism is not developed and justified; it remains an assertion. But even so it is the kind of assertion we in the systems movement need more of. We need to dispel the complacency which is typical of the systems movement, because it could be that attacks on a 'systems approach' are rare because it is devoid of content, as Monod implies, because it is not worth attacking. We ought to be disturbed that practically everybody is in favor of a 'systems approach', and especially so because most enthusiasts would find it difficult to explain just what a 'systems approach' consists of.

The fact that a large number of people are in favor of an intellectual abstraction which is very ill-defined gives rise to a marked gap between, on the one hand, the enthusiasts and, on the other, people like Jacques Monod who find it meaningless. This gap is a gap between what systems thinking is *said to be* and what it is *seen to do*. It is a gap we in the systems movement ought to try to close, for two reasons. Firstly, matching what the systems view *is* to what the systems method *does* potentially provides a program for the systems movement, a program whose execution will ensure that a 'systems approach' is more than a rallying cry. Secondly, it would be greatly to the advantage of the systems movement if we could persuade serious potential critics like Monod that the systems outlook is serious enough to be worth attacking seriously.

I believe we can make progress in closing the gap, in bringing together the systems view and the systems method, by examining *what a systems approach is*, which means examining where it comes from—its intellectual source—since this, I believe, shows us what systems thinkers ought to be *doing*.

II

I take it as indisputable that the systems movement is a part of the broad sweep of the science movement. Systems thinking is, or is supposed to be, a variety of scientific thought. This being so, we can best take a view of what systems thinking is by seeing it in the context of the history of science.

Science is characteristic of Western civilisation, and arose together with philosophy, from which it was at first indistinguishable, in 6th Century B.C. Greece. We now see Thales, Anaximander and the other Ionian philosophers as the founders of the science tradition because when they suggested that there must be a single component, a unitary stuff, from which the world was constructed they were founding a tradition of rational argument. Critical discussion, without recourse to myths or the supernatural, characterized Greek science, and has remained a prime characteristic of science as an organized human activity. But although Aristotle worked as a marine biologist, and founded a tradition of observation, the Greeks did not contribute in a major way to the experimental tradition that we see in science. That derives more directly from the medieval clerics who struggled with the problems of inductive argument—such as Grosseteste and William of Ockham—and, later and most dramatically with the experimentalists like Galileo and Gilbert who contributed to the scientific revolution of the 17th Century which culminated in Newton's world picture: rationally argued, experimentally verifiable and expressed through the generalising power of mathematics. Since the 17th Century, the exponential rise in the activity of science has virtually created our world. What we have learnt most clearly during this period of the exploitation of science is: firstly, that science is an unprecedentedly powerful means of finding things out, a highly successful 'learning system'; secondly, through the downfall of the Newtonian world picture and its replacement by Einstein's model in which space is no longer an absolute framework, we have learnt that all scientific knowledge is provisional, and that at any moment of time the scientific knowledge we have is simply the best-tested knowledge: it may be replaced by future conjectures which survive more stringent tests.

What the history of science teaches us is that science consists of rational thinking applied to experience, especially the kind of experience obtained in the special kind of controlled observations we call 'experiments'. We may characterize science, in fact, as a learning system, in terms of three crucial characteristics: reductionism, repeatability and refutation. [2] Science is *reductionist* in the sense that in experiments we isolate a small part of the world in the laboratory and investigate just a few variables under controlled conditions. And it is reductionist in the sense which derives from William of Ockham; using the principle known as Ockham's Razor ('do not multiply entities unnecessarily') we seek to explain the results in the most simple way, using as small a number of concepts as possible. *Repeatability* is the criterion which the results of experiments must satisfy if they are to be accepted as scientific knowledge; this is concisely expressed in Ziman's definition of science as 'public knowledge'. Finally, science makes progress by subjecting conjectures to experimental testing, retaining those which best survive the tests—Einstein's theory, for example, being preferred to Newton's precisely because it can encompass all Newton's results *and*, for example, some apparent anomalies in the motion of Mercury which defeat Newton's formulation. The aim of the experiments can thus be viewed as an attempt at the *refutation* of the hypotheses embodied in the experiments.

Science, as a learning system characterized by reductionism, repeatability and refutation, is manifestly successful. It can point to demonstrable successes, where consulting astrologers or reading the entrails of slaughtered goats cannot. But, nevertheless, it is out of the limitations of science that the systems movement arises.

Science is at its most powerful in the investigation of the physical regularities of the universe, phenomena such as the properties of light, magnetism, the laws of chemical combination, etc. The method is stretched to the limit as the complexity of the phenomena studied increases. It is clear, for example, that the bold program of the pioneers of social science—to establish the laws of society, to set alongside those of physics—has not been fulfilled; questions of methodology in the social sciences are still crucial issues. Perhaps it is not surprising that systems thinking, self-conscious thinking in terms of 'wholes', should arise in the science of biology which in complexity is intermediate between physics and chemistry on the one hand and the social sciences on the other. Systems thinking, thinking in terms of wholes and their properties, appears to be an appropriate weapon whenever the investigation is concerned with densely-connected whole entities which show properties described as 'emergent', that is to say, characteristic of the level of complexity being studied and without meaning in terms of lower levels. If I may quote a summarizing statement from a paper currently in the press [2]:

'Thus while the Weltanschauung of science is that the world consists of groups of phenomena which may be investigated by the method of science, the counter-Weltanschauung of the systems movement is that the world consists of a complex of wholes which we term 'systems'. The systems thinker assumes that the worl will exhibit emergent properties at virtually all levels of complexity and that it will be useful to examine the world in terms of the wholes which exhibit those properties, and to develop principles of 'wholeness'. The long-term program of the systems movement may be taken to be the search for the conditions governing the existence of emergent properties and the wholes which exhibit them.'

The systems movement can thus be seen as a part of the science movement, but one which hoped to make progress in understanding complex phenomena by trying not to be reductionist, by trying to analyze in terms of wholes which, compared with the components which comprise them, show emergent properties. One might say that the reductionism of the systems thinker consists of saying 'I will analyze and explain in terms of components which are themselves coherent wholes'. The intention is not to dispense with analytical reduction, but to create an intellectual weapon which is complementary to it.

Given this view of the nature of systems thinking, as a response to some problems in science, can we surmise why progress has been so slow? Can we obtain from the history of science any indications of what the systems movement ought to do?

It seems to me characteristic of science that its practitioners have on the whole shown a considerable lack of theoretical interest in its method and philosophy and that this has been a positive advantage to the development of science. Scientists have always been intensely interested in *problems, puzzles, or paradoxes,* and have been anxious to take for granted the philosophy of the activity. The systems literature, however, sags under a heavy load of untested assertions, models and conceptualizations. When the Aristotelian world picture was replaced by that due to Galileo and Newton, science took an immense step forward; it happened because motion was taken to be a problem. According to Aristotle motion could be sustained only if there were a local source of motion continuously at work. In the case of an arrow flying through the air Aristotelians had to assume that the air pushed out of the way by the front of the arrow rushed round to the back to provide the push. The flight of a projectile, they conjectured, was a straight line followed by a vertical fall to earth. Now, the point is that the new way of examining such things—experimental science—came to see that these conjectures were simply not true. Galileo discovered that the flight of a projectile was a parabola. Science took up motion *as a problem*, and by solving it changed mankind's way of looking at the universe. Science has always tackled problems; but where are the major problems with which the systems movement is currently engaged? And what kind of testable conjecture is emerging which will allow the problems to be solved? The answers to both questions are bleak ones.

Three hundred years ago Isaac Newton, in a letter to Roger Cotes, wrote:

> If, instead of sending the observations of able seamen to able mathematicians on land, the land would send able mathematicians to sea, it would signify much more to the improvement of navi-

gation and the safety of men's lives and estates upon that element.

In modern language this is a call to 'action research'; I think it is highly apposite today. Addressing this Conference two years ago James Miller suggested that what was needed in the systems movement was less talk and more data, a greater readiness to formulate hypotheses and to submit them to test [3]. That is still the challenge; and finally I would like briefly to indicate one way of trying to face it: the way in which the Department of Systems at Lancaster has been trying to meet it.

III

James Miller's group is concerned with one particular kind of system, namely 'living systems', and their approach is to formulate empirically testable hypotheses about the nature of such systems.

At Lancaster our concern has been with 'real-world problems'. By this phrase we mean problems of decision which arise unasked, which we find ourselves facing, as opposed to the laboratory problem which the scientist can select, define and constrain to suit his inclinations, interests and resources. The problem or puzzle at the foundation of our work is the fact that society is not good at tackling real-world problems, and that science, for all its power in other directions, seems not able to help. Real-world problems are frequently a perceived mis-match between 'what is' and 'what might be', and in this sense are 'management' problems defining this term broadly. The body of knowledge known as 'management science' has clearly not been very successful so far in providing ways of tackling real-world problems. Certainly that is the case if we judge by the low opinion of management science held by most real-world managers. So we took as a problem the fact that real-world problem-solving was astonishingly difficult; our approach was to try to use systems ideas in tackling such problems; and our anticipation, or hope, was that out of the work would come some system-based methods of problem-solving and, perhaps, a systemic base for an improved management science.

The choices of problem and means of approaching the problem has restricted our attention to systems of a particular kind. Where Miller has concentrated on 'living systems' our focus has been on a kind of system which has not always been recognized as such. We call a connected set of purposeful activities a 'human activity system'. Much of our effort has turned out to be an exploration of the nature of such systems.

In the method of approach adopted we hope to have avoided the danger of merely *talking about* real-world problem solving by making the research 'action research'—of the kind we hope Newton would have approved. We have tried to work within problem situations with problem owners who wanted problems solved. We try to be involved in action-taking, even though this means that research aims formulated before the event cannot always be followed. Action research has to follow where the action leads. The advantage of this has been that it provides in the long term—over a number of studies—a criterion by which the systems content of the work can be judged. If in a large number of studies we can persuade a number of problem owners of different types that problems have been alleviated or solved, or at least that insight has been gained, then we may surmise that some of the virtue lies in the systems thinking used in the problem-solving! This is by no means a sharp criterion, cut and dried; but we, like everyone else working in systems containing human beings, have found it extremely difficult to find a way of formulating testable systems hypotheses. By taking part in the action of problem-solving we have at least insured that we have been using systems ideas, not merely talking about using them.

Our problems tackled in systems studies have been varied. They have covered problems within organizations large and small, both user-supported and public-supported, and they have covered problems not organization-bound. Problem owners have ranged from chief executives in industry, to middle managers, to public authorities, to a charity organization, to a community action group.

All these varied experiences have only one thing in common: that they have encompassed problem situations in which we have tried to use systems ideas in problem solving. What every systems study has had in common with all the others, indeed the only thing they all have in common, is: methodology. So it is inevitable that the outcome of the work has been some principles of systems method—a systems-based *methodology* for problem-solving [4]. What has been most interesting about this outcome has been the fact that its emergence from an action research program has forced us to recognize that methodology is in itself of little value because it is, on its own, untestable. (Success might be due to extraneous reasons, and failure might be due simply to incompetence in applying the methodology!) In order to judge methodology it is necessary to take methodology plus problem; and the only test lies in the question: was the problem solved?—hence our criterion of success in this work. I do not beleive that we would have recognized this salutary fact so clearly had we been involved only in talking about using systems ideas rather than actually trying to use them.

I shall not discuss the methodology itself here (though some of the experiences of using it are the subject of a later paper to one of the symposia [N] of this meeting) beyond saying that it involves selecting and naming human activity systems possibly (hopefully) relevant to solving the problem or improving the problem, making systems models of those activity systems and comparing those models with what is happening in the real-world in order to structure debate about possible changes which meet two criteria: are they arguably *desirable* and are they actually *feasible* given prevailing attitudes, feelings, power structures and resource availability? The overall outcome of developing it in use in fifty-odd systems studies is a view of the special nature of human activity systems. What is especially interesting about such systems is that no single account of them can be given; it is not possible ever to describe something which without doubt *is* a particular human activity system. Such systems are inseparable from perceptions of them, and these perceptions, whether belonging to actors in them or to observers of them, will never be unitary. Only in the case of trivial human activity systems, about which nobody feels strongly or *cares*, will there ever be even a general consensus on what they consist of. Human activity systems are only meaningful, they only exist, in association with a particular *Weltanschauung*, and real-world problems are usually ill-structured debates about unstated *Weltanschauungen.* Systems thinking can improve the quality of such debates.

Such is the outcome of this work so far, and it has implications to be followed up both for management science and for methodological problems in the social sciences generally. But the content of this work is not here my main concern. The point I wish to make now is that I would have no confidence in its outcomes had I thought them up sitting at a desk in Lancaster. Only by trying to solve problems and by seeking ways of testing its assertions can the systems outlook link with systems method to make progress.

In summary, I have argued:

- —that the systems movement is much readier to talk about doing than to do;
- —that systems thinking can be seen as a response to certain problems within science;
- —that science is characterized by a readiness to concentrate on problems, puzzles or paradoxes, not on its philosophy;
- —that the systems movement ought to do the same, seeking out opportunities to try out systems thinking in systems action.

References

1. MONOD, J., 'On "Chance and necessity"', in Ayala, F.J. and Dobzhansky, T., *Studies in the Philosophy of Biology*, Macmillan (1974).
2. CHECKLAND, P.B., Science and the systems paradigm', *International Journal of General Systems* (1976), 3 (1972), pp.127-134.
3. MILLER, J.G., 'General systems theory—less talk and more data', 2nd European Meeting on Cybernetics and Systems Research, Vienna, Austria (1974).
4. CHECKLAND, P.B., 'Towards a systems-based methodology for real-world problem solving', *Journal of Systems Engineering* 3(2), (1972).

GENERAL SYSTEMS METHODOLOGY

The role of randomness in system theory

I.H. WITTEN and B.R. GAINES
Department of Electrical Engineering Science,
Colchester, Essex, UK

1. Introduction

The main role of the concept of randomness in system theory has always been that of subsuming residual phenomena not encompassed by the main model or theory, eg. 'we will assume that this signal is degraded by a channel noise of...'. In application studies the noise sources invariably play the role of defects, such as interference, unreliability, or imprecise measurement, and a major objective in system engineering is to reduce the effects of these defects to tolerable levels. This negative view of randomness is consistent with more fundamental concepts of knowledge and science where causality is taken as a basic postulate [1, pp.316-324] and randomness is introduced as a last resort, temporary 'explanation' when the assumption of causality fails.

In recent years, however, new foundations have been established for the concept of randomness that relate it very closely to that of *computational complexity* [2-7]. This relationship raises some interesting new system-theoretic questions about random behavior, notably: what form of world model will be obtained if we assume that the world is deterministic in circumstances where its behavior is really randomly generated?; can we use the 'complex' behavior of a random source in circumstances where we need complex behavior and would otherwise have to use substantial memory and/or computing power to generate it?

This chapter presents recent results that answer these two questions. It is shown, on the one hand, that a deterministic modeller of a world that exhibits the slightest degree of random behavior will obtain a model that becomes indefinitely complex with the number of observations. On the other hand, theoretical and practical examples are given of situations in which the complexity of random behavior may be used to solve problems that would otherwise be expensive in storage and computing power. A variety of examples and consequences of both these results are given.

The remainder of this introduction is a brief survey of the negative role of randomness in science followed by an outline of more constructive recent results in system theory.

A. Negative aspects of randomness in science

In his wide-ranging review of the emergence of the concept of probability Hacking [8] suggests and illustrates a peculiar symbiosis between causality and randomness in which the very stress on the former as a basis for scientific explanation itself necessitates the introduction of the latter to account for the gap between observation and theory. He remarks (p.2) that, "Europe began to understand concepts of randomness, probability, chance and expectation precisely at that point in its history when theological views of divine fore-knowledge were being reinforced by the amazing success of mechanistic models". Our scientific models are historically, and psychologically, causal. Random phenomena are thought of as those at the boundary between causality and acausality where some incomplete order can be extracted from nature. This boundary is regarded classically as that of our 'knowledge', and one that should be continually expanded outwards so that the inexplicable becomes random and eventually causal. As Hume[9] put it, "it is commonly allowed by philosophers, that what the vulgar call chance is nothing but a secret and concealed cause".

Recently Suppes [10] has been vehemently critical of this negative view of the role of randomness in science. He argues that, "the old theology has been replaced by a new theology of philosophy and science that is equally fallacious and mistaken in character", and that tenets of the new theology are, for example, "every event has a sufficient determinant cause", and "knowledge must be grounded in certainty". Suppes argues that, "the fundamental laws of natural phenomena are essentially probabilistic rather than deterministic in character". He goes on to draw a historic analogy between a refusal to accept probabilistic laws and Descartes' refusal to accept the concept of action at a distance which influenced Newton's own presentation of his laws of gravitation.

However, the negative view of randomness coupled with a positive one of causality appears to be more than an expression of scientific philosophy and to be rooted deeply in human psychology. For example, Michotte's [11] experiments in which certain moving visual patterns are immediately perceived as physical, cause-effect relationships between objects, illustrate that the assumption of causality may be inbuilt strongly at a comparatively low level of brain functioning. Gaines [12 and see section III] has reported the complex causal models generated by subjects performing extremely simple cognitive tasks which have a random component. These innate human factors rather than a reasoned scientific philosophy, appear to dominate the attitudes of many eminent scientists in their rejection of randomness as a legitimate part of scientific explanation. Charles Darwin [13] states, "I cannot think the world as we see it is the result of chance", and the conflict that he saw remains an issue in biology today [14,15,16]. Freud's [17] psychopathology of forgetting, mistakes and dreams originates from his refusal to accept explanations of these phenomena as chance occurrences and his resultant development of causal models for them. Finally, perhaps the most widely known example of this attitude towards randomness is that of Einstein in the controversy over the status of probability in quantum mechanics. In a letter to Born he remarks, "You believe in God playing dice and I in perfect laws" [18].

B. Positive aspects of randomness in system engineering

Even if the assumption of causality was not deeply embedded in the human nervous system, the attitudes of Darwin, Freud and Einstein would be justified by the very success of that assumption in the development of modern science. [1 p.316]. Hacking's argument above that the concept of randomness is complementary to, and subservient to, the determinism that led to modern physics, "this specific mode of determinism is essential to the formation of concepts of chance and probability" [8 p.3], also empasises the negative status of randomness. Thus it is not surprising that only in recent years have constructive applications of random phenomena begun to be investigated. To ask the question, 'can we make use of random phenomena', is itself an inversion of customary modes of thought. To look for practical advantages in so doing would seem foolish, and to expect there to be situations in which random phenomena are essential to achieve goals would seem akin to madness.

However, this decade has seen the growth of substantial research into, and practical application of, the constructive aspects of random phenomena. A key result for this work was Rabin's [19] 1963 paper on stochastic acceptors in which he demonstrated that there are regular events recognized by a two-state stochastic automaton that are only recognized by a deterministic automaton with an arbitrarily large number of states. This result is the theoretical basis for such engineering structures as the stochastic computer [20,21], and its commercial applications, eg. signal averaging [2], digital-analog conversion [23], and electronic [24] and fluidic [25] instruments. It is also the basis for Hellman and Cover's [25] result that for deterministic automata of arbitrary size there exist hypothesis-testing problems which they cannot solve but for which a two-state stochastic automaton has arbitrarily small error.

A comparable series of results have been developed for stochastic generators. Gaines [20,27] has shown that there are pattern classifications that cannot be learnt by an adaptive-threshold-logic element (ATLE) with discrete, bounded weights, unless the weight changes are probabilistic. Independently Clapper [28] discovered this phnomenon in a practical pattern recognizer. Gaines [29] generalized the result to show that there was a class of control problems insoluble for any deterministic automaton with a bounded set of states but universally soluble by a two-state stochastic automaton. Independently Gold [30] showed that this class of problems was soluble by a simple recursive automaton, thus establishing a correspondence in power between finite-state stochastic and infinite-state recursive controllers. It may be argued that this correspondence underlies the major role of chance in genetic development as emphasized by Monod [51] and illustrated in the models described by von Foerster [31] and Block [32].

The constructive role of randomness in many applications of electronic systems has been surveyed recently by Gupta [33]. In the remainder of this chapter I shall concentrate on some system-theoretic results that illustrate both the negative aspects of randomness discussed in the previous section and the positive aspects outlined above.

2. Some identification-theoretic results

The psychological force of the assumption that an observed system is causal is very strong as illustrated in section 1A. Some observations of human problem-solving made in the course of the STeLLA [34] project made apparent the importance of the assumption of deterministic causality in cognitive tasks. One test environment for the learning machine was a game of nim played against a partially random player of variable optimality. To demonstrate that the task was non-trivial for people also we built a nim automaton with four lamps (representing the number of matches left) and three keys (representing taking one to three matches) and a 'reward' light indicating that the game had been won. In practice human beings found the game virtually impossible to learn in these circumstances, whereas in its normal form nim is trivial.

Most of the problem in learning seemed to stem from the random nature of the opponent's play. Although this makes the game easier to win (the opponent makes fatal errors) it also makes the behavior of the automaton indeterminate, and a hypothesis of randomness, or indeterminism, was *never* introduced by human players, and the longer they played the more confused they became. To investigate this further we built a simpler automaton with two keys and two lamps such that one lamp came on for a short period whenever a key was depressed. The problem was to depress the key under whatever light came on with equal probability. Again in tests with a wide range of subjects this random element was never recognized and individuals were prepared to pit their wits against the automaton for long periods of time.

Records were kept of verbal introspection as subjects played against the automaton, and it was found that most people generated elaborate models which they felt were 'not quite correct'. One of the most interesting records is that of a behavioral scientist who played against the automaton for some thirty minutes and kept a tally of occasions correct less occasions wrong. After that period he was over twenty ahead on his count and announced he had a good strategy but felt he should explore others and seek a better one. His accumulated count then rapidly declined and when it went negative he announced that he was going back to his 'good strategy'. The count continued to decline, however, and he finally announced that the box contained a system which modelled his strategy and then structured itself in the opposing form to outwit him! The hypothesis of such a complex structure (a 'frustration automaton in the sense of Section (3)) on the basis of interaction with a simple two-state stochastic automaton suggested that the assumption of causality was a dangerous one in system identification and that it would be fruitful to investigate it more formally.

Gaines [12] analyzes a formal model of a *modeller* forming a causal model of some other system based on his observations of its behavior. The *observed behavior* is some sequence drawn from the union of input and output alphabets, and the *length* of a sequence of observations is the number of output symbols in it. For any finite sequence of observations there will be an unbounded number of finite-state machines whose input/output behavior, starting from a specified state, is identical to that observed. Out of these possible models there will be some such that no other machine has a smaller number of states, and an *optimal causal modeller* is defined as one who always chooses such a minimal state model.

Five results are established to describe the dynamic behavior of such a modeller, in particular, the relationship between the number of states in the model and the length of the observation sequence:

1. The number of states in the models formed by an optimal causal modeller of a given sequence of behavior is a montonic non-decreasing function of the length of that sequence;
2. The number of states in the models formed by an optimal causal modeller of a sequence of behavior generated by a finite-state deter-

ministic machine with M states cannot exceed M;

3. An optimal causal modeller of an observed behavior of length N will form a model with not more than N states;
4. There are observed behaviors of length N such that the model formed by an optimal causal modeller has N states;
5. The *expected* number of states in the model formed by an optimal causal modeller observing a sequence of behavior of length N may be at least $N\text{-}\log_2 N\text{-}2$.

The first two 'results' are just a formalization of what we, and the modeller, expect to happen when it is faced by a finite-state deterministic system—the number of states in the model will increase up to final value which cannot exceed the number in the observed system. The next two results show that whilst the number of states in the model cannot exceed the length of observation it can grow as fast as the number of observations. Hence non-causal systems may give rise to behavior requiring indefinitely complex models. However, if the system whose behavior is observed is relatively simple, eg. finite-state even if acausal, we might expect the modeller to behave in a similar way to when modelling a causal system. Clearly a stochastic source could generate isolated examples of the sequences noted in (4) requiring as many states in the model as the number of observations. However, it seems reasonable to suppose that such 'pathological' sequences might be generated only infrequently by a finite-state stochastic automaton.

Hence a more interesting characteristic of the observer's behavior would be, not the maximum state model he might generate, but rather the expected number of states in the model (the average complexity). This might have one of three possible behaviors as a function of the length of the observation sequence, N. The expected number of states in the model formed by an optimal causal modeller of the behavior of a finite-state stochastic automaton might be:

a. Asymptotic to a finite number, ie. closely similar to the situation when the behavior modelled is generated by a finite-state deterministic automaton;

b. Growing without limit but slower than the number of observations itself, eg. as log N. One might hypothesize that at least the ratio:

$$R_N = \frac{\text{expected number of states}}{\text{number of observations}}$$

would tend to zero with N.

c. Growing without limit at a rate similar to the maximum possible, N. This would imply that nearly all the sequences generated were 'pathological', requiring maximum-sized models growing as fast as the number of observations.

The main result, (5), shows that case (*c*) occurs and that the ratio R_N tends not to zero but to unity. The assumption of causality when modelling acausal systems can lead to ridiculously complex models which do no more than retain all past observations. The two-lamp, two-key, automaton we used with human subjects generates sequences leading to result (5), thus explaining the complexity of the models they described. One is tempted to paraphrase Einstein and say that if one assumes God does not play dice when he does, one will obtain an over-complex view of the universe—perhaps just what has happened in psychopathology.

This result indicates why some of the most attractive and apparently powerful techniques in automata theory are never seen in their obvious applications. The Nerode construction [35,36] gives a straightforward algorithm for determining the optimal causal model (minimum-state finite automaton) corresponding to any sample of observed behavior. Thus, epistemologically, we appear to be justified in assuming that knowledge of behavior can always be condensed into a canonical structure, and that general system-theoretic results are applicable if expressed in terms of the state-space of such structures. In practice, we should be able to resolve the problems of treating general non-linear systems, and in particular the whole of behavioral psychology, by feeding behavioral data into a canonical modeller. Such an approach seems to offer the possibility of a neutral behavioral science in which the structure of reality is derived without preconceptions, assumptions, or distortion. In a perfectly causal, closed universe in which all information was available to us this would be so, but, alas, the slightest acausality [12] destroys the validity of our modelling technique.

The failure of deterministic modelling with random sources has prompted a wealth of reearch on modelling techniques for discrete stochastic systems; Fu and Booth [37] have recently surveyed this work. However, until recently, there has been no formulation of the problem comparable to that for deterministic systems on which the elegant results of Nerode, Arbib and Goguen are based. What is an 'optimal model' of a stochastic system based on a finite sample of that system's behavior—is it a complex model exactly reproducing the behavior (in which case we ultimately reach the absurd deterministic models discussed above)—or is it a simple model that approximates the behavior (and then how simple for what level of approximation)?

Gaines [6,7] has recently given a precise formulation of the general identification problem that makes these possible trade-offs quite explicit in that the result is not an 'optimal model' but an *admissible set* of models ordered by *complexity* in one dimension and *poorness-of-fit* in the other. Models in the admissible set are such that no other model has both a less poor fit and a lesser complexity. For deterministic modelling the admissible set has only one member (up to an isomorphism) whereas for stochastic modelling it has a spectrum of members with clearly defined drops in 'poorness-of-fit' as real structure in the source is picked out by more complex models [7]. The formulation is applicable to general non-deterministic sources and, in the stochastic case, establishes a close link between the three formulations of probability theory [38] in terms of: relative frequency; subjective probability; and computational complexity. It demonstrates that the system-theoretic tools required for Suppes [10] 'probabilistic metaphsyics' can be made available.

There *is* a positive application of the results of the section in that there are occasions when we wish to simulate complexity in a fairly simple system, for example to enable an ELIZA-like system [39] to pass the Turing test. An interactive conversational computer program based on simple rules is rapidly dismissed as mechanical and its rules identified unless some random selection is introduced. This is particularly relevant in tutorial systems where one wishes to establish at least parity between (automatic) teacher and student. Albritton's comment that he is more likely to apply human terms to a robot if part of its program is random is one of the more telling points in the discussion following Putman's [40] paper on the role of automata in psychology and philosophy. A contrary example is the problem created by fluctuations in the response time at a terminal to an interactive time-shared computer. The variations with system loading are an acausal phenomenom to the interactive user who may attempt to ascribe them to variations in his own activity, eg. error inputs, and become very disturbed by them [41].

In conclusion, whilst the theory of (deterministic) automata is immediately applicable to digital computers, its powerful and elegant results and techniques are severely limited in the normal noisy world. We need equivalent results for non-determinate and stochastic automata, and must not fall into the trap of assuming that we can obtain them by assuming that our observations can be fitted into a causal framework—they can but only at the expense of diluting reality with triviality.

3. Some control-theoretic results

The results of section 2 illustrate some of the dangers of randomness if we assume that it is not present. However, as indicated in section 1*B*, these are also constructive aspects of randomness—advantages to be gained by its deliberate introduction in system design. For example, consider the following abstract formulation of the problem of regulating a discrete dynamical system to maintain its state within a prescribed region. Represent the system as an automaton, (I,P,S,σ,π), where I a finite input alphabet, $P\equiv\{0,1\}$ is a binary set of output, S is a set of states, $\sigma: S\times I\rightarrow S$ is the next-state function, and $\pi: S\rightarrow P$ is the output function. In this formulation π is a performance function and the problem is to regulate the inputs to the automaton to cause its output to become, and remain, 1.

Suppose that there is some distinguished element $\lambda\in I$ (the 'zero' input for the autonomous system), and consider the following sets of states:

$$W\equiv\{s:\pi(s)=1\}$$

$$A\equiv\{s:\ \forall n\geqslant 0,\ \sigma(s,\lambda^n)\in W\}$$

$$B\equiv\{s:\ \exists n:\ \sigma(s,\lambda^n)\in A\}$$

W is the subset of S in which it is desired that the state should reside; A is a weak attractor within W, and B is its region of attraction [42]. We assume that both B and S–B are non-empty so that the autonomous system has a region of local asymptotic stability but is not asymptotically stable in the large and consider the family of control automata whose inputs are from P and whose outputs are in I which induce global stability. For this family to be non-empty it is necessary that B be reachable from S, that is:-

$$\forall\, s\in S,\quad u\in I^*:\exists\sigma(s,u)\in B$$

where I^* is the free semigroup generated by I.

This problem is one of a class considered independently by Gold [30] in his paper on 'Universal Goal-seekers', and by Gaines [29] in a paper on stochastic automata. Both authors demonstrate that no finite-state deterministic automaton can act as a universal regulator for this class of problems. On the contrary it is shown that there is a finite-state dynamical system which is universally insoluble for all such regulators with less than a given number of states. Both authors give a construction for such a system which has the property that any action by the regulator other than that which it actually does would immediately solve the problem. Gaines calls such a system a *frustration automaton*, and Gold terms the resulting behavior *strongly worst*!

Having established that there is no universal finite-state deterministic regulator for this class of problems, the two papers diverge in an interesting fashion. Gold demonstrates that there is a universal primitive-recursive (and hence potentially infinite-state) deterministic regulator, and Gaines demonstrates that there is a universal two-state stochastic regulator. This equivalence between recursive and stochastic solutions, together with the non-existence of finite-state deterministic solutions, demonstrates the savings in memory and complexity resulting from the introduction of randomness. It suggests a role for random phenomena in both physical and biological systems in that they enable a simple, finite-memory structure to achieve the same control capability as a complex structure requiring unbounded memory.

A simple physical example of the principle involved is that of leveling the surface of a bowl of sand. A controller using a fine probe to push down grains on the surface could probably construct an appropriate trajectory through the state-space of the sand, but its storage requirements would be massive and one would have difficulty in constructing the control algorithm! Randomly tapping the bowl would be a rather more efficient means of achieving the same end. This example also illustrates a difference in informational requirements since the probe controller would require far more information as to the state of the sand (at least the locations of high spots) than the random controller (only a signal when the sand is level enough).

A psychological example of the principle in the context of instructional systems has been given by Gaines [43] who uses it to establish the existence of a universal trainer for any trainable system. The use of performance feedback only by the trainer without other information as to the structure or state of the trainees is given as an explanation of the effective operation of current instructional systems, including computer-based instruction, which do not build a model of the learner. The explanation of the observed phenomenon that many trainees are trainable is clearly outside these arguments, and requires postulates such as Pask's [44] characterization of "man as a system that needs to learn".

It might be supposed that the stochastic regulator is very much less effective than the recursive one in terms of the time taken to solve the problem. However, a simple example shows that this is not so. Moore's [45] 'combination lock' problem involves an automaton whose output is zero until the correct sequence of inputs is applied when it becomes one (the lock opens). At any stage an incorrect input causes the automaton to re-initialize. Moore shows that a sequence of length at least 2^{N-1} is necessary to guarantee to open a binary lock with N states. It may also be shown that the mean length of sequence to open the lock with a random generator is 2^N. Conway [46] emphasizes the complexity of the deterministic solution to this problem and recommends the stochastic one. Theoretical upper bounds on the length of the deterministic solution are ridiculously high, but simulation results suggest that the actual mean length is close to that for the stochastic solution [47].

Another interesting example is the deadlock problem in resource allocation. Suppose a number of automata are competing for the use of a single resource for a limited period. At each instant each automaton either claims, or does not claim, the resource. If there is only one claimant it is allocated the resource, otherwise all claims are rejected. It is clear that if the automata are identical in structure and commence in the same state then none of them will be allocated the resource (at each instant either none or all will claim), regardless of the complexity of their structure provided their behavior is deterministic. In a human situation typically that of two people stepping aside in the same direction to avoid one another, oscillations may occur but the deadlock is ultimately broken because their strategies are *not* identical. They do not have to communicate with one another or give way to a greater power as in the 'operating systems' solutions to such problems, eg. semaphores and mediation of claims through a central resource allocator. A simple solution to the general problem, that retains the independent, non-communicating, distributed control by individual automata is to make them claim resources probabilistically. In terms of Dijkstra's famous 'five philosophers' problem [48], one notes that each will get a fair share of the spaghetti if, on finding they have only one fork, they toss it in the air and, if it falls the right side up, put it back on the table for a little while so that another philosopher (whose toss has led him to keep his fork) may grab it!

The stochastic regulator cannot be frustrated because it does not show consistent behavior in those circumstances where it is not achieving its goal and hence cannot remain in a fruitless limit cycle. Ashby's [49] 'homeostat' is a beautiful example of a mechanism embodying this principle, and he suggests that limit cycles are the rule rather than the exception in complex coupled systems. Bremermann, Rogson and Salaff [50] and Clapper [28] have demonstrated the 'frustration' effect in deterministic, or insufficiently stochastic, learning automata. In particular, the results obtained formalize the requirement for an unconstrained random search mode in designs for universal learning automata such as Andreae's STeLLA [34].

In conclusion one may suggest that it is dangerous for the observer of a biological system to assume that stochastic behavior is 'noise' resulting from defects in the system or his imperfect observations— it may be playing a constructive and vital role. Equally the system engineer should consider that a stochastic element may provide a solution to problems otherwise requiring complex algorithms and associated storage. The pedagogue may take heart from the argument that the classic approach of presenting material in one way and, when it is not assimilated, trying another, etc., is not so 'unscientific' and inefficient as it may appear!

4. Conclusions

This chapter has been a wide-ranging, light-hearted and speculative survey of the role of randomness in science and system theory. It has shown both the destructive aspects of randomness if we reject it in modelling and the constructive aspects that made it an important factor in system design and function. If this paper has given some greater character, personality and life to the concept of randomness, and reduced the drabness of its role as a signal-destroying parasite called 'noise', then it will have achieved its objective. We need far richer results on random and general non-deterministic systems if system theory is to have wide application and automata theory is to deliver what it has always seemed to promise.

Acknowledgements

We are grateful to John Andreae and Judea Pearl for discussions that have contributed to the substance of this chapter.

5. References

1. NAGEL, E. *The Structure of Science*, London: Routledge and Kegan Paul (1961).
2. MARTIN-LOF, P. 'The definition of random sequences', *Information and Control* 9, pp.602-619 (1966).
3. KOLMOGOROV, A. 'Logical basis for information theory and probability theory', *IEEE Trans. Inf. Theor.* Vol.IT-14, pp.662-664 (1968).
4. WILLIS, D. 'Computational complexity and probability constructions', *JACM* 17, pp.241-259 (1970).
5. CHAITIN, G. 'A theory of program size formally identical to information theory', *JACM* 22, pp.329-340 (1975).
6. GAINES, B.R. 'Approximate identification of automata', *Electronics Letters* 11, pp.444-445 (1975).
7. GAINES, B.R. 'Behavior/structure transformations under uncertainty', EES-MMS-AUT-75, Department of Electrical Engineering, University of Essex, Colchester, UK (November 1975).
8. HACKING, I. *The Emergence of Probability*, Cambridge University Press (1975).
9. HUME, D. *A Treatise of Human Nature*, London (1739).
10. SUPPES, P. *Probabilistic Metaphysics*, Filosofiska Studier, Sweden: Uppsala University (1974).
11. MICHOTTE, A. *The perception of Causality*, London: Methuen (1963).
12. GAINES, B.R. 'On the complexity of causal models', IEEE Trans. Systems, Man and Cybernetics, January 1976, Vol.SMC-6.
13. DARWIN, F. *Life and Letters of Charles Darwin*, 1888, Vol.11, p.378, London: Collins (1972).
15. LEWIS, J. (ed), *Between Chance and Necessity*, London: Garnstone Press (1974).
16. AYALA, F.J., and DOBZHANSKY, T. (eds), *Studies in the Philosophy of Biology*, London: MacMillan (1974).
17. FREUD, S. *Psychopathology of Everyday Life*, London: Ernest Benn (1914).
18. SCHILPP, P. (ed), *Albert Einstein, Philosopher-Scientist*, Illinois: Evanston, p.176 (1949).
19. RABIN, M.O. 'Probabilistic automata', *Information and Control* 6, pp.230-245 (1963).
20. GAINES, B.R. 'Stochastic computing systems', *Advances in Information Systems Science* (ed. J.T. Tou), Vol.2, pp.37-172 (1969).
21. POPPELBAUM, W.J. 'Statistical processors', Department of Computer Science, University of Illinois, Urbana, Illinois, USA, (May 1974).

22. SCHUMANN, W.R. 'Method and apparatus for averaging a series of transients', United States Patent 3182181, May 5, 1965.
23. HIRSCH, J.J., CHEVALIER, G., and OLIVA, I. 'New types of digital to analog and analog to digital convertors using MOS/LSI technology', Proc. EPN Seminar, Paris (April 1972).
24. TUMFART, S. 'New instruments use probabilistic principles', *Electronics* **48**, pp.86-91 (July 1975).
25. MASSEN, R. 'Stochastic fluidic computing systems', Proc. 5th Cranfield Fuidics Conference, 1972, pp.G3 45-56.
26. HELLMAN, M.E., and COVER, T.M. 'On memory saved by randomization', *Annals of Mathematical Statistics*, 1971, **42**, pp.1075-1078.
27. GAINES, B.R. 'Techniques of identification with the stochastic computer', Proc. IFAC Symposium on Identification, Prague, June 1967.
28. CLAPPER, G.L. 'Machine looks, listens, learns', *Electronics*, 30 October 1967, pp.91-102.
29. GAINES, B.R. 'Memory minimization in control with stochastic automata', *Electronics Letters*, 1971, **7**, pp.710-711.
30. GOLD, E.M. 'Universal goal seekers', *Information and Control*, 1971, **18**, pp.395-403.
31. FOERSTER, H. von,'On self-organizing systems and their environments', *Self Organizing Systems* (eds. M.C. Yovits and S. Cameron) London: Pergamon Press, pp.31-50 (1960).
32. BLOCK, H.D. 'Simulation of statistically composite systems', *Prospects for Simulation and Simulators of Dynamic Systems',* (eds. G. Shapiro and M. Rogers), New York: Spartan Books, pp.23-68 (1967).
33. GUPTA, M.S. 'Applications of electrical noise', *Proc. IEEE,* 1975, **63**, pp.996-1010.
34. GAINES, B.R., and ANDREAE, J.H. 'A learning machine in the context of the general control problem'. Proc. 3rd International Congress IFAC, London 1966.
35. ARBIB, M.A. and ZEIGER, H.P. 'On the relevance of abstract algebra to control theory' *Automatica*, 1969, **5**, pp.589-606.
36. GOGUEN, J.A. 'Realization is universal'. *Mathematical System Theory*, 1973, **6**, pp.359-374.
37. FU, K.S. and BOOTH, T.L., 'Grammatical inference: introduction and survey—part II' *IEEE Transactions on Systems, Man and Cybernetics*, 1975, Vol.SMC-5, pp.409-423.
38. FINE, T.L. *Theories of Probability*, New York: Academic Press (1973).
39. WEIZENBAUM, J. 'ELIZA—a computer program for the study of natural language communication between man and machine.' *Comm. ACM, 1966,* **9** , pp.36-45.
40. PUTNAM, M. 'Robots: machines or artificially created life?' *Journal of Philosophy,* 1964 **61**, pp.668-691.
41. GAINES, B.R. and FACEY, P.V. 'Some experience in interactive system development and application' *Proc. IEEE* 1975, **63**, pp.894-911.
42. BHATIA, N.P. and SZEGO, G.P. *Dynamical systems: stability theory and applications.* Heidelberg: Springer-Verlag (1967).
43. GAINES, B.R. 'Training, stability and control'. *Instructional Science,* 1974, **3**, pp.151-176.
44. PASK, G. 'Man as a system that needs to learn' *Automaton Theory and Learning Systems* (ed. D.J. Stewart), London: Academic Press, pp.137-208 (1967).
45. MOORE, E.F. 'Gedanken experiments on sequential machines'. *Automata Studies* (eds. C.E. Shannon and J. McCarthy), Princeton University Press, pp.129-153 (1956).
46. CONWAY, J.H. *Regular Algebra and Finite Machines,'* London: Chapman and Hall, (1971).
47. WITTEN, I.H. 'Learning to control sequential and non-sequential environments', EES-MMS-CON-75, Department of Electrical Engineering, University of Essex, Colchester, UK, 1975.
48. DIJKSTRA, E.W. 'Hierarchical ordering of sequential processes', *Acta Informatica*, 1971, **1**, pp.115-138.
49. ASHBY,W.R. *Design for a Brain*, London: Chapman and Hall, (1960).
50. BREMERMANN, H.J. ROGSON, M., and SALAFF, S. 'Search by Evolution', *Biophysics and Cybernetic Systems* (eds. M. Maxfield, A. Callahan, and L.J. Fogel), Washington: Spartan Books, pp.157-167 (1965).
51. MONOD, J., *Chance and Necessity,* Collins, London (1972).

Procedure for generating hypothetical structures in the structure identification problem§

GEORGE J. KLIR and HUGO J.J. UYTTENHOVE
School of Advanced Technology,
State University of New York, Binghamton, New York, USA

1. Introduction

1.1 Epistemological levels of systems

During previous efforts to develop conceptual foundations for systems problem solving, a *hierarchy of epistemological levels of systems* has become inherent [1-5].

At the lowest level in this hierarchy, denoted as level 0, a system is characterized by a set of variables, set of potential states associated with each variable and descriptive mappings of these sets to real world attributes and their manifestations. This level is devoid of any knowledge pertaining to the relationship among the states of the variables. The term *source system* is used to refer to systems defined at this level.

Systems defined at different higher epistemological levels are distinguished from each other by the level of knowledge regarding the variables of the source system. A higher level system entails all knowledge of the corresponding systems at any lower level and contains some additional knowledge which is not available at the lower levels.

When a source system is supplemented by data, i.e., by actual arrays of states, either observed or desirable, which the involved variables produce, we consider the new system defined at epistemological level 1; systems defined at this level are called *data systems*.

Knowledge about some invariant properties through which the data can be generated, refers to higher levels. These properties may include time-invariant or space-invariant relations or generative relations among some variables invariant with respect to other variables.

At epistemological level 2, the invariant properties are described through a single generative relation among the variables of the source system and possibly some other variables. These other variables are obtained by shifting appropriate variables in the state subspace of those variables with respect to which the invariance is considered. Since systems at this level constitute a basis for generating data, they are called

§The research reported in this chapter was done at the Netherlands Institute for Advanced Studies in Wassenaar, The Netherlands.

generative systems.

A system at level 3 in the hierarchy consists of a set of generative systems (inferring lower level systems as well) which are regarded as elements of a larger system and some relation among them. At this level, the relations among the generative systems may be represented by direct couplings in terms of shared variables between the elements. Such variables prescribe a form of data exchange among the elements in addition to a manner in which the generative relations associated with the individual elements are composed. Systems at this level are called *structure systems.* Systems can be defined at still higher epistemological levels [3,4] but such systems are of no interest in this chapter.

1.2 Environment, neutral and directed systems

It is apparent that, regardless of the level at which a system is defined, states of some variables of the system may be determined by an outside source which is not part of the system under consideration. This outside source is referred to as the *environment.*

In fact, the environment of a system can be viewed as another system which is coupled to the system under consideration through the variables shared by these two systems. Although the environment can be recognized at any of the epistemological levels it is usually known only as a source system.

The task of identifying which variables of a system are determined by the environment (i.e., input variables) and which are determined through their relationship to other variables of the system (i.e. output variables), may prove difficult or impossible.

Systems which have some variables classified into input and output variables are called *directed systems,* while systems for which no such classification is known, are called *neutral systems.* Their validity in classification holds for each of the levels in the hierarchy of epistemological levels.

1.3 Structure identification problem

The structure identification problem, as introduced in [1], can loosely be described as follows: Given a neutral data system, identify the best representation of this system by a directed structure system whose elements are associated with subsets of variables of the data system.

The structure identification problem is clearly characterized by climbing up the hierarchy of epistemological levels, from the data system (level 1) to the structure system (level 3). It is a problem associated with empirical investigations; it encompasses the problem of transition from level 1 to level 2, which is the subject of [2]. Moreover, it may involve a transition from a neutral system to a directed system.

A procedure for solving the structure identification problem is suggested in [1]. It consists of the following three major subproblems:

1. An overall neutral generative system is determined for the given neutral data system by which the latter is best represented within given constraints.
2. All meaningful hypothetical structure systems are derived and conveniently ordered. Each of them includes all variables involved in the generative system. Elements of the individual structure systems are generative systems; each of them is identified by a subset of variables of the overall generative system and the empirical invariant relation derived for these variables from the given data.
3. Individual hypothetical structure systems are analyzed (relations of their elements are composed according to the configuration of couplings) and the resulting hypothetical relation is compared, using some convenient measure of conformation, with the corresponding empirical relation based directly on the data system.

Procedures described in [2] are directly applicable for solving subproblem (1); if desirable, they can be augmented by allowing an introduction of hypothetical internal (unobserved) variables.

Subproblems (2) and (3) are based on the assumption that a generative system representing the given data system has already been fixed. We shall refer to these two subproblems as the *structure identification problem proper.* While an overall approach to solving this problem is described in [1], no specific procedure is given for generating meaningful hypothetical structures. A derivation of such procedure is a subject of this chapter.

2. Hypothetical structure systems

2.1 Structure system

Let $E=\{e_1,e_2,...,e_m\}$ be a nonempty set of identifiers of elements involved in a structure system and let e_0 stand for the environment of the system. Let Γ_N be a nonempty set of neutral generative systems (see [1,4]). Each element of the structure system is one of these generative systems; the identification of a particular system for each element is given in terms of function

$$f_N : E \rightarrow \Gamma_N$$

Let X_i denote the set of external (observed) variables involved in the generative system associated with element e_i $(i=1,2,...,m)$ and let X_0 denote the set of variables of the environment. Then a neutral coupling $C_{i,j}$ between elements e_i and e_j is defined as the set of variables the two elements share, i.e.,

$$C_{i,j} = X_i \cap X_j \tag{1}$$

Let C denote the matrix $[C_{i,j}]$ of couplings between all pairs of elements as well as between the individual elements and the environment $(i,j=0,1,...,m)$. Variables in couplings of structure systems will be called *coupling variables*; matrix C will be called the *coupling matrix*.

The neutral structure system S_{3N} is defined as quintuple

$$S_{3N} = (E, e_0, \Gamma_N, f_N, C) \tag{2}$$

The following remarks attempt to clarify certain nuances of this definition which might be overlooked or misinterpreted:

1. The definition does not assign any system to the environment (f_N does not include e_0 in its domain). This is justified by taking the following position: *In case of neutral systems,* the concept of environment is reserved for the identification of variables in which the investigator is interested within a given context. As such, the environment serves as a reference with respect to which the analysis of the structure system is to be performed. No generative system is assigned to the environment because such system is determined when generative systems associated with the regular elements $(e_1,e_2,...,e_m)$ are composed so that all variables except those included in couplings with the environment are eliminated. If the environment were directly defined as a generative system independently of systems assigned to the regular elements and of couplings between the elements, the whole structure system would be overdetermined and a subject of likely inconsistencies.
2. Couplings of *neutral* structure systems are symmetrical, i.e., $C_{i,j}=C_{j,i}$ for all $i,j=0,1,...,m$.
3. For convenience, we shall define $C_{i,i}=\phi$ for every $i=0,1,...,m$.
4. If the set E contains only one element, the structure system degenerates to the generative system assigned to the element by f_N.
5. Sets X_i of variables associated with elements e_i $(i=0,1,...,m)$ can be determined from the couplings by the formula

$$X_i = \bigcup_j C_{i,j} \tag{3}$$

Directed structure systems differ from their neutral counterparts in two respects:

1. Each element is a directed generative system.
2. Couplings between elements are directed. A directed coupling $D_{i,j}$ between elements e_i, e_j is defined by

$$D_{i,j} = Z_i \cap Y_j, \tag{4}$$

where Z_i, Y_j stand for the set of output variables of element e_i and the set of input variables of element e_j, respectively. Clearly

$$Y_j = \bigcup_i D_{i,j}, \tag{5}$$

$$Z_i = \bigcup_j D_{i,j}$$

The *directed structure system* S_{3D} is defined as quintuple

$$S_{3D} = (E, e_0, \Gamma_D, f_D, D) \tag{6}$$

where Γ_D is a set of directed generative systems, f_D is a mapping $E \rightarrow \Gamma_D$, and D is a matrix $[D_{i,j}]$ of directed couplings between all pairs of elements including the environment; it will be called the *directed coupling matrix*.

Although sets Γ_N or Γ_D in S_{3N} or S_{3D}, respectively, may contain systems defined at levels 2, 1 or 0, we are interested in this chapter only when they are all defined at level 2 (generative systems).

2.2 Structure candidates

Let a neutral data system and a generative system representing the whole data system in terms of a particular sampling mask (determined previously [2]) be given. Furthermore, let a structure system be given. In order to be a meaningful candidate for a hypothetical structural process through which the data system can be implemented, within the constraints of the chosen mask, the structure system must be compatible with both the data system and the mask in four respects:

(i) The set of all coupling variables of the structure system is equal to the set of variables of the data system.

(ii) Each element of the structure system is identified by a nonempty subset of variables of the data system.

(iii) No element is identified by a subset of variables related by the other elements of the same structure system.

(iv) Subsets of variables of the data system attached to the individual elements of the structure system identify specific portions (submasks) of the overall sampling mask on the basis of which generative systems representing the elements are determined from the data system.

Structure systems which satisfy all of these requirements will be called *structure candidates* of the given data system subject to the constraints of the chosen mask. As no directions of variables are required, structure candidates are neutral structure systems. Each of them stands for a set of directed structure systems; they are obtained when coupling variables are assigned all possible directions.

Structure candidates $S_{3N} = (E, e_0, \Gamma_N, f_N, C)$ can be represented in a number of ways. The representation used in this chapter is characterized as follows: Elements of the structure system (elements of E) and the environment (e_0) are identified in terms of subsets of variables of the data system; generative systems in set Γ_N and mapping f_N are of no interest in this chapter; couplings follow directly from the identifiers of the elements.

2.3 *Ordering of structure candidates*

When n variables identify an element of a structure candidate, these variables are assumed to be related by a single relation; we say they are *directly related* in the structure candidate.

Applying this concept, structure candidates can be usefully ordered by partial ordering $\leqslant$ defined as follows:

Let C_i and C_j denote two structure candidates. Then, $C_i \leqslant C_j$ iff sets of directly related variables in C_i are all subsets of those directly related in C_j.

The relation is reflexive since the definition does not refer to proper subsets; it is antisymmetric because it is based on the relation 'being a subset of' which is antisymmetric; it is transitive for the same reason. Hence, it is indeed a partial ordering.

When $C_i \leqslant C_j$ we may use a suggestive phrase 'C_i is a more specialized structure candidate than C_j'.

The set of all structure candidates for a given set of variables together with the partial ordering 'being more specialized than' is a lattice.

The maximal and last candidate in the lattice is always candidate C_0, which consists of one element identified by all variables of the data system. It is the least specialized (or most general) structure candidate; it actually represents an interface between the overall mask identification (see [2]) and the structure identification proper.

The minimal and first element in the lattice is always the candidate whose number of elements is equal to the number of variables of the data system; each of its elements is identified by a single variable and there are no couplings between the elements.

2.4 *Lattices of structure candidate for three and four variables*

Let L_n denote the lattice of structure candidates for data systems consisting of n variables. To derive lattice L_n, we can make a list of all possible elements, each one associated with one nonempty subset of the variables involved. Then, we select all subsets of the elements which satisfy the requirements specified in Section 2.2.

Lists of elements for structure candidates of three and four variables are given in Tables I and II respectively; 1's in the tables indicate variables which are associated with the individual elements. Lattices

Table I List of elements for structure candidates of three variables

Elements	Variables x y z
0	1 1 1
1	1 1 0
2	1 0 1
3	0 1 1
4	1 0 0
5	0 1 0
6	0 0 1

Table II List of elements for structure candidates of four variables

Element	Variables x	y	z	w
0	1	1	1	1
1	1	1	1	0
2	1	1	0	1
3	1	0	1	1
4	0	1	1	1
5	1	1	0	0
6	1	0	1	0
7	1	0	0	1
8	0	1	1	0
9	0	1	0	1
10	0	0	1	1
11	1	0	0	0
12	0	1	0	0
13	0	0	1	0
14	0	0	0	1

L_3 and L_4 are defined in Tables III and IV respectively. The candidates in each lattice are partitioned into levels such that candidates at level k have immediate successors only at level $k+1$ $(k = 0,1,...)$.

Table III Lattice L_3 of structure candidates

Candidate	Level	Elements	Numbers of variables	Immediate successors
0	0	0	3	1, 2, 3
1	1	1, 2	2, 2	4, 5
2		1, 3		4, 5
3		2, 3		5, 6
4	2	1, 6	2, 1	7
5		2, 4		7
6		3, 5		7
7	3	4, 5, 6	1, 1, 1	none

Table IV Lattice L_4 of structure candidates

Candidate	Level	Elements	Numbers of variables	Immediate successors
0	0	0	4	1,2,3,4,5,6
1	1	1,2	3,3	7,8,10,11
2		1,3		7,9,13,14
3		1,4		8,9,16,17
4		2,3		10,12,13,15
5		2,4		11,12,16,18
6		3,4		14,15,17,18
7	2	1,7	3,2	19,23,26,31
8		1,9		19,24,28,33
9		1,10		19,25,29,34
10		2,6		20,23,24,32
11		2,8		20,26,28,36
12		2,10		20,27,30,38
13		3,5		21,23,25,27
14		3,8		21,31,34,37
15		3,9		21,32,35,38
16		4,5		22,28,29,30
17		4,6		22,33,34,35
18		4,7		22,36,37,38
19	3	1,14	3,1	42,44,47
20		2,13		43,45,49
21		3,12		46,48,50
22		4,11		51,52,53
23		5,6,7	2,2,2	42,43,46
24		5,6,9		40,42,45
25		5,6,10		39,42,48
26		5,7,8		41,43,44
27		5,7,10		39,43,50
28		5,8,9		44,45,51
29		5,8,10		39,44,52
30		5,9,10		39,45,53
31		6,7,8		41,46,47
32		6,7,9		40,46,49
33		6,8,9		40,47,51
34		6,8,10		47,48,52
35		6,9,10		40,48,53
36		7,8,9		41,49,51
37		7,8,10		41,50,52
38		7,9,10		49,50,52
39	4	5,10	2,2	54,59
40		6, 9		55,58
41		7, 8		56,57
42		5, 6,14		54,55
43		5, 7,13	2,2,1	54,56
44		5, 8,14		54,57
45		5, 9,13		54,58
46		6, 7,12		55,56
47		6, 8,14		55,57
48		6,10,12		55,59
49		7, 9,13		56,58
50		7,10,12		56,59
51		8, 9,11		57,58
52		8,10,11		57,59
53		9,10,11		58,59
54	5	5,13,14	2,1,1	60
55		6,12,14		60
56		7,12,13		60
57		8,11,14		60
58		9,11,13		60
59		10,11,12		60
60	6	11,12,13,14	1,1,1,1	none

3. Procedure for generating hypothetical structure systems

3.1 General discussion

The overall procedure for structure identification in empirical data described in [1] requires that either a catalog of lattices L_n for all desirable numbers n be available or, alternatively, that a procedure be available through which immediate successors are generated for an arbitrary structure candidate in L_n for each desirable n.

The number of structure candidates in L_n grows quite rapidly as n increases. As a consequence, the catalog of lattices is feasible only for small numbers n, say $n \leqslant 5$. Hence, we focus on the second alternative—a procedure for generating immediate successors of a given structure candidate in L_n.

Given a structure candidate, say C, it is obvious that its successors are derivable by reducing subsets of directly related variables in C. Consider now a successor of C, say C_r. Candidate C_r is an immediate successor of C iff there is no other candidate derivable from C, say C'_r, such that C_r is a successor of C'_r. Immediate successors of C are thus obtained by the smallest possible modifications of C.

A smallest possible modification is clearly a modification in a single element of C. Indeed, if two or more elements of C are modified, the obtained successors are derivable from other successors obtained by modifying only one of the elements.

It turns out that the smallest possible modifications of an element are its decompositions into two new elements each, associated with the largest possible subsets of the set of variables representing the decomposed element. This operation is described more precisely in the following section.

3.2 Decomposition

Consider a candidate lattice L_n based on a set $V=\{v_1,v_2,...,v_n\}$ of n variables. Assume that a particular candidate of L_n, say candidate C, consists of m elements identified by the identifier i $(i=1,2,3,...,m)$. Assume furthermore that each element of C is defined in terms of a nonempty subset, say E_i, of the variables involved, i.e., $E_i \subset V$. It is required (see Section 2.2) that

$$\bigcup_i E_i = V$$

and

$$E_i \not\subset E_j$$

for every pair i,j of distinct elements $(i \neq j)$. Due to the last property, we conveniently simplify phrases such as 'element i associated with the set E_i of vari-

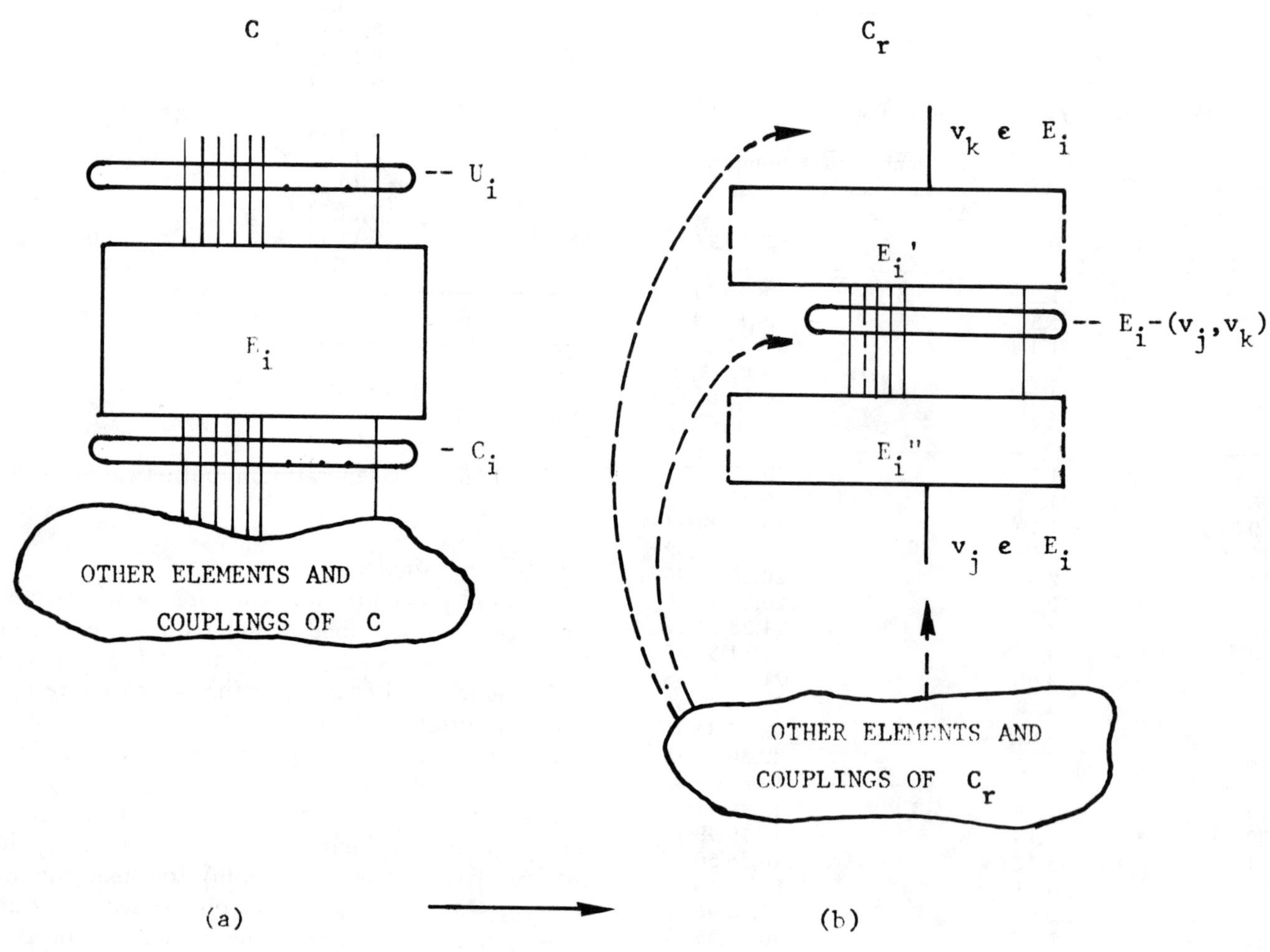

Fig.1 The Generative Rule

ables' by using the phrase 'element E_i' instead.

Consider now the general situation shown in Fig. 1*a* and assume that element E_i of structure candidate C is to be decomposed into two elements. Let the elements which replace E_i when the decomposition is performed be denoted as E_i' and E_i''. To obtain an immediate successor of C, say candidate C_r, it is required that elements E_i' and E_i'' contain as many variables taken from the set E_i as possible. This can be achieved by including as many variables of E_i as possible in the coupling between the elements of E_i' and E_i''. At the same time, however, E_i' must contain at least one variable which is not included in E_i'' and vice versa. To satisfy all these requirements the following simple rule can be used to generate an immediate successor C_r of structure candidate C, characterized in Fig. 1*a*, by decomposing its element E_i into elements E_i', E_i'':

Take two distinct variables of E_i, say v_j and v_k, and define

$$E_i' = E_i - \{v_j\}$$

and

$$E_i'' = E_i - \{v_k\}$$

The situation obtained by applying this rule is illustrated in Fig. 1*b*. Let the rule be referred to as the *generative rule.*

To generate all immediate successors of structure candidate C, the generative rule has to be applied to *each pair of variables* associated with *each element* of C.

Although we may be able to obtain a structure candidate C_r by the generative rule, it still must satisfy the requirements specified in Section 2.2 to be a meaningful structure candidate. It is clear that (i) is not violated because the generative rule does not exclude any variable. Since we select two distinct variables of E_i and assign one to E_i' and one to E_i'', the new elements of C_r are still identified by a nonempty set of variables of the data system; hence, (ii) is not violated. It is clear that when $|E_i|=1$ no selection of distinct variables v_j and v_k is possible and hence the generative rule is not applicable to such an element.

As it is not guaranteed that (iii) is satisfied, both elements obtained by the generative rule must be checked with respect to this property.

Element E_i' may be a subset of variables related (directly or indirectly) by the element

$$C_r - \{E_i'\}$$

If indeed it is such a subset, element E_i' must be deleted from C_r. This verification must also be carried out for elements E_i'' with respect to the variables related by the elements $C_r - \{E_i''\}$ where C_r may already be without E_i'. If E_i'' turns out to be a subset, it too must be deleted from the structure candidate C_r. In this manner, the validity of (iii) is retained by the final C_r.

Requirement (iii) can also be satisfied by making the appropriate adjustment in the set $E_i' \cap E_i''$, as shown in Section 3.3.

Requirement (iv) naturally holds as the generative rule only redefines elements and their identification of specific portions of the overall sampling mask.

3.3 *Procedure*

Given structure candidate $C=\{E_i | E_i \subset V, E_i \neq \emptyset, i=1,2,...,m\}$ where V is the set of variables of a given data system and the sets E_i are used as identifiers of the elements of C, the following procedure generates all immediate successors C_r of C:

(1) Let $i=0$ and $r=1$.
(2) Put $i+1 \rightarrow i$; if $i>m$, go to (19).
(3) Let $E_i=\{v_1,v_2,...,v_{|E_i|}\}$; if $|E_i|=1$, go to (2).
(4) Let $j=0$.
(5) Put $j+1 \rightarrow j$; if $j=|E_i|$, go to (2).
(6) Put $j \rightarrow k$.
(7) Put $k+1 \rightarrow k$; if $k>|E_i|$, go to (5).
(8) Replace E_i by $E_i'=E_i-\{v_j\}$ and $E_i''=E_i-\{v_k\}$ and let $C_r=\{C-\{E_i\}\} \cup \{E_i', E_i''\}$.
(9) Determine the set of variables related (directly or indirectly by the set

$$C_r - \{E_i'\}$$

let U denote this set of variables.
(10) If $E_i' \subset U$, put $\{C_r - \{E_i'\}\} \rightarrow C_r$.
(11) Redefine U as the set of variables related (directly or indirectly) by the set $C_r - \{E_i''\}$.
(12) If $E_i'' \subset U$, put $\{C_r - \{E_i''\}\} \rightarrow C_r$.
(13) Store C_r as a potential immediate successor of C.
(14) Put $r+1 \rightarrow r$.; if neither E_i' or E_i'' has been excluded from C_r continue with step (7).
(15) If E_i' (or E_i'') can be excluded from C_r in step (8), determine the set of variables of E_i'' (or E_i') related to v_k (or v_j) through the elements $C-E_i$. Denote this set by A.
(16) If $A \neq \emptyset$, redefine E_i'' (or E_i') as $\{E_i''-\{A\}\} \rightarrow E_i''$ (or $\{E_i'-\{A\}\} \rightarrow E_i'$) and store C_r as a potential

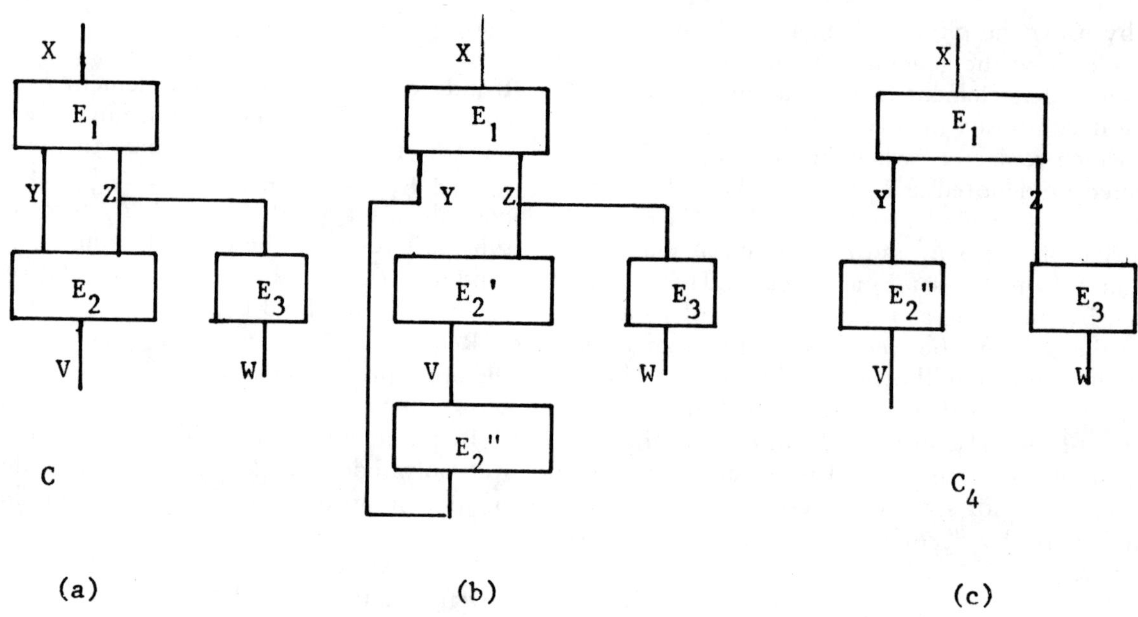

Fig.2 Illustration of generation of an immediate successor (C_4) from a structure candidate C in L_5

successor of C, else go to step (7).

(17) Put $r+1 \rightarrow r$ and go to (7).

(18) Reduce the set of all potential structure candidates to a set of pair-wise distinguishable candidates.

(19) End procedure.

3.4 Examples

Let structure candidate C defined by the block diagram in Fig. 2*a* be considered. Clearly,

$$C = \{E_i | E_i \subset V,\ E_i \neq \emptyset,\ i=1,2,3\},$$

where $V=\{x,y,z,v,w\}$, $E_1=\{x,y,z\}$, $E_2=\{y,z,v\}$ and $E_3=\{z,w\}$; C is taken from lattice L_5. When the procedure defined in Section 3.3 is applied to this candidate, its immediate successors in L_5 are generated as follows (detail description is used for $r=1$, followed by a simplified description).

(1) $i = 0$ and $r = 1$.
(2) $i = 1$; as condition $i > 3$ is not satisfied, go the next step.
(3) $E_1 = \{v_1,v_2,v_3 | v_1 = x,\ v_2 = y,\ v_3 = z\}$; as condition $|E_1| = 1$ is not satisfied, go the next step.
(4) $j = 0$.
(5) $j = 1$; as condition $j = |E_1|$ is not satisfied, go to the next step.
(6) $k = 1$.
(7) $k = 2$; as condition $k > |E_1|$ is not satisfied, go to the next step.
(8) Replace E_1 by $E_1' = \{v_2,v_3\}$ and $E_1'' = \{v_1,v_3\}$; $C_1 = \{E_1', E_1'', E_2, E_3\}$.
(9) $U = V$.
(10) As $E_1' \subset U$, put $\{E_1'', E_2, E_3\} \rightarrow C_1$.
(11) $U = \{y,z,v,w\}$.
(12) As $E_1'' \not\subset U$, C_1 does not change.
(13) C_1 is stored.
(14) $r = 2$, E_1' has been excluded and go to step (15).
(15) Since set $A = \emptyset$ go to step (7).
(7) $k = 3$; go to next step.
(8) $E_1' = \{x,z\}$; $E_1'' = \{x,y\}$; $C_2 = \{E_1', E_1', E_2, E_3\}$.
(9) $U = \{x,y,z,v,w\}$.

(10) $E'_1 \subset U$; put $\{E''_1, E_2, E_3\} \rightarow C_2$.
(11) $U = \{y,z,y,w\}$.
(12) $E'' \not\subset U$; C_2 is not changed.
(13) C_2 is stored.
(14) $r=3$; Since E'_1 has been excluded, go to step (15).
(15) Since $A = \emptyset$, go to step (7).
(7) $k = 4$; go to (5).
(5) $j = 2$; go to the next step.
(6) $k = 2$; go to the next step.
(7) $k = 3$; go to the next step.
(8) $E'_1 = \{x,y\}$; $E''_1 = \{y,v\}$; $C_3 = \{E'_1, E''_1, E_2, E_3\}$.
(9) $U = \{x,y,z,w,v\}$.
(10) $E'_1 \subset U$; put $\{E''_1, E_2, E_3\} \rightarrow C_3$.
(11) $U = \{y,v,z,w\}$
(12) $E''_1 \not\subset U$; C_3 does not change.
(13) C_3 is not stored since $C_3 = C_2$.
(14) $r = 4$; since E'_1 has been eliminated, go to step (15).
(15) Since $A = \emptyset$, go to step (7).
(7) $k = 4$; go to (5).
(5) $j = 3$; go to (2).
(2) $i = 2$; go to the next step.
(3) $E_2 = \{v_1, v_2, v_3 | v_1 = y, v_2 = z, v_3 = v\}$.
(4) $j = 0$; go to the next step.
(5) $j = 1$; go to the next step.
(6) $k = 1$; go to the next step.
(7) $k = 2$; go to the next step.
(8) $E'_2 = \{z,v\}$; $E''_2 = \{v,y\}$; $C_4 = \{E_1, E'_2, E''_2, E_3\}$ (see Fig. 2*b*).
(9) $U = \{x,y,z,v,w\}$.
(10) $E'_2 \subset U$; put $\{E_1, E''_2, E_3\} \rightarrow C_4$ (see Fig. 2*c*).
(11) $U = \{x,y,z,w\}$
(12) $E''_2 \not\subset U$; C_4 is not changed.
(13) C_4 is stored.
(14) $r = 5$; since E'_2 has been eliminated, go to step (15).
(15) Since $A = \emptyset$, to to step (7).
(7) $k = 3$; go to the next step.
(8) $E'_2 = \{v,z\}$; $E''_2 \{y,z\}$; $C_5 = \{E_1, E'_2, E''_2, E_3\}$.
(9) $U = \{x,y,z,w\}$
(10) $E'_2 \not\subset U$; C_5 is not changed.
(11) $U = \{x,y,z,v,w\}$.
(12) $E''_2 \subset U$; put $\{E_1, E'_2, E_3\} \rightarrow C_5$.
(13) C_5 is stored.
(14) $r = 6$; since E''_2 has been eliminated, go to step (15).
(15) Since $A = \emptyset$, go to step (7).
(7) $k = 4$; go to step (5).
(5) $j = 2$; go to the next step.
(6) $k = 2$; "
(7) $k = 3$; "
(8) $E'_2 = \{y,v\}$; $E''_2 = \{y,z\}$; $C_6 = \{E_1, E''_2, E_3\}$.
(9) $U = \{x,y,z,w\}$
(10) $E'_2 \not\subset U$; C_6 is not changed.
(11) $U = \{x,y,z,v,w\}$;
(12) $E''_2 \subset U$; put $\{E_1, E'_2, E_3\} \rightarrow C_6$
(13) C_6 is stored.

(14) $r = 7$; since E''_2 has been eliminated, go to step (15).
(15) Since $A = \emptyset$, go to step (7).
(7) $k = 4$; go to (5).
(5) $j = 3$; go to (2).
(2) $i = 3$; go to the next step.
(3) $E_3 = \{v_1, v_2 \mid v_1 = z, v_2 = w\}$.
(4) $j = 0$; go to the next step.
(5) $j = 1$; "
(6) $k = 1$; "
(7) $k = 2$; "
(8) $E'_3 = \{w\}$; $E''_3 = \{z\}$; $C_7 = \{E_1, E_2, E'_3, E''_3\}$
(9) $U = \{x, y, z, v, \}$.
(10) $E'_3 \subset U$; C_7 is not changed.
(11) $U = \{x, y, z, v, w\}$.
(12) $E''_3 \subset U$; put $\{E_1, E_2, E'_3\} \to C_7$
(13) C_7 is stored.
(14) $r = 8$; since E'_3 has been eliminated, go to step (15).
(15) Since $A = \emptyset$, to to step (7).
(7) $k = 3$; go to step (5).
(8) $j = 2$; go to step (2).
(2) $i = 4$; go to step (19) since $4 > m$.
(19) End.

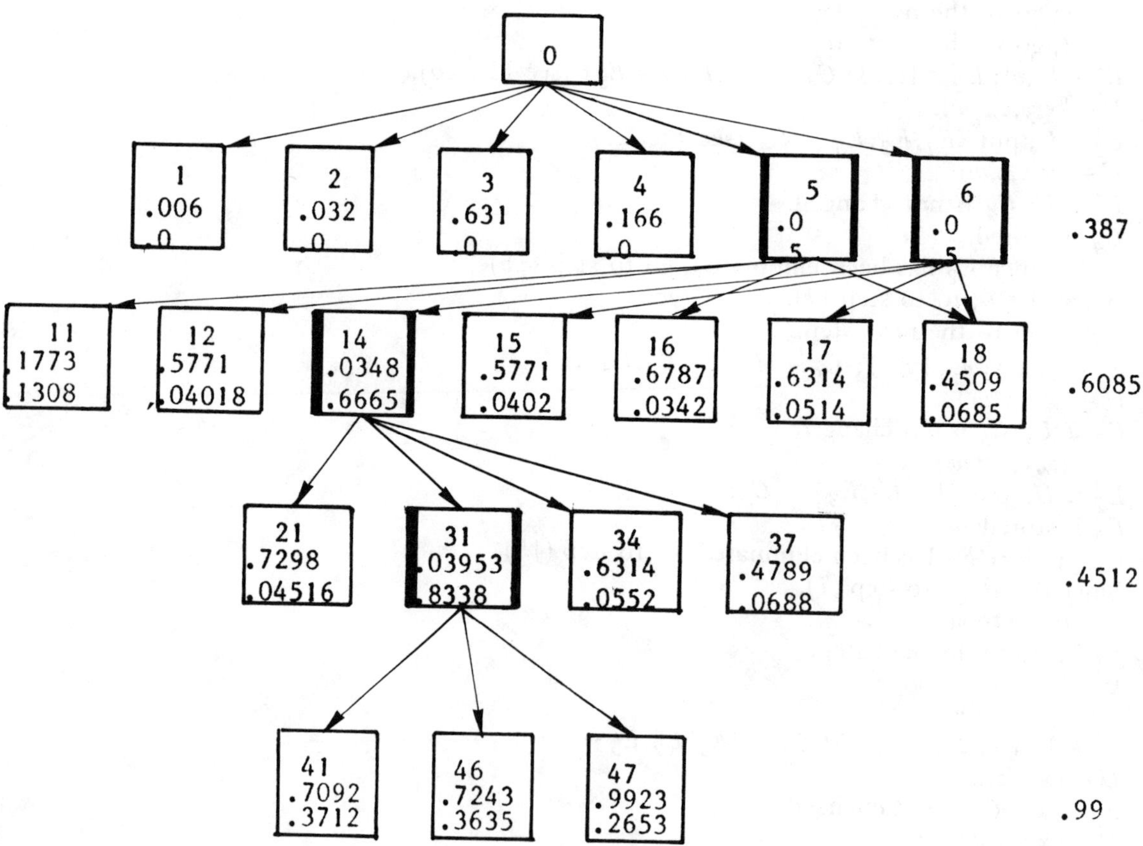

Fig.3 Example of structure identification in L_4

Figure 3 illustrates an example in which the procedure described in this chapter is incorporated in the overall procedure for structure identification described in Ref.[1], is associated with a data system consisting of four variables and 500 samples generated by a computer through structure candidate 31 in lattice L_4 (see Table IV). Blocks in Fig.3 indicate structure candidates in L_4 which have to be evaluated. The first number in each block is the structure candidate identifier, the second number is the distance [1] from the perfect candidate 0, the third number is the degree of confidence [1] associated with the candidate in relation to other potential candidates at the same level. The numbers attached to the individual levels are normalized entropies describing the uncertainty in the decision made at each level.

4. Conclusions

This chapter is the first step in refining the general ideas regarding the structure identification problem as described in Ref. [1]. It focuses on one particular partial problem involved in the overall procedure for structure identification—a problem of generating all immediate successors of a given structure candidate in lattice L_n for an arbitrary n.

References

1. KLIR, G.J., 'Identification of generative structures in empirical data', *International Journal of General Systems,* 3, No.2 (1976).
2. KLIR, G.J., 'On the representation of activity arrays', *International Journal of General Systems,* 2, No.3, pp.149-168 (1975).
3. KLIR, G.J., *An Approach to General Systems Theory,* Van Nostrand Reinhold, New York (1969).
4. ORCHARD, R.A., 'On an approach to general systems theory', in *Trends in General Systems Theory,* edited by G.J. Klir, John Wiley, New York (1972).
5. KLIR, G.J., 'Processing of fuzzy activities of neutral systems', in *Progress in Cybernetics and Systems Research,* Volume 1, edited by R. Trappl and F.R. Pichler, Hemisphere, Washington, D.C., pp.21-24 (1975).

Simplifying data systems: an information theoretic analysis

GERRIT BROEKSTRA
Graduate School of Management, Delft, The Netherlands

1. Introduction

Recently Klir [1] proposed a description of a general methodology of empirical investigation. Like any methodology it rests on a number of premises which, undoubtedly, may serve as a basis for fierce debates among philosophers of science. The issue at stake would be the possibility of a logic of scientific discovery. Although it is very tempting to enter a debate on topics which may revolutionize existing scientific paradigms, the purpose and content of this chapter are much more modest; it may be classified as a product of the ordinary puzzle-solving activities which, at least according to Kuhn, most scientists appear to be engaged in. For some background on a discovery approach to knowledge acquisition we refer to H.A. Simon's paper [2], and for an elaboration in terms of Klir's methodology to his paper [1].

A law-discovery process is, in Simon's terms, a process of finding a pattern in or recoding, in parsimonious fashion, sets of empirical data. Since information theory provides a number of measures of the amount or degree of variability, relation and structure, it is obvious to explore the application of these measures to the process. Important advantages of these measures are their non-negative nature and their applicability to both metric and non-metric variables. We are particularly referring to the concepts of multivariate uncertainty analysis as developed by McGill [3], Garner [4] and Ashby [5,6], et al. Specific earlier applications can be found, for example, in the field of psychology [7]. References to more recent applications in the field of biology can be found, e.g. in Conant [8]. Integration of these developments with Klir's methodology provides a useful generalization in the direction of establishing a 'systems methodology'.

2. Epistemological levels

Assuming an object of investigation and a purpose of investigating, the procedure of empirical investigation consists of a number of phases [1]. Each phase is characterized by a specific system definition. A system may be loosely defined as a 'list of variables' [9,10] plus some additional knowledge about these variables. A number of epistemological levels of increasing order may then be distinguished according to the nature and amount of additional induced knowledge. It has been pointed out [1,11] that, from this point of view, the procedure may be regarded as a process of uncertainty reduction. We will briefly summarize the several phases of the methodology on the basis of the corresponding system defintions. We will restrict the attention to well-defined discrete variables, which have not been classified in input and output variables, i.e. we will discuss non-oriented or neutral discrete-time systems. Further details are found in Klir [1,12].

2.1 *Source system*

A source system is defined on the object of investigation by a set of $n+1$ variables, each taking values of sets of states V_i, $i=0,1,\ldots,n$, and resolution levels L_i, $i=0,1,\ldots,n$. A general source system is defined isomorphically with the (empirical) source system by an ordered set of abstract variables X_i, each associated with a set of states X_i, $i=0,1,\ldots,n$ (we will use the same symbol for a variable and the set of states of the variable). A state of variable X_i is denoted by $x_i \in X_i$, or with time identifier $x_{i,t} \in X_i$, $t \in T$, where T stands for the set of abstract times $T=\{0,1,\ldots,r_t\}$. In general system notation both states and times will be represented by standard sets of symbols, e.g., non-negative integers.

At this stage of investigation (level 0) the investigator decides on an upper bound of uncertainty or potential information as to the states of the system. More specifically, when all possible states of the system $\{X_i \mid i=0,1,\ldots,n\}$ would occur independently and with equal probabilities, the investigator would be faced with the nominal uncertainty as to which state x of the system will occur

$$H_0(X_0,X_1,\ldots,X_n) = \log_2 |X|$$

where $X = X_0 \times X_1 \times \ldots \times X_n$, and $x \in X$. We note that this maximum uncertainty is directly related to the number of categories on the variables. We will demonstrate below that the choice of the number of states for each variable will show its effect throughout the consecutive phases of the procedure.

2.2 *Data system*

In the second phase of the investigation (level 1) data are gathered and may conveniently be organized in an $n+1$ by r_t+1 data matrix or activity array $[x_{i,t}]$, $x_{i,t} \in X_i$, $i=0,1,\ldots,n$; $t \in T$. We define a data system by the ordered pair

$$(\{X_i \mid i=0,1,\ldots,n\},\ [x_{i,t}])$$

Still assuming that the system variables are statistically independent, the investigator may now use the data matrix to obtain estimates of the marginal uncertainty measures for variables X_i, $i=0,1,\ldots,n$,

$$H(X_i) = -\sum_j p(x_j) \log_2 p(x_j)$$

in which $p(x_j)$ is the probability of the jth value of X_j. From a computational point of view it is more reliable to work with observed frequencies k_j of states x_j (summing to k),

$$H(X_i) = \log_2 k - \frac{1}{k} \sum_j k_j \log_2 k_j$$

The uncertainty as to which state of the system will occur at level 1 is given by the sum of marginal uncertainties

$$H_1(X_0,X_1,\ldots,X_n) = \sum_i H(X_i)$$

The uncertainty reduction with respect to level 0, due to the fact that in most cases actual probabilities are not distributed equally, is the difference $H_1 - H_0$, called distributional constraint or redundancy.

2.3 *Generative system*

The third phase of the investigation (level 2) is characterized by the processing of the data with the purpose of pattern detection. A set of sampling variables, $\{S_i \mid i=0,1,\ldots,q\}$, is defined, while states of a sampling variable are specified by $s_{k,t} = x_{i,t+a}$, such that $s_{k,t} \in S_k = X_i$, and k is identified by the pair (i,a). The set of pairs $\{i,a\}$ identifies a *sampling mask* M. In the mask we will distinguish a right-most sampling variable as one which has the largest a for a given i.

The ordered pair consisting of the set of sampling variables together with the mask will be called the *sampling system:*

$$(\{S_i \mid i=0,1,\ldots,q\}, M)$$

A state of the sampling system at reference time t_c $(a=0)$ is a sample of the data matrix $c \in S = S_0 \times S_1 \times \ldots \times S_q$. An exhaustive temporal sampling of the data matrix may result in a number of time-invariant relations for the specific sampling system. Klir [1,12] discerns basic behavior, generative behavior and state-transition relation. In this chapter we will restrict ourselves to generative behavior only. The time-invariant relation for this type of behavior is given by $R \subset S$, $c \in R$.

Let C_r denote the set of right-most sampling variables for a given mask, with a state c_r, and let $\overline{C}_r$ denote the set of sampling variables which is formed by the rest of the sampling system, with a state $\overline{c}_r$.

The estimated probability of state c_r conditional on $\bar{c}_r$ is denoted by $p(c_r|\bar{c}_r)$. *Generative behavior* is now defined by

$$GB = \{(c, p(c_r|\bar{c}_r)|c \equiv (\bar{c}_r, c_r),\ c \in R\}$$

The name is suggestive because generative behavior represents a rule of generating the data. A data matrix is a specific realization of all possible matrices that may be generated by the generative behavior for a specific sampling system. The ordered pair consisting of the sampling system and generative behavior is called the *generative system*

$$((\{S_i\,|i=0,1,...,q\}, M),\ GB)$$

We will associate generative systems with a specific order according to the number of elements in set $\bar{C}_r$. Also, each order specifies a particular sublevel of level 2. A generative system of first order is specified by $\bar{C}_r = \phi$, i.e. when $\bar{C}_r$ contains no elements. The set C_r, on the other hand, contains $n+1$ right-most sampling variables; one sampling variable S_i for each variable X_i, $i=0,1,...,n$, (not necessarily covering one column of the data matrix!). In fact, this type of behavior would be called basic [1].

The joint uncertainty as to which ordered $n+1$-tuple will occur, is given in this case by

$$H_{21}(S_0, S_1,..., S_n) = H(C_r)$$

$S_i = X_i$ $(i=0,1,...,n)$. The uncertainty reduction with respect to level *1* is equal to

$$\Sigma H(X_i) - H_{21}(S_0, S_1,..., S_n)$$

This equation is also the definition of an information measure called total constraint or transmission over $n+1$ variables: $T(S_0 : S_1 : ... : S_n)$, a concept, which we will discuss in more detail below. Since at level 2—or better sublevel 21—one takes into account the correlation between $n+1$ variables, the uncertainty reduction with respect to level *1* is also called correlational constraint or redundancy [4]. An investigator may either decide, for some reason, on the set C_r beforehand or he may search for those sampling variables which maximize the total amount of constraint among them, in other words, which minimize H_{21}. In the latter case, it is convenient to confine the search to an a priori specified rectangular *D*-mask [1], which gives the maximum number of columns of the data matrix one wants to cover with the mask at any instant of time.

Starting from a right-most side C_r, one may look for further reductions of uncertainty by gradually 'filling' the set $\bar{C}_r$ with sampling variables, possibly choosing among the remaining ones of a *D*-mask. A generative system of second order is thus defined by one sampling variable in set $\bar{C}_r$, of third order by two sampling variables in $\bar{C}_r$, etc. The uncertainty as to which state of C_r will occur, given $\bar{C}_r$, is defined by

$$H_{\bar{C}_r}(C_r) = H(\bar{C}_r, C_r) - H(\bar{C}_r) \tag{1}$$

where $H(\bar{C}_r, C_r)$ is the joint uncertainty as to a state of the sampling system, while $H(\bar{C}_r)$ is one's uncertainty about a state $\bar{c}_r$ alone. The information shared by $\bar{C}_r$ and C_r, or, in other words, the degree to which knowledge of $\bar{c}_r$ is helpful in predicting c_r is the well-known non-negative transmission measure defined thus:

$$T(\bar{C}_r : C_r) = H(\bar{C}_r) + H(C_r) - H(\bar{C}_r, C_r) \tag{2}$$

Inserting (1) into (2) gives the uncertainty reduction due to given $\bar{C}_r$ (at sub-level *2i*, when $\bar{C}_r$ contains *i-1* sampling variables) with respect to level 21:

$$T(\bar{C}_r : C_r) = H(C_r) - H_{\bar{C}_r}(C_r) \tag{3}$$

Since the transmission is strongly affected by the manner in which the quantizing of the variables has been performed, it is customary [13] to define the normalized transmission

$$A_{GB} = \frac{T(\bar{C}_r : C_r)}{H(C_r)} \tag{4}$$

Properties of A_{GB} have been discussed in another paper [11], so we summarize here

$$0 \leqslant A_{GB} \leqslant 1 \tag{5}$$

where it is unity if and only if $\bar{C}_r$ determines C_r completely, in which case a perfect generator of the data is obtained.

It is of some interest to consider the uncertainty reduction which may be obtained between two consecutive sublevels when $\bar{C}_r$ is extended with one variable S_k. For the transmission at the next higher-order level we have $T(\bar{C}_r, S_k : C_r)$, which can be partitioned in the following manner

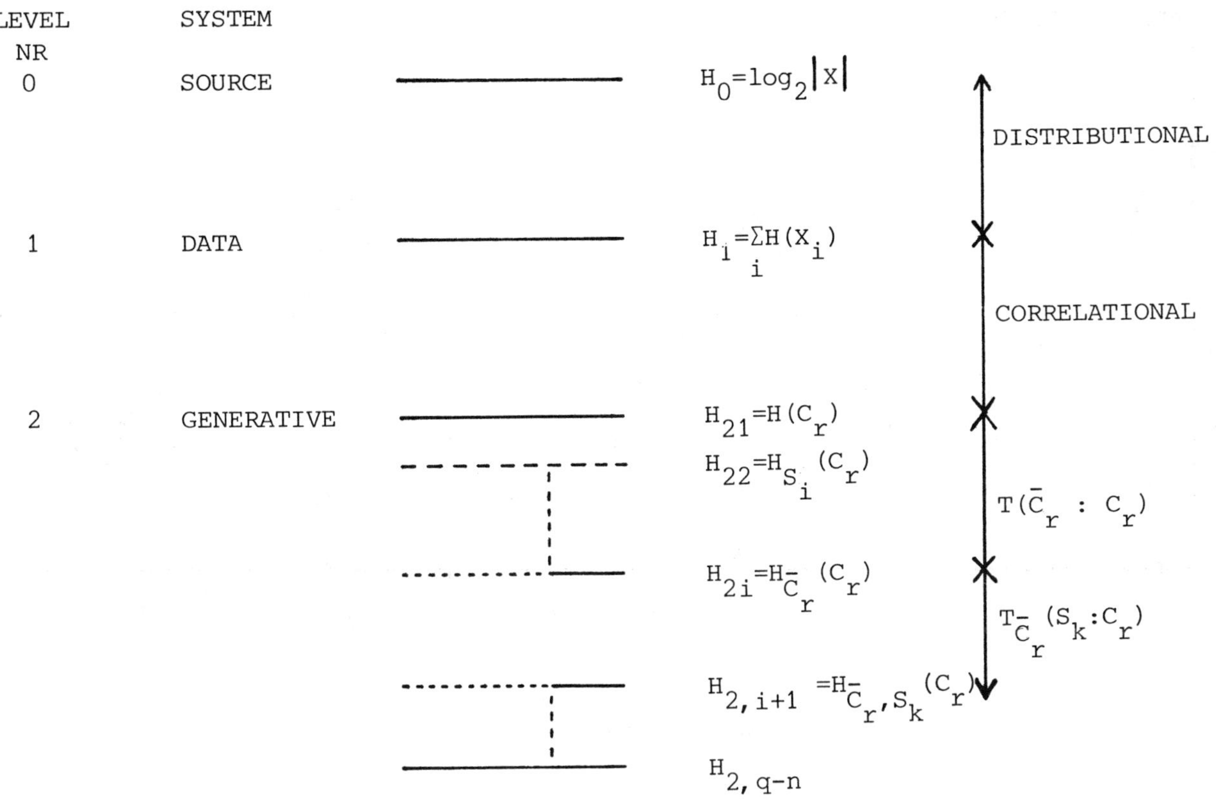

Fig.1 Epistemological levels and the process of uncertainty reduction for a generative behavior

$$T(\bar{C}_r,S_k:C_r) = T(\bar{C}_r:C_r) + T_{\bar{C}_r}(S_k:C_r) \qquad (6)$$

This equation shows that the transmission between $(\bar{C}_r,S_k)$ and C_r at one sub-level equals the transmission between $\bar{C}_r$ and C_r at the preceding sub-level plus the *extra* prediction of C_r which can be obtained from S_k after the effects of $\bar{C}_r$ are held constant. The latter quantity, which is called partial transmission, is also a non-negative quantity.

The process of uncertainty reduction has been summarized schematically in Figure 1.

2.4 *An example*

The previously discussed procedure will be applied to a concrete single variable example. It concerns the analysis of a time series consisting of observations on the amount of banknotes in circulation in the Netherlands. The time series consists of the 256 daily figures (5 days in a week) of the year 1970 as published by De Nederlandsche Bank N.V. Apart from the fact that information measures may also be applied in cases of non-metric variables, the value of a single variable example presumably does not go beyond that of a mere illustration. Well-developed techniques exist for the analysis of a time series, such as that of the banknotes, composed of a trend, a cycle, seasonal and accidental fluctuations [14]. However, it may be of some importance, and increasingly so when more than one variable are considered, to know merely which simultaneous or previous states of the variables involved give the strongest prediction of some other states. This may be of considerable help in the postulation of a stochastic model which underlies the realization of a time series.

A sample of a time series of the amount of banknotes in circulation is shown in Fig.2. Weekly and monthly patterns are clearly suggested by the series. It is also known [14] that the time series contains a slowly rising trend and a regular seasonal cycle with a basic period of one year. In an attempt to improve ergodicity of the stochastic process the

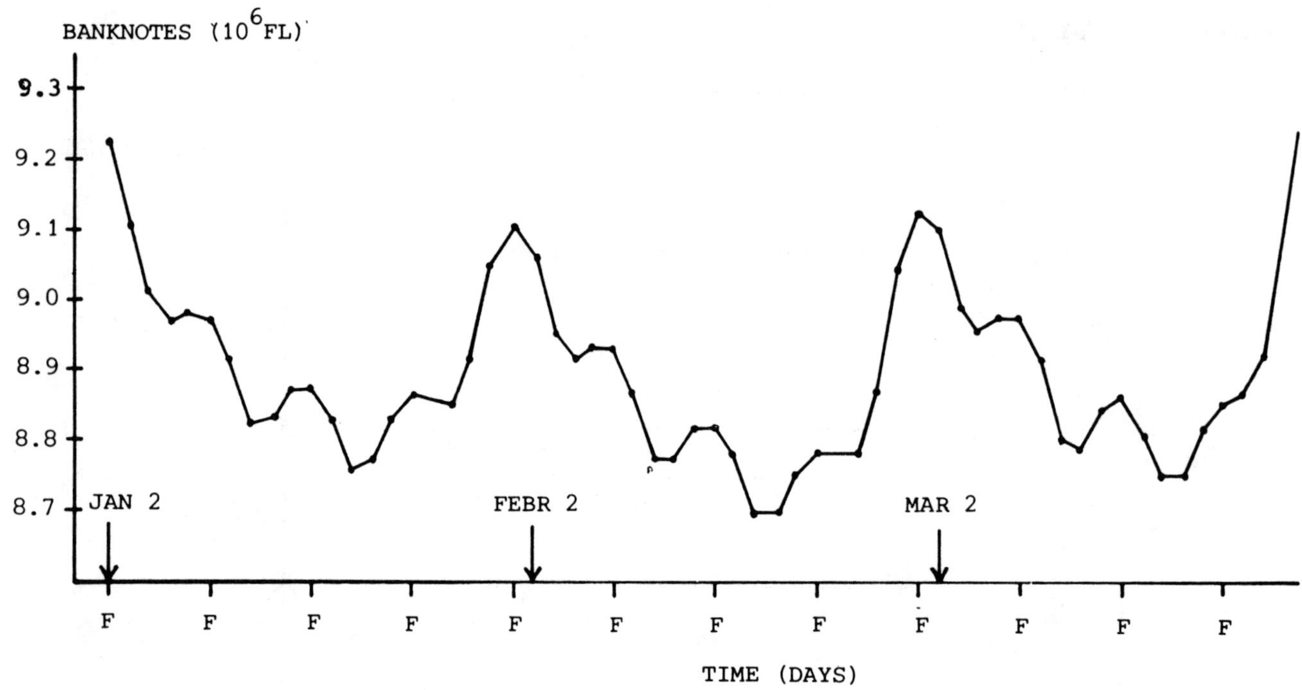

Fig.2 Banknotes in circulation (YEAR: 1970)

time series was filtered by taking first differences

$$\Delta a_t = a_t - a_{t-1}$$

where a_t stands for the state at day t of the variable A, representing the amount of banknotes in circulation. The resolution level of the source 'system' was determined by defining a number of categories by dividing the difference max (Δa_t) - min (Δa_t) by the appropriate number, and by classifying the differences Δa_t in the resulting classes. To obtain a general source 'system' $\{X\}$, these classes were labeled by integers, representing the states x_t. For example, in the case of 4 states $X = \{0,1,2,3\}$, the corresponding 4 classes were given by the intervals $-0.169 \leqslant \Delta a_t < -0.076$, $-0.076 \leqslant \Delta a_t < 0.017$, $0.017 \leqslant \Delta a_t < 0.110$, and $0.110 \leqslant \Delta a_t \leqslant 0.093$ (in 10^6 guilders).

The process of uncertainty reduction in the case of 4 categories on variable X is shown in Fig.3. The D-mask contained 6 consecutive sample positions $\{S_1,S_2,S_3,S_4,S_5,S_6\}$, where $C_r = \{S_6\}$ corresponds to the day to be predicted. At level 0 the uncertainty $H_0 = 2$ bits, i.e. 4 equal probabilities. Level 1 en 21 coincide, of course, with $H(S_6) = 1.62$ bits. At level 22 one searches for the single variable among S_i, $i = 1,2,...,5$, which maximizes $T(S_i : S_6)$. It is shown that the largest transmission is obtained by the same day one week earlier (S_1), while somewhat less prediction of the amount of banknotes on the 6th day is obtained by knowing the amount on the previous (5th) day. Given the transmission of S_1, one would like to know how much extra prediction can be obtained from one of the remaining sampling variables S_i, $i = 2,3,4,5$. It is shown that the third day gives a slightly higher contribution than the fifth day, although the difference is presumably not significant (significance tests have not been performed, see for example ref. [7]); in terms of the previous paragraph,

$$T(S_1,S_3 : S_6) > T(S_1,S_5 : S_6)$$

or, according to (6)

$$T_{S_1}(S_3 : S_6) > T_{S_1}(S_5 : S_6)$$

The procedure of adding one variable at a time to $\overline{C}_r$ suggests a decision-tree like algorithm: starting with $\overline{C}_r = \phi$, find the one variable which maximizes the transmission, add this one to $\overline{C}_r$; next, find the one variable which gives the highest extra (partial) transmission, add this one to $\overline{C}_r$, etc. The left-most path of Fig.3 demonstrates this procedure. Such a

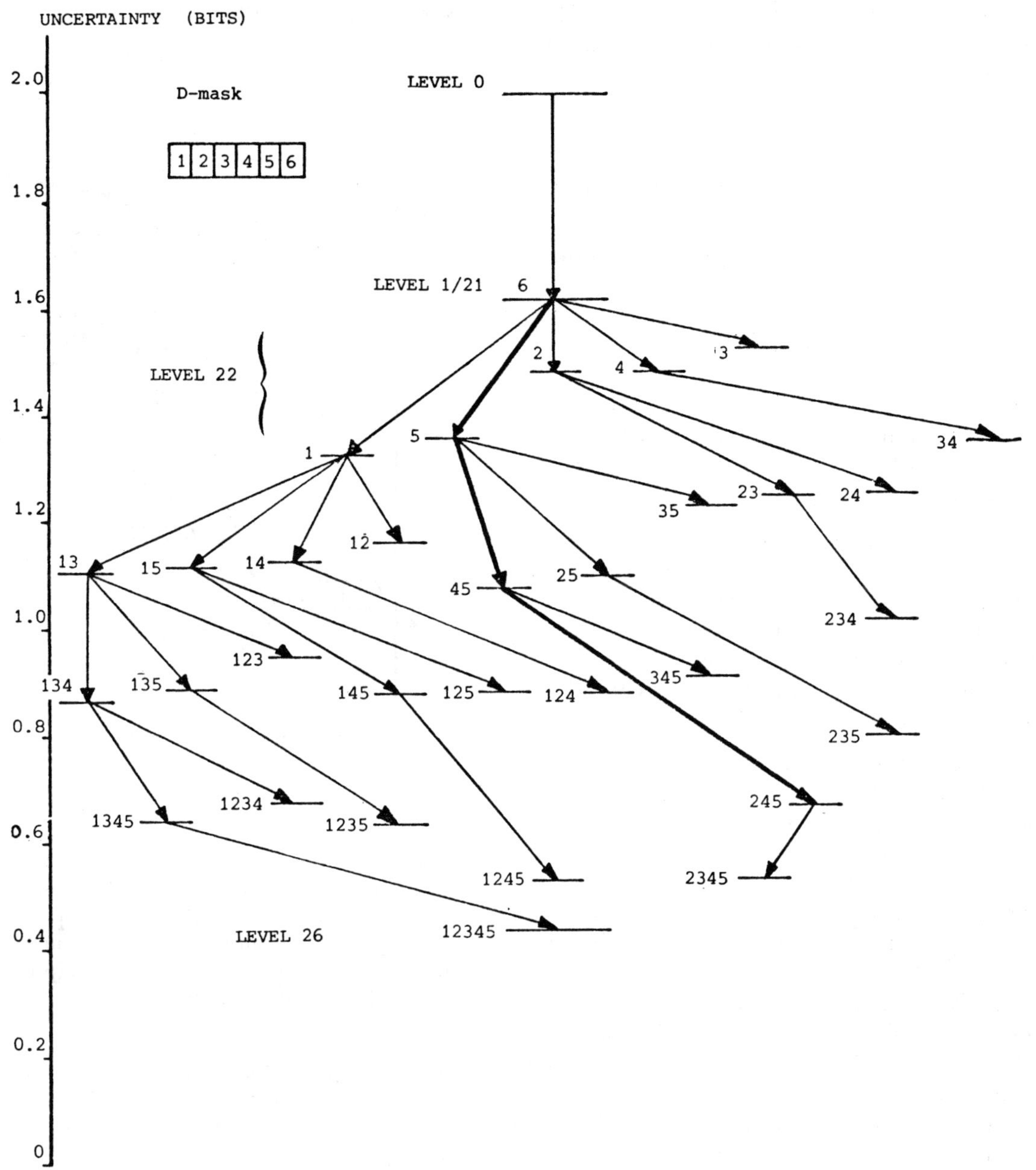

Fig.3 The process of uncertainty reduction for the example of banknotes (X={0,1,2,3,}). Arrows indicate some of the possible transitions

procedure could cut down tremendously the amount of computing time. Unfortunately, the path 6-5-45-245 shows that the suggested algorithm will not continue after the first stage to give the highest uncertainty reduction. For at level 23:

$$T(S_4,S_5:S_6) > T(S_1,S_3:S_6)$$

and even more pronounced at level 24:

$$T(S_2,S_4,S_5:S_6) > T(S_1,S_3,S_4:S_6)$$

Figure 3 suggests a modification of the algorithm: instead of continuing, after a single variable prediction has been obtained, with this one variable, one could also take into account the next best one in performing the calculations for two variables. In this particular case, it would have given the 'optimal path'. I am however convinced that this is a matter of mere chance. The argument is as follows. The transmission between the two variables S_4, S_5 and S_6 may also be partitioned thus

$$T(S_4, S_5 : S_6) = T(S_4 : S_6) + T(S_5 : S_6) + \\ + Q(S_4, S_5, S_6) \qquad (7)$$

where the interaction uncertainty may be defined by

$$Q(S_4, S_5, S_6) = T_5(4{:}6) - T(4{:}6) \qquad (8)$$

Here the interaction uncertainty is identified as that portion of the transmission which cannot be ascribed to any of the variables acting in pairs. Ashby states:

> 'It represents, in other words, the amount of transmission (constraint, law, entropy) that is ascribable only to the three variables acting as a unique triple. It thus measures the degree to which the system (here of three variables) is *irreducibly* complex, i.e. not to be treated by examination of the variables two at a time. [5]'

Applying the argument to our case, we have to admit that knowledge of the single variable transmissions at level 22 will not be able to indicate the magnitudes of transmissions involving pairs of variables. (For higher order transmissions, involving three or more predictor variables a similar argument, involving higher-order interactions, can be held). In our example, $T(S_4 : S_6)$ is relatively small, while $T_5(S_4 : S_6)$ is relatively large. Since $T(S_4 : S_6)$ measures the transmission with all other variables ignored, apparently $T(S_4 : S_6)$ gives a small prediction because to some extent the various positive effects have balanced and annulled each other. The more 'direct' transmission between S_4 and S_6, $T_5(S_4 : S_6)$, apparently does not suffer from these effects, resulting in a relatively high interaction term.

It would still be feasible to look for more sophisticated, presumably heuristic, algorithms in order to reduce computing time. However, these procedures may have to incorporate information about the details of the probability distributions of the variables, which, undoubtedly, will be reflected in the amount of computing time. We will leave the discussion on the possibility of an algorithm here.

We will make some additional remarks. Figure 4 shows the normalized transmission A_{GB} as a function of the number of sampling variables involved,

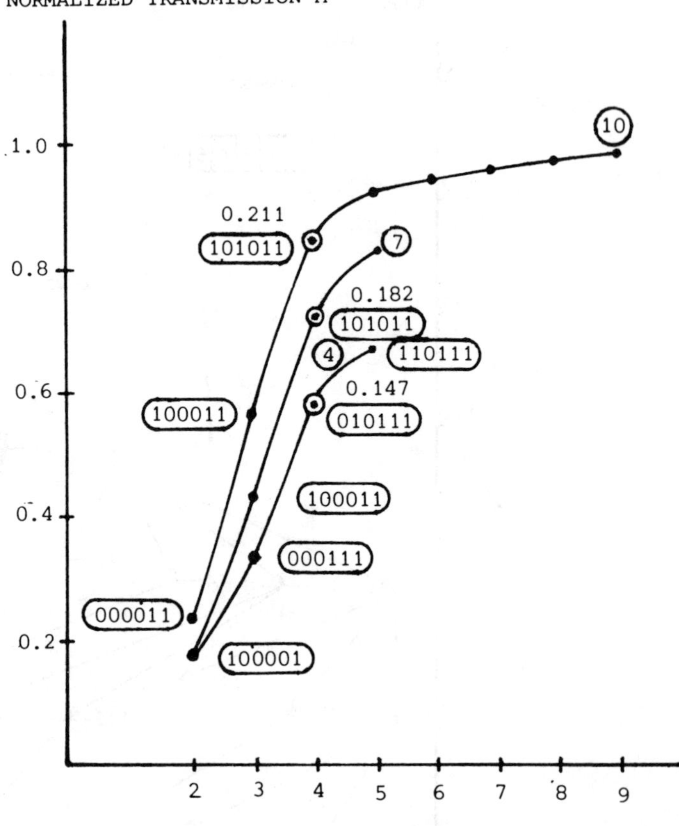

Fig.4 The A-measure as a function of the number of sampling variables; parameter is the number of states of X, resp. 4, 7 and 10. Encircled points with associated numbers give the highest quality mask and the quality. Series of 0's (absence of sampling variable) and 1's (presence of sampling variable) constitute the mask

where A_{GB} is the highest one for each sub-level. Also the number of states of X were varied, so that three different cases are presented, one for 4 categories (discussed above), one for 7, and one for 10 categories. The effects both on A and the optimal mask at each level is clearly shown. With 3 and 4 sampling variables the optimal masks for 7 and 10 categories are similar, but dissimilar to the ones for 4 categories.

It is also shown that the A-measure is still sensitive to the number of categories employed; however, much less sensitive than the corresponding T-measures. At 4 sampling variables we have the ratio's $T_7 : T_4 = 1.83$, $T_{10} : T_4 = 2.51$, while $A_7 : A_4 = 1.24$, $A_{10} : A_4 = 1.44$, where the index notation is assumed obvious.

Finally, the general trend of the curves suggests the usefulness of a measure, previously introduced by Klir [1], the so-called quality of the mask, defined by

$$Q = \frac{A}{m}$$

where m stands for the number of sampling variables in the mask. This quantity measures the uncertainty reduction per sampling variable (an alternative measure would involve only the elements of the predictor set $\overline{C}_r$). It is seen that all curves agree on the number of sampling variables in the highest quality mask.

3. Causal inferences

3.1 Correlation

Suppose we have decided on a specific mask that has been obtained by some procedure at epistemological level 3. This means that we have at our disposal a set of sampling variables. We may now proceed to apply information measures, such as transmissions and partial transmissions, in an attempt to detect the strength of intervariable relations. The ultimate purpose, of course, is that of making causal inferences, given a knowledge of these interrelations. The inference of the causal ordering or network underlying the interrelations is one of the major problems in all sciences.

The problem of making causal inferences from *correlational* data has been more or less systematically investigated by Simon [15] and Blalock [16, 17,18] in the social sciences. We strongly believe that both the generalization as to the location of variables, that seem to be most important in accounting for the variation in some other (dependent) variables, by the concept of the mask at level 3, and the relatively general applicability of information measures, in particular in non-metric domains, to determine the strength of intervariable relations, may prove a powerful additional methodological tool for theory building. Because of the analogies of correlation techniques with information techniques, however, we may utilize the former to guide the elaboration of the latter.

We will present an outline of the main essentials involved in the Simon-Blalock method. The affirmative answer to the question of the possibility of causal inference from correlational data is only conditional, that is, two kinds of assumptions, apart from the assumptions of interval scales and linear regression equations, need to be made. Firstly, the investigator has to introduce some postulates that several of the causal relationships do *not* hold between the variables (usually two-way causation is ruled out from the start). Secondly, environmental variables that are not involved in the causal scheme are admitted to exert their influence upon the system, but it is assumed that these 'error terms' have essentially random effects on the variables.

The induced knowledge or additional information involved in the assumptions cannot be derived from the data. It must come from theoretical considerations, common sense, etc. So, indeed, one may discern a fourth epistemological level of structural or causal inference, with an associated structural system definition. The study of the relationship between correlation and causality by the Simon-Blalock method involves a set of simultaneous equations expressing the relationships when each variable is taken as a possible dependent variable. We will confine the attention to the case of three variables X, Y and Z, measured about their respective means; thus:

$$X = a_{12}Y + a_{13}Z + U_1$$

$$Y = a_{21}X + a_{23}Z + U_2$$

$$Z = a_{31}X + a_{32}Y + U_3 \tag{9}$$

where the U's are error terms due to the effects of all outside variables. In this set of linear mechanisms each of the variables is directly influenced by the other two. Simon has demonstrated that this set of equations cannot be solved unless *a priori* assumptions are made about some of the unknown terms. We will consider three specific examples, which have been diagrammed in Fig.5. We will examine the three cases in some detail.

Case (a): $a_{12} = a_{13} = a_{21} = a_{32} = 0$. A change in U_1 will change the value of X directly, and values of Z and Y indirectly. A change in U_2 will change Y directly, but will leave X and Z unchanged. A change in U_3 will change Z directly, Y indirectly, and will not affect X. In Simon's terminology we may say that Z is causally dependent on X, and Y is causally dependent on Z. Because of the assumed uncorrelated error terms one may also expect the correlation between X and Y to be smaller in absolute magnitude than that between X and Z and between Z and Y. For it is easily shown that the following relation is predicted between the correlation coefficients:

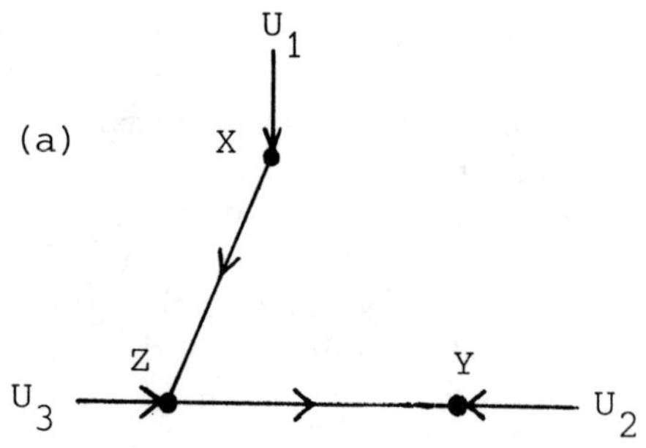

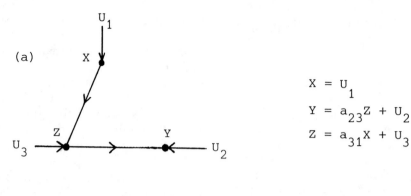

$$X = U_1$$
$$Y = a_{23}Z + U_2$$
$$Z = a_{31}X + U_3$$

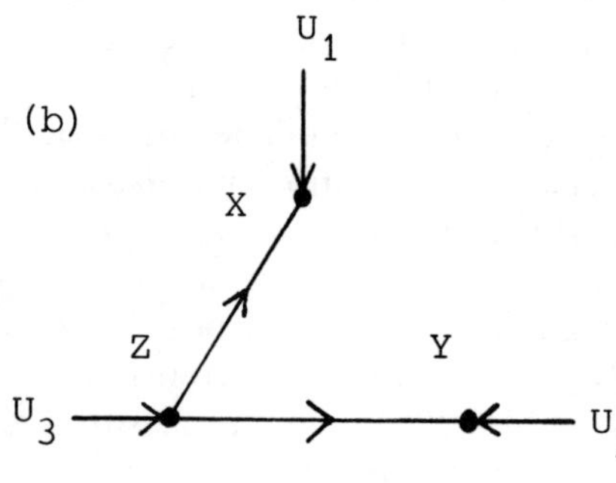

$$X = a_{13}Z + U_1$$
$$Y = a_{23}Z + U_2$$
$$Z = U_3$$

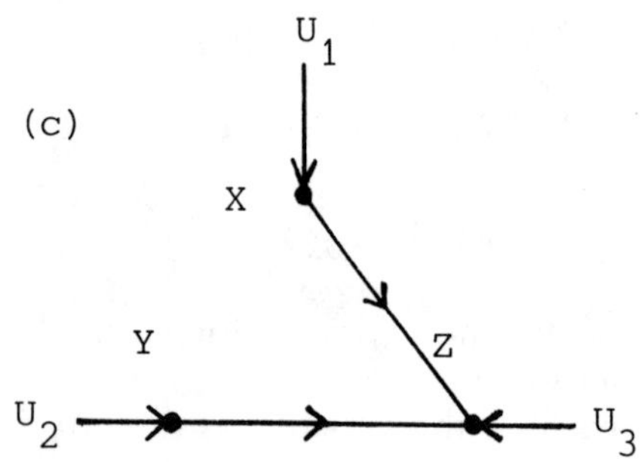

$$X = U_1$$
$$Y = U_2$$
$$Z = a_{31}X + a_{32}Y + U_3$$

Fig.5 Some causal models of X, Y and Z

$$r_{XY} = r_{XZ} r_{ZY}$$

which gives (disregarding signs)

$$r_{XY} < r_{XZ}, r_{ZY} \tag{10}$$

Furthermore, the analysis will predict the vanishing of the partial correlation between X and Y:

$$r_{XY.Z} = 0 \tag{11}$$

Case (*b*): $a_{12} = a_{21} = a_{31} = a_{32} = 0$. A similar analysis will reveal that both equations (10) and (11) are predicted also for this case. This unfortunate fact demonstrates that without going beyond the data one cannot possibly choose between these models. (Also, reversing the arrows in Case (*a*) gives the same empirical predictions, which emphasizes the general point that there may be several distinct causal models which all predict the same empirical relationships among correlation coefficients). Blalock states:

> "This means that it will usually not be possible to establish the correctness of any given model unless certain alternative models can be ruled out either theoretically or on empirical grounds other than the magnitude of the correlation coefficients (e.g. a knowledge of time sequences). We can proceed by eliminating false or inadequate models, however. The method can be used to choose between two existing theoretical models, where such theories lead to different empirical predict-

ions, or—perhaps more realistically—it can be used as an exploratory device through which we note those instances where the theory fails to predict correctly, making successive changes until more adequate predictions have been obtained [17]".

Case (*c*): $a_{12} = a_{13} = a_{21} = a_{23} = 0$. In this case, where both X and Y are independent causes of Z, the model predicts

$$r_{XY} = 0 \qquad (12)$$

Consequently, on the basis of the correlation between X and Y one may distinguish the latter model from the cases (*a*) and (*b*) (ignoring the question of sampling errors).

3.2 *Transmission*

The information theoretic counterpart of the correlation between variables is the transmission measure [19]. In an exploratory analysis we utilized the simultaneous equations of the three Cases (*a*), (*b*) and (*c*) to generate three variable data matrices (3×210 samples each). We did not bother about measuring the variables about their means; the equations used to generate the data were identical to (9), while, according to the model at hand, appropriate coefficients were set equal to zero; the remaining coefficients were set equal to one. The values zero or one of the error terms were generated randomly, and independently from each other, according to some predetermined probability distributions. From all cases we have studied, one representative example for each Case (*a*), (*b*) and (*c*) has been given in Table 1. Case (*d*) is a special case: the causal model is Case (*a*), but we took the probability distribution of U_2 conditional on the value of U_1. A first look at the Table reveals the following.

Case (*a*): $T(X{:}Y) < T(X{:}Z),\ T(Y{:}Z)$

$$T_Z(X{:}Y) = 0$$

Analogously to the results of correlation techniques, the transmission, as a measure of the strength of the relationships, between adjacent variables in the causal network is the largest, and becomes smaller the further the variables are removed in a causal chain (i.e. X and Y). Due to the symmetrical nature of the transmission the direction of causality cannot be concluded from the data. Furthermore, the partial transmission $T_Z(X{:}Y) = 0$, indicating that the transmission between X and Y is somehow mediated

Table 1

Case:	(*a*)*	(*b*)**	(*c*)*	(*d*)***
$T(X{:}Y)$	0.452	0.365	0	0.401
$T(X{:}Z)$	0.604	0.604	0.452	0.460
$T(Y{:}Z)$	0.894	0.646	0.248	0.925
$T_Z(X{:}Y)$	0	0.004	0.194	0.025
$T_Y(X{:}Z)$	0.152	0.243	0.646	0.084
$T_X(Y{:}Z)$	0.442	0.286	0.442	0.549
$Q(X,Y,Z)$	-0.452	-0.361	0.194	-0.376
$T(X{:}Y{:}Z)$	1.498	1.255	0.894	1.410

* $p(U_1=1)=0.5; p(U_2=1)=0.2; p(U_3=1)=0.2$
** $p(U_1=1)=0.2; p(U_2=1)=0.2; p(U_3=1)=0.5$
*** $p(U_1=1)=p(U_3=1)=0.5; p(U_2=1|U_1=0)=0.6; p(U_2=1|U_1=1)=0.8$

by the third variable Z.

Case (*b*): $T(X{:}Y) < T(X{:}Z),\ T(Y{:}Z)$

$T_Z(X{:}Y) = 0$ (disregarding sampling error)

Case (*c*): $T(X{:}Y) = 0$

The analogy with correlational predictions is clear. Also, again, on the basis of the transmission 'predictions' alone we are not able to distinguish between 'true' transmission [Case (*a*)] and 'spurious' transmission [Case (*b*)] between X and Y (compare Simon [15]).

3.3 *Exploratory development of causal models*

It has been argued [17] that the Simon-Blalock method is a technique that can be used not only to check the adequacy of any particular causal model, but also in an exploratory fashion to help develop alternative models. Although it presents a small basis, the results of the preceding paragraph suggest a similar use of information techniques. In particular, the partitioning approach of the total transmission between the variables of a system may guide in an exploratory way the search for underlying causal mechanisms. The total transmission is defined by

$$T(S_0{:}S_1{:}\dots{:}S_q) = H(S_0) + H(S_1) + \dots$$
$$\dots + H(S_q) - H(S_0, S_1, \dots, S_q)$$

According to Ashby,

> "It is perhaps the most important quantity of the system, for it measures the total constraint holding over the system (the entropies of the individual variables being regarded as given). For this reason it measures the total quantity of relationships that exist in the system—the total quantity of law, as one might put it. Once an actual system has yielded a primary body of factual data, the 'total transmission' computed from this data measures the total quantity of law that can be extracted from the data. Thus it is possible to measure how much law a given body of data contains *before* the particular details of the law (or laws) have been discovered. [5]"

One way of partitioning the total transmission is that in a sum of transmissions between pairs of variables and interaction terms. For the three variable case $\{X,Y,Z\}$ we may write

$$T(X{:}Y{:}Z) = T(X{:}Y) + T(X{:}Z) + T(Y{:}Z) + \\ + Q(X,Y,Z) \qquad (13)$$

Consider Case (*a*) and (*b*) of the previous paragraph. A negative interaction indicates a correction on the total constraint due to the fact that the transmission between X and Y is a consequence of the regression on the third variable Z.

Inserting

$$Q(X,Y,Z) = T_Z(X{:}Y) - T(X{:}Y)$$

into (12) gives

$$T(X{:}Y{:}Z) = T(X{:}Z) + T(Y{:}Z) + T_Z(X{:}Y)$$

Since the latter term is equal to zero (disregarding sampling errors) the equation, together with the considerations of the preceding section, suggests a rule of thumb for a preliminary exploration of the causal structure: *a)* a causal relation (the direction of which cannot be derived from the data) is assumed between the pair of variables which gives the highest transmission. In Case (*a*) and (*b*) we find $T(Y{:}Z)$; the relation is written as Y-Z. (*b*) Subtract the value of this quantity from the total transmission and check if the magnitude of the remaining total constraint allows the subtraction of the transmission between another pair of variables (the transmissions are all non-negative quantities); the choice may be guided by the magnitude of the interaction term. In Case (*a*) and (*b*) $T(X{:}Z)$ exhausts the available total constraint, postulating the causal relation X-Z. Hence, the suggested structure in both Case (*a*) and (*b*) is of type: X-Z-Y.

Case (*c*) suggests in addition to this rule of thumb: we may again derive the structure X-Z-Y from the data but also the 'non-correlation' of X and Y. Some extra total transmission, resulting in a positive interaction term, is obtained, because, for a given value of Z, there must be a constraint between the two variables X and Y 'causing' this value of Z. Equation (8) shows that the sign of Q is non-negative when the transmission between two variables equals zero. It is interesting to note that even in Case (*d*), where the condition of the essentially random effect of the environmental variables is violated, an indication may be acquired about the right structure.

We will give another illustration of the suggested rule of thumb for a four variable case. In Klir [12] an array of probabilities was deduced from a 500 memoryless sample activity matrix of four three-valued variables X,Y,Z,W:

X	Y	Z	W	$p(x,y,z,w)$
0	0	0	1	0.024
0	2	0	1	0.084
0	2	2	1	0.144
1	0	0	0	0.076
1	0	0	2	0.028
1	2	0	0	0.244
1	2	0	2	0.048
1	2	2	0	0.126
1	2	2	2	0.022
2	1	1	1	0.204

From this array one can compute all entropies, transmissions, etc., as shown in Table 2. The total transmission in this four variable case is partitioned as follows

$$T(X{:}Y{:}Z{:}W) = T(X{:}Y)+T(X{:}Z)+T(X{:}W)+T(Y{:}Z)+ \\ +T(Y{:}W)+T(Z{:}W)+Q(X,Y,Z)+ \\ +Q(X,Y,W)+Q(X,Z,W)+Q(Y,Z,W)+ \\ +Q(X,Y,Z,W)$$

Table 2

$H(X)$	1.447	$T(X:Y)$	0.724
$H(Y)$	1.221	$T(X:Z)$	0.777
$H(Z)$	1.485	$T(X:W)$	0.994
$H(W)$	1.365	$T(Y:Z)$	0.809
$H(X,Y)$	1.944	$T(Y:W)$	0.276
$H(X,Z)$	2.154	$T(Z:W)$	0.326
$H(X,W)$	1.817	$Q(X,Y,Z)$	-0.724
$H(Y,Z)$	1.900	$Q(X,Y,W)$	-0.272
$H(Y,W)$	2.310	$Q(X,Z,W)$	-0.325
$H(Z,W)$	2.524	$Q(Y,Z,W)$	-0.272
$H(X,Y,Z)$	2.566	$Q(X,Y,Z,W)$	0.271
$H(X,Y,W)$	2.310	$T(X:Y:Z:W)$	2.586
$H(X,Z,W)$	2.524		
$H(Y,Z,W)$	2.931		
$H(X,Y,Z,W)$	2.931		

Applying the above procedure the following relations are suggested in the order of decreasing transmissions 1. X-W, 2. Y-Z, 3. X-Z. The sum of the corresponding transmissions equals 2.580, which is almost equal to the calculated total transmission: 2.586. Consequently, the suggested structure is

$$W - X - Z - Y$$

which is indeed the structure by which the data matrix was generated originally [12].

Let us examine the example in some more detail. The magnitudes of $T(X:Y)$ and $Q(X,Y,Z)$ suggest that the transmission between X and Y is mediated by Z, which is indeed the case. Likewise, comparing $Q(X,Y,W)$ and $T(Y:W)$, $Q(X,Z,W)$ and $T(Z:W)$, $Q(Y,Z,W)$ and $T(Y:W)$, suggests, respectively that transmission between W and Y is mediated by X, between Z and W by X and between Y and W by Z. A further check on the correctness of the structure may be obtained by the calculation of some partial transmissions; for example $T_X(W:Z) = T_Z(X:Y) = 0$, while $T_Y(X:W) = 0.723$. It is also of interest to note that the magnitudes of the transmissions between adjacent variables, $T(W:X)$, $T(X:Z)$, $T(Y:Z)$, are all larger than the transmissions between pairs, which are mediated by one other variable, i.e. $T(Z:W)$ and $T(X:Y)$; the latter, in turn, are larger than $T(Y:W)$, where the transmission is mediated by two other variables.

The partitioning laws of multivariate information analysis may also be utilized for the detection of subsystems within the structural system [13]. For example, in the above four variable case we may write

$$T(X:Y:Z:W) = T(W,X:Y,Z) + T(W:X) + T(Y:Z)$$

where $\{W,X\}$ was taken as one subset, and $\{Y,Z\}$ as another. The equation expresses the fact that the total amount of relationships in the system is the sum of the amounts in the subsystems plus the amounts of relationship between the subsystems. Numerically,

$$2.586 = 0.783 + 0.994 + 0.809$$

In general, however, we would require the magnitude of the transmission between subsystems to be considerably smaller than those within the subsystems. So that, in this case, it is pointless to discern subsystems.

4. Concluding remarks

The foregoing does not constitute proof in any sense. It is an exploration guided by reasoning by analogy and, perhaps misguided, by some illustrative simple examples. Especially the potential utility of information measures in the explorative phase of causal model building is in need of a more thorough investigation. Furthermore, the connection between the investigations at level 3 in terms of a generative system and those at level 4 in terms of a structural system need to be made more explicit. One cannot deny, however, that the results look promising so far, at least for the recursive type of models which have been considered here.

One aspect of Klir's conception of a methodology of empirical investigation emerges as particularly appealing from the above sections. It clarifies exactly where in the process of investigation one has to introduce additional information or postulates which cannot be derived from the data. It also allows for the generalization and comparative study of methods and techniques which are being used in diverse, and sometimes alien disciplines, thus "heaping our science together".

Acknowledgments

I am very grateful to Prof. George J. Klir of the School of Advanced Technology, State University of New York, Binghamton, for his generous support in connection with this chapter. I also thank Mr. Geilenkirchen of the Graduate School of Management for dedicating his excellent computational skills to this research work.

References

1. KLIR, G.J., 'On the representation of activity arrays', *Int. J. General Systems* 2(2), 149-168 (1975).
2. SIMON, H.A., 'Does scientific discovery have a logic?', *Philosophy of Science,* 471-480 (December 1973).
3. McGILL, W.J., 'Multivariate information transmission', *Psychometrika,* 19, 97-116 (June 1954).
4. GARNER, W.R., *Uncertainty and Structure as Psychological Concepts,* John Wiley, New York (1962).
5. ASHBY, W.R., 'Measuring the internal informational exchange in a system', *Cybernetica,* 8, 5-22 (1965).
6. ASHBY, W.R., 'Two tables of identities governing information flows within large systems', *Comm. Amer. Soc. Cyb.* 1(2), 2-8 (1969).
7. ATTNEAVE, F., *Applications of Information Theory to Psychology,* Holt, New York (1959).
8. CONANT, R.C. and STEINBERG, I.B., 'Information exchanged in grasshopper interactions', *Information and Control,* 23, 221-233 (1973).
9. ASHBY, W.R., *Introduction to Cybernetics,* Chapman and Hall, London (1956).
10. BROEKSTRA, G., 'System definitions: an approach in the language of variables', *Annals of Systems Research,* 4, 141-157 (1974).
11. BROEKSTRA, G., 'Some comments on the application of informational measures to the processing of activity arrays', *Int. J. General Systems,* 3, 1-9 (1976).
12. KLIR, G.J., 'Identification of generative structures in empirical data', *Int. J. General Systems,* 3 (1976).
13. CONANT, R.C., 'Detecting subsystems of a complex system', *IEEE Trans. Syst. Man. Cyber.* SMC-2, No.4, 550-553 (September 1972).
14. ABRAHAMSE, A.P.J., 'Parametric decomposition of time series', (unpublished) Lecture Notes AA/2, Neth. School of Economics, Econometric Institute (1972).
15. SIMON, H.A., 'Spurious correlation: a causal interpretation', *I. Am. Stat. Ass.,* 49, 467-479 (1954).
16. BLALOCK, H.M., 'Correlational analysis and causal inferences', *American Anthropologist,* 62, 624-631 (1960).
17. BLALOCK, H.M., 'Correlation and causality: the multivariate case', *Social Forces,* 39, 246-251 (1961).
18. BLALOCK, H.M. and BLALOCK, A.B., *Methodology in Social Research,* McGraw-Hill, London (1971).
19. McGILL, W.J., 'Isomorphism in statistical analysis' in *Information Theory in Psychology* (Edit. by H. Quastler), The Free Press, Illinois (1954).

Predicting the behavioral effects of system size in a family of complex abstract systems

CRAYTON C. WALKER
The University of Connecticut, Storrs, USA

Introduction

Size is an important system characteristic, since both structural and behavioral complexity can expand exponentially as size increases. To aid the development of theory dealing with the behavior of complex systems in general, it would be useful to know what conceptual connections, if any, exist between size and other system characteristics. It also appears useful to examine the empirical relation between system behavior and size with the aim of finding procedures for predicting the behavior of large systems, even in the absence of definitive theory.

In this chapter I consider a recent suggestion (Walker, [1]) that size is essentially a structural variable, in the sense that size affects behavior in much the same way as 'structure' (this term will be defined below). I also consider an implication of that suggestion which would have it that predictions of system behavior for large systems can be had solely from the behavior of small systems. These two points are examined with respect to certain aspects of system behavior over time. In particular, the behaviors dealt with are, length of transient behavior, periodicity of long-term behavior, and their sum, a quantity which measures the minimum amount of time required for an observer to distinguish transient from long-term behavior (Walker and Aadryan [2])

Systems examined

The test-bed for these inquiries is a particularly basic family of abstract systems, namely, a subset of those systems that are autonomous, binary, fixed in structure, and built of elements having two input 'lines'. That is, I consider the behavior of a well defined family of systems which all (1) behave independently of the environment, (2) are logic networks with components or elements that have exactly two states, (3) have connections among these elements that remain unchanged once they have been made, and (4) are composed of elements receiving the 'signals' of two elements in the system.

It is the network of connections among elements in a given system that I refer to as the structure of that system. By 'structural effect' I mean the change in behavior produced by change in the network.

In the particular subset of systems that this chapter examines, the elements in a given system carry the various elements' internal states, and the elements' next internal state is a determinate function of both the internal state in each element and the states carried on the elements' inputs at the present instant of (discrete) system time. Finally, all functions computed by the elements of a given system are identical.

This family of systems is important in general systems theory for at least two reasons. First, the simplicity of the family inclines us to use it to test hypotheses, as any system theories which pretend generality must hold in such basic examples. Second, among fixed-structure systems composed of heterogeneous elements with the same number of input lines, those with two-input elements form a class which is at the same time the simplest class to allow genuine stuctural complexity, and the most complex class that would appear to have long-term behavioral periodicity and stability characteristics reasonable enough for the class to provide attractive models of empirical systems (Kauffman [3]). Thus, we have reason to consider few-input systems as theoretically pivotal. We are here discussing modeling efforts which proceed in the absence of detailed knowledge of system structure, but it should be remembered that model building and theorizing in such circumstances is often inescapable, and on occasion even desirable.

System size and system behavior

Size, as that term is used in this chapter, refers to the number of elements in a system. Is size essentially a structural variable? If it were so, it would be expected that the relationship between system behavior and size would be mirrored by a uniquely strong relationship between system behavior and structure. That is, we would expect variables other than structure to suffer by comparison as predictors of behavior across changes in system size. Unfortunately, such comparisons are not straightforward: size as a variable is both ordered and unbounded, while the variables I will define over systems fixed in size are neither. For example, structure at a fixed size is, with respect to our present knowledge, not clearly ordered, and it is certainly finite. Noting this difficulty in specifying any general effect of size on behavior in a way that is directly comparable to structural effects at a fixed system size, I will restrict attention to the effect of size (over an arbitrary span of system size) in terms of behavioral variability, since the effects of structure on behavior can also be quantified in terms of variability.

The reasonableness of the suggestion that size is a structural variable in the first place hinges on the observation that in specifying structure completely, size must be specified as well. Therefore, size is an aspect of structure. We then set up the strong hypothesis that size *is* a structural variable, and proceed to test; for granting the importance of size, if the strong form of the hypothesis can bear weight, we would expect strong results to follow.

The variable contrasted with structure is the state at which the system is started in its behavioral space, the 'initial state'. The strategy of the study is to select a particular behavior, and then to determine, for each function which can be computed by the elements, the components of observed variability in behavior associated with the various variables. The determination of the variability separately associated with initial state, with structure, and with their interaction, is carried out at a fixed, small system size. These components of variability having been estimated, the relationships between them and the variability associated with a change to a larger system size are compared.

The systems examined here are determinate in behavior. From any given initial state, a system proceeds stepwise through system states in its behavior space. The state of the system at any instant of system time is just the ordered set of internal states of the various elements at that time. At length, the system must re-encounter a state that had been passed through before. From that point on, the system will continue to cycle through a fixed sequence of states indefinitely. The number of states the system passed through before reaching the final cycle, the length of the 'run in', is the length of the transient behavior of that system in the given starting conditions. The number of distinct states in the cyclic behavior the cycle length, is the periodicity of the long-term behavior. The sum of the runin and cycle lengths of a given behavior trajectory is called the length of the 'disclosure'.

Procedure

Behavioral data for the family of systems were obtained by computer simulation. For a given system size and a given function (recall that each element in a system computes the same function) ten initial states and ten structures were chosen at random. One hundred behavioral trajectories were therefore examined, one for each initial state and structure pair. In each trajectory, runin and cycle lengths were determined and recorded.

In the family of systems examined here, there are 256 different functions computable by system elements. Behavioral identities among these functions allow a reduction to 88 without sacrificing information (Walker, [1]). Therefore, the basic data are runin and cycle lengths obtained using a fully factorial experimental design for each of the 88 functions, at two levels of system size. The sizes used are seven and seventeen. In the first part of the chapter I assess the relationship between behavioral variability across systems of size seven and seventeen and components of variance at size seven. The second part of the chapter examines the predictability of behavior at the larger size using data provided by systems of the smaller size. Seventeen was chosen for the size of the larger systems because, for some functions, extravagant computing effort would be required for any increased size. As it was, some cycle lengths exceeding 16,000 were observed. The dependent variable used in the first part of the study is an estimate of the variance between behavior means at the two system sizes, for each function. The quantity estimated can be easily shown to be $(\mu_2 - \mu_1)^2/4$, where μ_2 and μ_1 are the behavior population means at the larger and smaller system sizes respectively. As it happens, the analysis always involved taking logarithms, before using standard correlation techniques, so the factor of one quarter was dropped and the analysis continued working with an unbiased estimate of $(\mu_2 - \mu_1)^2$, namely $(\bar{X}_2 - \bar{X}_1)^2 - (s_2^2 + s_1^2)/100$ where $\bar{X}$ is an observed mean, and s^2 is an observed variance estimate, with subscripts as before.

Analysis

The analysis proceeds as follows, using runin length as an example of the behavior examined. For each of the 88 functions, standard components of variance technique was used to determine estimates of variance in runin lengths associated with initial state, structure, and interaction; this for data from the smaller system size. For each function the observed means of runin length, at both system sizes, was used along with total variance at each system size to generate an estimate of variance associated with system size. The size-variance estimates were then correlated separately with the variance components, over the 88 data points for each variable pair. This yields a description of the relatedness of the variables in the family of systems, as separated into the 88 behavioral equivalence classes. Similar procedure was used for cycle lengths and disclosure lengths.

The correlation data discussed is presented in Table 1, in the s_N^2 columns. In that table, correlations for logarithms of cycle length, and logarithms of disclosure length are also given. These transformed variables were introduced for reasons discussed below.

Correlations with a second dependent variable, the mean at the larger system size, are also presented in Table 1 under $\bar{X}_2$.

Table 1 Correlations between variance estimates across system size (s_N^2), means at the greater system size ($\bar{X}_2$); and estimated total variances, variance components and mean ($\bar{X}_1$) at the smaller system size. (Decimal points omitted.)

Smaller System Variances and Mean (log trans.)	Runin Lengths (log transformation) s_N^2	$\bar{X}_2$	Cycle Lengths (log transformation) s_N^2	$\bar{X}_2$	Log Cycle Lengths (log transformation) s_N^2	$\bar{X}_2$
Total	77	86	83	85	75	73
Initial state	18	22	63	67	48	37
Structure	65	70	82	85	72	65
Interaction	80	87	83	85	67	70
$\bar{X}_1$	61	84	81	88	55	88

	Disclosure Lengths (log transformation) s_N^2	$\bar{X}_2$	Log Disclosure Lengths (log transformation) s_N^2	$\bar{X}_2$
Total	81	87	70	64
Initial state	40	42	28	20
Structure	77	85	63	53
Interaction	80	83	54	57
$\bar{X}_1$	80	87	54	90

To rectify the observed data, which showed a strong positive skew for all variables, logarithmic transformations were used for all variables before correlation. Since zeros were observed, all variables except one were translated by *0·1*, yielding the final before-correlation transformation of log $(X+0.1)$. The one exception is the variance estimate of disclosure length due to system size which had to be translated by *+0·12* to make the estimate non-zero for one function.

Examination of scatter plots of the correlated data indicate that the logarithmic transformation did an acceptable job of rectifying the relationship, so that no important additional non-linear relationships need be considered.

Results and discussion

1. Is size essentially a structural variable?
Transient system behavior. It seems reasonable before the fact to argue that structure is most unlikely to be strongly related to size in the case of runin lengths. Temporary behavior would seem much more dependent on the initial state, and only rather indirectly related to structure, so that even if size were exclusively a structural variable, for some variables it would not appear so in an analysis of runin lengths. By this argument, a finding that no unique relationship existed between size and structure in temporary behavior would not be evidence contrary to the hypothesised structural nature of the system size variable. However, in analyzing the variance of runin lengths at the smaller system size there are 21 functions out of the 88 which show an F ratio for initial state variance significant at the 0·01 level, while 59 functions show, at the same level, a significantly non-zero variance component for structure. The data indicate that, contrary to expectation, structure can influence temporary behavior.

As to the observed correlations between variance components of runin length at the smaller system size and variance of runin length across system size, the log initial state-log size variance correlation is 0·18, the log structure-log size variance is 0·65, and the log interaction-log size variance is 0·80, or percent variances accounted for of 3%, 42%, and 64% respectively. The findings are that for transient system behavior, the variance due to change in system size is not ralated to the variance due to initial state, but is apparently related to structural variance, and to interaction variance. In normal, bivariate populations with zero correlation, coefficients between approximately ±0·22 can be expected in 95% of random samples of size 88. Since the populations in question are likely non-normal, the 0·22 figure is only a rough guide to what would be expected in the event no correlation exists in the population, but I will use it as at least suggestive.

Long-term system behavior. It seems evident that the periodicity of long-term system behavior would be influenced by the structure of a system. In fixed-structure systems the long-term behavior is that which continues after the transient states in a trajectory vanish, allowing more time for the influence of the starting state to be submerged under the effects of particularities of structural feed-back loops. Analyzing the variance of the logarithms of cycle lengths, a point further considered in the next paragraph, it is found that at the smaller system size, only six functions have F ratios for initial state that are significant at the 0·01 level, while at the same level there are 63 ratios significant for structure. For long-term behavior the expectation that structure can be influential is realized.

In analyzing the variance of cycle lengths one characteristic of this variable is worth noting. It is observed that starting the same system at different initial states often results in cycle lengths that are multiples or divisors of one another. This suggests that it is more appropriate to convert cycle lengths to logarithms before variance is analysed, owing to the additive nature of the model on which the analysis is based.

The results for logs of cycle lengths are as follows, using the transformations already described prior to computing correlations coefficients. There does not appear to be a moderate relationship between log structural variance and log size variance: $r=0{\cdot}72$, indicating that 52% of the variance in log size variance is linearly accounted for by log structure variance, while log initial state variance correlates *0.48,* accounting for only 23% of the variance in log size variance. Log interaction variance correlates *0·67* and accounts for 45% of the variance in log size variance.

Amount of time required to distinguish transient from long-term behavior. Since disclosure lengths are the sum of runin lengths and cycle lengths the analysis can appropriately follow that method used for cycle lengths, namely analyzing the variance of log disclosure lengths prior to correlating. The findings are that initial state variance accounts for 8% of the size-produced variance, structure accounts for 40%, and interaction accounts for 29%.

Summary. Considering runin, cycle, and disclosure lengths, structure accounts for at least 40% of size variance, while the initial state accounts for at most 23%. Looking solely at the mutually independent parts of the structure and initial state variables, as separated by the analysis of variance model, we find structure clearly linked to size effects while the initial state is either not related or is related to size effects at a low level.

Unfortunately, this clear picture is clouded by the finding that interaction variance rivals or, for runin lengths, exceeds structure in accounting for size effect variance. We are forced to conclude that one of two situations holds: either initial state and structure jointly account for roughly as much variance in size effects as does structure, or, the components of variance model is inappropriate in the case of all three dependent variables examined.

Lacking alternatives to the components of variance models, it appears most appropriate to conclude at this time that while an important linkage has been established between structure and size for runin, cycle, and disclosure lengths, the initial state can combine with structure to influence size effects, particularly in the case of runin lengths.

2. Are the effects of size predictable from behavior at a fixed system size?

To this point in the chapter I have found that while structure is indeed linked to size, it does not appear to be the only variable involved in explaining size effects—the initial state is also jointly important, at least for temporary behavior. Furthermore, if predicting size effects and not the comparative importance of pre-selected variables is at issue, then it is worth noting that the variance components isolated to this point are not very strongly related to size-produced effects. It also happens that the correlation structure which obtains is such as to provide no useful gain in linear predictability using the several components of variance as predictors, as contrasted with the use of their sum. These facts in mind, when it comes to predicting size related effects for larger systems, there appears to be no convincing reason to decompose variance. Attention can therefore be turned to the more straightforward question of what convenient means, if any, exist for predicting behavior in larger systems from smaller systems.

Here the findings appear simple and convincing. In predicting the mean behaviors of the larger systems, means for the small systems are quite useful. Table 1 shows that the (log) mean runin lengths of each function at the smaller system size correlate 0·84 with the (log) mean runin length of the function at the larger system size. The correlations for (log) mean cycle length, and (log) mean disclosure length are 0·88 and 0·87 respectively. Over the span of system sizes used here, at least 70% of the variance in (log) means at the greater system size is predictable using a linear prediction rule based on (log) means at the smaller size. While it is not known at present how a greater span in sizes would affect these results, the data at least suggest that small system data in certain circumstances can furnish useful predictions for large systems.

Acknowledgements

I thank Alan Gelfand of the University of Connecticut and John Hartigan of Yale University for helpful comments.

References

1. WALKER, C., 'Behavior of a class of complex systems: the effect of system size on properties of terminal cycles', *J. Cybern.*, 1, pp.55-67 (1971).
2. WALKER, C. and AADRYAN, A., 'Amount of computation preceding externally detectable steady-state behavior in a class of complex systems', *Int. J. Bio-Med. Comput.*, 2, pp.85-94 (1971).
3. KAUFFMAN, S., 'The organization of cellular genetic control systems', *Lectures on Math. in Life Sciences*, Vol.13, Providence, R.I.: Am. Math. Society (1972).

The arithmetic of closure

FRANCISCO J. VARELA
University of Colorado, Denver, Colorado, USA
and **JOSEPH A. GOGUEN**
University of California, UCLA

1. Introduction: the closure thesis

1.1 The idea of a system-whole

What is the proper notion of a *whole?* A major motivation for thinking in systems terms, stems from a need to deal with the coherence and interconnectedness with which whole units present us in every domain. It is a suspected universality, irrespective of the particularities of parts and specific processes, that lends its fascination to a general theory of such systems. Yet, we feel that little effort has been dedicated to arrive at a precise notion of this central intuition of 'wholeness' or 'systemness'.

There are, we shall argue, several important philosophical and historical reasons for this neglect. This chapter is our attempt to lay bare these issues and to offer an improvement. Our proposal hinges on a characterization of wholes which allows us to have a mathematical handle on their description. Let us start with their characterization.

The context in which we are considering wholes is that of systems theory. By stating this, we wish to direct the imagination of the reader mainly to what we may call natural systems, non-man made wholes, from cells to solar systems, from rocks to societies. We shall use the hybrid word *system-whole* for the purpose of this denotation, and to establish a distinction from the many other possible meanings that the word 'whole' has come to bear.

In such system-wholes we want to consider their constitutive processes: productions, regulations, descriptions, computations of any form and kind. And in the best spirit of a theory of systems, let us agree to abstract the specifics and consider just general interactions between unspecified parts. Let us agree to call an assembly of such interactions a system's *organization.* We can think of such an organization as a graph where nodes are the components and internode links represent processes and interactions. So much for vocabulary.

Now to the first important point. Question: What have we learned from the descriptions of system-wholes in the last decades? Answer: That in order to account for the coherence of the observed systems, their constitutive interactions must be *mutual* and *reciprocal,* so as to become an interconnected network.

There seems to be plenty of evidence to substantiate this view of system-wholes. The traditional source of examples has been living systems. Surely in them the circularity of interconnectedness is more striking than anywhere else, both topologically and functionally. But, although outstanding, biological systems are not unique in this respect, and the current interest in ecological wholeness and world models are a testimony. One could exhibit examples ad nauseam; we hope the reader agrees with us that this is not necessary here [1].

In terms of organization, what this empirical conclusion reveals is that system-wholes are *organizationally closed*: their organization is a circular network of interactions rather than a tree of hierarchical processes [2]. Conversely then, if we are trying to make more precise our notion of a whole, we propose to make these empirical results a guideline. That is, we propose to take the circular and mutual interconnectedness of organization, or organizational closure, as the *characterization* of system-wholes. In brief we propose the

Closure Thesis:

Every system-whole is organizationally closed.

Let us see now what this idea can do for us.

1.2 Distinction and stability

Behind the simplest idea of a system stands the basic act of splitting the world into what we consider separable and significant entities. To be sure, there are many ways to perform this subdivision of our experience. But some *criterion of distinction* is always present. Given some criterion, we distinguish and recognize things such as animals, galaxies or families. Some system-wholes appear to be quite universally distinguished, (i.e. persons); others (i.e. nations), seem more variable. Every culture will select quite specifically which are the predominant criteria of distinction; this selection need not concern us now. What we are considering is the common properties of system-wholes under *any* such criteria of distinction.

Next question is then: What is the common basis for a criterion of distinction to isolate system-wholes? Our answer: The specification of forms of interaction which identify a system-whole by its *stability*. That is, designation is possible only because we can enter in certain interactions which are repetitive enough. If a certain degree of repetitiveness exists, a system can be identified by its permanence or stability.

This is very interesting, for the stability of a system is a manifestation of its wholeness, insofar as the disruption of its organization will make it lose its stability. In the face of interactions that perturb it, a system-whole, asserts its individuality through compensations. But how is this stability achieved? We know: through the mutual balance and regulation of the processes that constitute it. We are back to the Thesis again! It is closeness in organization that ensures stability; organizational closure represents a universal mechanism for stabilization. The Closure Thesis appears to combine the manner in which we distinguish system-wholes, and our experience of how do they attain this stability and retain individuality in the face of interactions. The 'wholeness' of a system is embodied in its organizational closure. The whole is not the *sum* of its parts: it is the organizational closure of its parts.

Please note that when we speak of organizational closure, by no means do we imply *interactional* closure, i.e. the system in total isolation. We do assume that every system will maintain endless interactions with the environment which will impinge and perturb it. Lest this be so, we could not even distinguish them.

This openness to interaction, of course, raises an important objection to the Thesis: where do we draw the line to separate those interactions that participate in the system's organization, and those that are environmental disturbances? This is admittedly fuzzy. But this fuzziness rests in the many alternative criteria of distinction, and not in the notion of system boundary and organizational closure. Once a criterion of distinction (and hence some form of testing stability) is given and fixed, the boundaries of a system are perfectly clear. In many instances, however, we have more than one criterion of distinction, and we use them in sequence or alternatively. But this is another matter.

At a second level the Thesis can be suspect of another criticism: it seems to run counter to the classical approach in systems theory, of specifying inputs/outputs as part of the organization itself. This is an important point. We shall argue, however, that it is only an apparent objection, and that the classical treatment of systems is a particular form of a more inclusive approach for handling system-wholes, as proposed here. There is a lot more to say about this, but we will postpone this discussion for later in the chapter. It would distract too much from the central ideas we want to introduce in the first place.

In summary: we have the three interrelated notions of criteria of indication, systemic stability, and organizational closure. They appear related thus: given a criterion for distinction, system-wholes can be identified through their stable properties, and empirical experience tells us that such stability is due to organizational closure. Whence the Closure Thesis, that can be now restated in a less compact form.

Closure Thesis (second form):

Every (distinguishable) system-whole is (distinguishable through its stable properties arising from it

being organizationally closed.

2. The order of forms

2.1 Re-entry and the forms of system-wholes

What we need then is a more precise description of system-wholes in terms of their organizational closure. Such description should start from indicational grounds, that is, starting from the basic act of distinguishing against some indicational space.

The exploration of this indicational grounds is quite recent, and we owe it to the superb work of G. Spencer-Brown [3]. We are aware that most readers will not be familiar with this work, but we are forced to be very sketchy in this first part, and refer the puzzled reader to previous references [4]. Should we spend too much time on preliminaries now, we would not have a chance to introduce the new ideas!

Let us move then to a basic domain of distinctions, looking for the *forms* of indication irrespective of the processes, parts, and interactions they might stand for. In Brown's notation, let us indicate a distinction by a mark $\urcorner$ for the marked state, and let us indicate by the absence of such a mark. Brown's two axioms for these indicational constants are

$$\urcorner\,\urcorner = \urcorner$$

$$\overline{\urcorner}| = \quad .$$

From these axioms, we may calculate the value of any indicational form containing no variables, an *arithmetical* expression [5]. When variables are introduced, a complete algebra can be obtained with initials

$$\overline{\overline{p}|\ \overline{q}|}|\ r = \overline{\overline{pr}|\ \overline{qr}|}|$$

$$\overline{\overline{p}|\ p}| = \quad .$$

One of the many key innovations in Brown's work is that indications are relative to one another, as they all stand in relation to some indicational space or domain. This makes it very natural to ask: Can we have forms that connect with one another and with themselves through such a common indicational space? This is the case for the forms of system-wholes, since their organizational closure is not compatible with hierarchical forms such as

$$\overline{\overline{\urcorner\,\urcorner}|\ \ \urcorner}| \quad \text{or} \quad \overline{\overline{\overline{a}|\ b}|\ \overline{\overline{a}|\ c}|}|\ d\,,$$

but they resemble rather

$$\overline{\overline{\urcorner\,\urcorner}|\ \ \urcorner}|\ (\text{re-entrant}) \quad \text{or} \quad \overline{\overline{\urcorner\, b}|\ \ \overline{\urcorner\, c}|}|\ d\,(\text{re-entrant}),$$

where the form feeds-back on itself, in-forms itself.

In general, such forms arise when we contemplate expressions that are identical with parts of its contents,

$$f = \Phi(f) \tag{1}$$

where Φ is some indicational expression containing f as a variable. In another terminology, these are forms that are *fixed* points for some expression; as we shall see later, this is not an idle remark. For example

$$f = \overline{\overline{f}|}|$$

is one such expression. In studying them, Brown proposed the convenient convention of replacing the variable by an extension of the cross, i.e.

$$f = \overline{\urcorner}|\ (\text{re-entrant})\,.$$

Whence, in the example above we have

$$g = \overline{\overline{\urcorner\,\urcorner}|\ \ \urcorner}|\ (\text{re-entrant})$$

as being identical to

$$g = \overline{\overline{g\urcorner\,\urcorner}|\ \ g\urcorner}|\,.$$

All of these expressions have the peculiarity that, in re-entering their own indicational space, they in-form themselves, they are *self-referential.* Thus, in (1) we may take f to be asserting that Φ is the case for itself, to be self-indicatory.

Now, self-reference is really the nerve of the logic behind the description of wholes. This is so according to the Closure Thesis, since processes constituting their organization will be circular, mutually feeding back on each other, as in the simplest case of a feedback loop. Correspondingly, when considered in their bare form, stripped of all particularities of their processes, they will exhibit re-entry. At the level of indications, circularity of interactions, and self-reference is neatly captured as *re-entry*. Thus, the study of system-wholes (and for that matter, the study of anything where self-reference is central and not dispensable) requires that we can deal with re-entering forms in a satisfactory way.

The question is, can we in Brown's setting? The answer is, alas, no. Let us see why, formally and conceptually. Brown calls expressions like (1) Boolean equations of higher degree; equations of degree 1 being those with no re-entry, of degree 2 being those where a variable re-enters one time, and so on [6]. We prefer to speak of re-entering or fixed-point expressions, without distinctions of degree. Whichever the terminology the basic facts remain the same. We have that either (i) we cannot express every form of re-entry, or (ii) if we allow every form of re-entry the calculus is inconsistent. In other words, for certain expressions such as

$$f = \overline{f}\rceil$$

there is no value for f that can satisfy it: $f = \urcorner(f)$ has no fixed point.

This is a sudden let down, because we surely need access to all possible re-entering forms to represent the forms of system-wholes. Our initial proposal [4] was to take this at face value, that is, to look at this incompleteness as an indication that self-referential situations cannot be represented as non-self-referential ones. And to assign a specific value to the self-referential domain we suggested expanding the initial arithmetic to include a mark □ of self-indication or *autonomous* value. Then

$$f = \overline{f}\rceil = \square$$

has precisely this value, and in fact every expression of the form (1) has a fixed point in this Extended Calculus of Indications (ECI) [4].

We do not think now that this represents a really satisfactory solution to the handling of re-entering forms. What the ECI provides us with is a map of where things went wrong, rather than a map of the forms we wish to express. In introducing re-entry we see the collapse of the coherence on which the Primary Calculus of Indications (PCI) rests. Instead of letting it collapse by inconsistency, we extend it so that we can still contemplate what is happening when re-entry is allowed, a meta-standpoint. And what is it happening? That can be seen patently in the very nature of the autonomous value we had to introduce. It is a minimal self-referential situation, a self-oscillation. Taken from another point of view, it represents an extension of the grounds on which we are standing to *infinite* regress or *time* as Brown chooses to interpret it. Whichever way we wish to interpret it, the fact is that re-entry requires a more *extensive* ground than purely non-reentering expressions.

The introduction of a third arithmetic value to accomplish this necessary extension will not really do. First of all, because every re-entering form which is infinite (i.e. 'vitiously' self-referring) takes the same value, every one of these forms is compressed into the same state, and we can't see their differences, as we need to. Secondly, because in introducing a third value, certain forms with a great deal of intuitive meaning, lose this meaning, such as

$$\overline{\overline{p}\rceil \;\; p}\rceil = \quad ,$$

which is obscured to

$$\overline{\overline{p}\rceil \;\; p}\rceil\,\square = p\,\square \,.$$

Thirdly, we have tried to build a propositional logic out of ECI, as a test of its capacity to deal with closure in this particularly neat domain of logical expressions. The results, as reported elsewhere [4], are not really satisfactory. It can be done, but only through the introduction of certain ad-hoc rules that make the calculus awkward.

As we see it then, we must take the extension in ECI only as transitory stage to achieve a truly satisfactory ground for re-entry. What must be re-examined is the connection between re-entry and infinity or time. And this is much more an examination of the structure of re-entering forms, than the introduction of a value. The introduction of a new arithmetical value seems to do much violence to the initial indicational grounds.

In mathematics, infinity arises in most instances associated with the idea of approximation, and its companion notion of limit. We owe to the beautiful work of Dana Scott [7] this insight that we now regard as quite fundamental. His ideas were developed for the foundations of computer science; we see them as having a wider validity as we shall show below.

Thus we propose a change in strategy. Brown's suggestion and our own previous work can be made analogous to the extension of the real to the complex numbers, through the introduction of a new arithmetical value $\sqrt{-1}$. Scott's strategy, is to approach the problem in a fashion analogous to the extension of the rational to the real numbers, not by value introduction, but by approximation and limit, so that an entirely new ground emerges naturally from the initial one. Let us now see how this can be done.

2.2 Order and approximation of forms
The key idea to consider is really simple: forms have a very natural built-in order, in the sense that they can have relative degrees of *determination* or *approximation.*

More specifically, [8] let us denote by **I** the collection of all forms that can be constructed in the Primary Algebra of Indications. Let $\mathbf{I}_n$ denote the collection of forms of depth n or less. Naturally

$$\mathbf{I}_n \subseteq \mathbf{I}_{n+1}$$

where $\subseteq$ is used in its usual set-theoretic sense of inclusion. We assume that algebraic expressions contain variables from some initial list $X = \{a,b,c,...,z\}$. We also assume that an arbitrary, but fixed, assignment of values for the variables is given. Now as to the announced order relation in **I**.

Definition 1
Let f, g, be any expressions in **I**. Then we say that g is *at least as determined* as f, and write $f \sqsubseteq g$, if the contents of f coincide exactly with part of (or equal to) the contents of g when compared starting at the shallowest depth of both expressions ■

To see this more clearly let us convene to re-write a form as a *tree*, by spreading the form in two dimensions. Say for

$$f = \overline{\overline{\overline{d|}\; b|}\; \overline{c|}\,|}\; d$$

re-write it as

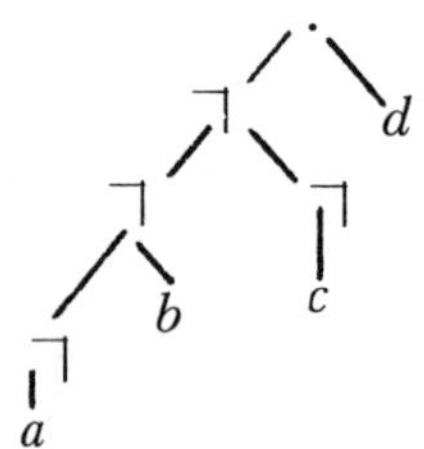

where '.' denotes the continence operation.

With this convention we can now re-formulate Definition 1 by saying that $f \sqsubseteq g$ iff when we take both trees and superimpose them starting at their roots, the branches of f will coincide exactly with part of the branches of g. Thus, at some points f will stop where g will continue to branch further. This partial coincidence can be made more precise by saying that, at the points where f stops and g continues, f has an *undetermined* value. In this sense, f is less determined than g, or f *approximates* g by a lesser degree of determination. We assume then, the existence of some undetermined value $\perp$, a *bottom*, which approximates everything.

Definition 2

$$\perp \sqsubseteq f \text{ for every } f \text{ in } \mathbf{I} \; ■$$

This idea of order in **I** is simple. For example, we have

$$\perp \sqsubseteq \overline{\perp\, b|} \sqsubseteq \overline{\overline{d|}\, b|}$$

and so on. Clearly

Proposition 1

$$f \sqsubseteq g \text{ and } g \sqsubseteq f \text{ if and only if } f = g \; ■$$

Note that not always two expressions can be compared; $\sqsubseteq$ is not total in **I**. For example, if the roots of two expression's trees are not the same, they cannot be compared. We want, however, to be able to construct an expression that a pair of forms could approximate.

Definition 3
An expression h is said to be an *upper bound* for f, g if and only if both $f \sqsubseteq h$ and $f \sqsubseteq h$ is the case ■

Now we can define a way to construct a least upper bound for pairs of forms

Definition 4
Let f, g be any two expressions. The *join* of f and g, written $f \sqcup g$, is that upper bound of f, g obtained by superimposing f and g at their roots, identifying identical branches, and dropping bottoms where some determined expression occurs at its place. $f \sqcup g$ is undefined whenever f, g cannot be thus superimposed ■

What we are doing here is taking the intuitive idea of order in **I**, and making it explicit as a 'poset' [8]. The order and the join are related very closely:

Proposition 2

$$f \sqcup g = g \text{ iff } f \sqsubseteq g \; ■$$

Consider now some sequence of expressions

$$f_1, f_2, ..., f_n = \langle f_i \rangle_n$$

We can inductively extend the idea of the join over such a collection.

Definition 5
The join of a sequence $\langle f_i \rangle_n$, when it exists, is a form f such that

$$f = \bigsqcup_{i=1}^{n} f_i = \left(\bigsqcup_{i=1}^{n-1} f_i \right) \sqcup f_n \quad \blacksquare$$

This allows us to consider a *limit* by extending joins over *infinite* sequences:

Definition 6
The *limit* of a sequence $\langle f_i \rangle_\infty = f_1,\dots,f_n,\dots$, when it exists, is a form f such that

$$f = \lim_{n \to \infty} \bigsqcup_{i=1}^{n} f_i = \bigsqcup_{i=1}^{\infty} f_i = \bigsqcup_{n} f_n \quad \blacksquare$$

Nothing has assured us that such a limit has any meaning as a form. We have begun to deal with infinite forms and our intuition about them is quite different from the finite ones. Remember that we had

$$\mathbf{I}_n \subseteq \mathbf{I}_m \quad \text{iff} \quad n < m\,,$$

and, in fact, the 'poset' **I** is

$$\mathbf{I} = \bigcup_{i=1}^{n} \mathbf{I}_i$$

for a large but finite n.

We can similarly consider the collection of infinite forms

$$\mathbf{I}_\infty = \bigcup_{n=1}^{\infty} \mathbf{I}_n.$$

What is this structure? What does it look like? Surely $\mathbf{I}_\infty$ still has a coherent partial ordering. Furthermore, we know what a join looks like for a finite collection of forms. But what does an f in $\mathbf{I}_\infty$ look like? Take any chain

$$f_1 \sqsubseteq f_2 \sqsubseteq \dots \sqsubseteq fn \sqsubseteq \dots$$

and its limit

$$\bigsqcup_{n} f_n = f.$$

Since for every n,

$$\bigsqcup_{i=1}^{n} f_i = f_n\,,$$

the limit is not as abstract as it seems, but is the nth member of the infinite sequences as we 'watch' it grow for unbounded values of n. Once a sequence is well specified, $\bigsqcup_{n} f_n$, is also well specified in the sense of being an effective construction for an unending form. Symbols like $\bigsqcup_{n} f_n$ do not denote objects that we can graphically display, but they are a well-defined mathematical construction.

This gives us an idea of what the elements of $\mathbf{I}_\infty$ look like. In fact, every element in $\mathbf{I}_\infty$ can be *defined* as the limit $\bigsqcup_{n} fn$ of a sequence

$$f_1 \sqsubseteq f_2 \sqsubseteq \dots \sqsubseteq f_n \sqsubseteq \dots$$

where we can take any f and chop it at some depth i

$$\text{Chop}_i(f) = f_i$$

and, of course,

$$f_i = \bigsqcup_{j=1}^{i} f_j$$

so that for any f we can construct a sequence $\langle f_i \rangle_n$ that approximates f with any desired degree of accuracy. Thus, we have

Proposition 3
$\mathbf{I}_\infty$ is a chain-complete 'poset', that is, every chain has a least upper bound ■

This is quite nice, because there is a neat correspondence between sequences and elements in $\mathbf{I}_\infty$. We do not need to assume anything at all about $\mathbf{I}_\infty$, we can construct its elements: They are limits of sequences. In $\mathbf{I}_\infty$, therefore, we have as nice a structure as we had in $\mathbf{I}_n$.

Definition 7
For any two forms, f,g, in $\mathbf{I}_\infty$

$$f \sqcup g = \bigsqcup_n (f_n \sqcup g_n),$$

$$f \sqsubseteq g \text{ iff } f_i \sqsubseteq g_i \blacksquare$$

$\mathbf{I}_\infty$ is also a 'poset' [8].

The operations of crossing and continence can be naturally extended to $\mathbf{I}_\infty$, by the convention of looking at every finite form as an infinite sequence of identical forms. That is, for any f in $\mathbf{I}_n$, write its sequence as $\{f_1, f_2, ..., f_n .....\}$, $f_1 = f_2 = ... = f_n = f$, so that

$$f = \bigsqcup_n f_n$$

Then for any form in $\mathbf{I}_\infty$ crossing and continence can be extended thus

$$fg = \bigsqcup_n (f_n g_n)$$

$$\overline{f|} = \bigsqcup_n (\overline{f_n|})$$

For example we have that, as expected,

$$f\urcorner = \bigsqcup_n f_n\urcorner = \bigsqcup_n \urcorner = \urcorner \sqcup \ \sqcup ... = \urcorner$$

since every f_i is in some $\mathbf{I}_i$.

2.3 *Fixed points of re-entering expressions*

This is as much as we need to know about the completed grounds of infinite indicational forms. Let us consider now re-entry in these completed grounds.

As we saw, a re-entering expression takes the form of fixed point of a general expression

$$g = \Phi(f) \tag{2}$$

where Φ is some algebraic expression. Even more generally, we may have multiple re-entry, and a system of interrelated equations

$$g_1 = \Phi_1(f_1), ..., g_n = \Phi_n(f_n) \tag{3}$$

Now, an equation like (2) is really a map between forms

$$\Phi: \mathbf{I}_\infty \to \mathbf{I}_\infty$$

and, similarly for (3),

$$\langle \Phi_1, ..., \Phi_n \rangle : (\mathbf{I}_\infty)^n \to (\mathbf{I}_\infty)^n \ ;$$

re-entry arises as the fixed points of these maps of $\mathbf{I}^i_\infty$ onto itself.

Consider now any algebraic expression Φ. Let $\Phi^n = \Phi(\Phi^{n-1})$, and $\Phi^o(f) = f$. Consider the chain, for some f,

$$\Phi^o(f) \sqsubseteq \Phi^1(f) \sqsubseteq ... \sqsubseteq \Phi^n(f) ... ;$$

surely $\Phi^n(f) \sqsubseteq \Phi^{n+1}(f)$. Then we have the

Proposition 3
For any algebraic expression Φ, and a given f, $\langle\Phi^n(f)\rangle$ is a chain and has a limit $\bigsqcup_n \Phi^n(f) = \Phi_f \blacksquare$

Next we introduce the notion of continuity [7].

Definition 8
A function $\varphi: \mathbf{I}_\infty \to \mathbf{I}_\infty$ is continuous if for every chain

$$\varphi(\bigsqcup_n f_n) = \bigsqcup_n \varphi(f_n),$$

(i.e., it preserves upper bounds) $\blacksquare$

Note that if φ is continuous then φ is also monotonous, i.e.

Proposition 4
If φ is continuous, then $f \sqsubseteq g$ implies $\varphi(f) \sqsubseteq \varphi(g)$.

Proof:
φ continuous implies

$$\varphi(f \sqcup g) = \varphi(f) \sqcup \varphi(g) \ ;$$

$f \sqsubseteq g$ implies $f \sqcup g = g$.

Then $\varphi(f \sqcup g) = \varphi(g) = \varphi(f) \sqcup \varphi(g)$,

hence $\varphi(f) \sqsubseteq \varphi(g) \blacksquare$

As it is clear from the definition of the join,

Proposition 5
Crossing and continence are continuous, i.e.

$$\overline{\bigsqcup_n f_n|} = \bigsqcup_n \overline{f_n|}$$

and

$$(\bigsqcup_n f_n)(\bigsqcup_n g_n) = \bigsqcup_n (f_n g_n) \ \blacksquare$$

Furthermore, since an algebraic expression Φ is composed only of repeated application of crosses we obtain the

Proposition 6
Every algebraic expression is continuous ■

With all of this, we can finally state the result that we were seeking all the way [9]:

Theorem 1
Every algebraic expression Φ has a fixed point, f_Φ, in $\mathbf{I}_\infty$. In fact

$$f_\Phi = \bigsqcup_n \Phi^n(\bot).$$

Proof:

$$\Phi(f_\Phi) = \Phi\Big(\bigsqcup_n \Phi^n(\bot)\Big) = \bigsqcup_n \Phi^{n+1}(\bot)$$

$$= \bigsqcup_n \Phi^n(\bot) = f_\Phi \; ■$$

This fixed point may or may not be the only one, but it is always available.

Theorem 2
Every system of n equations has at least one fixed point in $\mathbf{I}^n_\infty$.

Proof:
Consider first a system of two equations

$$\Phi = \langle \Phi_1, \Phi_2 \rangle : \mathbf{I}_\infty \times \mathbf{I}_\infty \rightarrow \mathbf{I}_\infty \times \mathbf{I}_\infty.$$

This is also a continuous function on $\mathbf{I}^2_\infty$, since it is composed of two continuous components Φ_1, Φ_2. Also a chain $\langle f_{1i}, f_{2i} \rangle$ in $\mathbf{I}^2_\infty$ will have a limit $\langle f_1, f_2 \rangle$ as the paired limit of its composing sequences

$$\langle f_1, f_2 \rangle = \bigsqcup_n \langle f_{1n}, f_{2n} \rangle = \langle \bigsqcup_n f_{1n}, \bigsqcup_n f_{2n} \rangle.$$

Then Φ has a fixed point

$$\Phi \langle f_{1\Phi}, f_{2\Phi} \rangle = \langle f_{1\Phi}, f_{2\Phi} \rangle,$$

since the proof of Theorem 1 can be carried out identically for any n-tuple of forms. Thus, the case with n equations is similarly proved ■

These two results permit us to be sure that

Theorem 3
Every system of equations has at least one solution in $\mathbf{I}^n_\infty$ ■

By considering finite forms as embedded in this much larger ground of $\mathbf{I}_\infty$ our problems seem to be over. Let us insist that $\mathbf{I}_\infty$ is really a mathematical construct that serves us to express and calculate. Most of the components being infinite, they are not, properly speaking 'objects'. This is not to say that we cannot *represent* them in quite intuitive ways, construct models.

To see this more clearly, let us convene to rewrite a form as an extended two-dimensional tree. In such a tree, re-entry can, especially in the complex cases, be indicated more explicitly.

Examples
Let us take the simple

$$f = \overline{f\,}| = \Phi_1(f).$$

The solution for this equation can be constructed as

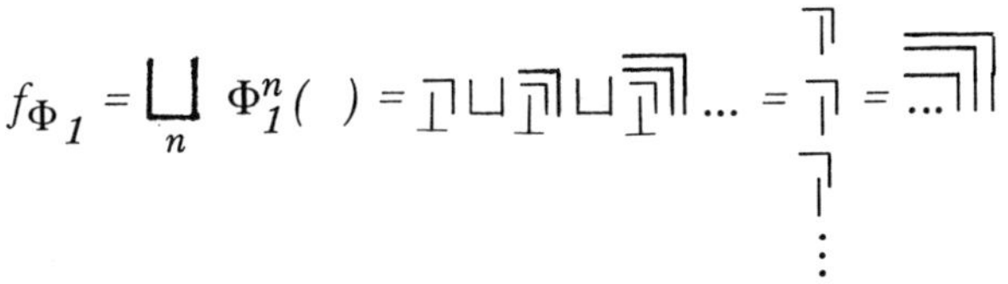

$$f_{\Phi_1} = \bigsqcup_n \Phi^n_1(\;) = \overline{\bot}| \sqcup \overline{\overline{\bot}|}| \sqcup \overline{\overline{\overline{\bot}|}|}| \ldots = \ldots$$

in $\mathbf{I}_\infty$. Yet it is clearly better to *represent* it as the minimal re-entry

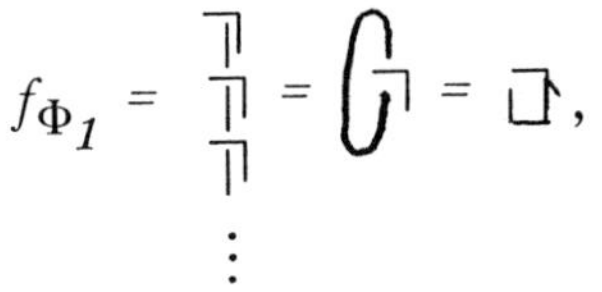

$$f_{\Phi_1} = \ldots$$

in the form of a closed circuit.

Next take

$$f = \overline{\overline{f\,}|}| = \Phi_2(f);$$

here

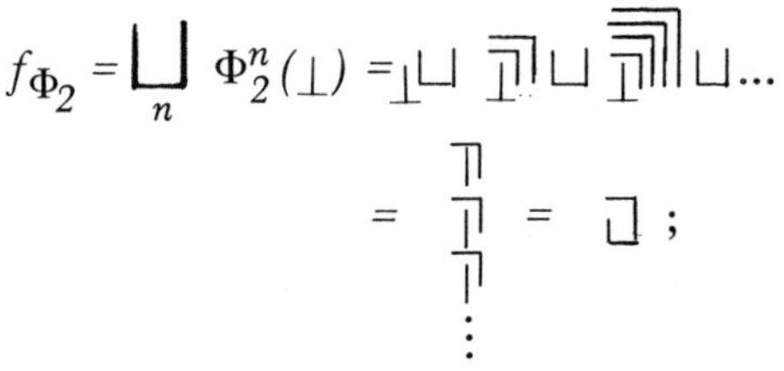

$$f_{\Phi_2} = \bigsqcup_n \Phi^n_2(\bot) = \bot \sqcup \overline{\overline{\bot}|}| \sqcup \overline{\overline{\overline{\overline{\bot}|}|}|}| \sqcup \ldots$$

$$= \ldots ;$$

so $\sqcup$ is also a fixed point for Φ_2. In fact, any expression f is a fixed point for Φ_2,

$$\Phi_2(f) = \overline{f\urcorner}\urcorner = f.$$

For a less simple case take

$$f = \Phi_3(f) = \overline{\overline{f\urcorner p}\urcorner\; q}\urcorner\;;$$

then

$$f_{\Phi_3} = \cdots \quad = \quad \overline{\lfloor p\urcorner\; q}\urcorner .$$

Or, in the case of simultaneous equations

$$\Phi_4(f)\begin{cases}\Phi_{41}(f_1) = \boxed{\overline{a}\urcorner\ \lfloor\overline{c}\rfloor\ \ d} = \overline{f_1\ a\urcorner\ \overline{f_1\ f_2\ c}\urcorner\ d}\urcorner\\ \Phi_{42}(f_2) = \boxed{f_1\ \ c} = \overline{f_2 f_1\ \ c}\urcorner\end{cases}$$

with the 'double' infinite tree

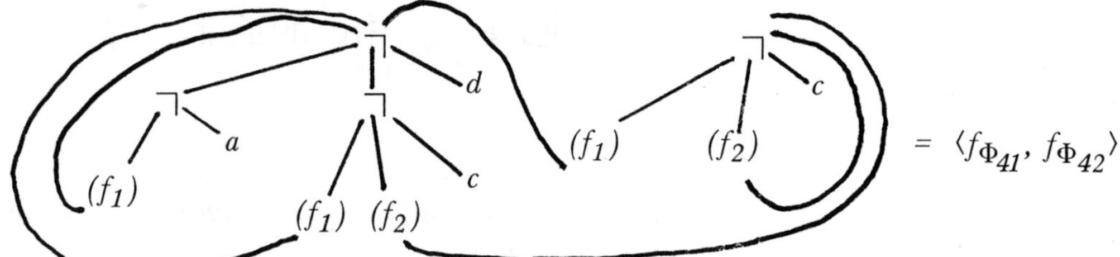

as a fixed point

$$\langle f_{\Phi_{41}}, f_{\Phi_{42}}\rangle = \bigsqcup_{n=0}^{\infty} \langle \Phi_{41}^{n}(\perp), \Phi_{42}^{n}(\perp)\rangle.$$

This last example points to the central question of the kinds of simple expressions that are available in $\mathbf{I}_\infty^n$ to which every expression can be shown equivalent or reduced to, once all variables are assigned a value. Such simple expressions are the arithmetical *values* for this algebra of infinite expressions. We now know that $\urcorner$, , $\sqcup$ belong to $\mathbf{I}_\infty$. The question is: are there more? Partial answer:

Definition 9
Let x in $\mathbf{I}_\infty^n$ be an *arithmetical expression* iff it contains no variables ■

Definition 10
Let an arithmetical expression be called *rational* [10] iff it is finite or is the fixed point of some finite algebraic expression ■

Definition 11
Let an arithmetical expression be called a *value* of the arithmetic iff it cannot be reduced any further by substitution using

$$\overline{x\urcorner}\urcorner = x$$

$$x\urcorner = \urcorner$$

$$x = xx \quad ■$$

Theorem 4
The only rational arithmetical values in $\mathbf{I}_\infty$ are $\urcorner$, , $\sqcup$.

Proof
Surely $\urcorner$, , $\sqcup$ are arithmetical values. Assume some other distinct rational arithmetical value x exists in $\mathbf{I}_\infty$. x cannot be finite for it would reduce to either $\urcorner$ or . So x must be the infinite solution of an equation

$$x = \Phi(x) \qquad (*)$$

for some finite algebraic expression Φ.

It must be the case that Φ contains no variables, for x is an arithmetical expression. Consider then the contents of $\Phi(x)$ at some depth i, and let n be the deepest space occurring in $\Phi(x)$. At any depth i, there are exactly three possibilities:

1. only the unmarked state occurs;
2. the marked states occur at least once;
3. only x occurs (perhaps more than once).

If for every i, $0 \leqslant i \leqslant n$, (1) and (2) are the case, then it is clear that (*) reduces to

$$x = \quad ,$$

or

$$x = \,\urcorner\, ,$$

contrary to the initial assumption.
Thus, (3) must be the case at least once, and at most n times. Let us simplify $\Phi(x)$ by (1) and (2), and suppose (3) is the case some $j \leqslant n$ times, i.e.

$$\Phi(x) = \underbrace{\overline{\overline{\overline{\overline{x}|\;\; x}|\ldots x}|\, x}|}_{j}$$

This, again, contradicts the initial assumption that x is not ⊔⌉. For let

$$⊔⌉ = x\,,$$

then

$$\begin{aligned} ⊔⌉ &= \overline{x}| \\ &= \overline{xx}| \\ &= \overline{\overline{x}|\; x}| \\ &\;\vdots \Big\}\, j \\ &= \underbrace{\overline{\overline{\overline{\overline{x}|\;\; x}|\ldots x}|\;\; x}|}_{j} \\ &= \Phi(x) \end{aligned}$$

Thus, in every case the assumption that x is not ⌉, , or ⊔⌉, leads to a contradiction. This completes the proof ■

This is very interesting for two reasons. First, it gives us something that we can relate to the Extended Calculus (more about this later). Secondly, and more important, it tells us that it is the simultaneity of re-entry that provides the source of diversity in arithmetic values. These we can draw and depict as multiple interconnections and nested loops. Mathematically, they are tuples of infinite forms. In other words it is the 'poset'

$$\mathbf{I}_\infty \times \mathbf{I}_\infty \times \mathbf{I}_\infty \times \ldots = \mathbf{CI}$$

the one that is really our *complete* (continuous) ground for indicational forms, and which we propose to call the *Complete Calculus of Indications* **CI**. We know **CI** is sufficient to solve any equations we wish. And not surprisingly

Theorem 5
There are infinitely many arithmetical values in **CI**.

Proof:
It is enough to exhibit some infinite collection of arithmetical values. Consider the pair of forms,

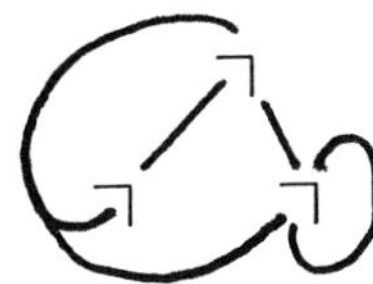

that is, the least fixed point of the system

$$\Phi_1 \begin{cases} f_0 = \overline{f_0 \; f_1}| \\ f_1 = \overline{\overline{f_1}|\; \overline{f_1 \;\; f_0}|}|\,, \end{cases}$$

or

$$f_{\Phi_1} = \overline{\urcorner\,|\,⊔⌉}|\,.$$

f_{Φ_1} cannot be reduced further to ⌉, , or ⊔⌉. Hence it is a simple expression.

Define Φ_2 as $f_{\Phi_2} = \overline{\urcorner\,|\,\overline{\urcorner\,|\,⊔⌉}|}|\,,$

that is

$$\Phi_2 \begin{cases} f_0 = \overline{f_1 \;\; f_0}| \\ f_1 = \overline{\overline{f_1}|\; \overline{f_1 \;\; f_0}|}| \\ f_2 = \overline{\overline{f_2}|\; \overline{f_2 \;\; f_1}|}| \end{cases}$$

and inductively

$$\Phi_n \begin{cases} \quad\vdots \\ f_n = \overline{\overline{f_n}|\; \overline{f_n \;\; f_{n-1}}|}|\,. \end{cases}$$

Then $f_{\Phi_1}, f_{\Phi_2}, \ldots, f_{\Phi_n}, \ldots$ is an infinite collection of irreducible arithmetic forms in **CI**. Hence the theorem follows ■

2.4 Arithmetics of closure
As a result of completing the indicational forms, we have reconstructed ECI. In fact, $\mathbf{I}_\infty$ has the same

arithmetic values, and the ECI axioms are also valid

$$\overline{\sqcup}\rceil = \sqcup,\ \sqcup\rceil = \rceil,\ \sqcup\sqcup = \sqcup,$$

This tells us how unnecessary it was to assume the autonomous state rather than construct it. In $\mathbf{I}_\infty$ we can see the autonomous value as representing the state of *self*-interaction or self-indication. But as we consider *two* mutually interrelated elements we have a value in $\mathbf{I}^2_\infty$, and so on. This opens up, as we saw by Theorem 4, an infinity of arithmetic values in **CI**.

This has to be evaluated properly. An arithmetic value is a constant that embodies certain properties of a domain close to our direct understanding. It is in the contemplation of what sorts of relations these arithmetic expressions can have, that a calculus for these arithmetic expressions emerges, and stands at the base for algebraic generalization [5]. As we learned from Brown, the true richness of a domain lies in its arithmetic, as it is closer to that which is directly accessible to us, rather than through algebraic calculation. It is this realization that allowed Brown to look for, and find, the proper arithmetic behind Boolean algebras and establish for the first time the arithmetic of indication (PCI). What is truly fascinating, then, is that, when this domain of indications is opened up for closed or self-referential forms, an entirely new variety of arithmetical values is revealed. Not only a paradigmatic autonomy, as a tail-chasing $\sqcup$, but, in fact, an infinity of mutually interrelated forms in every conceivable appearance and degree of intricacy [11]. This *arithmetic of closure* we regard as the truly novel and beautiful insight that we have to report here.

The full exploration of the complete indicational arithmetic is yet to be done. We can easily identify, of course, finite and infinite forms (i.e. of first degree and 'vitiously' self-referring forms). But there is much more than that. How do these arithmetical values relate to each other individually? Are these interrelations reducible to a finite list of axioms? Further, consider those forms which are *not* fixed-point solutions to a set of equations: do they exist?, how do they look?, how can they be represented? Also, we have only considered a finite number of mutually interrelated elements, but in **CI** we need not do so. We may contemplate a 'field' of indications, an infinite series of mutual interrelations. Where can these and other exploration of **CI** lead to? We do not know, but there it is, open.

We could have presented this extension of the laws of form just by itself. We believe, however, that we miss its true arithmetic substance, in the Brownian sense, unless we see that self-indication plays a key role in the description of any *universe*. For, any universe that has been conceived of is populated of system-wholes, and, accordingly, it is to the laws of indications and their forms that we can trace their source. This is our reason for the way this chapter is presented, and why we would like to finish it by returning to reconsider closure in system-wholes.

3. Re-evaluating the thesis

3.1 From a Fregean to a Brownian approach to systems

Closely considering the nature of system-wholes has revealed quite a number of implications. When we strip systems to their bones, to their indicational forms, this is patently revealed. That we had to complete our basic descriptive grounds suggests to me that whenever we consider elements interacting reciprocally, new properties emerge which cannot be described in terms of isolated components. In our indicational formalism, this is reflected as an infinity of arithmetic values. In our direct experience with systems, it represents the emerging of 'holistic' properties that systems exhibit as they acquire stability by and through closure.

Thus, the heated philosophical debate between holism and reductionism has, in **CI**, a very clear counterpart. But, here it is not a debate any more, but a complementarity. In a numerical analogy, there is no 'opposition' between the rational and the real numbers. If we stay only within the rational domain there are numerical relations that have no solution; but it is through the rationals that real numbers can be constructed. *Mutatis mutandi*, to consider hierarchical non-circular interactions is quite possible, but they cannot account for the re-entering ones, which can, instead, be seen to arise from them by an infinite approximation. Thus, the study of forms, open or closed, is a ground, on which there is a superation of the dichotomy holism/reductionism.

What is at stake is the interdependence between the need to consider whole systems, and the correlated necessary appearance of circular interrelations of processes. I do not feel that the full impact of this cognitive issue has been fully adsorbed. When Wiener brought to the foreground the feedback idea, not only did it become immediately recognized as a foundational concept, but it also raised major philosophical questions as to the validity of the cause-effect doctrine. The picture seemed closer to a circular causation, where one can deal only with the ensuing totality and its manifested stability. In other words, the nature of feedback is that it gives a mechanism, which is independent of particular properties of components, for constituting a stable unit. And from this mechanism, the appearance of stability gives a rationale to the observed purposive behavior of systems and understanding teleology [12]. Since Wiener, the analysis of various types of systems bears this same generalization: whenever a whole is identified, its interactions turn out to be circularly interconnected, and cannot be taken as linear cause-effect relationships if one is not to lose the system's characteristics [1]. This is reflected in the simultaneity of solutions in $\mathbf{I}^n_\infty$, when there are n forms mutually related. In a paraphrase:

> Principle of Inaccessibility of Circuit Elements:
> *In an organizational closure, no element is accessible in isolation.*

All of these considerations seem to us necessary if we are not to betray our deepest intuitions about systems and organization. It is surprising that there hasn't been more attention paid to the key role of closure. Even further, I suspect that many will be outraged. What about input/outputs and control? This is surely not closure! Let us spend some time with this objection.

We contend that the reluctance to concede a central role to circularity in system's organization is basically a heritage from empiricism, or what we would like to call a *Fregean* systems view-point. The basic assumption here is that we can look at a system and identify initial or atomic elements with which a larger system can be constituted and so on until an output is reached. The idealized form of this logic is the Whitehead-Russell theory of types, where some atomic elements are given, and do not affect operations of higher types. The mental picture is that of a tree with roots and branches. But, this view is awkward for describing system wholes, where the picture is more that of a closed network, where root and branches intertwine. It resembles the network of language that the late Wittgenstein was concerned with [13]. No type distinctions are possible in such a network. This kind of logic is the basis of what we wish to call a *Brownian* approach to systems.

Surely, a lot of contemporary cybernetics and system theory does recognize implicitly the relevance of circularity and in practice, (on an ad-hoc basis Brown would say) attempt to deal with their description via temporal and sequential descriptions, via delays or recursion (see Appendix). This is fine, and can take care of itself. Our point is, however, that in formulating such notions explicitly, there is return to a Fregean attitude, and this is what is involved in putting inputs and outputs, or reference points, or finiteness in the recursion, where, again, there is openness of organization. This reflects, I believe, the historical fact that the most sophisticated tools in systems theory have been generated in the context of engineering and computer science. There, the goal of the *design* is the motivating force, and hence the input/output approach is quite suitable. The system is quite definitionally open.

In contradistinction, in dealing with natural systems, the whole idea of i/o becomes muddled. Who and how are we to select a fixed set of input and output spaces? It is more adequate to talk about environmental perturbations/compensations. And this is quite different. For then, we explicitly start with a system's stability, coming from the closure of its organization, which is the basis for its capacity to confront a perturbation and compensate it. Of course, we may take a *fixed* set of such p/c's and treat the system *as if* it were organizationally open. However, useful this has been and is in engineering and·design, it misses a deeper insight about the system's organization proper (and for those systems of which man is a part, this has proven to be disastrous) [14]. For if we know what the closure of interactions is, and the limits of the stability to which it gives rise, then we can make sense of any observed pair of p/c's.

We are saying then that in our dealing with natural systems, we have carried a philosophical and methodological position adequate only for the domain of design, and that this is a conceptual inconsistency. A general science of organizations, and first of all of the organization of natural systems, has to effect a transition from a Fregean to a Brownian foundation. The Closure Thesis, I submit, is a methodological guideline: if you are to study a system, assume it has a closed closed organization, analyze individual pathways until a reconstruction of the network is obtained, and then putting all of these circuits together simultaneously, see what kinds of stability they can generate. The i/o approach is, in fact, a moment in this process, insofar as we fix certain modes of interaction with a purpose in mind. This is the only way of investigating particular paths within a given organizational network. The simultaneous performance of all of these pathways together, however, is a different matter.

The i/o view is like a Newtonian view of the universe where some absolute framework is needed; a closure view is more 'Einsteinian', where a variety of frames of reference are possible. And, to continue this analogy to physics, the Closure Thesis suggests something like a 'frictionless mobile', an idealized situation, which can be observed in several degrees of 'friction'. Similarly, only well defined p/c's can give us an idea of the basis for an observed stable behavior arising from closure.

The key point is then that the most fundamental *stuff* with which a general theory of systems has to deal with is organizationally closed, and that we can do so by (conceptually or actually) opening their network of interactions and perturbing stability. However useful this opening might be, it is only a stage in obtaining a fuller representation. Hence we propose the

Closure Principle:

A system-whole, if devoid of perturbations that that can interrupt its organizational closure, will maintain its stability indefinitely.

The last aspect of the Closure Thesis we wish to consider is that of self-organization. In fact, by adopting the Thesis we are *ipso facto* saying that the establishment of an organizational closure in a given domain is the mechanism for self-organization [15]. By self-organization we mean here the spontaneous assembly of a system's components to a stable unit of emergent properties. Surely every natural system is self-organizing. Thus the Thesis applies to them simply by implication. A clear paradigm for this is the origin of cells: only when a set of chemical reactions closes onto itself by self-production are stable systems attained that can distinguish themselves from their environment by boundaries, proto-cellular systems [16]. Quite in general, reciprocal interaction is the mechanism on which a ladder of stable systems can be ascended, from atoms to society. Or vice versa, if some system arises through self-organization and is stable, this is surely due, according to the Thesis, to the reciprocal interactions of the components. This means that we can take the Closure Thesis not only as a guideline to the description of system-whole, but *also* to the mechanism of their self-organization and evolution.

3.2 Second order closure

Our previous statement to the effect that we haven't yet absorbed fully the cognitive and epistemological impact of moving from a Fregean to a Brownian foundation should not be underestimated. We wish to close this chapter by laboring on this a bit more, that is, by considering the closure, not of observed systems, but of observing systems, a *second-order* closure [17].

The central point emerging from the study of closure and self-reference is that the structure on which we have to base our description has to be endowed with a new property that can be described as *time* or infinity, and its manifestation as sequential approximation and recursion. **CI** is a descriptive calculus which contains time *innately*. It is a logical ground in which time cannot be extirpated unless one loses most of it.

Conversely, the tradition in the west has been predominantly two-valued, avoiding self-reference like the plague, and accepting implicit self-reference only as non-vicious (finite) recursion only lately in this century. This is important, for this preference goes hand in hand with the western epistemology of excluding observer from observation: it's either subject or object. Gothard Günther [18] has been a lonely voice in pointing out that the problem of time, self-reference, and the observer/observed interrelations is one and the same, and that it looks so formidable a problem is essentially because of the grounds on which western logic has decided to stand. To incorporate the observer into our descriptions, Günther says, time has to be given a logical *locus*.

Günther's claims are precisely borne out when the re-entry of forms is considered. Not only because time is absolutely required, but because its introduction necessitated the Brownian foundation of indication, which is so obviously observer-bound. A distinction reveals where the observer stands and what he chooses among the many possible forms in which his reality can be dismembered. On indicational grounds, the world is made of proofs, not facts. More pictorially

... a distinction is done by an observer
which reveals his properties
which reveal what the contents of ...,

or, more succinctly,

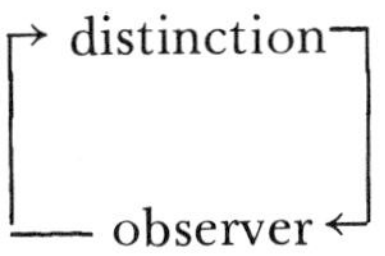

Since the form of this interaction is clearly one of re-entry, observer and observed are, by the Principle of Inaccessibility, meaningless if considered in isolation from each other. Reality, as we know it, is not separable from we, that know it; we, as knowers are not independent of the reality we know. Epistemologically, this is a shift, as von Foerster has said so neatly [19], from the classical paradigm of 'The properties of the observer shall not enter into the description', to a paradigm where 'The description shall reveal the properties of the observer'. This paradigm shift makes the world a different world. And it leaves us with the realization that the study of wholes leads directly to us and is a part of our own study, so that the theory illuminates the subject, and the subject is what makes theorizing possible. The science of organization we are pursuing is of this nature: it is done by us, but it contains us, and this is to be not implicit but explicit. It is not a simple transition in the history of science. But this is the kind of science required for a proper study of systems-wholes. The arithmetic of closure being, in the end, the closure of arithmetic.

Appendix—Forms as systems

Can we make more explicit the connection between what forms are and what specific systems are?

We will take, first of all, a flip-flop circuit, as a simple and well-known example, and examine it in some detail to get a flavor for how things work. A good reason to choose this example is that being a logical circuit, we can interpret it directly as a form. In fact the typical flip-flop

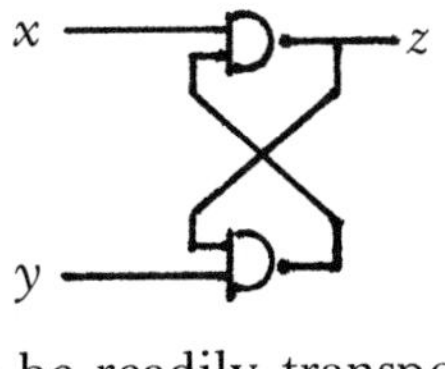

can be readily transposed into its corresponding form

$$z = \overline{\overline{z}\rceil\ \overline{y}\rceil}\rceil\ \overline{x}\rceil \qquad (*)$$

or, if you wish,

$$z' = \overline{z}\rceil = \overline{\overline{\overline{z}\rceil\ \overline{y}\rceil}\rceil\ \overline{x}\rceil}\rceil = \boxed{\overline{\overline{y}\rceil}\rceil\ \overline{x}\rceil}$$

with its tree

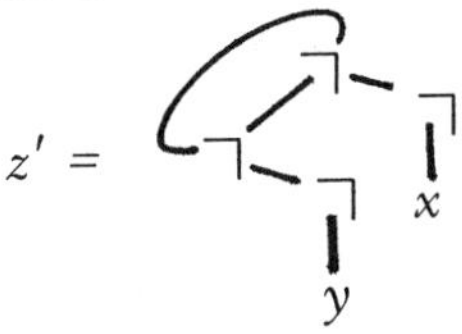

Now, z is the limit of an approximation

$$z = \bigsqcup_n z_n$$

where

$$z_1 = \overline{\overline{z_0}\rceil\ \overline{y_1}\rceil}\rceil\ \overline{x_1}\rceil ,$$

and, in general,

$$z_n = \overline{\overline{z_{n-1}}\rceil\ \overline{y_n}\rceil}\rceil\ \overline{x_n}\rceil \qquad (**) .$$

Clearly

$$z_n \sqsubseteq z_{n+1} .$$

This also specifies as sequence for x and y,

$$x = \bigsqcup_n x_n ,$$

$$y = \bigsqcup_n y_n .$$

All of this makes sense because, in an actual flip-flop, expression (*) is, of course, interpreted in time as a discrete step-by-step recursive function (**), for a given sequence of inputs x_i, y_i. In fact, we could have done that all along in **CI**, by interpreting z (in time) as a finite sequence, starting with some z_0, and under some finite of x_i's and y_i's, the following algebraic expression is valid (as can be easily verified by induction):

$$z_n = \overline{\overline{z_0}\rceil\ \alpha(n)}\rceil\ \beta(n) ,$$

with

$$\alpha(n) = \overline{y_1}\rceil\ \overline{y_2}\rceil\ \dots\ \overline{y_n}\rceil ,$$

$$\beta(n) = \overline{\overline{y_2}\rceil \dots \overline{y_n}\rceil\ x_1}\rceil\ \overline{\overline{y_3}\rceil \dots \overline{y_n}\rceil\ x_2}\rceil \dots$$

$$\dots \overline{\overline{y_n}\rceil\ x_{n-1}}\rceil\ \overline{x_n}\rceil .$$

This is a recursive expression that algorithmically determines z_n for every n, and this is what is normally done in representing these kinds of logical circuits with feedback.

We can see, however, that this approach fits hand in glove with our approximation to an infinite expression (*), which, as it were, embodies the self-referring quality of this re-entering circuit. It represents, formally and intuitively, the basic structure of the flip-flop as a logical design, rather than describes it as an ad-hoc sequential expression. The time/recursive expression shows how it can actually be operated; its re-entering forms show what it is and what it means.

This example says little as to how to represent specific system-wholes, where there is closure but where the interactions that occur are not simply transcribable into circuit logic. The extension of a simple and unique value of indication to a more diversified *domain of operators* can nevertheless, be done very elegantly due to the construction of initial continuous algebras [8]. But to make this extension minimally intelligible will take another chapter, which we hope to report on shortly.

Notes

1. A full discussion on this view of living systems appears in Maturana and Varela (1973, 1975), and Varela, Maturana, and Uribe (1975). Some of our favorite studies of closure in significant classes of systems are: the closed nervous system approach to cognition as in Maturana (1969, 1975), and also Powers (1973); the teacher/taught or dialogue interaction as in Pask (1975), and Bateson's view of schizophrenia and alcoholism (1972), part III.
2. Our feeling in talking about organizational closure is that it is too obvious to even argue about it. Yet, many conversations have convinced us that this is only our personal illusion. Perhaps being too obvious is what makes it so difficult to see self-reference in full focus, in spite of its appeal in several areas. This can be illustrated by contemplating the language used in several of these areas, as in the list below:

systems engineering, cybernetics	feedback
	⟨OE. fook-food+back-reverse⟩
logic, philosophy	self-reference
	⟨L. re-back+ferre-to bear⟩
philosophy, biology	autonomy
	⟨Gr. autos-self+nomos-law⟩
logic, computer science	recursion
	⟨L.re-back+currere-to run⟩
psychology	self-reflection, reflection
	⟨L.re-back+flectere-to bend⟩
informal	circular ⟨Gr. kirkos⟩ / loop ⟨ME. loup⟩

In considering their more original meaning, all these words seem to arise from the same basic intuition, surfacing in different roots for different disciplines.

3. SPENCER-BROWN, G. (1969). There is a second American edition by Julian Press (1971). Also see his book under the pen name of James Keys, which he claims as a complement of *Laws of Form*. We have also learned much from informal transcripts of a conference with Brown at Esalen, California, during April, 1973. We are grateful to J. Lilly for making those transcripts available to us.
4. Cf. VARELA (1975) and two related publications, in which the ideas are carried to logic, VARELA (1976 a,b).
5. Arithmetic is, according to Brown (Abramovitz *et al.*, 1975) p.136 "A calculus in which the constants operated on all have specific values..." Arithmetic relations are proved by theorems, rather than mechanical demonstrations, as in an algebra. We find this epistemology of mathematics very insightful. For other views on it see (Richardson, 1974; Kreisel, 1971).
6. We do not think that this is necessary, since any equation can be reduced to a degree less than or equal to 3, cf. Theorem 14 in Brown (1969) and Theorem 6 in Varela (1975).
7. The more rigorous treatment is to be found in (Scott, 1972). A very clear exposition of the basic ideas is Scott (1971). His general programme for a foundation of computer science is in Scott (1970).
8. In the mathematical presentation that follows, we have taken a compromise between conciseness and rigor, so that the whole exposition has an introductory and rather sketchy character. I hope, however, that enough detail is provided so that the reader can see the mathematical flavor of the pro-

posed developments.

Thus, for example, I have used the word 'poset' in quotation marks as a warning. Properly speaking (Birkhoff, 1967; Markowsky, 1976) **I** is not a poset unless the definitions are refined in the sense of a better specification of depth and order through some encoding of the form's contents. See Goguen *et al.* (1975) and the references therein, specially Thatcher (1973). I tried this more adequate definition in an earlier draft of this chapter, but found that to make it mathematically sound and clear takes more space than I can afford here.

Thus, we are aware, and the reader should be aware, that a more precise construction is needed for a more thorough exploration of the order of indicational forms. I have decided to compromise with precision in this presentation for the sake of making the basic ideas more clear. I ask the indulgence of the more rigorous-oriented reader for this choice. A full presentation will appear elsewhere.

9. This fixed point theorem was apparently first recognized in all its significance by Tarski (1955), and discussed in Birkhoff (1967) p.115. For other related applications see Kleene (1952) and Manna (1974), chap. 5, on the fixed point theory of computation.
10. The idea of rational here is that which can be, recursively or iteratively, specified. For all we know, there might be non-rational expressions and/or values in $\mathbf{I}^n_\infty$. Similarly, there are infinite algebraic functions which could be of the non-rational sort. The restriction the rational type is useful since these are the only expressions we *actually* deal with and *need.*
11. Can we help being reminded of the amazing discovery, that once one number such as $\sqrt{2}$ is constructed, the entire world of irrationals appears as well?
12. First discussion on this is in Wiener, Rosenblueth, and Bigelow (1943); see also von Foerster *et al.* (1968).
13. We owe this insight and its connections with the Fregean idea of elements, to an unpublished paper by I. Waldo B. presented during the Naropa Seminary, Fall of 1976.
14. We suspect this is the point, if at all, where cybernetics and a general theory of systems part ways. For, cybernetics is essentially concerned with design as the science of effective action; in this sense is a part of the more inclusive view of a general theory of systems. This is only a matter of christening, and we are aware that almost everybody will have a different proposal.
15. The converse: every conceivable closed organization is stable, seems false. It raises the question of viability in a given domain, of different degrees of stability for different forms of closure. Viability is determined by specific properties of components, and is not, properly speaking, inherent in the organization itself.
16. See Varela *et al.* (1975). For a biophysical discussion of the same ideas see the remarkable work of Eigen (1971), and also Fox and Dose (1972), chaps. 6 and 7.
17. Cf. von Foerster in Abramovitz *et al.* (1974), p.1.
18. Günther's work is not easy to read, and we have found his papers on time, of 1967, more illuminating than other references. For a more complete bibliography see Abramovitz *et al.* (1974) p.487. It is no accident that Günther found the origin of his interests in Hegel. The marxist version of the hegelian dialects is, in our opinion, the closest approximation to Günther's thinking and to the kind of epistemology espoused here. Unfortunately, dialectic epistemologies have been quite degraded by most popularizers such as F. Engels. For a very thoughtful study on dialectics see Kozik (1970).
19. Cf. von Foerster (1975).

References

ABRAMOVITZ *et al.* (Eds.), *Cybernetics of Cybernetics,* Biological Computer Laboratory, Univ. of Illinois, Urbana (1974).

BATESON, G., *Steps to an Ecology of Mind,* Ballantine Books, New York (1972).

BIRKHOFF, G., *Lattice Theory*, A.M.S. Colloquium Publications, Vol.XXV, Providence, Rhode Island (1967).

BIRKHOFF, G., What can lattices do for you?, in: J. Abbott (Ed.), *Trends in Lattice Theory*, Van Nostrad Reinhold, New York (1970).

EIGEN, M., Self-organization of matter, and the organization of biological macromolecules, *Naturwiss* **10**, p.466 (1971).

FOX, S. and DOSE, K., *Molecular Evolution and the Origin of Life,* Freeman, San Francisco (1972).

GOGUEN, J., THACHTER, J., WAGNER, E., and WRIGHT, J., 'Initial Algebra Semantics', IBM Research Report RC 5243, Yorktown Heights, New York (1975).

GÜNTHER, G., 'Time, timeless logic, and self-referential systems', *Ann. N.Y. Acad. Sci.* **138**, 396 (1967a).

GÜNTHER, G., *Logik, Zeit, Emanation und Evolution,* Heft 136, Westdeutscher Verlag, Köln und Opladen (1967b).

KEYS, J., *Only Two Can Play This Game,* Ballantine Books, N. York (1974).

KLEENE, S., *Introduction to Metamathematics,* Van Nostrand, N. York (1952).

KREISEL, G., Review of the Coll. Papers of G. Gentzen, *J. Philos.* **68**, 238 (1971).

MANNA, Z., *Mathematical Theory of Computation,* McGraw-Hill, N. York (1974).

MARKOWSKY, G., Chain-complete posets and directed sets with applications, *Algebra Universalis* (in press), (1976).

MATURANA, H., Neurophysiology of Cognition, in: P. Garvin (Ed.), *Cognition,* Spartan Books, N. York (1969).

MATURANA, H., Cognitive Strategies, in: *L'Unite de L'homme,* Plon Paris (1975).

MATURANA, H. and VARELA, F., *De Máquinas y Seres Vivos,* Editorial Universitaria, Santiago, Chile, (1973).

MATURANA, H. and VARELA, F., *Autopoietic Systems,* BCL Res. Report 9.4, University of Illinois, Urbana (1975).

PASK, G., *Conversation, Cognition and Learning,* Elsevier, N. York (1975).

POWERS, W.T., *Behavior: The Control of Perception,* Aldine, Chicago (1973).

RICHARDS, J., 'Epistemology and mathematical proof', Res. Report Mathemagenics Activities Prog., No.14, Univ. of Georgia, Athens (1974).

SCOTT, D., 'Outline of a mathematical theory of computation', Proc. IV Annual Princeton Conf. on Information Sci. and Systems, pp.169-176 (1970).

SCOTT, D., 'The lattice of flow diagrams' in: E. Engler (Ed.), Springer Lecture Notes in Mathematics 182, pp.311-366 (1971).

SCOTT, D., 'Continuous Lattices' in: Springer Lecture Notes in Mathematics 274, pp.97-136 (1972).

SPENCER-BROWN, G., *Laws of Form,* George Allen & Unwin, London (1969).

TARSKI, A., 'A lattice-theoretic fixed point and its applications', *Pacific J. Math.* **5**, 285 (1955).

THATCHER, J.W., 'Tree automata: an informal survey', in: *Currents in Computing,* Prentice-Hall, N. Jersey, pp.143-172 (1973).

VON FOERSTER, H., Notes on an epistemology for living things, in: *L'Unite de L'homme,* Plon, Paris (1975).

VON FOERSTER, *et al., Purposive Systems,* Spartan, N. York (1968).

VARELA, F., A calculus for self-reference, *Int. J. Gen. Syst.* **2**, 5 (1975).

VARELA, F., 'The extended calculus of indications interpreted as a three-valued logic', *Notre Dame J. Formal Logic* (in press) (1976a).

VARELA, F., 'The grounds for a closed logic (submitted for publication) (1976b).

VARELA, F., H. MATURANA, and URIBE, R., 'Autopoiesis: the organization of living systems, its characterization and a model', *Biosystems* **5**, 187 (1974).

WIENER, N., ROSENBLUETH, A., and BIGELOW, J., 'Behavior, purpose and teleology', *Phil. Sci.* **10**, 18 (1943).

Acknowledgements

We wish to thank the many friends who have helped and encouraged us in many sorts of ways: Juan Bulnes, Heinz von Foerster, Gordon Pask, Alia Carl, James Farned, Peter Boyle, Samy Frenk and George Klir. Thanks also to Chögyam Trungpa, Rinpoche and the Vajraddathu Seminary and the Naropa Institute for providing the environment in which some of these ideas developed. We acknowledge the influence of the long time collaborators J.W. Thatcher, E.G. Wagner and J.B. Wright.

This research was financially supported by Grant BMS 73-06766 from the National Science Foundation and the Alfred P. Sloan Foundation.

APL-AUTOPOIESIS: Experiments in self-organization of complexity

MILAN ZELENY
Columbia University, USA

Introductory notes

Most *organic systems*, e.g. biological and social organizations, are characterized by their *autopoiesis* (self-production or self-renewal) in a given environmental domain. An observer's perception and recognition of autopoiesis is fuzzily referred to as 'life' and autopoietic systems as 'living systems'.

Autopoietic organization is realized as an autonomous and self-maintaining unity through a network of component-producing processes such that the components, through their interaction, generate recursively *the same* network of processes which produced them.

For example, a cell produces cell-forming molecules, a group 'produces' group-maintaining individuals, etc. The product of an autopoietic organization is not different from the organization itself. In contrast, the product of *allopoietic organization* is different from itself; some spatially determined structures, like crystals or macromolecular chains, formal hierarchies, etc. are allopoietic. Allopoietic organization does not produce the components and processes which realize it as a distinct unity. Allopoietic systems are not perceived as 'living' and are often fuzzily referred to as mechanistic or contrived systems.

The first significant advance toward the theory of living organizations was made by Varela, Maturana and Uribe [1] in their seminal article on autopoiesis. Further extensions and generalizations of the basic model, especially in the direction of social systems and organizational behavior, appeared in the works of Zeleny and Pierre [2,3] and Zeleny [25,26]. A theory of the nervous system as a closed autopoietic organization was presented by Maturana [4]. The main purpose of this chapter is to report some new experimental results and experience and to establish autopoietic modeling as an eminently suitable tool for the study of complex organizations.

Before proceeding further let us attempt to define, in concordance with Maturana [4], the distinct concepts of systemic *structure and organization.*

A given system, observed as being distinct in its environmental domain, can be viewed, as a composite of interrelated and further unspecified components.

A network of interactions among the components, defining the system as a distinct unity, is referred to as a *systemic organization.*

Actual spatial arrangement of components and their interrelations, realized temporarily in a given physical milieu, constitutes a *systemic structure.*

The unity of systemic organization and structure is commonly referred to as a *system.*

Note that two distinct systems may have the same organization but different structures. Structural changes do not change a system as a unity as long as its organization remains invariant. A system cannot be explained by simply reproducing its structure. The structure of a system determines the space in which its components interact among themselves, with its environment and with the observer. It does not determine its properties, as a distinct whole and unity.

Returning to the definition of autopoiesis (a process of continuous self-renewal of a systemic whole) we refer to *autopoietic systems* as those resulting from the autopoietic organization of their components. Although the organization of an autopoietic system is independent of the properties of its components, its structure is actually realized through the components and their interactions.

Structural interaction of components and structural changes are driven by autopoiesis. One structural state is thus determined by the previous structural state and the history of structural changes of an autopoietic unity, ie. its dynamics, *is organizationally closed* according to Varela [5]. Structural changes are not triggered by an external and independent disturbance but arise internally from the structure itself. It would be a simplification to view autopoietic systems as 'Pavlov's dogs', ie. as being purely *reactive*, with apparent inputs and outputs. All living systems are *interactive*, ie. structurally coupled with their environments, *organizationally closed*, ie. structurally self-determined, and with *no apparent inputs and outputs,* ie. they are not mechanical contrivances.

Autopoietic model of a cell.

The simplest autopoietic system is that of a *living cell*. It exhibits the *minimal* organization of components which is necessary for autopoiesis. There is a *catalytic nucleus* capable of interaction with the medium of *substrate* so that the *membrane-forming components* are continually produced. The resulting structure displays a *membraneous boundary* that defines the system as a separate and autonomous unity in the space of its components. The changing medium of substrate creates perturbations which generate structural changes while the cell's autopoietic organization remains invariant. In order to achieve such dynamic equilibrium of structural changes, the cellular components must interact with the medium of substrate (and extracellular components) so that new autopoiesis-preserving structures can be generated. Without such spatio-temporal rapport between the cell and its medium either allopoiesis or disintegration would be observed.

In accordance with the basic organization of a cell, the simplest model of its autopoietic organization must consist of a medium of substrate, a catalyst capable of producing more complex elements (links), which are in turn capable of their own bonding, ultimately concatenating into a membrane around the catalyst. We designate the basic components of the model by the following symbols:

hole	(space)
substrate	○
free link	□
singly bonded link	□-
fully bonded link	-⊞-
catalyst	*

The original model by Varela *et al.* [1], is based on the following organization of components:

1. *Composition:* 2○ + * → □ + * + (space). A catalyst and two units of substrate produce a free link and a hole, while the catalyst is assumed to be unaffected by the operation.
2. *Disintegration:* □ + (space) → 2○
   -□ + (space) → 2○
   -⊞- + (space) → 2○
   A link, free or bonded, disintegrates into two units of substrate which will occupy the available holes.
3. *Bonding:* □-⊞- ··· -□ + □ → □-⊞- ··· -⊞-□
   A free link can be bonded with a chain of bonded links; two chains of bonded links can be bonded into one; two free links can be bonded together to start a chain formation.

Observe that the essential organization of components is given by the three *interaction rules* above. More detailed rules, guiding the movement of all components, specifying the necessary conditions for the three interaction rules, and allowing spontaneous encounters and bonding of components, as well as their subsequent disintegration, are formally presented by Zeleny and Pierre [2] and Zeleny [26]. Only a brief summary is provided here.

Each link is allowed to have at most two bonds: it can be either free, □, or single bonded, -□, or fully bonded, -⊞-. (Additional bonds would induce branching of chains and their introduction would only be meaningful in three-dimensional spaces.) Unbranched chains of bonded links can ultimately form a membrane around the catalyst, creating an enclosure which is not penetrable by either * or □. These components are thus forced to function for the benefit of the autopoietic unity. Substrate ○ can pass freely through the membrane and participate in the production of □ by the catalyst. Any disintegrated links, causing ruptures in the membrane, are being effectively repaired by the on-going production. The unity of the cell is thus recursively maintained through a series of minor structural changes.

A series of computer printouts in Fig. 1, provides a sample profile of a structural history of an autopoietic system driven by the above set of rules.

APL-AUTOPOIESIS: Experiments in self-organization of complexity

A computer program, 'APL-Autopoiesis', simulating autopoietic behavior of self-renewing systems, has been coded in APL (A Programming Language) and a series of computerized experiments performed on the IBM 360/75 system of Columbia University.

We define a two-dimensional tesselation grid of $n \times n$ positions. Each position is identifiable as (i,j), $i,j = 1,2,...,n$.

Recall that each position can be in one of the six possible states: hole, substrate, link, singly or fully bonded link and a catalyst.

Each component can either stay in place or move (be displaced). We allow only the moves in a rectangular fashion, i.e. four basic rectangular movements can be defined: north, east, south and west. Correspondingly, there are four basic movement operators in a given set of movement operators, M:

$$M = \{M_N = (-1,0),\ M_E = (0,1),\ M_W = (1,0),\ M_S = (0,-1)\}$$

To demonstrate the usage of movement operators, we can write:

$$(i,j) + M_N \rightarrow (i-1,j) + M_N \rightarrow (i-2,j)$$

$$(i,j) + M_W \rightarrow (i+1,j) \ + M_S \rightarrow (i+1,j-1)$$

etc.

A very important concept of a *neighborhood of* (i,j) can now be introduced. All positions reachable from (i,j) by a single move define the first neighborhood, N_1; all positions reachable by one or two moves define the second neighborhood, N_2; those reachable by up to three moves form N_3, etc. Complete neighborhoods of (i,j) are graphically represented in Fig. 2.

Other types of neighborhoods: asymmetric, incomplete or composite, can be defined as well. The neighborhoods are utilized to identify all adjacent positions into which a given component can move or be displaced, to simulate production power field or 'reach' of a catalyst, to identify bonding regions, to define overlapping power fields of multiple catalysts, and many other useful purposes.

Movement of components over the tesselation grid is guided by the following hierarchical rules:

1. Bonded links and holes are not subject to any motion at all.
2. Units of substrate, free links and catalysts can move into adjacent holes.
3. Free links can displace units of substrate into adjacent holes or exchange positions with them.
4. Catalysts can displace both free links and substrate into adjacent holes or exchange positions with them.
5. Units of substrate can pass through a bonded link segment. Neither free links nor catalysts can do so.

There are additional adjustable features of the simulation model. Whenever two adjacent neighbor positions of a catalyst are occupied by substrate, a free link can be produced. Every free link can be bonded with another free link or with a single bonded link. Each free or bonded link can disintegrate into two units of substrate, provided there is a hole in its neighborhood into which the additional substrate can sink.

The three interaction rules of composition, disintegration and bonding, supplemented by the above rules of movement and positioning, represent an autopoietic organization of components—its *bioalgebra*, a formal tool for studying organic systems.

1. *Form and function*

We have already introduced the distinction between systemic organization and structure. Their dialectics presents itself as an interaction between function and form (or content and shape).

In Fig. 3, observe that the size of a membrane can be simply encoded in system's bioalgebra. If only the first neighborhood is considered (a weak catalyst) only a narrow membrane will form, possibly coinciding with the neighborhood itself.

Autopoiesis can be considerably slowed down or even cease entirely. An allopoeitic structure, a crystal, might ultimately form. It can neither disintegrate nor expand or move. In the lower part of Fig. 3 please observe the turbulence which results from allowing the catalyst to move too rapidly over the space of autopoiesis, simulating the conditions of a tempesteous ocean for instance.

Similarly, by considering a larger neighborhood, ie. by simulating stronger catalytic action, one induces a more spacious membrane into its existence. Catalytic neighborhood cannot be increased without limits, a membrane might never form and autopoietic unity might not even be perceived. There is an optimal strength of catalytic field which gives rise

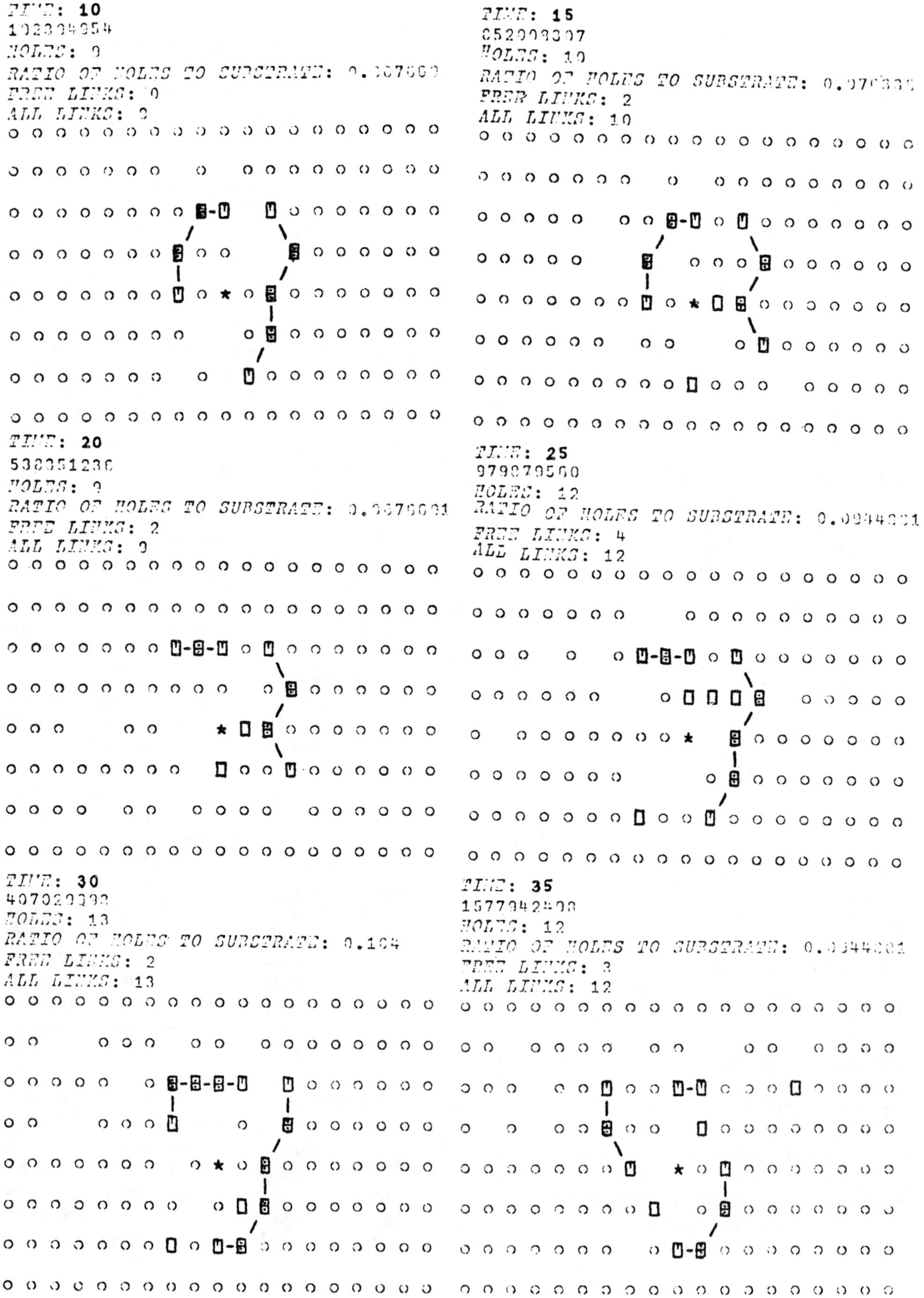

```
TIME: 10
102394954
HOLES: 9
RATIO OF HOLES TO SUBSTRATE: 0.087000
FREE LINKS: 0
ALL LINKS: 9
```

```
TIME: 15
052008307
HOLES: 10
RATIO OF HOLES TO SUBSTRATE: 0.076333
FREE LINKS: 2
ALL LINKS: 10
```

```
TIME: 20
538351236
HOLES: 9
RATIO OF HOLES TO SUBSTRATE: 0.0876001
FREE LINKS: 2
ALL LINKS: 9
```

```
TIME: 25
979879560
HOLES: 12
RATIO OF HOLES TO SUBSTRATE: 0.0944001
FREE LINKS: 4
ALL LINKS: 12
```

```
TIME: 30
407020332
HOLES: 13
RATIO OF HOLES TO SUBSTRATE: 0.104
FREE LINKS: 2
ALL LINKS: 13
```

```
TIME: 35
1577942408
HOLES: 12
RATIO OF HOLES TO SUBSTRATE: 0.0944001
FREE LINKS: 3
ALL LINKS: 12
```

Fig.1

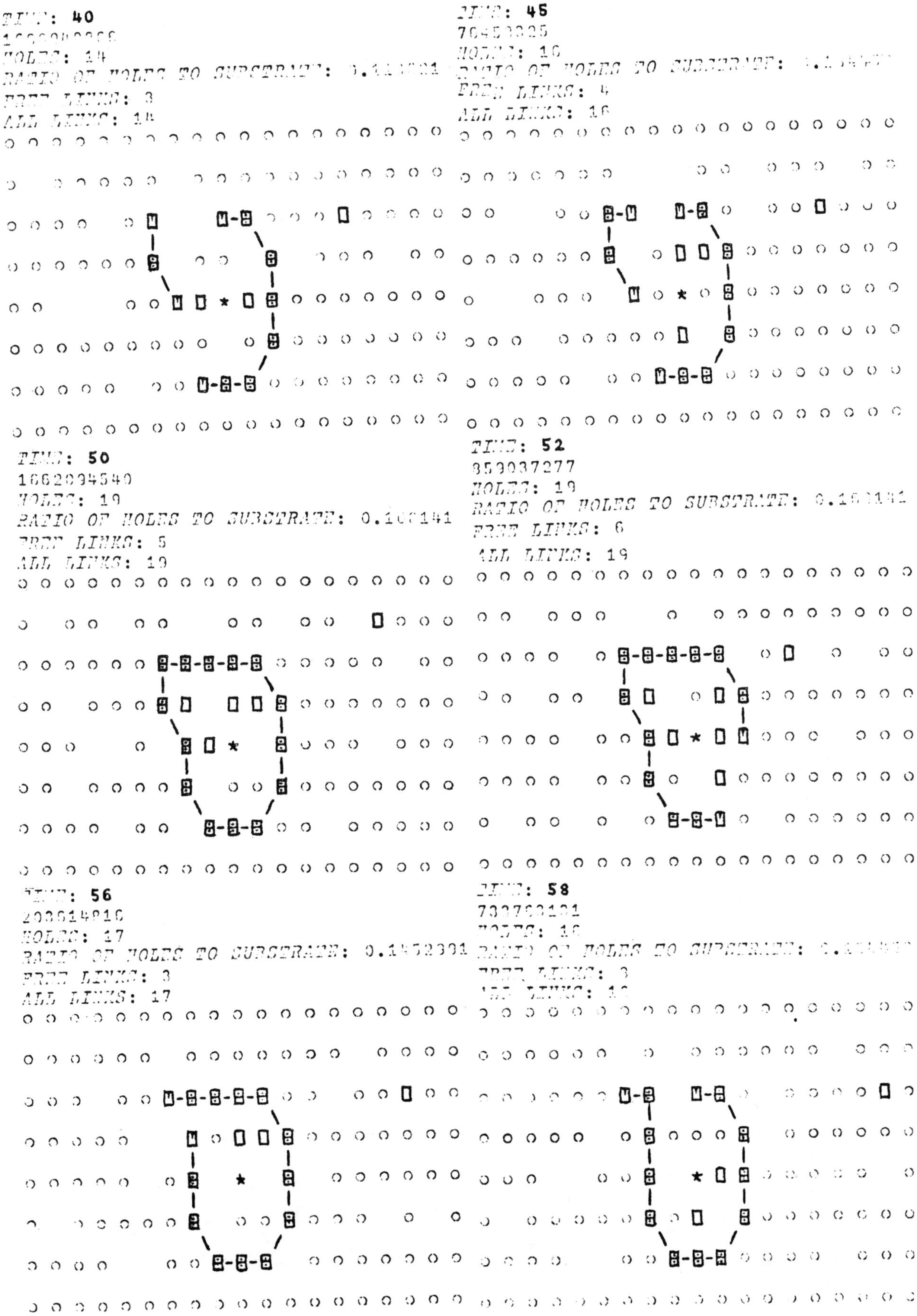

```
TIME: 40
[illegible]
HOLES: 14
RATIO OF HOLES TO SUBSTRATE: 0.[illegible]
FREE LINKS: 3
ALL LINKS: 14
```

```
TIME: 45
70450325
HOLES: 16
RATIO OF HOLES TO SUBSTRATE: [illegible]
FREE LINKS: 4
ALL LINKS: 16
```

```
TIME: 50
1662094540
HOLES: 19
RATIO OF HOLES TO SUBSTRATE: 0.160141
FREE LINKS: 5
ALL LINKS: 19
```

```
TIME: 52
359037277
HOLES: 19
RATIO OF HOLES TO SUBSTRATE: 0.160141
FREE LINKS: 6
ALL LINKS: 19
```

```
TIME: 56
203614816
HOLES: 17
RATIO OF HOLES TO SUBSTRATE: 0.1432301
FREE LINKS: 3
ALL LINKS: 17
```

```
TIME: 58
733769191
HOLES: 16
RATIO OF HOLES TO SUBSTRATE: [illegible]
FREE LINKS: 3
ALL LINKS: 16
```

Fig.1 (cont.)

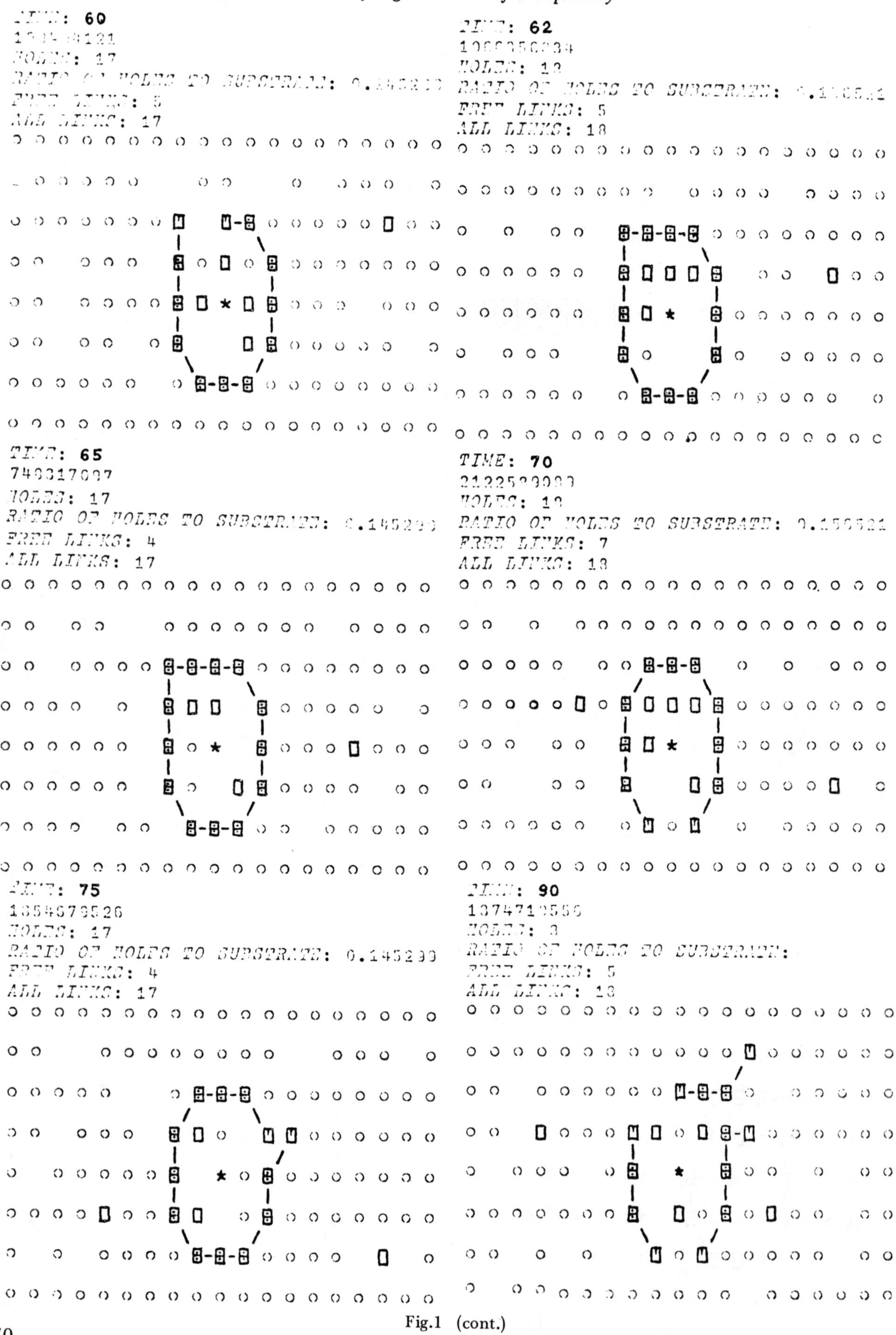

Fig.1 (cont.)

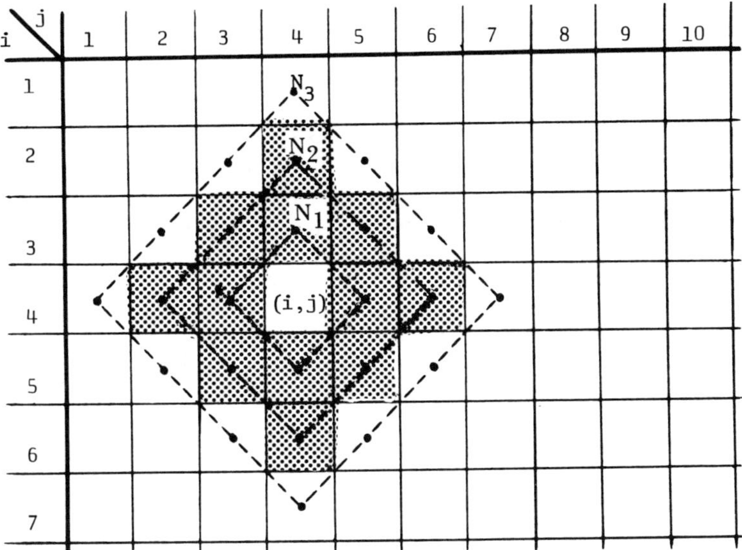

Fig.2

to a safe and fast self-assemblance of a cell, avoiding both the crystallization and structural turbulence. See Fig. 4.

It is now obvious that a growth of an organism can be simulated quite simply. One can establish a state-dependent *neighborhood change regime*, allowing transitions $N_1 \rightarrow N_2 \rightarrow N_3$..., and vice versa. Such regime of structural change can also be cyclical, continuous, irregular or in jumps. In any case, the regime is not pre-programmed but it is entirely state-dependent. A particular state of an autopoietic system, ie, a particular structure, determines the change of its state (eg. a larger membrane) while establishing a new rapport with the newly structured medium. Autopoietic unity and identity of the system is being maintained throughout.

Simple growth or shrinkage of an organism does not explain the actual shape or form. It can be shown that even very complex shapes and patterns can arise spontaneously from autopoietic organization. In all print-outs presented so far a fully symmetrical catalytic neighborhood has been assumed. The catalytic action, however, is not uniformly symmetrical. In the vicinity of boundaries only partial neighborhoods can be defined, environmental conditions can distort the field of catalytic function, the structure of a medium itself reinforces or weakens the catalytic 'reach' in different directions.

In Fig. 5 we observe a self-organization of an autopoietic system in the *shape of a cross*. All autopoietic rules are retained and there are no changes in cells' vital functions. There is a strong tendency that the membrane forms a closure of this particular shape. Theoretically any shape can be 'encoded' quite simply in the bioalgebra.

Again, state-dependent regimes, combining simultaneous changes in size and shape, can be established. A large variety of emergent structural properties can be observed and studied.

2. Biological clock

One deficiency of all previously modelled autopoietic systems is that there is no 'life cycle' to be observed. Yet, the *ageing phenomenon* is as closely connected with life as any other function. After the cell established its dynamic equilibrium in the given medium it would keep renewing itself through a series of oscillations between rupture and closure, practically forever, ie. until we turn the computer off. Such stability or permanence is excessive and does not reflect the actual behavior observed in living systems. There is no life without death.

The first rule of composition is obviously very approximative. It is unreasonable to assume that the catalyst is unaffected by its participation in the production of free links. Its catalytic power is more likely to be diminished, for example

$$2\bigcirc + * \rightarrow \square + (* - \Delta) + (\text{space})$$

would indicate that Δ has been subtracted from the power of *. It is then quite simple to relate the accumulation of Δ's to a particular change regime of catalytic neighborhoods. The relation between catalyst's power and its neighborhood 'reach' is thus

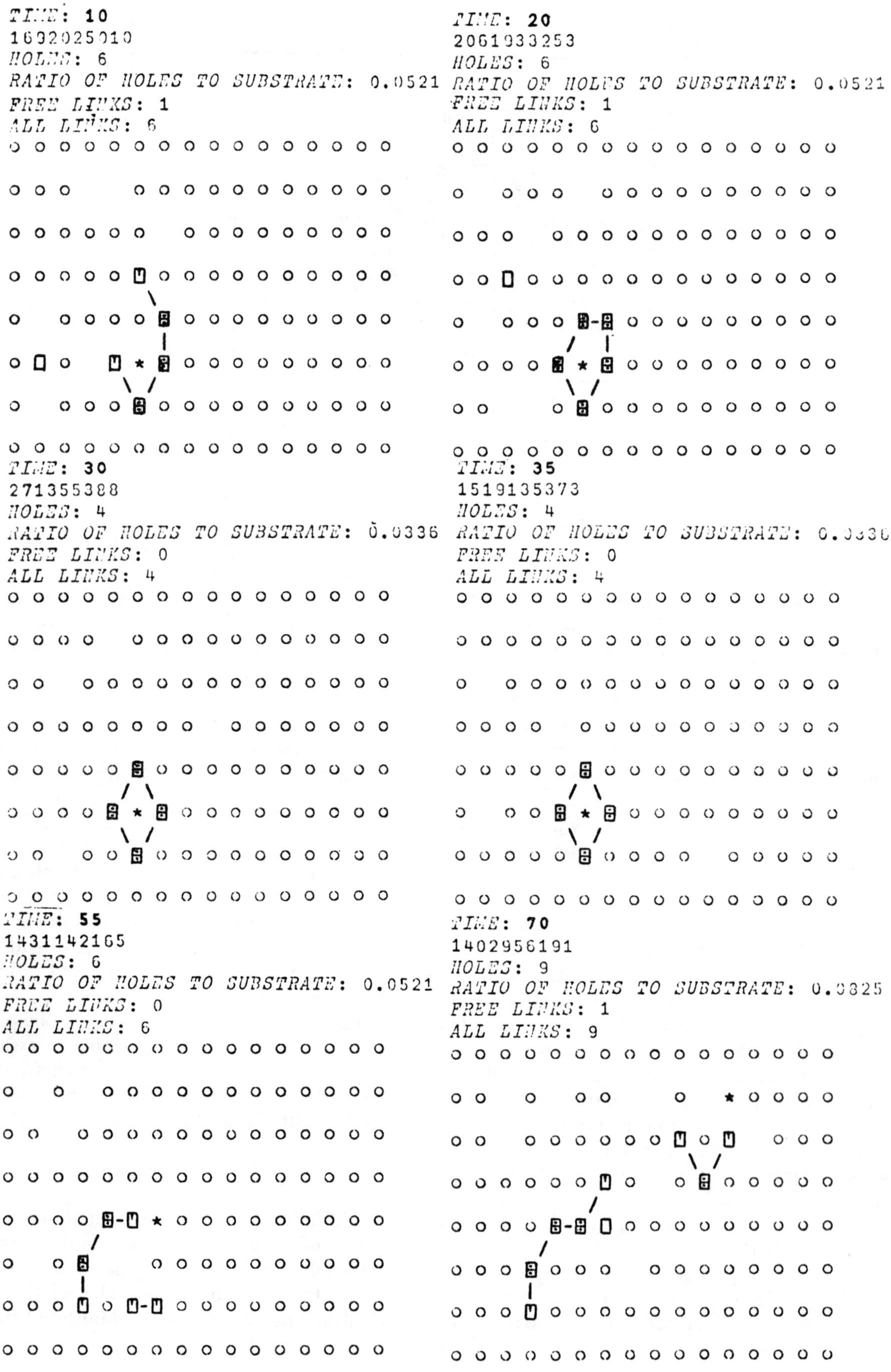

```
TIME: 10
1692025010
HOLES: 6
RATIO OF HOLES TO SUBSTRATE: 0.0521
FREE LINKS: 1
ALL LINKS: 6
o o o o o o o o o o o o o o o o

o o o     o o o o o o o o o o o

o o o o o o   o o o o o o o o o

o o o o o □ o o o o o o o o o o
           \
o   o o o o □ o o o o o o o o o
            |
o □ o   □ * □ o o o o o o o o o
         \ /
o   o o o □ o o o o o o o o o o

o o o o o o o o o o o o o o o o
```

```
TIME: 20
2061933253
HOLES: 6
RATIO OF HOLES TO SUBSTRATE: 0.0521
FREE LINKS: 1
ALL LINKS: 6
o o o o o o o o o o o o o o o o

o   o o o   o o o o o o o o o o

o o o   o o o o o o o o o o o o

o o □ o o o o o o o o o o o o o

o   o o o □-□ o o o o o o o o o
         /  |
o o o o □ * □ o o o o o o o o o
         \ /
o o     o □ o o o o o o o o o o

o o o o o o o o o o o o o o o o
```

```
TIME: 30
271355388
HOLES: 4
RATIO OF HOLES TO SUBSTRATE: 0.0336
FREE LINKS: 0
ALL LINKS: 4
o o o o o o o o o o o o o o o o

o o o o   o o o o o o o o o o o

o o   o o o o o o o o o o o o o

o o o o o o o o   o o o o o o o

o o o o o □ o o o o o o o o o o
         / \
o o o o □ * □ o o o o o o o o o
         \ /
o o   o o □ o o o o o o o o o o

o o o o o o o o o o o o o o o o
```

```
TIME: 35
1519135373
HOLES: 4
RATIO OF HOLES TO SUBSTRATE: 0.0336
FREE LINKS: 0
ALL LINKS: 4
o o o o o o o o o o o o o o o o

o o o o o o o o o o o o o o o o

o   o o o o o o o o o o o o o o

o o o o   o o o o o o o o o o o

o o o o o □ o o o o o o o o o o
         / \
o   o o □ * □ o o o o o o o o o
         \ /
o o o o o □ o o o o   o o o o o

o o o o o o o o o o o o o o o o
```

```
TIME: 55
1431142165
HOLES: 6
RATIO OF HOLES TO SUBSTRATE: 0.0521
FREE LINKS: 0
ALL LINKS: 6
o o o o o o o o o o o o o o o o

o   o   o o o o o o o o o o o o

o o   o o o o o o o o o o o o o

o o o o o o o o o o o o o o o o

o o o o □-□ * o o o o o o o o o
       /
o   o □     o o o o o o o o o o
      |
o o o □ o □-□ o o o o o o o o o

o o o o o o o o o o o o o o o o
```

```
TIME: 70
1402956191
HOLES: 9
RATIO OF HOLES TO SUBSTRATE: 0.0825
FREE LINKS: 1
ALL LINKS: 9
o o o o o o o o o o o o o o o o

o o   o   o o     o   * o o o o

o o   o o o o o o □ o □   o o o
                   \ /
o o o o o o □ o   o □ o o o o o
           /
o o o o □-□ □ o o o o o o o o o
       /
o o o □ o o o   o o o o o o o o
      |
o o o □ o o o o o o o o o o o o

o o o o o o o o o o o o o o o o
```

Fig.3

```
TIME: 24
755644608
HOLES: 36
RATIO OF HOLES TO SUBSTRATE: 0.2352
FREE LINKS: 9
ALL LINKS: 35
CUMULATIVE PRODUCTIONS: 45
PRODUCTIONS THIS CYCLE: 1
```

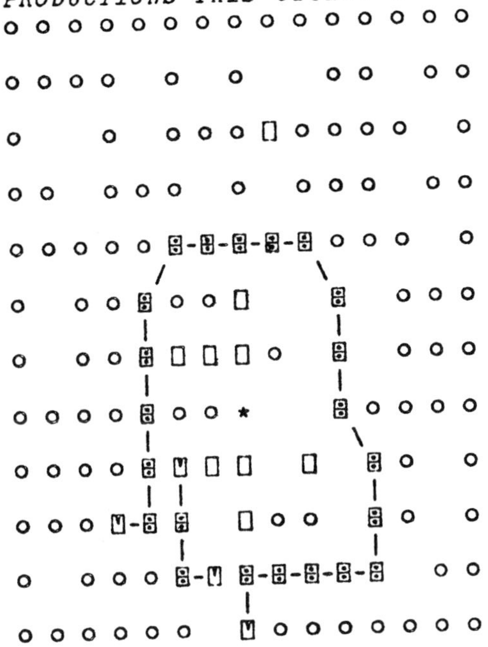

```
TIME: 34
1117321188
HOLES: 41
RATIO OF HOLES TO SUBSTRATE: 0.2867
FREE LINKS: 14
ALL LINKS: 40
CUMULATIVE PRODUCTIONS: 58
PRODUCTIONS THIS CYCLE: 1
```

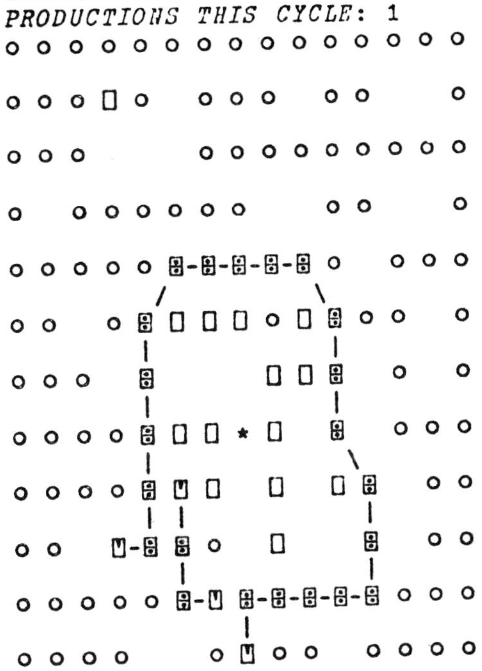

```
TIME: 16
1007693858
HOLES: 33
RATIO OF HOLES TO SUBSTRATE: 0.1333333333
FREE LINKS: 14
ALL LINKS: 33
```

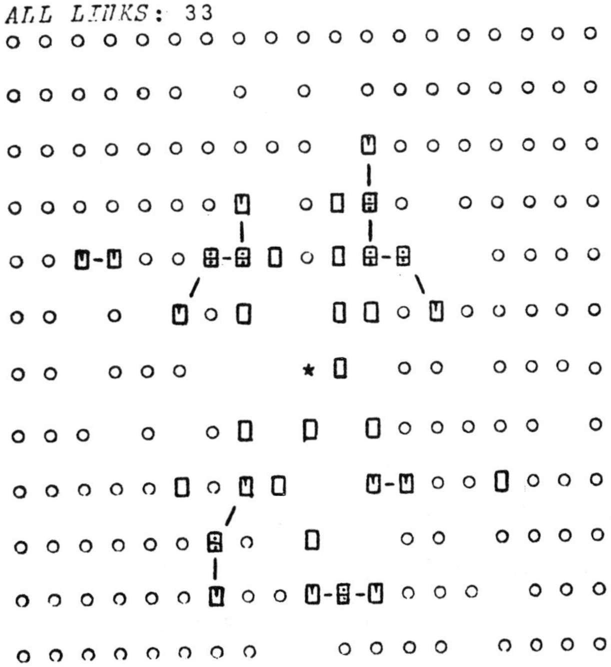

Fig.4

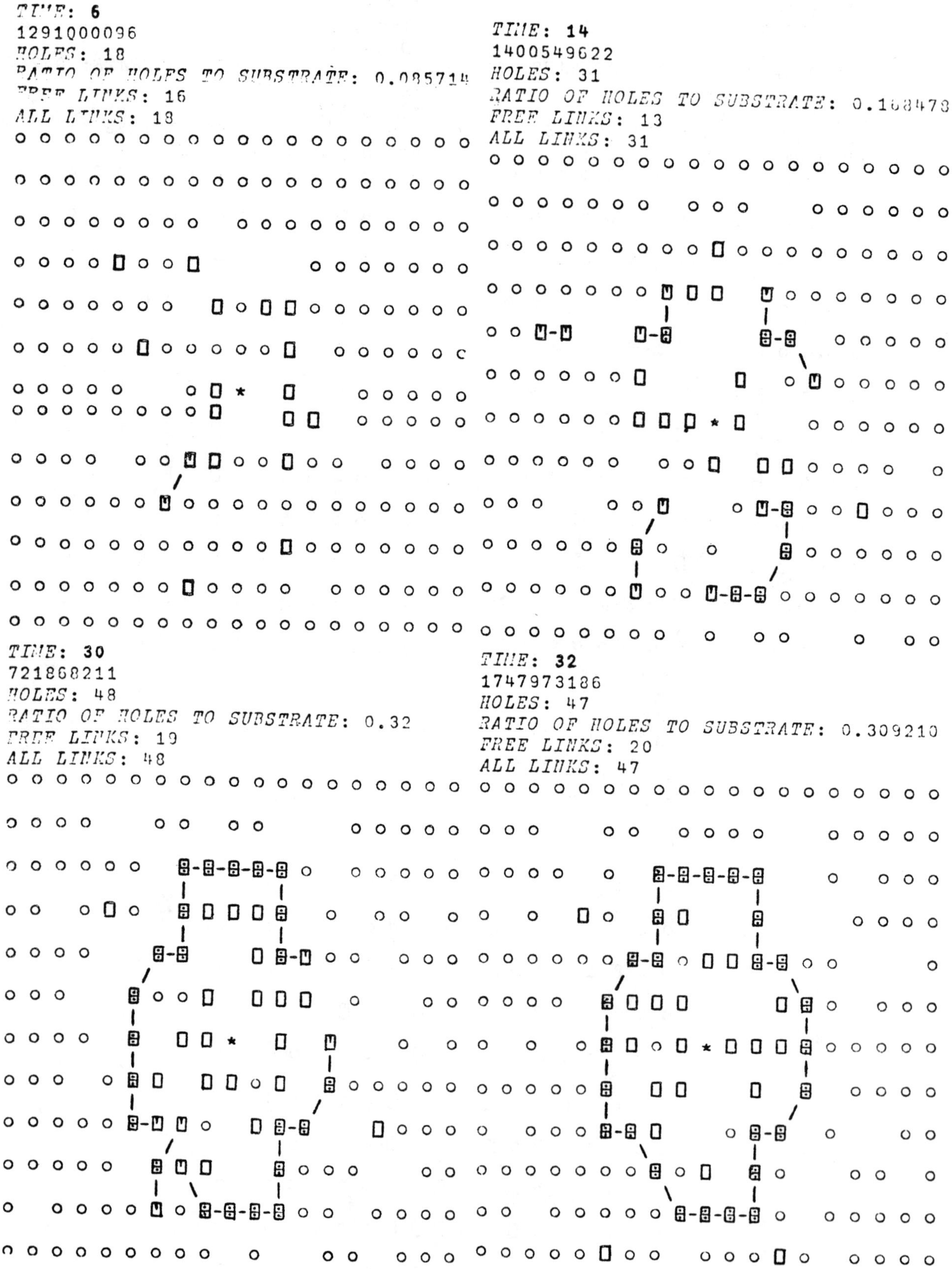

```
TIME: 6
1291000096
HOLES: 18
RATIO OF HOLES TO SUBSTRATE: 0.085714
FREE LINKS: 16
ALL LINKS: 18
```

```
TIME: 14
1400549622
HOLES: 31
RATIO OF HOLES TO SUBSTRATE: 0.168478
FREE LINKS: 13
ALL LINKS: 31
```

```
TIME: 30
721868211
HOLES: 48
RATIO OF HOLES TO SUBSTRATE: 0.32
FREE LINKS: 19
ALL LINKS: 48
```

```
TIME: 32
1747973186
HOLES: 47
RATIO OF HOLES TO SUBSTRATE: 0.309210
FREE LINKS: 20
ALL LINKS: 47
```

Fig.5

effectively simulated. Changes in catalysis are directly related to the changes in the size of an autopoietic membrane.

More visible can be the effect of catalytic changes on the *rate of composition* (or production). Initially we assume that only one link can be produced per each step even though more units of substrate might be present in the catalyst's neighborhood. That would simulate a steady and unchanging catalytic power.

Because individual computer steps do not correspond to actual elapsed time periods, we can simulate multiple link productions in each step without violating the temporal constraints. Thus we start with maximum of four links per step, then three, two, one etc., until, say, one per each ten steps. State-dependent regime is thus established: the stronger the catalyst the more acts of productions being engaged in. So, we experience the fastest 'ageing' of the catalyst in the initial stages of the most vigorous production activity. This 'ageing rate' becomes progressively slower, although it might never really stop. As a result, an *internally coded biological clock* is being effectively simulated. A cell, continually in a dynamic equilibrium with its environment, passes through a relatively fast stage of intensive production and structural build-up, a stable plateau of balanced rupture-closure pulsation, and, finally, through the declining rates of production, self-repair and internal mobility.

In Fig. 6 we present a sample profile of autopoietic structural history which exhibits a clearly perceivable life cycle. Actual behavior of the cell results from a composite impact of many natural biorhythms. Initially, there is a lot of free substrate and the number of produced links is naturally very high. Because the number of holes necessary for disintegration is initially very low, there is a large build-up of the organized matter in these initial stages. As the amount of substrate decreases and the number of holes increases, the two rates achieve a balance which is characteristic for a relatively stable period of membraneous enclosures. As the production rate decreases further, it is exceeded by disintegration rate and the total amount of organized matter is steadily declining. Because the holes become fewer again the rate of this decline is steadily declining as well. Observe how the rapid loss of catalytic power is initially offset by more available substrate, thus keeping the actual production rate steady and high for some time.

It is observed that the behavior of an autopoietic system is highly cyclical and a variety of emergent rhythms can be identified. None of them were purposefully designed or otherwise encoded in the model.

For example, assuming stable production and disintegration rates, even under the conditions of autopoietic stability of a cell, there is a natural cycle observed in the ratio of holes to substrate. More substrate leads to more links and higher incidence of bonding. Consequently, the actual amount of available substrate decreased as well as the actual number of produced links. The number of available holes is increased, however, thus allowing more links to disintegrate, creating more substrate and less holes. It would be a gross fallacy to interpret such structural rapport between the system and the medium as some kind of a feedback mechanism. There is none. It only appears as such to an observer.

3. Multiple catalysts

It should be apparent that multiple catalysts can function in the same medium of substrate. The simplest case is when the catalysts are distant enough so that they can enclose themselves independently and without any interference. One can then observe a group of independent autopoietic cells, each and all in a dynamic equilibrium with the medium.

The closer the catalysts are the more interesting their behavior becomes. In Fig. 7 we provide a sketch of two catalysts with their respective catalytic neighborhoods overlapping.

Observe that because no bonding is realized within the respective neighborhoods (because of catalytic turbulence), the two stationary, catalysts cannot be enclosed by a single membrane, if their catalytic neighborhoods overlap. If at least one of the catalysts is freely moving, it will ultimately float sufficiently apart so that two distinct autopoietic systems can be observed. Note that only if one of the catalysts is substantially weaker than the other, ie. one neighborhood can be a proper subset of the other, then their catalytic action is reinforced by a single membrane.

The most interesting case arises if we assume that at a certain stage the catalyst is allowed to divide itself into two identical replicas. For example, the first total closure of a membrane may be assumed to provide a trigger which causes such catalytic replication. The new catalyst occupies any immediately adjacent hole. If both catalysts are of comparable strength, ie. the power field of one cannot be a proper subset of the power field of the other, then we can sketch the situation as in Fig. 8.

Observe that a large portion of the original mem-

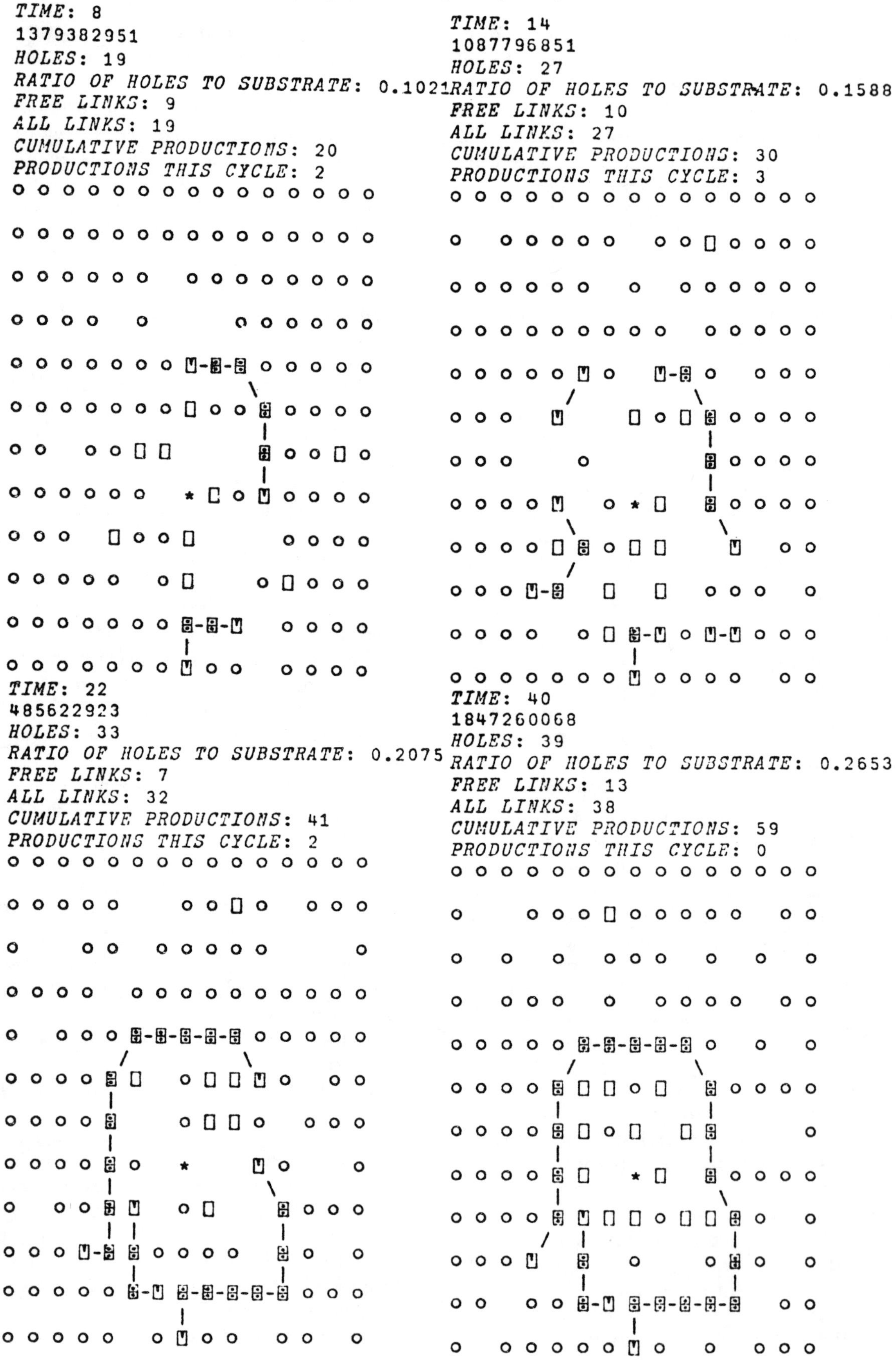

```
TIME: 8
1379382951
HOLES: 19
RATIO OF HOLES TO SUBSTRATE: 0.1021
FREE LINKS: 9
ALL LINKS: 19
CUMULATIVE PRODUCTIONS: 20
PRODUCTIONS THIS CYCLE: 2
○ ○ ○ ○ ○ ○ ○ ○ ○ ○ ○ ○ ○ ○ ○ ○ ○ ○

○ ○ ○ ○ ○ ○ ○ ○ ○ ○ ○ ○ ○ ○ ○ ○ ○ ○

○ ○ ○ ○ ○ ○   ○ ○ ○ ○ ○ ○ ○ ○ ○

○ ○ ○ ○   ○       ○ ○ ○ ○ ○ ○

○ ○ ○ ○ ○ ○ ○ ▯-▣-▣ ○ ○ ○ ○ ○
                  \
○ ○ ○ ○ ○ ○ ○ □ ○ ○ ▣ ○ ○ ○ ○
                    |
○ ○   ○ ○ □ □       ▣ ○ ○ □ ○
                    |
○ ○ ○ ○ ○ ○   * □ ○ ▯ ○ ○ ○ ○

○ ○ ○   □ ○ ○ □       ○ ○ ○ ○

○ ○ ○ ○ ○   ○ □     ○ □ ○ ○ ○

○ ○ ○ ○ ○ ○ ○ ○ ▣-▣-▯   ○ ○ ○ ○
                |
○ ○ ○ ○ ○ ○ ○ ○ ▯ ○ ○   ○ ○ ○ ○
```

```
TIME: 14
1087796851
HOLES: 27
RATIO OF HOLES TO SUBSTRATE: 0.1588
FREE LINKS: 10
ALL LINKS: 27
CUMULATIVE PRODUCTIONS: 30
PRODUCTIONS THIS CYCLE: 3
○ ○ ○ ○ ○ ○ ○ ○ ○ ○ ○ ○ ○ ○ ○ ○ ○ ○

○   ○ ○ ○ ○ ○   ○ ○ □ ○ ○ ○ ○

○ ○ ○ ○ ○ ○   ○   ○ ○ ○ ○ ○ ○

○ ○ ○ ○ ○ ○ ○ ○ ○   ○ ○ ○ ○ ○

○ ○ ○ ○ ○ ▯ ○   ▯-▣ ○   ○ ○ ○
          /          \
○ ○ ○   ▯     □ ○ □ ▣ ○ ○ ○ ○
                    |
○ ○ ○     ○         ▣ ○ ○ ○ ○
                    |
○ ○ ○ ○ ▯   ○ * □   ▣ ○ ○ ○ ○
          \           \
○ ○ ○ ○ □ ▣ ○ □ □     ▯   ○ ○
          /
○ ○ ○ ▯-▣   □   □   ○ ○ ○   ○

○ ○ ○ ○   ○ □ ▣-▯ ○ ▯-▯ ○ ○ ○
              |
○ ○ ○ ○ ○ ○ ○ ▯ ○ ○ ○ ○   ○ ○
```

```
TIME: 22
485622923
HOLES: 33
RATIO OF HOLES TO SUBSTRATE: 0.2075
FREE LINKS: 7
ALL LINKS: 32
CUMULATIVE PRODUCTIONS: 41
PRODUCTIONS THIS CYCLE: 2
○ ○ ○ ○ ○ ○ ○ ○ ○ ○ ○ ○ ○ ○ ○ ○ ○ ○

○ ○ ○ ○ ○     ○ ○ □ ○   ○ ○ ○

○     ○ ○   ○ ○ ○ ○ ○         ○

○ ○ ○ ○   ○ ○ ○ ○ ○ ○ ○ ○ ○ ○ ○

○   ○ ○ ○ ▣-▣-▣-▣-▣ ○ ○ ○ ○ ○
          /         \
○ ○ ○ ○ ▣ □   ○ □ □ ▯ ○   ○ ○
        |
○ ○ ○ ○ ▣     ○ □ □ ○   ○ ○ ○
        |
○ ○ ○ ○ ▣ ○   *     ▯ ○     ○
        |             \
○   ○ ○ ▣ ▯   ○ □     ▣ ○ ○ ○
        | |           |
○ ○ ○ ▯-▣ ▣ ○ ○ ○ ○   ▣ ○   ○
          |           |
○ ○ ○ ○ ○ ▣-▯ ▣-▣-▣-▣-▣ ○ ○ ○
              |
○ ○ ○ ○ ○   ○ ▯ ○ ○   ○ ○   ○
```

```
TIME: 40
1847260068
HOLES: 39
RATIO OF HOLES TO SUBSTRATE: 0.2653
FREE LINKS: 13
ALL LINKS: 38
CUMULATIVE PRODUCTIONS: 59
PRODUCTIONS THIS CYCLE: 0
○ ○ ○ ○ ○ ○ ○ ○ ○ ○ ○ ○ ○ ○ ○ ○ ○ ○

○     ○ ○ ○ □ ○ ○ ○ ○ ○ ○   ○ ○

○   ○   ○   ○ ○ ○   ○   ○   ○

○   ○ ○ ○   ○   ○ ○ ○ ○   ○ ○

○ ○ ○ ○ ○ ▣-▣-▣-▣-▣ ○   ○   ○
          /         \
○ ○ ○ ○ ▣ □ □ ○ □   ▣ ○ ○ ○ ○
        |           |
○ ○ ○ ○ ▣ □ ○ □   □ ▣       ○
        |           |
○ ○ ○ ○ ▣ □   * □   ▣ ○ ○ ○ ○
        |             \
○ ○ ○ ○ ▣ ▯ □ □ ○ □ □ ▣ ○   ○
      /   |           |
○ ○ ○ ▯   ▣   ○     ○ ▣ ○   ○
          |           |
○ ○   ○ ○ ▣-▯ ▣-▣-▣-▣-▣   ○ ○
              |
○   ○ ○ ○ ○ ○ ▯ ○   ○   ○ ○ ○
```


Fig.6

```
TIME: 46
1403767852
HOLES: 37
RATIO OF HOLES TO SUBSTRATE: 0.2450
FREE LINKS: 13
ALL LINKS: 36
CUMULATIVE PRODUCTIONS: 59
PRODUCTIONS THIS CYCLE: 0
o o o o o o o o o o o o o o o

o   o o o o o   o o o o o o o

o     o o o o o o o o o o o o

o   o o o o o   □ o o o     o

o   o o o ▣-▣-▣-▣-▣ o o o   o
         /         \
o o o o ▣ □ □ □ □ □ ▣     o o
        |           |
o o o o ▣ o o       ▣     o o
        |           |
o o   o ▣ □ □ * □ □ ▣ o     o
        |            \
o   o   ▣ ◫ □ o   □   ◫ o o o
       /  |
o o o ◫ o ▣ o □ o   o o   o o
          |
o o   o o ◫ o ▣-▣-▣-▣-◫     o
              |
o   o o o o o ◫   o o   o   o
```

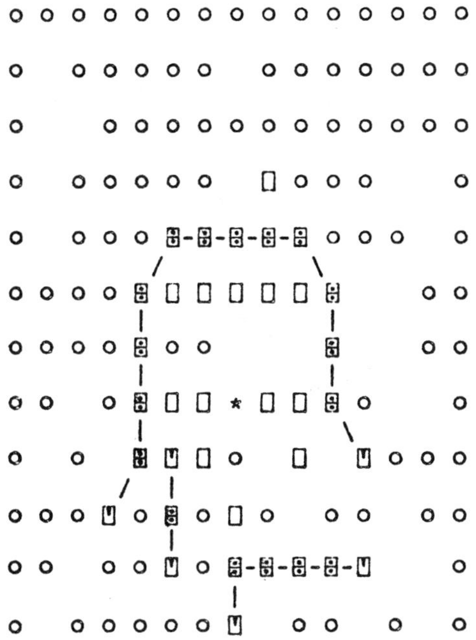

```
TIME: 50
1917167081
HOLES: 34
RATIO OF HOLES TO SUBSTRATE: 0.2165
FREE LINKS: 9
ALL LINKS: 33
CUMULATIVE PRODUCTIONS: 60
PRODUCTIONS THIS CYCLE: 0
o o o o o o o o o o o o o o o

o     o o o o o   o o o o o o

o   o o o o o o o   o o   o o

o o o o o o o o o ◫ o o o   o
                  |
o   o   o ▣-▣-▣-▣-▣ o o o o o
         /
o o o   ▣ o □ □   o o   o o o
        |
o   o o ▣ □ o     □ ◫ o o   o
        |           |
o o o o ▣ □ o * o   ▣ o o o o
        |            \
o o   o ▣ ◫   □ o □ □ ◫ o o o
       /  |
o o o ◫   ▣   o   □ o o o   o
          |
o o o o o ▣-◫ ▣-▣-▣-▣-◫   o o
              |
o o o o   o o ◫   o o o o   o
```

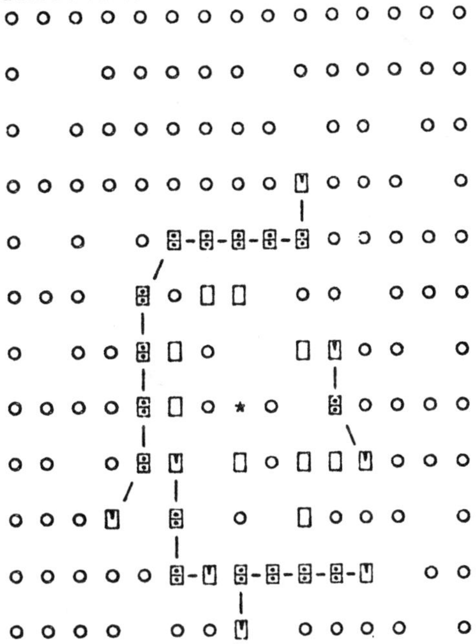

```
TIME: 64
1877143764
HOLES: 30
RATIO OF HOLES TO SUBSTRATE: 0.1818
FREE LINKS: 8
ALL LINKS: 29
CUMULATIVE PRODUCTIONS: 65
PRODUCTIONS THIS CYCLE: 0
o o o o o o o o o o o o o o o

o o     o     o o o o   o o o

o   o o o o o o o o o o o o o

o o o o o o o o o ◫ o o o   o
                  |
o o   o o ◫-▣-▣-▣-▣ □ o o o o

o o o o   o o   o o o o o o o

o o o o ◫   □   □ o ◫   o   o
        |           |
o o   o ▣ o o * □ □ ▣   o o o
        |            \
o   o o ▣ ◫ o         ◫ o o o
       /  |
o o o ◫ o ▣ □ □ o o       o o
          |
o o o   o ▣-▣-▣ ◫-◫ o o   o o
              |
o o o o o o o ◫ o o o   o   o
```

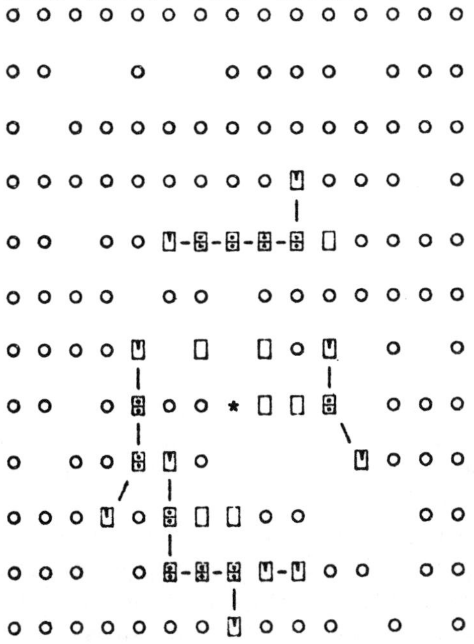

```
TIME: 70
385435594
HOLES: 28
RATIO OF HOLES TO SUBSTRATE: 0.1656
FREE LINKS: 7
ALL LINKS: 27
CUMULATIVE PRODUCTIONS: 67
PRODUCTIONS THIS CYCLE: 0
o o o o o o o o o o o o o o o

o     o o o   o o o o o o o o

o   o   o o o   o o o o o o o

o o o o o o   o o o ◫   o o o
                    |
o   o o o ◫-◫ o ◫-▣-▣ o o   o

o o o o o   o o o o o o o o o

o   o o ◫ o o □ □ o   □   o o
        |
o o o o ▣ o o * o o o o o o o
        |
o   o o ▣ ◫     □ o o ◫     o
       /  |             |
o o o ◫ o ▣ □ o   □   ◫ o o o
          |
o o o     ▣-▣-▣ ◫-◫ o o o o o
              |
o o o o o o   ◫ o o   o o □ o
```

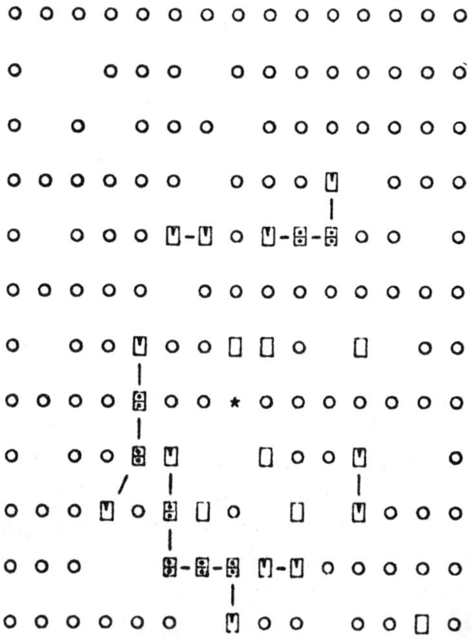

Fig.6 (cont.)

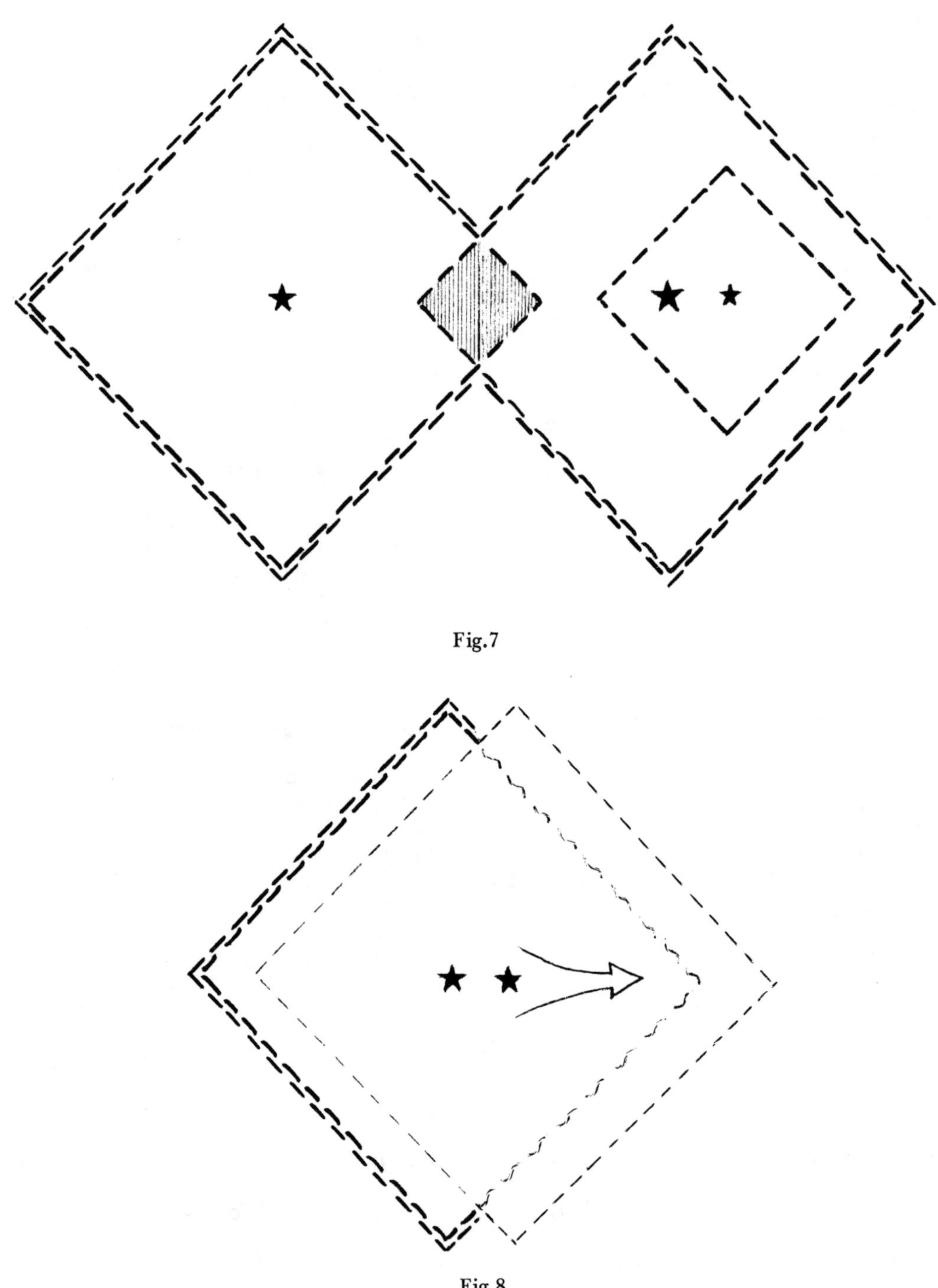

Fig.7

Fig.8

brane will disintegrate because no re-bonding is possible in the area of the overlap. Because a catalyst cannot pass through bonded links, it will naturally float toward the new opening. The two catalysts of equal power will thus float apart until they can enclose themselves independently by two separate membranes. The larger is the overlap of the two catalytic neighborhoods, the stronger 'pulling apart' is experienced by the two catalysts. The process can be accelerated considerably by allowing only the newly created catalyst to move while keeping the original one stationary.

In Fig. 9 we present a structural history of a very simple 'cellular division'. Imagine this as a film projected at higher speeds, providing a continuous impression of the process of the division.

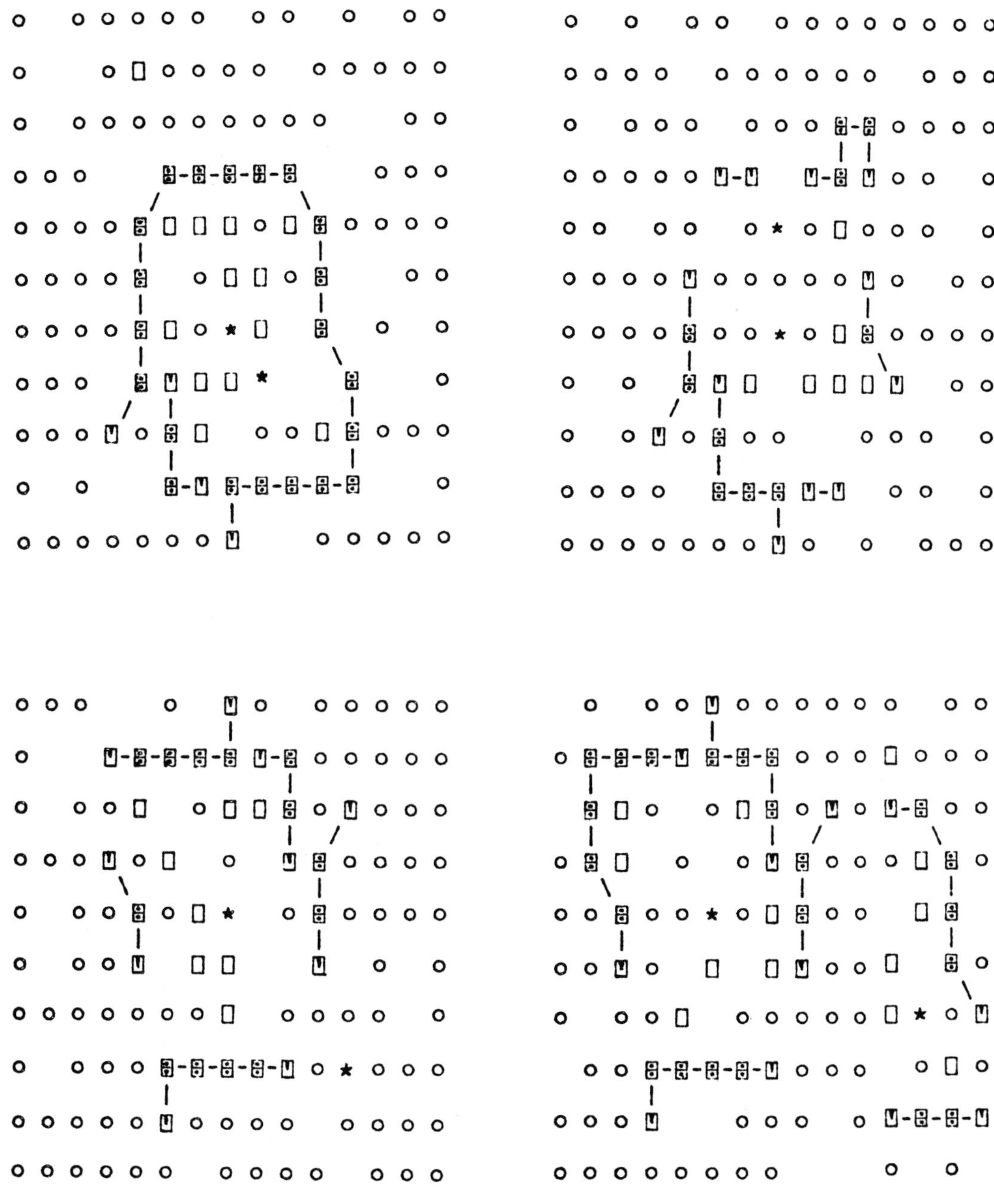

Fig.9

We note that a simple mechanical division of the catalyst leads to self-reproduction of an autopoietic cell. Approximate replica of a cell's organization and structure is obtained without any copying, coding or information processing mechanisms.

Some emerging theses of autopoiesis

Experiments with 'computer-driven' autopoiesis, as those presented in this report, enhance man's understanding of 'natural' autopoietic systems, or *A-systems* for brevity. Coupled with some reflective thought and a keen awareness of past accomplishments, they are also instrumental in unfolding a new and potent paradigm of illuminating the still murky complexity of the observable universe. The following theses appear to be consistent with human experience, though not necessarily with the scientifically interpreted experience.

1. Autopoietic systems

Autopoiesis is the main characteristic of organic systems, physical, biological or social. 'Life' represents a particular class of structural realizations of an autopoietic organization and all 'living' systems are autopoietic. A-systems exhibit self-renewal and autonomy in a given environment: they recursively 'produce' themselves by producing the same components that define them as concrete wholes or distinct unities.

According to Maturana [4]:

> There is a class of mechanistic systems in which each member of the class is a dynamic system defined as a unity by relations that constitute it as a network of processes of production of components which: (*a*) recursively participate through their interactions in the generation and realization of the network of processes of production of components which produced them; and (*b*) constitute this network of processes of production of components as a unity in the space in which they (the components) exist by realizing its boundaries.

2. Organization and structure of A-systems

Organization of an A-system refers to that essential set of interactions among its components which define the system as a recognizable unity. *Structure*, on the other hand, refers to a particular spatial and temporal arrangement of components. While the organization defines the properties of an A-system as a distinct whole, its structure determines the space in which it actually exists and can be perturbed. As Maturana [4] insists:

> ...when the organization of a unity is to be explained it is a mistake to reproduce its structure, it is necessary and sufficient to reproduce its organization...

Distinct A-systems may have identical organization but different structures. Individual structural histories are not sufficient to explain the underlying organization and its present functioning. Humberto R. Maturana was the first to discover that A-systems have a fundamentally simple organization and that their structural complexity is only a result of their evolutionary and ontogenic histories, and, therefore, irrelevant to the explanation of their organization.

3. Modeling of A-systems

A-systems generally exhibit high complexity of interactions between their components. An A-system cannot be understood by studying the properties of its components, its 'wholeness' is crucial for its comprehension. It is their organization which must be modeled, not their structure. The requisite complexity of structural modeling is not only prohibitive but a grossly misplaced effort as well.

Von Foerster [6] expressed his conviction that:

> The idea is to abandon the strategy of reformulating the problem into terms that smack of mathematical rigor but lack the contextual richness originally perceived, and to develop the algorithms that transform the descriptions of certain aspects of a system into paraphrases that uncover new semantic relations pertaining to the system as a whole.

Beer and Casti [7] remark that:

> One of the principal difficulties associated with attempts at mathematically modelling social and economic phenomena has been the natural tendency of modelling to slavishly, in fact to almost religiously adhere to the modelling apparatus which has served so well in physics and engineering.

We don't have to blame our tools but rather our eternal confusion of organization and structure as well as our inability to recognize that this confusion, although pardonable when dealing with allopoietic systems of mechanics and physics, becomes fatal when transplanted into the domain of A-systems.

Modeling of A-systems requires modeling of their organization, ie. the bioalgebra of their rules of interaction. Whether computer simulation, arithmetics, algebra or topology will prove to be the most useful tool, that is not essential.

4. A-systems and their environment

A-systems are not definable by drawing simple mechanical distinctions (or boundaries) which would separate them from their environment. A-systems are inseparable from their environment, ie. from the medium within which their components interact. Because an autopoietic unity is different from its components, even the components of the A-system itself become part of its medium or environment. The interaction is circular: structural changes of the A-system induce structural changes in the medium which constitute the perturbations under which further structural changes are realized. Maturana [4] talks about structural coupling, ie. "The effective spatio-temporal correspondence of changes of state of the organism with the recurrent changes of state of the medium while the organism remains autopoietic." This rapport between the structural dynamics of an A-system and the structural dynamics of the medium is often referred to as either adaptation or learning.

5. Closure of A-systems

A-systems are closed systems. All their components interact through the intervening environment in a closed and circular fashion. According to Varela, [5]: "Every system-whole is organizationally closed." Because of system's structural coupling with its environment there is no inside or outside and no 'input or output' points can be identified. System's bioalgebra defines a closed set of behavioral rules, a closed organization with no environmental interfaces explicitly encoded.

Open systems are those which are characterized by independent input and output points facilitating their interaction with the environment. Their inside and outside is well defined and their behavior is allopoietic. Man-designed systems are typically open: with well identified inputs and outputs, information processing channels, hierarchical relations between components, etc. All 'living' systems are closed and cannot be 'engineered' by man.

6. An observer

An observer is capable of recognizing a system as a separate entity by drawing its concrete or conceptual boundary. Structural interactions between A-systems and their environment include also the interactions with an observer. Observation represents a process of structural coupling between the observer and the observed, ie. between the system and its environment.

A-system is perceived as being allopoietic if its organization becomes open by the observer. If the observer, either experimentally or conceptually, describes an autopoietic system as having input and output surfaces, he can then describe its organization only in terms of hierarchical transfer functions. Such 'organization' is relevant only for the descriptive system of references induced by the observer.

Actual realization of an autopoietic system proceeds in the environment which includes the observer as well. Given a closed A-system, its inside or outside exists only in the eyes of the observer, not for the system itself. G. Spencer Brown [8] puts it in the following way:

> But in order to do so, evidently it [the world] must first cut itself up into at least one state which sees, and at least one other state which is seen. In this severed and mutilated condition, whatever it sees is only partly itself. We may take it that the world undoubtedly is itself (ie. is indistinct from itself), but, in any attempt to see itself as an object, it must, equally undoubtedly, act so as to make itself distinct from, and therefore false to, itself. In this condition it will always partially elude itself.

7. Aggregation of A-systems

The problem of an aggregation of autopoietic systems led Ricardo Uribe to propose the following conjecture: "There are no possible hierarchies of autopoietic systems." Clearly, there are no internal hierarchical relations, except those observed to fit a purposeful design of the observer. The components of A-systems can be A-systems themselves and thus the conclusion that there is no hierarchy among A-systems is reinforced. It seems plausible that an observer, in order to comprehend autopoiesis at all, must perceive the autopoietic components as behaving with apparent input and output with respect to the aggregate autopoietic system.

An observer can, however, within limits, displace himself. Changing his 'point of view', either concretely or conceptually, the observer perceives an organism as being autopoietic; equipped with a microscope he observes its components, the cells, as being autopoietic; from a tall tower he might observe the aggregates of organisms, eg. social groupings, as being autopoietic. One may argue that the observer becomes a different observer after any process of concrete or conceptual displacement described above.

8. Synthetic biology

Synthesis of molecular autopoiesis has been successfully accomplished through the hundreds of experiments performed by Leduc [9] and others. Because the detailed chemical properties of components, beyond those of catalysis, linkage, mobility and decay, are not essential for autopoiesis, Leduc and other synthetic biologists were capable of creating 'living forms' in the medium composed of essentially inorganic molecules.

Certain substances in concentrated solution (substrate) have the property of forming membranes when they come in contact with other chemical compounds (catalysts). According to Leduc:

> Most beautiful osmotic cells may be produced by dropping a fragment of fused calcium chloride into a saturated solution of potassium carbonate or tribasic potassium phosphate, the cal-

cium chloride becoming surrounded by an osmotic membrane of calcium carbonate or calcium phosphate.

Quite often the cells multiply and some elaborate growths result. Certain of these artificial cells may even be made to grow out of the solution into the air. Forms, resembling plants, mushrooms, shells, or even fish-like swimming organisms and undulating medusas, have repeatedly been reproduced by synthetic biologists. Leduc insists that: "The resemblance is so perfect that some of our productions have been taken for fungi even by experts."

9. Cells

It appears that cellular replication and hereditary transfer can be accomplished through a simple separation and distribution of cellular structure while keeping the organization invariant. Maturana [4] writes:

> ...a simple mechanical fragmentation of the autopoietic unity (self-division or self-reproduction) produces at least two new autopoietic unities that may have identical or different structures according to how uniform was the component's distribution in the original unity. Heredity of organization and structure with the possibility of hereditary structural change is, therefore, a necessary consequence of distributed autopoiesis.

No copying, no program, no coding, and no transmission of information are necessary. The molecular components become uniformly distributed throughout the cell in the course of its mitosis, two identical autopoietic unities emerge. If the distribution of the molecular content is less uniform the two emerging cells are differentiated cells. The fact that DNA's double helix replicates itself in the process of mitosis is a simple byproduct and not the cause of autopoietic self-division.

10. Nervous system

Maturana [4] defines the nervous system as an autopoietic unity, ie. a closed neuronal network with no input and output surfaces imbedded in its organization. He proposed:

> The sensory and the effector surfaces that an observer can describe in an actual organism, do not make the nervous system an open neuronal network because the environment (where the observer stands) acts only as an intervening element through which the effector and sensory neurons interact; completing the closure of the system.
>
> The nervous system is structurally coupled with the organism (its medium) and jointly with the environment. The rapport between the changes of state of the organism and the changes of state of the nervous system appear as an apparent input-output relationship. The actually neuronal activity is determined by the structure of the nervous system and not by independent environmental perturbations. Perturbations only provide triggering conditions for internally determined changes of state.

11. Social interactions

Two or more organisms and their nervous systems can become structuraly coupled with each other, and with the medium of their interaction as they represent a medium for each other. Structural rapport means that the internally determined structural changes of closed autopoietic systems appear in an interlocked order, of mutually complementary perturbations. Apparent complexity of human interaction results from observer's description of the observed behavior and not from its internal autopoiesis. Two structurally coupled A-systems form a closed network of internally triggered structural changes under the invariant underlying organization.

Maturana [4] provides an explanation of structural coupling between observers themselves:

> ...linguistic behavior is structurally determined behavior in ontogenically structurally coupled organisms, in which structural coupling determines the sequential order of the mutually triggering alternating changes of state. Semantics exists only in a metadomain of descriptions as a property projected upon the interacting systems by the observer, and valid only for him.

12. Social order

Von Hayek [10], co-recipient of the 1974 Nobel Prize in economics, distinguished two basic kinds of order in society: the kind of order achieved by arranging the relation between the components according to a preconceived plan, ie. man-designed social 'organization', and a 'spontaneous' order resulting from the action of individuals without their intending to create such an order. The ordering forces are the rules (organization, bioalgebra) governing the behavior of the components which form the orders. In this, Von Hayek has pre-conceived social autopoiesis:

> If we understand the forces that determine such an order, we can use them by creating the conditions under which such an order will form itself.

> This indirect method of bringing about an order has the advantage that it can be used to produce orders that are far more complex than any order we can produce by putting the individual pieces in their appropriate places.

Many theorists of 'organization design' still consider social complexity to be of much lower order than physical or biological complexity. It isn't. Although the conduct of the individuals is guided in part by deliberately enforced rules, the order is still a spontaneous order, corresponding to an organism rather than to a machine. Man-conceived designs often represent only a tinkering and interference with spontaneously recurrent activities of a complex society. According to Von Hayek [10]:

> It is thus a paradox, based on a complete misunderstanding of these connections, when it is sometimes contended that we must deliberately plan modern society because it has grown so complex. The fact is rather that we can preserve an order of such complexity only if we control it not by the method of planning, ie. by direct orders, but, on the contrary, aim at the formation of a spontaneous order based on general rules.

13. Mangement of human systems

Trentowski [11] is the father of Cybernetyka, the science and the art of management of complex human systems. He recognized that human systems, since they are not simple machines, cannot be designed or analyzed, but should be managed. This manager, Cyberneta, is not a designer but a catalyst of spontaneous social forces. Human systems management is analogous to the 'management' known from physics or biology. Crystals are not produced by directly arranging the individual molecules but by creating the conditions under which the crystals form themselves. Plants or animals are not created by mechanistic designs but by inducing conditions favorable to their growth. Their resulting shape and structure can be determined only within narrow limits. The same applies to spontaneous social orders.

According to Trentowski, the Cyberneta does not design a social system but helps to bring about the conditions, as a skilful social obstetrician, under which the social organization evolves on its own. A skill beyond that of an engineer and wisdom higher than that of a social philosopher are needed for management of human systems. Trentowski's Cybernetyka goes beyond feedback mechanisms of modern cybernetics, mainly because: "...people are neither mathematical numbers nor logical categories."

In conclusion, it is not presumptuous to regard autopoietic modeling as the first real opportunity to evolve an experimental laboratory for the study of social systems. Also, if one introduces a set of rules which would induce changes in system's bioalgebra in accordance with structural changes of the system itself, providing thus a higher closure of its organization, a rich variety of unexpected emergent behavioral patterns could be observed. The implications of such closure are enormous.

Acknowledgements

I am personally indebted to Mr. Norbert A. Pierre for his invaluable computing assistance as well as for his effective intellectual challenge during our working association.

In addition to my 'spiritual fathers' of the present and of the past, whose works I list in an attempt to share some of the excitement I experienced in studying their thought, I am also indebted to many colleagues and friends who provided their personal encouragement, support and enlightenment throughout this project. Although this particular experience cannot be shared, I still wish to mention their names, if only to show my sincere gratitude: Stafford Beer, Erich Jantsch, George Klir, Ladislav Minaru, Oskar Morgenstern, Gordon Pask, Robert Trappl, Ricardo Uribe, Francisco Varela, Heinz Von Foerster and Betka Zeleny.

This research has not been supported or funded by any official grant or foundation. It has been made possible by Faculty Research Fund of the Graduate School of Business at Columbia University.

References

In order to provide a more complete picture of the conceptual foundations from which this new paradigm of thought—Autopoiesis— has been emerging for more than a century, I endeavor to list also the works which are not explicitly quoted in the text. That does not subtract from their seminal importance but merely reflects my own inability to share this view in its totality and through this particular medium.

1. VARELA, F., MATURANA, H.R., and URIBE, R., 'Autopoiesis: The organization of living systems, its characterization and a model', *Bio-Systems* 5(4), 1974, 187-196.
2. ZELENY, M. and PIERRE, N.A., 'Simulation models of autopoietic systems', *Proceedings of the*

1975 Summer Computer Simulation Conference, Simulations Council, La Jolla, California, pp. 831-842 (1975).

3. ZELENY, M. and PIERRE, N.A., 'Simulation of self-renewing systems', in E. Jantsch and C.H. Waddington, eds. *Evolution and Consciousnes*, Addison-Wesley, Reading, Massachusetts (1976).
4. MATURANA, H.R., 'The organization of the living: a theory of the living organization', *International Journal of Man-Machine Studies* 7, 1975, 313-332.
5. VARELA, F., 'The arithmetics of closure', Third European Meeting on Cybernetics and Systems Research, Vienna, 1976. In this book, pp. 48-63
6. VON FOERSTER, Heinz, 'Computing in the semantic domain', *Annals of the New York Academy of Sciences*, 184, 239-241 (7 June 1971).
7. BEER, S. and CASTI, J., 'Investment against disaster in large organizations', *IIASA Research Memorandum*, RM-75-16 (April 1975).
8. SPENCER-BROWN, G., *Laws of Form*, The Julian Press, New York (1972).
9. LEDUC, Stéphane, *The Mechanism of Life*, Rebman Ltd., London (1911).
10. VON HAYEK, F.A., 'Kinds of order in society', *Studies in Social Theory*, No.5, Institute for Humane Studies, Menlo Park, California (1975).
11. TRENTOWSKI, Bronislaw, *Stosunek Filozofii do Cybernetyki czyli sztuki rzadzenia narodom* (On the Relationship of Philosophy to Cybernetyka or the art of managing a nation), J.K. Župański, Poznań (1843).
12. BEER, S., *Platform for Change*, John Wiley and Sons, New York (1975).
13. BOGDANOV, A., *Tektologiia: vseobshchaia organizatsionnaia nauka* (Tectology: general science of organization), Z.I. Grschebin Verlag, Berlin (1922).
14. JANTSCH, Erich, *Design for Evolution*, George Braziller, New York (1975).
15. MATURANA, H.R., 'Neurophysiology of cognition', in *Cognition: A Multiple View*, edited by Paul Garvin, Spartan Books, New York (1970).
16. MENGER, Carl, *Untersuchungen über die methode der Sozialwissenschaften und der Politischen Ökonomie insbesondere,* Duncker & Humblot, Leipzig (1883).
17. MORGENSTERN, Oskar, 'Thirteen critical points in contemporary economic theory: An interpretation', *Journal of Economic Literature,* 10, 1165-1166 (December 1972).
18. ORTEGA Y GASSET, José, *The Revolt of the Masses,* W.W. Norton & Co., Inc., New York (1960).
19. PASK, Gordon, *Conversation, Cognition and Learning,* Elsevier, New York (1975).
20. PRIGOGINE, Ilya, *Stability, Fluctuations, and Complexity,* Brussels (1974).
21. SMUTS, Jan Christiaan, *Holism and Evolution,* Greenwood Press, Westport, Connecticut (1973).
22. TEILHARD de CHARDIN, Pierre, *The Phenomenon of Man,* Harper Books, New York (1961).
23. THOMPSON, D'Arcy W., *On Growth and Form,* Cambridge University Press, London, p.346 (1971).
24. VARELA, F., 'A calculus for self-reference', *International Journal of General Systems* 2(1), 5-24 (1975).
25. ZELENY, M., 'Organization as an Organism', *Proceedings of the Annual North American Meeting of the Society for General Systems Research,* SGSR, Washington, D.C., pp.262-270 (1977).
26. ZELENY, M., 'Self-organization of living systems: a formal model of autopoiesis', *International Journal of General Systems* 4(1), 13-28 (1977).

Backward computers for cell spaces

U. GOLZE
Technische Universität Hannover
Germany (FRG)

1. Introduction and outline

As just one out of many applications, some deterministic cell spaces have been used to model the mammalian cerebellar cortex (Mortimer [1]). For a configuration c describing the present state of a nervous system, the cell space computes a successor configuration $F(c)$. For a special configuration 'nervous breakdown', for example, it might be interesting to compute possible predecessor configurations which will lead to the breakdown.

In the present chapter we will study the 'inversion' of global cellular transition functions F; a backward computer when given an input configuration c, outputs a predecessor b such that $F(b)=c$. Forward computation is a straightforward matter involving a well known amount of complexity; even when predecessors exist, backward computation may become arbitrarily difficult, often impossible.

To avoid searching for algorithms that cannot exist, we will try to encircle the class of unsolvable predecessor problems. We will consider various types of solvability induced by the stop-iff-undefined and output-iff-defined concepts of backward computers. We obtain different results for finite, rational, recursive and infinite configurations being defined as multi-dimensional generalizations of rational, recursive and arbitrary real numbers. There exist fundamental differences between one- and higher-dimensional spaces when computing backwards.

2. Configurations and recursiveness

For convenience of the reader, we briefly review some definitions introduced in an earlier paper [2]. A function e is represented by $e: A \to B$ or $a \mapsto b$ in the usual way. Domain and range of e are referred to by dom e and rg e, respectively. Induced are two functions $\tilde{e}$ and $\underset{\sim}{e}$ assigning to every set C the sets $\tilde{e}(C) = \{b \in \text{rg}\, e : \exists a \in C(e(a)=b)\}$ and $\underset{\sim}{e}(C)=\{a \in \text{dom}\, e: \exists b \in C(e(a)=b)\}$. When convenient, a is identified with $\{a\}$.

An *(infinite) cell space* Z consists of the *dimension* $d \in \mathbb{N}$, the finite set of *cell states* Q having at least two elements, the *local transition function* $g: Q^m \to Q$ and the *neighborhood template* $N \in (\mathbb{Z}^d)^m$ for some $m \in \mathbb{N}$. An infinite space is *finite* if there is a *quiescent state* $\theta \in Q$ such that $g(\theta,\ldots,\theta)=\theta$. Throughout this chapter, d, Q, g, N and θ, and F, F^κ, C^κ, S^κ, $G^{\kappa\lambda}$ and s^κ for $\kappa,\lambda \in \{f,q,r,i\}$ (see below) are reserved for the components associated with the cell space under consideration in each case. Let CS be the class of all cell spaces, let CS^f denote the class of finite spaces and let the subscript d refer to the d-dimensional spaces.

Given a space $Z \in CS$, any mapping $c: \mathbb{Z}^d \to Q$ is called an *(infinite) configuration.* These configurations cannot be coded by a natural number, that is, by a finite amount of information. Therefore, often only finite spaces and finite configurations have been considered. $c \in C^i$ is *finite* if the *support* $\sup c :=$ $\{x \in \mathbb{Z}^d : c(x) \neq \theta\}$ is finite, i.e., if all but a finite number of cells are in the quiescent state.

When trying to find configurations of which certain predecessors can (cannot) be computed or approximated effectively, the finite configurations turned out to be a rather unnatural class. The first generalization still permitting effective descriptions is achieved by considering periodic rather than constant environments. For $Z \in CS_1$, a configuration $c \in C^i$ is *rational* if there is an $\omega \in \mathbb{N}$ such that the sequences $(c(\omega+n))_{n\in\mathbb{N}}$ and $(c(-\omega-n))_{n\in\mathbb{N}}$ have period ω. We will only sketch out the generalization to higher dimensions. In 2-dimensional spaces, a

		⋮		
	C	B_1	C	
…	B_4	A	B_2	…
	C	B_3	C	
		⋮		

rational configuration c has a finite kernel $c|A$; it is double-periodic (periodic in every dimension) when restricted to C and single-periodic when restricted to any of the B_n for $n=1,\ldots,4$.

For $Z \in CS$, configuration $c \in C^i$ is *recursive* if the state of every cell can be computed effectively, that is, if the function (code number of $x \in \mathbb{Z}^d$) $\mapsto$ (code number of $c(x)$) is recursive. c can be described by a finite amount of information, namely the Gödel number of a machine computing c.

Denoting the sets of finite, rational and recursive configurations by C^f, C^q and C^r, respectively, one immediately obtains for every space the hierarchy $C^f \subset C^q \subset C^r \subset C^i$. Unless stated differently, κ and λ stand for any of the configuration type indices f, q, r, i. Whenever f occurs, we assume that the space considered is finite.

The cellular system described thus far becomes dynamic by introducing the *global transition function* $F: C^i \to C^i$ defined by $\forall c \in C^i\, \forall x \in \mathbb{Z}^d: F(c)(x) = g(c(x+x_1),\ldots,c(x+x_m))$ where $N=(x_1,\ldots,x_m)$. For every configuration c, F determines the *successor* $F(c)$ which never belongs to a higher class in the hierarchy, i.e. $\forall\kappa: \tilde{F}(C^\kappa) \subseteq C^\kappa$. Any $b \in \underline{F}(c)$ is called a *predecessor* of c. F^κ is defined as the restriction $F|C^\kappa$.

Intuitively, F is a recursive function; but its domain cannot be coded by natural numbers. Moreover, the stop-iff-defined concept of Turing machines is not directly applicable to such 'recursive functionals' (see Shoenfield [3, Ch.7.2]). For an easy visualization as well as a short abstract formulation, we will introduce the notion of a *cellular Turing machine*. This is a Turing machine M which in addition to its 1-dimensional computation tape is equipped with a read-only *input tape* of structure $\mathbb{Z}^m$ and a write-only *output tape* of structure $\mathbb{Z}^n$ $(m,n \in \mathbb{N})$. The alphabet X for all tapes contains a blank symbol. The normal instruction set is extended such that M may move its three tape arms in any direction possible as well as execute: P_x (print a non-blank $x \in X$ into the output tape if a blank is scanned), $T_x(\ell)$ (go to instruction ℓ if x is read in the input tape) and S (stop).

The partial function G from A into B is ε-*recursive* if the following holds:

1. There is a cellular Turing machine M as described above where every $a \in A$ is an input tape inscription $(a: \mathbb{Z}^m \to X)$ and every $b \in \operatorname{rg} G$ is a non-blank output tape inscription $(b: \mathbb{Z}^n \to X \setminus \{\text{blank}\})$.

2. Let an input tape inscription $a \in \operatorname{dom} G$ be given, the output and computation tape being blank and all tape arms scanning $(0,\ldots,0)$. Then, for every output cell $x \in \mathbb{Z}^n$, there is a $t \in \mathbb{N}$ such that at time t, M prints $G(a)(x)$ into the output tape.

3. If $a \in A \setminus \operatorname{dom} G$ is presented in the same format, then condition ε holds where $\varepsilon \in \{\alpha,\beta,\gamma,\delta\}$ and

   $\alpha :=$ (M never prints anything into the output tape),
   $\beta :=$ (M will eventually stop),
   $\gamma := \alpha \wedge \beta$,
   $\delta := (1=1)$.

If an ε-recursive G is defined for x, M will gradually write the result $G(x)$ into the output tape. If an α-recursive G is undefined, nothing will ever be written into the output tape ('output-iff-defined'). If a β-recursive G is undefined, M will stop ('stop-iff-undefined').

If an α-recursive G is defined for x, it can eventually be determined effectively that $G(x)$ is defined. For β- but not α-recursive functions, this is, in general, never possible. If on the other hand a β-recursive G is undefined for x, it can eventually be determined effectively that $G(x)$ is undefined. For α- but not β-recursive functions, this is generally never possible.

For γ-recursive functions, the definedness is decidable. Therefore, the γ-recursiveness is the strongest notion. The δ-recursiveness is the weakest notion and will mainly be used for characterizing non-δ-recursive functions. A configuration c is recursive iff the function G: {empty tape} $\rightarrow \{c\}$ is γ-recursive.

The following diagram summarizes the implications and independences of the various notions of recursiveness. Proofs can be found in [4].

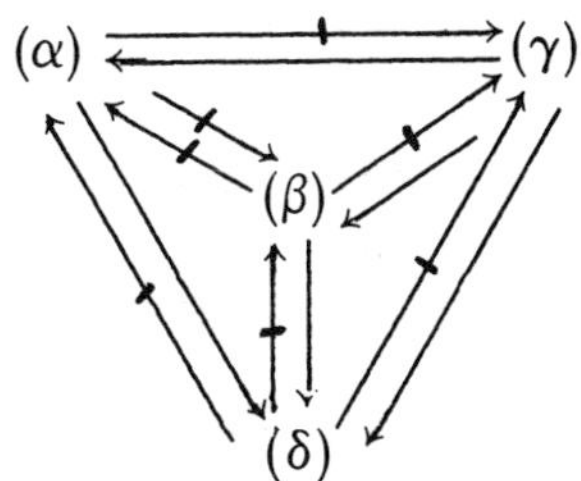

3. Predecessor functions

A type λ predecessor of a given type κ configuration c does not exist if c belongs to the *Garden-Eden* set $G^{\kappa\lambda} := C^{\kappa} \backslash \bar{F}(C^{\lambda})$. The various applications of the $G^{\kappa\lambda}$-framework have been discussed in [2]. In this chapter we will touch on the decidability of some of the $G^{\kappa\lambda}$ and try to compute λ-predecessors of κ-configurations not in $G^{\kappa\lambda}$.

Let $P(C^f, C^f)$ be the set of (partial) predecessor functions G from C^f into C^f where $C^f \backslash \text{dom}\, G = G^{ff}$ and $c \in \text{dom}\, G$ implies $G(c) \in \underset{\sim}{F}^f(c)$. For a given cell space there is in general no cellular Turing machine which δ-recursively computes finite predecessors of finite configurations:

Theorem 1

$$\forall d \in \mathbb{N}\ \exists Z \in CS^f_d\ \forall G \in P(C^f, C^f) : \neg\, (G\ \delta\text{-}recursive).$$

Proof. 1-dimensional spaces can be considered as degenerated d-dimensional spaces. A 1- and thus d-dimensional counter example is a finite space with state set $Q = \{0,1,2,3\}$, neighborhood template $N = (0,1)$ and a local transition function g where $\underset{\sim}{g}(0) = \{(0,0), (1,1), (2,0)\}$, $\underset{\sim}{g}(1) = \{(0,1), (0,2)\}$ and $\underset{\sim}{g}(2) = \{(1,0)\}$.

(1) The finite configuration c_1: ... 0 0 1 0 0 0 ... (cell 0 ↓ at the 1)
has only one finite predecessor: ... 0 0 0 2 0 0 ... .

(2) The finite configuration $c_2(x)$: ... 0 0 1 0 0 ... 0 0 2 0 0 ... (cell 0 ↓ at the 1, cell x ↓ at the 2)
has only one finite predecessor: ... 0 0 0 1 1 ... 1 1 1 0 0 ... .

(3) A cellular Turing machine M computing a predecessor function G from C^f into C^f would print a 2 in output cell 1 at some time t when given input c_1; however, at time t, M cannot have recognized that also the input $c_2(t+1)$ might have been given requiring the printing of a 1 rather than a 2. This contradiction proves Theorem 1.

Similar examples imply that we can only hope to compute predecessors if the inputs contain additional coded information like the size of a finite configuration or the algorithm for computing a recursive configuration, etc.

For $Z \in CS$, let $S^q \subseteq \mathbb{N}$ be a Gödelization of C^q such that $s \in S^q$ codes the size and state of the finite kernel of a rational configuration $c \in C^q$ as well as the sizes and states of the periods of c. Let $s^q : S^q \rightarrow C^q$, $s \mapsto c$ be the associated bijective, γ-recursive mapping, let $S^f \subseteq S^q$ code the finite configurations and define s^f as $s^q | S^f$. In an effective enumeration of all Turing machines we distinguish the set $S^r \subseteq \mathbb{N}$ of those numbers whose corresponding machines compute recursive configurations in the sense of their definition. This surjective, γ-recursive mapping $s^r : S^r \rightarrow C^r$ is naturally highly non-injective.

Let R^{κ} be a variable for C^{κ} or if $\kappa \neq i$, for S^{κ}. Define R^{λ} analogously. We will now study the sets $P(R^{\kappa}, R^{\lambda})$ of (partial) *predecessor functions* G from R^{κ} into R^{λ} being defined such that for any $x \in R^{\kappa}$ and the corresponding configurations

$$c := \begin{cases} x & \text{for } R^{\kappa} = C^{\kappa} \\ s^{\kappa}(x) & \text{for } R^{\kappa} = S^{\kappa} \end{cases},$$

$G(x)$ is undefined if the set $A := \underset{\sim}{F}^{\lambda}(c)$ of λ-predecessors of c is empty, and

$$G(x) \in \begin{cases} A & \text{for } R^{\lambda} = C^{\lambda} \\ \underset{\sim}{s}^{\lambda}(A) & \text{for } R^{\lambda} = S^{\lambda} \end{cases}$$

otherwise.

A generalization of the negative Theorem 1 is:

Theorem 2

$\forall\kappa\ \forall\lambda\ \forall d \in \mathbb{N}\ \exists Z \in CS_d$:

(1) $G \in P(C^\kappa, C^\lambda) \Rightarrow \neg$ (G δ-recursive),
(2) $\lambda \neq i \wedge G \in P(C^\kappa, S^\lambda) \Rightarrow \neg$ (G δ-recursive).

(1) is shown similarly as in the previous proof; (2) is implied by (1) and the γ-recursiveness of the code functions s^λ.

The first positive result states the effective computability of coded finite and rational predecessors of coded finite and rational configurations:

Theorem 3

$\forall\kappa \in \{f,q\}\ \forall\lambda \in \{f,q\}\ \forall Z \in CS\ \exists G \in P(S^\kappa, S^\lambda)$:

G partial recursive.

Because a successor never belongs to a higher configuration class than any predecessor, it is of little interest (but not meaningless) to search, for example, for finite predecessors of rational configurations ($G \in P(S^q, S^f)$).

Theorem 3 can be proved by 'brute force': S^f and S^q are recursively enumerable, for every enumerated code of a configuration c the 'finiteness' of the finite and rational configurations permits to check whether c's successor coincides with the configuration whose predecessors are wanted. More efficient algorithms exist in the 1-dimensional case [4] and would be of interest for higher dimensions.

In 1-dimensional spaces, the situation is more pleasant: also recursive and infinite predecessors of at most rational configurations are computable; the predecessor computation is γ-recursive; therefore, the existence of predecessors and thus the Garden-Eden sets $G^{f\lambda}$ and $G^{q\lambda}$ are decidable:

Theorem 4

$\forall\kappa \in \{f,q\}\ \forall\lambda\ \forall Z \in CS_1$:

(1) $\exists G \in P(S^\kappa, C^\lambda)$: G γ-recursive,
(2) $\lambda \neq i \Rightarrow \exists G \in P(S^\kappa, S^\lambda)$: G partial recursive.

Proof. The cases $\lambda \in \{r,i\}$ need not be considered because in 1-dimensional spaces we have $G^{fq} = G^{fr} = G^{fi}$: if a finite configuration has a predecessor at all, then there always exists a rational predecessor. This, by the way, is in sharp distinction to higher-dimensional spaces where finite successors can be found having not even recursive predecessors: $G^{fi} \subset G^{fr}$ (see [1] for details). Thus (2) follows from Theorem 3. Using (2) and the γ-recursiveness of s^λ, the proof is complete if we can find a cellular Turing machine stopping only for those 1-dimensional input configurations that have no predecessors. This is done in

Lemma 5

For all $Z \in CS_1$, the partial function $G : C^i \rightarrow C^i$,

$$c \mapsto \begin{cases} \text{undefined if } c \in G^{ii} \\ c \text{ else} \end{cases} \quad \text{is } \beta\text{-recursive.}$$

Proof. The cellular Turing machine M to be constructed starts at input cell 0 and operates symmetrically and simultaneously to the left and to the right. We will only recursively describe the operation from 0 to the right. Without loss of generality, Z has the neighborhood template $N=(-1,0,1)$ (for a justification see Yamada and Amoroso [5]). Let $c \in C^i$ be given. Base of recursion: $Q_0 := g(c(0))$. Q_0 contains all cell state triples mapped onto $c(0)$ by the local transition function. Recursion: Let Q_n be defined. If $Q_n = \emptyset$, stop. Otherwise copy $c(n)$ onto the output tape and construct $Q_{n+1} := g(c(n+1)) \cap \{(x,y,z) : \exists (w,x,y) \in Q_n\}$. Any predecessor b of c gives rise to a sequence $(\ldots,(b(n-1),b(n),b(n+1)),(b(n),b(n+1),b(n+2)),\ldots) \in$

$\in \ldots \times Q_n \times Q_{n+1} \times \ldots$ and vice versa. Therefore M stops if and only if c is Garden-Eden. This proves Lemma 5.

There are α-recursive computers of finite or rational predecessors of finite or rational configurations in spaces with $d \geqslant 2$—once they start outputting, the output is the initial piece of an existing predecessor. But those computers cannot always operate β-recursively:

Theorem 6

$\forall\kappa \in \{f,q\}\ \forall\lambda \in \{f,q\}\ \forall d \in \mathbb{N} \setminus \{1\}$:

(1) $\forall Z \in CS_d\ \exists G \in P(S^\kappa, C^\lambda)$: G α-recursive,
(2) $\exists Z \in CS_d\ \forall G \in P(S^\kappa, C^\lambda)$: $\neg$ (G β-recursive).

Proof.
(1) A suitable cellular Turing machine M simulates

a partial recursive G as in Theorem 3 thus computing a predecessor code s if existent; the output tape is used if and only if such an s has been found: M executes the γ-recursive transformation $s \mapsto s^{\lambda}(s)$.

(2) If (2) was false, then for some κ, λ and d there would be for all d-dimensional spaces an α- and β- hence γ-recursive $G \in P(S^{\kappa}, C^{\lambda})$. This would contradict the following undecidability result.

Theorem 7

(G^{ff}, G^{fq}, G^{qf} and G^{qq} are undecidable for $d \geqslant 2$.)

$\forall \kappa \in \{f,q\}\ \forall \lambda \in \{f,q\}\ \forall d \in \mathbb{N}\setminus\{1\}\ \exists Z \in CS_d$:

$(G : S^{\kappa} \to \{0,1\})$ is not recursive if

$$s \mapsto \begin{cases} 1 \text{ for } s^{\kappa}(s) \in G^{\kappa\lambda} \\ 0 \text{ else} \end{cases}$$

Proof. The case $\kappa = \lambda = f$ has been shown by Yaku [6]. The counter example used there covers also the more general case of Theorem 7.

For recursive configurations the predecessor question is not even decidable for 1-dimensional spaces:

Theorem 8

$\forall d \in \mathbb{N}\ \exists Z \in CS_d\ \forall \lambda$: *$(G : S^r \to \{0,1\})$ is not recursive if*

$$s \mapsto \begin{cases} 1 \text{ for } s^r(s) \in G^{r\lambda} \\ 0 \text{ else} \end{cases}$$

The proof needs only be given for $d = 1$. $\lambda = r$: Consider a space where $Q = \{0,1\}$, $N = 0$ and $g \equiv 0$. Then only the trivial configuration $c \equiv 0$ has a predecessor. For all natural n, define a recursive configuration c_n such that for all $x \in \mathbb{Z}$, $c_n(x) = 1$ if Turing machine number n with empty input will stop after x time steps, and $c_n(x) = 0$ else. The negation of Theorem 8 would solve the halting problem. $\lambda \neq r$: For the space just defined we have $G^{rr} = G^{r\lambda}$.

In any dimension one cannot expect, in general, to be able to compute a predecessor code of a recursive configuration:

Theorem 9

$\forall d \in \mathbb{N}\ \exists Z \in CS_d \forall G \in P(S^r, S^r) : \neg$ *(G partial recursive).*

Proof. Assume that $\exists d \in \mathbb{N}\ \forall Z \in CS_d\ \exists$ partial recursive $G \in P(S^r, S^r)$. Then the same is true for $d = 1$ because any 1-dimensional space Z with neighborhood template $N_Z = (x_\nu)_{\nu=1,\dots,n}$, local transition function g_Z and cell state set Q_Z behaves like the d-dimensional space Y where $N_Y = ((x_\nu, 0, \dots, 0)_{\nu=1,\dots,n})$, $g_Y = g_Z$ and $Q_Y = Q_Z$. We obtain for every 1-dimensional space Z a machine stopping exactly for inputs s where $s^r(s) \notin G^{rr}$. Lemma 5 yields the opposite machine stopping iff $s^r(s) \in G^{rr}$. G^{rr} becomes decidable contradicting Theorem 8.

With a similar argument we exclude the α-recursive computability of recursive or infinite predecessors of recursive configurations:

Theorem 10

$\forall d \in \mathbb{N}\ \exists Z \in CS_d\ \forall \lambda \in \{r, i\}\ \forall G \in P(S^r, C^{\lambda})$:

$\neg$ *(Gα-recursive).*

Proof. An α-recursive G would start outputting if and only if input $s \in S^r$ codes a successor; a second machine could stop in the same situation. As in the previous proof, we obtain the contradictory decidability of $G^{r\lambda}$.

Not even of finite and rational configurations can we hope to be always capable of computing recursive predecessors in a partial or α-recursive manner:

Theorem 11

$\forall d \in \mathbb{N}\setminus\{1\}\ \forall \kappa \in \{f,q\}\ \exists Z \in CS_d$:

(1) $G \in P(S^{\kappa}, S^r) \Rightarrow \neg$ *(G partial recursive),*
(2) $G \in P(S^{\kappa}, C^r) \Rightarrow \neg$ *(G α-recursive).*

For the rather long and tedious proof of (1) refer to [4]. The negation of (2) would contradict (1) as follows. Assume there are a $d \neq 1$ and a κ such that every d-dimensional space has a cellular Turing machine M α-recursively computing a $G \in P(S^{\kappa}, C^r)$. For $s \in S^{\kappa}$ we may construct a normal Turing machine R which when given the coordinates of any $x \in \mathbb{Z}^d$ will simulate M with input s and will print the number n and stop if and only if M would print n into output cell x. The effective constructibility of R can be done by a Turing machine T: for input s, T will print the code number of R and stop if configuration $s^{\kappa}(s) \notin G^{\kappa r}$. The latter condition can be determined because M operates α-recursively (output-iff-defined). Such a T contradicts (1).

We propose but have not proved yet that 'α' in Theorem 10 for $\lambda \neq i$ and Theorem 11 is replaceable by 'δ'. This would mean that certain recursive predecessors c although existent cannot be approximated, in other words, c's effective description cannot be found effectively.

The existence of finite successors without recursive predecessors [2] implies that the computation of infinite configurations may not even be δ-recursive:

Theorem 12

$$\forall d \in \mathbb{N}\backslash\{1\}\ \exists Z \in CS^f_d\ \forall \kappa \in \{f,q,r\}\ \forall G \in P(S^\kappa, C^i):$$

$$\neg\ (G\ \delta\text{-recursive}).$$

4. Summary

In the following matrix, the row index m gives dimension, type and coding mode (uncoded or coded) of a configuration class. The column index n determines type and coding mode of a predecessor class. Matrix element (m,n) informs which property some predecessor function $G \in P(m,n)$ has or, when '¬' is used, which all $G \in P(m,n)$ do generally not have. The number in parentheses points to the corresponding theorem. (pr = partial recursive, δ = δ-recursive, ...)

References

1. MORTIMER, J.A., 'A cellular model for mammalian cerebellar cortex', PhD thesis, 03296-7-T, Dept. of Comp. & Comm. Sc., Univ. of Michigan (1970).
2. GOLZE, U., 'Differences between 1- and 2-dimensional cell spaces', *Automata, Languages, Development* (edit. by A. Lindenmayer and G. Rozenberg) North-Holland (1976).
3. SHOENFIELD, J.R., *Mathematical Logic,* Addison-Wesley (1967).

Type and coding mode of a -

- configuration - predecessor

dimension		C^f	C^q	C^r	C^i	S^f	S^q	S^r
$d \geq 1$	C^f	¬δ (1)	¬δ (2)	¬δ (2)	¬δ (2)	¬δ (2)	¬δ (2)	¬δ (2)
	C^q	-	¬δ (2)	¬δ (2)	¬δ (2)	-	¬δ (2)	¬δ (2)
	C^r	-	-	¬δ (2)	¬δ (2)	-	-	¬δ (2)
	C^i	-	-	-	¬δ (2)	-	-	-
$d = 1$	S^f	γ (4)	γ (4)	γ (4)	γ (4)	pr(3)	pr(3)	pr(4)
	S^q	-	γ (4)	γ (4)	γ (4)	-	pr(3)	pr(4)
	S^r	-	-	¬α (10)	¬α (10)	-	-	¬pr(9)
$d > 1$	S^f	α (6) ¬β (6)	α (6) ¬β (6)	¬α (11)	¬δ (12)	pr(3)	pr(3)	¬pr(11)
	S^q	-	α (6) ¬β (6)	¬α (11)	¬δ (12)	-	pr(3)	¬pr(11)
	S^r	-	-	¬α (10)	¬δ (12)	-	-	¬pr(9)

4. GOLZE, U., 'Finite, periodic and recursive cellular configurations: computation of predecessors and Garden-Eden problems', Doct. Diss., Techn. Univ. Hannover, W. Germany (1975).
5. YAMADA, H., and AMOROSO, S., 'Structural and behavioral equivalences of tessellation automata', *Inf. & Contr.* 18 (1971).
6. YAKU, T., 'The constructibility of a configuration in a cellular automaton', *J. Comp. & Syst. Sc.* 7 (1973).

The subspace approach to pattern recognition

JOSEF KITTLER*
Department of Electronics, The University, Southampton, England

1. Introduction

The mode of functioning of many biological, social and engineering systems depends on a finite set of possible situations which are recognized by information processing and decision making channels of the systems. Thus, for instance, a particular behavioral pattern of a certain category of animals depends entirely on the class of visual patterns in the field of view; the feedback action of time optimal control systems is a two-valued function of the position of the state space vector with respect to the optimal switching surface, etc. It follows that a pattern recognizer forms an essential part of such systems. Its function is to extract from the abundance of available information features characterizing the patterns and subsequently, on the basis of this information, classify the patterns into their appropriate categories. The outcome of this decision making process then determines the subsequent modus operandi of the system.

There are various approaches to modelling and design of pattern recognition systems [1]. In our discussion, however, we shall concentrate on the subspace methods of pattern recognition which combine the role of feature selection and classification. In section 2 of this chapter the subspace approach to pattern classification is introduced. Subsequently, in section 3, existing techniques for determining class subspaces are reviewed and a new method based on the generalized Fukunaga-Koontz feature selection procedure is proposed. Finally, in the Appendix, two simple methods for determining the intersection of projection operators are suggested.

2. The subspace approach to pattern recognition

The process of recognizing representation patterns is usually viewed as a sequence of two independent functions, namely feature selection and classification. Thus, given an N-dimensional representation pattern $\mathbf{x}$, which belongs to one of m possible pattern classes ω_i, $i=1,2,...,m$, a system utilizing this two stage concept first maps the pattern vector into a lower dimensional feature space and subsequently makes a decision about the membership of the pattern on the basis of the information contained in the feature vector. Restricting the class of admissible mappings of pattern vectors from the representation space into the decision space to linear transformations, this two stage process can be formally described as follows: assign pattern vector $\mathbf{x}$ to class ω_i if the ith class discriminant c_i,

*This work was supported by the Science Research Council, England.

$$\mathbf{c}_i = [c_{i1},\ldots,c_{in}]^T \tag{1}$$

satisfies

$$\mathbf{c}_i^T\mathbf{y} = \mathbf{c}_i^T A^T \mathbf{x} > \mathbf{c}_j^T\mathbf{y} \qquad \forall j \neq i \tag{2}$$

where $\mathbf{y}$ denotes the feature vector and A is an $N \times n$ matrix which maps pattern $\mathbf{x}$ into the feature space.

This two stage concept is, however, quite arbitrary and has been proposed only in order to divide the complex problem of pattern recognition system design into manageable subproblems. It is, of course, quite reasonable to combine the process of feature selection and pattern classification into one mapping operation. This is apparent, for instance, from (1) if we denote the product of class discriminant $\mathbf{c}_i$ and matrix A by $\mathbf{d}_i$, i.e.

$$\mathbf{d}_i = A\,\mathbf{c}_i \tag{3}$$

Then vector $\mathbf{d}_i$ defines a linear transformation which maps pattern $\mathbf{x}$ directly into the decision space. It would be possible to interpret vector $\mathbf{d}_i$ as a direction in the pattern representation space which defines one dimensional subspace associated with class ω_i. Pattern $\mathbf{x}$ is then assigned to that class for which the projection of $\mathbf{x}$ into the corresponding subspace is maximum.

In these terms the problem of pattern recognition system design could be reformulated as one of determining the subspaces occupied by individual classes. In general, these class subspaces will be multi-dimensional but their dimensionality will be substantially smaller than that of the pattern representation space. The classification process is then simply implemented by measuring the magnitude of the projection of patterns with unknown membership into these class subspaces. In the following section we shall discuss various methods for determining class subspaces.

3. Methods for determining class subspaces

All the methods of determining the subspaces occupied by individual classes which are discussed in this chapter are based on the Karhunen-Loeve expansion and its information compression properties [1,2,3]. In this approach the constituent axes of the ith class subspace, A_i,

$$A_i = [\mathbf{a}_{i1}, \mathbf{a}_{i2}, \ldots, \mathbf{a}_{in_i}] \tag{4}$$

are defined as the eigenvectors of correlation matrix $E\{\mathbf{x}\mathbf{x}^T\}$, $\mathbf{x} \in \omega_i$, that are associated with the n_i largest eigenvalues of the correlation matrix. By virtue of the expansion the axes $\mathbf{a}_{ij}$, $j = 1,2,\ldots,n_i$ are orthonormal, i.e.

$$\mathbf{a}_{ij}^T\mathbf{a}_{ik} = \delta_{jk} \tag{5}$$

where δ_{jk} is the Kronecker delta function.

Before discussing individual subspace methods first it will be worthwhile introducing the concept of projection operators. Let q_i be the square of the magnitude of the projection of vector $\mathbf{x}$ into the space spanned by the column vectors of matrix A_i, i.e.

$$q_i = \mathbf{y}_i^T\mathbf{y}_i = \mathbf{x}^T A_i A_i^T \mathbf{x} \tag{6}$$

Now let us denote $A_iA_i^T$ by P_i, i.e.

$$P_i = A_i A_i^T \tag{7}$$

It is apparent that P_i is a real symmetric matrix, i.e.

$$P_i = P_i^T \tag{8}$$

Moreover P_i^2 satisfies

$$P_i^2 = A_iA_i^TA_iA_i^T = A_iA_i^T = A_iA_i^T = P_i \tag{9}$$

since, owing to (5), $A_i^TA_i = I$. Thus P_i is also an idempotent matrix and it is, therefore, a projection operator. Using (6) and (9) we can write

$$q_i = \mathbf{x}^T P_i^T P_i \mathbf{x} \tag{10}$$

Since the projection operator P_i is an idempotent matrix its eigenvalues must be unity or zero. This can be seen by expressing P_i in terms of its matrix of eigenvectors and eigenvalues B and Λ respectively. Then

$$P_i^2 = P_i = B^T \Lambda B \tag{11}$$

But P_i^2 is given as

$$P_i^2 = B^T \Lambda B B^T \Lambda B = B^T \Lambda^2 B \tag{12}$$

and it follows that

$$\Lambda^2 = \Lambda \tag{13}$$

Relationship (13) will be satisfied only if the elements of the diagonal matrix Λ are either 1 or 0.

Recalling (4) and (7) we can write P_i in terms of the columns of A_i as

$$P_i = \sum_{j=1}^{k} \mathbf{a}_{ij}\mathbf{a}_{ij}^T \tag{14}$$

The operator P_i, therefore, spans only n_i-dimensional space of the original pattern representation space, defined by vectors $\mathbf{a}_{ij}$. It can be readily seen that vectors $\mathbf{a}_{ij}$ are eigenvectors of P_i with unity eigenvalues since

$$P_i\mathbf{a}_{ij} = \mathbf{a}_{ij} \tag{15}$$

In addition matrix P_i will have N-n_i zero eigenvalues and the corresponding eigenvectors will be orthogonal to the space spanned by P_i since

$$P_i\mathbf{a}_{ij} = 0 \qquad j=n_i+1,\ldots,N \tag{16}$$

It thus follows that operator P_i uniquely defines a subspace of the pattern space. Any pattern vector $\mathbf{x}$ lying in this subspace will satisfy

$$P_i\mathbf{x} = \mathbf{x} \tag{17}$$

and is, therefore, an eigenvector of P_i with unity eigenvalue. Its projection by P_i is identical with the pattern vector. On the other hand the projection of a pattern orthogonal to the subspace defined by P_i will be zero.

Now suppose that the generating matrix A_i in (4) is composed of all N eigenvectors of the ith class correlation matrix. Then A_i is $N{\times}N$ nonsingular matrix and the corresponding projection operator P_i

$$P_i = A_iA_i^T = A_iA_i^{-1} = I \tag{18}$$

becomes the identity matrix. Thus projection operator associated with the complete pattern space is the identity matrix. It is interesting to note that when operator P_i defines only a subspace of the pattern representation space, the projection operator $\bar{P}_i$ defined by eigenvectors of P_i with eigenvalues equal to zero can be considered as the complement projection operator to P_i, i.e.

$$\bar{P}_i = I - P_i \tag{19}$$

From (19) any vector $\mathbf{x}$ in general must satisfy

$$I\mathbf{x} = P_i\mathbf{x} + \bar{P}_i\mathbf{x} \tag{20}$$

and thus it can be expressed, as expected intuitively, in terms of its respective projections into subspaces of the pattern space defined by the projection operators P_i and $\bar{P}_i$. But it has been emphasized that the pattern vectors from class ω_i will lie mainly in the subspace of the pattern space determined by projection operator P_i. This means that the projection of pattern vectors $\mathbf{x} \in \omega_i$ into this subspace will satisfy

$$P_i\mathbf{x} = \mathbf{x} \tag{21}$$

while the projection of these vectors into the complement space $\bar{P}_i$ is negligible, i.e.

$$\bar{P}_i\mathbf{x} \doteq 0 \tag{22}$$

3.1 *Clafic*

The simplest method of constructing class subspaces is known under the acronym Clafic [4]. The method utilizes directly the constituent axes A_i of the class subspace, as determined by the Karhunen-Loeve expansion, for determination of the ith class projection operator. Thus P_i is given as

$$P_i = A_iA_i^T = \sum_{j=1}^{n_i} \mathbf{a}_{ij}\mathbf{a}_{ij}^T \tag{23}$$

Once the projection operators are determined for all the classes a pattern vector with unknown membership is assigned to that class whose associated projection operator yields the maximal projection of the vector into the corresponding class subspace.

3.2 *Orthogonal subspace method*

The main disadvantage of Clafic is that the subspaces spanned by the eigenvectors of projection operators P_i of individual classes may overlap. Consequently, even a cautious use of Clafic procedure may result in high misrecognition rate. It follows that it is desirable to modify the projection operators so that the new class subspaces defined by the amended projection operators $\tilde{P}_i$, $i=1,\ldots,m$ are orthogonal [5], that is any eigenvector of $\tilde{P}_i$ (any vector lying in the subspace defined by $\tilde{P}_i$) is orthogonal to all eigenvectors of $\tilde{P}_j$, $j \neq i$. Thus $\tilde{P}_i, \tilde{P}_j$ must satisfy

$$\tilde{P}_i\tilde{P}_j = 0$$

It is apparent that $\tilde{P}_i$ will be orthogonal to $\tilde{P}_j$ if $\tilde{P}_i$ satisfies a stronger condition, namely that

$$\tilde{P}_i P_j = 0 \tag{24}$$

or, alternatively, if $\tilde{P}_i$ is included in the complement subspace $\bar{P}_j = I - P_j$, i.e.

$$\tilde{P}_i \to \bar{P}_j \tag{25}$$

In order to satisfy the orthogonality condition (24) for all j, $\tilde{P}_i$ must be included in the intersection of all the complement subspaces $\bar{P}_j$, $\forall j \neq i$, i.e.

$$\tilde{P}_i \to \bigcap_{j \neq i} \bar{P}_j \tag{26}$$

But since $\tilde{P}_i$ is also included in the original ith class subspace P_i it follows that $\tilde{P}_i$ is the conjunction of the operators P_i and $\bigcap_{j \neq i} \bar{P}_j$. Thus we finally have

$$\tilde{P}_i = P_i \cap \bigcap_{j \neq i} \bar{P}_j \tag{27}$$

and this expression can be solved using one of the methods discussed in the Appendix.

3.3 *Nonorthogonal retrenched subspace method*

While the subspaces yielded by the procedure Clafic provide no guarantee that adequate discrimination of classes will be achieved by simply measuring the amount of projections of pattern vectors into these subspaces, the orthogonal subspace method is too restrictive. A suitable compromise would be to amend these subspaces by removing only the overlapping subspaces of the projection operators [5]. More specifically, let the conjunction of projection operators P_i, P_j be nonzero, i.e.

$$P_i \cap P_j = 0 \tag{28}$$

Then the contribution to the total amount of projection of the pattern vectors into subspaces P_i and P_j due to the intersection subspace P_{ij} will be identical to both subspaces. The greater the overlapping subspace P_{ij} the more difficult it will be to discriminate between classes purely on the basis of the magnitudes of projection. However, if this overlapping subspace is removed from both projection operators P_i and P_j then the discriminatory potential of the resulting subspaces $\tilde{P}_i = P_i - P_{ij}$ and $\tilde{P}_j = P_j - P_{ij}$ will greatly increase.

It should be noted here that these reduced subspaces do not necessarily have to be orthogonal. This relaxation of the orthogonality condition allows us to find a solution (a set of class subspaces) with discriminatory potential in cases where the orthogonal subspace method fails.

In order to improve the discriminatory ability of the class subspaces it is necessary to remove all the overlaps P_{ij}, $\forall j \neq i$ of P_i with the subspaces of the other classes. Then the amended subspace $\tilde{P}_i$ will be given as

$$\tilde{P}_i = P_i - \bigcup_{j \neq i} P_i \cap P_j \tag{29}$$

Since each term of this union is included in P_i we can write

$$P_i \cap P_j = P_i (P_i \cap P_j) \tag{30}$$

Substituting for $P_i \cap P_j$ in (29) from (30) expression (29) can now be simplified to

$$\tilde{P}_i = P_i (I - \bigcup_{j \neq i} P_i \cap P_j) \tag{31}$$

Since the union in (31) can be expressed as

$$\bigcup_{j \neq i} (P_i \cap P_j) = I - \bigcap_{j \neq i} \overline{P_i \cap P_j} \tag{32}$$

we can finally write (29) as

$$\tilde{P}_i = P_i (\bigcap_{j \neq i} \overline{P_i \cap P_j}) \tag{33}$$

From our result it follows that the determination of the retrenched subspaces is much more difficult than acquisition of orthogonal subspaces since it involves calculation of the complements of all the pairwise conjunctions of projection operators. On the other hand this method allows for obtaining a solution when orthogonal class subspaces do not exist.

3.4 *Generalized Fukunaga-Koontz method*

Let us consider the mixture correlation matrix C,

$$C = \sum_{i=1}^{m} P(\omega_i) E\{\mathbf{x}_i \mathbf{x}_i^T\} \qquad \mathbf{x}_i \in \omega_i \tag{34}$$

Since this matrix is symmetric there exists a matrix B such that

$$B^T C B = I \tag{35}$$

In other words matrix B is a linear normalizing operator which transforms the correlation matrix into the identity matrix. This transformation implies that instead of pattern vector $\mathbf{x}$ we now work with the normalized vector $\mathbf{g}$

$$\mathbf{g} = B^T \mathbf{x} \tag{36}$$

whose components are uncorrelated and their variance is unity. The discriminatory ability in the pattern vector **g** is retained since the transformation involves a simple rotation and scaling of the coordinates which are identical to all the classes.

Suppose thate we now expand pattern vectors, $\mathbf{g}_j$, from one class alone, say ω_j, using the Karhunen-Loeve expansion to find the subspace occupied by class ω_j. Then the system of eigenvectors A_j

$$A_j = [\mathbf{a}_{j1},\ldots,\mathbf{a}_{jN}] \tag{37}$$

of the correlation matrix C_j

$$C_j = P(\omega_j)E\{\mathbf{g}_j\mathbf{g}_j^T\} \qquad \mathbf{g}_j \in \omega_j \tag{38}$$

which forms the system of candidate axes of the jth class subspace satisfies

$$C_jA_j = A_j\Lambda_j \tag{39}$$

where Λ_j is the diagonal matrix of eigenvalues of C_j. Utilizing (35) C_j can be written as

$$C_j = I - B^T \sum_{\substack{i=1\\ i\neq j}}^{m} P(\omega_i)E\{\mathbf{x}_i\mathbf{x}_i^T\}B \tag{40}$$

But now from (39) and (40) it follows that

$$B^T[\sum_{\substack{i=1\\ i\neq j}}^{m} P(\omega_i)E\{\mathbf{x}_i\mathbf{x}_i^T\}]BA_j = A_j(I - \Lambda_j) \tag{41}$$

We can conclude that the coordinate system A_j for class ω_j is also the system of eigenvectors of the sum of the correlation matrices of all the other classes $\tilde{C}_j$,

$$\tilde{C}_j = \sum_{\substack{i=1\\ i\neq j}}^{m} P(\omega_i)E\{\mathbf{x}_i\mathbf{x}_i^T\} \tag{42}$$

with the matrix of eigenvalues $\tilde{\Lambda}_j$

$$\tilde{\Lambda}_j = I - \Lambda_j \tag{43}$$

Thus the elements of matrices $\tilde{\Lambda}_j$ and Λ_j are related as

$$\tilde{\lambda}_{jk} = 1 - \lambda_{jk} \tag{44}$$

Owing to the symmetry and positive definiteness of matrices C_j and $\tilde{C}_j$ the eigenvalues of C_j and $\tilde{C}_j$ are positive and therefore less or equal to unity. Thus large eigenvalue λ_{jk} of class ω_j implies a small eigenvalue $\tilde{\lambda}_{jk}$ of all the other classes and vice versa. This result is, of course, very interesting. It means that the average projection of the pattern vectors from class ω_j into the coordinate axes $\mathbf{a}_{jk}$ is substantially greater than the corresponding projection of vectors from any other class. It is apparent that this property can be used advantageously for construction of class subspaces.

Suppose that matrix Λ_j has n_j eigenvalues equal to unity and that the corresponding eigenvectors $\mathbf{a}_{jk}$, $k = 1,\ldots,n_j$ are used to generate the jth class projection operator P_j, i.e.

$$P_j = \sum_{k=1}^{n_j} \mathbf{a}_{jk}\mathbf{a}_{jk}^T \tag{45}$$

By virtue of this process of class subspace construction all operators P_i, $i \neq j$ will be included in the operator $I - P_j$, i.e.

$$P_i \rightarrow I - P_j \tag{46}$$

But since P_j satisfies

$$P_j(I - P_j) = P_j - P_j = 0 \tag{47}$$

all the projection operators will be orthogonal to the jth class projection operators, i.e.

$$P_jP_i = 0 \qquad \forall i, i \neq j \tag{48}$$

It follows that the generalized Fukunaga-Koontz method can be used to find the orthogonal class subspaces. Simply by constructing the projection operators from the eigenvectors of the normalized class conditional correlation matrices whose corresponding eigenvalues are unity, the projection operators will be automatically orthogonal.

In section 3.3 it was pointed out that the condition of orthogonality imposed on class subspaces is too restrictive. The main advantage of the method described in this section is that it can be easily modified to yield class subspaces which are not necessarily orthogonal. If the projection operators are constructed not only from eigenvectors of class conditional correlation matrices whose associated eigenvalues are unity but also from those whose corresponding eigenvalues are less than unity we will have a better chance of finding a suitable solution. The resulting projection operators, of course, will not be orthogonal but the amount of overlap will be known since it is a function of the magnitude of the eigenvalues and thus it can be easily controlled.

4. Conclusions

This chapter discussed the subspace approach to the design and modelling of pattern recognition systems. The existing techniques of Watanabe and Pakvasa for determining class conditional subspaces have been reviewed. Two simple methods have been proposed for calculating the intersection of projection operators. The methods avoid the computationally demanding solution suggested by Watanabe and Pakvasa in terms of the infinite products of the projection operators. Finally a new procedure for determining class subspaces has been developed. The method is based on the generalized Fukunaga-Koontz method of feature selection. It allows the designer a greater control over the class subspaces and their overlap and in this way it enables him to find the best suboptimal set of projection operators.

References

1. FUKUNAGA, K., *Introduction to Statistical Pattern Recognition,* Academic Press, New York (1972).
2. WATANABE, S., 'Karhunen-Loeve expansion and factor analysis', Trans. 4th Prague Conf. on Information Theory (1965).
3. KITTLER, J. and YOUNG, P.C., 'A new approach to feature selection based on the Karhunen-Loeve expansion, Pattern Recognition, 5, pp.335-352 (1973).
4. WATANABE, S., *Knowing and Guessing*, Wiley, New York (1969).
5. WATANABE, S. and PAKVASA, N., 'Subspace method in pattern recognition', Proc. 1st Internat-Conf. on Pattern Recognition, Washington (1973).
6. FUKUNAGA, K. and KOONTZ, W.L.G., 'Application of the Karhunen-Loeve expansion to feature selection and ordering', IEEE Trans. Comput., C-19, pp.826-829 (1970).

Appendix

The solution of expressions (27) and (33) requires computation of the intersection of projection operators of the general form

$$\tilde{P} = \prod_{j=1}^{m} P_j \qquad \text{(A.1)}$$

Watanabe and Pakvasa suggested a solution in terms of the infinite product of the operators, i.e.

$$\tilde{P} = P_1 P_2,...,P_m P_1 P_2....P_m... \qquad \text{(A.2)}$$

It is apparent that in practice only a finite number of products in expression (A.2) can be calculated. Consequently the resulting projection operator will only approach the optimal solution. In the following two alternative methods of determining operator $\tilde{P}$ will be discussed.

Let us consider the constituent eigenvectors, $\mathbf{e}_k$, of operator $\tilde{P}$. Since $\tilde{P}$ is included in the projection operators P_j, $\forall j$, then $\mathbf{e}_k$ is an eigenvector of each of these operators. Thus

$$P_j \mathbf{e}_k = \mathbf{e}_k$$

Now premultiplying equation (A.3) by P_i we get

$$P_i P_j \mathbf{e}_k = \mathbf{e}_k$$

But the right hand side of (A.4) is again $\mathbf{e}_k$. By analogy, premultiplying (A.4) subsequently by P_i, $\forall i \neq j$ we get

$$P_1 P_2 ... P_m \mathbf{e}_k = \mathbf{e}_k \qquad \text{(A.5)}$$

We can conclude that the projection operator $\tilde{P}$ is defined by the system of those eigenvectors of matrix $\hat{P}$,

$$\hat{P} = P_1 P_2 .... P_m \qquad \text{(A.6)}$$

that are associated with unity eigenvalues.

An alternative method of determining projection operator P is to sum up respectively the left hand and the right hand sides of Eq.(A.3) for all $j = 1,...,m$. Then we get

$$\sum_{j=1}^{m} P_j \mathbf{e}_k = m \mathbf{e}_k \qquad \text{(A.7)}$$

Thus the eigenvectors $\mathbf{e}_k$ of $\tilde{P}$ can be obtained as eigenvectors of matrix $\hat{P}$, i.e.

$$\hat{P} = \sum_{j=1}^{m} P_j \qquad \text{(A.8)}$$

with eigenvalue m. This latter approach is computationally less demanding because the acquisition of the generating matrix $\hat{P}$ defined in (A.8) does not require matrix multiplications.

Mathematical modelling of mobile systems

GEORGE E. LASKER
University of Windsor,
Ontario, Canada

Mobile systems can be broadly viewed as natural or artificial dynamic systems that have the capability to move within a specific geographical environment from one physical position to another. Formally, these systems can be represented through various kinds of mathematical models [1,2,3,4,5,6]. In this article, a special mathematical structure is introduced through which it is possible to formally define different types of mobile systems at various levels of complexity. A mathematical construct is also proposed that allows us to formally represent the geographical environment(s) or setting(s) in whcih the mobile systems operate. The physical mobility and the kinetic behavior of dynamic systems can thus be studied within the context of environmental contingencies that can be formally expressed in well defined mathematical terms.

1. FORMAL REPRESENTATION OF MOBILE SYSTEMS

In a most general form, a mobile system **M** can be represented through a finite configuration of specified systems entities whose interrelationship(s) describes the functional, structural and behavioral properties of the system. Mathematically, a mobile system **M** can be defined as follows:

$$\mathbf{M} = \langle \Sigma, A, M, \Phi \rangle \qquad (1)$$

where

Σ = a configuration of interrelated entities $\Sigma_0, \Sigma_1, \Sigma_2, \ldots, \Sigma_n$ of which the system consists. Each entity may be visualized as an abstract element of the system, or as a principal quality or quantity of **M**. Each entity can be further specified through a set (or configuration) of parameters, descriptors, entity determinants or entity attributes. An entity may be an input/output alphabet of the system, a set of internal states of the system, a memory of the system, a program, a control unit, move generator, and so forth.

A = a finite set of attributes $a_0, a_1, a_2, \ldots, a_k$ that can characterize the individual entities of **M**. It is assumed that each entity $\Sigma_i \in \Sigma$ may be associated with a specific configuration $A_i \subseteq A$ of entity attributes. The association of Σ_i with A_i may be specified through a special function or a program ϕ_i that is mapping $\Sigma_i \rightarrow A_i$, where $\phi_i \in \Phi$.

M = a non-empty set of all possible moves $m_0, m_1, m_2, \ldots, m_z$ that the system **M** can make within a specified environment under given conditions. These moves may be viewed as spatial transitions of **M** in its geographical environment.

Φ = a set or family of functions (or programmes) $\phi_0, \phi_1, \phi_2, \ldots, \phi_r$ interrelating individual entities of the system **M**. These functions/programmes determine the structure, the organization and the behavior of the system **M** in its respective environment. In general it is assumed that

$$(\exists \phi_i \in \Phi_i | \Phi_i \subset \Phi)(\exists A_i \subseteq A)(\exists \Sigma_i \in \Sigma) \quad (2)$$

$$\phi_i : \Sigma_i \rightarrow A_i \qquad \text{for all } i \in I | I = \{0,1,2,...\}$$

where ϕ_i is a particular function that associates the individual systems entity $\Sigma_i \in \Sigma$ with a set A_i of specific attributes.

Further it is assumed that

$$(\exists \phi_j \in \Phi_j | \Phi_j \subseteq \Phi)(\exists A_k \subseteq A)(\exists \Sigma_j \in \Sigma)(\forall \Sigma_i \in \Sigma' | \Sigma' \subseteq \Sigma) \quad (3)$$

$$\phi_j : A_k \times \left\{ \mathop{\mathsf{X}}_{i=1}^{n} \Sigma_i \right\} \rightarrow \Sigma_j \qquad \text{for all } i,j,k \in I$$

where $\phi_j \in \Phi_j$ is a specific function that determines a particular entity $\Sigma_j \in \Sigma$ through a configuration of A_k and

$$\mathop{\mathsf{X}}_{i=1}^{n} \Sigma_i \quad ,$$

where $\Sigma_i \in \Sigma' | \Sigma' \subseteq \Sigma$ and where $\mathop{\mathsf{X}}_{i=1}^{n} \Sigma_i$ is a cartesian product of all entity sets under consideration, i.e. $\left\{ \mathop{\mathsf{X}}_{i=1}^{n} \Sigma_i \right\} = \{\Sigma_1 \mathsf{x} \Sigma_2 \mathsf{x} ... \mathsf{x} \Sigma_n\}$.

By analogy, a special function $\phi_m \in \Phi$ can be defined as follows:

$$(\exists \phi_m \in \Phi)(\exists A_k \subseteq A)(\forall \Sigma_i \in \Sigma' | \Sigma' \subseteq \Sigma) \quad (4)$$

$$\phi_m : A_k \times \left\{ \mathop{\mathsf{X}}_{i=1}^{n} \Sigma_i \right\} \rightarrow M$$

The function ϕ_m determines here the moves that the system **M** will make under given conditions. Subsequently, this function will be called a *move* function or a spatial transition function and it will be denoted by the symbol μ. Thus $\phi_m \equiv \mu$.

The above defined algebraic structure forms a general base from which a large variety of kinematic systems models can be derived. The individual models can be derived from this structure by selecting and separately identifying those entities that are assumed to adequately characterize the system(s) that we wish to represent. In principle though, each kinematic model should incorporate only those entities that are actually relevant to the study objective(s). Once the systems entities have been identified and formally defined, we have to determine the compatible sets A, M and Φ. When this is accomplished, we have to explicitly define the interrelationship amongst the selected entities and to determine the conditions under which the mobile system is assumed to operate in its respective environment.

Let us consider a situation in which we want to represent mobile system as an input-output finite-state sequential system that can move in discrete steps in specified directions to successively enter and occupy a finite set P of physical positions $p_1,p_2,...,p_n$ within an arbitrarily defined environment E. Formally, such a system can be defined by the following configuration of systems entities.

$$\mathbf{M} = (X,Q,Y,A,M,F) \quad (5)$$

where

X = a finite non-empty set of inputs $x_1,x_2,...,x_m$
Q = a finite non-empty set of states $q_1,q_2,...,q_n$
Y = a finite non-empty set of outputs $y_1,y_2,...,y_r$
A = a finite non-empty set of attributes $a_1,a_2,...,a_s$
M = a finite non-empty set of moves $m_1,m_2,...,m_z$
F = a finite non-empty set of functions $\alpha,\beta,\delta,\lambda$ and μ

where

α = an input function that maps $X \times Q \times A \rightarrow X$
β = an assignment function that maps $Q \rightarrow A$
δ = a state transition function that maps $X \times Q \times A \rightarrow Q$
λ = an output function that maps $X \times Q \times A \rightarrow Y$
μ = a move function that maps $X \times Q \times A \rightarrow M$

In defining the above functions, a variety of alternative mappings may be also used. Thus for instance we may wish to define the input function α by the mapping $Q \times A \rightarrow X$ or even $Q \rightarrow X$. Similarly, we may wish to define the output function by the mapping $Q \times A \rightarrow Y$ or $Q \rightarrow Y$. Also the function β, δ and μ can be modified.

The individual sets of parameters through which the system is described in Eq.(5) are conceptually viewed as *basic* systems entities. By definition, the system **M** is assumed to operate in discrete steps within a time continuum **T**. At any time $t \in \mathbf{T}$, the mobile system **M** is in its environment E entirely characterized by a specific state $q(t) \in Q$ and by a specific geographical position $p(t) \in P$. Thus

$$\forall t \in \mathbf{T} [\exists q(t) \in Q \;\&\; \exists p(t) \in P \;\Rightarrow\; \mathbf{M} | q(t),p(t)] \quad (6)$$

where $\mathbf{M} | q(t),p(t)$ indicates that at time t the mobile system **M** assumes the state $q(t)$ and occupies the position $p(t)$.

Provided that an initial state $q(0)$ of **M** has been specified, the internal state $q(t)$ can be defined recursively through the *state transition* δ-function as follows:

$$\begin{cases} q(0) = q_i & \text{for } q(0),q(t),q(t+1) \in Q \\ & x(t) \in X \\ q(t+1) = \delta[x(t),a(t),q(t)] & a(t) \in A \end{cases} \quad (7)$$

where

$q(0) = q_i$ is an initial state of **M** such that $q_i \in Q$
$x(t)$ = a present configuration of input quantities
$a(t)$ = a present configuration of attributes
$q(t)$ = a present state of **M**
$q(t+1)$ = a next state of **M**

Equation (7) essentially indicates that the next state $q(t+1)$ of **M** is determined through δ-function by the present input pattern $x(t)$, by the present configuration $a(t)$ of specific attributes and by the present internal state $q(t)$ of **M**.

A situation may be considered in which the state $q(t+1)$ can be also defined through δ-function in terms of $x(t)$, $q(t)$, $a(t)$ and $p(t)$, where $p(t) \in P$ represents a position that **M** currently occupies. In that case Eq.(7) must be modified accordingly. Sometimes, however, it is more convenient to treat $p(t)$ as an integral part of the input pattern $x(t)$.

It is assumed that at any time t each state $q(t)$ of **M** can be associated with a configuration $a(t)$ of specific attributes. The association is formally defined through the assignment function β as follows:

$$a(t) = \beta[q(t)] \qquad \text{for } a(t) \in A,\; q(t) \in Q \quad (8)$$

Sometimes the function β may be rewritten as a production function. In that case

$$\beta : q(t) \Rightarrow a(t) \quad (9)$$

which means that the state $q(t)$ of **M** generates a specific configuration $a(t)$ of attributes. These individual attributes may be defined in various ways. For instance $a(t)$ may represent a configuration of thresholds and input weights, or it may represent an entropy of $\mathbf{M}|q(t),p(t)$, etc. (see Lasker [7]).

The *input* function α determines for each input pattern $x(t)$ whether or not the system $\mathbf{M}|q(t),p(t)$, will accept the pattern $x(t)$ at time t. Formally, this function can be formulated in a variety of ways. Thus, for instance we may define α-function as follows:

$$\alpha[x(t),q(t),a(t)] = \begin{cases} 1 \text{ iff } \delta[x(t),q(t),a(t)] \in Q_1 | Q_1 \subset Q \\ 0 \text{ iff } \delta[x(t),q(t),a(t)] \notin Q_1 | Q_1 \subset Q \end{cases} \quad (10)$$

The input pattern $x(t) \in X$ is said to be accepted by $\mathbf{M}|q(t),p(t)$ at time t if and only if the system would transit under $x(t)$ from $q(t)$ into $q(t+1) \in Q_1 | Q_1 \subset Q$, where Q_1 is a specific subset of Q. Otherwise, the input pattern $x(t)$ is said to be rejected by $\mathbf{M}|q(t),p(t)$

Another possible formulation of α-function may be devised in which the input pattern $x(t)$ is defined as a combination of binary value inputs $x_1(t),x_2(t),..,x_r(t)$ and in which each internal state $q(t) \in Q$ is associated with a specific threshold vector

$$a(t) = [T(t);w_1(t),w_2(t),...,w_r(t)]$$

where $T(t)$ is a threshold of $\mathbf{M}|q(t),p(t)$ and $w_1(t),w_2(t),...,w_r(t)$ are inputs weights such that each input weight $w_i(t)$ is associated with a specific input $x_i(t)$ for $i \in I | I = \{1,2,...,r\}$. In this case the α-function can be defined as follows:

$$\alpha[x(t),q(t),a(t)] = \begin{cases} 1 \text{ iff } \sum_{i=1}^{r} w_i(t)x_i(t) \geqslant T(t) \\ 0 \text{ iff } \sum_{i=1}^{r} w_i(t)x_i(t) < T(t) \end{cases} \quad (11)$$

The input pattern $x(t) \in X$ is said to be accepted by $\mathbf{M}|q(t),p(t)$ if and only if the value of input function α is 1, otherwise $x(t)$ is said to be rejected by $\mathbf{M}|q(t),p(t)$. This means that $x(t)$ is accepted if and only if the sum $\sum_{i=1}^{r} w_i(t),x_i(t)$ of weighted inputs $x_1(t),x_2(t),...,x_r(t)$ is larger than or equal to a specific threshold $T(t)$. If this sum is smaller than $T(t)$ the input pattern $x(t)$ is not accepted.

Sometimes it may be convenient to define input function α as a function that selectively determines *what part(s)* of input pattern $x(t)$ would be 'perceived and processed by $\mathbf{M}|q(t),p(t)$. In this case the α-function would 'filter out' specific parts of the total pattern and would accept only those parts that can meet specified criteria.

The *output* function λ determines what kind of output will be produced by $\mathbf{M}|q(t),p(t)$ at time t.

In general, this function can also be defined in various ways. The two most common forms of representation can be expressed as follows:

$$y(t) = \lambda[x(t),q(t),a(t)] \quad \text{where} \quad \begin{matrix} y(t) \in Y, \ q(t) \in Q \\ x(t) \in X, \ a(t) \in A \end{matrix} \tag{12}$$

$$y(t) = \lambda[q(t)] \quad \text{where} \quad \begin{matrix} y(t) \in Y \\ q(t) \in Q \end{matrix} \tag{13}$$

The Eq.(12) determines the output $y(t)$ as a function of the present input $x(t)$, present state $q(t)$ and a present configuration $a(t)$ of specific attributes. The Eq.(13) defines the output $y(t)$ in terms of the present state $q(t)$ of **M** only. Essentially, the Eq.(12) employs the modelling concept of Meally [8] and Lasker [7], while Eq.(13) uses the modelling concept of Moore [9].

It is assumed that the mobile system **M** can make in its environment a series of distinct moves (or motions) $m_1, m_2, \ldots, m_z$. Each of these moves is *a priori* specified by the direction in which the system should travel and by the distance that the system should traverse within a specific period of time. Each move can be also specified by the spatial transition of **M** from one geographical position to another, or by the route that the system should take in order to reach a certain position in its environment, to carry out a certain task or to achieve a specific goal. In general, there are many alternative ways through which the moves of **M** can be defined.

Formally, a move $m(t) \in M$ to be made by the system **M** at time t can be defined by the *move* function μ as follows:

$$m(t) = \mu[x(t),q(t),a(t)] \quad \text{where:} \quad \begin{matrix} m(t) \in M, \ q(t) \in Q \\ x(t) \in X, \ a(t) \in A \end{matrix} \tag{14}$$

Equation (14) indicates that the move $m(t)$ of **M** is determined by the configuration of $x(t)$, $q(t)$ and $a(t)$. Sometimes, an alternative form of representation can be used in which the move $m(t)$ is defined in terms of the state $q(t)$ as follows:

$$m(t) = \mu[q(t)] \tag{15}$$

which means, that if **M** is in a state $q(t)$ at time t it will realize a specific *move* $m(t)$. This move will then be realized (repeatedly) as long as the system assumes this state.

Each move $m(t) \in M$ that **M** can realize may be also defined by a spatial transition of the system from one geographical position $p_i \in P$ to another position $p_j \in P$. Provided that $p(t)$ represent a specific geographical position $p_i \in P$ that **M** currently occupies and that $p(t+1)$ represent the next position $p_j \in P$ to be entered by **M** in the next period of time, the move $m(t)$ can be represented by the transition $\tau_{ij} = \overline{P_i P_j} = \tau(i,j)$ or equivalently by $\tau(t,t+1) = \overline{p(t),p(t+1)}$. In this respect an option is open to incorporate the set P of geographical (or spatial) positions $p_1, p_2, \ldots, p_n$ into a structure (1) as one of the basic systems entities† and to replace the above-defined move function μ by a *spatial transition* (or space-transition) function π, that maps $X \times Q \times A \times P \rightarrow P$. In that case, the spatial transition or positional function π can be defined recursively as follows:

$$\begin{cases} p(0) = p_k \\ p(t+1) = \pi[p(t),x(t),q(t),a(t)] \end{cases} \quad \text{where:} \quad \begin{matrix} p(0),p(t),p(t+1) \in P \\ x(t) \in X \\ q(t) \in Q \\ a(t) \in A \end{matrix} \tag{16}$$

Provided that the initial geographical position $p(0)$ is specified by p_k, the next position $p(t+1)$ of **M** is determined through by the configuration of the present input pattern $x(t)$, present state $q(t)$, present combination $a(t)$ of attributes and presently held position $p(t)$. The above-defined positional function π can be very useful in situations in which we do not wish to explicitly define the moves of **M** *a priori* or in which we want to make moves contingent on specific environmental positions.

Mobile systems that are defined through the model structure (1), (5) or through any of their derivations can be analytically represented by transition tables and transition diagrams.

The *transition table* representation portrays the properties of selected mappings defined on set Φ or F in tabular form. The columns and rows of the table can be designed to correspond to the specific system entities used in the mapping(s) so that the table entries that are found at the intersections of these columns and rows would define the resultant images of the mapping(s). For instance, if we wish to represent a specific mobile system defined by a simplified version of structure (5), then the columns of the table could correspond to the possible inputs of the system and the rows could correspond to the possible states of the system (see Fig. 1).

†In this case the set P would replace the set M in the model structure (5).

Q \ X	x_i	x_j
q_i	q_j, y_ℓ, m_r	q_i, y_k, m_r
q_j	q_k, y_ℓ, m_s	q_j, y_k, m_z
q_k	q_i, y_ℓ, m_z	q_k, y_k, m_s

Fig. 1 Transition table representation of system **M**

In this case the table entry found at the intersection of each column and row would indicate the next state of the system, its output and its move. Thus, for example, the table entry at the intersection of the *i*th column and *j*th row would define:

1. the next state $q(t+1) = \delta[x(t),q(t)] = \delta[x_i,q_j] = q_k$
2. the output $y(t) = \lambda[x(t),q(t)] = \lambda[x_i,q_j] = y_l$
3. the move $m(t) = \mu[x(t),q(t)] = \mu[x_i,q_j] = m_s$.

If, however, a positional or spatial transition function π has been defined for **M** instead of a move function μ, the row descriptors would correspond to the possible geographical positions that the system **M** can enter and the column would correspond to the possible combinations of input patterns and states that can occur under given circumstances. In that case the entry found at the intersection of the *l*th column and *h*th row would indicate the next geographical position $p(t+1) = p_{lh}$ to be entered by **M** at time $t+1$. It is assumed that $p_{lh} \in P$, where $p_{lh} = \pi(c_l,p_h)$, provided that c_l is a combination of x_i and q_j and that p_h is a presently held position $p(t) \in P$. The table representation of such positional or spatial transition function π is illustrated in Fig. 2. This table essentially indicates all the spatial or geographical transitions that the system **M** can make under given conditions.

In certain situations it is possible to translate the table representation form of the π-function into a functionally equivalent table representation form of the μ-function and *vice versa.* The translation techniques that are suitable for this purpose have been introduced by Lasker (7), who also examined the conditions under which the functional equivalence of π and μ can be established.

Transition diagrams (or kinematic graphs) provide a graphical description of the kinetic behavior of mobile systems in their respective geographical environment(s). In general, various versions of transition diagrams can be constructed, depending on the definition of specific functions in the set Φ or F.

Formally, a transition diagram of kinematic graph consists of a set of vertices/nodes whose labels correspond to the states of the system. The individual vertices/nodes are connected by directed edges/arrows indicating the temporal and functional relationship between the considered states. The edges also indicate the moves (or positional geographical transitions) that the system can realize under given conditions. If there exists an input $x_i \in X$ and states $q_j, q_k \in Q$ such that $\delta(x_i,q_j) = q_k$, then a directed edge will connect the vertex q_j to the vertex q_k for each ordered pair of states q_j and q_k from Q. If there also exists an output $y_l \in Y$ and *move* $m_s \in M$ such that $\lambda(x_i,q_j) = y_l$ and $\mu(x_i,q_j) = m_s$, then the directed edge will be associated with a label x_i, y_l, m_s indicating (i) the present input $x_i \in X$, (ii) the present output $y_l \in Y$, and (iii) the present move $m_s \in M$. To illustrate this idea, Fig. 3 presents a simple example of a transition diagram (or kinematic graph) of a mobile system that has been originally defined in transition table form in Fig. 1.

Sometimes it is convenient to represent the moves of the system either as separate diagram entities or in the association with the specified outputs. For this purpose a box can be inserted into each edge-line to accommodate the label(s) that indicates the considered move(s) separately (see Fig. 4*a*) or in association with an appropriate system's output (Fig. 4*b*). All

Present Configuration

present positions

P \ C	c_1	. . .	c_ℓ	. . .	c_m
	(x_1,q_1)	. . .	(x_i,q_j)	. . .	(x_m,q_n)
p_1	p_{11}	. . .	$p_{\ell 1}$	. . .	p_{m1}
⋮	⋮		⋮		⋮
p_h	p_{1h}	. . .	$p_{\ell h}$	. . .	p_{mh}
⋮	⋮		⋮		⋮
p_n	p_{1n}	. . .	$p_{\ell n}$	. . .	p_{mn}

Fig.2 General table form representation of positional function π

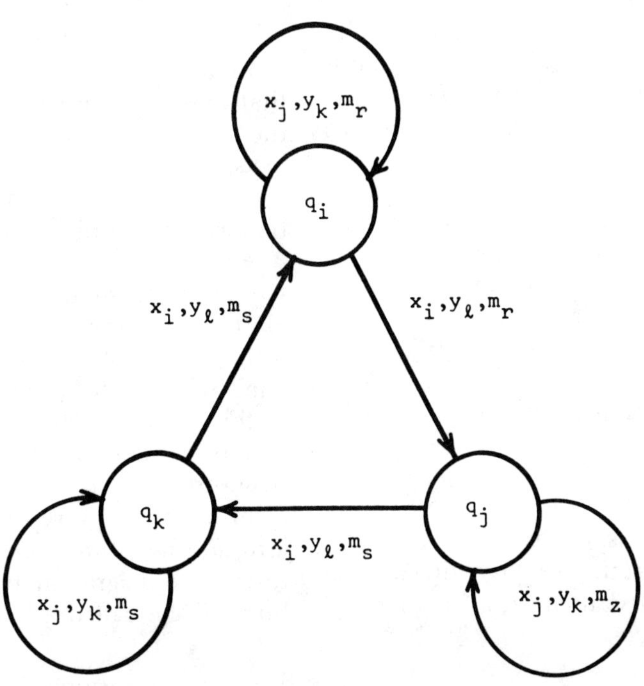

Fig.3 Transition diagram representation of system **M**

three types of functionally equivalent representation forms illustrated by Figs. 3 and 4 can be interpreted as follows: For instance, if the system is initially in the state $q_i \in Q$ and the input $x_i \in X$ is applied, then the system will (1) produce output $y_l \in Y$, (2) realize move $m_r \in M$, and (3) it will transit into a new next-state $q_j \in Q$. If the system is presently in the state $q_j \in Q$ and input x_i is applied then the system will (1) produce output $y_l \in Y$, (2) realize move $m_s \in M$, and (3) take on a next-state $q_k \in Q$; and so forth.

2. MATHEMATICAL REPRESENTATION OF SYSTEMS ENVIRONMENT

The environment $\mathbf{E_M}$, in which a mobile system **M** operates will be called an *operation* space or operation environment of the system. In general, this environment can be defined by a finite configuration of environmental entities that allow us to formally describe the relevant properties of a geographical setting in which the mobile system is assumed to travel. In a most general form, the operation environment $\mathbf{E_M}$ can be defined as follows:

$$\mathbf{E_M} = \langle E, A \rangle \tag{17}$$

where E represents a configuration of various environmental entities $E_1, E_2, ..., E_n$ such that each entity $E_j = \{e_i | i \in I\}$. It is assumed that $E_i \cap E_j = 0$ for each $E_i, E_j \in E$, where $i \neq j$ and $i,j \in I$. Under certain conditions E may be also defined as $E = \langle\{\mathbf{x}\, E_i : i \in I\}\rangle$, where **x** denotes a Cartesian product and I is the index set.

A represents a configuration of entities attributes $A_1, A_2, ..., A_m$ such that $A_j = \{a_i | i \in I\}$ for each $A_j \in A$. It is assumed that $A_i \cap A_j \neq 0$ for some $A_i, A_j \in A$, where $i \neq j; i,j \in I$. It is further assumed that

$$(\exists E_\alpha \in E)(\exists E_i \in E)(\exists A_i \in A)(\exists \alpha_i \in E_\alpha) \tag{18}$$

$$\alpha_i : E_i \rightarrow A_i \qquad \text{for all } i \in I$$

where α_i is a function that associates a particular configuration of environmental entity E_i with a specific configuration A_i of entity attributes. Furthermore it is assumed that

$$(\exists E_j \in E)(\forall E_i \in E' | E' \subseteq E)(\exists E_\alpha \in E)(\exists \alpha_j \in E_\alpha) \tag{19}$$

$$\alpha_j : \{\mathop{\mathbf{X}}_{i \in I} E_i\} \mathbf{X} A \rightarrow E_j \qquad \text{for all } i,j \in I$$

where α_j is a function that interrelates various environmental entities and their attributes. In certain situations we may assume that

$$(\exists E_j, E_k, E_l \in E)(\forall E_i \in E' | E' \subseteq E)(\exists E_\alpha \in E)(\exists \alpha_j^* \in E_\alpha) \tag{20}$$

$$\alpha_j^* : \{\mathop{\mathbf{X}}_{i \in I} E_i\} \mathbf{X} \{E_k \cup E_l\} \mathbf{X} A \rightarrow E_j \qquad \text{for } i,j,k,l \in I$$

where α_j^* is a function that interrelates operational configurations of various environmental entities and their attributes.

The environmental entities used in the model are specified by the properties of the setting(s) that they are supposed to represent. In general, a large variety of entities may be used to describe the operation environment(s) of a particular mobile system. Essentially though, only a small number of well defined entities may suffice to portray all those environmental characteristics that are considered relevant to the study objective. Amongst the most important environmental entities may be counted:

a. a structure σ of the environment upon which the system operates

b. a set P of geographical positions $p_1, p_2, ..., p_n$ that the system can enter and occupy

c. a set R of routes $r_1, r_2, ..., r_z$ through which the system can travel.

In addition, a long list of entities can be compiled that may include for instance: a set of facilities that the system can use during its operation, a set of objects that may obstruct certain moves of the system, a set of special devices that are used to control the traffic flow within the environment, etc.

A structure of the operation environment E, subsequently denoted by σ, can be viewed as a diagrammatic projection of the geographical setting into an orthogonal two- or three-dimensional coordinate system. Schematically this structure can be represented in the form of a directed graph, cellular array, maze structure, geographic map, web structure, network structure, and so forth. Under certain conditions one form of structure representation can be translated into another, functionally equivalent representation form. The diagrammatic description of the σ-structure will be called a map. This map essentially indicates:

i. the location of physical positions within the environment $\mathbf{E_M}$

ii. the location of routes that connect the designated positions in $\mathbf{E_M}$

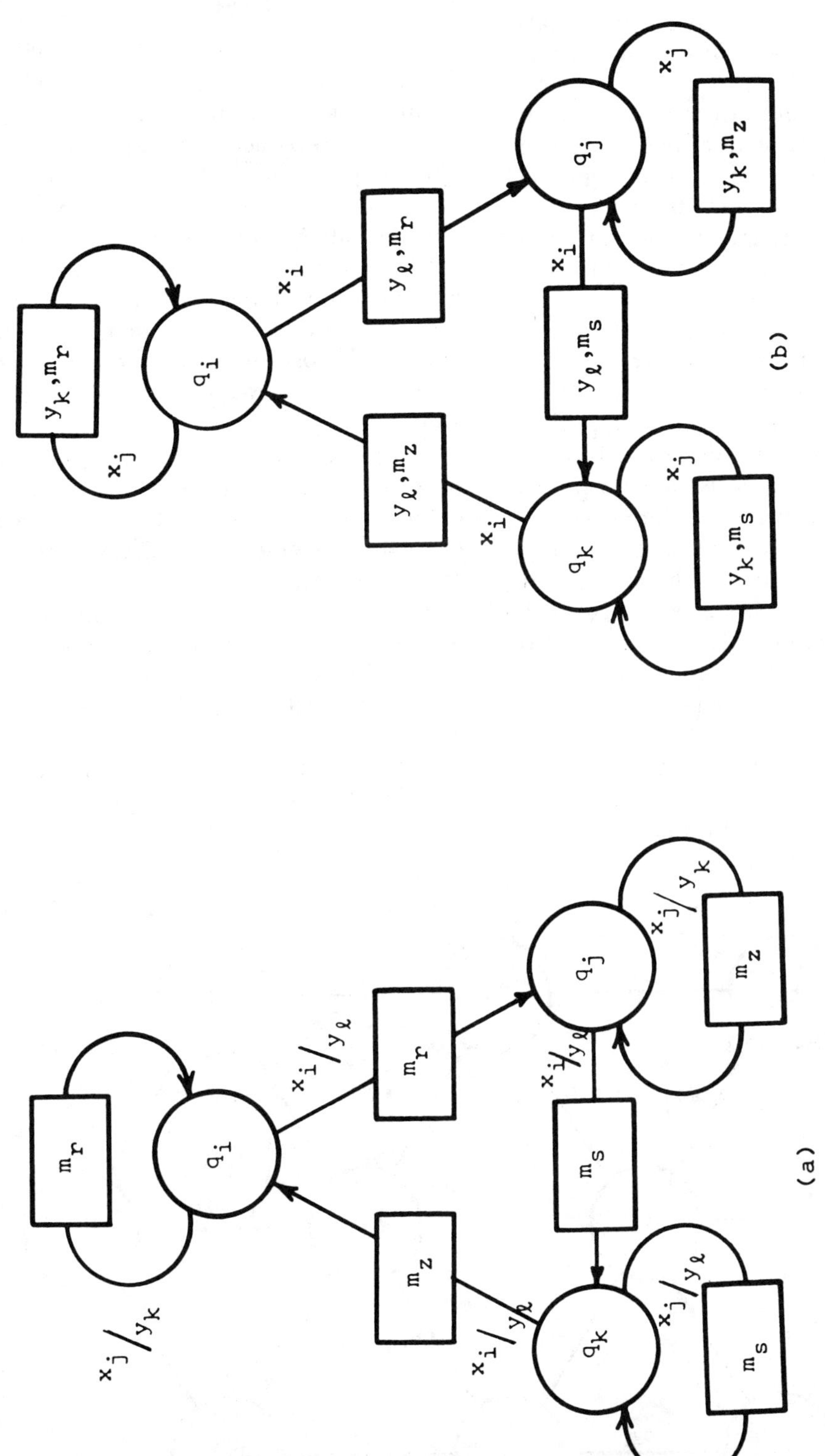

Fig.4 Kinematic graph representation of system **M**

iii. the location of the territorial boundaries of the environment $\mathbf{E_M}$.

In addition, the map may also indicate the location of various objects, devices, facilities and other elements of $\mathbf{E_M}$.

Figure 5 illustrates a structure of an environment $\mathbf{E_M}$ that is schematically represented by a map of a directed graph. The nodes or vertices of this graph represent the physical positions that the system can enter. The arcs or edges of the graph represent the routes through which the system can move in order to reach a specific position.

Each route connects a pair of positions. Diagrammatically, each route is depicted by an arc or arrow that is directed from one position to another. If an arc/arrow is directed from a position p_i to a position p_j, then the position p_j is said to be a successor of the position p_i and the position p_i is called a predecessor of the position p_j. When two positions are successors of each other, the pair of directed arcs or arrows may be replaced by an edge. The arrowhead on each arc or edge points out the direction in which the system is allowed to travel under given conditions. Thus arcs/arrows may be visualized as one-way traffic routes, while edges may be visualized as two-way traffic routes. In general it is assumed that a pair of positions may be connected with more than one alternative route. This gives a system an option to select a route which is deemed to be the most appropriate under given circumstances.

A sequence of connected positions $p_{i_1}, p_{i_2}, \ldots, p_{i_k}$, where each p_{i_j} is a successor of $p_{i_{j-1}}$ for $j \in I$ is said to form a path or trajectory of length 'k' traveled from position p_{i_1} to position p_{i_k}. If a path exists from a position p_a to a position p_c then the position p_c is said to be accessible from the position p_a. A specific position $p_i \in P$ which the mobile system initially occupies is said to be the initial position of the system. A position that the system strives to reach or is guided towards is said to be a target or goal position of the system. This position may be also viewed as a terminal or final position of the system. A position that is not connected by a route with any other position is said to be an isolated position.

Formally, each geographical position $p_i \in P$ may be defined as a point in two- or three-dimensional orthogonal space, as a cellular compartment, as an elementary segment of the environment or as a special ecological system (eco-system). The location of each position in the environment can be specified by the configuration of appropriate spatial coordinates and/or by other relevant indicator(s). Conceptually, each position may be visualized as a generator, carrier or a source of information or energy and it may be assumed that this information or energy can serve as an input to a system that enters the position.

Each geographical position can be formally charac-

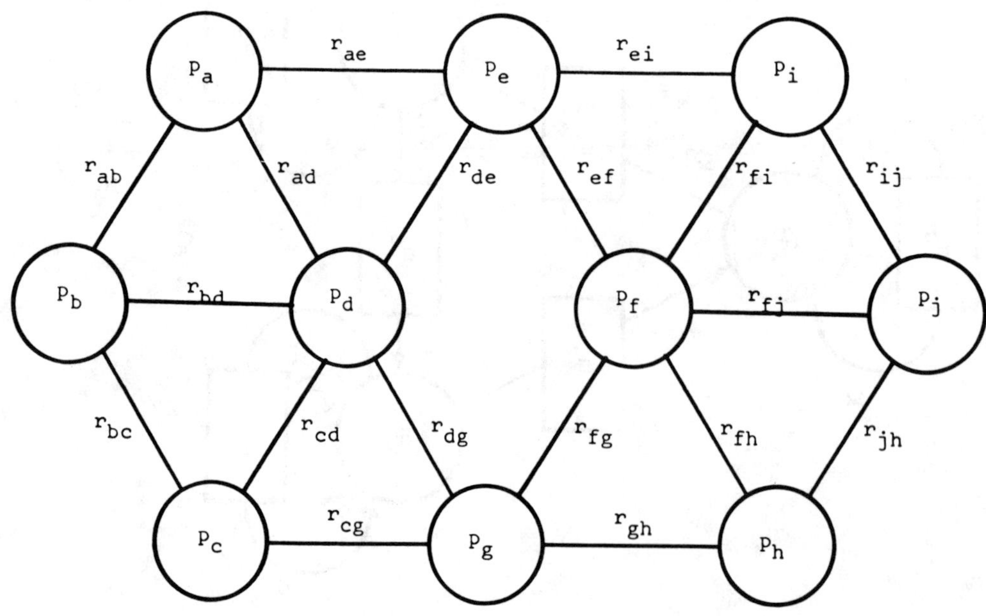

Fig.5 Graph structure representation of operation environment of M

terized by a configuration of attributes or parameters, each of which defines a particular property of the position. These attributes may be alphabetical, numerical, alphanumerical or logical. They may indicate the address of the position (i.e. its name and location), the information content, the storage capacity, the present status, the utility value, the structural ordinacy, etc. As an example let us consider a situation in which each position $p_i \in P$ is defined by its address attribute $\mathscr{A}$ and a content attribute C. Formally

$$p_i \Rightarrow p\langle \mathscr{A}, \mathrm{C} \rangle \quad \text{where } \mathscr{A}, \mathrm{C} \in \mathrm{A} \tag{21}$$

The address $\mathscr{A}$ of the position p_i incorporates: (1) the name of the position (that may be defined by a separate name attribute or 'tag', N, and (2) the location of the position (that may be specified by the location attribute or location identifier, L. The content or load attribute C indicates the quality and/or quantity of information or energy that this position contains and that can be eventually passed on to the mobile system for further processing. In this case Eq.(21) can be rewritten through the following equivalent representation forms.

$$p\langle A_i \rangle \equiv p\langle \mathscr{A}, \mathrm{C} \rangle \equiv p\langle N, L, \mathrm{C} \rangle \equiv p_i \langle \mathrm{C} \rangle \tag{22}$$

In general, the index subscript $i \in I$ may serve as an address identifier of the position $p_i \in P$, so that the bracket $\langle\ \rangle$ may be reserved to indicate the information content or information load of the position p_i.

It may be assumed that some of the considered attributes are constants, while others may change with circumstances. Sometimes, it may be useful to design special register or counter attributes, that could continually indicate for instance: the number of visits that the system made to the position, the length of time the system spent there, the amount of information or energy that the system picked up from or deposited to the position during each visit, etc.

The routes through which the system can travel may also be conceptually defined in a variety of forms. For instance, they may be defined as single lines, curves or connectives between two or more positions, they may be visualized as finite/or infinite sequence(s) of ordered adjoint points, cells or elementary ecological systems, they may be viewed as communication or transportation channels, corridors, or special segments of the environment, etc. Each route can be defined by a configuration of route attributes or route parameters. These attributes specify various properties of the route such as: the name and the location of the route, its length, shape and width, the number of lanes, noise level, capacity, utility value, and so forth. These attributes/parameters provide an information, on the basis of which some mobile systems can make a decision as to which route or trajectory to select under given conditions in order to reach a specified target or homing position within the shortest possible time, at minimum cost and/or with maximum pay-off.

All other environmental entities that we may further wish to consider for modelling of the environment $\mathbf{E_M}$ can be treated in a similar manner as the entities described above.

Example 1

To illustrate the way in which it is possible to study the kinetic behavior of mobile systems within the context of their environment and to clarify the meaning of the theoretical concepts and expressions that have been introduced in the previous sections of this article, let us consider an example of a specific mobile system $\mathbf{M}_1$ that operates within a certain cellular environment E_c. This system is defined as follows:

$$\mathbf{M}_1 = (X, Q, Y, M, F) \tag{23}$$

where

X = input set $\{0,1\}$

Q = state set $\{q_1, q_2, q_3\}$

Y = output set $\{0,1\}$

M = move set $\{m_1, m_2, \ldots, m_6\}$

F = function set $\{\delta, \lambda, \mu\}$

$$\delta: \begin{cases} q(0) = q_1 \\ q(t+1) = \delta[x(t), q(t)] \end{cases}$$

$$\lambda: \ y(t) = \lambda[x(t), q(t)]$$

$$\mu: \ m(t) = \mu[x(t), q(t)]$$

The functions δ, λ and μ of $\mathbf{M}_1$ are defined by the transition table in Fig. 6 and the kinetic behavior of

Q \ X	0	1
q_1	q_1, 1, m_1 = NE	q_2, 0, m_4 = SE
q_2	q_2, 0, m_2 = S	q_3, 0, m_5 = SW
q_3	q_3, 1, m_3 = NW	q_1, 1, m_6 = N

Fig.6 Table representation of system $\mathbf{M}_1$

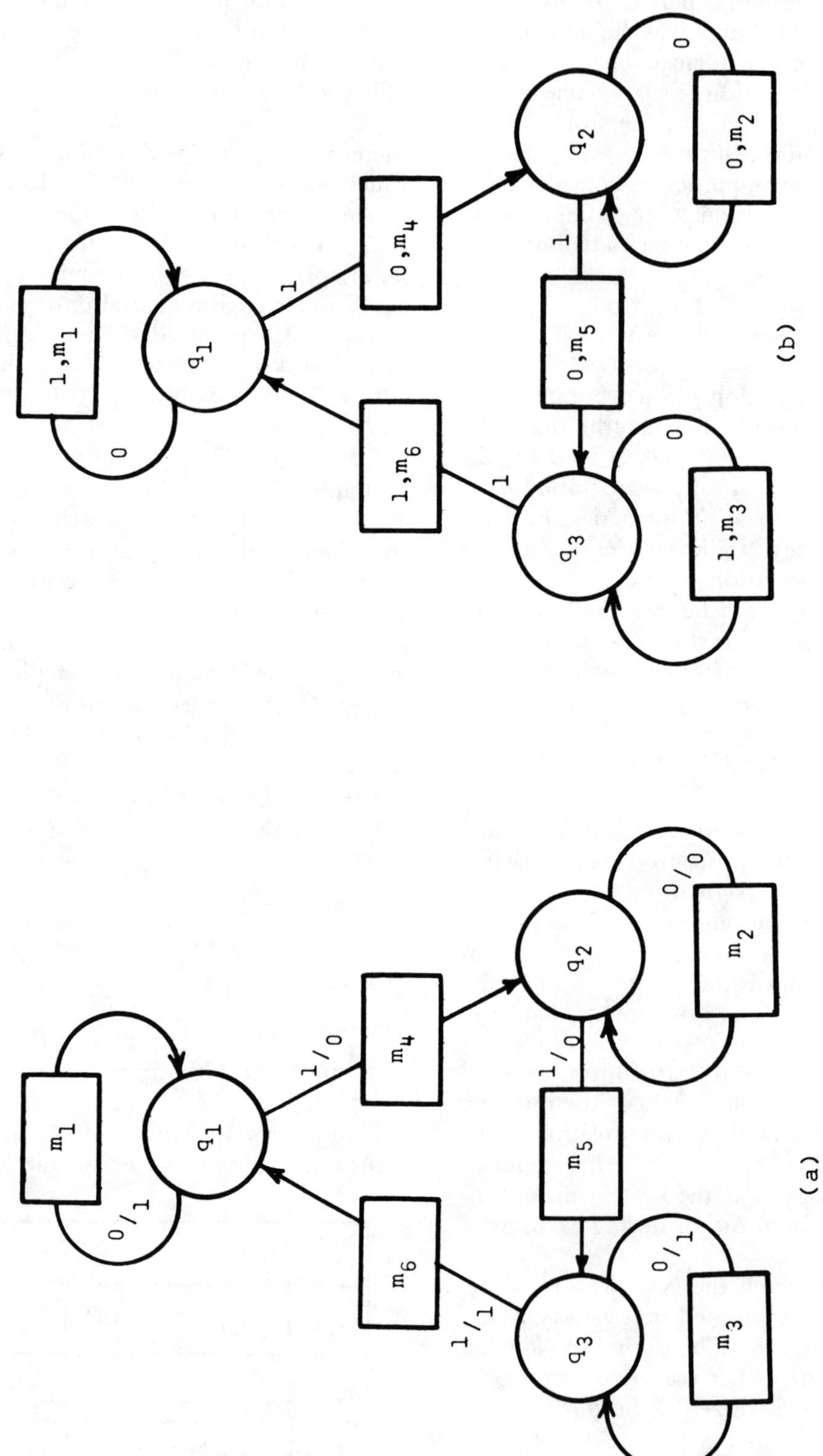

Fig.7 Kinematic graph representation of system $\mathbf{M}_1$

the system is schematically illustrated by the kinematic graph, displayed in Fig. 7.

The mobile system $\mathbf{M}_1$ is assumed to operate in an environment, whose structure σ is schematically described by an array of thirteen adjacent hexagonal cells depicted in Fig. 8. Each cell represents here a distinct physical position that the system under certain conditions can enter. Each position is labelled by two attributes. One attribute indicates the address of the position and the other attribute indicates the information content or 'the information load' of the position. Thus for instance the cell descriptor $p_{11}\langle 0\rangle$ indicates that this particular position has address *11* and carries the message (information content) *0*. Each cellular position acts as a carrier or a source of information. It is assumed, that the mobile system upon entering a specific position can pick up and process the information content of that position as an input and deposit in its place the currently produced output.

The system $\mathbf{M}_1$ can make six distinct moves $m_1,\ldots,m_6$. Each move is defined as a single step spatial transition to the adjacent cellular position in one of the six designated directions. The move m_1 defines the transition in north-east direction, m_2 in southern direction, m_3 in north-west direction, m_4 in south-east direction, m_5 in south-west direction and m_6 in northern direction. It is assumed that $\mathbf{M}_1$ moves in discrete unit steps and that it makes one step at a time.

Let us consider a situation in which the system $\mathbf{M}_1$ is initially in the state $q(0)=q_1$ and let us assume that it moves first into the position $p_{11}\langle 0\rangle$, whereupon $\mathbf{M}_1$ will pick up the message $\langle 0\rangle$ as its input $x\langle 0\rangle$. Under the effect of this input, $\mathbf{M}_1|q(0)$ will produce an output $y(0)=\lambda[x(0),q(0)]=\lambda[0,q_1]=1$ which will be immediately deposited (imprinted) as a new message (information content) in the cellular position p_{11}, replacing the original information content $\langle 0\rangle$ by the content $\langle 1\rangle$. Under the effect of $x(0)$, the system $\mathbf{M}_1|q(0)$ will also realize a move

$$m(0) = \mu[x(0),q(0)] = \mu[0,q_1] = m_1 = \text{NE}$$

transiting from the initial position p_{11} in north-east direction to a new position p_{21}. Subsequently the system will assume a new internal state

$$q(1) = \delta[x(0),q(0)] = \delta[0,q_1] = q_1$$

which in this case is identical with the initial state. Upon arrival to the new position $p_{21}\langle 1\rangle$, the system $\mathbf{M}|q(1)$ will pick up the message that occurs in the brackets of $p_{21}\langle 1\rangle$ and it will process it as its new

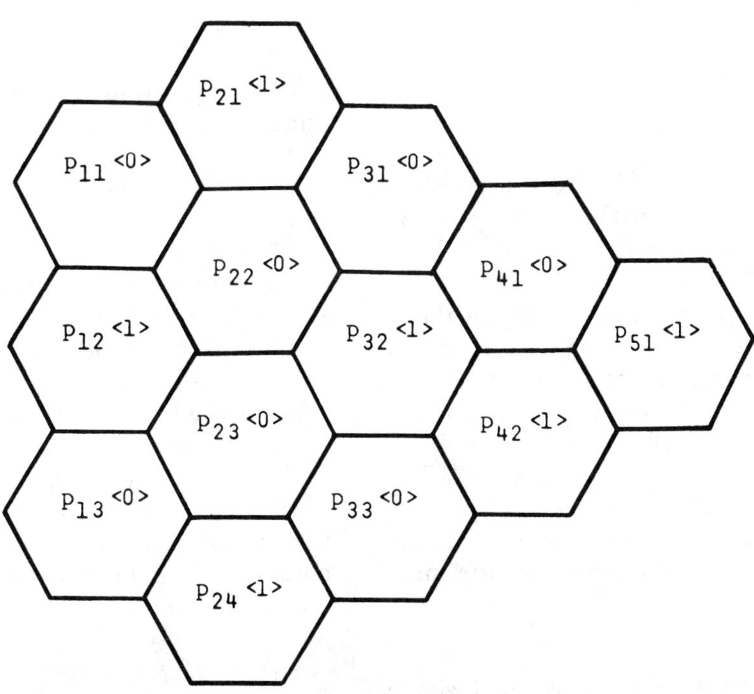

Fig.8 Operation environment of system $\mathbf{M}_1$

input $x(1)$. This input $x(1) = 1$ will then act upon $\mathbf{M}_1 | q(1)$ producing a new instantaneous output

$$y(1) = \lambda[x(1), q(1)] = \lambda[1, q_1] = 0$$

which in turn is deposited as a new information content into the position p_{21}. Subsequently, the system $\mathbf{M}_1$ will realize the move

$$m(1) = \mu[x(1), q(1)] = \mu[1, q_1] = m_4 = \mathrm{SE}$$

i.e. it will transit from the visited 'processed' position p_{21} to a new position $p_{31}\langle 0 \rangle$ and it will assume a new internal state

$$q(2) = \delta[x(1), q(1)] = \delta[1, q_1] = q_2$$

Upon arrival to the new position p_{31}, the system will pick up the message (information content) contained in the brackets of $p_{31}\langle 0 \rangle$ as its new input $x(2)$ and it will produce an output

$$y(2) = \lambda[x(2), q(2)] = \lambda[0, q_2] = 0$$

that will be put in place of the original message carried by p_{31}†. Subsequently, the system will realize a new move

$$m(2) = \mu[x(2), q(2)] = \mu[0, q_2] = m_2 = \mathrm{S}$$

and it will assume a new internal state

$$q(3) = \delta[x(2), q(2)] = \delta[0, q_2] = q_2$$

Upon entering the position $p_{32}\langle 1 \rangle$ the system $\mathbf{M}_1 | q_2$ will pick up the information carried by the position p_{32} and will use it as its new input $x(3)$ with the consequences outlined above.

By analyzing the kinetic behavior of $\mathbf{M}_1$ in the above-described fashion, we can define the complete trajectory of the system within its specified cellular environment. In our example, we may notice, that a specific trajectory of $\mathbf{M}_1$ will be determined partially by the internal state that the system will initially assume and partially by the cellular position that the system will initially enter, or more correctly by the message (information content) that the initial position may carry.

†Note that in this case, the value of $y(2)$ is identical with the current information content of p_{31} and that the information exchange will not affect the value of the information content of this position.

Sometimes we may wish to translate the move function μ of $\mathbf{M}_1$ into a position transition function π. To accomplish that we have to use the table representation form of μ-function depicted in Fig. 6 and the cellular structure representation of operation space $\mathbf{E}_{\mathbf{M}_1}$ depicted in Fig. 8. From the table in Fig. 6 we shall look up the move that the system under given conditions would make and then, from the cellular structure in Fig. 8 we shall determine the position that the system under specified conditions would enter. The found position can be subsequently marked and placed as an entry at the intersection of an appropriate row and column of the table, that defines the positional function π in Fig. 9

Thus, for instance we may wish to find a position that the system $\mathbf{M}_1$ can enter under conditions that it presently assumes the internal state q_1, occupies the position p_{11} and is exposed to the input 0, i.e. under conditions that $q(t) = q_1$, $p(t) = p_{11}$ and $x(t) = 0$ In such case we have to find out from Fig. 6 the move $m(t)$ that $\mathbf{M}_1$ would make under the given conditions, i.e. under the combination of input $x(t) = 0$ and state $q(t) = q_1$. Figures 6 and 8 indicate, that if the system $\mathbf{M}_1 | q_1$ is exposed to the input $x(t) = 0$ at the position $p(t) = p_{11}$, then the system will realize the move $m(t) = m_1$, transiting in north-east direction from the present position p_{11} to the next available position p_{21}. This is due to the fact that

$$m(t) = \mu[x(t), q(t)] = \mu[0, q_1] = m_1 = \mathrm{NE}$$

The next position is then defined through the π-function as follows:

$$p(t+1) = \pi[x(t), q(t), p(t)] = \pi[0, q_1, p_{11}] = p_{21}$$

This position can be subsequently placed as an entry at the intersection of the first row and first column of the table in Fig. 9.

If we, for instance, wish to find the next position that $\mathbf{M}_1$ can enter under conditions $q(t) = q_2$, $p(t) = p_{12}$ and $x(t) = 0$, then we shall proceed in the same fashion as outlined above, i.e. from Fig. 6 we look up the move $m(t)$ to be realized under the given circumstances and from Fig. 8 we shall find out the next position $p(t+1)$ to be entered by $\mathbf{M}_1$ upon the completion of the move $m(t)$. In this case, the Fig. 6 indicates that

$$m(t) = \mu[x(t), q(t)] = \mu[0, q_2] = m_2 = \mathrm{S}$$

and Fig. 8 indicates that the spatial transition is realized from the position $p(t) = p_{12}$ in a southern

C \ P	c_1 $(0,q_1)$	c_2 $(0,q_2)$	c_3 $(0,q_3)$	c_4 $(1,q_1)$	c_5 $(1,q_2)$	c_6 $(1,q_3)$
p_{11}	p_{21}	p_{12}	*	p_{22}	*	*
p_{12}	p_{22}	p_{13}	*	p_{23}	*	p_{11}
p_{13}	p_{23}	*	*	p_{24}	*	p_{12}
p_{21}	*	p_{22}	*	p_{31}	p_{11}	*
p_{22}	p_{31}	p_{23}	p_{11}	p_{32}	p_{12}	p_{21}
p_{23}	p_{32}	p_{24}	p_{12}	p_{33}	p_{13}	p_{22}
p_{24}	p_{33}	*	p_{13}	*	*	p_{23}
p_{31}	*	p_{32}	p_{21}	p_{41}	p_{22}	*
p_{32}	p_{41}	p_{33}	p_{22}	p_{42}	p_{23}	p_{31}
p_{33}	p_{42}	*	p_{23}	*	p_{24}	p_{32}
p_{41}	*	p_{42}	p_{31}	p_{51}	p_{32}	*
p_{42}	p_{51}	*	p_{32}	*	p_{33}	p_{41}
p_{51}	*	*	p_{41}	*	p_{42}	*

Fig.9 Table representation of positional function for system π for system M_1

direction to the position $p(t+1)=p_{13}$. Thus in terms of π-function the position

$$p(t+1) = \pi[x(t),q(t),q(t)] = \pi[0,q_2,p_{12}] = p_{13}$$

This position can then be placed as an entry at the intersection of the second row and second column of the table in Fig. 9. All remaining table entries can be filled in analogously, i.e. in the same manner as described above.

Under certain configurations of conditions, however, we may find out that the system may be forced to move outside the boundary of the specified cellular space E_{M_1} and to leave the structure of its present environment. When this happens and no specified position is available for $\mathbf{M}_1$ to move to and the system is forced to leave its environment, the relevant table entry in Fig. 9 would be marked by the star, indicating that at that point the system $\mathbf{M}_1$ would exit the operation space $\mathbf{E}_{\mathbf{M}_1}$. Thus, for instance, under condition that $q(t)=q_1$, $p(t)=p_{21}$ and $x(t)=0$, the system $\mathbf{M}_1$ would realize a transition

$$m(t) = \mu[x(t),q(t)] = \mu[0,q_1] = m_1 = \text{NE}$$

moving from the position $p(t)=p_{21}$ in north-eastern direction outwards and leaving the boundary of the defined cellular structure $\mathbf{E}_{\mathbf{M}_1}$. This would be indicated in Fig. 9 by an appropriate entry at the intersection of the fourth row and first column.

Example II

Let us consider a mobile system $\mathbf{M}_2$ given as follows:

$$\mathbf{M}_2 = (B,Q,A,F,P) \tag{24}$$

where

B = an alphabet of the system defined by the set $\{1,2,3\}$

Q = a state set $\{q_1,q_2,q_3\}$ $\quad \beta: Q \rightarrow A$

$\quad \delta: B\times Q \rightarrow Q$

A = an attribute set $\{1,2,3\}$ $\quad \lambda: B\times Q \rightarrow B$

F = a function set $\{\beta, \delta, \lambda, \pi\}$ $\quad \pi: B\times A\times Q\times P \rightarrow P$

P = a position set $\{p_1,p_2,p_3,p_4,p_5\}$

The assignment function β associates each internal state $q_i \in Q$ of the system $\mathbf{M}_2$ with a specific configuration A_i of numerical attributes. Each configuration A_i of attributes represents a combination of threshold T_j and input weight w_j. Thus

$$\beta:\ q_i \Rightarrow A_i \qquad \text{where } A_i \subset A \text{ for } i \in I \tag{25}$$

and

$$A_i = (T_j;w_k) \qquad \text{where } T_j,\, w_k \in A \text{ for } i,j,k \in I \tag{26}$$

In our example:

$$q_1 \Rightarrow A_1 = (T_1;w_1) = (1;3)$$

$$q_2 \Rightarrow A_2 = (T_2;w_2) = (2;2)$$

$$q_3 \Rightarrow A_3 = (T_3;w_3) = (3;1)$$

The state transition function δ and the output function λ of $\mathbf{M}_2$ are defined in table form in Fig. 10. The system $\mathbf{M}_2$ operates in an environment $\mathbf{E}_{\mathbf{M}_2}$, that is schematically depicted by the directed graph structure, presented in Fig. 11. This environment contains

δ, λ :

Q \ X	1	2	3
q_1	q_1,2	q_2,1	q_3,3
q_2	q_2,3	q_3,2	q_1,1
q_3	q_3,3	q_1,2	q_2,1

Fig.10 Table representation of δ and λ-functions for system $\mathbf{M}$

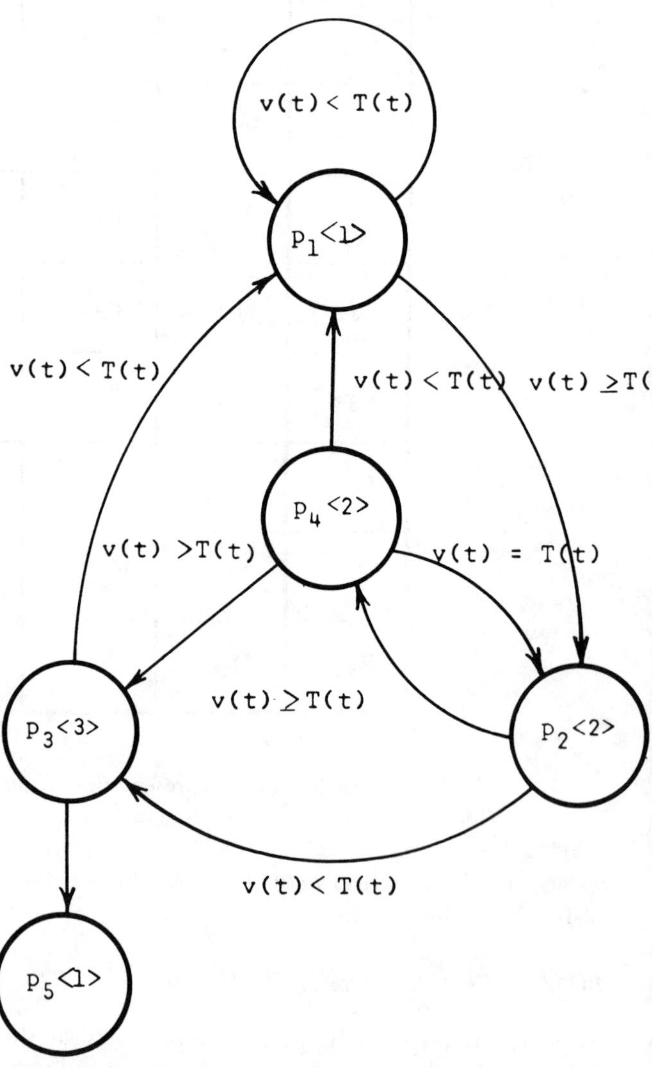

Fig.11 Operation environment of mobile system $\mathbf{M}_2$

five distinct positions $p_1,...,p_5$ that the system can enter through the designated routes. Each position is associated with an attribute that indicates the current information content (message) of the position. It is assumed that during the visit at the individual positions the system will pick up the available information content of the position as its input and it will replace it by the ensuing output. The recorded input information is assigned a specific weight by the system and is compared to a specific threshold. The result of this comparison determines then the next position to which the system will move. The individual moves that the system can make are defined as positional transitions through the function π.

The position transition function π is conceptually defined as a threshold function. In general form, this function can be defined as follows:

$$\pi_i[x(t),q(t),A(t),p_i(t)] = p(t+1) = \begin{cases} p_j(t+1) & \text{if } x(t)w(t)<T(t) \\ p_k(t+1) & \text{if } x(t)w(t)\geqslant T(t) \end{cases} \quad (27)$$

where $p_i(t)$ is a specific position that the system presently occupies and $p(t+1)$ is a position to be entered by the system in the next period of time. Equation (27) indicates that the system $\mathbf{M}_2|q(t),p_i(t)$ will enter the position $p_j(t+1)$ if and only if the value of weighted input $x(t)w(t)=v(t)$ is smaller than the value of the threshold $T(t)$, currently associated with the present state $q(t)$. Otherwise, the system will enter position $p_k(t+1)$. It is assumed that $p_i(t),p_j(t+1),p_k(t+1)\in P$, $x(t)\in B$, $q(t)\in Q$, $A(t)\subset A$ and that $i,j,k\in I$.

In our example, five component functionals of the position transition function π will be used to define all the possible transitions that the system $\mathbf{M}_2$ can realize under given conditions.

$$\pi_1[c(t),p(t)=p_1] = p(t+1) = \begin{cases} p_1 & \text{if } v(t)<T(t) \\ p_2 & \text{if } v(t)\geqslant T(t) \end{cases} \quad (28)$$

$$\pi_2[c(t),p(t)=p_2] = p(t+1) = \begin{cases} p_3 & \text{if } v(t)<T(t) \\ p_4 & \text{if } v(t)\geqslant T(t) \end{cases} \quad (29)$$

$$\pi_3[c(t),p(t)=p_3] = p(t+1) = \begin{cases} p_1 & \text{if } v(t)<T(t) \\ p_5 & \text{if } v(t)\geqslant T(t) \end{cases} \quad (30)$$

$$\pi_4[c(t),p(t)=p_4] = p(t+1) = \begin{cases} p_1 & \text{if } v(t)<T(t) \\ p_2 & \text{if } v(t)=T(t) \\ p_3 & \text{if } v(t)>T(t) \end{cases} \quad (31)$$

$$\pi_5[c(t),p(t)=p_5] = p(t+1) = p_5 \text{ for all } v(t) \quad (32)$$

In the above-defined formulas the expression $c(t)$ represents the combination $x(t)$, $q(t)$ and $A(t)$ and the expression $v(t)$ represents the value of the weighted input. Thus $c(t)=[x(t), q(t), A(t)]$ and $v(t)=x(t).w(t)$. The symbol $T(t)$ represents the value of the threshold, that is presently associated with $q(t)$. It is assumed that $q(t) \Rightarrow A(t)|A(t) = [T(t);w(t)]$ where $A(t)\subset A$. The position p_5 is considered to be the terminal or target position of $\mathbf{M}_2$ from which no further transition is possible.

The position transition function π of $\mathbf{M}_2$ can be also defined in a matrix form, as illustrated in Fig. 12.

This figure indicates the conditions under which the individual positional transitions are realized. When the system presently occupies a specific position in $\mathbf{E}_{\mathbf{M}_2}$, the table in Fig. 12 will show us what the next position of the system should be under given circumstances.

To illustrate the moves of $\mathbf{M}_2$ within the environment $\mathbf{E}_{\mathbf{M}_2}$, let us consider a situation in which the system in state q_1 enters initially (at time $t=0$) the position $p_1\langle 1\rangle$. The information content of this position is picked up by $\mathbf{M}_2$ as an input $x(t)$. Under the effect of this input, the system $\mathbf{M}_2|q_1$ will produce an output $y(0)=\lambda[x(0),q(0)]=\lambda[1,q_1]=2$ whose value is deposited in place of the original information content of the position p_1. (The value of the present output is indicated by an entry at the intersection of the first row and first column of the table in Fig. 10.) Since the initial internal state $q(0)=q_1$ is associated with a configuration $A(0)=[T(0);w(0)]$ of the threshold $T(0)=1$ and input weight $w(0)=3$, the initially recorded input $x(0)$ will be assigned the weight $w(0)=3$ and the value $v(0)$ of weighted input will be compared with the value of the threshold $T(0)$ in order to determine the positional transition of $\mathbf{M}_2$. In this case the value $v(0)=x(0).w(0)=3$ and the value of $T(0)=1$. Immediately we can see that $v(0)>T(0)$ and hence according to Eq.(28) the system $\mathbf{M}_2$ shall move from the present position p_1 to a new position p_2, changing its initial internal state

present positions

next positions

p(t+1) \ p(t)	p_1	p_2	p_3	p_4	p_5
p_1	if $v(t) < T(t)$	—	if $v(t) < T(t)$	if $v(t) < T(t)$	—
p_2	if $v(t) \geq T(t)$	—	—	if $v(t) = T(t)$	—
p_3	—	if $v(t) < T(t)$	—	if $v(t) > T(t)$	—
p_4	—	if $v(t) \geq T(t)$	—	—	—
p_5	—	—	if $v(t) \geq T(t)$	—	for $v(t) \lesseqgtr T(t)$

Fig.12 Matrix representation of π-function for system $\mathbf{M}_2$

$q(0)$ into a new state

$$q(1) = \delta[x(0),q(0)] = \delta = [1,q_1] = q_1$$

Here we may notice that under the given conditions the internal state $q(0)$ is coincidentally the same as the state $q(1)$.

As soon as the system enters a new position $p_2\langle 2\rangle$ it will pick up the information content $\langle 2\rangle$ of this position as its new input $x(1)$. Under the effect of this input the system will produce a new output

$$y(1) = \lambda[x(1),q(1)] = \lambda[2,q_1] = 1$$

whose value will be deposited in place of the present information content of p_2. Since the internal state $q(1)$ that $\mathbf{M}_2$ assumes upon the arrival in p_2 is associated with a configuration $A(1) = [T(1), w(1)]$ of the threshold $T(1) = 1$ and input weight $w(1) = 3$, the recorded input $x(1) = 2$ will be assigned the corresponding weight and the value of $v(1) = x(1).w(1) = (2).(3) = 6$ will be compared with the value of $T(1) = 1$. Since $v(1) > T(1)$, the system will move from its present position p_2 to a new position p_4, transforming its present internal state $q(1) = q_1$ into a new state

$$q(2) = \delta[x(1),q(1)] = \delta[2,q_1] = q_2$$

When $\mathbf{M}_2$ enters the position $p_4\langle 2\rangle$ it will pick up the information content $\langle 2\rangle$ as its new input $x(2)$, replacing it with the output†.

$$y(2) = \lambda[x(2),q(2)] = \lambda[2,q_2] = 2$$

†Note that also in this case the value of output $y(2)$ is identical with the current information content of the position p_4.

Since the present internal state $q(2) \Rightarrow [T(2), w(2)] = = [2,2]$, the input $x(2)$ will be associated with the weight $w(2)$ and the value of $v(2) = x(2).w(2) = = (2).(2) = 4$ will be compared to the value of $T(2) = 2$. The result of this comparison indicates that $\mathbf{M}_2$ will move from p_4 to the position $p_3\langle 3\rangle$. Further examination of this case would reveal that under given circumstances $\mathbf{M}_2$ will subsequently move from the position p_3 to the terminal position p_5. In this position the system $\mathbf{M}_2$ may continue to operate in the above described manner, but all its future operations would be confined to the terminal position p_5 only.

SUMMARY

A special algebraic structure is introduced that enables us to formally represent various types of mobile systems within the conceptual framework of General System Theory. The proposed structure functions essentially as a general-purpose model generating devise from which a large variety of kinematic systems models can be derived. The structure permits us to define any mobile system through a finite configuration of selected systems entities whose functional and structural interrelationships describe the operation properties and kinetic behavior of the system. The individual systems entities are defined by sets of relevant parameters and their interrelationships are expressed analytically through appropriate functions.

A new systems-theoretical framework is also introduced that allows us to formally represent the environment in which mobile systems operate.

Various forms of model representation are discussed throughout the text and two examples of kinematic system models are presented to illustrate the meaning of various theoretical terms and concepts that have been introduced in the first two sections of the article. One example illustrates a finite state model of an elementary mobile system that operates within a specified cellular environment. This system has the capability to move in discrete steps in designated directions from one cellular position to another exchanging with each visited cell various amounts of information. The moves of this system are defined through a move function μ. The example also describes and interprets the conditions under which the individual moves of the system can be realized.

The second example presents a threshold model of a mobile system. The moves of this system are determined through a position transition function π that is formally defined as a threshold function. The system is assumed to operate within an environment which is schematically represented by a directed graph structure. The system can move within this environment from one position to another and exchange with each visited position various quantities of information. The example also describes the individual moves that the system can make under various environmental contingencies. It also illustrates the physical trajectory that the system would follow under specific prearranged conditions.

References

1. AMOSOV, N.M., KASATKIN, A.M., KASATKINA, L.M., KUSSUL, E.M., TALAEV, S.A., and FOMENKO, V.D., 'Intelligent behavior systems based on semantic networks', *Kybernetes* 2(4), pp.211-126 (1973).
2. LASKER, G.E., *Theory of Mobile Automata*, Lecture Scriptum, School of Computer Science, University of Windsor, Windsor, Ontario, Canada (1969).
3. LASKER, G.E., 'Theory of mobile automata', in *Advances in Cybernetics and Systems* (ed. J. Rose) Gordon & Breach, London, Vol.2, pp.897-910 (1975).
4. PELIKAN, P., 'Dévélopment du modèle de l'instinct de conservation', *Information Processing Machines,* Collection of Papers, Vol.10, pp.313-320, Czechoslovak Academy of Sciences Press, Prague (1964).
5. RAICHL, J., 'An attempt to simulate some simple behaviors of lowest organisms on a computer', *Information Processing Machines*, Collection of Papers, Vol.12, pp.121-126, Czechoslovak Academy of Sciences Press, Prague (1962).
6. SVOBODA, A., 'Un modèle d'instinct de conservation', *Information Processing Machines,* Collection of Papers, Vol.7, pp.147-155, Czechoslovak Academy of Sciences Press, Prague (1960).
7. LASKER, G.E., 'Threshold automata', in *Advances in Cybernetics and Systems* (ed. J. Rose), Gordon & Breach, London, Vol.2, pp.911-921 (1975).
8. MEALY, G.H., 'A method of synthesizing sequential circuits', *Bell System Techn. J.* **XXXIV**, pp.1045-1079 (1955).
9. MOORE, E.F., 'Gedanken-experiments on sequential machines', *Automata Studies*, Annals of Mathematical Studies, No.34, pp.129-153, Princeton University Press, Princeton N.J. (1956).
10. ARBIB, M., *Theories of Abstract Automata*, Prentice-Hall, Englewood Cliffs, N.J. (1969).
11. HARTMANIS, J. and STEARNS, R.E., *Algebraic*

Theory of Sequential Machines, Prentice-Hall, Englewood Cliffs, N.J. (1966).
12. KITAGAWA, Tosio, 'Cell space approaches in bio-mathematics', *Proc. 1st Symposium on Uniformly Structured Automata and Logic,* pp.81-85, Tokyo (August 1975).
13. KLIR, G.J., *An Approach to General Systems Theory,* Van Nostrand Reinhold, New York (1969).
14. KLIR, G.J. and VALACH, M., *Cybernetic Modelling,* D. Van Nostrand, Princeton, N.J. (1967).

Unsolvability in systems of constructing automata

ALEXANDER LEITSCH
Interfak. Rechenzentrum d.
Universität Wien, Austria

Introduction
Let us consider the following system (a similar construction can be found in [1]):

We fix a construction-universal cellular space ([2],[3]) and embed an automaton consisting of a Turing machine and a builder-controller.

Instructions given by the Turing machine enable the builder-controller to build another automaton of the same type. We consider a recursive enumeration of all Turing machines, such that the number of any Turing machine can be found effectively (i.e. a function $t:N \rightarrow TM$ with the property:

$M \in TM$ we can compute effectively an integer k with $t(k) = M$).

Now the Turing machine of the automaton gets a tape with an input representing an integer n (if you consider a 2-symbol Turing machine you can take an unbroken string of I-symbols). If the Turing machine comes to a halt at a standard-configuration corresponding to a number k, the builder-controller constructs the automaton consisting of the same builder-controller and of the Turing machine $t(k)$. If the Turing machine does not halt at a standard-configuration, then the automaton is 'sterile with respect to the input n'. Unlike [1] the machines are not generated by a fixed input. We omit therefore the research of computing-ability, which is carried out in [1]. The definition of the constructing system in this chapter is related to the definition of an universal system of construction-algorithms in [4].

Because the input varies, the question naturally arises, whether an automaton is potentially able to construct descendants of a certain type. At first sight the Turing machines computing the identity seem to be the 'best', because the next generation consists of all automata. But such Turing machines produce all automata, that are sterile with respect to every input (totally sterile). This follows from the definition of the underlying enumeration.

There is an analogy with biological systems (see also [1] and [5]). The tape of the Turing machine can be interpreted to be the environment of an organism (the organism is now the automaton). The sexual ability of an organism depends naturally on its environment. When the conditions for the organism are not convenient, it can not produce descendants, or the descendents are genetically degenerated; i.e. in automata-theoretic interpretation: The first Turing machine does not halt at a standard-configuration or further generations have such a property. In one part of the chapter we shall deal with automata generating only fertile descendants (independent of the input).

Instead of the recursive enumeration of Turing machines we consider a Gödel numbering of partial recursive functions, because it is more practical to deal with partial recursive functions than to deal with Turing machines (it is well known that both ways are equivalent). In [6] one can find a research about the constructing-ability of automata with different inputs not using a Gödel numbering. But this way is more general and axiomatic and not so adequate to the research of unsolvability problems. In [1] the recursion-theoretic method is used to demonstrate undecidability and also to determine the degree of unsolvability (see [7]). Similar methods are used to classify the set of integers appearing in the problems concerned in this chapter. For the most part the degree of unsolvability in the arithmetical hierarchy is not at a lower level than the second or even the third. The unsolvability (at least the undecidability) appears in all interesting cases, because the type of automaton is very 'general'. In this chapter especially self-reproduction and fertility will be found to be unsolvable (at least undecidable) properties.

1. Preliminaries

Let P^k_1 be the set of partial recursive functions in k variables and R^k_1 the set of (total) recursive functions in k variables.

$\phi \in P^2_1$ is called a Gödel numbering of P^1_1, if for all $\psi \in P^2_1$ there is an $h \in R^1_1$ such that $(\forall x)(\forall y)\phi(h(x),y) = \psi(x,y)$. Obviously there is $\{\phi_i / i \in N\} = P^1_1$, where ϕ_i is the function $j \to \phi(i,j)$. Let $A,B \subseteq N$; if there is an $h \in R^1_1$ such that $x \in A \Leftrightarrow h(x) \in B$ we write $A \underset{m}{\leqslant} B$ ([7] chapter 6,7); if h is one-one, we write $A \underset{1}{\leqslant} B$. $A \underset{m}{\equiv} B$ means $A \underset{m}{\leqslant} B$ and $B \underset{m}{\leqslant} A$, $A \underset{1}{\equiv} B$ means $A \underset{1}{\leqslant} B$ and $B \underset{1}{\leqslant} A$. Note that $\underset{m}{\leqslant}$ and $\underset{1}{\leqslant}$ define a partial ordering in the set of all subsets of N. In [7] (Chapter 7.4) it is demonstrated that $A \underset{1}{\equiv} B \Leftrightarrow A \equiv B$, i.e. there is a $g \in R^1_1$, g one-one and onto with $g(A) = B$.

$T(x,y,z)$ denotes the predicate: The universal program computing ϕ_x with input y comes to a halt at a standard configuration after z steps. If $T(x,y,z_0)$ we define $T(x,y,z)\,(\forall z)\, z > z_0$. Then $\phi(x,y)$ is defined $\Leftrightarrow (\exists z)T(x,y,z)$. Furthermore $T(x,y,z)$ is a recursive predicate (see [7], [4]).

Consider the set $\{(x_1,\ldots,x_m)/(Q_1 y_1) \ldots \ldots (Q_n y_n) R(x_1,\ldots,x_m, y_1 \ldots y_n)\}$ where $R(x_1,\ldots,x_m,y_1,\ldots,y_n)$ is a $(m+n)$-ary recursive relation and $Q_1,\ldots,Q_n$ is a sequence of $\exists$ and $\forall$ quantifiers. If the sequence of quantifiers begins with $\exists (\forall)$and there are n quantifiers, except the first $\exists(\forall)$ which have a quantifier of the other type in their immediate neighborhood, then we say that the set is in $\Sigma_{(n+1)}(\Pi_{(n+1)})$. For example the set $\{x(\exists y)/\phi(x,y) \text{ defined}\} = \{x/(\exists y)(\exists z)T(x,y,z)\}$ is contained in $\Sigma_{(1)}$. Every recursive predicate can be considered as $\Sigma_{(0)}$-predicate (there is no quantifier). For example $\{x/(\forall y)(\exists z)(\forall w)[T(x,y,z) \Rightarrow T(x,w,z)]\}$ is in $\Pi_{(3)}$. It holds $\Sigma_{(n)} \cup \Pi_{(n)} \subseteq \Sigma_{(n+1)} \cap \Pi_{(n+1)}$ (hierarchy-property; for a more exact information use [7] chapter 14, [4] chapter 6.5).

We call a set $x \in \Sigma_{(n)}(\Pi_{(n)})$ $\Sigma_{(n)}(\Pi_n)$—complete if for all $B \in \Sigma_{(n)}(\Pi_{(n)}) B \underset{1}{\leqslant} X$ holds. If a set is $\Sigma_{(n+1)}(\Pi_{(n+1)})$—complete it is not contained in $\Sigma_{(n)}(\Pi_{(n)})$. Therefore the demonstration of completeness is a very useful method to prove that certain sets are not in $\Sigma_{(0)}$ or not in $\Sigma_{(1)}$.

If the Turing machine $t(i)$ with input m halts with output n, we write: $\phi(i,m) = n$ or $\phi_i(m) = n$. This defines a function $\phi \in P^2_1$. The properties of the enumeration guarantee that ϕ is a Gödel numbering. $\phi(i,m) = n$ means: The ith automaton with input m constructs the nth automaton. If $\phi(i,m)$ is undefined, the ith automaton with input m is sterile. If $\phi(\phi(i,m), n)$ exists, it is the number of the automaton in the third generation with progenitor number i, where the input to the first automaton is m and to its descendent u. In the nth generation we must consider expressions of the form: $(x_1,\ldots,x_n) \to \phi(\ldots\phi(\phi(x_1,x_2)x_3)\ldots x_n)$. They denote sequences of generations beginning with x_1.

2. Self-reproduction

First we define three types of 'weak' self-reproduction.

1. There is a descendent in the $(k+1)$th generation which is equal to its progenitor in the first generation (in the first generation there is only one automaton).
2. There is such a descendent in the jth generation with $j \leqslant k+1$.
3. There is such a k, that 1) is fulfilled. More exactly: let $k \geqslant 1$.

1) $SP_k = \{x/(\exists m_1)\ldots(\exists m_k)\,\phi(\ldots\phi(\phi(x,m_1),m_2),\ldots$

$\ldots m_k) = x\}$

2) $SP_{\leqslant k} = \{x/(\exists j \leqslant k)(\exists m_1)..(\exists m_j)\phi(\ldots\phi(\phi(x,m_1),$

$m_2)\ldots m_j) = x\}$

3) $SP_E = \{x/(\exists k)(\exists m_1)\ldots(\exists m_k)\phi(\ldots\phi(\phi(x,m_1),m_2)$

$\ldots m_k) = x\}$

Because the predicate $m_1,\ldots,m_k[\phi(\ldots\phi(\phi(x,m_1),m_2)\ldots m_k) = x]$ is a $\Sigma_{(1)}$-predicate, the Tarski-Kuratowski algorithm (see [7] chapter 14) yields immediately that SP_k and $SP_{\leqslant k}$ are in $\Sigma_{(1)}$. In the case SP_E we note that $x \in SP_E \Leftrightarrow (\exists k)x \in SP_k$. Because $SP_k \in \Sigma_{(1)}$ it follows $SP_E \in \Sigma_{(1)}$.

Let $K = \{x/\phi(x,x)$ defined$\}$ ('halting problem').

Theorem 2.1

$SP_E \equiv SP_k \equiv SP_{\leqslant k} \equiv K$. That means: All these sets are $\Sigma_{(1)}$-complete (where $k \geqslant 1$).

Proof: It is sufficient to demonstrate: $K \underset{m}{\leqslant} SP_E$, SP_k, $SP_{\leqslant k}$ $(\forall k \geqslant 1)$. Because K is $\Sigma_{(1)}$-complete ([7] chapter 7.2) it follows $K \underset{m}{\equiv} SP_E, SP_k, SP_{\leqslant k}$. But the $\Sigma_{(n)}$-complete sets (which form an $\underset{1}{\leqslant}$ degree) form also an $\underset{m}{\leqslant}$ degree ([7] p.332). Therefore $K \underset{1}{\equiv} SP_E, SP_k, SP_{\leqslant k}$ and $K \equiv SP_E, SP_k, SP_{\leqslant k}$ (see Chapter 1). Consider $f: f(x,y) = y$ if $x \in K$

= undefined otherwise

Then $f \in P^1_1$. The *s-m-n* theorem ([7] chapter 1.8) implies the existence of an $h \in R^1_1$ with:

$\phi_{h(x)}(y) = y$ if $x \in K$

= undefined otherwise

This implies $x \in K \Rightarrow \phi_{h(x)}$ onto $\Rightarrow (\exists j)\phi_{h(x)}(j) = h(x) \Rightarrow$
$\Rightarrow h(x) \in SP_1 \Rightarrow h(x) \in SP_E, SP_{\leqslant k}\ (\forall k \geqslant 1)$
and $h(x) \in SP_E, SP_{\leqslant k} \Rightarrow (\exists y)\phi(h(x),y)$ defined $\Rightarrow$
$\Rightarrow x \in K$.

Therefore $(\forall k \geqslant 1)\, x \in K \Leftrightarrow h(x) \in SP_E, SP_{\leqslant k}$; that means: $K \underset{m}{\leqslant} SP_E, SP_{\leqslant k}$.

Let us show that $h(x) \in SP_k\ (\forall k \geqslant 2)$ if $x \in K$. Because $\phi_{h(x)}$ is onto for $x \in K$ there is a y_1: $\phi(h(x),y_1) \in \{x/\phi_x$ onto$\}$. Therefore there is a y_2: $\phi(\phi(h(x),y_1),y_2) \in \{x/\phi_x$ onto$\}$ and so on. We get $y_1,\ldots y_k$: $\phi(\ldots\phi(h(x),y_1)\ldots y_k) \in \{x/\phi_x$ onto$\}$. That means $(\exists y_{k+1})\phi(\phi(\ldots\phi(h(x),y_1),\ldots y_k),y_{k+1}) =$
$= h(x)$ and $h(x) \in SP_{k+1} (\forall k \geqslant 1)$. It follows $K \underset{m}{\leqslant} SP_k$ for all $k \geqslant 1$ □

Now we define stronger forms of self-reproduction:

1) $SP_A = \{x/(\forall y)\phi(x,y) = x\}$ (self-reproducing with every input)

2) $SP^*_A = \{x/(\forall y)\phi(x,y) \in SP_A\}$ (for every input 'total' self-reproduction begins in the second generation).

Theorem 2.2

a) $SP_A \notin \Sigma_{(0)}$; *b)* SP^*_A is $\Pi_{(2)}$-complete.

Proof: a) Let $\mathbf{C} = \{\phi_x/(\forall y)\phi_x(y) = x\}$. The recursion theorem yields $\mathbf{C} \neq \emptyset$. Obviously $\mathbf{C} \neq P^1_1$. Now the theorem of Rice ([7] p.34) can be used to demonstrate:

$\{x/\phi_x \in \mathbf{C}\} \in \Sigma_{(0)} \Leftrightarrow \mathbf{C} = \emptyset \vee \mathbf{C} = P^1_1$. This implies in our case $\{x/\phi_x \in \mathbf{C}\} = SP_A \notin \Sigma_{(0)}$.

b) Because $y[\phi(x,y) = x]$ is a $\Sigma_{(1)}$-predicate one can see immediately: $SP_A, SP^*_A \in \Pi_{(2)}$.

Let $T = \{x/\phi_x$ total$\}$. T is $\Pi_{(2)}$-complete ([7] p.264); it is enough to show that $T \underset{m}{\leqslant} SP^*_A$. Then it follows $T \underset{m}{\equiv} SP^*_A$ and $\overline{T} \equiv \overline{SP^*_A}$. Because $\overline{T}$ is $\Sigma_{(2)}$-complete, the remark in the beginning of theorem 2.1 yields, that $\overline{SP^*_A}$ is $\Sigma_{(2)}$-complete and therefore SP^*_A is $\Pi_{(2)}$-complete.

Let j be such that $(\forall m)\phi(j,m) = j$.

Then there is an $h \in R^1_1$ such that:

$\phi_{h(x)}(y) = j$ if $\phi(x,y)$ defined

= undefined otherwise

then $x \in T \Rightarrow (\forall y)\phi_{h(x)}(y) = j \Rightarrow (\forall y)\phi_{h(x)}(y) \in SP_A \Rightarrow$

$$\Rightarrow h(x) \in SP^*_A.$$

$h(x) \in SP^*_A \Rightarrow \phi_{h(x)}$ defined everywhere [otherwise there is a y_0 with $\phi(\phi(h(x),y_0),j)$ undefined for all j and therefore $\phi(h(x),y_0) \notin SP_A$]. That means:

$h(x) \in SP^*_A \Rightarrow \phi_{h(x)}$ defined everywhere $\Rightarrow (\forall y)\phi(x,y)$ defined $\Rightarrow x \in T$. Therefore $T \underset{m}{\leqslant} SP^*_A$ □

Remark: Theorem 2.1 and theorem 2.2 yield: SP_k, $SP_{\leqslant k}$, SP_E and SP_A are undecidable. SP^*_A and $\overline{SP^*_A}$ are not recursively enumerable.

It follows immediately from the proof of theorem 2.2 that $SP_A \neq SP^*_A$ [There are infinitely many x with $\phi_x(y) = j\,(\forall y)$; at most one of the x can be equal to j.] $SP_A \subseteq SP^*_A$ is trivial. It holds also $SP_k \neq SP_j \;\; k \neq j$.

Take x such that $(\forall m)\phi(x,m) = i_0$, $x \neq i_0$ and $i_0 \in \{x/\phi_x$ onto$\}$; then $x \notin SP_{(1)}$ and $x \in SP_{(2)}$. Consider now y: $(\forall m)\phi_y(m) = x$, $y \neq x$ and $y \neq i_0$. It follows $y \in SP_{(3)} - (SP_{(2)} \cup SP_{(1)})$. For every n this method yields a $z \in SP_{(n+1)} - \bigcup_{k=1}^{n} SP_{(k)}$.

3. Sterility

In this section we investigate the potential ability to generate sterile descendents. We consider two types of sterility:

1) Sterility does not appear before a delay of k generations or: $V_k = \{x/(\forall m_1)\ldots(\forall m_k)\phi(\ldots$ $\ldots\phi(\phi(x,m_1),m_2)\ldots m_k)$ defined$\}$ for $k \geqslant 1$.

2) Sterility appears never, or one can also write:

$$V = \{x/(\forall k \geqslant 1)(\forall m_1)\ldots(\forall m_k)\ \phi(\ldots\phi(\phi(x,m_1), m_2)\ldots m_k) \text{ defined}\}$$

The Tarski-Kuratowski algorithm yields that the prefix of V_k is of the form $\forall\exists$. Therefore $V_k \in \Pi_{(2)}$. Because $V = \{x/(\forall k \geqslant 1) x \in V_k\}$ we see immediately that also $V \in \Pi_{(2)}$ holds.

Theorem 3.1

V_k (for all $k \geqslant 1$) and V are $\Pi_{(2)}$-complete.

Proof: The fixpoint-theorem yields a y_0 with $\phi(y_0,x) = y_0$ for all x. Then $y_0 \in V_k$ and $y_0 \in V$ holds (compare with the proof of theorem 2.2). There is an $h \in R^1_1$ such that:

$$\phi_{h(x)}(y) = y_0 \text{ if } \phi(x,y) \text{ defined}$$
$$= \text{undefined otherwise}$$

Now T (as in theorem 2.2) is $\Pi_{(2)}$-complete and V_k, $V \in \Pi_{(2)}$ for all $k \geqslant 1$. Therefore it is sufficient to show $T \underset{m}{\leqslant} V_k, V$.

Now: $x \in T \Rightarrow (\forall y)\varphi_x(y)$ defined $\Rightarrow (\forall y)\phi_{h(x)}(y) = y_0 \Rightarrow$ $\Rightarrow h(x) \in V$.

$h(x) \in V \Rightarrow (\forall y)\phi_{h(x)}(y)$ defined $\Rightarrow (\forall y)\phi(x,y)$ defined $\Rightarrow x \in T$.

The same steps yield that $x \in T \Leftrightarrow h(x) \in V_k$ for every $k \geqslant 1$. Therefore $(\forall k \geqslant 1) T \underset{m}{\leqslant} V_k$ and $T \underset{m}{\leqslant} V$.

It follows $T \equiv V_k \equiv V$ □

When the progenitor-automaton has a number belonging to V, a special kind of 'evolution' can be established: The automaton reads the first input and the Turing machine of the automaton halts after finitely many steps (because of the definition of V) at a standard-configuration. Then the builder-controller builds the automaton, whose number is printed out by the Turing machine. If the descendent is built, the progenitor-automaton gets the second input and simultaneously the descendent gets the first input. The second automaton constructs the third, whereas the first automaton constructs the fourth. Now the progenitor gets the third input, the second automaton gets the second input and so on... If we assume that the time used for building an automaton is always T, at time TN 2^N automata are constructed.

If we consider for example a Turing machine, which computes a function which is onto, the automaton belonging to this Turing machine can produce all automata (of this type). In the second generation there are sterile automata and the procedure for automata belonging to V must fail in this case. (If a Turing machine does not halt, no further descendents can be produced.) Naturally there are methods to construct all machines in all generations, but such a construction will need more time, because the halting problem is not decidable.

Let $U_k = \{x/(\exists m_1) \ldots (\exists m_k)\phi(x,m_1),\ldots m_k) \in T\}$ ($T = \{x/\phi_x$ total$\}$) and $U = \{x/(\exists k \geqslant 1)(\exists m_1) \ldots$ $\ldots(\exists m_k)\phi(\ldots\phi(x,m_1),\ldots m_k) \in T\}$

U_k: After a delay of k generations 'total fertility' appears.

U: There is such a delay.

Because $T \in \Pi_{(2)}$ it is obvious that U_k, $U \in \Sigma_{(3)}$.

Theorem 3.2

U_k, $U \in \Sigma_{(3)} - (\Sigma_{(2)} \cup \Pi_{(2)})$ (for all $k \geqslant 1$).

Proof:

1) $T \underset{m}{\leqslant} U$; if $f \in P_1^3$ there is a $g \in R_1^2$: $\phi_{g(x,y)}(z) = f(x,y,z)$ (*s-m-n* Theorem of Kleene [7] p.23) and there is an $h \in R_1^1$ $\phi_{h(x)}(y) = g(x,y)$. Therefore $f(x,y,z) = \phi(\phi(h(x),y),z)$.

Let $i_0 \in \{x/\text{domain } (\phi_x) = \emptyset\}$ and

$f(x,y,z) = i_0$ if $\phi(x,z)$ defined
= undefined otherwise

then there is an $h \in R_1^1$: $\phi(\phi(h(x),y),z)$
$= i_0$ if $\phi(x,z)$ defined
= undefined otherwise

Therefore: $x \in \overline{T} \Rightarrow (\exists z)\phi(x,z)$ undefined $\Rightarrow (\exists z)(\forall y)\phi(\phi(h(x),y),z)$ undefined ($\phi(\phi(h(x),y),z)$ is independent of y) $\Rightarrow (\forall y)\phi(h(x),y) \in \overline{T}$. But the choice of i_0 implies $(\forall z)(\forall y)\phi(\phi(h(x),y),z) \in \overline{T}$, even $(\forall z)(\forall y)\phi(\phi(h(x),y),z) \in \overline{U}$ [if $\phi(i,j)$ is undefined we write $\phi(i,j) \in \overline{T}$ because $(\forall z)\phi(\phi(i,j),z)$ is undefined and therefore $\phi_{\phi(i,j)}$ is not total.]

This implies: $h(x)$ can not have total descendents in the second or in a higher generation, when $x \in \overline{T}$. Therefore $h(x) \in \overline{U}$; $h(x) \in \overline{U} \Rightarrow h(x) \in \overline{U}_1 \Rightarrow (\forall y)\phi(h(x),y) \notin T \Rightarrow (\exists z)(\forall y)\phi(\phi(h(x),y),z)$ undefined (because $\phi(\phi(h(x),y),z)$ is independent of y) $\Rightarrow (\exists z)\phi(x,z)$ undefined $\Rightarrow x \in \overline{T}$.

Therefore $\overline{T} \underset{m}{\leqslant} \overline{U}$ and $T \underset{m}{\leqslant} U$.

2) *If* $k < e$ $T \underset{m}{\leqslant} U_k \underset{m}{\leqslant} U_e$

First we show $T \leqslant U_{(1)}$. Consider $h \in R_1^1$: $\phi(h(x),y) = x$ $(\forall y)$.

Then $x \in T \Leftrightarrow (\exists y)\phi(h(x),y) \in T \Leftrightarrow (\forall y)\phi(h(x),y) \in T$ (because $\phi_{h(x)}$ is a constant function).

That means $T \underset{m}{\leqslant} U_{(1)}$

Now: $(\exists m_1) \ldots (\exists m_k)\phi(\ldots\phi(x,m_1),\ldots m_k) \in T \Leftrightarrow$
$\Leftrightarrow (\exists m_1) \ldots (\exists m_k)(\exists y)\phi(\ldots\phi(\phi(h(x),y),m_1) \ldots m_k) \in T$;

therefore $U_k \underset{m}{\leqslant} U_{k+1}$ and $(\forall k)(\forall e)[k < e \Rightarrow U_k \underset{m}{\leqslant} U_e]$

3) $\overline{T} \underset{m}{\leqslant} U_1$ and $\overline{T} \underset{m}{\leqslant} U$. Let i_o as before and h such that:

$\phi(\phi(h(x),y),z)$ = undefined if $T(x,y,z)$
$= i_o$ if $\neg T(x,y,z)$

(compare with point 1) of the proof; $T(x,y,z)$ is defined in Section 1) $x \in \overline{T} \Leftrightarrow (\exists y)(\forall z)\neg T(x,y,z) \Leftrightarrow$ $\Leftrightarrow (\exists y)(\forall z)\phi(\phi(h(x),y),z) = i_o \Leftrightarrow (\exists y)\phi(h(x),y) \in T \Leftrightarrow$ $\Leftrightarrow h(x) \in U_1$ (and $h(x) \in U_1 \subseteq U$).

That means $\overline{T} \underset{m}{\leqslant} U_1$.

$x \in \overline{T} \Rightarrow h(x) \in U$ is demonstrated and:

$h(x) \in U \Rightarrow (\exists y)(\forall z)\phi(\phi(h(x),y),z)$ defined, because $h(x) \notin U_1$ otherwise; but further generations can not be total because ϕ_{i_o} is undefined everywhere, what would imply $h(x) \notin U$. Therefore $h(x) \notin U_1$. It follows $\overline{T} \leqslant U$ because $h(x) \in U_1 \Leftrightarrow x \in \overline{T}$.

4) $U, U_k \notin \Sigma_{(2)} \cup \Pi_{(2)}$: Suppose $U \in \Pi_{(2)}$; then T is $\Pi_{(2)}$-complete and therefore $U \underset{m}{\leqslant} T$. We showed $\overline{T} \underset{m}{\leqslant} U$. It would follow $\overline{T} \underset{m}{\leqslant} T$ which is wrong (see [7] Chapter 14.5). Similarly the assumption $U \in \Sigma_{(2)}$ leads to a contradiction.

The same method yields $U_1 \notin \Sigma_{(2)} \cup \Pi_{(2)}$. Because $U_1 \leqslant U_k$ for all $k \geqslant 1$: $U_k \notin \Sigma_{(2)} \cup \Pi_{(2)}$ □

At the end of the chapter we shall investigate automata which can generate descendants with the ability to generate infinitely many pairwise different descendants. We consider a 'weak' property:

$B_{\chi_o} = \{x/(\exists k \geqslant 1)(\exists m_1) \ldots (\exists m_k)\phi(\ldots\phi(x,m_1) \ldots \ldots m_k) \in I\}$

where

$I = \{x/\text{card}(\text{range}(\phi_x)) = \chi_o\}$.

Theorem 3.3
$B_{\chi_o} \in \Sigma_{(3)} - (\Sigma_{(2)} \cup \Pi_{(2)})$.

Proof: 1) $T \underset{m}{\leqslant} B_{\chi_o}$; Let W_x = domain(ϕ_x) and $\{x/W_x = \phi\} = Y_o$. Then there is a one-one $h \in R_1^1$ with $\phi_{h(x)}(y) =$ undef. $(\forall x)(\forall y)$ (see [4] p.82); furthermore there is a $g \in R_1^1 : \phi(\phi(g(x),y),z) =$
$= h(z)$ if $\neg T(x,y,z)$
$=$ undefined if $T(x,y,z)$

Then $x \in \overline{T} \Rightarrow (\exists y)(\forall z)\neg T(x,y,z) \Rightarrow (\exists y)(\forall z)\phi(\phi(g(x),y),z) = h(z) \Rightarrow (\exists y)\phi(g(x),y) \in I \Rightarrow g(x) \in B_{\chi_o}$
$g(x) \in B_{\chi_o}$ must imply $\phi(g(x),y) \in I$ because $\phi(\phi(g(x),y),z) \in Y_o$ for all z and for all y. In the third generation the machines are completely sterile. Therefore if $\phi(g(x),y) \notin I$, it follows $g(x) \notin B_{\chi_o}$.
Therefore $(\exists y)\phi(g(x),y) \in I \Rightarrow (\exists y)(\overset{\infty}{\forall} z)\phi(\phi(g(x),y),z)$ is defined ($(\overset{\infty}{\forall} z)$ denotes: for infinitely many z) $\Rightarrow (\exists y)(\overset{\infty}{\forall} z)\neg T(x,y,z) \Rightarrow (\exists y)(\forall z)\neg T(x,y,z) \Rightarrow$ (thanks to the definition of the predicate $T(x,y,z)$). That means $\overline{T} \underset{m}{\leqslant} B_{\chi_o}$.

2) $\{x/\text{card}(W_x) = \chi_o\} = L \underset{m}{\leqslant} B_{\chi_o}$; Let h be as in part 1) of the proof. Then there is a $g \in R_1^1 : \phi(\phi(g(x),y),z) = h(z)$ if $\phi(x,z)$ defined
$=$ undefined otherwise.
Then $x \in L \Rightarrow (\overset{\infty}{\forall} z)\phi(x,z)$ def. $\Rightarrow (\overset{\infty}{\forall} z)\phi(\phi(g(x),y),z) =$ $= h(z)$ independent of $y \Rightarrow g(x) \in B_{\chi_o}$ because h is one-one. $g(x) \in B_{\chi_o} \Rightarrow (\overset{\infty}{\forall} z)\phi(\phi(g(x),y),z) = h(z)$ because the third generation is completely sterile, and $\phi(g(x),y) \notin I$ would imply $g(x) \notin B_{\chi_o}$. But then $(\overset{\infty}{\forall} z)\phi(x,z)$ defined and $x \in L$. That means $L \underset{m}{\leqslant} B_{\chi_o}$. Because $L \equiv T$ (see [7] p.264) it follows $T \underset{m}{\leqslant} B_{\chi_o}$. This implies $T \underset{m}{\leqslant} B_{\chi_o}$ and $\overline{T} \underset{m}{\leqslant} B_{\chi_o}$. As in the proof of Theorem 3.2 we conclude $B_{\chi_o} \notin \Sigma_{(2)} \cup \Pi_{(2)}$. One can see immediately that $B_{\chi_o} \in \Sigma_{(3)} (I \in \Pi_{(2)})$ and therefore $B_{\chi_o} \in \Sigma_{(3)} - (\Sigma_{(2)} \cup \Pi_{(2)})$ □

4. Mixed properties

Because the input varies, different phenomena belonging to the descendants may appear. For example, an automaton is self-reproducing with one input and sterile with another input. Now we investigate two such phenomena.

Theorem 4.1
$K \underset{m}{\leqslant} SP_E \cap \overline{V}$ (SP_E as in theorem 2.1, V as in theorem 3.1).

Proof:
Let $i_o \in \{x/\phi_x \text{ onto}\}$. Then there is an $h \in R_1^1$:

$\phi_{h(x)}(y) = i_o$ if $T(x,x,y)$
$=$ undefined if $\neg T(x,x,y)$

$x \in K \Rightarrow (\exists y)T(x,x,y) \Rightarrow (\exists y)\phi(h(x),y) = i_o$. This implies: $(\exists z)(\exists y)\phi(\phi(h(x),y),z) = h(x)$ and $(\exists w)(\exists y)\phi(\phi(h(x),y),w) \in \{x : W_x = \phi\}$ (then $(\exists v)(\exists w)(\ \ y)\phi(\phi(\phi(h(x),y),w),v) =$ undefined) because ϕ_{i_o} is onto.
That means: $x \in K \Rightarrow h(x) \in SP_E \cap \overline{V}$
$h(x) \in SP_E \cap \overline{V} \Rightarrow h(x) \in SP_E \Rightarrow (\exists y)\phi_{h(x)}(y)$ defined $\Rightarrow (\exists y)T(x,x,y) \Rightarrow x \in K$. That means $K \underset{m}{\leqslant} SP_E \cap \overline{V}$ □

Theorem 4.2
$B_{\chi_o} \cap U \in \Sigma_{(3)} - (\Sigma_{(2)} \cup \Pi_{(2)})$. ($U$ as in Theorem 3.2, B_{χ_o} as in Theorem 3.3.)

Proof:
Let $i_o \in SP_A$ (SP_A as in Theorem 2.2) and $h \in R_1^1$, h one-one and $(\forall x)(\forall y)\phi(h(x),y) = i_o$ (compare with the proof of Theorem 3.3.)
consider $g \in R_1^1 : \phi(\phi(g(x),y),z) = h(z)$ if $\neg T(x,y,z)$
undefined if $T(x,y,z)$
a) $x \in \overline{T} \Rightarrow (\exists y)(\forall z)\neg T(x,y,z) \Rightarrow (\exists y)(\forall z)\phi(\phi(g(x),y),z) = h(z) \Rightarrow (\exists y)\phi(g(x),y) \in I \Rightarrow g(x) \in B_{\chi_o}$ (I as in theorem 3.3).
because $h(z) \in T$, $g(x) \in U$ if $x \in \overline{T}$.
$g(x) \in B_{\chi_o} \cap U \Rightarrow g(x) \in B_{\chi_o}$. It must hold: $(\exists y)\phi(g(x),y) \in I$, because in higher generations the descendants are always constant if they exist [$\phi_{h(z)}$ constant for all z and ϕ_{i_o} constant] and therefore not in I. But then $(\exists y)(\overset{\infty}{\forall} z)\phi(\phi(g(x),y),z) = h(z) \Rightarrow$ $\Rightarrow (\exists y)(\overset{\infty}{\forall} z)\neg T(x,y,z) \Rightarrow x \in \overline{T}$. That means: $\overline{T} \underset{m}{\leqslant} B_{\chi_o} \cap U$.

b) Let h be as in part *a)*, L as in Theorem 3.3 and g such that: $\phi(\phi(g(x),y),z) = h(z)$ if $\phi(x,z)$ defined
= undefined otherwise

Then $x \in L \Rightarrow (\overset{\infty}{\forall} z)\phi(x,z)$ defined $\Rightarrow (\overset{\infty}{\forall} z)\phi(\phi(g(x),z) = h(z)$ (independent of y) $\Rightarrow g(x) \in B_{\chi_o} \cap U$ (because $h(z) \in T$, and h one-one). $g(x) \in B_{\chi_o}$. The same argument as in *a)* yields $(\exists y)\phi(g(x),y) \in I$. Therefore $(\exists y)(\overset{\infty}{\forall} z)\phi(\phi(g(x),y),z) = h(z) \Rightarrow (\overset{\infty}{\forall} z)\phi(x,z)$ def. $\Rightarrow x \in L$. That means: $L \underset{m}{\leqslant} B_{\chi_o} \cap U$; as in Theorem 3.3 we conclude $B_{\chi_o} \cap U \in \Sigma_{(3)} - (\Sigma_{(2)} \cup \Pi_{(2)})$. □

If we define $X = \{x | (\exists k)(\exists x_1) \ldots (\exists x_k)\phi(\ldots \phi(x,x_1),\ldots \ldots x_k) \in V\}$, ($V$ as in Chapter 3) we can show with the same method of reduction (even with the same reducing functions), that also $X \cap U \in \Sigma_{(3)} - (\Sigma_{(2)} \cup \pi_{(2)})$.

5. Conclusions

In Sections 2 and 3 we mainly investigated self-reproduction and fertility and we showed, that all these properties are undecidable. Moreover, in many cases the degree of unsolvability to the index-sets of the considered subclasses of automata was found to be greater than 1. This phenomenon is based on the enumerations of the automata considered in this chapter, which are equivalent to Gödel numberings. The existence of self-describing functions in a Gödel numbering [i.e. $(\exists i)(\forall n)\phi_i(n) = i$] guarantees the existence of automata, which are self-reproducing with every input. This fact, based on the recursion theorem of Kleene, is used in the proofs of the Theorems 2.2, 3.1 and 4.2.

Theorem 2.2 shows, that the ability of an automaton to reproduce itself with every input (we called the index set belonging to these automata SP_A) or the ability to produce only descendants belonging to SP_A (i.e. the automaton belongs to SP_A^*) are undecidable properties. Furthermore we found that the degree of SP_A^* coincides with the degree of $\Pi_{(2)}$-complete sets. The set V (index set of automata producing only fertile descendants) considered in Theorem 3.1 has the same degree as SP_A^*. Automata whose index-set belongs to V admit a special method to construct their descendants. This method, which yields an automatic production of descendants of a high 'fertility-degree' is carried out in Section 3.

The ability of an automaton to produce at least one descendant, which has the property to produce only non-sterile descendants, was studied in Theorem 3.2 (corresponding index set: U). This property defines a degree of unsolvability, that is greater than 2. A method applied in the proof of Theorem 3.2 is useful in the research of other phenomena; i.e. for example the property 'B_{χ_o}'. B_{χ_o} is the index-set of automata, which are able to produce at least one descendant, producing infinitely many pairwise different automata. As in Theorem 3.2 (for U) we found, that the degree of B_{χ_o} is greater than 2.

In Section 4 we investigated the ability of automata to produce descendants of two different types; for example: a generation-sequence leads to a sterile automaton and another leads to a self-reproducing automaton. Naturally it is possible, that two properties are incompatible; that means, for example: $SP_A \cap \overline{V} = \phi$ (it is impossible that an automaton reproduces itself with every input and a sterile descendant as well). If we consider only compatible properties, it is conceivable, that the set of automata, which can produce two different types of descendants, is decidable, although the set of automata belonging to one of these types is undecidable (i.e. for example: Is $SP_E \cap V$ decidable, although SP_E and V are undecidable?).

Setting two examples we showed generally, that the above assumption is wrong. If K denotes the 'halting problem' and SP_E is defined as in Section 2 we proved: $K \underset{m}{\leqslant} SP_E \cap \overline{V}$ and $U \cap B_{\chi_o}$ has a degree higher than 2.

Collecting all results of this chapter we get the following sequences of degrees:

$$SP_E,\ SP_A \underset{m}{\leqslant} SP_A^* \equiv V \equiv V_k \underset{m}{\leqslant} U,\ U_k,\ B_{\chi_o},$$

$$B_{\chi_o} \cap U \text{ and } K \equiv SP_E \underset{m}{\leqslant} SP_E \cap V.$$

If we set: $A \underset{m}{<} B \Leftrightarrow A \underset{m}{\leqslant} B$ and not $B \underset{m}{\leqslant} A$

then: $SP_E \underset{m}{<} SP_A^* \equiv V \equiv V_k \underset{m}{<} U,\ U_k,\ B_{\chi_o} \cap U.$

References

1. CASE, J., 'Periodicity in generations of automata', *Math. Systemtheory* 8, 15-31 (1974).
2. CODD, E.F., *Cellular Automata,* Academic Press, New York (1968).
3. NEUMANN, J. von, *Theory of Selfreproducing*

Automata (edited and completed by A.W. Burks), University of Illinois Press, Urbana, Ill. (1966).

4. SCHNORR, C.P., 'Rekursive Funktionen und ihre Komplexität', *Teubner Studienbücher–Informatik,* Stuttgart (1974).
5. CASE, J., 'Recursion theorems and automata which construct', Proceedings of the 1974 Conf. on Biol. Math. Automata Theory, McLean, Virginia (1974).
6. MYHILL, J., 'Abstract theory of selfreproduction', *Essays on Cellular Automata,* University of Illinois Press (editor A.W. Burks), Urbana Ill., pp. 204-218 (1970).
7. ROGERS, H., *Theory of Recursive Functions and Effective Computability,* McGraw-Hill, New York (1967).

Formalisation du concept de système décomposable

MAURICE MILGRAM
Université de Technologie de Compiègne, France

I. Introduction

Dans de nombreux problèmes, on dispose d'un système de grande taille, soit par le nombre de ses éléments, soit par le nombre des relations entre ses éléments. Le système peut être à caractère concret, par exemple une unité industrielle ou un service de transport; le système peut avoir un caractère abstrait: résolution d'un vaste système d'équations linéaires par exemple. Les problèmes qui sont posés concernant un système S ne peuvent pas toujours être résolus globalement. Si on dispose d'une décomposition de S en deux sous-systèmes S_1 et S_2, on souhaite que le problème posé sur S puisse être résolu sur S_1 et S_2 séparément; dans certains cas, on pourra retrouver par 'recollement' une solution pour S. Les méthodes de Dantzig-Wolfe en programmation linéaire illustrent bien cette approche. Il arrive que la décomposition possède plus de 2 niveaux, c'est par exemple le cas dans les solutions hiérarchisées de certains problèmes de commande [1,2].

Dans chaque type de problème, on définira un concept spécifique de décomposition. Si on cherche à faire une synthèse de ces différents concepts, on rencontre une difficulté majeure: les différents problèmes font appel à des structures variées: espaces vectoriels dans le cas de la programmation linéaire, graphes values dans les problèmes d'implantation, etc.,... Un moyen pour tourner cette difficulté est d'adopter un langage plus général, ceci conduit à présenter une définition des systèmes décomposables dans le cadre de la Théorie des Catégories. Nous allons rappeler brièvement les propriétés fondamentales des Catégories [3,4].

II. Rappel sur les catégories

Une catégorie C est donnée par une classe d'objets $Ob(C)$ et une classe de morphismes entre ces objets. Si x, $y \in Ob(C)$ et si f est un morphisme de x vers y (On note $f: x \to y$) on écrira que $f \in \mathrm{Mor}_C(x,y)$. De plus on exige de pouvoir composer les morphismes c'est à dire qu'à tout couple $(f,g) \in \mathrm{Mor}_C(x,y) \times \mathrm{Mor}_C(y,z)$ on sait associer un morphisme $h \in \mathrm{Mor}_C(x,z)$ et on écrit $h = g.f$. Cette opération sur les morphismes doit vérifier:

1. $\forall f,g,h \quad (f.g).h = f.(g.h)$
   dès que $f.g$ et $(f.g)h$ sont définis
2. $\forall x \in Ob(C) \qquad \exists e \in \mathrm{Mor}_C(x,x)$

   $\forall f \in \mathrm{Mor}_C(x,y) \qquad \forall g \in \mathrm{Mor}_C(y,x)$
   $f.e = f$ et $e.f = g$

On notera ce morphisme Id_x et on l'appellera 'unité de x'. On peut résumer ce qui précède en disant qu'une catégorie C possède des objets et des morphismes. Il existe une loi de composition non partout définie mais associative entre morphismes, de plus il existe un élément neutre (unité) par objet de C.

Exemples

1. La catégorie de tous les ensembles, Ens, est la catégorie fondamentale. Les morphismes sont les applications entre ensembles, les unités sont les applications identiques.
2. Soit G la catégorie des objets appelés groupes. Un groupe est un couple $G=(\hat{G}, *)$ où $\hat{G}$ est l'ensemble sous-jacent à G et $*$ une loi de composition ayant certaines propriétés.

On dira que $f \in \mathrm{Mor}_{\mathsf{G}}(G,G')$ si:

a. $f \in \mathrm{Mor}_{\mathrm{Ens}}(\hat{G},\hat{G}')$ autrement dit f est une application entre les ensembles sous-jacents.

b. $\forall (u,v) \in \hat{G}\times\hat{G} \qquad f(u * v) = f(u) \circ f(v)$
(On a posé: $G' = (\hat{G},\circ))$

Les homomorphismes de groupe ne sont pas des applications entre groupes mais entre ensembles sous-jacents. En général en confond G et $\hat{G}$ ce qui est gênant lorsqu'un ensemble peut être muni de plusieurs lois distinctes.

3. La catégorie θ des couples $(E,\leqslant)$ d'ensembles ordonnés; un morphisme f de $(E,\leqslant)$ vers $(E',\leqslant')$ est une application de E vers E' telle que:

$$\forall (u,v) \in E \times E \qquad u \leqslant v \Rightarrow f(u) \leqslant' f(v)$$

Ces trois exemples montrent la généralité du concept de catégorie. On définit la notion de sous-catégorie en restreignant la classe des objets et/ou des morphismes. La catégorie θ_t des ensembles totalement ordonnés est une sous-catégorie de θ. La catégorie G_C des groupes commutatifs est une sous-catégorie de G. Nous dirons que deux objets $x,y \in \mathrm{Ob}(C)$ sont isomorphes s'il existe $f \in \mathrm{Mor}_C(x,y)$ et $g \in \mathrm{Mor}_C(y,x)$ tels que:

$$f.g = \mathrm{Id}_y \quad \text{et} \quad g.f = \mathrm{Id}_x$$

et f et g sont appelés isomorphismes de x à y.

Exemples

1. Les objets isomorphes de Ens sont des ensembles de même cardinal; les isomorphismes sont des bijections.
2. Les isomorphismes de θ sont des bijections croissantes dont la bijection réciproque est aussi croissante. Remarquons que dans θ_t les bijections croissantes sont toutes des isomorphismes ce qui explique l'intérêt des ensembles totalement ordonnés.

Dans une catégorie C, un objet x_0 est dit terminal si pour tout $x \in \mathrm{Ob}(C)$, $\mathrm{Mor}_C(x,x_0)$ ne possède qu'un seul élément.

Exemples

1. Dans Ens, les objets terminaux sont les singletons.
2. Dans G, les objets terminaux sont les groupes réduits à leur élément neutre.

On a le résultat suivant:

Lemme: Tous les objets terminaux d'une catégorie C sont isomorphes entre eux.

Preuve: Soit x_1 et x_2 deux objets terminaux et les f_i uniques:

$$f_1 : x_1 \to x_1 \quad f_2 : x_1 \to x_2 \quad f_3 : x_2 \to x_1$$

$$f_4 : x_2 \to x_2$$

Puisque $\mathrm{Id}_{x_1} \in \mathrm{Mor}(x_1,x_1)$ on a $f_1 = \mathrm{Id}_{x_1}$ et de même $f_4 = \mathrm{Id}_{x_2}$.

De plus: $\quad f_3 \cdot f_2 = \mathrm{Id}_{x_1}$

$$f_2 \cdot f_3 = \mathrm{Id}_{x_2}$$

donc f_2 est un isomorphisme de x_1 à x_2.

Dans une catégorie C, les morphismes permettent les comparaisons entre les objets de $\mathrm{Ob}(C)$. Un moyen de construire de nouveaux objets à partir d'autres est le suivant:

Soient $x,y \in \mathrm{Ob}(C)$ un objet $p \in \mathrm{Ob}(C)$ et deux morphismes $\Pi_1 : p \to x$ et $\Pi_2 : p \to y$. On dira que (p,Π_1,Π_2) est un produit de x et y et on notera

$$p = x \times y$$

dans C si:

$\forall(p',\Pi'_1,\Pi'_2) \qquad \Pi'_1 : p' \to x \qquad \Pi'_2 : p' \to y$

$\exists!\ h : p' \to p$ tel que:

$\Pi'_1 = \Pi_1 . h \qquad \Pi'_2 = \Pi_2 . h$

Ces propriétés qui caractérisent $p = x \times y$ sont dites 'universelles' et nous verrons plus loin pourquoi. Le diagramme suivant résume ce qui précède.

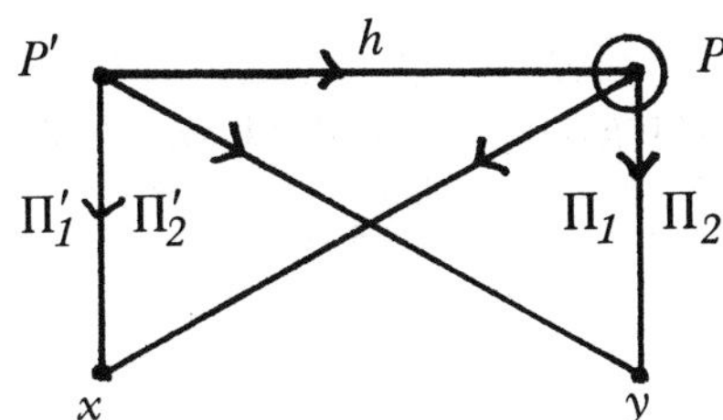

Un triangle du type:

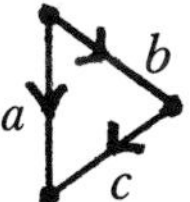

signifie: $a = c . b$

Exemples

1. Dans la catégorie Ens le produit est tout simplement le produit cartésien et Π_i est la i-ième projection

$P = (A \times B) \qquad \Pi_1(a,b) = a \qquad \Pi_2(a,b) = b$

2. Dans la catégorie G le produit de $G_1 = (\hat{G}_1, \circ_1)$ et $G_2 = (\hat{G}_2, \circ_2)$peut être pris comme $P = (\hat{G}_1 \times \hat{G}_2, \circ)$ avec:

$$(x_1,x_2) \circ (y_1,y_2) = (x_1 \circ_1 y_1,\ x_2 \circ_2 y_2)$$

et les projections Π_1 et Π_2 sont les mêmes que dans Ens. On vérifie aisément que c'est un produit et en particulier que Π_1 et Π_2 sont des homomorphismes de groupes.

Nous allons montrer que la construction du produit dans une catégorie obéit à un schèma général, celui d'objet universel. Considérons C une catégorie et $C(a,b)$ la catégorie des Triplets (x,Π_1,Π_2) où $\Pi_1 : x \to a$, $\Pi_2 : x \to b$ et $x \in \mathrm{Ob}(C)$. Les morphismes de $C(a,b)$ entre (x,Π_1,Π_2) et (x',Π'_1,Π'_2) sont des morphismes h de C entre x et x' tels que

$\Pi'_2 . h = \Pi_2$ et $\Pi'_1 . h = \Pi_1$

On vérifie que $C(a,b)$ est bien une catégorie (notamment que si h et h_1 sont des morphismes composables de $C(a,b)$, alors $h . h_1$ est un morphisme de $C(a,b)$.) Si on reprend maintenant la définition du produit dans C des objets a et b, on voit qu'elle coincide avec celle d'objet terminal de $C(a,b)$. Puisque, d'après le lemme précédent, tous les objets terminaux sont isomorphes, on a:

Lemme: Soit C une catègorie et $a,b \in \mathrm{Ob}(C)$, tous les produits de a et b dans C sont isomorphes.

On peut maintenant étendre la définition du produit à une famille indexée d'objets de C . Soit $(a_i \mid i \in I)$ une telle famille. Considérons la catégorie $C(a_i \mid i \in I)$ dont les objets sont de la forme $[x\,;(\Pi_i \mid i \in I)]$ avec $x \in \mathrm{Ob}(C)$ et $\Pi_i \in \mathrm{Mor}_V(x,a_i)$ $\forall i \in I$ et dont les morphismes entre un objet $[x\,;(\Pi_i \mid i \in I)]$ et un objet $[x'; (\Pi'_i \mid i \in I)]$ sont les morphismes $h : x \to x'$ de C tels que:

$\forall i \in I \qquad \Pi_i = \Pi'_i . h$

Par définition, un produit de la famille $(a_i \mid i \in I)$ sera un objet terminal de $C(a_i \mid i \in I)$.

Nous allons maintenant introduire un nouveau mode de construction des objets; en inversant le sens des morphismes dans les définitions nous définissons les coproduits appelés aussi sommes dans une catégorie quelconque.

Précisément soit $(a_i \mid i \in I)$ une famille d'objets de C. Une somme de $(a_i \mid i \in I)$ est un couple $[s\,;(\sigma_i \mid i \in I)]$ avec $s \in \mathrm{Ob}(C)$ $\sigma_i : a_i \to s$ $\forall i \in I$ et tel que pour tout autre $[s'\,;(\sigma'_i \mid i \in I)]$ vérifiant la même relation, il existe un unique $h : s \to s'$ tel que:

$\forall i \qquad \sigma'_i \doteq k . \sigma_i$

(Il suffit d'inverser le sens des flèches dans la définition du produit.)

Exemples

1. Dans Ens la somme d'une famille $(E_i \mid i \in I)$ d'ensembles est $(s\,;(f_i \mid i \in I)$ avec $s = \bigcup_{i \in I} E_i \times \{i\}$

et $\quad f_i(x) = (x, \{i\})$

C'est donc une réunion disjointe des ensembles E_i, même s'il existe i et j avec $E_i \cap E_j \neq \emptyset$ on aura bien

$(E_i \times \{i\}) \cap (E_j \times \{j\} = \emptyset \qquad \forall i \neq j$

2. Il existe également une somme dans G et on

montre que si l'on fait la somme ou le produit d'un nombre fini d'objets de G, on retrouve deux objets isomorphes. Comme avec les produits, les sommes peuvent être présentées en tant qu'objets terminaux dans une catégorie 'bien choisie'. Ceci établit que toutes les sommes sont isomorphes entre elles.

Les sommes et produits permettent déjà de fabriquer de nombreux objets dans une catégorie, néanmoins nous allons encore présenter une nouvelle construction. Comme solution d'un nouveau schèma universel (objet terminal) cette construction va nous conduire au concept d'objet décomposable.

Soit C une catégorie; considerons un système noté ainsi:

$\tau = [b\,; (a_i, t_i) \mid i \in I]$ avec:

$b \in \mathrm{Ob}(C) \quad \forall i \in I \quad a_i \in \mathrm{Ob}(C)$ et $t_i \in \mathrm{Mor}_C(b, a_i)$

la famille $((a_i, t_i) \mid i \in I)$ est appelée le corps du système et l'objet b le résidu du système.

Considérons la catégorie $C(\tau)$ dont les objets sont de la forme $(s\,; v_i \mid i \in I)$ avec: $s \in \mathrm{Ob}(C)$ et $v_i : a_i \to s$ et tels que: $\forall i \in I$

$$v_i . t_i = v_j . t_j$$

et dont les morphismes entre un objet $(s\,; v_i \mid i \in I)$ et un autre $(s'\,; v'_i \in I)$ est un morphisme $h : s \to s'$ de C tel que $\forall i \in I$

$$v'_i = h . v_i$$

Le diagramme ci-dessous correspond à $I = \{1,2\}$

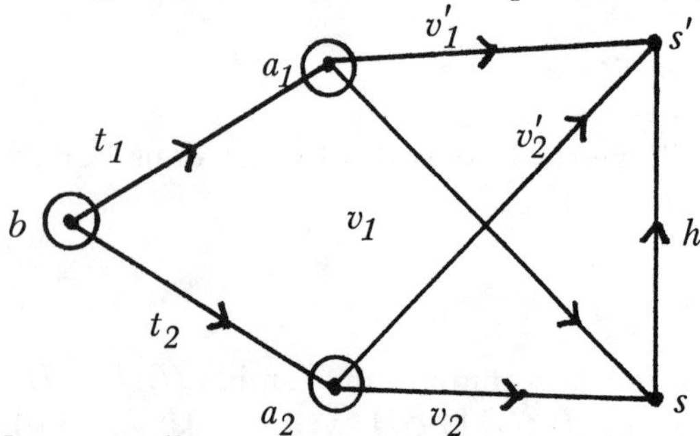

Nous appellerons somme amalgamée du système τ un objet terminal de $C(\tau)$. Si on supprime l'objet b du système, on retrouve la définition d'une somme dans C. L'intérét d'une somme amalgamée est de permettre des modifications internes à l'objet construit independamment de la structure en cause.

Exemple
Prenons encore la catégorie Ens. Soit deux ensembles A et B, Soit (K, t_A, t_B) avec:

$$t_A : K \to A \qquad t_B : K \to B$$

Considérons l'ensemble $S = A \cup B/\mathbf{R}$ où $\mathbf{R}$ est la relation d'équivalence engendrée par la relation binaire $\mathbf{R}_1$ définie par:

$$x \mathbf{R}_1 y \Leftrightarrow \exists b \in K \quad x = t_A(b) \text{ et } y = t_B(b)$$

Considérons les applications $v_A : A \to S$ et $v_B : B \to S$ définies par:

$$\begin{cases} v_A(x) = \mathbf{R} \text{ —classe de } x \text{ dans } S \\ v_B(x) = \mathbf{R} \text{ —classe de } x \text{ dans } S \end{cases}$$

On peut vérifier que (S, v_A, v_B) est bien une somme amalgamée du système de résidu K et de corps la famille $[(A,B); (t_A, t_B)]$.

III. Concept d'objet decomposable

III.1 Définition
Nous disposons maintenant des outils permettant de définir le concept d'objet décomposable dans une catégorie quelconque sur laquelle est définie la somme amalgamée.

Soit C une catégorie. Considérons 2 ensembles $A \subseteq \mathrm{Ob}(C)$ et $R \subseteq \mathrm{Ob}(C)$. Nous dirons qu'un objet $x \in \mathrm{Ob}(C)$ est (A,R)-décomposable si il existe un système

$S = [r\,; (a_i, t_i) \mid i \in I]$ tel que:

1. x est une somme amalgamée du système
2. $\forall i \quad a_i \in A$ (A est appelée classe des atomes)
3. $r \in R$ (R est appelée classe des résidus)

Nous définirons aussi la dimension de la décomposition comme étant le cardinal de I:

$$\mathrm{Dim}(S) = \mathrm{Card}\, I$$

Un objet peut éventuellement avoir des décompositions de dimensions différentes.

Exemple

C = catégorie θ des ensembles ordonnés
A = ensembles totalement ordonnés de θ
R = ensembles discrets, un élément n'est comparable qu'à lui-même.

On démontre que tout ensemble ordonné est (A,R)-décomposable'

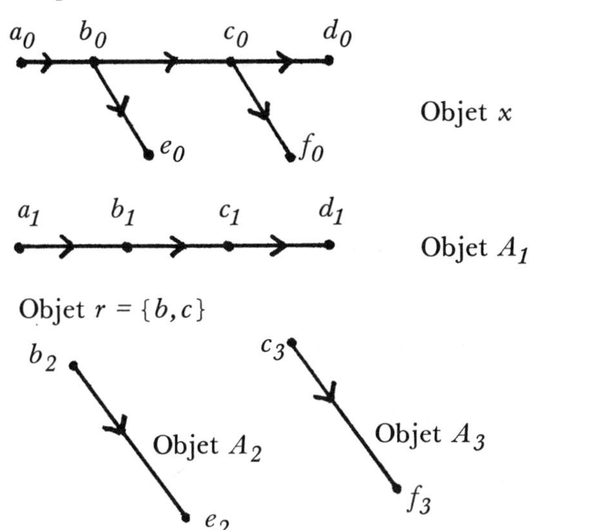

$i = 1,2 \quad t_i(b) = b_i$

et

$t_i(c) = c_i$ $\quad r, t_1, t_2$ recollent les 3 objets A_1, A_2, A_3 pour faire x

Les morphismes des A_i vers x sont définis par:

$$v_i : A_i \to x \qquad v_i(x_i) = x_0 \qquad \begin{cases} x = a,b,c,d,e,f \\ i = 1,\ 2 \end{cases}$$

Les dimensions minimales d'une décomposition de x sont égales au nombre d'éléments de x deux à deux incomparables d'après le théorème de Dillworth.

III.2 Application à la catégorie des graphes valués

III.2.1 Construction

Parmi toutes les structures mathématiques utilisées pour formaliser le concept de système, la structure de graphe valué est une desplus riche. Que le système soit un réseau de transport d'énergie, une organisation humaine, un calculateur électronique, une unité de production industrielle, on peut en avoir une image sous forme d'un graphe valué. La pertinence de la représentation est liée aux choix des sommets, des arcs et de la valuation du graphe.

Si on considère un système d'équations linéaires ou non, par exemple:

$$f_i(x_1, x_2, \ldots, x_n) = 0 \qquad \text{pour } i = 1, \ldots, k$$

On peut lui associer le graphe G dont les sommets sont les variables x_i, deux sommets x_i et x_j constitueront une arête si x_i et x_j figurent simultanément dans une même équation f_r.

On pourra pondérer l'arête (x_i, x_j) par un nombre relié à la sensibilité de f_r pour de petites perturbations de x_i et x_j. On aurait pu aussi considérer le graphe dont les sommets sont les équations f_i, deux équations étant reliées si elles comportent une même inconnue.

Exemple:

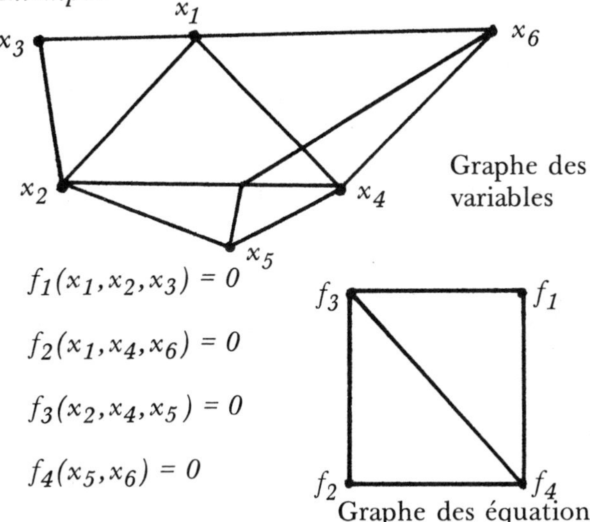

Graphe des équations

Nous considérons d'abord la catégorie des graphes puis nous étendrons sans difficulté nos constructions à celle des graphes valués.

Soit donc **G** la catégorie des graphes. Ob(**G**) est constitué des objets $G = (X,U)$ où U est une partie de $X \times X$. Un morphisme f de l'objet $G = (X,U)$ vers l'objet $G' = (X', U')$ est une application de X vers X' telle que:

$$\forall (x,y) \in U \qquad (f(x), f(y)) \in U'$$

Soit $G_i = (X_i, U_i)$ pour $i \in I$ une famille de graphes et $R = (X,U)$ un résidu. Soit $t_i : R \to G_i$ une famille de morphismes.

On obtient un système dans **G**:

$$\tau = [R\,;(G_i, t_i) \mid i \in I]$$

Ce système admet une image dans Ens, catégorie des ensembles, si on associe à tout graphe l'ensemble de ses sommets.

$$\hat{\tau} = [X\,;(X_i, t_i) \mid i \in I]$$

Soit Y la somme amalgamée ensembliste de $\hat{\tau}$ et

$\sigma_i : X_i \to Y$ les morphismes qui lui sont attachés.

Notons $\sigma_i \times \sigma_i$ l'application de U_i dans Y^2 définie par:

$$\sigma_i \times \sigma_i(a,b) = (\sigma_i(a), \sigma_i(b))$$

Posons alors:

$$V = \bigcup_{i \in I} \sigma_i \times \sigma_i(U_i)$$

On peut alors vérifier que $S = (Y,V)$ muni de morphismes $\sigma_i : G_i \to S$ est bien une somme amalgamée de τ dans la catégorie des graphes **G**. Nous allons étendre cette construction aux graphes valués. Nous noterons $\mathbf{G}^V$ la catégorie des graphes valués. Un objet de $\mathbf{G}^V$ est un graphe valué:

$G^V = (X, U, f, g)$ où $\mathbf{G} = (X,U)$ est le graphe sous-jacent

et $f : X \to R^+$, $g : U \to R^+$ sont deux fonctions de pondération respectivement pour les sommets et pour les arcs de G.

Un morphisme h de $G^V = (X,U,f,g)$ vers $G_1^V = (X_1,U_1,f_1,g_1)$ est un morphisme de G vers G_1 qui vérifie de plus:

1. $f_1 \circ h = f$
2. $\forall (a,b) \in U \qquad g_1[h(a), h(b)] = g(a,b)$

La construction de la somme amalgamée S d'un système

$$\tau = [R^V, (G_i^V, t_i) \mid i \in I] \text{ avec } G_i^V = (X_i, U_i, f_i, g_i)$$

est rigoureusement identique à celle des graphes non valués.

Il faut définir sur $S = (Y,V)$ des valuations f_S et g_S. Ces valuations doivent être telles que les morphismes de graphe σ_i de G_i vers S deviennent des morphismes valués de G_i^V vers S^V.

Hors, si des éléments $x^i \in X_i$ et $x^j \in X_j$ ont la même image y dans Y, on peut montrer qu'ils sont imagés par t_i et t_j d'un même élément $r \in R$.

Donc on posera:

$$f_S(y) = f_i(x_i) = f_j(x_j) = f_R(r)$$

On définit de même:

$$g_S[(y,y')] = g_i(x_i, x_i')$$

III.2.2 Exemples

Voici un exemple de somme amalgamée de 2 graphes (non valués).

$G_1 = (X_1, U_1)$ et $G_2 = (X_2, U_2)$ avec un résidu $R = (Z, W)$ $\; t_1 : R \to G_1 \quad t_2 : R \to G_2$

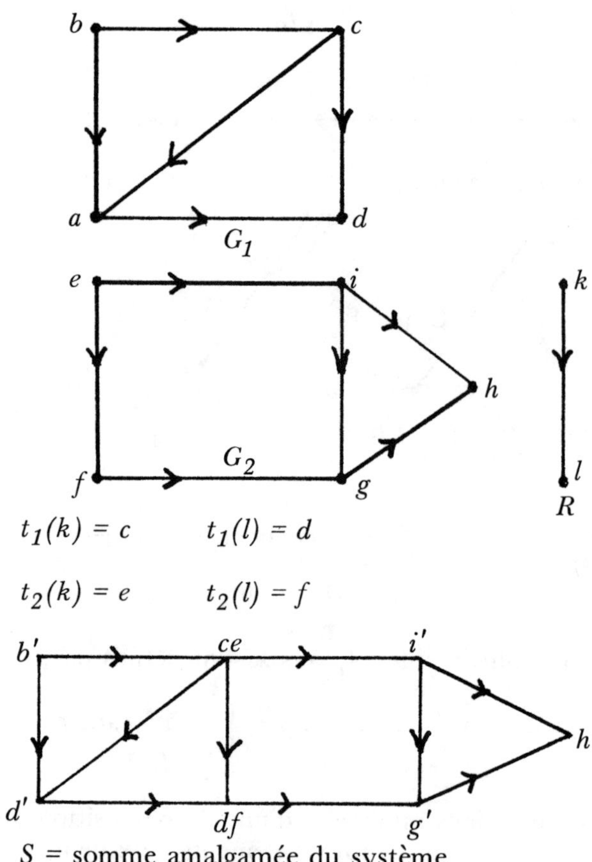

S = somme amalgamée du système

$= [R\,; (G_i, t_i) \mid i = 1,2]$

$\sigma_1(a) = a' \quad \sigma_1(b) = b' \quad \sigma_1(c) = ce \quad \sigma_1(d) = df$

$\sigma_2(e) = ce \quad \sigma_2(d) = df \quad \sigma_2(i) = i' \quad \sigma_2(g) = g'$

$\sigma_2(h) = h'$

En choisissant convenablement l'ensemble des atomes et celui des résidus, on peut reformuler en terme décomposables des problèmes classiques. Par exemple, si la classe des atomes est celle des cliques (graphes tels que 2 sommets quelconques sont liés par un arc) et celle des résidus constituée des graphes sans arcs ($U = \emptyset$), on débouche sur une question classique en théorie des graphes qui est le recouvrement des arcs par un nombre minimum de cliques.

D'autres problèmes, tels que la décomposition d'un graphe en sous-graphes connexes ayant moins de N sommets et rendant minimum le nombre d'arcs joignant deux tels sous-graphes, peuvent s'exprimer en prenant pour atomes les graphes connexes de moins de N sommets et pour résidus des graphes ayant le moins d'arcs possibles.

III.3 Applications aux sysèmes dynamiques linéaires
Nous allons présenter la construction de la somme amalgamée dans une catégorie définie par Arbib et Manes [5].

Nous nous contenterons dans cette construction de manipuler les systèmes dynamiques sans entrées ni sorties. La généralisation aux systèmes dynamiques avec entrée et sortie est aisée. Cette construction permet alors, en choisissant les atomes et les résidus de définir diverses classes d'objets décomposables pour les systèmes dynamiques.

On peut aussi, dans une autre direction, supprimer l'hypothèse de linéarité en se restreignant toutefois aux catégories 'à somme'.

Un système dynamique linéaire (SDL) est un couple $\hat{E} = (E, U)$ où E est un espace vectoriel sur un corps K et u une application linéaire de $E \to E$. Nous noterons Vect_K la catégorie des espaces vectoriels sur K dont les morphismes sont les applications linéaires. Nous noterons S_K la catégorie des SDL dont les objets ont été introduits ci-dessus. Soit $\hat{E}_1 = (E_1, u_1)$ et $\hat{E}_1 \to \hat{E}_2$ est une application linéaire $h : E_1 \to E_2$ telle que $u_2 \circ h = h \circ u_1$

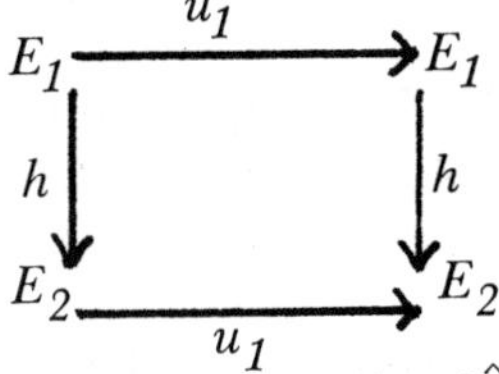

Soit un système de S_K $\hat{\tau} = [\hat{R} ; (\hat{Q}_i, t_i) | i \in I]$ avec:

$$\hat{R} = (R, f) \qquad \hat{Q}_i = (Q_i, f_i)$$

On lui associe un système dans Vect_K le système:

$$\tau = [R ; (Q_i, t_i) \mid i \in I]$$

Nous allons construire la somme amalgamée dans Vect_K.

Considérons $S_1 = \underset{i \in I}{\oplus} Q_i$ la somme directe des espaces vectoriels Q_i munie des injections canoniques $\tau_i : Q_i \to S_1$.

Soit $[N]$ le sous-espace de S_1 engendré par l'ensemble des vecteurs $\{(\tau_i \circ t_i - \tau_j \circ t_j)(r)$ pour $r \in R$ et $i,j \in I$.

On pose $S = S_{1/[N]}$ et $\pi : S_1 \to S$ est la projection canonique d'un espace vectoriel vers son quotient.

Le schèma suivant illustre cette construction:

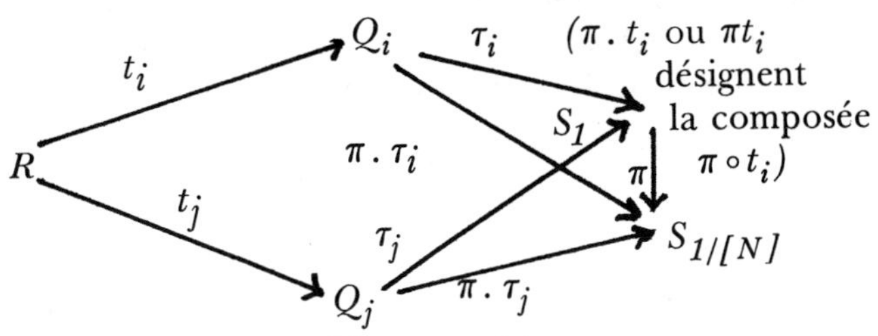

On a: $\forall r \in R \quad \pi \tau_i t_i(r) = \pi \tau_j t_j(r)$

Puisque: $\tau_i t_i(r) - \tau_j t_j(r) \in [N]$

Et donc: $\pi[\tau_i t_i(r) - \tau_j t_j(r)] = 0$

$$\pi \tau_i t_i(r) = \pi \tau_j t_j(r)$$

On peut montrer que l'objet $S = S_{1/[N]}$ muni des morphismes $s_i = \pi . \pi_i$ est bien la somme amalgamée du système τ dans la catégorie Vect_K.

Il faut maintenant passer à la somme amalgamée de $\hat{\tau}$ que nous noterons $\hat{S}$ dans S_K.

Posons $\hat{S} = (S, g)$.

On doit avoir: $\forall i \in I \qquad g . s_i = s_i . f_i$

Puisque $\hat{Q}_i = (Q_i, f_i)$ et $S_i : \hat{Q}_i \to \hat{S}$

Si $x = \pi(y)$ on a $y = \sum_{i \in I} \tau_i(q_i)$

Posons $g(x) = \sum_{i \in I} s_i . f_i(q_i) = \pi \{ \sum_{i\, I} \tau_i . f_i(q_i) \}$

Supposons que

$$x = \pi(y') \quad \text{avec} \quad y' = \sum_{i \in I} \tau_i(q_i')$$

La construction de $g(x)$ à partir de y' donnera le même élément de S puisque $y - y' \in [N]$.

Il faut vérifier que:

$$\forall i \quad \forall q_i \in Q_i \qquad g . s_i(q_i) = s_i f_i(q_i)$$

On a: $s_i(q_i) = \pi \tau_i(q_i)$

Donc $g s_i(q_i) = s_i f_i(q_i)$ par construction de $g(x)$ avec $x = s_i(q_i)$ et $y = \tau_i(q_i)$.
Nous avons donc construit un objet $\hat{S} = (S, g)$ muni de morphismes $S_i : \hat{Q}_i \to \hat{S}$ dans S_K.
Nous admettrons que S est encore solution du problème universel, ce résultat pouvant être vérifié par inspection méthodique des conditions à satisfaire. Nous avons ainsi montré que la catégorie S_K

ayant des sommes amalgamées, permet de définir les objets décomposables.

Un exemple de choix d'atomes consiste à prendre les systèmes linéaires mono-dimensionnels. Nous ne pouvons néanmoins aborder ici de manière approfondie le problème du choix des atomes et des résidus.

IV Conclusion

Si nous résumons le contenu du travail qui vient d'être exposé, nous pouvons dire que nous avons présenté une méthodologie pour définir, quelque soit la structure sous-jacente, le concept de système décomposable. Il nous semble, d'autre part, que l'on peut essayer de prolonger cette recherche dans plusieurs directions:

1. Approfondir le support mathématique c'est à dire, chercher en quoi les axiomes fondamentatux peuvent être améliorés.
2. Préciser, pour des structures particulières, la classe des objets décomposables lorsqu'on se fixe une classe d'atomes et une classe de résidus. Dans ce sens, les systèmes dynamiques avec entrée et sortie peuvent offrir un premier terrain d'exploration, en particulier, il semble intéressant d'examiner la notion d'interaction entre sous-systèmes dans l'optique des catégories. Si on choisit pour atomes des sous-systèmes stables (resp. observables, commandables) il peut être instructif de construire la classe des systèmes décomposables.
3. Examiner ce que devient la classe des objets décomposables lorsqu'on modifie la catégorie de référence. Par exemple, si on passe des morphismes de graphes valués (catégorie des graphes valués) à des morphismes de graphes planaires. (Utilisés dans les problèmes de connexions électroniques). Plus généralement, préciser les modifications de concept de décomposition lorsqu'on enrichit ou appauvrit la catégorie en question, du point de vue des morphismes et des objets.

Le travail que nous avons présenté pourra sembler très abstrait et donc éloigné de la méthodologie de l'analyse des systèmes.

En fait, il nous semble souhaitable de pouvoir, parfois, replacer certaines notions dans un cadre plus général.

Notre définition des objets (systèmes) décomposables permet d'aborder les modèles linéaires et les modèles combinatoires. En réalité, nous pensons tenir ici une méthode permettant une approche de la décomposition des systèmes pour une classe très large de systèmes, tous ceux qui sont modélisables à l'aide d'une structure 'riche'. Par structure riche, nous entendons une structure produisant une catégorie où se laisse définir la somme amalgamée. C'est le cas des structures de graphes valués et de systèmes dynamiques linéaires (de dimension finie ou non). Pour chaque problème concret, si on a défini un modèle dans une catégorie à somme amalgamée, il reste à choisir les atomes et les résidus. Ce choix doit refléter la spécificité du problème, c'est à travers ce choix que l'on essaie de retrouver les notions, très floues, de simplicité et de complexité.

Bibliographie

1. MESAROVIC, MACKO, TAKAHARA, *Theory of hierarchical multilevel systems,* Academic Press (1970).
2. DANTZIG, WOLFE, 'Decomposition principle for linear programs', *Operat. Res.,* **8**, 101-111 (1960).
3. ERHESMANN, C., *Cours d'algèbre C3,* C.D.U. Paris (1969).
4. LANG, S., *Algebra,* Addison Wesley.
5. ARBIB, MANES, 'Foundations of system theory: decomposable systems', *Automatica,* 10, 285-302, Pergamon Press (1974).

Abstract

A formalisation of the concept of decomposable systems

Maurice Milgram

Many authors have proposed different methods of system decomposition. The various definitions are not always compatible. I would like to put forward the formalisation of the concept of the decomposable system.

For this reason I would like to use the language and methods of the theory of categories. Certain authors such as Arbib have already made use of the theory of categories in order to obtain a general theory of dynamic systems, linear or not.

I would like to study two particular classes of objects. The first class is made up of atoms, simple objects which will serve as parts to construct more complex ones. The second class is made up of objects called residues, these objects serve to assemble the atoms in between them. In order to give a precise meaning to this construction, one can make use of the concept of 'blended sum'. The decomposable objects of a category are these which one can construct as 'blended sum' of a family of atoms and of residues of that particular category.

I would like then to deal with one category in particular: that of valued graphs. I would like to take up once again some accepted results of the theory of graphs, but viewed under a different light. I would like to show how atoms and residues can be chosen and what type of valued decomposable graph can be so produced.

I would like then to examine various criteria which lead respectively to particular decompositions of valued graphs and I would try to show how one can present them in a more unified fashion.

Observability, controllability and duality of general input-output systems

E. PFAFFELHUBER
Institute for Information Sciences,
University of Tübingen, Germany (FRG)

I. Introduction and summary

The concepts of observability, controllability and duality are powerful notions in linear systems theory, the main reason being Kalman's duality theorem which says that a linear system is controllable (observable) iff the dual system is observable (controllable) (see e.g. [1-3]). It is very surprising therefore that, as far as the author is aware, no serious attempt has yet been made to carry this theorem over to the case of non-linear systems, or to the framework of general systems theory (cf. [4-7]). The book by Bucy and Joseph [8] contains a definition of observability and controllability for the case of non-linear differential systems; however, this definition is not used subsequently in the book nor does the latter contain any concept of duality for the non-linear case. A similar comment applies to the papers by Brockett [9] and by Sussmann and Jurdjevic [10,11] which contain definitions and also useful criteria for observability and controllability in the non-linear or bilinear case, but again the notion of duality, and, therefore, an analog of Kalman's duality theorem, is missing. Of course, the full generalization of the above concepts to the general non-linear case is difficult and perhaps impossible, and will not be attempted here. Instead we consider the simple case of a static functional input-output system only. This restriction is possibly not too severe, since, in a certain sense, not much more has been achieved in linear systems theory. Namely, in the latter, the concept of duality is defined by means of certain matrices, which represent, from a basic viewpoint, linear static mappings from certain vector spaces into others. Also it is known that (at least if no restrictions are imposed on the inputs, outputs and system states) linear systems are either observable (controllable) in an arbitrarily short time (instantaneously, so to speak), or not at all. Hence, the notions of controllability and observability are concerned, from a basic viewpoint, with instantaneous (i.e. static) functional relations between two sets, namely the input set and the state space, and the state space and the output set, respectively. Thus it seems justified to study, as a first approach to the non-linear case, static functional systems which involve two sets only, and thus may be considered as input-output systems.

To this end, an apparently quite natural approach

seems the following. Call the triple

$$\Sigma = (X,Y,R)$$

a system iff X,Y are sets, and R is a relation on, i.e. a subset of, the cartesian product $X \times Y$ (see]4-6]), the idea being that $(x,y) \in R$ iff y is a possible system output generated by the input x. It is natural to call the system observable iff to every $y \in Y$ there is at most one $x \in X$ such that $(x,y) \in R$ (so that the system input can be uniquely determined from its output), and to call the system controllable iff to every $y \in Y$ there exists at least one $x \in X$ such that $(x,y) \in R$ and $(x,y') \in R$ implies $y = y'$ (so that to each possible system output there is at least one system input which generates solely that output). In other words, the system would be called controllable iff the relation R is functional and surjective, and observable iff the inverse relation

$$R^{-1} = \{(y,x) : (x,y) \in R\} \tag{2}$$

is functional. It is also natural to call the triple

$$\Sigma^d = (Y,X,R^{-1}) \tag{3}$$

the system dual to (1). If we use the above concepts we can easily prove that the dual system (3) is controllable iff the system (1) is observable and the domain of R is all X. For all those purposes where the concept of a 'system' need not involve a *functional* relation between inputs and outputs, the above definitions and results are quite useful. However, as pointed out before, we are interested in the case of functional systems. Now even if the relation R in (1) is functional, the dual system (3) need not be functional since (2) need not be so. If we attempt to remedy the situation by restricting ourselves from the very beginning to the case where both R and R^{-1} are functional we end up with the rather uninteresting special case where observability holds generally, and controllability holds whenever the range of R is all Y.

Therefore it is clear that the above approach will not yield a satisfactory answer to our problem. This explains why in the following we use a slightly refined definition of a functional input-output system in which the system output is considered itself as a function on a certain set. (A similar situation is encountered in the generalized-function approach to linear systems theory (cf. [12,13]) where both input and output are treated as functionals on certain test function spaces.) This will allow us to introduce the concept of dual systems in such a way that both the system and its dual are automatically functional; also the concepts of observability and controllability can be introduced in a natural way. The main result then is to show that, if certain completeness conditions are satisfied, the observability (controllability) of a system implies (or is equivalent to) the controllability (observability) of the dual system. The case of systems generated by intput-output relations which are functional or have a functional inverse (so-called systems of type $\mathcal{N}$ and $\mathcal{S}$) are considered as an illustration.

II. Observability, controllability and duality of general input-output systems

In the following, if M' is a set of K-valued partial functions on a set M, then to each $m \in M$ there is, in a natural way, assigned a K-valued partial function on the set M', which, at the point $m' \in M'$, assumes the value $m'(m)$, and is undefined if $m'(m)$ is undefined. This partial function on the set M' generated by $m \in M$ will be denoted by $\hat{m}$.

We start with the following

Definition 1
The 5-tuple

$$\Sigma = (K,X,Y,Y',f) \tag{4}$$

is an input-output system (in short, a system) iff

a) K,X,Y are sets (X is called the input set),
b) Y' is a set of K-valued partial functions on Y (Y' is called the output set),
c) f is a partial function (called the system function) from X into Y'.

Consider now the following so-called completeness conditions.

C1) If $y_1, y_2 \in Y$ are such that $\hat{y}_1$ and $\hat{y}_2$ are identical (two functions are identical, if they are defined on the same set, and assume the same values on this set), then $y_1 = y_2$.
C2) If Range$(f) \neq Y'$, then there are two different elements $y_1, y_2 \in Y$ such that $\hat{y}_1$ and $\hat{y}_2$ are identical on Range(f).

Condition *C1)* guarantees that there are sufficiently many functions in Y' in the sense that, if the values of all these functions at the points $y_1, y_2 \in Y$ are identical (or undefined), then this can happen only if the points y_1, y_2 are identical. Condition *C2)*, on the other hand, requires, that there are also sufficiently many elements in Y; namely, if the functions $f(x)$ (which are elements of Y') where x varies over X, do not exhaust the whole set Y', then this should be testable by two appropriate points y_1, y_2 of Y, at which solely the functions $y' \in \text{Range}(f)$ (but not all possible $y' \in Y'$) assume the same value (or are undefined).

Let us next define the notions of observability and controllability of systems.

Definition 2
A) The system (4) is input observable (in short, observable), iff to every $y' \in Y'$ there is at most one $x \in X$ such that $f(x)$ is defined and $f(x) = y'$.
B) The system (4) is output controllable (in short, controllable) iff to every $y' \in Y'$ there is at least one $x \in X$ such that $f(x)$ is defined and $f(x) = y'$.

Part *A)* of the definition says that Σ is observable iff the system input can be uniquely determined from its output, which means that f is injective. Similarly, according to part *B)* of the definition, Σ is controllable iff by a suitable choice of the input any element of the output set can actually be generated as an output, which means that f is surjective. Note that the above concepts of observability and controllability are dual notions in the sense of category theory since the concepts of injectivity and surjectivity are (cf. [14]).

It seems worthwhile to point out the following. If, for the system (4) we set

$$h(x,y) = f(x)(y) \tag{5}$$

then h is a partial function from $X \times Y$ into K. Conversely, if we are given such a function h, and if we define a system function f by (5), i.e.

$$f(x) = h(x,.) \tag{6}$$

then f is a partial function from X into the set of functions

$$\{h(x,.) : x \in X\} \tag{7}$$

or any set Y' that contains (7). In this case we shall say that Σ is generated by h. Our above considerations may now be summarized by saying that any system can be looked upon as being generated by an appropriate function h, namely the function h from (5).

Next we introduce the concept of dual systems according to

Definition 3
The systems $\Sigma = (K,X,Y,Y',f)$ and $\Sigma^d = (L,U,V,V',g)$ are dual to each other iff

a) $L = K$, $U = Y$, $V = X$,
b) for $x \in X$, $y \in Y$, we have

$$f(x)(y) = g(y)(x)$$

in the sense, that, whenever one side is defined the other is defined, too, and assumes the same value.

If Σ and Σ^d are dual systems generated by h and h^d, respectively, then it is immediately clear from (5) that

$$h(x,y) = h^d(y,x) \tag{8}$$

whenever one side is defined. Thus, if h and h^d are considered as abstract matrices with components $h(x,y)$ and $h^d(y,x)$, respectively, then the dual matrix h^d is simply the transpose of h.

It is easy to see that if we take K equal to the real or complex numbers; X and Y as finite dimensional vector spaces over K with basis elements e_i and f_j, respectively; X', Y' as the corresponding dual spaces with dual basis vectors r^i and t^j, respectively; and, furthermore, f as a linear function from X into Y',

$$f = \sum_{i,j} a_{ij} t^j \otimes r^i$$

(where the a_{ij} are elements of K), then the system function of the dual system is given by

$$g = \sum_{i,j} a_{ji} r^j \otimes t^i$$

It appears that it is this linear case which is at the basis of the duality concept in linear systems theory. (note that finite dimensional vector spaces are automatically self-dual, which introduces an additional simplification in the linear case).

Now we state

Theorem 1
Let $\Sigma = (K,X,Y,Y',f)$ and $\Sigma^d = (K,Y,X,X',g)$ be dual systems, which both satisfy the completeness conditions C1) and C2). Then Σ is controllable (ob-

servable) iff Σ^d is observable (controllable).

Proof: It is obviously sufficient to prove the equivalence between the controllability of Σ and the observability of Σ^d. The following statements are equivalent.

a. Σ is not controllable.
b. Range $(f) \neq Y'$.
c. $\exists y_1, y_2 \in Y$ such that $y_1 \neq y_2$ and $\hat{y}_1, \hat{y}_2$ are identical on Range (f).
d. $\exists y_1, y_2 \in Y$ such that $y_1 \neq y_2$ and $\hat{y}_1 \circ f, \hat{y}_2 \circ f$, are identical on X.
e. $\exists y_1, y_2 \in Y$ such that $y_1 \neq y_2$ and $g(y_1), g(y_2)$ are identical on X.
f. Σ^d is not observable.

Here, the equivalence between *a)* and *b)* and between *e)* and *f)* follows from definition 2, that between *b)* and *c)* from the completeness conditions C1) and C2), and that between *d)* and *e)* from condition *b)* in definition 3. [The symbol $\circ$ in the above statement *d)* denotes the composition of functions.] The proof is complete.

A more careful examination of which kind of completeness condition is needed at what part of the proof reveals that we have

Theorem 2
We have the following implications.

a. Σ controllable, Σ satisfies C1) $\Rightarrow \Sigma^d$ observable.
b. Σ observable, Σ^d satisfies C2) $\Rightarrow \Sigma^d$ controllable.
c. Σ^d controllable, Σ^d satisfies C1) $\Rightarrow$
$\Rightarrow \Sigma$ observable
d. Σ^d observable, Σ satisfies C2) $\Rightarrow \Sigma$ controllable.

Let us consider a special case of an input-output system, namely a system generated by an input-output characteristic R. By this we mean the following. K should contain the two elements 0 and 1 (the decisive thing, of course, is that K possesses at least two elements), R is a relation on $X \times Y$, and we have

$$f(x)(y) = h(x,y) = \chi_R(x,y) = \begin{cases} 1 \text{ for } (x,y) \in R \\ 0 \text{ for } (x,y) \notin R \end{cases}$$

(χ_R is the characteristic function of R), and

$$Y' = \{\chi_R(x,.) : x \in X\}$$

It is easy to see from (2), (8) that the corresponding dual system is generated by the inverse relation R^{-1}.

Two interesting special cases arise if the graph of the relation R is of the 'generic' form as in Figs 1 or 2, respectively.

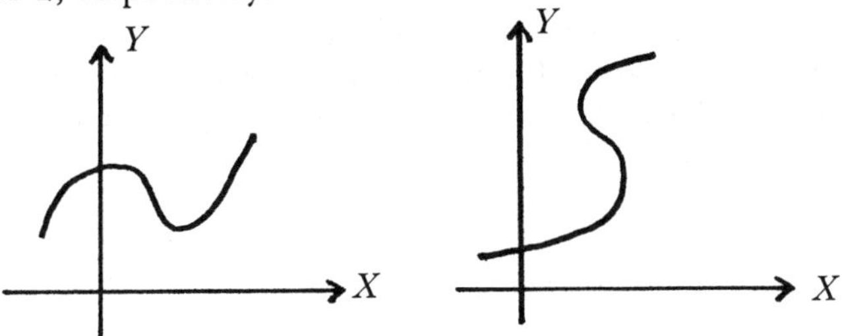

The first case is characterized by the condition that R be functional, so that

$$R = \{(x,\alpha(x)) : x \in X\}$$

where α is a partial function from X into Y. Such a system will be called of type **∿** and generated by α. Here (5), (6) may be written as

$$h(x,y) = \delta(y,\alpha(x))$$

or, equivalently, as

$$f(x) = \delta(.,\alpha(x))$$

where the generalized Kronecker symbol is

$$\delta(\xi,\eta) = \begin{cases} 1 \text{ if } \xi \text{ and } \eta \text{ are both defined and } \xi = \eta \\ 0 \text{ otherwise} \end{cases}$$

In the second case the inverse relation R^{-1} is functional, so that

$$R = \{(\beta(y),y) : y \in Y\}$$

where β is a partial function from Y into X. Such a system will be called of type **S** and generated by β. Here (5), (6) may be written as

$$h(x,y) = \delta(x,\beta(y))$$

or, equivalently, as

$$f(x) = \delta(x,\beta(.))$$

It is an easy matter to verify the following. If $\Sigma = (K,X,Y,Y',f)$ is a system of type **∿** generated by α in the sense of (6), and $\Sigma^d = (K,Y,X,X',g)$ is a system of type $\mathcal{S}$ generated by α according to

$$g(y) = \delta(y,\alpha(.))$$

then Σ and Σ^d are dual systems. Also it is easy to see that a system of type $\mathcal{N}$ generated by α is always controllable. In addition, it satisfies condition C1) iff the range of α essentially exhausts the set Y, with the possible exception of one element, i.e. iff

$$Y - \text{Range}(\alpha) \text{ has at most one element}$$

This can be seen as follows. Condition C1) is satisfied iff the statement

$$\delta(y_1,\alpha(x)) = \delta(y_2,\alpha(x)) \text{ for all } x \in X$$

where $y_1, y_2 \in Y$, implies $y_1 = y_2$. Obviously, this implication holds anyway if y_1 or y_2 belong to the range of α. Thus C1) holds iff there is at most one element in Y which does not belong to Range(α), as was to be shown.

Invoking now the duality between systems of type $\mathcal{N}$ and of type $\mathcal{S}$ and the statement *c)* of Theorem 2, we conclude that a system $\Sigma = (K,X,Y,Y',f)$ of type $\mathcal{S}$ generated by β is observable iff Range(β) essentially exhausts X, i.e. iff

$$X - \text{Range}(\beta) \text{ has at most one element}$$

Of course, in this simple case the above assertion could have been proven directly, without invoking Theorem 2. However, the above considerations indicate that our definitions of duality, controllability and observability appear to be natural ones, since in the special case of systems of type $\mathcal{N}$ and $\mathcal{S}$ they reduce to familiar results which, however, have apparently not yet been formulated in a rigorous way within the framework of general systems theory. (It is, of course, well known to engineers that systems of type $\mathcal{N}$ and $\mathcal{S}$ are typically controllable and observable, respectively.)

Acknowledgements
The author is grateful to Dr. H. Hahn for some helpful and stimulating discussions.

References

1. KALMAN, R.E., *Proc. Nat. Acad. Sci.* **48**, 596 (1962).
2. WEISS, L. and KALMAN, R.E., *Int. J. Engng. Sci.* **3**, 141 (1965).
3. KALMAN, R.E., FALB, P.L., and ARBIB, M.A., *Topics in Mathematical System Theory,* McGraw-Hill (1969).
4. MESAROVIC, M.D., *Math. Systems Theory* **2**, 203 (1968).
5. SALOVAARA, S., *Acta Polytechnica Scandinavica, Mathematics and Computing Machinery Series* No.15 (1967).
6. PICHLER, F., *Mathematische Systemtheorie,* De Gruyter (1975).
7. PFAFFELHUBER, E., 'A note on reversibility and causality of general time systems', in *Progress in Cybernetics and Systems Research,* Vol.I, Hemisphere Publishing Corporation (Proc. Second Meeting on Cybernetics and Systems Research, Vienna, 1974) (1975).
8. BUCY, R.S. and JOSEPH, P.D., *Filtering for Stochastic Processes with Applications to Guidance,* Interscience-Wiley (1968).
9. BROCKETT, R.W., *SIAM J. Control* **10**, 265 (1972).
10. SUSSMANN, H.J. and JURDJEVIC, V., *J. Differential Eq.* **12**, 95 (1972).
11. JURDJEVIC, V. and SUSSMANN, H.J., *J. Differential Eq.* **12**, 313 (1972).
12. PFAFFELHUBER, E., *IEEE Trans. Circuit Theory* **18**, 218 (1971).
13. PFAFFELHUBER, E., *IEEE Trans. Information Theory* **21**, 605 (1975).
14. HERRLICH, H. and STRECKER, G.E., *Category Theory,* Allyn and Bacon (1973).

Networks of finite automata

DIETER[†] AND WALBURGA[‡] RÖDDING
in cooperation with HANSJÜRGEN BRÄMIK[‡], HANS KLEINE BÜNING[†], LUTZ PRIESE[‡]
and communicated by EGON BÖRGER[†]
[†]*University of Münster, FRG*
[‡]*University of Dortmund, FRG*

1. Introduction

1.1
The following is a short report about some work on the aggregation of finite automata to form a system which can be regarded as an automaton of just the same kind as that of its component automata. This work has been done by a team at the universities of Münster (FRG) and Dortmund (FRG). Some results are published but all of them are available by xerox. The work reported here certainly is not finished.

1.2
Up to this moment it has been concentrated on the following problems listed briefly (details will be given later):

1. Determining one (some) mathematically precise concept(s) of *aggregating automata* (of a certain type) *to a network* (of the *same* type).
2. Determining a '*basis*' of components as simple as possible such that, for each automaton, a network can be constructed using only those components and simulating the given automaton.
3. If one calls a basis like the foregoing 'universal' there is also the problem of showing, for some possible sets of components, that they do *not* form a *universal* basis.
4. Studying the *effect* that there are *not allowed arbitrary* constructions, but rather those which obey well-defined laws of construction, especially the construction principles of two-dimensional or one-dimensional iterative arrays (cellular spaces, chains).
5. Studying some *logical* problems arising from a theory of aggregation, problems concerned with the determination of *complexity;* to be more precise: the theory has been used to give examples for some very simple undecidable cellular spaces and secondly it has been shown that a certain first order theory describing chain constructions is undecidable.
6. It is possible to make precise the concept that an *automaton realizes a preference relation* (partially) ordering its states (or transitions): there are some results concerning the aggregation of preferences of components to a global preference of a network built up from the components; these results will have immediate consequences as to *developing a sort of combinatorial approach to understanding the aggregation of social phenomena.*
7. It is possible to have a mathematically precise definition of a 'fault' in a network and then to discuss the concept of *self-correction in a purely combinatorial manner,* that means, without using statistics; there are some theorems concerning the existence of a basis which is universal (in the above sense) and also self-correcting in those constructions which are needed to prove universality: as to the

question how simple the components to be used may be, there are theorems concerning a basis which is universal, but not universal *and* self-correcting.

8. The *von Neumann-problem of self-reduplication* of automata which is regarded usually as the problem of reduplication of statepatterns in a cellular space, can be discussed again for the case of general nets instead of cellular arrays, if there is a mathematically precise version of the possibility of changing the interconnections between the automata of a network along the utilization of this network.

9. If as components one allows not only finite automata, but also 'registers' which can store a natural number, add '1', subtract '1' and perform zero-test, one can make precise not only once more the concept of recursive function, but also in a quite natural manner a concept of *computable functional of finite type;* for example, it is possible to show that primitive recursion and the partial recursive μ-operator can be represented each of them by a network consisting of finite automata and registers.

1.3

At the end of this report it is emphasized that the above presented aggregation technique and theory are applicable not only as a special handling of modular decomposition of automata, but also as a *tool for the combinatorial treatment of topics in cybernetics, social sciences, theoretical biology and systems science.*

2. Automata

2.1

With regard to the concept to be developed later on, the usual definition of a MEALY-automaton should be modified. A MEALY-automaton is a tuple (X,Y,Z,δ,λ) where X is the (finite, non-empty) set of inputs, Y is the (finite, non-empty) set of outputs, and Z is the (finite, non-empty) set of states; δ is a function defined on $X \otimes Z$ with values in Z (transition-function) and λ is a function defined on $X \otimes Z$ with values in Y. The problem as to this concept is that it is presupposed that each input signal gives rise to an output signal; our intention, however, is rather to formalize that there can be a sequence of accepted input signals without any resulting outputs, and also a sequence of output signals without the acceptance of an input signal. Further, we have to assume that there are spontaneous changes of state without the participation of any input or output signal. This leads to the following definition of a

2.2 Generalized automaton:

It is a tuple $(X,Y,Z,\alpha,\beta,\gamma)$, where:
X,Y,Z as above,
$\alpha \subset (X \otimes Z) \otimes Z$,
$\beta \subset (Z \otimes Z)$
$\gamma \subset Z \otimes (Z \otimes Y)$.

We speak of an α-transition, if an input signal is accepted and state is changed simultaneously; a β-transition means a spontaneous change of state; a γ-transition consists of disseminating an output and simultaneously changing the state.

Remark: Clearly, neither a determinacy nor a completeness assumption is presupposed (there may be signals which cannot be accepted in certain states).

3. Channels

Our intention is to make precise that an automaton has not only one but finitely many input channels each of which bears one symbol taken from a finite set of symbols which may vary from channel to channel. (There is no real loss of generality in the assumption that each channel only can bear one possible signal (we shall call it 'unit signal') but it is essential for the concept of network to be introduced later that we have the distinction between different channels). Now we can interpret the set X of 2.2 as consisting of tuples: the components of each tuple design a signal, or indicate that there is no signal, and the index of the component in the tuple designs the channel of the signal. An α-transition consists of the acceptance of a single signal on a single channel, the constellation of other signals on the other channels remaining unchanged. A similar consideration applies to outputs.

4. Experiments

An experiment is a sequence of triples (x_n, z_n, y_n), $0 \leqslant n \leqslant n_0$, $x_n \in X$, $z_n \in Z$, $y_n \in Y$, where the interconnection between the triples with index n and $n+1$ is given by one of the following five cases:

1. α-transition: $y_n = y_{n+1}$, $((x_n, z_n), z_{n+1}) \in \alpha$, x_{n+1} results from x_n by taking into account (cf. 3) that an α-transition removes one input signal from one input channel, that means, x_{n+1} describes the constellation of the remaining signals of x_n after α

has been applied.

2. β-transition: $x_n = x_{n+1}$, $y_n = y_{n+1}$, $(z_n, z_{n+1}) \in \beta$.

3. γ-transition: $x_n = x_{n+1}$, $(z_n, (z_{n+1}, y_n)) \in \gamma$, and the connection between y_n and y_{n+1} is similar to that described in 1., but respective to γ.

4. Input: $y_n = y_{n+1}$, $z_n = z_{n+1}$; x_{n+1} is like x_n, except that one channel, which is empty for x_n, bears a signal for x_{n+1} (remember that a component of $x \in X$ can indicate that there is no signal).

5. Output: $x_n = x_{n+1}$, $z_n = z_{n+1}$, y_{n+1} is like y_n except that one channel, which is empty for y_{n+1}, bears a signal for y_n.

Remark: The last two cases formalize the direct interaction between the automaton and an (assumed) 'experimentor'.

5. Protocol

This is a sequence consisting of the information about the input-output-behavior of an automaton generated by an experiment (cf. No.4) with this automaton. We shall not give the precise mathematical formalization (cf. Rödding and Nachtkamp, 1975*c*).

6. Simulation

6.1

To each state of an automaton and initial constellation of input and output signals there corresponds the set of possible experiments, resp. protocols. We define that A is simulated by B, if (assuming identical input and output devices belonging to A resp. B) to each initial state of A there is a state of B such that to each initial choice of input and output signals the corresponding sets of protocols are identical.

6.2

It should be remarked shortly that not all aspects of the intention of simulation are covered by this definition because the case of an experiment that can be continued with A, but cannot be continued with B is possibly contained in the case that A is simulated by B.

7. Networks

7.1

The intention of the following definition of network is to describe the fact that for a certain collection of automata interconnections between these automata are introduced by identifying output channels of some automata with input channels of some (not necessarily different) other automata. One has, in a natural way, the concept of an 'inner' channel, and of 'outer' input- and output-channels.

7.2

One can interpret such a network as a generalized automaton (cf. 2.2) assuming X to be built up from the possible signal constellations of the outer inputs, Y built up in the same manner from the outer outputs, and Z built up from the description of the states of the component automata together with the description of the distribution of signals on the inner channels of the network. An α-transition of the net is now an α-transition of one of its component automata concerning an outer input of the net. A γ-transition of the net is similarly an 'outer' γ-transition of one of its components. A β-transition of the net is an α-transition of one of its components concerning an inner channel of the net or likewise an 'inner' γ-transition of one component or a β-transition of a component automaton. This explication has been given instead of a precise mathematical definition which is passed over here.

7.3

Alternatively to the above there is a formalization which disregards the possibility of a signal 'waiting' for an indefinite time on an inner channel of the net. In this case it is rather assumed that a γ-transition of a component which concerns an 'inner' output of this component is followed immediately by a corresponding α-transition of that component automaton the input channel of which has (during the construction of the net) been identified with this output channel. One can formalize this by defining β-transitions of the net to be either β-transitions of a component automaton or pairs of corresponding γ- and α-transitions of (not necessarily different) component automata. But we shall not do so.

8. Basis (Introduction)

With the concept of network one has a decomposition principle for automata and it is natural to ask how simple components can be from which one can build up networks which are able to simulate arbitrary automata. The general structure theory of those nets has been researched by Priese (Priese,

1975). Several finite sets of automata of which the nets under consideration have to be compounded have been compared. The relations to the theory of Petri-nets are of some canonical interest as our general nets over (α, β, γ)-automata obey no synchronizing device, i.e. are concurrent, asynchronous nets of abstract automata. It should be emphasized that in the work of Priese not the just mentioned simulation-concept but a much more strong type of 'hang-up free' realization–similar to the concepts of Hack 1975–has been used. However, for mathematical simplicity and smoothness of results the chapter will be mainly concerned with so-called 'normed networks', introduced now. Thus, in order to have a good mathematical handling for answering the just mentioned and related questions, it is advisable to study the problem under some restrictive conditions: So it seems quite natural to restrict the attention to automata and networks which use one unit signal only (cf. Ottmann 1973). One also can approximate the classical case of a Mealy-automaton as far as possible by postulating that only those experiments are allowed in which the experimentor restricts his actions to processing an input after an output and vice versa. Of course in this case the possible component automata should be approximately chosen. This leads to the concept of normed networks.

9. Normed network

9.1

For the sake of a lucid mathematical treatment it is useful to start with only slight modifications of the classical automata concept. Further we shall first treat the special case of determinacy. We generalize the concept of Mealy-automaton only in that we allow the functions δ, γ to be undefined (simultaneously) for some argument pairs (x,z). In this special case we repeat the definition of network and of network action in a manner in which the whole action of a network between the acceptance of an input signal and the (not necessarily possible) production of an output signal is regarded as one step. We assume that, at each moment there is at most one (unit-) signal on the channels of the net and similarly we interpret the behavior of the component automata.

9.2 Automata

We consider (cf. 9.1) tuples (X,Y,Z,δ,γ) like in No.2.1 except that we allow δ and γ to be (simultaneously) undefined for certain argument pairs (x,z), $x \in X$, $z \in Z$. This is done with respect to the fact that a unit-signal may enter a net and never leave it. The input set X will be regarded to consist of a disjoint union of one-element sets $X = \{x_1\} \cup \ldots \cup \{x_n\}$. An input to an automaton thus no longer is a vector–i.e. referring to several 'unit-impulses'–but is one element x_i, $1 \leqslant i \leqslant n$. Thus, in what follows only one 'impulse' or 'unit-signal' runs through our automata and normed networks–this explains the notation of 'normed'–whereas in the general theory–compare chapter 14 –an arbitrary amount of concurrent unit-signals is allowed.

9.3 Normed networks

Our intention is to give an inductive definition of a normed network by regarding its construction as a repeated application of two basic steps, namely parallel composition and feedback. Given $A_i = (X_i, Y_i, Z_i, \delta_i, \lambda_i)\ (i=1,2)$ the *parallel composition* A of A_1 and A_2 is simply defined as follows: Without loss of generality we can assume that $X_1 \cap X_2 = \emptyset$, $Y_1 \cap Y_2 = \emptyset$. Now

$$A = (X_1 \cup X_2,\ Y_1 \cup Y_2,\ Z_1 \oplus Z_2,\ \delta,\ \lambda) \text{ where}$$

$$\delta(x,(z_1,z_2)) = \begin{cases} (\delta_1(x,z_1),z_2), & \text{if } x \in X_1 \\ (z_1,\ \delta_2(x,z_2)), & \text{if } x \in X_2 \end{cases}$$

$$\lambda(x,(z_1,z_2)) = \begin{cases} \lambda_1(x,z_1), & \text{if } x \in X_1 \\ \lambda_2(x,z_2), & \text{if } x \in X_2 \end{cases}$$

Given $A = (X,Y,Z,\delta,\lambda)$ and $x \in X$, $y \in Y$, by the *feedback* (of A resp. x,y) A_x^y we mean the result of connecting the output y with the input x, that is $A_x^y = (X - \{x\},\ Y - \{y\},\ Z,\ \delta',\ \lambda')$ where δ', λ' are defined as follows: For $x' \in X - \{x\}$, $y' \in Y - \{y\}$, $z, z' \in Z$ we have $\delta'(z,x') = z'$ and $\lambda'(z,x') = y'$, if either $\lambda(z,x') = y'$ and $\delta(z,x') = z'$ or $\lambda(z,x') = y$ and there is a finite sequence of states, starting with $\delta(z,x')$, corresponding to repeated x-y-actions of A and afterwards one has output y' and state z'. (It is trivial to give a precise mathematical formalization of this.) We shall assume that in all other cases $\delta'(z,x')$ and $\lambda'(z,x')$ simultaneously are undefined. Because it is possible that for everywhere defined δ and λ the functions δ' and λ' may be undefined for certain arguments, we had to modify (cf. 9.2) the definition of a Mealy-automaton. A network construction can be achieved by repeated application of the two steps described above, and clearly the exact succession in the performance of those steps is not essential. Given automata $A_1,...,A_n$ $NN(A_1,...$ $...,A_n)$ let be the class of all networks which can be constructed from $A_1,...,A_n$ by finite number of parallel compositions and feedback constructions. Especially $A_i \in NN(A_1,...,A_n)$ for $1 \leqslant i \leqslant n$.

9.4 Basis (for normed networks)
We call $\{A_1,...,A_n\}$ a *basis,* if for each automaton A there exists an automaton $B \in NN(A_1,...,A_n)$ simulating A. Of course the concept of simulation used here must be adapted to the concept of normed network: Especially we regard only experiments which start with only one unit signal and underly the restriction that there must be always a change between input and output signal applications. We shall give a formal definition: An automaton C (or a normed network C) simulates or realizes an automaton B, in signs $B \rightarrow C$, iff

a) $X_B = X_C,\ Y_B = Y_C$ and

b) $\forall s \in Z_B:\ \exists K_s \subseteq Z_C:\ K_s \neq \emptyset:\ \forall s' \in K_s:$

$\forall x \in X_B:\ \delta_C(x,s') \in K_{\delta_B(x,s)}$ and

$\lambda_C(x,s') = \lambda_B(x,s)$.

If in addition for all s of Z_B holds that the set K_s consists of exactly one element we say that C simulates B isomorphically and write $B \xrightarrow{\text{iso}} C$. In both cases C has the same input-output-behavior as B if C starts with states that correspond to states of B. Most of the following results hold for isomorphical simulation, too.

10. Basis theorems

10.1 Some examples of specially normed automata
For the purpose of getting a basis of quite simple automata we shall describe some special automata and later on we shall use sketches to describe network constructions and sketches for the special automata used in these constructions. So we shall give accompanying sketches to the special automata which we shall speak about now. They also shall get symbols as names for future reference:

1. K has two input channels and one output channel, further one state. Each input symbol causes one output symbol. The formal description of K is as follows:

$$K = (X_K, Y_K, Z_K, \delta_K, \lambda_K)$$

$$X_K = \{1,2\},\quad Y_K = \{3\},\quad Z_K = \{s\}$$

$$\delta_K(1,s) = \delta_K(2,s) = s,\quad \lambda_K(1,s) = \lambda_K(2,s) = 3$$

(See Fig.1[†] or more simply Fig.2.)

2. F has one input-channel, two output channels and two states. Each input signal changes the state and the output channels are used alternatively (and therefore corresponding to the actual date). Formal description:

$$F = (X_F, Y_F, Z_F, \delta_F, \lambda_F)$$

$$X_F = \{1\},\quad Y_F = \{2,3\},\quad Z_F = \{l,r\}$$

$$\delta_F(1,l) = r,\ \delta_F(1,r) = l,\ \lambda_F(1,l) = 2,\ \lambda_F(1,r) = 3$$

(see Fig.3).

3. D has two input channels, two output channels and two states. One input channel can be used as a sort of test for the actual state because the resulting output is corresponding to the state if this input is used. The second input changes the state; the output is fixed for this input. Formal description:

$$D = (X_D, Y_D, Z_D, \delta_D, \lambda_D)$$

$$X_D = \{1,2\},\quad Y_D = \{3,4\},\quad Z_D = \{l,r\}$$

δ_D	l	r
1	l	r
2	r	l

λ_D	l	r
1	4	3
2	3	3

(See Fig.4.)

4. H has three input channels (called u,v,w), four output channels (called u', u'', v', w') and two states (called l,r). Input v always causes state l and output v'. Similarly input w always causes state r and output w'. Input u is a test because according to the state (which is not changed when using this input), the resulting output is u' or u''. Formal description:

$$H = (X_H, Y_H, Z_H, \delta_H, \lambda_H)$$

$$X_H = \{u,v,w\},\quad Y_H = \{u',u'',v'w'\},\quad Z_H = \{l,r\}$$

[†] Figure 1 and the following figures can be found in the Appendix of this article.

δ_H	l	r
u	l	r
v	l	l
w	r	r

λ_H	l	r
u	u'	u''
v	v'	v'
w	w'	w'

(See Fig.5.)

10.2
It can be shown that the automata *K, F* and *D* form a basis. Instead of a proof for this we shall indicate with some sketches how to construct a network from *K, F* and *D* which can simulate *H*. This may be motivated by the fact that several constructions have been shown explicitly (Cohors-Fresenborg, 1976; Priese and Rödding, D., 1974; Ottmann, 1975*b* and 1973) that prove *K* and *H* being sufficient for isomorphically simulating an arbitrary automaton. The last mentioned paper (Ottmann, 1973) in addition tries to minimize the necessary number of *H*-component-machines in general realizations (but therefore operates with *K, H* and *F*-components).

Our intention for showing this construction in detail is firstly to acquaint the reader with this specific sequential technique used in working with such networks and secondly we have introduced the automaton *H* because it plays a crucial role in some later considerations. Now the details: First we define an automaton *E:*

$$E = (X_E, Y_E, Z_E, \delta_E, \lambda_E)$$

$$X_E = \{t,c\}, \quad Y_E = \{t^l, t^r, c'\}, \quad Z_E = \{l,r\}$$

δ_E	l	r
t	l	r
c	r	l

λ_E	l	r
t	t^l	t^r
c	c'	c'

(See Fig.6.)

Simulation of *E* by a network constructed from *K,F,D:* (See Fig.7.)

Z can easily be composed as a network by using 3 examplars of *F:*

$$Z = (X_Z, Y_Z, Z_Z, \lambda_Z, \delta_Z)$$

$$X_Z = \{0\}, \quad Y_Z = \{1,2,3,4\}, \quad Z_Z = \{5,6,7,8\}$$

$$\delta_Z(0,i) = \begin{cases} i+1 \text{ for } 5 \leqslant i < 8 \\ 5 \text{ for } i = 8 \end{cases}, \quad \lambda_Z(0,i) = i-4$$

Simulation of *H* by a network constructed from *K,E:* (See Fig.8.)

Now one can combine these two constructions getting a simulation of *H* by a network constructed from *K,F,D.*

10.3
Naturally the question arises if the basic automata *K,F* and *D* form a *minimal* basis. Indeed, the following can be shown: If one defines the complexity of $A = (X,Z,Y,\delta,\lambda)$ simply to be the number $(|X| \dot{-} 1)\cdot(|Y| \dot{-} 1)\cdot(|Z| \dot{-} 1)$, then no finite set of automata of complexity 0 is a basis. (Indeed, *D* has complexity 1.) It is mentioned that the class *NN(K,F)* is certainly not universal for Mealy-automata (more precisely, one cannot simulate *D* in it). Thus, questions about intermediate (not universal) classes of networks arise. (For results, cf. Ottmann 1973 and 1975*a*, Koerber 1976.)

10.4
Another way to get intermediate classes is to study the structure of networks. We will list some results (Kleine Büning 1975): If we have networks without closed paths (cascade-networks formed only by parallel and serial composition) no finite basis can be found and there arises a hierarchy of intermediate classes.

For networks with closed paths we have the following results: There exists a finite basis such that to each Mealy-automaton a simulating network can be found with the additional property that all closed paths possess a common edge.

This basis has complexity 2. (It is an open problem whether a basis can be found, in the above sense, with complexity 1.) As an interpretation *one* delay is sufficient for general construction.

If we restrict the length of all closed paths in the networks, no finite basis exists and, again, we will find a hierarchy of intermediate classes.

10.5
Another question (stimulated by electronics) asks whether it is possible to get a basis such that every Mealy-automaton *A* can be simulated by a normed network *N* with the additional property that each of its component-machines can be used at most once during each 'period of using' of *N* (see Section 13). As an interpretation, in principle no delay is necessary if nets are used strictly sequentially. This holds for the definition of simulation with basis *H, F,* and *K* but fails for isomorphical simulation.

10.6
Another question is: Is it possible to get a basis which is even more simple than the above-mentioned one, if one allows the construction of networks from components which have no normed behavior but in such a manner that the network as a whole *has* a normed behavior. In this case, on intends to simulate at least all Mealy-automata. According to this aim, there indeed a further reduction of the basic automata is possible (cf. Koerber 1971, Winckelmann 1971).

10.7
By the way, the above-mentioned basis theorem concerning normed networks involves determinism. In order to have a corresponding theorem for indetermined Mealy-automata one simply has to introduce the following automaton *I:*

$$I = (X_I, Y_I, Z_I, \alpha_I, \beta_I, \gamma_I)$$

$$X_I = \{1\}, \quad Y_I = \{2,3\}, \quad Z_I = \{4,5,6\}$$

$$\alpha_I = \{((1,4),5), ((1,4),6)\}, \quad \beta_I = \emptyset,$$

$$\gamma_I = \{(5,(4,2)), (6,(4,3))$$

Now it can be proved that each indeterminate Mealy-automaton can be simulated by a network built up from *K,F,D,I.*

11. Normed constructions

11.1
If one interprets the network constructions as a sort of hardware technology (maybe in electronics), one should be interested in the case that no arbitrary constructions are allowed. One should be interested in an approach analogous to that one in the theory of cellular spaces. If one interprets a cellular space as a normed network construction, namely a two-dimensional iterative array, one can go one step further and also consider one-dimensional iterative arrays, namely chain constructions of the following form: (see Figs 9 and 10).

If one uses a restriction to a unit signal, the capacity of channels between the single automata of a chain can be described trivially by the number of interconnections between them (in both directions). It appears natural to define the *type* of an automaton in this context by a quadruple *k,l,m,n* so describing the respective number of input-output channels corresponding with the following sketch. (see Fig.11.)

For the sake of simplicity we shall confine ourselves to the case $k = l = m = n = 2$ [if there is no or at most one channel in one direction, at each side, the combinatorial theory is trivial (cf. Koerber and Ottmann 1973/74)]. Now one has two versions of the basis problem: Firstly to find simple automata which form a basis for chain construction and secondly to find *one* ('universal') automaton which only by repetition in a chain construction, already guarantees universality, but of course, with various initial states in its occurrencies in the chain.

In Koeber and Ottmann 1973/74 a set of 13 basic-automata is specified each of which has at most four states. This set of automata is universal and using two end-automata, every finite Mealy-automaton can be simulated by chain-construction. Koeber 1976 has shown that one has also a basis with respect to concatenation with state-number 2.

Kleine Büning 1976 specifies universal automata with respect to two-dimensional finite iterative arrays (cellular spaces) which allow to find very simple two-dimensional universal Turing-Machines. Especially there are simple universal one-element bases (for example with 2 states and 5 inputs and outputs).

11.2
It may be permitted to add some remarks concerning related logical problems: The general question could be: How complicated is a theory of network constructions? There are at least two possible methods to make this question more precise. In both cases we shall confine ourselves to the context of chains. Firstly one can take the operation of joining two chains as fundamental; then one can use constants for certain special chain-automata [that means of type (2,2,2,2)]. Now one can build up a first-order language using these fundamental constants (and maybe some other). The complexity problem reduces to the question of (recursive) complexity of that language: It has been shown that for an adequately argumented language of the above type the definable predicates over automata are precisely the arithmetical predicates. [One should not confuse this result with the fact that *semiotic* concatenation is a basis for arithmetic (cf. Ottmann 1971, 1974*a*).] Secondly one can make precise the concept of an arithmetic treatment of chain constructions and then ask if there exist simple axioms and rules such that one would get a sort of 'calculus' of chain-constructions. The answer is also positive (cf. Koerber 1976).

11.3
There is a connection between the theory of cellular spaces and a certain generalization of the theory of these systems. Namely, if one first interprets words over a finite alphabet as finite patterns on 1-dimensional 'Turing'-tape which is filled up with blanks at those places which are not covered with the word, and secondly restricts the interest to such a sort of Thue-systems in which all rules have words of equal length on the right and left side (and use also the blank symbol), then one can clearly simulate arbitrary Thue-systems by such systems.

Now, one can interpret the rules of this type as a sort of command to replace a part of a one-dimensional pattern by a part of equal length or, as one could say, of equal shape. Now, let us generalize this to the two-dimensional case: Rules are given to replace parts of a pattern by patterns of a shape equal to the part. Now, one can simulate a cellular space with local action by the above-sketched combinatorial system of a finite number of rules. One may be interested in such systems with, on the one hand, rules as simple as possible, and with an alphabet as small as possible, on the other hand, with undecidable stopping problem or something like that. It has been shown by using more refined versions of the above basis theorems that there exists a system over an alphabet with only three letters, which has only three rules, the premises and conclusions of which have only three letters, such that this system has an undecidable stop-problem (cf. Priese 1973, 1976*a*, 1976*b*). It is a purpose of this remark to show that the combinatorial analysis with networks leads to fields related to some very simple but undecidable systems. We shall give an idea of how to refine the above-mentioned bases for dealing with Thue-systems:

In order to apply our theory to Thue-systems our automata and normed networks have been generalized to 'reversible' automata and a 'reversible feedback-operation' for network-constructions is introduced (Priese 1973, 1976*a*). These results could be strengthened by applying the concept of 'reversibility' to the general nets over (α, β, γ)-automata, this leading to a concept of reversible Petri-nets. With this technology some very simple computation-universal, reversible, asynchronous, concurrent, growing, self-duplicating systems could be found (Priese 1973, 1976*b*), solving the von-Neumann-problem of self-reproduction and, in addition, obeying the chemic law of microscopic reversibility.

12. Automata with preferences
This denomination is an abbreviation for finite, indeterminate automata which have their sets of states ordered by a transitive, but not necessarily complete ordering. The underlying idea–depicted in broad outline–is that an automaton realizes its preference structure if, in every case, it chooses such a transition from those open to it that will give it a better state than it has, if it can improve its state in a concrete situation (given by an input x and a state z) by means of a transition. This concept of automata with preferences is dealt with, because–in the realm of social sciences–it seems appropriate for the modelling and theoretical treatment of mutual interactions among units in society,

– it permits an *exact* theory in the social sciences starting from the analysis of *actions* in society which are grasped as transitions from one situation to another,

– it is extremely flexible with regard to the social situations to be grasped,

– it makes possible a solution of the well-known aggregation problem by aggregating automata to form networks, using information exchange as an aggregation mechanism.

For these reasons, the concept of automata with preferences, used as a model of an action unit, serves as a basic concept for an exact theory of social interactions. Let us summarize the results we could deduce up to now, which are concentrated in four theorems proved after having formulated all essential concepts in a mathematically precise manner. Here, they are presented not in their exact mathematical form (cf. Rödding, W. and Nachtkamp 1975*c*), but in a way that already takes into account their interpretation:

First conclusion: Observation exclusively of the external behavior of an action unit (represented by the input-output-behavior of the corresponding automaton) which does not take into account its internal organization, does not permit inferences to be made concerning the preferences on which this behavior is based.

Second conclusion: It is not true that, if an action unit dispenses with certain of its originally possible reactions, a preference structure can always be fixed, the realization of which would lead to the relevant restriction of behavior.

The first conclusion affects the scientific bases of revealed preference theory, widely known in economic fields (Morgenstern 1948; Rödding, W. and

Nachtkamp 1975*c*), the second one shows that the realization of a preference ordering is a non-trivial concept to which not all kinds of behavior can be ascribed.

Third conclusion: The aggregation of action units modelled by finite, indeterminate automata *with* preferences does not necessarily lead to networks with preferences.

Fourth conclusion: If the possibilities of behavior of a given but freely elective indeterminate automaton are restricted in a freely elective way, then its now restricted behavior can be interpreted as that of another indeterminate automaton, originally equipped with more possibilities of behavior, which now realizes preferences.

The third conclusion (Rödding, W. 1975*a*; Rödding, W. and Nachtkamp 1975*c*) is a statement akin to Arrow's impossibility theorem. This last, however, is based on a theory of preference which dispenses with any notion of action, which is static from its origin and assumes complete preference ordering, and it makes use of a special aggregation mechanism, a voting procedure, in contrast to our more general principle of information exchange. So, to a certain degree, the third conclusion can be interpreted as a dynamic generalization of Arrow's theorem.

The fourth conclusion may seem surprising in connection with the second conclusion, and we will outline in a few words how it comes about:

A principle can be found for the construction of networks whereby the aggregation of preferences constantly leads again to preferences. This principle is, on the other hand, sufficiently general for the simulation of chosen automata, and in such a fashion that a network of similar structure, the components of which have preferences, undergoes a restriction of its behavior due to these preferences which permits the simulation of any chosen, prescribed behavioral restriction. Thus, for every automaton A and every sub-automaton $A' \subseteq A$ there exist an automaton B and a sub-automaton $B' \subseteq B$, and also a preference ordering P with the following property: B realizing P produces B' in such a way that B simulates A and B' simulates A'.

We wish to confine ourselves here to this brief outline of conclusions and to proceed with two further developments of the argument:

The first concerns the question of the possibility of conceiving, in a precise way, by the use of models, for example the existence of power relationships in social systems. Indeed, an exact recursive definition of the presence of a power relationship can be established, with the result that computers, in principle, are enabled to recognize power relationship in prescribed simulated social structures (Rödding, W. 1973).

The second development concerns analoga to the concept of the stability of systems in the framework of the line of argument here put forward: An opportunity to approach a stability problem of this nature exists in the investigation of the question, to what extent networks are capable of recognizing autonomous changes of states of their components, and of correcting them on their own initiative, without external intervention, on the basis of their special construction. For this question a number of conclusions could be discovered which are dealt with in the next paragraph.

13. Self-correction

We shall treat the following problem: Is it possible, to have a basis of simple automata such that each automaton can be simulated by a network built up from this basis, but with the following restrictive postulate: Whenever there is a 'fault' in the network, this fault is recognized by the network and repaired during the use of the network. There should be no restrictions according to the possible location of such a fault, but there should be, of course, restrictions concerning the number of occurrences of faults during a period of using the network. Now to make precise the notion of a 'fault', let us assume it to be a spontaneous (that is: not caused by signals) change of state of a component in a network. Clearly, one could have had also another sort of 'fault', namely e.g., spontaneous appearance of a signal on a channel in the network. In order to make precise the concept of a 'period of using' (see above), let us assume that we are dealing with a normed network. For some of the below-mentioned results this is no restriction. We define 'period of using' to be the time interval between a (unit)-signal entering the network and leaving it, and we shall introduce the concept of resting-interval by denoting the time interval between the last signal leaving a network and the next signal entering this network. It is of some importance for the considerations below to distinguish two possible cases for the appearance of faults in the network:

In the first case, we shall allow the appearance of a fault only during the resting-interval and, in the simplest case, we shall allow the appearance of at most one fault during each resting interval. In the second case we shall have only a restriction concerning the maximum number of faults appearing in a network during the time-interval between the entering of a unit-signal and the next unit-signal. We reiterate that we shall treat here normed networks, that means, especially, that each input-signal is followed by an output-signal and vice versa. The above introduced modules H and E (cf. Priese and Rödding, D. 1974) play a crucial role in our considerations concerning self-correction.

On the one side, it is true (cf. Ottmann 1973) that the modules K, F and D form a basis, also K and E, and also K and H. One can say that the module H can be replaced by the simpler module E, and E can again be replaced by the simpler modules F and D. Now, let us call a basis self-correcting, if each automaton can be simulated by a network built up from the modules of this basis but having the additional property of self-correction (with respect to some of the above sketched precise definitions of fault and resulting definition of self-correction). We shall not give the exact mathematical definition of self-correcting behavior here, but our intention is, of course, to formalize the fact that the appearance of faults (an appearance obeying some restrictions; see above) does not disturb the factual simulation. [A precise definition of simulation and simulation in self-correction can be found in an article of Priese (Priese 1976*e*) in this proceedings. There, a complete, more detailed, characterization of self-correcting networks can be found.] Now, it is possible to prove that K and H form a self-correcting basis, also in the above-mentioned second case of precise definition of fault, while there is a theorem saying that it is not possible to have the corresponding result for the modules K and E even if one restricts to the first case of fault appearance (see above). A proof can be found in Priese 1976*c*. As each F and D can be substituted by *one* module E it follows easily that the basis K, F, D is not self-correcting. There is some work pursuing the matter further in studying some generalizations of E and it is possible to distinguish various generalizations of E by the possibility of forming a self-correcting basis (together with K) with respect to various concepts of fault-implementation. A complete characterization of these generalizations of the automaton E sufficient for self-correction is given by Priese 1976*c* for two classes of self-correction (the first type of self-correction and one slight generalization). For the second, more restrictive, type of self-correction some results concerning these so-called E^1_m-automata can be found in Priese 1976*d*.

We have not given proofs in this survey, but we shall give a simple example for the self-correcting property of basis K, H, namely in the case of simulating the automaton H itself, and under the restriction that at most one fault may appear during the resting period (and no fault during the period of using). Let H^0 be a shorthand description of the following network formed from 5 modules H (see Figs 12 and 13). Let H^* (Fig.14) be a shorthand description of the following network formed from H^0 and 4 modules H (Fig.15). The network shown in Fig.16 built up from four exemplars of the network H^* simulates H and is self-correcting (in the above mentioned restricted sense. For the notation of H see Section 10).

It is somewhat surprising that if we allow the (at most one) fault to appear also during the period of using we only have *to replace* in the above construction each module H by a corresponding network H^* and to interconnect the 'correcting' inputs and outputs c, c' (see sketch of H^*) in the same way chain-like as one has done in the construction above.

14. Concurrent nets

14.1

As we mentioned in Section 8 the concept of normed networks should be regarded as a specialization of the general nets over (α,β,γ)-automata. With the language developed for the theory of normed networks we will briefly introduce some results of our general theory. As the theory of Petri-nets is commonly well-known, and well-researched, let us first compare both concepts of nets.

A Petri-net consists of a graph with directed arcs between so-called 'events' and 'conditions' and a distribution, marking, of so-called 'tokens' on the conditions. The dynamic behavior of a Petri-net is given by a calculus that operates—in a non-determined, non-synchronous manner—on the markings according to the connection of the conditions via some events. Thus, the atomar objects—events and conditions—of a Petri-net are no action units as they are unable to run a calculation themselves (e.g. they cannot solve any situation of commonly shared input-conditions). This shows the inherent difference of both concepts: The nets over (α,β,γ)-automata receive their dynamic behavior by the—non-synchronized—activities of their action units, their component-machines, that are able to react on some received information independently of some structural phenomenas.

14.2
However, as Petri-nets are widely accepted to model a large class of concurrent situations both concepts have to be compared.

We regard the following (α,β,γ)-automata V and W defined by

$$V := (\{1\}, \{2,3\}, \{a,b,c,d\}, \alpha, \beta, \gamma), \text{ with}$$
$$\alpha = \{((1,a),b), ((2,a),c), ((2,b),d), ((1,c),d)\}$$
$$\beta = \emptyset$$
$$\gamma = \{(d,(a,3))\};$$
$$W := (\{1,2\}, \{3\}, \{a,b,c,d\}, \alpha,\beta,\gamma), \text{ with}$$
$$\alpha = \{((1,a),b)\}$$
$$\beta = \emptyset$$
$$\gamma = \{(b,(c,2)), (b,(d,3)), (c,(a,3)), (d,(a,2))\}$$

W is a waiter that needs two impulses to react and V doubles impulses.

With some quite canonical notation of 'equivalence' it could be shown (Priese 1975) that the following classes of nets are equivalent—here $N(A_1,...,A_n)$ denotes the class of all nets over the (α,β,γ)-automata $A_1,...,A_n$; P is an automaton equivalent to an unbounded shift-register; R is a counter-machine with a zero-testing possibility (see Section 15):

$$\{\text{finite } (\alpha,\beta,\gamma)\text{-automata}\} = \{\text{safe Petri-nets}\} = N(K,E,V,W),$$
$$N(K,E,V,W,P) = \{\text{Petri-nets}\},$$
$$N(K,E,V,W,R) = \{\text{combinatorial Petri-nets}\}.$$

Some more classifications have been shown regarding the possibility of finding finite bases of modular Petri-nets that are able to simulate arbitrary Petri-nets.

14.3
It should be noted that there are some hints that no finite base (as K,E,V,W in the first equation) exists that simulates any finite (α,β,γ)-automaton, if some additional properties—such as 'promptness' —of the simulation-concept are required. But, dealing with automata with preferences such a base could be found.

14.4
If nets over (α,β,γ)-automata are researched in order to simulate normed networks some simple bases could be found (cf. Koerber 1971, operating with a concept of simulation that bases on the principle of protocols) for arbitrary Mealy-type automata, very similar to Petri-net constructions. Some research (Winkelmann 1971) has been made into properties of the general nets simulating normed networks whilst obeying some 'synchronizing'-like restrictions of allowed transition steps.

15. Representations of computable functionals of finite types by networks

This section is devoted to a problem which seems to be (at first glance) of a purely mathematical-logical character. But we shall give some concluding remarks which shall indicate that the representation of functionals which we propose below can serve as a sort of general approach to hardware representation of constructive softwâre concepts.

We shall be concerned with functionals of finite type over the natural numbers. Types are used only as a notational instrument to distinguish between the various sorts of functionals. We shall introduce types of inductive definition:

o is a type, and if σ, τ are types, then also $(\sigma \rightarrow \tau)$. Now, we shall introduce universes u_τ (τ a type) by induction over types: μ is a set of natural numbers, and $u_{\sigma\rightarrow\tau}$ is the set of all (total or partial: both cases are considered) functions, defined on u_σ, with values in u_τ. Our intention is to represent the functionals of finite type (that is, members of one of the u_σ) by normed networks built up from a certain basis, and we shall for this purpose introduce 'automata with types'. Here, the term 'type' means only a certain ordering of the input and output channels of an automaton. Firstly, we introduce datalines: A dataline is simply a finite sequence of $\uparrow$ and $\downarrow$. We shall introduce an identification of types with certain data-lines. To this end, we introduce the dualization of a data-line d called $\bar{d}$: $\bar{d}$ is got from d by simply replacing each $\uparrow$ by $\downarrow$ and vice versa. Now, an automaton A together with a certain ordering of its input- and output-channels, indicated by d, can be written simply as in Fig.17 or as in Fig.18. In order to introduce automata with types, we shall identify types with certain special data-lines, defined by type-induction: we identify type o with $\downarrow\uparrow\uparrow$, and if σ and τ are data-lines, then $\bar{\sigma}\tau$ is to be identified with $\sigma \rightarrow \tau$. An automaton A of type τ is simply as in Fig.19.

Now we shall introduce the crucial concept of representation of functionals of type τ by automata of type τ, and in order to do this, we have to make two preparational steps:

Step 1 is to single out a certain class N of automata, namely the networks built up from K,F,D and R where R is the following infinite automaton:

$$R = (X_R, Y_R, Z_R, \delta_R, \lambda_R)$$

$$X_R = \{a,s\}, \quad Y_R = \{e,i\}, \quad Z_R = \mathbb{N}_0$$

(*a:* addition, *s:* subtraction, *e:* executed, *i:* impossible to execute)

$$\delta_R(n,a) = n+1$$

$$\delta_R(n+1,s) = n \;, \qquad \delta_R(0,s) = 0$$

$$\lambda_R(n,a) = e \;, \quad \lambda_R(n+1,s) = e \;, \quad \lambda_R(0,s) = i$$

Step 2. If we have two automata A, B, A of type σ and B of type $\sigma\rightarrow\tau$, we shall define a composition of A and B to a new automaton (called AB). This composition is of course a network construction (see Fig. 20) that means that we connect simultaneously the corresponding channels in the part $\bar{\sigma}$ of the data-line of B, and the data-line σ of A. Clearly, the resulting automaton is of type τ.

Now we can define the representation of a functional x of some type by an automaton A (of the same type) by type-induction:

1. *Representation of numbers:* Here and for the following we assume automata to be 'initial', that is, to be defined with a specified initial state.

The type of numbers is ↓↑↑ and we shall denote an automaton A of this type as follows: (see Fig. 21).

We define: A represents number x iff the experiment with protocol

$$\underbrace{ab \ldots ab}_{x \text{ times}}\; ac$$

causes the automaton to start and to end with its initial state.

2. *Representation of x of type $\sigma\rightarrow\tau$:* Let A be an automaton (taken from N) of type $\sigma\rightarrow\tau$. We define A to represent x if for each y of type σ and each B (B taken from N) of type τ such that B represents y, the automaton AB represents xy (xy denotes the application of x to y).

This is the inductive definition of representation.

Remark: Clearly, in the case of 'full' functionals of finite type, the fact of being representable or not is established by using only little part of the 'course of value' of such a functional. One can escape this difficulty by using more 'meager' universes u_σ instead of the full introduced here. In this case, one has to formalize a more intricate connection between the automata of N on the one side, and the universes to be looked at on the other side. But we have not done so for the sake of simplicity.

It is quite trivial to see that all numbers are representable. It can easily be proved that the representable (one-place) functions are precisely the (total or partial) recursive functions, and we shall give some examples of representable higher-type functionals: If one takes the partial approach, it can be shown by some more or less complicated constructions that functionals are representable which describe substitution and iteration of functions, primitive recursion or more general types of recursion and μ-operator. For example we shall give some details in the last case, that is for the functional $\lambda fx.\, \mu y.fxy = 0$ of type $(o\rightarrow(o\rightarrow o))\rightarrow(o\rightarrow o)$ (λ-notation as usual). In constructing a network representing this functional we make use of automata H (see Section 10), R^* and Z (see below), which can easily be seen as replaceable by network constructions in N.

R^*: (see Fig.22)

(*t:* test, *a:* addition, *z:* zero)

$$X_R^* = \{t,a,z\} \qquad Y_R^* = \{t_0,t',a',z'\} \qquad Z_R^* = \mathbb{N}_0$$

$$\delta_{R^*}(n,t) = n \qquad \delta_{R^*}(n,a) = n+1 \qquad \delta_{R^*}(n,z) = 0$$

$$\lambda_{R^*}(0,t) = t_0 \;, \quad \lambda_{R^*}(n+1,t) = t'$$

$$\lambda_{R^*}(n,a) = a' \;, \quad \lambda_{R^*}(n,z) = z'$$

Z: (see Fig.23)

$$Z = (X_Z, Y_Z, Z_Z, \delta_Z, \lambda_Z)$$

$$X_Z = \{a,d,s,t\}, \quad Y_Z = \{a',d',d'',s',t',t_0\}$$

$$Z_Z = \{\langle i,j\rangle : i,j \in \mathbb{N}_0,\; i \leqslant j\}$$

$$\delta_Z(\langle i,j\rangle, d) = \langle i+1 \bmod j+1,\, j\rangle$$

$$\delta_Z(\langle i,j\rangle, a) = \langle i, j+1\rangle$$

$$\delta_Z(\langle i,j\rangle, s) = \begin{cases} \langle i,j-1\rangle & \text{if } i<j \\ \langle i,j\rangle & \text{otherwise} \end{cases}$$

$$\delta_Z(\langle i,j\rangle, t) = \langle i,j\rangle$$

$$\lambda_Z(\langle i,j\rangle, d) = \begin{cases} d', & \text{if } i<j \\ d'', & \text{otherwise} \end{cases}$$

$$\lambda_Z(\langle i,j\rangle, a) = a' \;, \qquad \lambda_Z(\langle i,j\rangle, s) = s'$$

$$\lambda_Z(\langle 0,0\rangle, t) \;= t_0 \;, \quad \lambda_Z(\langle i,j{+}1\rangle, t) = t'$$

(all states in the sense of condition $i \leqslant j$ for Z_Z, which is preserved by these definitions).

Now a representing network for the given functional is: (see Fig. 24)

H being in the state indicated by a dot, R^* and Z being in state 0 and $\langle 0,0\rangle$ respectively for the initial condition.

We also have indicated how networks, representing f and x, have to be composed with the given one for representation of the least y such that $fxy = 0$; so it is easy to verify that this construction will work.

Now let us come back to the introductory remarks about a sort of hardware programming of constructive software concepts. While it is of little use to give a further precise definition of recrusivity of functions (which has been done often enough since the time of the formulation of Church's thesis), there are alternatives for the formalization of the concept of computable functionals. If one assumes constructive concepts to be represented by computable functionals, the above given interconnections between functionals and networks give rise to a hardware-oriented treatment of such concepts. There is no difficulty to getting canonical representations for logical junctors and also (restricted) quantifiers. So one can handle a mathematical description by a direct translation into a hardware construction, and a similar approach is possible, if one uses the tool of combinatorial logic and the specific sort of mathematical analysis, given in that discipline.

But these more speculative remarks belong to our last paragraph of this text in which we shall give some comments concerning the possible usefulness and relevance of the further development of the technique and theory of networks sketched up to now.

16. Comments

We hope that the further development of the matter can have some influence on topics like *cybernetics, theoretical biology, social sciences* and, quite generally, a more combinatorial sort of approach to questions of aggregation. We think that the proposed approach allows both a rigorous mathematical treatment and an adequate modelling technique for a rather wide class of systems. Without having given any details. we are also sure that a formalization including the possibility of change of the structure of the interconnections between the components of a network is possible in the same style and would allow the expansion of the specified mathematical treatment also to questions of development of systems in whichever of the above cited topics that would be of interest.

Literature

COHORS-FRESENBORG, Elmar, *Mathematik mit Kalkülen und Maschinen,* Vieweg Wiesbaden (1976).

HACK, Michael, 'Petri-net languages', MIT, Project MAC, Computation Structure Group, Memo 124 (1975).

KLEINE BÜNING, Hans, 'Netzwerke endlicher Automaten für Potentialautomaten', Diplomarbeit am Institut für mathematische Logik und Grundlagenforschung der Universität Münster (1975).

KLEINE BÜNING, Hans, 'Endliche Flächen von Zellen', Working Paper (1976).

KOERBER, Peter, 'Simulation endlicher Automaten durch Netzwerke', Diplomarbeit am Institut für mathematische Logik und Grundlagenforschung der Universität Münster (1971).

KOERBER, Peter, Dissertation, Münster, not yet published (1976).

KOERBER, Peter, and OTTMANN, Thomas, 'Simulation endlicher Automaten durch Ketten aus einfachen Baustein-Automaten', Bericht 15, Forschungsberichte des Instituts für angewandte Informatik und formale Beschreibungsverfahren, Hg.: H.A. Maurer, W. Stucky, Universität Karlsruhe (1973)
und in
Elektronische Informationsverarbeitung und Kybernetik **EIK 10**, 2/3, 133-148 (1974).

MORGENSTERN, Oskar, 'Demand theory reconsidered', *Quarterly Journal of Economics* **6** (1948), pp.165, reprinted in R.B. Eklund, Jr. et al. (eds.), *The Evolution of Modern Demand Theory: A Collection of Essays,* Lexington, Mass., Heath Lexington Books (1972).

MORGENSTERN, Oskar, 'Thirteen critical points in contemporary economic theory: an interpretation', *Journal of Economic Literature* **10**, p.1163 (1972).

NACHTKAMP, Hans Heinrich, and RÖDDING, Walburga, see RÖDDING, W. and NACHTKAMP, H.

OTTMANN, Thomas, 'Darstellungen endlicher Automaten', Wiss. Haus-arbeit am Institut für mathematische Logik und Grundlagenforschung der Universität Münster (1969).

OTTMANN, Thomas, 'Eine Theorie sequentieller Netzwerke', Dissertation, Münster (1971).

OTTMANN, Thomas, 'Über Möglichkeiten zur Simulation endlicher Automaten durch eine Art sequentieller Netzwerke aus einfachen Bausteinen', *Zeitschrift für math. Logik und Grundlagen der Mathematik,* Bd. 19 (1973).

OTTMANN, Thomas, and KOERBER, Peter, see KOERBER, P. and OTTMANN, Th.

OTTMANN, Thomas, 'Arithmetische Prädikate über einem Bereich endlicher Automaten', *Archiv für math. Logik und Grundlagenforschung* **16** (1974*a*).

OTTMANN, Thomas, 'Some classes of nets of finite automata', Bericht 29, Forschungsberichte des Instituts für angewandte Informatik und formale Beschreibungsverfahren, Hg.: H.A. Maurer, W. Stucky, Universität Karlsruhe (1975*a*).

OTTMANN, Thomas, 'Einfache universelle mehrdimensionale Turingmaschinen', *Habilitationsschrift,* Karlsruhe (1975*b*).

OTTMANN, Thomas, 'Eine universelle Turingmaschine mit zweidimensionalem Band, 7 Buchstaben und 2 Zuständen,
Bericht 22, Forschungsberichte des Instituts für angewandte Informatik und formale Beschreibungsverfahren, Hg.: H.A. Maurer, W. Stucky, Universität Karlsruhe (1974)
und in
Elektronische Informationsverarbeitung und Kybernetik EIK 11, 1/2, 20-38 (1975).

PRIESE, Lutz, 'Über einfache unentscheidbare Probleme: Computational- und Constructional-universelle asynchrone cellulare Räume', Diss. am Institut für mathematische Logik und Grundlagenforschung der Universität Münster (1973).

PRIESE, Lutz, and RÖDDING, Dieter, 'A combinatorial approach to self-correction', *Journal of Cybernetics,* 4, 3, 7-25 (1974).

PRIESE, Lutz, 'Eine Erweiterung normierter Netze zu asynchronen, parallel arbeitenden Netzen abstrakter Automaten, Vortrag 5. Jahrestagung GI Dortmund (1975), not yet published.

PRIESE, Lutz, 'Reversible Automaten und einfache universelle 2-dimensionale Thue-Systeme', to appear in *Zeitschrift für mathematische Logik und Grundlagen der Mathematik* (1976*a*).

PRIESE, Lutz, 'On a simple combinatorial structure sufficient for sublying non-trivial self-reproduction', in preparation (1976*b*).

PRIESE, Lutz, 'On the minimal complexity of component-machines for self-correcting networks', in preparation (1976*c*).

PRIESE, Lutz, 'On fault-tolerant networks', in preparation (1976*d*).

PRIESE, Lutz, 'On stable organizations of normed networks', to appear in Proceedings of the Third European Meeting on Cybernetics and Systems Research (1976*e*).

RÖDDING, Dieter, 'Some Logical Problems connected with Networks of Finite Automata', Manuskript eines Vortrags an der Universität Oxford (1974).

RÖDDING, Dieter, 'Funktionale endlicher Typen und Netzwerke von Automaten', Working Paper (1975).

RÖDDING, Dieter, and PRIESE, Lutz, see PRIESE, L. and RÖDDING, D.

RÖDDING, Walburga, 'An algorithm for constructing compromises', Göttingen (1975*a*), (the manuskript in German language was finished in 1972).

RÖDDING, Walburga, 'Macht: Präzisierung und Meßbarkeit', in H.K. Schneider, Chr. Watrin (eds.), *'Macht und ökonomisches Gesetz',* Schriften des Vereins für Sozialpolitik, Gesellschaft für Wirtschafts- und Sozialwissenschaften, Neue Folge Band 74/I, Macht und ökonomisches Gesetz, Berlin (1973).

RÖDDING, Walburga, 'Netzwerke endlicher Automaten als Modelle wirtschaftlicher und sozialer Systeme,' Schriftenreihe der Österreichischen Studiengesellschaft für Kybernetik (Austrian Society for Cybernetic Studies) (1975*b*).

RÖDDING, Walburga, and NACHTKAMP, Hans Heinrich, 'On the aggregation of preferences to form a preference of a system', *Naval Research Logistics Quarterly,* Washinton D.C. (1975*c*).

WINKELMANN, Monika, 'Untersuchungen zur Theorie der endlichen Automaten und simulierenden Netzwerke,' Diplomarbeit am Institut für mathematische Logik und Grundlagenforschung der Universität Münster (1971).

Appendix

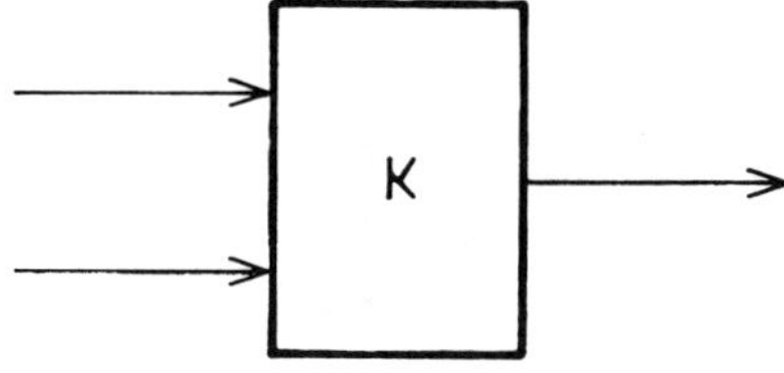

Fig.1

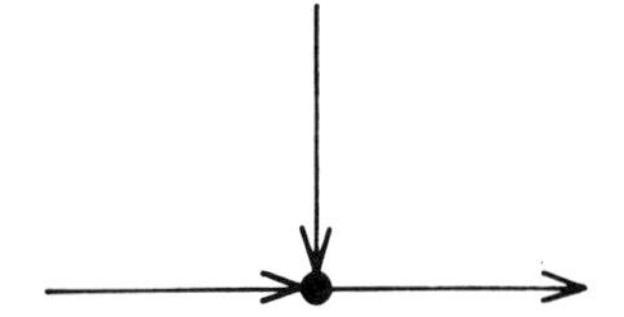

Fig.2

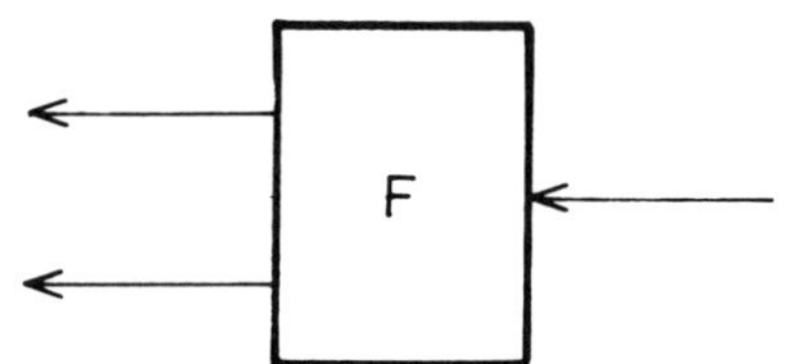

Fig.3

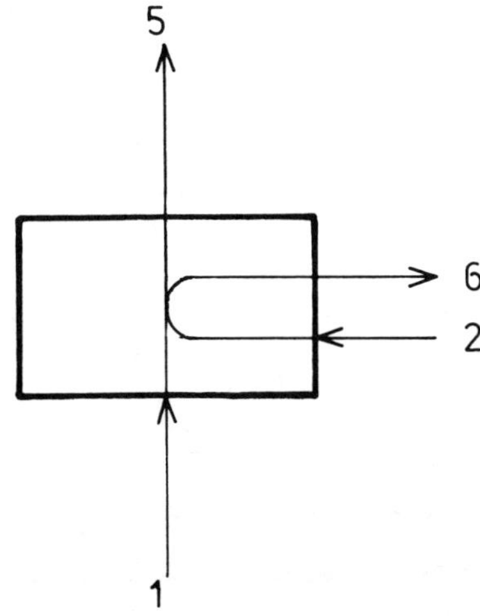

Fig.4

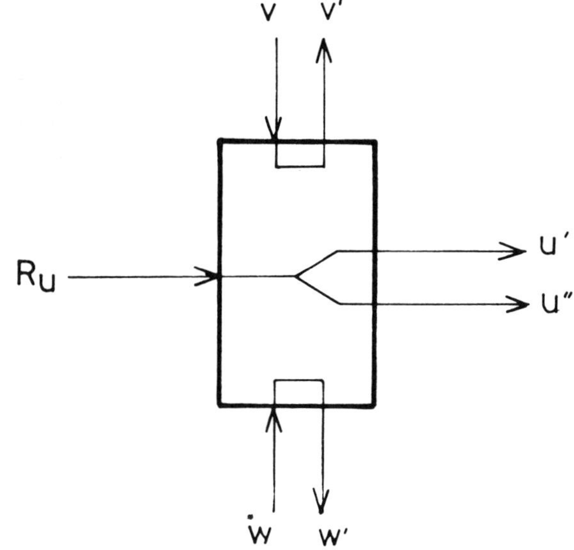

Fig.5

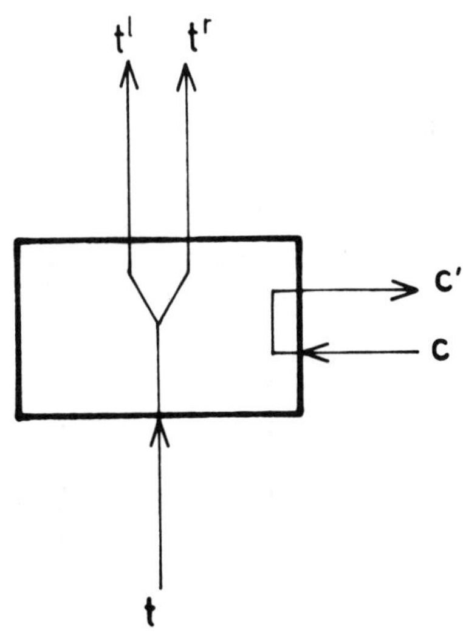

Fig.6

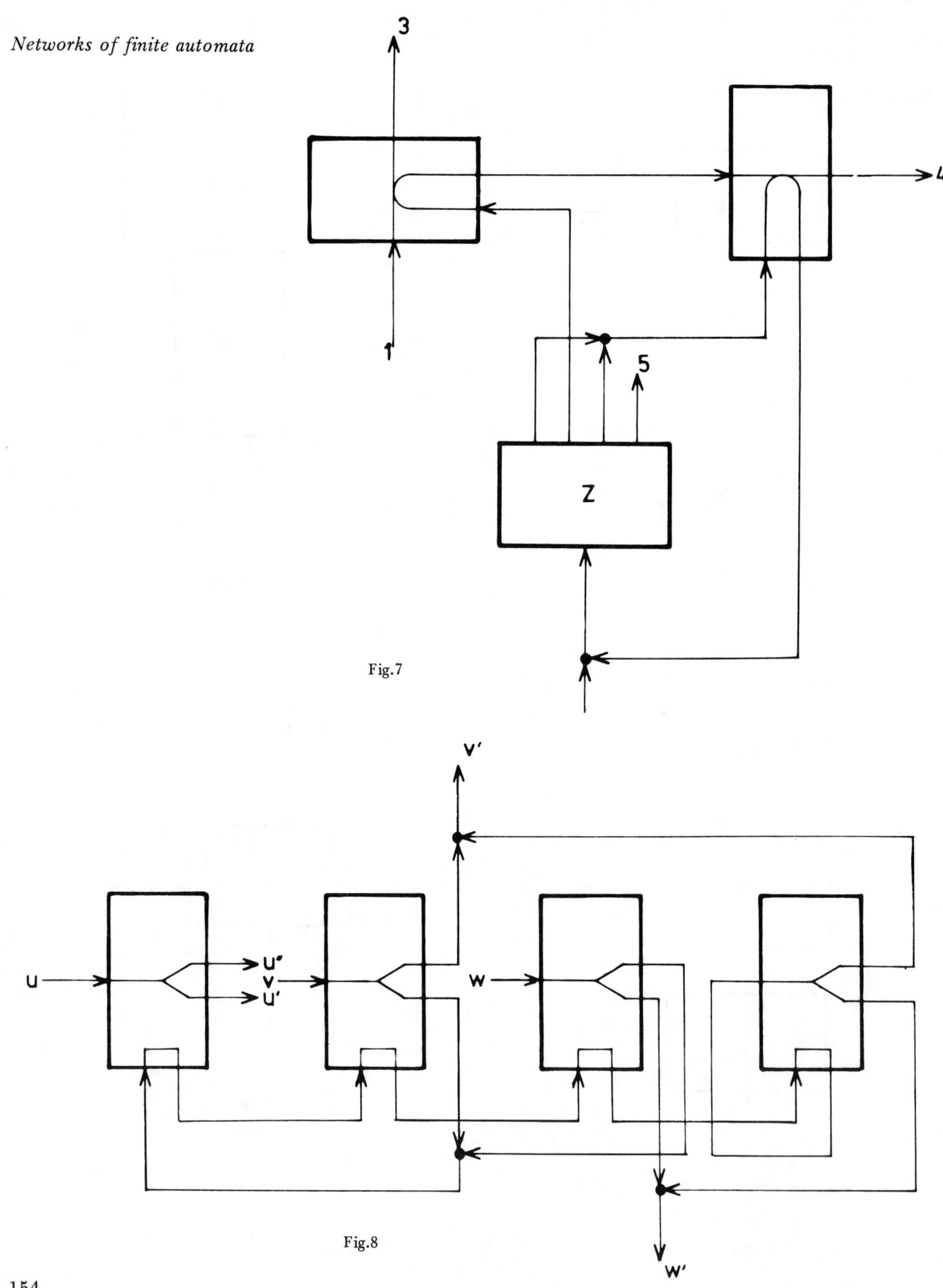

Fig.7

Fig.8

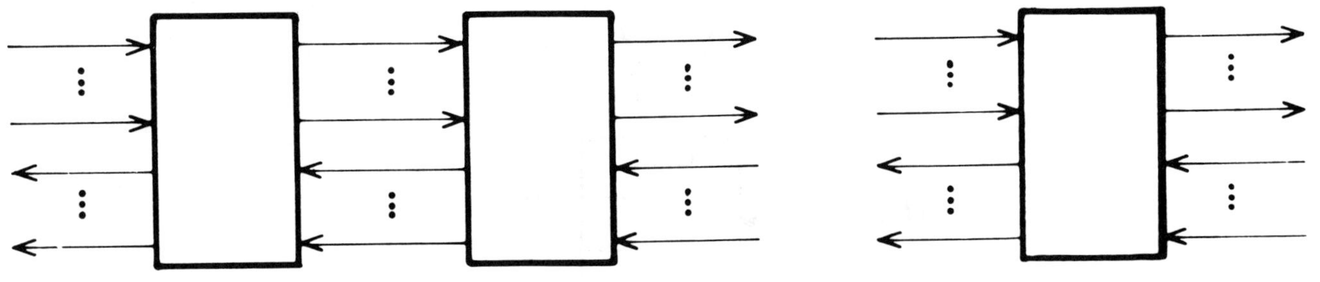

Fig.9

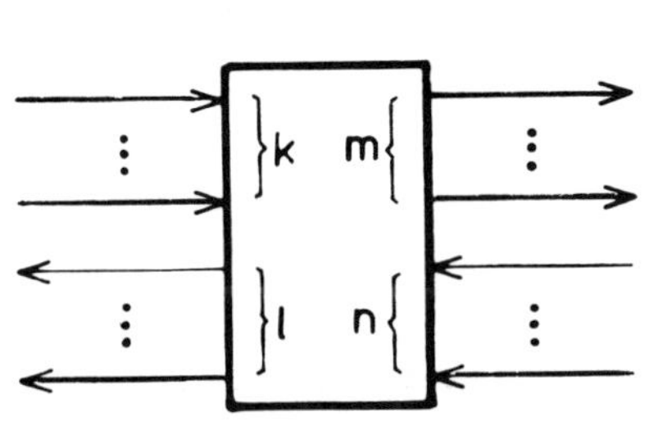

Fig.10

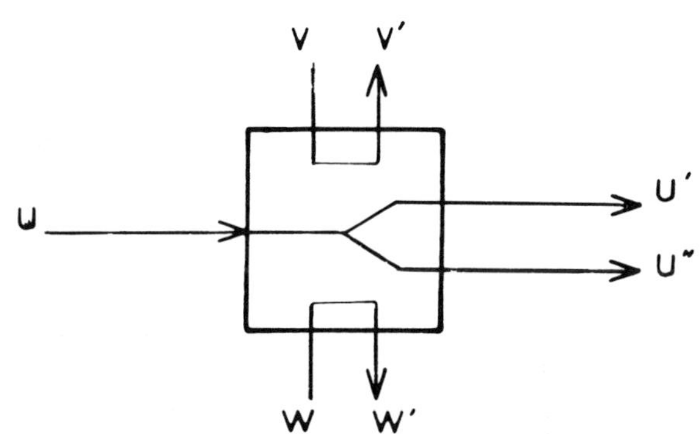

Fig.11

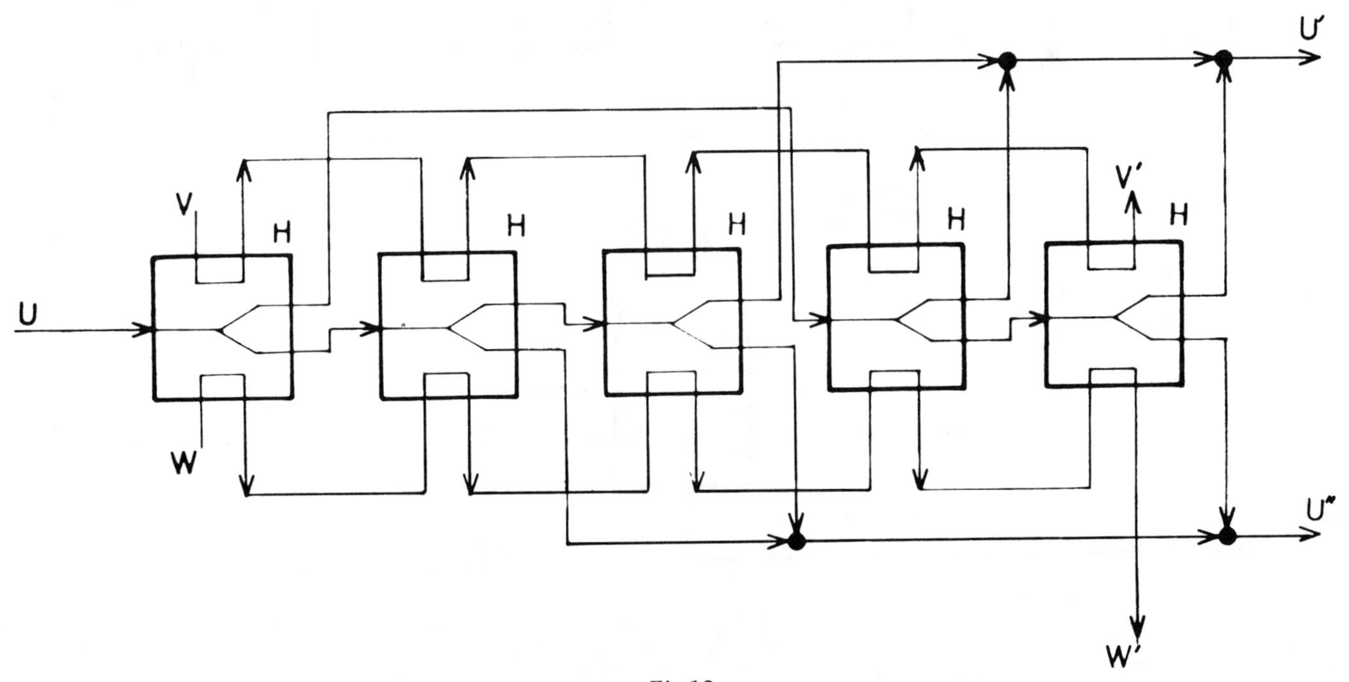

Fig.12

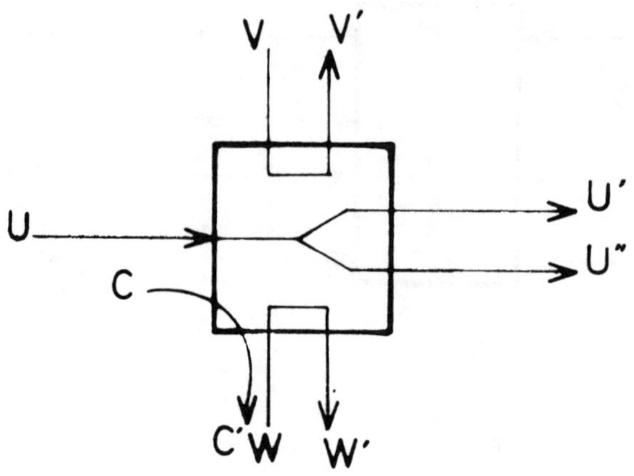

Fig.13

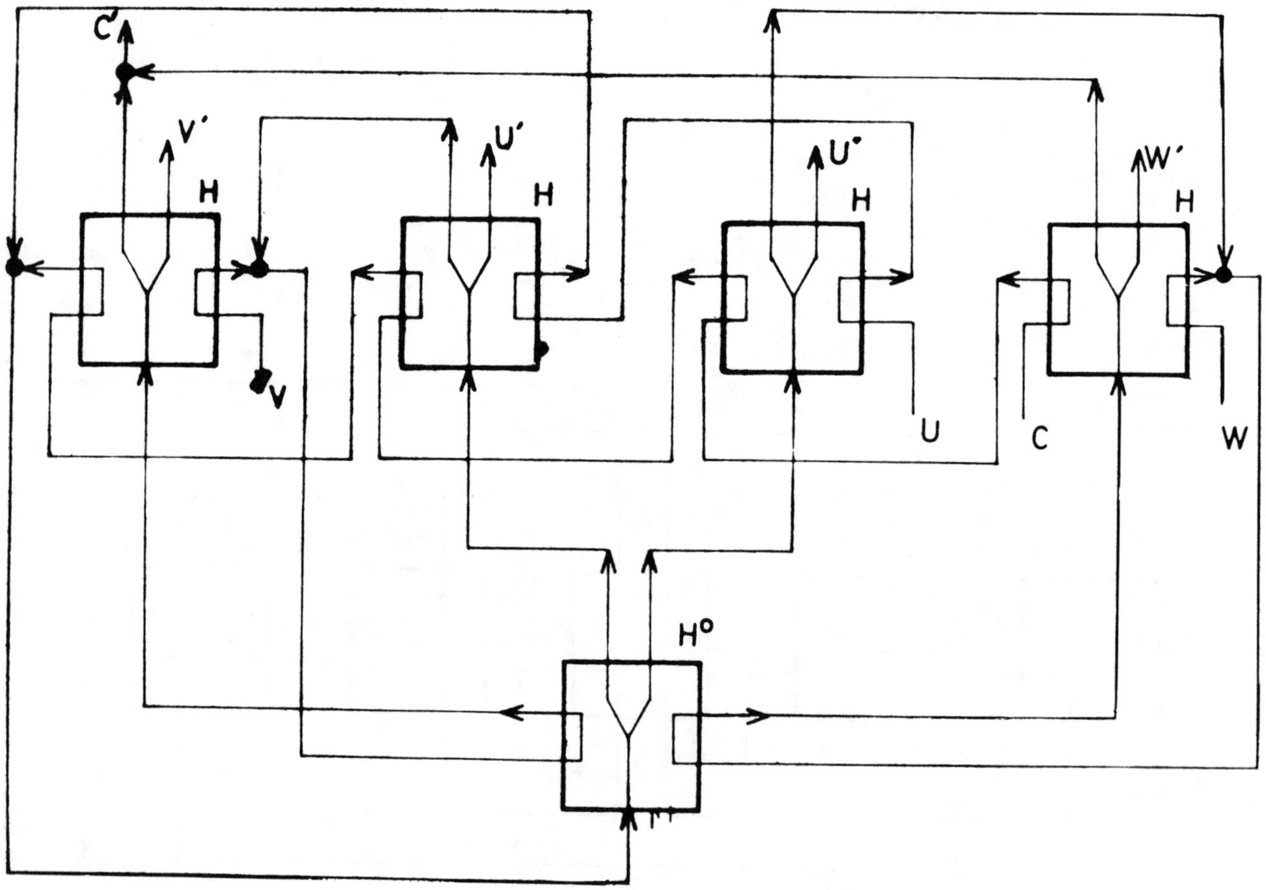

Fig.14

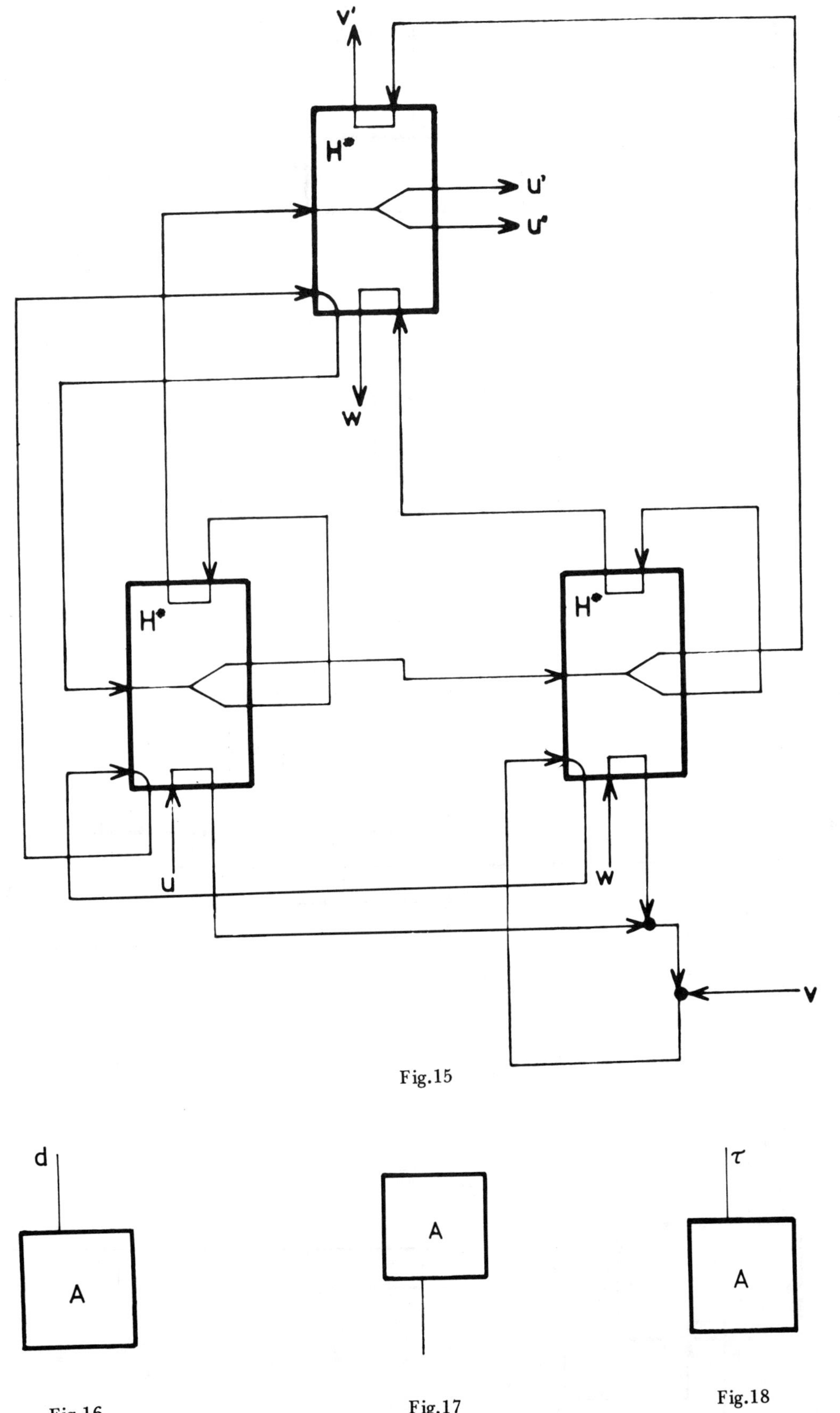

Fig.15

Fig.16

Fig.17

Fig.18

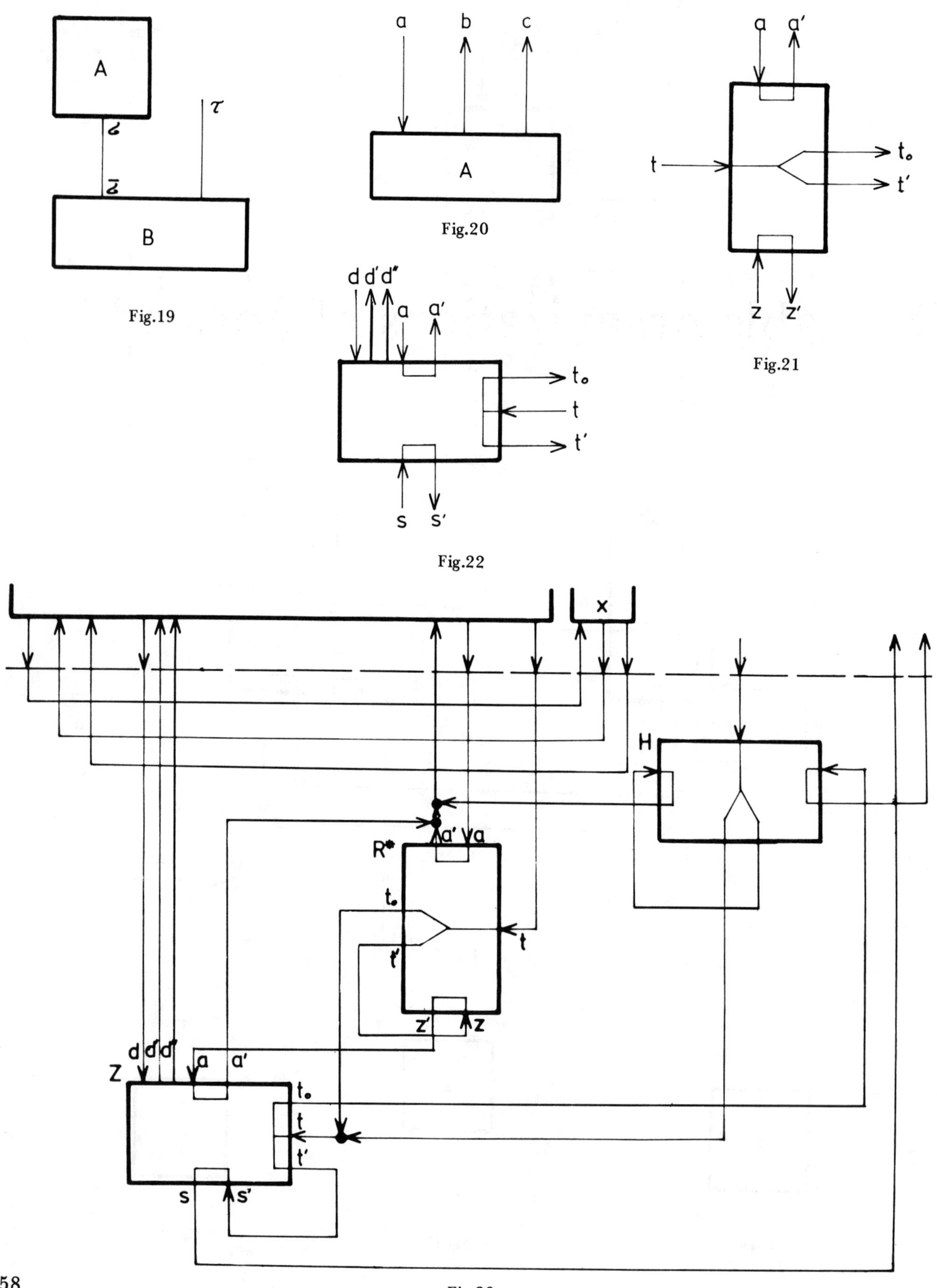

Fig.19

Fig.20

Fig.21

Fig.22

Fig.23

On stable organizations of normed networks

LUTZ PRIESE
Fachgebiet Systemtheorie und Systemtechnik,
Universität Dortmund, FRG

Normed networks

If one accepts automata—of, for example, the Mealy-type—as an interesting cybernetical tool, decomposition of automata into simpler elements and compounding of automata to more complex systems almost immediately prove to be important techniques. Whereas most techniques of net-building leave the concept of formal Mealy-automata the concept of Normed Networks developed by D. Rödding and his co-workers operates entirely on the principle of Mealy-automata.

Consider a (Mealy-)automaton A defined as a tuple $A=(I_A,O_A,S_A,R_A)$ of mutually disjoint finite sets of inputs I_A, outputs O_A, states S_A, and a functional transition-relation $R_A \subseteq (I_A \times S_A)\times(O_A \times S_A)$.

The product $A//B$ of two automata A and B simply is the automaton

$$A//B := (I_A \dot{\cup} I_B, O_A \dot{\cup} O_B, S_A \times S_B, R_{A//B})^{\dagger}$$

with

$$R_{A//B} := \{((x,(s_1,s_2)),(y,(s'_1,s'_2))) \,/\, ((x,s_1),(y,s'_1)) \in R_A$$

$$\& \; s_2=s'_2 \;\vee\; ((x,s_2),(y,s'_2)) \in R_B \;\&\; s_1=s'_1\}.$$

†Without any restriction I_A and I_B, O_A and O_B, shall be disjoint sets as we will not distinguish isomorphic automata.

Applying an output y of A to an input x of A the *feed-back* A^x_y can be defined as the automaton

$$A^x_y := (I_A-\{x\}, O_A-\{y\}, S_A, R'_A), \quad \text{with}$$

$$R'_A := Cl^o(R_A \cup N^x_{Ay}) \cap E^x_{Ay},$$

$$N^x_{Ay} := \{((y,s),(x,s)) \,/\, s \in S_A\},$$

$$E^x_{Ay} := \{((e,s),(o,s')) \,/\, s,s' \in S_A,\ \{e,o\}\cap\{x,y\}=\emptyset,$$

$$e,o \in I_A \cup O_A\}$$

For any relation $R \subseteq M\times N$ between two sets M and N the closure $Cl^\circ(R)$ of R is inductively defined by:

$$R \subseteq Cl^\circ(R)$$

$$(a,b),(b,c) \in Cl^\circ(R) \rightarrow (a,c) := (b,c)^\circ(a,b) \in Cl^\circ(R).$$

For the automata $A_1,\ldots,A_n$ the class $NN(A_1,\ldots,A_n)$ of *Normed Networks over* $A_1,\ldots,A_n$ is inductively defined by:

$$A_1,\ldots,A_n \in NN(A_1,\ldots,A_n) \quad ,$$

$$A,B \in NN(A_1,\ldots,A_n),\ x \in I_A,\ y \in O_A$$

$$\rightarrow A//B,\ A^x_y \in NN(A_1,\ldots,A_n).$$

Thus, any Normed Network is an automaton by definition and any automaton may be regarded as a Normed Network (i.e. over itself).

Let us note that, for both operations // and ^x_y, the associative and commutative laws do hold. We may thus drop the parentheses in $(A//B)//C$ and $(A^{x_1}_{y_1})^{x_2}_{y_2}$, e.g. The following elementary but important lemma will ease working with Normed Networks:

Lemma: For any Normed Network N over $A_1,\ldots,A_n$ there exist occurrences $B_1,\ldots,B_t$ of the automata $A_1,\ldots,A_n$, inputs $x_i \in \bigcup_{1\leqslant\tau\leqslant t} I_{B_\tau}$, and outputs $y_i \in \bigcup_{1\leqslant\tau\leqslant t} O_{B_\tau}$, $1\leqslant i\leqslant n$, such that

$$N = (B_1//\ldots//B_t)^{x_1,\ldots,x_n}_{y_1,\ldots,y_n}\ ,$$

and for all permutations $\sigma \in \gamma_t$, $\pi \in \gamma_n$:

$$N = (B_{\sigma(1)}//\ldots//B_{\sigma(t)})^{x_{\pi(1)},\ldots,x_{\pi(n)}}_{y_{\pi(1)},\ldots,y_{\pi(n)}}.$$

For a proof we refer to Priese [1].

The feed-back operation and the product operation differ from commonly known operations on automata in that they can dispense with any synchronization.

Using the notations as applied in the above lemma, we note that a state of N is a t-vector (as

$$S_N = \bigtimes_{1\leqslant\tau\leqslant t} S_{B_\tau}).$$

Any transition of R_N is the result of a composition $c=(f_l \circ n_{l-1} \circ \ldots \circ n_1 \circ f_1)$, where all f_λ, $1\leqslant\lambda\leqslant l$, describe a transition of a relation R_{B_τ}, $1\leqslant\tau\leqslant t$, and all n_λ, $1\leqslant\lambda\leqslant l$, describe a transition 'through some wire of N' as they are elements of some relation N^x_{Ay}. Because all component-machines B_τ are compounded by the product-operation to a larger automaton the transitions f_λ, $1\leqslant\lambda\leqslant l$, cannot be elements of some B_τ, but they give rise to a projection that is exactly a transition of a component-machine. Thus each f_λ is essentially a transition of a component-machine.

With these terms we may describe the intuitive idea of how a Normed Network operates by introducing the concept of an 'impulse' and a 'utilization':

As any transition T of N is the result of a sequence c—as defined above—we may regard such a sequence as a path and describe a movement along that path by an impulse that passes through the net N and thus fulfils the requirements for transition T. To be more concise we will use the term 'impulse' to uniquely describe such a sequence and, thus, a transition of N.

In this terminology: a sequence $c=f_l \circ n_{l-1} \circ \ldots \circ n_1 \circ f_1$ is the result of an impulse entering the Normed Network N via the input of a component-machine B_{i_1}, changing the state of B_{i_1}, and leaving B_{i_1} via an output according to the transition of B_{i_1} that is essentially f_1. The impulse then has to pass a wire according to a feed-back described by n_1 and, applying the same mechanism to a component-machine B_{i_2}, one obtains f_2 and so forth. This process obviously needs no synchronization.

By a utilization (or utilization-phase) of N we imply the fact that an impulse is entering—or has already entered—the net N, but has not left it. As a Normed Network is an automaton it operates sequentially on strings of inputs: A new impulse cannot enter N before the previous impulse has left it. This is a consequence by the definition of R_N and is the description of a transition of N by the action of an impulse.

If no impulse operates on N, we say that N is in a pause-phase. Thus, if a string of inputs operates on a Normed Network N, N is alternatively in a utilization-phase and a pause-phase. It is noteworthy that no time-factor in these phases is known. Indeed, according to the algebraic definition there is no necessity in dealing with time.

For a more detailed introduction to Normed Networks we refer to Th. Ottmann [2], L. Priese [1], and W. Rödding, H. Nachtkamp [3].

Bases

Two types of automata proved to be convenient and important for the analyses of Normed Networks:

a. The switch $H_{l,m,n}$ defined by

$$H_{l,m,n} := (\{t_1,\ldots,t_l,u_1,\ldots,u_m,d_1,\ldots,d_n\},\{t^u_1,\ldots,t^u_l,\ t^d_1,\ldots,t^d_l,u'_1,\ldots,u'_m,d'_1,\ldots,d'_n\},\{\text{up},\text{down}\},R),$$

where R is defined by the following transitions:

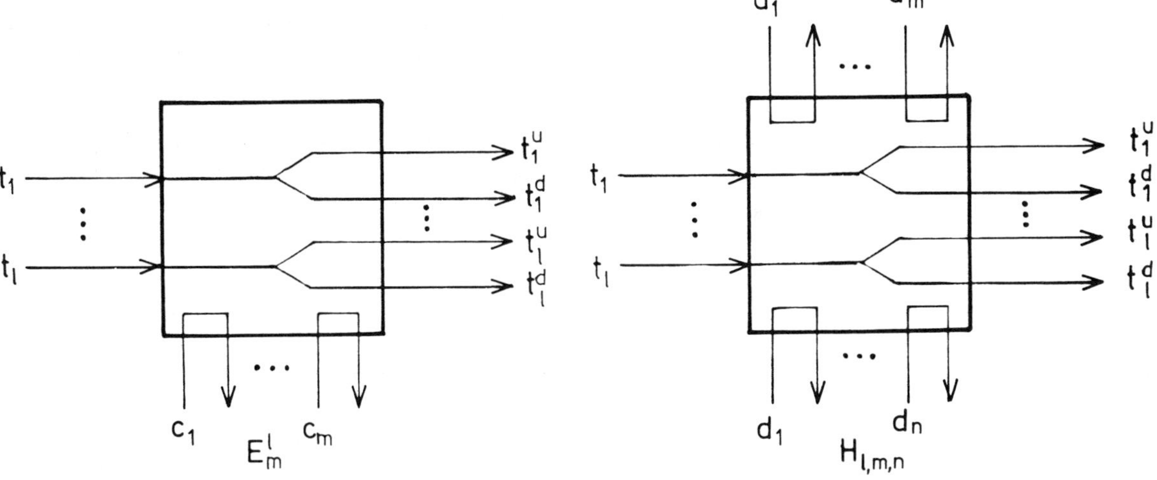

Fig.1

t_i, up	$\rightarrow$	t_i^u, up †		$1 \leqslant i \leqslant l$
t_i, down	$\rightarrow$	t_i^d, down		$1 \leqslant i \leqslant l$
u_i, up	$\rightarrow$	u_i', up		$1 \leqslant i \leqslant m$
u_i, down	$\rightarrow$	u_i', up		$1 \leqslant i \leqslant m$
d_i, up	$\rightarrow$	d_i', down		$1 \leqslant i \leqslant n$
d_i, down	$\rightarrow$	d_i', down		$1 \leqslant i \leqslant n$

The switch $H_{l,m,n}$ possesses two states, l independent test-lines t_i, and m, n independent switch-lines u_i, d_i, that always change the state to either up or down.

b. The switch E_m^l defined by

$$E_m^l := (\{t_1,...,t_l, c_1,...,c_m\}, \{t_1^u,...,t_l^u, t_1^d,...,t_l^d, c_1',...,c_m'\}, \{\text{up}, \text{down}\}, R),$$

where R is defined by the transitions

t_i, up	$\rightarrow$	t_i^u, up	$1 \leqslant i \leqslant l$
t_i down	$\rightarrow$	t_i^d, down	$1 \leqslant i \leqslant l$
c_i up	$\rightarrow$	c_i', down	$1 \leqslant i \leqslant m$
c_i down	$\rightarrow$	c_i', up	$1 \leqslant i \leqslant m$

The switch E_m^l possesses two states, l independent test-lines t_i, and m independent switch-lines c_i that bring about alternative changes in the states.

As a utilization of a switch-line in an E_m^l-automaton doesn't always result in the same state—as a utilization of a u- or d-switch-line in the $H_{l,m,n}$-case will do—operating with E_m^l-modules will prove more difficult than operating with $H_{l,m,n}$-modules.

As we don't possess output-identifying operations on Normed Networks we will need the automaton K defined as

$$K := (\{1,2\}, \{3\}, \{s\}, \{i, s \rightarrow 3, s \,/\, i = 1,2\}).$$

K, $H_{l,m,n}$, and E_m^l are given the diagrams of Fig. 1.

In the following we will define automata and Normed Networks in terms of diagrams. A network-diagram uniquely defines an automaton by means of composing all occurrent automata applying the product and feed-back operations according to the structure of the network-diagram. By our first lemma the resulting automaton is independent of the order the operations are applied in.

The following theorem can be proved with little difficulty:

Theorem 1: For all automata A there exist integers l,m,n and a Normed Network N^1 over K and $H_{l,m,n}$ such that

$$A \rightarrow N^1$$

holds.

† Instead of $((e,s), (e',s'))$ we sometimes write $e,s \rightarrow e',s'$.

For automata A and B we say that B *simulates* A, $A \to B$, iff $I_A = I_B$, $O_A = O_B$, and for all $s \in S_A$ there exists an equivalent state $s' \in S_B$.

Theorem 2: For all automata A there exist integers l,m and a Normed Network N^2 over K and E^l_m such that

$$A \to N^2$$

holds.

Theorem 3: For all integers l,m,n there exists a Normed Network H' over K and $H_{1,1,1}$ such that

$$H_{l,m,n} \to H'$$

holds.

Theorem 4: For all integers l,m there exists a Normed Network E' over K and E^1_1 such that

$$E^l_m \to E'$$

holds.

Thus we conclude:

Theorem 5: For all automata A there exists a Normed Network $N^{5,1}$ over K and $H_{1,1,1}$ and a Normed Network $N^{5,2}$ over K and E^1_1 such that

$$A \to N^{5,1} \quad \text{and} \quad A \to N^{5,2}$$

holds.

Thus, $\{K,E^1_1\}$ and $\{K,H_{1,1,1}\}$ form bases for all automata.

As an exercise and a demonstration we will construct a Normed Network N_H over K and E^1_1 that simulates $H_{1,1,1}$. Just choose N_H to be the Normed Network defined by Fig. 2.

Define $S_{N_H} := \times_{1 \leq i \leq 4} S_{B_i}$, then it can be easily seen:

$$S_{H_{1,1,1}} \ni \text{up} \sim (\text{up}, \text{up}, \text{up}, \text{up}) \in S_{N_H}$$

$$S_{H_{1,1,1}} \ni \text{down} \sim (\text{down}, \text{down}, \text{down}, \text{down}) \in S_{N_H}.$$

Thus, we know $H \to N_H$, q.e.d.

The above construction can even prove the state-correspondence to define an S-homomorphism. Thus, H is isomorphic to a sub-automaton of N_H.

The concept of self-correction

If, in a Normed Network N over H $(=H_{1,1,1})$ and some other automata, we substitute some occurrences of H by the subnet N_H the input-output-behavior of N will not be changed. But this construction N_H somehow appears to be 'unstable':

Suppose in N the state of one occurrence of H is switched to an incorrect state. If this component-machine H of N is now used on its u- or d-line the first time after this fault-implementation, this fault-implementation doesn't harm the correct behavior of N. But if H is substituted by N_H and such a fault-implementation occurs in N_H the state of one component-machine of N_H is incorrectly changed. That is, a first utilization of N_H on its u- or d-line will correct only a fault inside B_1 or B_3. A fault-implementation inside B_2 or B_4 will, in any case, yield an incorrect behavior.

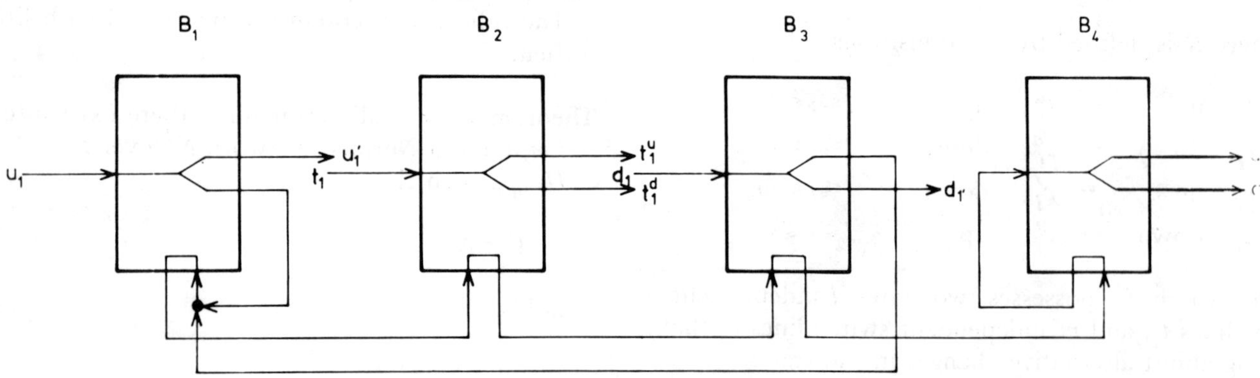

Fig.2

This informal discussion leads one to feel that the organization of N_H does simulate H, but doesn't possess the same 'stability'. We have to make this concept a precise one–in terms of 'fault-tolerance' or 'self-correction'. Our aim is to deal with organizations that are fault-correcting.

Given a network N, we will confine ourselves to cases in which faults can only occur inside the component-machines of the network. Consequently, rather than the structure of a network being destroyed, only the elements of a network may become defective. As we do not wish to deal with a repairing-device we will put forward an arrangement whereby a fault inside a component-machine will not mean that this element will stop working but that it will work incorrectly. This can be subsumed by switching an element to an incorrect state. Obviously, the number of allowed fault-implementations has to be limited as there can be no possible correct behavior if all component-machines of a net are allowed to change their state: i.e. a state

$$s \in S_N = \bigtimes \{S_B / B \text{ component-machine of } N\}$$

could be changed to any other state $s' \in S_N$ in this manner. To facilitate further discussion we first require still another standardizing of fault-implementation: those fault-implementations only occurring between successive utilizations of a network are valid.

As a state of a Normed Network

$$N = (B_1 // \ldots // B_t)^{x_1,\ldots,x_n}_{y_1,\ldots,y_n}$$

is a t-vector, we introduce on S_N the Hamming-distance by putting $d(s,s') := k$ iff s and s' differ in exactly k components. We regard S_N as the normed space (S_N, d) and use the common topological notations such as $U^k_N(S) := \{s \in S_N / \exists s' \in S : d(s,s') \leqslant k\}$. Giving an implementation of k faults to a state s of N results in s changing to s', with $d(s,s') = k$.

Consider the following:

A Normed Network N is k-self-correcting, iff all states s,s' of N with $d(s,s') \leqslant k$ are equivalent.

But this definition doesn't make sense as, even for $k = 1$, k-self-correcting Normed Networks can only consist of equivalent states. The reason for this is that all states of N are accepted as starting-states in the above 'definition'. But, usually in the theory of automata, nets have to fulfil certain 'tasks'; only a small number of states are needed as starting-states. To solve a 'task' simply implies simulating another automaton. Let us repeat the definition of *realization* or *simulation:*

An automaton C realizes (or simulates) an automaton A, $A \to C$, iff

a. $I_C = I_A$, $O_C = O_A$

b. *For all states $s \in S_A$ there exists a nonempty set K_s of S_C such that for all inputs x of I_A and states s' of K_s it holds:*

$$\lambda_C(x,s') = \lambda_A(x,s) \quad \text{and}$$
$$\delta_C(x,s') \in K_{\delta_A(x,s)} .$$

λ denotes the output-function and δ the next-state-function of automata. Thus, only the states of

$$\bigcup_{s \in S_A} K_s$$

are starting-states of C. We can give a precise notation of self-correction now by:

A Normed Network N simulates an automaton A in k-self-correction, $A \underset{k-sc}{\to} N$, iff

a. $I_A = I_N$, $O_A = O_N$

b. *For all states $s \in S_A$ there exists a nonempty set K_s of S_N such that for all inputs x of I_A and states s' of $U^k_{S_N}(K_s)$ it holds:*

$$\lambda_N(x,s') = \lambda_A(x,s) \quad \text{and}$$
$$\delta_N(x,s') \in K_{\delta_A(x,s)} .$$

We will introduce some different definitions for self-correction later.

Self-correcting bases

Two theorems about self-correction, similar to theorems 1 and 2, can be proved straightforwardly:

Theorem 6: For all integers k and all automata A there exist integers l,m,n and a Normed Network N^6 over K and $H_{l,m,n}$ such that

$$A \underset{k-sc}{\to} N^6$$

holds.

Theorem 7: For all integers k and all automata A there exist integers l,m and a Normed Network N^7 over K and E^l_m such that

$$A \rightarrow_{k\text{-}sc} N^7$$

holds.

Thus, $\bigcup_{l,m,n \in N} \{H_{l,m,n}\} \cup \{K\}$ and $\bigcup_{l,m \in N} \{E^l_m\} \cup \{K\}$ form two infinite bases for self-correction. But we are interested in finite sets of automata that form universal bases for self-correction. We will define:

A finite set $\{A_1,...,A_r\}$ of automata is a k-self-correcting base, iff for all automata A there exists a Normed Network N over $A_1,...,A_r$ that simulates A in a k-self-correction.

A finite set $\{A_1,...,A_r\}$ of automata is a self-correcting base, iff it is a k-self-correcting base for all integers k.

We will restrict ourselves to the case of E^l_m-component-machines first. In an initial step we will develop a method for combinatorial constructions that gives a short but elaborated proof for K and E^2_2 forming a 1-self-correcting base.

Theorem 8: $\{K,E^2_2\}$ forms a 1-self-correcting base.

Outline of proof:

We will call a Normed Network N *dense*, iff any utilization of N yields a utilization of each component-machine with two or more states. In other words, whenever an impulse passes a dense net N it must pass each component-machine B of N with $|S_B|>1$.

Given an automaton A we will construct a Normed Network N over K, E^1_1, and $F^\wedge$ with

$$F^\wedge := (\{1,4\},\{2,3,5\},\{\text{up},\text{down}\},R),$$

where R is defined by the following transitions:

1, up $\rightarrow 2$, down
1, down $\rightarrow 3$, up
4, up $\rightarrow 5$, up
4, down $\rightarrow 5$, down ,

such that

a. $A \rightarrow N$.
b. N is dense.
c. During each utilization of N, whenever an im-

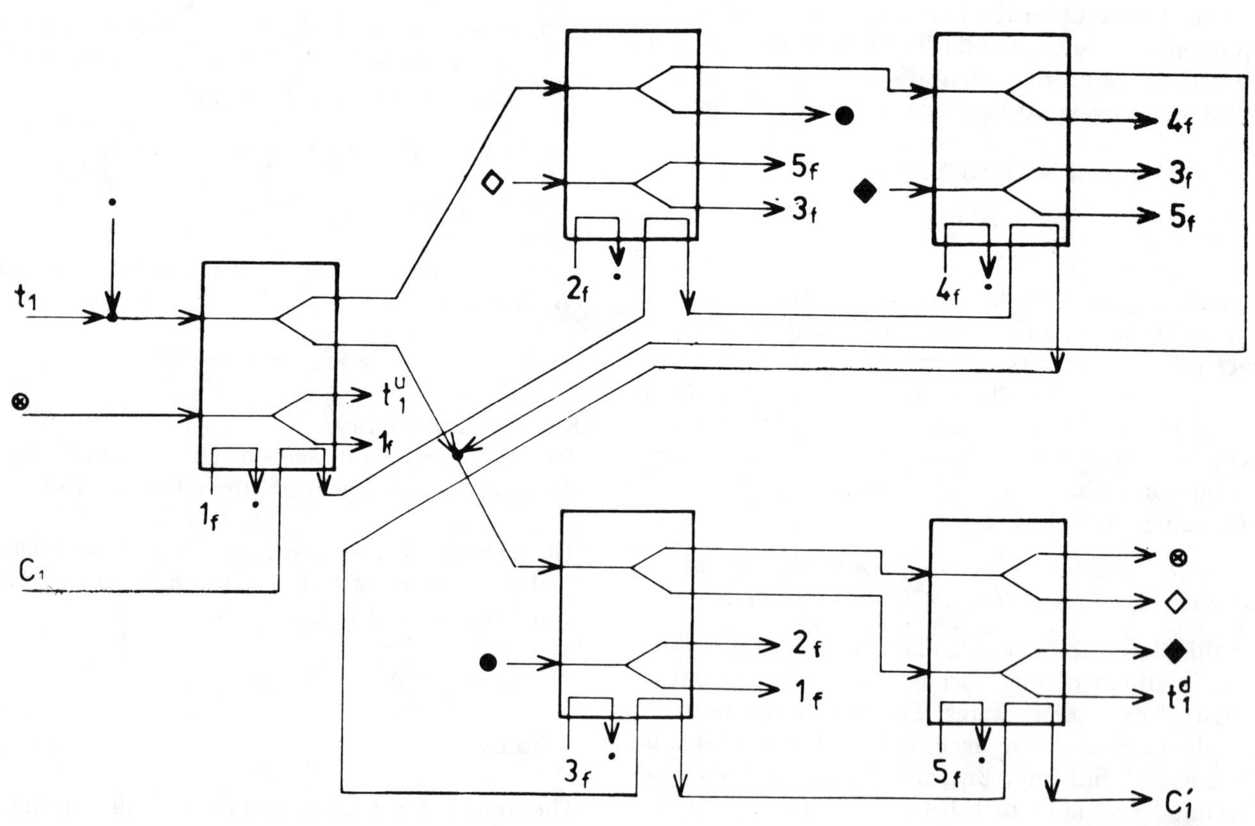

Fig.3

pulse passes an $F^{\wedge}$-module for the first time, it will use the 4-input of $F^{\wedge}$, the latter being in the state up.

d. During each utilization of N the impulse passes each E_1^1-module in the state up by its t_1-t_1^u-line at least once.

We, thus, obtain a Normed Network N simulating A where we, in addition, have some information about the way its component-machines have to be used. We now only need a Normed Network F^+ over K and E_2^2 that simulates $F^{\wedge}$ and that corrects one fault if used on its 4-input, and a Normed Network E^+ over K and E_2^2 that simulates E_1^1 correctly even if one possible fault occurs and that will correct that fault if used on its t_1-input in the state up. Now substitute in N each $F^{\wedge}$ by F^+ and each E_1^1 by E^+. The resulting net obviously simulates the given automaton A in 1-self-correction and only consists of K- and E_2^2-component-machines.

E^+ may be chosen as the Normed Network as defined by Fig. 3. If all components of E^+ are correct they shall have the same state, and E^+ has the state up (down) iff all components are in the state up (down, respectively). For details we refer to L. Priese and D. Rödding [4].

We will need the same idea of proof later again. Theorem 8 can be further improved:

Theorem 9: $\{K, E_1^2\}$ forms a self-correcting base.

Hence, we have to construct for all integers k and all automata A a Normed Network N^9 over K and E_1^2 that simulates A in k-self-correction. Such a construction will be successful by developing a new reduction technique:

In the theory of Normed Networks it is a main but hidden problem to find an organization consisting, say, of E_m^l-modules for some fixed integers l and m that allows a line to be used in different situations without losing information about the context that has led to the utilization of that line. To be more concrete:

Define two Multiliners M_1^n, M_2^n by:

$$M_1^n := (\{1,\dots,n,e\},\{1',\dots,n',o\},\{s_1,\dots,s_n\},R),$$

where R is defined by the transitions

$$i,s_j \rightarrow o,s_i \qquad 1\leqslant i,j\leqslant n$$
$$e,s_i \rightarrow i',s_i \qquad 1\leqslant 1\leqslant n\ ,$$

$$M_2^n := (\{1,\dots,n,e_1,e_2\},\{1',\dots,n',1'',\dots,n'',o\},$$
$$\{s_1,\dots,s_n\},R),$$

where R is defined by the transitions

$$i,s_j \rightarrow o,s_i \qquad 1\leqslant i,j\leqslant n$$
$$e_1,s_i \rightarrow i',s_i \qquad 1\leqslant i\leqslant n$$
$$e_2,s_i \rightarrow i'',s_i \qquad 1\leqslant i\leqslant n\ .$$

M_1^n and M_2^n receive the diagrams of Fig. 4.

Now the Normed Network of Fig. 5 obviously simulates E_m^l.

So the test-input of E_1^1 in Fig. 5 can be used in l different situations and the switch-input in m different situations. The overall organization is managed by the two Multiliners M_2^l and M_1^m.

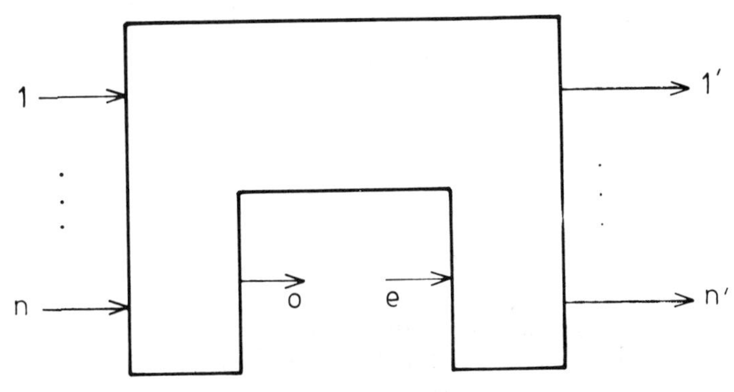

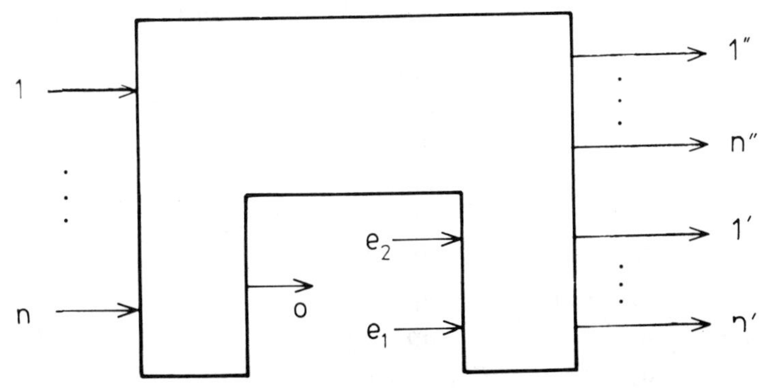

Fig.4

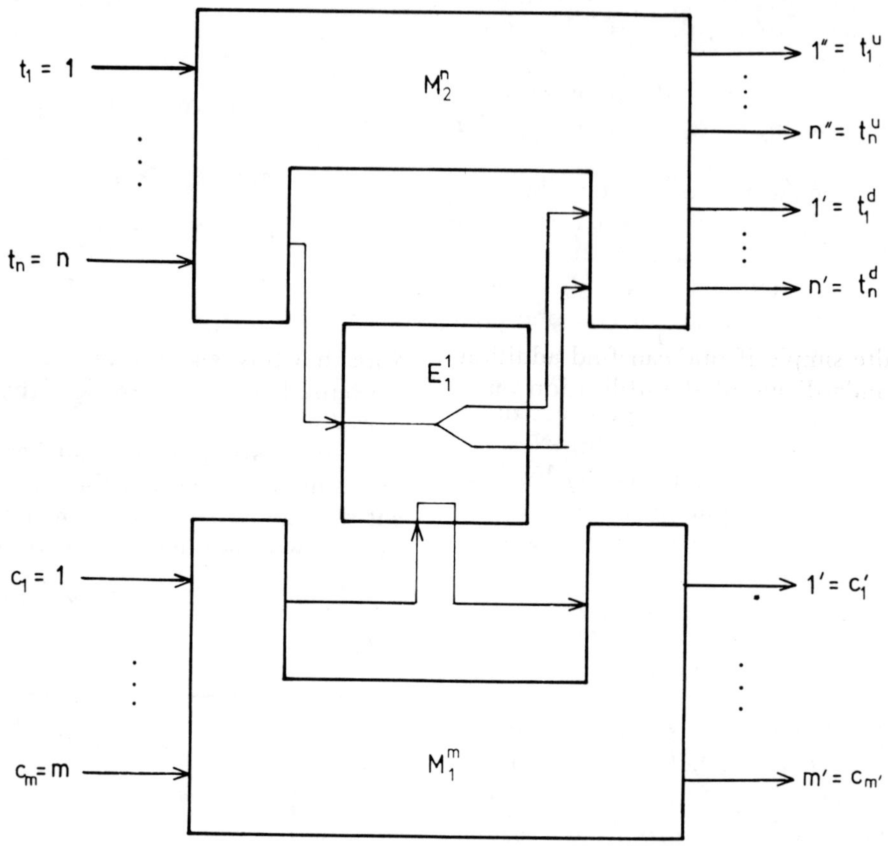

Fig.5

The reduction of theorem 2 to theorem 4, e.g., may be interpreted as being that organization of both kinds of Multiliners that can be compounded in Normed Networks over K and E_1^1. We will try to reduce theorem 7 to theorem 9 by looking for a stable organization for the Multiliners in terms of K and E_1^2 components!

In the first step we will operate with automata

$$M := (\{t,u,d\}, \{t^u, t^d\}, \{\text{up}, \text{down}\}, R),$$

where R is defined by the transitions

t, up $\rightarrow t^u$, up
t, down $\rightarrow t^d$, down
u, up $\rightarrow t^u$, up
u, down $\rightarrow t^u$, up
d, up $\rightarrow t^d$, down
d, down $\rightarrow t^d$, down ,

and

$$F := (\{c,u\}, \{c^u, c^d, u'\}, \{\text{up}, \text{down}\}, R),$$

where R is defined by the transitions

c, up $\rightarrow c^d$, down
c, down $\rightarrow c^u$, up
u, up $\rightarrow u'$, up
u, down $\rightarrow u'$, up .

We can compound a stable organization $\hat{M}_i^n$ for both Multiliners M_i^n, $1 = 1,2$, by means of M and F component-machines only, such that the states of all component-machines M and F may be switched to an arbitrary state but $\hat{M}_i^n$ will still behave correctly as M_i^n if used in the net of Figure 5! Thus there holds:

Theorem 10: For all integers k and all automata A there exists a Normed Network N^{10} over K, E_1^1, M, and F such that it holds:

a. $A \underset{k-sc}{\rightarrow} N^{10}$

b. N^{10} simulates A correctly even if before each utilization of N^{10} all M- and F-component-machines are allowed to be switched to an arbitrary state.

The first application is to drop E in theorem 10:

Lemma: $\{K,F,M\}$ is a self-correcting base.

It thus remains that a stable simulation for F and M by Normed Networks over K and E_1^2 be found. This should become quite simple if one can find additional properties of standardizing of the utilization of the M- and F-component-machines in N^{10}. Such an additional standardizing will be that each first utilization of a component-machine M and F will be on their u-input. Now, in terms of K and E_1^2, it is not too hard to simulate M and F with this restricted behavior.

Theorem 8 thus results. For details we refer to L. Priese [5].

On E^1_m-component-machines

Up to now $\{K,E^l_m\}$ is recognized as forming a self-correcting base for all integers $l \geqslant 2$ and $m \geqslant 1$. There remains the case of the E^1_m-component-machines.

This part is solved by the following theorem.

Theorem 11: For all integers m there exists no Normed Network N over K and E^1_m that simulates the automaton E^1_1 in 1-self-correction.

A proof can be found in L. Priese [5]. Our theory thus has the acceptable property of giving a sharp boundary of complexity for self-correction:

Lemma: For all integers k,l,m $\{K,E^l_m\}$ forms a k-self-correcting base iff $l \geqslant 2$ and $m \geqslant 1$.

Distributed fault-implementation

We have already mentioned that the amount of fault-implementation has to be limited in some way. Instead of fixing an upper bound of allowed faults one might try to standardize fault-implementation in such a way that there shall be no local accumulation of defects.

Let us call two component-machines of a net *directly connected* if there exists a line that connects one output of one component-machine to one input of the other without passing any further component-machine except some K-modules eventually. We now state a definition:

A Normed Network N simulates an automaton A in distributed self-correction, $A \to_{dsc} N$, *iff N simulates the input-output-behavior of A correctly despite the fact that between any two utilizations of N all component-machines may be switched to an arbitrary state according to the rule that no two directly connected component-machines will be both switched to an incorrect state.*

Note that this is a kind of fault-implementation that can be self-corrected by biological systems, whilst an increasing of local defects generally leads to severe damage.

If one analyses the proof of theorem 8 one obtains:

Theorem 12: For all automata A there exists a Normed Network N^{12} over K and E_2^2 such that

a. $A \to_{dsc} N^{12}$.

b. N^{12} is dense.

c. N^{12} consists of only $10|I_A||O_A|+2$ component-machines E_2^2.

d. N^{12} corrects up to 60% of its component-machines if the faults occur in the distributed manner.

Analyzing theorem 9 gives rise to a similar result. Together with theorem 11 we thus know:

Theorem 13: For all integers l,m: $\{K,E^l_m\}$ forms a distributed self-correcting base iff $l \geqslant 2$ and $m \geqslant 1$.

Stable self-correction

The most restrictive standardizing of both definitions on k-self-correction and distributed self-correction is the requirement that faults shall not occur during a utilization-phase at all. We will now drop this requirement.

The behavior of a Normed Network is as follows: An impulse enters N and, after generally changing the state of N, leaves N (utilization-phase). After that nothing happens inside N (pause-phase) until a new utilization-phase is started. We will call a pair (pause-phase, utilization-phase) a *unit-phase*. If we want to allow fault-implementations during a utilization-phase, too, we thus have to limit the fault-implementations during a unit-phase:

A Normed Network N simulates an automaton A in k-stable-self-correction, $A \to_{k\text{-}st\text{-}sc} N$, iff N simulates the input-output-behavior of A correctly, even if up to k component-machines of N may be switched to an incorrect state during each unit-phase.

Obviously this type of fault-implementation is much more severe—and more interesting. Now it is impossible to ensure that after a utilization-phase no faults inside the network are left uncorrected.

As far as can be seen there seems to be no possibility of translating the proof of the self-correcting universality of E^2_1—theorem 9—to this kind of fault-implementation. This is because the organization of K- and E^2_1-modules to compound Multiliners is not capable of remembering to correct inputs if faults are allowed during a utilization.

But the technique of theorem 8 can be used to construct k-stable-self-correcting organizations built up with K- and E^2_2-component-machines! It is not hard to use this technique and further prove:

Theorem 14: For all integers k and all automata A there exists a Normed Network N^{14} over K and E^2_2 that simulates A in k-stable-self-correction.

Together with theorem 11 this yields:

Theorem 15: For all integers *k,l,m:*

$l \geqslant 2,\ m \geqslant 2 \to \{K, E^l_m\}$
forms a k-stable-self-correcting base

$l = 1 \to \{K, E^1_m\}$
doesn't form a k-stable-self-correcting base.

The E^2_1-case is an open problem.

Stable distributed self-correction

We will try to introduce the concept of distributed fault-implementation for stable self-correction, too:

A Normed Network N simulates an automaton A in stable distributed self-correction, $A \to_{st\text{-}d\text{-}sc} N$, iff N simulates the input-output-behavior of A correctly, even if during each unit-phase all component-machines of N may be switched to an arbitrary state according to the rule that no two directly connected component-machines will be both switched to an incorrect state.

If dealing with distributed fault-implementation during a unit-phase up to two faults in each directly connected subnet† of N can occur: one of the last unit-phase that couldn't be corrected (because the fault was implemented after that part of the net was utilized last) and one of the unit-phase under consideration.

With regard to E^2_2-component-machines the directly connected subparts of a Normed Network must be quite small—but still up to two faults are allowed. It is thus somewhat surprising that E^2_2-modules can handle this kind of fault-implementation. But with some 'tricky' constructions it can be proved:

Theorem 16: For all automata A there exists a Normed Network N^{16} over K and E^2_2 that simulates A in stable-distributed-self-correction.

For details we refer to L. Priese [6]. Nothing is known about the E^2_1-case.

On $H_{l,m,n}$-component-machines

It is easier to handle $H_{l,m,n}$-component-machines than E^l_m-modules, as a utilization of a u_i-, $1 \leqslant i \leqslant m$, or d_i-, $1 \leqslant i \leqslant n$, input results in the unique states up and down, respectively. Nevertheless, it is surprising that there exists one technique developed by D. Rödding, that manages the proofs of all following results:

Theorem 17: For all integers *k,l,m,n:*

a. $\{K, H_{l,m,n}\}$ forms a k-self-correcting base.

b. $\{K, H_{l,m,n}\}$
forms a distributed-self-correcting base.

c. $\{K, H_{l,m,n}\}$
forms a k-stable-self-correcting base.

(a) and *(b)* are proved in L. Priese, D. Rödding [4], and the same technique as developed there will also prove *(c)*.

References

1. PRIESE, L., 'Reversible Automaten und einfache universelle 2-dimensionale Thue-System', *Zeitschrift für math. Logik und Grundlagen d. Math.*, Vol.22, 76.

†A subnet is called directly connected if each pair of components of this subnet is directly connected.

2. OTTMANN, Th., 'Über Möglichkeiten zur Simulation endlicher Automaten durch eine Art sequentieller Netzwerke aus einfachen Bausteinen', *Zeitschrift f. math. Logik u. Grundlgen. d. Math.*, Vol.19, 73.
3. RÖDDING, W. and NACHTKAMP, H., 'On the Aggregation of preferences to form a preference of a system', to appear in *Naval Research Logistic Quarterly,* Washington.
4. PRIESE, L. and RÖDDING, D., 'A combinatorial approach to self-correction', *Journal of Cybernetics* 4(3), 74.
5. PRIESE, L., 'On fault-tolerant networks', in preparation.
6. PRIESE, L., 'On the minimal complexity of component-machines for self-correcting networks', *Journal of Cybernetics* 5(4), 75.

Aggregation in system dynamics models

J.A. SHARP
System Dynamics Research Group,
University of Bradford, England

Introduction

Most systems that are the subject of a System Dynamics study usually contain hundreds or even thousands of variables that potentially might be included in the model. Since it is widely accepted, however, that the difficulty of constructing a working model increases exponentially with the number of variables involved it is of considerable practical importance, as the budget for an S.D. study is limited, to reduce the number of variables to a manageable figure. In part this can be done by the omission of variables that are deemed to be unimportant, but the main approach adopted is generally to make extensive use of aggregation. At the same time this aggregation fits in well with the aim of using an S.D. model to examine policies that will improve system performance. In general it is far simpler to set up broad policies to deal with aggregates, e.g. company investment, rather than individual components of these aggregates such as investment in milling machines. Normally devising improved control policies for a socioeconomic system is difficult enough without further complicating the problem by attempting to design different control policies for each component of an aggregate. Indeed it is clear that were this to be attempted a great deal of the power of S.D. methods would be lost because of the consequent complexity of the redesign problem, and the fact that this very complexity would inevitably reduce the number of redesign options that could be examined. At the same time practical policy making for the higher level management and political problems to which System Dynamics is usually applied does in fact require just such a design of broad policies rather than the focus on individual detail [1].

There are thus excellent practical reasons why S.D. models should be aggregated and few, if any, models do not make implicit use of aggregates. Thus the production-distribution model [2] deals with an 'aggregate' product made up of a number of individual product items, as does the Sprague study in the same work. The use of aggregation in World Dynamics [3] and Urban Dynamics [4] is sufficiently obvious for comment to be unnecessary.

The importance of aggregation has generally been recognized in the System Dynamics literature [4] but there appear to be no published systematic studies of where it is justified, though recently there has been considerable interest in the construction of partially disaggregated models as a result of the World Dynamics Project [5].

Such lack of interest in the theory of aggregation in S.D. models is rather surprising since the subject has aroused considerable interest in Econometrics and the resulting work would tend to suggest that aggregation for multi-equation systems is difficult to justify except under rather special circumstances [6].

The Econometric studies that have been carried out have, however, been oriented toward the study of regression [6] or Production Functions [7]. The aim of this chapter is, therefore, to examine the problem of aggregation for S.D. models.

The aggregation problem

There would seem to be three broad approaches to the construction of aggregate equations for an S.D. model. The first, which we may dub *Optimality Aggregation* relies on assumptions of optimal behavior on the part of those individuals who determine the individual variables that make up the aggregate. There is an extensive economic literature that deals with such aggregates cf. [8,9]. The second type we may term *Statistical Aggregation* in that it relies on observed regularities in the behavior of the aggregates independent of their individual components. Such behavior may be found, for example, in connection with the distribution of incomes [10]. The third type of aggregation we may call *Aggregation by Analogy*, in that aggregates are assumed to follow the same laws as that for individual components. As an example of such a process we consider equation 123 (p.203 [4]).

TRNP.K=TRNP.J+(DT/TRNPT)(TRN.J-TRNP.J)

This equation may be said to represent the mechanism by which individuals adjust their perception (TRNP) of the TAX RATIO NEEDED (TRN) but adjusted for aggregation purposes by taking as TRNPT–the perception time–some average of the perception time across all voters. In the model's own terms, however, it would seem likely that these perception times would be different for the managerial, labor and under-employed groups and, since the numbers in each group are continually changing as a result of the dynamics of the system, the average perception time on which the aggregation is based may be expected to change continually also. Furthermore, in any practical application, it would seem unlikely that the three different worker groups could in fact be considered homogeneous so that we might expect the basis of aggregation of most of the model equations, e.g. the demographic equations to be continually changing. Moreover even if the groups were homogeneous such an aggregation process needs justification.

Similar comments of course apply to the World Model [3] where aggregation takes place across the population and countries of the World, and the Dynamic effects of differing economic and population growth rates may be expected to be considerable, and the model has been criticized on this basis [11].

In general aggregation by analogy may be said to be the most common type of aggregation in S.D. models and, since it lacks either the theoretical or empirical justification of the first two types of aggregation, we shall devote most of this chapter to its study. In considering it three questions arise which we shall attempt to answer:

a) Under what circumstances may aggregate equations be formed on the basis of equations describing the behavior of individual system components?

b) How should the structure of these aggregate equations be related to the structure of the individual equations?

c) How should the parameters of these aggregate equations be related to that of the parameters of the individual equations?

Satisfactory and perfect aggregation

Since the reasons for aggregation are essentially pragmatic, it seems sensible to define adequate aggregation in a similar spirit. We accordingly define *Satisfactory Aggregation* as follows. Aggregation is said to be satisfactory if no significant inaccuracies are introduced into the study by the use of an aggregate model.

Such a definition though acceptable is rather difficult to define mathematically so we shall first examine the possibility of a more rigorous type of aggregation which can be so defined. This type of aggregation we shall call 'perfect' after Theil [6, p.140]. Aggregation is said to be perfect if the results obtained for the behavior of the aggregates are the same whether they are derived from the aggregate model or by working with models of individual components and deriving the aggregates from these micro-models.

'Perfect' Aggregation if attainable is obviously very desirable in that there is no conflict between information concerning the aggregates derived from the individual models and the aggregate models. Where Perfect Aggregation is not possible we may expect a number of problems to arise.

Firstly, the properties of the aggregate model and

the individual models may differ with regard to such basic characteristics as Stability.

Secondly, we shall not be sure how far errors that are found during validation of the Aggregate model are due to errors introduced by virtue of the Aggregation process, rather than weaknesses in the model formulation.

Thirdly, we shall not be sure that redesigned policies based upon the aggregate model will in fact have the same effect when applied to the actual system.

These are obviously serious problems so it is clearly necessary to assume that as far as our weaker definition is concerned, that aggregation can only be deemed to be satisfactory if they are of negligible importance for the system under consideration.

A general aggregation model

In order to make any progress in the mathematical examination of the Aggregation problem it is necessary to formulate a general model of the process of aggregation by analogy in System Dynamics. For simplicity we shall ignore the effects of noise and deal with deterministic models only. In order to reduce the notational complexity of the discussion we shall deal with the continuous formulation of an S.D. model [12] though similar results can be derived for the discrete form.

We assume, therefore, that an S.D. model exists describing the behavior of the individual components of the system, e.g. companies, income groups, families or even individual people of the form

$$\dot{w}_{iI} = f_{iI}(w_{jJ}, x_{kK}) \tag{1}$$

$$x_{qQ} = g_{qQ}(w_{rR}, x_{sS}) \tag{2}$$

$$\begin{aligned} &1 \leqslant i,j,r \leqslant m_1 \\ &1 \leqslant k,q,s \leqslant m_2 \\ &1 \leqslant I,J,R \leqslant M(i) \\ &1 \leqslant K,Q,S \leqslant N(q) \end{aligned} \tag{3}$$

The variables w_{iI} we shall call state variables and the variables x_{qQ} non-state variables. The lower case letters are assumed to refer to distinct variable types, e.g. Family Size, Family Income, whilst the Upper Case subscripts refer to individual representations of these variables. Thus the variable x_{qQ_1} for a particular value of q might represent the Income of the Q_1th family, whilst x_{qQ_2} would then represent the income of the Q_2th family. We assume that equations with the same lower case subscripts are similar in general form but that the parameters depend on the upper case subscript. This can always be achieved by putting some parameters in certain of the equations equal to zero if necessary.

We note that in some cases, e.g. in the demographic sector of a model, the equations in as far as they relate to actual individuals may need to be interpreted as expectations in the probability sense, e.g. expected birth rate.

The model (1) to (3) is said to be capable of perfect aggregation if there exist functions $W_i(w_{iI})$, $X_q(x_{qQ})$, $F_i(W_j, x_k)$, $G_q(W_r, X_s)$ such that

$$\dot{W}_i = F_i(W_j, X_j) \tag{4}$$

$$X_q = G_q(W_r, X_s) \tag{5}$$

By allowing aggregation to take place over different numbers of variables depending on the value of i or q we obtain a very general model. Thus if we have c consumers and f firms we might wish to write

x_{1CF} = amount purchased by consumer C from firm F

If we define $x_{1K} = x_{1CF}$

where $K = c \times (F-1) + C \quad 1 \leqslant K \leqslant cf$

then our new variable is of the required form as in equation (4) and can clearly be aggregated to X_1 which represents total consumption. Similarly if x_{2K} = production of Kth firm ($1 \leqslant K \leqslant f$) we can form X_2 representing total production by simple summation. Our formulation then enables us to envisage the aggregation of cf variables in the former case and f in the latter.

Necessary conditions for perfect aggregation

We shall now investigate the necessary conditions for perfect aggregation. Since we shall show that these are most unlikely to be fulfilled in practice we shall ignore the question of what constitute sufficient conditions. The method we shall use is essentially a generalization of [7]. Since our approach is basically exploratory we assume that the second derivatives of all the functions involved exist.

To simplify the notation we write

$$W_{i,jJ} = \frac{\partial W_i}{\partial w_{jJ}}$$

$$W_{i,iIiI} = \frac{\partial^2 W_i}{\partial w_{iI}^2}$$

etc.

Furthermore we employ a summation convention to avoid a plethora of summation signs.

Variables which are to be summed over are, therefore, enclosed in angle brackets beneath the relevant equation.

Thus $f_{iI,jJ}dw_{jJ} + f_{iI,kK}dx_{kK}$ (6)

$\langle I, j, J, k, K\rangle$ denotes

$$\sum_I \left\{ \sum_{j=1}^{m_1} \sum_{J=1}^{M(j)} \frac{\partial f_{iI}}{\partial w_{jJ}} w_{jJ} + \sum_{k=1}^{m_2} \sum_{K=1}^{N(k)} \frac{\partial f_{iI}}{\partial x_{kK}} \right\} \quad (7)$$

where m_1, m_2 $M(j)$, $N(k)$ are as defined in (3).

We therefore have from (4)

$$d\dot{W}_i = F_{i,j}W_{j,jJ}dw_{jJ} + F_{i,k}X_{k,kK}\,dx_{kK} \quad (8)$$

$\langle j, J, k, K\rangle$

Also from the definition of W_i

$$\dot{W}_i = W_{i,iI}\,\dot{w}_{iI}$$

$\langle I\rangle$

$$= W_{i,iI}f_{iI} \quad (9)$$

$\langle I\rangle$

Thus

$$d\dot{W}_i = W_{i,iIiI}\,f_{iI}\,dw_{iI} + W_{i,iI}f_{iI,jJ}\,dw_{jJ} + f_{iI,kK}\,dx_{kK} \quad (10)$$

$\langle I, j, J, k, K\rangle$

An extension of Leontieff's approach would now lead us to equate the coefficients of corresponding differentials in Eqs (8) and (10) which, of course, is only valid if all the differentials can vary independently. In our case, however, the x_{kK} depend on w_{jJ} through Eq.(2) so this assumption is not obviously valid. In fact in a normal S.D. system the variables will normally all depend on one or two exogenous variables so that the assumption is, on the face of it, even more questionable. We can, however, justify the assumption that the differentials may be regarded as independent by the following arguments:

a) The individual relations in an S.D. model are considered to be valid for wide ranges of the values of the variables they contain and not merely for the values which they are constrained to take by the behavior of the rest of the system.

b) The fact that variables are influenced by random noise not included in (1) and (2). The influence of this noise is akin to that of differential changes in the variable concerned and these changes are at least in part indpendent of each other. Since we do not wish our aggregation to be affected by noise and because of argument *a)* we therefore conclude that the differentials in Eqs (8) and (10) must be regarded as capable of independent variation and it is, therefore, in order to equate corresponding coefficients in the two equations. This gives:

Equating coefficients of dw_{jJ}

$$F_{i,j}W_{j,jJ} = \delta_{ij}W_{j,jJjJ}f_{iJ} + W_{i,iI}F_{iI,jJ} \quad (11)$$

$\langle I\rangle$

where $\delta_{ij} = 1 \quad j = i$
$\qquad\quad = 0 \quad j \neq i$

Equating coefficients of dx_{kK}

$$F_{i,k}X_{k,kK} + W_{i,iI}\,f_{iI,kK} \quad (12)$$

$\langle I\rangle$

Applying the same approach to (2) and (5) gives

$$G_{q,r}W_{r,rR} = X_{q,qQ}\,g_{qQ,rR} \quad (13)$$

$\langle Q\rangle$

$$G_{q,s}S_{s,sS} = X_{q,}x_{qQ,sS} \quad (14)$$

$\langle Q\rangle$

Two trivial consequences of Eqs (11) and (12) are

a) If for every I and J, $\dot{w}_{iI}$ does not depend on w_{jJ} $j \neq i$ then $\dot{W}_i$ does not depend on W_j.

b) If for all I and K, $\dot{w}_{iI}$ does not depend on x_{kK} then $\dot{W}_i$ does not depend on X_k.

Similar results can be derived from (13) and (14).

These results show that if Perfect Aggregation is possible it must be intuitively reasonable. If, for example, sales are considered to be independent of investment for individual companies in an industry then likewise industry sales must be independent of industry investment. In this respect then aggregation by analogy is not incompatible with perfect aggregation.

Application of the perfect aggregation rules

We now consider the application of the above rules to some common equation types.

We first consider the aggregation of a number of individual smoothing equations given by

$$\dot{a}_I = \alpha_I(b_I - a_I) \quad (15)$$

where α_I is a parameter.

To fix ideas we assume that the b_I are non-state variables. Setting $A \equiv A(a_I)$, $B = (b_I)$ and assuming the aggregate relationship is $\dot{A} = F(A,B)$ (16) we have from (12)

$$w_{iI} \equiv a_I$$

$$x_{kK} \equiv b_K$$

$$f_{iI} \equiv \alpha_I(b_I - a_I)$$

Applying (12) we have

$$F_{i,k} \equiv \frac{\partial F}{\partial B}$$

$$X_{k,kK} \equiv \frac{\partial B}{\partial b_K} \quad (17)$$

$$W_{i,iI} \equiv \frac{\partial A}{\partial a_I}$$

$$f_{iI,kK} = \alpha_I \qquad K = I$$
$$= 0 \qquad K \neq I$$

$\therefore$ (12) gives

$$\frac{\partial F}{\partial B}\;\frac{\partial B}{\partial b_K} = \alpha_K \frac{\partial A}{\partial a_K} \quad (18)$$

Taking the ratio of (18) for two different values of K, K_1 and K_2 we find

$$\frac{\partial B}{\partial b_{L_1}} \Big/ \frac{\partial B}{\partial b_{K_2}} = \alpha_{K_1} \frac{\partial A}{\partial a_{K_1}} \Big/ \alpha_{K_2} \frac{\partial A}{\partial a_{K_2}} \quad (19)$$

Since the l.h.s. of (19) is a function only of the b_K and the r.h.s. only of the a_K it follows that for (19) to be true for arbitrary a_K and b_K both sides must be constant

whence [8, p.20] $A \equiv P(\sum_I \beta_I a_I)$

$$B \equiv Q(\sum_I \gamma_I b_I)$$

where P and Q are arbitrary functions and $\alpha_I \beta_I \equiv c\gamma_I$

where c, β_I and γ_I are constants.

Clearly since $F(P(\Sigma\beta_I a_I), Q(\Sigma\gamma_I b_I)) \equiv F^*(\Sigma\beta_I a_I, \Sigma\gamma_I b_I)$

we can without loss of generality take

$$A \equiv \Sigma\beta_I a_I$$

$$B \equiv \Sigma\gamma_I b_I$$

Following a similar procedure but using (11) gives

$$\frac{\partial F}{\partial A}\;\frac{\partial A}{\partial a_I} = \frac{\partial^2 A}{\partial a_I^2} + \frac{\partial A}{\partial a_I} \cdot -\alpha_I \quad (21)$$

By virtue of (20) we have from (21)

$$\frac{\partial F}{\partial A} \cdot \beta_I = -\alpha_I \cdot \beta_I \quad (22)$$

or taking the ratio of (22) for two arbitrary values of I we have

$$\frac{\alpha_{I1}}{\alpha_{I2}} = 1$$

Thus even for such a commonly occurring equation as the smoothing equation it would appear that perfect aggregation is not possible for arbitrary values of the individual variables, unless all the smoothing parameters are the same, in which case of course aggregation by summation is perfectly adequate.

As a further example of the type of problems

that may arise we consider a typical level equation

$$\dot{x}_I = y_I - z_I$$

where y_I and z_I are non-state variables. This leads to vid. Eqs (15)-(19)

$$\frac{\partial F}{\partial Y} \frac{\partial Y}{\partial y_K} = \frac{\partial Z}{\partial z_K}$$

which leads as above to $Y = \sum_I p_I y_I$ where

$$\frac{p_I}{q_I} = \text{constant}$$

$$Z = \sum_I q_I z_I$$

and thus

$$X = \sum_I p_I x_I \tag{23}$$

If we now assume that x_I appears in 2 other equations, say

$$m_I = \alpha_I x_I$$
$$n_I = \beta_I x_I \tag{24}$$

where m_I and n_I are non-state variables and α_I, and β_I are parameters, it is easy to show that unless $\alpha_I = k\beta_I$ no aggregation of (24) is possible that is compatible with (23).

As another illustration of the problem, let us assume that the variables x_I defined above appear in the non-state equations.

$$u_I = \frac{x_I}{w_I}$$

Applying Eq.(14) with $x_{ss} \equiv x_I$ and $U \equiv U(X,W)$ gives

$$\frac{\partial X}{\partial x_{I1}} \Big/ \frac{\partial X}{\partial x_{I2}} = \frac{w_{I2} \dfrac{\partial U}{\partial u_{I1}}}{w_{I1} \dfrac{\partial U}{\partial u_{I2}}}$$

which by virtue of (23) gives

$$\frac{w_{I1}}{w_{I2}} = k \frac{\dfrac{\partial U}{\partial u_{I1}}}{\dfrac{\partial U}{\partial u_{I2}}} \tag{25}$$

Now the l.h.s. of (25) is a function of w_{I1} and w_{I2} only whilst the r.h.s. is through u_{I1} and u_{I2} a function of x_{I1}, x_{I2}, w_{I1} and w_{I2} and therefore (25) cannot hold in general.

The arguments of the previous section are sufficient to indicate that in general Perfect Aggregation is unlikely to be possible in S.D. models. Clearly the root of the trouble is that we have assumed all the individual variables and parameters can vary independently which leads to conditions for Perfect Aggregation that are impossible to fulfil. The question then arises as to whether if we allow dependencies between the individual variables or the individual parameters whether aggregation will become possible and if so what type of dependence will do the trick.

It has been known for some time in Economics (cf. [6]) that in simple economic relations of the form

$$y_I = \sum_j \sum_J a_{jJ} x_{jJ}$$

aggregation is possible if x_{jJ_1}/x_{jJ2} = constant all J_1, J_2

or equivalently $x_{jJ} = k_{jJ} X_j$ where k_{jJ} = constant and $\sum_J k_{jJ} = 1$.

Before considering this type of aggregation, however, it seems worthwhile–since we are imposing restrictions on the variables–to impose some restrictions on the type of aggregation we will accept. We therefore impose the condition that the aggregation functions for the individual variables are not changed by a change in the parameters of the system equations where those parameters are not fixed a priori, i.e. the aggregates do not depend explicitly on the system parameters. There are a number of reasons why this condition appears desirable in S.D. models.

a) It makes possible general aggregate models in that the same model can be applied to different systems with the same method of aggregation.
b) When system parameters are changed since the basis of aggregation is unchanged it is easy to measure the improvements in performance generated by the change.
c) Given the uncertainty of many of our parameter estimates it is clearly undesirable for the aggregates to depend explicitly on the parameters. Indeed if the aggregates did depend on the parameters it would be a complex task to determine whether a change in a parameter would improve

the correspondence of the model to reality since the basis of aggregation would continually be changing.

We further assume that aggregates are simple weighted sums as is frequently the case in practice, e.g., $X(x_I) = \sum_I \beta_I x_I$ so, that aggregates are simple to interpret and measure.

We shall also make one other assumption that the aggregate equation has the same form as the individual equations. This again as remarked earlier is conventional S.D. practice.

Armed with these assumptions we now proceed to demonstrate, in a heuristic way, that aggregation is only likely to be possible where individual members of an aggregate are proportional to each other, or all the individual equations have the same parameters.

We consider an equation of the form

$$x_I = \alpha_I y_I \tag{26}$$

where α_I is a parameter.

By virtue of our assumptions the aggregate equation is of the form

$$X = \alpha Y \tag{27}$$

One solution of (26) and (27) is $\alpha_I = \alpha^*$, $X = \Sigma x_I$, $Y = \Sigma y_I$, $\alpha = \alpha^*$ (28)

Let us now consider the case where the α_I are not equal to each other. We have from (26)

$$dX = \frac{\partial X}{\partial \alpha_I} dI = y_I \frac{\partial X}{\partial x_I} d\alpha_I \tag{29}$$

whilst (27) gives since the aggregates are assumed not to depend explicitly on the parameters.

$$dX = Y \frac{\partial \alpha}{\partial \alpha_I} d\alpha_I \tag{30}$$

Thus

$$\frac{\partial X}{\partial x_I} = \frac{Y}{y_I} \frac{\partial \alpha}{\partial \alpha_I} \tag{31}$$

Since by assumption $X \equiv \sum_I \beta_I x_I$ we have

$$\frac{Y}{y_I} \frac{\partial \alpha}{\partial \alpha_I} = \beta_I$$

or

$$\frac{y_I}{Y} = \frac{1}{\beta_I} \frac{\partial \alpha}{\partial \alpha_I} \tag{32}$$

Since α is a function of the α_I only it follows that the r.h.s. of (32) must be independent of the y_I whence

$$\frac{y_I}{Y} = \text{constant}$$

or

$$y_I = k_{Y,I} Y \quad \text{where } \sum_I k_{Y,I} = 1$$

and

$$Y \equiv \sum_I y_I \tag{33}$$

It follows therefore that

$$\alpha = \sum_I \alpha_I k_{Y,I} \tag{34}$$

and

$$\frac{x_I}{X} = k_{X,I} = \frac{\alpha_I k_{Y,I}}{\alpha}$$

and

$$\sum_I k_{X,I} = 1 , \qquad \text{i.e. } X \equiv \Sigma x_I \tag{35}$$

Let us now consider an equation of the form

$$x_I = \alpha_I \frac{y_I}{z_I}$$

with corresponding aggregate equation $x = \alpha \dfrac{Y}{Z}$ (36)

Following the method used above we find that possible solutions are

$$x_I = k_{X,I} X \quad y_I = k_{Y,I} Y \quad z_I = k_{Z,I} Z$$

$$x = \sum_I x_I , \quad Y = \sum_I , \quad Z = \sum_I Z_i$$

$$\alpha = \frac{\Sigma \alpha_I k_{Y,I}}{k_{Z,I}}$$

$$k_{X,I} = \frac{\alpha_I k_{Y,I}}{\alpha k_{Z,I}} \tag{37}$$

or

$$\alpha = \alpha_I = \alpha^*_, \; X = \sum_I x_I \,, \; Y = \sum_I \frac{y_I}{k_{Z,I}}, \quad Z = \sum_I z_I \tag{38}$$

and

$$z_I = k_{z,I} Z$$

The considerations of this section to date, though they can have no pretence to rigor, suggest the following rules of thumb

a) Aggregation is only likely to be possible for linear systems if either all the parameters in the individual equations are the same or the individual variables move in proportion to each other.
b) Aggregation of nonlinear relations is only likely to be possible if some at least of the individual variables in those equations move proportionately to each other†, cf. Eq.(38).
c) In either case the aggregates concerned will be simple weighted sums so that their interpretation and measurement will be straightforward.

The case of individual linear equations with identical parameters needs no further discussion since the basis of aggregation is obvious. It is, however, of considerable practical interest since the individual model variables can usually be grouped in such a way that this is a reasonable assumption, e.g. by dividing the population into reasonably homogeneous social groups.

The case where individual variables move in proportion to each other is, however, rather more complex so we now consider it further.

In this situation the relationship between the individual variables is affected by the dynamics of the system itself, so the constant proportionality can only be expected to be maintained under certain special circumstances which need to be established. Of course, for linear models without exogenous inputs such proportionality can be shown to be inevitable at a certain stage of system evolution [13] but, since such systems are rare, we need to study the problem in greater generality.

We assume that $w_{iI} = k_{iI} W_i$

$$x_{kK} = k_{kK} X_k \quad \text{etc.} \tag{39}$$

i.e. proportionality between the variables exists.

†This can only be interpreted as a guideline since it is not always sufficient. Consider for example the aggregation of a set of functions $x_I = e^{a_I y_I}$.

We consider first the aggregation of a linear state equation of the form

$$\dot{w}_{iI} = A_{iIjJ}\, w_{jJ} + B_{iIkK} x_{kK} \qquad \langle j, J, k, K \rangle \tag{40}$$

which by virtue of (39) can be written

$$k_{iI} \dot{W}_i = A_{iIjJ}\, k_{jJ}\, W_j + B_{iIkK} k_{kK} X_K \qquad \langle j, J, k, K \rangle$$

or

$$\dot{W}_i = \frac{A_{iIjJ} k_{jJ}}{k_{iI}} W_j + \frac{B_{iIkK} k_{kK}}{k_{iI}} X_K \qquad \langle j, J, k, K \rangle \tag{41}$$

In order for an aggregate equation to exist it is clearly necessary that for any I

$$\frac{A_{iIjJ} k_{jJ}}{k_{iI}} W_j + \frac{B_{iIkK} k_{kK}}{k_{iI}} X_k \qquad \langle j, J, k, K \rangle \tag{42}$$

is independent of I.

The case where this holds by virtue of the fact that the terms (42) are identical for different values of I may be christened *equation similarity*. The case where (42) holds by virtue of the relationships

$$\frac{A_{iIjJ} k_{jJ}}{k_{iI}} = A_{ij} \tag{43}$$

$$\frac{B_{iIkK} k_{kK}}{k_{iI}} = B_{ik} \qquad \langle J \rangle$$

we may call term by term similarity.

Similar reasoning applies to linear non-state equations. The notions of equation and term by term similarity are easily generalized to nonlinear equations in that

$$\frac{1}{k_{iI}} f_{iI}(w_{jJ}, x_{kK}) \tag{44}$$

$$\frac{1}{k_{qQ}} g_{qQ}(w_{rR}, \dot{x}_{sS})$$

must evince the appropriate type of similarity.

In order to examine how (44) can be applied in practice we now consider the aggregation of the set of equations

$$x_I = \frac{y_I}{z_I + c_I} \tag{45}$$

where c_I is a parameter.

Writing $x_I = k_{X,I}X$, etc., we have that for term by term or equation similarity we must have

$$\frac{1}{k_{X,I}} \frac{k_{Y,I}Y}{k_{Z,I}Z + c_I} \tag{46}$$

is independent of I, or

$$\frac{Y}{\frac{k_{X,I}k_{Z,I}}{k_{Y,I}} Z + \frac{k_{X,I}c_I}{k_{Y,I}}} \tag{47}$$

is independent of I whence we must have

$$\frac{k_{X,I}k_{Z,I}}{k_{Y,I}} = a \quad \text{and} \quad \frac{k_{X,I}c_I}{k_{Y,I}} = b \tag{48}$$

where a and b are constants in which case

$$X = \frac{Y}{a\ Z+b} \tag{49}$$

Parameters of the aggregate equation

The assumption that variables are interrelated though it enables us to aggregate the equations can lead to rather curious aggregate parameters.

Let us consider (45) with $c_I = 0$ and interpret y_I as target stock, z_I as actual stock and hence x_I as the stock cover ratio for the Ith company. Clearly, unless a is fortuitously equal to 1, (49) cannot be interpreted as the stock cover ratio for the industry. If, however, we assume that x_I is required to calculate the production for the Ith company via the relationship

$$p_I = d_I x_I \tag{50}$$

where d_I is some parameter then similarity considerations dictate

$\frac{d_I k_{X,I}}{k_{p,I}}$ is independent of I so we arrive at the equation for industry production $P = dX$ (51)

$$\text{where } d \equiv \frac{d_I\, k_{X,I}}{k_{p,I}}$$

Thus X whilst it is required in (51) and will actually be calculated by (47) is not, unless $a = 1$, the inventory cover ratio for the industry.

We thus have two choices

a) Use equations (49) and (51) but abandon the interpretation of X as the industry stock cover ratio.

b) Modify the definition of X so that it is the industry stock cover ratio, by using the equation

$$X = \frac{Y}{Z}$$

and change the parameter d in (51) to d/a.

In either case the interpretation of the equations is no longer straightforward. The problem posed by the discussion of this section is of especial interest when considered in the light of actual practice in building aggregate models, e.g. the equation for Manufacturing Delay in the Electronics Industry Model [2, p.227, eqn.17-50].

To show that the problem is not illusory we take a very simple example of the above equations. We consider an industry with only two companies with

$$x_1 = x_2 = 1$$

$$y_1 = y_2 \equiv 1$$

$$z_1 = z_2 = 1$$

$$d_1 = d_2 = 1 = d$$

$$c_1 = c_2 = 0$$

We then have from (50)

$$P = p_1 + p_2 = 1+1 = 2$$

The industry stock cover ratio (ISCR) is $\frac{1+1}{1+1} = 1.$

Hence $P \neq d \times$ ISCR.

Thus even in a very favorable case where both the parameters d_I are the same calculation of Industry Production on the basis of strict analogy is invalid.

Aggregation with varying parameters

The conditions for equation or term by term similarity are clearly strict and it is unreasonable to assume they will always be satisfied. It is, however, possible to produce aggregate equations that are formally correct. If we allow the constants k_{iI} to depend on *time.* For this purpose it is convenient to employ the discrete form of the system equations, which is used for simulation [12].

We accordingly consider the discrete system

$$w_{n+1,iI} = F_{iI}(w_{n,jJ}, x_{n,kK}) \qquad (53)$$

$$x_{n,qQ} = G_{qQ}(w_{n,rR}, x_{n,sS}) \qquad (54)$$

where $W_{n,iI}$ is the value of W_{iI} at the nth time step, etc. We now write $\sum_I w_{n,iI} = W_{n,i}$ etc.

$$\frac{w_{n,iI}}{W_{m,i}} = k_{n,iI}, \text{ etc.} \qquad (55)$$

We then have from (53), (54) and (55)

$$W_{n+1,i} \equiv \sum_I F_{iI}(k_{n,jJ} W_{n,j}, k_{n,kK} X_{n,k}) \qquad (56)$$

$$X_{n,q} \equiv \sum_Q G_{qQ}(k_{n,rR} W_w, k_{n,sS} S_s) \qquad (57)$$

Thus perfect aggregation is always possible provided the $k_{n,iI}$ are allowed to vary with time. The effect of this is, of course, to cause the aggregate parameters themselves to vary with time. This fact in itself causes no great problem conceptually, though it contrasts interestingly with the usual S.D. practice of assuming constant parameters. There is, however, a considerable practical problem in that the evolution of the aggregate parameters can only be calculated if the $k_{n,iI}$ are known which, of course, implies that $w_{n,iI}$, etc., are known. Since the whole point of an aggregate model is to avoid calculating the individual variables it is clearly impossible to *calculate* how the aggregate parameters will evolve. Nevertheless equations (53) to (57) do suggest various approaches to the aggregation problem, which we shall now discuss.

Aggregation in practice

If we wish to construct an aggregate model on the basis of equations (53) and (54) it would seem that the best way in which we could derive the aggregate parameters would be to estimate the values of $k_{o,iI}$, etc., at the time when the model is to be started and then calculate the aggregate parameters in the light of the discussion of the last section but one.

There would then seem to be two approaches to the running of the aggregate model

a) to use as aggregate parameters throughout the model run their initial values,

b) to modify the aggregate parameters as the simulation progresses on the basis of what can reasonably be forecast about their evolution from knowledge of the system being studied. Thus in a production planning system model, for example, it may well be that long range market forecasts and corporate plans, provide sufficient information about the likely evolution of product mix for the evolution of parameters aggregate parameters that depend on it, e.g. average production time, to be estimated reasonably well. It is unlikely, however, that such an approach can be employed with any great confidence where the time scale is long, e.g. {3].

Method *a)* is likely to be preferred unless it seems reasonable to assume that the evolution of the aggregate parameters can be predicted. It also has the advantage of being simpler. It has, however, the great disadvantage (at least in theory) that the aggregate equations may not behave in the same way as the individual sub-systems. For instance, the stability properties of the aggregate system may differ from those of the individual subsystems as shown by the following example.

Consider the system of individual equations

$$\dot{x}_{11} = -x_{11} + 2x_{21}$$

$$\dot{x}_{21} = -x_{11} - x_{21}$$

$$\dot{x}_{12} = -x_{12}$$

$$\dot{x}_{22} = 2x_{12} - x_{22}$$

where the second subscript refers to the subsystem.

Obviously both individual subsystems are stable since the eigenvalues of equations are $-1 \pm i\sqrt{2}$ and of the second -1.

We now aggregate with $k_{11} = \frac{1}{4}$, $k_{12} = \frac{3}{4}$; $k_{21} = \frac{3}{4}$, $k_{22} = \frac{1}{4}$ assuming these values are initially correct. This gives

$$k_{11}\dot{X}_1 + k_{12}\dot{X}_1 = \dot{X}_1 = -\tfrac{1}{4}X_1 - \tfrac{3}{4}X_1 + 2 \times \tfrac{3}{4}X_2$$

$$= -x_1 + \frac{3}{2}X_2$$

$$k_{21}\dot{X}_2 + k_{22}\dot{X}_2 = \dot{X}_2 = -\tfrac{1}{4}X_1 + 2\times\tfrac{3}{4}X_1 - \tfrac{3}{4}X_2 - \tfrac{1}{4}X_2$$
$$= \frac{5}{4}X_1 - X_2$$

This system has characteristic equation

$$\begin{vmatrix} -1 & -\lambda & \frac{3}{2} \\ \frac{5}{4} & & 1-\lambda \end{vmatrix} = 1 + 2\lambda + \lambda^2 - \frac{15}{8} = \lambda^2 + 2\lambda - \frac{7}{8}$$

and therefore has a positive eigenvalue and is hence unstable. The implication of this then is that an unfortunate choice of the aggregation constants may well result in the aggregate system having very different properties to that of the individual systems.

Given the possibility that aggregation by analogy could conceivably lead to incorrect results there is clearly a need to develop methods that allow the aggregation error to be assessed. In order to be useful such criteria need to be couched in terms of the aggregate model since an approach that relied on the calculation of the individual equations would be pointless in that we should expect it to lead to as much computational effort as working with a disaggregated model.

On the other hand aggregation causes a loss of information so any method for testing the accuracy of the aggregation via the aggregate equations must inevitably be fairly crude. With this proviso we now examine two such methods.

Aggregation and sensitivity analysis

If we use method *a)* of the last section to construct the aggregate equations then equations (56) and (57) show that we are in fact ignoring the variations of the aggregate parameters over time. This immediately suggests that the acceptability of the aggregation can be reduced to a sensitivity problem in that it seems reasonable to conclude that if the aggregate model is insensitive to changes in the aggregate parameters that may occur due to changes in the constants $k_{x,I}(t)$, etc., then we may reasonably conclude that the aggregation is satisfactory.

We, therefore, require estimates of how the aggregate parameters might be changed by a change in the constants $k_{x,I}(t)$, etc., so that sensitivity analysis [12] can be applied.

In the case of a linear relationship, e.g.

$$x_I = a_I y_I \tag{58}$$

we shall have the aggregate relationship

$$X = a\,Y \tag{59}$$

where

$$a \equiv \sum_I k_{y,I}\, a_I \tag{60}$$

Rough estimates of the likely maximum variation in the $k_{y,I}$ (δk) and the maximum (a_{max}) and the minimum (a_{min}) values of the a_I can be made whence the maximum change in the aggregate parameter a can be estimated as

$$a = \delta k\,(a_{max} - a_{min}) \tag{61}$$

In the case of non-linear equations such as

$$x_I = \frac{y_I}{z_I} \tag{62}$$

which leads to the aggregate equation

$$X = a\,\frac{Y}{Z} \tag{63}$$

where

$$a \sum_I \frac{k_{y,I}}{k_{z,I}} \tag{64}$$

rough estimates of the maxima and minima might be obtained by assuming all the $k_{y,I}$ increase by $p\%$ and all the $k_{z,I}$ decrease by $q\%$ (since $\sum_I k_{y,I} = 1$, etc., this is not possible in practice but we require simple estimates) which leads to an estimate to first order of the maximum change

$$\delta a = a\,\frac{(p+q)}{100} \tag{65}$$

Of course such estimates of aggregate parameter changes are crude and likely to overstate the actual changes. Furthermore they are not necessary if sensitivity analysis shows that the model is insensitive to large changes in all the aggregate parameters. On the other hand this rarely seems to be the case in practice. It will thus normally be necessary to examine the effects of a change in the basis of aggregation on the aggregate parameters, particularly since such examination will often show that changes will be negligible.

Thus (60) is unaffected by changes in the basis

of aggregation if all the a_I are equal and (64) is unaffected if $k_{y,I} = k_{z,I}$. Both conditions may often be expected to hold approximately in practice in the former case because we are, for example, aggregating across similar companies in the latter, for example because the various variables in a company are all proportional to its size relative to the industry as a whole.

Aggregation error as noise

We have already shown that where term by term similarity exists, aggregation without error is possible. In this case the equations can be put in the form

$$\dot{w}_{iI}\, k_{iI}\, \dot{W}_i = k_{iI}\, f_i\, (\underline{W}, \underline{X}, \underline{\lambda}) \tag{66}$$

$$x_{qQ} = k_{qQ} X_q = k_{qQ}\, g_q\, (\underline{W}, \underline{X}, \underline{\mu}) \tag{67}$$

where $\underline{\lambda}$ and $\underline{\mu}$ are vectors of parameters.

In using method *a)* of the section 'Aggregation in Practice' to aggregate an equation of the form (45) we encounter the problem that, although the equation can theoretically be aggregated by assuming fixed proportions, the aggregate equation assumes a very awkward form, viz.

$$X = \sum_I \frac{k_{y,I}\, Y}{k_{z,I}\, Z + c_I} \tag{68}$$

which cannot easily be simplified.

This suggests that in cases such as this the aggregate model might better be assumed to be based on term by term similarity so that Eq.(68) can then be written following (47) and (66) and (67) as

$$X = \frac{Y}{aZ + b} \tag{69}$$

where a and b are constants as defined by (48).

In practice however we cannot expect exact similarity to hold so we must expect that if we calculate a and b from (48) their values will depend on I, i.e. we can write

$$a_I\, \frac{k_{x,I}\, k_{z,I}}{k_{y,I}} = a + \delta a_I \tag{70}$$

and similarly $b = b + \delta b_I$.

By suitable definition of a and b we can set

$$\sum_I k_{x,I}\, \delta a_I = \sum_I k_{x,I}\, \delta b_I = 0 \tag{71}$$

In more general terms we can put any model with fixed relationships between the variables in the form

$$\dot{w}_{iI} = k_{iI}\, \dot{W}_i = k_{iI}\, f_i\, (\underline{W}, \underline{X}, \underline{\lambda} + \underline{\delta\lambda}_i) \tag{72}$$

$$x_{qQ} = k_{qQ} X_q = k_{qQ}\, g_q (\underline{W}, \underline{X}, \underline{\mu} + \underline{\delta\mu}_Q \tag{73}$$

where

$$\sum_I k_{iI}\, \underline{\delta\lambda}_I = \sum_Q k_{qQ}\, \underline{\delta\mu}_Q = \underline{0} \tag{74}$$

and the $\underline{\delta\lambda}_I$ and $\underline{\delta\mu}_Q$ represent the deviations from the values required for similarity for the individual parameters.

The condition (74) is chosen because a natural definition of λ_r is

$$\sum_I k_{iI} (\lambda_r + \delta\lambda_{r,I}) = \lambda_r \tag{75}$$

i.e. λ_r is defined to be the weighted mean of the individual $\lambda_{r,I}$. It has the further advantage that equations such as $x_I = a y_I$ then lead to an aggregate parameter a^* given by

$$a^* = \sum_I k_{X,I}\, \frac{a k_{Y,I}}{k_{X,I}} \equiv a$$

Now we have already shown that if similarity exists the basis of aggregation will remain unchanged, and the aggregate parameters $\underline{\lambda}$ and $\underline{\mu}$ will remain unchanged. The question of whether aggregation is justified can, therefore, be transformed into the question of whether the errors $\underline{\delta\lambda}_I$ and $\underline{\delta\mu}_Q$ cause significant changes in system behavior.

Thus the question of the accuracy of aggregation in this case is again reduced to a problem of sensitivity analysis.

Optimality and statistical aggregation

So far we have concentrated on Aggregation by Analogy because of its importance in System Dynamics and the lack of attention it has received in the past. Both optimal and statistical aggregation do, however, have a role within S.D. modelling and indeed the latter form of aggregation has been extensively used.

Optimality aggregation

In the past optimality assumptions have not been common within the System Dynamics field, on the grounds that the model equations should represent the actual decision-making processes within the system and that these are rarely optimal (cf. [2, Ch.10]). We have seen, however, that when equations representing the individual decision-making process are aggregated, errors may arise by virtue of the aggregation process itself. Since all models involve a degree of approximation it would seem possible that the use of decision functions that can be aggregated without error based on optimality assumptions, which we know do not describe the actual decision-making process but which enable the variables concerned to be predicted reasonably accurately, may be prefereable to the use of equations that describe the process more accurately but which lead to an aggregation error that is difficult to estimate.

Statistical aggregation

Statistical aggregation divides into two types.

a) Aggregation based on observed regularities in distributions that enable the proportionality between individual variables that make up an aggregate to be established. As we have already seen aggregation is always possible where such fixed proportions exist.
Such regularities occur frequently in economic contexts, e.g. in income distributions [10].

b) The use of relationships that are found to hold between the aggregates themselves rather than the individual variables comprizing the aggregates. Such relationships apparently enable the aggregation problem to be sidestepped. In fact, however, a number of problems arise unless the relationship found can be justified by optimality assumptions in that the usual way of establishing such relationships will be by means of regression (cf. [14]).

The normal way of determining the structure of the regression equation will be by analogy with the individual equations.† In such cases the basis of aggregation will continually be changing so it is difficult to be confident that such empirical relationships will hold for any considerable length of time [6]. Furthermore the regression process, since it makes use of past data, has an inevitable bias towards relationships that have existed in the past.

There is also a further complication in that such relationships between aggregates may well be improved as predictors by the introduction of variables that are irrelevant to the particular process being modelled because the addition of these variables may, to some extent, adjust for the changing basis of aggregation, [15]. This, of course, is contrary to normal S.D. practice.

Thus the consideration of Optimality or Statistical aggregation of type *b)* suggests that from the aggregation point of view it may not necessarily be best to base the model equations on a description of the actual decision-making process, if we wish to avoid aggregation problems.

Rules for satisfactory aggregation

The discussion of the last section shows that there are difficulties associated with both Optimality Aggregation—which involves assumptions about decision-making that are rarely acceptable—and Statistical Aggregation which, in as far as it is based on regression, is open to various objections.

The problems of aggregation by analogy have been discussed at length in the light of the Leontieff conditions. It has been shown [16] for aggregation to be approximately correct these conditions must be satisfied approximately so there is little doubt that the problems we have raised in connexion with this type of aggregation are genuine ones.

Given the difficulties that we have discussed in attaining acceptable aggregation it seems appropriate to summarize our conclusions to date as a 'Code of Practice' for models. Such a code may be considered as making more explicit the somewhat general precepts of [2, Ch.11].

A suitable procedure for the construction of an aggregate model would seem to be:

i) Decide the basis on which each individual aggregate relationship is to be constructed, i.e. Optimality, Statistical or Analogy. Consideration will obviously have to be given to the availability of data, the appropriateness of optimality assumptions, etc.

ii) If aggregation is to take place on the basis of analogy the individual elements should be grouped together into sets which have equations with approximately identical parameters. In the case of linear relationships this will allow perfect aggregation. It will, however, often mean that a number of sub-aggregates have to be constructed, e.g. instead of aggregating over

†There is clearly a strong resemblance between this case and that of aggregation by analogy.

an industry we aggregate over groups of similar firms in the industry, say large firms and small ones.

iii)Where non-linear relationships are aggregated by analogy we shall also normally require that the individual variables that make up the aggregates are as far as possible in fixed proportion to each other. Thus such relationships need to exhibit reasonable equation or term by term similarity as discussed in this section.

iv)Where aggregation is by analogy proper attention should be devoted to the calculation of aggregate parameters, particularly in the case of non-linear relationships, since, as shown earlier, the aggregate parameters may not have the 'obvious' values.

v) Where aggregation is by analogy and we do not have equation or term by term similarity the sensitivity of the aggregate model to the resulting errors in the aggregate parameters should be checked by methods such as those discussed above. These tests would also seem useful where Statistical Aggregation is used.

These criteria though simple are by no means easy to satisfy, as can be seen by examining a highly aggregated model such as [3]. It is interesting to note that in their discussions of the World Dynamics models, the University of Sussex group make several references to points at which one or other of the models conflicts with the above criteria. Thus the discussion of capital/output ratios [17] and pollution [18], all suggest that in these respects and no doubt others the models implicitly conflict with criteria ii) and iv) above. Similarly Appendix A of the discussion of the Capital sub-system [17] raises certain queries with regard to the optimality assumptions implicit in the use of an aggregate production function. The discussion of the relationship between Yield and Fertilizer inputs [19] provides an example of a statistical aggregation that is questionable.

It seems, therefore, that the above criteria, though doubtless capable of improvement, are nonetheless useful for judging the acceptability of aggregate models.

References

1. BARNETT, A.B., 'System dynamics applied to oilfield development', PhD Thesis, University of Bradford, Management Centre (1973).
2. FORRESTER, J.W., *Industrial Dynamics,* MIT Press (1961).
3. FORRESTER, J.W., *World Dynamics*, Wright Allen (1971).
4. FORRESTER, J.W., *Urban Dynamics*, MIT Press (1969).
5. PESTEL, E.C. and MESAROVIC, M.D., *Kybernetes I,* pp.79 (1972).
6. THEIL, H., *Linear Aggregation of Economic Relations,* North Holland (1965).
7. LEONTIEFF, W.W., *Bulletin of American Mathematical Society,* Vol. 53, 343 (1947).
8. GREEN, H.A., *Aggregation in Economic Analysis,* Princeton University Press (1964).
9. FISHER, F.M., *Review of Economic Studies* 35, 417 (1968).
10. MOGRIDGE, M., IFAC/IFORS International Conference on Dynamic Modelling and Control of National Economies, IEE Conference Publication No.lOl, l20.
11. NORDHAUS, W.D., *Economic Journal*, 83, 1156 (1973).
12. SHARP, J.A., 'A study of some problems of system dynamics methodology', PhD Thesis, University of Bradford (1974).
13. SIMON, H.A. and ANDO, A., *Econometrica*, 29, pp.111 (1961).
14. HAMILTON, H. *et al., Systems Simulation for Regional Analysis,* MIT Press (1969).
15. GUPTA, K., *Aggregation in Economics,* Rotterdam University Press (1969).
16. FISHER, F.M., *Econometrica,* 37, 457 (1969).
17. JULIEN, P.A., FREEMAN, C., and COOPER, C.M., *Futures,* 5, 66 (1973).
18. MARSTRAND, P.K. and SINCLAIR, T.C., *Futures*, 5, 80 (1973).
19. MARSTRAND, P.K. and PAVITT, K.L.R., *Futures,* 5, 56 (1973).

A methodology for the evaluation of model adequacy

FLAVIO ARGENTESI
Commission of the European Communities,
JRC—Euratom, Establishment of Ispra

Introduction

The term model is at present used in two different senses, although some important attempts have been made to establish a unified concept (preferably to incorporate one sense within the other).

Nevertheless a unified theory of models does not seem to be near at hand. Dálla Chiara Scabbia [1].

The first meaning of the term model, is the one used in formal logic, that is the model of a formal system. In fact, upon the basis of Tarski's semantic, after 1930, we can follow the development of a very important branch of formal logic, the so called 'theory of models'; as detailed references see Shoenfield [2] and Addison, Henkin and Tarski [3].

For an analysis of the attempts to extend this theory of models to fit within the framework of empirical science, especially physics, see Sneed [4], Suppes [5] and Przelecki [6].

In this context our reference will be made exclusively to the second meaning of the term model, that in use in the empirical science. In empirical science a model is intended to be a theory for the 'explanation' of a given framework of experience.

This definition and some aspects of its weakness will be enlightened by the following formal assessment, based on the relevant work of Reggiani and Marchetti [7,8].

Let θ_0 be an object pertaining to the external 'reality'. If we make an investigation either starting from a detailed series of well defined hypotheses about θ_0 or simply from a naive and intuitive framework, we always only ever achieve a model θ of θ_0, that which emerges from our experimental investigation. We can never have a complete knowledge of any 'real object' because 'real objects' are too complex for man to model. In other words, the 'reality' cannot be exhausted by scientific investigation; only a continuous improvement of scientific theories seems to be possible. Bellone [9] and Popper [10]. Therefore the starting point of the modelling process is a previous less defined model θ of θ_0.

This process normally produces a more clearly defined counterpart M of θ, which contains a relevant and significant hypothesis about the inner structure of θ_0. M differs in character from θ, because it is set up with the aim of 'describing' and 'explaining' θ.

The relevance and significance of M are clearly related to the aims followed by the investigator. At this point, M has to be considered as a new starting point for further experimental investigations on θ_0, which will produce a new θ, that could give rise to a new M, and so on.

Now we will try to define the concept of adequac

of a model, to represent a given object, that is a given θ of θ_0. The term adequacy is used to avoid the philosophical problems related to terms like verification or validation; for a discussion on this subject see Mirham [11].

Let a model M be characterized by a set of k functions $T_k M$.

We will now define the most important problem of the scientific approach, the so called 'inverse problem'.

Let $\{M_t\}$ be a set of models and θ a given object, that is the result of one or more experimental investigations into θ_0 and the previous general knowledge of θ_0.

The inverse problem is that of finding in $\{M_t\}$ the most adequate M to represent θ (giving a series of criteria to establish the adequacy).

It has to be noticed that this kind of problem cannot be reduced simply to a reproduction by the model outputs of some past measured behaviors of θ_0 but is more general in character. Adequacy cannot be reduced to fitting. The $t_k\theta$ and $T_k M$ have not to be only thought of as some experimental measurements on θ_0 and as the correspondent model outputs, but they also contain other information of a more general character about θ and M. We shall consider θ 'weak' when there are few or no measurements of the 'relevant $t_k\theta$' for some past behavior of θ_0. With the term 'relevant $t_k\theta$' we mean the information related to quantitative measurements on θ_0 that have to be reproduced by the model. We state the adequacy criteria as a series of rules for finding the distance (in a sense that will be defined later) between θ and the models proposed to represent it, and for ordering the models in relation to this distance.

Different kinds of distance between 'relevant $t_k\theta$' and the corresponding $T_k M$ are already largely in use. All these are related to the concept of distance between functions as defined in functional analysis, see for instance Collatz [12]. As previously noticed, the criterion of fitting alone, does not seem sufficient to establish a proper methodology of adequacy. Other properties of the models have to be taken into account, for instance simplicity of the model, ability of the model to preserve its nearness to θ when its parameters are subject to perturbations, etc.

Therefore a series of indices seem necessary for a proper characterization of the relationship between M and θ.

We consider this series of indices as a vector, the characterizing vector of M and θ, the adequacy will be established as a proper distance between vectors. Having a criterion to evaluate the distance between an object and its models, and between models, the problem of ordering models in relation to their adequacy can be approached.

In the next section we give some considerations to the problem of characterization and ordering of deterministic models. Some preliminary ideas in relation to the problem of adequacy of stochastic models will be presented later.

Models characterization and models ordering

The production of a vector V characterizing a model M of θ can be achieved in many different ways and by many different criteria. In fact, the choice of the elements of V is a subjective decision of the researcher; also the criteria, by which a distance between vectors can be evaluated, are always selected in a subjective way. Therefore every scale of adequacy, that can be achieved is, at least in part, inevitably a subjective issue as suggested by Hermann [13]. Nevertheless, the general framework of the assessed scientific methodology has a strong influence on this subjectivity. Actually a series of elements for a characterization of a model are already well defined and of practical use.

In the following points we summarize those which in our opinion are the more relevant:

1. Global fitting index, evaluated on the basis of the global comparison between the relevant $t_k\theta$ and the corresponding $T_k M$.
2. Partial fitting indices, related to the comparison between each couple of relevant $t_k\theta$ and $T_k M$.
3. A simplicity index that can be posed, equal to the total number of parameters present in the model.
4. Sensitivity analysis and error analysis indices; their number is variable depending on the way in which the sensitivity and error analysis are performed. In the simplest case, they can be reduced to two, see Miller [14].
5. Relevant and significant characteristic indices that are related to the occurrence in the model of important feature of θ. For instance asymptotic properties, common statistical properties like those established by time series analysis, forecasting attitude of the model, etc. The indices expressing qualitative correspondence can be binary in nature, conveniently 0, if the correspondence is present, and 1 under the opposite conditions.

Under these conditions the characterizing vector of

of the object θ is composed only by zeros, because we assume that the object has no parameters and that it is equal in every respect to itself.

A proper convention, in the choice of the values of the characterizing indices, is needed in order to put the vector of the object ($\emptyset$) to zero.

Before going through a discussion on the distance between vectors characterizing models, some comments are needed about the vector composition suggested.

Points 1) and 2) can be approached by the common statistical and numerical methods, see for instance Bard [15] and Box, Davies and Swann [30]. Point 4) is a complex one; it cannot be fully discussed, in this context. Relevant references are Miller [14,15], Burns [16] and Garrat [17]. A simple approach to sensitivity analysis is discussed by Miller [14]; Burns [16] gives a more general approach to both sensitivity and error analysis. In any case, sensitivity and error analysis methods are well established in many fields. Therefore, in general, the estimation of the indices related to sensitivity or to error transmission properties of models, can be properly achieved. Point 5) is the more heuristic one; these kinds of indices have to be determined in relation to any particular problem under investigation. Generalizations do not seem easy at present.

When the set $\{V_t\}$ of the characterizing vectors of $\{M_t\}$ has been produced we must establish some criteria in order to assess the adequacy of the $\{M_t\}$ to represent θ and therefore θ_0. The simplest ordering criteria is that of measuring some kind of distance between V and $\emptyset$ and to take the inverse value of this distance as a measure of adequacy. It also seems advisable to measure the distance between the models themselves and to check the relative distances. For instance:

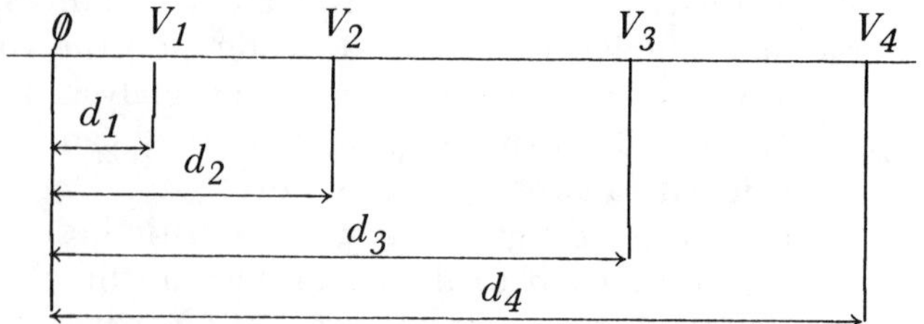

If $d_{12} = d(V_1, V_2)$, $d_{13} = d(V_1, V_3)$, etc.

then

$$d_{12} \cong d_2 - d_1, \quad d_{13} \cong d_3 - d_1, \quad \text{etc.}$$

The adequacy could be assessed as:

$$\text{ad}\ (M_1) = 1/d_1$$

$$\text{ad}\ (M_2) = 1/d_2$$

We will refer now to the notion of distance between vectors. In the present context, this notion will be assimilated into the measure of similarity between objects. The measure of similarity is a well known area of the classification methods, cluster analysis and multivariate analysis. Several indices of similarity can be used, as relevant references, see Bijnen [31], Cormack [18] and Hartigan [19]. The choice of a proper index of similarity is related to the nature of the elements of V (quantitative measurements, scores, binary data, etc.).

For the proper criteria to apply in the choice of a proper distance index, see the references given above. In any case, it seems advisable to try the assessment of the adequacy on the bases of different distance indices, in order to improve the significance of the ordering achieved. When the cardinality of $\{V_t\}$ is high, the introduction of the techniques for forming clusters of V and therefore of M might be useful. As references on the clustering techniques see Hartigan [19], Cormick [18], Fisher [20], Bijnen [31], Sokal and Sneath [21] and Jardine and Sibson [22].

The clustering methodology can be useful in two important ways. First, if the elements of $\{V_t\}$ can be ordered in clusters (different kinds of clustering have to be attempted) every cluster produced has to be considered as a new object. This object will be composed of a given number of vectors. It is convenient to arrange these vectors in a matrix C_i of $l \times p$ dimensions, where l is the number of vectors aggregated in the cluster and p is the number of elements present in each vector. Clearly at this point all the methods of the multivariate analysis can be applied to the C_i. It is possible that the aggregation of model achieved in the C_i yields meaningful results in some respects. The models associated to each cluster might have common relevant and significant properties, that differ notably from those of the other clusters and that have not appreciated before the analysis. In this way, some kind of model taxonomy can be achieved and, by the techniques of the multivariate analysis, it will be possible to measure the distance between clusters, that is types of models, and $\emptyset$. The final results of such a process could then decide what kind of modes seems more appropriate for the modelling process under investigation. Second, the clustering methodology can also be useful for possible feedback to the modelling process itself. Supposing a relevant clustering of models has been achieved, then we can study the changes in a given model, that remove it from its cluster, and if the configuration of cluster is significant, the modeller might reach a better understanding of the modelling process he is producing. For multivariate analysis references see Anderson [23], Morrison [24] and Roy [25].

Some preliminary considerations about the problem of the adequacy of stochastic models

We could consider θ as stochastic, when a repetition of the same experimental investigation of θ_0, maintaining a general regularity, leads to remarkably different realizations of some $t_k\theta$. The random behavior of some elements of θ_0 can be seen in two convenient ways. First, we can think of θ_0 as a collective (population); a realization sample of dimension m, of the relevant $t_k\theta$, can be observed. Second, the source of the random behavior of θ can be thought of as an inner property of θ_0 and the $t_k\theta$ have only one observable realization. In fact, the two points of view are the same but we distinguish between them mainly to show that in one case we have the observation of m realizations and in the other, only one can be observed. In these preliminary considerations, we will mainly refer to the former situation; the latter seems to be more complex for an adequacy assessment. When several realizations of θ_0 are observed for relevant $t_k\theta$, the characterizing vector for the stochastic models of θ, $\{MS_t\}$ can be achieved in a way very similar to that proposed for the deterministic case. Nevertheless, some differences have to be noticed:

a) The fitting indices as in 1) and 2) have to be achieved, minimizing the distance between m realizations of relevant $t_k\theta$ and n realizations of T_kMS. This can be done as follows:
Let X_1 and X_2 be the vectors of the means for $t_k\theta$ and T_kMS respectively, and S, the estimation of the variance-covariance matrix of $t_k\theta$ and T_kMS. We can take for instance as distance index, the Hotelling's T^2:

$$T^2 = nm/(n+m)(X_1 - X_2)'S^{-1}(X_1 - X_2)$$

Let P_j be the parameters of MS, then the fitting problems as at the points 1) and 2) can be solved as:

$$\min_{\langle P_j \rangle} T^2$$

b) The error analysis (and the sensitivity analysis) as in point 4) can be achieved in a manner very similar to those indicated in *a)*. On the basis of P_j we can produce n_1 realizations of T_kMS, then P_j can be perturbed until the parameters vector P_j^1 and n_2 realizations of T_kMS can be achieved. The distance D between the two groups of realizations can be evaluated as in *a)*. The distances D can already be estimated in *a)* but we perform a further evaluation, in order to check if the order of our vector of sampling functions for the stochastic process is appropriate. This process can be repeated several times until a proper sample of the function

$$D = f(P_j)$$

has been achieved.
$f(P_j)$ might be approximated by proper base functions, for example polynomials, spline functions, etc. Myers [26] and Duffy and Franklin [27]. The number of realizations n_1 and n_2 have to be such that the variance associated to D is as close as possible to zero. The sensitivity functions can now be estimated performing the partial derivatives

$$H(P_j) = \partial f(P_j)/\partial P_j$$

Finally a proper error analysis can be done, performing an analysis of the hypersurface D.

The case where only one realization of the $t_k\theta$ can be observed (for example a unique object as, for instance, a lake) has to be considered 'weak' in a sense similar to that given for the deterministic case. 'Weak' fitting indices can be achieved, following the approach given in *a)*. Sensitivity and error analysis are still possible.

An example of application of the adequacy methodology proposed to the modelling of a biological process

In this example we show briefly how in principle the proposed methodology could be applied. The process that we are considering is the interspecific competition between populations. A general deterministic model for interspecific competition can be formulated as follows:

$$dN_i/dt = D_i(N_i(t), N_j(t); r_i, K_i, \alpha_{ij}, \beta_{ij}, \ldots)$$

where D_i is an unspecified function, giving the population change per unit time, N_i and N_j are respectively the number of species i and j at time t, and r_i, K_i, α_{ij}, β_{ij}, etc., are certain parameters, whose values are independent of time. Ayala, Gilpin and Ehrenfeld [28], reviewing the literature on the subject have found 11 significant formulations for D_i. The following table shows these formulations:

Model	D_i	num. of par.
1	$(r_iN_i/K_i)(K_i-N_i-\alpha_{ij}N_j)$	3
2	$(r_iN_i/\log(K_i))(\log(K_i)-\log(N_i)-\alpha_{ij}\log(N_j))$	3
3	$(r_iN_i/K_i^{1/2})(K_i^{1/2}-\alpha_{ij}N_j/K_i^{1/2})$	3
4	$(r_iN_i/K_i)(K_i-N_i-\alpha_{ij}N_j-\beta_{ij}N_iN_j)$	4
5	$(r_iN_i/K_i)(K_i-N_i-\alpha_{ij}N_j-\beta_iN_i^2)$	4
6	$(r_iN_i/K_i)(K_i-N_i-\alpha_{ij}N_j-\beta_jN_j^2)$	4
7	$(r_iN_i/K_i^{\theta i})(K_i^{\theta i}-\alpha_{ij}N_j/K_i^{q-\theta_i})$	4
8	$(r_iN_i/K_i)(K_i-N_i-\alpha_{ij}N_j-\beta_i(1-\exp(-\gamma_iN_i)))$	5
9	$(r_iN_i/K_i)(K_i-N_i-\alpha_{ij}N_j-\beta_j(1-\exp(-\gamma_jN_j)))$	5
10	$(r_iN_i/K_i)(K_i-N_i-\alpha_{ij}N_j-\beta_{ij}N_iN_j-\delta_iN_i^2)$	5
11	$(r_iN_i/K_i)(K_i-N_i-\alpha_{ij}N_j-\beta_{ij}N_iN_j-\delta_iN_i^2-\gamma_jN_j^2)$	6

In their paper the authors indicated above try to check the models in the case of the competition between two species of *Drosophila*. Therefore two 'relevant $t_k\theta$' are measured experimentally, $N_1(t)$ and $N_2(t)$, and a detailed series of further information about θ_0 are collected. Moreover several significant characteristics of the models are pointed out. We will summarize the most important ones:

Total fitting index and the two partial fitting indices for $N_1(t)$ and $N_2(t)$.

Differences between the theoretical and experimental equilibrium densities.

Different characteristics of the models, such as: relationship with the Lotka-Volterra model, linearity in the parameters and biological meaning of the parameters.

No error or sensitivity analysis are given by Ayala, Gilpin and Ehrenfeld but this point could be achieved without severe difficulties.

As can be seen, the set $\{V_t\}$ for $t = 1,...,11$ may be properly formulated on the basis of the given information and performing some further analysis. For instance, a composition of V could be:

1. General fitting index.
2. Partial fitting index for $N_1(t)$.
3. Partial fitting index for $N_2(t)$.
4. Total number of parameters.
5. Sensitivity and error analysis indices.
6. Comparison between carrying capacities and equilibrium densities.

For the only comparison between models, other indices, such as linearity in the parameters, relationship to the Lotka-Volterra model, etc., can be taken into account. The distances between $\emptyset$ and V and between the V can now be estimated, preferably taking several different indices of distance. On the basis of these distances, an adequacy ordering of the models can be achieved. It has to be noticed that the same methodology can be repeated for several competition experiments with a different number and kinds of species. In this way, it seems to be possible to identify the most adequate model to represent competition in general.

Stochastic versions of the 11 deterministic models, given above, are possible. They can be formulated conveniently following, for instance, the approach of Bartlett [29]. We will show the procedure for the model 1 in the case of two populations. The table of the transition probabilities can be formulated as:

Transition	Probability
$N_1 \rightarrow N_1+1$	$r_1 N_1 dt$
$N_1 \rightarrow N_1-1$	$r_1/k_1(N_1^2 - \alpha_{12}N_1N_2)dt$
$N_2 \rightarrow N_2+1$	r_2N_2dt
$N_2 \rightarrow N_2-1$	$r_2/K_2(N_2^2 - \alpha_{21}N_1N_2)dt$
no change	$1 - (r_1N_1dt + r_2N_2dt + r_1/K_1(N_1^2 - \alpha_{12}N_1N_2)dt + r_2/K_2(N_2^2 - \alpha_{21}N_1N_2)dt)$

The rate of occurrence of events per unit time is:

$$R = r_1N_1 + r_2N_2 + r_1/K_1(N_1^2 - \alpha_{12}N_1N_2) + + r_2/K_2(N_2^2 - \alpha_{21}N_1N_2))$$

Therefore the stochastic process is a Markov chain in two dimensions and so the density function of the time between events is exponential with parameter R. Hence a Monte Carlo method can be used to generate the time to the next event and to determine what that event will be. It seems clear that for a given set of parameters by Monte Carlo simulation several realizations of the given stochastic model can be achieved. In this way, minimizing the Hotelling's T^2 between the realizations of θ_0 and those of our stochastic model, r_1, r_2, K_1, K_2, α_{12} and α_{21} can be estimated and the various fitting indices evaluated. Sensitivity and error analysis can be performed, following the outline, previously given. Therefore the set $\{V_t\}$ can also be produced for the stochastic versions of the 11 deterministic models and by the methodology discussed above, an adequacy ordering can be achieved.

As can be seen by this example, a practical solution of the adequacy problem is a difficult task, although in principle a solution for the various items can be related to well implemented statistical techniques.

Conclusion

The methodology outlined above seems to be powerful enough to tackle the adequacy ordering for both deterministic and stochastic models. The criteria are given only in principle but they fall into the area of well defined statistical and mathematical techniques. It seems clear that the methodology discussed can be further specified to deal with every relevant and significant class of problems of interest, for example population dynamics, dynamics of socio-economic systems, statistical models of computer systems, etc. The advantage of having well defined criteria for model evaluation in times, of which there is a large production of models of doubtful significance has not to be underestimated. Many points, only briefly outlined in this chapter, have to be developed; in our opinion, we have summarized the most important ones here:

1. Further investigation on the general criteria, upon which the elements of the characterizing vectors have to be chosen.
2. Detailed analysis of the similarity indices for relevant kinds of problems.
3. Development of operating techniques for the sensitivity and error analysis, both for deterministic and stochastic models.
4. Development of the parameter estimation and fitting evaluation techniques for stochastic models.
5. Evaluation of the relevance of the clustering techniques for the improvement of the modelling process and of the models understanding.

Finally, it has to be mentioned that the value of a continuous investigation into the modelling process, from the point of view of the philosophy of science, is a critical and vital element, that cannot be neglected.

References

1. DALLA CHIARA SCABBIA, M.L., *Logica*, Isedi, Milano (1974).
2. SHOENFIELD, J.R., *Mathematical Logic*, Addison-Wesley, London (1967).
3. ADDISON, J.W., HENKIN, L., and TARSKI, A., *The Theory of Models*, North Holland, Amsterdam (1965).
4. SNEED, J., *The Logical Structure of Mathematical Physics,* Reidel, Dordrecht (1971).
5. SUPPES, P., *Introduction to Logic,* Van Nostrand, New York (1957).

6. PRZELECKI, *The Logic of Empirical Theory,* Routledge and Kegan Paul, London (1969).
7. REGGIANI, M.G. and MARCHETTI, F.E.,'Proposta di un nuovo criterio per la valutazione della adeguatezza dei modelli', *Alta Frequenza,* Vol.XLI, 257-260 (1973).
8. REGGIANI, M.G. and MARCHETTI, F.E., 'On assessing model adequacy', *IEEE Trans. Syst., Man and Cyber.,* Vol.SMC-5, 322-330 (1975).
9. BELLONE, E., *I Modelli e la Concezione del Mondo nella Fisica Moderna da Laplace a Bohr,* Feltrinelli, Milano (1973).
10. POPPER, K., *Objective Knowledge, An Evolutionary Approach,* Oxford at the Clarendon Press (1972).
11. MIRHAM, G.A., 'Some practical aspects of the verification and validation of simulation models', *Operational Research Quarterly,* 23, 17-29 (1971).
12. COLLATZ, L., *Functional Analysis and Numerical Mathematics,* Academic Press (1966).
13. HERMANN, C.F., 'Validation problems in games and simulations with special reference to models of international politics', *Behavioral Science* 12, 216-230 (1967).
14. MILLER, D.R., 'Sensitivity analysis and validation of simulation models', *J. Theor. Biol.* 48, 345-360 (1974).
15. MILLER, D.R., 'Model validation through sensitivity analysis', Proc. 1974 Summer Computer Simulation conf., Houston, Tex, pp.911-914 (1974).
16. BURNS, J.R., 'Error analysis of non-linear simulations: Application to world dynamics', *IEEE Trans. Syst., Man and Cyber.,* Vol.SMC-5, 331-340 (1975).
17. GARRAT, M., 'Statistical validation of simulation models', Proc. 1974 Summer Computer Simulation Conf. Houston, Tex. pp.915-926, (1974).
18. CORMACK, R.M., 'A review of classification', *Journal of the Royal Statistical Society,* Series A, 134, 321-353 (1971).
19. HARTIGAN, J.A., *Clustering Algorithms,* John Wiley and Sons, London (1975).
20. FISHER, W.D., *Clustering and Aggregation in Economics,* John Hopkins Press, Baltimore (1969).
21. SOKAL, R.R. and SNEATH, P.H.A., *Principles of Numerical Taxonomy,* Freeman, London (1963).
22. JARDINE, N. and SIBSON, R., *Mathematical Taxonomy,* John Wiley and Sons, London (1971).
23. ANDERSON, T.W., *Introduction to Multivariate Statistical Analysis,* John Wiley and Sons, New York (1958).
24. MORRISON, D.F., *Multivariate Statistical Methods,* McGraw Hill, San Francisco (1967).
25. ROY, S.M., *Some Aspects of Multivariate Analysis,* John Wiley and Sons, New York (1959).
26. MYER, R.H., *Response Surface Methodology,* Allyn and Bacon, Boston (1971).
27. DUFFY, J.J. and FRANKLIN, M.A., 'A learning identification algorithm and its application to an environmental system', IEEE Trans. Syst., Man and Cyber., Vol.SMC.-5, 345-352 (1975).
28. AYALA, F.J., GILPIN, M.E. and EHRENFELD, J.G., 'Competition between species: Theoretical models and experimental tests', *Theoretical Population Biology,* 4, 331-356 (1973).
29. BARTLETT, M.S., *Stochastic Population Models,* Methuen and Co., London (1960).
30. BOX, M.J., DAVIES, D., and SWANN, W.H., *Non-linear Optimization Techniques,* Oliver and Boyd, London (1967).
31. BIJNEN, E.J., *Cluster Analysis, Survey and Evaluation of Techniques,* Tilburg University Press, The Netherlands (1973).
32. BARD, Y., *Non-linear Parameter Estimation,* Academic Press, London (1975).

Network traps

J. DENNIS WILLIGAN
University of Utah,
Salt Lake City, USA

Introduction

The possibility of a scientific sociology is based upon the assumption that human societies can be empirically demonstrated to possess identifiable structures or networks. The discovery of parameters establishes the framework for the macrosociological analysis of social structure in empirical and theoretical terms (Blau [1]: 619). Structures and their parameters function as constraints upon societal activity. From a systems analysis perspective one may say that the specification of the set of constraints associated with a given phenomenon is equivalent to its definition. In this sense, the relevant information which can be extracted from a system resides in the nature of its constraints. This carries the important implication that the analysis of a policy problem should start from an explicit recognition of relevant constraints.

Population system problems are often assumed to be inherently solvable. Realistically, however, the question of solvability must, in practice, be judged against the nature of the constraints present in each situation under investigation. The discoveries of Abel, Galois, Lindemann and others which led to the development of modern algebra and group theory were significantly stimulated by the revoltionary idea that certain problems (e.g. the determination of the roots of equations of degree $\geqslant 5$, by means of rational operations and radicals) were inherently unsolvable. We intend to propose that analogous states of 'unsolvability' may exist for types of population systems which characterise human societies. While the foundations of social technology rest upon the knowledge of how to manipulate 'operators' (Heise [2]: 18-23) within the objective limits to population system modification, the discovery and theoretical explanation of those limits constitutes the core subject matter for macrosociology.

Societal networks

Using the expression 'population systems' to refer to the structural configurations constitutive of human societies, we may in turn consider these systems to be made up of concatenations of networks whose nodes are key societal subsystems, each of which possess distinctive topologies of material and information flow. Models of human societies as stochastic networks may prove extremely valuable for integrating concepts of structure with concepts of process in macrosociological theory.

Harrison White has drawn attention to the fact that "in the last decade it has become common to accept networks as the natural metaphor for describing how people fit together in social organization" ([3]:45). Here we are considering networks at a macrosocial level. Whereas in White's formulations [4], persons are nodes in networks, this discussion focuses upon societal subsystems as nodes within networks.

No theory in sociology can be fully a theory unless

it is mathematical. Scientific theories when perfected, are inevitably mathematical in nature. The idea that mathematics is exclusively a science of quantity is a nineteenth-century notion, and sociologists who pursue it or else attack the inclusion of mathematics in the domain of theoretical sociology are immersing themselves in obsolete ideas (Howard [5] :2). Contemporary mathematics (e.g., symbolic logic, modern algebra, set theory, and topology) primarily discuss *relations* rather than numbers. Qualitative analysis, no matter how mathematical in nature, is of course never wholly equivalent to actual quantitative calculations. Both approaches are necessary, complement each other and are required to recognize the essential factors which govern (societal) processes and to get direct insights into the 'laws' of (population) systems (Weisskopf [6] : 605).

Just as structural linguists such as Noam Chomsky have endeavored to establish that there are 'deep level' universals which underlie the diversity of human languages, the type of structural sociology espoused by the author seeks to discover 'deep level' universals which underlie the diversity of human societies. Sociologists have been searching for such structures since the turn of the century and have met with only meagre success. Sociological structuralism, while neither strictly a theory nor a method in the traditional sense but 'a way of looking at things', may provide the key to this problem. It offers a new approach which can, in principle, attain exactness, systemicity and testability, and it can strive for genuine depth in detecting the basic patterns beneath societal appearances (Bunge [7] : 142).

Structures or networks are basically mathematical ideas. To communicate an intuitive sense of their essential nature, we shall quote an example given by Edmund Leach which is in turn derived from Bertrand Russell:

> If I listen to a broadcast version of a piano sonata the music has gone through a whole series of transformations. It started out as a score written on a piece of paper; it was interpreted in the head of the pianist and then expressed by movements of the pianist's fingers the piano then produced a patterned noise imposed on the air which was converted by electronic mechanisms into grooves on a phonograph record; subsequently other electronic devices converted the music into radio frequency vibrations and after a further series of transformations it eventually reached my ears as a patterned noise. Now it is perfectly clear that *something* must be common to all the forms through which the music has passed. It is that common something, a patterning of internally organized relationships which I refer to by the word *structure*. [8] : 40-41.

Structures can be expressed in multiple forms which are transformations of one another, no one of which is a more correct expression of the latent structure than any other.

Homomorphisms, simulation, and structural models

Lorrain and White [9] have shown how patterns of global relations present in social psychological networks can be derived algebraically through mappings which are generalized homomorphisms. A prime tenet of the type of macrosociological structuralism proposed by the author is that the structures which human societies create are far from arbitrary and have a high degree of order; and that by investigating this order we can uncover latent structures which hold for populations systems in general. The concept of analogy as a search-guiding technique lies at the heart of both endeavors. We shall now indicate how homomorphisms are related to simulation and how these in turn are related to structural models. This will provide a theoretical foundation for the subsequent structural analysis in the chapter. The discussion draws upon material given in Bunge [7: 114-30].

The universe of discourse will be the whole set W of real and conceptual elements. W will be partitioned into three subsets: [1] the set S of human societies, [2] the set N of system networks, and [3] the set of macrosociological theories, hypotheses, and concepts. S, N, and T are mutually disjoint and collectively exhaust the content of W.

An element of the universal set W, w_i, will be said to be *analogous* to another element w_j, if: (1) w_i and w_j share several objective properties, or (2) there exists a correspondence between the properties of w_i and those of w_j. If w_i and w_j are analogous, this will be expressed as $w_i \simeq w_j$ and each will be said to be an analog of the other. If w_i and w_j satisfy condition (1) above they will be said to be *substantially* analogous, and *formally* analogous if condition (2) holds. In the case where both conditions are met, the analogy will be called a *homology*.

We shall distinguish two types of formal analogy: (1) *plain*, if only some elements of w_i are paired off to some elements of w_j, and (2) *injective*, if every element in w_i is paired to an element in w_j. Injective analogy may be divided into two categories: (1) *weak*, when every element of w_i is maped onto some element of w_j, and (2) *homomorphic*. when there exists a correspondence that maps every ele-

SYMBOL	DESCRIPTION
$\simeq_1 \subset S \times S$	human society - human society similarity
$\simeq_2 \subset S \times N$	human society - system network similarity
$\simeq_3 \subset S \times T$	human society - macrosociological theory similarity
$\simeq_4 \subset N \times S$	identical to $\simeq_2$
$\simeq_5 \subset N \times N$	system network - system network similarity
$\simeq_6 \subset N \times T$	system network - macrosociological theory similarity
$\simeq_7 \subset T \times S$	identical to $\simeq_3$
$\simeq_8 \subset T \times N$	identical to $\simeq_6$
$\simeq_9 \subset T \times T$	macrosociological theory - macrosociological theory similarity

Table 1 Types of analogy derivable from the relata made possible by the tripartition of W into S, N, and T. Extensions to higher order types can easily be derived from those listed here

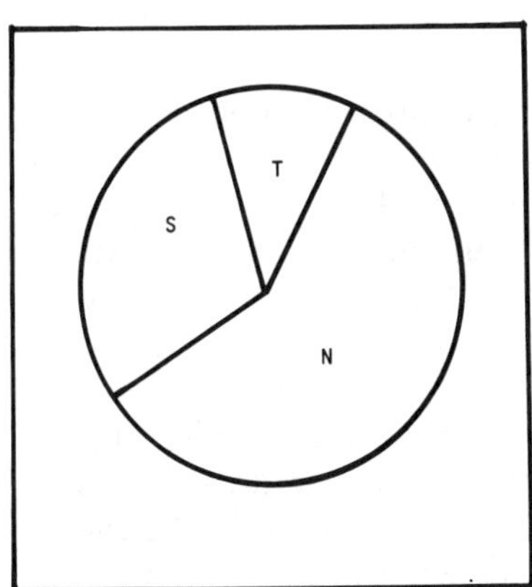

Fig.1 The universe set W and its subsets

ment of w_i onto some element of w_j while also preserving the relations and operations in w_i. An *isomorphism* exists between w_i and w_j if there is a homomorphism from w_i into w_j and also one from w_j into w_i such that the two morphisms compensate each other. The relations between the categories just discussed are shown in Fig. 2.

The analogy relation $\simeq$ is a binary relation on W but it is not connected since no two elements in W are necessarily analogous. The relation $\simeq$ is symmetric (i.e., if $w_i \simeq w_j$, then $w_j \simeq w_i$), reflexive (i.e., $w_i \simeq w_i$ and $w_j \simeq w_j$) but neither transitive nor intransitive (i.e., $w_i \simeq w_j$ and $w_j \simeq w_k$ do not jointly imply $w_i \simeq w_k$). If a similarity is transitive it is called a *contagious* analogy ($\cong$) and is an equivalence relation. But if all the elements in a subset of W are pairwise similar (i.e., $\simeq$ is connected in that subset), the elements are equivalent and will be said to consitute an equivalence class. Identity implies equality; equality implies equivalence; and equivalence implies similarily (i.e., $\equiv \subset = \subset \cong \subset \simeq$, where '$\subset$' refers to the concept of subrelation).

An element w_i belonging to N or T *simulates* an element w_j in W if: [1] w_i is contagiously analogous to w_j (i.e., $w_i \cong w_j$), and [2] this analogy is 'valuable' to w_i itself or to a third element w_k in S that controls w_i in some way. If w_i simulates w_j, this will be symbolized as $w_i \triangleq w_j$. $\triangleq$ is a binary relation on W, with domain $N \cup T$ and codomain W; it is symmetrical, reflexive, transitive, and therefore an equivalence relation. The set $w_i = \{w_k \mid w_k \triangleq w_i\}$ of all the

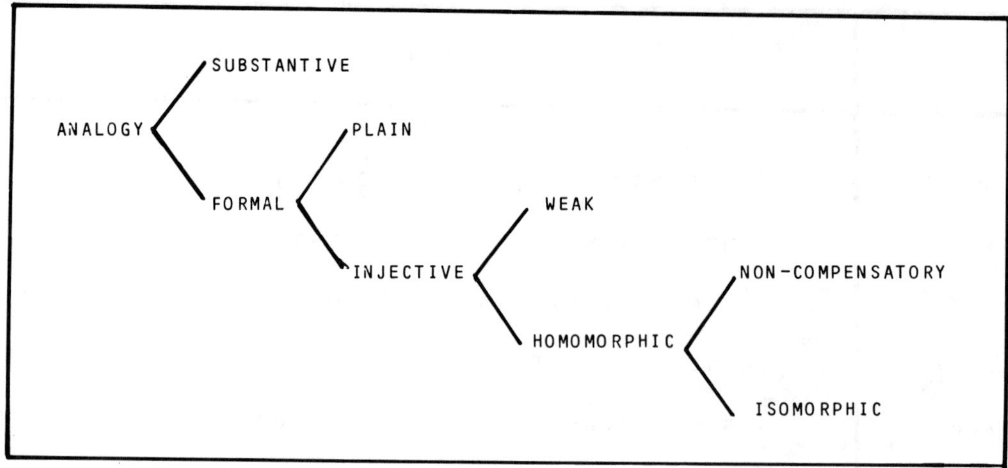

Fig.2 A tree diagram of the fundamental analogic categories

simulates of w_i constitutes an equilvalence class in regard to the simulation relation. The discovery of equivalence classes in the universe set W can be said to consitute the formal object of all macrosociological inquiry.

Some system networks represent possible societal elements, others do not. The set N will therefore be partitioned into the subset M of model-able system networks and its complement, $N-M$. Similarly, we shall partition the set T into the subset of model-able macrosociological theories Q (i.e., theories which can be operationalized), and its complement, $T-Q$. We can now say that an element w_i in $M \cup Q$ *models* an element in W if and only if w_i is a simulate of w_j. This relation will be written $w_i \triangleq w_j$. A model is a subrelation of simulation, $\triangleq \subset \hat{=}$. The model relation restricts the simulation relation to the subset $M \cup Q$ of $N \cup T$.

Let us assume that we are given two concrete population systems, P_1 and P_2, and that we introduce a third relational system or conceptual structure, Y, which is regarded for heuristic purposes as a type of Platonic idea. Now we start from the formal structure of Y and descend to empirical human societies. P_1 and P_2 are regarded as two (among infinitely many) potential empirical realizations or concrete models of the formal system Y. All such concrete models of Y will be said to be formally analogous to each other and with respect to Y. This exemplifies the meaning of the concept of isomorphic representation: the goal of all theorizing in macrosociology. To further clarify this idea we shall propose a definition of system isomorphism in model-theoretic terms: two population systems (represented empirically or symbolically), P_1 and P_2, will be isomorphic with respect to a third system, the relational system Y, if P_1 and P_2 are models of Y. Hence, to determine whether two systems are analogous, we must produce their theories. Judgements of analogy are always theory-dependent and contingent upon the nature of some relational system. Without analogy and the far stronger (transitive) relation of equivalence there would be no sociological 'knowledge'.

The concepts presented above provide an intuitive foundation for understanding the possibility of a mathematical sociology. A few decades ago it appeared that human societies were inexpressible in mathematical terms. Sociologists are now beginning to realize that their discipline has ceased in many aspects to be methodological different from the 'hard' sciences: while the elements which these disciplines deal with are different and as a consequence they must devise specialized methods, their general methodological goals are the same, namely to uncover objective (natural or social) laws and to systematize such laws into theories or hypothetico-deductive systems (Bunge [7, 132]).

Traps and countertraps

To explore more concretely the concepts presented up to this point, we shall turn to a relatively new area of theoretical discussion which has grown around the subject of 'traps'. John Platt has defined *social traps* as "situations in society that contain traps formally like a fish trap, where men or organizations or whole societies get themselves started in some direction or some set of relationships that later prove to be unpleasant or lethal and that they see no easy way to back out of or to avoid" [10: 641]. Conversely, *countertraps* act like barriers in situations where consideration of the advantages accruing to a member of a group of entities prevents that member from acting in ways that might be of benefit to the whole group.

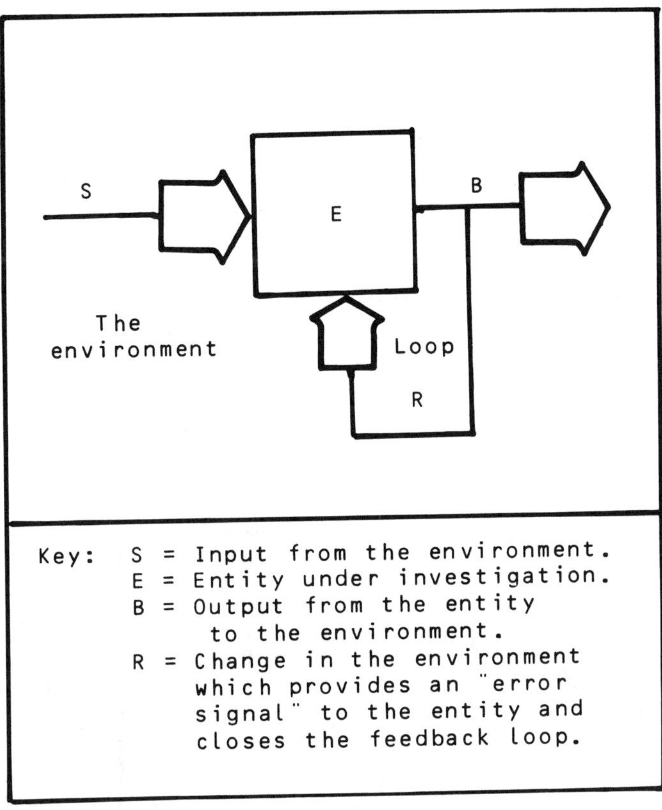

Fig.3 The fundamental Skinnerian feedback loop

Trap and countertrap situations can be formalized in terms of a Skinnerian feedback loop in which an entity *E* is subject to an input *S* from its environment, as a result of which *E* emits some output *B* back to the environment which gives a changed 'error signal' *R* to the entity and closes the circuit (Platt [10:642]. If *R* is an environmental consequence which makes *B* more probable when the entity is subjected to *S* again, we shall indicate this by the symbol *R[+]*. If the environmental consequence makes a given *B* less probable when *S* occurs again, we shall use the symbol *R[-]*. From a Skinnerian viewpoint (Skinner [11]), a trap would occur when there is an opposition between high intensity short-term *R[+]* or *R[-]*, and long-term consequences *R[+]* or *R[-]*. A countertrap takes place when a short-term *R[-]* tends to block *B* even though there would be a long-term, high intensity *R[+]*. If we turn aside from time considerations, an alternative criterion can be whether a member of a network of enitities *R[+]* or *R[-]* is in opposition to the collective network's *R[+]* or *R[-]*. Traps or countertraps can arise depending upon whether a network member's initial environmental consequence is *[+]* or *[-]*. This type of analysis may prove very illuminating in tracking the dissolution of social networks present in the type of groups considered by White [4].

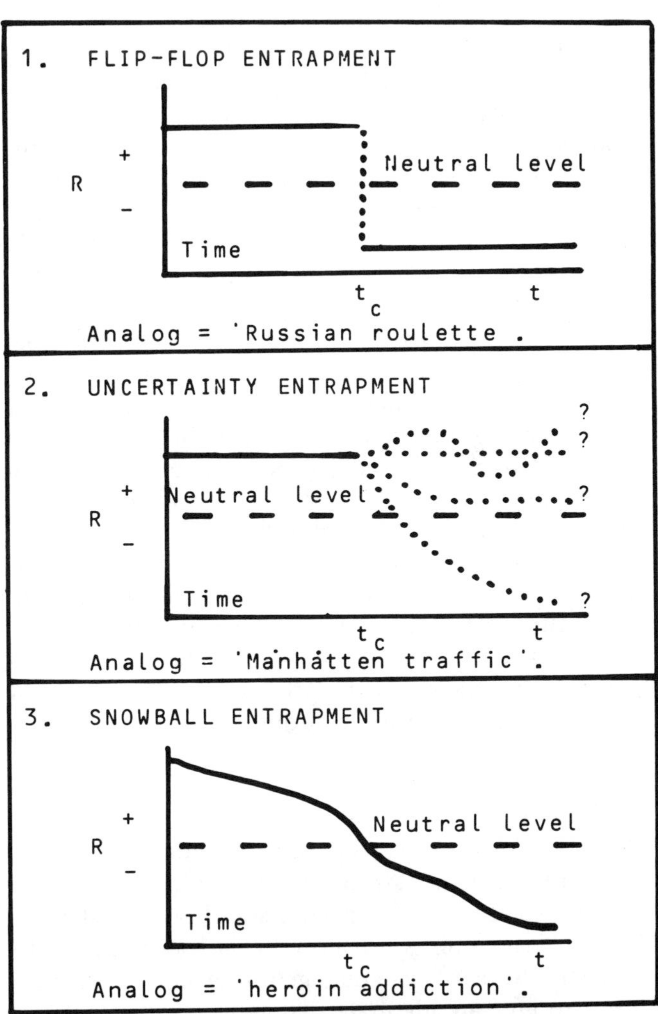

Fig.4 Three types of 'entrapment'. Combination or variations of these traps may be easily conceptualized

We may distinguish three fundamental types of traps into which an entity may move: [1] flip-flop entrapment where after a given period of time t_c an *R[+]* 'flips over' to an *R[-]*; [2] uncertainty entrapment where the problem is not centered upon temporal delay as in [1] but rather ignorance of the nature of the outcome after time t_c; and [3] snowball entrapment in which an *R[+]* gradually attenuates and then becomes a steadily increasing *R[-]*. In each case *E* is locked into a fixed network of relations whose logic is played out until the trap is 'sprung'.

Extricating Durkheim

Emile Durkheim's sociology viewed human societies as totalities emerging on a higher level from geographically circumscribed aggregates of individuals and reacting upon them by imposing a variety of 'constraints'. Durkheim considered the process of constraint imposition to be the most general social fact: social behavior is governed by rules. Jean Piaget claims that Durkheim's sociology has "died a natural death for the lack of a relational structuralism which might have supplied some laws of composition or construction instead of referring unremittingly to a totality conceived as ready made" [12: 22].

By presenting a structuralist approach to a particular type of relational network, the trap, this chapter attempts to exemplify a new perspective toward human societies which offers a provisional solution to the critical problem accounting for the fossilization of Durkheimian and Parsonian sociology. Adapted from recent work by Piaget, the structuralism proposed here is relational and posits netwroks of interactions or transformations as the primary reality; network elements or nodes are therfore subordinated from the start to the relations in which they are embedded and, reciprocally, the networks as a whole are conceptualized as the products of the composition of these generative interactions.

Unfolding traps: Examples

Relational analysis, as opposed to reductionism, attempts to demonstrate the affinity between general mental structures and general social system structures. Piaget points out how Levi-Strauss resorts to the structures of general algebra (groups, lattices, etc.) to give an adequate expression to his anthropological structuralism so that the sociological explanation coincides with a qualitative mathematization "similar to that which occurs in the setting up of logical structures, the process of which can be followed in the spontaneous thinking of children and adolescents, though not in their school learning" [13: 32]. The discovery of homomorphisms leads us to much deeper explanatory tendencies than simple reductionism. The examples which follow unfold over time a special class of networks, traps, which possess certain interdisciplinary 'deep structures', each of which may assist in a *constructivist* (as opposed to a reductionist) *understanding* of the others. It is our position that the analogic capabilities of structuralism make possible an entirely new, constructivist approach to sociological theorizing.

A. Topological catastrophe

R. Thom defines a topological catastrophe as the destruction of a structurally stable 'regime' (attractor) by the variation of a vector field [14: 323]. The intuitive notion underlying Thom's models of ordinary topological catastrophes and which corresponds to the concept of a trap, is that a catastrophe may occur in a structurally stable way, according to a fixed algebraic model given by theoretical consideration. An example of such a catastrophe is given in Fig. 5.

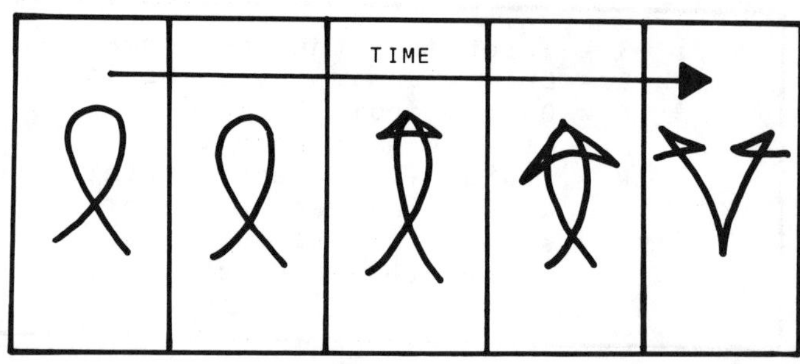

Fig.5 Some plane sections of the universal unfolding of the parabolic umbilic ($V = x^2y + y^4 + wx^2 + ty^2 - ux - vy$) on four-dimensional space-time, illustrating a trap-like catastrophe of internal dimension two (Adapted from Thom, 1969:327)

Thom proposes a 'semantic model' in which the decomposition of a structural process taking place on an Euclidean space E^m can be considered as a kind of generalized m-dimensional language. The everyday language we use is a semantic model of dimension one (the time) whose 'islands of determinacy' or chreods (Waddington [15: 32]) are the *words.* This 'semantic model' gives rise to two problems: [1] the classification of all types of chreods, and the understanding of the dynamic processes which insure their stability; and [2] ensembles of chreods (associations of words which appear more often than others) which give rise to a multi-dimensional syntax directing the semantic model and need to be described and formalized as one may formalize rules in linguistics (Thom [14: 322]). This idea suggests how Chomsky's constructivist grammar laws might be related to certain topological structures.

The task proposed above requires that one build a dictionary of chreods and then create what linguists refer to as the 'corpus' of a language. The goal of scattering experiments in elementary particle physics is precisely of this nature. The problem of interpreting the data arising from such situations and constructing a formal theory is equivalent to deciphering an unknown language. Plato's Cratylus poses the question of how to understand the manner in which the phonetic structure of a word proceeds from its meaning. Analogously in sociology, we often seek to infer elemental structures or chroeds from the qualitative dynamics which insure the stability of human societies. This quest is inevitably associated with the detection of 'traps' or potential structural destabilizers inherent in the very design of population systems. A deterministic network may exhibit, in a 'structurally stable way', a complete indeterminism in the qualitative prediction of the final outcome of its evolution (Thom [14: 317]). This situation corresponds to the uncertainty entrapment illustrated in cell 2 of Fig. 4.

B. Famine in the Sahelian zone

Jule Chaney *et al.*, employing a computer model of atmospheric circulation, have recently demonstrated the plausibility of a biogeophsical feedback mechanism which may have a substantial effect upon climate [16: 434-5]. Lester R. Brown reports that as of the summer of 1974, "several hundred thousand people had perished in the Sahelian zone just south of the Sahara Desert in West Africa and Ethiopia in a famine of disastrous proportions, which has been spreading since the early 'seventies as a result of years of drought and the accelerating southward spread of the desert" [17: 28].

It now appears that the tens of thosuands of people who have recently died in the Sahel were victims of a biogeophysical trap. The key mechanism in the trap was the dependence of the surface albedo (reflective power) on plant cover. A decrease in plant cover tends to increase the surface albedo which in turn leads to a decrease in the net incoming radiation. This produces radiative cooling of the air. In order to restore thermal equilibrium, the air is adiabatically compressed, yielding cumulus convection and consequent curtailment of rainfall. The drop in rainfall reinforces the original decrease in plant cover and thus closes the positive feedback loop in the biogeophysical trap. Chaney *et al.* also indicate this positive feedback loop may be initiated by overgrazing. Cattle serve as a major focal point in the culture of many Sahelian people. Tragically, their presence may be the critical link in the causal netwrok associated with current famine conditions.

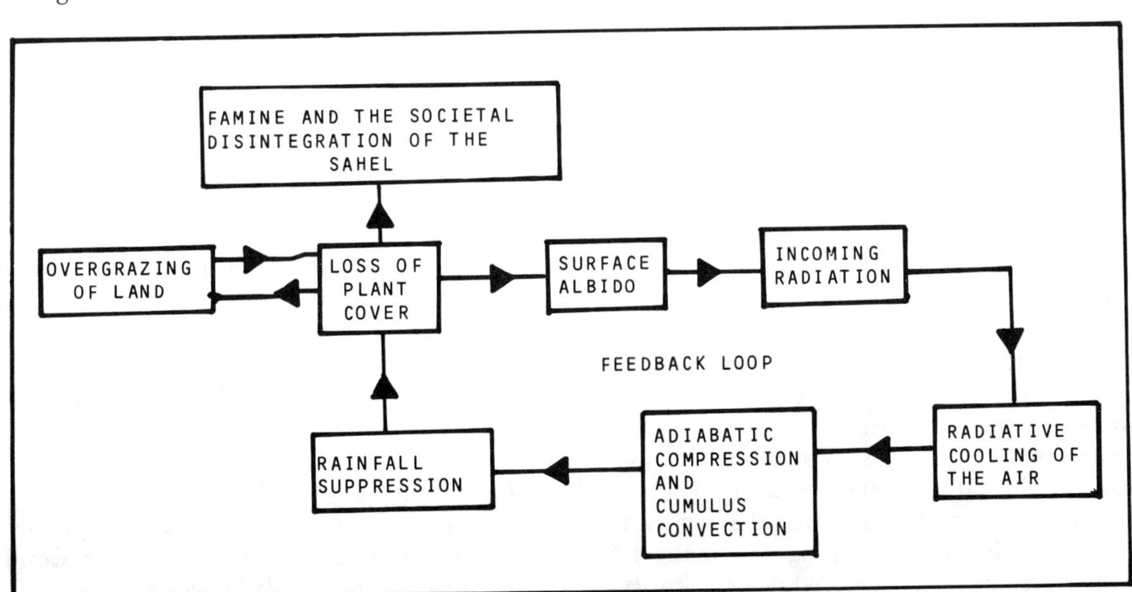

Fig.6 Famine and societal disintegration as a biogeophysical trap in the Sahelian zone

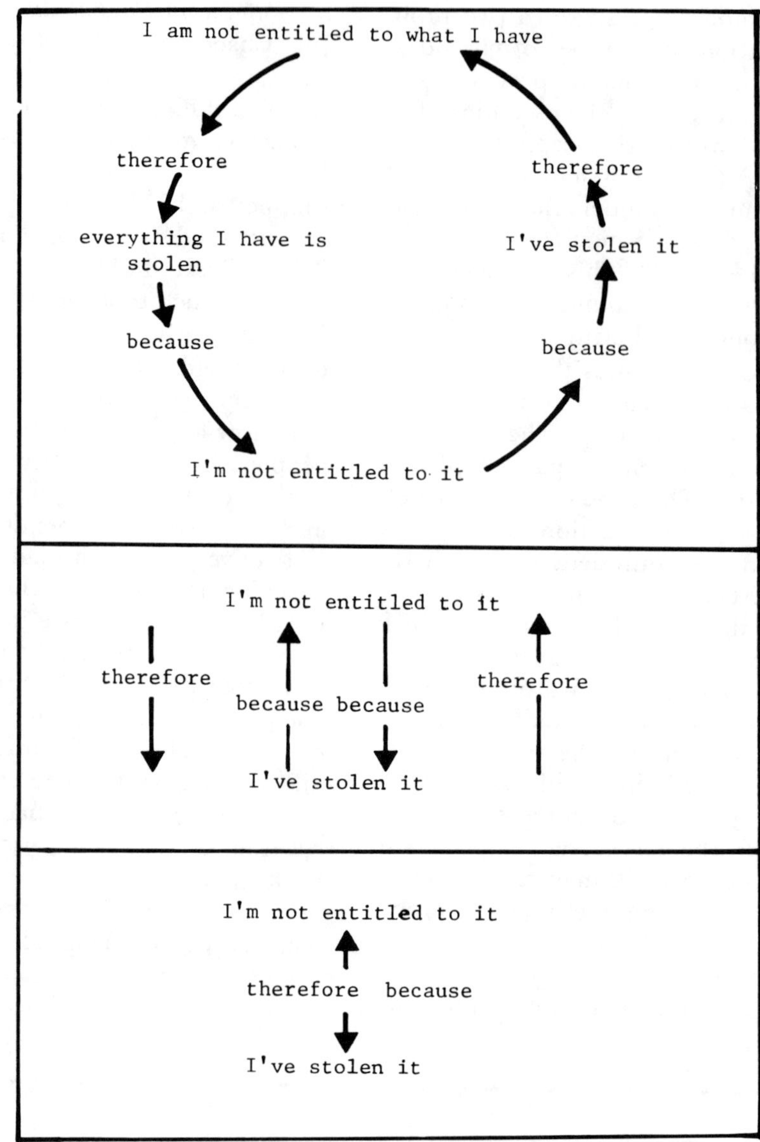

Fig.7 A pathological 'knot' (Adapted from Laing, 1970:36-7)

C. Pathological knots

Impasses in cognitive patterns can create cyclic traps paralyzing an individual's behavior, locking it into what R.D. Laing refers to as a 'knot' [18]. While he does not attempt to do so, Laing proposes that such knots are capable of distillation into an abstract logico-mathematical calculus. A structuralist approach to 'knots' would naturally seek to derive such a calculus, the knowledge of which would serve as a necessary prerequisite for developing systematic techniques for psychiatric therapy. While the 'knots' discussed by Laing are restricted to one and two person traps, the structure of these 'knots' can readily be found in group behavior. This suggests that a structuralist approach to Laingian knots would provide an easy transit to the analysis of social psychological pathologies, linking together and classifying individual and social traps.

D. Arrow's impossibility theorem

In the eigtheenth century the Marquis de Condorcet developed a paradox known as 'the Condorcet effect' (Granger [19: 94-129]) which later was to be generalized into 'Arrow's impossibility theorem' (Arrow

[20: 92-120]). Arrow's theorem states that if there are at least *three* alternatives which the members of a network are free to order in any way, every network decision rule satisfying some reasonable properties, and yielding a network ordering A that is connected (i.e. for all alternative network states n_1 and n_2, $n_1 A n_2$ or $n_2 A n_1$) and transitive (i.e. for all n_1, n_2, n_3, $n_1 A n_2$ and $n_2 A n_3$ implies $n_1 A n_3$), is either imposed or dictatorial. This important theorem illustrates a type of decision-making trap to which certain types of networks are vulnerable. *Regardless* of the chosen rule of aggregation of the preferences of the network members, network decisions do not necessarily satisfy even some minimal requirements of rationality (Majone [21: 66]). This type of trap has obvious relevance to policy analyses associated with the nature of the controllability of complex systems.

E. Unraveling an 'integrated' network

The structure of social networks can be vulnerable to what at first appear to be only minor stochastic perturbations. This type of trap produces an unraveling effect which results in a fundamentally new, undesirable network. Thomas Shelling [22] has drawn up an intriguing example of just such a trap. Shelling's model will now be recast into network language and presented below.

We shall start with equal numbers of α's and β's and impose the 'moderate' constraint that each α and β should have more than 1/3 neighboring nodes like itself. This 'integrated' network is shown in cell 1 of Fig. 8. There are 60 α's and β's, 20 of which will now be removed, using a table of random numbers to designate the empty nodes. Five of the empty nodes are now chosen at random and an α or β is inserted into each with equal probability. Such a result is illustrated in cell 2 of Fig. 8. The original 'integrated' network has now been systematically perturbed but still appears to be stably 'integrated'. There are only nine 'unintegrated' α's and β's out of a total of 45. Their nodes are given in cell 3 of Fig. 8.

The problem now is one of trying to 'integrate' the network by allowing the nine 'unintegrated' members to move somewhere among the 19 empty nodes. The trap now unfolds once we recognize that any of these nine α's and β's which move to new nodes (seeking an 'integrated' node in the network), affect the nodal relations that existed at their old node and create additional relations at their new node. Cells 4 and 5 in Fig. 8 show two possible outcomes which from a network perspective yield undesirable 'segregated' patterns. The original 'integrated' pattern has been unraveled and the network, trapped by the 'integrationist demands' of minority members, is now more 'segregated' than ever. In cell 2 of Fig. 8, the α's have a ratio of α-neighboring nodes to β-neighboring nodes of 1.1, with a small cluster of α's in the upper left corner and six 'unintegrated' α's widely distributed throughout the network. In cell 4 of Fig. 8 the average ratio of similar to dissimilar neighboring nodes for α's and β's together increases to 2.3, and in cell 5 it increases to 2.8 (almost 3 times the initial ratio and 4 times the minimum required for an 'integrated' network). The analysis of traps of this nature have obvious relevance for the study of residential segregation patterns by demonstrating how sorting processes aimed at increasing integration can be trapped into a segregated network.

F. A global trap

John Knowles, President of the Rockefeller Foundation, recently remarked that we are now "in the midst of the greatest and most profound cultural discontinuity and contradiction in history" [23: 3]. A total world population of 4 billion humans was anticipated in 1975 and 7 billion expected within the next 25 years. D. Pimentel has argued that since most of the arable land of the earth is already in production, "the only means of increasing production will be to intensify production on the available arable land using fossil fuel inputs [24: 561]. 'Green revolution' agriculture requires large fuel inputs. Since fossil fuel energy is now widely recognized as a finite environmental resource by OPEC suppliers and others, its price value has and will continue to significantly increase (barring some drastic change in the current balance of power). Hence, many of the world's socieities seem to be moving deeper and deeper into what may prove to be a lethal trap for large sectors of their populations. This crucial problem may possess no feasible technical or political solution.

Some human ecologists, such as Amos Hawley, have presented an optimistic case for believing that "although the environment may be finite, organizational determinants will come into force long before the environment itself operates as a restraint on population" (Hawley [25: 1200]). This position is quite similar to Wynne-Edwards' theory that elementary social processes in animal life act as homeostatic mechanisms tending to maintain ideal population size and dispersion [26]. A more ominous note has been sounded by others, such as Beryl Crowe, who believes that "there is evidence of structural as well

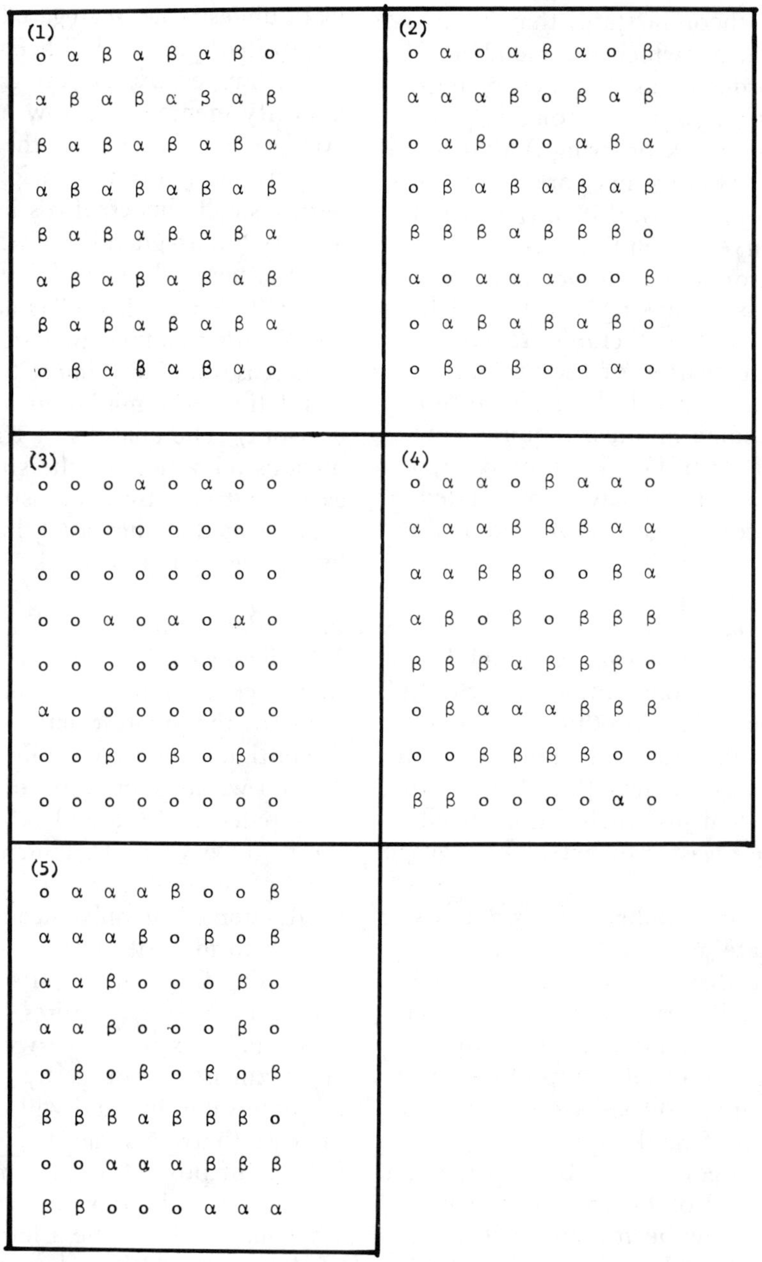

(1)

o	α	β	α	β	α	β	o
α	β	α	β	α	β	α	β
β	α	β	α	β	α	β	α
α	β	α	β	α	β	α	β
β	α	β	α	β	α	β	α
α	β	α	β	α	β	α	β
β	α	β	α	β	α	β	α
o	β	α	β	α	β	α	o

(2)

o	α	o	α	β	α	o	β
α	α	α	β	o	β	α	β
o	α	β	o	o	α	β	α
o	β	α	β	α	β	α	β
β	β	β	α	β	β	β	o
α	o	α	α	α	o	o	β
o	α	β	α	β	α	β	o
o	β	o	β	o	o	α	o

(3)

o	o	o	α	o	α	o	o
o	o	o	o	o	o	o	o
o	o	o	o	o	o	o	o
o	o	α	o	α	o	α	o
o	o	o	o	o	o	o	o
α	o	o	o	o	o	o	o
o	o	β	o	β	o	β	o
o	o	o	o	o	o	o	o

(4)

o	α	α	o	β	α	α	o
α	α	α	β	β	β	α	α
α	α	β	β	o	o	β	α
α	β	o	β	o	β	β	β
β	β	β	α	β	β	β	o
o	β	α	α	α	β	β	β
o	o	β	β	β	β	o	o
β	β	o	o	o	o	α	o

(5)

o	α	α	α	β	o	o	β
α	α	α	β	o	β	o	β
α	α	β	o	o	o	β	o
α	α	β	o	o	o	β	o
o	β	o	β	o	β	o	β
β	β	β	α	β	β	β	o
o	o	α	α	α	β	β	β
β	β	o	o	o	α	α	α

Fig.8 Network scenarios illustrating a 'segregation' trap. The "o's" indicate vacant nodes in the network. The α's and β's can be given any interpretation desired

as value problems which make comprehensive solutions impossible and these conditions have been present for some time" (Crowe [27: 1106]).

It seems clear from historical evidence that life-support mechanisms may be removed so rapidly and massively from a population system that organizational factors will simply be overwhelmed (e.g. the recent famine in the Sahel and the mid-nineteenth century Irish famines). The post-disaster or survivor population (if one exists) will adaptively redesign its system features but only after a period of virtually total organizational fragmentation. Population system catastrophes, by definition, are characterized by a major perturbation which either poses a serious risk to the continued existence of the system or actually exterminates it. A necessary prerequisite for a population system catastrophe, then, is that the system lack the adaptive potential to absorb or neutralize certain types of perturbations.

Crowe argues that catastrophic types of population system perturbations are visible on the time horizon which demand urgent attention from the technical and social sciences. Crowe takes the position, however, that the technical and social sciences have cut themselves off from each other and have *trapped* themselves in a structure of insularity and specialization which insures that the truly critical problems facing contemporary population systems will not be attacked with the socio-technical expertise required to solve them:

> Perhaps the major problems of modern society have, in large part, been allowed to develop and intensify through this structure of insularity and specialization because it serves both psychological and professional functions for both scientific communities. Under such conditions, the natural sciences can recognize that some problems are not technically soluble and relegate them to the nether land of politics, while the social sciences recognise solutions and then postpone a search for solutions while they wait for new technologies with which to attack the problem. Both sciences can thus avoid responsibility and protect their respective myths of competence and relevance, while they avoid having to face the awesome and awful possibility that each has independently isolated the same set of problems and given them different names. Thus, both never have to face the consequences of their respective findings. Meanwhile, due to the specialization and insularity of modern society, man's most critical problems lie in limbo, while the specialists in problem-solving go on to less critical problems for which they can find technical or political solutions (Crowe [27: 1103]).

The social sciences cannot afford to defer considering critical population system problems on the basis of false expectations regarding future technological advances, while the natural sciences simultaneously defer the same problems because of false expectations concerning socio-political change. Adaptation is necessarily an organizational process (Hawley [25: 1200]). Without the requisite organisation, adaptation cannot occur in a population system. This is the fundamental dilemma.

Garrett Hardin's article 'The Tragedy of the Commons' has been reprinted in so many edited anthologies that it has almost assumed the status of a 'classic'. We shall use Hardin's model of the commons to illustrate certain population system traps as well as possible escape mechanisms. The Hardin model assumes that energy acquisition is the core problem facing population systems. Starting with a brief metaphorical discussion of the depletion of common-property resources associated with the social institution of pasturage set aside for public use by English common law, Hardin extends the metaphor to population system problems in general. Hawley has attacked Hardin's analogy as "a view that has been informed by neither a historic perspective nor a competent assessment of the existing informational and institutional resources for change" [25: 1197]. Nonetheless, the Hardin model can serve as a facile representation of the basic trap-like mechanism which many scientists believe threatens population systems at the present time. We shall explore the logic of the Hardin model by means of computer simulation. Technological and organizational change will be introduced into the model and the system outcomes assessed.

Jay Anderson [28,29] has operationalized some of the ideas presented in the Hardin article [30] into a formal computer model. This model lays Hardin's assumptions and conclusions regarding potential population system traps open to critical examination. Anderson translates Hardin's conceptualization of the commons problem into an ecosystem model with two principal components: capital and commons. 'Commons' can represent any renewable common-property resource (e.g. plants, water, wilderness, etc.) and 'capital' is any means of production of goods (e.g. cattle, factories, humans, etc.). Figure 9 shows the basic feedback loops present in the systems causal network. Capital is controlled by 'investment' and 'depreciation', and commons by 'natural regeneration' and 'usage'. The model is now in a form which can be translated into DYNAMO computer language†. The symbols used in the DYNAMO equations are listed in Table 2, a DYNAMO flow diagram of the model in Fig. 10, and the DYNAMO equations themselves along with the graphical output for the unmodified Hardin model are given in Fig. 11.

The graphical output in Fig. 11 reproduces the essential qualitative dynamics predicted by Hardin's verbal theory (i.e., the population system degenerates). To evaluate the logic of the causal network posulated by Hardin and to test his claim that the

† The reader who may be unfamiliar with system dynamics modeling and DYNAMO is advised to consult J.W. Forrester's *Principles of Systems* [34] or M.R. Goodman's more recent *Study Notes in System Dynamics* [35].

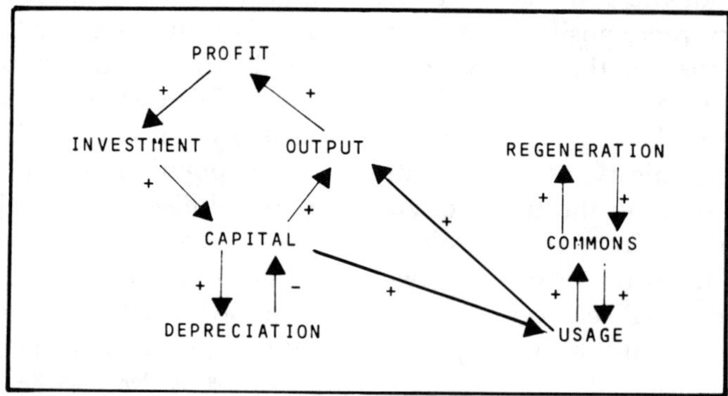

Fig.9 The causal network underlying Hardin's 'The Tragedy of the Commons', as operationalized by Anderson (1974)

DYNAMO SYMBOLS	DEFINITIONS AND UNITS
CAP	Capital, $
CCOST	Cost parameter, $/unit
CCR	Common-capital ratio, units/$
CDR	Capital depreciation rate, $/year
CIR	Capital investment rate, $/year
COM	Commons, units
COR	Capital-output ratio, dimensionless
CRR	Commons regeneration rate, units/year
CUR	Commons usage rate, units/year
EXCOST	External cost, $
EXCSW	External cost switch, 1 or 0, dimensionless
IT	Investment time, years
LC	Lifetime of capital
LIMIT	Limit on use of commons, units/year
OUT	Output, $
PROF	Profit, $
RT	Regeneration time, years

Table 2 Definitions and units of measurement for symbols in DYNAMO equations listed in Table 3

network is a member of the class of problems which possess "no technical solution", two types of technological change were simulated' [1] the common-capital ratio CCR was lowered from 0.75 to 0.50, the computer results of which are given in Fig. 12, and [2] the commons regeneration rate was modified to 2.0*COMI/(COMI-COM.K+1), yielding the graphical output in Fig. 13. Once again Hardin's predictions are confirmed. The DYNAMO output shows that realistic and feasible technological changes fail to alter significantly the qualitative dynamics of the unmodified model.

An "organizational". change is now introduced by charging for the use of the common. This is simulated by switching the parameter EXCSW from 0 to 1, and increasing the common's cost CCOST from 0.01 in the unmodified model to 0.20 $/commons unit. Figure 14 shows that this new condition also fails to substantially modify the original dynamics. At this point we might give up and accept as an established fact that Hardin's proposed network constitutes an inextricable population system trap. However, we shall now implement another 'organizational' change by restricting the use of the commons from 100 (full use) to 10. Figure 15 reveals quite dramatic results. The populations system fails to

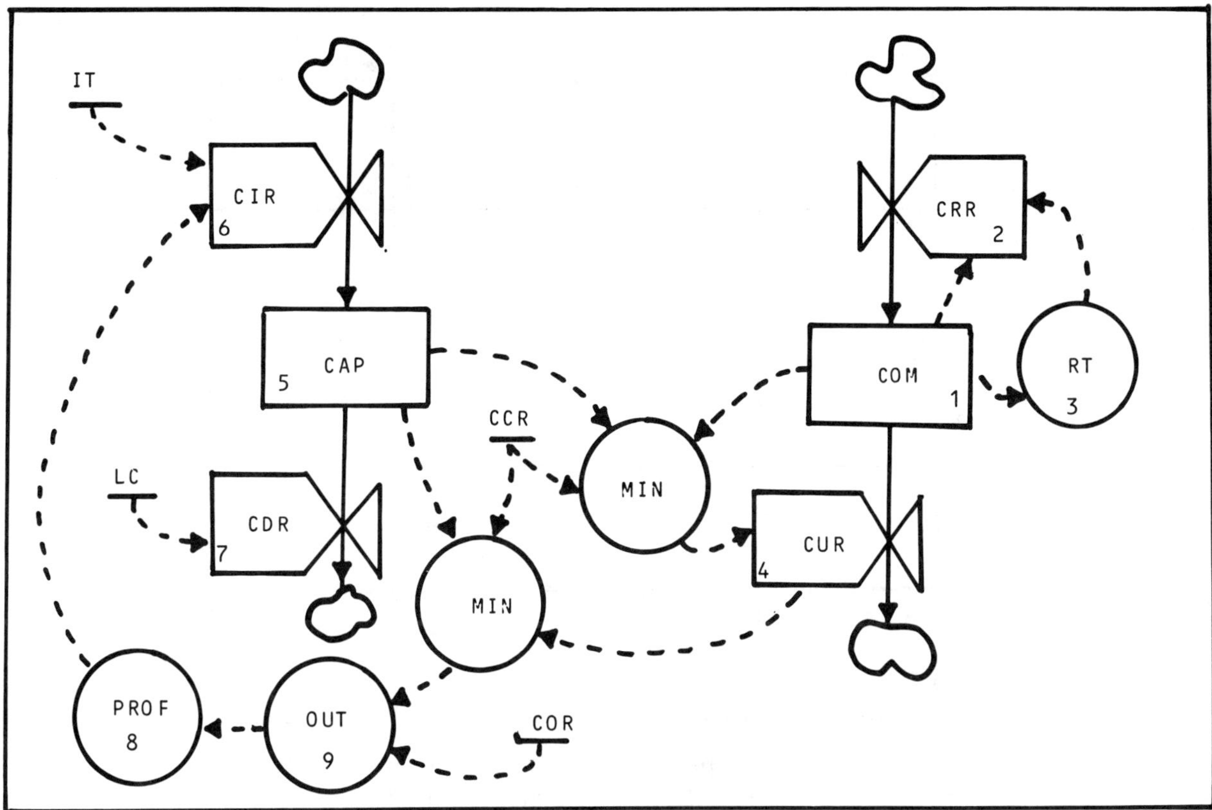

Fig.10 DYNAMO flow diagram for Hardin's 'The Tragedy of the Commons' (Adapted from Anderson, 1975)

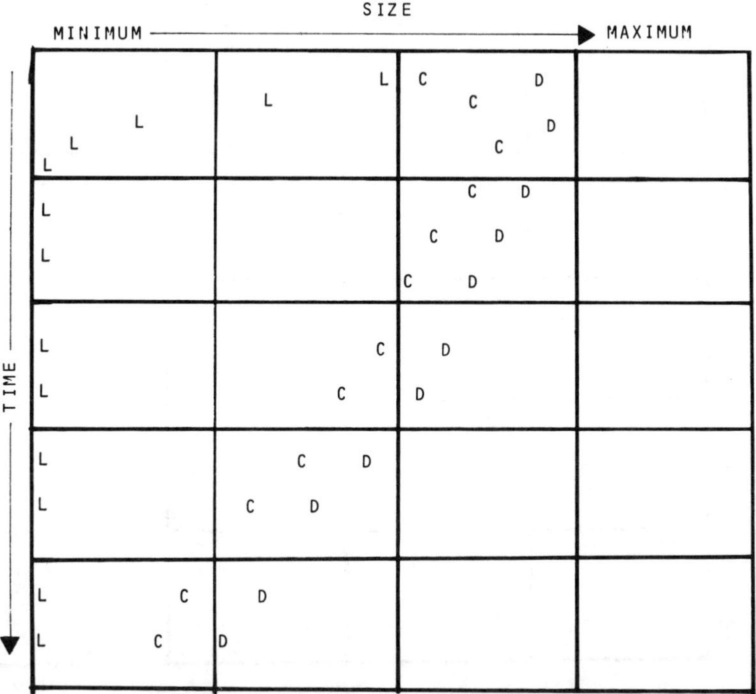

Fig.11 Basic characteristics of the DYNAMO output for the unmodified simulation of Hardin's 'The Tragedy of the Commons'

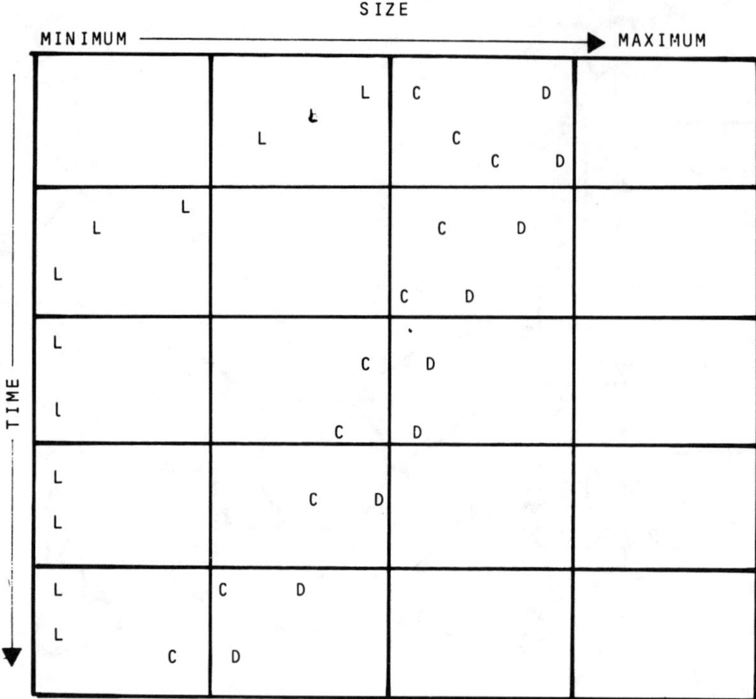

Fig.12 Basic characteristics of the DYNAMO output for the Hardin model, given the technological innovation represented by changing the value of CCR in the unmodified model from 0.75 to 0.50

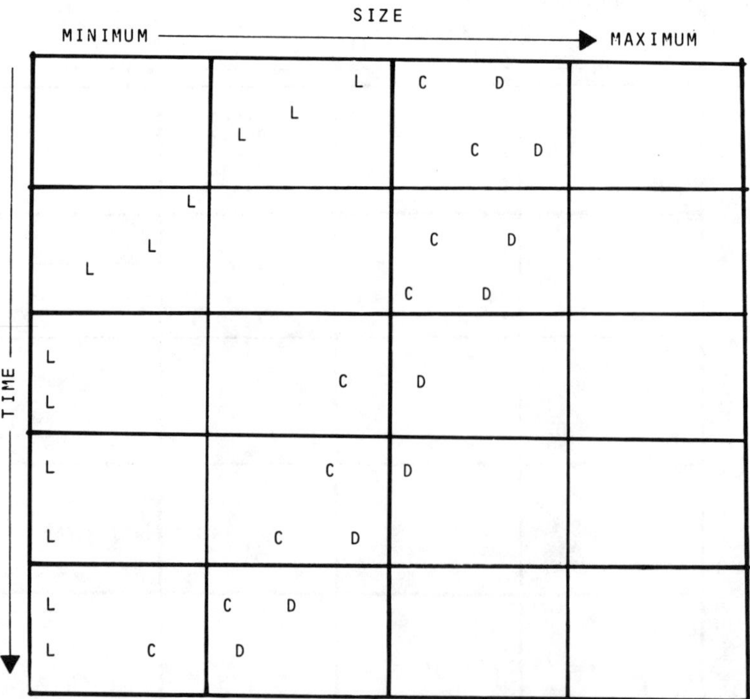

Fig.13 Basic characteristics of the DYNAMO output for the Hardin model, given the technological innovation represented by changing the value of RT in the unmodified model to 2.0*COMI/(COMI-COM.K)+1)

```
*    TRAGEDY OF THE COMMONS
L    COM.K=COM.J+DT*(CRR.JK-CUR.JK)
N    COM=COMI
C    COMI=100
R    CRR.KL=COM.K/RT.K
A    RT.K=2.5*COMI/(COMI-CCM.K+1)
R    CUR.KL=MIN(MIN(CAP.K*CRR.JK,LIMIT),COM.K)
C    CCR=.75
C    LIMIT=100
L    CAP.K=CAP.J+DT*(CIR.JK-CDR.JK)
N    CAP=10
R    CIR.KL=MAX(0,PROF.K)/IT
C    IT=10
R    CDR.KL=CAP.K/LC
C    LC=20
A    PROF.K=OUT.K-EXCOST.K
A    OUT.K=MIN(CAP.K,CUR.JK/CCR)/COR
C    COR=1
A    EXCOST.K=SWITCH(0,CCOST*(COMI-COM.K),EXCSW)
C    CCOST=.01
C    EXCSW=0
C    DT=.25
PLOT   COM=L/CAP=C/CUR=U/CIR=I/CRR=R
C    LENGTH=50
RUN   UNMODIFIED MODEL
```

Table 3 DYNAMO equations used to operationalize Hardin's theoretical model of 'The Tragedy of the Commons'. The simplified adaptations of the DYNAMO output which appear in Figs 11-15 have been based on these equations

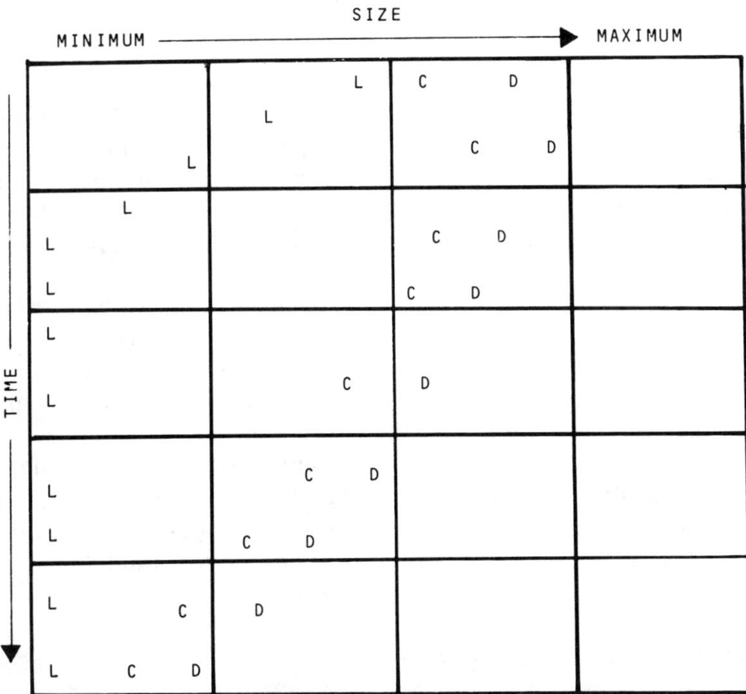

Fig.14 Basic characteristics of the DYNAMO output for the Hardin model, given the 'organizational' constraint represented by changing the value of the parameter EXCSW to 1 and raising CCOST to 0.20 $/commons-unit

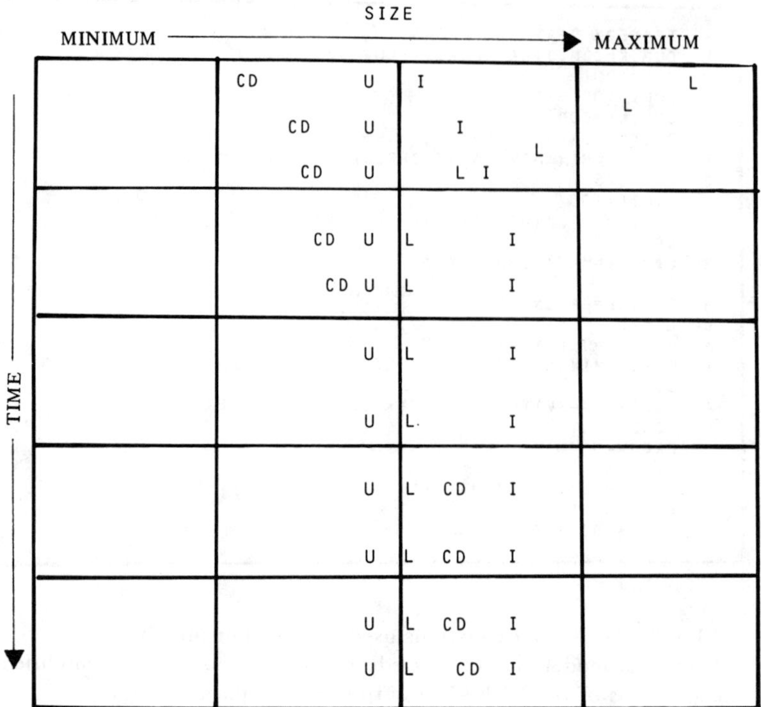

Fig.15 Basic characteristics of the DYNAMO output for the Hardin model, given the 'organizational' constraint represented by decreasing the parameter LIMIT from 100 to 10

degenerate and stabilizes at an acceptable level. This interesting result illustrates how one might utilize computer simulation to design an 'organizational interface' between a population and its environment to prevent them from being progressively drawn into a trap of mutual destruction.

Conclusion

Network traps all tend to exhibit similar structure, they are homomorphic. Relational structures within a given class, such as traps, can be exploited in an interdisciplinary manner to amplify knowledge in a discipline where structural information is at a low level by drawing upon other disciplines possessing relatively high informational levels relevant to the same class of structures.

Figure 16 suggests how this idea would apply to the six examples of trap-like structures already presented in this chapter. The approach presented here also points to the potential value of developing a synthetic social science such as human ecology (Hawley [25: 1200]). A structuralist social science will inevitably be led to break down the barriers between technical and social sciences which Crowe maintains are impeding research on critical human problems. Applications to social forecasting (Meadows [31]) and crisis management (Kupperman *et al.* [32]) should also prove to be particularly fruitful in terms of critical information acquisition.

The strategic idea presented in this chapter has been to demonstrate how a *depth level* (Hartmann [33: 39]) may be introduced (e.g. network traps) that can serve as a structure around which one can draw together and synthesize information from disparate fields. The choice of a depth level is ultimately a matter of informed decision or ingenuity. A structuralist sociology will require a wholly new repertoire of theoretical models and methods. This chapter has provided some glimpses of their probable nature.

References

1. BLAU, P.M., 'Parameters of social structure', *American Sociological Review* 39(5), pp.615-35 (October 1974).
2. HEISE, D.R., *Causal Analysis*, New York: Wiley-Interscience (1975).

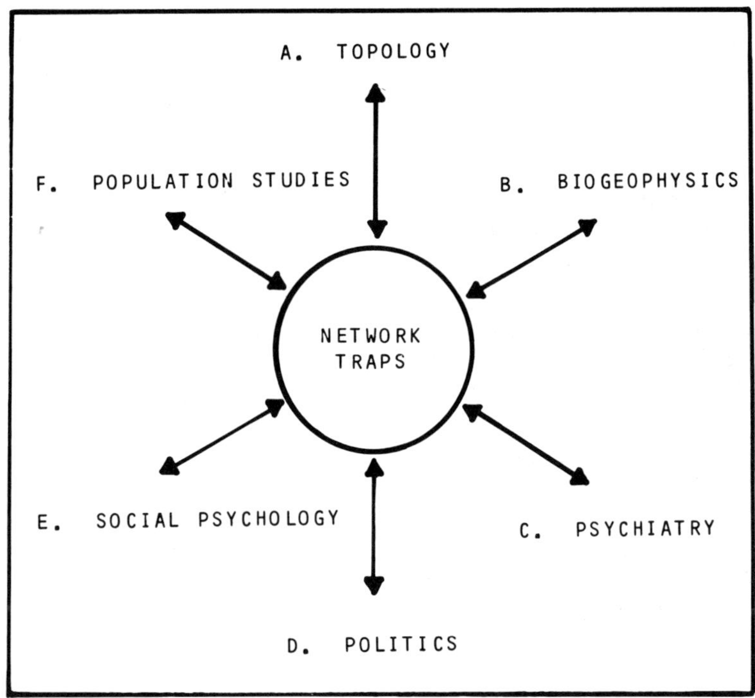

Fig.16 A schematic representation of how network traps serve as a structuralist 'depth level' for the examples cited in this article

3. WHITE, H.C., 'Everyday life in stochastic networks', *Sociological Inquiry* **43**, pp.43-9 (Spring 1973).
4. WHITE, H.C., 'Patterns across networks', Paper presented at the Annual Meeting of the American Association for the Advancement of Science, New York City (28 January 1975).
5. HOWARD, N., *Paradoxes of Rationality: Theory of Metagames and Political Behavior,* Cambridge, MA: M.I.T. Press (1971).
6. WEISSKOPF, V.F., 'Of atoms, mountains, and stars: a study in qualitative physics', *Science* **187** (4177), pp.605-12 (February 1975).
7. BUNGE, M., *Method, Model and Matter,* Dordrecht, Holland: D. Reidel Publishing Company (1973).
8. LEACH, E., 'Structuralism in social anthropology', in (D. Robey: editor) *Structuralism: an Introduction,* pp.35-56, Oxford: Clarendon Press (1973).
9. LORRAIN, F. and WHITE, H.C., 'Structural equivalence of individuals in social networks', *Journal of Mathematical Sociology* **1**(1), pp.49-80 (1971).
10. PLATT, J., 'Social traps', *American Psychologist* **28**(8), pp.641-51 (August 1973).
11. SKINNER, B.F., *Contingencies of Reinforcement: A Theoretical Analysis,* New York: Appleton (1969).
12. PIAGET, J., *Main Trends in Inter-Disciplinary Research,* New York: Harper Torchbooks (1973).
13. PIAGET, J., *Main Trends in Psychology,* New York: Harper Torchbooks (1973).
14. THOM, R., 'Theoretical models in biology', *Topology* **8**(3), pp.313-35 (July 1969).
15. WADDINGTON, C.H., *The Strategy of the Genes,* London: Allen & Unwin Ltd. (1957).
16. CHARNEY, J. *et al.*, 'Drought in the Sahara: a biogeophysical feedback mechanism', *Science* **187**(4175), pp.434-5 (7 February 1975).
17. BROWN, L.R., *By Bread Alone,* New York: Praeger Publishers (1974).
18. LAING, R.D., *Knots,* New York: Pantheon Books (1970).
19. GRANGER, G.-G., *La Mathématique Social du Marquis de Condorcet,* Paris: Presses Universitaires de France (1956).
20. ARROW, K.J., *Social Choice and Individual Values,* 2nd ed., New York: John Wiley & Sons, Inc. (1963).
21. MAJONE, G., 'The role of constraints in policy analysis', *Quality and Quantity* **8**(1), pp.65-76 (March 1974).

22. SHELLING, T.C., 'On the ecology of micromotives', *The Public Interest* **25**, pp.59-98 (Fall 1971).
23. KNOWLES, J.W., 'Food, energy, and American austerity', *Boston University Journal* **22**(2), pp.3-9 (Spring 1974).
24. PIMENTEL, D., 'Food production and the energy crisis: reply to a comment by V.W. Ruttan', *Science* **187**(4178), p.561 (14 February 1975).
25. HAWLEY, A., 'Ecology and population', *Science* **179**(4079), pp.1196-201 (23 March 1973).
26. WYNNE-EDWARDS, V.C., *Animal Dispersion in Relation to Social Behavior,* London: Oliver and Bond, Ltd. (1962).
27. CROWE, B.L., 'The tragedy of the commons revisited', *Science* **166**(3909), pp.1103-7 (28 November 1969).
28. ANDERSON, J.W., 'A model for "The tragedy of the commons"', *IEEE Transactions on Systems, Man, and Cybernetics* SMC-4(1), pp.103-4 (January 1974).
29. ANDERSON, J.W., Personal communication to the author (1975).
30. HARDIN, G., 'The tragedy of the commons', *Science* **162**(3859), pp.1243-8 (13 December 1968).
31. MEADOWS, D.L., 'Toward a science of social forecasting', *Proceedings of the National Academy of Sciences* **69**(12), pp.3828-31 (December 1972).
32. KUPPERMAN, R.H., WILCOX, R.H., and SMITH, H.A., 'Crisis management: some opportunities', *Science* **187**(4175), pp.404-10 (7 February 1975).
33. HARTMAN, K., 'Lévi-Strauss and Sartre', *Journal of the British Society for Phenomenology* **2**(3), pp.37-45 (October 1971).
34. FORRESTER, J.W., *Principles of Systems*, Cambridge, MA: Wright-Allen Press, Inc. (1968).
35. GOODMAN, M.R., *Study Notes in System Dynamics,* Cambridge, MA: Wright-Allen Press, Inc. (1974).

On control of multilevel active systems

RYSZARD WAŚNIOWSKI
Technical University of Wrocław, Poland

Classes composed of persons or a group of persons are among the most important classes of systems. There are various examples of those systems, e.g. sociological, economical, and political ones.

A characteristic feature of the class mentioned above is the existence of subsystems that have their own objective functions and, in general, are not identical with those of the whole system. The man including subsystems are, moreover, active in such a sense that in order to maximize their objective functions they tend to draw not only on available resources but also on information channels by means of which they receive information on the activities of other subsystems, in general, and on control-subsystems with respect to the subsystems considered, in particular.

Approaches to the control of organization systems may be arbitrarily divided into three groups. The first one includes those approaches which are based on the principles of coordination worked out by Mesarovic [1].

The basic idea of that work can be expressed in the following way: the objective function of a system is decomposed into N functions, corresponding to the goals of N subsystems. Each subsystem optimizes its function and a higher level coordinate solutions of lower subsystems so as to reach the optimum of global objective function. The coordination is realized by means of certain auxiliary variables being the parameters for lower subsystems.

The second group includes the approaches based on the game theory. A typical method of that group lies in consideration of a system of two levels consisting of 'Center' and 'Performers'. The objective function of the Center is defined by the results of the Performers' function and the Performers may independently undertake their optimum programmes. Reserves are at the disposal of the Center. A typical game situation occurs and the solutions are sought within the unantagonizing games theory [2].

The third group includes the approaches which take into account the goals of adequate subsystems and their active behavior. What is to be mentioned is the definition of the activity of a system. By the activity of a system we mean, that:

- the system tries to gain its goals,
- it has potentials to realize the programmes,
- it knows its potentials better than the controlling organs of higher levels,
- it is informed on the principles of decision-making on higher levels, and makes use of that information for its own goals.

The scheme of the function of an active system may be described in the following way. A system of higher level, i.e. Central Organ (CO), charges N active subsystems with N tasks. Starting with its own goals each sybsystem formulates a vector with components characterizing the degree of interest it takes in a given task and presents this vector to the CO. When allocating the tasks the CO tries to satisfy maximally the interests of lower subsystems, i.e. it solves the task allotted. When the Central Organ receives the results of the tasks realized it issues rewards or punishments. The approach below is based on the paper [3].

1. Formulation of the problem of control in active systems

To simplify, we shall consider first a two-level system comprising the CO and n active elements functioning independently. Control algorithms are generally based on control techniques adopted in the system: e.g. decentralized and centralized planning. In the class of centralized planning the following methods can be distinguished:

- firm centralization method, its main characteristic is that in the process of planning CO takes account mainly of the benefit of the CO,
- 'fair play' planning method, in which CO applies the so called priority function characterizing the scope of interests of subsystems in the adoption of a given plan,
- coordinated planning, in which CO takes equally into account the benefit of subsystems.

The problem of control in a two-level active system consisting of a CO and n subsystems is formulated as follows: does there exist a decision CO at which a certain permissible rigor of subsystems work provides the required functioning of a system as a whole?

Formally the problem is denoted by

$x = (x_1, x_2, \ldots, x_n)$ – decision made by CO,

$y = (y_1, y_2, \ldots, y_n)$ – decisions of subsystems,

$\Phi(x,y) \to \max_x$ – objective function of CO,

$\varphi(x_i, y_i) \to \max_{y_i}$ – objective function of subsystems

It is assumed that CO has the priority in decision making. Now we shall introduce three special cases of that problem leading to mathematical programming.

1.1 Problems of tasks assignment [4]

n performers and the CO assigning m tasks are given. We denote by

$$\tau_i = (\tau_{i1}, \ldots, \tau_{ij}, \ldots, \tau_{im}) \qquad i = 1,2,\ldots,n \quad j = 1,2,\ldots,m$$

the characteristic of the performer corresponding to the real level of 'remunerativeness' for a given task performer; $_{ij}$-variable mapping the acceptance or rejection of the task, $X_{ij} = 1$ if the task j is designed for the performer i; $X_{ij} = 0$ if the contrary case occurs, λ_j is the reward for the realization of the task j; $S_i = (S_{i1}, \ldots, S_{ij}, \ldots, S_{im})$, $i = 1,2,\ldots,n$, $j = 1,2,\ldots,m$. Evaluations of characteristics τ_i are transmitted by the performers to the CO at the stage of formulation of data.

The goal function of the performer is presented in the form:

$$\varphi_i = \sum_{j=1}^{m} (\tau_{ij} + \lambda_j) X_{ij} \; ; \quad i=1,2,\ldots,n \tag{11}$$

The goal function of the whole system is defined by the expression

$$\Phi = \sum_{i=1}^{n} \sum_{j=1}^{m} \tau_{ij} X_{ij} \tag{12}$$

which represents the total 'profit' of realization of all the tasks designed.

On the stage of planning, CO solves the following problem:

$$\sum_{i,j} S_{ij} X_{ij} \to \max \tag{13}$$

$$\sum_{i=1}^{n} X_{ij} = 1 \; ; \quad j=1,2,\ldots,m \tag{14}$$

$$\sum_{j=1}^{m} X_{ij} \leqslant 1 \; ; \quad i=1,2,\ldots,n \tag{15}$$

The above problem corresponds to the case when the system applies the principle of firm centralization and the size of the rewards λ_j is being settled (may also equal 0). The principle of co-ordinate control requires the necessity of adjusting the goals of the whole system to the goals of performers. The following restrictions are added to tasks (13-15)

$$[\max_1 (S_{i1} + \lambda_1) - (S_{ij} + \lambda_j)] X_{ij} = 0 \tag{16}$$

obliging the CO to assign only rational programmes for the performers, that is, the programmes warranting the performer i a 'foreseen' reward

$$(S_{ij} + \lambda_j)X_{ij}$$

not smaller than the highest of all possible values conforming to the evaluations transmitted to the performers.

1.2 Problems of resource allocation

Let

- n be the number of organizations belonging to the system,
- R the resources being at the disposal of the CO,
- u_i funds allocated for the institution i, $i=1,2,...,n$,
- $f(u_i)$ a function defining the size of resource demand in particular institutions (the same for all of them and known to the CO),
- τ_i the coefficient of effectiveness of resources utilization by the ith institution,
- $\tau_i f(u_i)$ the effect of utilizing by the ith institution the funds in quantity u_i,
- S_i the evaluation of the size τ_i by managers of the ith institution transmitted to the CO at the stage of formulation of the data.

The goal function of the whole system (presented by CO is defined by the expression:

$$\Phi = \sum_{i=1}^{n} \tau_i f(u_i) \tag{17}$$

At the planning stage, CO solves the problem

$$\Phi = \sum_{i=1}^{n} S_i f(u_i) \rightarrow \max \tag{18}$$

$$\sum_{i=1}^{n} u_i \leqslant R \tag{19}$$

and obtains the values of u_i.

At the stage of plan realization the performance of the institution is evaluated by the CO on the basis of comparison of the effects $\tau_i f(u_i)$ of performance of the institution and resources utilized by it.

Formally, a goal function of the institution can be written in the following form:

$$\varphi_i = \tau_i f(u_i) - \lambda u_i \tag{20}$$

where λ is the coefficient of bringing the resources to effective utilization (under certain conditions, it can be called the price of a resource unit).

In the case of analysis of the principle of firm centralization, the 'price' of a resource unit is stable and determined arbitrarily by the Central Organ; it is not directly concerned with the distribution of resources: we have to take into account the order of magnitude $S_i f_i(u_i)$ and u_i such that the numbers of φ_i be positive.

While utilizing the principle of coordinate control in the system, the central organ should take into consideration not only condition (19) but also the condition of warranting the highest possible foreseen value of reward to each active element (that is, the highest value at a fixed quantity of s_i transferred). This condition can be written in the form

$$S_i f_i(u_i) - \lambda u_i = \max_{0<z_i<\infty} [S_i f_i(z_i) - \lambda z_i] \tag{21}$$

The right choice of 'prince' of resource unit warrants the fulfilment of this condition, λ is determined by the set of equations: $df(u_i)/du_i = \lambda/S_i$, $i=1,2,...,n$; at $\sum_{i=1}^{n} u_i = R$. To simplify the calculation, it is recommended to assume $f(u_i) = \sqrt{u_i}$. Then the solution of task CO at the planning stage has the form

$$u_i = \frac{R}{\sum_{j=1}^{n} S_j^2} S_i^2 \tag{22}$$

and the value λ (in the case of coordinate planning) is obtained from the formula

$$\lambda = \frac{1}{2} \sqrt{\frac{\sum_{j=1}^{n} S_j^2}{R}} \tag{23}$$

1.3 Problems of synthesis of multilevel active systems

Let us assume a group consisting of n people be transmitted a plan $V^0 = (V_1,..., V_m)$ including m items. The plan of each individual work is described by two vectors:

I $\quad S^j = (S_1^j,..., S_m^j)$

the vector of individual goals or interests referring to each item of the plan. The individual goals are directly associated with his needs which can vary in time.

II $\quad W^j = (W_1^j,..., W_m^j)$

the vector of the individual potentials of a man to perform each item of the plan (the number W_i^j denotes the level of potentials of the jth realizing the ith work). The work realized by a given individual may be described by means of the following vector:

$$Q^j = (q_1^j, \ldots, q_m^j) \tag{24}$$

where: $q_i^j = f_i(W_i^j, S_i^j)$; $i=1,2,\ldots,m$; f_i is a given function.

A subordinate takes into account the goals of his superior. We may say that the goals of the superior deform those of the subordinate. For the jth individual who is subordinate to l superior this deformation will be generally written in the following way:

$$S'^j_i = F(S_i^j, S_i^l) \tag{25}$$

The F function may be given, e.g., in the following way:

$$S_i^{j'} = \mu_{jli} S_i^j + \gamma_{jli} S_i^l, \quad \mu_{jli} + \gamma_{jli} = 1 \tag{26}$$

where $\mu_{jli} \leqslant 1$ is the coefficient of the loss of the ith goal by the jth goal individual due to influence of the lth superior and γ_{jli} is the coefficient of importance of the jth individual, ith goal from the direct superior l. The coefficients may be defined, e.g., in the following way:

$$\gamma_{jl} = \frac{t_{jl}}{T} \tag{27}$$

where t_{jl} is the time given to the jth individual by l superior, T the total work time.

If p people are subordinate to one superior then some natural restrictions appear.

$$\sum_{j=1}^{p} t_{jl} \leqslant T \tag{28}$$

We shall assume that the system should have hierarchical levels with CO at the top, different levels being responsible for the realization of different items of the plan. We assume that vectors of goals are introduced for each level.

$$U^k = (U_1^k, \ldots, U_m^k); \quad k=1,2,\ldots,r \tag{29}$$

The coordinates of the vectors of goals range from 0 to 1.

$$\sum_{k=1}^{r} U_i^k = 1; \quad i=1,2,\ldots,m \tag{30}$$

We shall assume that we know how to transform the vector of plan V^0 into the vector of goals of the superior of a higher level.

$$S^0 = q(V^0), \quad \text{e.g.} \quad S^0 = kV^0 \tag{31}$$

where k is a certain coefficient of proportionality. Then, in accordance with (25) or (26), we can define the individual deformation of goal for each man, the superior of the system included. Now the work realized by one man is on a k level of the system, according to the ith item of the plan.

$$q_i^j = S_i^{j'} W_i^j U_i^k \tag{32}$$

Summing up the work realized by all people, according to each item of the plan, the following vector of the realization of the task by a given system is obtained

$$Q = (q_1, \ldots, q_m) \tag{33}$$

The task consists in finding such a structure of the system (the number of levels being responsible) given that the following condition be fulfilled.

$$q_i \geqslant V_i; \quad i=1,2,\ldots,m \tag{34}$$

It can be seen that for several cases it is relatively difficult to reduce a multi-criterion task to a singe-criterion one. We can do it by introducing the criterion of the assessment of effectiveness of a structure, i.e.:

$$E = \sum_{i=1}^{m} \alpha_i |q_i - V_i| \rightarrow \min \tag{35}$$

where

$$\alpha_i = \begin{cases} 1, & |q_i - V_i| < 0 \\ 0, & |q_i - V_i| \geqslant 0 \end{cases}$$

It should be noticed that the minimization of (35) is with respect to two extra parameters of weights and to the vector of goals of level.

2. Conclusion

The idea of 'active system' enables us to describe essential features of systems including groups of people, e.g. the existence of their 'own' functions of goals, transmission of information concerning their potentials with regard to these functions, usage of information concerning controlling systems, etc.

Further investigation should be concerned with:

- coalitions against the central organ, constituted by joined active subsystems,
- system control by means of resource and punishment function,
- definition of the optimum for active systems.

References

1. MESAROVIC, M.D., MACKO, D., TAKAHARA, Y., *Theory of Hierarchical, Multilevel Systems,* Academic Press (1970).
2. GEIMIER, Ju.B., 'Igrowyj koncepcji w issledowanii sistiem', *Tiechnicze skaja kibiernietika* No.2 (1970).
3. BURKOV, V.N. and LERNER, A.Ya., 'Fair Play' in Control of Active Systems, *Avtomatika i telemekhanika,* No.8 (in Russian) (1970).
4. BURKOV, V.N., MAKAROV, I.M., and SOKOLOV, V.B., Centralized Control of Active Systems, *Avtomatika i telemekhanika,* No.8 (in Russian) (1973).
5. WAŚNIOWSKI, R., 'On some concepts of active systems', *Prace Naukowe Ośrodka Badań Prognostycznych Politechniki Wrocławskiej* No.4, Wrocław (1975).

A systems approach to adaptive, multilevel, multigoal control

CHRISTER CARLSSON
Åbo Swedish University School of Economics, Finland

1. Introduction

The scope of this chapter is to give an outline and present a few steps toward a solution of the problem of implementing an adaptive, multigoal control in a hierarchical, multilevel system environment.

The first position we will take in this chapter is excellently formulated by R.L. Ackoff [1]:

> "Eulogies are delivered in which accounts are given about how 'messes' (~ systems of problems) were murdered by reducing them to problems, how problems were murdered by reducing them to models, and models were murdered by exposing them excessively to mathematics."

Ackoff's concept 'messes' is motivated by and based upon the following discussion, (cf. R.L. Ackoff [1]):

> "...problems exist only as *abstract subjective constructs,* not as concrete objective states. Furthermore, I will argue that, even if they were objective states, they *would not have solutions* if by 'solutions' we mean actions that extinguish a problem or put it to rest... Problems have traditionally been assumed to be *given* or *presented* to an actor, much as they are to students at the end of chapters in text books... problems are *taken up by*, not *given* to decision makers. William James argued that problems are extracted from unstructured states of confusion. John Dewery referred to such states as *indeterminate* or *problematic.* I prefer to call them 'messes'."

Our *first* point is thus that an excessive reduction in complexity of a studied context will rather efficiently distort that context. Then, consequently, as we choose a 'mess' as context for the present discussion we will be able to keep it rather close to reality and to outline our results under not too rigid assumptions about the structure of the relevant context of analysis.

Secondly, we find general systems theory intuitively well suited to serve as a conceptual framework for the study of 'messes', because

- states can be reproduced by elements, or sets of elements, forming a system,
- an ill-structured set of states can be given a well-defined structure in terms of a system's structure,
- interactive or interdependent states correspond to interactive or interdependent system's elements (or sets of system's elements),
- concepts of different dimension—or even different levels of abstraction—can be given varying degrees of interdependence,

- new, well-structured concepts can be formed from existing, but ill-structured concepts in terms of well-defined sets of system's elements.

We will then formulate our *second* point as follows: in an ill-structured and complex context the system's concept can form a conceptual framework for at least a preliminary structuring of the context. Furthermore, as a complex context is most efficiently sorted out by aggregation of some concepts and disaggregation of others, and as this gives the studied context a hierarchical structure, a system's concept forming a hierarchical, multilevel system seems to be most appropriate for the present purpose.

The problem *taken up* by the present author can be formulated as follows: to find and formulate a few operational principles for a program implementing and making some set of goals operational in a context that can be characterized as a 'mess'—which, due to our second point, can be specified as: to find and formulate an analogous program in a hierarchical, multilevel system.

The problem of implementing and coordinating a set of goals in some chosen environment has been tackled with a variety of methods: goal programming, problem-solving techniques, game theory, etc. (cf., for example, J.L. Cochrane—M. Zeleny [2]). Here we will try out an approach based on an algorithm coordinating a set of adaptive control functions, which are implemented in a hierarchical, multilevel system, toward a state where all the goals implemented in that multilevel system are simultaneously attained.

The relevance of this problem is based on three facts:

- the synergistic effects of an interaction of two or more adaptive control functions in a multigoal, multilevel system environment are not too well known, and
- we do not have too many operational guides for problem-solving and decision-making in environments of the complexity of a 'mess', and
- the existing techniques for making a set of goals operational are somewhat unsatisfactory as an operational tool for modelling, problem-solving and decision-making.

As the approach chosen in this chapter is developed in a rather general environment, i.e. that of a hierarchical, multilevel system (resembling a 'mess'), and as it is tested in a numerical, simulation model of the same structure as that environment, it is believed that the approach may offer a few hints toward a solution of the stated problem.

The problem thus stated we will give our discussion the following structure: in Section 2 an approach to a formulation of a hierarchical, multilevel system and a multigoal concept is presented and discussed; Section 3 describes a principle for implementing an adaptive, multigoal control in a hierarchical, multilevel system environment; in Section 4 results from a test of heuristics implementing the principle are presented, and Section 5 finally, holds a summary of the main points of our discussion.

2. An approach to a hierarchical, multilevel system

> "Classical science in its diverse disciplines, ..., tried to isolate the elements of the observed universe—chemical compounds and enzymes, cells, elementary sensations, freely computing individuals, as the case may be—expecting that by putting them together again conceptually or experimentally, the whole or system—cell, mind, society—would result and be intelligible. Now we have learned that for an understanding not only the elements but their interrelations as well are required: say, the interplay of enzymes in a cell or of many mental processes conscious and unconscious, the structure and dynamics of social systems, and so forth" (Ludwig von Bertalanffy [3])

The *system's* concept conveys essentially the idea that a set of two or more elements forms an *entity* as a result of interactions or interdependence between the elements. This statement is an analogy to Ackoff's [4] formulation, that

> "a system viewed structurally is a divisible whole, but viewed functionally it is an indivisible whole in the sense that some of its essential properties are lost in taking it apart."

The same emphasis on *wholeness* can be found in Ervin Laszlo's [5] introduction to system's philosophy:

> "An *ordered whole* is a non-summative system in which a number of constant constraints are imposed by fixed forces, yielding a structure with mathematically calculable parameters.
>
> The concept wholeness defines the character of the system as such, in contrast to the character of its parts in isolation. A whole possesses characteristics which are not possessed by its parts singly. Insofar as this is the case, therefore, the whole is other than the simple sum of its parts."

For the present purpose we will, however, concentrate upon an analysis, carried out in rather operational terms, of the elements that form such a wholeness—rather than taking the philosophical discussion any further.

According to M.D. Mesarovic [6] a *system* can be defined as a relation in the set-theoretic sense; if it is assumed that a family of sets is given,

$$\bar{A} = \{A^{(h)} : h \in H\} \tag{2.1}$$

where H is the index set, a system defined on $\bar{A}$ is a proper subset of $x\bar{A}$, i.e.

$$A \subset x\{A^{(h)} : h \in H\} \tag{2.2}$$

"A system is defined as a set (of a particular kind, i.e. a relation). It stands for the collection of all appearances of the object of study rather than for the object of study itself. This is necessitated by the use of mathematics as the language for the theory in which a 'mechanism' (a function or a relation) is defined as a set, i.e. as a collection of all proper combinations of components. Such a characterization of a system ought not to create any difficulty since the set relation, with additional specifications, contains all the information about the actual 'mechanism' we can legitimately use in the development of formal theory." (M.D. Mesarovic [6])

If a system thus is assumed to represent all the appearances of an object of study, OS, then a system's element, defined by

$$a_j \in A^{(i)}, \quad i \in H \text{ and } j \in N(i) \tag{2.3}$$

corresponds to some subset of relevant appearances. In order to state this point more precisely, it should be observed, that

- a subset of appearances can be classified as relevant only in relation to some active subject,
- the assumption of finite sets, cf. (2.3), is also based on the fact that a subject usually masters only a finite set of elements, and that
- a system's element is mostly a set of observations carried out in one or more dimensions and related to a point or a period of time.

This is our *first* step toward an operationalization of Mesarovic's general system's concept.

On the basis of (2.2-2.3) we have,

$$A \subseteq x\{A_k^{(h)} : h \in H \text{ and } k \in K(h)\} \tag{2.4}$$

i.e. a system may be defined as a subset of a relation on finite sets. Furthermore, we have,

$$A_k^{(h)} \subseteq \{A_k^{(h)} \times B_k^{(h)} \times A_l^{(h)} : k, l \in K(h); k \neq l\} \tag{2.5}$$

i.e. subset k, which is one appearance of the studied object OS (which may be is described in several dimensions), is defined as a subset of a relation on three sets which are characterized as follows:

- $A_k^{(h)}$ represents a finite set of appearances of one part of OS at a point or over some interval of time,
- $A_k^{(h)} \times B_k^{(h)} \times A_l^{(h)}$ represents a *coordination* or a *simultaneous appearance* of any two subsets $A_k^{(h)}$ such that any pair of elements for which there exists an element of $B_k^{(h)} \neq 0$ is an element of $A_k^{(h)}$.†

This sort of classification is of course based on the subject's knowledge of OS and its environment at some chosen point or over some interval of time.

Along the same lines we also have,

$$A \subseteq \{A_k^{(h)} \times C \times A_l^{(h')} : h, h' \in H_j\ h \neq h' \text{ and } k \in K(h),\ l \in K(h')\} \tag{2.6}$$

i.e. the system A is defined as a subset of a relation on three subsets which may be characterized as follows:

- $A_k^{(h)}$ represents a finite set of relevant appearances of one part of OS, as described above,
- $A_k^{(h)} \times C \times A_l^{(h')}$ is an *aggregation or a disaggregation* so that any subset $A_k^{(h)}$ for which there exists an element of $C \neq 0$ either contains or is a subset of another subset $A_l^{(h')}$.

The set B will be referred to as the set of *intrarelations* (or *horizontal* relations) and C as the set of *interrelations* (or *vertical* relations). If we then specify a_j as (cf. (2.3)),

$$a_{kj}^{(h)} \subseteq A_k^{(h)}, \quad k \in K(h) \text{ and } j \in N_k(h) \tag{2.7}$$

it can be characterized as a *basic system element;* this element will be called an *activity unit*. As cor-

† $B \supset B^{(h)} : h \in H$ and $B^{(h)} \supset B_k^{(h)} : k \in K(h)$

responding elements can be singled out in the sets B and C they should also be referred to as basic system elements. Consequently a system is built of three types of basic elements: *activity units, intrarelations* and *interrelations.*

When used in an operational systems formulation it would be necessary for the operational definitions of these elements to satisfy at least the claims George J. Klir [7] formulated for system's definitions:

- they should be based only on *constant* traits,
- they should be based on *characteristic* traits that are supposed to be completely known,
- the definitions should not be underdetermining, i.e. it should be possible to uniquely determine their consistency with the traits on which they are based,
- the definitions should not be *overdetermining,* i.e. they should not contain redundant traits.

These claims should be regarded as minimum requirements which are to be supplemented with situation-specific claims in order to give the elements the wanted operational formulation.

On the basis of our discussion, and the information in (2.1-2.7), the following formulation of the system's concept seems appropriate for the present purpose:

> A *system* is an entity formed by a relation on sets of *activity units, intra-* and *inter-*relations. The entity is given a *hierarchical structure* by a hierarchical ordering rule H, so that at least two levels, with at least two activity units on the lower level and at least one activity unit on the higher level can be identified. The activity units are joined horizontally by *intrarelations* and vertically by *interrelations* so that each activity unit is joined to at least one other activity unit, both horizontally and vertically. (2.8)

The reasons for giving a system a hierarchical structure are mainly two:

- from experience hierarchical classification is outstanding as a method for organizing information, observations and human knowledge,
- a hierarchical structure is necessary if concepts of different levels of aggregation should be applied.

As these observations most probably are valid for most cases when the system's concept is applied for structuring a 'mess', we regard the choice of system's structure motivated.

We propose that the hierarchical ordering rule is implemented in a system by giving the index set H a hierarchical structure. This can be carried out on the basis of, e.g., the following principles:

- the elements are organized according to some asymmetrical relation–mostly a dominance-relation (cf., for example, M. Bunge [8]),
- the organization is based on a relation of causality (cf., for example, A. Ando [9])
- the elements, or subsets of elements, are organized according to some rule of aggregation/disaggregation (cf., for example, R. Rosen [10]).

Here the third of these principles, *a rule of aggregation,* will be applied, because aggregation/disaggregation was the function given the interrelations. Then the set H is organized into a series of levels of aggregation and the system is correspondingly given a hierarchical structure, which differs, e.g., from Bunge's definition in that the hierarchy will not necessarily have a single beginner. This is explained by the fact that a rule of aggregation/disaggregation does not a priori represent any relation of power or dominance.

As these principles are applied a system will get the structure shown in Fig.2.1 (a four-level hierarchical system; interrelations are represented as ⟶, intrarelations as ——).

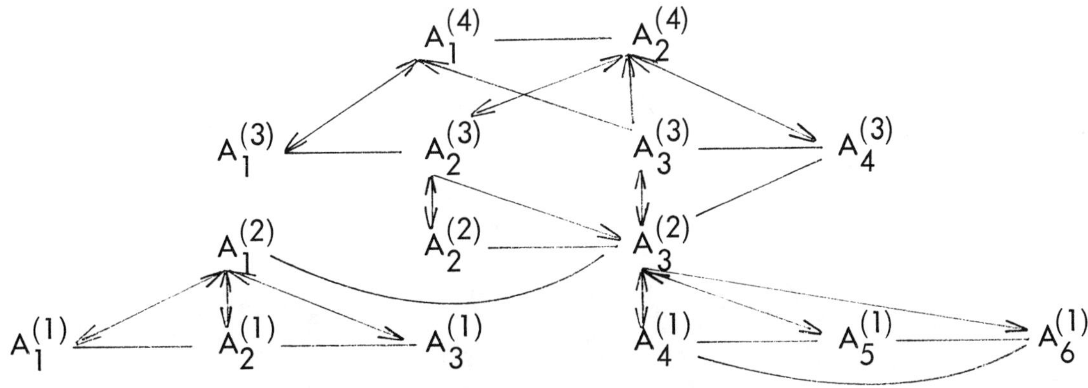

Fig.2.1 The structure of a hierarchical system

For the present purpose we will need a classification of the subsystems of the hierarchical system:

> A system of at least two activity units joined only by intrarelations forms an *elementary system* (which is not exactly a system, cf. 2.8). (2.9)

> A system of at least three activity units, joined by at least one intra- and one interrelation forms a *subsystem*. If the most extended interrelation covers two levels in a hierarchical system the subsystem is said to be of the 1st degree, if three levels are covered the subsystem is of the 2nd degree, etc. If all of the activity units of a subsystem belong to the highest level of hierarchy it is called a *suprasystem*. (2.10)

> Every subsystem in A (cf. 2.4-2.6) in which at least one subsystem is of the 2nd degree is a *multilevel system*. (2.11)

As the system's concept was given a hierarchical structure the goal concept should be given an analogous form in order to facilitate an implementation of a set of goals for interacting activity units of different levels of aggregation. Here we will apply the following formulation:

> A goal $G^{(h)}_{o}$ ($\subseteq G$, a given set of goals; $o \in S(h)$) is defined as a subset of one or more activity units, i.e.: (2.12)
>
> $$G^{(h)}_{ko} \subseteq a^{(h)}_{ki}, \text{ where } o \in (S_k(h) \cap N_k(h));$$
>
> $$S_k(h) \subseteq S(h) \text{ and } k \in K(h),\ i \in N_k(h)$$

A straightforward interpretation of (2.12) is thus that a goal represents a set of wanted appearances of the studied object. From (2.12) it is found, that

- activity units and goals are expressed in compatible units,
- due to the hierarchical structure goal elements could be aggregated and/or disaggregated,
- the elements of a set of goals—even of different levels of aggregation—can be made simultaneously operational if it is possible to implement the corresponding, necessary combination of activity units.

The aggregation/disaggregation of goal elements can be carried out with interrelations—goals were defined as subsets of the corresponding $a^{(h)}_{kj}$-elements. Here we will however utilize a special class of relations, *goal relations*, in order to be able to carry out independent operations of aggregation/disaggregation in G. Functionally these relations will be a combination of intra- and interrelations; for *coordination* of the elements of a (sub-) set of goals without aggregation the goalrelations will be used as intrarelations, otherwise as interrelations.

> A system is said to be a *hierarchical, multilevel, multigoal system* if it satisfies (2.8) and (2.11) and there are minimum two elements in an implemented goal set. (2.13)

As a conclusion to this Section the resulting structure of a hierarchical, multilevel and multigoal system will be shown in Fig.2.2 (a hierarchical, three-level system; – – – –→ represents aggregating goal-relations, – – – – represents coordinating goal relations and $\underline{a}^{(h)}_{kj}$ activity units for which goals have been implemented).

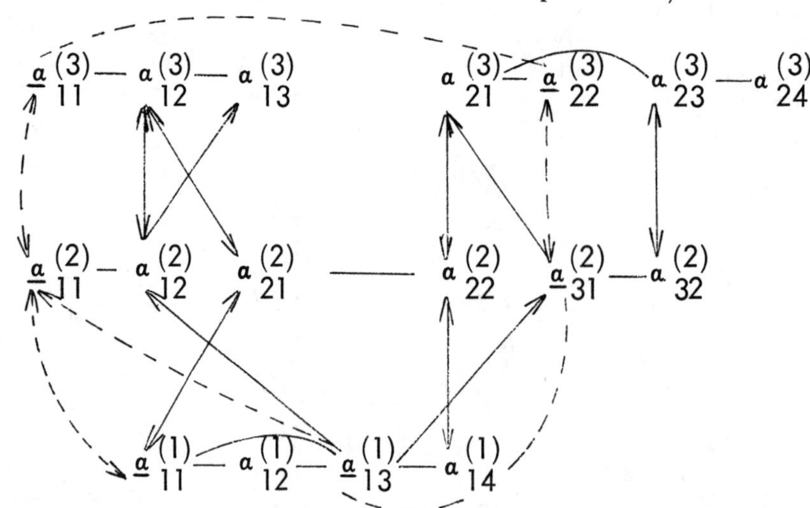

Fig.2.2 The structure of a hierarchical, multilevel and multigoal system

We suggest that the system's concept developed in this Section is well suited for a fairly close reproduction of the essential features of a 'mess':

- the states of a studied object (and its environment) can be reproduced in terms of activity units, intra- and interrelations,
- an ill-structured whole can fairly efficiently be turned into a well-structured entity when modelled as a hierarchical system,
- interactions are reproduced with intra- and interrelations,
- concepts of different levels of abstraction are reproduced as subsystems of different levels of hierarchy, as an aggregation of activity units can be made consistent with a transformation of levels of abstraction,
- a set of goals can be both implemented and transformed through aggregation/disaggregation and coordination—in a hierarchical, multilevel system.

3. A principle for implementing adaptive, multigoal control in a hierarchical, multilevel system environment

The problem we set out to solve was formulated as this: to find and formulate a few operational principles for a program implementing and making some set of goals operational in a context that can be characterized as a 'mess'. As we found in the previous section that a hierarchical, multilevel system reasonably well reproduces the essential features of a 'mess', we suggest that principles developed in such a system, with some care, could be generalized to the wider context of a 'mess'.

The problem-solving approach chosen in this Section is rather tangible: to answer the following question:

- how should a program be constructed that implements and makes a set of goals operational in a hierarchical, multilevel system environment?

and in the process of searching for an answer to formulate the principles applied in the various phases of the process.

By now we know something about human problem-solving behavior; we have, e.g. the following characteristic by A.J. Newell and H.A. Simon [11]:

> "Human problem solving is to be understood by describing the task environment in which it takes place; the space the problem solver uses to represent the environment, the task and the knowledge about it that he gradually accumulates, and the program the problem solver assembles for approaching the task."

The problem-solving process thus results from an interaction of a set of concepts—representing the problem-solver's knowledge of the essentials of the problem—and a task environment.

The overall organization of a problem-solving process could, e.g. have the following structure (cf. Newell-Simon [11]):

> "1. An initial process ... *input translation* ... produces inside the problem solver an internal *representation* of the external environment, at the same time selecting a problem space.
> 2. Once a problem is represented internally, the (problem solving-) system responds by selecting a particular problem solving *method.*
> 3. The selected method is applied: which is to say, it comes to control the behavior, both internal and external, of the problem solver.
> 4. When a method is terminated: ... (*a*) another method may be attempted, (*b*) a different internal representation may be selected and the problem reformulated, or (*c*) the attempt to solve the problem may be abandoned.
> 5. During its operation, a method may produce new problems ..."

This structure suggests that one of the basic principles for the searched program should be a form for *functional isomorphism* with the problem or (task) environment. Such an isomorphism can be found in applications of the *control* concept:

The problem of *controllability* is essentially a problem of how a certain performance should be achieved by use of a finite set of inputs for some defined system. Almost the same meaning is outlined in Ackoff's [12] definition of the concept, which however is slightly more operational:

> "An element or a system controls another element or system (or itself) if its behavior is either necessary or sufficient for the subsequent behavior of the other element or system (or itself), and the subsequent behavior is necessary or sufficient for the attainment of one or more of its goals." (3.1)

This definition seems both sufficiently general and explicit enough to serve as a basis for a further development of the suggestion on a functional isomorphism. A key principle for this isomorphism seems to be to make the control *adaptive.*

Adaptivity implies essentially an ability to react efficiently on changing conditions of both a functional and a structural character; here we will apply a formulation given by R.L. Ackoff—F.E. Emery [12]:

> "...an individual or a system is adaptive if, when there is a change in its environmental and/or internal state that has reduced its efficiency in performing its functions, it reacts or responds by changing its own state and/or that of its environment so as to increase its efficiency with respect to its functions. (3.2)

On the basis of (3.1-3.2) it should be possible to define the meaning of an adaptive multigoal control: the definitions suggest that the concept could be built around the following elements:

- an *incentive* that is either necessary or sufficient to start the control process; such an incentive could for instance be the event that one or more elements of a given set of goals are not attained at some moment of time,
- a *purpose* for the control process, in which an aspiration to react efficiently on the incentive is a necessary element,
- a *multigoal* characteristic, which implies that a set of goals should be satisfied.

Thus the following definition of the concept is suggested:

> An element, or a system, exerts an *adaptive, multigoal control* over another element or system (or itself) if its behavior is an efficient reaction to a defined incentive, and is either necessary or sufficient for the establishment of a control that satisfies a set of objectives. (3.3)

Now this concept conveys the idea of a functional isomorphism: a program implementing or making a set of goals operational should be adapted (or adapt itself) to the problem (or task) environment, and should furthermore provide either necessary or sufficient conditions for an attainment of a set of goals—in the next phase, or in some finite series of phases of the program.

An implementation of adaptive, multigoal control in a hierarchical, multilevel system could then be carried out along the following lines:

In order to exert an adaptive, multigoal control in an elementary system $A_i^{(1)}$ by t_k, a control function should have the following elements of information at t_{k-1}:††

$$a_{kj_1}^{(1)}(t_{k-1}) = b_{kl}^{(1)}(a_{kj_2}^{(1)}(t_{k-1}), t) \quad \text{where } j_1 \neq j_2, \text{ but } j_1, j_2 \in N_k(1) \text{ and } b_{kl} \in B_k^{(1)},\ l = f(j_1, j_2) \tag{3.4}$$

$$a_{kj_1}^{(1)}(t_k) = b_{kl}^{(1)}(\Delta_{kj_2}^{(1)}, t) \tag{3.5}$$

where $\Delta_{kj_2}^{(1)}$ is a change in the state of activity of $a_{kj_2}^{(1)}$ so that,

$$a_{kj_1}^{(1)}(t_k) \subseteq G_{ko}^{(1)}(t_k) \quad \text{for } o \in (S_k(1) \cap N_k) \tag{3.6}$$

and $\forall j_1, \forall j_2$ in (3.4-3.6)

If the control function determines $\Delta_{kj_2}^{(1)}$ so that (3.6) is fulfilled it is an adaptive control function, which is multigoal if there are at least two elements in $G_k^{(1)}$ (as $G_{ko}^{(1)} \subseteq G_k^{(1)}$). Necessary and sufficient elements for an implementation of the control are $G_{ko}^{(1)}(t_k)$, $\Delta_{kj_2}^{(1)}$ and (3.5).

If, correspondingly, a control function should exert an adaptive, multigoal control in a *subsystem* $A_k^{(h)}$ ($\overset{*}{h}$ for the suprasystem level) by some moment of time t_k, the control function should have the following elements of information at t_{k-1}:†

$$a_{kj_1}^{(h)}(t_{k-1}) = b_{kl}^{(h)}(a_{kj_2}^{(h)}(t_{k-1}), t) + C(A_{k'}^{(\bar{h})}, t), \tag{3.7}$$

where $\bar{h} = 1,...,h-1$; $j_1 \neq j_2$; but $h, \bar{h} \in H$ and $j_1, j_2 \in N_k(h)$; furthermore $k' \in K(h)$ and the addition is symbolical if the elements are defined in incommensurable concepts (l is defined in 3.4).

$$a_{kj_2}^{(h)}(t_k) = C(A_{k'}^{(\bar{h})}(t_{k-1}), t_k) \tag{3.8}$$

$$\Delta_{kj_2}^{(h)}(t_k) = \Delta_{kj_2}^{(h)}(a_{kj_2}^{(h)}, b_{kl}^{(h)}, t_k) \tag{3.9}$$

a change in a $a_{kj_2}^{(h)}$ so that (3.10) and (3.11) hold;

†The functions are rather symbolical; operational formulations should be based on the task environment.

††The functions are again rather symbolical.

$$a_{kj_1}^{(h)}(t_k) = b_{kl}^{(h)}(\Delta_{kj_2}^{(h)}(t_k)) \quad (3.10)$$

$$a_{kj_1}^{(h)}(t_k) \subseteq G_{ko}^{(h)}(t_k) \text{ for } o \in (S_k(h) \cap N_k(h)), \quad (3.11)$$

so that (3.7-3.11) will hold for all $h \in H$ and all $j_1, j_2 \in N_k(h)$.

If the control function determines $\Delta_{kj_2}^{(h)}(t_k)$, so that (3.11) is fulfilled, it represents an adaptive, on line control which is multigoal if $G_k^{(h)}$ has at least two elements (as $G_{ko}^{(h)} \subseteq G_k^{(h)}$). But if (3.9) is rewritten as,

$$\Delta_{kj_2}^{(h)}(t_k) = \Delta_{kj_2}^{(h)}(a_{kj_2}^{(h)}, b_{kl}^{(h)}, C, A_k^{(h)}, t_k) \quad (3.12)$$

$$\Delta_k^{(\bar{h})}(t_{k-\epsilon}) = \Delta_k^{(\bar{h})}(A_k^{(h)}(t_k), C, A_k^{(\bar{h})}(t_{k-1}), t_{k-\epsilon}): \forall h \quad (3.13)$$

where ϵ is assumed to be the reaction time,

the control function will simultaneously be an off line control (as it is now operating on the whole of the system, not only on a subsystem or an elementary system) for the h-levels of the hierarchy. As is indicated by (3.12-3.13) the control function will preadjust the activity units of lower level subsystems, or make the lower level control functions adapt, by t_k, to the wanted states on level h, if (3.11) is found not to hold as a function of $A_k^{(\bar{h})}(t_k)$. It should be observed that $b_{kl}^{(h)}$ and C are assumed to be independent of time—giving the multilevel system a *static* structure; if a more complex system structure is needed they can of course be made time functions.

Necessary elements for an implementation of the control are $G_{ko}^{(h)}(t_k)$, $\Delta_{kj_2}^{(h)}(t_k)$ and (3.10); these elements will also be sufficient elements for an implementation if they are supplemented with C and $A_k^{(\bar{h})}(t_k), \forall h$.

Now then, if we compare adaptive, multigoal control with the overall organization of problem-solving behavior, we have:

1. The representation of the environment is given by $A_k^{(h)} \subseteq A^{(h)}$, where $h \in H$ and $k \in K(h)$: K is delimited by the relevant environment, $B \subseteq A^{(h)}$, $C \subseteq A^{(h)}$ $(A^{(h)} \subseteq A)$ and G, the relevant set of goals.
2. An algorithm, or some set of heuristic rules, that makes the relations (3.6) and (3.11) hold by some chosen point, or over some interval, of time is an equivalent to a problem-solving method.

3-4. An application of the algorithm is initiated as soon as (3.6) and (3.11) do not hold, and is terminated as soon as the relations are established.

5. A central feature of an adaptive, multigoal control is that it detects inconsistency in the goal set, intra- and interrelations that are contradictory, activity units which are incompatible, etc.

Consequently, implementing an adaptive multigoal control introduces a process that at least in some but essential respects resembles the behavior of a problem solver.

The problem of overall control of a hierarchical, multilevel, multigoal system is complex in the sense that the elements of the system and the goal set are inter- and intrarelated. A consequence of this will be that an adaptive action in an h-level subsystem will have effects in all lower-level subsystems interrelated with that system; sometimes it will have effects also in higher-level subsystems if a lower-level adaption process is allowed to influence higher levels. As a consequence of these interinfluences a rather complex set of inter- and intrarelations should be taken into consideration when the adaptive multigoal control is implemented.

The problem of coordination is essentially that the goal attainment should be made simultaneous for all elements of G (or made to follow some predefined pattern). This could be achieved by coordinating the (local) adaptive control functions of the subsystems so that they form a global adaption process. This process can, e.g., be implemented as a series of successive choices of relevant intra- and interrelations (cf. Section 4) if the structure of the system is at least approximately static and the elements of the goal set are consistent.

If the structure of the hierarchical, multilevel system is *dynamic* it is possible for the set of relevant intra- and interrelations to change before the adaption process has succeeded. Then some other method for coordination should be applied; here we will suggest that the system should be made *ultrastable* (cf. W.R. Ashby, C.C. Carlsson [13]):

"A multilevel, multigoal system is said to be ultrastable if an implemented control function (3.14) is able to establish necessary and/or sufficient conditions for the system to shift from any instable state (at t_1) to a stable one, by some chosen moment of time (t_2), and this process could be repeated at any $t > t_2$ (within reasonable changes of environment and system structure)."

A subsystem $A_k^{(h)}$ is said to be in a *stable state* if all the elements of an implemented set of goals are attained (cf. (3.11)).

Here it should be observed that an ultrastable system not necessarily establishes and maintains (or restores) a state of system equilibrium, but a state of *stationary* or *quasi-stationary non-equilibrium*. E. Laszlo [5] argues for this point with reference to natural systems:

> "The reasons are potent for discarding the concept of the equilibrium state in favor of that of a non-equilibrium steady state in natural systems: (i) equilibrium states do not have available, usable energy whereas natural systems of the widest variety do; (ii) equilibrium states are 'memoryless', whereas natural systems behave in large part in function of their past histories. In short, an equilibrium system is a dead system—more 'dead' even that atoms and molecules. Thus, although a machine may go to equilibrium as its preferred state, natural systems go to increasingly organized non-equilibrium states."

In the present context this means that the state where every element of an implemented set of goals is attained should not be defined as a state of equilibrium but as a stationary state, and that the methods for attaining such a state should be more oriented toward a dynamic coordination of the goals than toward efforts to find some optimal combination of goals (cf., for example, goal programming). A set of heuristic rules representing an approach to dynamic coordination of a set of local adaptive control functions are presented in Section 4.

As a conclusion we will then answer the question we used as a framework for this Section:

- a program that implements and makes a set of goals operational in a hierarchical, multilevel system environment, can be realized with a program that implements an adaptive, multigoal control in that environment.

We found the following principles central for that program:

- it should be functionally isomorphic with the problem (or task) environment,
- it should provide either necessary or sufficient conditions for a goal attainment,
- it should resemble the behavior of a human problem solver,
- it should provide at least the necessary conditions for ultrastability.

4. Heuristic rules and test results

In order to get a preliminary test of the applicability of the principles outlined in the previous sections an experimental system's model was developed, programmed in FORTRAN and made operational on an IBM 1130 computer. Structurally the model is a hierarchical, multilevel, multigoal system based on the principles developed in Section 2; although still an embryo it is functionally an implementation of the principles for adaptive multigoal control.

The structure of the model is shown in Fig.4.1; the entities (x_{11}, x_{12}, x_{13}), (x_{23}), (x_{24}), etc., represent *elementary systems* (cf. (2.9)), the complexes (x_{11}-x_{14}), (x_{21}-x_{24}), (x_{31}-x_{34}) are *subsystems* (cf. (2.10)) and the complex (Y_1-Y_9) is a *suprasystem* (cf. 2.10). Interrelations are represented as ⟶, intrarelations as ——, aggregating goalrelations as -------→ and coordinating goalrelations as -----. The constructions (~log norm, ~norm, etc.) are random generators with log normal-, normal-, Poisson- and Erlang-distributions, which are applied to generate stochastic disturbances in the system.

The elements C_1-C_7 are *goals* of the hierarchically highest level, the elements B_{11}-B_{101} correspondingly *goals* of hierarchically lower levels. A closer study reveals that the construction of both the subsystems and the set of goals is aimed at a rather varying structure.

The subsystems are programmed as system-simulating models with an implicit time axis. *Adaptive control function* (based on (3.7-3.13)) of enough capacity to make the subsystems adapt both to influences generated by the disturbance generators and to influences communicated from other subsystems were constructed for the models. The task of the control functions is, as a test run normally started from an initially stable state, to either *retain* or *regain* a state of stability (if the system has a dynamic structure, which is the case here), which of course changes numerically during the run of the model. A sample on the functional characteristics is given on pp. 225-227. In order to obtain stability in the whole of the system complex these local adaption processes should be coordinated into

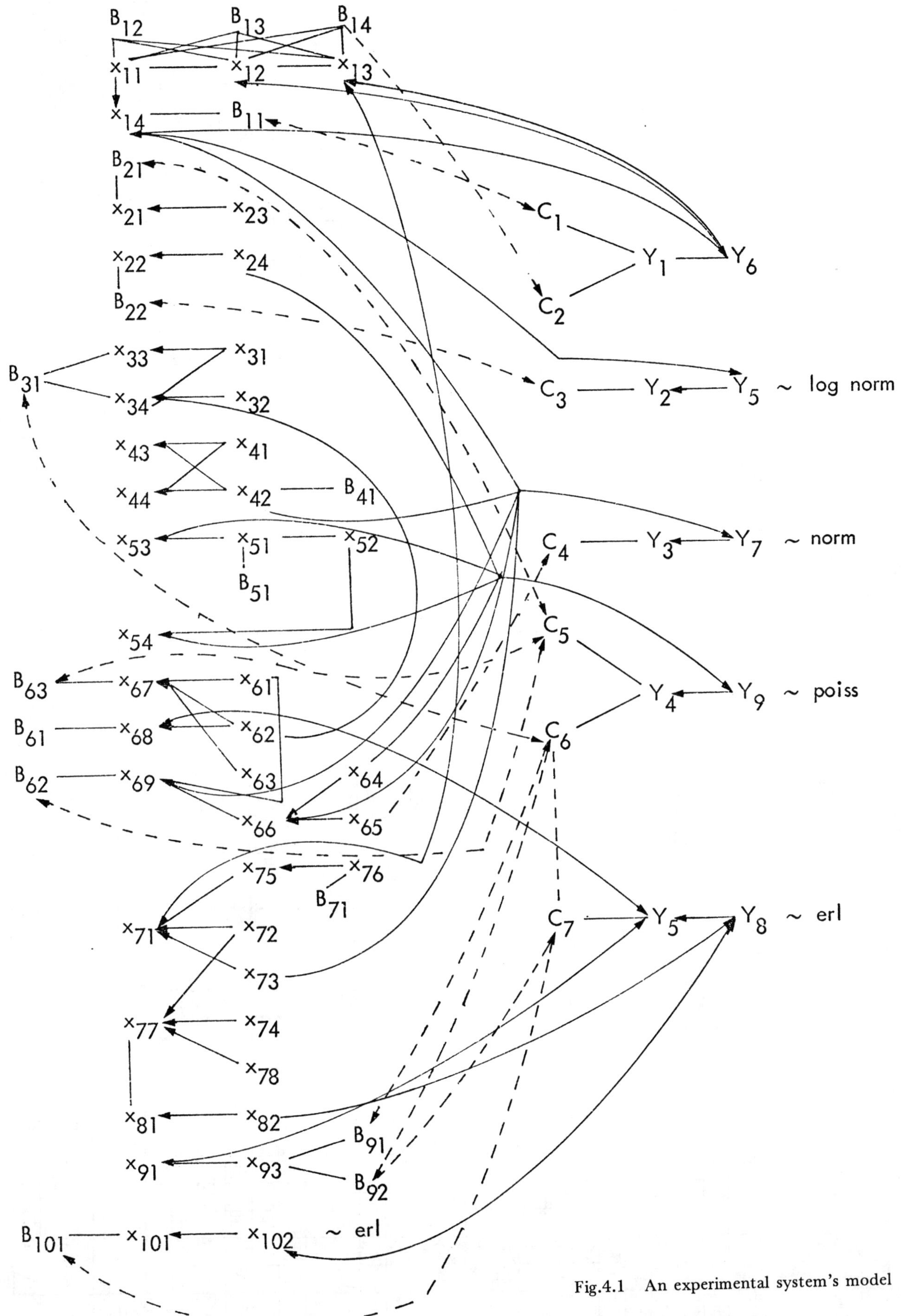

Fig.4.1 An experimental system's model

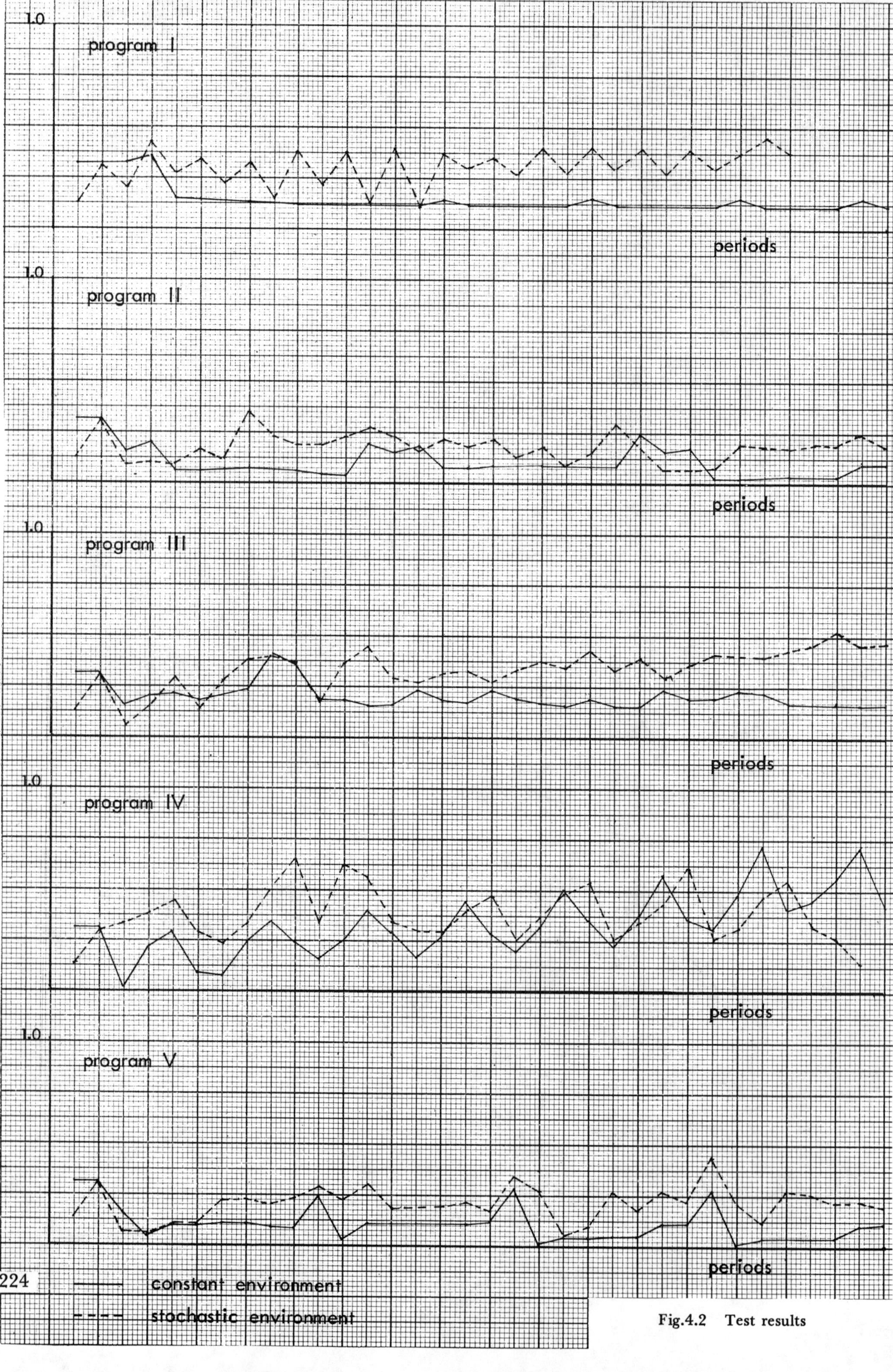

Fig.4.2 Test results

a global adaption process.

In the present study, which should be regarded as preliminary and exploratory, a heuristic method for coordination was developed and tested. Because local adaption processes in part neutralize the instability of a system, it is assumed that an appropriate communication pattern, which represents a series of instability-reducing adaption processes, will eventually take the system into a *globally stationary state*. If some communication pattern can be proved always to result in a globally stationary state the corresponding system will be an *ultra-stable* system (cf. Section 3). Here the following patterns were tested (the test variable, shown in Fig.4.2, represents the sum of the relative goal deviations of the suprasystem).

Program I

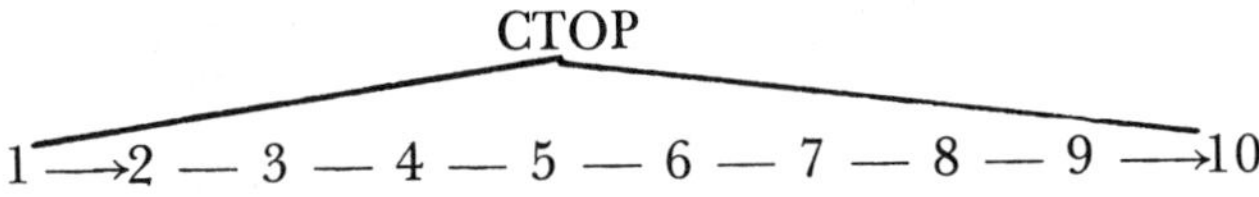

(CTOP is the suprasystem; the numbers 1-10 refer to subsystems)

RANDOM ORDER: the communication pattern is not based on any property of the system.

Comments: the fairly stable states in the constant environment have too large deviations and the structure causing these deviations could not be broken by the communication pattern; the oscillations and growing deviations in the stochastic environment show that a global adaption process was not achieved.

Program II

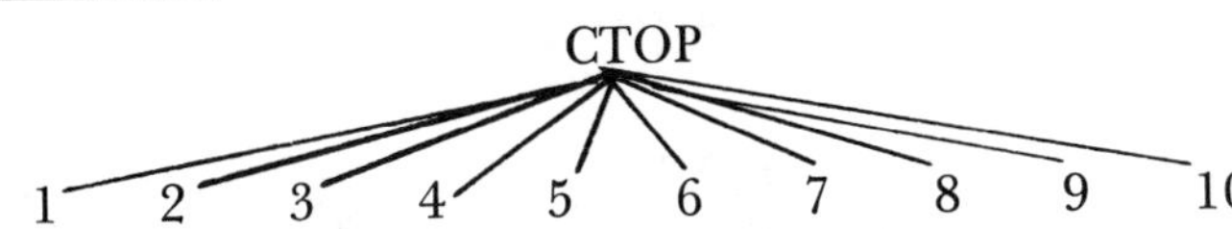

SUBOPTIMIZATION: the subsystems are handled as independent parts of the system.

Comments: the results are quite good in the constant environment, but they are nevertheless not satisfactory because a cyclic movement in the deviations could not be broken under the communication pattern; no global adaption process in the stochastic environment.

<u>CTOP</u>	Y_6	Y_1	
	7.50	18.60	C_1 10.00 (≥) C_2 19.70 (≤)
LOG NORM (0.53, 21.29)	~ Y_5	Y_2	
		15.57	C_3 10.63 ± 5.49
NORM (12.00, 1.07)	~ Y_7	Y_3	
		8.32	C_4 13.00 (≤)
POISS (1.43, 0.38)	~ Y_9	Y_4	
		4.34	C_5 2.28 (≥) C_6 14.00 (≤)
ERL (16.00, 0.96)	~ Y_8	Y_5	
		0.09	C_7 19.99 (≤)

Program III

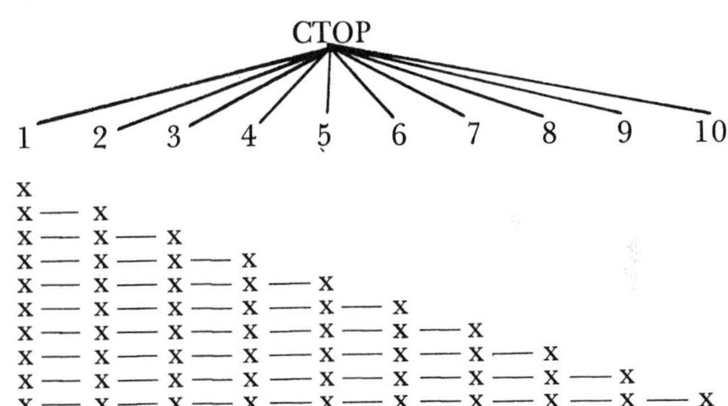

EXPANDING SUBOPTIMIZATION: the subsystems form complexes in a random order.

Comments: a rather unfortunate construct: goal deviations are not eliminated—rather then reinforced—presumably through cyclic effects generated by the inter-relations.

Program IV

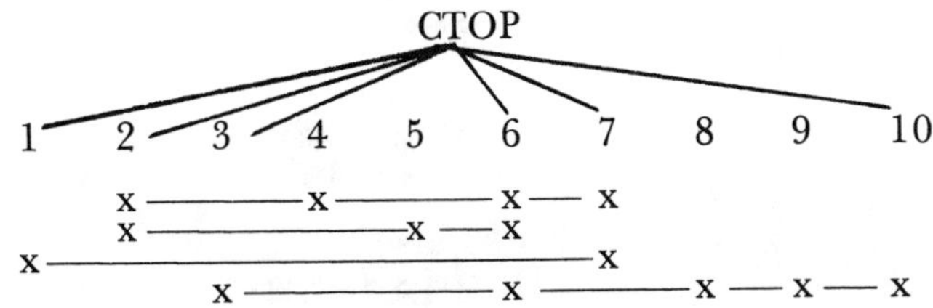

FUNCTIONAL CHARACTERISTICS OF THE SUBSYSTEMS

CDS 1

X_{11}	X_{12}	X_{13}	X_{14}
5.32	4.37	6.08	217.3
4.32	4.37	7.15	217.3
3.32	4.37	8.22	217.3
6.12	3.37	6.08	217.3
7.92	2.37	6.08	217.3
6.25	4.37	5.08	217.3
7.18	4.31	4.08	217.3

B_{11} 217.3 (≥)

B_{12} 34.0 (≤)

B_{13} 30.0 (≤)

B_{14} 78.9 (≤)

CDS 2

X_{23}	X_{21}	X_{24}	X_{22}
6.35	10.19	8.85	8.84
6.85	11.00	7.85	8.84
5.85	9.39	9.85	8.84
7.30	11.73	6.85	8.84
5.40	8.67	10.85	8.84

B_{21} 10.20 ± 1.53

B_{22} 8.85 ± 1.32

CDS 3

X_{31}	X_{33}	X_{32}	X_{34}	X_{35}
4.00	24.31	6.00	9.72	2.50
4.13	25.93	5.87	9.70	2.67
3.87	22.76	6.13	9.75	2.33
4.25	27.45	5.75	9.69	2.83
3.75	21.37	6.25	9.77	2.19

$B_{31} = 2.50 \pm 0.37$

$X_{35} = X_{33} / X_{34}$

CDS 4

X_{41}	X_{43}	X_{42}	X_{44}
3.20	43.23	4.90	0.02
3.45	49.66	5.21	0.02
2.95	37.28	4.59	0.03
3.70	56.57	5.52	0.01
2.70	31.79	4.28	0.04

B_{41} 4.89 ± 0.62

CDS 5

X_{52}	X_{51}	X_{53}	X_{54}
0.56	3.88	1.36	1.96
0.50	4.12	1.38	1.85
0.61	3.70	1.35	2.04
0.44	4.32	1.40	1.74
0.38	4.51	1.41	1.61

B_{51} 4.28 ± 0.60

CDS 6

X_{61}	X_{62}	X_{63}	X_{67}	X_{68}	X_{64}	X_{65}	X_{66}	X_{69}
0.64	0.96	5.00	25.41	4.23	4.30	8.25	14.19	2.64
0.73	1.01	5.50	26.24	4.45	4.80	8.50	15.00	2.40
0.68	0.91	5.20	24.59	4.13	3.80	8.00	13.40	2.62
0.60	1.06	6.00	27.06	4.68	5.30	8.75	15.80	2.64
0.55	0.86	5.70	22.20	4.10	3.30	7.75	12.60	3.04

B_{61} 4.55 ± 0.68 B_{62} 2.64 ± 0.39 B_{63} 23.85 ± 1.65

CDS 7

X_{76}	X_{75}	X_{72}	X_{73}	X_{71}	X_{74}	X_{78}	X_{77}
11.00	2000.00	15.00	1420.00	1359.98	0.08	0.02	842.61
12.04	2100.00	16.13	1520.00	1001.67	0.18	0.12	904.02
10.36	1900.00	13.87	1320.00	1160.98	0.09	0.06	1104.79
12.88	2200.00	17.25	1620.00	879.63	0.28	0.22	965.14
9.52	1800.00	12.75	1220.00	836.62	0.14	0.11	1018.33

B_{71} 11.20 ± 1.68

CDS 8

X_{82}	X_{81}
14.00	0.25
15.05	0.18
12.95	0.27
16.10	0.10
11.90	0.23

CDS 9

X_{93}	X_{91}
1.80	0.25
1.63	0.18
1.93	0.30
2.23	0.41
2.53	0.51

B_{91} 1.78 ± 0.15

B_{92} 2.20 ± 0.33

CDS 10

ERL ~ (20.45,614) X_{102}	X_{101}
0.95	-0.53
0.75	-0.54
1.15	-0.52
0.55	-0.55
1.35	-0.53

B_{101} 22.00 (≤)

COOPERATIVE ADAPTION: interrelated subsystems are treated as complexes within which an interadaption of the adaption processes takes place.

Comments: supports the impression from Program III that rather randomly formed complexes of subsystems easily produce cyclic effects, presumably through interactions in the interrelations, which makes goal deviations grow.

Program V

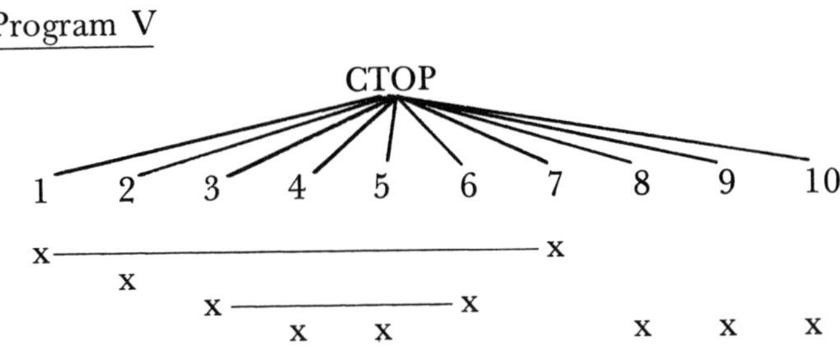

SELECTIVE ADAPTION: subsystems in interaction through one or more interrelations are treated as complexes with a corresponding interadaption of the adaption processes; the rest of the subsystems are treated in random order.

Comments: suggests that building complexes of subsystems on the basis of interacting interrelations might be a good idea; the communication process as such is however not sufficient to constitute a global adaption process.

In conclusion then, the communication patterns tested were not efficient enough for the system to reach a state of global stationarity, but the basic principles seem operational enough. More elaborate communication patterns, perhaps combined with an algorithm for dynamic global coordination of the local control functions, might result in the wanted global adaption process.

5. Conclusion

The main results that could be claimed on the basis of this chapter are the following:

1. A hierarchical multilevel, multigoal system has the true complexity to fairly closely reproduce the structure of a 'mess'.
2. A program designed to implement an adaptive, multigoal control in a hierarchical, multilevel system environment represents an approach to a solution of the multigoal problem.
3. In a static system environment a set of goals could be implemented by coordinating local adaptive control functions to a global adaption process; in a dynamic system environment the corresponding implementation could be achieved through adaptive control functions establishing necessary and/or sufficient conditions for the system to become ultrastable, i.e. restoring or maintaining states of stationary or quasi-stationary non-equilibrium.
4. A global adaption process is not easily realized in a hierarchical multilevel, multigoal system environment; a sufficiently elaborate pattern of interaction between relevant subsystems, in which local adaptive control functions are implemented, combined with an algorithm for dynamic global coordination of the local control functions could produce the wanted adaption process.

(Devil's Dictionary: *Faith:* belief without evidence in what is told by one who speaks without knowledge, of things without parallel.)

References

1. ACKOFF, R.L., 'Beyond problem solving', *General Systems,* Vol.XIX (1974).
2. COCHRANE, J.L. and ZELENY, M. (editors), *Multiple Criteria Decision Making,* University of South Carolina Press, Columbia (1973).
3. von BERTALANFFY, L., *Foreword* to *Introduction to Systems Philosophy* by E. Laszlo (1972).
4. ACKOFF, R.L., 'The systems revolution', *Long Range Planning,* Vol.17, No.6 (December 1974).
5. LASZLO, E., *Introduction to Systems Philosophy,* Harper Torchbooks, New York (1972).
6. MESAROVIC, M.D. and TAKAHARA, Y., *General Systems Theory: Mathematical Foundations,* Academic Press, New York (1975).
7. KLIR, G.J., *An Approach to General Systems Theory,* van Nostrand Reinhold Company, New York (1969).
8. BUNGE, M., 'The Metaphysics, Epistemology and Methodology of Levels', in Whyte-Wilson-Wilson (ed.), (1969).
9. ANDO, A., FISCHER, F., and SIMON, H.A., *'Essays on the Structure of Social Science Models,* MIT Press, Cambridge (1963), quoted in Whyte-Wilson-Wilson (ed.), (1969).
10. ROSEN, R., 'Comments on the Use of the Term Hierarchy', in Whyte-Wilson-Wilson (ed.), (1969).
11. NEWELL, A.J. and SIMON, H.A., *Human Problem Solving,* Prentic Hall, Englewood Cliffs (1972).

12. ACKOFF, R.L. and EMERY, F.E., *On Purposeful Systems,* Aldine-Atherton, Chicago (1972).
13. CARLSSON, C.C., 'On the Principles for Problem-solving Heuristics in a Multilevel, Multigoal Control Problem', Memo-Stencil, No.18, Åbo (1975).
14. CARLSSON, C.C., 'On Heuristics for Adaptive, Multigoal Control', NOAK-75, Åbo (1975).
15. WHYTE, L.L., WILSON, A.G., and WILSON, W., *Hierarchical Structures,* American Elsevier, New York (1969).

Conflicting goals, pareto optima in systems analysis

THOMAS GAL
Aachen

1. Introduction

Defining objectives for a system to be designed and modelled it may happen that an over-all objective cannot be found. The goal conflict may arise, e.g. from an inconsistency of goals in the system and/or its subsystems. In such cases trials to introduce utility functions are made or some ways are sought, such as optimizing with regard to preference orderings.

Another possible way of overcoming difficulties in conflicting goals is to determine all pareto-optimal solutions (also known as efficient solutions) in the appropriate model. This means that an improvement of an individual value of one of the objectives implies deterioration of an individual value of at least one of the remaining ones.

When the set E of all pareto-optimal (efficient) solutions is found, then there are again several possibilities of selecting one 'optimal' or 'compromise' solution from the set E. One of them is to optimize a utility function over the set E; another one is to apply the so-called interactive approach by using the knowledge of the set E.

Therefore a procedure for determining the set E of all efficient solutions is of interest. Procedures have been published describing how to determine efficient vertices, edges and faces of the restriction set X *[2, 3, 4, 12, 13]*. However, these methods either need separate subprograms for each of the three mentioned cases, or they can be used only under quite strict conditions on the restriction set, such as boundedness, non-degeneracy, etc.

In this chapter we shall show the principles of a method which finds the set E of all efficient solutions (i.e. efficient vertices, edges and faces) to a linear vector maximum problem (LVMP) via one subprogram and which does not depend on special properties of the restrictions set X (e.g. compactness, degeneracy, etc.). The procedure bases on two multiparametric algorithms [6,7]. The proofs of the theorems in Section 2 are found in [8]. Also, in order to give more insight into the procedure, we shall use in this chapter rather geometrical comments. The special cases are discussed only concisely.

2. Basic results

Let

$$z_k = (c^k)^T x\,, \quad k=1,\ldots,K\,, \quad c^k \in \mathbb{R}^n, \quad x \in \mathbb{R}^n, \tag{2.1}$$

be K linear objective functions. The linear vector maximum problem (LVMP) is then to

$$\text{'max'}\ Z = C^T x\,, \quad C=(c^1,\ldots,c^K)\,, \quad Z=(z_1,\ldots,z_K)^T, \quad x \in X \tag{2.2}$$

where

$$X = \{x \in \mathbb{R}^n \mid Ax = b,\ x \geqslant 0\} \tag{2.3}$$

A being an (m,n) matrix, C an (n,K) matrix, $b \in \mathbb{R}^m$. The LVMP is in fact to find the set of all efficient solutions, i.e. the set

$$E = \{\bar{x} \in X \mid \text{there is no } x \in X \text{ such that } C^T x \geqslant C^T \bar{x} \text{ and } C^T x \neq C^T \bar{x}\} \tag{2.4}$$

Consider the multi-parametric problem

$$\max_{x \in X} z = (Ct)^T x\ ,\quad t \geqslant 0\ ,\quad t \in \mathbb{R}^K. \tag{2.5}$$

This is in fact problem *(H)* in [6] enlarged by $t \geqslant 0$. As it is well known (cf. [9,10,11,12]) the following theorem is true:

Theorem 1
$x^0 \in X$ is an efficient solution to (2.2) iff there exists $t^0 > 0$ such that $x^0(t^0) \in X$ is an optimal solution to (2.5) with $t = t^0$.

Now let $x^s \in X$ be a vertex of X, i.e. a basic feasible solution to (2.3) with basis B_s and basis index ρ_s. Suppose $X \neq \emptyset$ and compact.

Denote P a supporting hyperplane to X. Let x^s for $s=1,\ldots,S$ be some vertices of X [basic feasible solutions to (2.2)], and

$$\mathop{\gamma}_{s=1}^{S}(x^s) = \gamma(x^1,\ldots,x^S) = \{x \in X \mid x = \sum_{s=1}^{S} \lambda_s x^s,\ \sum_{s=1}^{S} \lambda_s = 1,\ \lambda_s \geqslant 0\,\forall s\}$$

the convex hull of $x^1,\ldots,x^S$.

Define for our purposes a face F of X as follows.

Definition 1
Let

$$F = \mathop{\gamma}_{s=1}^{S}(x^s),\ S \text{ is the number of all vertices of } F.$$

This convex polyhedron with dimension $1 \leqslant \dim F \leqslant n-1$ is said to be a face F of X iff there exists P such that $F \subset P$.

Note: By definition an edge between two neighboring vertices of X is also a face.

Theorem 2
All $x \in F$ are efficient solutions to (2.2) iff there exists $t^0 > 0$ such that $x^s(t^0)$ for all $s=1,\ldots,S$ are optimal solutions to (2.5) with $t = t^0$.

Corollary
$x \in \operatorname{Int} F$ is an efficient solution to (2.2) iff all $x \in F$ are efficient solutions to (2.2).

Let

$$x^s = (x^s_{j_1},\ldots,x^s_{j_m})^T,\quad x_{j_i} > 0\ ,\quad i=1,\ldots,m\ ,$$

$$\rho_s = \{j_1,\ldots,j_m\}.$$

Let

$$\Delta c_j^k = (c_B^k)^T B_s^{-1} a^j - c_j^k,\quad k=1,\ldots,K\ ,\quad j \notin \rho_s$$

(cf. also [6,7]). Denote

y_{ij} the elements of matrix $B_s^{-1} A$

y_{io} the values of the basic variables x_{j_i}.

The region

$$R_s = \{t \in \mathbb{R}^K \mid -\sum_{k=1}^{K} \Delta c_j^k t_k \leqslant 0\ ,\ j \notin \rho_s\ ,\ t_s > 0\,\forall k\}$$

defines the set of $t \in \mathbb{R}^K$ such that for all $t \in R_s$ the solution x^s is optimal to (2.5) or efficient to (2.2) (cf. Theorem 2 and [6] and [7]).

Definition 2
Two basic solutions x^s, $x^{s+1} \in X$ are said to be efficient neighbors iff: 1) x^s and x^{s+1} correspond to neighboring vertices of X in the usual sense (cf. also [6,13,7]), and 2) x^s and x^{s+1} are both efficient basic solutions to (2.2).

Lemma: Let $w > 0$ be an arbitrary but fixed real number, and

$$M_s = \{t \in \mathbb{R}^K \mid \sum_{k=1}^{K} t_k = w\ ,\ t_k > 0\,\forall k\}$$

Then, for every $w > 0$,

$$R_s \neq \emptyset \leftrightarrow R_s \cap M_s \neq \emptyset$$

As in [7], assign to (2.5) a graph G such that each node $\rho_s \in G$ corresponds to an efficient basic solution to (2.2), and between two nodes ρ_s and ρ_{s+1} there exists an arc iff the corresponding basic solutions are efficient neighbors in the sense of Definition 2.

Note: Such a graph has all the properties of the graph defined in [7].

By this the principles of the required theory are given. Let us describe the procedure.

3. The method

In order to find the set E of all efficient solutions to (2.2) we proceed in two phases:

Phase 1: Determine an efficient basic solution x^s to (2.2) with basis ρ_s.

Phase 2: Part 1: Starting with ρ_s determine all nodes of graph G.

Part 2: Based on the results of Part 1 determine the efficient faces.

Let us describe some details:

Phase 1: This is carried out as described by J.G. Ecker and I.A. Kouada [2] or by H. Isermann [4].

Note: Since [2] does not necessarily provide an efficient vertex (it can be an efficient interior point of an edge or of a face) the method by Isermann [4] is rather used.

Phase 2: Part 1:

1. Suppose that x^1 is the first efficient basic solutions to (2.2) as found in Phase 1. The according simplex tableau called the main tableau for basis ρ_1, is thereby generated. Due to Theorem 1 we test the efficiency of x^1 by solving the system

$$\min \xi_j \quad \text{all} \quad j \notin \rho_1 \tag{3.1}$$

s.t.

$$-\sum_{k=1}^{K} \Delta c_j^k t_k + \xi_j = 0 \tag{3.2}$$

$$\sum_{k=1}^{K} t_k + p = 1, \tag{3.3}$$

$t_k > 0 \forall k$, $\xi_j \geqslant 0 \forall j \in \rho_1$, $p \geqslant 0$ artificial variable

The system (3.2) defines the set R_1, (3.3) defines the set M_1 with $w = 1$ from the Lemma.

Since, by assumption, $x^1 \in E$, there must exist $t^0 > 0$ which solves (3.2), (3.3).

Note: The procedure for solving (3.1)-(3.3) is found in [6,13,7].

Evidently, the min $\xi_j \geqslant 0$. If min $\xi_j > 0$ for some fixed j, there does not exist a neighbor regarding the pivot in the jth column of the according main simplex tableau, and, at the same time, the jth condition in (3.2) is redundant regarding R_s [14].

Note: The redundant constraints can be cancelled in the course of the procedure in order to save computer time (cf. [14]).

2. Denote $t_B^u \in R_1$, $u = 1,...,U$, all basic feasible 't-solutions' in (3.2), (3.3). Determine t_B^u for all u. Let $u^0 \in \{1,...,U\}$ be arbitrary but fixed and suppose that, e.g., $\xi_{j_1} = 0$, $j_1 \notin \rho_1$, is a non-basic variable in the according solution to the subprogram (3.1)-(3.3). That means, min $\xi_{j_1} = 0$ is found.

   Look at the main tableau for basis ρ_1 and determine the pivot element in column j_1, provided that $y_{ij_1} > 0$ for at least one i.

   By this a new neighboring (efficient) basis ρ_2 with the solution x^2 to (2.5), and hence to (2.2) is generated; due to Theorem 2 and the Corollary there is also found that the edge between x^1 and x^2, i.e. the convex hull $\gamma(x^1, x^2)$ is efficient.

   Find in the same manner all efficient neighbors to ρ_1. List all t_B^u and use the lists V_0, W_0 as described in [7] for listing the efficient bases.

3. Choose a recognized efficient basis, e.g. ρ_2, for which there does not yet exist the main tableau. Carry out a simplex step to get the main tableau for basis ρ_2 with the solution x^2. As follows from [6,13,7], regarding ρ_2 there exists $t_B^u \in R_2$ such that for at least one u, say u^*

$$t^{u^*} \in R_1 \quad \text{and} \quad t^{u^*} \in R_2$$

   Go back to step 2, i.e. determine the efficient neighbors to ρ_2 by solving (3.1)-(3.3) with ρ_2.

4. Assume, $h+1$ main tableaus are already generated. Using the lists V_h, W_h (cf. [6,13,7]) proceed until $W_h = \emptyset$ for some h. By this all efficient basic solutions to (2.2) are found.

Part 2:

5. Look at the subprograms (3.1)-(3.3) for all generated efficient bases. Compare $t_B^u \in R_s$ for all s and collect those which are identical. The corresponding vertices generate an efficient face, due to Theorem 2 and Corollary. The union of all efficient faces defines the set E of all efficient solutions to (2.2) (see also [3]).

Theorem 2 provides the possibility of describing the principle of the method as follows: Map X into Z by the linear relation $Z(x)$. The efficiency Theorem 1 says that in the region of dominancy regarding an efficient point $z(x)$ there exists a vector $t > 0$. This is based on the proof of the efficiency Theorem 1 [10], saying that to each efficient point there exists a supporting hyperplane, which separates the region of dominancy and the whole region $Z(x)$. Such a separating hyperplane is also that one which is parallel with a face of $Z(x)$. From the theory of multi-parametric programming, it then follows that it is possible to find normal vector t^0 to such a separating hyperplane. And this t^0 is provided by the method.

Hence, after having found t^0 regarding a vertex x^s, the same t^0 must result regarding all efficient neighbors of x^s. This implies that the according edges are efficient as well.

Having determined the efficient vertices (and edges) it suffices to compare $t^i > 0$ belonging to the vertices in order to find out:

1. Which of the vertices generates a face, and
2. Whether or not this face is efficient.

This fact provides also the theoretical background for the case, X is not compact. If, namely, there exists a $t^0 > 0$ which is normal to the supporting hyperplane P, such that the investigated face $F \subset P$, then F is efficient. This is due to the corollary.

Degeneracy has evidently no influence on the procedure.

References

1. EVANS, J.P. and STEUER, R.E., 'A revised simplex method for linear multiple objective programs', *Math. Program.* **5**, pp.54-72 (1973).
2. ECKER, J.G. and KOUADA, I.A., 'Finding efficient points for linear multiple objective programs', *Math. Program.* **8**, pp.375-377 (1975).
3. ECKER, J.G. and KOUADA, I.A., 'Generating all efficient faces for multiple objective linear programs', Working Paper: DP CORE (1975).
4. ISERMANN, H., 'Lineare Vektoroptimierung', Dissertation, Regensburg (1974).
5. ZELENY, M., 'Linear multi-objective programming', *Lecture Notes in Econ. and Math. Syst.* No.95, Springer-Verlag, Berlin/New York (1974).
6. GAL, T., 'Homogene mehrparametrische lineare Programmierung', *Z. f. OR* **16**, pp.115-136 (1972).
7. GAL, T., and NEDOMA, J., 'Multiparametric linear programming', *Management Science* **18**, pp.406-422 (1972).
8. GAL, T., 'A method for determining the set of all efficient solutions to a linear vector maximum problem', *Europ. J. of OR* **1**, pp.307-322 (1977).
9. DINKELBACH, W., *Sensitivitätsanalysen und parametrische Programmierung,* Chapter 6, Springer-Verlag Berlin/New York (1969).
10. FOCKE, J., Vektormaximumprobleme und parametrische Optimierung, *Math. Oper.-Forsch. u. Statist.* **4**, pp.365-369 (1973).
11. GEOFFRION, A.M., 'Proper efficiency and the theory of vector maximization', *J. of Math.-Anal. & Appl.* **22**, pp.618-630 (1968).
12. ISERMANN, H., 'Existence and duality in multiple objective linear programming', presented at: 'Multiple Criteria Decision-Making', Jouy-en-Josas, Paris (21-23 May 1975).
13. GAL, T., *Betriebliche Entscheidungsprobleme, Sensitivitätsanalyse und parametrische Programmierung,* De Gruyter-Verlag, Berlin/New York (1973), English: McGraw-Hill Int. (to appear 1978)
14. GAL, T., 'Zur Identifikation redundanter Nebenbedingungen in linearen Programmen', *Z. f. OR.* **19**, pp.19-28 (1975).

Game-theoretic treatment of perturbations

O. HÁJEK
Case Western Reserve University, Ohio, USA

In many real systems the situation arises that

a. There are available certain controls which are to be chosen so as to achieve a desired outcome.
b. There are present more or less easily identifiable factors (errors, disturbances) which are, in some sense, small.
c. If entirely absent, the analysis and behavior prediction would be profoundly simplified, but
d. Cannot be ignored, since they essentially or even qualitatively affect the process.

We propose to use the interplay of (*a*) and (*b*) in such a manner that the simplified model of (*c*) can still be used while the perturbing effects (*d*) are accurately accounted for. A somewhat over-simplified description of the method is as follows: one sets up an auxiliary game, in which the disturbances are interpreted as the action of a fictional second player, with unpredictable but observable behavior, and possibly antagonistic intent. It may be noted that all of (*a*) - (*d*) are essential here; in contrast, the van der Pol equation exhibits (*b*) - (*d*) for small gain, but not (*a*).

The idea is certainly not new, and probably belongs to game-theoretic lore; however, it remained rather impractical until solution methods had been developed sufficiently. The point of this chapter is to show that it works, and even surprisingly well. In the example there is the added complication that, while the given controls are external, the perturbations are not: they are internal, of typical feedback type.

The present account is a slightly improved version of that in [1], Section 3.2; the numerical results are taken from there. The conclusions were previously announced in [2] (and, more informally, elsewhere). The reader may, according to his interests, omit either the Example and Numerical Results, or the General Theory. Further background material appears in [1].

The Example. Consider a mathematical pendulum, which it is desired to bring to rest, in least time, by the action of an external torque. For simplicity assume that the motion is planar and frictionless, the gravitational field constant; that the kinematic equation, and also the bounds on the control torque τ are normalized,

$$\ddot{\theta} + \sin\theta = \tau , \quad -1 \leqslant \tau(t) \leqslant 1 \tag{1}$$

and that the termination condition is $\theta = 0 = \dot{\theta}$ (disregarding the values $\theta = \pm 2\pi, \pm 4\pi, \ldots$ appropriate to large or rapid initial motion).

The problem belongs to nonlinear control theory,

and the governing principles are well known (see, for example, Chap. 7 in [3], even with a linear friction term allowed); we shall merely summarize the results. In the phase plane, with coordinates $x = \theta$ and $y = \dot{\theta}$, there is a 'switching curve'

$$y = -\operatorname{sgn} x \cdot \sqrt{2(|x| + \cos x - 1)}$$

(see Fig.1). The optimal controls τ always have extreme values, 1 below and -1 above this locus. The optimal trajectories follow the curves

$$y^2 = 2(\tau x + \cos x + c) \tag{2}$$

with constant c appropriate to the initial phase, until they meet the switching curve, which they then follow into the origin (see Fig. 2 and 3).

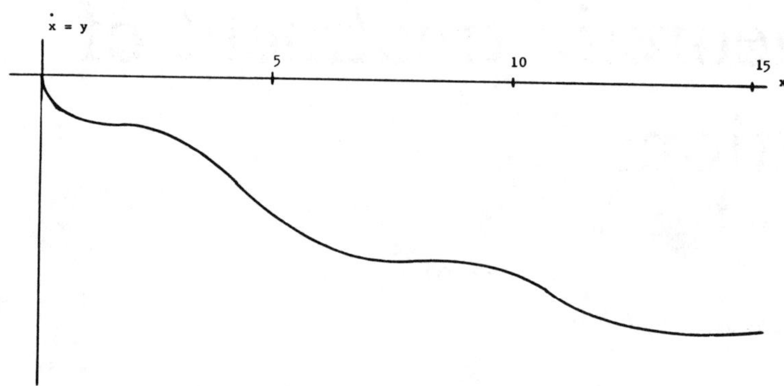

Fig.1 Switching curve (lower part) $y = -\sqrt{2(x + \cos x - 1)}$, in phase plane of torque-controlled pendulum

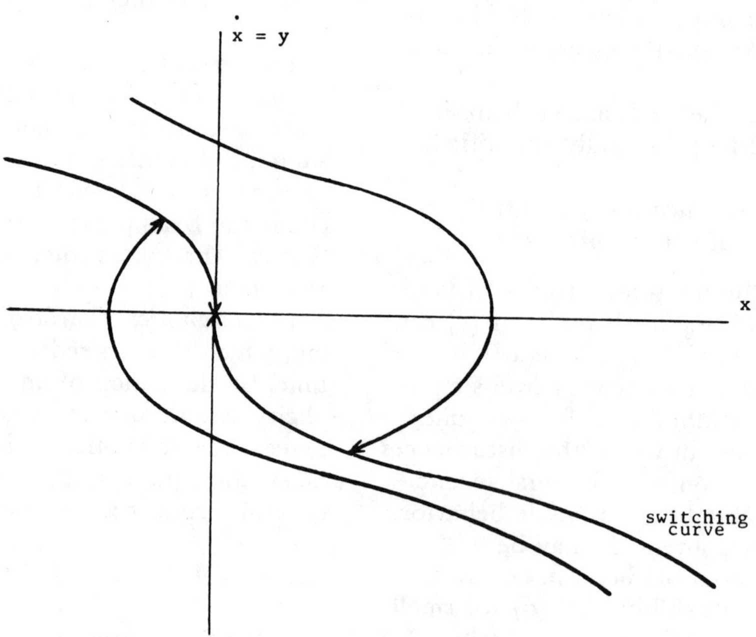

Fig.2 Optimal trajectories and switching regime (schematic) in phase plane of torque-controlled pendulum

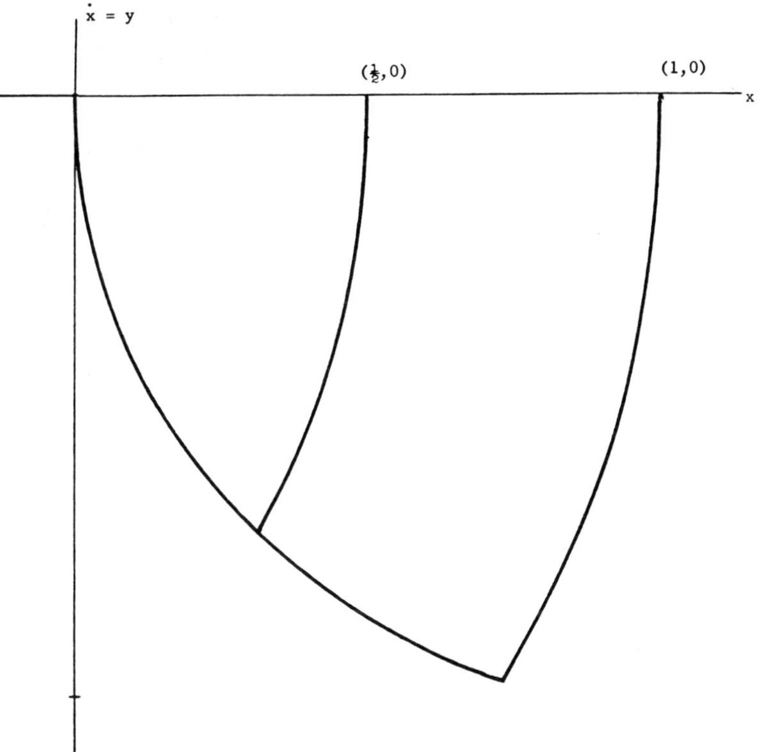

Fig.3 Torque-controlled pendulum: optimal trajectories for initial data (½,0) and (1,0) (for the former also see Table 1)

Qualitatively, this is clear sailing; a contributing effect being the fortuitous direct integrability, (2), of the first-order phase equation

$$\frac{dy}{dx} = \frac{\dot{y}}{\dot{x}} = \frac{\tau - \sin x}{y}$$

(For example, this is not the case for the controlled van der Pol equation $\ddot{x} + \epsilon(x^2-1)\dot{x} + x = u$.)

However, from the point of view of numerical computation, this is still distinctly unpleasant; the nonlinearity in (1) being the culprit. To be specific, let us determine the optimal time T for bringing the point $(x,y) = (\theta,\dot{\theta}) = (½,0)$ to rest. The 12-step integration procedure of Table 1 yields the result

$$T = 1{,}32219 \pm 0{,}02492 \tag{3}$$

the error is much too large for our pains (and even a 4-step integration readily dissuades one from using this approach).

Unorthodox linearization

The proposed method begins by re-writing (1) as

$$\theta + \theta = \tau - (\sin\theta - \theta) \text{ ,}$$

interpreting the bracketed term as the action of a notional opponent, and 'solving' the resulting game,

$$\ddot{\theta} + \theta = \tau - \rho \tag{4}$$

The last step consists in finding 'winning strategies' for the controller τ, enabling him to steer to the target and forcing the outcome $\theta = 0 = \dot{\theta}$ against all possible action of his opponent ρ. The game belongs to the class of linear pursuit games (see [1]; but the reader will, however, readily solve this game without having recourse to the book). Actually, it is more efficient to replace (4) by

$$\ddot{\theta} + \omega^2\theta = \tau - \rho \text{ ,} \quad \rho = \sin\theta - \omega^2\theta \tag{5}$$

with a free parameter ω. Here one instinctively takes $\omega = 1$ as in (4), and 'obviously' $\omega = 0$ would be a bad choice; two further values for ω are treated below.

The method for solving (5) proposed in 3.1 of [1] is deceptively simple. Having control bounds

$$|\tau| \leqslant 1 , \quad |\rho| \leqslant \rho_0 \tag{6}$$

(with ρ_0 still to be determined), one always neutralizes the opponent completely, by taking

$$\tau = u + \rho \tag{7}$$

the control u is chosen, independently of ρ, so as to steer the initial point to termination within the control system

$$\ddot{\theta} + \omega^2\theta = u \tag{8}$$

The crux is whether, for each control ρ, the τ in (7) is admissible: i.e., whether $|\rho| \leqslant \rho_0$ implies $|\tau| = |u + \rho| \leqslant 1$ to satisfy (6). Obviously this is so if $\rho_0 < 1$, whereupon the constraint on u is $|u| \leqslant 1 - \rho_0$. Note that then (8) is globally controllable.

To recapitulate, one chooses ω a priori (determining ρ_0). Then u is taken so as to steer the given initial point to the origin within (8); finally, the strategy of (7) is used within the game (5). For the special opponent control ρ used there, the control τ will steer as required within the original system (1).

The problem reduction (1) → (5) → (8) appearing here is

$$\begin{matrix}\text{nonlinear} & & \text{linear} & & \text{linear} \\ \text{control system} & \rightarrow & \text{game} & \rightarrow & \text{control system}\end{matrix} \tag{9}$$

The results, for the same initial phase, are in Table 2. The final result, that $1{,}32450 > T$, is well toward the center of the confidence interval of (3); it provides a better upper bound on T, obtained with far less effort.

One interpretation is that, in the first transition of (9) one passes from T to an upper estimate on choosing to lose information by treating the deterministic behavior of $\theta(\tau)$ in $\rho = \sin\theta - \omega^2\theta$ as the unpredictable (but bounded) action of an opponent ρ. Possibly a more satisfactory one is that, while the time-optimal controls in (1) are quite rigid, the sub-optimal ones are not: and one may efficiently choose the latter so that the state variable tracks along the optimal trajectories of the linear system (8).

Orthodox linearization

This is rather disappointing. Suppose that one 'linearizes' (1) to

$$\ddot{\theta} + \theta = \tau , \quad -1 \leqslant \tau \leqslant 1$$

and then applies the optimal controls of this within (1). Specifically, let the initial data be $\theta = ½$, $\dot{\theta} = 0$ again. Then the control has value -1 initially; it switches on an appropriate circle through the origin —an easy calculation shows that this occurs above the true switching curve of Fig. 1 or 3; and then drives the state variable to miss the target entirely, and on to ∞ within the first quadrant.

Numerical results

Referring first to the original control system (1), and initial values

$$\theta = x = ½ , \quad \dot{\theta} = y = 0 ,$$

the optimal trajectories were examined: (2) with $c = -0{,}37758$ and the portion $y = -\sqrt{2}(x + \cos x - 1)$ of the switching curve after intersection (see Fig.4). For chosen values of x, the corresponding y appear in Table 1. Then, referring to the state equation $\dot{y} = \sin x - 1$ above the switching curve, the time taken to move from (x_1, y_1) to (x_2, y_2) satisfies

$$\frac{y_1 - y_2}{1 + \sin x_1} \leqslant t \leqslant \frac{y_1 - y_2}{1 + \sin x_2}$$

similarly, on the switching curve,

$$\frac{y_2 - y_1}{1 - \sin x_2} \leqslant t \leqslant \frac{y_2 - y_1}{1 - \sin x_1}$$

The extreme values, labelled t_- and t_+, appear in Table 1. Finally, T in (3) is the mean of Σt_- and Σt_+.

Now let us turn to unorthodox linearization. The first task is to specify ω. There is the obvious first guess $\omega = 1$, and a bad one $\omega = 0$; one naturally seeks to make $|\sin\theta - \omega^2\theta|$ small. For the initial point (½,0), the least squares method yields $\omega^2 = 0{,}97536$; the best uniform value is $\omega^2 = 0{,}9691$. Actually, only the former is put forward seriously as an improvement on $\omega = 1$; the best uniform value requires a complicated computation, and is only included for comparison.

Having ω, one finds

$$\rho_0 = \max_{0 \leqslant \theta \leqslant ½} |\sin\theta - \omega^2\theta|$$

for (6), determining the control system (5) completely; see Table 2. The optimal trajectories are made up of elliptic arcs,

TABLE 1 Points (x,y) on optimal trajectory of (1) through (½,0); upper portion is on $y^2 = 2(-x + \cos x + c)$ with $c = -0{,}37758$, lower on switch curve. Lower and upper estimates t_-, t_+ on transition times.

x	$-y$	t_-	t_+
0,5	0		
		0,20049	0,20415
0,47	0,29661		
		0,05755	0,05860
0,45	0,38176		
		0,10727	0,11078
0,4	0,53569		
		0,08287	0,08574
0,35	0,65083		
		0,05560	0,05721
0,31121	0,72549		
		0,08472	0,09175
0,25	0,66173		
		0,07689	0,08187
0,2	0,60012		
		0,08617	0,09147
0,15	0,52682		
		0,10101	0,10691
0,1	0,43589		
		0,13014	0,13635
0,05	0,31225		
		0,11556	0,11921
0,02	0,19900		
		0,19900	0,20306
0	0		
		Σ: 1,29727	1,34710

$$\omega^2 x^2 \pm 2(1-\rho_0)x + y^2 = \text{const.} \tag{10}$$

(parabolas for $\omega = 0$), and the switching curve consists of translates of the semi-ellipses with const = 0. In the case of a single switch one readily determines the constant in (10); then the intersection (x_1, y_1) with the switching curve; and finally the times t_1, t_2 needed to reach first this intersection and then the origin. For the same initial point as before, the results appear in Table 2.

The results of Tables 1 and 2 are summarized in Fig.3; this contains, in addition, intermediate points on the trajectory obtained by unorthodox linearization, and also treats the initial point (1,0). The optimal and approximate trajectories are indistinguishable on the scale of this drawing.

General theory

Obviously there is a more general underlying principle. We shall formulate and comment on it, and then suggest possible extensions.

Consider the problem of steering to the origin within the nonlinear control system (vector notation in R^n)

$$\dot{x} = f(x) - p, \quad p(t) \in P, \quad \text{end: } x = 0 \tag{11}$$

We rewrite the differential equation as $\dot{x} = Ax - p + (f(x) - Ax)$, and treat the differential game this suggests:

$$\dot{x} = Ax - p + q, \quad p(t) \in P, \quad q(t) \in Q, \quad \text{end: } x = 0 \tag{12}$$

(the new data of (12) are as yet unrestricted).

Theorem

In (11) let $f: R^n \to R^n$ be continuous, with closed nonvoid $P \subset R^n$. Choose a (n,n) matrix A and subsets Q, Θ of R^n such that

$$Q \subset P, \quad f(x) - Ax \in Q \quad \text{for } x \in \Theta \tag{13}$$

If, in the game (12), there is a memory-less winning strategy for pursuer p (forcing an initial point x_0 to 0 against all action of quarry q within time-interval $[0,\theta]$) such that the state variable remains in Θ throughout the play, then there also exists an admissible control p for the control system (11) (steering x_0 to 0 within $[0,\theta]$).

Proof:

Refer first to the linear game. Once pursuer has forced to 0, he can use the strategy $p = q$ (since $Q \subset P$) to overcome future attempts of quarry and hold the state response at 0 until the end of the time interval. Thus we may assume that forcing to 0 occurs at time θ for all quarry action. According

TABLE 2 Minimal time T for initial point (½,0) under unorthodox linearization: four values of ω.

Method for ω	ω^2	ρ_0	x_1	$-y_1$	ωt_1	ωt_2	T
Bad guess	0	0,47943					1,96009
First guess	1	0,02057	0,31381	0,71850	0,50710	0,82357	1,33067
Least squares	0,97526	0,00825	0,31147	0,72331	0,50389	0,80419	1,32450
Best uniform	0,96911	0,00512	0,31088	0,72451	0,50304	0,79935	1,32298

to the First Reciprocity Theorem [1, Sec.3.1], the strategy can then be taken as follows. There is a mapping $u: R^1 \to R^n$ such that, for any quarry control $q(\cdot)$, pursuer takes $p = u + q$ and wins. Here u is any control which steers x_0 to 0 at time θ within an auxiliary control system,

$$\dot{x} = Ax - u, \quad u(t) \in U,$$

where the constraint set

$$U = \{y \in R^n: \; y + Q \subset P\} \tag{14}$$

(the so-called Pontrjagin difference $P \underline{*} Q$). Since $p = u + q$, the intermediate values of the state variable,

$$x(t) = e^{At}(x_0 - \int_0^t e^{-As}(p(s) - q(s))\, ds)$$

$$= e^{At}(x_0 - \int_0^t e^{-As}\, u(s)\, ds)$$

are independent of $q(\cdot)$.

Return to the system (11). Having $u(\cdot)$ and $x(\cdot)$, set

$$q(t) = f(x(t)) - Ax(t)$$

(note that these q-values are indeed in Q); and let $p = u + q$. The latter values are in $U + Q \subset P$ [see (14)], so that p is an admissible control for (11).

To see that p steers x_0 to 0, note that for $x = x(t)$, etc.,

$$f(x) - p = Ax - p + q = Ax - u = \dot{x}$$

and $x(\theta) = 0$ by assumption on u.

Remarks: In the assumption (13), the nonlinear portion $f(x) - Ax \in Q \subset P$; e.g., the method would not apply to $\ddot{x} + \sin\dot{x} = \tau$, since the nonlinearity is athwart the scope of compensatory controls τ.

It is unfortunate that, at present, the method is restricted to targets which are single points (one often wishes to steer to zero only some observed component of the 'full' state variable). The problem is not with the available apparatus in linear games, see [1, Section 3.3], but rather in one step of the preceding proof; in the case treated, this is avoided by having the state response independent of quarry controls. It would be highly desirable to remove this restriction.

Finally, more extensive numerical experimentation (other initial points, or targets, or even dynamics) would surely be instructive.

References

1. HÁJEK, O., *Pursuit Games,* Academic Press, New York (1975).
2. HÁJEK, O., 'Applications of first duality theorem' in *The Theory and Application of Differential Games* (ed. by J.D. Grote), pp.288-289, Reidel, Dordrecht (1975).
3. LEE, E.B. and MARKUS, L., *Foundations of Optimal Control Theory,* Wiley, New York (1967).

Practical problem. A tentative formalization (the praxiological-systemic approach)

WOJCIECH W. GASPARSKI
Polish Academy of Sciences,
Warsaw, Poland

1.

While accepting the concept of the state of affairs as intuitively understandable let us introduce the notion of event $e \in E$ as a transition from a certain state s_i at moment t_i to another state s_j at moment t_j, with t_j being later than t_i or simultaneous with t_i. We write this in the following form:

$$e = Def(s_i|_{t_i}, s_j|_{t_j}) \quad (1)$$

In particular: if $s_i \neq s_j$, event e is called a kinetic event in interval t_{i-j}, but if $s_i = s_j$, event e is called a static event or state[†] in interval t_{i-j}.

Facts which occur in the real world, Fig.1, and, which we shall identify with events e are subject to evaluation by various criteria $k \in K$. If for some event e the result of this evaluation is positive: $v(e/k) = 1$, we say that event $e \in E$, being evaluated, is satisfactory from the point of view of criterion (criteria) $k \in K$. This we write as follows:

$$SAT(e/k) = D\acute{e}f\{\exists e \in E \wedge \exists k \in K \rightarrow [v(e/k) = 1]\} \quad (2)$$

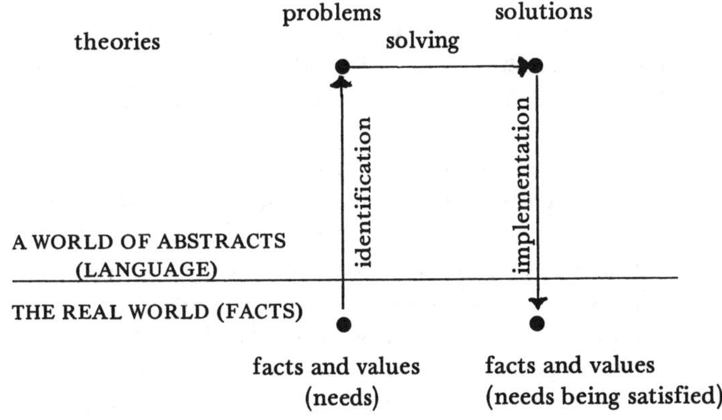

Fig.1 General scheme of practical (design) problem-solving procedure (on a base of Kemeny's scheme of scientific procedure)

† "Events are either changes (kinetic events), such as, for instance, the flight of a bullet, or a state (static events), such as, for instance, a key being in a lock. Every event occupies a certain time interval which we call the moment of that event". (T. Kotarbiński: Pojecie czynu [The Concept of Action], Warszawa 1953.)

If, however, the result of evaluation of event e is negative: $v(e/k) = 0$, we say that event e, being evaluated, is unsatisfactory from the point of view of criterion (criteria) $k \in K$ at moment t. This we write as follows:

$$\overline{SAT}(e/k) = Def\{\exists e \in E \wedge \exists k \in K \rightarrow [v(e/k) = 0]\} \tag{3}$$

If for same event e, for which $\overline{SAT}(e/k)$, it is possible to distinguish components $a_1, a_2, \ldots, a_q$ in such a way that a change of any component, except the two of them: an aim and a criterion, causes $SAT(e/k)$, we say that expression (3) denotes a need, while expression (2) denotes its fulfilment. We also say that the expression:

$$!(a_1, a_2, \ldots) \rightarrow ?(\ldots, a_{q-1}, a_q) \tag{4}$$

the left-hand side of which is composed of unchanging components, and the right-hand side is composed of changing components, denotes a practical problem suited to the need $\overline{SAT}(e/k)$.

In expression (4) "!" functor "is given", "?" functor "is sought".

Expression (4) reflects well, it seems, intuition concerning problems being solved in practical activities if we interpret polynomial $e = (a_1, a_2, \ldots, a_q)$ as an action. It refers to some earlier theoretical proposals in this field (Leniewicz [1]).

Let us note that practical problems are questions concerning what should be done in order to achieve the objective with reference to given components of activity.

Some practical problems are solved at once or almost at once. They are standard problems which occur repeatedly, or problems which approximate them. Other practical problems are non-trivial questions which require some processing in consequence of which their solution is obtained. Among those non-standard problems there are also practical problems, whose processing is possible only after they have been clearly formulated in the form of an utterance in a certain language. The practical problems which are uttered in language $l \in L$ we shall call design problems. The general structural form of design problems is the same as that of practical problems†, i.e. it is consistent with formula (4).

†The structural form of practical problems does not depend upon the content carried by these problems.

2.
Components of any action A may be distinguished, in principle, by any number, depending upon the objective set by the researcher. For instance: E. Leniewicz distinguishes four components: the objective, the conditions, the action, the evaluation criterion [1]; R.L. Ackoff distinguishes four components: the decision-maker, the objective, the method of action, the context (Ackoff [2]).
For example, for design purposes, I think, it is worth while to distinguish a somewhat greater number of components of an action, namely $q = 7$. These components are:

$$A(C,T,O,R,E) \tag{5}$$

and $T\ (N,M,Z)$, where C denotes the evaluation criterion, T the method ("technique") of action including: the apparatus (tools) of action N, the method of action M, the resources of action Z, O the objective of action, R the realizer of action, E the environment of action (Gasparski [3]).

Substituting appropriate components of activity in formula (4); given on the left-hand side and sought on the right-hand side, we obtain an expression for one kind of design problems. Here is an example:

$$!(C,M,Z,O,E) \rightarrow ?(N,R) \tag{4.1}$$

which we read, "If the criterion, the method, the resources, the objective of action A and the environment E are given, which tools and which realizer of this action should be selected, if for event e, consisting in the occurrence of action A in the interval t_{i-j} we want to have $SAT(e/k)\, t_{i-j}$?"

A full list of practical design problems contains 31 kinds of problems" (Table 1). The design problems in the table are divided into five classes in accordance with the number of sought components of action. It is worth noting here that the classification table of design problems may provide a basis for the construction of a multidimensional morphological box which makes possible further divisions of design problems into subkinds of various degrees (Fig.2). It is to be expected that these subdivisions may be helpful in indicating clearly the kind and subkind of design problems being typical to the particular kind of design activity. For instance, design problems in which a certain type of tool is sought for the instrumentalization of activity we are inclined to include in engineering designing, while problems in the selection of a realizer are

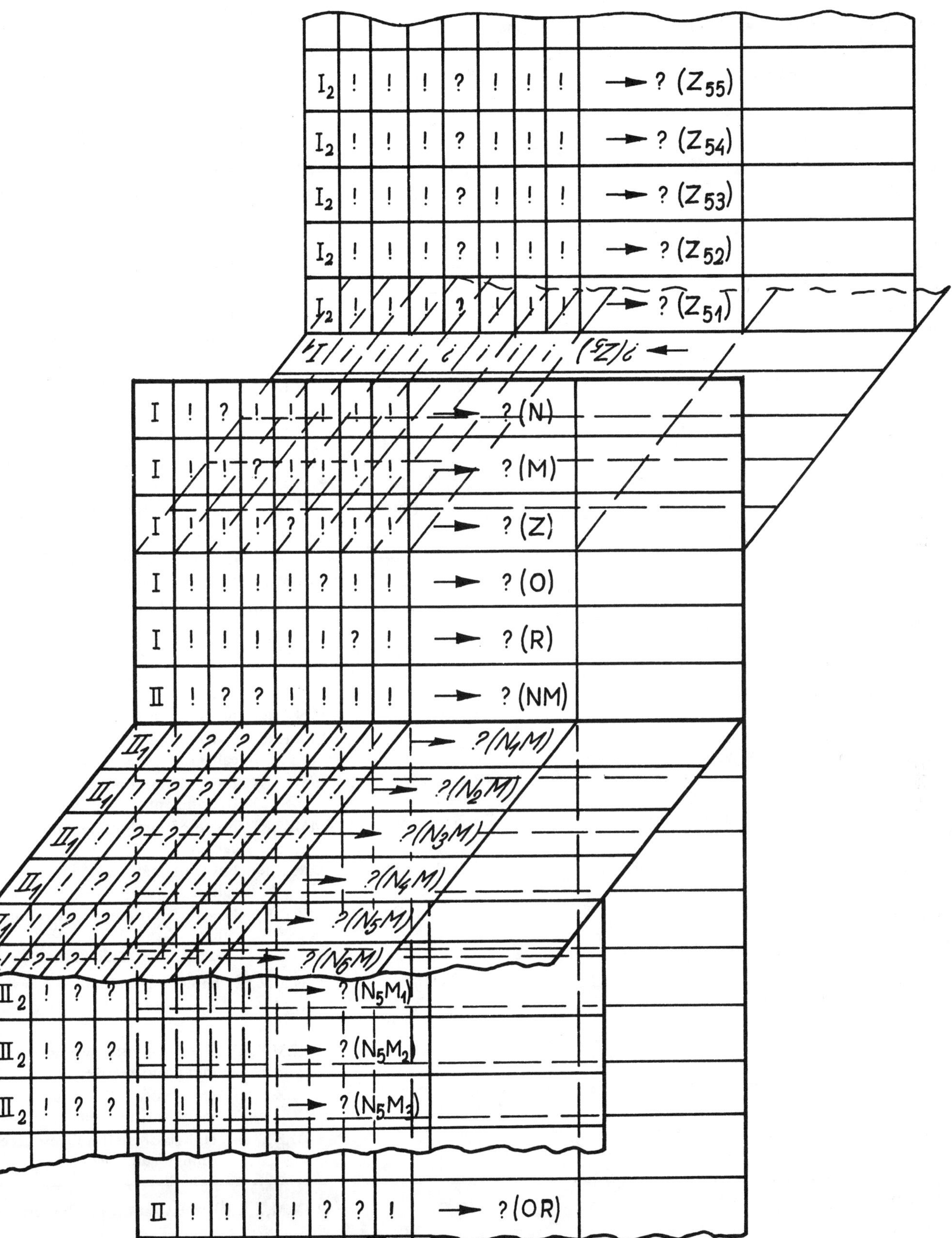

I2 ! ! ! ? ! ! ! → ? (Z55)
I2 ! ! ! ? ! ! ! → ? (Z54)
I2 ! ! ! ? ! ! ! → ? (Z53)
I2 ! ! ! ? ! ! ! → ? (Z52)
I2 ! ! ! ? ! ! ! → ? (Z51)
I1 ! ! ! ? ! ! ! → ? (Z5)
I ! ? ! ! ! ! ! → ? (N)
I ! ! ? ! ! ! ! → ? (M)
I ! ! ! ? ! ! ! → ? (Z)
I ! ! ! ! ? ! ! → ? (O)
I ! ! ! ! ! ? ! → ? (R)
II ! ? ? ! ! ! ! → ? (NM)
II1 ! ? ? ! ! ! ! → ?(N1M)
II1 ! ? ? ! ! ! ! → ?(N2M)
II1 ! ? ? ! ! ! ! → ?(N3M)
II1 ! ? ? ! ! ! ! → ?(N4M)
II1 ! ? ? ! ! ! ! → ?(N5M)
II1 ! ? ? ! ! ! ! → ?(N6M)
II2 ! ? ? ! ! ! ! → ? (N5M1)
II2 ! ? ? ! ! ! ! → ? (N5M2)
II2 ! ? ? ! ! ! ! → ? (N5M3)
II ! ! ! ! ? ? ! → ? (OR)

Fig.2

Practical problem. A tentative formalization (the praxiological-systemic approach)

TABLE 1 Classification of (practical) design problems

Class of problem	Given (!) Sought (?)							Problem	
	O	N	M	Z	E	R	C	Formula	Name
FIRST CLASS I	!	?	!	!	!	!	!	!(O,M,Z,E,R,C) →? (N)	choice of tool
	!	!	?	!	!	!	!	!(O,N,Z,E,R,C) →? (M)	choice of method
	!	!	!	?	!	!	!	!(O,N,M,E,R,C) →? (Z)	choice of resources
	!	!	!	!	?	!	!	!(O,N,M,Z,R,C) →? (E)	choice of environment
	!	!	!	!	!	?	!	!(O,N,M,Z,E,C) →? (R)	choice of realizer
SECOND CLASS II	!	?	?	!	!	!	!	!(O,Z,E,R,C) →? (N,M)	choice of tool and method
	!	?	!	?	!	!	!	!(O,M,E,R,C) →? (N,Z)	choice of tool and resources
	!	?	!	!	?	!	!	!(O,M,Z,R,C) →? (N,E)	choice of tool and environment
	!	?	!	!	!	?	!	!(O,M,Z,E,C) →? (N,R)	choice of tool and realizer
	!	!	?	?	!	!	!	!(O,N,E,R,C) →? (M,Z)	choice of method and resources
	!	!	?	!	?	!	!	!(O,N,Z,R,C) →? (M,E)	choice of method and environment
	!	!	?	!	!	?	!	!(O,N,Z,E,C) →? (M,R)	choice of method and realizer
	!	!	!	?	?	!	!	!(O,N,M,R,C) →? (Z,E)	choice of resources and environment
	!	!	!	?	!	?	!	!(O,N,M,E,C) →? (Z,R)	choice of resources and realizer
	!	!	!	!	?	?	!	!(O,N,M,Z,C) →? (E,R)	choice of environment and realizer
THIRD CLASS III	!	?	?	?	!	!	!	!(O,E,R,C) →? (N,M,Z)	choice of method ("technique")
	!	?	?	!	?	!	!	!(O,Z,R,C) →? (N,M,E,)	choice of tool, method and environment
	!	?	?	!	!	?	!	!(O,Z,E,C) →? (N,M,R)	choice of tool, method and realizer
	!	?	!	?	?	!	!	!(O,M,R,C) →? (N,Z,E)	choice of tool, resources and environment
	!	?	!	?	!	?	!	!(O,M,E,C) →? (N,Z,R)	choice of tool, resources and realizer
	!	?	!	!	?	?	!	!(O,M,Z,C) →? (N,E,R)	choice of tool, environment and realizer
	!	!	?	?	?	!	!	!(O,N,R,C) →? (M,Z,E)	choice of method, resources and environment
	!	!	?	?	!	?	!	!(O,N,E,C) →? (M,Z,R)	choice of method, resources and realizer
	!	!	?	!	?	?	!	!(O,N,Z,C) →? (M,E,R)	choice of method, environment and realizer
	!	!	!	?	?	?	!	!(O,N,M,C) →? (Z,E,R)	choice of resources, environment and realizer
FOURTH CLASS IV	!	?	?	?	?	!	!	!(O,R,C) →? (N,M,Z,E)	choice of method and environment
	!	?	?	?	!	?	!	!(O,E,C) →? (N,M,Z,R)	choice of method, ('techniques') and realizer
	!	?	?	!	?	?	!	!(O,Z,C) →? (N,M,E,R)	choice of tool, method, environment and realizer
	!	?	!	?	?	?	!	!(O,M,C) →? (N,Z,E,R)	choice of tool, resources, environment and realizer
	!	!	?	?	?	?	!	!(O,N,C) →? (M,Z,E,R)	choice of method, resources, environment and realizer
FIFTH CLASS V	!	?	?	?	?	?	!	!(O,C) →? (N,M,Z,E,R)	relatization of objective

typical to organization design, or to work design, as it is sometimes called (Nadler [4]).

The classification table of practical (design) problems could also suggest the usefulness of certain solution procedures for some classes of problems and their uselessness for other classes (e.g. usefulness of systems techniques for the fifth class problems).††

References

1. LENIEWICZ, H., *Dyrektywy praktyczne. Konstrukcja i uzasadnienie (Practical Directives. Construction and Justification)*, PWN, Warszawa (1971).
2. ACKOFF, R.L., *Scientific Method: Optimizing Applied Research Decisions*, J. Wiley, New York (1962).
3. GASPARSKI, W., 'Projektowanie w systemie działań (Designing in the System of Actions), a paper delivered to the First Conference on Design Methodology, Warszawa (1971), published in *Metodologia projektowania inzynierskiego (Methodology of Engineering Design)*, W. Gasparski *et al.* (ed.), PWN, Warszawa (1973).
4. NADLER, G., *Work Design. A Systems Concept*, Irwin, Homewood Illinois (1970).
5. KEMENY, J.G., *A Philosopher Looks at Science*, D. Van Nostrand Co. Inc. (1959).

††The enriched version of this chapter was delivered at the "Changing Design" DRS Conference, Portsmouth 1976 on the title "Problem Situation in Designing. (Toward a Semiotic and a Logic of Design)". It was partially published in *Design Methods and Theories*, Vol.10, No.2, Apr.-June (1976).

Scientific model building and experimenting — a cybernetic analysis of their interaction

HELMUT HIRSCH
Federal Ministry for Trade, Commerce and Industry, Vienna, Austria

1. Introduction

The process of model formation in science has been considered from many different angles, from the standpoint of the philosopher, statistician, physicist, etc. Although the original purpose of this chapter was to develop a synthesis of the different approaches, limited space does not allow us to achieve so ambitious a goal; we shall mainly restrict ourselves to presenting simple schemes which, it is hoped, will help to clarify the overall picture, and develop qualitatively the concept of the two states (active and passive) of information. There are many philosophical questions linked with the problem of how exactly a scientific theory is formed; they will not be considered here.

Cybernetic models for processes of scientific cognition have been constructed by, among others, Klaus and Steinbuch [1,2]; their approach centers on the interaction of a single individual with its environment, possibly in symbiosis with an automaton. The basic pattern of scientific research is also discussed by Klir [3]. Unlike this author, however, we shall employ throughout the concept of a multilevel, hierarchical system as developed, for example, by Mesarović *et al* [4]; a very appropriate tool in the description of many realistic structures.

It should be noted that the discussion is restricted to sciences that can be instrumentalized, ie. used as a basis for a technology. 'Technology' is meant in the widest possible sense, including social technology as practised by some psychologists and sociologists, as well as biological and medical technology. It certainly would be interesting to specialize our schemes for natural and social sciences, respectively, and consider the differences; but again, the space is too limited for this.

2. Active and passive information

The simplest possible scheme representing the process of scientific development is given in Fig. 1. This is a hierarchical three-level system; the levels are linked by flow of information in both directions. The character of this exchange, clearly showing the hierarchical character of the system, will be discussed later. The limits of this system were chosen in a very restrictive manner; for a full picture, addit-

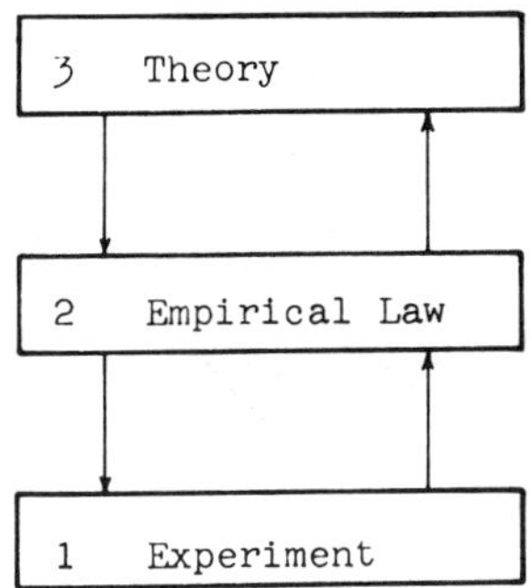

Fig.1 Basic scheme of scientific development

ional levels and a more complicated structure is required, as shown in section 4.

The lowest, or experimental, level contains only the actual gathering of results according to pre-set rules. On the following level, empirical laws, ie. laws expressed in a mathematical form that is suitable, but of no deeper meaning (eg. regression polynomials whose coefficients are purely formal terms) are fitted to the data and their validity is tested. On the highest level, empirical laws are integrated towards a theory, ie. a family of functions (which might be characterized by means of a differential equation) whose form and parameters are 'meaningful' in terms of a model of the system under study and are interpreted as certain observable quantities. We do not distinguish between a model in the sense of Stachowiak [5] and a theory; in fact, if the presented scheme were more detailed, it might be appropriate to refer as model to what is called theory above, and reserve the term 'theory' for larger systems, embracing a great range of phenomena (like relativity theory or quantum theory). This distinction, however, is basically only a semantic question.

The situation is different, and less clear, when empirical laws and theories are not expressed mathematically; as we consider a mathematical formulation as most convenient and most advanced, our attention will center around it in this section.

The arrows pointing upwards in Fig. 1 stand for the flow of experimental results, properly grouped and structured according to the design of the experiment, and of the characteristics of empirical laws, being relevant in the context. So far, a conventional description of the progress data→ empirical law→ theory has emerged. This static picture, however, needs to be enlarged by the two arrows closing the feedback-circuits and thus to be converted into a dynamic one. From level (3) to (2), the relevant variables, the interdependence of which has to be mathematically described, the parameters to be taken into account, and their range, are communicated; all this information is chosen from the need to test a hypothesis, gain evidence for or against it, discriminate between several rival theories or establish a new one. Accordingly, level (2) communicates to level (1) the optimal experimental design for the task set by level (3), the choice of optimal values of the given parameters within their given range. The circuit enclosing (1) and (2) will work independently for several 'periods' (measurement–evaluation–design–measurement...), until some criterion for stopping set by level (3) is fulfilled and the higher-order circuit is entered.

Figure 2 gives, for illustration, a more detailed scheme for a special, although quite important, case of measurements. For the functional dependence of a variable, subject to statistical fluctuations, on another one, several regression models appear possible; the measurements have to be planned to achieve optimal discrimination. Such problems are extensively treated in the recent statistical literature [6-10] (among others); in the figure, a Bayesian approach is assumed.

In the interaction between levels (1) and (2), two criteria are employed–a stopping criterion, which is relevant for the decision to enter level (3), and a design criterion for the quantitative planning (ie. the choice of regressor values), which is regularly tested for its adequacy. After the stopping criterion is fulfilled, it is attempted to test or establish theories (in the sense as defined above) on the basis of the different weights (probabilities π_i) the regression models have been given.

So far, we have used the concept of information without any further differentiation. However, it appears advisable, for the purpose of this short presentation as well as for more general application, to distinguish between two states of information: the active and the passive one. Both can be mathematically expressed in the same way (if a quantitative definition is possible at all) in terms of a probability distribution. The passive information refers to a-posteriori data, to knowledge about some event that has already taken place, thus reducing uncertainty (like the message that the result of a throw of dice was two sixes, say), whereas the active information represents a-priori knowledge, data which are the basis for planning and executing future steps, which do not directly come from a system where an event took place, but which are destined to become effective in a system in the future (like the

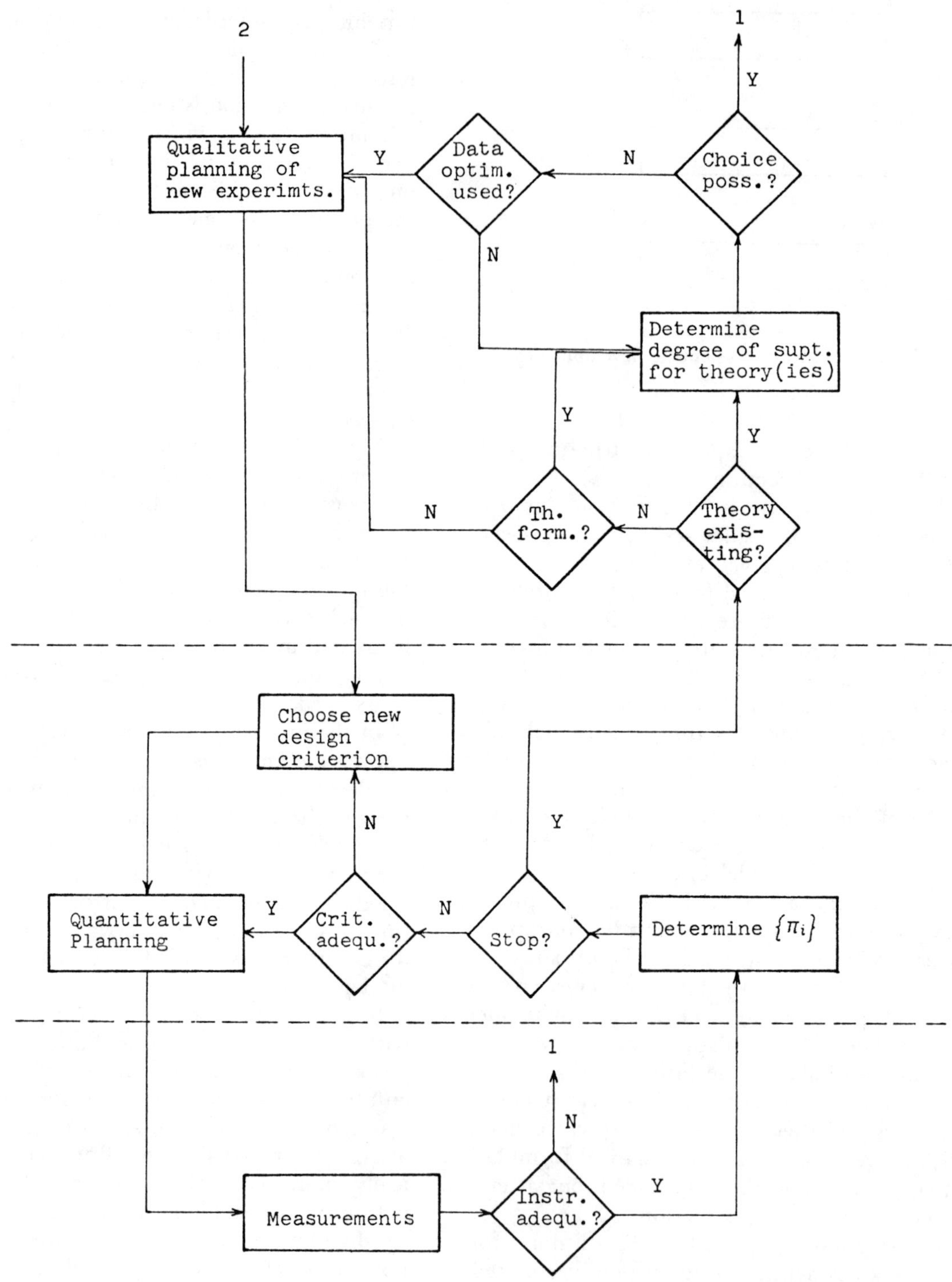

Fig.2 Theory formation and testing based on model discriminating experiments.
$\{\pi_i\}$......... set of probabilities of regression models
1 .... Impulses to other theories, technology, etc.
2 .... Impulses from other theories, technology, etc.

order that two dice will have to be laid down with the sixes on top). Clearly, this distinction is senseless in the realms of pure mathematical information theory—it only gains significance when systems are considered that consist of one material component to be explored and changed, and one spiritual component doing the exploring and changing; in other words, in science, taken as a system of an interacting object and subject. The character of the exchange of information in Fig. 1 is now clearly revealed: the upwards-pointing arrows stand for flow of passive information, the downwards-pointing ones for flow of active information. This is a general property of multilevel hierarchical systems with information feedback: passive information flows from lower levels to higher ones, and active information in the opposite direction.

It should be well noted that the terms 'active' and 'passive' are chosen from the viewpoint that the former can be regarded as orders or rules governing activity, whereas the latter 'has to be acted on' and is more like a sort of raw material. The terms should carry no other connotations or associations.

A short remark seems appropriate at this point on the role of creative thinking which is involved, especially in level (3), in converting passive information into active one. Clearly, information of some characteristics not considered so far enters here; an analysis of this problem would require a separate chapter. So we only note that 'creative' information must be regarded as meta-information relative to our system and serves as an encoder.

The amounts of active and passive information flowing in and out of a certain level are closely linked together. When planning an experiment, eg. by choosing values of the independent variables for which the measurements have to be performed, it is hardly reasonable to determine them (the active information) with a significantly greater, or smaller, accuracy than that of the measurements themselves (the passive information), ie. the amounts of information will be approximately equal. One level higher, if the validity of a theory is still very uncertain, due to only a small amount of passive information 'condensed' into it, its power for planning and prediction—producing active information—is small and vice versa.

Thus, if A stands for the amount of active, P of passive information, and the subscript i and o denote input and output respectively of the level under consideration, the hypothesis that approximate equality between active and passive information flows is necessary for the functioning of a multilevel hierarchical system with information feedback, can be expressed as

$$\Delta A_o \simeq \Delta P_i$$

(and

$$\Delta A_i \simeq \Delta P_o)$$

where the amounts of information flowing during a time Δt, which must be greater than one 'period' of the system, are entering the approximate equations. (At a certain instant, only one type of information might be flowing; but there will be some sort of periodic performance as indicated above over longer intervals; also, it has to be taken into account that passive information might be stored for later use, or brought forward from such a store to be transformed into active information, so that the 'law' will only be meaningful over longer periods of time.) One of those can be regarded as redundant, as they hold for all levels and the output of one is the input of another.

Especially in regard to the higher levels, this 'approximate equality' hypothesis must be regarded as the symbolic statement of a qualitative relation, and not as a mathematical relation, as quantitative measures for amounts of information can only be given in very rare cases. This, however, is due more to the degree of complexity of most real systems than to principal impossibilities; therefore, the 'equations' are not meaningless. A considerable amount of additional work will be required, however, to put them to the test in any realistic situation.

The 'approximate equality' hypothesis is the dynamic formulation of the fact that, seen statically, the 'degree of structuredness' of a theory must be approximately equal to that of the underlying data.

In the communication between the lowest levels [(1) and (2) in Fig. 1], $\Delta P > \Delta A$ means suboptimal, inaccurate planning, and $\Delta P < \Delta A$ senseless accuracy in planning, equivalent to waste of resources. In the criterion-controlling circuit (Fig. 2), $\Delta P > \Delta A$ corresponds to the prolonged use of an inadequate criterion, whereas $\Delta P < \Delta A$ will lead to changes that are not necessary, again involving waste of resources (computing time and others). In the interaction between the theory and the empirical law level, $\Delta P > \Delta A$ brings either a loss in the applicability of the theory or suboptimal possibilities for linking up theories into greater structures due to a formulation less clear or less certain than is possible in principle. $\Delta P < \Delta A$ means that theories are formulated, based too much on personal intuition, consensus among experts, prejudice and fixed thought-patterns, and not on data (Fig. 2). This situation, unfortunately, arises frequently in practice. According to personal prejudices, passive information is filtered with the aim of biasing it in the desired direction and reducing it in the process where this is not possible. Convention and publication habits can play a large role here.

As there is no room to discuss this point at any length, a short example must suffice here; an example of more general significance is given in section 3. As was shown by Sterling [11], in fields where tests of significance are commonly used (eg. psychology) research yielding nonsignificant results is not published. Such research, being unknown to other investigators, may be repeated until eventually by chance a significant result occurs—an 'error of the first kind'—and is published. Significant results published in these fields are seldom verified by independent replication. The possibility thus arises that the literature of such a field consists in substantial part of false conclusions resulting from errors of the first kind in statistical tests of significance. The filtering of passive information by unwritten rules proves disastrous, especially when the wrong theories are used to produce active information. This active information will appear better founded than it really is, if the wrong results are not challenged; (wrong) passive information has been simulated by the lack of critics giving rise to the production of an unduly large amount of active information.

The approximate equalities stated above have to be modified, when costs of experimentation and planning are taken into account. If the latter is of an order of magnitude comparable to that of the advantages gained by it (or even bigger), it will be economical to simplify the planning process to the point where the savings in plannings costs minus the additional experimental costs (to reach a given accuracy, eg. with the suboptimal planning) are a maximum.

The same applies on all levels where the use of passive information involves processes creating costs of the same order of magnitude as the benefits from this use. Thus, in general, not all passive information will be used, and we can write, as a modification of the approximate equality, simplified and symbolically

$$\Delta P \gtrless \Delta A$$

This inequality is again valid only for the accumulated amounts over several 'periods' of the system. This addition is important as, eg. active information may be produced before the acquisition of passive information, as in block-wise design of experiments with non-Bayesian methods [7-9].

Compared with other attempts to further develop the concept of information, as, for example, Oeser's 'erkenntnistheoretische' information [12], that "contains the message for its own further processing", we have, by the distinction between active and passive information, gone a step further in describing information in a way relevant in real hierarchical multilevel systems. In abstract algorithms, as considered by Oeser, this distinction is meaningless, as there is no interaction with an unknown external world, only the dialogue between man and the extension of his brain, the program-controled automaton. Also, in algorithmic processing, there are no feedback-circuits, only a one-way road (which can have branches) from beginning to end.

3. An example

Whereas the interaction between levels (1) and (2) (Fig. 1) is comparatively simple, that between (2) and (3) remains somewhat schematic, vague and abstract. To clarify this point to some extent, but mostly to show that further work in this direction might prove highly relevant in many fields, an example of inadequate theory formation is briefly discussed below.

In water vapor in thermal equilibrium, not only single molecules, but also clusters of several molecules are present. When condensation is stimulated by, for example, lowering the temperature, the clusters assume an important role in serving as condensation nuclei (if no other such nuclei are present—'homogenous condensation'). The number of clusters above a certain size under given thermodynamic conditions can thus be determined by inducing condensation in a defined way and counting the number of emerging droplets. This is the main method of acquiring information about cluster formation. The theories, however, based on this passive information, are highly involved structures. The 'classical', comparatively simple model, treats each cluster in bulk and considers bulk energy and surface energy. Its predictions on the cluster formation rate (ie. the active information forthcoming from it) are in the right order of magnitude, compared to experiment [13]. Allen and Kassner [14], not satisfied with the rough correspondence, but without relevant additional information, constructed a model including various terms such as translational and rotational cluster energy and its changes from the impact of single molecules. The disagreement between experiment and theory reached a factor between 10^{17} and 10^{21}; perhaps the most impressive incidence of this sort ever known in the history of science. It should be mentioned, though, that later work by Hale and Plummer [15,16], incorporating new parameters like terms arising from the de-activation of the six degrees of freedom of the condensing molecule achieved an error factor of only one or two orders of magnitude; still no gain compared with the early 'classical', and simple, model. It remains to be seen if these highly 'over-sophisticated' models based on too little passive information will be of great practical value.

In a closely related field, the study of droplet growth well above the cluster size, similar troubles have been avoided elegantly. Several authors, using different models, calculated highly different growth rates until the 'mass-accommodation coefficient' (the probability that a molecule touching the droplet surface will actually condense and not be scattered away again) was introduced. Values of this coefficient vary, in the literature, from about 10^{-2} to 1 ([17], further references given there). Needless to say, all theories now fit the data fairly well.

This was a short example to demonstrate the disastrous consequences of overloading a theory with parameters when only little data is available, or, expressed in terms of our dynamic scheme, creating too much active information from too little passive information.

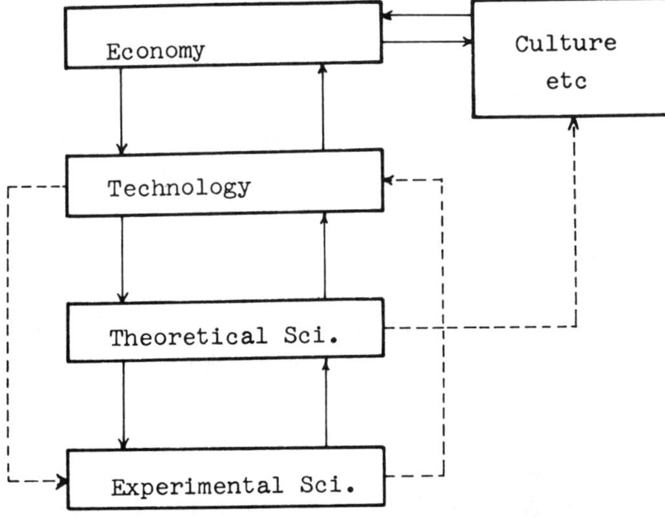

Fig.3 Basic interactions of scientific development with other areas of society; main information flows

4. A generalization

In order to get a complete picture of the process of scientific development, it has to be seen in its socio-economic context. For the purpose of a very condensed discussion, a simple scheme (Fig. 3) of the now complete system will be sufficient. Note that interactions are now not strictly confined to neighbouring levels, while the main streams of information still follow the basic pattern of the hierarchy. Scientific knowledge has the greatest impact on society at large by becoming operative in technology (in the widest sense—including planning and social 'techniques' as well), thus invading the life of every person, changing the structure of society and thus culture, ethical standards, etc. To a considerable extent, however, scientific theories have direct impact in these areas, directly changing the 'Weltbild' of a society, as, for example, Darwin's theory of evolution, Einstein's theory of relativity and, last but not least, cybernetics itself. In addition, direct impulses to technology may come from the experimental level, arising from measurement problems, and vice versa.

Any attempt to describe information flows in this system quantitatively is futile because of its complexity and the present standard of mathematics; the 'approximate equality' hypothesis in its modified version $\Delta A \lessapprox \Delta P$ must again be taken as symbolic. However, the effects of drastic disturbances in the equilibrium between active and passive information are manifesting themselves everywhere.

In our society, based to a large extent on individual prestige as the motivating force for all activities, the policy of each individual or group is to keep his output ΔA and ΔP as high, and his input ΔA and ΔP as low as possible. In a system of interaction between equals, these tendencies are checked by the simple fact that they are common practice and the input of one is the output of another. In a hierarchical system, however, the higher level has the power to follow its own policy at the expense of the lower ones. The result is that the upward flow of passive information is consistently filtered and reduced, while the downward flow of active information grows out of proportion to it. Thus, too much will be based on too little (too much planning and decision based on too little actual data from 'below') and all the shortcomings, that are only too clearly seen by a neutral observer, arise.

In big industrial enterprises, for example, decisions by the top management are often taken for reasons of prestige; afterwards, a rational basis is invented, thus 'simulating' the passive information actually lacking in the process, from either being disregarded or not collected at all. The decisions, in turn, influence technological and scientific development.

Professor George Wald, the Nobel Prize winner, was referring to the same problem when he recently wrote to the Colloquium on Environmental Aspects of Nuclear Energy [18], that the general belief that governments make their policy for the benefit of the public, has to be challenged. In his opinion, the contrary is the case; governments decide on a policy (for political or economical reasons) and then search for the information to support it: that means that information, provided by research, often follows the decisions it should rightfully precede. In our terms, we can add that this produces a heavy bias in the passive information acquired, as well as a filtering as only data suitable to the decision already made will be considered.

In the subsystem considered in section (2) (Fig. 1), there were at least some factors counteracting this alarming tendency—a theory has to be applicable, it cannot be wholly without merit if it is to be accepted. In addition, the contact between experimenter and theorist is quite close and mostly on equal terms. But even there personal prestige and authority as well as group consensus disturb the equilibrium, as well as the tendency to develop theories for their own sake with little regard for experimental data (section 3). The more complicated the structures become, the further up in the hierarchical system we get, the larger the number of interactions becomes—the stronger are the disturbances. In this way, no real progress is possible; there is only a development which can lead, by continuous reinforcement of the established patterns by socialisation techniques, to an actual breakdown.

The consequence of this is the necessity of establishing a parallel control-structure to the multilevel system in Fig. 3 (since the original structure cannot simply be destroyed—this would lead to destruction, not progress, of society), which creates the possibility that active information may flow from the lower levels to higher ones as well, to guarantee an overall equilibrium (or approximate equilibrium). The establishment of such a control would again create a host of new problems. These, however, will have to be faced to avoid the short-term danger of an irreversible malfunction of the system. The necessity of increasing control power of lower levels very likely applies to other sectors of human society as well. An analysis of this lies outside the scope of this chapter.

5. Conclusion

Due to the high complexity of the topic under discussion, it could only be simplified and treated very sketchily. However, it was shown that some ideas about the way large systems function can be acquired by considering analogies in smaller ones.

For a more thorough treatment, two directions of research need to be further followed up:

1. Empirical research: In order to be able to conceive a more realistic, fully detailed scheme for the system under study, various case studies are necessary to reveal the structure of concrete multilevel systems; these can then be merged by a synopsis to arrive at a general theory (following the pattern described in section 2).

2. Theoretical research: The concept of numerical measures of amount of information is far less developed than required for these studies. The optimism of the late forties and early fifties, expecting, on the basis of Shannon's theory, an imminent breakthrough regarding measures for information in physics, semantics and various other fields [19-21], has been thoroughly frustrated, leaving the mathematical theory of information strongly in need of new impulses.

Section (4) was of a highly speculative character; however, we think that these speculations serve a real purpose as all too often it is the scientist who gets impulses (active information) from the rest of society (mainly its higher levels) and sends back only passive information, to be converted into active information elsewhere. If the hierarchical structures are to be challenged, it is necessary to reverse this situation and also to produce large amounts of active information. This insight, luckily, is slowly becoming recognized by scientists today; still, there is yet a long way to go to achieve the postulated 'approximate information' in information flow.

Acknowledgement

The author wishes to express his thanks to Dipl.Ing. M. Schmutzer, PhD, for his constructive criticism and helpful suggestions concerning the topic of this chapter.

References

1. KLAUS, G., *Kybernetik und Erkenntnistheorie*, VEB Deutscher Verlag der Wissenschaften, Berlin (1966).
2. STEINBUCH, K., *Automat und Mensch*, Springer

Verlag, Berlin (1971).
3. KLIR, G.J., *An Approach to General Systems Theory,* Van Nostrand Reinhold Company, New York (1969).
4. MESAROVIĆ, M.D., MACKO, D., and TAKAHARA, Y., *Theory of Hierarchical, Multilevel, Systems,* Academic Press, New York (1970).
5. STACHOWIAK, H., *Allgemeine Modelltheorie,* Springer Verlag, Wien (1973).
6. BOX, G.E.P. and Hill, W.J., *Technometrics* **9**(1), 57-71 (1967).
7. FEDOROV, V.V., *Theory of Optimal Experiments,* Academic Press, New York (1972).
8. ATKINSON, A.C. and COX, D.R., *J. Roy. Stat. Soc. B* **36**(3), 321-348 (1974).
9. ATKINSON, A.C. and FEDOROV, V.V., *Biometrika* **62**(2), 289-303 (1975).
10. BORTH, D.M., *J. Roy. Stat. Soc. B* **37**(1), 77-87 (1975).
11. STERLING, T.D., *J. Amer. Stat. Ass.* **54**(1), 30-34 (1959).
12. OESER, E., *Informationsprozesse,* Gesellschaft für Mathematik und Datenverarbeitung, Bonn (1974).
13. LOTHE, J. and POUND, G.M., *Nucleation* (Ed. A.C. Zettlemoyer), 119-120, Dekker, New York (1969).
14. ALLEN, L.B. and KASSNER, J.L., *J. Coll. Interface Sci.* **30**(1), 81-86 (1968).
15. PLUMMER, P.L.M. and HALE, B.N., *J. Chem. Phys.* **56**(9), 4329-4334 (1972).
16. HALE, B.N. and PLUMMER, P.L.M., *J. Atmosph. Sci.* **31**(6), 1615-1621 (1974).
17. WAGNER, P., 'Untersuchung des Tröpfchenwachstums in einer schnellen Expansionsnebelkammer', 22f, Dissertation, Universität Wien (1974).
18. WALD, G., 'Message to the Colloquium on Environmental Aspects of Nuclear Energy', August 28 and 29, Vienna (1975) (published at press conference
19. SHANNON, C.E. and WEAVER, W., *The Mathematical Theory of Communication,* University of Illinois Press, Urbana (1949).
20. BRILLOUIN, L., *Science and Information Theory,* Academic Press, New York (1956).
21. BAR-HILLEL, Y. and CARNAP, R., 'Semantic information', in: *Proceedings of a Symposium on Applications of Information Theory* (Ed. W. Jackson), Butterworth, London (1953).

Progress in structural modelling — a biased review

J.M. McLEAN, P. SHEPHERD, and R.C. CURNOW
Science Policy Research Unit,
University of Sussex, UK

1.1 Background to the research

The Science Policy Research Unit has recently had, and continues to maintain, a considerable interest in the methodology of modelling and forecasting. Firstly, the Unit has an overall concern with *all* forecasting techniques through its research program into 'Social and Technical Alternatives for the Future' (STAFF). Secondly, as a particular commitment of the STAFF program the Unit keeps a watching brief on major developments in the modelling of social and economic systems. Thirdly, there is a small group within the Unit which is researching into alternative techniques for the investigation of the structural and behavioral properties of large dynamic systems. This chapter presents a review of the 'state of the art' in what we have chosen to term 'structural modelling'.

In the first section we attempt to establish the argument that the development of conventional dynamic modelling techniques has shifted attention from the important task of producing a model structure which adequately represents a complex system. We then go on to review ongoing and recently completed modelling work in which the concept of model structure has been predominant. Finally, we present a survey of some automatic procedures which we have developed to provide what we believe to be an essential balance in the methodology of modelling. We have described elsewhere our ongoing research in greater detail [1, 2,3]. The principal aim of this chapter is to set our work firmly in the context of the international modelling research effort.

1.2 The case for structural modelling

Existing dynamic modelling methodology focuses attention on a single aspect of the modelling process. When a model is built to study a complex situation there are two definite, if not distinct, phases to the research. Firstly, there is the stage of identifying the problem and specifying it in terms of a structure, this structure usually being a set of mathematical equations which purport to describe the system under study. The second stage of the process then takes this model structure and uses it to derive the expected behavior of the system. Modelling methodology centered on the use of continuous system simulation languages (CSSL) thus focuses attention on the second part of the modelling process. This is not to say that the first phase is often omitted or understated since the construction of a structure always occurs; what is missing, however, is a systematic methodology to get from the statement of a problem in terms of a complex situation to a structure which represents that situation. This process is usually performed by

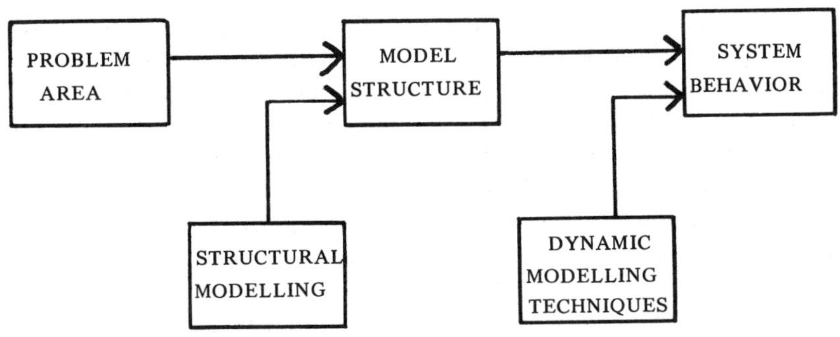

Fig.1 The model-building process

pragmatic procedures based upon common sense and conventional statistical methods. Whereas sophisticated computational techniques have been exploited to derive the system behavior from system structure, the process of obtaining the system structure does not usually benefit from the use of such techniques and from the available computational resources. Computational techniques which we have termed tools for *structural* modelling might play an analogous role to conventional dynamic modelling methods but at an earlier stage of the modelling process (as illustrated in Fig. 1).

It is important at this stage to say what we mean by a *structural* model. One possible definition of the term 'structure' is 'the way in which the component parts of the complex whole are interrelated; that which is made up of many component parts'. Structure in this sense is inherent in any mathematical model; all such models consist of components with interactions specified between those components. Thus, the essence of structural modelling is one of emphasis only; that is to say that a structural model focuses the interest of the modeller on the task of selecting the components of a model and explicitly stating the interactions between them.

What then are the areas in which we feel automatic techniques could contribute to this phase of the modelling process? To begin with, we feel that the attention of the modeller should be focused on the nature of the feedback processes which are built into a dynamic model. A second key area is the way in which these feedback processes combine and conflict to produce dominant patterns of model behavior. The final area is concerned with the division of model structure into sub-systems. The process of applying these techniques to a structural model should be essentially iterative; after each analysis the original specification of both the components and the interactions of the model can be modified according to the changed perceptions brought about by the analysis.

1.3 Summary of methodological background

In common with several 'systems theorists' we feel that dynamic models are fundamentally statements of system structure. It is also commonly asserted that our knowledge of the structure and internal motivation of complex systems is in general more reliable than our understanding of the implied behavior of such systems.

Whilst we agree with the assertion that the correct identification of a systems structure is an essential prerequisite to the modelling of that system; we also feel that our ability to discern system structure has been greatly overestimated. This conjecture is greatly reinforced by the quantity of criticism leveled at what many critics felt to be 'structural' inconsistencies and omissions in the early 'World Models' produced by the M.I.T. team. [4]

The overall objective of our work is thus the development of tools and techniques which improve our capacity to 'think about systems'. Our work is, in fact, directed towards two major application areas. Firstly, we feel that the examination of a simplified 'map' of a system would be a useful preliminary task in the process of building a full-scale dynamic model. The inclusion of this 'pre-modelling' phase would, we feel, assist a modelling group in making fundamental decisions about such structural features of the model as aggregation levels, choice of sub-systems structure, and inclusion of feedback mechanisms.

In addition, we feel that our methods could be used *directly* to gain knowledge about a system. That is to say that the modelling process illustrated in Fig. 1 could be terminated with the completion of a structural model. Such a model of a system is perhaps the most appropriate for the discussion of alternative policies involving structural changes in the system being studied. In fact, two major projects have been completed (described in greater detail in the next section) the OECD R & D [5], and the Roberts energy study [6] in which structural modelling was the sole methodology employed.

These projects were hampered by the lack of automatic techniques for the analysis of the model structures. Our development of such techniques will, we hope, revive interest in building structural models of complex systems.

2. Current state of the art

2.1 Background

It is hard to comment directly on the 'state of the art' since to some extent our research project is unique both in content and objectives. Nevertheless, there are a number of earlier pieces of research which have had considerable influence on the project, and a number of ongoing research projects which have much in common with our approach.

Much of the motivation for our earlier work on stability was derived from the field of ecological modelling. From the work of Gardner and Ashby [7] we first obtained the concept of an *interaction* matrix as a simplified 'map' of system structure and the idea that one can derive general system properties from such a 'map'.

In practice we have found that an interaction matrix is the most convenient method of representing the structure of a model. This matrix can thus be used as the basis of a structural model. Such an interaction matrix can be derived in a number of forms from a conventional dynamic model, and can, as mentioned in section 1.2, be used as a model in its own right.

Alternatively, we may represent the structure of a model by a *directed graph,* such a graph may often be directly derived from the 'causal diagram' or 'system flow chart' which is usually produced as a natural step in the modelling process. The graph presents model variables as nodes with the functional relationships between the variables indicated by directed links between the nodes. There is a one-to-one correspondence between such a graph and a matrix in which the interactions between system components or variables are displayed as non-zero element values.

Both the 'directed graph' and the 'interaction matrix' method of describing a system have been found to be extremely adaptable to a wide variety of levels of system representation. At one extreme when the interactions are specified by real numbers the matrix and graph correspond to a set of linear differential equations which completely determine the *local* behavior of the system. At the opposite extreme we may construct merely a 'binary connection matrix' or 'signed digraph' which specifies only the existence, non-existence and sign of links. Such a representation is, of course, an extremely simplified version of both the model and the real-world system that the model purports to represent. Nevertheless, such a system description embodies essential information concerning the structural features of the system. We should also mention a limitation inherent in our structural model representations that is that they are both only capable of describing the structure of *linear* models. This drawback is, however, far from serious for the following reasons. Firstly, many of the structural features of models are independent of the form of the functional relationship between variables. Secondly, most models can be reproduced in a linear form if a sufficiently small time domain is selected. Furthermore, the latter technique enables us to compare and contrast structural analyses of linear 'snapshots' of models at different time phases. Finally, we feel that the construction and analysis of a simplified linear 'map' of a model might be a useful first step in the eventual construction of a more sophisticated non-linear model.

We should also mention that one aspect of non-linearity *can* be incorporated into our feedback loop and behavior mode analyses, there we allow a separate time delay to be associated with each interaction in the model; these delays are stored separately in a *delay matrix.*

Before we go on to describe the 'state of the art' in this field, we ought to clear up a possible point of confusion by listing in Table 1 the equivalent terms, as defined above, used in the latter parts of the chapter:

Table 1 Equivalence between terms in various contexts

Equation set	Interaction or connection matrix	Graph
Variable	Component	Node
Functional Relationship	Matrix Element	Edge
Partial Derivative	Element Value	Edge Weight
Feedback Loop	Feedback Loop	Cycle

Following the pioneering work of Gardner and Ashby many ecologists have followed up their attempt to derive a simple relationship between the stability properties and the degree of inter-connection of large dynamic systems — a relationship with particular significance for the theoretical basis of ecology.

Many ecologists are still working in this field, the work of May [8] and Holling [9] being of particular interest. Both our early work and the continuing work of the ecologists has leant heavily on the body of classical stability theory and we continue to draw inspiration from this area.

With regard to the present phase of our research project the fields of study upon which we have drawn are more diverse. We can categorize them into theoretical work, general application areas and particular projects. These are discussed in the following sections.

2.2 Theoretical approaches

2.2.1 There is a considerable body of mathematical theory dealing with the abstract notion of 'structure'. Its preparation has been only partially motivated by the belief that the knowledge of abstract structure will eventually be of value to investigators interested in various kinds of empirical structure; much of the work has remained 'pure' mathematics. This work is epitomized by the book on structural models by Harary, Norman and Cartwright [10]; here the authors develop a vast number of theorems derived from the theory of directed graphs; many of these theorems, however, seem to lack any interpretation in the field of empirical analysis. Nevertheless, the book is an extremely useful reference work for the utilisation of graph theory to analyze the structure of models.

More recent work in this field has been directed towards empirical and operational ideas. In particular, the work of Atkin [11] and Schofield [12] is concerned with the application of mathematical ideas — simplices and cliques respectively, to real world problems. These authors have, however, proceeded to the opposite extreme of using one particular simple technique on problems of great complexity in contrast with the above authors' approach of using techniques of great sophistication to analyze very simple problems.

Whilst we have borrowed much of the work of these mathamaticians we have been very careful to choose several operational techniques and to match each technique to a particular problem area. This eclectic approach avoids the pitfalls of theoretical sterility and the dangers of over-use of a single conceptual tool.

2.2.2. The above techniques have, of course, not escaped the attention of computer scientists, there have been numerous attempts to operationalize the above mathematical ideas and many published algorithms are available.

Most algorithms produced in the computer science literature are far from being immediately usable, they are concerned with a technique — leaving the problems of data input and provision of output in an intelligible form to the user. We have accordingly re-structured most of the algorithms which we have incorporated into our research program in a form suitable for both standard data input and more easily interpreted results.

2.3 Application areas

2.3.1 One area in which structural modelling has been developed to a considerable extent is geography. Geographers have long been interested in, for example, the structure of natural watercourses and have utilized graph-theoretic concepts as tools for understanding such structures. There has also been much work on man-made networks, such as road and rail systems which has relied on network theory and matrix analysis. Chorley and Haggert [13] have recently produced an excellent survey of work in this area and one of our techniques, matrix powering, has been derived from this source. More recently, geographers have also become interested in the problem of sub-system identification in physical networks (see Harvey and Auwerter [14]).

2.3.2 Whilst not explicitly stated as such, statisticians have long been concerned with the problems of sub-system identification. In fact a fundamental problem of statistics is the identification of natural clusters or groupings in objects according to certain criteria which determine the similarity of one object to another. Attempts to solve this problem have resulted in a large body of literature on the subject of 'cluster' analysis, Everitt [15] provides a useful recent survey of this field. The determination of the sub-system structure of a model is a particular case of this more general problem; in which the objects to be grouped are the components of a system and the similarity between components is determined by their degree of interaction.

Although the principles of cluster analysis may be presented in such a simple fashion, there are many practical difficulties involved in its use in any area. There are two major reasons for this:

1. the many alternative measures of similarity which can be derived from a set of observations
2. the absence of a definition of a cluster

the latter being more crucial to our work.

At this stage of the program we have selected one of the many available clustering algorithms which seems to be most suited for our problem. As with other available algorithms we have modified it to allow for a standard input/output interface which is consistent with our overall methodology.

Part of our ongoing research involves the monitoring of recent developments in cluster analysis in order to select and adapt any new techniques which prove to be of interest.

2.4 Other research projects

2.4.1 Historically there has only been one other research program which has shared our objectives and methodology to a large extent. This was the work of Geradin [16] whose report (in French) on 'Cybernetic Studies of Complex Systems by means of Topological Structural Analysis' did much to stimulate interest in the field, particularly in France. Geradin, like us, was concerned with structural information which could be derived from a simple interaction matrix representation of a complex system. Unfortunately, the only automated method of analysis which he developed was the use of matrix powering for the determination of critical components of a system. This work, whilst somewhat limited in scope, nevertheless encouraged the development of two other studies of relevance to our work. These are described in the following sections.

2.4.2 Duperrin and Godet [17] combined Geradin's concept of an interaction matrix with a well established forecasting technique – cross impact matrices (see Gordon and Hayward [18]). The first step in their procedure is to build an interaction matrix (see Section 2.1) of the variables required for forecasting in a particular problem area. This matrix is specified in terms of unsigned binary relationships, 1 representing the existence of a relationship between two variables, 0 signifying no relationship. The MICMAC method of Duperrin and Godet then uses matrix powering (see Section 3) to rank variables in order of the importance of their indirect effects on the other variables of the system. The method then selects an arbitrary number, say 20, of the most 'critical' variables selected by this procedure and these variables are sorted into a smaller number of major categories (say 6). Next an event is selected which represents in some way the development of the variables in each major category. The next step is to derive, by expert concensus, the probability of the occurrence of each event both in isolation and with reference to the occurrence or non-occurrence of each other event. Duperrin and Godet then use the SMIC 74 method to render this collection of probabilities consistent. Once this has been achieved the probability of 'scenarios' based on the six events can be calculated and the most probable scenarios analyzed.

This technique has obtained considerable popularity in France and has been applied to forecasting in the field of nuclear energy and air transport. (See [19], [20]). The principal weaknesses of their approach are basically the lack of information contained in the original unsigned binary interaction matrix and the number of arbitrary assumptions which need to be introduced at each stage of the method.

In contrast, the results produced by the Duperrin and Godet method appear to be positive and definite. The chief danger with these techniques is that the computational methodology can be used to conceal the inherently arbitrary nature of the method and the lack of adequate theoretical understanding.

2.4.3 Another large scale study to some extent generated by the Geradin report was the OECD [5] attempt to build a structural model of the United States Research and Development System. The model-building was preceded by a phase of exhaustive conventional analysis and the final structural model in the form of an interactive matrix encapsulated a great deal of both theory and data. Unfortunately, the OECD team were hampered by the lack of tools for guiding the analysis of the completed model, their analysis being confined to a discussion of various feedback loops identified by inspection. Thus the model has never really been adequately analyzed. As part of our research we have produced some limited and tentative analyses of the OECD model and we hope that our work will eventually revive the interest of the OECD in extending and developing the model.

2.4.4 Roberts [6] has for some time now been developing structural models to analyze future demand for energy. Unlike our work he pays a lot of attention to the problem of deriving an interaction matrix representation of a system from a collection of expert judgements; for him this problem is separate from the problem of analyzing the model once it has been constructed. For us the first problem does not exist since we believe that the construction of a model is an ongoing process synchronous with the analysis of that model.

Roberts' analyses depend on a rather cumbersome use of various mathematical theorems (some taken from [10]) concerning various stability properties of models. We describe the techniques as cumbersome since little use is made of automatic analyses and mathematical sophistication and extreme ingenuity substitutes for computational power.

Since his analysis relies on rigorous mathematical theorems, in his early work he limited his model structures to 'signed digraphs', that is, models specified in terms of signed binary numbers only. The limitations of this kind of model for behavior mode analysis are pointed out in another of our publications [3]. We have accordingly automated his procedures so that we can deal with models in which weights can be given to various links in a system. He has recently extended his methodology to include quantitative data but still relies on manual (non-computerized) techniques of analysis. (See [21]).

2.4.5 The work of Kane [22] is very similar in both content and purpose to that kind of structural modelling for which we have argued in the first section of this chapter. Kane has produced a combined methodology/programming language to deal with cross-impact matrices which represent the structural dynamics of a system. His computer language, called KSIM, is designed as an *alternative* to existing dynamic simulation languages with the emphasis placed on ease of use and the ability to quickly assess the implications of structural changes. KSIM, however, provides no facilities for structural *analysis* to guide the improvement of models under construction. Furthermore, KSIM has another fundamental limitation built in to the method by which system behavior is calculated. Kane makes the somewhat arbitrary assumption that all functional relationships expressed in the model take the form of a logistic curve; that is to say that all state variables in a system described by a set of KSIM equations are only allowed to vary between an explicit upper and lower bound. Now whilst it is true that as Kane asserts; "Such growth and decay patterns are characteristic of many economic, technological, and biological processes" [23] it is also undoubtedly true that many of the variables commonly included in models of social and economic systems behave as if no such bounds existed.

Whilst we are in full agreement with Kane's diagnosis of the need for a systematic methodology for structural modelling we also feel that KSIM, due to the inherent limitations mentioned above, is not really adequate for the task.

2.5 Summary of 'State-of-the-art' review

In the above review we have detected what we believe to be all the ingredients for a systematic structural modelling methodology. We have seen, however, that these necessary ingredients have rarely been brought together by a single research project or within a single discipline. The wide dispersal of the techniques is especially tragic in view of the potential for a synergistic application of several techniques together in conjunction with a single application. In the following section we go on to describe how we have attempted to remedy this situation in our own research program.

3. Techniques for the analysis of system structure

Lack of space in this chapter prevents us from providing a full technical discussion of each of our techniques. We can do little more than give a brief outline of the fundamental principles and capabilities of some of the available forms of analysis. Further information is available in the publications referenced in the first section. [1,2,3].

Several of the basic concepts of many of our techniques have been derived from the work discussed in the previous section. We feel, nevertheless, that we have fulfilled a useful function in such circumstances by bringing together various methods from a variety of disciplines in a standard and operational format.

The various techniques described here have been implemented as segments of a fully operational user-oriented computer program. The program is designed to operate in interactive mode in conjunction with a visual display unit. Each type of analysis uses a common data set representing the model structure and can be initiated by a simple command and parameter entry from a terminal. The program is currently able to run on both IBM 370 and ICL 1900 series machines. This computer program has been developed at the Science Policy Research Unit by a team consisting of the authors. We would like to acknowledge the financial support and intellectual stimulus of the IBM United Kingdom Scientific Centre, Peterlee, Co. Durham.

3.1 Identification and analysis of feedback structure

Even in relatively simple models it is not easy to identify all of the feedback processes even from a causal loop diagram representation of the model. We have, therefore, implemented a program to extract these feedback loops from an interaction matrix representation of a system, this program was based on an algorithm by Syslo (see [24]).

Using a graph theoretic approach, a feedback loop can be regarded as a sequence of variables or path,

$$P = i_1, i_2, \ldots, i_k, \qquad \text{where } i_1 = i_k$$

that is, a closed path. The graph G, see Fig. 2, contains the simple circuits:

$$C_1 = x, y, x \qquad C_2 = y, z, y$$

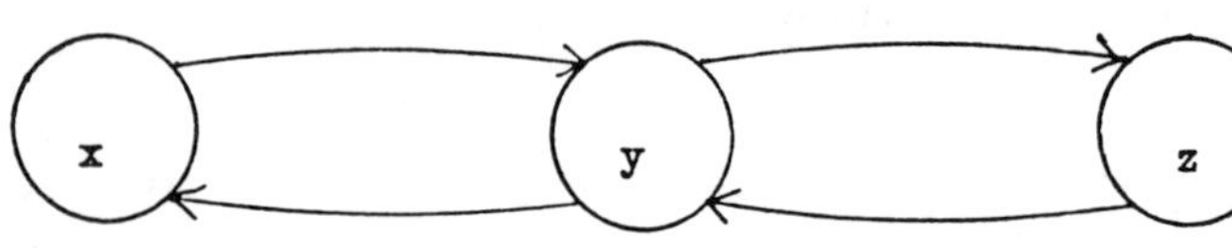

Fig.2

A cycle can be said to possess four properties:

- length – number of variables on the path,
- strength – product of the edge weights or values,
- delay – sum of the edge delays,
- sign – the sign (+ or -) of the strength.

Because of the large number of feedback loops present even in small models it is necessary to establish a filter on those listed as an aid to interpretation. This is done by means of user defined, optional, constaints which can include the four properties mentioned above and statements concerning the variables or relationships in which one is interested. Using combinations of these constraints it becomes possible to identify precisely the feedback processes under study.

The dominant behavior mode of dynamic models is determined by the conflict between the various feedback processes built into the model, continuous simulation languages provide one method of determining this behavior. Dynamic models based on the use of CSSLs are, however, fairly cumbersome and difficult to alter. Consequently it is often difficult to determine changes in behavior mode which can be brought about by minor alterations to the feedback structure of the model. We have developed a technique, called pulse analysis, for which the work of Roberts [6] has provided a theoretical framework.

Our pulse analysis program enables the modeller to regard a weighted interaction matrix, A, as a representation of a set of dynamic equations which govern the behavior of the system.

Thus, we treat the interaction matrix as a simplified form of a dynamic simulation model. The program allows the user to interactively establish the initial conditions and the pulses to be applied, the model is then run for a specified time period and the results produced in a graphical form on the terminal. Thus, we can tell whether a pulse in a certain input variable leads to exponential growth in a certain output variable, to exponential decline, to a relatively stable or cyclic pattern of behavior. It is possible, therefore, not only to view the effects of internal and external perturbations but also the effects of changes in the model structure.

Our pulse analysis technique is thus very similar in use and concept to KSIM (see 2.4.5); we can, however, claim two advantages for our technique. Firstly, we do not assume that all variables are bounded, the matrix elements in our interaction matrix in fact correspond to true partial derivatives which means that classical stability criteria can be applied to mathematical analyses of the matrix properties. Furthermore, for us pulse analysis is just *one* of a series of techniques for the analysis of system structure; used in conjuction with the other techniques (especially feedback analysis and the identification of critical components) described here, its power is considerably enhanced. This technique of pulse analysis is also a considerable advance over the earlier work of Roberts in which his analysis was limited to the determination of all or nothing stability properties or inherently unrealistic models due to his reliance on manual computational techniques.

Even Roberts' later work which can deal with interaction specified in terms of real numbers still relies on strict mathematical methods of analysis. Thus, pulse analysis combines the ability of KSIM to display the principal behavior modes of models with the flexibility and mathematical rigor of the work of Roberts.

3.2 Critical components

A second key area is the existence of critical components within the model; whilst some components of the model may obviously be critical in the sense that their quantitative interactions with other components are highly significant, other critical components may be quantitatively insignificant but qualitatively and interactionally very significant. Matrix powering is one tool which is capable of providing such an insight.

Using the interaction matrix, A, a *total connection matrix* (TCM) may be constructed by summing the powers of A up to m, (where m is the longest chain of connection within the model). The row sums indicate the importance of the components as a source of change whereas the column sums represent the component's sensitivity to changes elsewhere in the model. The detailed contents of the TCM can also be used as an indication of the importance of the *overall* interaction between pairs of components within the model. These properties of the TCM have also been exploited in a different context by Duperrin and Godet [17]. As we mentioned above the use of matrix powering to locate the critical components of a model can be a useful guide for the use of the pulse analysis technique mentioned in the previous section.

As a first step a pulse analysis can be set up with perturbations being applied to those components which are the most significant sources of change in the model and the components which act as sinks can be output as the most sensitive indicators of the behavior mode of the system.

3.3 Sub-system identification

The final area in which we believe that the use of computational techniques can help to focus attention of the adequacy of model structure is concerned with the division of a complex model structure into sub-systems. It is often convenient both for the planning and implementation of the model to divide the structure into smaller units. A large complex model divided into several sub-systems is far more easy to communicate than a complex model taken as a whole. Although most dynamic modellers have in fact divided their models into sub-systems there has been very little work on evolving a rigorous method of achieving this division.

When a system is represented in the form of an interaction matrix the numbering order associated with the variable set is purely arbitrary. One approach to finding the sub-system structure of a model is to re-arrange the rows and columns of the interaction matrix so that groups of components which interact strongly with each other are placed close together. We have developed an algorithm which exploits this property of the matrix which has already demonstrated some success in the identification of syb-system structure.

Alternative approaches can be devised based on the work of Atkin [11] in which the mathematical concept of a 'simplex' can be associated with a subsystem or on the work of Schofield [12] in which an analogous concept, the 'clique', can be similarly utilized. Schofield's 'clique' corresponds closely to a concept of Roberts' which he terms a 'strong component', a notion which plays an important role in his own methodology.

As mentioned in section 2.3.2 there seems to be considerable potential in the development of cluster analysis techniques to attack this problem. So far we have been unable to identify an algorithm which is entirely satisfactory for this purpose. A full account of our research in this area is given in section 4.4 of our working paper [3].

4. Results and conclusions

During the course of their development we have repeatedly tested our techniques on the structures of actual models. Whilst our detailed findings are reported elsewhere, we can say here that we have been sufficiently satisfied with the results obtained so far as to embark on a full scale program of further development of our ideas combined with a commitment to apply our existing techniques to a number of models at present under construction.

In conclusion, we would also like to emphasize the underlying rationale behind our choice of objectives. We feel strongly that many types of simulation modelling have a vast potential contribution in the field of planning and policy making. Nevertheless, many planners and policy makers are, perhaps justifiably, suspicious of such techniques. We feel that structural modelling can contribute to this problem in two ways.

Firstly, we feel that using structural modelling in parallel with the building of a dynamic model can demonstrate openly a systematic approach to devising the structure of the model. Often at present potential users of such models are presented with the system structure as a 'fait accompli'. The use of interactive techniques to build an interaction matrix means that the decision maker could be involved at an early stage of model building. Constructing an interaction matrix requires little technical skill; the planner can suggest components for the model in terms of his own priorities and objectives, and interactive programs can highlight the structural consequences of the inclusion of specified variables and interactions.

Secondly, the use of structural models in certain situations as an *alternative* to dynamic modelling can certainly improve the chances of such a model being successfully communicated. A systemic structural model is easy to understand because of its simplicity, it forces the model builder to justify the relationships expressed in the model; he cannot hide in the convenient jungle of complex numerical relationships incomprehensible to all but experts on models. A structural model compels open discussion of the crucial relationships relating to understanding the future – the *structure* of the system.

References

1. McLEAN, M., SHEPHERD, P., and CURNOW, R.C., 'Systematics—The Study of the Behavior of Large Complex Systems', Proceedings of the 2nd International Symposium on Mathematical Modelling, Jablonna, Poland (1974), in press.
2. McLEAN, M. and SHEPHERD, P., 'The importance of system structure', *Futures*, **8**(1), (February 1976).
3. McLEAN, M., SHEPHERD, P., and CURNOW, R.C., 'Techniques for the Analysis of System Structure', Occasional Paper Series of the Science Policy Research Unit, No.1, University of Sussex (February 1976).
4. FORRESTER, J.W., *World Dynamics*, Wright-Allen, Cambridge, Mass. (1971).
5. Organisation of Economic Co-operation and Development, *The Slowdown in R & D Expenditure and the Scientific and Technical System,* Paper No. SPT (74) 1, Paris, OECD (1974).
6. ROBERTS, F., *Signed Digraphs and the Growing Demand for Energy,* The Rand Corporation, R-756-NSF (1971).
7. GARDNER, M.R. and ASHBY, W.R., 'Connectance of large dynamic (cybernetic) systems: critical values for stability,' *Nature* **228** (21 November 1970).
8. MAY, R.M., *Stability and Complexity in Model Ecosystems,* Princeton University Press, Princeton, N.J. (1973).
9. HOLLING, C.S., 'Resilience and stability of ecological systems', *Annual Review Ecological Systematics* **4**, 1-23 (1973).
10. HARARY, F., NORMAN, R.Z., and CARTWRIGHT, D., *Structural Models—An Introduction to the Theory of Directed Graphs,* John Wiley, New York (1965).
11. ATKIN, R.H., *Mathematical Structure in Human Affairs,* Heinemann, London (1974).
12. SCHOFIELD, N., 'On structure in an organization: an empirical example', mimeo, Department of Government, University of Essex, undated.
13. HAGGETT, P. and CHORLEY, R.J., *Network Models in Geography,* Methuen (1969).
14. HARVEY, M.E. and AUWERTER, J., 'Derivation and decomposition of hierarchies from interaction data', Department of Geography, Discussion Paper, No.31, Ohio State University (1969).
15. EVERITT, B., *Cluster Analysis*, Heinmann (1974).
16. GERADIN, L.A., 'To forecast decision-making with system analysis—systemic analysis to forsee alternative futures', Thompson-CSF, Report No. 523/74 (1974).
17. DUPERRIN, J.C. and GODET, M., 'System Forecasting using a new method of Cross-Impacts', METRA Report, Vol.XII, No.4 (1974).
18. GORDON, T.J. and HAYWARD, H., 'Initial Experiments with the Cross-Impact Method of Forecasting', *Futures* **1**(2), (1968).
19. DUPERRIN, J.C. and GODET, M., 'Methode de Hierarchisation des Elements d'un Systeme: Essai de Prospective due Systeme de l'Energie Nucleaire dans son contexte societal', Rapport Economique C.E.A. R 4541 (December 1973).
20. GODET, M., 'Scenarios for Air Transport up to 1990: strategic analysis using the SMIC 74 method' *Metra*, Vol.XIV, No.1 (1975).
21. ROBERTS, F.S., 'Weighted digraph models for the assessment of energy use and air pollution in transportation systems', *Environment and Planning*, A 7, pp.703-724 (1975).
22. KANE, J., 'A Primer for a New Cross-Impact Language—KSIM', *Technological Forecasting and Social Change,* Vol.4, pp.129-142 (1972).
23. IBID., p.133.
24. SYSLO, M.M., 'The elementary circuits of a graph, Algorithm 459', *CACM,* **16**(10), pp.632-633.

FUZZY MATHEMATICS AND FUZZY SYSTEMS

Introduction

H.J. ZIMMERMANN
University of Aachen, FRG

The European Working Group on Fuzzy Sets was founded in January 1975 as one of the working groups of the Association of the European Operational Research Societies within IFORS. It meets approximately twice a year in order to promote the advancement of theory and applications of Fuzzy Sets and to improve the communication in this area.

It was highly appreciated that the organizers of the Third European Meeting on Cybernetics and Systems Research accommodated the meeting of the working group, which at the present has about 100 members.

The scope of the papers presented in the framework of the third meeting of the group—at the same time Symposium K of the congress—was very wide spread. This seems to be characteristic of this area at the present time. While theoretical developments are still going on in mathematical areas such as topology, logic, optimization, control-theory, clustering, etc., applications can be found in the areas such far apart as medicine and approximate reasoning or fuzzy document retrieval and the application of fuzzy controls to production processes.

For obvious reasons the discussion about the relationships between Probability and Fuzziness are still of great interest. Therefore a penal discussion on this topic was arranged at the end of the meeting. This discussion—based on two principal contributions by Dr. Arigoni (Italy) and Dr. Sagaama (France)—seemed to be of considerable interest judged by the large number of participants.

After more than two hours of very fruitful discussions the following conclusions had to be drawn:

1. Neither a definition of a fuzzy set nor that of 'probability' is unique.
2. Fuzzy sets have to be compared with probabilities on several levels: On an axiomatic level, on a semantic level and probably on a philosophical level.

The participants agreed that the problem could not be solved in the framework of the penal discussion but that a great deal of clarification was reached.

It was also noticed that the audience of the Vienna meeting (similar to the meeting in Bucharest) was very heterogeneous with respect to knowledge in the area of fuzzy sets. While some persons are mainly interested in an introduction to the area of fuzzy sets others are already specialists in this area and looking for original and new contributions. Because of this difference in aspirations it was decided that the next meeting of the European Working Group on Fuzzy Sets, which will take place in the framework of the EURO II conference in Stockholm, from November 29-December 1, 1976, shall have the following structure:

(*a*) 1/2 tutorial-type contributions to acquaint persons interested in this area with the basics of fuzzy sets.

(*b*) Special original contributions concerning advancements of theory and applications in this area.

(*c*) Another–well-prepared–penal discussion on fuzzy sets vice versa probabilities.

It was hoped that this penal discussion could achieve a final clarification of the interdependencies and the differences of these two areas.

Membership characteristic function of fuzzy elements: fundamental theoretical basis

ANIO O. ARIGONI
Istituto di Matematica, Università di Bologna, Italy

1. Introduction

The membership to a subset, of the elements forming the set which includes the subset itself, usually has no binary value. In the ordinary meaning of set, often such a membership's value ranges in the real interval [0, 1].

This fact was enucleated by J. Lukasiewicz in discussing its '3-valued logic', in 1920 [1]. The same line of thought had been followed in Zadeh's work 'Fuzzy sets' issued in 1965 [2].

By the latter, a general theory analyzing the particular type of sets which may be formed by the aforesaid elements was introduced.

Subsequently, the theory in subject was developed by receiving contributions from Zadeh himself and from other workers [3, 4, 5].

Such a theory has now assumed elegant aspects and extended possibilities of application could be prospected to it [6, 7]. The first applications of it are, for instance, in Systems Theory [8], in many valued Logics [9], and in other disciplines.

This, notwithstanding in the same theory the values of membership of the single elements, to the subset including the, are left to subjective determination [1, 3]. However the informal character deriving to it from the anthropomorphic aspect, just as subjectivity is, reduces the acceptability of it.

Essentially, the aim of introducing 'fuzzy sets' is, in fact, to allow the evaluation of the measure of a subset of a given set, when the inclusion of the single elements in such a subset derives from having defined them by natural language. Thus, in those branches of Mathematics such as Probability, Theory of Information, Theory of Dynamic Systems, and others which, in applying the theory of the function of a real variable, utilize the concept of 'measure of a set', the application of fuzzy sets theory results to be censurable. This, in that, the last may lead to a measure which, substantially, is subjectively determined.†

The subject of the present chapter represents the basic part of a more extended work. In this last, the aim of evaluating quantitatively the membership of the single elements of a set to a specific subset of the former is pursued.

†An analogous case exists: the censure, substained by the logicians, which was made to the Bourbakian definition of set. In this last, reference to the possession of the properties by the elements of the set itself is made conceiving the same properties as qualities of the elements, which are subjectively interpreted [10].

The above stated argument concerns the fundamental criteria which the entire work is developed upon. The same criteria consist as follows: analysis of a set's single elements, hinged on considering the properties characterizing the elements themselves; evincing those of such properties which result to be irrelevant to a specific classification of the elements; then, those of these last forming a set which results equivalent to that given, for the classification itself, are evinced.

In papers which will follow, the concept of 'content' of an equivalent set, as referred to the given one, will be introduced as succeeding to that of cardinality. Finally, the characteristic membership of the elements function will be determined. This will derive from the mentioned content of a set which will be, in fact, consequent: to the relationship between the content of each single element of a set and the content contribution yielded by this last to the set equivalent to that given.

In so doing, the membership functions in subject will be calculated by keeping into account objective data exclusively; thus, any ambiguity which is involved in subjective evaluations will be avoided.

2. A preliminary consideration

As already hinted in the introduction, the sets of elements related to the natural linguistic expressions are, in general, infinite and uncountable. These characteristics derive from the practically infinite quantity of details which can be described by the natural language. These, intersecting each other, become shaded to such a degree that they constitute a 'descriptive continuum'. The rational representation of this last can be pursued by considering an infinity **X** of 'capsulated' elements; elements countable and differentiating one from the other for at least one of the properties thereof. Such a representation is analogous to that which can be made for real numbers through the rational ones.

In so doing the elements of **X** are characterized by a countable infinity **P** of properties susceptible of a definition in **X**; i.e. the possession of which can be *decided* for every element of **X** itself.

In the case that the considered properties form the finite set $\{P_h : h = 1,2,...,l\}$, as for rational reasons it can be required, then every P_h partitions **X** into two classes: that which includes the elements having the property P_h, and that including the elements which the property P_h itself has not. Thus, as a function of the l properties, a finite set X of subsets is derived from **X**. Each element of X is formed by an infinity of elements of **X** itself. Such elements are joined in common by a specific disposition with repetition of the logical values of the properties P_h $(h = 1,2,...,l)$†. In addition, the same elements differentiate one another for one or more values that the $|\mathbf{P}| - l$ infinite properties, which are not definable in X, take in them. X is the set $\{x_\alpha : \alpha = 1,2,...,m;\ m = 2^l\}$.

From that it derives that each single element $x_\alpha \in X$ may also be considered as a set of finite whatever quantity q_α of elements; these are said to be *single components* and denoted by $x_{i\alpha}$; thus, $x_\alpha = \{x_{i\alpha} : i = 1,2,...,q_\alpha\}$.

Any strictly proper subset $x_{a'} \subsetneqq x_\alpha$, $\{x_{i\alpha} : i = 1,2,..., q_{a'} \lneqq q_\alpha\}$ is said to be a *virtual element* of X.

Each element $x_\alpha \in X$, and likewise the single components forming it, $x_{i\alpha}$, may be analyzed in the logical values of the properties P_h characterizing them; in such a case, the same values are denoted by $P_{h\alpha}$ and $p_{hi\alpha}$ respectively. Therefore, it results that $x_\alpha = (P_{1\alpha}, ..., P_{h\alpha}, ..., P_{l\alpha})$ and $x_{i\alpha} = (p_{1i\alpha}, ..., p_{hi\alpha}, ..., p_{li\alpha})$.

3. The partition of X

Let the infinite countable set **X** be given, some *conditions* E_ν $(\nu = 1,2,..)$ may be stated on its elements. That is, propositions resulting from connecting, by the usual logical operators, atomic statements on the inherence of specific properties to the elements $x_\alpha \in X$ may be enunciated. Those elements which correspond with one, two, ... conditions form as many not necessarily disjoint sets S, T, ... finite, if finite is the number of the aforesaid atomic propositions.

Let us now consider the relation between each set S, T, ... and that including all the elements x_α which may be formed by combining the logical values of the properties P_h $(h = 1,2,..., l)$, i.e. the set $X = \{x_\alpha : \alpha = 1,2,...,m;\ m = 2^l\}$. From the analysis in subject, those sets which, as a consequence of the conditions determining them, equal to set X are left out.

†It is recalled that 'logical value of a property', in an element of a set, is agreed the denotation of either the inherence or not of the same property to the element in subject.

Lemma 3.1 Let the set X and a set S which is determined by a condition E_ν in **X** be given; iff the properties relative to the condition are among those P_h, S is a subset of X either proper or not.

Proof: The necessary and sufficient condition, so that X includes S is that $\forall x_\alpha \in S$ the value of the characteristic function $\psi_X(x_\alpha)$ can be decided. This happens only if the properties included in the condition in subject are definable in X, i.e. if they are identical to either a part or all those P_h. This last concept can be found, stated in different terms, in the Axiom of Specification of Set Theory (see, e.g., [11]). On the other side, if one or more condition's properties are not included among those P_h, the inherency of them to the single elements $x_\alpha \in S$ is not decidable, then neither the value of $\psi_X(x_\alpha)$ may be calculated, i.e. it can not be ascertained if either $S \subset X$ or $S \not\subset X$.

Theorem 3.2 A condition E_ν relative to properties included among those P_h determines a set S in **X** which is an equivalence class $C_j \subset X$ $(j = 1,2)$.

Proof: For Lemma 3.1 the set S is a subset of X. Therefore, $\forall x_\alpha \in S$ the function $\psi_X(x_\alpha)$ is not only decidable, its value, in addition, invariably is 1. As concerns the characteristic function $\psi_S(x_\alpha)$ related to every $x_\alpha \in X$, the same statement is not valid. This is because: for some x_α $\psi_S(x_\alpha) = 1$; for other elements, that is not the case and $\psi_S(x_\alpha) = 0$, i.e. $\psi_{\bar{S}}(x_\alpha) = 1$. Since $\forall x_\alpha \in X$ either $x_\alpha \in S$ or $x_\alpha \notin S$, as it is intuitive, then $S \cup \bar{S} = X$. Definitively, the condition determining $S \subset$ **X**, partitions X into two subsets S and $\bar{S}$; and these constitute two equivalence classes C_1 and C_2 of X.

Corollary 3.2.1 The condition E_ν determining S in **X**, introduces an equivalence relation $\mathbb{E}_\nu$ $(\nu = 1,2,..)$ in X.

The equivalence among the elements $x_\alpha, x_\beta, \ldots$ $\ldots \in X$ included in the same class C_j, is denoted by $x_\alpha \equiv x_\beta \equiv \ldots \in C_j$.

4. Irrelevance of the properties

Considerations on the logical values the single properties P_h take in the elements $x_\alpha \equiv x_\beta \equiv x_\gamma \equiv \ldots$ $\in X$ are now made. Let us have a set X in which an equivalence relation $\mathbb{E}_\nu$ has been introduced. The said logical values can be a significant matter as related to their relevance with respect to relation $\mathbb{E}_\nu$. Such a matter forms the subject of the Axiom of Irrelevance that follows.

Axiom 4.1 Let us have two equivalent elements of X, $x_\alpha \equiv x_\beta \in C_j$, such that $\exists!h : x_{h\alpha} \neq x_{h\beta}$ and $\forall k \neq h\, x_{k\alpha} = x_{k\beta}$; the hth property is *irrelevant* in them, with respect to $\mathbb{E}_\nu$.

Two elements $x_\alpha \equiv x_\beta \in C_j$ in which the hth property results irrelevant are said *h-coinciding*; this is denoted by $|\Delta(\alpha, \beta)| = 1$. Definitely, $\mathbb{E}_\nu$ makes the same elements to be equal to another element, $x'_\alpha \notin X$, such that: $\forall k \neq h\, x'_{k\gamma} = x_{k\alpha} = x_{k\beta}$ and $x'_{h\gamma} = *$ (* being the symbol signifying the irrelevance of the corresponding property).

The derivation of an element x'_γ, from two h-coinciding elements, is a summing operation allowed because the algebraic structure of the set including all elements of X united to those which can be obtained from X itself, just by the derivation in subject.

The derivation from all possible couples of h-coinciding elements in X, of the elements of the type 'x'_γ', is here called *representation* of X with respect to $\mathbb{E}_\nu$.

5. The equivalent set X'

The derivation of the elements x'_γ, the union of which forms a new set $X' = \mathbb{E}_\nu(X)$, consists in performing an analysis and a subsequent synthesis of the represented set's elements.

We call *analysis* of a set X, the disjoint tabular representation of all the logical values $p_{hi\alpha}$ which, respectively, form the elements included in the classes C_1 and C_2 of X. That is:

$$\{(p_{hi\alpha}), \ldots, (p_{li\alpha}) : i = 1,2,\ldots, q_\alpha;\ \alpha = 1,2,\ldots, |C_1|\}$$

and

$$\{(p_{hi\alpha}), \ldots, (p_{li\alpha}) : i = 1,2,\ldots, q_\alpha;\ \alpha = 1,2,\ldots, |C_2|\}.$$

What we call *synthesis* of the set X's elements, consists in deriving, from every couples of h-coinciding single components of the elements themselves which are included in the class either C_1 or, disjunctively, C_2, after the last has been analysed as it is described previously. In it an equivalent single element is derived from each of the said couples $x_{i\alpha} \equiv x_{i\beta}$: $|\Delta(\alpha, \beta)| = 1$; then, by grouping all the identical single components-images, new elements are obtained. These last since result by synthesizing etherogeneous elements—eventually virtual elements—are called *synthetical elements.* They are the images, $x'_\alpha \in X'$, of *h*-coinciding couples of elements $x_\alpha \in X$.

The process by which the representation of X is performed and the set X' is obtained, is not necessarily exhausted in a unique phase. The set X' may in fact include new couples of *h*-coinciding elements. In such an event, the process itself must be iterated on the elements x'_α of the classes C'_1 and, separately, of C_2, on their analogous forming the so obtained set X'', and so on. This, up to a complete dissipation of all possible *h*-coincidences. Of course, in the elements of X', of X'', ..., one, two, ... properties may respectively result irrelevant.

The logical values by which the *l*-tuples that each element $x'_\gamma \in X'$ consist of—which represent $x_\alpha \equiv x_\beta \in C_j$: $|\Delta(\alpha,\beta)| = 1$—those of the elements $x'' \in X''$—which represent $x'_\alpha \equiv x'_\beta \in C'_j$: $|\Delta(\alpha,\beta)| = 1$— and so on, form the set $\{0, 1, *\}$. The combinatorial product of l times the last set, $\{0, 1, *\}_h$, constitutes the set of all *l*-tuples forming the set $X = \{x_\alpha : \alpha = 1,2,...,m;\ m = 2^l\}$, that $X' = \{x'_\alpha : \alpha = 1, 2,...,m';\ m' < m\}$, that X'', ..., and, finally, X^{l-1}. About this last, it may be observed that its classes C_i^{l-1} and C_2^{l-1} both are singleton and in their unique element only one property P_h is relevant [13].†

The product-set above seen as indicated by $\vec{X}$; this to denote the structure, of particular linear space, that the product itself has, as it has been shown [13, 14].

The representation of X by a set $\vec{X}^h$ $(h = ', '', ..., ..., {}^{l-1})$ no elements of which is *h*-coinciding to others, is called *equivalent set* of X, with respect to $\mathbb{E}_\nu$; it is denoted by $\vec{X}^h$ to outline that it includes elements of the space $\vec{X}$. Formally: $\vec{X}^h = \mathbb{E}_\nu(X) \to \vec{X}$. The $\vec{X}^h$'s elements also are denoted by arrowing x', ...: $\vec{x}'_\alpha$, $\vec{x}''_\alpha$, and so on.

†As a matter of fact, the combinatorial product $\prod_{h=1}^{l} \{0,1,*\}_h$ includes an element in which all the components are '*'. This last would be obtained in a set X in which a class C_1 equals X itself and that $C_2 = \phi$: because such a partition of X is very particular and not deriving from an equivalence relation, this case is not considered here.

The described synthesis on the elements of X, which may be performed exclusively on elements included in the same class C_j of X itself, can be of three different types: 1st) simple elements synthesis (SES), it is that which is performed on two elements $x_\alpha \equiv x_\beta \in C_j$: $|\Delta(\alpha, \beta)| = 1$; 2nd) multiple elements synthesis (MES), it is that performed on more than two elements $x_\alpha \equiv x_\beta \equiv x_\gamma \equiv ... \in C_j$ such that one of them, let us say x_α, is *h*-coinciding to the others; 3rd) complex elements synthesis, is that performed on more than two elements $x_\alpha \equiv x_\beta \equiv x_\gamma ... \in C_j$ such that each of them is *h*-coinciding to one or more of the others.

The synthetical elements resulting from a SES, as the last is performed on real elements, are composed as denoted by: $x_{(\alpha \cup \beta)}$.

Those resulting from a MES, as the last necessarily is performed on virtual and real elements, are composed as denoted by: $x_{(a' \cup \beta)}$, $x_{(a'' \cup \gamma)}$, ... Finally, synthetical elements resulting from a CES, as the last is performed on any kind of elements, are composed either as the previous ones or as those denoted by: $x_{(a' \cup b')}$, $x_{(a'' \cup c')}$, $x_{(b'' \cup c'')}$, ...

7. Conclusion

By this chapter, the basis for an objective evaluation of the membership characteristic function of an element of a set X to a subset $S \subset X$ are set.

An analysis concerning the properties of a set's elements constitutes the preliminary argument. Through this, any set X is shown that can be regarded as a set of hypothetical sets: those formed by elements which differentiate one from another for properties not definable in X.

The partition of X by an equivalence relation $\mathbb{E}_\nu$ is then considered; this, to the end of individuating those elements which, being equivalent, may be regarded as identical if their objective differentiation results irrelevant with respect to $\mathbb{E}_\nu$ itself.

The above said consistence of X's elements offers the possibility of decomposing singularly the elements themselves and to form 'virtual elements'. This is utilized in synthesizing new elements: one for every couple of *h*-coinciding elements; these are identical in the sense specified previously. In so doing, a set X' which is equivalent to X—relatively to $\mathbb{E}_\nu$—is finally obtained; it includes all the synthetical elements derived from X, by representing X in $\vec{X}$. The inclusion in X' of elements in which one or more of their components—logical values of the properties—are not relevant, causes the 'cardinality' of X' itself to be succeeded by the 'content' of the same set X'. Such a content can result from the contribution that each single element $x_\alpha \in X$ gives to this last; it is that contribution which expresses the membership of the corresponding element, to the subsets of X, the element itself is caused to be included by the partition actuated by introducing $\mathbb{E}_\nu$. The determination of a set's content and of membership characteristic function of the elements will be subjects of future papers.

List of symbols used

X infinite denumberable set of elements-intervals

P infinite denumberable set of properties; those by which the **X**'s elements are characterized

X finite denumberable set of elements which are subsets of **X**; it is denoted by

$\{x_\alpha : \alpha = 1,2,...,m;\ m = 2^l\}$

x_α, x_β elements of X which are characterized by a finite set of properties

$x_{i\alpha}, x_{i\beta}$ single components of $x_\alpha, x_\beta, ...$ respectively

$x_{a'}, x_{a''}, ...$ virtual elements forming an element $x_\alpha \in X$

l quantity of properties by which the elements of X are characterized

$P_{h\alpha}$ logical value of the hth property ($h = 1,2,...,l$) in the element x

$p_{hi\alpha}$ logical value of the hth property in the single component $x_{i\alpha}$

$q_\alpha, q_\beta, ...$ quantity of single components of, respectively, the elements $x_\alpha, x_\beta, ...$

$q_{a'}, q_{a''}, ...$ quantities of single components forming the virtual elements $x_{a'}, x_{a''}, ...$ such that

$q_{a'} + q_{a''} + ... = q_\alpha$

$\mathbb{E}_\nu$ equivalence relation (ν—greek letter 'nu'— = 1,2,...)

C_j equivalence class ($j = 1, 2$) in which X is partitioned by $\mathbb{E}_\nu$

$|\Delta(\alpha, \beta)| = 1$ h-coincidence between the two equivalent element x_α and x_β of X

X^h set representing, respectively, $X, X', ..., X^{l-2}$ $(h = ', '', ..., ^{l-2})$

S canonical subset of X

S fuzzy subset of X

$\psi_S(x_\alpha)$ membership characteristic function of x_α to S

$\varphi_{\mathbf{S}}(x_\alpha)$ membership characteristic function of x_α to **S**

$\vec{X}$ particular type of linear space

$\vec{X}^h$ equivalent set to X

$\vec{x}_\alpha$ elements of $\vec{X}$ ($\alpha = 1,2,...,m^h;\ m^h = 2^{l-h};\ h = 1,2,..., l-1$)

$\vec{x}_{(\alpha \cup \beta)}, \vec{x}_{(\gamma \cup \delta)}, ...$ synthetical elements from a SES

$\vec{x}_{(a' \cup \beta)}, \vec{x}_{(a'' \cup \gamma)}, ...$ synthetical elements from a MES

$\vec{x}_{(a' \cup b')}, \vec{x}_{(a'' \cup c')}, ...$ synthetical elements from a CES

References

1. LUKASIEWICZ, 'Bermerkungen zu mehrwertigen Systemen des Aussa *genkalkuls*', Comptes rendus de la Soc. des Sciences et lettres de Varsovie, Classe III, vol.xxiii, pp.51-77 (1930).
2. ZADEH, L.A., *Fuzzy Sets*, Information and Control **8**, 338-353 (1965).
3. ZADEH, L.A., *Probability Measures of Fuzzy Events, J. Math. An. and Appl.* **23**, 421-427 (1968).
4. KAUFMAN, A., *Introduction a la theorie des sous-ensembles flous,* Masson et C., Paris (1973).
5. GOGUEN, J.A., *L-Fuzzy Sets, J. Math. An. and Appl.* **23**, 421-427 (1968).
6. CHANG, L., *Fuzzy Topological Spaces, J. Math. An. and Appl.* **24**, 182-190 (1968).
7. ZADEH, L.A., *Quantitative Fuzzy Semantics, Information Sciences* **3**, 159-176 (1971).
8. ZADEH, L.A., *Fuzzy Sets and Systems*, Proc. of Symposium on Theory, Pol. Inst. of Brooklin, N.Y., 29-39 (1969).
9. LAKOF, G., *Hedgs: A Study in Meaning Criteria and the Logic of Fuzzy Sets, J. Phil. Logic* **2**, 458-508 (1973).
10. BOURBAKI, N., *Theorie des ensembles-Fascicule des resultats,* Masson et C., Paris (1939).
11. HALMOS, P.R., *Naive Sets Theory*, D. Van Nostrand Co., N.Y. (1960).
12. ARIGONI, A.O. and BALBONI, E., *S-Spaces* (submitted to: *Information Sciences*).
13. ARIGONI, A.O., *Algebraic Structure of the Semiotic Dimension of Information,* Int. Congr. for Semiotical Studies, Milano (1974).

General fuzzy logics

B.R. GAINES
University of Essex, Colchester, UK

1. What is a fuzzy logic?

One aspect of the tremendous growth of interest in fuzzy systems and fuzzy reasoning [1, 2] is the development of *fuzzy logics* [3-12] and their relationships [11, 12] to standard multivalued logics (MVLs) [13]. I have noted previously [12] the varying usage of the term 'fuzzy logic' which may be classified in three broad categories:

(a) A basis for reasoning with vague statements. This general definition is consistent with the colloquial meaning of 'fuzzy', and also with its use in a technical sense different from, but related to [14, 15] that of Zadeh's 'fuzzy sets theory'.

(b) A basis for reasoning with vague statements using fuzzy set theory for the fuzzification of logical structures. This restricted form of (*a*) seems closest to Zadeh's own usage of the term 'fuzzy logic' [3] and his general use of the term 'fuzzy' as a qualifier.

(c) A multivalued logic in which truth values are in the interval [0, 1] and the valuation of a disjunction is the maximum of the disjuncts, whilst that of a conjunction is the minimum of the conjuncts. This narrow definition encompasses the population stereotype of a 'fuzzy logic' [4-9]. It is interesting that most infinitely-valued MVLs [13] have min/max connectives for conjunction/disjunction and hence are 'fuzzy logics' on this definition.

One may note some scope for confusion between these three definitions because:

(i) There is some disagreement about the basic conjunction/disjunction connectives of fuzzy logic [8, 9], i.e. neither (*b*) and (*c*) are necessarily accepted;

(ii) even if the min/max connectives of fuzzy logic are accepted in the sense of (*c*), the further connectives of equivalence/implication/negation are left undefined [7], or defined [3, 4, 5] or assumed [6], in different forms;

(iii) fuzzification of the classical propositional calculus (PC) in the sense of (*b*) gives a fuzzy logic in the sense of (*c*) but one with an inappropriate form of implication [10, 12, 6] in which the assertion of $A \supset B$ does not necessitate the degree of membership of B being greater than that of A;

(iv) for his models of truth values in human verbal reasoning Zadeh [3] fuzzifies in terms of (*b*) Lukasiewicz infinitely-valued logic [13] (here abbreviated to L_1), a logic which is itself a 'fuzzy logic' in the sense of (*c*).

There is no sense in which one would wish to legislate in favor of one of the three defintions—all are appropriate in their proper contexts. However, one may also note that at the level of definition (*a*), there has been much previous work on practical reasoning with vague data under the auspices of *probability theory*. I have previously suggested [10-12] both formal and semantic links between fuzzy and probability logics that provide foundation for a general *logic of uncertainty* encompassing these logics, many classical MVLs, and some modal logics.

The following section briefly presents such a logic of uncertainty, a general fuzzy logic, or a *basic probability logic* as I have previously termed it. Section 3 elucidates the effects of imposing the constraint of definition (*c*) upon the logic, and Section 4 gives a semantics for the general logic which illuminates the formal relationship between fuzzy and probability logics.

2. A general logic of uncertainty

To integrate together the various logics of uncertainty, including fuzzy and probability logics, it is essential to make a clear initial distinction between the (algebraic) structure of *propositions* and the ascription to these propositions to *truth values* (making them into *statements*). Indeed, it will be shown that fuzzy logics may be uniquely distinguished from other logics of uncertainty by the irrelevance, only in their case, of propositional structures when assigning truth values to compound statements.

The natural and conventional algebraic semantics for a propositional calculus is a lattice structure: $L(X, F, T, \vee, \wedge)$, generated by a set of elements, X, under two (idempotent, commutative) monoid operations, $\vee$ (disjunction), $\wedge$ (conjunction), with maximum element, T, and minimum element, F, i.e. L satisfies:

$$\forall\ x \in L \qquad x \vee x = x \wedge x = x \tag{1}$$

$$\forall\ x, y \in L \qquad x \vee y = y \vee x,\ x \wedge y = y \wedge x \tag{2}$$

$$\forall\ x,y,z \in L \qquad x \vee (y \vee z) = (x \vee y) \vee z,\ x \wedge (y \wedge z) = (x \wedge y) \wedge z \tag{3}$$

$$\forall\ x,y \in L \qquad x \wedge (x \vee y) = x,\ x \wedge (x \wedge y) = x \tag{4}$$

$$\forall\ x \in L \qquad x \vee F = x,\ x \wedge F = F,\ x \vee T = T,\ x \wedge T = X \tag{5}$$

the idempotent, commutative, associative, and adsorption postulates, together with a definition of the minimal and maximal elements [16]. The usual order relation is also defined:

$$\forall\ x,y \in L \qquad x \leqslant y \Leftrightarrow \exists z \in L : y = x \vee z \tag{6}$$

It is possible to make a case for weaker structures (e.g. dropping idempotency) but, for present purposes, this will be taken as an unreasonably wide generalization of our concepts of conjunction and disjunction. Now suppose that every element of L is assigned a 'truth-value' (for different applications different terminologies might be more appropriate, 'probability', 'degree of knowledge', 'level of belief', etc.) in the closed interval, [0, 1] by a continuous, order-preserving function $p: L \rightarrow [0, 1]$, with the constraints:

$$p(F) = 0,\ p(T) = 1 \tag{7}$$

$$\forall\ x, y \in L,\ p(x \vee y) + p(x \wedge y) = p(x) + p(y) \tag{8}$$

i.e. p is a continuous, order-preserving, *valuation* [16] on L. Note that, for p to exist, the lattice must be modular, and that we have:

$$p(x \wedge y) \leqslant \min(p(x), p(y)) \leqslant \max(p(x), p(y)) \leqslant p(x \vee y) \tag{9}$$

To complete the definition of an MVL one needs values for equivalence, $\equiv$, implication, $\supset$, and negation, $\bar{\ }$, of propositions. These may be defined naturally by noting that the equivalence relation on L, $\equiv$, defined by:

$$x \equiv y \Leftrightarrow p(x \vee y) = p(x \wedge y) \tag{10}$$

is a congruence on L, and that the deviation from equality in (10) defines a *metric* on L under this congruence [16]. Hence it is reasonable to define:

$$p(x \equiv y) = 1 - p(x \vee y) + p(x \wedge y) \tag{11}$$

as a measure of the degree of equivalence between x and y.

Implication may be defined in terms of equivalence by noting that, according to the usual lattice semantics, we require:

$$\forall\ x, y \in L,\ x \leqslant y \Leftrightarrow p(x \supset y) = 1 \tag{12}$$

but that for x, y satisfying this we have: $x \vee y = y$, $x \wedge y = x$. Thus it is natural to define implication as the degree to which these equivalences hold:

$$p(x \supset y) = p((x \vee y) \equiv y) = 1 - p(x \vee y) + p(y)$$
$$= 1 - p(x) + p(x \vee y) = p((x \vee y) \equiv x) \quad (13)$$

Negation may also be defined in terms of equivalence in the usual way:

$$p(\bar{x}) = p(x \equiv F) = p(x \supset F) = 1 - p(x) \quad (14)$$

I have previously called the logic thus defined a *basic probability logic* (BPL) because it satisfies the usual definitions of a probability logic (PL) [13], or probability over a language [17], except for the law of the excluded middle (LEM). In a BPL the law of contradiction and LEM are not necessarily theses, but if one is then so is the other, i.e. we have (from (8) and (14)):

$$p(x \vee \bar{x}) + p(x \wedge \bar{x}) = 1 \quad (15)$$

so that: $p(x \vee \bar{x}) = 1 \Leftrightarrow p(x \wedge \bar{x}) = 0 \quad (16)$

The form of implication in a BPL has the property, from (13), that:

$$p(y) = p(x \vee y) - 1 + p(x \supset y) \geqslant p(x) - (1 - p(x \supset y)) \quad (17)$$

which enables a lower bound to be placed on the truth value of y given those of x and $x \supset y$. Thus it satisfies the normal requirement [5, 6] that the assertion of $x \supset y$ may be used to infer that $p(y) \geqslant p(x)$, and hence also that $p(y) \geqslant \max(p(x_i))$ where y is constrained by 'rules' of the form $x_i \supset y$, a common pattern of inference in applications of fuzzy reasoning [18].

3. Truth functionality in BPLs

A BPL is not truth-functional (TF) in that the truth-values of the binary connectives, conjunction/disjunction/equivalence/implication, are not uniquely defined in terms of those of the two connected propositions. Note, that there is only one degree of freedom in that fixing the value of any of these connectives fixes that of all of them. There have been many debates in philosophical logic about truth-functionality but, particularly in the context of a logic of uncertainty, there seems to be no fundamental basis on which to demand truth-functionality, quite the contrary. However, a great deal may be learned, and many interesting logics derived, by considering various ways in which a BPL may be made TF:

(i) *A BPL with binary truth values is precisely PC.* If the truth values in a BPL are restricted to the end points of [0, 1] then it becomes TF with truth tables for all the connectives precisely as in the classical propositional calculus. Thus a BPL, and all its derivatives, are proper extensions of PC.

(ii) *A strongly truth-functional BPL is the 'fuzzy logic' L_1.* I have called a logic *strongly TF* [12] if there is a single equational definition for each of its binary connectives giving their truth values in terms of those of the connected propositions *regardless of the propositional structures* (e.g. having generating elements in common). The arguments of Bellman and Giertz [7] may be used to show that a strongly TF BPL is necessarily a 'fuzzy logic' (in sense (*c*) of Section 1), with the min/max bounds of (*a*) being attained.

Thus a fuzzy logic is a limiting case of a BPL and has the important computational property of allowing one to drop, without loss of information, the propositional structure of a statement and retain only its truth value. This property also holds for PC and hence it is a natural assumption in an MVL. However, its very strength is also its weakness because it implies that for any two statements, x and y, either $p(x \supset y) = 1$, or $p(y \supset x) = 1$, i.e. the lattice, L, reduces to a *chain* of propositions mutuall connected by implication. This very strong requirement is unlikely in general, although there are situations in which it becomes a reasonable hypothesis (see Section 4).

(iii) *A BPL with LEM is Rescher's probability logic.* Adding the law of the excluded middle (and hence also the law of contradiction) to a BPL gives a classical probability logic [13]. The PL is still not truth-functional but the demand for LEM makes it impossible for it to be strongly TF. In many practical cases the semantics require LEM and one is led to consider weaker forms of truth-functionality in which the computation of truth values of the binary connectives requires both the values *and* the propositional structure of the connected propositions:

(iv) *A BPL with LEM may be made truth-functional by an equational definition of the connectives for conjunction or disjunction which is commu-*

tative, associative and such that LEM or the law of contradiction applies, and is applied to pairs of propositions with no common elements. Notice that it is now essential to retain propositional structures in order to use the lattice laws and definition of a valuation to evaluate connectives in terms of pairs of components that have no common element. However, the resultant logics can now be made consistent with a far wider range of semantics, essentially now applying to the generating set of basic propositions, X, e.g.:

(v) *Assuming the truth-value of conjunction in X is the minimum of the truth-values of the conjuncts gives a 'fuzzy logic' in which the truth-value of a disjunction in X is the maximum of those of the disjuncts.* Thus min/max connectives are not incompatible with LEM. They imply that the generating propositions form a chain but that their negations form a separate chain, thus enabling LEM to apply.

(vi) *Assuming the truth-value of conjunction in X is the product of the truth-values of the conjuncts gives a logic of statistical independence.* Thus is the common assumption made by system analysts and engineers in order to make an uncertain system truth-functional.

(vii) *Assuming the truth-value of disjunction in X is the sum of the truth-values of the disjuncts gives a logic of mutual exclusion.* This is another common assumption, justified when the generating elements represent events, such as being in different states, that cannot occur together.

Thus a BPL may be made TF in a variety of ways of which only a few 'pure' examples have been given. In practice different propositions in the generating set may be connected in different ways and it makes more sense to reverse the definitions and consider *which* propositions are mutually exclusive, statistically independent, fuzzily related, etc., i.e. to classify the structure of the particular propositional calculus encountered in each practical situation. This concept will be further clarified in the semantic examples of the next section.

4. Semantics for BPLs

The close relationship established in Sections 2 and 3 between fuzzy and probability logics may evoke suspicion since we know that in many applications a fuzzy 'degree of membership' is most definitely *not* associated with a (physical) probability. For example, the man who has a degree of membership (dm) .5 to the class of short men or the woman who has dm .7 to the class of beautiful woman do not necessarily represent samples from a population (they may be the *only* people). Neither is there a sampling distribution in our own measurements that makes the man appear smaller than 5 feet on 50% of the occasions we measure his height—indeed, for beauty we possess no physical measuring rule!

Thus the formal relationship of probability theory to fuzzy logic may appear as spurious. However, this would be to adopt too narrow a view of probability theory, taking a strict 'physical frequentist' interpretation when there are well-established alternative semantics for probability in terms of 'subjective probability' [19-21], 'belief' [22, 23], etc., that are closely related to both classical and computational-complexity-based probability [24, 25]. There is a common semantics for all these interpretations of a BPL in terms of the binary responses of a population that shows that the formal relationships established are more than a mathematical artifice.

Consider a population each member of which can 'respond' to certain questions with a binary, yes or no, reply. The forms of question will involve evaluating a statement which belongs to the generating set, X, of a lattice, L, as defined in Section 2. For example, 'is this statement, $x \in X$, true or false, or reasonable or unreasonable, or generally believed, etc.'. The valuation of x is defined to be the proportion of the population replying yes to the question. A compound statement in L is given a valuation in terms of the proportion of the population who say yes to each of x and y for terms of the form, $x \wedge y$, or who say yes to either x or y for terms of the form, $x \vee y$, and similarly for more complex combinations of conjunction and disjunction.

This is essentially a set-theoretic model for L as a lattice of sub-sets of the population and (1) through (10) are clearly valid. A distance measure and hence valuations of logical equivalence, implication, and negation, may be defined as in (11) through (14). Thus, for any given population whose members are able to give one of two responses to a question about each element of X, there is a simple and well-defined procedure for ascertaining the valuation of any arbitrary statement in L, involving, conjunction, disjunction, equivalence, implication and negation, which is consistent with (1) through (14). Thus such a population is a model for a (distributive)

BPL.

Returning to the initial examples, one may now suppose that 50% of some test 'population' agree that the man is 'short' whilst 70% agree that the woman is 'beautiful'. If the 'population' was one of measuring instruments then the results express the effects of physical 'noise'. If the 'population' is one of people then this is a social acceptance model of linguistic usage, a reasonable model of Zadeh's 'fuzzy reasoning' based on human linguistic behavior. If the 'population' is one of 'neurons' then this is a model of individual decision-making. If we allow metalinguistic statements about the value of $p(x)$ to be made by members of the population then this is a model of 'subjective probability' or 'belief', and so on.

Consider now the additional constraints that must be placed upon the behavior of the population to correspond to result (iii) and (ii) of Section 3. Rescher's probability logic is obtained if someone who says 'yes' to x must say 'no' to $\sim x$. Lukasiewicz's L_1 is obtained if members of the population each evaluated the evidence for x in the same way but applied differing thresholds of acceptance. The member with the lowest threshold would then always respond with 'yes' when any other member did, and so on up the scale of thresholds, thus giving the required relation of implication between propositions. This model, although unusual, has its intuitive attractions, e.g. Reason [26] has shown that the threshold applied by human being in coming to a binary decision on an essentially analog variable seems to be associated with personality factors and a trait of the individual. If so, human populations would tend to show more a fuzzy, than a stochastic, logic in their decision making.

Similarly populations showing the 'statistical independence' of (vi) or the 'mutual exclusion' of (vii) may be defined. However, rather than argue the case for one type of population or another, one can now envisage that logics based on a real population will be of mixed type and hence it is more interesting to insert the concepts and talk in terms of a 'fuzzy', 'probabilistic', 'independence', 'exclusion', etc. *relationship* between propositions. Such relationships are mainly of interest to the extent that they are *necessary* and hence would appear as modal operators over a family of possible p's or L. In terms of our example so far it seems unlikely that anyone would argue for the logical necessity of semantics that make it possible to compute the truth value of 'the man is short and the woman is beautiful' on a truth-functional basis. However there would be reasonable grounds for the fuzzy TF of 'the man is short and he is not heavy'. A BPL and associated population semantics can encompass all these possible variants on a general logic of uncertainty.

5. Conclusions

In conclusion, there is no one logical system that stands out clearly as *the* logic of vagueness, uncertainty or fuzzy reasoning. It has been shown that a non-functional basic probability logic provides a formal foundation for a general logic of uncertainty encompassing both fuzzy and probability logics. Classical probability logic is obtained by adding the law of the excluded middle. The fuzzy logic L_1 is obtained by demanding strong truth-functionality. Since both LEM and TF have been subject to philosophical debate over many centuries one is unlikely to choose between them on general grounds!

However, it has also been shown that LEM is consisten with weaker forms of TF leading to partially 'fuzzy logics' (with min/max connectives between primitive propositions), logics of statistical independence, mutual exclusion, etc. The 'population model' semantics given show that these formal relationships between various logics of uncertainty carry over to an intuitively satisfying model of uncertain reasoning. The model also clarifies the distinction between fuzzy 'degree of membership' and conventional 'probability', showing it to be one of detailed semantic interpretation rather than one of logic or basic semantics.

Finally, the characterization of fuzzy logic as bieng *strongly TF* highlights its unique computational advantages. They are not so much ones of numerical simplicity (of min/max operations) as ones that stem from the *memory-reduction* possible through the irrelevance of propositional structures when computing truth-values. In any other logic of uncertainty it is necessary to know the actual structure (in terms of primitive propositions in the generating set) of propositions, x and y, when computing $x \wedge y$, $x \supset y$, etc., *whereas in fuzzy logic it is necessary only to remember the truth values, $p(x)$, $p(y)$.* Thus, regardless of whether fuzzy logic is *correct* in a given application, it is *easy to apply,* requiring a substantially lower memory load in generating or following complex arguments. This is not only practically important but may also be very relevant to the role of fuzzy logic in modelling human reasoning where memory resources are notoriously weak. It may, for example, explain Edwards [27] results that humans, whilst being good probability estimators, do not use the information efficiently in Bayesian computations (requiring a logic of statistical independence). A fuzzy logic is easier to apply, but is equivalent in this context to throwing away information.

Thus, the wider framework for logics of uncertainty described in this chapter establishes a close link between fuzzy logic and probability theory, to the mutual advantage of both fields. It also makes clear the unique computational advantage of fuzzy logic derived from its strong truth-functionality.

Acknowledgements

I am grateful to Joe Goguen, Ladislav Kohout, Abe Mamdani, and Lofti Zadeh for discussions relating to the theme of this chapter.

References

1. ZIMMERMANN, H.J., 'Bibliography: Theory and application of fuzzy sets', Lehrstuhl fur Unternehmensforschung, RWTH Aachen, West Germany (October 1975).
2. GAINES, B.R. and KOHOUT, L.J., 'The fuzzy decade: a bibliography of fuzzy systems and closely related topics', *International Journal of Man-Machine Studies* **9**, pp.1-68 (1977).
3. ZADEH, L.A., 'Fuzzy logic and approximate reasoning', *Synthese* **30**, pp.407-428 (1975).
4. GOGUEN .J.A., 'The logic of inexact concepts', *Synthese* **19**, pp.325-373 (1969).
5. LEE, R.C.T. and CHANG, C.L., 'Some properties of fuzzy logic', *Information and Control* **19**, pp.417-431 (1971).
6. LEE, R.C.T., 'Fuzzy logic and the resolution principle', *JACM* **19**, pp.109-119 (1972).
7. BELLMAN, R. and GIERTZ, M., 'On the analytic formalism of the theory of fuzzy sets', *Information Sciences* **5**, pp.149-156 (1973).
8. KOCZY, L.T. and HAJNAL, M., 'A new fuzzy calculus and its application as a pattern recognition technique', Proc. 3rd Int. Congr. WOGSC, Bucharest (1975).
9. SANFORD, D.H., 'Borderline logic', *American Philosophical Quarterly* **12**, pp.29-39 (1975).
10. GAINES, B.R., 'Multivalued logics and fuzzy reasoning', Lecture Notes of AISB Summer School, Cambridge, UK, pp.100-112 (1975).
11. GAINES, B.R., 'Stochastic and fuzzy logics', *Electronics Letters* **11**, pp.188-189 (1975).
12. GAINES, B.R., 'Fuzzy reasoning and the logics of uncertainty', Proc. 6th Int. Symp. Multiple-Valued Logic, IEEE 76CH1111-4C, pp.179-188 (1976).
13. RESCHER, N., *Many-valued logic,* McGraw-Hill, New York (1969).
14. GOGUEN, J.A., 'Concept representation in natural and artificial languages: axiom, extensions and applications for fuzzy sets', *International Journal of Man-Machine Studies* **6**, pp.513-561 (1974)
15. ARBIB, M.A. and MANES, E.G., 'A category-theoretic approach to systems in a fuzzy world', *Synthese* **30**, pp.381-406 (1975).
16. BIRKHOFF, G., *Lattice Theory*, Rhode Island: American Mathematical Society (1948).
17. FENSTAD, J.E., 'Representations of probabilities defined on first order languages', in J.N. Crossley, *Sets, Models and Recursion Theory,* Amsterdam: North-Holland, pp.156-172 (1967).
18. MAMDANI, E.H. and ASSILIAN, S., 'An experiment in linguistic synthesis with a fuzzy logic controller', *International Journal of Man-Machine Studies* **7**, pp.1-13 (1975).
19. SAVAGE, L.J., 'Elicitation of personal probabilities and expectations', *J. American Statistical Association* **66**, pp.783-801 (1971).
20. FINETTI, B. de, *Probability, Induction and Statistics,* London: John Wiley (1972).
21. MENGES, G., 'On subjective probability and related problems', *Theory and Decision* **1**, pp.40-60 (1970).
22. GROFMAN, B. and HYMAN, G., 'Probability and logic in belief system', *Theory and Decision* **4**, pp.179-195 (1973).
23. MITROFF, I.I., 'On the problem of representation in Lockean and dialectical belief systems: a systems approach to policy analysis', *International Journal of General Systems* **4**, pp.75-85 (1975).
24. FINE, T.L., *Theories of Probability,* NY: Academic Press (1973).
25. GAINES, B.R., 'Behavior/structure transformations under uncertainty', *International Journal of Man-Machine Studies* **8**, pp.337-365 (1976).
26. REASON, J.T., 'Motion sickness—some theoretical considerations', *International Journal of Man-Machine Studies,* **1**, pp.21-38 (1969).
27. EDWARDS, W., PHILLIPS, L.D., HAYES, W.L., and GOODMAN, B.C., 'Probabilistic information processing systems: design and evaluation', *IEEE transactions on Systems Science and Cybernetics,* Vol.SSC-4, pp.248-265 (1968).

Uber logische Verknüpfungen unscharfer Aussagen und deren zugehörige Bewertungs-funktionen

HORST HAMACHER
Lehrstuhl für Unternehmensforschung, RWTH Aachen, FRG

Zeichenerklärung

X, K, M	Mengen
$X \times K$	kartesisches Produkt von X und K
$[0,1]^X$	Menge aller Abbildungen von X in das Einheitsintervall
$\mathbb{R}, \mathbb{N}$	Menge der reellen, natürlichen Zahlen
$\mathbb{R}^-, \mathbb{R}^+$	Menge der nicht-positiven, nicht-negativen reellen Zahlen
$I \subset R$	reelles Intervall I
$(a, b]$	linksoffenes, rechtsabgeschlossenes Intervall
inf X, sup X	größe untere Schranke–, kleinste obere Schranke von X
$(X_n) \in \mathbb{R}^{\mathbb{N}}$	reelle Zahlenfolge
ran f	Bildbereich von f
dom f	Definitionsbereich von f
$\bigwedge$	Allquantor
$\bigvee$	Existenzquantor
$\dot{\bigvee}$	Existenzquantor bei Eindeutigkeit

Einleitung

Ein Handelsunternehmen habe nach eingehender Kostenkalkulation und erfolgter Preisfixierung für ein in seinem Sortiment befindliches Produkt seine maximale Rabattgewährung festgelegt, sagen wir 30% des ausgezeichneten Preises. Es ist klar, daß dieser maximale Rabatt nicht jedem Kunden eingeräumt werden kann. Vielmehr läßt man sich in dem Unternehmen bei der Rabattgewährung vom dem simplen Grundsatz leiten:

(+) 'Gute' Kunden erhalten mehr Rabatt als 'weniger gute' Kunden.

Ein Entscheidungsfäller sieht sich dann vor die Frage gestellt:

(++) Welcher Rabattprozentsatz r soll dem Kunden x gewährt werden?

Der Beantwortung dieser Frage gehen offensichtlich folgende Problemstellungen voraus:

(*a*) Welches 'Gütemaß' μ soll dem Kunden x zugeordnet werden?

und

(*b*) Wie soll dieses 'Gütemaß' die Rabattgewährung beeinflussen?

Unter (*a*) ist also nach einer 'Wahrheitsbewertung' oder 'Wahrheitsfunktion' $\mu(x)$ auf der Menge aller

Kunden x unseres Unternehmens bezüglich der unscharfen Aussage:

'x ist ein guter Kunde'

gefragt. Bezeichne K die Menge unserer Kunden, dann ist (a) mit der Angabe einer Abb. $\mu : K \rightarrow I$, wobei I ein kompaktes Intervall im $\mathbb{R}^+$ ist, gelöst. (b) hingegen ist wegen (+) mit der Angabe einer streng monoton steigenden Abb. $r : I \rightarrow [0, 30]$ (wir wollen r Rabattfunktion nennen) gelöst.

Die Entscheidung für eine gewisse Rabattfunktion r_0 hängt natürlich von den Unternehmenszielen ab.

Die Fragestellung unter (b) soll im Folgenden nicht mehr betrachtet werden. Vielmehr wollen wir uns mit der Lösung von (a) beschäftigen. Drei Vorgehensweisen sind naheliegend:

1. Wir verlangen vom Entscheidungsfäller ad hoc, jedem Kunden x bzgl. der unscharfen Aussage 'x ist guter Kunde'—einen Wahrheitswert der vorgegebenen Bewertungsskala I zuzuordnen.
2. Wir versuchen, über statistische Verfahren (Rangordnungsverfahren) die Einschätzung eines jeden Kunden x bzgl. seiner 'Güte' vom Entscheidungsfäller zu erfahren.
3. Wir verfeinern 2. und erfragen vom Entscheidungsfäller einen Katalog von, möglicherweise unscharfen Kriterien, die er bei seiner Einschätzung der Kunden bzgl. ihrer 'Güte' zu beachten gedenkt, sowie der logischen Operatoren, über die er die von ihm gewählten Kriterien miteinander verknüpft, so daß die so entstandene Verknüpfung die 'Güte' eines Kunden ausreichend beschreibt.

Jedoch:
Der unter 1. vorgeschlagene Weg ist offensichtlich nicht sinnvoll. Der Befragte ist überfordert.

Der unter 2. vorgeschlagene Weg ist natürlich. 3. soll hier dennoch vorgezogen werden, weil wir dadurch zur Analyse unscharfer Aussagen gezwungen sind und deshalb das Problem der W-Bewertung unscharfer Aussagen möglicherweise realitätsnäher lösen konnen, nachdem wir über den Inhalt der 'und'—bzw. 'oder'—Verknüpfung klarer sehen.

Ein Entscheidungsfäller halte einen Kunden x dann für einen 'guten' Kunden, wenn x

(i) in der Vergangenheit unsere Forderungen 'stets in angemessener Frist' beglichen hat (Zuverlässigkeitsindikator)

und

(ii) in der Vergangenheit 'regelmäßig' Warenkontingente 'gewisser Größe' von uns bezogen hat (Mengenindikator)

und

(iii) schon 'längere Zeit' unserem Kundenstamm angehört (Treueindikator)

Sicherlich sind weitere Kriterien notwendig (bzw. denkbar), um die 'Güte' eines Kunden zu beschreiben. Unsere Problemstellung wird jedoch mit (i), (ii), (iii) bereits hinreichend verdeutlicht. Nämlich:

Keines der Kriterien (i), (ii), (iii) sondert in *eindeutiger* Weise aus der Menge K aller Kunden eine Teilmenge von Kunden aus, da auf Grund der unscharfen Begriffe: 'angemessene Frist', 'regelmäßig' —'gewisse Größe', 'längere Zeit'— für die meisten Kunden keine zweiwertige Antwort (ja-nein) gegeben werden kann.

Wir unterstellen deshalb für das Folgende, daß bereits über 2. auf K I-wertige Funktionen $\mu_{(i)}$, $\mu_{(ii)}$, $\mu_{(iii)}$ ermittelt worden sind, die für jeden Kunden x die W-Bewertung des Entscheidungsfällers bzgl. (i), (ii), (iii) wiederspiegeln.

Die W-Bewertung $\mu_G : K \rightarrow I$ (G : · ist guter Kunde) finden heißt dann, diejenige Operation O ermitteln, die die 'und'—Verknüpfung der Kriterien (i), (ii), (iii) beschreibt. Bezeichne $\wedge$ das logische Symbol für 'und', dann suchen wir einen Operator O mit der Eigenschaft:

$$O\,(\mu_{(i)}, \mu_{(ii)}, \mu_{(iii)}) = \mu_G = \mu_{(i) \wedge (ii) \wedge (iii)}$$

1. Terminologie und grundlegende Vereinbarungen

Von unserem bisherigen Beispiel abstrahierend betrachten wir im folgenden statt K irgendeine nicht leere Menge X und statt G: 'x ist guter Kunde' irgendeine F—Aussage A bzgl. der Elemente von X. Dann heißt jede Abb. $\mu_A : X \rightarrow I$ (I Bewertungsskala im $\mathbb{R}^+$) *W'bewertung* (oder *Membershipfunction*) *auf X bzgl. A*. Das Paar (X,A) bzw. (X, μ_A) heißt *F-Menge in X*. Es sei (X, μ_A) eine F-Menge in X, dann heißt die Abb. $\mu'_A : X \rightarrow [0, 1]$

$$\mu'_A = \frac{\mu_A - \inf\limits_{x \in X} \mu_A(x)}{\sup\limits_{x \in X} \mu_A(x) - \inf\limits_{x \in X} \mu_A(x)}$$

normalisierte W-Bewertung auf X bzgl. A.

Für das Folgende unterstellen wir *stets* das Vorliegen normalisierter *W*-Bewertungen!

Der klassischen Mengenlehre liegt das *Extensionalitätsprinzip* zugrunde. Es besagt:

> Der *W*'Wert einer durch logische Verknüpfung von n Aussagen entstandenen Aussage ist *nur* von den *n W'*-Werten der beteiligten Aussagen abhängig. Er ist *weder* vom *sprachlichen Inhalt* der Aussagen *noch* von dem *Individuum,* welches die verknüpfte Aussage macht, *abhängig.*

Um für unsere Untersuchungen eine möglichst übersichtliche und sinnvolle Situation zu gewährleisten, legen wir auch für die 'und'-bzw. 'oder'-Verknüpfung von *F*-Aussagen das Extensionalitätsprinzip zugrunde. Dann lautet das uns gestellte Problem:
Gesucht sind:

> *W*'bewertungen (verallgemeinerte Wahrheitstafeln) für die *F*-Aussageformen '•und•' bzw. '•oder•',

also eine *binäre* Operation

(*a*) $D : [0,1] \times [0,1] \rightarrow [0,1]$, die die 'und'-Verknüpfung charakterisiert, d.h.

$$D(\mu_A, \mu_B)(x) = \mu_{A \wedge B}(x)$$

und eine *binäre* Operation

(*b*) $V : [0,1] \times [0,1] \rightarrow [0,1]$, die die 'oder'–Verknüpfung charakterisiert, d.h.

$$V(\mu_A, \mu_B)(x) = \mu_{A \vee B}(x)$$

In Analogie zur klassischen Mengenlehre fassen wir *D* auch als *Durchschnittsbildung, V* auch als *Vereinigungsbildung* von *F*-Mengen auf.

2. *Ein minimales Axiomensystem für die 'und'– bzw. 'oder'– Verknüpfung*

2.1 Zunächst werden wir uns eingehend mit der 'und'–Verknüpfung beschäftigen und dann in 2.2 die gewonnenen Erkenntnisse auf die 'oder' –Verknüpfung in analoger Weise ausdehnen.

Drei Mindestanforderungen an die 'und'–Verknüpfung sind unmittelbar einsichtig:

A1 *D* ist assoziativ

Diese Eigenschaft von *D* ist notwendig und sinnvoll, da sie die Ausdehnbarkeit der 'und' –Verknüpfung auf mehr als nur 2 *F*-Aussagen gewährleistet und außerdem der intuitiven Vorstellung, daß den *F*-Aussagen

$$A \wedge (B \wedge C) \text{ bzw. } (A \wedge B) \wedge C$$

gleicher Wert zukommt, Rechnung trägt.
Ferner:

A2 *D* ist stetig

Die Stetigkeit von *D* erweist sich für das Folgende als ebenso nützlich wie sinnvoll; sie bewirkt nämlich eine gewisse 'Stabilität' des *W*'Wertes einer 'und'-Verknüpfung gegenüber 'kleinen' *W*'wertänderungen der an der Verknüpfung beteiligten *F*-Aussagen.
Und:

A3 *D* ist in jedem Argument injektiv

Es scheint plausibel, daß Individuen die *F*-Aussagen $A \wedge B$ bzw. $A \wedge C$ verschieden bewerten, falls *B* und *C* verschieden bewertet worden sind. Gleiches soll für die *F*-Aussagen $A \wedge C$ bzw. $B \wedge C$ gelten.†

Aus beweistechnischen Gründen betrachten wir *D* zunächst als Abb.: $D : (0, 1]^2 \rightarrow (0, 1]$ und leiten später stetige Ergänzbarkeit zum Rande hin her.
Die Axiome A1-A3 erlauben bereits einige bemerkenswerte Erkenntnisse für *D*, nämlich:

Satz 1: Es sei $D : (0, 1]^2 \rightarrow (0, 1]$, *D* genüge A1-A3. Dann gilt:

(i) $\bigvee_{n \in (0,1]} D(n,n) = n \Rightarrow \dot{\bigvee}_{n \in (0,1]} D(n,n) = n$

(ii) $\bigwedge_{x,y \in (0,1]} \left\{ \begin{array}{l} D(x,y) = x \Leftrightarrow D(y,y) = y \\ D(y,x) = x \Leftrightarrow D(y,y) = y \end{array} \right\}$

(iii) *D* ist in jedem Argument streng monton wachsend.

Satz 1 besagt:
Soll mittels einer A1-A3 genügenden 'und'–Verknüpfung eine 'identische' Bewertung zugelassen sein, so kann dies nur in *eindeutiger* Weise geschehen!

†Diese Forderung unterscheidet unser Axiomensystem wesentlich von dem in [2] vorgestellten.

Dies Ergebnis kommt uns sehr gelegen!
Da uns intuitiv richtig zu sein scheint, daß genau wie im zweiwertigen Fall) die *F*-Aussage '$A \wedge B$' jedenfalls dann mit *W*'Wert 1 zu versehen ist, wenn jeweils *A* und *B* mit *W*'Wert 1 versehen sind, können wir sinnvollerweise zusätzlich noch

$$A4: \bigwedge_{x \in (0,1]} D(x,x) = x \Leftrightarrow x = 1$$

annehmen.

Dann impliziert Satz 1(ii) weiter:
Einer *F*-Aussage '$A \wedge B$' kommt dann und nur dann der *W*'Wert einer der beteiligten Aussagen zu, wenn die andere beteiligte Aussage mit *W*'Wert 1 versehen ist. Ferner gilt unter Einbeziehung von A4 über Satz 1 hinausgehend:

Satz 2: (iv) $\bigwedge_{y \in (0,1]} \lim_{x \to 0} D(x,y) = \lim_{x \to 0} D(y,x) = 0$

(v) $\bigwedge_{x \in (0,1)} D(x,x) < x$

(vi) $\bigwedge_{x,y \in (0,1)} D(x,y) < \min\{x,y\}$

Bemerkungen zu Satz 2:
2(iv) gestattet stetige Fortsetzung von *D* auf den Rand von $(0,1]^2$ in einer durchaus wünschenswerten Weise, nämlich: $D(x,0) = D(0,x) = 0$.
Der 'und'-Verknüpfung zweier *F*-Aussagen kommt *W*'Wert Null zu, wenn eine der *F*-Aussagen mit *W*'Wert Null versehen ist. Eine kleine Überlegung zeigt übrigens, daß schwächere Voraussetzungen als A1-A4 o.g. Ergebnis bereits implizierten, und zwar die Forderungen:

(*a*) Die Restriktion der Operation *D* auf zwei-wertige Aussagen ergibt die *W*'tafel des dual-logischen 'und'.
(*b*) *D* ist in jedem Argument monton nichtfallend.

Die 'und'-Verknüpfung '$\wedge$' im klassischen Fall ist sowohl über das Minimum als auch das Produkt der beteiligten *W*'Werte erzielbar. (Vgl. *W*'tafel für '$\wedge$'). Es ist klar, daß dies für die Erweiterung von $\wedge$ auf $[0,1]^X \times [0,1]^X$ nicht mehr gelten muß.
Daher ist es naheliegend zu fragen, ob und unter welchen Umständen (d.h. unter Zugrundelegung welcher Anforderungen) beide Operationen $\cdot$ bzw. min oder eine von beiden oder eine wie immer geartete Kombination von beiden, die hier gesuchte 'und'-Verknüpfung darstellt.
Auf Grund von A3 ist unmittelbar klar, daß die Operation min, wie sie etwa Zadeh in [1] vorschlägt, *ausscheidet.* Darüberhinaus präzisiert Satz 2(vi):

$$\bigwedge_{x,y \in (0,1)} D(x,y) < \min\{x,y\}$$

Man prüft leicht, daß die Operation 'Produkt' den Axiomen A1-A4 genügt. *Eine* der im klassischen Fall verwendbaren Operationen genügt also unseren Mindestanforderungen für die 'und' –Verknüpfung von *F*-Aussagen. Damit drängt sich sofort die Frage auf:

Ist etwa *D* durch A1-A4 eindeutig festgelegt, nämlich

$$D(x,y) = x \cdot y$$

oder gibt es weitere Funktionen, die A1-A4 genügen und somit für uns als die 'und'-Verknüpfung charakterisierende Funktionen in Frage kommen?

Die folgenden Sätze beantworten diese Frage!

Satz 3: (s. [5], pp.176 *ff.*)
Sei $D : (0,1]^2 \to (0,1]$, *D* genüge A1-A3.
Dann gilt:
Es gibt eine reelle, stetige, streng monton wachsende Funktion *f* mit ran *f* = ran *D*, dom $f \subset \mathbb{R}$, so daß

$$D(x,y) = f(f^{-1}(x) + f^{-1}(y))$$

Der folgende Satz 4 *präzisiert:*
Zu vorgegebenem *D* ist *f* bis auf Dilatation eindeutig bestimmt.

Satz 4: Sei *g* reelle, stetige, streng monoton wachsende Funktion mit ran *g* = ran $D = (0,1]$ und

$$D(x,y) = g(g^{-1}(x) + g^{-1}(y))$$

dann gilt:
Es gibt eine reelle Konstante *c*, so daß

$$\bigwedge_{x \in \text{dom } g} g(x) = f(cx) .$$

Zu Satz 3 und 4 kann bemerkt werden:

(i) Die Umkehrung von Satz 3 ist richtig.

(ii) Im Hinblick auf die gesuchte 'und'-Verknüpfung haben wir nur die stetigen, streng monoton wachsenden Lösungen f der Funktionalgleichung

$$D(f(x), f(y)) = f(x+y)$$

zu ermitteln und dann D nach

$$D(x,y) = f(f^{-1}(x) + f^{-1}(y))$$

zu konstruieren.

(iii) Die zur Konstruktion von D heranzuziehenden Funktionen sind (Satz 4) bis auf Dilatation eindeutig bestimmt.

(iv) Jede durch A1-A3 beschriebene 'und'-Verknüpfung ist *kommutativ*, d.h. Individuen, die für die 'und'-Verknüpfung von F-Aussagen A1-A3 akzeptieren, müssen dann die F-Aussagen

'$A \wedge B$' bzw. '$B \wedge A$'

mit gleichem W-Wert versehen. Dies impliziert, daß von A1-A3 beschriebene 'und'-Verknüpfungen invariant gegenüber der Reihenfolge der beteiligten F-Aussagen sind. Ein ebenfalls wünschenswertes Ergebnis!

(v) Wird für D noch A4 akzeptiert, so gilt für das Definitionsintervall $(a, b]$ von $f : (a, b] = (-\infty, 0]$ (weshalb für c aus Satz 4 sofort $c > 0$ folgt), denn:

$$f^{-1}(1) = f^{-1}(D(1,1)) = f^{-1}(1) + f^{-1}(1)$$

impliziert $f^{-1}(1) = 0$ d.h. $f(0) = 1$, wegen der streng steigenden Monotonie muß deshalb $(a,b]$ Teilintervall der nicht positiven reellen Achse sein.

Sei aber $(a, 0] \subset \mathbb{R}^-$ beschränkt, und (x_n) eine monotone Nullfolge in $(0,1]$, dann gilt auf Grund der stetigen Ergänzbarkeit von $D(x,y)$ und der strengen Monotonie von $D(x,x)$:

$D(x_n, x_n)$ ist ebenfalls monotone Nullfolge! Dies zieht dann wegen der Beschränktheit von $(a, 0]$ die Konvergenz der monotonen Folgen $f^{-1}(D(x_n, x_n))$ und $f^{-1}(x_n)$ mit *gleichem* Grenzwert nach sich, da ferner

$$\bigwedge_{n\in\mathbb{N}} f^{-1}(D(x_n, x_n)) = f^{-1}(x_n) + f^{-1}(x_n)$$

gilt, sind die Grenzwerte auf beiden Seiten der obigen Gleichung gleich, was sofort $\lim_{n\to\infty} f^{-1}(x_n) = 0$ impliziert und deshalb im Widerspruch zur Monotonie von f steht.

Also gilt $(a, b] = (-\infty, 0]$ q.e.d.

Wir haben also im Folgenden die Lösungsgesamtheit der Funktionalgleichung

$$D(f(x), f(y)) = f(x+y),\ f : (-\infty, 0] \to (0, 1]$$

zu ermitteln. Dies geschieht sinnvollerweise, indem über A1-A4 hinausgehende Anforderungen an D gestellt werden, nämlich:

1) D sei algebraische Funktion, d.h. D genüge auf $(0, 1]^2$ der Gleichung:

$$P_0(x,y) + P_1(x,y)D(x,y) + \ldots + P_n(x,y)D^n(x,y) \equiv 0$$

für geeignete Polynome $P_0, \ldots, P_n$, die *nicht* sämtlich Nullpolynome sind.

2) D sei rationale Funktion, d.h.

$$D(x,y) = \frac{P_n(x,y)}{P_m(x,y)},\ \text{für geeignete Polynome } P_n, P_m .$$

3) D sei Polynom in x und y, d.h.

$$D(x,y) = \sum_{\substack{i_1 \leq n \\ i_2 \leq m}} a_{i_1 i_2}\, x^{i_1} \cdot y^{i_2} .$$

Den Fall, daß D transzendente Funktion ist, wollen wir hier nicht betrachten. Bereits für den algebraischen Fall sind wir gezwungen, recht komplexe Ergebnisse aus der Weierstraßschen Theorie der elliptischen Funktionen anzuwenden, so daß wir uns auch hierfür an dieser Stelle mit einem Zwischenergebnis begnügen. Nämlich

Satz 5: Sei $D : (0,1]^2 \to (0,1]$, D genüge A1-A3 und sei algebraisch. Dann kommen zur Konstruktion von D höchstens rationale Funktionen in x bzw. e^x in Frage (bis auf Dilatation) (s. [5] pp.61)

Für rationales D hingegen besagen Untersuchungen von Alt (1940) und Kuwagaki (1951) (s. [5] pp.61):

Lemma 1: Die stetigen Lösungen f der Funktionalgleichung $D(f(x), f(y)) = f(x+y)$ sind notwendig vom Typ:

$$f(x) = \frac{ax+b}{cx+d} \quad \text{oder} \quad f(x) = \frac{ae^{tx}+b}{ce^{tx}+d}, \text{ wobei}$$

a, b, c, d, t geeignete Konstanten sind.

Lemma 1 gibt uns die Möglichkeit, die gesuchte 'und'-Verknüpfung im rationalen Fall unmittelbar anzugeben:

Satz 6: Sei $D : (0,1]^2 \to (0,1]$, D genüge A1-A4 und sei rationale Funktion. Dann gilt:

$$D(x,y) = \frac{dxy}{a+(d-a)(x+y-xy)}, \text{ für geeignete}$$

Konstanten a, d.

Beweis: Zunächst betrachten wir Lösungen f vom Typ

$$f(x) = \frac{ax+b}{cx+d}$$

Da f das Intervall $(-\infty, 0]$ stetig und streng monoton auf $(0, 1]$ abbildet, muß $a = 0$ gelten. Wegen $f(0) = 1$ gilt $b = d \neq 0$. Außerdem ist dann $c \neq 0$. Eine kleine Rechnung zeigt dann

$$D(x,y) = \frac{x \cdot y}{x+y-x \cdot y} \quad (= f(f^{-1}(x)+f^{-1}(y)))$$

Sei nun $f(x) = \dfrac{ae^{tx}+b}{ce^{tx}+d}$. Dann gilt (gleiche Argumentation wie oben):

$b = 0, \quad a = c+d, \quad d \neq 0$

Dies ergibt:

$$D(x,y) = \frac{dxy}{a+(d-a)(x+y-xy)}$$

$$= \frac{dxy}{a(1-x)(1-y)+d(x+y-xy)}$$

$$= \frac{dxy}{a(1-(x+y)+(xy))+d(x+y-xy)}$$

q.e.d.

Das für rationale D vollständige Ergebnis unserer Untersuchungen lautet:

Satz 7: Sei $D : (0,1]^2 \to (0,1]$

D genügt A1-A4 und ist rationale Funktion dann und nur dann, wenn *vorzeichengleiche* Konstanten a, d existieren, so daß gilt:

$$D(x,y) = \frac{dxy}{a+(d-a)(x+y-xy)}$$

Das zugehörige f lautet

$$f(x) = \frac{b}{cx+b} \quad \text{im Fall} \quad f(x) = \frac{ax+b}{cx+d}, \text{ und}$$

$$f(x) = \frac{ae^{tx}}{(a-d)e^{tx}+d} \quad \text{im Fall}$$

$$f(x) = \frac{ae^{tx}+b}{ce^{tx}+d}$$

Beweis: '⇒' Die Existenz der Konstanten ist mit Satz 6 klar. Der Fall

$$f(x) = \frac{ax+b}{cx+d}$$

ergibt für $a = 0$ und beliebiges $d \neq 0$ das gewünschte Ergebnis. Im Fall

$$f(x) = \frac{ae^{tx}+b}{ce^{tx}+d}$$

gilt $b = 0$ und $a = c + d$, $d \neq 0$. Dann ist aber $a \cdot d \geqslant 0$ gerade notwendig und hinreichend für die wachsende Monotonie von f auf $(-\infty, 0]$

'⇐' trivial.

q.e.d.

Der Beweis zu Satz 6 liefert noch weitere Darstellungen der mittels A1-A4 beschriebenen 'und'-Verknüpfung, nämlich:

$$D(x,y) = \frac{dxy}{a(1-(x+y)+xy)+d(x+y-xy)}$$

$$= \frac{dxy}{a(1-x)(1-y)+d(x+y-xy)}$$

Diese Darstellungen werden sich möglicherweise bei späteren Untersuchungen im Zusammenhang mit 'oder'- und 'Komplement'-Verknüpfung bei Betrachtung der 'De-Morgan'schen Regeln' noch als nützlich und interessant erweisen.

Bevor wir uns weiteren Untersuchungen über logische Verknüpfungen zuwenden, wollen wir uns kurz mit dem polynomialen Fall für D beschäftigen.

Satz 8: $D : (0,1]^2 \rightarrow (0,1]$, D genügt A1-A4 und ist Polynom in x und y.
Dann und nur dann gilt:

$$D(x,y) = x \cdot y$$

Beweis: Nicht trivial ist nur '$\Rightarrow$' des Beweises. Sei also $D(x,y)$ wie vorausgesetzt; dann ist auf Grund von Satz 3 D kommutativ, d.h. D ist in x und y Polynom vom gleichen Grad, sagen wir n. Dann steht aber auf Grund von

$$\text{A1:} \quad D(x,D(y,z)) = D(D(x,y),z)$$

auf der linken Seite ein Polynom vom Grad n in x auf der rechten Seite aber ein Polynom vom Grad n^2 in x. Also ist $n = n^2$.
Dies ist nur für $n = 0$ bzw. $n = 1$ richtig.
$n = 0$ scheidet aus, D wäre konstant!
Also gilt:

$$D(x,y) = Ax \cdot y + Bx + Cy + D$$

Die stetige Ergänzbarkeit von $D(x,y)$ zum Rand hin mit Wert Null ergibt sofort.

$$D = 0 \qquad (\text{wegen } D(0,0) = 0)$$

$$B = C = 0 \qquad (\text{wegen } D(0,1) = D(1,0) = 0)$$

Schließlich ergibt $D(1,1) = 1 : A = 1$ also gilt:

$$D(x,y) = x \cdot y$$

q.e.d.

Als Lösung der zu polynomialem D gehörigen Funktionalgleichung $D(f(x), f(y)) = f(x+y)$ ergibt sich:

$$f : (-\infty, 0] \rightarrow (0, 1] \qquad \text{mit:}$$

$$f(x) = e^{tx} \quad , \qquad t > 0$$

Außerdem sehen wir, was nicht anders zu erwarten war, daß diese Lösung ein spezielles Ergebnis des rationalen Falls ist. Wir brauchen dort nämlich nur $a = d$ zu wählen!

2.2

Die gleiche Argumentation wie im vorigen Kapitel bzgl. D, führt dazu, daß wir auch für die 'oder'-Verknüpfung $V : [0,1)^2 \rightarrow [0,1)$ die Axiome A1-A3 (selbstverständlich unter Beachtung des V zugrundeliegenden Definitions- und Bildbereichs!) fordern.
Das Satz 1 entsprechende Ergebnis lautet dann:

Satz 9: Es sei $V : [0,1)^2 \rightarrow [0,1)$, V genüge A1-A3.

Dann gilt:

(i) $\bigvee_{n \in [0,1)} V(n,n) = n \Rightarrow \dot{\bigvee}_{n \in [0,1)} V(n,n) = n$

(ii) $\bigwedge_{x,y \in [0,1)} \{V(x,y) = x \Leftrightarrow$
$\Leftrightarrow V(y,y) = y$ und
$V(y,x) = x \Leftrightarrow V(y,y) = y\}$

(iii) V ist in jedem Argument echt monoton wachsend.

Auch zu diesem Sachverhalt drängen sich zu Satz 1 analoge Bemerkungen auf.
Deshalb fordern wir für V noch:

$$A'4 \quad \bigwedge_{x \in [0,1)} V(x,x) = x \Leftrightarrow x = 0$$

Damit können wir ähnlich wie im Beweis zu Satz 2 schließen und erhalten

Satz 10: Sei V wie in Satz 9 vorausgesetzt und genüge außerdem A'4. Dann gilt:

(iv) $\bigwedge_{y \in [0,1)} \lim_{x \to 1} V(x,y) = \lim_{x \to 1} V(y,x) = 1$

(v) $\bigwedge_{x \in (0,1)} V(x,x) > x$

(vi) $\bigwedge_{x,y \in (0,1)} V(x,y) > \max\{x,y\}$

Dies impliziert auch für V die Möglichkeit der stetigen Ergänzbarkeit zum Rand von $[0,1)^2$ hin, und zwar wie folgt:

$$\bigwedge_{x \in [0,1)} V(x,1) = V(1,x) = V(1,1) = 1$$

Wiederum stellt sich nun die Frage:
Wie sehen auf Grund von A1-A'4 beschriebene 'oder'-Verknüpfungen aus?
Die gleichen Überlegungen (Modifikation von Satz 3, im Folgenden mit Satz 3' bezeichnet) wie im vorigen Kapitel liefern:

Satz 11: Sei $V: [0,1)^2 \to [0,1)$, V genügt den Axiomen A1, A2, A3, A'4 und ist rational. Dann gibt es geeignete reelle Konstanten a', d', so daß gilt:

$$V(x,y) = \frac{(d'-a') \cdot xy \; + \; a' \cdot (x+y)}{d' \cdot xy \; + \; a'}$$

$$= \frac{d' \cdot xy \; + \; a' \cdot (x+y-xy)}{d' \cdot xy \; + \; a'}$$

Das für rationale V vollständige Ergebnis lautet:

Satz 12: Sei $V: [0,1)^2 \to [0,1)$, V genügt A1, A2, A3, A'4 und ist rational.
Dann und nur dann gibt es reelle Konstanten a', d' mit $d' \cdot a' \geqslant -a'$ $(a' \neq 0)$, so daß gilt:

$$V(x,y) = \frac{(d'-a') \cdot xy \; + \; a' \cdot (x+y)}{d' \cdot xy \; + \; a'}$$

Das zugehörige f lautet:

$$f(x) = \frac{a' \cdot x}{a' \cdot x + d'} \quad \text{bzw.} \quad f(x) = \frac{a' \cdot (e^{tx} - 1)}{a' \cdot e^{tx} + d'}$$

Beweis: Auf Grund von Satz 11 ist nur die Relation für die Konstanten zu verifizieren. Das Ergebnis für $f(x) = \dfrac{a' \cdot x}{a' \cdot x + d'}$ ist leicht unter den Fall

$$f(x) = \frac{a' \cdot e^{tx} - a'}{a' \cdot e^{tx} + d'} \qquad \text{subsummierbar}$$

$(a' = 1,\ d' = -1$ gesetzt!). Für diesen Fall ist die Relation trivialerweise richtig. Für den anderen Fall $(a' \neq -d'!)$ ergibt sich die Relation $d' \cdot a' \geqslant -a'$ als notwendige und hinreichende Bedingung für streng steigende Monotonie von f.

q.e.d.

Für polynomiales V kommen wir über zu Satz 8 analoge Überlegungen auf das Ergebnis:

$$V(x,y) = x + y - x \cdot y$$

3. Ein reduziertes Bellman-Giertz'sches Axiomensystem zur Beschreibung der 'und'-bzw. 'oder'-Verknüpfung

Im Gegensatz zu unserem Vorgehen betrachten Bellman und Giertz in [2] die Operationen D und V als in gewissem Sinne zueinander duale Operationen. Konsequenterweise enthält deshalb ihr Axiomensystem die in der dualen Logik wesentlichen Forderungen der *Distributivität* der 'und' *über* der 'oder'-Verknüpfung und umgekehrt. (Eine andere Möglichkeit die Dualität von D, V zu verdeutlichen, besteht übrigens in der Vorgabe einer 'Komplementoperation' K und der Forderung der De-Morgan'schen Regeln für D, V, K.) Die Plausibilität dieser Forderungen ist sicherlich strittig; jedenfalls trägt die Distributivität wesentlich, wie wir noch sehen werden, dazu bei, daß Bellman und Giertz ein *eindeutiges* Ergebnis für D und V erhalten, nämlich:

$$D(x,y) = \min\{x,y\} \quad \text{und} \quad V(x,y) = \max\{x,y\}$$

Außerdem betrachten die o.g. Autoren D und V sofort als Abb. von $[0,1]^2$ nach $[0,1]$, wohingegen wir uns aus beweistechnischen Gründen (Anwendung des Satzes 3!) zunächst auf halboffene Einheitsintervalle beschränken, dann jedoch stetige Ergänzbarkeit herleiten und stetig zum Rand ergänzen.

Bellman und Giertz haben in [2] gezeigt:

Satz 13: Es seien D, V: $[0,1]^2 \to [0,1]$
D, V genügen:

BG1 $$\bigwedge_{x,y,z \in [0,1]} D(x,V(y,z)) = V(D(x,y),D(x,z))$$

BG2 $$\bigwedge_{x,y,z \in [0,1]} V(x,D(y,z)) = D(V(x,y),V(x,z))$$

BG3 $\bigwedge_{x,y,z\in[0,1]} \{D(x,D(y,z)) = D(D(x,y),z)$ und $V(x,V(y,z)) = V(V(x,y),z\}$

BG4 D, V sind monoton wachsend in jedem Argument.

BG5 D, V sind auf der Diagonalen *streng* monoton wachsend.

BG6 $\bigwedge_{x,y\in[0,1]} D(x,y) \leqslant \min\{x,y\} \wedge V(x,y) \geqslant \max\{x,y\}$

BG7 $D(1,1) = 1$ und $V(0,0) = 0$

BG8 $\bigwedge_{x,y\in[0,1]} D(x,y) = D(y,x) \wedge V(x,y) = V(y,x)$

BG9 D und V sind stetig.

Dann und nur dann gilt:

$D(x,y) = \min\{x,y\}$ und $V(x,y) = \max\{x,y\}$

Bevor wir die Axiomensysteme A1-A4(A'4) und BG1-BG9 miteinander vergleichen, sollen BG1-BG9 kurz diskutiert werden:

1. BG1 und BG2 sind aus der dualen Logik übernommen und sollen die Dualität der Operatoren D und V beschreiben.
2. BG3 ist genau A1. (Des eigentlich nötigen Vermerks: 'bis auf gewisse Randpunkte', bedarf es hier, wie in noch folgenden Fällen wegen der stetigen Ergänzbarkeit von D, V, nicht!)
3. A1-A3 implizieren BG4 und BG5 (s. Sätze 1(iii), 9(iii)). Der W'Wert einer 'und'–bzw. 'oder'-Verknüpfung soll *nicht* fallen, falls der W'Wert auch nur einer der beteiligten Aussagen steigt und der W'Wert der anderen Aussage konstant bleibt. Der W'Wert einer 'und'–bzw. 'oder'-Verknüpfung zweier gleicher F-Aussagen soll steigen, falls der W'Wert der beteiligten F-Aussage steigt.
4. BG6 stellt eine leichte Abschwächung des aus Satz 2(vi) und 10(vi) bekannten Sachverhalts dar. Also: A1-A4(A'4) implizieren BG6. Der W'Wert der 'und'-Verknüpfung soll nicht größer sein als der W'Wert jeder beteiligten F-Aussage. Dadurch mag ausgedrückt sein, daß eine F-Aussage über die 'und'-Verknüpfung mit einer weiteren F-Aussage 'straffer' oder jedenfalls *nicht* 'weicher' wird. Für die 'oder'-Verknüpfung kann BG6 analog interpretiert werden.
5. BG7 ist genau A4 bzw. A'4. Der W'Wert für eine 'und'-Verknüpfung von F-Aussagen soll jedenfalls dann 1 sein, wenn der W'Wert *jeder* der beteiligten Aussagen 1 ist. Für die 'oder'-Verknüpfung mag BG7 analog interpretiert werden.
6. BG8 ist (über die Sätze 3, 3') direkte Folgerung aus A1-A3. Es soll für den W'Wert der 'und'–bzw. 'oder'-Verknüpfung von F-Aussagen nicht auf die Reihenfolge der beteiligten Aussagen ankommen.
7. BG9 ist genau A2. Der W'Wert der 'und'–bzw. 'oder'–Verknüpfung soll stabil gegenüber 'kleinen' W'Wertänderungen der beteiligten F-Aussagen sein.

Das Bellman-Giertz'sche Axiomensystem beschreibt in durchaus bejahenswerter Weise (bis auf vielleicht BG1 und BG2) die Eigenschaften der gesuchten Verknüpfungen und dennoch mag sein *eindeutiges* Ergebnis manchem unbefriedigend erscheinen. Denn über die mit $D(x,y) = \min\{x,y\}$ beschriebene 'und'-Verknüpfung werden zwei zusammengesetzte F-Aussagen

'A und B' bzw. 'A' und B'

mit dem *gleichen* W'Wert versehen, wenn z.B. in beiden Verknüpfungen die beteiligten F-Aussagen mit dem *kleineren* W'Wert gleich sind (hier: 'B'), völlig unabhängig davon, ob und wie stark sich die W'Werte der jeweils 'höher' bewerteten F-Aussagen (hier: 'A' bzw. 'A'') voneinander und/oder von der niedriger bewerteten F-Aussage (hier: 'B') unterscheiden. Analoges kann natürlich auch über $V(x,y) = \max\{x,y\}$ angemerkt werden.

Diese Kritikpunkte veranlaßten Rödder, ausgiebige statistische Tests druchzuführen und in der Tat ist er in [3] in der Lage, unter gewissen Regularitätsvoraussetzungen die Hypothese: $D(x,y) = \min\{x,y\}$ mit über 95%-iger Wahrscheinlichkeit zu verwerfen! Wenn aber das Bellman-Giertz'sche Ergebnis als nicht wünschenswert oder realitätsnah einzustufen ist, andererseits aber das Bellman-Giertz'sche Axiomen system einer ersten, spontanen Diskussion, wie wir gesehen haben, durchaus standhält, so ist zu fragen, welche der Axiome BG1-BG9 fallengelassen werden müssen, um ein anderes, dann eventuell nicht mehr eindeutiges Ergebnis zuzulassen.

Wir werden sehen, daß die Axiome BG7, BG8, BG9 bereits in den Axiomen BG1-BG6 enthalten sind, d.h. durch weglassen von BG7, BG8, BG9 wird das Bellman-Giertz'sche Axiomensystem *qualitativ* nicht geändert. Es gilt nämlich der

Satz 14: Es seien $D, V : [0,1]^2 \to [0,1]$

D, V genügen BG1-BG6.
Dann und nur dann gilt:

$$D(x,y) = \min\{x,y\}, \; V(x,y) = \max\{x,y\}$$

Beweis: Nicht trivial ist nur '⇒' des Beweises.
Zunächst zeigen wir:

(i) $\bigwedge_{x,y \in [0,1]} (D(x,x) = y \Leftrightarrow V(y,y) = x)$

Sei $x \in [0,1]$ und $D(x,x) = y$, und zwar gilt dann auf Grund von BG6 und BG2:

$$D(x,x) = y \leqslant V(y,D(y,y)) = D(V(y,y), V(y,y))$$

und damit wegen BG5:

$$x \leqslant V(y,y)$$

Andererseits gilt ebenfalls wegen BG6 sowie BG1:

$$x \geqslant D(x,V(x,x)) = V(D(x,x), D(x,x)) = = V(y,y)$$

also insgesamt:

$$\bigwedge_{x,y \in [0,1]} (D(x,x) = y \Rightarrow V(y,y) = x)$$

Analoge Argumentation ergibt:

$$\bigwedge_{x,y \in [0,1]} (V(y,y) = x \Rightarrow D(x,x) = y)$$

Damit ist (i) bereits nachgewiesen.
(i) und BG1, BG2 ergeben sofort:

(ii) $\begin{cases} \bigwedge_{x \in [0,1]} D(x, V(x,x)) = V(D(x,x), D(x,x)) = = V(y,y) = x \\ \bigwedge_{x \in [0,1]} V(x, D(x,x)) = D(V(x,x), V(x,x)) = = D(y,y) = x \end{cases}$

Sei nun $y \in [0,1]$ und $V(y,y) = x$, dann ergeben (i) and (ii):

$$V(y,y) = x = V(x,D(x,x)) = V(x,y) = V(V(y,y),y)$$

Analog: $D(x,x) = y = D(y,V(y,y)) = D(y,x) = D(D(x,x),x)$, also:

(iii) $\bigwedge_{x \in [0,1]} (D(x,x) = D(D(x,x),x) \wedge V(x,x) = = V(V(x,x),x))$

Demnach trivialerweise nach Anwendung von (iii) und BG3:

(iv) $\begin{cases} \bigwedge_{x \in [0,1]} D(x,x) = D(D(x,x), D(x,x)) \\ \bigwedge_{x \in [0,1]} V(x,x) = V(V(x,x), V(x,x)) \end{cases}$

Danach gilt dann auf Grund von BG5 (impliziert nämlich Injektivität auf der Diagonalen):

(v) $\bigwedge_{x \in [0,1]} D(x,x) = x \wedge V(x,x) = x$

BG1-BG6 ziehen also Idempotenz von D und V nach sich!
Deshalb gilt:

(vi) $\bigwedge_{x,y \in [0,1]} \min\{x,y\} \leqslant D(x,y), V(x,y) \leqslant \leqslant \max\{x,y\}$

Sei nämlich $x,y \in [0,1]$, o.B.d.A. $x \leqslant y$ dann gilt:

$$\min\{x,y\} = x = D(x,x) \leqslant D(x,y) \leqslant D(y,y) = = y = \max\{x,y\}$$

und

$$\min\{x,y\} = x = V(x,x) \leqslant V(x,y) \leqslant V(y,y) = y = \max\{x,y\}$$

also ist (vi) richtig!
(vi) und BG6 implizieren dann:

$$D(x,y) = \min\{x.y\}, \quad V(x,y) = \max\{x,y\}$$

q.e.d.

4. Vergleich der Axiomensysteme
Es sei mit D_A die Klasse aller A1-A4 genügenden Funktionen, mit V_A die Klasse aller A1-A'4 genügenden Funktionen bezeichnet.

Sowohl D_A als auch V_A enthalten *überabzählbar viele* Funktionen! Teilklassen von D_A und V_A, nämlich die in D_A, V_A enthaltenen rationalen Funktionen wurden bereits ermittelt.
Die Klasse **BG** der BG1-BG6 genügenden Funktionenpaare (BG7, BG8, BG9 sind wegen Satz 14 verzichtbar) ist *einelementig.*

Es ist klar, daß *kein* Paar $(D,V) \in D_A \times V_A$ BG1 und BG2 erfüllt, also *distributiv* übereinander ist. Jedes Paar (D,V) erfüllt nämlich BG3-BG6, wie die Diskussionspunkte 2)-7) gezeigt haben. Andererseits genügen sowohl $D = \min$ als auch $V = \max$ den Axiomen A1, A2, A4 bzw. A'4. D.h. BG1-BG6 schließen A3 (Injektivität in jedem Argument) aus. Dies wiederum bedeutet, daß durch folgende Verschärfung BG'4 von BG4:

BG'4: *D, V* sind *streng* monoton wachsend in jedem Argument

das System BG1-BG3, BG'4, BG6 (BG5 ist dann in BG'4 enthalten!) inkonsistent wird d.h. es gibt *keine* BG1-BG3, BG'4, BG6 genügenden Funktionen.

Da aber die Forderungen BG3, BG'4, BG6 in den Axiomen A1-A3 enthalten (also schwächer!) sind lehrt dies die 'Stärke' der Distributivitätsforderungen:

> Selbst bei Abschwächung des minimalen Axiomensystems A1-A3 zu BG3, BG'4, BG6 ist die Existenz distributiver Funktionen *D* und *V* ausgeschlossen!

Das Gewicht der Distributivitätsforderungen wird ganz deutlich, wenn A1-A3 durch noch schwächere Anforderungen als BG3, BG'4, BG6 ersetzt werden:

M: D und *V* seien *kommutativ* und in jedem Argument *monoton nicht fallend.*

In diesem Fall können wir nämlich zeigen: Bereits zu harmlosen Funktionen *D* existiert *keine* (für uns akzeptable) distributive Funktion *V*.

Satz 15: Sei $D(x,y) = x \cdot y$

Dann gibt es zu *D* außer der identisch-Null Funktion *keine M* genügende, distributive binäre Operation $V : [0,1]^2 \to [0,1]$.

Beweis: *V* genüge *M* und sei zu $D(x,y) = x \cdot y$ distributiv, dann gilt sofort:

(i) $$\bigwedge_{x,y,z \in [0,1]} x \cdot V(y,z) = V(x \cdot y, x \cdot z)$$

(ii) $$\bigwedge_{x,y,z \in [0,1]} V(x,y \cdot z) = V(x,y) \cdot V(x,z)$$

(i) ergibt ($x = 0$ gesetzt!):

$$V(0,0) = 0$$

(ii) ergibt ($x = y = z = 1$ gesetzt!):

$$V(1,1) = V(1,1) \cdot V(1,1)$$

was natürlich nur für $V(1,1) = 0$ oder $V(1,1) = 1$ richtig sein kann. $V(1,1) = 0$ impliziert wegen der nicht fallen den Monotonie bereits: $V \equiv 0$.

Sei also $V(1,1) = 1$!

Dann ergibt (i) ($y = z = 1$ gesetzt!):

$$\bigwedge_{x \in [0,1]} x = x \cdot V(1,1) = V(x,x)$$

also die Idempotenz von *V*!
Ferner ergibt (ii) ($y = z = 1$ gesetzt!):

(iii) $$\bigwedge_{x \in [0,1]} V(x,1) = V(x,1) \cdot V(x,1)$$

Angenommen es existiert ein $x_0 \in (0,1)$ mit der Eigenschaft:

$$V(x_0, 1) = 0$$

dann gilt wegen der nicht fallenden Monotonie in beiden Argumenten für $y = x_0$:

$$V(x_0, y) = V(x_0, x_0) \leqslant V(x_0, 1) = 0$$

im Widerspruch zur Idempotenz und $0 < x_0 < 1$, also gilt wegen (iii):

$$\bigwedge_{x \in (0,1]} V(x,1) = 1 = V(1,x)$$

Sei nun $x \neq 0 \neq y$, dann gilt mit (ii), (i) und $V(x,1) = 1$ ($y = z$ gesetzt!):

$$V(x,y^2) = V(x,y) \cdot V(x,y) = x \cdot V(1, \frac{y^2}{x}) = x$$

im Fall $x \geqslant y^2$; und

$$V(x,y^2) = V(x,y) \cdot V(x,y) = y^2 \cdot V(\frac{x}{y^2}, 1) = y^2$$

im Fall $y^2 \geqslant x$.

Insgesamt also:

$$V(x,y) = \begin{cases} \sqrt{x} & , x \geqslant y^2 \\ y & , y^2 \geqslant x \end{cases}$$

dann jedoch impliziert $x = y = 1/2$ (also $x > y^2 !$):

$$V(\frac{1}{2}, \frac{1}{2}) = \frac{1}{\sqrt{2}}$$

im Widerspruch zur Idempotenz von V. Also muß $V(1, 1) = 1$ falsch sein.

q.e.d.

Dies zeigt insgesamt, daß ersatzlose Streichung der Axiome BG1, BG2 aus BG1-BG6, die Mächtigkeit der zugelassenen Funktionenklasse schlagartig erhöht und zwar ist die Menge der mittels BG3-BG6 beschriebenen Funktionen umfassender als die Menge der mittels A1-A4 (A'4) beschriebenen Funktionen.

Literaturverzeichnis

1. ZADEH, L.A., 'Fuzzy Sets', *Information and Control* 8, pp.338-353 (1965).
2. BELLMAN, R. and GIERTZ, M., 'On the Analytic Formalism of the Theory of Fuzzy Sets', *Information Sciences* 5, pp.149-156 (1973).
3. RÖDDER, W., 'On 'and' and 'or' Connectives in Fuzzy Set Theory, EURO I, 1975, (WP) 75/07, Lehrstuhl für Unternehmensforschung, RWTH Aachen.
4. FUNG, L. and FU, K.S., 'An Axiomatic Approach to Ratoinal Decision Making Based on Fuzzy Sets, (WP) November 1973, Purdue University, West Lafayette, Indiana 47907.
5. ACZÉL, J., 'Vorlesungen über Funktionalgleichungen und ihre Anwendungen', Birkhäuser Verlag Basel (1961).

ON LOGICAL CONNECTIVES OF FUZZY STATEMENTS AND THEIR AFFILIATED TRUTH-FUNCTIONS

H. Hamacher, Lehrstuhl f. Unternehmensforschung, RWTH Aachen, Templergraben 64, FRG

In the first publication on Fuzzy Sets [1], ZADEH defined 'and' ($\wedge$) respectively 'or' ($\vee$) operators on multivalued predicates (i.e. $A, B : X \to [0,1], X \neq \emptyset$) as follows:

(i) $\bigwedge_{x \in X} A \wedge B(x) := \min\{A(x), B(x)\}$

(ii) $\bigwedge_{x \in X} A \vee B(x) := \max\{A(x), B(x)\}$. Alternatively he suggested:

(iii) $\bigwedge_{x \in X} A \wedge B(x) := A(x) \cdot B(x)$

(iv) $\bigwedge_{x \in X} A \vee B(x) := A(x) + B(x) - A(x) \cdot B(x)$.

Meanwhile, only few authors discussed fundamental questions related to these operators. BELLMAN/GIERTZ presented in [2] a system of axioms for which (i) and (ii) are necessary and sufficient. RÖDDER doubted in [3] the 'reality' of (i) resp. (ii) and presented an empirical test, which enables him to reject the hypothesis:

$A \wedge B(x) = \min\{A(x), B(x)\}$ by a 99% probability.

Our researches are directed as follows:

1. By a slight modification of BELLMAN/GIERTZ'S proof, we show that a subsystem of axioms out of [2] is necessary and sufficient for (i) and (ii).
2. For $\wedge$ resp. $\vee$ we present a 'poor' system of axioms, namely:

   (*a*) $\wedge, \vee$ are associative

   (*b*) $\wedge, \vee$ are strictly monotone increasing in each of their arguments

   (*c*) $\wedge, \vee$ are continuous

   (*d*) $\wedge(1,1) = 1, \vee(0,0) = 0$

By help of some special results of the theory of functional equations, we show the following theorem:

Be $\wedge : (0,1]^2 \to (0,1]$, $\vee : [0,1)^2 \to [0,1)$ rational functions

then *(a)-(d)* hold if and only if there exist two reals $\gamma \geqslant 0$, $\gamma' \geqslant -1$, such that the following holds:

$$\wedge(x,y) = \frac{xy}{\gamma + (1-\gamma)(x+y-xy)}, \vee(x,y) = \frac{(\gamma'-1)xy + x + y}{xy + 1}$$

Subjective probabilities, fuzzy sets and decision making

S. SAGAAMA
Ecole Nationale Supérieure des Mines de Saint-Etienne, France

Introduction

Fuzzy sets theory enables us to valuate (1) non-measurable (2) subjective vagueness. From a first point of view, the common point between fuzzy sets theory and subjective probabilities, is the subjective valuation.

Our purpose in this chapter is to distinguish fuzzy sets from subjective probabilities.

1. The basis of subjective probabilities

De Finetti, in [4], thinks it is reasonable to admit the additivity criterion of subjective probabilities.

Raifa, in [9], proved the additivity of subjective probabilities, basing his proof on two principles concerning substitution by subjective equivalence and transitivity by subjective equivalence (or preference), taking a lottery (game) as an example. These two principles (called P_1 and P_2) are:

1) P_1: Substitution principle [9]

If a lottery is modified by substituting one of the prizes by another, all other gains being unchanged, and if the first prize and its substitute are equivalent for the gambler (decider), then, the first lottery and the modified one are equivalent.

2) P_2: Transitivity principle: [9]

Let *A*, *B*, *C*, be three possible situations.

If the decider prefers one of the situations, then logically, his preferences must satisfy the following propositions.

a) If (*A* and *B*) and (*B* and *C*) are equivalent for the decider then *A* and *C* are equivalent for him.

b) If he prefers *A* to *B* and *B* to *C* then, he prefers *A* to *C*.

c) If he prefers *A* to *B* and if *B* and *C* are equivalent for him then he prefers *A* to *C*.

Note that these two principles are based on subjective preference and subjective equivalence.

Through these principles, subjective probabilities' additivity is proved in [9].

Our purpose is to discuss the principles of substitution and transitivity of subjective equivalence.

Let us first examine a construction of a subjective probability (or its distribution function).

In the sequel, we note SP = subjective probability.

II. Construction of a subjective probability

Let $\mathbb{R}$ be the real space.

Let $[a, b] \subset \mathbb{R}$, be the construction domain of the subjective probability.

[1] Instead of 'measure'; a term suggested by Pr. Kaufmann

[2] In the sense of Lebesgue measure.

Let x_0 be the point precised by the decider, signifying his indifference between the intervals $[a, x_0[$ and $[x_0, b]$.

This means that the decider bets 0.5 for $[a, x_0[$ and 0.5 for $[x_0, b]$, and we write $p_{0.5} = x_0$, which means that SP $\{X < x_0\} = 0.5$, when X is his 'subjective random variable'.

After that, we then ask the decider to give his indifference point in the interval $[a, x_0[$, if x_1 is this point, we write SP $\{X < x_1\} = 0.5/2 = 0.25$.

Repeating the same process with $[a, x_1[$, we find a point x_2 and we note SP $\{X < x_2\} = 0.25/2 = 0.125$, and so on...

Then, we consider the interval $[x_0, b]$. Let x_3 be the indifference point, we note SP$\{X < x_3\} = 0.5 + 0.5/2 = 0.75 =$ SP$(\{X < x_0\} \cup \{x_0 \leqslant X < x_3\})$, and so on...

We proceed by dichotomy, and we take care to note that if A and B are two intervals in $[a, b]$, and if $A \cap B = \phi$ then SP(A) + SP(B) = SP$(A \cup B)$. This means that we must rationalize the answers of the decider, according to the principles P_1 and P_2, which leads to a probabilistic distribution which has no subjectivity (in its construction).

We are now going to discuss the two principles P_1 and P_2.

III. Remarks about the transitivity principle

Let us consider the case of a subjective equivalence in the following example:

Let M and A, be two goods. Let us suppose that their prices are rather high, but they (the goods) are equivalent for the decider. Let us take for example: A = apartment, M = a car (Mercedes 450 SL).

First we have:

$M \sim A$	($\sim$: equivalent)
$M \sim M + \$1$	(reason: $1 is insignificant)
$M + \$1 \sim A$	(same reason)
$A \sim A + \$1$	(" ")
$A + \$1 \sim M + \$2$	(" ")
.	
.	
.	
$A + \$9999 \sim M + \$10{,}000$	(" ")
$M + \$10{,}000 \sim A + \$10{,}000$	(" ")

If we accept the subjective equivalence transitivity [9], we have $A \sim A + \$10{,}000$, and this conclusion is out of the question for the decider!

So, the subjective equivalence is not transitive.

In order to make the decider transitive, the solution proposed, is that he determines a sum d, such as below $A + d$, his equivalences are transitive.

But if we ask him to find this sum of money (d), he will be at a loss to answer, because:

1. He is unable to find it with precision; and
2. If he finds a sum of money d, and admits $A, A + \$1, ..., A + \$d$, to be equivalent, then $A + \$(d+1), ..., A + \$2d$, might also be equivalent, then his equivalences are rather blocks of elements than one only block.

Let I_1 be the set of following goods:

$I_1 = \{M, M + \$1, ..., M + \$n\}$, and J_1 the following set of goods: $J_1 = \{A, A + \$1, ..., A + \$m\}$.

Let $I = I_1 \cup J_1$.

1. We consider the following (particular) case: Let $(x,y) \in I \times I$. First we decide if x and y are equivalent or not.

We note $u(x,y) = 1$ if x and y are equivalent and $u(x,y) = 0$ if not.

Particularly, if $m = n = \$10{,}000$ and $(M, A + \$10{,}000)$ are not equivalent for the decider, we note that $u(M, A + \$10{,}000) = 0$. This means that the weight of the arc $(M, A + \$10{,}000) = 0$. We say that the equivalences are transitive (fuzzy—transitive), if $u(M, A + \$10{,}000) \geqslant \geqslant \max_Z (\text{MIN}\ \{u(M, Z), u(Z, A + \$10{,}000\})$. This means that the weight of the arc $(M, A + \$10{,}000)$ is greater than the weight of any way joining M to $A + \$10{,}000$ [6].

This means also that the 'degree of equivalence between M and $A + \$10{,}000$' (or the weight of equivalence between M and $A + \$10{,}000$) is greater than or equal to the 'degree of equivalence of any equivalence chain joinging M to $A + \$10{,}000$'. In our particular case, $u(M, A + \$10{,}000) = 0$, but nevertheless, there is a way whose weight is 1, for example:

$$(A, M+\$1), (M+\$1, M+\$2), ..., (M+\$9999, A+\$10{,}000)$$

because

$$\text{MIN}\{u(A, M+\$1), ..., u(M+\$9999, A+\$10{,}000)\} = 1$$

we conclude that the equivalences of the decider are not transitive (in the sense of fuzzy transitivity), but they are boolean transitive.

We can pursue this further:

By considering the matrix obtained on $I \times I$ (reflexive and symmetric), using the Malgrange method, we can decompose it in maximal sub-relations of similitude [6], e.g.: reflexive symmetric and transitive submatrixes.

Remark:
In these sub-matrixes, and in this particular case (e.g.: the decider's equivalence of couples is yes (or) no), the fuzzy transitivity and the boolean one are the same (e.g.: a relation R processes boolean transitivity if: $R(x,y) \leqslant R(x,Z) + R(Z,y)$, $(+)$ is the boolean operator), and the sub-matrixes which we obtain are in fact classes of equivalences.

Example:
Let

$$I_1 = \{x_1 = M,\ x_2 = M + \$100,\ x_3 = M + \$200\}$$

$$J_1 = \{x_4 = A,\ x_5 = A + \$100,\ x_6 = A + \$200\}.$$

Let us give the weight 1 to $(x,y) \in I \times I$ if x and y are equivalent, and 0 if not.

Suppose we have $I = \{x_1, x_2, x_3, x_4, x_5, x_6\}$ and the following matrix:

	x_1	x_2	x_3	x_4	x_5	x_6
x_1	1	1	0	1	1	0
x_2	1	1	1	1	1	0
x_3	0	1	1	0	1	1
x_4	1	1	0	1	1	0
x_5	1	1	1	1	1	0
x_6	0	0	1	0	0	1

Using Malgrange method [6], we have:

$$S = (x_1 + x_3 x_6)(x_2 + x_6)(x_3 + x_1 x_4)(x_4 + x_3 x_6)(x_5 + x_6)(x_6 + x_1 x_2 x_4 x_5)$$

$$= (x_1 x_2 + x_1 x_6 + x_3 x_6)(x_3 x_4 + x_3 x_6 + x_1 x_4)(x_6 + x_1 x_2 x_4 x_5)$$

$$= (x_1 x_2 x_4 + x_1 x_4 x_6 + x_3 x_6)(x_6 + x_1 x_2 x_4 x_5)$$

$$= x_1 x_4 x_6 + x_3 x_6 + x_1 x_2 x_4 x_5 \Rightarrow S' = x_2 x_3 x_5 + x_1 x_2 x_4 x_5 + x_3 x_6$$

Thus, the maximal sub-relations of similitude are:

	M_1	M_2	A_1
M_1	1	1	1
M_2	1	1	1
A_1	1	1	1

	M_0	M_1	A_0	A_1
M_0	1	1	1	1
M_1	1	1	1	1
A_0	1	1	1	1
A_1	1	1	1	1

	M_2	A_2
M_2	1	1
A_2	1	1

N.B. The different sub-matrixes are not necessarily separated, hence the equivalence partitions are:

1. Couples $(M+d, A+d')$ with $0 < d \leqslant \$200$, $d' = \$100$.
2. Couples $(M+d, A+d')$ with $d \leqslant \$100$, $d' \leqslant \$100$.
3. (M_2, A_2) are equivalent because the relation on $I \times J$ is reflexive.

2. *Case of fuzzy equivalence*

If we question some one about the equivalences of the following couples, he can say that:

(M_0, M_0) are certainly equivalent
(M_0, M_1) are almost equivalent
(M_0, A_2) are sufficiently equivalent

This means that the message of equivalences is fuzzy.

Let μ be the degree of fuzziness regarding the equivalence for each couple.

Let us suppose for simplicity, that μ is symmetric and suppose we have the following matrix reflecting the fuzzy equivalence of the couples.

R =

	M_0	M_1	M_2	A_0	A_1	A_2
M_0	1	0.9	0.7	1	0.9	0.6
M_1	0.9	1	0.9	0.9	1	0.7
M_2	0.7	0.9	1	0.6	0.9	1
A_0	1	0.9	0.6	1	0.9	0.7
A_1	0.9	1	0.9	0.9	1	0.7
A_2	0.6	0.7	1	0.7	0.7	1

If the decider considers that the couples (x,y) are equivalent at the threshold $u(x,y) \geqslant 0.9$, then from the matrix R, we obtain a matrix R_1 such that:

$u_1(x,y) = 1$ if $u(x,y) \geqslant 0.9$ and
$u_1(x,y) = 0$ if not.

Then, R_1 =

	M_0	M_1	M_2	A_0	A_1	A_2
M_0	1	1	0	1	1	0
M_1	1	1	1	1	1	0
M_2	0	1	1	0	1	1
A_0	1	1	0	1	1	0
A_1	1	1	1	1	1	0
A_2	0	0	1	0	0	1

We have already seen this matrix and we give the decider his classes of equivalence. He is transitive only in the classes.

If he decides that his threshold is 0.7 for example, this means that he considers (x,y) equivalent if $u(x,y) \geqslant 0.7$. Then repeating the same process as for the matrix R, we obtain a matrix R_2 and his classes of equivalence.

3. Fuzzy case.
Here, the decider is not obliged to be symmetric. Then, the matrix R is only reflexive. However, the transitive closing of R is a relation of pre-order (transitive and reflexive). To employ the Malgrange method, we construct a matrix R_3 such that:

$u_3(x,y) = 0$ if $u(x,y) \neq u(y,x)$ or if

$u(x,y) = u(y,x) = 0$

$u_3(x,y) = 1$ if not.

We employ the Malgrange method and determine the classes of equivalence or his maximal sub-relations of similitude. We can conclude that the transitivity problem is solved: if the decider's equivalences are not transitive (fuzzy transitive) we must determine the classes of alternatives in which he is reflexive symmetric and transitive (e.g.: we determine his maximal sub-relations of similitude).

IV. Remarks about the substitution principle
Let us consider the following lottery:

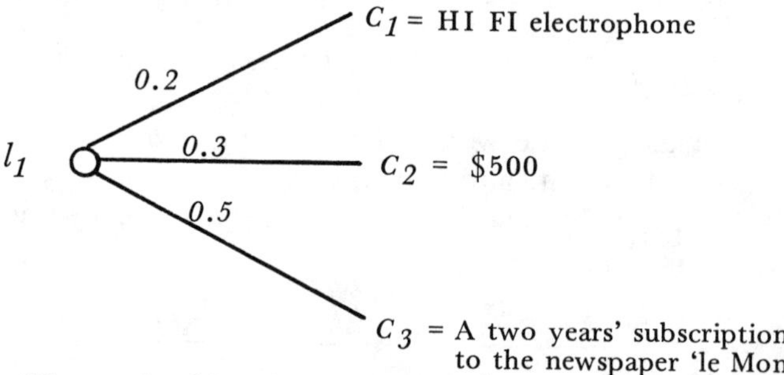

We are looking for a lottery l'_1 equivalent (by substitution) to l_1, according to the principle P_1. Let us state this principle again.

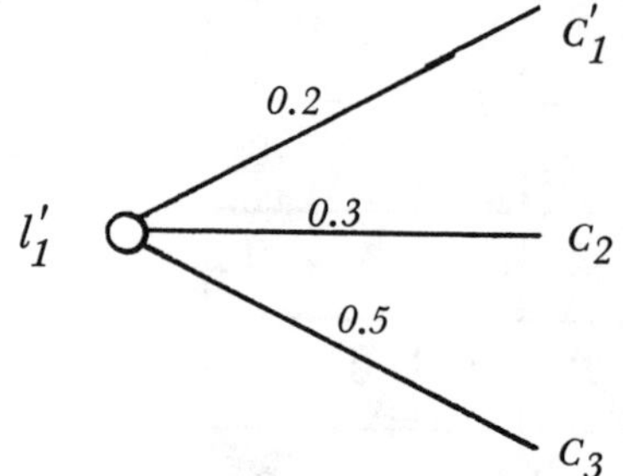

If a lottery is modified by changing a prize (C_1) to an equivalent prize (C'_1), without changing the other winnings (C_2, C_3) and the winning probabilities, the first lottery (l_1) and the modified one (l'_1) are equivalent for the decider (gambler).

This means that we are preoccupied by looking for an equivalent to the winning C_1 and only C_1; disregarding the other winnings.

Let us take an example of a decider who reasons as follows:

Considering the lottery l_1: for this decider (gambler), a Hi-Fi is equivalent to a five years' subscription to *'Le Monde'*, his basis for this decision is that he already has a Hi-Fi for one, and secondly, a five years' subscription to the newspaper will distinguish him from his colleagues (psychological advantage).

Without changing either the probability of winning, or C_2 and C_3 we have:

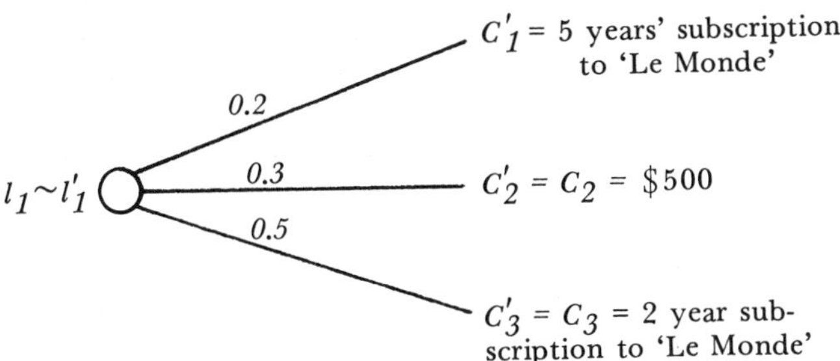

Let us continue:

The same decider, considering C_2 and only C_2 in l'_1 estimates that C_2 is equivalent to a one year subscription to the newspaper *'Le Monde'*. He thinks that his wife would look up to him if he subscribed to such a serious newspaper as *'Le Monde'*. Then by substitution (P_1) we have:

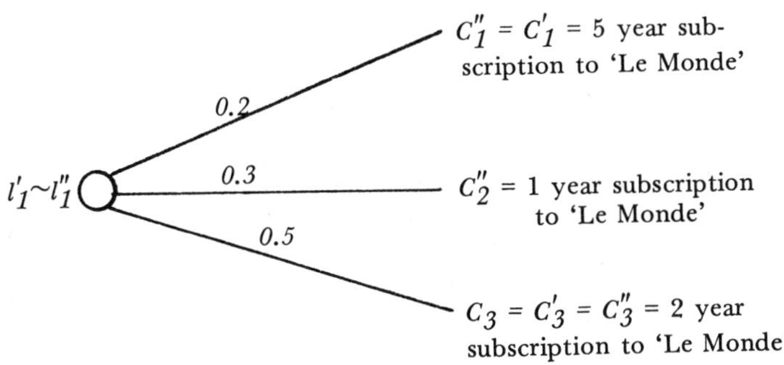

Thus, according to the principles P_1 and P_2 we have:

$$l_1 \sim l'_1 \sim l''_1$$

Let us analyse what l''_1 represents:

l''_1 is equivalent to an urn in which there are 2 red balls (5 years subscription to *'Le Monde'*), three white balls (one year subscription to *'Le Monde'*) and five yellow balls (2 years subscription to *'Le Monde'*). Then the only possible result: a subscription to the newspaper *'Le Monde'*.

However, the decider, comparing l_1 to l''_1, can conclude that, in l_1, he has five chances to win a subscription to the newspaper, three chances to win $500 which permit him to buy the promised bicycle for his son if he passes his exams (another criterion), and two chances to win a Hi-Fi which will give great pleasure to his daughter for her birthday (different criterion).

So, comparing l_1 to l''_1, he certainly prefers l_1. He thinks that l_1 is more 'balanced' than l''_1, and estimates that he has been selfish in l''_1. This is contrary to the substitution principle. That is, in the principle P_1, the decider judges a winning, apart from the other prizes of the lottery. If he reasons like this, in every substitution case, he can have a different criterion of equivalence substitution, when he must consider the winnings all together. In this manner, he can judge with the maximum number of preference or equivalence criteria. The subjective probabilities are based on the two principles P_1 and P_2. If the substitution and transitivity principles are based only on one criterion all along the substitution by equivalences, then we can employ subjective probabilities. If not, we think that we end to contradictions, without assigning these contradictions to the decider. Then, what can we do if we reason with many criteria at once?

Let us take an example:

Let us consider the set of prizes of the lottery l_1, and the set of goods which would be equivalent to the prizes.

Let us consider the substitution (multi-criteria now) of the Hi-Fi, and consider the goods which would be equivalent to it.

Let us note

H = Hi-Fi
D = Diamond
$AB5$ = Five year subscription to the newspaper *'Le Monde'*
P = Skin diving equipment

Let us consider the fuzzy equivalence message if H is the initial prize in l_1: we note it **H**.

Suppose we have **H**: $\{(H, 1); (D, 0.9); (AB5, 0.1); (P, 0.5)\}$

This means that:

H is certainly equivalent to H: $u(H,H) = 1$ (reflexivity)

H is nearly equivalent to D : $u(H,D) = 0.9$; because a diamond will give as much pleasure to his daughter as a Hi-Fi.

There is a very little equivalence between H and $AB5$, the decider estimates that he has an important probability to win a subscription to the newspaper in $C_3 \Rightarrow u(H,AB5) = 0.1$. He estimates also, that H is not very equivalent to P, because if so, he seems to be selfish (he likes skin diving and wants to have a subscription to *'Le Monde'*).

If we continue to reason like that, the decider proceeds alike if D is the initial state, next if $AB5$ is the initial state and also if P is the initial prize.

Let us suppose we have the following messages:

H : $\{(H, 1); (D, 0.9), (AB5, 0.1); (P, 0.5)\}$

D : $\{(H, 0.9); (D, 1); (AB5, 0.5); (P, 0.4)\}$

AB5 : $\{(H, 0.9); (D, 0.9), (AB5, 1), (P, 0.8)\}$

P : $\{(H, 0.8), (D, 0.9), (AB5, 0.3), (P, 1)\}$

1. First method

We shall determine the nearest transitive messages.

For example, if H and $AB5$ are the nearest transitive messages, and if we are obliged to substitute another prize to H, this winning must be $AB5$.

We shall note, for simplicity in the sequel, **H** : *H;* **D** : *D,* **AB5** : *AB5,* **P** : *P.*

Let us then, compute the relative Hamming's distance [6], of these fuzzy messages (example: $\delta(H,D) = \frac{1}{4} d(H,D) = \delta(D,H)$).

We have:

$\delta(H,D) = 0.175$
$\delta(H,AB5) = 0.325$
$\delta(H,P) = 0.225$
$\delta(D,AB5) = 0.25$
$\delta(D,P) = 0.25$
$\delta(AB5,P) = 0.25$

We find the following matrix, which we call R:

$R =$

	H	D	AB5	P
H	0	0.175	0.325	0.225
D	0.175	0	0.25	0.25
AB5	0.325	0.25	0	0.25
P	0.225	0.25	0.25	0

Let us look for the transitive closure min-max of this relation R we have: $R \circ R = R^2$:

$R^2 =$

	H	D	AB5	P
H	0	0.175	0.25	0.225
D	0.175	0	0.25	0.225
AB5	0.25	0.25	0	0.25
P	0.225	0.225	0.25	0

We find $R^3 = R^2$. Let us note $\check{R}$ the min-max transitive closure of R. We have $\check{R} = R \cap R^2$.

$\check{R} =$

	H	D	AB5	P
H	0	0.175	0.25	0.225
D	0.175	0	0.25	0.225
AB5	0.25	0.25	0	0.25
P	0.225	0.225	0.25	0

and we find the following hierarchy of the fuzzy messages:

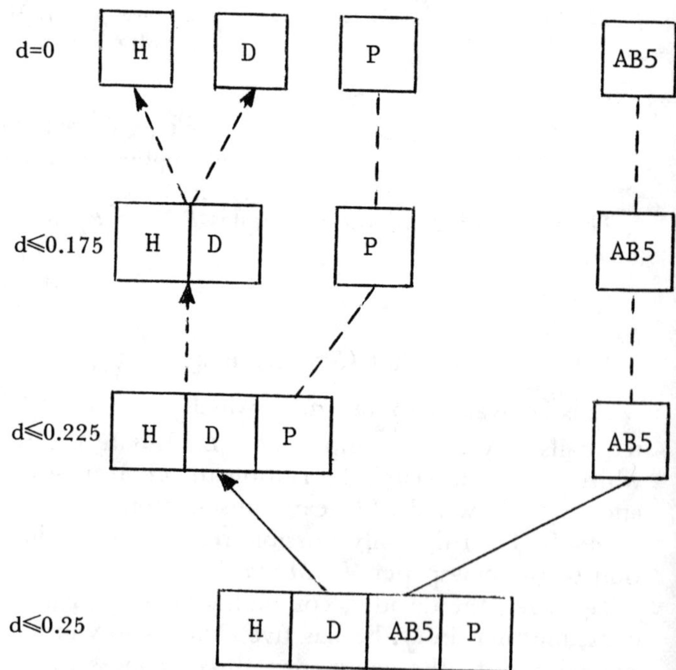

Thus, we see that the nearest transitive messages are H and D. That is, if the initial stage is H, the decider considers that H and D are equivalent and vice-versa. If he could risk more, he might admit that H, D and P are equivalent (for $d \leqslant 0.225$).

2. *Second method*

Let us consider the initial messages **H, D, P, AB5.** When we have noted these messages, we have noted these messages, we have noted in fact the product: $\{H,D,AB5,P\} \times \{H,D,AB5,P\}$. Then we have the following matrix R_1.

$R_1 =$

	H	D	AB5	P
H	1	0.9	0.1	0.5
D	0.9	1	0.5	0.4
AB5	0.9	0.9	1	0.8
P	0.8	0.9	0.3	1

Let us look for the transitive max-min closure of R_1.

Let $\hat{R}_1$ we have:

$R_1^2 =$

	H	D	AB5	P
H	1	0.9	0.5	0.5
D	0.9	1	0.5	0.5
AB5	0.9	0.9	1	0.8
P	0.9	0.9	0.5	1

We find $\hat{R}_1 = R_1 \cup R_1^2$. Then, $\hat{R}_1$ is transitive and reflexive. Let us look for the maximal sub-relations of similitude in $\hat{R}_1$ then we consider the matrix R_3 such that:

$$u_{R_3}(x,y) = \begin{cases} 1 & \text{if } u_{\hat{R}_1}(x,y) = u_{\hat{R}_1}(y,x) \neq 0 \\ 0 & \text{if not} \end{cases}$$

We obtain:

$R_3 =$

	H	D	AB5	P
H	1	1	0	0
D	1	1	0	0
AB5	0	0	1	
P	0	0		1

and the only maximal sub-relations of similitude are: $\{H,D\}$, $\{AB5\}$, $\{P\}$

Then, the decider can substitute D to H in the lottery l_1, and if the degree of equivalence $(u(H,D) = 0.9)$ satisfies him.

V. Conclusion

The examples seen above are not a proof that subjective probabilities are invalid, and this was not our objective. But some remarks might be made.

1) The substitution principle is rarely connected to a single criterion of judgement about equivalence (or preference). The substitution of a piece of goods to another, must be done (as in our example), by considering all the goods which result from a lottery.

We note that subjective probabilities are valid if we have one substitution criterion.

2) We now have a method to solve the problem of multicriterium substitution: we select the nearest transitive messages of the winnings.

3) If we don't consider the 'classical' substitution principle, the subjective probabilities will not be additive; then we are not obliged to admit that $SP(E_1) + SP(E_2) = SP(E_1 \cup E_2)$ if $E_1 \cap E_2 = \phi$. AND instead of the additivity operator we must employ other operators (MAX,MIN,...), which make the subjective probabilities lose their characteristics, such as probability of random subjective events, then we are obliged to deal with non-random events.

4) We observe that if the 'classical' transitivity is invalid, the fuzzy transitivity (example max-min) is valid and if the classical substitution is invalid, 'fuzzy substitution is valid'. This might help us to solve decision problems, in which we have to deal with non-random imprecisions.

Bibliography

1. CAPORELLI and DE LUCA, *Fuzzy sets and decisions theory infor. & control,* pp.446-473 (1973).
2. CHANG, C.L., 'Fuzzy topological spaces', *Jour. math. analysis and applications,* **24**, pp.182-190.
3. GLUSS, B., 'Fuzzy multistage decision making', *International jour. control UK* **17**, no.1, pp.177-192 (1973).
4. DE FINETTI, *Probability, Induction and Statistics*, John Wiley & Sons, New York.
5. FUNG & FU, 'On the *K*th optimal policy algorithms for decision making in a Fuzzy environment (1973).
6. KAUFMANN, A., 'Introduction à la théorie des sous-ensembles flous,' Tome 1 (Masson).
7. KAUFMANN, A., 'Introduction à la théorie des sous-ensembles flous', Tome 2 (Masson).
8. KYOJI & TANAKA, 'On the Fuzzy mathematical programming', (1973).
9. RAIFA, H., 'Analyse de la décision. Introduction aux choix en avenir incertain' (Dunod).
10. ZADEH & BELLMANN, 'Decision making in a Fuzzy environment', *Manag. Science* **17**, 4 (December 1970).
11. ZADEH, L.A., 'Outline of a new approach to the analysis of complex systems and decision process', ERL MEMO M342, University of California, Berkeley (July 1972).

A fuzzy algorithm for nonlinear systems identification

E.F. CAMACHO, C.G. MONTES, and J. ARACIL
Esc. Tec. Sup. Ing. Industriales, University of Seville, Spain

I. Introduction

The identification problem consists in the determination, on the basis of observation of input and output, of a system within a specified class of systems to which the system under test is equivalent [1]. The identification of a system is based on the optimization of a functional that represents the difference existing between the behavior of the model and the observed output of the system, such a functional cannot be obtained for most systems in a deterministic way as these behavior criteria are too subjective and too fuzzy to be expressed in such a way.

In this work a fuzzy formulation of the functional is presented, it permits a very effective convergence of the most significant characteristics of the observed output of the system, as maximum and minimum values, final value, etc.; topics which are not directly attended by the functionals normally used.

For the minimization of the fuzzy functional the simplex method is proposed; this method is shown to be effective and computationally compact. Spendley [2] introduced the simplex method for tracking optimum conditions by evaluating a function at a set of points forming new simpleces by reflecting the worst point of the simplex in the hyperplane formed by the remaining points. This idea has been successfully applied to the problem of minimization of a function of several variables by Nelder and Mead [3].

The simplex method is clearly applicable to the problem stated above: an analytical formulation of the criterion is not necessary and it is possible to insert fuzzy statements in the identification algorithm.

The proposed method is applied to the identification of a nonlinear system describing the growth and regulation of animal population. Finally the results are compared with the ones obtained by the application of the least square method.

II. Formulation of the fuzzy criterion

The specific formulation of a fuzzy criterion is bound to the characteristic behavior of the system considered; therefore the general ideas of the formulation will be made on the following system.

A continuous dynamic model of the prey-predator system is described by

$$dN_1/dt = p_1N_1 - p_3N_1N_2 \qquad (1a)$$

$$dN_2/dt = -p_2N_2 + p_4N_1N_2 \qquad (1b)$$

where t is time, N_1 and N_2 are the instantaneous populations of the prey and predators, respectively, and p_i $(i=1,...,4)$ are all positive constants. This is known as the Lotka-Volterra model [4].

The identification problem in this system is that of evaluating the constants p_i from the output data $N_1(t)$ given in Fig.1, that is estimating the set of values $\hat{p}_i$ so that the corresponding dynamic behavior $\hat{N}_1(t)$ is *as similar as possible* to $N_1(t)$.

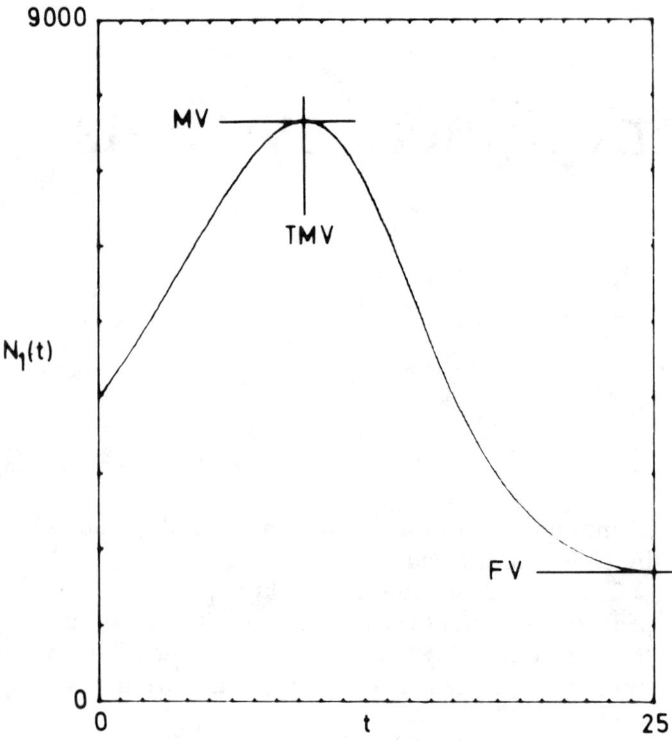

Fig.1 Output data of the system

The usual form of quantising the term as similar as possible is by using the functional:

$$J(\hat{p}_i) = \int_0^{TFV} (N_1(t) - \hat{N}_1(t))^2 \, dt \qquad (2)$$

which evaluates *all* the deviations of $\hat{N}_1(t)$ in respect to $N_1(t)$ in a non selective way.

The proposed formulation of $J(\hat{p}_i)$ permits the estimations $\hat{p}_i$ to be manoeuvered according to the most significant characteristics of the behavior of the system, that is: the maximum value (MV), the correspondent value of t (TMV) and the final value (FV); see Fig.1.

It is clear that if the structure of the model (1) previously attributed to the system is exact, an identification algorithm based on the minimization of $J(\hat{p}_i)$ formulated as in (2) will proportion a good fit of MV, TMV, FV, as well as all the other values that integrate $N_i(t)$. However if the given structure of the system is not exact, which is the most normal case in ecological systems, the functional (2) does not ensure an identification to cover satisfactorily MV, TMV, and FV, although the global solution $\hat{N}_1(t)$, $t=0,...,TFV$, is the best one possible from the point of view of minimizing the errors in every t.

Zadeh's development of fuzzy sets [5] and fuzzy algorithms [6] provides a means of expressing linguistic characteristics of $\hat{N}_1(t)$ in a form suitable for processing using a computer.

An index J formulated by the variables

$$EMV = (M\hat{V} - MV)/MV \qquad (3)$$

$$ETMV = (T\hat{M}V - TMV)/TMV \qquad (4)$$

$$EFV = (F\hat{V} - FV)/FV \qquad (5)$$

permits the fit $\hat{N}_1(t)$ to be quantised by using the most significant characteristics of $N_1(t)$.

The goodness of $\hat{N}_1(t)$ can be evaluated by the grade of membership to fuzzy sets of the variables EMV, ETMV and EFV. For that the fuzzy set **MV** of good estimations $\hat{p}_i$ in respect to MV, that is MV is closed to zero, is defined. $f_{\mathbf{MV}}(EMV)$ is the membership function corresponding to the fuzzy set **MV**. **TMV** and **FV** are defined in a similar way.

In accordance with these definitions the fit of $\hat{N}_1(t)$ to $N_1(t)$ can be quantised by the algebraic product of the sets **MV**, **TMV** and **FV**. denoted by **G** and defined by

$$f_{\mathbf{G}}(EMV, ETMV, EFV) = $$

$$= f_{\mathbf{MV}}(EMV)\, f_{\mathbf{TMV}}(ETMV)\, f_{\mathbf{FV}}(EFV) \qquad (6)$$

Remembering that the membership function is expressed in real positive numbers of the closed interval [0,1], the index $J(\hat{p}_i)$ to be minimized by the identification algorithm can be formulated as follows:

$$J(\hat{p}_i) = 1 - f_{\mathbf{MV}}(\hat{p}_i)\, f_{\mathbf{TMV}}(\hat{p}_i)\, f_{\mathbf{FV}}(\hat{p}_i) \qquad (7)$$

The minimizing set of (7) is a fuzzy set **P** in $\hat{p}_i$ such that the grade of membership $f_{\mathbf{G}}$ of $\hat{p}_i$ in **P**

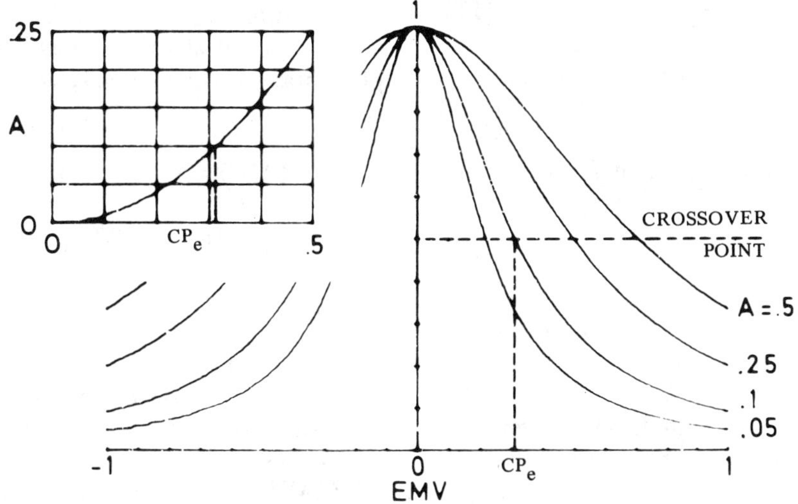

Fig.2 Grade of membership functions

represents the degree to which (7) is close to zero, that is the degree to which each variable EMV, ETMV, and EFV are null.

For the formulation of the function $f_{\mathbf{MV}}(EMV)$ the following expression has been used

$$f_{\mathbf{MV}}(FMV) = A_{MV}/(A_{MV} + EMV^2) \qquad (8)$$

where A_{MV} is a constant positive real value, $f_{\mathbf{TMV}}$ and $f_{\mathbf{FV}}$ are formulated in the same way. The value of A_{MV} determines the precision of $M\hat{V}$ corresponding to the crossover point of the fuzzy set **MV**, see Fig.2.

III. Minimization algorithm

Considering the type of index to be optimized $J(\hat{p}_i)$, the simplex algorithm proposed by Nelder *et al.* [3] with some slight modifications has been used. The Nelder and Mead version of the simplex method is widely considered as a very compact method for unconstrained minimization, it also permits the introduction of fuzzy statements in the process of optimization.

Consider the optimization of an index of n variables. A proper simplex is defined as a set of $n+1$ points $P_0, P_1, ..., P_n$ which span the space, that is: the rank of the $n \times (n+1)$ matrix P form as follows

$$P = [P_0 \mid P_1 \mid ... \mid P_n] \qquad (9)$$

is n. The simplex is denoted as regular if the points are mutually equi-distant (a tetrahedron in a 3-dimensional space). The following terminology is used:

P_i , $i = 0,1,...,n$	are the vertices of the simplex, each of them having coordinates p_{ij}, $j = 1,2,...,n$
$J(p_i)$	is the index to be optimized
w	is the suffix such that P_w is the *worst* point of the simplex
b	is the suffix such that P_b is the *best* point of the simplex
P_c	is the centroid of the points P_i with $i \neq w$
(P_i, P_j)	is the distance between the points P_i and P_j

The method works as follows: At each stage in the process the *worst* point P_w is replaced by a new point. For the determination of the new point three operations are used:

(i) Reflection: The reflection of P_w is denoted by P_{w1} and it is defined by the relation

$$P_{w1} = (1+a)P_c - aP_w \qquad (10)$$

where a is a positive constant called the reflection coefficient. It has been fixed to one in the present application.

(ii) Expansion: The expansion of P_{w1} is denoted by P_{w2} and defined by the following relation

$$P_{w2} = b\,P_{w1} + (1-b)P_c \qquad (11)$$

where b is a constant greater than the unity and called the expansion coefficient. In this case $b = 1.5$.

(iii) Contraction: The contraction of P_w is denoted by P_{w2} and defined by

$$P_{w2} = c\,P_w + (1-c)P_c \qquad (12)$$

The contraction coefficient c is a positive constant smaller than unity. It has been fixed to 0.5 in this application.

The principal features of the method consist in applying a reflection to P_w and then an expansion or a contraction if P_{w1} is better or *worse* than P_b or P_w respectively. P_w is replaced by the best point obtained through these transformations. If a point *better* than P_w is not obtained all the points of the simplex are replaced by $P_i = (P_i + P_b)/2,\ i = 0,1,\ldots,n$, and the process is restarted.

The evaluation of the relationship P_i *better* than P_j is made through the membership functions $f_{\mathbf{G}}(P_i)$ and $f_{\mathbf{G}}(P_j)$ where **G** is the fuzzy set which has been defined above.

IV. Results

The proposed algorithm has been tried with the output data, shown in Fig.1, with membership functions:

$$f_{\mathbf{MV}}(EMV) = 0.01/(0.01 + EMV^2) \qquad (13)$$

$$f_{\mathbf{TMV}}(ETMV) = 0.005/(0.005 + TMV^2) \qquad (14)$$

$$f_{\mathbf{FV}}(EFV) = 0.01/(0.01 \div EFV^2) \qquad (15)$$

In Fig.3 the evolution of EMV is represented against the number of transformations applied to the original simplex; in the same way the evolution of EFV is illustrated in Fig.4. The different values of ETMV are not shown as they are negligible after the first few transformations.

The results obtained using a combined formulation of $J(\hat{P}_i)$, described as follows, are also shown.

$$J(\hat{p}_i) = [\int_0^{TFV} (N_1(t) - \hat{N}_1(t))^2\,dt] \,/\, p_1 p_2 p_3 \qquad (16)$$

It can be seen that the combined index gives, as is to be expected, an intermediate convergence between those proportioned by the square index (2)

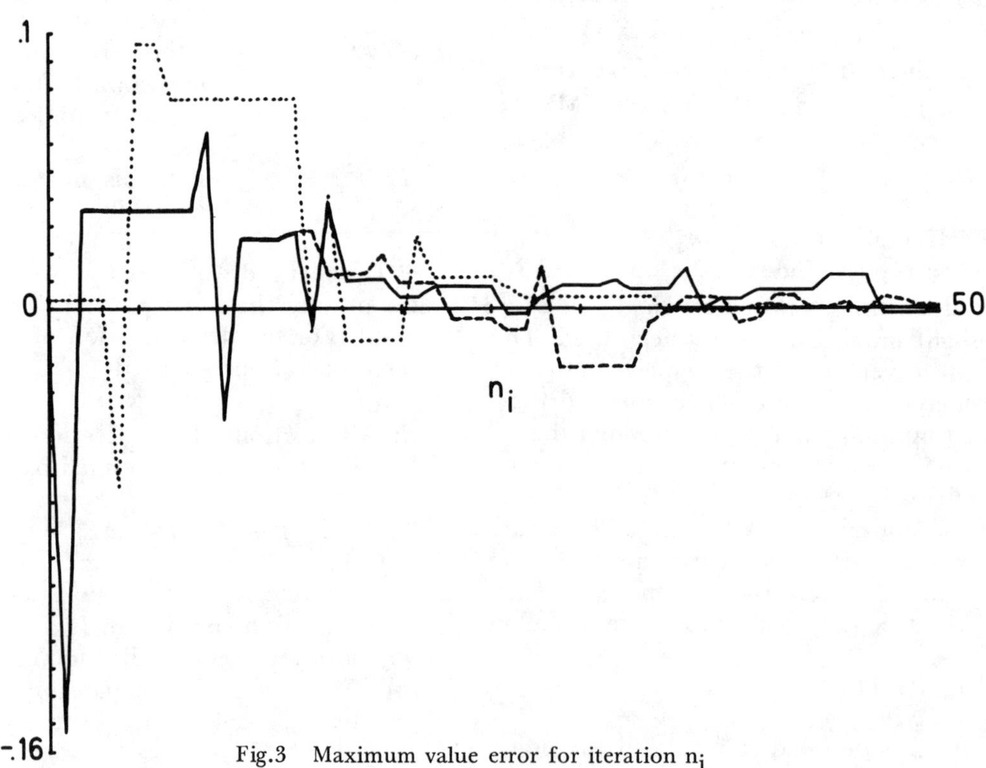

Fig.3 Maximum value error for iteration n_i

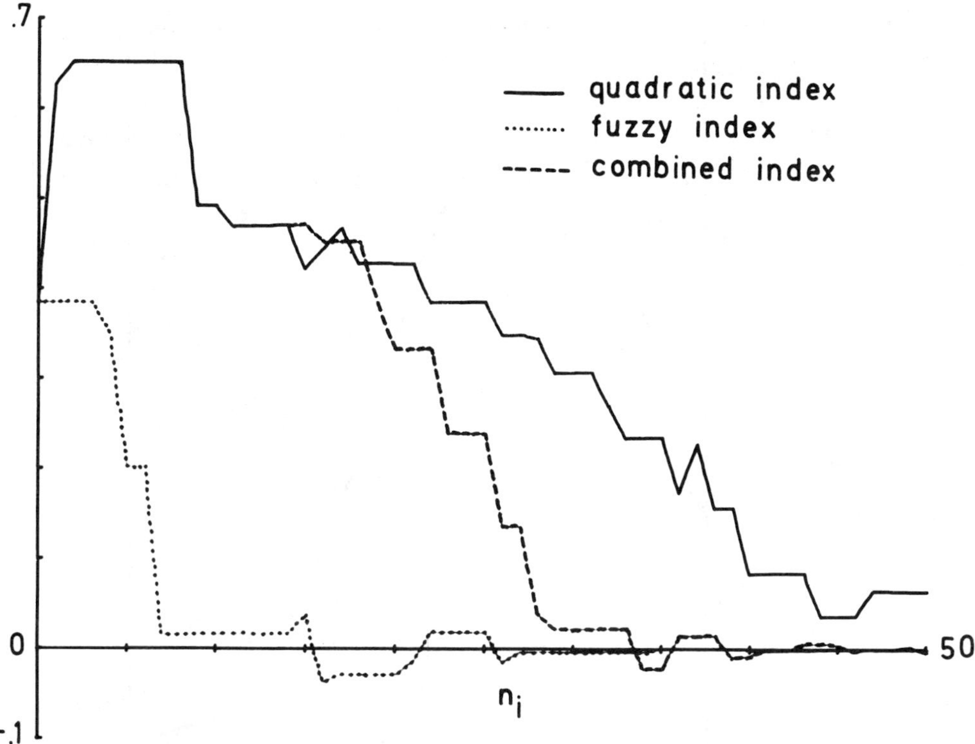

Fig.4 Final value error for iteration n_i

and the fuzzy one (7). However, there is a fact that makes the use of the combined index (16), useful; it minimizes the sum of the square errors (SE) with a greater speed of convergence than the pure square index (2), see table I.

No. iter	SE with the expression (2)	SE with the expression (16)
0	0.4301×10^8	0.4501×10^8
10	0.3963×10^8	0.3104×10^7
20	0.2023×10^7	0.3680×10^6
30	0.1587×10^6	0.3109×10^5
40	0.4256×10^5	0.7985×10^3
50	0.1099×10^5	0.2552×10^3

Finally, in Fig.5 the global estimation $\hat{N}_1(t)$ obtained with fuzzy and square indexes can be compared after the thirtieth iteration.

V. Conclusions

The results obtained by the fuzzy algorithm of identification are comparable to those given by square index methods, with the advantage of ensuring a fit of $\hat{N}_1(t)$ to the 'singularities' of the known output data $N_1(t)$ (maximum, minimum, final value, etc.).

The approach described here is not proposed as an alternative to conventional identification algorithms where these are effective; however in systems with structures not exactly known, fuzzy algorithms may be a non trivial alternative approach.

In spite of the fact that the proposed algorithm is not the most obvious choice for systems with exactly known model structures, as in the case of the presented application, it has been possible to prove that the combined index is more effective than the square from all points of view.

References

1. ZADEH, A., 'From circuit theory to system theory', Proceedings of the IEE, pp.856-865 (May 1962).

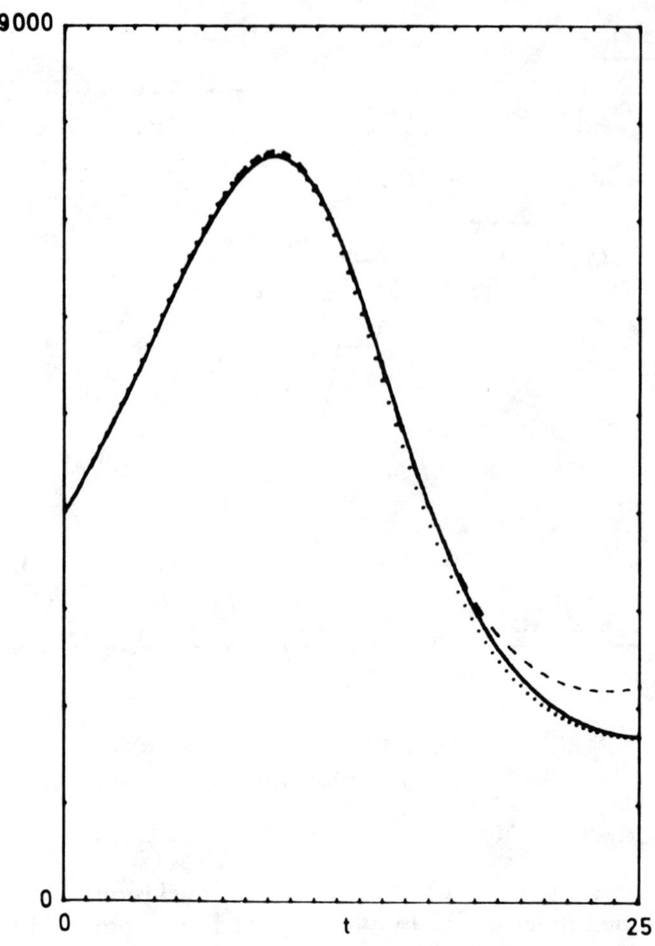

Fig.5 Comparison of $\hat{N}_1(t)$ corresponding to the iteration 30
—— given output data
.... using the fuzzy index
- - - - using the quadratic index

2. SPENDLEY, W., HEXT, G.R. and HIMSWORTH, F.R., 'Sequential application of simplex designs in optimization and evolutionary operation', *Technometrics* 4, pp.441-461 (1962).
3. NELDER, J.A. and MEAD, R., 'A simplex method for function minimization', *Comput. J.* 7, pp.308-313 (1965).
4. VOLTERRA, V., 'The general equations of biological strite in the case of historical actions', Proc. Edinburgh Math. Soc. 6 (1939).
5. ZADEH, L.A., 'Fuzzy sets', *Information and Control* 8, pp.338-353 (1965).
6. ZADEH, L.A., 'Fuzzy algorithms', *Information and Control* 12, pp.94-102 (1968).

A fuzzy programming approach to an air pollution regulation problem

G. SOMMER[†] and M.A. POLLATSCHEK[‡]

[†] *Lehrstuhl für Unternehmensforschung (Operations Research), Aachen, FRG*
[‡] *Operations Research Area, Faculty of Management Science, Technion, Haifa, Israel*

A remark on optimization of abatement policies
In this article we present a refinement of the mathematical model of abatement policies as were proposed in numerous publications of Gorr, Gustafson and Kortanek *et al.* (see in particular [2-8] and the references given there).

They considered a finite number of sources in a given control area and investigated the reduction of emission rates of controllable sources so as to comply with given standards in the entire control area. Their problem was to choose the reduction factors, which entail minimum overall cost, they also obtained those locations of the control-area in which the pollutant's concentration is maximal. The techniques developed by them for one specific pollutant may be extended to the general case of a number of pollutants. Their assumptions which led to the optimization model were:

1. There is a formula for each source which gives the concentration of the pollutant in a given location (determined by its coordinates) called transfer function. It is constant in time and depends only on the source and the location.
2. The reduction is feasible but limited.
3. The air-quality standard is given as a fixed concentration which should not be surpassed in a given location.

Although these assumptions are very convenient in the mathematical model, they are not very realistic: changes of weather and emission rates in the source imply transfer functions which depend on time. Moreover, air-quality standards are of 'long-run' and 'short-run' types, the former referring to average and the latter to peak concentration, in some form or other. Therefore, no single limit can be given for it as under 3). The same is true for assumption 2).

In short, the difference between reality and a model is that reality is vague or fuzzy, whereas a model normally is clear-cut. This is typical not only of the model under consideration, but of any mathematical model to some extent. Recently the gap between reality and a model has become smaller by the introduction of fuzzy mathematics and decision-making or optimization in fuzzy environment [1, 9-17].

We shall show that it is possible to introduce fuzziness into the described air pollution regulation problem above and therefore to obtain a truer representation of reality.

For keeping the presentation brief and simple we shall consider explicitly the problem of emission-reduction of a few controllable sources when the sampling stations are fixed in known locations. The same ideas can be applied to more involved models. Our aim is, however, to introduce only the idea and method of Fuzzy Programming in pollution abatement policies and to show its usefulness.

In Section 1 we describe the air pollution control problem that is to be solved by fuzzy programming.

In the Section 2 and 3 we consider the elements of fuzzy optimization and construct a special fuzzy optimization model for the air pollution regulation problem in discussion, which will be illustrated in Section 4 by the numerical example from Gustafson and Kortanek [7].

The last Section 5 contains some general remarks on optimization in fuzzy environments, including the handling of the logical 'and' by the introduced method [13].

1. Description of the air pollution control problem

We intend to improve the quality of life as concerns the purity of air in a given control area, in which each point is expressed by the coordinates (x_1, x_2) of a cartesian coordinate system.

In concreto: The present pollutant's concentration known to us is to be modified so that

- given air quality standards are strictly complied to, and
- (*) - people's desire to reach an even better concentration rate than the adopted standard is taken into consideration, and
- (*) - the reduction of the emission does not entail too big a decrease of the plants' production
- (*) - by 'minimizing' the reduction cost.

Because of the requirements (*), this leads to a fuzzy optimization problem which we want to solve in sections below.

At first we make the following assumptions to simplify the mathematical handling of this task. These assumptions do not change the model essentially.

- We consider a single kind of pollutant which is released from several sources in the control area.
- The given air quality standards are to be complied with at scattered so-called receptor-points in the control area.
- There is no change in weather and climate conditions. This implies that we do not need parameters expressing quantities like wind-speed and direction, pressure change, moisture, temperature, etc. This can be considered in a more elaborate model.

A known 'transfer function' s_j exists for each source j $(j = 1, \ldots, J)$ which indicates the concentration of the pollutant at the ground level with the coordinates (x_1, x_2). This function may be of generalized Gauss type

$$s_j(x_1, x_2) = \alpha_j e^{\beta_j}$$

where the sources' special parameters α_j and β_j are known functions of x_1 and x_2. The concentration rate at a receptor-point i $(i = 1, \ldots, I)$ caused by the source j is

$$s_{ji} = s_j(x_1^i, x_2^i)$$

where (x_1^i, x_2^i) are the coordinates of the receptor-point i. s_{ji} is one of the given data and expresses the actual concentration rates.

Next, we introduce the decision variable $E_j \in [0, 1]$ as an indicator for the reduction of the pollutant at the source j, so that

E_j	is the reduction rate, and
$(1 - E_j)s_{ji}$	is the concentration rate after reduction.

Finally, assuming that the total concentration at one receptor-point is the sum of the pollutant's concentrations that are released from all sources, we determine

$$\sum_{j=1}^{J} (1 - E_j)s_{ji}$$

as the receptor-point's total concentration (after reduction, if $\exists j : E_j > 0$).

Now, the first type of fuzzy restrictions is:

'Let, for each receptor-point i, $\sum_{j=1}^{J} (1 - E_j) s_{ji}$ not exceed the standard d but try to go below e_i which is the desirable quantity at this receptor-point i!'

This, we write as:

$$\sum_{j=1}^{J} (1 - E_j)s_{ji} \lesssim e_i \,;\, d \qquad \forall i \qquad (1)$$

The second type of fuzzy restriction is:

'Ensure, for each source j, that the reduction rate does not exceed $\bar{E}_j$ and even try to go below w_j!'

This, we write as:

$$E_j \lesssim w_j \,;\, \overline{E}_j \qquad \forall j \qquad (2)$$

$(\overline{E}_j, w_j \in [0, 1])$ Finally, we want to formulate the fuzzy objective function as:

$$\sum_{j=1}^{J} c_j \quad E_j \quad \text{'min'} \qquad (3)$$

where c_j is the total reduction cost at the source j, assuming (or having approximately determined) that the cost function is linear on $[0, \overline{E}_j]$.

How to treat this fuzzy programming problem (FPP) mathematically, will be discussed in the section below.

2. Developing linear programming problems as substitutes for fuzzy optimization problems

Section 2.1 (The peculiarity of fuzzy restrictions) In ordinary mathematical programming problems we formulate equations and inequations in order to express requirements such as

→ do not exceed ..., or
→ exceed (or comply with) ..., or
→ comply exactly with ... .

We call them 'common restrictions' and write them (in Linear Programming Problems):

$$\forall i \in N_4 {}^{*)} \quad a_i' x \leqslant \alpha_i \qquad \text{Type } `\leqslant' \qquad (4)$$

$$\forall i \in N_5 \quad a_i' x \geqslant \alpha_i \qquad \text{Type } `\geqslant' \qquad (5)$$

$$\forall i \in N_6 \quad a_i' x = \beta_i \qquad \text{Type } `=' \qquad (6)$$

where $a_i \in R^m$ is the vector of the coefficients of the decision variables $x \in R^m$ and α_i, $\beta_i \in R$ is the quantity which is (not) to be exceeded or complied with.

*) $N_1 = \{i \mid i = 1, \ldots, n_1\}$
$N_2 = \{i \mid i = n_1+1, \ldots, n_1+n_2\}$
$N_3 = \{i \mid i = n_1+n_2+1, \ldots, n\}$
$N_4 = \{i \mid i = n+1, \ldots, n+n_4\}$
$N_5 = \{i \mid i = n+n_4+1, \ldots, n+n_4+n_5\}$
$N_6 = \{i \mid i = n+n_4+n_5+1, \ldots, n+\overline{n}\}$

This is the well-known notation of restrictions in common LP.

In our case, we must formulate restrictions like these, but it is also necessary to express, that *it would be desirable* to reach even a better value of $a_i' x$, namely

$$a_i' x \leqslant \beta_i \quad (\beta_i \leqslant \alpha_i \text{ in case } `\leqslant'), \text{ or}$$

$$a_i' x \geqslant \beta_i \quad (\beta_i \geqslant \alpha_i \text{ in case } `\geqslant') .$$

This leads to the formulation of 'fuzzy' restrictions.

(*a*) Do not exceed α_i but try to go below β_i will be the first type of a fuzzy restriction, which we write as

$$\forall i \in N_1 \quad a_i' x \lesssim \beta_i \,;\, \alpha_i \qquad \text{Type } `\lesssim' \qquad (7)$$

(*b*) Otherwise, do not go below α_i but try to exceed β_i is written as

$$\forall i \in N_2 \quad a_i' x \gtrsim \beta_i \,;\, \alpha_i \qquad \text{Type } `\gtrsim' \qquad (8)$$

(*c*) Finally, the 'fuzzyfization' of a '='-restriction would be:

$$\forall i \in N_3 \quad a_i' x \eqsim \beta_i ;\; \alpha_i \,;\, \gamma_i \qquad \text{Type } `\eqsim' \qquad (9)$$

which means: Try to comply with β_i, but ensure, that $\alpha_i \leqslant a_i' x \leqslant \gamma_i$.

A so-called *membership-function*

$$\forall i \in \bigcup_{j=1}^{3} N_j \qquad \mu_i : \{a_i' x\} \rightarrow [0, 1]$$

expresses the decision-maker's degree of satisfaction about the actual value $a_i' x$ of the restriction i. We define these functions as:

$$\forall i \in N_1 \quad \mu_i = \begin{cases} 1 & \text{if } a_i' x < \beta_i \\ \dfrac{\alpha_i - a_i' x}{\alpha_i - \beta_i} & \text{if } a_i' x \in [\beta_i, \alpha_i] \end{cases}$$

$$\forall i \in N_2 \quad \mu_i = \begin{cases} 1 & \text{if } a_i' x > \beta_i \\ 1 - \dfrac{\beta_i - a_i' x}{\beta_i - \alpha_i} & \text{if } a_i' x \in [\alpha_i, \beta_i] \end{cases}$$

$$\forall i \in N_3 \quad \mu_i = \begin{cases} 1 - \dfrac{\beta_i - a_i'x}{\beta_i - \alpha_i} & \text{if } a_i'x \in [\alpha_i, \beta_i] \\[2ex] \dfrac{\gamma_i - a_i'x}{\gamma_i - \beta_i} & \text{if } a_i'x \in (\beta_i, \gamma_i] \end{cases}$$

and illustrate them in Fig.1-3.

We discuss each kind of FPP and formulate it as a linear programming problem. In general, the type of the special objective function will be the distinguishing mark for the FPP.

At first, we deal in FPPs without any fuzzy objective function. The next task will be the discussion of FPPs with an objective function which leads to the description of the FPP which we need in order to solve the air pollution regulation problem.

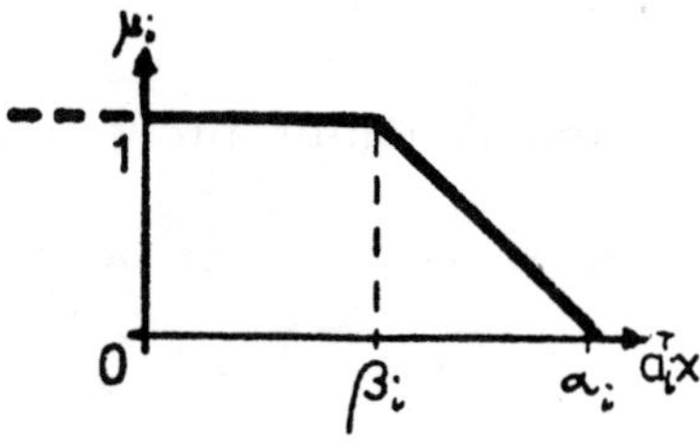

Fig.1 (Type '$\lesssim$')

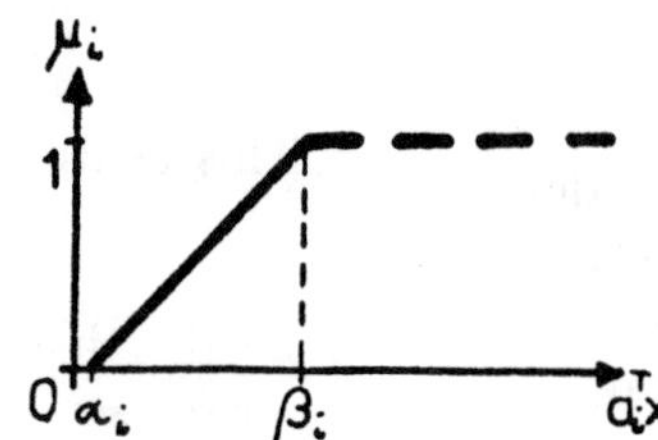

Fig.2 (Type '$\gtrsim$')

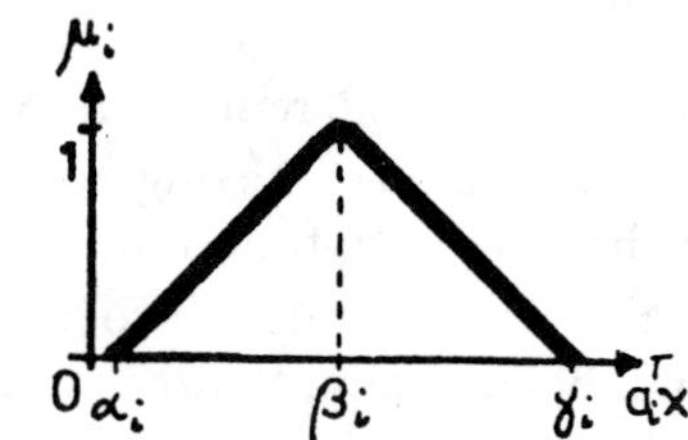

Fig.3 (Type '$\eqsim$')

Section 2.2 (The peculiarity of fuzzy objective functions)
Contrary to common LP we interpret in our approach a so-called fuzzy objective function

$c'x \rightarrow$ 'max (min)'

as: 'Maximize (Minimize)' $c'x$, $c \in \mathbb{R}^m$, but take the μ_i of all fuzzy restrictions into consideration. And, in fuzzy decision problems, we also have to consider the case that there is no objective function at all. Similar to the section above we want to introduce a symbol for each type:

$c'x \rightarrow$ 'max (min)'	Type '$\rightarrow$'
no objective function	Type '0'

Section 2.3 (Special types of derived optimization problems).
Combining the different types of objective functions with the types of restrictions leads to a set of various fuzzy programming problems (FPP) that are listed below.

	at least one of '$\leqslant$', '$\geqslant$' or '=' and '$\lesssim$', '$\gtrsim$' or '$\eqsim$'	'$\lesssim$' and/or '$\gtrsim$' and/or '$\eqsim$'
'0'	mixed FPP '0'	pure FPP '0'
'$\rightarrow$'	mixed FPP '$\rightarrow$'	pure FPP '$\rightarrow$'

Table 1 Typology of Fuzzy Programming Problems

Section 2.4 (LP as an FPP's substitute in case of no objective function). We consider the pure FPP '0':

(*a*) A restriction of type '$\lesssim$' (see (7)) is formulated as two inequalities by means of a slack variable $s_i^0 \geqslant 0$:

$$\forall i \in N_1 \begin{cases} a_i'x - s_i^0 \leqslant \beta_i, & \text{while adding} \\ s_i^0 \leqslant \alpha_i - \beta_i \end{cases} \tag{10}$$

to ensure that $a_i'x \leqslant \alpha_i$, $\forall i \in N_1$.

(*b*) An equivalent substitute for a restriction of type '$\gtrsim$' is found with $s_i^u \geqslant 0$:

$$\forall i \in N_2 \begin{cases} a_i'x + s_i^u \geqslant \beta_i, \text{ and} \\ s_i^u \leqslant \beta_i - \alpha_i \end{cases} \tag{11}$$

(*c*) Finally, type '$\eqsim$' can be written with s_i^0, $s_i^u \geqslant 0$:

$$\forall i \in N_3 \begin{cases} a_i'x - s_i^0 + s_i^u = \beta_i, \text{ and} \\ s_i^0 \leqslant \gamma_i - \beta_i, \text{ and} \\ s_i^u \leqslant \beta_i - \alpha_i \end{cases} \tag{12}$$

The membership-functions for each type imply:

$$\forall i \in N_1 \qquad \mu_i(s_i^0) = 1 - \frac{s_i^0}{\alpha_i - \beta_i} \qquad (13)$$

$$\forall i \in N_2 \qquad \mu_i(s_i^u) = 1 - \frac{s_i^u}{\beta_i - \alpha_i} \qquad (14)$$

$$\forall i \in N_3 \qquad \mu_i(s_i^0, s_i^u) = 1 - \frac{s_i^u}{\beta_i - \alpha_i} - \frac{s_i^0}{\gamma_i - \beta_i} \ {}^{*)} \qquad (15)$$

Adding

$$(NN) \begin{cases} \forall x \in R^m & x \geqslant 0 \text{ componentwise} \\ \forall i \in N_1 \cup N_3 & s_i^0 \geqslant 0 \\ \forall i \in N_2 \cup N_3 & s_i^u \geqslant 0 \end{cases}$$

(requesting the non-negativity of each variable) the set of the feasible solutions is given by (10, 11, 12 and NN).

Our intention is the maximization of the satisfaction functions as follows:

$$\sum_{i \in N_1} \mu_i(s_i^0) + \sum_{i \in N_2} \mu_i(s_i^u) + \sum_{i \in N_3} \mu_i(s_i^0, s_i^u) \rightarrow \max \qquad (16)$$

(The problem of handling the logical 'and' in combining such membership-functions is briefly described in Section 5.)

The complete linear substitute for (7, 8 and 9) is now (10, 11, 12, NN and 16).

An example (see [13]) is given to illustrate this approach: The FPP '0'

$$4x_1 + x_2 \lesssim 16;\ 20$$

$$x_1 + x_2 \gtrsim 10;\ 9$$

$$-x_1 + x_2 \approx 4;\ 3.5\ ;\ 5 \qquad x_1, x_2 \geqslant 0$$

*) An annotation: The necessary condition that $s_i^0 = 0$ if $a_i'x \in [\alpha_i, \beta_i]$, and $s_i^u = 0$ if $a_i'x \in [\beta_i, \gamma_i]$ is always fulfilled by the term $\sum_{i \in N_3} \mu_i \rightarrow \max$ (see (16) below) which can easily be proved.

leads to the linear substitute:

$$0.25\, s_1^0 + s_2^u + 2 s_3^u + s_3^0 \rightarrow \min$$

subject to

$$\begin{array}{llr} 4x_1 + x_2 - s_1^0 & & \leqslant 16 \\ s_1^0 & & \leqslant 4 \\ x_1 + x_2 + s_2^u & & \geqslant 10 \\ s_2^u & & \leqslant 1 \\ -x_1 + x_2 - s_3^0 + s_3^u & & = 4 \\ s_3^0 & & \leqslant 1 \\ s_3^u & & \leqslant 0.5 \\ x_1, x_2, s_1^0, s_2^u, s_3^0, s_3^u & & \geqslant 0 \end{array}$$

Its optimal solution is:

$$(x_1, x_2, s_1^0, s_2^u, s_3^0, s_3^u)_{opt\,1} = (3, 7, 3, 0, 0, 0)$$

The linear substitute for a *mixed* FPP '0' is the above program enlarged by adding the restrictions of type '$\leqslant$', '$\geqslant$' or '=' (see (4, 5, 6).

Section 2.5 (LP as an FPP's substitute in case of a fuzzy objective function) We consider the pure FPP '$\hookrightarrow$' (see list in Section 2.3) in case of 'maximization'. I.e. we add

$$c'x \rightarrow \text{'max'} \qquad (17)$$

to (7, 8, 9), respectively to (10, 11, 12, NN). Similar to the formulation of the membership function above we define

$$\bar{\mu}_0 = \begin{cases} 1 & \text{if } c'x > \beta_0 \\ 1 - \dfrac{\beta_0 - c'x}{\beta_0 - \alpha_0} & \text{if } c'x \in [\alpha_0, \beta_0] \\ 0 & \text{if } c'x < \alpha_0 \end{cases}$$

as the expression of the decision-maker's degree of satisfaction about the objective function's value $c'x$ (see Fig.4). Contrary to the above discussion, where the β_i of all fuzzy restrictions are given data, we have the problem to determine a priori the—let us call them lower and upper limits—α_0 and β_0.

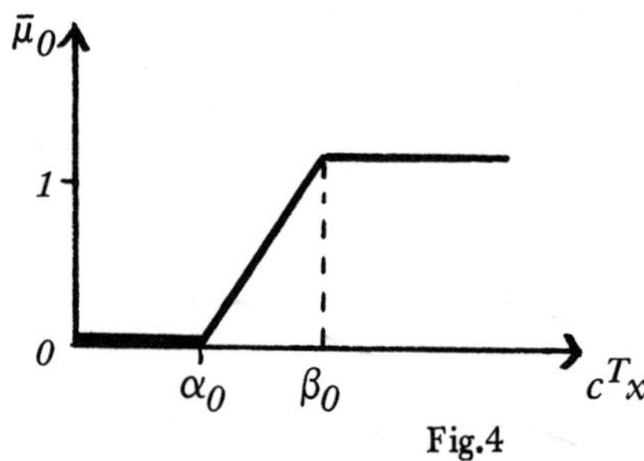

Fig.4

(Determination of α_0). The value of the objective function will not go below $c'\underline{x}$ where $\underline{x}$ is the optimal vector of the decision variables given by the linear substitute mentioned last (10, 11, 12, NN and 16).

Hence, we get $\alpha_0 = c'\underline{x}$ by solving the FPP's substitute without regarding the fuzzy objective function (respectively its membership function).

(Determination of β_0). The value of the objective function will not exceed $c'\bar{x}$ where $\bar{x}$ is the optimal solution of:

$$c'x \rightarrow \max$$

so that

$$\begin{aligned} a_i'x &\leqslant \alpha_i && \forall i \in N_1 \\ a_i'x &\geqslant \alpha_i && \forall i \in N_2 \\ \left.\begin{aligned} a_i'x &\leqslant \gamma_i \\ a_i'x &\geqslant \alpha_i \end{aligned}\right\} & && \forall i \in N_3 \\ x &\geqslant 0 && \text{componentwise} \end{aligned}$$

Hence, we get $\beta_0 = c'\bar{x}$ by solving the problem without regarding its fuzziness.

Now we add the fuzzy objective function's membership function to the substitute's objective function (16):

$$1 - \frac{\beta_0 - c'x}{\beta_0 - \alpha_0} + \sum_{i=1}^{n} \mu_i \rightarrow \max \qquad (18)$$

An example (see [13]): We add the fuzzy objective function

$$x_1 + x_2 \rightarrow \text{'max'}$$

to the numerical example above. At first we determine

$$\alpha_0 = 10$$

$$\beta_0 = 11$$

which leads to its satisfaction function

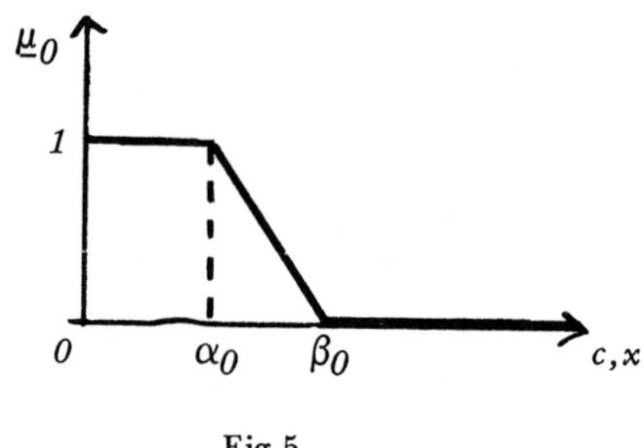

Fig.5

$$\bar{\mu}_0 = 1 - \frac{11 - (x_1 + x_2)}{11 - 10}$$

The optimal solution of the linear substitute

$$0.25\, s_1^0 + s_2^u + 2\, s_3^u - x_1 - x_2 \rightarrow \min$$

subject to the same set of feasible solutions above is

$$(x_1, x_2, s_1^0, s_2^u, s_3^0, s_3^u)_{opt\ 2} = (3\tfrac{1}{5}, 7\tfrac{1}{5}, 4, 0, 0, 0)$$

The pure FPP '$\rightarrow$' in case of a fuzzy objective function which is to be *minimized*, is treated in a similar way. We define its satisfaction function $\underline{\mu}_0$ as (see Fig.5):

$$\underline{\mu}_0 = 1 - \bar{\mu}_0 = \begin{cases} 0 & \text{if} \quad c'x > \beta_0 \\ \dfrac{\beta_0 - c'x}{\beta_0 - \alpha_0} & \text{if} \quad c'x \in [\alpha_0, \beta_0] \\ 1 & \text{if} \quad c'x < \alpha_0 \end{cases}$$

We determine the lower limit α_0 by solving:

$$c'x \rightarrow \min$$

so that

$$\begin{aligned} a_i'x &\leqslant \alpha_i && \forall i \in N_1 \\ a_i'x &\geqslant \alpha_i && \forall i \in N_2 \\ \left.\begin{aligned} a_i'x &\leqslant \gamma_i \\ a_i'x &\geqslant \alpha_i \end{aligned}\right\} & && \forall i \in N_3 \\ x &\geqslant 0 && \text{componentwise} \end{aligned}$$

Let $\underline{\underline{x}}$ be the optimal solution, then $\alpha_0 = c'\underline{\underline{x}}$.

We determine the upper limit $\beta_0 = c'\bar{\bar{x}}$ by solving the linear substitute without regarding the fuzzy objective function. Let $\bar{\bar{x}}$ be its optimal solution.

The linear substitute for a *mixed* FPP '→' is the above program enlarged by adding the restrictions of type '⩽', '⩾', or '='.

3. Solving the air pollution control problem

We get the linear substitute for the FPP described by (1, 2 and 3) as follows:

$$\forall i \begin{cases} \sum_{j=1}^{J} (1 - E_j) s_{ji} - s_i \leqslant e_i \\ s_i \leqslant d - e_i \end{cases} \quad (19)$$

$$\forall j \begin{cases} E_j - t_j \leqslant w_j \\ t_j \leqslant \bar{E}_j - w_j \end{cases} \quad (20)$$

The membership functions are:

$$\forall i, \text{ see (19)} \quad \mu_i = 1 - \frac{s_i}{d - e_i}$$

$$\forall j, \text{ see (20)} \quad \mu_j = 1 - \frac{t_j}{\bar{E}_j - w_j}$$

for the fuzzy objective funct. see (3)

$$\mu_0 = \frac{\beta_0 - \sum_{j=1}^{J} c_j E_j}{\beta_0 - \alpha_0}$$

which leads to the substitute's objective function

$$\sum_{i=1}^{I} \frac{s_i}{d - e_i} + \sum_{j=1}^{J} \frac{t_j}{\bar{E}_j - w_j} + \sum_{j=1}^{J} \frac{c_j E_j}{\beta_0 - \alpha_0} \rightarrow \min \quad (21)$$

Given data is s_{ji}, e_i, d, w_j, $\bar{E}_j$, c_j. Data to be determined a priori is α_0, β_0.

Variables are E_j (structural decision variables) and s_i, t_j.

A numerical example will conclude our discussion.

4. A numerical example

Referring to the worked example in [7], p.13 we obtain the transfer functions

$$s_1(x_1, x_2) = \frac{1}{x_1} e^{-\frac{1}{x_1}(1+(x_2-1)^2)} \qquad x_1 > 0$$

$$s_2(x_1, x_2) = \frac{1}{x_1} e^{-\frac{1}{x_1}(2+x_2^2/4)} \qquad x_1 > 0$$

$$s_3(x_1, x_2) = \frac{1}{x_1 - 2} e^{-\frac{1}{x_1-2}(1+(x_2+1)^2)} \qquad x_1 > 2$$

for the three sources and make the same assumptions as in [7]. Beyond this we assume that we only have six scattered receptor points as in the picture below. The control area is divided into three different parts, so that the desirable concentration rate e_i depends on the location but the maximum permitted concentration d is a constant standard over the entire control area.

We assume the desirable concentration e_i and the maximum permitted d to be:

area	e_i	d
recreation	.42	.5
urban	.44	.5
industrial	.49	.5

Table 2 Desirable concentration rates and the standard

We get the actual concentration rates listed in the next column.

The receptor points characterized by (*) are not taken into account because they already have a smaller sum of the plant's concentration rates than the desirable rates.

The data of the sources' assumed maximum permitted reduction rate $\bar{E}_j$, the desirable rate w_j and the reduction cost c_j are:

j	c_j	$\bar{E}_j$	w_j
1	2	.2	.04
2	4	.35	.15
3	1	.55	.3

Table 4 Data concerning the reduction

receptor points and its coordinates			actual concentration rates (before reduction)				desirable concent. rate	
i	x_1^i	x_2^i	s_{1i}	s_{2i}	s_{3i}	$\sum_{i=1}^{3} s_{ji}$	e_i	
1	1	1	.3679	.1054	-- --	.4733	.42	
2	3	1	.2388	.1575	.0067	.4030	.44	(*)
3	5	1	.1638	.1275	.0630	.3543	.44	(*)
4	1	-1	.0067	.1054	--	.1121	.49	(*)
5	3	-1	.0630	.1575	.3679	.5884	.49	
6	5	-1	.0736	.1275	.2388	.4399	.42	

Table 3 Actual situation of pollution

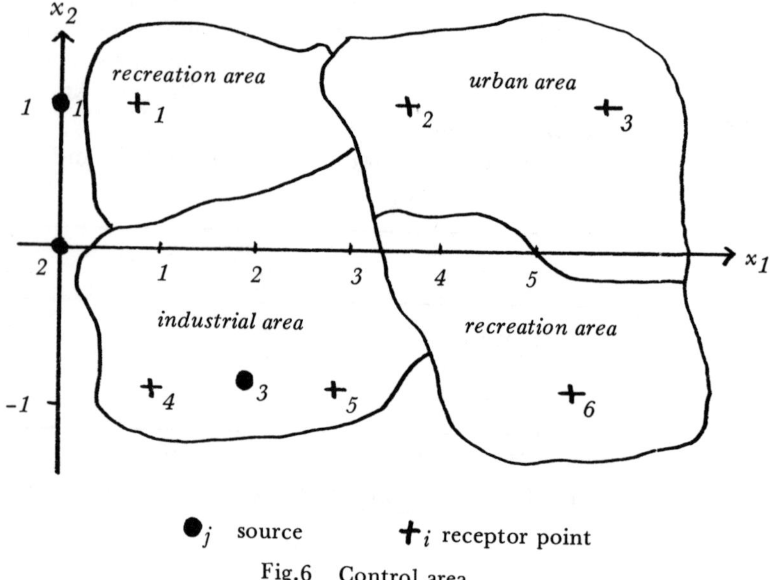

Fig.6 Control area

For the numerical treatment we have to formulate (19) as follows:

$$\left.\begin{aligned} \sum_{j=1}^{J} E_j s_{ji} + s_i &\geqslant \sum_{j=1}^{J} s_{ji} - e_i \\ s_i &\leqslant d - e_i \end{aligned}\right\} \qquad (22)$$

with the above data this leads to the following set of feasible solutions:

$$
\left\{
\begin{array}{ll}
\left.
\begin{array}{rl}
.3679\,E_1 + .1054\,E_2 \qquad\qquad + s_1 & \geqslant .0533 \\
s_1 & \leqslant .08 \\
.0630\,E_1 + .1575\,E_2 + .3679\,E_3 + s_5 & \geqslant .0984 \\
s_5 & \leqslant .01 \\
.0736\,E_1 + .1275\,E_2 + .2388\,E_3 + s_6 & \geqslant .0199 \\
s_6 & \leqslant .08
\end{array}
\right\} & \text{requirements for the receptor points, see (19), (22)} \\
\left.
\begin{array}{rl}
E_1 \qquad\qquad\qquad - t_1 & \leqslant .04 \\
t_1 & \leqslant .16 \\
E_2 \qquad - t_2 & \leqslant .15 \\
t_2 & \leqslant .2 \\
E_3 - t_3 & \leqslant .3 \\
t_3 & \leqslant .25
\end{array}
\right\} & \text{requirements for the sources, see (20)}
\end{array}
\right. \qquad (23)
$$

$E_j,\ s_i,\ t_j$ nonnegative

(Determination of the limits α_0 and β_0)
The lower limit α_0 (the best objective function's value at all) is found by solving:

$$\sum_{j=1}^{J} c_j E_j \rightarrow \min$$

so that
$$\left.\begin{array}{l} \sum_{j=1}^{J} (1 - E_j)\, s_{ji} \leqslant d \\ \left(\text{resp. } \sum_{j=1}^{J} E_j\, s_{ji} \geqslant \sum_{j=1}^{J} s_{ji} - d\right) \end{array}\right\} \text{for all relevant receptor points } i$$

$$E_j \leqslant \bar{E}_j \qquad \text{for all sources}$$

E_j nonnegative

In our case:

$$
\begin{array}{l}
2\,E_1 + 4\,E_2 + E_3 \rightarrow \min \\
.0630\,E_1 + .1575\,E_2 + .3679\,E_3 \geqslant .0884
\end{array}
$$

(the only relevant receptor point is No. $i = 5$)

$$
\begin{array}{rl}
E_1 & \leqslant .2 \\
E_2 & \leqslant .35 \\
E_3 & \leqslant .55
\end{array}
$$

E_j nonnegative

The optimal solution is:

$$\underline{\underline{E}}_1 = .0$$
$$\underline{\underline{E}}_2 = .0 \qquad \underline{\underline{E}}_3 = .2403$$

which leads to the lower limit α_0:

$$\sum_{j=1}^{J} c_j \underline{\underline{E}}_j = .2403$$

The upper limit β_0 (the worst objective function's value at all) is found by solving:

$$\sum_{i \in \{1,5,6\}} \frac{s_i}{d - e_i} + \sum_{j=1}^{J} \frac{t_j}{\bar{E}_j - w_j} \rightarrow \min$$

under the restrictions (19, 20) and

$s_i,\ t_j$ nonnegative

In our case:

$$12.5\,s_1 + 100\,s_5 + 12.5\,s_6 + 6.25\,t_1 + 5\,t_2 + 4\,t_3 \rightarrow \min$$

under consideration of the set of feasible solutions (23).

In the optimal solution we get:

$$\bar{\bar{E}}_1 = .04$$

$$\bar{\bar{E}}_2 = .15$$

$$\bar{\bar{E}}_3 = .19643$$

which leads to the upper limit β_0:

$$\sum_{j=1}^{J} c_j \bar{\bar{E}}_j = .87643$$

The distance between the limits is:

$$\beta_0 - \alpha_0 = .63613$$

Having determined the limits α_0 and β_0 we are now able to solve the given air pollution control problem. Its objective function (see (21)) is:

$$12.5\,s_1 + 100\,s_5 + 12.5\,s_6 + 6.25\,t_1 + 5\,t_2 + 4\,t_3 + 3.144\,E_1 + 6.288\,E_2 + 1.572\,E_3 \quad (24)$$

Minimizing (24) under consideration of (23) leads to the optimal solution:

$$E_1^{opt} = .04$$

$$E_2^{opt} = .0$$

$$E_3^{opt} = .26063$$

5. A remark on—and combining of—fuzzy statements

Since '65 when Zadeh published his basic concept of fuzzy sets and fuzzy statements [15] a number of authors in the meantime started a fundamental discussion whether or not combining two fuzzy statements by the logical 'and' can be represented by taking the minimum of two membership-functions. Even if Zimmermann in [16] and Rödder/Zimmermann in [10] discussed a fuzzy optimization approach underlaying the minimization concept they could not decide that handling the logical 'and' by using the minimum operator is always a true representation of reality.

Empirical experiments on man's behavior [11, 12] turned out to falsify the minimum hypothesis. Hamacher [9] did some research on the axiomatic side of this question.

All these results seem to justify a different concept of 'and'-combining fuzzy statements than minimum.

The research work about handling the logical 'and' has just started. We propose the concept of adding membership functions rather than taking the minimum not only because the latter leads to a much too high loss of information [13,14], but also because our aim was to 'solve' such a fuzzy programming problem in the most simple way conceivable. This is not to say that using our procedure is the only way constructing a mathematical model for the fuzzy reality.

References

1. BELLMAN, R. and ZADEH, L.A., 'Decision-making in a Fuzzy Environment', *Management Science* 17, No.4, pp. B 141- B 164 (1970).
2. BIRD, C.G. and KORTANEK, K.O., 'Game theoretic approaches to some air pollution regulation problems', *Socio-Econ. Plan: Sci.* 8, pp. 141-147, Pergamon Press (1974).
3. CARBONE, R., GORR, W.L., GUSTAFSON, S.-A., KORTANEK, K.O. and SWEIGART, J.R., 'Environmental-energy-equity models for resource management in an airshed', Series in Numerical Optimization and Pollution Abatement, Report No.23, Carnegie-Mellon University, Pittsburgh, Pennsylvania (1975).
4. GORR, W.L., GUSTAFSON, S.-A., KORTANEK, K.O., 'Optimal control strategies for air quality standards and regulatory policy', *Envirment and Planning* 4, pp.183-192 (1972).
5. GORR, W.L. and KORTANEK, K.D., 'Numerical aspects of pollution abatement problems: Optimal control strategies for air quality standards', Proceedings in Operations Research, Eds. M. Henke, A. Jaeger, R. Wartmann, H.-J. Zimmermann, Physica-Verlag, Würzburg-Wien, pp.34-58 (1972).
6. GORR, W.L., 'A theoretical and empirical investigation of the design of optimal air pollution regulations', Ph.D. thesis, School of Urban: and Public affairs, Carnegie-Mellon University, Pittsburgh, Pennsylvania (1973).
7. GUSTAFSON, S.-A., and KORTANEK, K.O., 'Mathematical models for air pollution control: Numerical determination of optimizing abatement policies', Series in Numerical Optimization and Pollution Abatement, Technical Report No.6, Carnegie-Mellon University, Pittsburgh, Pennsylvania (1972).
8. GUSTAFSON, S.-A. and KORTANEK, K.O., 'Mathematical models for optimizing air pollution abatement policies: numerical treatment',

Kolokvium, Ochrana, Zivotuiho, 1974, resp. Reprint No.132, School of Urban and Public Affairs, Carnegie-Mellon University, Pittsburgh, Pennsylvania (1975).

9. HAMACHER, H., 'Über logische Verknüpfungen unscharfer Aussagen und deren zugehörige Bewertungsfunktionen', Progress in Cybernetics and Systems Research, Vol.III.

10. RÖDDER, W., and ZIMMERMANN, H.-J., 'Analyse, Beschreibung und Optimierung von unscharf formulierten Problemen', Zeitschrift für Operations Research, pp.1-18 (1977).

11. RÖDDER, W., 'Ein Beitrag zur Verknüpfung unscharfer Mengen', presented at EURO I, Brussels, Paper No.54 (January 1975).

12. RÖDDER, W., 'On 'and' and 'or' connectives in fuzzy set theory', Working Paper No.75/07, Institut für Wirtschaftswissenschaften, Lehrstuhl für Unternehmensforschung (Operations Research).

13. SOMMER, G., 'Lineare Ersatzprogramme für unscharfe Entscheidungsprobleme', *Zeitschrift für Operations Research* (1977).

14. SOMMER, G., 'Zur Diskussion über optimale Entscheidungen bei unscharfer Problembeschreibung', *Zeitschrift für Betriebswirtschaft,* pp.921-923 (1976).

15. ZADEH, L.A., 'Fuzzy Sets', *Information and Control* 8, pp.338-353 (1965).

16. ZIMMERMANN, H.-J., 'Description and optimization of fuzzy systems', *International Journal of General Systems*, pp.209-215 (1976).

17. ZIMMERMANN, H.-J., 'Optimale Entscheidungen bei unscharfen Problembeschreibungen', *Zeitschrift für betriebswirtschaftliche Forschung,* pp.785-795 (1975).

A fuzzy approach to residential location theory

B. BONA[†], D. INAUDI,[‡] and V. MAURO[†]

[†] *Centro Elaborazione Numerale dei Segnali (CNR)*
Istituto di Elettrotechnica, Politecnico di Torino
Istituto Elettrotecnico Nazionale Galileo Ferraris, Torino

[‡] *Gruppo Nazionale Informatica Matematica (CNR)*
Istituto Scienza dell'Informazione, Università di Torino

Introduction
The residential location theory plays an important role on planning, as it can serve for correct prevision of the development of demand and offer in a given area (as a result of government intervention, changes in economics, population, services, etc.). Many different models for territorial development, land use, urban spatial structure, etc., have been presented in the past years. See, for instance [1, 2, 3, 4, 5, 7, 8, 9, 10, 11, 12].

The considerably high level of complexity of the problem leads the different authors to many simplifications which, on the other hand, influence strongly the results obtained.

In models of spatial interaction, or in land use studies, which are similar to the case introduced in this chapter, a common assumption is the validity of the 'maximum utility criterion' for the consumer [1], along with the 'equilibrium between demand and offer' [3, 7] for the market.

The models of development are, in turn, based on very simple economic laws [8].

From these assumptions it is possible in general to obtain precision, both on the residential location and market, and on the area development.

These results are not useful when considering short-term planning strategies in developed areas; the effect of changes in services, for instance, depends strongly on the existing situation. Moreover, the previsions of land use must take into account the limited offer which is presented to the single consumer in a given social and economic context.

Finally, there exists an interaction, in a longer period, between the consumer aptitude and the structure of the offer.

In this chapter, it is suggested that the problem can be studied by means of several partial models. In particular, it is presented in detail the model for short-term consumer choice for residences.

1. Outline of a general model
The location of residences in a given area depends on many different factors, which can be divided into

four main classes; i.e.

- the consumer aptitude in a given context,
- the residential market,
- the government intervention and, finally
- the general economic situation.

An important role in the consumer choice is played by such criteria as: *a*) closeness of residence to the work place, *b*) cost and quality of facilities, *c*) cost and quality of residences, *d*) cost and quality of transportation system.

These criteria, at their turn, depend on factors influenced by economic, social and historical context. The amount of residences offered to the consumer in a district as well as their cost, is related to the general economic, as well as market, situation; the same is true for services and transportation, which depend also on the public intervention.

So a complete model for the residential location should take into account all these factors; a suitable, very schematic block diagram could be as in Fig.1.

The block *A* represents the model of the offer for services, residences, transportation facilities, which is driven by the demand (r) and the external input i_1 which summarizes the effects due to the general economic situation and to the government intervention;

the block *B* represents the model of the consumer's choice: its inputs are the offer situation (s) and the number of individuals to be allocated; its output is the number of individuals allocated in each district;

the block *C* represents the model of the variation in the consumer 'level of satisfaction' due to variation in the residential situation. It operates on the vector of residents (r) and generates the vector of individuals which are looking for new residences;

the block *D* represents the model of the variation in time of the labor force situation (vector of employed); it depends strongly on the external

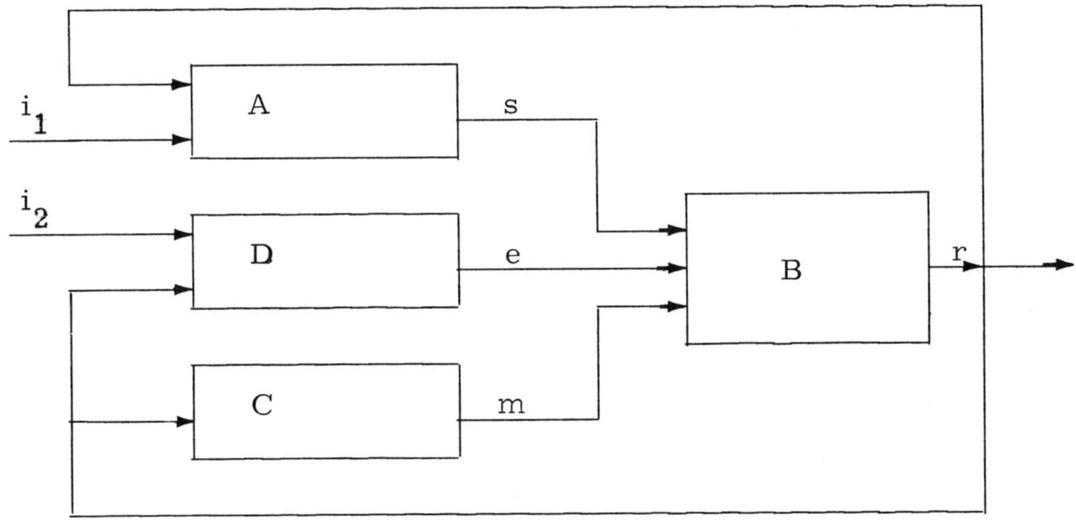

Fig.1

The letters in the figure assume the following meaning:

s represents the vector of offered services (amount, quality and cost) in every district of the area;

e is the vector of new employed in every district (amount and salaries);

m is the vector of inhabitants looking for new residences (amount, income and job location);

r is the vector of residents (amount, income and job location);

i_1 and i_2 represent the external inputs;

input i_2, but also on the residential situation in the area.

While block *B* is essentially a static one, the others are essentially dynamics; there is indeed an inherent delay between variation in demand and in offer *(A, D)*, between variation in distribution of inhabitants and level of satisfaction (*C*).

The model in Fig.1 is a very simple representation of residential dynamics, but it can show the general framework in which the short term consumer choice operates and the strong interactions between all the different factors listed above.

The aim of this chapter is to deal with the block B; i.e. the model of the consumer choice for residences in a given environment.

The detailed analysis of block B could be useful as a part of a general work, which would deal also with the study of the other models; but it is surely useful as it stands, due to the previsions it gives of the short term changes in a particular environment. As an example, it would allow the study of variation of demand for residences due to changes in the labor force, or in location of factories.

2. The fuzzy model of consumer choice

The choice of the residence by the consumer is accomplished under many different criteria which cannot be put in an exact mathematical form. Indeed, one chooses a residential place provided that it is 'good enough', 'not too far from work place', 'not too expensive', 'in a nice site', etc. Moreover, the relative weight of these criteria varies from individual to individual, from region to region. Finally, the information given to the consumer at choice time is limited and partially wrong: so, all these criteria are applied with a considerably high level of uncertainty.

The above considerations suggest that a correct framework for the formulation of a model is given by the 'fuzzy set theory'.

A fuzzy model of consumer choice would provide, for any given distribution of population over the area under analysis, its membership, relative to the set of 'satisfactory' distributions. Indeed, it is assumed here that an hypothetical configuration is more likely if it is more 'satisfactory' for the population interested.

To build the fuzzy model which has been outlined, let us introduce the following defintions and notations:

a) every consumer has a finite number of possible choices.

b) there is a finite number of criteria for the choice (i.e. transportation cost, transportation services, residences cost, etc.).

c) there is a finite number of homogeneous classes in which the population can be divided, relatively to social and economic attributes (salaries, social position, etc.).

d) there is a finite number of residential districts in which the whole area is divided.

e) all the 'basic' activities† in the area can be separated in a finite number of districts; the number of people working in 'basic' activities in every district is a priori known.

f) the consumers which, belonging to the same class, make the same choice have the same 'level of satisfaction' for the criterion involved in that choice.

A distribution of population is defined by the set of quantities

$$x_h^{k,l} \qquad 1 \leqslant h \leqslant H;\ 1 \leqslant k \leqslant K;\ 1 \leqslant l \leqslant L_k \tag{2.1}$$

where $x_h^{k,l}$ is the amount of people in class h (see assumption *c*), which, with respect to the criterion k (see assumption *b*), makes the choice l.

The sequence of quantity $x_h^{k,l}$ must verify some constraints (i.e. all of them must be non-negative, the total of commuters to a work district must equal the number of workers in that district, etc.).

All these constraints define an admissible set for the sequence

$$\left\{ x_h^{k,l} \right\} \in X \tag{2.2}$$

For every $x_h^{k,l}$ it is defined a 'level of satisfaction', which from now on will be given by the following membership function, relative to the set of 'satisfactory ones':

$$\mu_h^{k,l}\,(x_1^{k,l}, \ldots, x_h^{k,l}, \ldots, x_H^{k,l}) \tag{2.3}$$

Notice that:

- the membership depends on the number of people making the same choice. This appears as a very important feature of the model here suggested, as it makes the model closed to the real world. Indeed, the 'goodness' of a given service (school, for instance), depends on the number of users. This assumption apparently makes more complicated the application of this model; but, in authors' opinion, it cannot be

†The definition of 'basic' activities is given in [5]. In the present approach, the basic activities must include also all the services which do not vary in short period and whose presence or absence in a district is taken as an element of choice by the consumer (example: schools, hospitals, banks, etc.). All the other services are generated by the model itself.

avoided. In section 2 an attempt will be made in order to simplify the development of the model.
- the formulation given in (2.3) allows us to consider the cross-influences of different social groups in the choice process.

The second logical step of the modeling procedure, is the computation of the membership of the whole sequence as a composition of the single memberships as given by (2.3).

This composition could be done by different rules, which would be chosen accordingly to the particular case. In order to obtain a relatively simple algorithm, it will here be assumed that the membership of the whole distribution is obtained by the product of the single membership. The model, in this form, gives to the user an indication of 'likelihood' of every hypothetic distribution (remember that 'likelihood' follows from 'level of satisfaction'); this indication alone is not very useful, as the model should compare too many different distributions in order to have a correct idea of the level of 'likelihood' of the given area. So, the model suggested here computes also the distribution to which it corresponds the maximum value of the membership. To do this, the following problem must be solved:

$$\max_{\{x_h^{k,l} \in X\}} \prod_{h,k,l} \mu_h^{k,l}(x_1^{k,l}, \ldots, x_h^{k,l}, \ldots, x_H^{k,l}) \quad (2.4)$$

In the following section, problem (2.4) will be solved in a particular case, showing that, by making use of a suitable algorithm, its complexity can be considerably reduced.

3. Maximal membership distribution

For the sake of simplicity, it will be assumed in this section that the various individual criteria can be summarized in three different memberships: transportation, residential cost and services. Moreover, it is assumed that the population of the area is homogeneous. The algorithm resulting from these assumptions can be adapted, with minor modifications, to deal with the general case.

For clarity, the notations from now on will be the following:

- e_{ij} is the amount of people community from district i to district j
- r_i is the number of residents in district i.

There will be a membership for transportation, one for residences, and one for services; let them be:

$$\overset{t,ij}{\mu}(e_{ij}), \quad \overset{r,i}{\mu}(r_i), \quad \overset{s,i}{\mu}(r_i)$$

The set X defined in (2.2) is given by the following set of constraints:

$$e_{ij} \geqslant 0 \quad (3.1)$$

$$r_i = \sum_{j=1}^{N} e_{ij} \quad (3.2)$$

$$a_j = \sum_{i=1}^{M} e_{ij} \quad (3.3)$$

where N is the number of residential districts, M is the number of work districts, a_j is the amount of labor force in district j.

The membership functions have the following form:

$$\begin{aligned} \overset{t,ij}{\mu}(e_{ij}) &= \exp[-Q_{ij}(e_{ij})] \\ \overset{r,i}{\mu}(r_i) &= \exp[-P_i^r(r_i)] \\ \overset{s,i}{\mu}(r_i) &= \exp[-P_i^s(r_i)] \end{aligned} \qquad \begin{aligned} i &= 1,2,\ldots,M \\ j &= 1,2,\ldots,N \end{aligned} \quad (3.4)$$

where $Q_{ij}(e_{ij})$, $P_i^r(r_i)$, $P_i^s(r_i)$ are suitable convex functions.

Notice that for the proposed algorithm to be useful it is essential that the single membership functions have an exponential form, as well as the membership composition rule is the product one.

So the general membership function of the entire area is:

$$\mu = \mu(E) = \exp\left\{-\left[\sum_{i=1}^{N}\sum_{j=1}^{M} Q_{ij}(e_{ij}) + \sum_{i=1}^{M} P_i(r_i)\right]\right\} = \exp[-f(E)] \quad (3.5)$$

where

$$E = \begin{bmatrix} e_{11} & e_{12} & \cdots\cdots & e_{1N} & r_1 \\ e_{21} & e_{22} & \cdots\cdots & e_{2N} & r_2 \\ \vdots & \vdots & & & \\ e_{M1} & & & e_{MN} & r_M \end{bmatrix} \qquad (3.6)$$

$$P_i(r_i) = P_i^r(r_i) \cdot P_i^s(r_i)$$

Our purpose is to obtain the optimal matrix E_0 which maximizes the above general membership function $\mu(\cdot)$. We have to solve the following non-linear problem (NLP)

$$\max_E \mu(E) = \max_E \ln \mu(E) = \min_E f(E) \qquad (3.7)$$

subject to the following constraints:

$$E \in C \cap D$$

where

$$C = \{E \mid e_{ij} \geqslant 0; \; r_i \geqslant 0\}$$

$$D = \left\{ E \,\middle|\, \sum_{i=1}^{M} e_{ij} = a_j \;\; \sum_{j=1}^{N} e_{ij} = r_i \right\} \qquad (3.8)$$

The above NLP problem can be solved by applying Fenchel's theorem [6]:

$$\min_{E \in C \cap D} [f(E) - g(E)] = \max_{E^* \in C^* \cap D^*} [g^*(E^*) - f^*(E^*)] \qquad (3.9)$$

where

- $g^*(E^*)$, $f^*(E^*)$ are the conjugate functionals of $g(E)$, $f(E)$;
- C^*, D^* are the conjugate sets of C, D.

Let us now introduce the following notations:

$$E_j = \begin{bmatrix} 0 & 1 & 0 \\ 0 & 1 & 0 \\ \vdots & \vdots & \vdots \\ 0 & 1 & 0 \end{bmatrix}$$ $M \times (N+1)$ matrix where the ones appear in the jth column.

$$A_i = \begin{bmatrix} 0 & 0 & \cdots & 0 \\ \vdots & \vdots & & \vdots \\ 1 & 1 & \cdots & 1 \\ \vdots & \vdots & & \vdots \\ 0 & 0 & & 0 \end{bmatrix}$$ $M \times (N+1)$ matrix where the ones appear in the ith row

$$(E, E_j) = \text{trace } E^T E_j$$

The constraints (3.8) can be rewritten as follows:

$$(E, E_j) = a_j$$

$$(E, A_i) = 0$$

The set E^* can be defined as:

$$E^* = \sum_{j=1}^{N} \lambda_j E_j + \sum_{i=1}^{M} \nu_i A_i \qquad (3.10)$$

The expression of $f^*(E^*)$ is

$$f^*(E^*) = \max_{E \in C} \left\{ \sum_{j=1}^{N} \lambda_j \cdot \sum_{i=1}^{M} e_{ij} + \sum_{i=1}^{M} \nu_i \cdot \sum_{j=1}^{N} e_{ij} - \sum_{i=1}^{M} \nu_i r_i - f(E) \right\} \qquad (3.11)$$

Introducing in (3.11) in place of $f(E)$ its representation (3.5), we obtain that the given NLP can be solved by the simultaneous solution of:

$N \times M$ independent problems

$$\max_{e_{ij} \geqslant 0} [(\lambda_j + \nu_i) e_{ij} - Q_{ij}(e_{ij})] \qquad (3.12)$$

M independent problems

$$\max_{r_i \geqslant 0} [-\nu_i r_i - P_i(r_i)] \qquad (3.13)$$

where the quantities λ_j and ν_i must be found by satisfying the constraints (3.2), (3.3), i.e. by solving a system of $N+M$ equations.

In this way, the problem is reduced to the solution of a set of equations in $N+M$ variables (λ_i and ν_i), where the computation of residuals requires the solution of $N \times M+M$ independent maximization problems in one variable.

The resulting procedure, by making use of general iteration schemes, could be time consuming:

in the following section an 'ad hoc' rapidly convergent iteration is suggested.

4. Optimization algorithm

In this section we present a procedure which simplifies the problem (3.7) shown in the previous section, and gives a fast minimization algorithm when the number of variables is large (> 1000).

We write the minimum problem again

$$\min_E f(E) = \min_E \left[\sum_{i=1}^{M} \sum_{j=1}^{N} Q_{ij}(e_{ij}) + \sum_{i=1}^{M} P_i(r_i) \right] \tag{4.1}$$

subject to

$$\begin{aligned} & e_{ij} \geqslant 0 \\ & \sum_{i=1}^{M} e_{ij} = a_j \qquad i = 1,2,\ldots,M \\ & \qquad\qquad\qquad\quad j = 1,2,\ldots,N \\ & \sum_{j=1}^{N} e_{ij} = r_i \end{aligned} \tag{4.2}$$

The function $P_i(\cdot)$ is iteratively approximated in the following way

$$P_i\left(\sum_{j=1}^{N} e_{ij}\right) = \sum_{j=1}^{N} \overline{P}_i^k(e_{ij}) \tag{4.3}$$

where:

$\overline{P}_i^k(\cdot)$ is a convex quadratic function in e_{ij}

k represents the value of current iteration.

The function $\overline{P}_i^k(e_{ij})$ must satisfy the following properties

a) $$\operatorname{grad}\left[P_i \sum_{j=1}^{N} \right]_{\overline{e}_{ij}^{k-1}} = \operatorname{grad}\left[\sum_{j=1}^{N} \overline{P}_i^k(e_{ij}) \right]_{\overline{e}_{ij}^{k-1}} \tag{4.4}$$

where $\overline{e}_{ij}^{k-1}$ is the point obtained at the $(k-1)$th iteration

b) assuming that $\min\limits_{r_i \geqslant 0} P_i(r_i) = r_i^0$ the minimum of $\sum\limits_j \overline{P}_i^k$, defined as

$$e_{ij,0}^k = \min_{e_{ij}} \sum_{j=1}^{N} \overline{P}_i^k(e_{ij})$$

must satisfy:

$$\sum_{j=1}^{N} (\overline{e}_{ij}^{k-1} - e_{ij,0}^k)^2 = \min_{e_{ij}^k} (\overline{e}_{ij}^{k-1} - e_{ij}^k)^2 \tag{4.6}$$

(condition (4.6) can be substituted by any suitable minimum-distance criterion).

The iteration scheme is then defined as:

$$E^k = \min_E \left[\sum_{i=1}^{M} \sum_{j=1}^{N} Q_{ij}(e_{ij}) + \sum_{i=1}^{M} \sum_{j=1}^{N} \overline{P}_i^k(e_{ij}) \right] \tag{4.7}$$

subject to (4.2).

We can decompose problem (4.7) in N problems of the type:

$$\min_E \left[\sum_{i=1}^{M} Q_{ij}(e_{ij}) + \sum_{i=1}^{M} \overline{P}_i^k(e_{ij}) \right] \tag{4.8}$$

subject to:

$$\begin{aligned} & e_{ij} \geqslant 0 \\ & \sum_{i=1}^{M} e_{ij} = a_j \end{aligned} \qquad \text{for } j = 1,2,\ldots,N$$

Each problem (4.8) can be easily reduced by duality to M one-dimensional minimizations. We can now describe the iterative minimization algorithm:

STEP 1) - Compute

$$E^1: \min_E \sum_{i=1}^{M} Q_{ij}(e_{ij})$$

subject to the constraints (4.2)

- Solve M problems

$$\min_{r_i} P_i(r_i)$$

and obtain

$$r^0 = [r_1^0 \, r_2^0 \ldots\ldots r_M^0]^T$$

STEP 2) - Construct $\overline{P}_i^k(e_{ij})$ using properties (4.4), (4.5), (4.6).

STEP 3) - Solve problems (4.8) and obtain E^k.

STEP 4) - If $\|E^k - E^{k-1}\| < \epsilon$ with $\epsilon > 0$ small the convergence is satisfied and the minimum found else set $k = k+1$ and return to step 2.

The given algorithm has the following convergence properties:

- it is surely convergent in the given context (i.e. for function $P_i(r_i)$ and $Q_{ij}(e_{ij})$ strictly convex). The proof is omitted for brevity.
- it can be easily implemented also on small computers and very fast for large problems. It has been successfully tested in a 40 district problem (1600 variables), on a minicomputer. The convergence was reached with a suitable precision after 10-20 iterations; the total computing time has been of few (2-3) minutes.

Conclusions

In this chapter a model of the consumer for residences in a given environment has been presented. The main features of the model are:

- it is conceived as a part of a more general model of area development.
- it takes into account the 'saturation' of offered facilities; i.e. the 'appeal' of a given facility is assumed to be dependent on the number of its users. This assumption is not often made in models of this type, as it makes more complicated the model application.
- a rapidly convergent algorithm is given for the computation of the 'maximal membership' distribution. This algorithm can solve on a small computer very large problems.

The above described features make the model as it stands useful for practical applications and further investigations on the outlined general model of area development.

Acknowledgements

This work has partially been supported by C.N.R. (C.E.N.S. and G.N.I.M.).

References

1. ALONSO, W., *Location and Land Use,* Harvard University Press, Cambridge, Mass. (1964).
2. BATTY, M., BOURKE, R., CORMODE, P. and ANDERSON-NICHOLLS, M., 'Experiments in urban modelling for country structure planning: the area 8 pilot Model', *Environment and Planning* 6 (1974).
3. BECKMANN, M.J., 'On the distribution of urban Rent and residential density', *Journal of Economic Theory* 1 (1969).
4. ECHENIQUE, M., FEO, A., HERRERA, R. and RIQUEZES, J., 'A disaggregated model of urban spatial structure: theoretical framework', *Environment and Planning A*, vol. 6 (1974).
5. LOWRY, I.S., 'A Model of Metropolis' RM-4125-RC, Rand Corporation, Santa Monica (1964).
6. LUENBERGER, D.G., *Optimization Theory by Vector Space Methods,* Wiley (1969).
7. MONTESANO, A., 'A restatement of Beckmann's model on the distribution of urban rent and residential density', *Journal of Economic Theory* 4 (1972).
8. OXLEY, M.J., 'Economic Theory and Urban Planning', *Environment and Planning A,* vol.7 (1975).
9. SENIOR, M.L., 'Approaches to residential location modelling 1 urban ecological and spatial interaction models (a review)', *Environment and Planning A,* vol. 5 (1973).
10. SENIOR, M.L., 'Approaches to residential location modelling n. 2', *Environment and Planning A,* vol. 6 (1974).
11. WILSON, A.G., *Entropy in Urban and Regional Modelling,* Pion, London (1971).
12. WILSON, A.G., *Urban and Regional Models in Geography and Planning,* Wiley (1974).
13. ZADEH, L.A., 'Fuzzy sets', *Information and Control* 8 (1965).
14. ZADEH, L.A., 'Outline of a new approach to the Analysis of Complex Systems and decision Processes', *IEEE Trans. System, Man and Cybernetics,* vol. SMC-3, No.1 (gennaio 1973).

On the representation of the results of measurements by fuzzy sets

PAULETTE FEVRIER
Université Pierre et Marie Curie, Paris

1. Notion of result of measurement
Measurements, inquiries, surveys results are mostly expressed by numbers and estimates about the precision of results. Hence the result of a measurement for an observable A is expressed by a pair of numbers $(a, \Delta a)$. According to custom, one usually writes the result of a measurement $a \pm \Delta a$.

In the conventional conception, the Δa which affects the result a is not considered an imprecision but an upper limit of the error in the value of a. The traditional theory of measurement (Stevens, Campbell) is presently in dispute. Both Bridgeman's operational conception and the technical progress due to the use of intermediary electronical devices bringing in the signal theory and allowing the measurement of quantities that vary rapidly, and considerations on microphysical measurements, have shown that the conventional measurement theory is not sufficient [1]. The notions of 'real value' and 'error in real value' have no operational defintion, no positive content. Every result of measurement is basically imprecise. Hence a 'real value' cannot be found by experiment.

A measurement provides a more or less determined interval, from which is extracted an estimated value a; tests upon the method of measurement and the apparatus used provide an evaluation of the imprecision Δa of the measurement result. Then, from that point of view, the result of a measurement is always a pair of two estimated numbers $(a, \Delta a)$, one of them giving the result, the other its imprecision. But nothing allows us to assert that an objective and intrinsic value a_1 of a exists, whose a would be a false evaluation, or that

$$a - \Delta a < a_1 < a + \Delta a \ ,$$

which would be a metaphysical hypothesis. (However such an hypothesis cannot be eliminated, since a_1 cannot be reached by experiment. But if it is in accordance with the conceptions in favor at the beginning of the present century, it is in disagreement with the present technology.) [1]

2. Expressing the result of a measurement
From a mathematical point of view, it is more convenient to replace a pair of numbers $(a, \Delta a)$ by an interval $[a - \Delta a, a + \Delta a]$. Moreover, the numbers a and Δa being known only approximately, they can

be supposed to be rational. Thus a measurement result can be expressed by an interval with rational ends, denoted $I_{a,\Delta a}$.

$$I_{a,\Delta a} = [a - \Delta a,\ a + \Delta a]$$

The statement expressing such a result concerning an observable A can be written in the following form:

$$\text{Re Mes } A \subseteq I_{a,\Delta a}$$

As a and Δa are not quite determined, the interval $I_{a,\Delta a}$ is not well determined but belongs to the family of intervals with rational ends. If we denote by φ the characteristic function of $I_{a,\Delta a}$, we cannot determine two numbers a_1 and a_2 such that

$$a_1 = a - \Delta a \quad \text{and} \quad a_2 = a + \Delta a$$

and

$$\varphi(x) = 0 \quad \text{iff} \quad x < a_1 \quad \text{or} \quad x > a_2$$

$$\varphi(x) = 1 \quad \text{iff} \quad a_1 \leqslant x \leqslant a_2$$

Then it is more convenient to consider a function defining a fuzzy set such that $\varphi(x) = 0$ iff x surely does not belong to $I_{a,\Delta a}$, and $\varphi(x) = 1$ iff $x \in I_{a,\Delta a}$, and $0 < \varphi(x) < 1$ when it is possible but not sure that $x \in I_{a,\Delta a}$.

If $\int_{-\infty}^{+\infty} \varphi(x)\,dx = 1$ the function φ can be interpreted as a density probability for x belonging to $I_{a,\Delta a}$; $\varphi(x)$ must be a summable function whose non zero values constitute a rational interval.

So $I_{a,\Delta a}$ must not be considered as a usual rational interval, but as a fuzzy set.

3. Result of the measurement of a vectorial or tensorial observable

Instead of scalar observables, one can as well consider observables with components like vectorial observables or tensorial observables. In the conventional conception, the result of the measurement is represented by a point P, end of an O-originated vector, with an upper limit of the error expressed by a scalar r defining a sphere around P; or, if we use a vectorial basis, the result of the measurement is expressed by means of components on that basis, that is by an n-uple of numbers a_i with their upper error limit Δa_i. In that case, the conception can be changed in the same way as above: every result of measurement along a component is a scalar result and will be expressed by an interval with rational ends $I_{a_i,\Delta a_i}$ and the result of the measurement on the vector will be expressed by a Cartesian product of such intervals, that is

$$I_{a_1,\Delta a_1} \times I_{a_2,\Delta a_2} \times \dots \times I_{a_n,\Delta a_n}$$

which constitutes a so called n-interval with rational apexes. The volume of the n-interval is fixed by the imprecision of the measurement. This n-interval is not quite defined, since the numbers a_i and Δa_i are only estimated but not univocally determined. Then it is represented by a direct product of fuzzy sets of the same type as for the measurement of a scalar. The characteristic function is replaced by a function of point $\varphi(P)$ whose non-zero values are included into an n-interval with rational apexes, that is

$$0 \leqslant \varphi(P) \leqslant 1$$

$\varphi(P) = 0$ if surely P does not belong to the n-interval

$\varphi(P) = 1$ if surely P belongs to the n-interval

$\{P \mid 0 < \varphi(P) \leqslant 1\}$ is an n-interval with rational apexes.

Let Π be such an n-interval with rational apexes; then, an elementary statement expressing the result of the measurement of a vectorial observable V will take the following form:

$$\text{Re Mes } V \subseteq \Pi$$

4. Using logical connectors

As some measurement can be reiterated, several elementary statements can be connected by logical conjunctions 'and', 'or', definable by means of operations on sets, namely intersection and union; the same applies to the negation. More precisely, if

$$p_1 = \text{Re Mes } A \subseteq I_1, \qquad p_2 = \text{Re Mes } A \subseteq I_2$$

we write

$$p_1 \,\&\, p_2 =_d \text{Re Mes } A \subseteq I_1 \cap I_2$$

$p_1 \vee p_2 =_d$ Re Mes $A \subseteq I_1 \cup I_2$

$\neg p_1 =_d$ Re Mes $A \subseteq \complement I_1$

If we let these connectors be applied a finite number or enumerably infinite number of times, the corresponding tribe of sets is the borelian tribe, and in the case of such an extension, the empirical statements will be expressed by

Re Mes $A \subseteq E$

where E is a borelian fuzzy set.

Such a set can be defined by a function of point $\varphi(P)$ such that

$0 \leqslant \varphi(P) \leqslant 1$
$\varphi(P) = 0$ if surely P does not belong to E
$\varphi(P) = 1$ if surely P belongs to E
$\{P \mid 0 < \varphi(P) \leqslant 1\}$ is a borelian set
$\{P \mid \varphi(P) = 1\}$ is a borelian set

Then $\{P \mid 0 < \varphi(P) \leqslant 1\} \supseteq E \supseteq \{P \mid \varphi(P) = 1\}$ and $\varphi(P)$ can be interpreted as a probability density.

5. Structural consequences

The main advantage of this method of expressing results of measurements is that it uses the same tribe of sets, namely the borelian sets tribe, to express the results of the measurements and to express predictions in probabilities form, since the borelian sets tribe constitute the probabilisable sets tribe.

In the case above I took under consideration only a single observable A; there the set of the empirical statements is a boolean algebra. If we take the set of all the measurable observables of the studied system, two cases are to be distinguished:

a) All the observables are simultaneously measurable: in this case a compounding operation for the observables can be defined, and the set of the empirical statements is a boolean algebra.

b) There are at least two observables non-simultaneously measurable: in this case the set of the empirical statements is a non distributive lattice including, for each observable, a boolean sub-lattice. This case occurs in quantum mechanics (lattice of the closed linear manifolds of a Hilbert space), and in some models in socio-economy.

Thus the algebraical structure introduced by the logical connectors induces some structural properties on the system itself [2]. Then, according to the type of the structure introduced by the empirical statements, three classes of systems must be distinguished (see J.L. Destouches [3]).

1. boolean algebra
2. geometrical lattices
3. non-geometrical lattices

In the two first cases the system can be represented by a point in a suitable representative space, but not in the third case.

6. Discontinuous results of measurements

The possible values of some observables are integers and are measured by means of counters. In that case, when the measurement is precise enough, the result is univocally fixed and expressed by an integer. The formulation proposed above can be applied here, but to express the result of a measurement we have, instead of an interval, a single which is an integer, that is

Re Mes $N \subseteq \{n\}$

We have no fuzzy set occuring, but in the case of an imprecise measurement we shall have a finite fuzzy set. This holds in the case where, according to a theory, it can be asserted a priori that the result of the measurement belongs to a set Q (as in quantum mechanics). When this set Q contains a discontinuous subset, it can be asserted on the basis of the theory that the result is included into the set $I_{a,\Delta a} \cap Q$. If the measurement has been performed with a sufficient precision, this last set contains only one element, as it happens in the case of integers. But if the measurement is less precise, $I_{a,\Delta a} \cap Q$ is a finite fuzzy set, and it can even be enumerably infinite if $I_{a,\Delta a} \cap Q$ contains a limit point of Q.

Conclusion

That way of using fuzzy sets to express results of measurements seems much more adequate to the experimental data than the classical notion of a 'true value' represented by a number or a vector and an 'upper bound of the error'. It has the advantage of using the same tribe of sets for the results and the predictions in the form of probabilities.

Bibliography

1. DESTOUCHES, J.L., FINKELSTEIN, and GONELLA, Communications and discussions, IMEKO Colloquium, Enschede, December 1975; J.L. DESTOUCHES, Cours de 3ème cycle on Théorie des prévisions, 1975-1976, Paris; FINKELSTEIN and GONELLA, Lectures delivered at the Université P. et M. Curie, Paris, mars 1976, to be published in Revue philosophique, Paris, 1978.
2. FEVRIER, P., 'Structure des raisonnements expérimentaux et prévisionnels en Physique, Thèse de Mathematiques statistiques, Paris (1967).
3. DESTOUCHES, J.L., Communication EMCSR 76.

A model of learning based on fuzzy information

M. SUGENO† and T. TERANO‡
† *Department of Electrical and Electronic Engineering, Queen Mary College, University of London, UK*
‡ *Department of Control Engineering, Tokyo Institute of Technology, Tokyo, Japan*

1. Introduction

Since the concepts of fuzzy measures and fuzzy integrals were proposed, several applications have been reported [3,4,5,6]. Here fuzzy measures are set functions with monotonicity which are interpreted as subjective scales for fuzziness. These are similar to probability measures for randomness. Fuzzy integrals are functionals defined by using fuzzy measures and correspond to probability expectations, i.e. Lebesque integrals. We can evaluate fuzzy objects subjectively by using fuzzy integrals. Further there is a useful concept 'conditional fuzzy measures' [4,5] which of course corresponds to conditional probabilities.

This chapter presents a learning model based on a fuzzy information and discusses its application.

One of the essential features of human learning is a human's ability to learn through a fuzzy information. Generally speaking, a human being can learn and decide in a fuzzy environment. As a matter of fact, most of the available information in the actual world is fuzzy. Therefore it is very important to develop a model of learning which can work by accepting a fuzzy information.

The presented model is formulated by using the concept of conditional fuzzy measures, which is similar to Bayesian learning model in a stochastic environment. The model is compared with Bayesian model and applied to find an extremum of an unknown multimodel function.

2. Fuzzy measures and fuzzy integrals

In this section let us briefly explain about fuzzy measures and fuzzy integrals which are used in the next section.

Let X be an arbitrary set and **B** be a Borel field of X.

[Definition 1] A set function g defined on **B** which has the following properties is called a fuzzy measure.

1. $g(\phi) = 0,\ g(X) = 1$
2. If $A, B \in \mathbf{B}$ and $A \subset B$, then $g(A) \leqslant g(B)$
3. If $F_n \in \mathbf{B}$ and $\{F_n\}$ is monotone, then $\lim_{n\to\infty} g(F_n) = g(\lim_{n\to\infty} F_n)$

In the definition, if further g has finite additivity, then g becomes a probability measure. Note that a probability measure is one of fuzzy measures. A fuzzy measure can be interpreted in several ways. Here let us interpret it abstractly as follows. A more concrete interpretation is seen in [3,4,6].

First, suppose that a person picks up an element x out of X, but does not know which one he has picked up. Next, suppose that he guesses if x belongs to a given subset A. It is uncertain and fuzzy for him whether $x \in A$ or not. His guess would become subjective when there are few clues for guessing.

Assume in general that a human being has a subjective quantity called the grade of fuzziness measuring fuzziness such as stated above. If it is assumed in this case that he can consider a quantity $\mathrm{Gr}(x \in A)$ which expresses the grade of fuzziness of a statement '$x \in A$', then $g(A)$ is interpreted as $\mathrm{Gr}(X \in A)$. It will easily be accepted that if $A \subset B$, then $\mathrm{Gr}(x \in A) \leqslant \mathrm{Gr}(x \in B)$.

Let $h : X \rightarrow [0,1]$ be a **B**-measurable function.

[Definition 2] A fuzzy integral of h over A with respect to g is defined as follows

$$⨍_A h(x) \circ g(\cdot) = \sup_{\alpha \in [0,1]} [\alpha \wedge g(F_\alpha)] \tag{1}$$

where $F_\alpha = \{x \mid h(x) \geqslant \alpha\}$.

Fuzzy integrals correspond to probabilistic expectations and are also called fuzzy expectations. A triplet $(X, \mathbf{B}, g)$ is called an F-measure space. Here g is called a fuzzy measure of the measurable space $(X, \mathbf{B})$ or merely that of X. In the above definition, the symbol $⨍$ is an integral with a small bar and also shows a symbol of the letter f. The small circle is the symbol of the composition used in the fuzzy sets theory.

Hereafter, it is assumed that all the integrands including constants, have the range [0,1]. For simplification, a fuzzy integral is written as $⨍_A h \circ g(\cdot)$ or $⨍_A h \circ g$. In the case of $A = X$, it is written briefly as $⨍ h \circ g$.

By using fuzzy integrals, it is possible to express one's subjective evaluation of fuzzy objects. Fuzzy integrals have the following properties.

Let $a \in [0,1]$, then

$$⨍ a \circ g(\cdot) = a \tag{2}$$

$$⨍ (a \wedge h) \circ g(\cdot) = a \wedge ⨍ h \circ g(\cdot) \tag{3}$$

$$⨍ (a \vee h) \circ g(\cdot) = a \vee ⨍ h \circ g(\cdot) \tag{4}$$

If $h \leqslant h'$, there holds

$$⨍ h \circ g(\cdot) \leqslant ⨍ h' \circ g(\cdot) \tag{5}$$

If $A \subset B$, then there holds

$$⨍_A h \circ g(\cdot) \leqslant ⨍_B h \circ g(\cdot) \tag{6}$$

If $\{h_n\}$ is a monotone sequence of **B**-measurable functions, then

$$⨍ \lim_{n \to \infty} h_n \circ g = \lim_{n \to \infty} ⨍ h_n \circ g \tag{7}$$

If $\{h_n\}$ is a monotone decreasing (increasing) sequence of **B**-measurable functions and $\{a_n\}$ is a monotone increasing (decreasing) sequence of real numbers, then

$$⨍ [\bigvee_{n=1}^{\infty} (a_n \wedge h_n)] \circ g = \bigvee_{n=1}^{\infty} [a_n \wedge ⨍ h_n \circ g] \tag{8}$$

There holds $⨍_A h \circ g = M$ if and only if $g(A \cap F_M) \geqslant M \geqslant g(A \cap F_{M+O})$, where $F_M = \{x \mid h \geqslant M\}$ and $F_{M+O} = \{x \mid h > M\}$.

The fuzzy integrals are very similar to the Lebesgue integrals in their definition. Let $h(x)$ be a simple function such that

$$h(x) = \sum_{i=1}^{n} \alpha_i \chi_{E_i}(x)^{\mathrm{T}} \tag{9}$$

where $X = \sum_{i=1}^{n} E_i$, $E_i \in \mathbf{B}$, and $E_i \cap E_j = \phi (i \neq j)$.

Let μ be a Lebesque measure. In the measure space $(X, \mathbf{B}, \mu)$, the Lebesque integral of h over A is defined as

$$\int_A h \, d\mu = \sum_{i=1}^{n} \alpha_i \mu(A \cap E_i) \tag{10}$$

Here assume $0 \leqslant \alpha_i \leqslant 1$ $(1 \leqslant i \leqslant n)$ and $\alpha_1 \leqslant \alpha_2 \leqslant \ldots\ldots \leqslant \alpha_n$. Let further $F_i = E_i + E_{i+1} + \ldots$ $\ldots + E_n$ $(1 \leqslant i \leqslant n)$. Then a simple function $h(x)$ can also be written as

$$h(x) = \bigvee_{i=1}^{n} [\alpha_i \wedge \chi_{F_i}(x)] \tag{11}$$

and two expressions are identical. With respect to a simple function h on X, there holds

$$⨍_A h \circ g(\cdot) = \bigvee_{i=1}^{n} [\alpha_i \wedge g(A \cap F_i)] \tag{12}$$

The similarity of Lebesque and fuzzy integrals is clarified by comparing Eq.(9) with Eq.(11) and Eq.(10) with Eq.(12), respectively.

Next a quantitative comparison is tried. Let h be a **B**-measurable function. Then both integrals, fuzzy and Lebesque, with respect to a probability measure P can be defined and the following inequality is obtained. Let $(X, \mathbf{B}, P)$ be a probability space and $h : X \rightarrow [0,1]$ be a **B**-measurable function, then there holds

$$|\int_X h(x)\,dP - ⨍_X h(x) \circ P(\cdot)| \leqslant 1/4 \tag{13}$$

Since the operations of fuzzy integrals include only comparisons of grades, the above inequality implies that using only ∨ and ∧, a value different by at most ¼ from a probabilistic expectation can be obtained.

Now let $\phi : X \rightarrow Y$, then both the Borel field $\mathbf{B}^{(\phi)}$ and the fuzzy measure $g^{(\phi)}$ are induced from X into Y. That is:

$$F \in \mathbf{B}^{(\phi)} \text{ iff } \phi^{-1}(F) \in \mathbf{B}$$

$$g^{(\phi)}(F) = g(\phi^{-1}(F)).$$

A fuzzy measure space $(y, \mathbf{B}^{(\phi)}, g^{(\phi)})$ is interpreted in the following way. If Y is related to X by a mapping ϕ, then a fuzzy measure of Y by which grade of fuzziness in Y is measured should be also related to that of X.

[Definition 3] Let $E \in \mathbf{B}$ and $F \in \mathbf{B}^{(\phi)}$. By $\rho(E|\phi=y)$, denote the representative of all functions equivalent to $h(y)$ with respect to g† such that

$$g(E\ \phi^{-1}(F)) = ⨍_F h(y) \circ g^{(\phi)}(\cdot) \tag{14}$$

Here $\rho(\cdot\,|\,\phi = y)$ is called a conditional fuzzy measure under the condition of $\phi = y$.

Let $F = Y$ in the definition, then we obtain

$$g(E) = ⨍_Y \rho(E\,|\,\phi = y) \circ g^{(\phi)}(\cdot) \tag{15}$$

Conditional fuzzy measures have the following properties.

1. For a fixed $E \in \mathbf{B}$, $\rho(E\,|\,\phi = y)$ is, as a function of y, a $\mathbf{B}^{(\phi)}$-measurable function.
2. For fixed y, $\rho(\cdot\,|\,\phi = y)$ is a fuzzy measure of $(X, \mathbf{B})$ in the sense of $g^{(\phi)}$ – *a.e.*‡

Even if two fuzzy measure spaces, $(X, \mathbf{B}_X, g_X)$ and $(Y, \mathbf{B}_Y, g_Y)$, are related to each other, a mapping ϕ may not be explicit in general. We then write $\rho(\cdot\,|\,\phi = y)$ as $\rho_X(\cdot\,|\,y)$ which is called a conditional fuzzy measure from Y to X. In this case, there must hold similarly

$$g_X(\cdot) = ⨍_Y \rho_X(\cdot\,|\,y) \circ g_Y \tag{15'}$$

Under the above consideration, it is possible to give first $\rho_X(\cdot\,|y)$ and g_Y instead of g_Y and g_X.

Let us next show how to calculate fuzzy integrals. For simplicity we restrict the set X to a finite set, i.e., $X = \{x_1, x_2, \ldots, x_n\}$. First we give a construction rule of a fuzzy measure on 2^X which is called λ-rule.

$$\text{Let } 0 \leqslant g^i \leqslant 1, \quad 1 \leqslant i \leqslant n \tag{16}$$

and

$$\frac{1}{\lambda}[\prod_{x_i \in E}^{n} (1 + \lambda g^i) - 1] = 1, \quad -1 < \lambda < \infty \tag{17}$$

Define for $E \subset X$

$$g_\lambda(E) = \frac{1}{\lambda}[\prod_{x_i \in E} (1 + \lambda g^i) - 1] \tag{18}$$

Then g_λ satisfies all the conditions of fuzzy measures. From Eq.(18), it is obtained that

$$g_\lambda(\{x_i\}) = g^i, \quad 1 \leqslant i \leqslant n \tag{19}$$

and that if $A \cap B = \phi$, then

$$g_\lambda(A \cup B) = g_\lambda(A) + g_\lambda(B) + \lambda g_\lambda(A)\, g_\lambda(B) \tag{20}$$

When $\lambda = 0$, g_λ becomes additive and, hence, equal to a probability measure. In comparison with a probability density, g^i is called a fuzzy density.

†If $⨍_A h(x) \circ g = ⨍_A h'(x) \circ g$ for any $A \in \mathbf{B}$, then $h(x)$ is said to be equivalent to h' with respect to g.

$^T\chi_E(x) = 1$ if $x \in E$ and $\chi_E = 0$ if $x \notin E$.

‡When a proposition holds except on a null set E such that $g(E) = 0$, it is said to hold almost everywhere with respect to g. We write it as g - *a.e.*

Now let $h : X \to [0,1]$ and assume that $h(x_1) \geqslant h(x_2) \geqslant \ldots \geqslant h(x_n)$, then the fuzzy integral of h can be expressed as follows

$$⨍_X h(x) \circ g_\lambda(\cdot) = \bigvee_{i=1}^{n} [h(x_i) \wedge g_\lambda(E_i)] \qquad (21)$$

where $E_i \triangleq \{x_1, x_2, \ldots, x_i\}$.

Finding the maximum of $h(x_i) \wedge g_\lambda(E_i)$ with respect to i can be easily performed, since $h(x_i) \wedge g_\lambda(E_i)$ is convex with respect to i.

Concerning a conditional fuzzy measure, we can also apply λ-rule. Let $Y = \{y_1, y_2, \ldots, y_m\}$ and $\sigma_y(\cdot|x)$ be a conditional fuzzy measure from X to Y.

Let $0 \leqslant \sigma^{ij} \leqslant 1$ for $1 \leqslant i \leqslant n,\ 1 \leqslant j \leqslant m$ (22)

and

$$\frac{1}{\mu}[\prod_{j=1}^{m}(1+\mu\sigma^{ij}) - 1] = 1 \quad \text{for } 1 \leqslant i \leqslant n \qquad (23)$$

where $-1 < \mu < \infty$.

Define for $F \subset Y$

$$\sigma_y(F|x_i) = \frac{1}{\mu}[\prod_{y_j \in F}(1+\mu\sigma^{ij}) - 1] \qquad (24)$$

$$\text{for } 1 \leqslant i \leqslant n$$

Actually μ should be determined dependent on x_i, but we may choose it independently of x_i [4].

3. Learning model

Let us consider a problem of estimating causes through a fuzzy information when there are many results and causes. Let $X = \{x_1, x_2, \ldots, x_n\}$ be a set of causes and $Y = \{y_1, y_2, \ldots, y_m\}$ a set of results. Here g_X and g_Y are assumed to be fuzzy measures of X and Y. Further assume that g_Y is related to g_X by a conditional fuzzy measure $\sigma_Y(\cdot|x)$, that is,

$$g_Y(\cdot) = ⨍_X \sigma_Y(\cdot|x) \circ g_X \qquad (25)$$

Here g_X corresponds to a priori probability and $\sigma_Y(\cdot|x)$ to a conditional probability. For this reason, g_x may be called a priori fuzzy measure by which an individual estimates the grade of fuzziness of the statement 'one of the elements of E ($\subset X$) has been caused'. By $\sigma_Y(F|x)$ where $F \subset Y$, he measures also the grade of fuzziness of the statement 'one of the elements of F results because of x'.

Now let us consider the improvement of g_X by obtaining an information on actual results. This information can be expressed generally by a subset F of Y. The information F implies that one of the elements of F has resulted. Here an information is classified into three types. If F has only a single element, the information is deterministic. It is non-deterministic when F consists of some elements. Further it is fuzzy when F is a fuzzy set.

Hereafter we consider a fuzzy information A with a membership function $h_A(y)$: this is the most general. As is well known, a human being can learn through a fuzzy information. We may say that actually most of the available informations are fuzzy ones.

The fuzzy measure g_Y for a fuzzy set A is defined [4] by

$$g_Y(A) \triangleq ⨍_Y h_A(y) \circ g_Y \qquad (26)$$

Here $g_Y(A)$ can be interpreted to express the grade of fuzziness of the fuzzy information A when one estimates it beforehand. From Eqs (25) and (26) it follows that

$$\begin{aligned} g_Y(A) &= ⨍_Y h_A(y) \circ [⨍_X \sigma_Y(\cdot|x) \circ g_X] \\ &= ⨍_X \sigma_Y(A|x) \circ g_X \end{aligned} \qquad (27)$$

where

$$\sigma_Y(A|x) \triangleq ⨍_Y h_A(y) \circ \sigma_Y(\cdot|x) \qquad (28)$$

The derivation of the above Eq. is seen in [4].

After having the information A, g_X should be improved so that $g_Y(A)$ increases. This method of improvement seems to be quite reasonable. Now we assume that $g_X(\cdot)$ and $\sigma_Y(\cdot|x)$ satisfy λ-rule. If it is assumed that $\sigma_Y(A|x_i)$ is arranged decreasingly, we have

$$g_Y(A) = \bigvee_{i=1}^{n} [\sigma_Y(A|x_i) \wedge g_X(F_i)] \qquad (29)$$

where $F_i \triangleq \{x_1, x_2, \ldots, x_i\}$.

From this we have for an integer l that

$$g_Y(A) = \sigma_Y(A|x_l) \wedge g_X(F_l) \tag{30}$$

where l is the largest integer for which Eq.(30) holds, i.e., l satisfies the following inequality.

$$\sigma_Y(A|x_{l-1}) \wedge g_X(F_{l-1}) \leqslant \sigma_Y(A|x_l) \wedge g_X(F_l)$$
$$> \sigma_Y(A|x_{l+1}) \wedge g_X(F_{l+1})$$

Our learning can be accomplished by increasing fuzzy densities of g_X which contribute the increase of $g_Y(A)$ and by decreasing those which don't.

Let g^i, $1 \leqslant i \leqslant n$, be fuzzy densities of g_X, then as is easily seen, only g^i, $1 \leqslant i \leqslant l$ affects the value of $g_Y(A)$. Therefore the proposed learning algorithm is the following:

$$g^i = \alpha g^i + (1 - \alpha)\sigma_Y(A|x_i) \quad \text{for } 1 \leqslant i \leqslant l \tag{31}$$

$$g^i = \alpha g^i \quad \text{for } l < i \leqslant n \tag{32}$$

where $0 < \alpha < 1$.

In the above algorithm, it is not necessary to increase g^i for $1 \leqslant i \leqslant l$ more than $\sigma_Y(A|x_i)$ since such excessive increase of g^i does not affect the value of $g_Y(A)$. The parameter α is related to the speed of learning, i.e., the speed of convergence of g^i. The smaller the α, the more is the change of g^i.

Now we can derive analytically some characteristics of the learning model.

Property 1 When the same information is repeatedly given:

First we have new $g^i >$ old g^i for $1 \leqslant i \leqslant l$,
new $g^i <$ old g^i for $l < i \leqslant n$,

at a result, new $g_Y(A) \geqslant$ old $g_Y(A)$.

Therefore new $g_Y(A)$ is written such that

$$\text{new } g_Y(A) = \sigma_Y(A|x_k) \wedge g_X(F_k),$$
$$k \leqslant l.$$

From the fact $k \leqslant l$, by assuming $\sigma_Y(A|x_1) > \sigma_Y(A|x_2)$, we can conclude that g^1 finally converges to $\sigma_Y(A|x_1)$ and g^i to 0 for $2 \leqslant i \leqslant n$.

Property 2 In particular when the same information, such that $h_A(y) = C$ for all y, is repeatedly given then:

For simplicity let us call it a constant information. In this case we have

$$\sigma_Y(A|x) = f_Y\, C \circ \sigma_Y(\cdot|x)$$
$$= C$$

$$g_Y = C \wedge g_X(X).$$

From this we have $l = n$. Therefore it follows that g^i converges to C for all i.

Property 3 The limiting values of g^i's are independent of the initial values when the same information is repeatedly given.

Now let us have numerical examples. Table 1 shows conditional fuzzy densities $\sigma_Y(\{y_j\}|x_i)$ which obey the same λ-rule for different x_i. As is seen in the Table, a result y_j comes mainly from a cause x_j for $1 \leqslant j \leqslant 5$. Figures 1-3 and 5 show the improvements of a priori fuzzy densities under the condition that the same information has been given repeatedly. In Figure 4, two informations have been given alternatively.

	y_1	y_2	y_3	y_4	y_5
x_1	.70	.23	.16	.08	.39
x_2	.40	.64	.32	.16	.08
x_3	.16	.49	.57	.24	.16
x_4	.08	.16	.04	.64	.32
x_5	.17	.34	.25	.42	.50

Table 1 $\sigma_Y(\{y_j\}|x_i)$

Comparing Figure 2 with Figure 1, we can know the presented learning model works well with a fuzzy information. Of course in case of a fuzzy information, the speed of convergence becomes slower. Figures 2 and 3 show that the limiting values of g^i become the same for different initial conditions (see Property 3). It is interesting that the results shown in Figure 4 are almost the same as in Fig.1. Figure 5 shows the case when a constant information has been given (see Property 2).

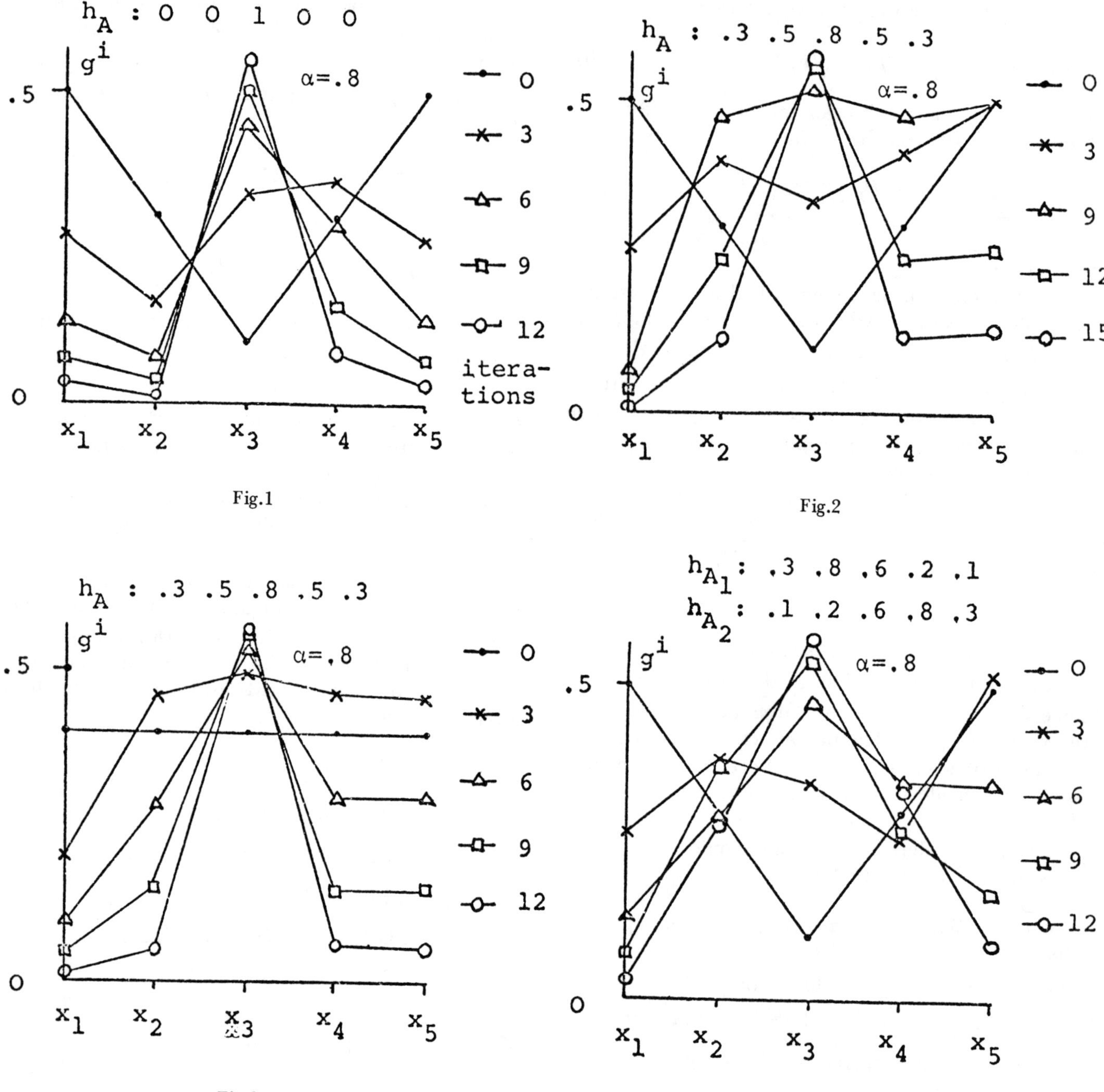

Fig.1

Fig.2

Fig.3

Fig.4

The most interesting and, perhaps, important characteristics of the learning model can be seen especially in Fig.2. In Fig.2, g^i's for $i = 2, 4, 5$ increase at the first several stages of learning and after that decrease. This implies that the model misunderstands at early stages the true cause because only a fuzzy information has been given. Note that this tendency is also seen in Fig.1 but not to a remarkable extent. After the same information has been given many times, the model can find finally the true cause, i.e., x_3.

This sort of misunderstanding seems to be quite natural particularly when only a fuzzy information is available. In other words, we could say that this sort of misunderstanding is even necessary in learning based on a fuzzy information because if a learning model reaches immediately one conclusion, it may lead us to a wrong conclusion. Because of the above characteristics, the presented model works well even in the case shown in Fig.4.

As has been mentioned the speed of the convergence is controlled by the parameter α. It seems that there exists a reasonable value of α in case by case.

Now let us compare our model with Bayesian learning model. What we should do first is to extend Bayesian learning model to a fuzzy case. Probabilities of fuzzy events have been already defined by L.A. Zadeh [1]. So we omit the details concerning those.

Let p_x^i, $1 \leqslant i \leqslant n$, be a priori probability densities of X and $\rho_Y(y_j | x_i)$, $i \leqslant j \leqslant m$, be conditional probability densities with respect to x_i. Then a posteriori probability of X after having an information $\{y_j\}$ is obtained as follows

$$\rho_X(x_i | y_j) = \frac{p_X^i \rho_Y(y_j | x_i)}{\sum_{i=1}^{n} p_X^i \rho_Y(y_j | x_i)} \tag{33}$$

Concerning a fuzzy event A, we have

$$\rho_Y(A | x_i) = \sum_{j=1}^{m} h_A(y_j) \rho_Y(y_j | x_i) \tag{34}$$

From this, a posteriori probability densities after having a fuzzy information A can be written as

$$\rho_X(x_i | A) = \frac{p_X^i \rho_Y(A | x_i)}{\sum_{i=1}^{n} p_X^i \rho_Y(A | x_i)} \tag{35}$$

In particular when a constant information such that $h_A(y) = C$ for all y is given, we have

$$\rho_Y(A | x_i) = \sum_{j=1}^{m} C \rho_Y(y_j | x_i)$$

$$= C.$$

From this it follows that

$$\rho_X(x_i | A) = p_X^i \quad \text{for} \quad 1 \leqslant i \leqslant n \tag{36}$$

That is, in Bayesian learning, to obtain a constant information means to learn nothing or to obtain no information. However, we should distinguish clearly between obtaining an information and not learning. As has been stated, we can distinguish between the above two cases in our learning model (see Property 2). Property 2 is very important when we deal with a fuzzy information because a fuzzy information is somewhat nearer to a constant one.

Now let us examine how Bayesian learning works with a fuzzy information. Table 2 shows conditional probability densities the values of which are chosen similar to those of conditional fuzzy densities. Figures 6, 7 and 8 are corresponding to Figures 2, 3 and 4, respectively. Those Figures show almost

	y_1	y_2	y_3	y_4	y_5
x_1	.45	.15	.10	.05	.25
x_2	.25	.40	.20	.10	.05
x_3	.10	.30	.35	.13	.19
x_4	.07	.13	.03	.52	.26
x_5	.10	.20	.15	.25	.30

Table 2 $\rho_Y(y_j | x_i)$

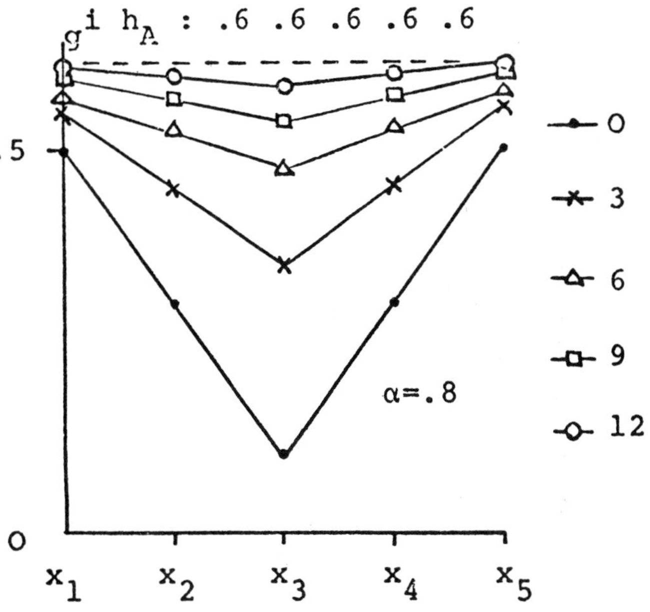

Fig.5

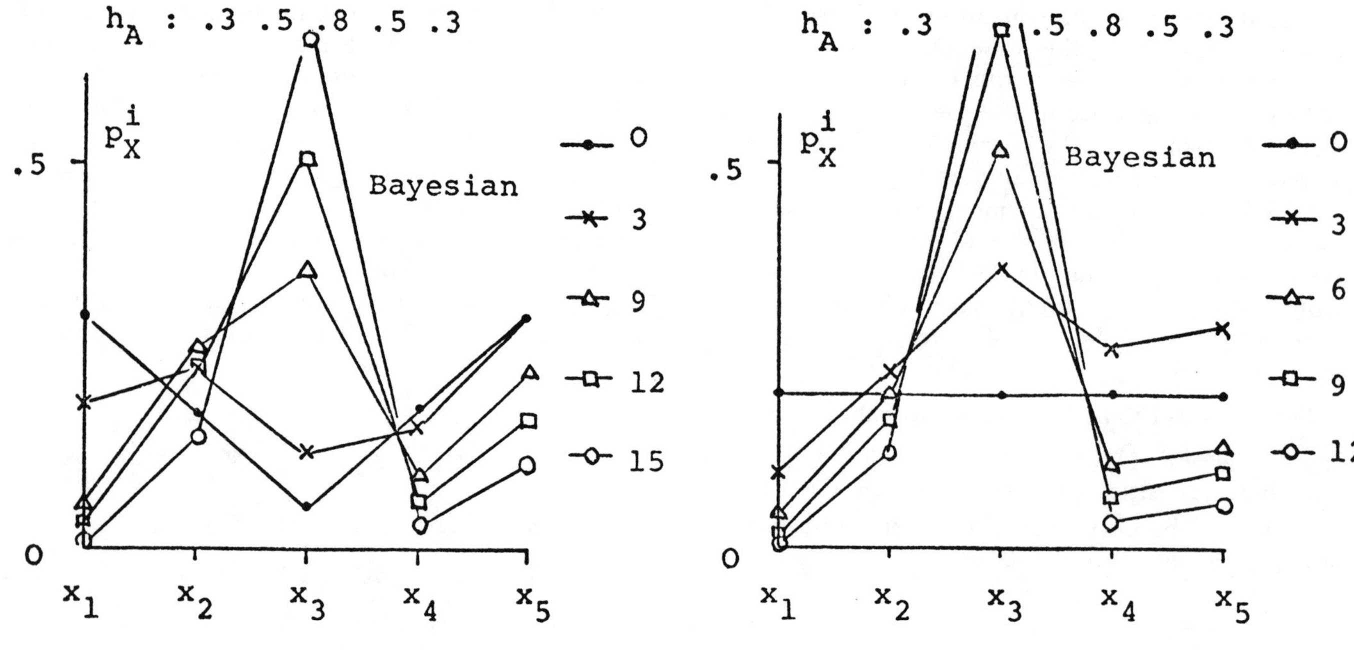

Fig.6

Fig.7

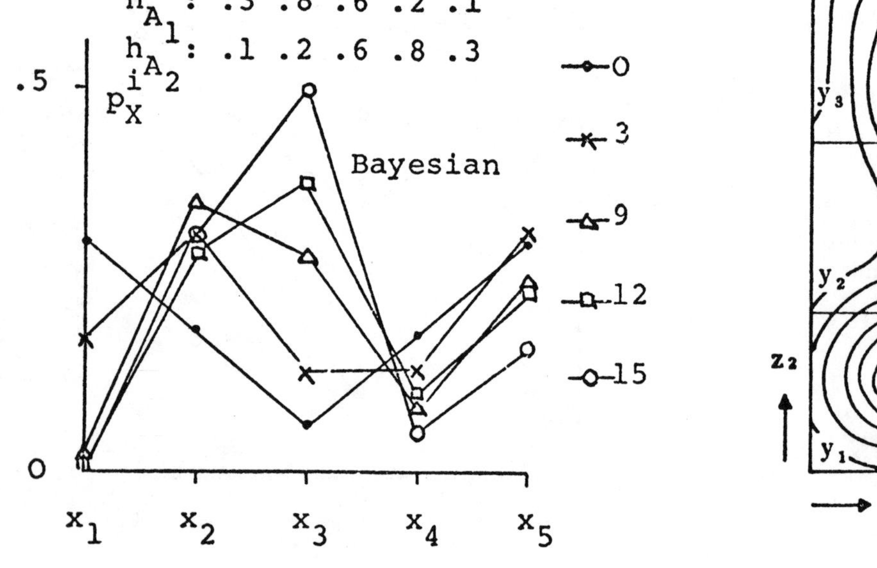

Fig.8

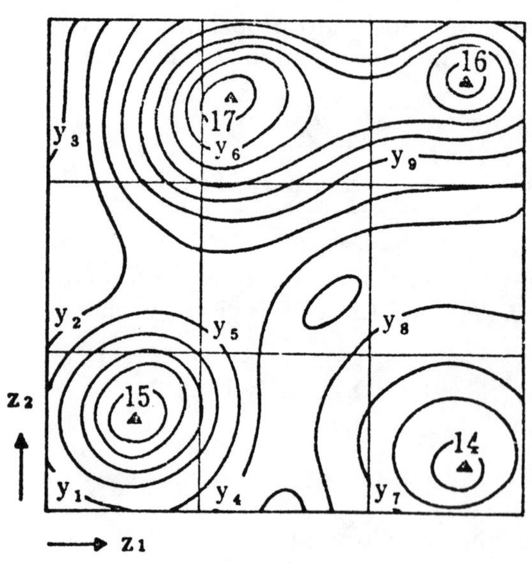

Fig.9 $J(z_1, z_2)$

the same results. As far as these numerical examples are concerned, the presented learning model can learn through a fuzzy information the same as Bayesian learning model.

However there are two remarkable differences between them.

1. The speed of convergence can be controlled by a parameter in the presented model.
2. The presented model can be expected to work through a fuzzy information more effectively than Bayesian learning model which could be concluded by the preceding discussions.

4. Application

In this section, the presented learning model is applied to finding an extremum of unknown multimodal function. From the view-point of engineering, it is more important to draw macroscopic map of an objective function covering the searched area and to know an approximate value of the maximum quickly than to obtain the exact value and the precise position of the maximum. Further, some experiments of the optimum seeking by a human being show us that he never tries to search locally before knowing the global feature of an objective function [2]. Therefore let us try to find an extremum in a macroscopic sense.

Our objective function is shown in Fig.9, which has four peaks and is generated by the following formula.

$$J(z_1, z_2) = 15 \exp[-20(z_1 - 0.22)^2 - 22(z_2 - 0.24^2] + 17 \exp[-19 \times (z_1 - 0.4)^2 - 15(z_2 - 0.85)^2] + 14 \exp[-23(z_1 - - 0.87)^2 - 18(z_2 - 0.15)^2] + + 16 \exp[-20(z_1 - 0.85)^2 - 20(z_2 - 0.85)^2] \quad (37)$$

The function is, of course, assumed to be unknown. The searched area is divided into some blocks.

Now let Y be a set of blocks. In this problem, $Y = \{y_1, y_2, \dots, y_9\}$ as shown in Fig.9. Let X be a set of criteria or clues by which one guesses if a block includes the true maximum. Further let g_X be a fuzzy measure which expresses the grade of importance of a criterion or, more precisely, a subset of X.

The following criteria are considered:

- x_1 concerning the number of the points examined in the past searches,
- x_2 concerning the average of the function obtained in the past searches,
- x_3 concerning the number of points the values of which belong to the best ten of those in the whole area,
- x_4 concerning the estimated maximum through the past searches,
- x_5 concerning the gradient of the function.

Next let $\sigma_Y(\cdot | x)$ be a conditional fuzzy measure by which one evaluates the grade of finding an extremum in a block or in a subset of Y from the view-point of a criterion x. To construct $\sigma_Y(\cdot | x)$, let us define the following functions $C_j(x)$ where j is the number of a block.

$C_j(x_1) = \exp[-m_j/3]$, where m_j is the number of the past searched points in the jth block. If m_j is small, the likelihood of finding the maximum in this block may become large.

$C_j(x_2) = [\tan^{-1}\{6(M_j - M)/M\}]/\pi + 0.5$, where M_j and M are the mean values obtained in the past searches in the jth block and in the whole area, respectively.

$C_j(x_3) = 0.3N_j + 0.1$, where N_j is the number of the searched points in the jth block which are classified into the best ten of those in the whole area.

$C_j(x_4) = 2[\tan^{-1}(3O_j/M_j)]/\pi$, $Q_j \triangleq M_j + (P_j - R_j)$, where P_j and R_j are the maximum and minimum obtained in the past searches in the jth block. Q_j means the estimated maximum from the past variance of the objective function.

$C_j(x_5) = [\tan^{-1}\{6(P_j - M_j\}]/\pi + 0.5$. This is the criterion related to the gradient of the objective function.

Here $\sigma_Y(\{y_j\} | x_i)$ is obtained by normalizing $C_j(x_i)$ so that $\sigma_Y(\cdot | x)$ satisfies λ-rule.

Choose β_i so that

$$\frac{1}{\mu}[\prod_{j=1}^{9}(1 + \mu\beta_i C_j(x_i)) - 1] = 1, \quad -1 < \mu < \infty \quad (38)$$

where μ is a given constant.

Define

$$\sigma_Y(\{y_j\}|\, x_i) \triangleq \beta_j C_j(x_i) \tag{39}$$

In this example, $\sigma_Y(\cdot|\, x)$ is not fixed but varies dependently on the stage of the search (see the definition of $C_j(x)$).

If g_X is given, we can estimate, by using g_Y obtained from Eq.(25), the grade of finding the maximum in a block. However we do not know which g_X is good. Therefore we first have to assume g_X with λ-rule a priori and afterward we can improve g_X according to the presented learning procedure.

An available information A is given as follows:

$$h_A(y_j) \triangleq (p_j - \min_j p_j)/(\max_j p_j - \min_j p_j) \tag{40}$$

where p_j is the maximum in the jth block which has been found so far.

Note that an information is essentially fuzzy in this example because we will never be able to know which block includes the true maximum. Therefore we have to learn through a fuzzy information.

Generally we may say that a suitable g_X depends on the shape of an objective function as well as the stages of the search, i.e., time. For example, at early stages of the search, we should attach higher importance to the criterion x_1.

The procedure of the heuristic search is as follows.

[step 1] In each block, two points are searched at random and $\sigma_Y(\{y_j\}|\, x_i)$ are calculated from $C_j(x_i)$. $g_Y(\{y_j\})$ are also calculated from Eq.(25), where g_X^i are given a priori. Next the information $h_A(y_j)$ are calculated from Eq.(40). The grade of fuzziness $g_Y(A)$ of total information can be known from Eq.(26), and g_X^i is corrected according to the learning rule Eqs. (31), (32).

[step 2] In proportion to $g_Y(\{y_j\})$, another eighteen search points are newly assigned as follows. Three points are assigned to each three blocks of the top class and two points to each three of the middle class, and one point to each three of the lowest. Then random search is done as same as step 1 and σ_Y, g_X, $g_y(\{y_j\})$, $g_Y(A)$ are calculated. g_X^i are corrected again according to the results.

[step 3] Step 2 is repeated until $g_Y(\{y_j\})$ converges. In our example, the repetition is stopped at the fourth iteration, because the total number of search points reaches 72.

Some results of the experiment are shown in the following figures and tables. Figures 10, 11, 12 show the process of the convergence of g_X^i on the different initial conditions (a priori measures). Tables 3, 4, 5 show the change of macroscopic evaluation $g_Y(\{y_j\})$ of each block at each step. From these figures, we can see the speed of convergence is most rapid in the case C. In Figs 10 and 11, g_X^i is expected to become similar to those of Fig.12.

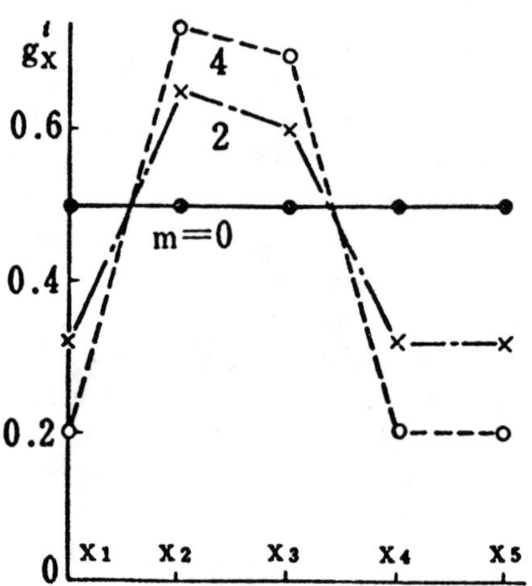

Fig.10 Improvement of g_X^i (Case A)

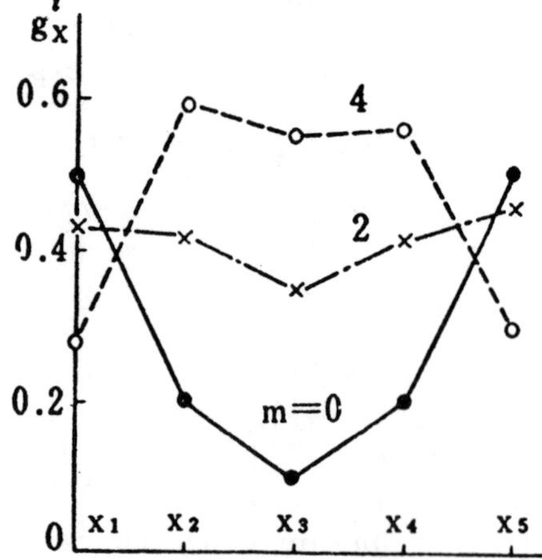

Fig.11 Improvement of g_X^i (Case B)

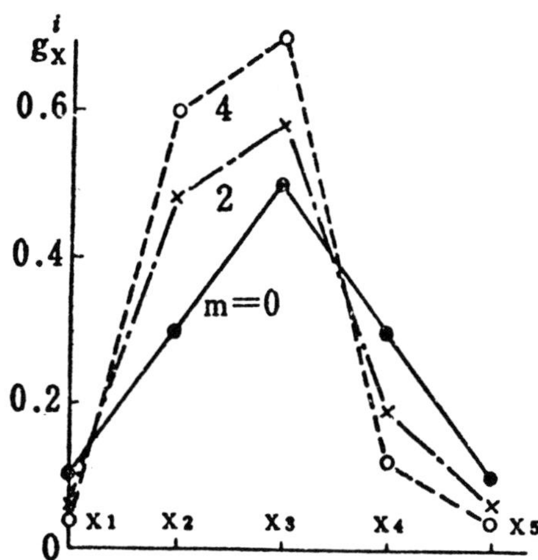

Fig.12 Improvement of g_X^i (Case C)

The locations of the real extreme are y_6, y_9, y_1, y_7 as shown in Fig.9. The evaluation $g_Y(\{y_j\})$ of each block does not always indicate the correct order, but the results are acceptable. We also understand from these figures that, among the many criteria, x_3 is most important and x_2 is next. It may be better to eliminate other criteria in this example.

The results are satisfactory. These results have been compared with those without learning. The comparison shows that learning scheme has been effective.

m \ y_j	y_1	y_2	y_3	y_4	y_5	y_6	y_7	y_8	y_9
1	.36	.41	.38	.25	.36	.50	.27	.31	.50
2	.30	.38	.39	.33	.26	.57	.33	.33	.59
3	.31	.32	.32	.32	.31	.60	.31	.31	.65
4	.26	.26	.32	.27	.30	.57	.26	.26	.67

Table 3 $g_Y(\{y_j\})$
(Fuzzy Information Case A)

m \ y_j	y_1	y_2	y_3	y_4	y_5	y_6	y_7	y_8	y_9
1	.33	.26	.50	.25	.36	.27	.31	.38	.30
2	.44	.40	.40	.33	.23	.33	.38	.22	.44
3	.44	.42	.36	.43	.27	.42	.42	.23	.42
4	.45	.34	.33	.37	.32	.50	.34	.32	.51

Table 4 $g_Y(\{y_j\})$
(Fuzzy information Case B)

m \ y_j	y_1	y_2	y_3	y_4	y_5	y_6	y_7	y_8	y_9
1	.36	.27	.17	.28	.16	.36	.36	.22	.36
2	.30	.24	.15	.26	.18	.53	.30	.21	.53
3	.39	.24	.19	.25	.16	.48	.39	.19	.53
4	.30	.20	.19	.20	.21	.59	.30	.20	.56

Table 5 $g_Y(\{y_j\})$
(Fuzzy Information Case C)

5. Conclusions

This chapter presents a learning model which works through a fuzzy information. It also applies the model to search an extremum of an unknown multimodal function.

It seems particularly important to deal with a fuzzy information. We cannot talk about human learning without mentioning a fuzzy information.

The authors hope that the method, suggested in this chapter, is not only effective for optimum seeking but also has wider applications as a decision model with a learning scheme based on a fuzzy information.

References

1. ZADEH, L.A., 'Probability measures of fuzzy events', *Journal of Mathematical Analysis and Applications* 23, 421/427 (1968).
2. NAKAMURA, K. and ODA, M., 'Heuristic and learning control' included in *Pattern Recognition and Machine Learning* (edited by K.S. Fu), p.297 Plenum Press (1971).
3. SUGENO, M. *et al.*, 'Subjective evaluation of fuzzy objects', IFAC Symposium on Stochastic Control, Budapest (1974).
4. SUGENO, M., 'Theory of fuzzy integrals and its applications', Ph.D. thesis, Tokyo Institute of Technology (1974).
5. TERANO, T. and SUGENO, M., 'Conditional

fuzzy measures and their applications', included in *Fuzzy Sets and their Applications to Cognitive and Decision Processes,* Academic Press (1975).

6. TSUKAMOTO, T., 'A subjective evaluation on attractivity of sightseeing zones', included in Summary of Papers on General Fuzzy Problems—Report No. 1, published by the Working Group on Fuzzy Systems, Tokyo, Japan (1975).

About the fuzziness in the analysis of information systems

MARKKU. I. NURMINEN
Turku School of Economics, Finland

1. Introduction

An organization can be considered as a socio-economic system, which includes human components. The participation of men in the activity of the organization implies some more or less fuzzy features in the system. The fuzziness in the system can be divided into three categories:

1. The information processed and transmitted is fuzzy.
2. The algorithms for processing the information are fuzzy.
3. The structure of the system is fuzzy.

The information and the rules used in the human decision making are often fuzzy. The structure is fuzzy if the couplings for transmitting information between the decision makers are fuzzy. In practice, it is not always exactly defined whether particular information should be regularly transmitted from one manager to another. This categorization is suitable especially for the information systems and the decision making systems.

There is a tendency to delegate a part of information processing and programmed decision making to a separate computer system. This system, on the other hand, has clearly a non-fuzzy character. The interest in using the computer systems is partly based on the properties such as reliability and accuracy, which are expressly non-fuzzy objectives. The computer has, if any, only a clumsy ability to process fuzzy data or process the data with fuzzy programs. I have not even found a reasonable interpretation for the fuzzy structure of a computer system. We have to accept the view, that the computer is essentially designed for non-fuzzy data processing and the computer systems are essentially non-fuzzy.

The man-machine interface between the fuzzy decision system and non-fuzzy computer system can be accomplished only on the conditions of the computer. Only programmed (i.e. non-fuzzy) decisions (or algorithms) which process non-fuzzy data are delegated. The couplings between the computer system and its environment are non-fuzzy, too. This means that the requirement of data to be produced by the computer system must be determined in detail, before the computer system can be implemented.

The method for the design of computer systems used in this chapter consists of two parts, analysis and synthesis†. In the analysis the object of inquiry

† Langefors: Theoretical Analysis of Information Systems.

is abstract information represented by information sets, independently of its future representation by data. The analysis is started from a given output information set. The processes and information sets preceding each of them are determined. The determination is repeated recursively until the input information sets are attained after the finding of all intermediate information sets and processes. The description of the system is presented at several levels of detailing. This gives a hierarchical, multilevel subsystem structure. The analysis is terminated when the elementary description is reached.

In the synthesis phase the information sets are interpreted as data files. Now the future structure of the computer system can be outlined and designed.

Logically, the decision of the information to be produced by the computer system must not be made later than the exact time of entering the synthesis phase. In practice, the decision has often been made when entering the analysis phase. In this chapter, the effect of the postponement of this decision is considered. Firstly, the character of the ambiguity in the requirement of information is considered in section 2. In sections 3 and 4 a quantitative measure of necessity is introduced and some manipulations with it as an instrument of information analysis are presented. Some points of view about the usefulness of this approach are discussed in section 5.

2. The requirement of information

In this chapter the requirement of information is considered in connection with the design of information systems and computer systems. The existence of the requirement of information is obviously needed to motivate the producing of it by the systems. Of course, the requirement exists—if it exists —independently of the information and data systems; I consider it as such in this section.

The analysis of the requirement of information is initiated with the information required by the interested parties of the enterprise. Much information is produced with no other purpose than to be transmitted and used outside of the borders of the firm. This kind of requirement can be taken as given when the design of the information system is considered. The determination of it is by no means problematic nor interesting. Thus, I exclude it in the rest of this section.

In this chapter I have supposed that the main purpose of the information inside the firm is the decision making. If the decision making is defined broadly, also the other possible purposes can be interpreted as decision making[†]. It is also important to see the structural similarity of the decision making and information processing. In both of them, input information is transformed to output information according to some more or less specified rules. The result of the decision making, the decision, is seen as information, too. This view of the decision making requires the distinguishing between the concepts 'problem solving' and 'decision making': Problem solving is a preceding phase of the decision making. In the decision making the results of problem solving are applied in the *actual* decision situation. After accepting this view no extra problems arise, although similar programmed decisions are made both automatically in a computer system and by human workers.

The requirement of information is thus fundamentally associated with the decision making. In order to better understand the character of this requirement of information for different types of decision, a classification of decision making is presented as follows:

- 1 Programmed decision making.
- 2 Non-programmed decision making
  - 21 Structured decision making
  - 22 Non-structured decision making
    - 221 Externally initiated
    - 222 Internally initiated

The decision making is programmed if the result can be obtained from the input information algorithmically. The algorithms can be programmed for a computer. All other decisions are non-programmed[‡].

The decision making is structured, if the decision situation is previously known and if the preceding problem solving has been performed. Programmed decisions are always structured. The problem solving does not always give programmed decision rules. Because the decision situation is known in advance, the decision making is usually delegated to a certain point in the organization.

The non-structured decision making can be initiated externally or internally. The environment of the firm can change so much that new types of decisions have to be made, which are unforeseeable, and therefore the firm cannot be well prepared to meet them. The internal initiation is closely related

[†]Other possible purposes of information are the production of new information and the operative use of information (see Nurminen: Information System and Control System).

[‡] Simon: The New Science of Management Decision.

with the strategic planning and other voluntary adaptive activities in order to develop the firm itself. The non-structured decision making is often unclear. The problem itself can be unidentified, the possible alternatives have to be searched for, as their possible consequences. When the problem is defined, a suitable method for solving it can be searched for and found.

The classification presented resembles the classification into strategic planning, management control, and operative control. The classification above is not based on the content or the organizational level, but on the type of the decision only. There is a tendency to structure the non-structured decisions and to program the non-programmed decisions. The change in the type of the decision does not change the content or the domain of the decision†. The type of the decision is most essential in determining the need for information. This gives the motivation for the classification presented.

The requirement of information for programmed decisions is clear. In a given situation the input information for the programmed decisions can readily be listed in as detailed a manner as is desired. The changes in technology or programming a previously non-programmed decision can change the set of programmed decisions. This kind of change is, however, the result of a non-programmed decision at a higher level of the organization. Henceforth the requirement of information for programmed decisions is supposed to be known.

For non-programmed, but structured decisions the decision maker can usually be identified. It is often possible to determine what information cannot be omitted in a structured decision making situation. Every occurrence of the structured decision making is unique; it is one of the reasons why it has not been programmed. Every time a decision is made, the actual requirement of additional information can be weighed. In fact, a decision at a new level has to be made: What information is exactly needed for this decision? How much is one willing to pay in money and in time to get additional information? This problem could be solved by comparing the costs and benefits of the additional information. Unfortunately, neither can be expressed exactly, or even quantitatively in terms of money. It is very difficult to show the marginal improvement in the decision caused by additional information, at least in advance. Rather it seems obvious, that the manager can only approximate what information is important and what is not. The approximation is based on intuition, on the insight into essential and non-essential facts, rather than algorithmic estimates of expected values.

Information can be necessary, useful, or useless. Given two useful information sets, a decision maker can say which of them is more important in a given structured decision. Thus, a measure with ordinal scale can be associated with each piece of information possibly useful in a structured decision. The measure is not a probability measure—either subjecttive or objective—because it is not a question of uncertainty but of the degree of importance. The final solution of whether to use this additional information cannot, of course, be obtained from this measure alone, but it can be a useful instrument. The measure has the character of a fuzzy membership function‡.

The requirement of information in a non-structured decision situation is far more problematic, because not even the situation is known in advance. Of course, we can consider the set of all possible decisions, estimate the subjective probability for the occurrence of each, and proceed with them as with structured decisions with associated probabilities. The number of all possible decisions is very great, however. Only a small number of the most probable decisions could be processed with this approach. The corresponding instrument for manipulations would consist of one measure of subjective probability and one fuzzy membership function, or one measure with properties of each.

The other approach for preparing to meet unexpected decisions is to create an information system, which can serve us in most thinkable situations. Instead of numbering the situations one by one, a holistic view of the firm as an entity is demanded. If we have all the essential information about ourselves and our environment, we are as ready as possible to meet any new and unexpected decision situation. Here again, we have a problem of measuring the essence of the information. Here it is not coupled to any specific decision, but to the knowledge of the firm as an entity. If this measure can be found, it has the character of the fuzzy membership function.

The design of an information system is usually initiated by the determination of the information which it is supposed to produce. There is no problem in stating the indispensable information needed

†The methods used for achieving the change from one class to another are often called problem solving.

‡Zadeh: Outline of a New Approach to the Analysis of Complex Systems and Decision Processes.

in the programmed or structured decisions and for the external information requirements. The problem is, which information is to be included in the system in order to improve the non-programmed decision making as additional information, or because it is considered essential enough, though it cannot be seen to be useful for any structured decision now in sight.

We cannot make a non-programmed decision on behalf of the decision maker, but we can provide him with relevant information to assist the decision making. The determination of the output information of the information system is not a programmed decision. The manager cannot be replaced in this decision making either. Instead we can assist him by giving him more supporting information.

In this chapter a method for obtaining additional information for this decision is presented. It is suggested that the performing of the information analysis should take place before the final selection of the information to be produced by the system. A substantially large amount of information should be taken in the analysis. It will be seen that the result of information analysis gives additional information which can help the final determination of the output information.

The problem of getting relevant approximations of the degree of necessity of different information sets is not treated in this chapter. In the next section it is assumed that with every output information set there is associated a measure of necessity which in the first place is a fuzzy membership function but which can also have properties of a probability measure.

3. Fuzziness in the elementary information system

Before the measure of necessity can be introduced in the information analysis, some basic definitions are required. We first consider fuzziness in an elementary information system. It is a one-level system with (elementary) information sets and (elementary) processes† as its components. The structure consists of input and output relations (or couplings) and is restricted as follows:

Restriction 1 Every information set has exactly one input relation and exactly one output relation.

Restriction 2 Every process has at least one input relation and at least one output relation.

Restriction 3 Directed loops are not allowed, except in the update cycle of standing information sets.

Restriction 4 The system is unified, i.e. it cannot be divided into two or more separate parts with no relation with each other.

The fuzziness in the information system originates from the judgment of the degree of necessity of the information which the system is supposed to produce. Thus, the output information set of the whole information system can be considered as a fuzzy set. To each element of the output information is assigned a value of the fuzzy membership function, which describes the degree of necessity of that element. The fuzziness originates externally; no internal source of fuzziness is accepted.

The fuzziness of the output information set has no effect in the coupling structure. The measure of necessity can be used for measuring the necessity of the information sets and processes, not that of their relations. The input or output relation between an information set and a process is meaningful only if both of its members exist. The existence of, for example, the input relation can be verified or falsified by giving the answer to the question: Is particularly this information set (say *a*) necessary as input information for the process *(A)* in order to produce another information set *(b)*? As restricted above, only programmed decisions can be delegated to the future computer system. The algorithm of this decision (process *A*) gives an unambiguous answer to this question at the same time as the solution to the existence problem of the input relation. The input relation itself is not fuzzy. The debatability of an output information set can cause the debatability of the preceding process and information set, but the input and output relations are valid or invalid indisputably.

The fuzziness of the output information set does not affect the input and output relations, whereas it certainly has an effect on the necessity of the preceding information sets and processes. The fuzziness enters the system through its output information set and is transferred over it through the network of input and output relations. The rules of transference depend on the character of the measure of necessity. In the first place the measure is interpreted as a fuzzy membership function. Sometimes, it can have some properties of subjective probability

† For simplicity, the word 'elementary' is omitted in this section. Thus, it has to be imagined before every occurrence of terms 'information set' and 'process'.

too. Therefore, an alternative rule (*1p*) for the transference is presented for use in this case.

Rule 1 The necessity $\mu_{Pe}(A)$ of the process A depends on the necessities $\mu_{Ie}(b_i)$ of its output information sets b_i as follows:

$$\mu_{Pe}(A) = \max_i \mu_{Ie}(b_i), \quad A \to b_i$$

Thus, a process is as necessary as the most necessary of its output information sets. This rule is in accordance with the general rules for operating with the fuzzy membership function.

The corresponding rule for the probability measure has the form:

Rule 1p The necessity $\mu_{Pe}(A)$ of the process A depends on the necessity $\mu_{Ie}(b_i)$ of its output information sets b_i as follows:

$$\mu_{Pe}(A) = 1 - \prod_i [1 - \mu_{Ie}(b_i)], \quad A \to b_i .$$

This rule gives an increase in the value of the measure for a process with several output information sets. The motivation for this rule can clearly be seen in a special case, when several duplicates of the same information are needed for different purposes, but the actual requirement of each is uncertain. The rules 2 and 3 can be used unmodified for the probability measure.

Rule 2 The necessity $\mu_{Ie}(a_j)$ of an information set a_j equals the necessity $\mu_{Pe}(A)$ of the process A, which uses it as its input information:

$$\mu_{Ie}(a_j) = \mu_{Pe}(A), \quad A \to a_j .$$

This rule takes advantage of the restriction 1 above. Because one information set is used as input information in exactly one process, all of the input information sets of one process have the same value of the measure of necessity.

Rule 3 The necessity of an information set in an update cycle is the same for each of its occurrences and equals the necessity of the first updating process of the cycle.

The starting point of the transference process is an elementary description of the information system with a measure of necessity assigned to each of its elementary output information sets. The process proceeds by applying each of the three rules in turn and thus finding the measure for the preceding processes and information sets. The measure of necessity of an information set can be determined immediately after the determination of the measure of necessity of the succeeding process. The determination of the measure of necessity of a process can be carried out first when the measure of necessity of all the succeeding information sets are determined. If an update cycle is identified, rule 3 is applied and necessary adjustments are performed. Because there are no other kinds of loops in the structure, the transference process can be completed with a finite number of applications of the rules.

It is important, that the description of the information system is elementary. If not, the process should be divided into (finally elementary) subsystems, because it gives a more detailed description of the necessity.

The probability measure (with rule 1p) is greater than or equal to the measure of fuzziness (with rule 1).

Remark: The measure of necessity is non-decreasing in every path, considered in the direction opposite to the flow of the network. This is a natural consequence of the absence of the internal sources (and sinks) of fuzziness.

Remark: If rule 1 is applied instead of rule 1p, no new numerical values of the measure of necessity are generated inside the system. Obviously the corresponding equivalence classes can be formed among the elements of the system.

The transference process outlined above means roughly speaking that for every information set and process all the paths ending in an elementary output information set are found. The maximum value of the measure of necessity of these elementary output information sets is assigned to the value of the measure of necessity of this particular information set or process. This can be performed alternatively by matrix operations using the precedence matrix of the system†. The Boolean transitive closure of the precedence matrix gives an elegant tool for determining the necessity of information sets. The number of elements in a precedence matrix of the entire system is very great, however—of course, this method fails if rule 1p is used.

†Langefors: Theoretical Analysis of Information Systems.

4. Fuzziness in information analysis

The elementary description of the information system used in the previous section is the result of information analysis. The method called information analysis consists of successive detailings, i.e. the subsystems are divided into smaller subsystems until finally the elementary level is reached.

A non-elementary subsystem, like the whole information system, is defined as a set of elementary processes. The internal information sets of a subsystem are included in the subsystem. The subsystem has a process-like character: It transforms input information into output information. The input and output information of a subsystem is presented in non-elementary information sets, which are collections of elementary information sets. The couplings between the subsystems are realized by means of the input and output relations between the information sets and subsystems. The relations form a structure with the same properties as that of the elementary information system (see section 2).

Sometimes it is desirable to introduce the measure of necessity already in a non-elementary description during the information analysis. Because of the structural identity with the elementary description rules 1, 2, and 3 for transference can be used in the non-elementary description with only slight revisions. Firstly, the initial state of the transference process requires the measure of necessity assigned to the output information sets, which are here non-elementary.

Definition 1 The *local measure of necessity* of a non-elementary information set (subsystem) is defined as the maximum value of the measures of necessity assigned to its elementary information sets (elementary processes):

$$\mu_{IL}(a) = \max_{a_j \subset a} \mu_{Ie}(a_j)$$

$$(\mu_{PL}(A) = \max_{A_j \in A} \mu_{Pe}(A_j)\ )\ ,$$

where the distinction between information sets *(I)* and processes or subsystems *(P)* is denoted by the first subscript. The second subscript is used to express the locality *(L)*, globality *(G)*, or the definition at the elementary level *(e)*.

Definition 2 The *global measure of necessity* of an information set or subsystem is determined by using rules 1, 2, and 3 below. The process is initiated by assigning the local measures of necessity to the values for the output information sets:

1. $\mu_{PG}(A) = \max_j \mu_{IG}(a_j), \quad A \rightarrow a_j$;

2. $\mu_{IG}(a) = \mu_{IL}(a)$, if a is output information set
   $\mu_{PG}(A), \quad a \rightarrow A$, otherwise;

3. Rule 3 of section 3 unmodified.

During the information analysis the elementary structure is not generally known; therefore the local measure is not usable in this phase. When the information analysis is completed, the local measures can be determined. The local measure should be considered as primary compared with the global measure.

Theorem The global measure of necessity for an information set or subsystem is greater than or equal to its local measure of necessity.

The proof is not given. It can easily be inferred as valid when considering the applications of the three rules above starting from the output information sets. At the beginning the two measures are equal. The equality remains in the application of rule 2, but rules 1 and 3 can cause an inequality for some elements. This inequality has the direction stated in the theorem.

If the description is non-elementary so that it consists of large information sets and subsystems with numerous components, the measure of necessity has great values. This is natural, because the local measure is obtained as the maximum of the measures of its elementary components, and the global measure is still greater. The measure of necessity for some parts of the system has a tendency to decrease in value, when the information analysis proceeds.

The information analysis is performed in several successive stages. At each stage every subsystem (starting with the whole system itself) is divided into smaller subsystems, until the resulting subsystems are elementary processes. Altogether four tasks have to be carried out when a division into subsystems is performed[†]

[†]Nurminen: Definition of Information System and Information Analysis.—For simplicity, in the list of tasks the (sub)-system to be divided is called a system and the resulting subsystems subsystems.

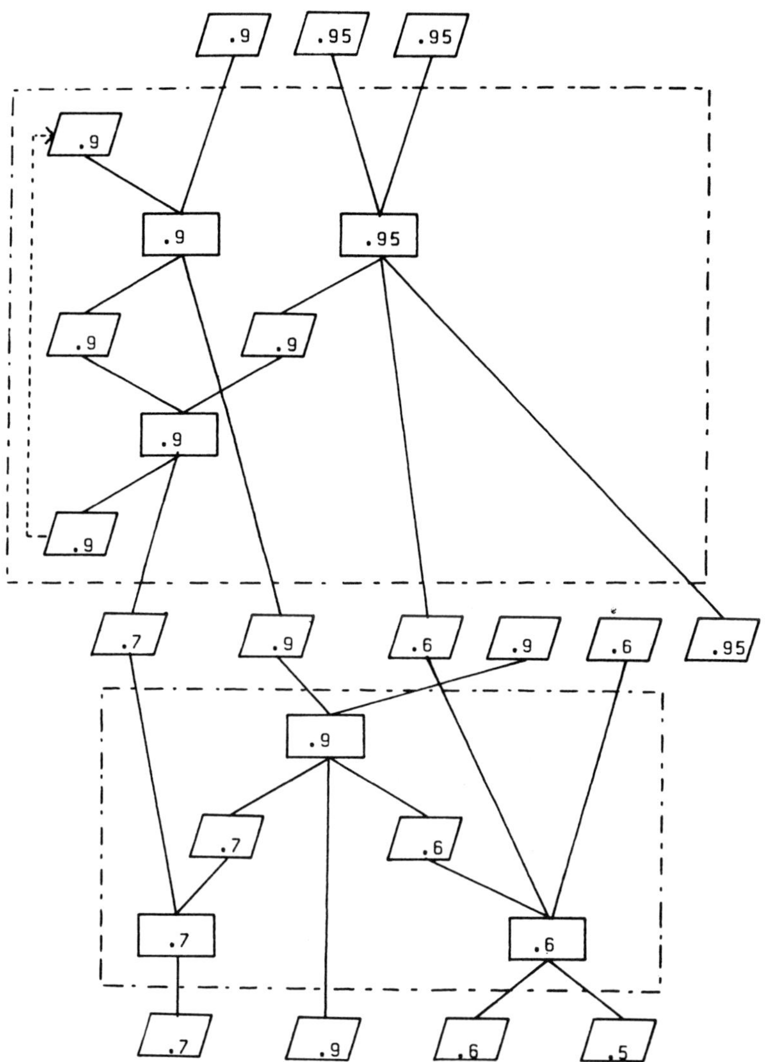

Fig.1 Description at an elementary level with subsystem boundaries and the values of the measure of necessity

1. The system is divided into subsystems. Because the system is defined as a set of elementary processes (which are not yet known in detail), the subsystems are subsets of it.
2. The couplings between the subsystems are introduced.
3. The new internal information sets of the system are introduced. These can be standing information stored inside the system or intermediate information between the subsystems.
4. The input and output information sets of the system are divided into components. This is obligatory in order to maintain the structural restriction 1. If one information set is found to have couplings with several subsystems, it must be divided into components, one for each subsystem. If the same component is needed in several subsystems, it has to be duplicated.

The four tasks presented form an indivisible operation. All of them have to be completed before the measure of necessity can be treated. Only the global measure can be determined. Even this can be difficult if the total output information is not analyzed at an elementary level and assigned detailed values of the measure of necessity. In the absence of this detailing only subjective approximations of the measure of necessity of the non-elementary output information sets can be used as the source of transference.

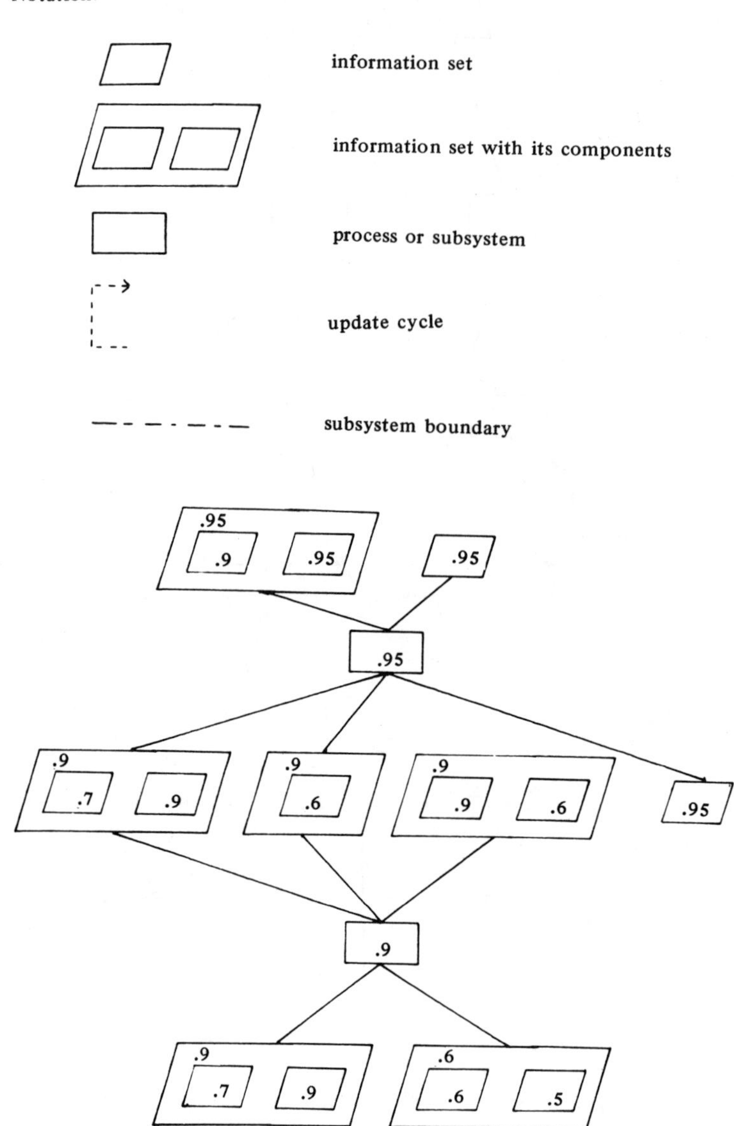

Fig.2 Description of the same system as in Fig.1 at a non-elementary level with the values of the global measures of necessity

At every stage, the consistency of the description as an entity has to be preserved. At least the following checkings are necessary:

When an output information set is divided into components, no component can be more necessary than their union, but the measure of necessity for at least one component must have the same value as that of the union. If the measures have originally been based on intuition only, some corrections may be needed.

Both horizontal and vertical consistency between all descriptions is required: The measure of necessity of a subsystem (information set) is greater than or equal to that of its subsystems (components). The measure of necessity of an information set is determined by the measure of necessity of the succeeding subsystem, not the preceding one.

After the information analysis the local measure must not be greater than the global measure.

During the information analysis the subsystems are perhaps not known in detail. This can compel the analyzer to make corrections to earlier, less elementary descriptions. Of course, the measures of necessity must be adjusted correspondingly.

The transference process is illustrated in Figs 1 and 2.

5. Discussion

The approach presented in this chapter suggests that the information analysis should include additional information, which is then eliminated before the synthesis phase, i.e. the designing of the computer system. This causes additional cost. So does the approximating of the measures of necessity and the transference of them through the coupling network. Some benefits must be achieved to justify these costs. Three partly interrelating views on this are outlined briefly:

Firstly, the final decision on the information which the computer system is to produce can be postponed until the information analysis is completed. This is advantageous especially in questionable cases. The conditions for this decision can clearly be better after the analysis than before. This is because the structure of the system is known in detail after the analysis. The elementary description gives access to all precedents of any output information set. This can be used to determine how big a part of the system is unnecessary if some less necessary output information set is cut off. This kind of examination can lead to rough estimates of the marginal cost of output information: one information set can, for example, be produced as a by-product with minor marginal costs, whereas the production of another can provide much processing or collecting of new input information to be used for this purpose only. It seems obvious, that the comparison of the requirement of information and the costs incurred in its production can be made on a slightly better basis than before the analysis. Sometimes it can be helpful if some less necessary information can be eliminated already at an earlier, non-elementary stage of information analysis. This means that the original high level of aspiration is lowered gradually when more knowledge of the structure of the system is achieved. This opportunity motivates the introduction of the measure of necessity already during the information analysis. It can also help the decision maker to give better approximations for the elementary information sets, if he has first considered the information as larger entities.

Secondly, the measurė of necessity assigned to the elements of the system can be used to find suggestions for the future subsystem structure of the computer system. It is important to notice this, especially if the final decision of the output information is based on the threshold method, where all elements with a value of the measure of necessity less than a given threshold value are cut off. It is not desirable to include elements with a large variation in the measure of necessity in the same subsystem of the computer system. The elements with low necessity will not be cut off by the threshold value which is greater than their measure of necessity, if they are included in a subsystem with high necessity. The equivalence classes often form unified subsets in the network, and thus give an elegant tool for this purpose.

Thirdly, useful suggestions for the order of implementation of the subsystems can be obtained if the measures of necessity are used. This can be important because usually all subsystems cannot be designed, programmed and implemented simultaneously. The realization of the subsystems with low necessity can readily be postponed to a later phase. Meanwhile, their functions must be compensated with, for example, manual systems. This kind of compensation routine can be simplified by eliminating some less important elements of the subsystem by means of a threshold applied to the measure of necessity. It is not necessary to realize all subsystems with a value of the measure of necessity greater than the threshold value. Especially if the measure has a strong character of subjective probability, the realization can wait until the actual need arises, provided that sufficient resources for rapid realization are available.

References

LANGEFORS, B., *Theoretical Analysis of Information Systems,* Studentlitteratur, Lund, Sweden (1973).

LUNDEBERG, M., and ANDERSEN, E.S., *Systemering - Informationsanalys,* Studentlitteratur, Lund, Sweden (1974).

NURMINEN, M.I., *Information System and Control System,* In Turun Kauppakorkeakoulu, Tutkielmi-a, Publications of the Turku School of Economics: Series A II-1:75, Turku, Finland, pp.273-284 (English translation in preparation) (1975).

NURMINEN, M.I., *Definition of Information System and Information Analysis,* Publications of the Turku School of Economics: Series A I-2: 75, Turku, Finland (English translation in preparation), (1975).

SIMON, H.A., *The New Science of Management Decision* (excerpts from), in Management Decision Making, Penguin Books, Middlesex, England, pp.13-29 (1970).

ZADEH, L.A., *Outline of a New Approach to the Analysis of Complex Systems and Decision Processes,* IEEE Transactions on Systems, Man, and Cybernetics, Vol.SMC-3, No.1, pp.28-44 (1973).

Application of fuzzy logic to controller design based on linguistic protocol

E.H. MAMDANI and T.J. PROCYK
Queen Mary College, University of London, UK

1. Introduction

The fact that mathematics as a whole is taken to be synonymous with precision has caused many scientists and philosophers to show considerable concern about its lack of application to real world problems. This concern arises because in logic as well as in science there is constantly a gap between theory and the interpretation of results from the inexact real world. Many eminent thinkers have contributed to the discussion on vagueness, occasionally holding human subjectivity as the culprit.

In an excellent analysis of the subject Black [1] says... 'that with the provision of an adequate symbolism the need is removed for regarding vagueness as a defect of language'. In his paper he strongly argues that vagueness should not be equated with subjectivity. Briefly, his argument may be summarized by noting that the color 'Blue', say, is vague but not subjective since its sensation among all human beings is roughly similar. It is possible to deal with color precisely by considering the e.m. radiation producing it but in doing so the important human sensation of color, as it happens to be vague, has to be sacrificed. Furthermore, it may be argued that vagueness is not a defect of language but also important source of creativity. Analogies are extremely important to creative thinking and vagueness plays a dominant role in such thought process.

Black's motivation to symbolise vagueness appears to be at the back of all investigations of 'Deviant Logics' [2]. An important contribution in the past 10 years has been that of Zadeh's fuzzy-set-theory and fuzzy-logic [3]. In his recent writings Zadeh [4, 5] states clearly his motivation which is to use fuzzy sets to symbolize Approximate Reasoning (AR). Whereas there are many applications of fuzzy-set-theory, this chapter describes one of the first results in the application of AR and linguistic synthesis.

1.1 An outline of the Chapter's content: The intention in this chapter is to review the whole program of investigation concerning the application of Fuzzy-logic to controller design and to analyse the findings in order to offer insightful comments and conclusions. The original work in this program was done in early 1974 [6] and first published later that year [7,8]. This was the control of a pilot scale steam-engine using fuzzy-logic to interpret linguistic rules which qualitatively express the control strategy. This work is briefly reviewed in the next section of this chapter.

Since the publication of the above work several researchers, elsewhere have also implemented the approach using different pilot scale plants. This together with the continuing work as part of this programme has produced results which throw more light on the usefulness of applying fuzzy-logic to linguistic synthesis. Section 3 below offers comments on some of the key findings of these studies.

One of the comments that has been made about fuzzy-logic is that in its present form it is essentially descriptive and does not offer a prescriptive approach to reasoning. In the first place, it should be noted that fuzzy-logic, like any other form of logic, can only be a system for inferring consequences from previously stated premises and only from these premises. A prescriptive system is possible, however, if a hierarchical decision-making approach is used so that the strategy at a lower level is derived as a consequence of a description at a higher level. Two early implementations of such a prescriptive method (some might term this a learning or an adaptive approach) are discussed in Section 4 of this chapter. To conclude this chapter the last section examines the future trend in this field in the light of experience being gained from current investigations described here.

2. An experiment in linguistic synthesis

2.1 A brief review of fuzzy-logic

The point of view adopted here is that the variables are equated to universes of discourse which are non-fuzzy sets. These variables take on specific linguistic values. These linguistic values are expressed as fuzzy subsets of the universes.

Given a subset A of X $(A \subset X)$ A can be represented by a characteristic function: $X_A: X \rightarrow \{0,1\}$. If the above mapping is from X to a closed interval $[0,1]$ then we have a fuzzy subset. Thus if A were a fuzzy subset of X it could be represented by a membership function: $\mu_A: X \rightarrow [0,1]$.

Note that X is a non-fuzzy support set of a universe of discourse, say, height of people. A can then be equated to a linguistic value such as tall people. Figure 1 shows two linguistic values A_1 and A_2 and their logical combinations $\bar{A}_1$; $A_1 \wedge A_2$; $A_1 \vee A_2$; where:

$\bar{A}_2$ is formed by taking $(1 - \mu_{A_2})$ as membership value at each element of support set,

$A_1 \wedge A_2$ is formed by taking min (μ_{A_1}, μ_{A_2}) at each element of support set, and

$A_1 \vee A_2$ is formed by taking max (μ_{A_1}, μ_{A_2}) at each element of support set.

It is in the definition of implication that this logic may be found to differ from other logics. Given $A \rightarrow B$ (If A then B), then it can happen that

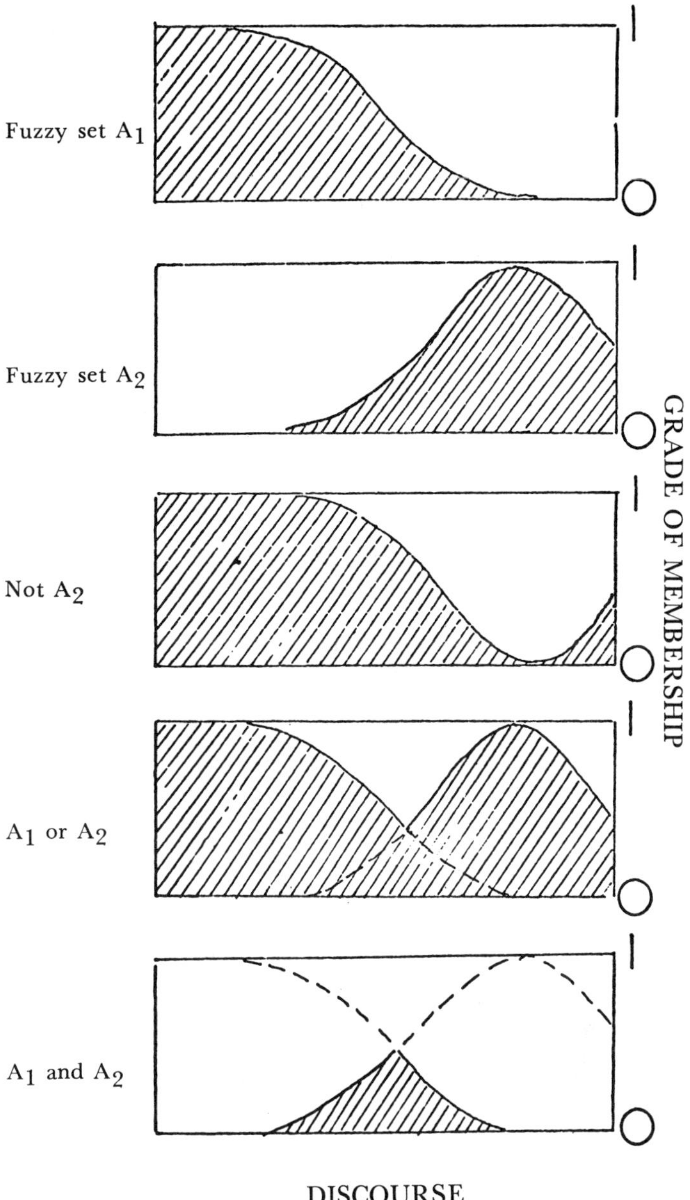

Fig.1

A and B are linguistic values of two disparate universes of support say X and Y. Note that here the implication is between individual values and not the underlying variables. Thus the relation R between A and B is a fuzzy subset of the universe of support $X \times Y$, the cross-product of X and Y. $\mu_R: X \times Y \rightarrow [0,1]$. $\mu_R(x \times y)$ is related to $\mu_A(x)$ and $\mu_B(y)$ (in the present application) by the following:

$$\mu_R(x \times y) = \min(\mu_A(x), \mu_B(y))$$

If the relation R represents a 'nested' implication (i.e. If A then (If B then C) or $A \rightarrow B \rightarrow C$), then R will have a corresponding higher order cross-product support set.

Now if some relation R between A and B is known and so is some value A^1 then the idea is to infer B^1 from R and A^1; $B^1 = A^1 \circ R$, where A^1 is composed with R. This has the effect of reducing the dimensionality of the support set of R to that of B^1. In this work, the compositional rule of inference used to relate μ_{B^1} to μ_R and μ_{A^1} is:

$$\mu_{B^1}(y) = \max_x \min (\mu_{A^1}(x), \mu_R(x \times y))$$

These definitions are themselves a matter of much discussion but that concern is outside the scope of this chapter. The setting up of relations R from stated implications between fuzzy values and the subsequent use of the rule of inference are the chief mechanism used in decision-making in the application described below.

2.2 *Application to fuzzy-controllers*

As stated earlier the linguistic synthesis approach was initially applied to control a pilot scale steam-engine, a more detailed description of which is given elsewhere [6,7,8]. A concise summary of this work is presented here. The overall control system is shown in Fig.2. One aspect of control in this system is the regulation of pressure in the boiler around a prescribed set-point. The control is achieved by measuring the pressure at regular intervals and inferring from this the heat setting to be used during that interval. The essence of this work is simply that if an experienced operator can provide the protocol for achieving such a control in qualitative linguistic terms, then fuzzy-logic as described above can be used to implement successfully this strategy.

The protocol obtained from the operator in this case considers *pressure error (PE)* and *change in the pressure error (CPE)* to infer the amount of *change in the heat (HC)*. The protocol consists of a set of rules in terms of specific linguistic values of these variables and is shown in Fig.3*. Now it can be seen that these rules are in the form of If...Then statements (implications) and thus, from above, each rule i will translate into a relation R_i. The overall protocol is then a relation R formed by 'oring' together the R_i's:

$$R = R_1 \vee R_2 \ldots \vee R_i \ldots \vee R_n$$

Let us say now that each rule R_i represents an implication $A_i \rightarrow B_i \rightarrow C_i$. The decision-making algorithm that is implemented contains two phases:

a. The initial setting up phase when the protocol R is formed from two sets of data:

(i) The individual linguistic values A_i, B_i, C_i given as fuzzy subsets.

(ii) The rules as in Fig.3 which specify the actual combination of these values to form each R_i.

b. The decision-making phase is invoked at each sampling instant during run-time with the exact measured values A^1 and B^1 supplied to it. This phase then is nothing but the use of compositional rule of inference to derive C^1 as follows:

$$C^1 = A^1 \circ (B^1 \circ R).$$

Note that A^1, B^1 can be non-fuzzy, whereas since C^1 is a fuzzy subset of the set of all possible actions,

*The abbreviations used for these linguistic values here are: ZE-zero; PZ-positive zero; PS-positive small, PM-positive medium; PB-positive big and the same for negative values NZ, NS, NM and NB. Change in Error negative is taken as movement towards set-point and positive as away from set-point.

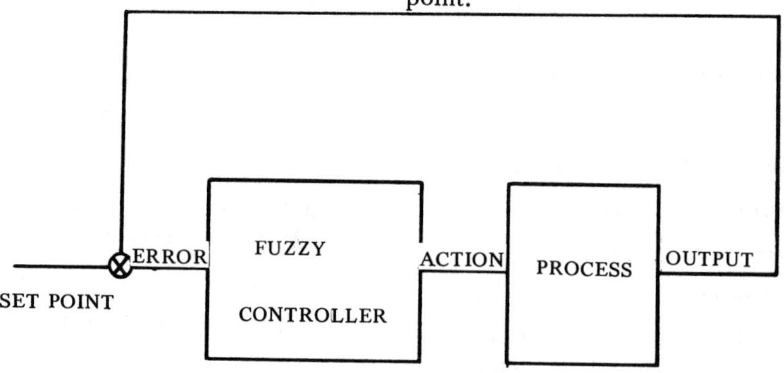

Fig.2 Fuzzy logic control system

PRESSURE ERROR = PE, CHANGE IN PRESSURE ERROR = CPE
and HEAT INPUT CHANGE = HC

IF PE = (NB OR NM) THEN IF CPE = NS THEN HC = PM
OR
IF PE = NS THEN IF CPE = PS THEN HC = PM
OR
IF PE = NO THEN IF CPE = (PB OR PM) THEN HC = PM
OR
IF PE = NO THEN IF CPE = (NB OR NM) THEN HC = NM
OR
IF PE = PO OR NO THEN IF CPE = NO THEN HC = NO
OR
IF PE = PO THEN IF CPE = (NB OR NM) THEN HC = PM
OR
IF PE = PO THEN IF CPE = (PB OR PM) THEN HC = NM
OR
IF PE = PS THEN IF CPE = (PS OR NO) THEN HC = NM
OR
IF PE = (PB OR PM) THEN IF CPE = NS THEN HC = NM

Fig.3

a procedure is required to determine the actual action to be taken from the knowledge of C^1. Also there is a certain advantage in deferring the computation of R until the second phase. Because then this provides a means of altering the control strategy on-line by altering the data structures containing the rules during run-time. However, what need concern us at present is the results obtained from the application of this method to the pilot scale plan. In repeated trials it was found that the results compared favorably with those from applying classical methods from control engineering practice (i.e. 2 or 3 term controllers).

3. Comments on fuzzy-logic controller studies

Two main conclusions have been drawn from this work.

First, that the results vindicate the approach advocated by Zadeh and demonstrate its potential. Second, it can be asserted that the method can easily be applied to many practical situations. This assertion is supported by considering a practical instance, that of cement kiln operation, in which a similar control protocol obtains. In a book on cement kilns, Peray and Waddell [9] list a collection of rules for controlling a kiln. Examples of these rules are shown in Fig.4. From this it is immediately apparent that the method as described, can be used for translating these rules. Furthermore, this method has also been tested on plants such as batch chemical reactors, heat-exchangers and so on. Some key feature emerging from these studies are mentioned here for the sake of interest.

In many of the studies, rules exactly as those given in Fig.3. are used with only minor changes. This is not surprising as the rules indicate the relationship between *error, change in error* and *control action* that exists in most dynamical plants. This relation is mainly one of monotonicity between the outputs of a plant and the input applied to it. What is more of interest is that in most studies it is found that this form of controller is far less sensitive to parameter changes within the plant than the classical 2-term controller. At this stage only a qualitative explanation can be offered for this. It appears that the former is a *reasonable* controller as it relies on the underlying relationships between the plant outputs and inputs whereas the latter is a *pedantic* controller in which the action is computed as a linear combination of the measurements and thus more susceptible to parameter changes.

It is the first conclusion above, however, which is more important. Approximate Reasoning approach outlined here is obviously applicable to other areas as well. The one that has been considered is the design of traffic signal controllers. Application to more obvious areas of decision-making in complex

BACK-END TEMPERATURE = BE, BURNING ZONE TEMPERATURE = BZ
PERCENTAGE OF OXYGEN GAS IN THE KILN EXIT GAS = OX

CASE	CONDITION	ACTION TO BE TAKEN
1	BZ LOW	WHEN BZ IS DRASTICALLY LOW
	OX LOW	A. REDUCE KILN SPEED
	BE LOW	B. REDUCE FUEL
		WHEN BZ IS SLIGHTLY LOW
		C. INCREASE I.D. FAN SPEED
		D. INCREASE FUEL RATE
2	BZ LOW	A. REDUCE KILN SPEED
	OX LOW	B. REDUCE FUEL RATE
	BE OK	C. REDUCE I.D. FAN SPEED
3	BZ LOW	A. REDUCE KILN SPEED
	OX LOW	B. REDUCE FUEL RATE
	BE HIGH	C. REDUCE I.D. FAN SPEED

TOTAL OF 27 RULES

Fig.4

and humanistic system will no doubt be attempted in the future. If the method described above is applied to these other areas then the likely sources of difficulties to be encountered can be attributed to one main factor. This is that the quality of decision is only as good as the relation R from which it is inferred. R in turn is affected by three factors.

First, it is affected by the set of rules in the protocol. With more complex situations a good protocol is not easy to derive. A great deal of investigatory effort normally referred to as human factors in control is devoted to exactly such matters. Unlikely as it may seem, the human being does not always find it easy to verbalise his considerations during decision-making. The only mitigating factor here is that it is far more difficult to determine the decision heuristics in a form amenable to treatment by a branch of precise mathematics than it is to derive rules for linguistic synthesis.

Second factor affecting the quality of decision (though not R itself explicitly) is the underlying range of elements in the support set which provides the context for interpreting the linguistic rules. This can be illustrated by noting that 'tall people' in a land of pygmies are likely to have the support set of range of height from 3 to 5 ft. 6 in. say whereas the more normal range of height may be say from 4 ft. to 7ft. Such considerations are implicit in any application and are equivalent to what a control engineer would term the gains applied to each variable.

Finally, R is affected by the membership values in the fuzzy subsets defining the linguistic values. This is perhaps the least important of all the factors because the degree of change permitted here is limited as too much change in the membership values of a fuzzy subset is likely to affect the linguistic meaning ascribed to it. This is illustrated in Fig.5 in which the effect of given linguistic value (bold line) is altered by using a different linguistic value (as in *a*), increasing the gain thus decreasing the range of the support set (as in *b*) and lastly in a minor way by adjusting the defining values of the fuzzy subset.

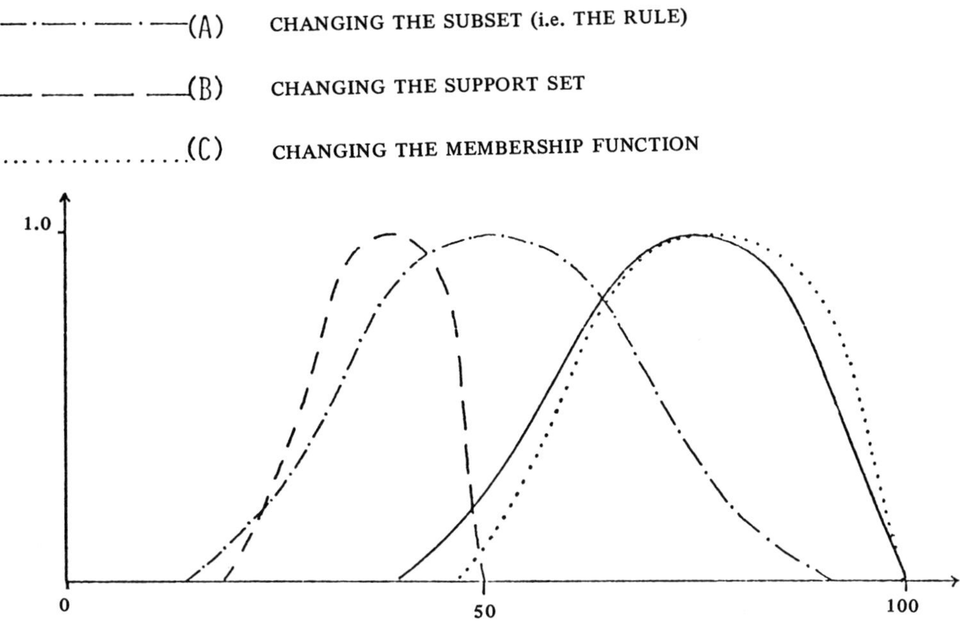

Fig.5 Modification of fuzzy control

4. A recipe for a prescriptive approach

4.1 An early implementation

As mentioned earlier, the main difficulty that arises is that a good decision requires that a good set of rules are described at the beginning. In any application of reasonable complexity this is not easy to achieve. Indeed it is quite possible that for some reason a protocol cannot be obtained at all. This may be due to the complexity of the plant (e.g. non-linearities) or to the fact that the operator cannot verbalize his decision process adequately or no consistent protocol can be found. However, the goal in any application and a set of assumptions regarding that application can often be much easier to state. This fact motivates investigations into so-called learning or adaptive systems.

In the control situation the goal is simply to bring the output to the set-point and keep it there, the only assumption being that the plant input and output are monotonically related. This monotonicity relation enables wrong control actions to be corrected. If the output is too high then too much input was applied and vice versa and so the proper amount of input required can usually be inferred backwards from the stated goal. An early attempt at implementing one such prescriptive system [11] is described here.

The overall schematic diagram for the control system is shown in Fig.6. The whole system consists of two hierarchical levels. The higher contains the goal which is effectively a bound within which the output is to be maintained. This band, Fig.7, is specified by a set of fuzzy rules whose input is

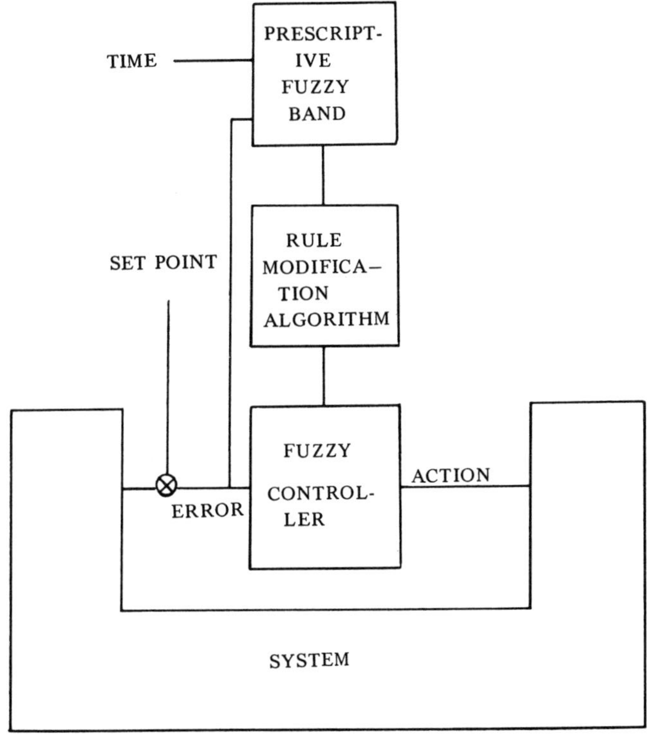

Fig.6

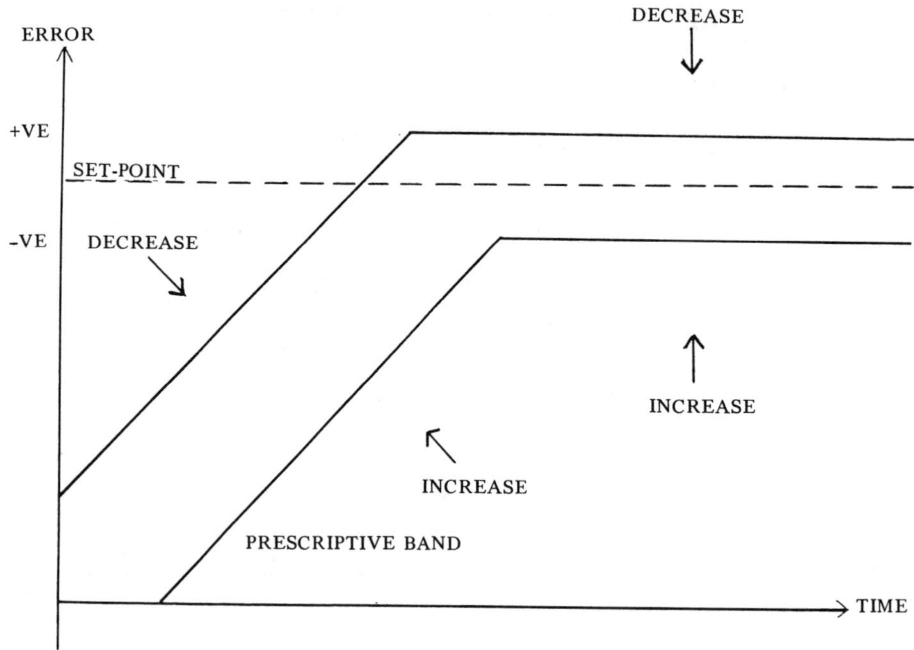

Fig.7

time from the start of control and set point deviation, i.e. the error signal. The output from the rules specify the changes to be made in the controller. For example:

a) IF Time is Small AND Error is Negative Big THEN Desired change is Big.

b) IF Time is Big AND Error is Positive Zero THEN Desired change is Zero.

The band can therefore be viewed as a set of local performance criteria which the response must satisfy.

The output from these 'teaching' rules alters the lower level control rules appropriately. Since the control rules are of the form $A_i \rightarrow B_i \rightarrow C_i$, the modification is effected by first finding the linguistic values A_i and B_i which best describe the plant state for which a change in action is desired. This search is simply carried out by a supremum operation over the range of linguistic values. The action, C_i, corresponding to that control rule is altered by the amount given by the 'teaching' algorithm. If no such rule exists then one is generated.

Figure 8 shows results obtained from applying this scheme. The tables are a method of displaying all the linguistic rules of the controller. The measurement error and change in error are given on the axes and the entries indicate the actions applied. The abbreviations are as stated in the footnote on p.348.

The rules in Fig.2(*a*) (without the asterisk) are the best designed rules of an experienced operator. On applying the above procedure the controller converges (i.e. there are no more requests for modification of rules) by creating the extra rules marked with an asterisk. Convergence in this sense means that if the system under question is controllable within the prescribed band, then the rules will converge to a solution in a finite number of training steps. On starting with no rules at all and then applying this procedure, convergence takes place to the set of rules shown in Fig.2(*b*). The output trajectory was observed to be marginally better in the second case. The output response of both these policies fit the prescribed band. When the band is narrowed then no convergent policy is found but the response tends to remain within the band. This lack of convergence could be attributed to the 'credit assignment' problem which could be tackled by the 'bootstrapping' technique. Furthermore, lack of convergence could also be attributed to the failure in including sufficient state variables of the plant in the controller.

These modifications are currently being included and are the subject of further experiments.

4.2 *An alternative approach*

The prescriptive approach described above is very

CHANGE IN ERROR	ERROR							
	NB	NM	NS	NZ	PZ	PS	PM	PB
NB	PB*	PB*	NS	NM	PM	PS		NB*
NM	PM*		NS	NM	PM	PS		
NS	PM	PM	ZE	NS	PS	ZE	NM	NM
ZE	PB	PB	PM	ZE	ZE	NM	NB	NB
PS	PB	PB	PM	PS	NS	NM	NB	NB
PM			PB	PM	NM	NB		
PB	PB	PB	PB	PM	NM	NB	NB	NB

(A)

CHANGE IN ERROR	ERROR							
	NB	NM	NS	NZ	PZ	PS	PM	PB
NB	PB	PB	NB	PB	PS	PB		
NM	PB		ZE	ZE	PB	NS		NB
NS	PS		PS	ZE	PM	NM		NB
ZE	PB		ZE	ZE	PS	NM		
PS		PS	PS	PB	ZE	NB		
PM			ZE	ZE	NM	NS		
PB	PB	PS	PB	PS	NS	NB	NB	NB

(B)

Fig.8

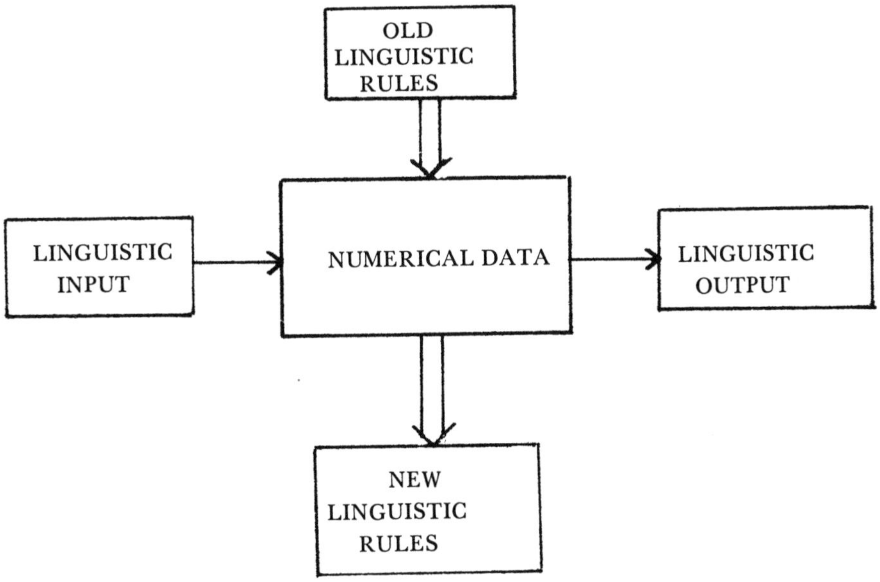

Fig.9

much an *ad hoc* implementation. It serves to illustrate what needs to be done to go beyond a simply descriptive system. What is desired is that such an approach should appear naturally in a suitably improved fuzzy logic theory itself. This is especially relevant to the way in which a change demanded by the higher or 'teaching' algorithm is transmitted to the lower one.

The general philosophy of this approach is shown in Fig.9, which depicts a very general learning system. The concept of the membership function enables the set of rules to be expressed as a data base and the linguistic input value as input data. All operations that are required are carried out on this numerical data and are retranslated into a linguistic output value or a new set of rules only when it is necessary to present the results. The idea of the membership function interfaces between the imprecise heuristics and the exact numerical data which describes them.

In the early implementation described above the method of transmitting the change from the higher to the lower algorithm was achieved by reverting to the linguistic rule text and substituting certain linguistic names by others or generating new ones. A much more direct approach is to perform numerical operations on the overall relation matrix which describes the controller according to the change demanded by the teaching algorithm.

The controller realtion matrix, R_{ABC}, is changed as follows. If for an input A^1 and B^1 and action $\hat{C}$ is required instead of C^1 then the relation $A^1 \times B^1 \times C^1$ is removed from R_{ABC} and a new relation $A^1 \times B^1 \times \hat{C}$ is included. To effect this the theory of fuzzy logic is extended to include operations between relation matrices as well as fuzzy sets.

Once the controller has converged and no more learning takes place it is necessary to develop the new rules as Fig.9 indicates. The new relation together with the old spreads and the new ones generated are input to a program which outputs the sets belonging to each rule having first performed a minimization.

This new approach appears more general and direct than the earlier one and experimental results (with a batch reactor plant) have so far proved encouraging.

5. Conclusions

The two prescriptive approaches described above are the first steps in an attempt to advance further than a purely descriptive system using fuzzy-logic theory. If, as is suggested here, hierarchical statements are a main requirement of such a theory then this means that fuzzy-logic should have an auto-descriptive property found in multiple valued logics [10]. From the application point of view both a learning situation described here as well as decision-making in complex systems are best framed in terms of hierarchical structures. This is very much the direction in which the theory of fuzzy-logic and approximate reasoning is likely to go. The work described in this chapter demonstrates the great potential of applying fuzzy-logic theory not only to control engineering problems but also to management and other humanistic systems.

References

1. BLACK, M., 'Vagueness', *Philosophy of Science* **4**, 472-455 (1937).
2. HAACK, S., *Deviant Logic,* Cambridge University Press (1974).
3. ZADEH, L.A., 'Fuzzy sets', *Inf. Contr.* **8**, 338-353 (1965).
4. ZADEH, L.A., 'Outline of a new approach to the analysis of complex systems and decision processes', *IEE Trans. Systems, Man and Cybernetics, SMC-3* (1973).
5. ZADEH, L.A., 'Calculus of fuzzy restrictions', *ERL Memorandum M474* (October 1974).
6. ASSILIAN, S., *Artificial intelligence in the control of real dynamical systems,* PhD. Thesis, London University (1974).
7. MAMDANI, E.H., 'Application of fuzzy algorithms for the control of a dynamic plant', *Proc. IEE, 121,* pp.1585-1588 (December 1974).
8. MAMDANI, E.H. and ASSILIAN, S., 'An experiment in linguistic synthesis with a fuzzy logic controller', *Int. J. Man-Machine Studies, 7,* pp.1-13 (1974).
9. PERAY, K.E. and WADDELL, J.J., *The Rotary Cement Kiln,* The Chemical Publishing Co., New York (1972).
10. RESCHER, M., *Many-valued Logics*, McGraw Hill, New York (1969).

Fuzzy document retrieval

P. BOLLMANN and E. KONRAD
Technische Universität Berlin

1. Introduction

We consider document retrieval systems which provide the user with references to documents. The term document retrieval is often used as synonymous with information retrieval.

The database of a document retrieval system consists of documents. The answer to a query is a subset of the database.

Document retrieval systems have been described by a variety of mathematical models: vector spaces, lattices, probability spaces, topological spaces, e.g.

Some of these models implicitly handle fuzzy concepts.

There are very few approaches—as known to the authors—which have explicitly modelled fuzzy concepts.

Using probability theory Maron and Kuhns [6] have introduced the concept of probabilistic indexing, while Zunde and Dexter [16] have developed a measure of indexing consistency based on fuzzy sets.

In this chapter we apply fuzzy mathematics as a general formal framework for document retrieval.

The fundamental concept of fuzzy mathematics is the membership function. On the other hand the concept of weighted items is often used in document retrieval systems.

Combining ideas of both areas we get fuzzy interpretations of document-descriptor-matrices, similarity matrices, retrieval vectors, and similarity measures. Documents are considered as fuzzy subsets of descriptors where the membership function is represented by the document-descriptor-matrix. Associations between objects (documents, descriptors, search requests, classes) are fuzzy relations whose membership functions are given by similarity measures.

The thesaurus of a document retrieval system can be structured by fuzzy clustering. The same technique is applicable to the document space. Fuzzy concepts also appear in connection with the evaluation of document retrieval systems: system's proposal, relevance judgement of the user, recall and precision.

2. Standard retrieval models

Let $\bar{D}$ be a set of documents in natural language. The information content of a document $\bar{d} \in \bar{D}$ can be described by terms (descriptors) of a term set T_0. In the same way a query of a user can be represented by terms. The transformation of documents and queries into term descriptions is called *indexing.* A document retrieval system compares the descriptions of a query S with the descriptions of documents and delivers a proposal $R(s)$, a subset of D. Let S be the set of the query descriptions. The function

$$R : S \rightarrow P(D)$$

is called *retrieval function; $P(D)$* is the power set of D [7].

Documents which are important to the user are called *relevant.* The evaluation of the system consists in comparing the relevant documents with the system's proposal.

The process of indexing is illustrated by the following example, the other concepts are illustrated later on.

Example 2.0.1

$\bar{d}$ = 'The concepts of equivalence, similarity, partial ordering, and linear ordering play basic roles in many fields of pure and applied science'. (From [13])

$\bar{d}$ can be represented by the following set of terms:

d = {concept, equivalence, similarity, partial, ordering, linear, pure, applied, science}

2.1 *Boolean retrieval*

The documents of $\bar{D}$ are represented as sets of terms, i.e. $D \subset P(T_0)$. The set of query descriptions S is defined as follows [5]:

i) If $t \in T_0$, then $t \in S$.

ii) If $s \in S$, then $(\neg s) \in S$.

iii) If $s_1, s_2 \in S$, then $(s_1 \vee s_2) \in S$.

iv) If $s_1, s_2 \in S$, then $(s_1 \wedge s_2) \in S$.

v) These are all query descriptions.

We define a retrieval function $R : S \to P(D)$:

i) Let $t \in T_0$, then $R(t) = \{d \in D \mid t \in d\}$

ii) $R(\neg s) = D \backslash R(s)$

iii) $R(s_1 \vee s_2) = R(s_1) \cup R(s_2)$

iv) $R(s_1 \wedge s_2) = R(s_1) \cap R(s_2)$

Example 2.1.1

$T_0 = \{a,b,c,d,e,f\}$

$D = \{d_1,..., d_5\}$

$d_1 = \{a,b\}$

$d_2 = \{b,d,f\}$

$d_3 = \{a,c,d\}$

$d_4 = \{c,d,f\}$

$d_5 = \{e,f\}$

Let $((\neg e) \wedge f) \in S$, then $R((\neg e) \wedge f) = \{d_2, d_4\}$.

This model which is used in many applications (e.g. [3], [11]) has some deficiencies.

In Salton's opinion [8] the main disadvantage is that an unexperienced user is not able to manage the system's proposal.

2.2 *Weighted retrieval*

This model was mainly developed by Salton [8].

The document collection of the system is described using a document-descriptor-matrix

$$\mathbf{D} = (d_{ij})_{\substack{i=1,...,n \\ j=1,...,m}}$$

d_{ij} is the weight of the descriptor (term) j with respect to the document i.

Every document $\bar{d}_i$ in natural language is mapped to a document vector $d_i = (d_{i1},...,d_{im})$. Differently from Salton we presuppose $0 \leqslant d_{ij} \leqslant 1$. This can always be achieved by special indexing techniques [10].

Furthermore we do not allow the zero vector to be a column or row in the document-descriptor-matrix.

Example 2.2.1

Let $T_0 = \{a,b,c,d,e,f\}$

$D = \{d_1,...,d_5\}$

and

$$\mathbf{D} = \begin{pmatrix} 1/2 & 1/2 & 0 & 0 & 0 & 0 \\ 0 & 1/2 & 0 & 1/2 & 0 & 1/3 \\ 1/2 & 0 & 1/2 & 0 & 0 & 0 \\ 0 & 0 & 1/2 & 1/2 & 0 & 1/3 \\ 0 & 0 & 0 & 0 & 1 & 1/3 \end{pmatrix}$$

A search request S of a user is represented as a vector $s = (s_1,...,s_m)$, $0 \leqslant s_j \leqslant 1$, $s \neq (0,0,...,0)$.

The search request s is matched against D using a similarity measure α. In the area of document retrieval the following correlation measures are used:

i) *Linear weighted retrieval [2, 8]*

$$\alpha_I(s,d_i) = \sum_{j=1}^{m} s_i d_{ij}$$

Strictly speaking α_I is not a similarity measure,

because $\alpha_I(s,d_j) = 1$ for $s = d_j$ is not always true.

ii) Tanimoto measure [2, 8, 9]

$$\alpha_T(s,d_i) \equiv \frac{\alpha_I(s,d_i)}{\alpha_I(s,s) + \alpha_I(d_i,d_i) - \alpha_I(s,d_i)}$$

It is $0 \leqslant \alpha_T(s,d_i) \leqslant 1$

iii) Cosine measure [2, 8, 9]

$$\alpha_c(s,d_i) = \frac{\alpha_I(s,d_i)}{\sqrt{\alpha_I(s,s) \cdot \alpha_I(d_i,d_i)}}$$

$$0 \leqslant \alpha_c(s,d_i) \leqslant 1$$

Example 2.2.2

Let **D** be defined as in 3.2.1 and $s = (½,0,1,0,0)$. Then we obtain for α_I:

$$\alpha_I(s,d_1) = ¼$$

$$\alpha_I(s,d_2) = 0$$

$$\alpha_I(s,d_3) = ¾$$

$$\alpha_I(s,d_4) = ½$$

$$\alpha_I(s,d_5) = 0$$

The vector $A = (\alpha(s,d_i))_{i=1,...,n}$ is called *response vector*. In the case of linear weighted retrieval the response vector is a linear transformation of the query vector, i.e.

$$A = \mathbf{D} \cdot s$$

A is the product of **D** and s using ordinary matrix multiplication. The retrieval function is defined as follows:

$$R(s) = \{d_i \mid \alpha(s,d_i) \geqslant c\}$$

c is a real number and called *cutoff*.

3. Fuzzy retrieval models

3.1 Fuzzy sets

We introduce some concepts from the theory of fuzzy sets which are useful in our approach to document retrieval. This work is based on [12, 13, 14].

Let U be the universe. A fuzzy subset $A \subset U$ is determined by its *membership function*

$$\mu_A : U \rightarrow [0,1]$$

$\mu_A(u)$, $u \in U$ tells us to what degree u belongs to A. Using the notation in [15] we have

$$A = \sum_{u \in U} \mu_A(u) \mid u$$

The set operations on fuzzy sets are defined using the membership function.

$$A \subseteq B \text{ iff } \mu_A(u) \leqslant \mu_B(u) \text{ for all } u \in U$$

The *complement* $\neg A$ of A has the membership function $\mu_{\neg A} = 1 - \mu_A$.

The *union* $A \cup B$ has the membership function $\mu_{A \cup B} = \max(\mu_A, \mu_B)$.

The *intersection* $A \cap B$ is defined by $\mu_{A \cap B} = \min(\mu_A, \mu_B)$.

The *product* $A \cdot B$ is defined by $\mu_{A \cdot B} = \mu_A \cdot \mu_B$.

The *scalar multiplication* aA, a real, $0 \leqslant a \leqslant 1$, is defined by $\mu_{aA} = a\mu_A$.

The *power* A^a of A, a real, $a \leqslant 0$, is defined by $\mu_{A^a} = \mu_A^a$.

$\sup \mu_A(u)$ is called the *height* $h(A)$ of A.

For a finite universe the *cardinality* $|A|$ of A is defined by $|A| = \sum_u \mu_A(u)$.

The set $A_c = \{u \mid \mu_A(u) \geqslant c\}$, $0 \leqslant c \leqslant 1$, is called *c-level-set* of A.

If $U_1,..., U_n$ are universes with the cartesian product $U_1 \times ... \times U_n$, then a *n-ary fuzzy relation* R is a fuzzy subset of $U_1 \times ... \times U_n$ with the membership function

$$\mu_R : U_1 \times ... \times U_n \rightarrow [0,1]$$

Example 3.1.1:

The document-descriptor-matrix **D** of example 2.2.1 describes a fuzzy relation on $D \times T_0$ with $\mu_\mathbf{D}(d_i, t_j) = d_{ij}$.

Remark: If a binary relation is represented by a matrix, then we do not distinguish them notationally.

A binary fuzzy relation $R \subset U \times U$ is called *resemblance relation* [4], if the following holds:

i) $\mu_R(u,u) = 1$

ii) $\mu_R(u,v) = \mu_R(v,u)$

These conditions are satisfied if the membership function is defined by one of the similarity measures.

Example 3.1.2:

Let $U = D$ be defined as in example 2.2.1.

$$\mu_R(d_i, d_j) = \frac{\alpha_I(d_i,d_j)}{\alpha_I(d_i,d_i) + \alpha_I(d_j,d_j) - \alpha_I(d_i,d_j)}$$

R is represented by the following matrix:

$$\begin{pmatrix} 1 & 9/31 & 1/3 & 0 & 0 \\ 9/31 & 1 & 0 & 13/31 & 2/29 \\ 1/3 & 0 & 1 & 9/31 & 0 \\ 0 & 13/31 & 9/31 & 1 & 2/29 \\ 0 & 2/29 & 0 & 2/29 & 1 \end{pmatrix}$$

Let $R \in U_1 \times U_2$ be a fuzzy relation with the membership function μ_R. The *inverse relation* $R^T \subset U_2 \times U_1$ is defined by $\mu_{R^T}(v,u) = \mu_R(u,v)$,
Remark: The matrix of an inverse relation is the transposed matrix. For a resemblance relation R holds $R = R^T$.

Let $R_1 \subset U_1 \times U_2$ and $R_2 \subset U_2 \times U_3$ be two fuzzy relations. The *'max-mult-product'* $R = R_1 \circ R_2$, $R \subset U_1 \times U_3$ is defined by

$$\mu_{R_1 \circ R_2}(u_1,u_3) = \sup_{u_2 \in U_2} (\mu_{R_1}(u_1,u_2) \cdot \mu_{R_2}(u_2,u_3))$$

The *max-min-product* is constructed in a similar way: the product is substituted by the minimum. For both products the following rules hold [13]:

i) $(R_1 \circ R_2) \circ R_3 = R_1 \circ (R_2 \circ R_3)$

ii) If $R_1 \subset R_2$, then $R_1 \circ R \subset R_2 \circ R$

iii) $R_1 \circ (R_1 \cup R_3) = R_1 \circ R_2 \cup R_1 \circ R_3$

The operations $\subset$ and $\cup$ are defined for relations and fuzzy sets. In addition it is supposed that the operations are defined on the universes.

Now we show how similarity measures can be applied to relations. Using the cosine measure we define $R = \alpha_c(R_1, R_2)$ by

$$\mu_R(u_1,u_3) = \frac{\sum_{u_2 \in U_2} \mu_{R_1}(u_1,u_2)\, \mu_{R_2}(u_2,u_3)}{\sqrt{\sum_{u_2 \in U_2} \mu^2_{R_1}(u_1,u_2) \sum_{u_2 \in U_2} \mu^2_{R_2}(u_2,u}}$$

For the Tanimoto measure the definition is analogous.

In both cases we have resemblance relations. Examp 3.1.2 shows how $\alpha_T(\mathbf{D}, \mathbf{D}^T)$ has been constructed. Using the cosine and Tanimoto measure these operations are not associative. The max-mult- resp. max-min-product does not define a resemblance relaion in general.

Analogously the composition of a binary fuzzy relation $R_1 \subset U_1 \times U_2$ and an unary fuzzy relation (fuzzy set) $R_2 \subset U_2$ is defined. The result is an unary fuzzy relation (fuzzy set) $R_3 \subset U_2$.

Example 3.1.3:

Let $\mathbf{D} \subset D \times T_0$ be as in example 2.2.1 and $s = (1/3, 0, 1, 0, 0)$. Using the max-mult-product we have

$$\mathbf{D} \cdot s = \frac{1}{6} \mid d_1 + ½ \mid d_3 + ½ \mid d_4$$

or in matrix form

$$\mathbf{D} \cdot s = (½, 0, ½, ½, 0)$$

The max-mult- resp. max-min-product of two fuzzy sets A and B is defined by

$h(A \cdot B)$ resp. $h(A \cap B)$.

3.2 Generalized Boolean retrieval

Let T_0 be the set of terms (descriptors). According to Zadeh [12] we define a query language S and its interpretation (Zadeh's concept of the linguistic variable [14, 15] is used here implicitly).

i) If $t \in T_0$, then $t \in S$

ii) If $s \in S$, then $(\neg s) \in S$

iii) If $s_1, s_2 \in S$, then $(s_1 \vee s_2) \in S$

iv) If $s_1, s_2 \in S$, then $(s_1 \wedge s_2) \in S$

In addition we need operators in order to express term weights.

v) If $0 \leqslant a$, $s \in S$, then $s^a \in S$

vi) If $0 \leqslant a \leqslant 1$, $s \in S$, then $as \in S$

vii) These are all queries.

The interpretation R is defined using the document descriptions D.

Let $t_j \in T_0$ $(j=1,...,m)$ be a descriptor and d_i $(i=1,...,n)$ be a document description with the membership function μ_i $(i=1,...,n)$. Now we define a fuzzy set of documents $A(t_j)$ by the membership function

$$\mu_{t_j}(d_i) = \mu_i(t_j)$$

This representation corresponds to an inverted file. Using the document-descriptor-matrix **D** (fuzzy relation) we have

$$\mu_{t_j}(d_i) = d_{ij}$$

Example 3.2.1:

The following document descriptions are given (comp. ex. 2.2.1).

$$d_1 = \tfrac{1}{2} \mid a + \tfrac{1}{2} \mid b$$

$$d_2 = \tfrac{1}{2} \mid b + \tfrac{1}{2} \mid d + 1/3 \mid f$$

$$d_3 = \tfrac{1}{2} \mid a + \tfrac{1}{2} \mid c$$

$$d_4 = \tfrac{1}{2} \mid c + \tfrac{1}{2} \mid d + 1/3 \mid f$$

$$d_5 = 1 \mid e + 1/3 \mid f$$

Then we have

$$A(b) = \tfrac{1}{2} \mid d_1 + \tfrac{1}{2} \mid d_2$$

The interpretation of the queries is defined recursively:

i) If $t \in T$, then $R(t) = A(t)$

ii) $R(\neg s) = \neg R(s)$

iii) $R(s_1 \vee s_2) = R(s_1) \cup R(s_2)$

iv) $R(s_1 \wedge s_2) = R(s_1) \cap R(s_2)$

v) $R(s^a) = R^a(s)$, $0 \leqslant a$

vi) $R(a\, s) = a\, R(s)$, $0 \leqslant a \leqslant 1$

The interpretation $R(s)$ of a query s corresponds to the response vector.

Example 3.2.2:

Take $s = ((\neg e) \wedge f^2)$ together with the document descriptions of example 2.2.1.

Then we have

$$R(s) = \frac{1}{9} \mid d_2 + \frac{1}{9} \mid d_4$$

3.3 Linear weighted retrieval

We restrict the query language of the generalized Boolean retrieval.

S is defined as follows:

i) If $t \in T_0$, then $t \in S$

ii) If $0 \leqslant a \leqslant 1$, $t \in T_0$, then $at \in S$

iii) If $0 \leqslant a \leqslant 1$, $s \in S$, then $(s \vee at) \in S$ (t not in s).

The definition of the retrieval function R is straightforward.

i) $R(t) = A(t)$

ii) $R(at) = aA(t)$

iii) $R(s \vee at) = R(s) \cup a\, A(t)$

R is a restriction of the function defined in 3.2.

Two queries are called equivalent, if they only differ in the order of occurrences of the $a_i t_i$.

If s_1 and s_2 are equivalent, then we have $R(s_1) = R(s_2)$. The converse does not hold. The equivalence classes are isomorphic to the fuzzy subsets $F(T_0) \setminus \{\phi\}$ with $F(T_0)$ being the set of all fuzzy subsets of T_0. The isomorphism

$$\phi : S(N) \rightarrow F(T_0) \setminus \{\phi\}$$

is defined as follows for the representatives of the equivalence classes.

i) $\phi(t) = 1 \mid t$

ii) $\phi(at) = a \mid t$

iii) $\phi(s \vee at) = \phi(s) + a \mid t$

Example 3.3.1:
If $s = c \vee \frac{1}{3} a$, then $\phi(s) = \frac{1}{3} | a + 1 | c$.

It follows from the above considerations that the non-empty fuzzy subsets of T_0 can be considered as the queries of the weighted retrieval model. Therefore they can be represented as unary relations resp. as matrices (vectors).

Let s be a query with the membership function μ_s. Then we have

$$s = \sum_{j=1,\dots,m} \mu_s(t_j) \mid t_j$$

$$R(s) = \bigcup_{j=1,\dots,m} \mu_s(t_j) \cdot A(t_j)$$

$$\mu_{R(s)}(d_i) = \max_{j=1,\dots,m} \mu_s(t_j) \cdot \mu_{A(t_j)}(d_i)$$

$$= \max_{j=1,\dots,m} \mu_s(t_j)\, \mu_i(t_j)$$

$$= h(s \cdot d_i)$$

$h(s \cdot d_i)$ is the height of the product set $s \cdot d_i$. This is exactly the max-mult-product of queries and documents. This means $R(s)$ can also be defined by $R(s) = \mathbf{D} \circ s$, i.e. the max-mult-product of the relations $\mathbf{D}$ and s.

Since the queries are fuzzy sets, we can consider queries of the form $s_1 \cup s_2$ and (as), $0 \leqslant a \leqslant 1$. We have

$$R(s_1 \cup s_2) = \mathbf{D} \circ (s_1 \cup s_2)$$

$$= \mathbf{D} \circ s_1 \cup \mathbf{D} \circ s_2$$

$$= R(s_1) \cup R(s_2)$$

and

$$R(as) = \mathbf{D} \circ (as)$$

$$= a(\mathbf{D}\, s)$$

$$= aR(s)$$

According to the definitions from 2.3 the correlation between s and d_i can be defined by the vector product $|s \cdot d_i|$. This case is not directly extendable to the fuzzy model because it is possible that $|s \cdot d_i| > 1$, but a modified vector product $\frac{1}{m} |s \cdot d_i|$ can be used here.

We have shown that both the Boolean Retrieval and the linear retrieval using the max-mult-product are special cases of the generalized Boolean retrieval. This is not the case if we use the max-min-product or the modified vector product, although they have many properties in common with the max-mult-product.

3.4 Weighted Retrieval with similarity measures
Let S be defined as in 3.3. The retrieval function is defined by operations on relations:

i) $R(s) = \alpha_T(\mathbf{D}, s)$ (Tanimoto measure)

ii) $R(s) = \alpha_c(\mathbf{D}, s)$ (cosine measure)

The membership functions are

i) $$\mu_{R(s)}(d_i) = \frac{|s \cdot d_i|}{|s \cdot s| + |d_i \cdot d_i| - |s \cdot d_i|}$$

ii) $$\mu_{R(s)}(d_i) = \frac{|s \cdot d_i|}{\sqrt{|s \cdot s| \cdot |d_i \cdot d_i|}}$$

In this case we do not have the same natural interpretation as in the preceding two models. In general we have

$$\mu_{R(t_j)}(d_i) \neq d_{ij}$$

Example 3.4.1
Let $\mathbf{D}$ be defined as in Example 2.2.1 and

$$s = \frac{1}{3} | a + 1 | c$$

Using the Tanimoto measure we have

$$R(s) = \frac{3}{26} | d_1 + \frac{12}{17} | d_3 + \frac{9}{23} | d_4$$

4. Extensions of the models

4.1 Classification
Classification procedures play an important role in

document retrieval systems.

The classification of terms establishes synonymy and hierarchical relations between concepts. Many procedures start from a descriptor-descriptor-matrix.

$$\Delta = (\delta_{ij})_{\substack{i = 1,\dots,m \\ j = 1,\dots,m}}$$

This matrix can be constructed from the relations **D** and $\mathbf{D}^T$ using similarity measures. Thus $\Delta \subset T_0 \times T_0$ is a resemblance relation.

Example 4.1.1

We construct $\Delta = \alpha_T(\mathbf{D}^T, \mathbf{D})$ from example 2.2.1. Then we have

$$\Delta = \begin{pmatrix} 1 & 1/3 & 1/3 & 0 & 0 & 0 \\ 1/3 & 1 & 0 & 1/3 & 0 & 1/4 \\ 1/3 & 0 & 1 & 1/3 & 0 & 1/4 \\ 0 & 1/3 & 1/3 & 1 & 0 & 2/3 \\ 0 & 0 & 0 & 0 & 1 & 1/3 \\ 0 & 1/4 & 1/4 & 2/3 & 1/3 & 1 \end{pmatrix}$$

Applying a cut-off c, $0 \leqslant c \leqslant 1$, to Δ we get the reduced similarity matrix

$$\Delta_c = (\delta'_{ij})_{\substack{i = 1,\dots,m \\ j = 1,\dots,m}}$$

$$\text{where } \delta'_{ij} = \begin{cases} 1, & \text{if } \delta_{ij} \geqslant c \\ 0, & \text{otherwise} \end{cases}$$

Δ_c is the c-level set of Δ and represents an unweighted undirected graph. There are different ways of defining classes on Δ_c [1, 8, 9], e.g. cliques, stars, and connected components.

Example 4.1.2:

Take Δ from example 6.1.1 and construct $\Delta_{0.2}$. Then we have

$$\Delta_{0.2} = \begin{pmatrix} 1 & 1 & 1 & 0 & 0 & 0 \\ 1 & 1 & 0 & 1 & 0 & 1 \\ 1 & 0 & 1 & 1 & 0 & 1 \\ 0 & 1 & 1 & 1 & 0 & 1 \\ 0 & 0 & 0 & 0 & 1 & 1 \\ 0 & 1 & 1 & 1 & 1 & 1 \end{pmatrix}$$

This matrix can be represented by the following graph

Classifying by cliques (maximally complete subgraphs) we have the classes:

$$\{1, 2, 3\};\ \{2, 3, 6\};\ \{4, 6\};\ \{5, 6\}$$

Using $\alpha(\mathbf{D}, \mathbf{D}^T)$ instead of $\alpha(\mathbf{D}^T, \mathbf{D})$ the same procedures can be applied to classify documents.

What is the role of classified terms in a document retrieval system? Experiments have shown that the systems effectivity can be improved by substituting terms by their classes [7, 9]. This implies that document descriptions are fuzzy sets of classes.

If K is the set of classes from T_0, then a classification **K** is a subset of $T_0 \times K$. The substitution of terms is managed by concatenating the relations $\mathbf{D} \subset D \times T_0$ and $\mathbf{K} \subset T_0 \times K$ into $\mathbf{D}' = \mathbf{D} \cdot \mathbf{K}$. The concatenation can be conducted using correlation measures. The effect is that classes containing many terms of the document d get a high weight in d [7, 9].

Example 4.1.3

Representing the classification of Example 4.1.2 as a relation $\mathbf{K} \subset T_0 \times K$ we have

$$\mathbf{K} = \begin{pmatrix} 1 & 0 & 0 & 0 \\ 1 & 1 & 0 & 0 \\ 1 & 1 & 0 & 0 \\ 0 & 0 & 1 & 0 \\ 0 & 0 & 0 & 1 \\ 0 & 1 & 1 & 1 \end{pmatrix}$$

We extend our approach to fuzzy classifications. $\mathbf{K} \subset T_0 \times K$ is a fuzzy relation. This procedure is useful if we want to give a lower weight to terms with a fuzzy meaning, i.e. terms of T_0 which occur in different classes [7]. The concatenation $\mathbf{D} \cdot \mathbf{K}$ is the same as in the case of fuzzy sets.

Example 4.1.4:

Diminishing the weight of terms that occur in different classes we get a classification $\mathbf{K}'$ from **K**.

$$\mathbf{K}' = \begin{pmatrix} 1 & 0 & 0 & 0 \\ ½ & ½ & 0 & 0 \\ ½ & ½ & 0 & 0 \\ 0 & 0 & 1 & 0 \\ 0 & 0 & 0 & 1 \\ 0 & 1/3 & 1/3 & 1/3 \end{pmatrix}$$

Taking **D** from Example 3.2.1 and applying the max-mult-product we have

$$\mathbf{D} \circ \mathbf{K}' = \begin{pmatrix} 1/2 & 1/4 & 0 & 0 \\ 1/4 & 1/4 & 1/2 & 1/9 \\ 1/2 & 1/4 & 0 & 0 \\ 1/4 & 1/4 & 1/2 & 1/9 \\ 0 & 1/9 & 1/9 & 1 \end{pmatrix}$$

In order to interpret a query s, it is necessary to apply a transformation $s' = \mathbf{K}^T \circ s$.

Example 4.1.5:

Let $s = \frac{1}{3} \mid a + 1 \mid c = (\frac{1}{3}, 0, 1, 0, 0)$

We take $\mathbf{K}'$ from the preceding example. Using the max-mult-product we have

$$\mathbf{K}'^T \circ s = (½, ½, 0, 0)$$

$$R(s) = (\mathbf{D} \circ \mathbf{K}') \circ (\mathbf{K}'^T \circ s)$$

$$= \frac{1}{4} \mid d_1 + \frac{1}{8} \mid d_2 + \frac{1}{4} \mid d_3 + \frac{1}{8} \mid d_4 + \frac{1}{18} \mid d_5$$

The associativity of the max-mult-product, the max-min-product or the modified vector product can be used to modify queries only:

$$R(s) = \mathbf{D} \circ ((\mathbf{K} \circ \mathbf{K}^T) \circ s)$$

or documents only:

$$R(s) = (\mathbf{D} \circ (\mathbf{K} \circ \mathbf{K}^T)) \circ s$$

4.2 Evaluation

The evaluation of a document retrieval system consists in comparing a set of documents $\bar{R}(\bar{s})$ which correspond to the user's need and a set $\bar{R}(s)$ he actually retrieves, $\mu_{\bar{R}(s)}(\bar{d}) = \mu_{R(s)}(d)$. $\bar{R}(\bar{s})$ is called the *set of relevant documents*, $R(s)_c$ the *system's proposal.* Usually $\bar{R}(\bar{s})$ is not considered as a fuzzy set, but there are proposals distinguishing degrees of relevance.

The most important evaluation measures are *recall* and *precision,* which are defined as follows:

Recall: $$r_c(\bar{s}) = \frac{|\bar{R}(\bar{s}) \cap \bar{R}(s)_c|}{|\bar{R}(\bar{s})|}$$

Precision: $$p_c(\bar{s}) = \frac{|\bar{R}(\bar{s}) \cap \bar{R}(\bar{s})_c|}{|\bar{R}(s)_c|}$$

Example 4.2.1:

Starting from Example 3.4.1 we choose a cut-off $c = 0.3$ such that $\bar{R}(s)_{0.3} = \{\bar{d}_3, \bar{d}_4\}$. Let $\bar{R}(\bar{s})$ be $\{\bar{d}_1, \bar{d}_3\}$. Then we have $r_{0.3}(\bar{s}) = ½$ and $p_{0.3}(\bar{s}) = ½$.

Using the concept of cardinality of fuzzy sets we can define recall and precision for fuzzy relevance judgements as follows:

$$r(\bar{s}) = \frac{|\bar{R}(\bar{s}) \cap \bar{R}(s)|}{|\bar{R}(\bar{s})|}$$

and

$$p(\bar{s}) = \frac{|\bar{R}(\bar{s}) \cap \bar{R}(s)|}{|\bar{R}(s)|}$$

We see that evaluating a document retrieval system cannot be a decision between 'good' or 'bad', it is a fuzzy problem too. There are a lot of criteria besides recall and precision for evaluation [8], each reflecting another aspect of the system and each fulfilled by the system more or less good.

For example

generality: $$\frac{|\bar{R}(\bar{s})|}{|D|}$$

fallout: $$\frac{|\bar{R}(s) \setminus \bar{R}(\bar{s})|}{|D \setminus \bar{R}(\bar{s})|}$$

etc.

Thus the evaluation of the system with respect to a query $\bar{s}$ can be considered as a fuzzy set $E(\bar{s}) \subset F(C)$, where C is the set of the used evaluation criteria.

5. Conclusion

Zadeh [12, 14] has mentioned the possibility of applying fuzzy sets to information retrieval problems. In this chapter we have shown how the structure of document retrieval systems can be described in terms of fuzzy mathematics. We have got two kinds of results: On the one hand we have unified separate approaches using implicitly fuzzy concepts. On the other hand we have got a generalization of Boolean retrieval and a fuzzification of clustering.

The generalized Boolean retrieval is being implemented in the FAKYR information retrieval system

[2]. We expect that the concept of weighted items, which led to improvements in retrieval with simularity measures [8] will show itself to be superior in this case also.

References

1. AUGUSTSON, J.G. and MINKER, J., 'An analysis of some graph theoretical cluster techniques', *J.A.C.M.* **17**, 4, 571-588 (1970).
2. BOCK, M., HAUSEN, H.L., KONRAD, E., and ZUSE, H., 'FAKYR—an on-line information retrieval system', Proc. 1975 Conference on Information Sciences and Systems, Baltimore, pp.364-369 (1975).
3. GOLEM2; Siemens-Schreiftenreihe data praxis.
4. KAUFMANN, A., *Introduction à la Theorie des sous-ensembles flous,* Paris (1973).
5. MAREK, W. and PAWLAK, Z., 'Information storage and retrieval system', *Mathematical Foundations,* Warszawa (1974).
6. MARON, M.E. and KUHNS, J.L., 'On relevance, probabilistic indexing, and information retrieval', *J.A.C.M.* 7, No.3, pp.216-244 (July 1960).
7. SALTON, G., *Automatic Information Organization and Retrieval,* New York (1968).
8. SALTON, G., *Dynamic Information and Library Processing,* Prentice Hall (1975).
9. SPARCK-JONES, K. and JACKSON, D., 'The use of automatically obtained keyword classification for information retrieval', *Inf. Stor. Retr.* **5**, 175-201 (1970).
10. SPARCK-JONES, K., 'Index term weighting', *Inf. Stor. Retr.* **9**, 619-633 (1973).
11. TELDOK 440, Telefunken Computer, No.31.60.01 (1972).
12. ZADEH, L.A., 'Quantitative fuzzy semantics', *Information Science* 3, 159-176 (1971).
13. ZADEH, L.A., Similarity relations and fuzzy ordering', *Information Sciences* 3, 177-200 (1971).
14. ZADEH, L.A., 'The concept of a linguistic variable and its application to approximate reasoning I', *Information Sciences* **8**, 199-249 (1975).
15. ZADEH, L.A., 'The concept of a linguistic variable and its application to approximate reasoning II', *Information Sciences* **8**, pp.301-357 (1975).
16. ZUNDE, P. and DEXTER, M.E., 'Indexing consistency and quality', *American Documentation* **20**(3), 259-257 (1969).

BIOCYBERNETICS AND THEORETICAL NEUROBIOLOGY

Introduction

LUIGI M. RICCIARDI
Department of Information Sciences,
University of Turin, Italy

I was 18 when I ran across the word 'cybernetics' for the first time. As an undergraduate at the University of Naples working for my bachelor's degree in physics, I happened to spot a notice announcing that Prof. Norbert Wiener, a guest at the Institute for Theoretical Physics at the time, was going to give a series of lectures on cybernetics. Much too confident of my high school mastery of the Greek classics, I thought it would be instructive for me to learn about the physics underlying the steering of a ship. To my dismay I instead ended up by learning some generalized harmonic analysis and a lot of Gibbsian statistical mechanics. I thought that I had not learned any cybernetics. But then, sometime after Wiener was gone, I managed to find a copy of the book *Cybernetics* and finally I got the message.

Ever since, I have tried to learn more about cybernetics and perhaps I have done a little cybernetics myself. Simultaneously I have come to realize how often Wiener's ideas have been misunderstood and how often the word 'cybernetics' has been purposely and dangerously misused. I have also developed a strong personal resentment toward certain blabbers and impostors (by now well known and confined to limited and much despised circles) that in the past kept wandering from country to country eloquently talking nonsense and aiming at extracting attention and money from unaware or much too trusting politicians and company executives: all, quite naturally, in the name of cybernetics and, therefore, under the robust shelter provided by Wiener's genius. This is why I felt a little apprehensive when the president of the Österreichische Studiengesellschaft für Kybernetik (O.S.G.K.) invited me to organize and chair a symposium on biocybernetics and theoretical neurobiology on the occasion of the Third European Meeting on Cybernetics and Systems Research. My hesitation, however, did not last long as I rationalized a little upon my dilemma. The O.S.G.K. was indeed known to me as an active and reliable organization very much in the best Austrian cultural and scientific tradition. Furthermore, I felt that by accepting this invitation I could have perhaps contributed to bringing together scientists *doing* rather than *talking* cybernetics, to the benefit of the ones authentically interested in the progress of scientific knowledge in a field where the topics under investigation are rather interdisciplinary and of a less conventional nature.

It is my conviction that the Symposium on Biocybernetics and Theoretical Neurobiology has been useful and stimulating. About 40 scientists from 13 countries (including Canada, Japan, and the USA, which indicates that it was not actually a 'European' congress) lectured and discussed on a variety of topics of current interest ranging from the foundations of model building in biocybernetics to the anatomy and the mathematics of the nervous systems. Once again I watched psychologists talk to engineers, physiologists to physicists and mathematicians to everybody. And once again I could not refrain from thinking that if all this passes unnoticed nowadays, being quite naturally accepted as an obvious consequence of the type of problems focused in such symposia, it is perhaps because over a third of a century ago a highly respected and appreciated mathematician was already actively supporting and practising such an unorthodox attitude in the name of cybernetics. Now, just as then, particularly in the USA, some scientists, quite differently from their European and Asian colleagues, still feel reluctant to use the word cybernetics. However, the message that emerged from the symposium is that cybernetic is alive and well. Whether or not scientists working in fields such as theoretical neurobiology care to cal what they do 'cybernetics' is a different and certainl unimportant question.

Tentative contributions of neuroanatomy to nerve net theories

G. PALM and V. BRAITENBERG
Max-Planck-Institut für biologische Kybernetik, Tübingen

Introduction

A great deal of information on the structure of the brain is contained in microscopical preparations. However, this information is widely ignored in brain theories. The theorist usually does not have direct access to microscopical preparations, and if he has, the inspection of the preparations is sometimes confusing to him. No systematic attempt has yet been made to extract this information in a form useful for theories.

We will review some well known anatomical facts on mammalian brains and try to point out their relevance for theories.

1. General framework

Whenever we look into a mammalian brain (Fig. 1), we see a discrete fibrous structure. The discreteness is essential because it is functional in the following sense: If you move a recording microelectrode approximately by the distance between neighboring elements seen in this picture, you sometimes record completely different signals. Therefore, these elements can be viewed as independent information-processing units. They are called neurons. As far as is known today, other constituents of the brain tissue (eg. blood vessels, glia cells) do not play an important role in the information handling but seem to serve mainly the supply of the neurons. So we can say that the brain is built essentially of one kind of element (the *neurons*) which *are the units of neuroanatomy as well as of electrophysiology. Therefore they should be the units of brain theory.*

Why are neurons fibrous? Imagine they were spherical. Then their possible connections would be very restricted by their arrangement. By building neurons fibrous one can, to a certain extent, override these metric restrictions to possible connections. *Fibers can override metric.*

These first few points indicate the framework we shall use: brain theories should be *nerve net theories*, ie. they should deal with discrete networks built of a finite number of one kind of element, namely the nerve cells or neurons.

In the following we will deal with the relations between anatomical shape and function on three levels: individual neurons; local arrangement of neurons; global organization. Finally we shall mention the plasticity of the brain and close this talk with some theoretical remarks.

2. Individual neurons

Let us give a rough description of the shape and functioning of a single neuron. It has an input tree (dendrites) and an output tree (axon). Incoming

Fig.1 From a section through a human brain stem, stained with silver nitrate. Many fibers can be seen, variously arranged in bundles and loose networks.

signals in the input tree (post-synaptic potentials) are weighted and added en route to the origin of the axon where an output signal (a spike or burst of spikes) is generated, the intensity of which is a function of the weighted sum of incoming signals. The output signal runs practically unchanged through all axonal branches, reaching the synapses which connect the axon to dendritic trees of other neurons. Then it passes the synapse and is changed to the new input signal in the adjacent neuron. This new input signal can be positive or negative (excitatory or inhibitory) but this feature seems to be constant in all axonal synapses and throughout the life of every single neuron.

Thus in principle (apart from the distinction between excitatory and inhibitory neurons) all neurons function the same way. The different roles they play in the network are mainly determined by the way they are connected to other neurons. The possible connections are grossly restricted by their shape. Therefore, the *shape of neurons* must be *correlated to their function in the network.* This is an important point for the relevance of neuroanatomy.

Let us look at the shapes of dendritic and axonal trees that actually occur (Fig. 2). The dendritic branches and most of the axonal branches are contained in a certain region near the cell body (within 1 mm). This region may be filled densely or loosely and may have widely varying shapes, but it is possible to define different *types* of neurons according to these shapes. One of the most frequent types (about 80% of cortical neurons) is the pyramidal cell. Its dendritic tree has two components: the apical dendrites, which fill a conical region, and the basal dendrites, which fill a more spherical one. Some axonal branches of the pyramidal cells (and in fact of most neurons in the mammalian brain) travel for some centimeters without ramifying and finally break up into a terminal arborization similar to the short range arborizations. This distinction between *short and long range connections* is an essential feature of the brain and is reflected microscopically in the separation of grey and white matter.

3. Local arrangement of neurons

Now let us focus our attention on the short range connections. These form a considerable portion of the total interneural contacts indicating that the metric is still important (though fibers can in principle override it). In fact *metric is used* in some parts of the brain: in input and output regions distances between neurons can represent spatial or temporal distances in the environment.

The importance of metric raises our interest in the spatial arrangement of neurons. Wherever we look into a mammalian brain, we can distinguish different regions by different patterns of cell arran-

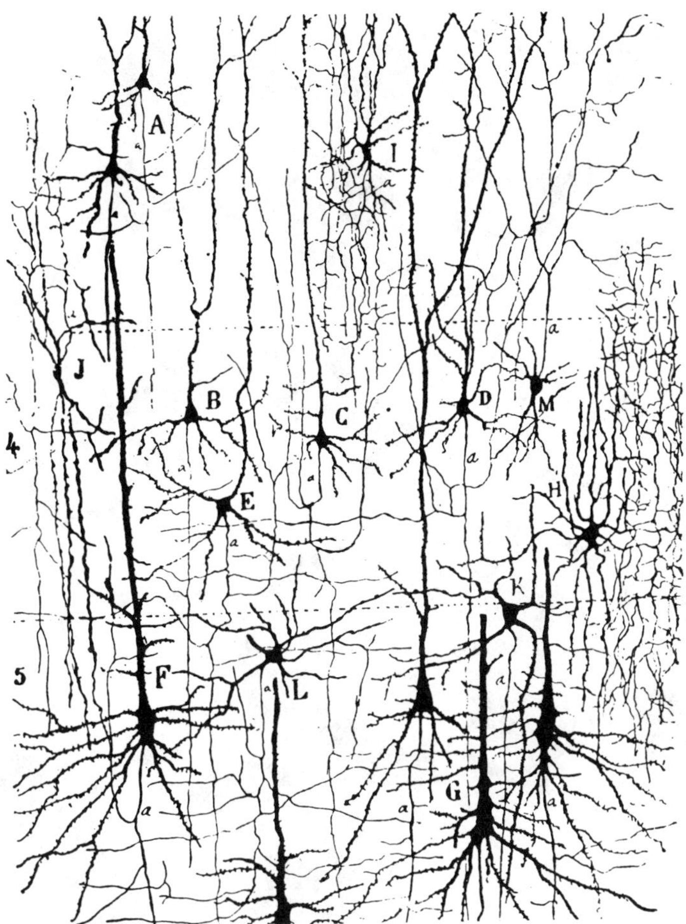

Fig.2 (from Ramon y Cajal, 1911). Various kinds of neurons in the cerebral cortex are shown, stained with the Golgi method which selects at random about one per cent or one per mille of the neurons present, and drawn with a 'camera lucida' microscope. The cell processes marked *a* are the axons, the others the dendrites.

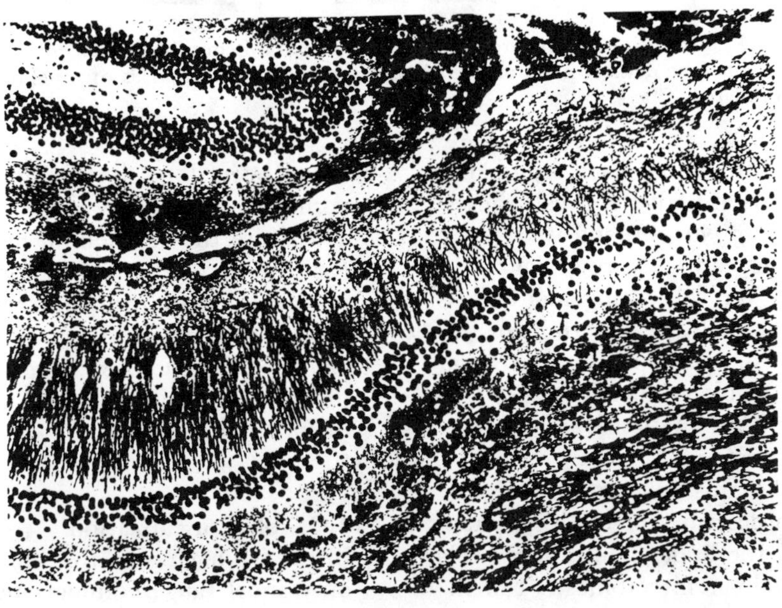

Fig.3 From a section through a mouse brain, Bodian stain. Parts of the hippocampus, fascia dentata and fimbria are shown.

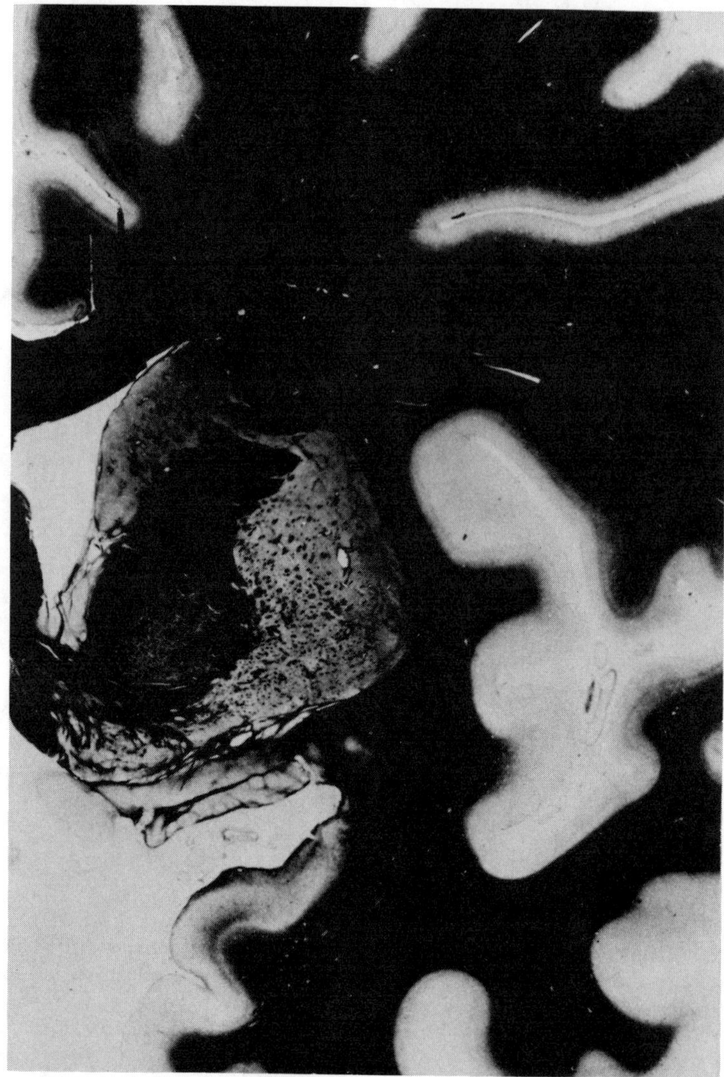

Fig.4 Part of a section through a human cerebral hemisphere, stained for myelin. The darkest parts of the picture represent dense masses of myelinated (insulated) fibers.

gement (Fig. 3). But throughout each region this pattern is relatively constant. We will call these *regions of uniform wiring pattern*, and we use the term 'wiring pattern' instead of 'wiring diagram' to indicate that (1) by the cell arrangement not every single connection is clearly determined, but only certain restrictions – perhaps of a statistical kind – are given to possible connections, (2) these restrictions occur in a highly repetitive fashion throughout each such region.

We think it would be a useful anatomo-mathematical work to describe these wiring patterns in different regions of the brain (and perhaps relate them to cyto- and myelo-architectonics). For instance, in the cerebellum this kind of work has already been done, but in this case the wiring pattern comes comparatively close to a wiring diagram. Therefore the cerebellum has been a favorite object for theories. The reason for that might also be that it is possible there to describe the function of the whole in terms of the unitary mesh. This method can probably be used in any well described region of of uniform wiring pattern, and we call it the method of *knitting theories.*

4. Global organization

In (2) we have seen that most neurons have not only short range connections, for which the metric is important, but also long range fibers. In other words, *most neurons have metric and ametric connections.* Now we will turn to the gross organization of the brain, which is provided by the long-range fibers (Fig. 4).

In this picture we can identify regions of grey matter, separated by a black mass. Actually they are connected by it since it consists of cables of long range fibers. This black mass corresponds to what goes by the name of white matter. The distinction between *regions of grey matter and connecting white matter* provides a natural articulation of the brain into *subunits, for which subtheories are possible.*

Only a very low percentage of the long range fibers serves for input-output relations of the brain as a whole. Most of them provide internal connections between different subunits or sometimes between distant parts of the same subunit, as in the case of the corticocortical connections. Therefore, all sorts of *loops are possible* (especially cortico-cortical ones) and signals may take complicated routes of widely varying length inside the brain. This is reflected in the fact that reaction time in man may vary from milliseconds to years, perhaps according to the difficulty of the task.

It is quite well known which of these subunits are connected by long distance fibers, but how they are connected is completely unclear. This, by the way, is the main reason why it is still impossible to give a meaningful interpretation, in terms of the significance for the whole organism, to the operation of the cerebellum, which is otherwise quite well understood.

5. Plasticity

All of the gross structure, seen for example in the last picture, and depicted in an atlas of the human brain, is already present at birth. Thus, the structure of the brain at birth is not at all random. This means that brain theories should incorporate *meaningful, non-trivial initial conditions* (eg. on the coupling coefficients in the network).

During life the reaction of the nerve net to the same inputs may change through experience. Our view that the functioning of single neurons cannot vary much throughout life (see 2) leads to the conclusion that the *structure of the network* (the coupling coefficients) *is changed by experience.* The important question concerning the relevance of anatomical observations for learning theories, clearly is: to what extent does the shape of neurons change, when they change their connections?

There is some evidence that the growth of neurons is dependent on their experience, ie., that *learning occurs with growth* (Valverde, [1]; Diamond, Krech, Rosenzweig, [2]). Our working hypothesis on this question is as follows: beyond the gross structure, the dendritic and axonal regions of individual neurons are also genetically predetermined. Only the way the neuron 'explores' its region and makes contacts within it depends on experience.

6. Theoretical remarks

Let us summarize this talk in terms of the italicized passages in the various sections, which might bear some importance on brain theories.

We think theorists have widely accepted the general framework of nerve nets (1), which 'learn' by changing the coupling coefficients (5). Our most important point, which occurs in several statements is the distinction between metric and ametric connections [see eg. (2), (4)]. If one tries to build a theory for the whole brain, one should be aware of this point which is in our opinion in favor of discrete rather than continuous theories. In fact one certainly can impose some metric on a discrete model, but we cannot imagine how one could break up the metric in a continuous model without involving discontinuities which would destroy the main advantages of continuous models.

There is an approach to a theory of the brain which avoids this difficulty: one can start with theories for very small parts and then connect them—at first perhaps by knitting—in regions of uniform wiring pattern (3), then one can start knitting together different meshes, when there is a juxtaposition of two such regions, and finally one can try to connect whole subtheories (for subregions of grey matter) (4) to a brain theory. One might believe that some of these problems of synthesis can be avoided if one builds theories only for subregions (eg. certain cortical regions, cerebellum); however, in these theories the problem of interpretation usually arises. What does a certain qualitative behavior of a solution of some system of differential equations 'mean', eg. at the next level of computation or for the behavior of the whole animal? What are the observables, in terms of which the meaningful aspects of qualitative behavior can be described (think of the analogy of statistical mechanics)? It seems to us that these questions lead again to the problem of connecting subtheories to a theory of the whole brain.

References

1. VALVERDE, F., *Exp. Brain Res.* 3, 337 (1967).
2. DIAMOND, M.C., KRECH, D., ROSENZWEIG, M.R., *J. comp. Neurol.* 123, 111 (1964).
3. CAJAL, S.R., *Histologie du système nerveux de l'homme et des vertébrés.* Paris: Maloin, 1911.

Adaptive pattern classification and universal recoding: parallel development and coding of neural feature detectors

STEPHEN GROSSBERG
Boston University, USA

1. Introduction

This chapter announces a model for the development and adult coding of neural feature detectors. The model is developed in Grossberg [1] and Grossberg [2]. The model shows how experience can retune feature detectors to respond to a prescribed convex set of spatial patterns. In particular, the detectors automatically respond to average features chosen from the set even if the average features have never been experienced. Using this procedure, any number of arbitrary spatial patterns can be recoded, or transformed, into any other spatial patterns (universal recoding). The scheme therefore embodies an adaptive pattern recognition network whereby reinforcements can generate complex pattern discriminations. The network is built from short term memory (STM) and long term memory (LTM) mechanisms, including mechanisms of adaptation, filtering, contrast enhancement, and tuning. These mechanisms capture various experimental properties of plasticity in the kitten visual cortex.

A context-dependent code can be synthesized in which no feature detector need uniquely characterize an input pattern; yet unique classification by the pattern of activity across feature detectors is possible. This property uses learned expectation mechanisms whereby unexpected patterns are temporarily suppressed and/or activate nonspecific arousal. A particular case describes reciprocal interactions via trainable synaptic pathways between two recurrent on-center off-surround networks undergoing mass action (or shunting) interactions. This unit is capable of establishing an *adaptive resonance*, or reverberation, between two regions if their coded patterns match, in an appropriate sense, and of suppressing the reverberation if their patterns do not match. This concept is used to analyse data on olfactory coding in the prepyriform cortex. Other special cases of the resonance idea include reverberation between conditioned reinforcers and generators of contingent negative variation when available external cues and internal drive demands match; and a search and lock mechanism whereby the disparity between two patterns (say to two eyes) can be minimized and the minimal disparity images locked into position.

A main theme in the model is: how can parallel processing of patterned data be accomplished in the presence of noise and saturation? Because of the generality of this problem, the model's developmental mechanism is formally analogous to data in non-neural developing structures, such as regeneration of *Hydra's* heads, sea urchin gastrulation, slime mold aggregation and slug motion, and the folds in the cuticle of *Rhodnius*; see Grossberg [3].

Another important theme in the model is: how can STM be reset in response to temporal sequences of spatial patterns? To do this, mechanisms are noted for suppressing uniform activity patterns in neuronal fields, yet generating outputs in response to differences in activity; subliminal search mechanisms for seeking the correct solution of a coding problem by comparing test patterns with expected patterns; mechanisms for shifting the focus of attention; and mechanisms for stabilizing in LTM the coding of a pattern after a critical developmental period.

Nonspecific arousal is implicated as a tuning, search, and attentional mechanism. Its use here generalizes its use in the theory of attention and discrimination learning that is suggested in Grossberg [4]. We suggest that nonspecific arousal is gated by a slowly varying transmitter system—for example, norepinephrine—whose relative states of accumulation and depletion at antagonistic pairs of on-cells and off-cells can shift the pattern of STM activity across a field of neuronal populations. A sudden increment in arousal can reverse, or rebound, the relative activities in on-cell and off-cell pairs across the field. A similar rebound mechanism, operating at midbrain sites, has been suggested to account for reversals in positive-incentive and negative-incentive in certain experiments on reinforcement (Grossberg [5]). This rebound mechanism has formal properties analogous to negative afterimages and spatial frequency adaptation. Such illusions add to others which formally arise in recurrent on-center off-surround shunting networks (Levine and Grossberg [6]), such as hysteresis, line neutralization, tilt aftereffect, and angle expansion.

2. The point of departure

The model evolved from earlier experimental and theoretical work. Various data showed that there is a critical period during which experimental manipulations can alter the patterns to which feature detectors in the visual cortex are tuned (e.g. Barlow and Pettigrew [7]; Blakemore and Cooper [8]; Blakemore and Mitchell [9]; Hirsch and Spinelli [10,11]; Hubel and Wiesel [12]; Wiesel and Hubel [13,14]. This work led Von der Malsburg [15] and Perez, Glass, and Schlaer [16] to construct models of the cortical tuning process, which they analysed using computer methods. Their models are strikingly similar. Both use a mechanism of long term memory (LTM) to encode changes in tuning. This mechanism learns by classical, or Pavlovian, conditioning (Kimble [17]) within a neural network. Such a concept was qualitatively described by Hebb [18] and was rigorously analysed in its present form by Grossberg (e.g. [19,20,21,22]). The LTM mechanism in a given interneuronal pathway is a plastic synaptic strength which has two crucial properties: (*a*) it is computed from a time average of the product of presynaptic signals and postsynaptic potentials; (*b*) it multiplicatively gates, or shunts, a presynaptic signal before it can perturb the postsynaptic cell.

Given this LTM mechanism, both models invoke various devices to regulate the retinocortical signals that drive the tuning process. On-center off-surround networks undergoing additive interactions, attenuation of small retino-cortical signals at the cortex, and conservation of the total synaptic strength impinging on each cortical cell are used in both models. Grossberg [1] realized that all of these mechanisms for distributing signals could be replaced by a minimal model for parallel processing of patterns in noise, which is realized by an on-center off-surround recurrent network whose interactions are of shunting type (Grossberg [23]). Three crucial properties of this model are: (*a*) normalization, or adaptation, of total network activity; (*b*) contrast enhancement of input patterns; and (*c*) short term memory (STM) storage of the contrast-enhanced pattern. Using these properties, Grossberg [1] eliminates the conservation of total synaptic strength—which is incompatible with classical conditioning—and shows that the tuning process can be derived from *adult* STM and LTM principles. It describes the interaction via trainable synaptic pathways from network region V_1 to network region V_2. V_1 and V_2 are separately capable of normalizing patterns, but V_2 can also contrast enhance patterns and store them in STM. In the original models of Von der Malsburg and Pérez, Glass, and Schlaer, V_1 was interpreted as a 'retina' and V_2 as 'visual cortex'. An analogous anatomy for V_1 as 'olfactory bulb' and V_2 as 'prepyriform cortex' can be noted. A more microscopic analysis of the model leads to a discussion of V_1 as a composite of retinal receptors, horizontal cells, and bipolar cells, and of V_2 as a composite of amacrine cells and ganglion cells. Such varied interpretations are possible because the same functional principles seem to operate in various anatomies.

3. The tuning process

This section reviews properties of the model that will be needed below. Suppose that V_1 consists of n states (or cells, or cell populations) v_{1i}, $i=1,2,\ldots,n$, which receive inputs $I_i(t)$ whose intensity depends on the presence of a prescribed feature, or features, in an external pattern. Let the population response (or activity, or average potential) of v_{i1} be $x_{i1}(t)$. The relative input intensity $\theta_i = I_i I^{-1}$, where $I = \sum_{k=1}^{n} I_k$, measures the relative importance of the feature coded by v_i in any given input pattern. If the θ_i's are constant during a given time interval, the inputs are said to form a *spatial pattern.* How can the laws governing the $x_{1i}(t)$ be determined so that $x_{1i}(t)$ is capable of accurately registering θ_i? Grossberg [23] showed that a bounded, linear law for x_{1i}, in which x_{1i} returns to equilibrium after inputs cease, and in which neither input pathways nor populations v_{1i} interact, does not suffice; cf. Grossberg and Levine [6] for a review. The problem is that as the total input I increases, given *fixed* θ_i values, each x_{1i} saturates at its maximal value. This does not happen if off-surround interactions also occur. For example, let the inputs I_i be distributed via a nonrecurrent, or feedforward, on-center off-surround anatomy undergoing shunting (or mass action, or passive membrane) interactions (Fig.1). Then

$$\dot{x}_{1i} = -A x_{1i} + (B - x_{1i})I_i - x_{1i} \sum_{k \neq i} I_k \qquad (1)$$

with $0 \leqslant x_{1i}(0) \leqslant B$. At equilibrium (namely, $\dot{x}_{1i} = 0$),

$$x_{1i} = \theta_i \frac{BI}{A+I} \qquad (2)$$

which is proportional to θ_i no matter how large I becomes. Since also $BI(A+I)^{-1} \leqslant B$, the total activity $x_1 \equiv \sum_{k=1}^{n} x_{1k}$ also never exceeds B; it is normalized,

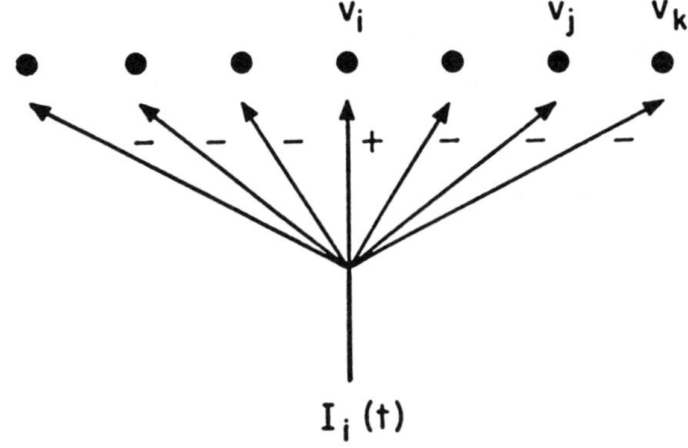

Fig.1

or adapts, due to automatic gain control by the inhibitory inputs. The normalization property in (2) shows that x_{1i} codes θ_i rather than instantaneous fluctuations in I.

To store patterns in STM, recurrent or feedback pathways are needed to keep signals active after the inputs cease. Again the problem of saturation must be dealt with, so that some type of recurrent on-center off-surround anatomy is suggested. The minimal solution is to let V_2 be governed by a system of the form

$$\dot{x}_{2j} = -A x_{2j} + (B - x_{2j})[f(x_{2j}) + I_{2j}] - x_{2j} \sum_{k \neq j} f(x_{2k}) \qquad (3)$$

where $f(w)$ is the average feedback signal produced by an average activity level w, and I_{2j} is the total excitatory input to v_{2j} (Fig.2). In particular, v_{2j} excites itself via the term $(B - x_{2j})f(x_{2j})$, and v_{2k} inhibits v_{2j} via the term $-x_{2j} f(x_{2k})$, for every $k \neq j$.

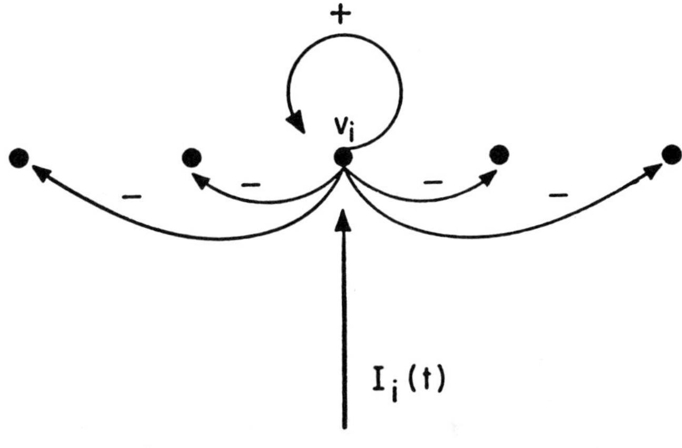

Fig.2

The choice of $f(w)$ dramatically influences how recurrent interactions within V_2 transform the input pattern $I^{(2)} = (I_{21}, I_{22}, \dots, I_{2N})$ through time. Grossberg [23] shows that a sigmoid, or S-shaped, $f(w)$ can reverberate important inputs in STM after contrast-enhancing them, yet can also suppress noise.

Various generalizations of recurrent networks have been studied, such as

$$\dot{x}_{2j} = -Ax_{2j} + (B - x_{2j}) \left[\sum_{k=1}^{N} f(x_{2k}) C_{kj} + I_{2j} \right] - (x_{2j} + D) \sum_{k=1}^{N} f(x_{2k}) E_{kj} \tag{4}$$

$D \geqslant 0$, where the excitatory coefficients C_{kj} ('on-center') decrease with the distance between populations v_{2k} and v_{2j} more rapidly than do the inhibitory coefficients E_{kj} ('off-surround'); cf. Ellias and Grossberg [24], Grossberg and Levine [27] and Levine and Grossberg [6].

Normalization in V_1 by (1) occurs gradually in time, as each x_{1i} adjusts to its new equilibrium value, but it will be assumed below to occur instantaneously with x_{1i} approaching θ_i rather than $\theta_i BI(A+I)^{-1}$. These simplifications yield theorems about the tuning process that avoid unimportant details. The assumption that normalization occurs instantaneously is tenable because the normalized pattern at V_1 drives slow changes in the strength of connections from V_1 to V_2. Instantaneous normalization means that the pattern at V_1 normalizes itself before the connection strengths have a chance to substantially change.

Let the synaptic strength of the pathway from v_{1i} to the jth population v_{2j} in V_2 be denoted by $z_{ij}(t)$. Let the total signal to v_{2j} due to the normalized pattern $\theta = (\theta_1, \theta_2, \dots, \theta_n)$ at V_1 and the vector $z^{(j)}(t) = (z_{1j}(t), z_{2j}(t), \dots, z_{nj}(t))$ of synaptic strengths be

$$S_j(t) \equiv \theta \cdot z^{(j)}(t) \equiv \sum_{k=1}^{n} \theta_k z_{kj}(t) \tag{5}$$

that is, each $z_{kj}(t)$ *gates* the signal θ_k from v_{1k} on its way to v_{2j}, and these gated signals combine additively at v_{2j} (cf. Grossberg [19,20,21,22]). Since $z^{(j)}(t)$ determines the size of the input to v_{2j}, given any pattern θ, it is called the *classifying vector* of v_{2j} at time t. *Every* v_{2j}, $j = 1,2,\dots,N$, in V_2 receives such a signal when θ is active at V_1. In this way, θ creates a pattern of activity across V_2.

Given any activity pattern across V_2, it can be transformed in several ways as time goes on. Two main questions about this process are: (*a*) will the *total* activity of V_2 be suppressed, or will some of its activities be stored in STM? and (*b*) which of the *relative* activities across V_2 will be preserved, suppressed, or enhanced? Several papers (Ellias and

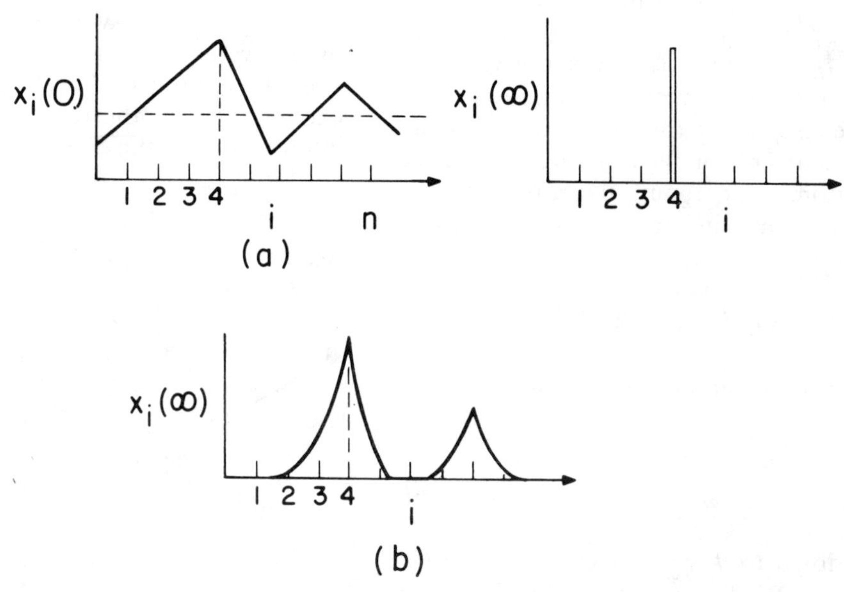

Fig.3

Grossberg [24], Grossberg [23], Grossberg and Levine [27], Levine and Grossberg [6]), analyze how the parameters of a reverberating shunting on-center off-surround network determine the answers to these questions. Below some of these facts are cited as they are needed. In particular, if all the activities are sufficiently small, then they will not be stored in STM. If they are sufficiently large, then they will be contrast enhanced, normalized, and stored in STM. Given suitable parameters, if some of the initial activities exceed a quenching threshold (QT), then V_2 will *choose* the population having maximal initial activity for storage in STM (Fig.3*a*). Under other circumstances, all initial activities below the QT are suppressed, whereas *all* initial activities above QT are contrast enhanced, normalized, and stored in STM; that is, *partial* contrast in STM is possible (Fig.3*b*). Grossberg [23] shows that partial contrast can occur if the signals between populations in a recurrent shunting on-center off-surround network are sigmoid (S-shaped) functions of their activity levels. Ellias and Grossberg [24] show that partial contrast can occur if the self-excitatory signals of populations in V_2 are stronger than their self-inhibitory signals, and moreover if the excitatory signals between populations in V_2 decrease with interpopulation distance faster than the inhibitory signals.

The enhancement and STM storage processes also occur much faster than the slow changes in connection strengths z_{ij}; hence, it is assumed below that these processes occur instantaneously in order to focus on the slow changes in z_{ij}.

The slow changes in z_{ij} are assumed to be determined by a time averaged product of the signal from v_{1i} to v_{2j} with the cortical response at v_{2j}; thus

$$\dot{z}_{ij} = -C_{ij}z_{ij} + D_{ij}x_{2j}$$

where C_{ij} is the decay rate (possibly variable) of z_{ij}, and D_{ij} is the signal from v_{1i} to v_{2j}. For example, if $C_{ij} = 1$, the V_1 and V_2 patterns are normalized, and V_2 chooses only the population v_{2j} whose initial activity is maximal for storage in STM, then while v_{2j} is active,

$$\dot{z}_{ij} = -z_{ij} + \theta_i, \quad \text{for all } i = 1,2,\ldots,n$$

It remains to determine how these z_{ij} and all other z_{ik}, $k \neq j$, change under other circumstances. To eliminate conceptual and mathematical difficulties that arise if z_{ij} can decay even when V_1 and V_2 are inactive, we let *all* changes in each z_{ij} be determined by which populations in V_2 have their activities chosen for storage in STM. In other words, all changes in z_{ij} are driven by the *feedback* within the excitatory recurrent loops of V_2 that establish STM storage. Then

$$\dot{z}_{ij} = (-z_{ij} + \theta_i)x_{2j} \tag{6}$$

where $\sum_{k=1}^{N} x_{2k}(t) = 1$ if STM in V_2 is active at time t, whereas $\sum_{k=1}^{N} x_{2k}(t) = 0$ if STM in V_2 is inactive at time t.

If V_2 *chooses* a population for storage in STM, then

$$x_{2j} = \begin{cases} 1 \text{ if } S_j > \max\{\epsilon, S_k:\ k \neq j\} \\ 0 \text{ if } S_j < \max\{\epsilon, S_k:\ k \neq j\} \end{cases} \tag{7}$$

where as in (5), $S_j = \theta \cdot z^{(j)}$ with $\theta_i = I_i(\sum_{k=1}^{n} I_k)^{-1}$.

Equation (7) omits the cases where two or more signals S_j are equal, and are larger than all other signals and ϵ. In these cases, the x_{2j}'s of such S_j's are equal and add up to 1. Such a normalization rule for equal maximal signals will be tacitly assumed in all the cases below, but will otherwise be ignored to avoid tedious details. Equation (6) shows that z_{ij} can change only if $x_{2j} > 0$. Equation (7) shows that V_2 chooses the maximal activity for storage in STM. This activity is normalized ($x_{2j} = 0$ or 1), and it corresponds to the population with largest initial signal ($S_j > \max\{S_k:\ k \neq j\}$). No changes in z_{ij} occur if all signals S_j are too small to be stored in STM (all $S_j \leqslant \epsilon$).

If partial contrast in STM holds, then the dynamics of a reverberating shunting network can be approximated by a rule of the form

$$x_{2j} = \begin{cases} f(S_j)[\sum_{S_k > \epsilon} f(S_k)]^{-1} & \text{if } S_j > \epsilon \\ 0 & \text{if } S_j < \epsilon \end{cases} \tag{8}$$

where $f(w)$ is an increasing nonnegative function of w such that $w = 0$; e.g. $f(w) = w^2$. In (8), the positive constant ϵ represents the QT; the function $f(w)$ controls how suprathreshold signals S_j will be contrast enhanced; and the ratio of $f(S_j)$ to $\Sigma\{f(S_k):\ S_k > \epsilon\}$ expresses the normalization of STM.

4. Ritualistic pattern classification

After developmental tuning has taken place, the above mechanisms describe a model of pattern classification in the 'adult' network. These mechanisms will be described first as interesting in themselves, and as a helpful prelude to understanding the tuning process. They are capable of classifying arbitrarily complicated spatial patterns into mutually nonoverlapping, or partially overlapping, sets depending on whether (7) or (8) holds. These mechanisms realize basic principles of pattern discrimination using shunting interactions. An alternative scheme of pattern discrimination using a mixture of shunting and additive mechanisms has already been given (Grossberg [25, 28]. Together these schemes suggest numerous anatomical and physiological variations that embody the same small class of functional principles. Since particular anatomies imply that particular physiological rules should be operative, intriguing questions about the dynamics of various neural structures, such as retina, neocortex, hippocampus, and cerebellum, are suggested.

First consider what happens if V_2 chooses a population for storage in STM. After learning ceases (that is, $\dot{z}_{ij} \equiv 0$), all classifying vectors $z^{(j)}$ are constant in time, and equations (6) and (7) reduce to the statement that population v_{2j} is stored in STM if

$$S_j > \max \{\epsilon, S_k : k \neq j\} \qquad (9)$$

In other words, v_{2j} *codes* all patterns θ such that (9) holds; alternatively stated, v_{2j} is a *feature detector* in the sense that all patterns

$$P_j = \{\theta : \theta \cdot z^{(j)} > \max(\epsilon, \theta \cdot z^{(k)} : k \neq j)\} \qquad (10)$$

are classified by v_{2j}. The set P_j defines a *convex cone* C_j in the space of nonnegative input vectors $J = (I_1, I_2, ..., I_n)$, since if two such vectors $J^{(1)}$ and $J^{(2)}$ are in C_j, then so are all the vectors $\alpha J^{(1)}$, $\beta J^{(2)}$, and $\gamma J^{(1)} + (1-\gamma) J^{(2)}$, where $\alpha > 0$, $\beta > 0$, and $0 < \gamma < 1$. The convex cone C_j defines the *feature* coded by v_{2j}.

The classification rule in (10) has an informative geometrical interpretation in n-dimensional Euclidean space. The signal $S_j = \theta \cdot z^{(j)}$ is the inner product of θ and $z^{(j)}$ (Greenspan and Benney [26]). Letting $\|\xi\| = \sqrt{\sum_{k=1}^{n} \xi_k^2}$ denote the Euclidean length of any real vector $\xi = (\xi_1, \xi_2, \ldots, \xi_n)$, and $\cos(\eta, \omega)$ denote the cosine between two vectors η and ω, it is elementary that

$$S_j = \|\theta\| \, \|z^{(j)}\| \cos(\theta, z^{(j)}) .$$

In other words, the signal S_j is the length of the projection of the normalized pattern θ on the classifying vector $z^{(j)}$ times the length of $z^{(j)}$. Thus if all $z^{(j)}$, $j = 1,2,...,N$, have equal length, then (10) classifies all patterns θ in P_j whose angle with $z^{(j)}$ is smaller than the angles between θ and any $z^{(k)}$, $k \neq j$, and is small enough to satisfy the ϵ-condition. In particular, patterns θ that are *parallel* to $z^{(j)}$ are classified in P_j. The choice of classifying vectors $z^{(j)}$ hereby determines how the patterns θ will be divided up. Section 7 will show that the tuning mechanism (6)-(7) makes the $z^{(j)}$ vectors more parallel to prescribed patterns θ, and thereupon changes the classifying sets P_j. In summary,

i. the number of populations in V_2 determines the maximum number N of pattern classes P_j;
ii. the choice of classifying vectors $z^{(j)}$ determines how different the sets P_j can be; for example, choosing all vectors $z^{(j)}$ equal will generate one class that is redundantly represented by all v_{2j}; and
iii. the size of ϵ determines how similar patterns must be to be classified by the same v_{2j}.

If the choice rule (7) is replaced by the partial contrast rule (8), then an important new possibility occurs, which can be described either by studying STM responses to all θ at fixed v_{2j}, or to a fixed θ at all v_{2j}. In the former case, each v_{2j} has a *tuning curve,* or *generalization gradient;* namely, a maximal response to certain patterns, and submaximal responses to other patterns. In the latter case, each pattern θ is *filtered* by V_2 in a way that shows how close θ lies to *each* of the classifying vectors $z^{(j)}$. The pattern will only be classified by v_{2j}—that is, stored in STM—if it lies sufficiently close to $z^{(j)}$ for its signal S_j to exceed the quenching threshold of V_2. The existence of tuning curves in a given cortex V_i increases the discriminative capabilities of the next cortex V_{i+1} in a hierarchy; cf. Grossberg [1].

5. Arousal as a tuning mechanism

The recurrent networks in V_2 all have a quenching threshold (QT); namely, a criterion activity level that must be exceeded before a population's activity can reverberate in STM. Changing the QT or, equivalently, changing the size of signals to V_2, can retune the responsiveness of populations in V_2 to prescribed patterns at V_1. For example, suppose that an unexpected, or novel, event triggers a nonspecific arousal input to V_2, which magnifies all the signals from V_1 to V_2. Then certain signals, which could not otherwise be stored in STM, will exceed the QT and be stored. For example, if V_2 is capable of partial contrast in STM and also receives a nonspecific arousal input, then (8) can be replaced by

$$x_{2j} = \begin{cases} f(\phi S_j)[\sum_{\phi S_k > \epsilon} f(\phi S_k)]^{-1} & \text{if } \phi S_j > \epsilon \\ 0 & \text{if } \phi S_j < \epsilon \end{cases} \tag{11}$$

where ϕ is an increasing function of the arousal level. Note that an increase in ϕ allows more V_2 populations to reverberate in STM; cf. Grossberg [23] for mathematical proofs. In a similar fashion, if an unexpected event triggers nonspecific shunting inhibition of the inhibition interneurons in the off-surrounds of V_2, then the QT will decrease (Grossberg [23]; Ellias and Grossberg [24]), yielding an equivalent effect.

Reductions in arousal level have the opposite effect. For example, if (11) holds, and arousal is lowered until only one population in V_2 exceeds the QT, then a choice will be made in STM. Thus a choice in STM can be due either to *structural* properties of the network, such as the rules for generating signals between populations in V_2 (cf. the faster-than-linear signal function in Grossberg [23]), or to an arousal level that is not high enough to create a tuning curve. Similarly, if arousal is too small, then all functions x_{2j} in (11) will always equal zero, and no STM storage will occur.

Changes in arousal can have a profound influence on the time course of LTM, as in (6), because they change the STM patterns that drive the learning process. For example, if during development arousal level is chosen to produce a choice in STM, then the tuning of classifying vectors $z^{(j)}$ will be sharper than if the arousal level were chosen to generate partial contrast in STM.

The influence of arousal on tuning of STM patterns can also be expressed in another way, which suggests a mechanism that is utilized when universal recoding is discussed.

6. Arousal as a search mechanism

Suppose that arousal level is fixed during learning trials, and that a given pattern θ at V_1 does not create any STM storage at V_2 because all the inner products $\theta \cdot z^{(j)}$ are too small. If arousal level is then increased in (11) until some $x_{2j} > 0$, STM storage will occur. In other words, changing the arousal level can facilitate *search* for a suitable classifying population in V_2.

Why does arousal level increase if no STM storage occurs at V_2? This is a property of the expectation mechanism that is developed in Grossberg [2]. A pattern θ at V_1 that is not classified by V_2 can use this mechanism to release a subliminal search routine that terminates when an admissible classification occurs.

7. Development of an STM code

System (6)-(7) will be analyzed mathematically because it illustrates properties of the model in a particularly simple and lucid way. The first result describes how this system responds to a single pattern that is iteratively presented through time.

Theorem 1 (One Pattern)
Given a pattern θ, suppose that there exists a unique j such that

$$S_j(0) > \max\{\epsilon, S_k(0) : k \neq j\}. \tag{12}$$

Let θ be practised during a sequence of nonoverlapping intervals $[U_k, V_k]$, $k = 1,2,\ldots$ Then the angle between $z^{(j)}(t)$ and θ monotonically decreases, the signal $S_j(t)$ is monotonically attracted toward $\|\theta\|^2$, and $\|z^{(j)}\|^2$ oscillates at most once as it pursues $S_j(t)$. In particular, if $\|z^{(j)}(0)\| \leqslant \|\theta\|$, then $S_j(t)$ is monotone increasing. Except in the trivial case that $S_j(0) = \|\theta\|^2$, the limiting relations

$$\lim_{t\to\infty} \|z^{(j)}(t)\|^2 = \lim_{t\to\infty} S_j(t) = \|\theta\|^2 \tag{13}$$

hold if and only if

$$\sum_{k=1}^{\infty} (V_k - U_k) = \infty. \tag{14}$$

Remark: If $z^{(j)}(0)$ is small, in the sense that $\|z^{(j)}(0)\| \leqslant \|\theta\|$, then by Theorem 1, as time goes on, the learning process maximizes the inner product signal $S_j(t) = \theta \cdot z^{(j)}(t)$ over all possible choices of $z^{(j)}$ such that $\|z^{(j)}\| \leqslant \|\theta\|$. Otherwise expressed, learning makes $z^{(j)}$ parallel to θ, and normalizes the length of $z^{(j)}$.

What happens if several different spatial patterns $\theta^{(k)} = (\theta_1^{(k)}, \theta_2^{(k)}, \ldots, \theta_n^{(k)})$, $k = 1,2,\ldots,M$, all perturb V_1 at different times? How are changes in the z_{ij}'s due to one pattern prevented from contradicting changes in the z_{ij}'s due to a different pattern? The choice-making property of V_2 does this for us; it acts as a sampling device that prevents contradictions from occurring. A heuristic argument will now be given to suggest how sampling works. This argument will then be refined and made rigorous. For definiteness, suppose that M spatial patterns $\theta^{(k)}$ are chosen, $M \leq N$, such that their signals at time $t = 0$ satisfy

$$\theta^{(k)} \cdot z^{(k)}(0) > \max\{\epsilon, \theta^{(k)} \cdot z^{(j)}(0) : j \neq k\} \quad (15)$$

for all $k = 1,2,\ldots,M$. In other words, at time $t = 0$, $\theta^{(k)}$ is coded by v_{2k}. Let $\theta^{(1)}$ be the first pattern to perturb V_1. By (15), population v_{21} receives the largest signal from V_1. All other populations v_{2j}, $j \neq 1$, are thereupon inhibited by the off-surround of v_{21}, whereas v_{21} reverberates in STM. By (6), none of the synaptic strengths $z^{(j)}(t)$, $j \neq 1$, can learn while $\theta^{(1)}$ is presented. As in Theorem 1, presenting $\theta^{(1)}$ makes $z^{(1)}(t)$ more parallel to $\theta^{(1)}$ as t increases. Consequently, if a different pattern, say $\theta^{(2)}$, perturbs V_1 on the next learning trial, then it will excite v_{22} more than any other v_{2j}, $j \neq 2$: it cannot excite v_{21} because the coefficients $z^{(1)}(t)$ are more parallel to $\theta^{(1)}$ than before; and it cannot excite any v_{2j}, $j \neq 1, 2$, because the v_{2j} coefficients $z^{(j)}(t)$ still equal $z^{(j)}(0)$. In response to $\theta^{(2)}$, v_{22} inhibits all other v_{2j}, $j \neq 2$. Consequently none of the v_{2j} coefficients $z^{(j)}(t)$ can learn, $j \neq 2$; learning makes the coefficients $z^{(2)}(t)$ become more parallel to $\theta^{(2)}$ as t increases. The same occurs on all learning trials. By inhibiting the postsynaptic part of the learning mechanism in all but the chosen V_2 population, the on-center off-surround network in V_2 samples one vector $z^{(j)}(t)$ of trainable coefficients at any time. In this way, V_2 can learn to classify as many as N patterns if it contains N populations.

This argument is almost correct. It fails, in general, because by making (say) $z^{(1)}(t)$ more parallel to $\theta^{(1)}$, it is also possible to make $z^{(1)}(t)$ more parallel to $\theta^{(2)}$ than $z^{(2)}(0)$ is. Thus when $\theta^{(2)}$ is presented, it will be coded by v_{21} rather than v_{22}. In other words, practising one pattern can recode other patterns. This property can be iterated to show how systematic trends in the sequence of practised patterns can produce systematic drifts in recoding. Moreover, if there are many patterns relative to the number of populations in V_2, and if the statistical structure of the practice sequences continually changes, then there need not exist a stable coding rule in V_2. This is quite unsatisfactory.

By contrast, if there are few, or sparse, patterns relative to the number of populations in V_2, then a stable coding rule does exist, and the STM choice rule in V_2 does provide an effective sampling technique. Such a situation is approximated, for example, when the network is exposed to a 'visually deprived' environment, in imitation of experiments on young animals. A theorem concerning this case will now be stated, if only to suggest what auxiliary mechanisms will be needed to establish a stable coding rule in the general case. This theorem shows how populations learn to code convex regions of features. In particular, if v_{2j} learns to code a certain set of features, then it automatically codes *average* features derived from this set.

The following nomenclature will be needed to state the theorem. A *partition* $\oplus_{k=1}^{K} \mathsf{P}_k$ of a finite set P is a subdivision of P into nonoverlapping and exhaustive subsets P_j. The *convex hull* $\mathsf{H(P)}$ of a finite set P is the set of all convex combinations of elements in P; for example, if $\mathsf{P} = \{\theta^{(1)}, \theta^{(2)}, \ldots, \theta^{(M)}\}$ then

$$\mathsf{H(P)} = \{\sum_{k=1}^{M} \lambda_k \theta^{(k)} : \text{each } \lambda_k \geq 0 \text{ and } \sum_{k=1}^{M} \lambda_k = 1\}.$$

Given a set P with subset Q, let $\mathsf{R} = \mathsf{P}\backslash\mathsf{Q}$ denote the set of elements in P that are not in Q. If the classifying vector $z^{(j)}(t)$ codes the set of patterns $\mathsf{P}_j(t)$, let $\mathsf{P}_j^*(t) = \mathsf{P}_j(t) \cup \{z^{(j)}(t)\}$. The *distance* between a vector P and a set of vectors Q, denoted by $\|P - \mathsf{Q}\|$, is defined by

$$\|P - \mathsf{Q}\| = \inf\{\|P - Q\| : Q \in \mathsf{Q}\}.$$

Theorem 2 (Sparse patterns)

Let the network practise any set $\mathsf{P} = \{\theta^{(i)} : i = 1,2,\ldots,M\}$ of patterns for which there exists a partition $\mathsf{P} = \oplus_{k=1}^{N} \mathsf{P}_k(0)$ such that

$$\min\{u \cdot v : u \in \mathsf{P}_j(0), v \in \mathsf{P}_j^*(0)\} >$$

$$> \max\{u \cdot v : u \in \mathsf{P}_j(0), v \in \mathsf{P}^*(0)\backslash\mathsf{P}_j^*(0)\} \quad (16)$$

for all $j = 1,2,\ldots,N$. Then $\mathsf{P}_j(t) = \mathsf{P}_j(0)$ and the functions

$$D_j(t) = \|z^{(j)}(t) - \mathsf{H}(\mathsf{P}^{(j)}(t))\| \quad (17)$$

are monotone decreasing for $t \geq 0$ and $j = 1,2,...,N$. If moreover the patterns in $P^{(j)}(0)$ are practised in intervals $[U_{jm}, V_{jm}]$, $m = 1,2,...$, such that

$$\sum_{m=1}^{\infty} (V_{jm} - U_{jm}) = \infty \tag{18}$$

then

$$\lim_{t \to \infty} D_j(t) = 0 \tag{19}$$

Remarks: In other words, if the $z^{(j)}(t)$ vectors are taught to code the patterns into sparse classes, in the sense of (16), then this code persists through time, and the classifying vectors approach a convex combination of their coded patterns. As (17) and (19) show, learning permits each v_{2j} to respond as vigorously as possible to its class of coded patterns.

The above results indicate that, given a fixed number of patterns, it becomes easier to establish a stable code for them as the number of populations in V_2 increases. Once V_2 is constructed, however, it is not possible to increase its number of populations at will. Moreover, *in vivo*, an enormous variety of patterns typically barrages the visual system. How can a stable code be guaranteed no matter how many patterns perturb V_1? The papers (Grossberg [2,29]) go on to analyze this problem and related concepts, such as adaptive resonance, and rebound in antagonistic populations due to the influence of temporal changes in nonspecific arousal on the spatial distribution of slow transmitter accumulation-depletion.

References

1. GROSSBERG, S., *Biol. Cybernetics,* **21**, 145-159, (1976).
2. GROSSBERG, S., 'Adaptive pattern classification and universal recoding: parallel development and coding of neural feature detectors, *Biol. Cybernetics,* **23**, 121-134, 187-202 (1976).
3. GROSSBERG, S., 'Communication, memory, and development', *Progress in Theoret. Biol.* (Eds.: R. Rosen and F. Snell), in press (1976).
4. GROSSBERG, S., In "International Review of Neurobiology", Vol.18 (Ed.: C.C. Pfeiffer), 263-327 (1975).
5. GROSSBERG, S., *Math. Biosci.* **15**, 39, 253 (1972).
6. LEVINE, D.S. and GROSSBERG, S., 'Visual illusions in neural networks: line neutralization, tilt aftereffect, and angle expansion', *J. Theor. Biol.,* **61**, 477-504 (1976).
7. BARLOW, H.B. and PETTIGREW, J.D., *J. Physiol,* **218**, 98P-100P, (1971).
8. BLAKEMORE, C. and COOPER, G.F., *Nature,* **228**, 477-478 (1970).
9. BLAKEMORE, C. and MITCHELL, D.E., *Nature,* **241**, 467-468 (1973).
10. HIRSCH, H.V.B. and SPINELLI, D.N., *Science,* **168**, 869-871 (1970).
11. HIRSCH, H.V.B. and SPINELLI, D.N., *Exp. Brain Res.,* **12**, 509-527 (1971).
12. HUBEL, D.H. and WIESEL, T.N., *J. Physiol. (Lond.),* **206**(2), 419-436 (1970).
13. WIESEL, T.N. and HUBEL, D.H., *J. Neurophysiol.,* **26**, 1003-1017 (1963).
14. WIESEL, T.N. and HUBEL, D.H., *J. Neurophysiol.,* **28**, 1029-1040 (1965).
15. VON DER MALSBURG, C., *Kybernetik,* **14**, 85-100 (1973).
16. PÉREZ, R., GLASS, L., and SCHLAER, R., *J. of Math. Biol.* (1974).
17. KIMBLE, G.A., *Foundations of Conditioning and Learning,* Appleton-Century-Crofts, New York (1967).
18. HEBB, D.O., *The Organization of Behavior*, Wiley, New York (1949).
19. GROSSBERG, S., *Proc. Natl. Acad. Sci. U.S.A.,* **58**, 1329-1334 (1967).
20. GROSSBERG, S., *Studies in Applied Math.,* **49**, 135-166 (1970).
21. GROSSBERG, S., *Proc. Natl. Acad. Sci. U.S.A.,* **68**, 828-831 (1971).
22. GROSSBERG, S., In "Progress in Theoretical Biology" (Eds.: R. Rosen and F. Snell), p.51-141, Academic Press, New York (1974).
23. GROSSBERG, S., *Studies in Applied Math.,* **52**, 213-257 (1973).
24. ELLIAS, S.A. and GROSSBERG, S., *Biol. Cybernetics,* **20**, 69-98 (1975).
25. GROSSBERG, S., *J. of Theoret. Biol.,* **27**, 291-337 (1970).
26. GREENSPAN, H.P. and BENNEY, D.J., *Calculus,* McGraw-Hill, New York (1973).
27. GROSSBERG, S. and LEVINE, D.S., *J. Theoret. Biol.* **53**, 341-380 (1975).
28. GROSSBERG, S., *Kybernetik,* **10**, 49-57 (1972).
29. GROSSBERG, S., In: *Progress in Theoretical Biology* (eds.: R. Rosen and F. Snell), Academic Press, New York (1977).

Dynamics of reverberation cycle and its implications to linguistics*

TOSIO KITAGAWA
Kyushu University and
Institute for Advanced Study of Social Information Science, Japan

1. Introduction

The purpose of the present chapter is to propose a neural dynamical approach aiming to establish a certain mathematical formulation which yields a link between reverberation phenomena in neurological activities and linguistics. Our proposal of such an approach has been influenced and stimulated by various works of many authors in different scientific fields. In particular we may and we should mention here the following definite and apparent indebtedness: (1°) the brain model to Zeeman [1], (2°) the neuronic equation systems to Caianiello [2], (3°) catastrophy theory in linguistics to Thom [3-5], (4°) classification of verbs to Tamachi-Okada [6], (5°) semantic theory in natural language to various authors including Yoshida [7], and (6°) biological foundations of languages to Lenneberg [8].

In spite of the apparent indebtedness which the present author owed in his pursuit for finding the relationship between brain activities and linguistics, there is a unified standpoint from which the whole aspect of our approach can be observed and by which the whole master plan of our approach can be organized and scheduled, as we shall explain in Section 2.

There is a set of five research strategies I-V which will constitute the keynote of our approach, as we shall explain in Section 2. In Section 3 we shall explain the research strategy III in more detail in connection with the Thom approaches [3 and 4] in linguistics. In Section 4 we shall propose a set of new problems of automaton theory which are required in order to make a further step in the direction suggested in Section 3. Section 5 is devoted to showing the application of informative logics, which was proposed by the present author in a series of the papers, Kitagawa [9,10,11,12], to the present problem. Section 6 is devoted to evaluation schemes in linguistic information with reference to three subspaces called control, eizon and creation. In short our present approach is a combination of the

*This work was done as a member in the research group "C-2: The Structure of data base and theory of information retrieval" belonging to the Special Research Project "Advanced Information Processing of Large Scale Data over a Broad Area" supported by the Research Grant of the Ministry of Education of Japan.

researches by the author on two different areas, one in the neuronic equations given in a series of papers, Kitagawa [13-18] and, the other in the logic of information science accumulating in what we call informative logics.

In conclusion of this short introduction we have to refer to our general standpoints for treating natural languages with an intention of reference to neural dynamics. Broadly speaking we are seeking biological foundations of languages under general five premises proposed by Lenneberg [8], Chapter Nine, "toward a biological theory of language development". Instead of going into a search for an explanation for the phylogenic development of language, which might be required for biological foundations of languages, however, we start with the present topics since there are a lot of indications to the effect that cerebral function is a determining factor for language behavior.

In conclusion of this Introduction, it ought to be remarked also that the mother tongue of the present author is Japanese which may give influence in a deep but implicit way to the thinking of the author, although he treats the subject rather universally without reference to any specific language. For instance the English noun 'time' is translated into the Japanese noun 'toki' which in turn is closely connected with the Japanese 'töku'. The Japanese verb 'töku' represents just indisintegration of reverberation cycle which corresponds to the Japanese abstract noun 'mono'.

2. Five research strategies

Our approach which we shall present in this chapter is based upon the consolidation of the following five research strategies.

Research strategy 1
We shall set up a system of neuronic equations regarding N neurons with time lag $(n-1)$ according to the formulation due to Caianiello [2] consisting of

I. neuronic (or decision) equations (NE),
II. mnemonic (or evolution) equations (ME), and
III. adiabatic learning hypothesis (or rule) (ALH).

The mathematical model of brain activities is given by the transition behaviors of the state configurations regarding N neurons with their time lag situations. Specific emphasis should be paid to considerations of (i) convergence to stable configurations, (ii) convergences to reverberation cycles, and (iii) disintegration of reverberation cycles into other types of configurations. It is the aim of research strategy I to give a mathematical formulation describing the brain activities in such a way that there are some definite functional equations which can directly refer to the dynamics of brain activities. Also for the present purpose we shall confine ourselves here to the neuronic equations (NE) without entering into (II) and (III).

Research strategy II
We shall pick up the roles of verb as determinating element in semantic analysis of any sentence. This attitude is entirely following that of many linguists as well as the characteristic considerations due to Thom [3-5] in his approach to linguistics by catastrophy theory. In this connection the roles of nouns, cutting principles and the set-up of subjectivity and objectivity are the indispensable frameworks for our considerations of semantics.

Research strategy III
We shall present an interpretation scheme of some fundamental verbs as transition phenomena to be observed among state configurations of neurons with time lag. Our presentation can be connected with the work of Thom [3-5] which is based upon his catastrophy theory, through a certain systematic correspondence between his approaches and ours.

Research strategy IV
In order to realize the transitions phenomena of neural state configurations which can be interpreted as corresponding to some fundamental verbs under our interpretation, we shall attempt to give an automata theoretic formulation to describe the transition phenomena of neural state configurations.

Research strategy V
We shall appeal to a general framework of informative logics introduced in our monograph [19] and paper [9] in which three subspaces called (1) control, (2) eizon, and (3) creation are introduced to describe three different feasibility aspects of existences. These three feasibility aspects will be used in order to give a semantic classification of certain set of fundamental verbs. Another role of informative logics to be expected in semantic analysis of any sentence is to appeal to general principles that should be adopted in any system formation, which is one of the indispensable components of informative logics. The terminology of informative logics was introduced by the author in his paper [12].

The details of each of these five research strategies will be either explicitly described or implicitly observed from our discussions in the following Sections. It is however imperative to give here a more definite illustration of research strategy III before ending this Section, because it will constitute a core of our methodology which is based upon a certain mathematical model for discussing the interrelationship between brain activities and linguistic phenomena. There is also another reason why we have to refer just at this moment. The reason is this. We have published already a series of papers [13-16] for discussing dynamical behaviors of solutions of N neuronic equations with time lag $(n-1)$, and we have started with a systematic investigation of them, which is under a certain research strategy, still now being undertaken in a progressive way step by step.

The idea of the present chapter is in fact suggested by such a systematic investigation. Nevertheless what is needed for explaining the substantial content of the chapter so far as Section 3 is concerned is not the total outcome given by our previous papers [13-16], but it is rather a general feature of neuronic equations which is required to be known.

There are five principal points (i)-(v) in this connection, as we shall explain. Before going into the explanation, let us start with some mathematical preparations. A system of N neuronic equations regarding N neurons with time lag $(n-1)$ is given by Caianiello [2] by the following vector equation, where $x_i(t)$ denotes the state of the ith neuron at time point t whose value is either 1 or 0.

$$\chi(t+1) = 1[A\chi(t) - \Theta(t)] \qquad (2.01)$$

where

$$\chi(t) = \begin{pmatrix} x_1(t) & x_2(t)\ldots\ldots & x_j(t)\ldots\ldots & x_N(t) \\ x_1(t-1) & x_2(t-1)\ldots\ldots & x_j(t-1)\ldots\ldots & x_N(t-1) \\ \ldots\ldots & \ldots\ldots & \ldots\ldots & \\ x_1(t-i) & x_2(t-i)\ldots\ldots & x_j(t-i)\ldots\ldots & x_N(t-i) \\ \ldots\ldots & \ldots\ldots & \ldots\ldots & \ldots\ldots \\ x_1(t-n+1) & x_2(t-n+1)\ldots & x_j(t-n+1)\ldots & x_N(t-n+1) \end{pmatrix} \qquad (2.02)$$

with

$$A = (a_{ij}(k);\ ij = 1,2,\ldots,N;\ k = 0,1,2,\ \ n\text{-}1) \qquad (2.03)$$

$$\Theta(t) = (\theta_1(t),\ \theta_2(t),\ \ldots\ldots,\ \theta_N(t)) \qquad (2.04)$$

and the vector threshold function $\mathbb{1}[u]$ for $\mathbf{u} = (u_1, u_2,\ldots, u_N)$ is defined by

$$1 : \mathbf{u} = (u_1, u_2,\ldots, u_N) \rightarrow (1[u_1], 1[u_2],\ldots, 1[u_N]) \qquad (2.05)$$

where

$$1[uj] = \begin{cases} 0 & \text{for } u_j \leqslant 0 \\ 1 & \text{for } u_j > 0 \end{cases} \qquad (2.06)$$

for each scalar value u_i. $(j=1,2,\ldots,N)$.

The minimal sufficient aspects of the neural dynamics which should be secured for our discussions in what follows by a general framework of (2.01) are the following five (i)-(v):

i. The multitudes of the systems of neuronic equations (2.01) and their solutions which can be secured by (*a*) the multitudes of A and/or $\Theta(t)$ and (*b*) the plasticity of the matrix A.
ii. Digraph representations of solutions of the equations (2.01) as discussed in our papers [13-16].
iii. Automaton theoretic approaches (See Section 4).
iv. Application of informative logics (See Section 5).
v. Deep structural digraphes, that is to say, deep structures associated with a dynamical system to be considered in connection with the system of neuronic equations (2.01). (See also our papers [13-16].)

3. Explanation of research strategy III

[1] The Thom theory of fundamental verbs

In connection with our research strategy III, we shall start with a certain version of the Thom theory of fundamental verbs, which can be summarized as follows:

i. The existence of one image or some images is assumed, and each image will be denoted by a point in some topological space.
ii. There is one coordinate axis which denotes the time.
iii. In combination of (i) and (ii), each image which continues to exist can be denoted by a line, which we may call image line.
iv. The genesis, growths, confluences, disintegration, and extinctions of these images are expressed as a mutual relationship among these image lines.
v. The various transitory phenomena among image lines can be classified according to the catastrophy theory due to Thom [3-5].

[2] The essential features of research strategy III

Our interpretation scheme of some fundamental verbs as transition phenomena to be observed among state configurations of neurons with time lag is based upon the following principal attitudes:

1° Each image in the Thom theory is now understood as corresponding to a deep structural digraph Q_ω which we have defined in our paper [14].

2° Each verb is understood as corresponding to transitionary behavior among state configurations of N neurons induced by a system of neuronic equations with time lag.

3° In order to secure the correspondences suggested in (2°) a new aspect of automaton theory will be cultivated, as we shall explain in Section 3.

[3] Examples

The purpose of the present paragraph is to give a set of examples by which to explain our principles (1°) and (2°) enunciated just now in the previous paragraph [2]. In each example, a verb is given with the Thom [3-5] expression by means of image lines. Then our interpretation appealing to a set of reverberation cycles and their transitionary phenomena is illustrated.

(1°) Être

Fig.3.1(a) The topological expression of 'être' by Thom

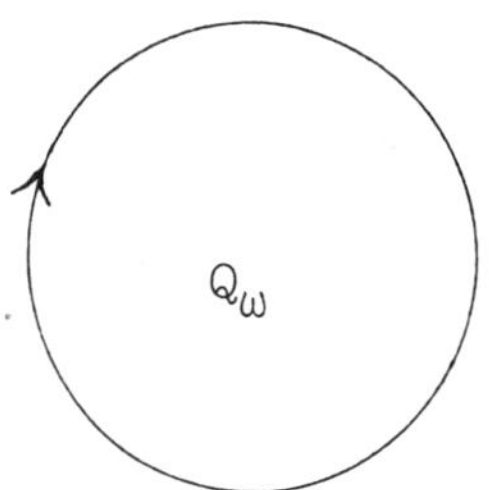

Fig.3.1(b) Interpretation of the verb 'être' by means of the existence of a reverberation cycle

(2°) Finir

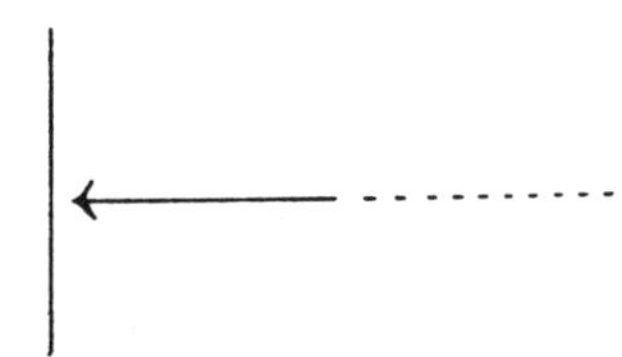

Fig.3.2(a) The topological expression of 'finir' by Thom

Fig.3.2(b) Interpretation of the verb 'finir' by means of convergence to a stable point

(3°) Commencer

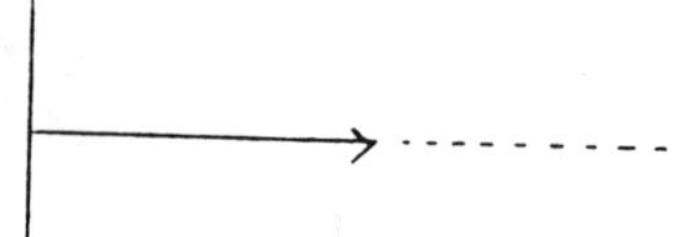

Fig.3.3(a) The topological expression of 'commencer' by Thom

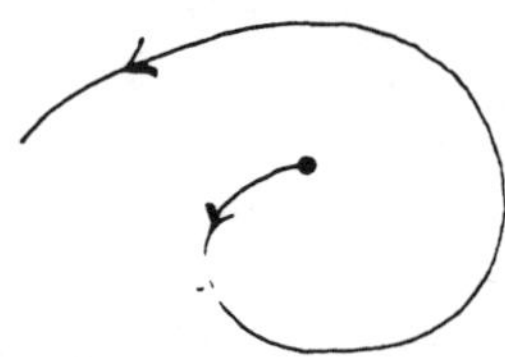

Fig.3.3(b) Interpretation of the verb 'commencer' by means of departure from a stable point or from a point belonging to Eden garden

(4°) Unir

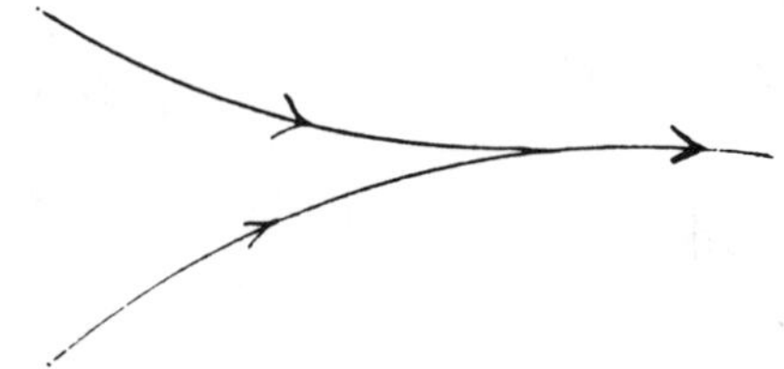

Fig.3.4(a) The topological expression of 'unir' by Thom

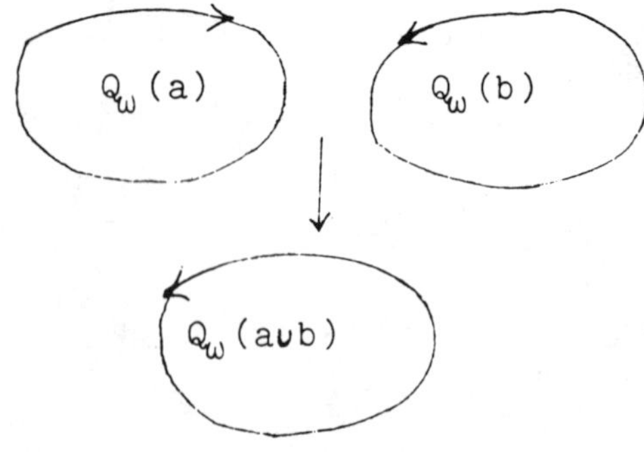

Fig.3.4(b) Interpretation of the verb 'unir' by means of amalgamation of two reverberation cycles

The verb 'unir' is interpreted as an amalgamation of two or more separated reverberation cycles into one reverberation cycle.

(5°) Séparer

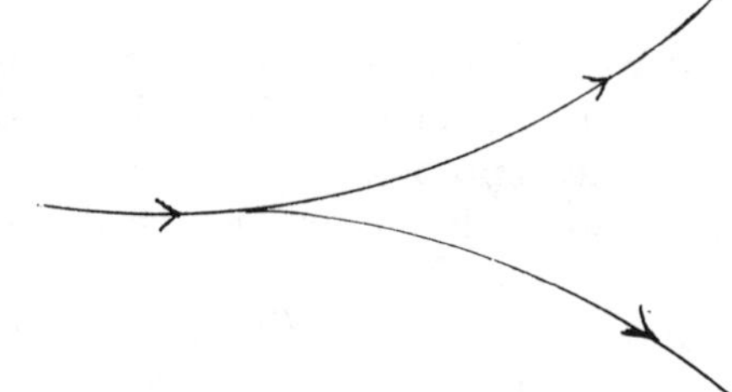

Fig.3.5(a) The topological expression of 'séparer' By Thom

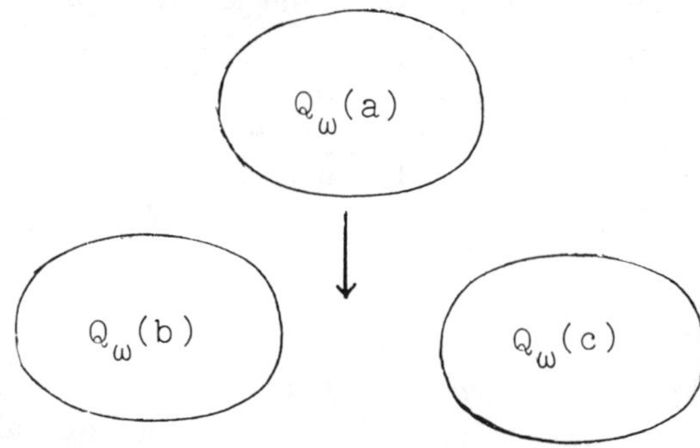

Fig.3.5(b) Interpretation of the verb 'séparer' by means of decomposition of one reverberation cycle into two cycles

The verb 'séparer' is interpreted as a decomposition of a reverberation cycle into two or more separated reverberation cycles.

(6°) Devenir

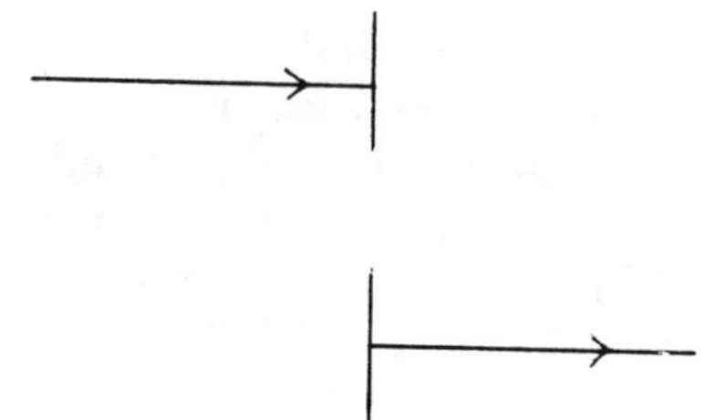

Fig.3.6(a) The topological expression of 'devenir' by Thom

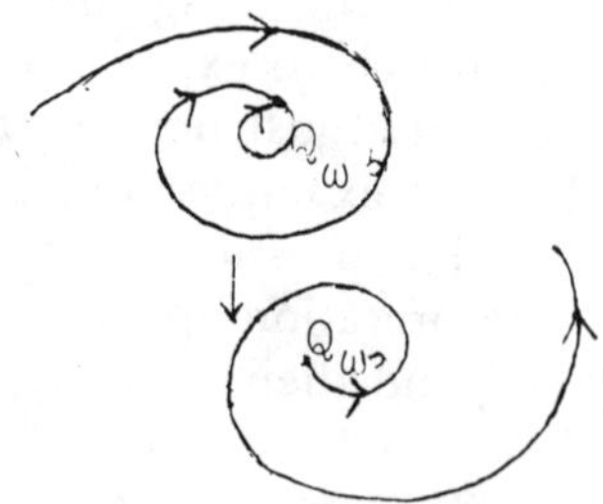

Fig.3.6(b) Interpretation of the verb 'devenir' by means of two stages of transitory behaviors

The verb 'devenir' is interpreted as consisting of two stages of transitory behaviors. The first stage is 'finir' in the sense of (2°), while the second one is 'commencer' in the sense of (3°).

(7°) Capturer

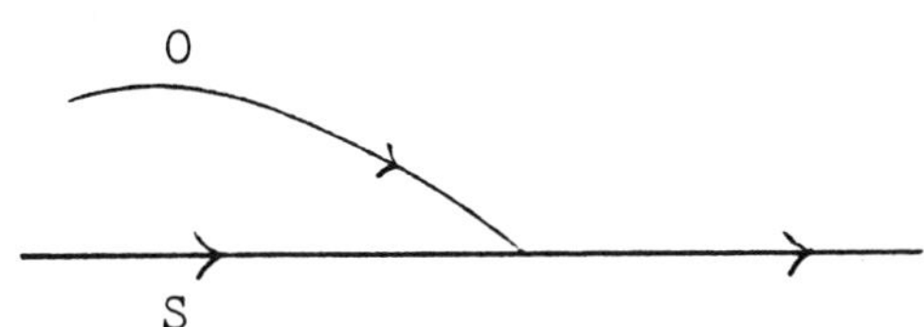

Fig.3.7(a) The topological expression of 'capturer' by Thom

(i) At the initial stage there are two reverberation cycles $Q_\omega(a)$ and $Q_\omega(b)$.

(ii) There is an introduction of cutting principle, which we have introduced in system formation of informative logics in our papers [9,12] and by which we shall assign $Q_\omega(a)$ and $Q_\omega(b)$ to the subject and the object respectively.

(iii) The verb 'capturer' is interpreted as the process of absorption of $Q_\omega(b)$ into $Q_\omega(a)$, which yields us a new reverberation cycle $Q_\omega(a')$ and an extinction of the cutt which distinguishes $Q_\omega(a)$ and $Q_\omega(b)$.

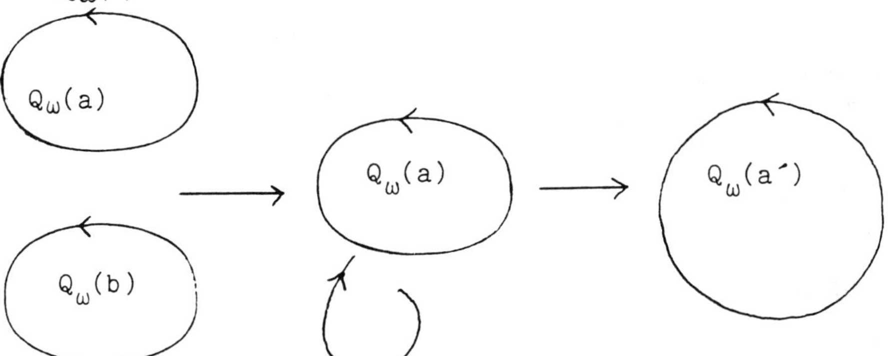

Fig.3.7(b) Interpretation of the verb 'capturer' by means of an absorption of another reverberation cycle into the original one

(8°) Emettre

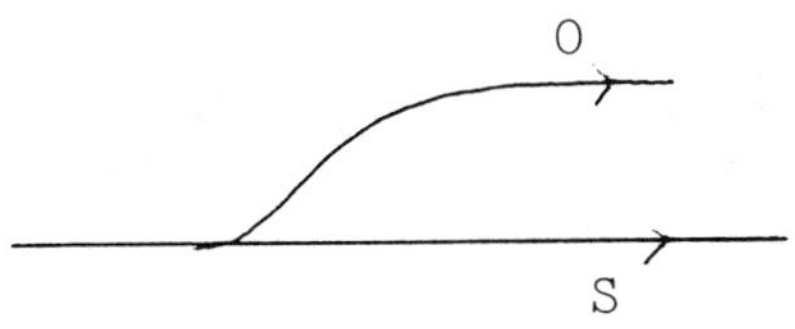

Fig.3.8(a) The topological expression of 'emettre' by Thom

(i) At the initial stage there is one reverberation cycle $Q_\omega(a)$.

(ii) The verb 'emettre' is interpreted as the process of decomposition of the reverberation cycle $Q_\omega(a)$ into two reverberation cycles $Q_\omega(b)$ and $Q_\omega(c)$ under the introduction of the cutting principle explained in (iii).

(iii) The cutting principle is used to identify $Q_\omega(b)$ with $Q_\omega(a)$ as the subject and to recognize $Q_\omega(c)$ as the object emitted from the subject.

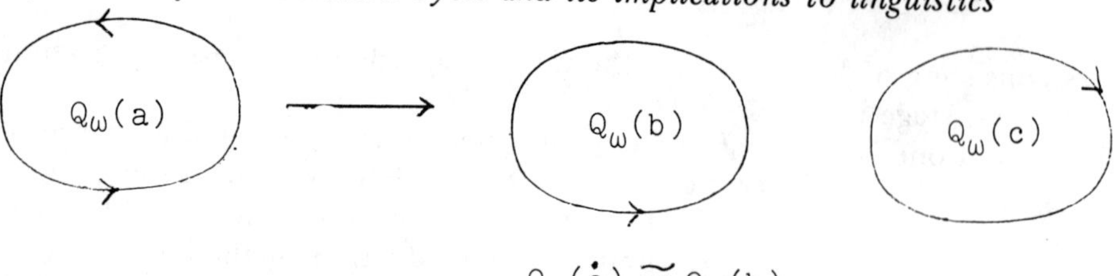

Fig.3.8(b) Interpretation of the verb 'emettre' by means of a creation of a new reverberation cycle with the original one by its decomposition

(10°) Suicide

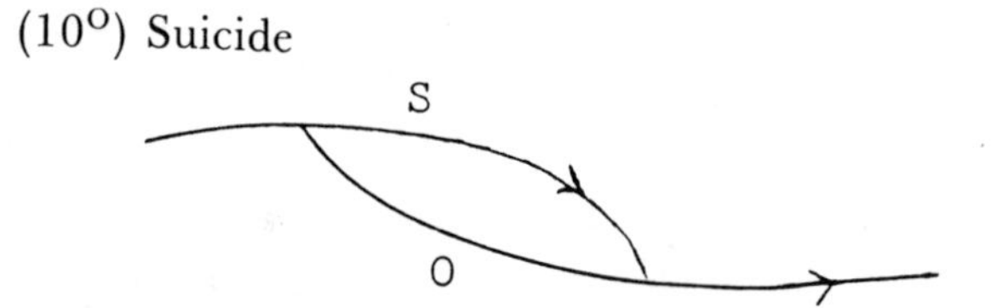

Fig.3.10(a) The topological expression of 'suicide' by Thom

The verb 'suicide' is interpreted as a combination of two verbs 'emettre' and 'capture' in the sense that *S* from which *O* has been emitted will be absorbed by *O*. This brilliant idea of Thom [3-5] can be represented in the realm of transitionary behaviors of reverberation cycles in the following way.

(i) A reverberation cycle $Q_\omega(a)$ is assumed to consist of two mutually disjoint parts *B* and *C*: $Q_\omega(a) = B + C,\ B \cap C = \phi$.

(ii) A series of transitionary phenomena happens in which *B* shrinks into a new reverberation cycle $Q_\omega(b)$, while *C* comes to be a set of transient states associated with $Q_\omega(b)$.

(iii) Then there follows a series of transitions by which the part *C* will be separated from $Q_\omega(b)$ and then will become another reverberation cycle called $Q_\omega(c)$.

(iv) The fourth step is a change of the reverberation cycle $Q_\omega(b)$ into a set of transient states associated with $Q_\omega(c)$.

(v) The last fifth stage is an extinction of the transient states given in (iv).

(11°) Agiter

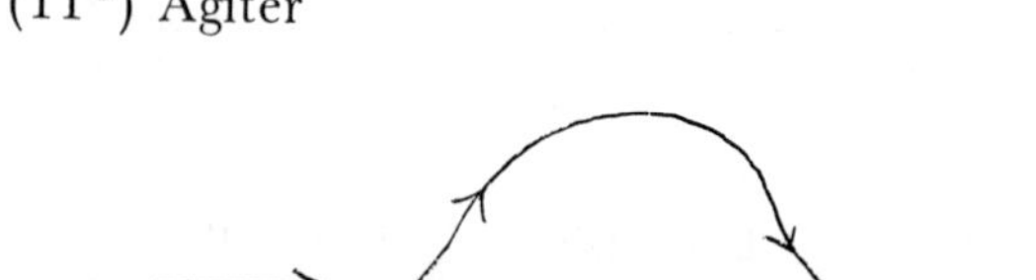

Fig.3.11(a) The topological expression of 'agiter' by Thom

The verb 'agiter' can be interpreted as a combination of two verbs 'separer' and 'unir' which we have discussed already in (5°) and (6°). In the consequence the representation of the verb 'agiter' in terms of the reverberation cycles is simply given by the illustration.

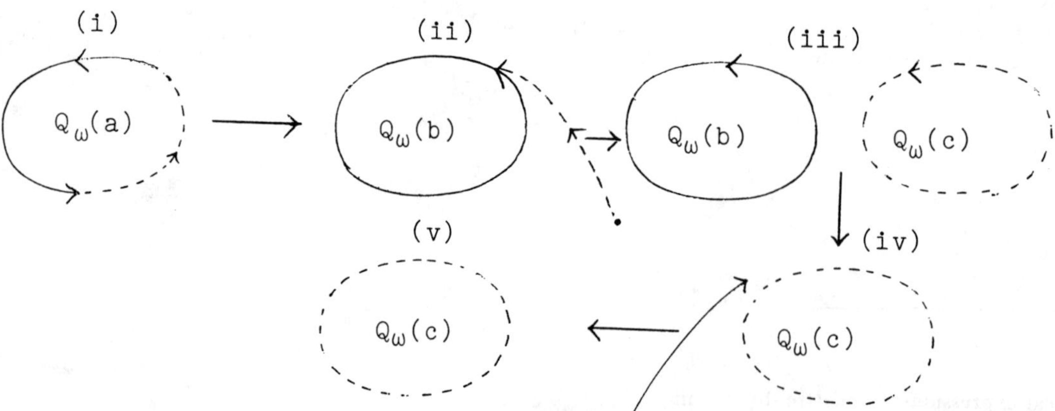

Fig.3.10(b) Interpretation of the verb 'suicide' by transitory behaviors between two reverberation cycles one of which has been emitted by the other one

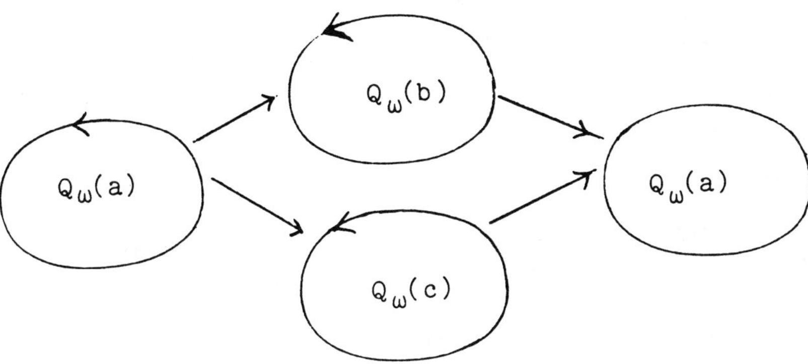

Fig.3.11(b) Interpretation of the verb 'agiter' by reverberation cycle phenomena

(12°) Repousser

Fig.3.12(a) The topological expression of 'repousser' by Thom

A representation of the verb 'repousser' is given by the following transitory phenomena (i) → (ii) → (iii) of two reverberation cycles, which can be denoted symbolically by $(a, b) \to a(b) \to (a, b)$.

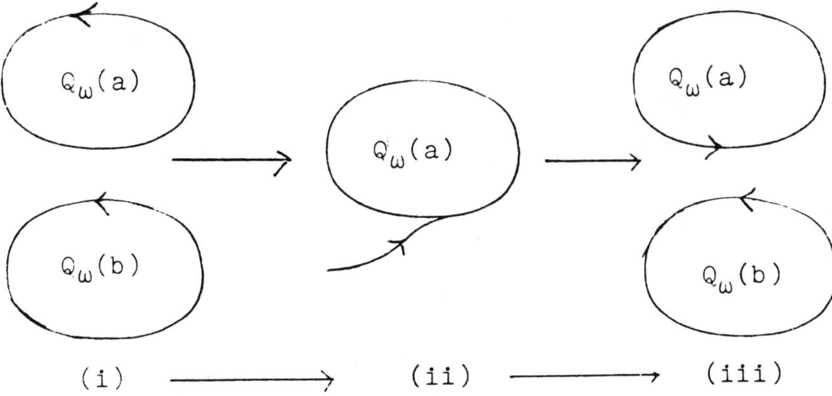

Fig.3.12(b) Interpretation of the verb 'repousser' by means of reverberation cycle phenomena

(13°) Traverser

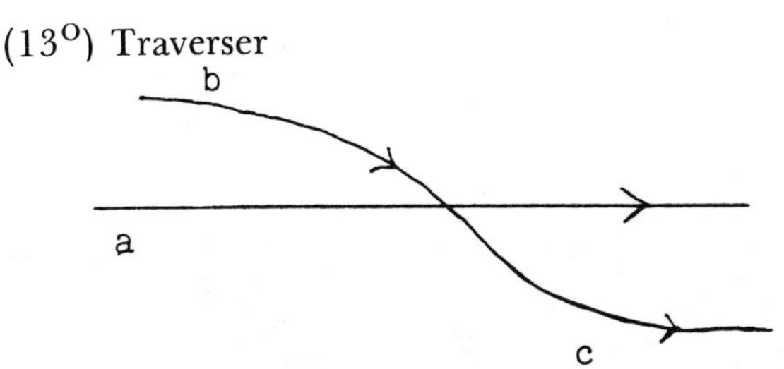

Fig.3.13(a) The topological expression of 'traverser' by Thom

In comparison with the verb 'reprousser' discussed in (12°), the verb 'traverser' is now represented by a sequence of transitory phenomena (i) → (ii) → → (iii) → (iv) of two reverberation cycles which will be denoted by $(a, b) \to a(b) \to a(c) \to (a, c)$ and illustrated by a sequence of figures:

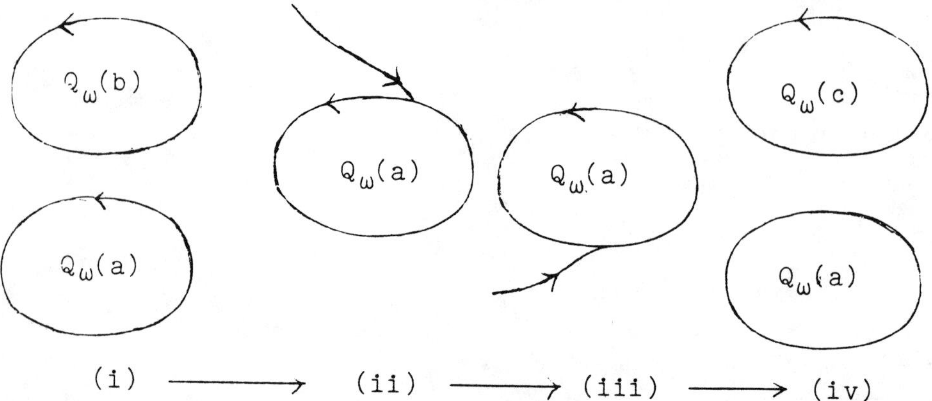

Fig.3.13(b). Interpretation of the verb 'traverser' by means of reverberation cycle phenomena

(14°) Donner

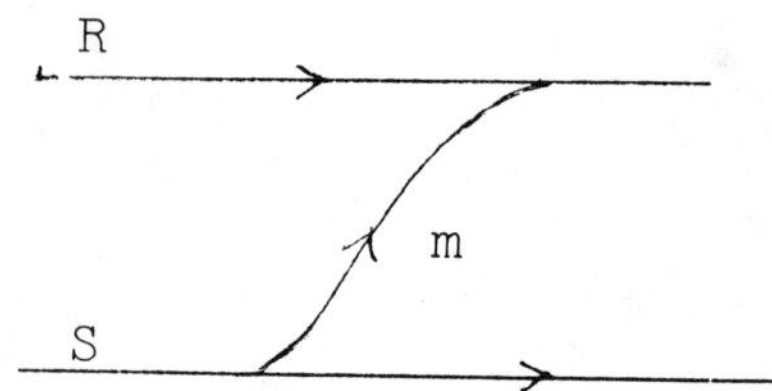

Fig.3.14(a) The topological expression of 'donner' by Thom

The cutting principle is applied to set up three existences S(= sender), R(= receiver) and m(= message). An interpretation of the verb 'donner' in our framework of reverberation cycles can be observed from the illustration of a sequence of transitory phenomena (i) → (ii) → (iii) → (iv) → (v) among three reverberation cycles to the following effect.

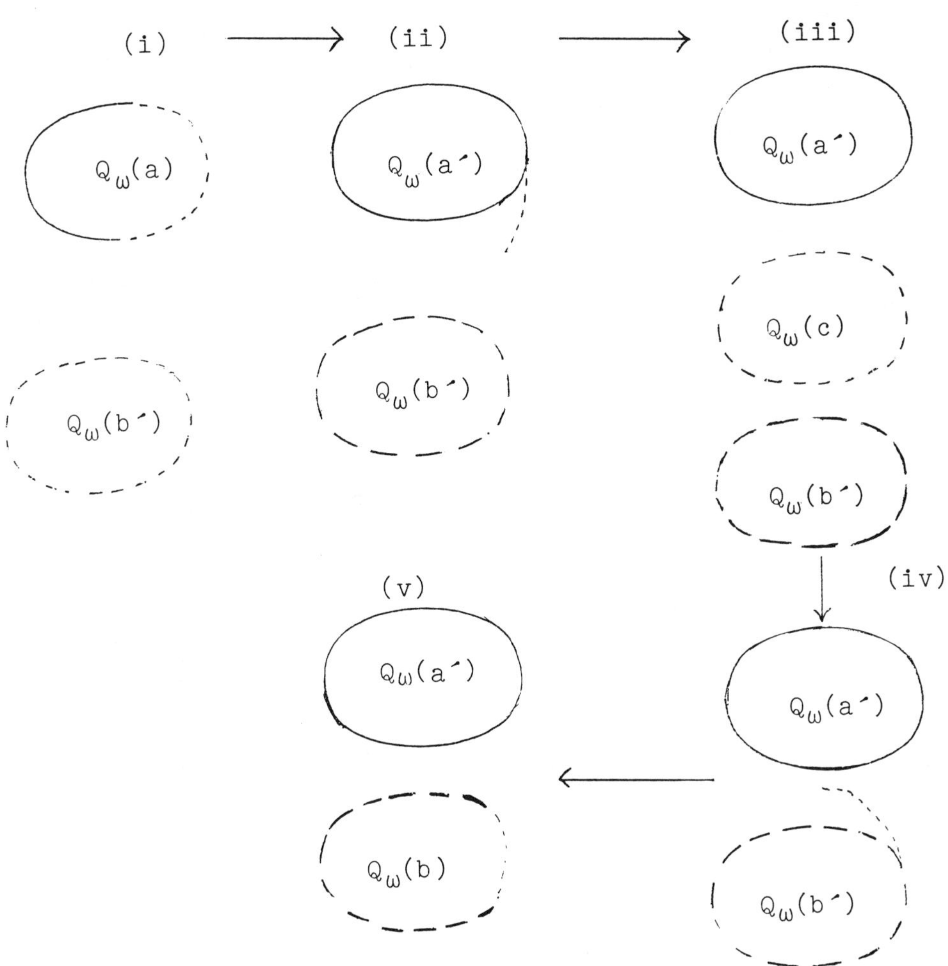

Fig.3.14(b) Interpretation of the verb 'donner' by means of reverberation cycle phenomena

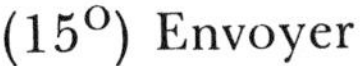

(15°) Envoyer

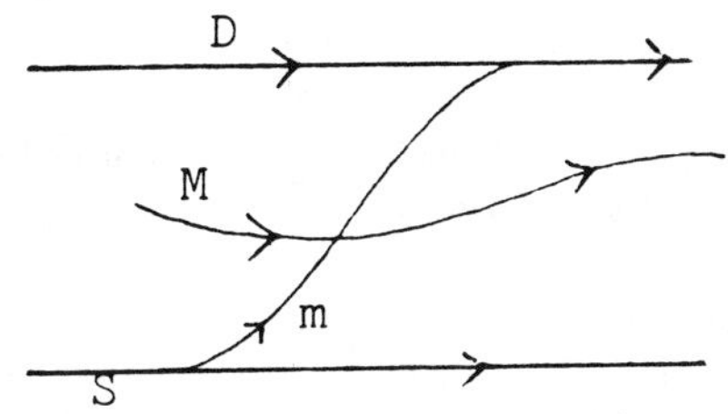

Fig.3.15(a) The topological expression of 'envoyer' by Thom

The verb 'envoyer' in the Thom theory is illustrated by the following French sentence.

'Jacques va de Paris à Strassbourg par le train.'

The correspondence between this sentence and the illustrative figure given by Thom [3-5] is given by

M = le train	m = Jacques
S = Strassbourg	D = Paris

under the interpretation of the figure by a sequence of phenomena among the images M, S, m and D.

(i) M approaches to S.
(ii) S is decomposed so as to emit its component m.
(iii) m is caught by M.
(iv) m is being carried by M and approaches to D.
(v) m becomes separated from M.
(vi) m is caught by D.

Now our interpretation of the verb 'envoyer' within the framework of reverberation cycles is given by a sequence of the transitory phenomena (i) → (vi):

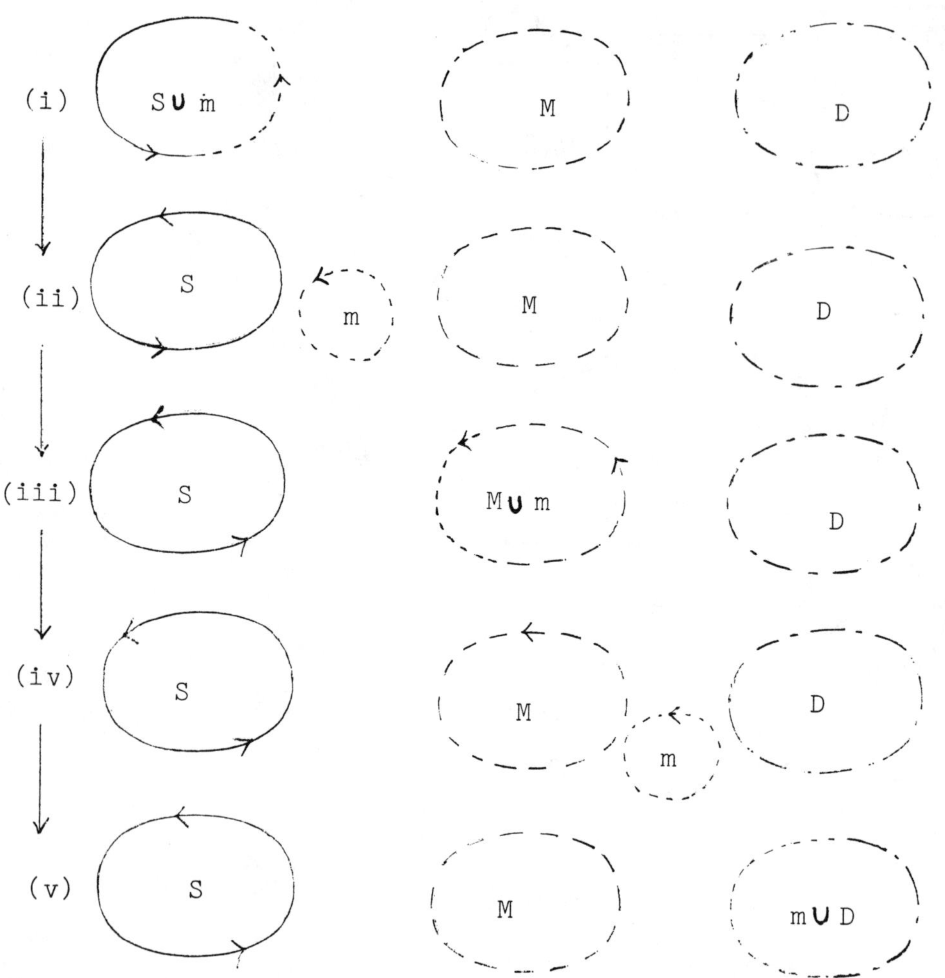

Fig.3.15(b) Interpretation of the verb 'envoyer' by means of reverberation cycle phenomena

(16°) Prendre

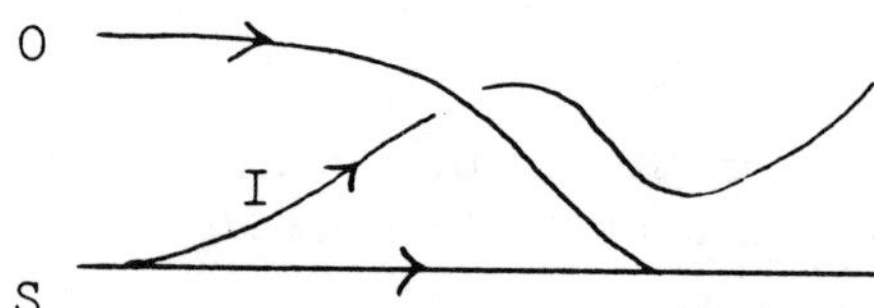

Fig.3.16(a) The topological expression of 'prendre' by Thom

The verb 'prendre' is illustrated in the Thom theory by the following French sentence:

'Jean met son chapeau sur sa tête.'

The correspondence between this sentence and the illustrative figure given by Thom [3-5] is given by

S = the head of Jean I = the hand of Jean
O = the hat of Jean

Under the interpretation of Fig.3.16(a) by a sequence of phenomena among the image the images S, I and O:

(i) S emits I.
(ii) I approaches to O.
(iii) A mixture of O and I approaches to S.
(iv) There arises a configuration in which the mixture of O and I, denoted $O \cup I$, is connected with S.
(v) The configuration obtained in the stage (iv) is unstable, and the connection between O and I is separated so that $S \cup I$ becomes stable and O is separated from $S \cup I$.

Now our interpretation of the verb 'prendre' within the framework of reverberation cycles is given by a sequence of the transitory phenomena

$(i)_1 \rightarrow (i)_2 \rightarrow (ii) \rightarrow (iii) \rightarrow (iv) \rightarrow (v) \rightarrow (vi)$:

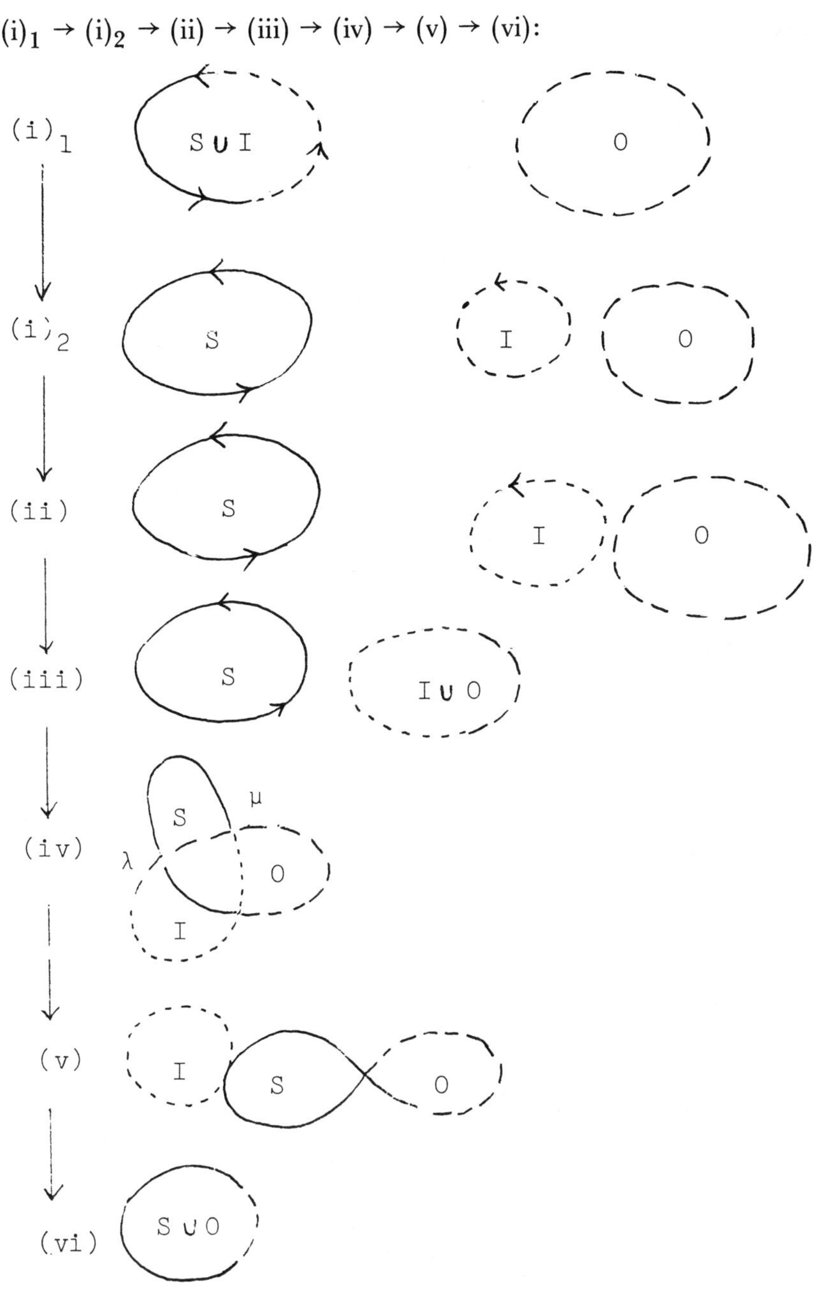

Fig.3.16(b) Interpretation of the verb 'prendre' by means of reverberation cycle phenomena

(17°) Lier

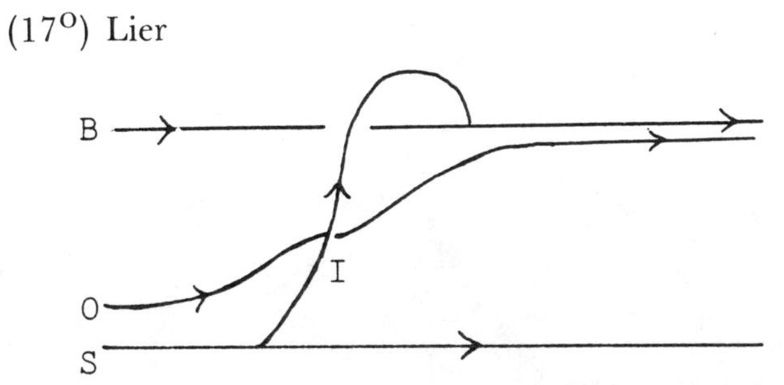

Fig.3.17(a) The topological expression of 'lier' by Thom

The verb 'lier' is illustrated in the Thom theory by the following French sentence.

'Pierre attache sa chevre à l'arbre avec une corde.'

The correspondence between this sentence and the illustrative figure given by Thom [3-5] is given by

S = Pierre, O = sa chèvre
B = l'arbre, I = une corde

under the interpretation of the figure by a sequence of transitory phenomena among the images *S, O, I* and *B*.

(i) *S* emits *I*.
(ii) *I* approaches to *O*.
(iii) A stable combination $I \cup O$ is formed.
(iv) $I \cup O$ is combined with *B*.
(v) A stable combination $I \cup O \cup B$ is formed.

Now our interpretation of the verb 'lier' within the framework of reverberation cycles is given by a sequence of the transitory phenomena $(i)_1 \rightarrow (i)_2 \rightarrow$ $\rightarrow (ii) \rightarrow (iii) \rightarrow (iv) \rightarrow (v)$:

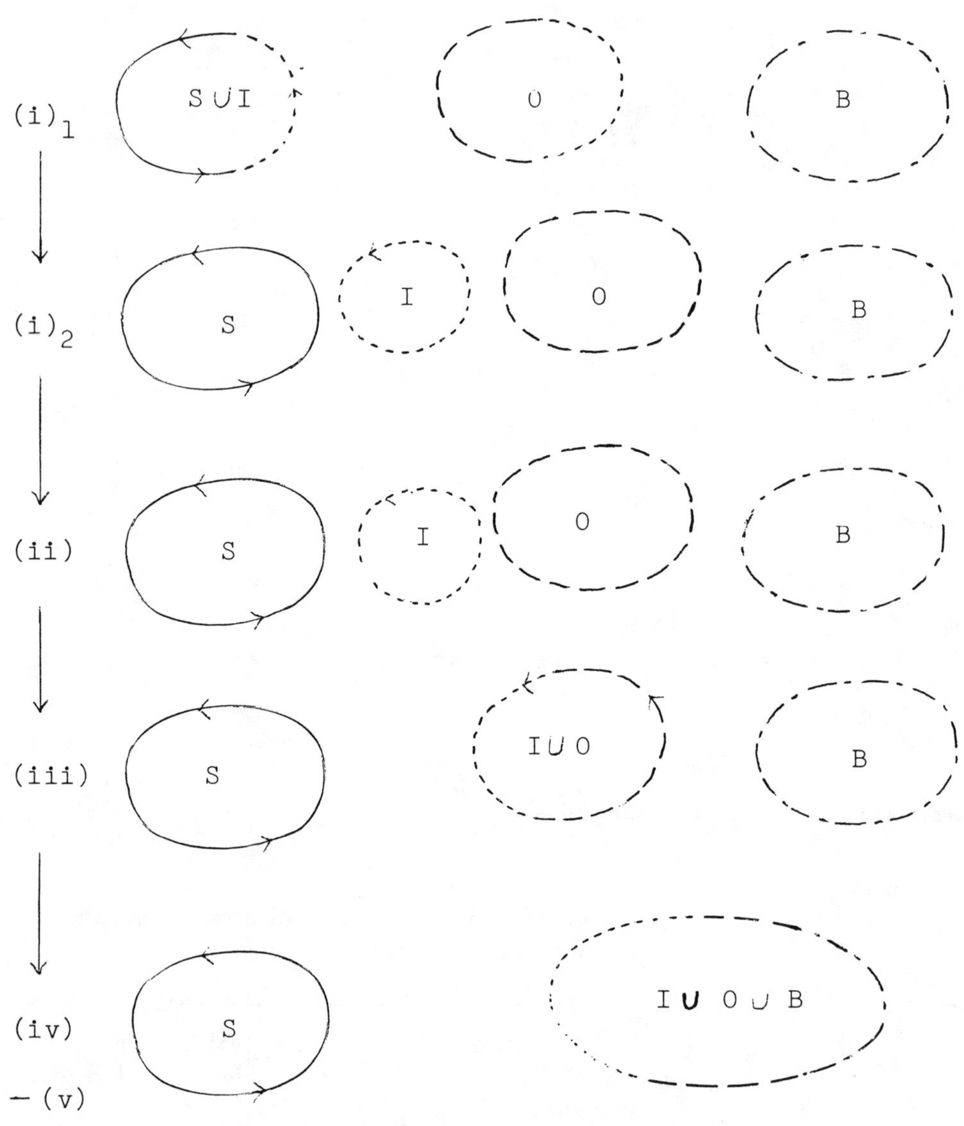

Fig.3.17(b) Interpretation of the verb 'lier' by means of reverberation cycle phenomena

(18°) Couper

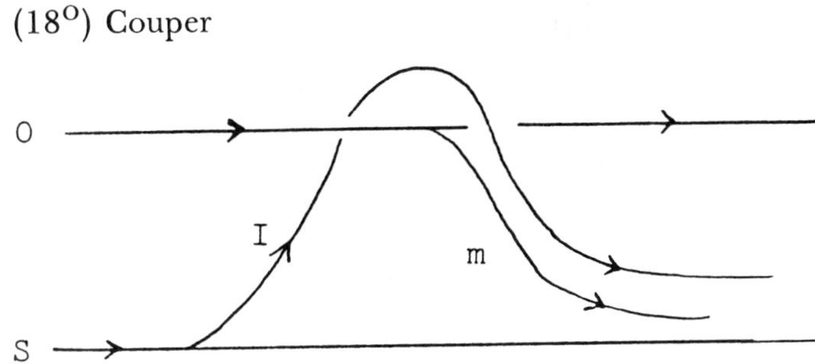

Fig.3.18(a) The topological expression of 'couper' by Thom

The verb 'couper' is illustrated in the Thom theory by the following French sentence.

'Jean m'a extrorque d'argent avec son revolver.'

The correspondence between this sentence and the illustrative figure given by Thom [3-5] is given by

S = Jean, I = son revolver
O = me, m = argent

under the interpretation of the figure by a sequence of phenomena among the images S, I, O and m.

(i) S emits I.
(ii) I approaches to O.
(iii) The approach of I to O emits m from O.
(iv) m is captured by I.
(v) $I \cup m$ is captured by S.

Now our interpretation of the verb 'couper' within the framework of reverberation cycles is given by a sequence of the following transitory phenomena (i) → (ii) → (iii) → (iv) → (v):

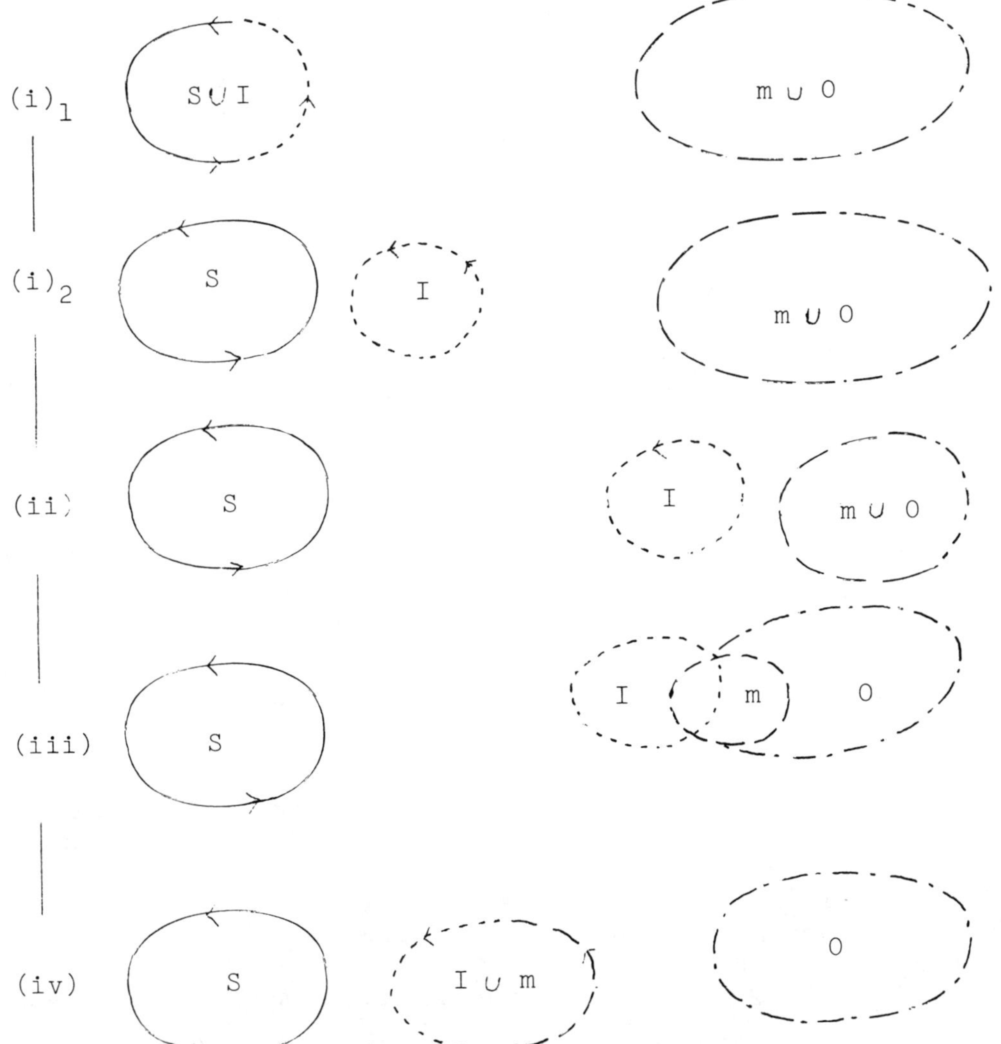

Fig.3.18(b) Interpretation of the verb 'couper' by means of reverberation cycle phenomena

4. Neural network approach to typological analysis in linguistics

Returning to the system of neuronic equations defined in (2.01), let us consider the set Q consisting of 2^{nN} nN-dimensional vectors δ such that

$$\boldsymbol{\delta} = \begin{pmatrix} \delta_{0,1} & \delta_{0,2}\cdots & \delta_{0,j}\cdots & \delta_{0,N} \\ \delta_{1,1} & \delta_{1,2}\cdots & \delta_{1,j}\cdots & \delta_{1,N} \\ \cdots & \cdots & \cdots & \\ \delta_{i,1} & \delta_{i,2}\cdots & \delta_{i,j}\cdots & \delta_{i,N} \\ \cdots & \cdots & \cdots & \\ \delta_{n-1,1} & \delta_{n-1,2}\cdots & \delta_{n-1,j}\cdots & \delta_{n-1,N} \end{pmatrix}$$

where each δ_{ij} is either 1 or 0 for $i=0,1,...,n-1$ and $j=1,2,...,N$.

The system of neuronic equations (2.01) induces the transition

$$L : \boldsymbol{\delta} \rightarrow L\{\boldsymbol{\delta}\},$$

where

$$L\{\boldsymbol{\delta}\} = \begin{pmatrix} g_1(\boldsymbol{\delta}) & g_2(\boldsymbol{\delta})\cdots & g_j(\boldsymbol{\delta})\cdots & g_N(\boldsymbol{\delta}) \\ \delta_{0,1} & \delta_{0,2}\cdots & \delta_{0,j}\cdots & \delta_{0,N} \\ \delta_{1,1} & \delta_{1,2}\cdots & \delta_{1,j}\cdots & \delta_{1,N} \\ \cdots & \cdots & \cdots & \\ \delta_{i,1} & \delta_{i,2}\cdots & \delta_{i,j}\cdots & \delta_{i,N} \\ \cdots & \cdots & \cdots & \\ \delta_{n-2,1} & \delta_{n-2,2}\cdots & \delta_{n-2,j} & \delta_{n-2,n} \end{pmatrix}$$

$$\equiv (L\{\boldsymbol{\delta}\}_1, \quad L\{\boldsymbol{\delta}\}_2, \ldots \quad L\{\boldsymbol{\delta}\}_j, \ldots \quad L\{\boldsymbol{\delta}\}_N)$$

and each $gj(\boldsymbol{\delta})$ $(j=1,2,...,N)$ may take either 1 or 0 respectively.

The set of all the possible combinations of N values $\{g_j(\boldsymbol{\delta})\}$ $(j=1,2,...,N)$ can be arranged in an array of the ascending order from 1 to 2^N with the inverse lexical ordering denoted by $\{\sigma_i\}$ $(i=1,2,...,2^N)$ such that

$$\left.\begin{array}{l} \sigma_1 \;;\; (0, 0, 0, ..., 0, ..., 0) \\ \sigma_2 \;;\; (1, 0, 0, ..., 0, ..., 0) \\ \sigma_3 \;;\; (0, 1, 0, ..., 0, ..., 0) \\ \sigma_4 \;;\; (1, 1, 0, ..., 0, ..., 0) \\ \sigma_5 \;;\; (0, 0, 1, ..., 0, ..., 0) \\ \sigma_6 \;;\; (1, 0, 1, ..., 0, ..., 0) \\ \sigma_7 \;;\; (1, 0, 1, ..., 0, ..., 0) \\ \sigma_8 \;;\; (1, 1, 1, ..., 0, ..., 0) \\ \cdots\cdots \\ \sigma_{2}N\text{-}1: (1, 1, 1, ..., 1, 0) \\ \sigma_{2}N\text{-}1_{+1}: (0, 0, 0, ..., 0, 1) \\ \sigma_{2}N\text{-}1_{+2}: (1,0, 0, ..., 0, 1) \\ \sigma_{2}N\text{-}1_{+3}: (0, 1, 0, ..., 0, 1) \\ \sigma_{2}N\text{-}1_{+4}: (1, 1, 0, ..., 0, 1) \\ \cdots\cdots \\ \sigma_{2^N-2} : (1, 0, 1, ..., 1, 1) \\ \sigma_{2^N-1} : (0, 1, 1, ..., 1, 1) \\ \sigma_{2^N} \;\; : (1, 1, 1, ..., 1, 1) \end{array}\right\} \quad (4.04)$$

Let us denote the set of all these σ_k $(k=1,2,...,2^N)$ by Σ:

$$\Sigma \equiv \{\sigma_k \mid k=1,2,...,2^N\} \quad (4.05)$$

In a similar way we may and we shall give an inverse lexical ordering of N-dimensional component vector $\boldsymbol{\delta}j$, which is defined by

$$\boldsymbol{\delta}_j = (\boldsymbol{\delta}_{0,j}, \boldsymbol{\delta}_{1,j}, \ldots \boldsymbol{\delta}_{n-1,j}) \quad (4.06)$$

for $j=1,2,...,N$. The set of all the possible vectors for $\boldsymbol{\delta}j$ whose size is equal to 2^n for each fixed j can be arranged in an inverse lexical ordering from 1 to 2^n, starting with $(0, 0, 0, .., 0)$ and ending with $(1, 1, 1, .., 1)$. In the consequence to each $\boldsymbol{\delta} = (\boldsymbol{\delta}_1, \boldsymbol{\delta}_2, ..., \boldsymbol{\delta}_N)$ there is one and only one corresponding form such that

$$\mathbb{P}(\delta) = (p(\boldsymbol{\delta}_1), p(\boldsymbol{\delta}_2), \ldots, p(\boldsymbol{\delta}_n)), \quad (4.07)$$

where $p(\boldsymbol{\delta}_i)$ ranges from 1 to 2^n for each $i=(1,2,...,N)$.

Now let us make use of some terminologies currently adopted in automaton theory. (See Eilenberg [21] and also Salomaa [20])

(i) States: Each $\boldsymbol{\delta}$ of Q is called state.

(ii) Initial states: a subset I of Q whose elements are called initial.

(iii) Terminal states: a subset I of Q whose elements are called terminal.

(iv) Edges: a subset E of $Q \times \Sigma \times Q$ whose element is a triplet $(\boldsymbol{\delta}, \sigma, \boldsymbol{\delta}')$ which begins at $\boldsymbol{\delta}$, ends at $\boldsymbol{\delta}'$, and carries the label σ, and is denoted by $\sigma : \boldsymbol{\delta} \to \boldsymbol{\delta}'$. An automaton A over a finite alphabet Σ consists of these four data (i)-(iv). A path c in A is a finite sequence

$$c = (\boldsymbol{\delta}^0, \sigma_1, \boldsymbol{\delta}^1)(\boldsymbol{\delta}^1, \sigma_2, \boldsymbol{\delta}^2) \dots (\boldsymbol{\delta}^{k-1}, \sigma_k, \boldsymbol{\delta}^k) \tag{4.08}$$

of consecutive edges. An abbreviated notation for a path is written as

$$c = (\boldsymbol{\delta}^0, \sigma_1, \sigma_2, \dots, \sigma_k, \boldsymbol{\delta}^k) \tag{4.09}$$

which is also written as

$$\boldsymbol{\delta}^0 \xrightarrow{\sigma_1} \boldsymbol{\delta}^1 \xrightarrow{\sigma_2} \boldsymbol{\delta}^2 \longrightarrow \cdots\cdots \to \boldsymbol{\delta}^{k-1} \xrightarrow{\sigma_k} \boldsymbol{\delta}^k \tag{4.10}$$

or simply $\boldsymbol{\delta}^0 \overset{c}{\to} \boldsymbol{\delta}^k$ or $c: \boldsymbol{\delta}^0 \to \boldsymbol{\delta}^k$.

The element $s = \sigma_0 \sigma_1 \dots \sigma_k \in \Sigma^*$ is called the label of C and is denoted by $|C|$. A path $C: i \to t$ with $i \in I$, $t \in T$ is called successful. The labels carried by the successful paths in A form a subset $|A|$ of Σ^* called the behavior of A. A subset A of Σ^* is said to be recognizable if there exists a Σ-automaton A such that $|A| = A$.

These terminologies just explained are being currently used in the automaton theory of the present time. Now there are needs for introducing a few important modifications of the standpoints in such a way that we can adequately make use of these terminologies in solving the problems represented in Section 3.

The problem in front of us is to find out transitory phenomena of reverberation cycles being enunciated in Section 3 in connection with our interpretation of fundamental verbs within the general framework of neural dynamics defined by systems of N neuronic equations with time lag $(n-1)$.

For this purpose there are a lot of freedoms to which we may appeal within the general framework of neural dynamics. It will be convenient to classify our situations into the following three cases according to the degree of freedom in this connection.

(1°) Case I. This is the case when the matrix A of inner connectivity coefficients are being fixed while the outer vector time function $\Theta(t)$ can be chosen from a certain broad class of functions.

(2°) Case II. This is the case when the matrix A of inner connectivity coefficients can be chosen from a certain broad class of matrices, while the outer vector time function $\Theta(t)$ is being fixed.

(3°) Case III. This is the case when both the matrix A and the time function $\Theta(t)$ can be chosen from a certain broad class of matrices and that of time functions respectively.

In view of these assumptions we think it worth while to investigate whether and how we can find each specific set of conditions for A and $\Theta(t)$ under which transitory phenomena among reverberation cycles corresponding to each fundamental verb given in Section 3 can be realized within the framework of our neural dynamical systems. It is quite natural that we may be interested in finding the set of the conditions on A and $\Theta(t)$ with the preference order of the pairs of N and n as small as possible. In order to proceed to such an investigation, more extensive studies should have been performed which can provide us with an extensive stock of digraph representation of the solutions of a system of N neuronic equations with time lag $(n-1)$. So far as the case when $N=1$ and $n=4$ is concerned, there is no theoretical difficulty to give all the possible digraphs among which 104 digraphs are given in the Appendix of our paper [15]. Before we propose to set up a methodology how to appeal to these three cases in order to make an approach to the problem in front of us, let us start with the following two preparatory considerations.

1. Observations on digraph representation of dynamical behaviors of solutions of systems of N neuronic equations with time lag $(n-1)$.

Let us remember here some of the results which we have obtained in a series of our papers Kitagawa [10, 12-15], in which we have given various examples of digraph representation of dynamical behaviors of the solutions of a system of N neuronic equations with time lag $(n-1)$. Besides general results concerning reverberation cycles and their disintegrations for specific types of allied operators given in our papers [10] and [12], exhaustive studies have been given merely for the cases (i) $N=1$, $n=4$ and (ii) $N=1$, $n=5$, as shown in Kitagawa [14] and [15] respectively.

Example 4.1. The following nine Figures 4.1

(a)-(f) are given to show an illustration of how the verb 'unir' can be realized among six reverberation cycles given in Figure 4.1 through the change of one letter σ (or τ) to another letter τ (or σ) at two mutually conjugate states respectively.

Figure 4.1(a) gives all the possible transitions among 2^4 states induced by a single neuronic equation with time lag 3 corresponding to the special case of (2.01) when $N=1$ and $n=4$ which was systematically treated in our previous paper [14]. In fact Figure 4.1(a) is the same as Fig.6.5 in our previous paper [14] apart from the indication of letters σ and τ in our present context referring to automaton theory. Figure 4.1(b) gives six reverberation cycles in the digraph representation of the operator $\mathcal{L}\beta(\bar{3})$ (= $\mathcal{L}\bar{\beta}(3)$) which corresponds to the case 9[$\bar{3}$] discussed in our paper [14], Section 7.

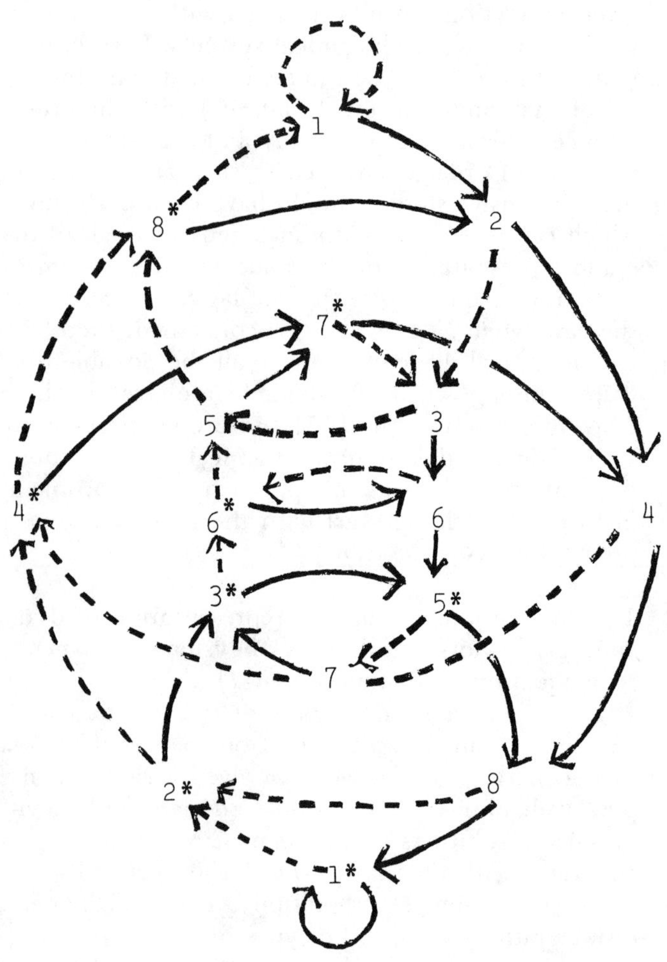

Fig.4.1(a) All the possible transitions among 2^4 states induced by a single neuronic equation with time lag 3

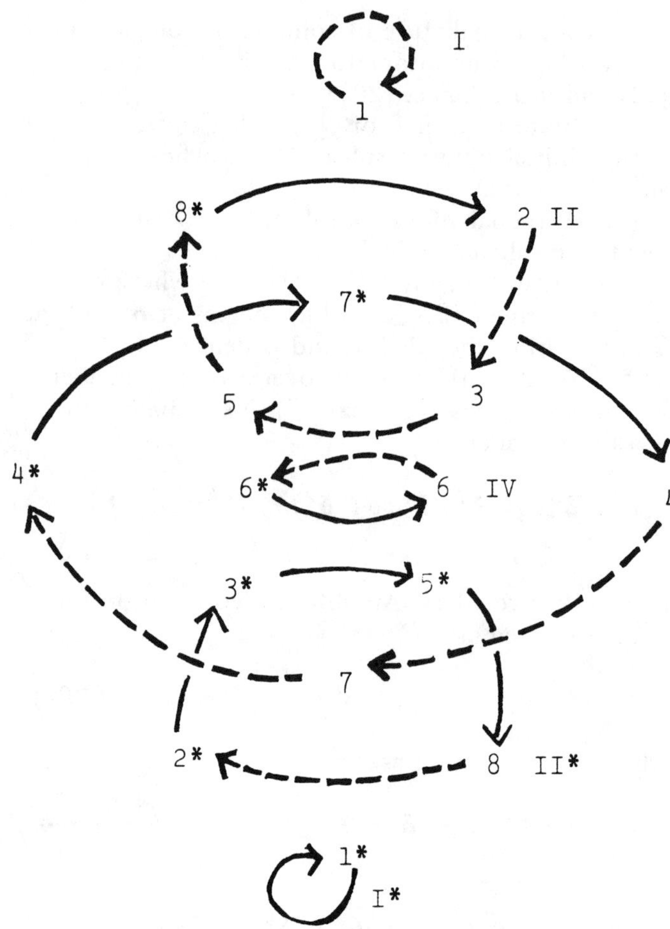

Fig.4.1(b) Digraph associated with L_9

In Fig.4.1(b) there are six reverberation cycles which we denote by I, II, III, IV, II*, and I* in the following way:

$$\text{I} = \{1\}, \quad \text{II} = \{2,3,5,8^*\}, \quad \text{III} = \{4,7,4^*,7^*\},$$

$$\text{IV} = \{6,6^*\}, \quad \text{I}^* = \{1^*\}, \quad \text{II}^* = \{2^*,3^*,5^*,8\}.$$

The verb 'unir' can be observed in Fig.4.1(c)-4.1(f).

Similar arrangements do not hold for the consolidation of two reverberation cycles II and III. There are various approaches which can be recognized to present the function of the verb 'unir' with respect to these two reverberation cycles II and III. One of them is a partial consolidation of the two reverberation cycles II and III in which some parts of II and III are joined together while the remaining other ones will not belong to the deep structure.

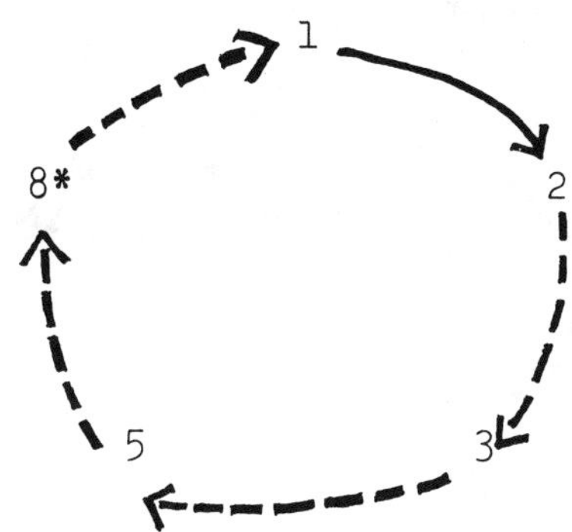

Fig.4.1(c) Consolidation of the reverberation cycles I and II representing the verb 'unir'

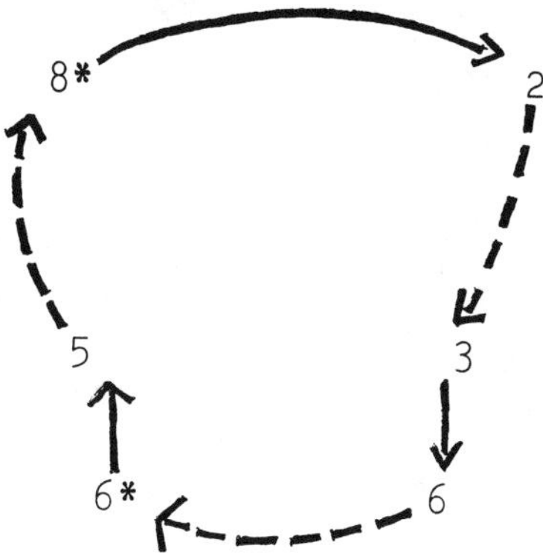

Fig.4.1(d) Consolidation of the reverberation cycles II and IV representing the verb 'unir'

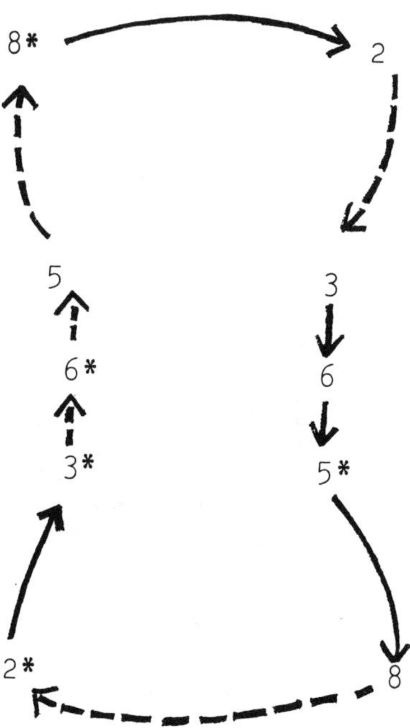

Fig.4.1(e) Consolidation of the reverberation cycles II, IV and II* representing the verb 'unir'

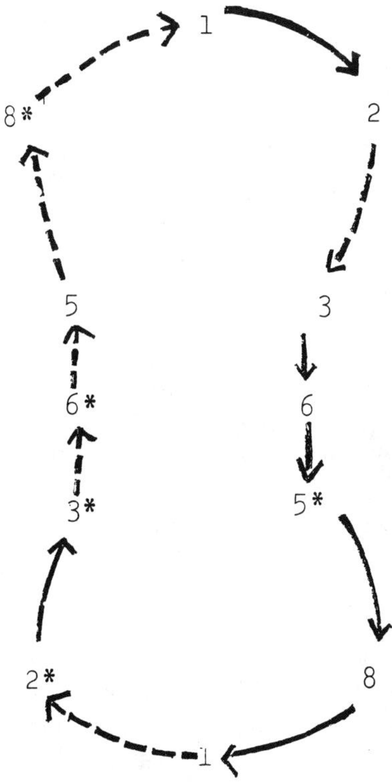

Fig.4.1(f) Consolidation of the reverberation cycles I, II, IV, II* and I* representing the verb 'unir'

An example of such a partial consolidation is given in Fig.4.1(*g*). Another approach is to make use of the complicated path which will pass through the same nodes more than once, as shown in Fig.4.1(*h*).

2. Dynamics of reverberation cycles. In spite of the superficial appearance of automaton theoretical formulation of neural network dynamics which we have proposed in this Section, there is a distinction between the current automaton theory and what we want to realize through an establishment of neural network dynamics in terms of automation oriented expressions.

First of all, in view of our idea explained in Section 3, it is natural to investigate how to construct reverberation cycles which should yield the

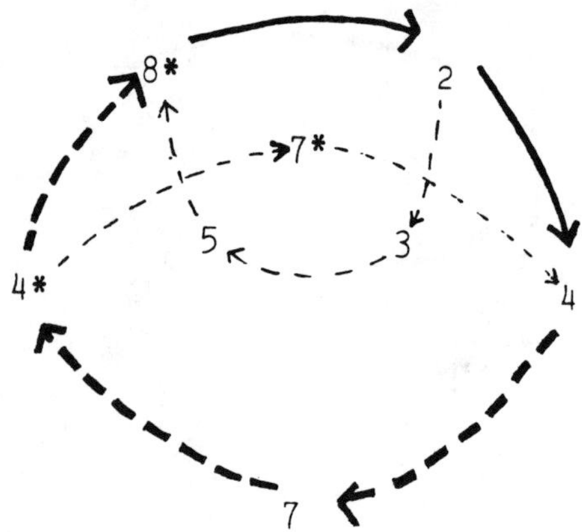

Fig.4.1(g) Partial consolidation of two reverberation cycles II and III

C_1: $2 \rightarrow 4 \rightarrow 7 \rightarrow 4^* \rightarrow 8^* \dashrightarrow 2^*$

(2474^*8^*2)

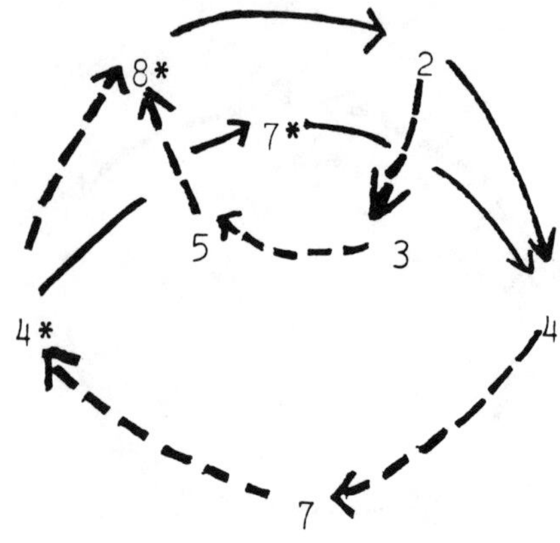

Fig.4.1(h) Complete consolidation of two reverberation cycles II and III

C_2: $((2358^*2)(474^*7^*4)$
$74^*8^*2)$

basis of the transitory phenomena corresponding to the fundamental verbs.

In the second place we have to seek for the neural content with some illustrative examples which will serve to enable us to understand the implication of our problem.

Example 4.2. The neural dynamics associated with *N=1* and *n=3*. The set of all possible states consists of the following eight with their respective inverse lexical orderings $Q = \{1,2,3,4,5,6,7,8\}$ where

$$\left.\begin{matrix} 1 = 0\,0\,0 & 3 = 0\,1\,0 & 5 = 0\,0\,1 & 7 = 0\,1\,1 \\ 2 = 1\,0\,0 & 4 = 1\,1\,0 & 6 = 1\,0\,1 & 8 = 1\,1\,1 \end{matrix}\right\} \tag{4.11}$$

We have the alphabet $\Sigma = \{\sigma.\ \tau\}$. The totality of edges are given in Fig.4.2. It is interesting to observe that various sorts of reverberation cycles can be obtained from the languages of the form

$$L = \{(\varphi(\sigma,\tau))^k \mid k \geqslant 1\} \tag{4.12}$$

where

$$\varphi(\sigma,\tau) = \sigma^{a_1}\tau^{b_1}\sigma^{a_2}\tau^{b_2} \ldots \sigma^{a_r}\tau^{b_r} \tag{4.13}$$

with any assigned initial state *x*. Figures 4.3-4.14 are given for illustration.

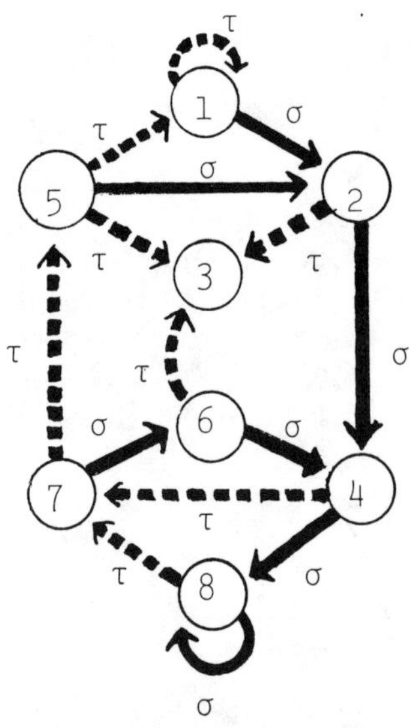

Fig.4.2 2^3 state configuration with $\Sigma = \{\sigma, \tau\}$

(1°) $L_1 = \{(\sigma\tau)^k \mid k \geq 1\}$

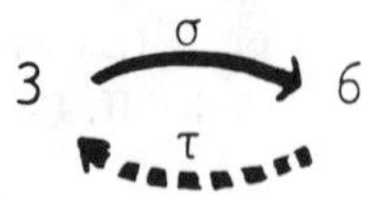

Fig.4.3 L_1

(2°) $L_2 = \{(\sigma^2\tau^2)^k \mid k \geq 1\}$

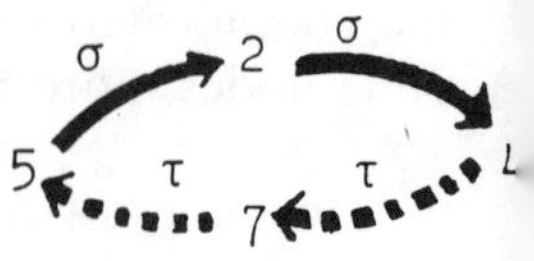

Fig.4.4 L_2

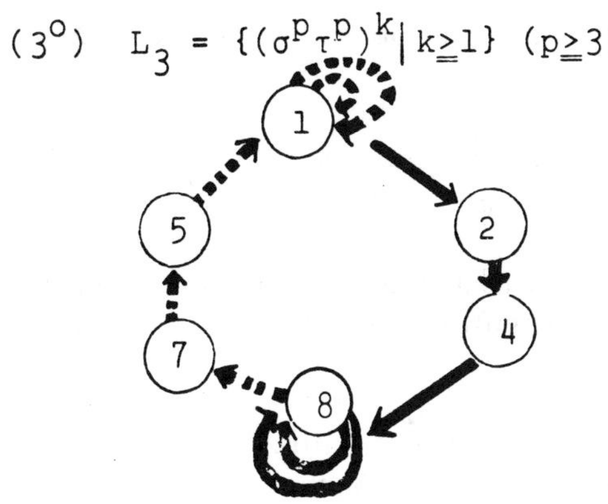

Fig.4.5 L_3

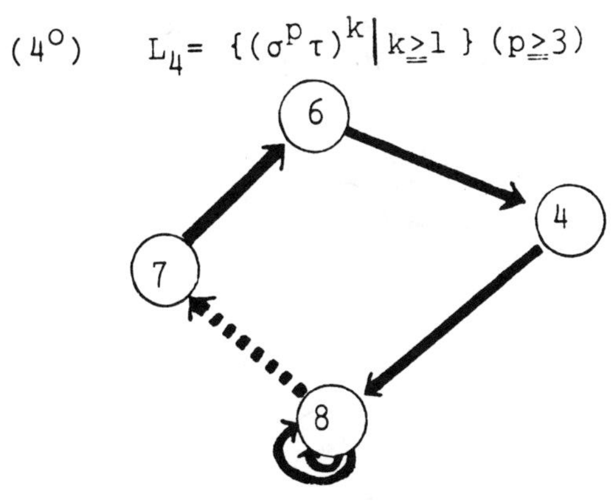

Fig.4.6 L_4

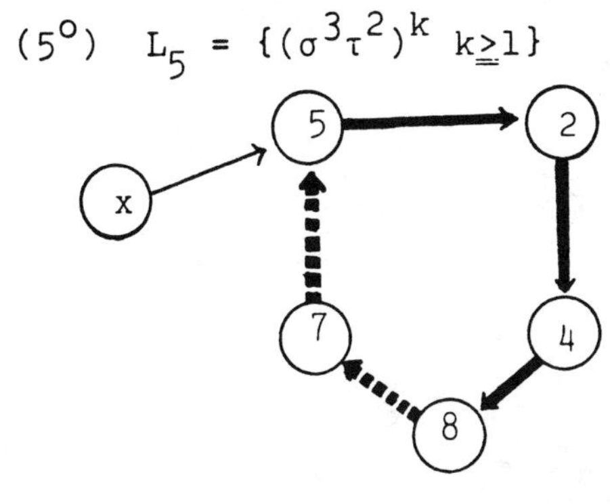

Fig.4.7 L_3

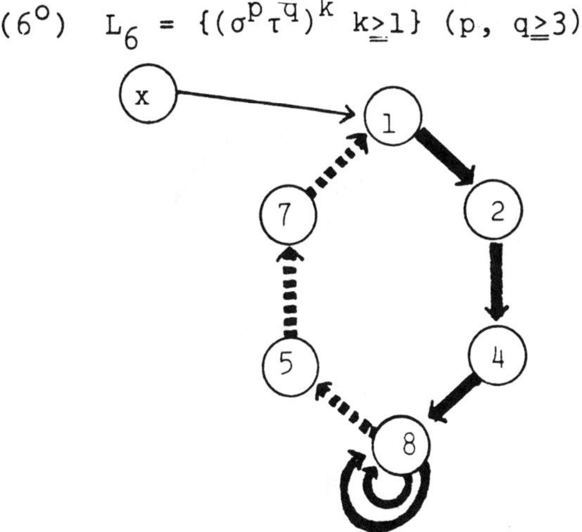

Fig.4.8 L_6

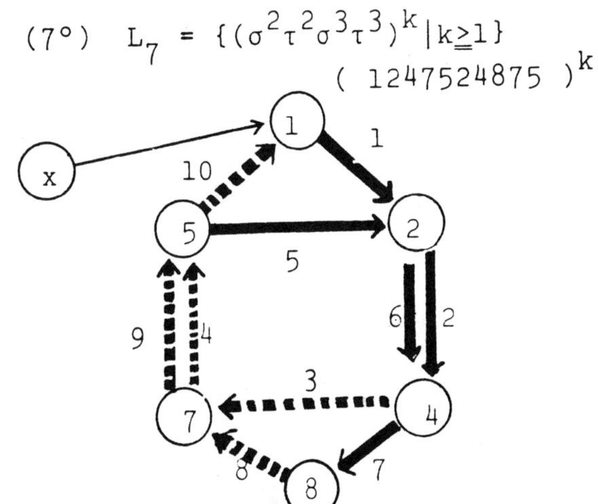

Fig.4.9 L_7

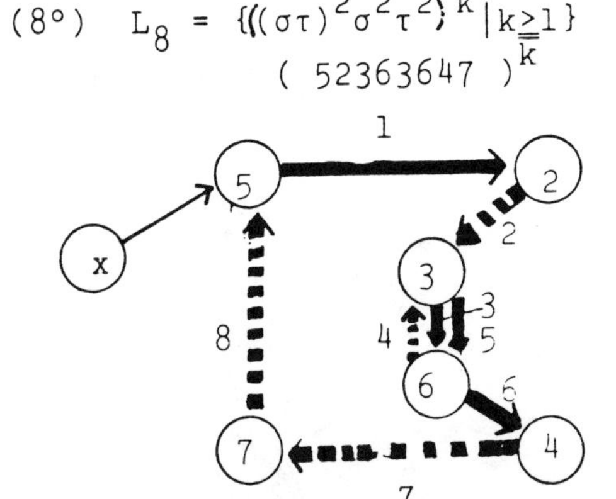

Fig.4.10 L_8

(9°) $L_9 = \{(\sigma\tau\sigma)^k \mid k \geq 1\}$

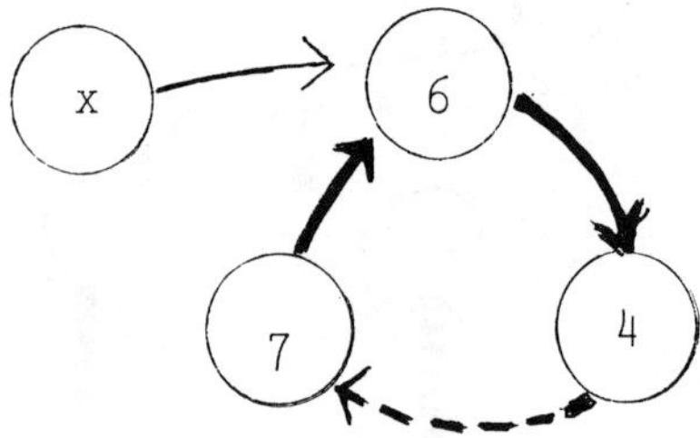

Fig.4.11 L_9

(10°) $L_{10} = \{(\sigma\tau^2\sigma)^k \mid k \geq 1\}$

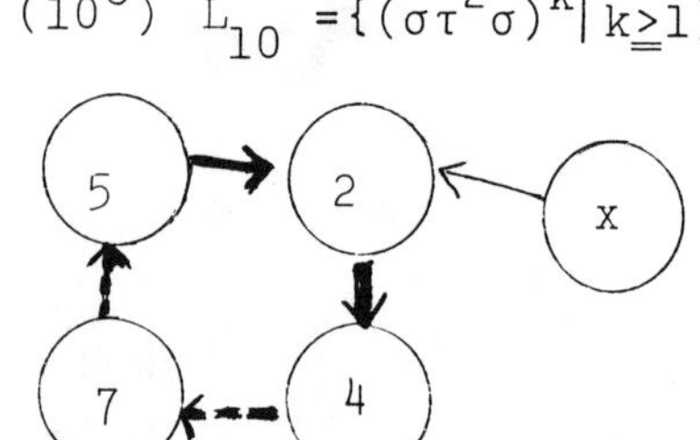

Fig.4.12 L_{10}

(11°) $L_{11} = \{(\sigma\tau^3\sigma)^k \mid k \geq 1\}$

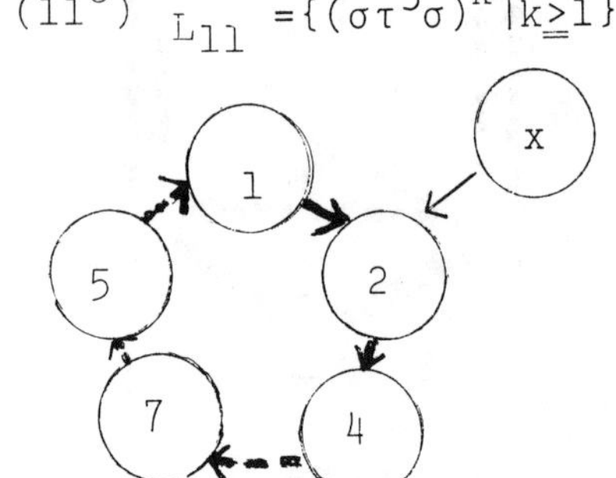

Fig.4.13 L_{11}

(12°) $L_{12} = \{(\sigma^2\tau\sigma)^k \mid k \geq 1\}$

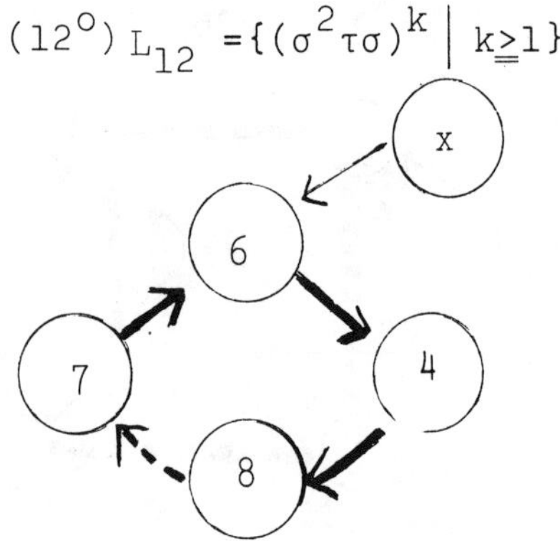

Fig.4.14 L_{12}

A neuronic equation with $N=1$ and $n=3$ is given by

$$x(t+1) = 1[\sum_{k=0}^{3-1} a_k x(t-k) - \Theta(t)] \tag{4.12}$$

It is noted that σ and τ can occur when the value $\Theta(t)$ is sufficiently small and when it is sufficiently large respectively, irrespective of the coefficient matrix $A=(a_0, a_1, a_2)'$. The implication of the independence of σ and τ with A is important. The automata having the alphabet $\Sigma = \{\sigma, \tau\}$ can be realized for any network so far as the outside input $\Theta(t)$ is sufficiently large in its absolute value and violates between positive and negative sides in a certain way.

Example 4.3. The neural dynamics associated with $N=1$, $n=3$ and $\Sigma = \{\sigma, \tau, \rho\}$. The neuronic equation is the same with Example 4.2, but we have now three letters σ, τ and ρ to the following effect as shown in Fig.4.15(*a*)-(*c*).

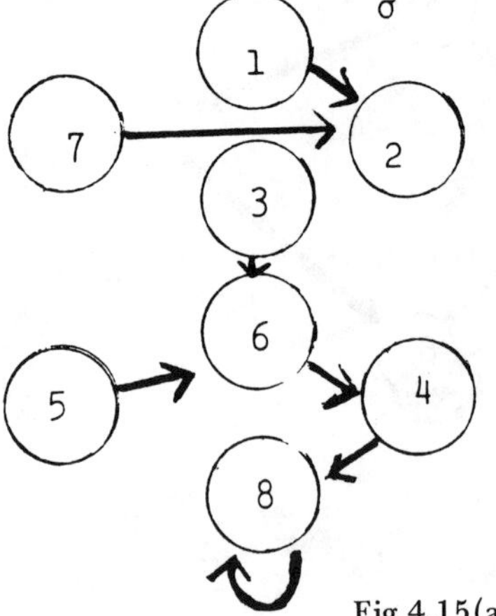

Fig.4.15(a) The letter σ in 2^3 state configurations

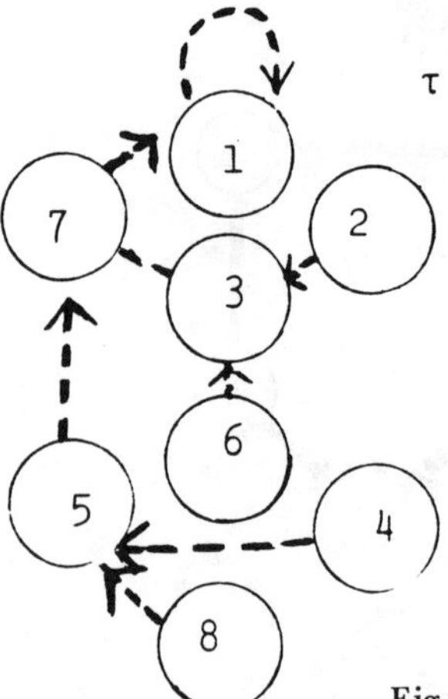

Fig.4.15(b) The letter τ in 2^3 state configurations

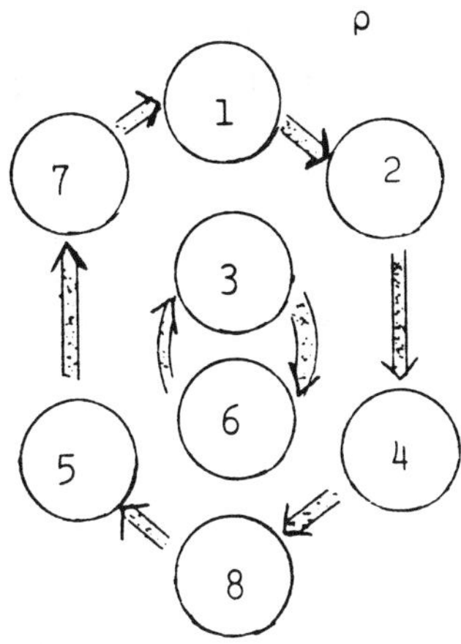

Fig.4.15(c) The letter ρ in 2^3 state configurations

It is now our concern to observe reverberation cycles generated by the languages of the form

$$L = \{\varphi(\sigma,\tau,\rho)^k \mid k \geqslant 1\} \tag{4.13}$$

where

$$\varphi(\sigma,\tau,\rho) = \sigma^{a_1}\tau^{b_1}\rho^{c_1}\,\sigma^{a_2}\tau^{b_2}\rho^{c_2}\,\ldots\,\sigma^{a_r}\tau^{b_r}\rho^{c_r} \tag{4.14}$$

We are content here with giving the following three illustrations. (Figs 4.16-4.18)

$(13^{\circ})\ L_{13} = \{(\tau\rho\sigma)^{2k} \mid k \geq 1\}$

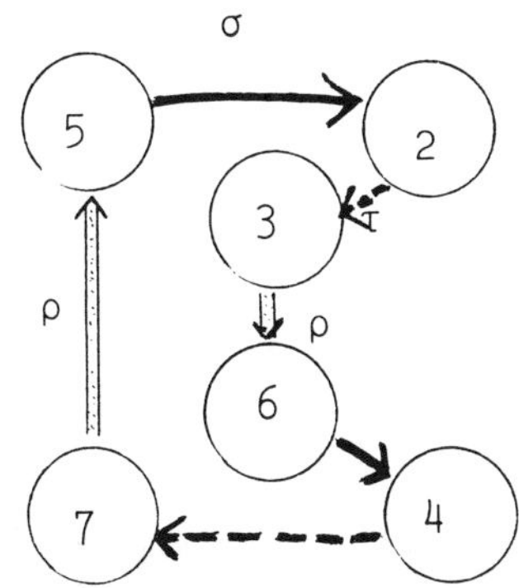

Fig.4.16 L_{13}

$(14^{\circ})\ L_{14} = \{(\tau^2\rho^2\sigma^2)^k \mid k \geq 1\}$

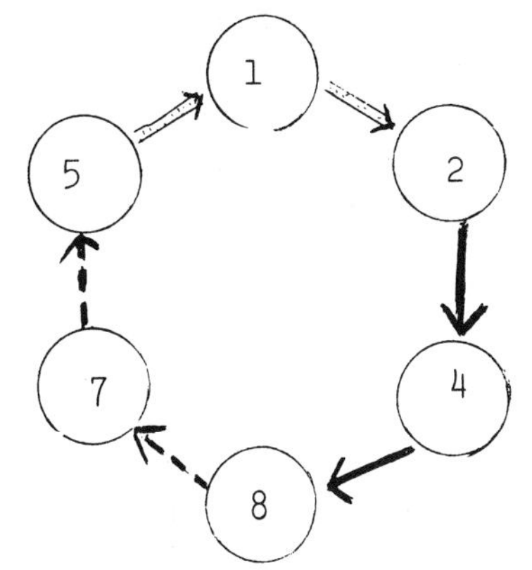

Fig.4.17 L_{14}

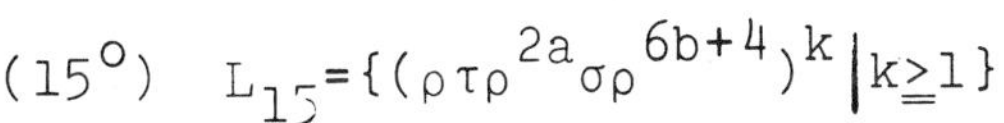

$(15^{\circ})\ L_{15} = \{(\rho\tau\rho^{2a}\sigma\rho^{6b+4})^k \mid k \geq 1\}$

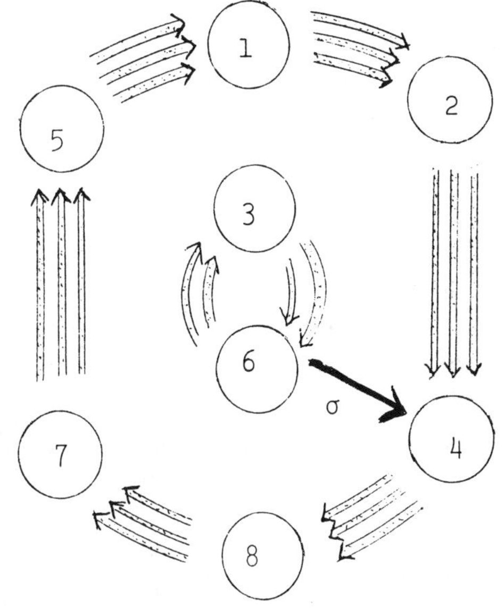

Fig.4.18 L_{15} with a = 2 and b = 3

Such neural dynamics can be realized for the neuronic equation (4.12) with $N=1$ and $n=3$ when the set of three coefficients $\{a_0, a_1, a_2\}$ satisfies a certain condition and $\Theta(t)$ remains within a certain interval such that $L\{\boldsymbol{\delta}\}$ is equivalent to operator $L_{\beta(2)}$ in the sense enunciated in our paper Kitagawa [12,14,15]. It implies that our neural dynamics can only be realized for a certain set of the neuronic equation with $N=1$ and $n=3$ with a certain type of outside input function $\Theta(t)$.

Example 4.4. The neural dynamics associated with $N=2$, $n=3$, and $\Sigma=\{\sigma,\rho,\lambda,\tau\}$. A system of two neuronic equations with two neurons and time lag 3 is given by

$$x_i(t+1) = 1[\sum_{j=1}^{2}\sum_{k=0}^{2} a_{ij}\, x_j(t-h) - \Theta_i(t)] \quad (i=1,2) \tag{4.15}$$

The letters are defined by

$$\nu\sigma = \begin{pmatrix} 1 & 1 \\ x_2 & y_2 \\ x_1 & y_1 \end{pmatrix}, \quad \nu\rho = \begin{pmatrix} 1 & 0 \\ x_2 & y_2 \\ x_1 & y_1 \end{pmatrix}$$
$$\nu\lambda = \begin{pmatrix} 0 & 1 \\ x_2 & y_2 \\ x_1 & y_1 \end{pmatrix}, \quad \nu\tau = \begin{pmatrix} 0 & 0 \\ x_2 & y_2 \\ x_1 & y_1 \end{pmatrix} \tag{4.16}$$

for any

$$\nu = \begin{pmatrix} x_2 & y_2 \\ x_1 & y_1 \\ x_0 & y_0 \end{pmatrix} = (x\ y)$$

where each of x_i and y_i $(i=0,1,2)$ is either 1 or 0. Each of x and y has the 8 possible states which we shall arrange in the inverse lexical ordering.

Consequently all the possible states of ν can now be denoted by (h, k) where $h, k=1,2,...,8$.

The four examples are given as shown in Figs 4.19-4.22 respectively.

(16°) $L_{16} = \{(\sigma\rho\lambda\tau)^k | k\geq 1\}$

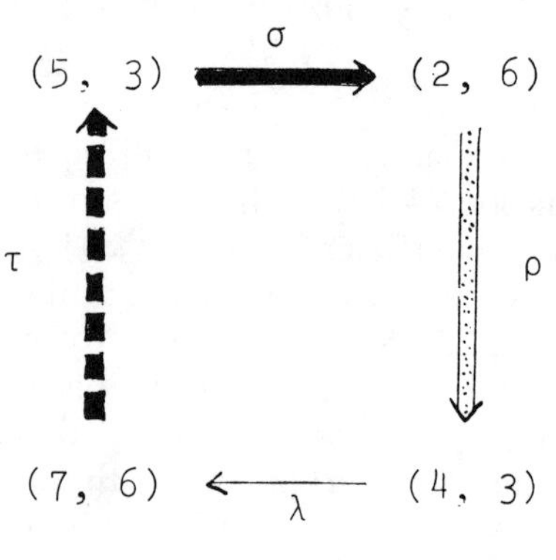

(5, 3) —σ→ (2, 6)

τ ↑   ↓ ρ

(7, 6) ←λ— (4, 3)

Fig.4.19 L_{16}

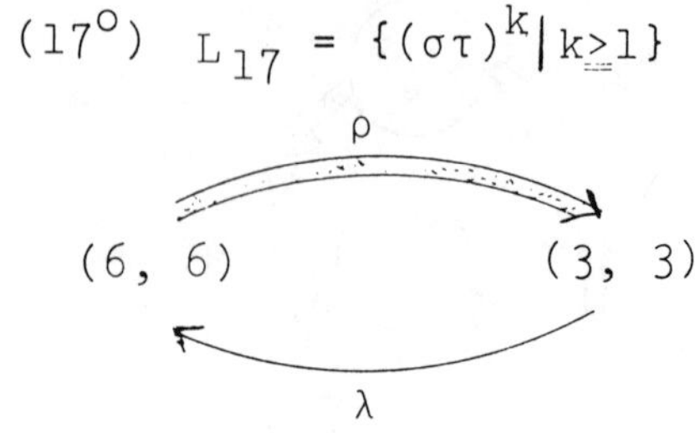

Fig.4.20 L_{17}

(18°) $L_{18} = \{(\sigma^2\rho^2\lambda^2\tau^2)^k | k\geq 1\}$

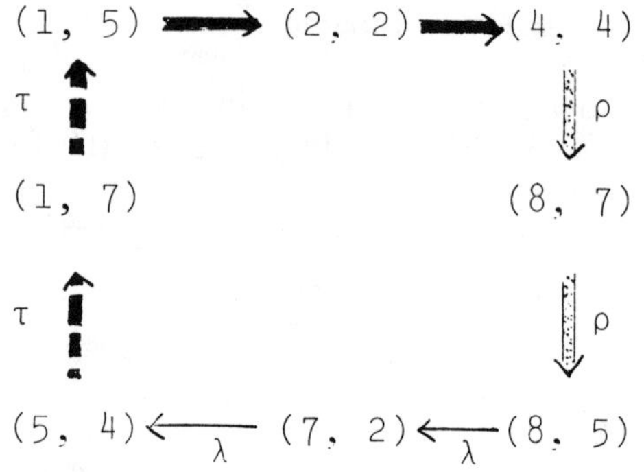

(1, 5) → (2, 2) → (4, 4)

τ ↑   ↓ ρ

(1, 7)   (8, 7)

τ ↑   ↓ ρ

(5, 4) ←λ— (7, 2) ←λ— (8, 5)

Fig.4.21 L_{18}

(19°) $L_{19} = \{(\sigma^3\rho^3\lambda^3\tau^3)^k | k\geq 1\}$

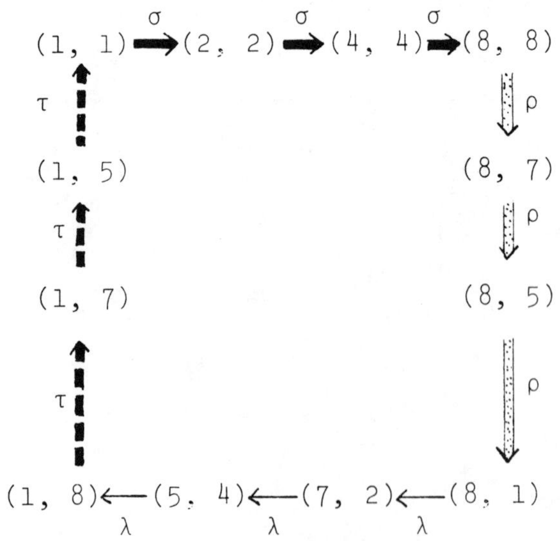

(1, 1) —σ→ (2, 2) —σ→ (4, 4) —σ→ (8, 8)

τ ↑   ↓ ρ

(1, 5)   (8, 7)

τ ↑   ↓ ρ

(1, 7)   (8, 5)

τ ↑   ↓ ρ

(1, 8) ←λ— (5, 4) ←λ— (7, 2) ←λ— (8, 1)

Fig.4.22 L_{19}

5. Fundamental features of linguistic informations viewed from the informative logic

In a series of our papers [9,10,11] we have attained at a unified standpoint upon which we can organize the fundamental logical features of information science and which leads us to introduce a certain scheme of logics called informative logics in our paper [22] as an indispensable basis upon which information science should be built. In particular we have introduced three aspects (subspaces) of feasibility of existence, which are called (i) control, (ii) eizon, and (iii) creation aspects (subspaces) respectively. Now it is the purpose of this Section 5 to provide us with an analysis of the meanings of the fundamental verbs in connection with these three aspects of feasibility of existence. The main conclusion which we are going to explain in what follows is an assertion which reads as follows.

"Each fundamental verb belongs to either one of the three subspaces called (I) control subspace, (II) eizon subspace and (III) creation subspace in the sense that each fundamental verb can be understood as carrying their respective informative concept which should belong to either one of three subspaces of feasibility existences."

We shall explain our assertion in the following Subsections 5.1-5.2.

5.1 Three aspects of feasibility of existence

[1] Eizon subspace. In this eizon subspace, either or both of the subject and the object may not be necessarily defined in a clear cut way and hence it may happen that a formal scheme that the subject applies its operation to the object does not hold true. In this subspace the fundamental aspect of feasibility of existence is rather to be understood as that which can be observed in an ecosphere where an existence in our concern is surrounded by an environment which is not clearly separated (cut) from the existence. So far as this fundamental aspect is concerned, our primary interests with transitory features should be concerned with the following features:

(i) changes of configuration relationship between existence and environment.

(ii) changes of mutual configuration relationship among various existences.

(iii) changes of environment without influence of existences.

There are various verbs which are concerned with some of these features (i), (ii) and (iii).

[2] Control subspace. In this control subspace the cutting principle, which is one of the fundamental principles in system formation as explained in our papers [9 and 12], is clearly applied so as to set up both the subject and the object, and the subject is concerned with the following three features:

(i) Recognition of the object.
(ii) Recognition of operation by the subject on the object.
(iii) Process of application of operations by the subject on the object.

[3] Creations subspace. It is the characteristic feature of creation subspace to emphasize the subjectivity more thoroughly than it would be in the control subspace in the sense that the subject would not adhere to a prescribed scheme of subject-object-operation. Namely the subject may have a freedom to appeal to a rearrangement of constituent schemes in control subspace so as to introduce a newly organized scheme from which the subject wants to realize what he aims at.

5.2 Descriptions in information science

A description of existing state is required to set up either one of subject, object and environment, and it has a certain common feature regardless whether it is concerned with subject, object and environment. In this connection it may be said to be rather an authentic course to appeal to the Kant theory of categories. Nevertheless it does seem to the present author to be most essential to point out three epistemological points which are indispensable in developing information science.

(i) The distinction between two procedures (*a*) setting up an existing state, and (*b*) understanding of an existing state is crucially important.

(ii) The procedure of cutting which we consider to be one of the most fundamental procedures in the formulation of information science is concerned with (*a*) but not with (*b*). It is noted that the logical action of cutting leads us eventually to the notion of space, while the continual existence of the parts having been cut in the space eventually to the notion of time. That is to say, the procedure of cutting in the mature form does imply two fundamental notions, space and time, as its necessary consequences. It is our standpoint to consider the operational procedure of cutting is more fundamental than the transcendental categories of space and time, so far as the formulation of information science is concerned.

(iii) In grasping and/or understanding of existing states, we have to appeal to various senses including (*a*) visual, (*b*) auditory, (*c*) olfactory, (*d*) gastatory and others. Some fundamental verbs are concerned with these senses.

In combination of these three observations (i)-(ii)-(iii), it is important to observe that the constituent components of information contents being given by each verb have the reference to the triplet (space) × x × (time). It is essential to investigate what is meant by the component x. There are two basic aspects regarding the component x.

(1°) Sensation aspect. This is basically due to biological information processings as observed in (iii) just mentioned.

(2°) Matter and energy aspect. This is due to the reduction of sensation to physical phenomena which are concerned with light, sound, heat, force and so on.

The following observations are useful for classification of the fundamental verbs.

(*a*) The standpoint of (2°) matter and energy provides us with an objective description of existing states.

(*b*) A fundamental verb can be understood as belonging to the framework (space) × (information scheme) × (time) where an information scheme is obtained as the result of organization of informations whose origins are either one or both of the standpoints (1°) and (2°).

(*c*) There are various applications of equivalent transformation principle in the sense of Ichikawa [19] which makes us able to generalize the uses of the fundamental verbs beyond the original aspects provided by either (1°) or (2°). In fact most of the usages of the verbs appeal to the application of equivalent transformation principle in creation subspace. This problem is deeply connected with the design aspect of language formation.

6. Evaluation schemes in linguistic information

Evaluation schemes in linguistic information are essential for our purpose of penetrating into the semantic implications of linguistic information. We shall discuss them from each of three features of feasibility of existence.

[1] Evaluation schemes in eizon subspace. There are two standpoints, constructive and transformational, whose combination is essential to understanding the semantic implication of verbs, as we shall explain in what follows.

(1°) Constructive standpoint. An analysis of existences is given to apply the cutting principle so as to have a set of the subjects

$$S = \{S_1, S_2, \ldots, S_m\} \tag{5.01}$$

and a set of the respective environments

$$E = \{E_{11}, E_{12}, \ldots, E_{1,n_1}; E_{21}, E_{22}, \ldots, E_{2,n_2}; \ldots; E_{m1}, E_{m2}, \ldots, E_{m,n_m}; E_1, E_2, \ldots, E_p\} \tag{5.02}$$

Evaluation schemes are then classified into the following cases regarding their concerns with S and E such as:

(i) with E exclusively
(ii) with S exclusively
(iii) with the surface of contract between S and E
(iv) with the relative situation between S and E
(v) with the change of a system (S, E).

(2°) Transformational standpoint

(i) We are concerned with a transformation from the state at initial time point s to the state at the terminal time point t:

$$T_{s,t} : (S, E_s) \rightarrow (S_t, E_t).$$

(ii) We are concerned with the state at an intermediate time point τ, $s < \tau < t$ or with the set of all the intermediate states for the time interval $s \leqslant \tau \leqslant t$.

[2] Evaluation schemes in creation subspace. Any understanding of linguistic phenomena can not be obtained unless we have some reference scheme with creation subspace. Even in the ages of childhood when inner word phenomena are so frequently observed, we can point out some sort of creation process by which children can obtain a sequence of image formations corresponding to words. In this connection we are rather naturally inclined to admit the following assertions, provided that we take into consideration the sensation standpoints which are essential in information science approach.

Assertion 1. Any form is not recognized as a continuous reflection of the objective existence, but rather as a result of construction by human ability of construction as enunciated by the Kant philosophy.

Assertion 2. Any form stores in itself movement as we have emphasized in reverberation cycle formulation in Section 3.

Assertion 3. Recognition has two distinct bases. In our one hand recognition is based upon sequential or successive process in which the framework of time is indispensable, while in the other hand recognition is based upon simultaneous covering of objects in concern in which the framework of space is indispensable.

Assertion 4. Existence of states is always subject to the whole process of genesis, growth, disintegration, breakdown, and extinction.

Assertion 5. A conception of an existence includes clear or vague consciousness of various attributes of the existence, which is sometimes understood as association process of memory in connection with storage of information. It is however to be noted clearly that an emphasis should be given to the fact that such an association process originates from our subjective concerns which are based upon our evaluation schemes.

[3] Evaluation schemes in control subspace. The prototype of evaluation schemes in control subspace is given by the universal human activities called labors with which the structure of control subspace regarding subjectivity, objectivity and practices is intrinsically connected, as can be observed from the correspondence of operation to subjectivity, pattern to objectivity, and optimalization to practices in informative logics of structure description of control subspace developed in a series of our papers Kitagawa [9,10,11]. Evaluation schemes in control subspace are based upon efficiency.

In conclusion of this Section we want to point out that our conception is always subject to the deformation process based upon our evaluation schemes belonging to the one feasibility aspects of existence, called creation. Among others there is always a sequence of applications of equivalent transformations of conceptions. In this sense without appealing to creation aspects no one can understand any semantic implication of verbs in which simile, metaphor, and allegory play indispensable roles.

References

1. ZEEMAN, E.C., 'The topology of the brain and visual perception', *Topology of 3-Manifolds* (Ed. M.K. Fort), Prentice-Hall, pp.240-256 (1960).
2. CAIANIELLO, E.R., 'Outline of a thought-processes and thinking machines', *Journ. Theoret. Biology* 2, 204-235 (1961).
3. THOM, R., *Stabilité Structurelle et Morphogénèse*, W.A. Benjamin, New York (1972).
4. THOM, R., 'Topologie et Signification', *L'Age de la Science*, No.4, Dunod, Paris, pp.1-24 (1968).
5. THOM, R., 'Topologie et Linguistique', *Essay on Topology and Related Topics,* Springer, Berlin (1970).
6. TAMACHI, T. and OKADA, N., 'Automatic description of meaning in picture patterns by natural languages', IIAS-Symposium on Natural Languages (7-8 October 1974).
7. YOSHIDA, S., 'On the natural language systems', IIAS-Symposium on Natural Languages (7-8 October 1974).
8. LENNEBERG, E.H., *Biological Foundations of Language,* John Wiley & Sons, New York/London/Sydney (1967).
9. KITAGAWA, T., *The Logic of Information Science* (in Japanese), Tokyo, Kodansha, Contemporary Series, Vol.200 (1969).
10. KITAGAWA, T., 'Three coordinate systems for information science approaches', *Information Sciences* **5**, 157-169 (1973).
11. KITAGAWA, T., 'Biorobots for simulation studies of learning and intelligent control', in *Learning Systems and Intelligent Robots*, edit. by K.S. Fu and J.T. Tou, Plenum Press, 33-46 (1974).
12. KITAGAWA, T., 'Fuzziness in informative logics', *Fuzzy Sets and Their Applications to Cognitive and Decision Processes,* edit. by L.A. Zadeh, K.S. Fu, K. Tanaka, and M. Shimura, Academic Press, 97-124 (1975).
13. KITAGAWA, T., 'Dynamical systems and operators associated with a single neuronic equation', *Math. Biosciences* **18**, 191-244 (1973).
14. KITAGAWA, T., 'Dynamical behaviors associated with a system of *N* neuronic equations with time lag,' (I), Research Institute of Fundamental Information Science, Kyushu Univ., Research Report No.46 (1975).
15. KITAGAWA, T., 'Dynamical behaviors associated with a system of *N* neuronic equations with time lag, (II)', Research Institute of Fundamental Information Science, Kyushu Univ., Research Report No.49 (1975) to appear.
16. KITAGAWA, T., 'Dynamical behaviors associated with a system of *N* neuronic equations with time lag, (III),' Research Institute of Fundamental Information Science, Kyushu Univ., Research Report No.54 (1975) to appear.
17. KITAGAWA, T., 'A mathematical formulation of intelligent and integrated information system and its implications to artificial intelligence', Research Institute of Fund. Artificial Intelligence, Research Institute of Fund. Information Science,

Kyushu Univ., Research Report No.57 (1975), presented to the 3rd International Congress of Cybernetics and System, Bucharest (25-29 August 1975).

18. KITAGAWA, T., 'Statistics and brainware in intelligent and integrated information system', Research Institute of Fund. Information Science, Kyushu Univ., Research Report No.58, 1-13 (1975); presented to the 40th Session of International Statistical Institute, Warsaw (1-9 September 1975).
19. ICHIKAWA, K., 'Science of creativity–Illustration and introduction to the equivalent transformational thinking' (in Japanese), Nippon Hoso-Shuppan Kyokai, Tokyo (1970).
20. SALOMAA, A., *Theory of Automata,* Pergamon Press, England (1969).
21. EILENBERG, S., *Automata, Languages, and Machines,* Vol.A, Academic Press, New York and London (1974).
22. KITAGAWA, T., 'Brainware concept in intelligent and integrated information system', Research Institute of Fundamental Information Science, Kyushu Univ., Research Report No.39, 1-20 (1974).

Synthesis of fragmented data on neuronal systems: a computer model of cerebellum

ANDRÁS PELLIONISZ

1st Department of Anatomy,
Semmelweis University Medical School,
Budapest, Hungary

1. INTRODUCTION

Neuroscience, when addressing itself to the revealing of basic ideas of the functioning of the brain, encounters a most formidable obstacle. Namely, this relatively young branch of science has to deal with by far the most complex system ever investigated. [Just for a numerical example (Llinás [1]) the nervous system is composed of a multitude of nerve cells in the range of 10^{11-12} and sometimes a single cell of these, the so-called Purkinje cell in the cerebellum, could possibly be connected with about 400,000 other neurons, so-called parallel fibers, the number of actual contacts ranging in the 70-80 thousands (Palkovits *et al.* [2]).] In addition, the system to be understood is readily, quite conveniently and relatively cheaply available for any conceivable experimental scrutiny in a wide variety of species.

Therefore it is understandable—though staggering to learn at first—that after the pioneering period of neuroscience (including the last decades of 'big boom' in basic sciences which accompanied the technological revolution) today on one hand an amazing wealth of experimental data is available on the nervous system both as a whole and on its special formations in particular, while on the other hand we are still lacking even the most primitive understanding on what principles are the neurons organized into subsystems and systems or how these neuronal organs work.

As a result of the escalating process of piling up bits and pieces of various kinds of evidence on the neuronal machinery under scrutiny there emerged an alarmingly diversified neuroscience (see, for example, Abstracts of Society for Neurosciences Vth New York meeting in 1976 with 4000 participants) widespread in the following main aspects: *(a) Targets of investigations:* neuroscience sets a set of simultaneous primary goals: basic understanding of different special subsystems (such as the retina, visual system, cohlea, auditory system, spinal cord, hippocampus or cerebellum) also those in various species of animals. *(b) Methods of investigations:* stemming from the traditional gross physiology and gross anatomy now the research is dominated by micro-physiological (microelectrode-electrophysiological) and micromorphological (light- and electronmicroscopical) methods. These methods were more recently joined by such complex and sophisticated techniques as computer-aided stereological quantitative histology, not to mention the field of histochemistry of the nervous tissue that reaches well into molecular biology. *(c) Levels of approaches:* a central problem of neuroscience is if the basics of nervous *system* are ever to be understood without having revealed first the fundamental function of a single nerve cell or vice versa: can one ever get a heuristic insight into the functioning of one single neuron without reaching an understanding first on the principles on the basis of which the elements are organized into a (unit of) nervous system. While this dilemma creates a hesitation in some theoreticians sometimes almost on a schizophrenic level the practice of experimental neuroscience ignores this dilemma by attacking the brain in *all conceivable levels* simultaneously: ie. both by holistic approaches (such as empirical interpretation of of EEG) and also digging into the deepest levels of approaches (such as molecular biology of membranes, in trying to reveal the how's of the electrogenesis of membranes, ultimately responsible for the activity of nerve cells). Within these extremes spans a continuous spectrum of the applied levels of approaches (eg. membrane biophysics, synapthology, biochemistry of transmitters, single neuron physiology, physiological and morphological investigations of localistic neuronal networks, etc.).

Therefore, not only is it quite unlikely that the fragmented data of all sorts would respect the obvious limits of integrating capacity of our brain in handling such immense degree of complexity but also already there are warning signs that some heuristic approaches are cutting off ties with the monstrous body of data called neuroscience, simply because it reached into the stage where the widely scattered holding of data can no longer be accounted for in its total complexity by any available means. This flag was first raised by the fact that artificial intelligence research had virtually abandoned neuroscience, its natural background, and so do some branches of mathematical neuronal modeling.

Although the amount of available fragmented data today certainly justifies and calls for it, a due synthesis has not been attempted heretofore in a rigorous manner because of lacking a systematic methodology which could render it possible to ascertain the relative significance of the particular experimental findings in a coherent self-consistent framework. Computer modeling was proposed just as a possible candidate to meet this demand (Pellionisz [3]) in the case of the organ or motor-coordination: the cerebellum. Cerebellum, of course, was elected not by chance as a target for such 'crystallization' of not always ultra-hard, fragmented experimental biological data into a kind of synthesis (first into a loose, but then gradually tightening self-consistent model). This organ excels in its remarkably simple and obviously highly specialized regular structure and rock-bottom functioning, lacking for example any circuits for reverberation. (For detailed morphological and physiological description see Eccles *et al.* [4], Palay & Palay [5], and Llinás [6]). Based on the great body of data available on this neuronal circuitry the search for a higher order structure-functional correlation started early (Szentágothai [7,8]) first relying solely on paper and pencil extrpolation of the functioning of a small portion of the neuronal network. Then, the early simulations (Pellionisz [3], Mortimer [9]) created a demand for a bi-directional flow of information between the experimental morphology, physiology and computer modeling. In this way, while prompting a series of systematic quantitative morphological studies (Palkovits *et al.* [10,11,2,12]) the computer model evolved in a simulation of the quantitatively realistic neuronal network of the cat cerebellum (Pellionisz & Szentágothai, [13,14]). These models, on the one hand, triggered such important experimental findings as the ratio of parallel fibers that make synaptic contact with the Purkinje cell they cross (Palkovits *et al.* [2]), and on the other, helped to reveal the possible functional consequences of the variation of morphological parameters (such as the length of paallel fibers that is subject to functional deprivation in kittens, (Pellionisz *et al.* [15]).

II. METHODS

The present software model was developed in the form of packages of FORTRAN programs (each assigned to a different job: generating separate neuronal fibers, dendritic trees, modeling the electrical activity of a single neuron, etc.). By pooling several packages together with auxiliary programs capable of displaying the modeled structure and function, any desired spatial or temporal segment of the model could be generated and analyzed, limited only by the available memory or required computation time. The programs were run partly on an IBM 360/67 (1250K) computer at the Stanford University Computation Center. Mostly, however, the PDP 15 computer setup (32K) of the Division of Neurobiology, University of Iowa was used, the latter equipped a dynamic, TV type, and a static, storage scope graphic display and with a versatile mass-storage system: a fix head- and a cartridge DECPACK disk along with two DECTAPE units. The different programs frequently communicated through the body of data, created by one program, stored on disk, or taped, and used by the adjoining program package.

For the morphogenesis of the network of cerebellar cortex (for details see Pellionisz *et al.* [16]) the neuronal elements: locations of cell bodies and starting and ending points of dendritic branches and axons were considered as situating at lattice points of a three dimensional rectangular grid, with 5 micron spacing in the granular- and Purkinje layers and 10 micron spacing in the molecular layer. For the parameters determining the arborization of dendritic and axonal appendages probability distributions were set up based on quantitative judgements derived from morphological investigations (Hillman 1969, personal communications). Individual elements (cell bodies, axonal and dendritic branches) were then generated with values drawn randomly from these distributions.

For the methodology of multicompartmental Hodgkin-Huxley models of the electrical activity of a single cell (for details see Pellionisz and Llinás [17]) some basics are given in Fig. 7. As shown there, the Hodgkin-Huxley equivalent circuitry (after Frankenhaeuser and Huxley [18] is applied to each of the sixty-two spatial compartments, each compartment constituting a single or three coupled cylindrical section and the compartments are pooled together by the cable equations. Digital simulation is then applied simply by successive numerical integration of the equations. The numerical integration was chosen as the classical Euler (explicit) integration formula which requires suitably limited Δx and Δt to avoid instabilities of the computation. Therefore for finite Δt-*s* for the total set of compartments an individually and asynchronously variable value was selected by the program, within the range of 200 nsec to 25 μsec. Results of the simulation were presented on the graphic display for hard copies taken directly, or (for a continuous time display of the single cell activity) single frames were taken, controlled by the program, to compose a cinematographic animation for demonstrative purposes.

III. RESULTS

1. Computer morphogenesis of the neuronal network of cerebellar cortex of frog

While earlier attempts at building models that retain the identity of single neurons (Pellionisz [3], Pellionisz & Szentágothai [13.14]) in the case of the cat's cerebellum had to be limited to a rather restricted localistic piece of the network, the frog's cerebellum with its (i) single-laminar overall structure, (ii) limited cell numbers, totalling only less than two million neurons, (iii) apparent lack of sophisticated interneuron-systems (especially basket-inhibition, which is a dominant complicating factor in cat's cerebellum) make it altogether feasible aiming at creating a holistic model of the neuronal network of cerebellar cortex. Such an overall model seems ultimately inevitable because otherwise it might not be possible to relate the single neuron recordings (that sometimes show unmistakable correlation with spatial localization; Precht and Llinás [19, 20]) to the known micro-morphological features of the neuronal structure.

Figure 1, an outline of the model, is to explain the mentioned features. As seen in *A* and especially in *B* the cerebellum of frog is a slab of tissue behind the optic tecta, arching over the fourth ventricle. Its simple, unfolded layers (in sharp, refreshing contrast to the highly multiplex, convoluting laminae of cerebella of higher species) lie in dorsoventral direction as shown in *C* and *D* of Fig. 1. As mentioned, the structure is largely stripped of interneurons, built only of four kinds of neurons (therefore it may be defined as 'basic cerebellar circuit'; Llinás [21]). These kinds of cells are: (i) the main cerebellar output cells: *Purkinje cells*, E, totalling less than 10^4; (ii) the diffuse and numerous input lines, so-called *mossy fibers*, F (as seen on F, these fibers, and *eo ipso* all fibers enter and leave the body of cerebellum through the so-called cerebellar peduncles which anchor this slab of tissue at its lateral ends to the brain stem; peduncles are marked by X and W on central part of Fig. 1; (iii) the scattered 'clouds' of mossy fiber endings are connected to tiny *granule cells* (the most numerous cells in the nervous system, about 1·68 million in frog's cerebellar granular layer) which then send their axons (see G) directly up to the molecular layer where their T-shaped axons are called *parallel fibers;* (iv) every Purkinje cell is overwhelmed by 200-300 contacts (Hillman [22]) of one and only one so-called *climbing fiber,* which, as being a kind of a 'shadow' of Purkinje cells, are

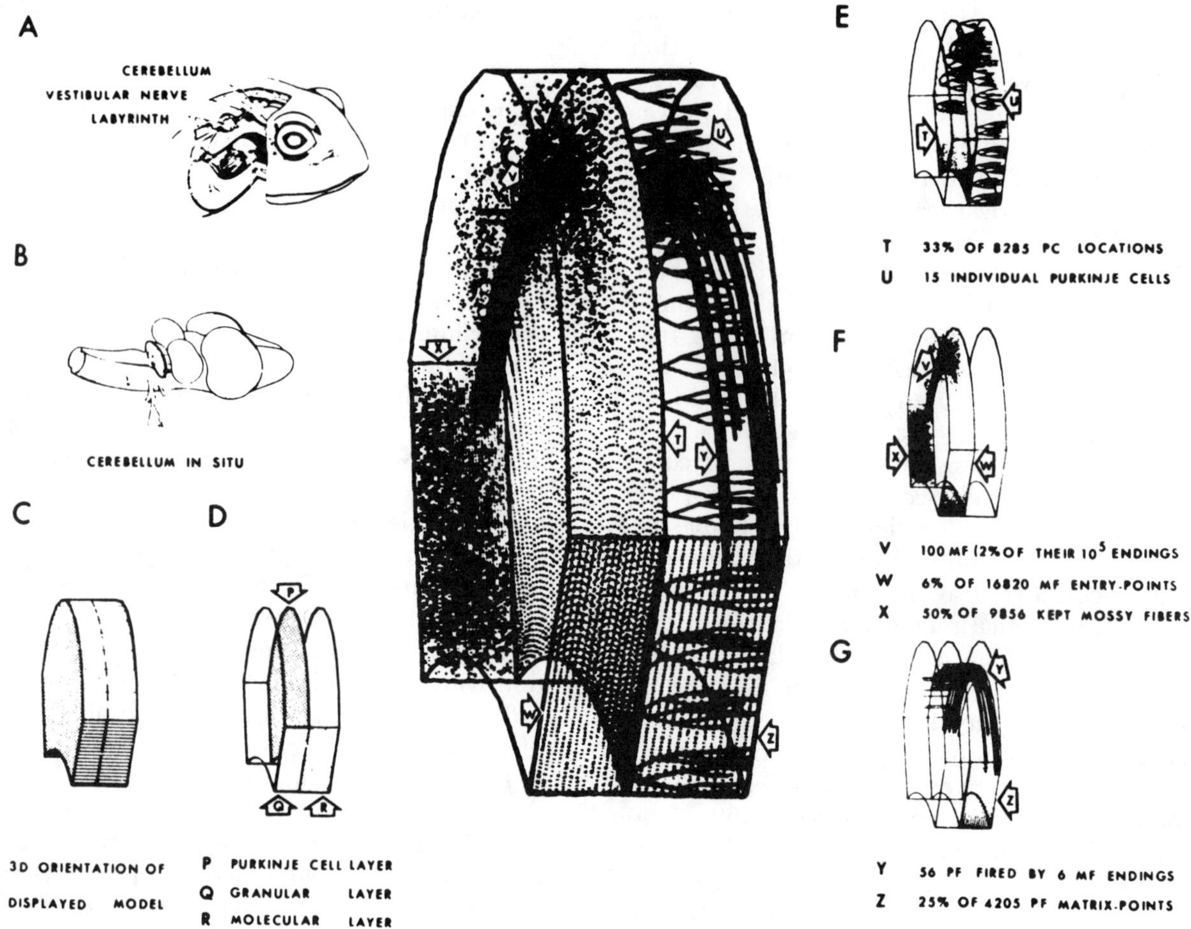

Fig.1 Morphological structure of the computer model of the frog cerebellum (central inset). The overall features of the cerebellum are shown in the insets at left. The neuronal elements of which the central model is assembled are shown at right: Purkinje cells (E), mossy fibers (F) and parallel fibers (G). Further description in text.

not shown separately in the circuitry model. In the center of Fig. 1 some representative elements (shown separately on the insets E-G) are assembled together giving an impression of the three-dimensional complexity of this simple cerebellar cortex. The morphogenesis of these elements was implemented as follows (for details see Pellionisz *et al.* [16]).

Morphogenesis of Purkinje cells

Purkinje cells are represented by 'skeletons' of their dendritic trees (see Fig. 2). Dichotomous trees with four generations of bifurcations (altogether 31 branches) were generated on a 29 by 29 lattice, with pilot values for the angular and dendritic branch-lengths in mind. [ie. that the skeleton of representative Purkinje cells would be a matching abstraction of the morphological appearance of the Pukinje cells observed in Golgi staining (Hillman [22])]. As seen, these trees allow for a variation regarding *(a)* a symmetrical (see cell 1) or asymmetrical (2) arborization, *(b)* an overlapping or non-overlapping or partially overlapping character of the trees (cf. 3,25,20), *(c)* different sizes of Purkinje cell dendritic trees (3,27). The dotted lattice points represent locations on the dendritic tree at which the crossing parallel fibers may establish synaptic contacts with the cells. As shown in Fig. 3 the different types of generated Purkinje cell arborization are implanted to the cerebellar molecular layer, where the dots represent the locations of the 8285 Purkinje cell bodies, arranged in a staggered pattern, similar to that deduced in cat (Palkovits *et al.* [10]), while the numbers of dendritic trees is reduced towards one end, since otherwise the picture would be incomprehensible to the viewer.

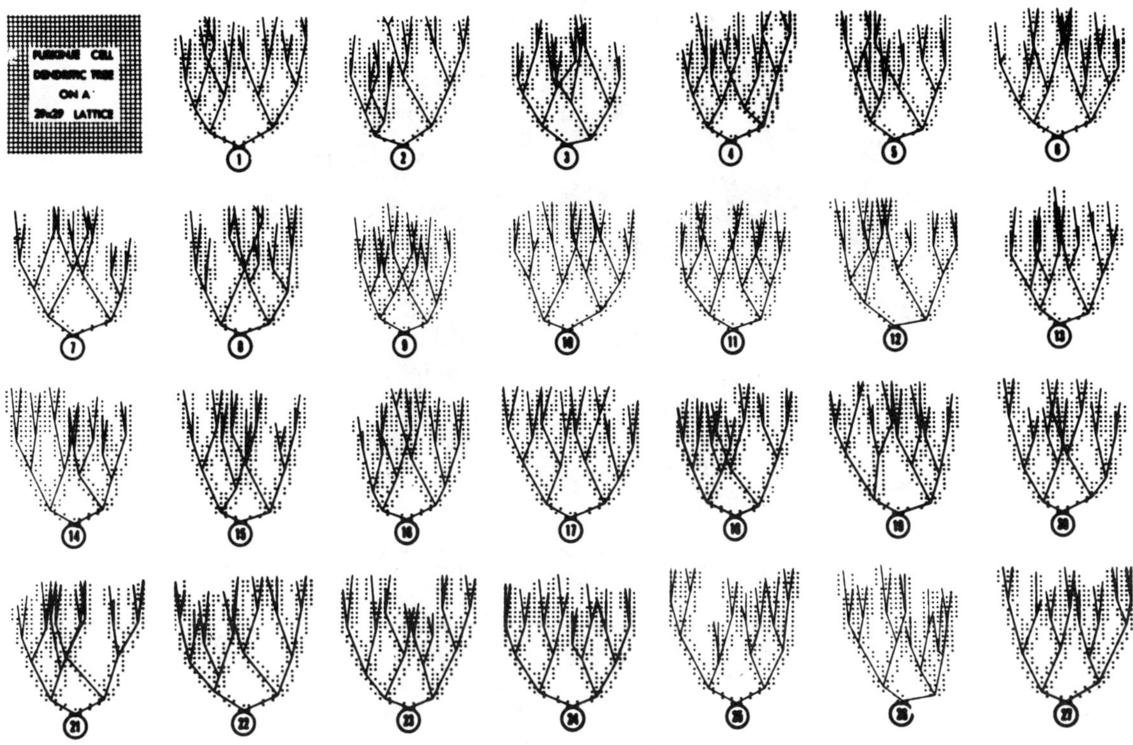

Fig.2 Purkinje cell branching patterns. Each modeled Purkinje cell dendritic tree is shown on a 29 x 29 lattice. Dots show locations where the crossing parallel fibers may contact the cell. Note the variation of tree-patterns in symmetry, overlapping and size. For details see text.

Morphogenesis of mossy fibers

Of the only two kinds of afferent fibers to the cerebellar cortex, the mossy fibers attract the most interest from an anatomical point of view. This is not because they far outnumber the other (climbing fiber) input but also because the mossy fiber input goes through an apparent spatial 'preprocessing' (Llinás [23]) before the signals reach their destination, the Purkinje cells. These fibers, branching diffusely in the granular layer, distribute the excitation on to a population of granule cells (and, therefore, to their axons, the parallel fibers) by a multitude of endings on every mossy fiber. As a result, it can be expected that given a certain locatlized set of activated mossy fibers, the structure of granular layer will relay this into a widely scattered pattern of input to the Purkinje cells via the parallel fibers.

Although both physiological observations (Precht & Llinás [19,20]) and theoretical considerations (Llinás [23]) are available for the analysis of the mossy fiber ending-granule cell transfer, very little is known wbout its real spatial features. [For example, an attractive speculation on the activity of a column of Purkinje cells along an activated strip of parallel fibers (Szentágothai [7,8]) is based on an implicit assumption that such strips do, indeed, exist. While they can certainly be created by electrophysiological local stimulation in cat (Eccles *et al.* [24]) because of the convoluted structure of the laminae and diffuse branching of mossy fibers in cat it is unknown how likely it is that such strips are created by physiological circumstances.]

Therefore, in generating the mossy fiber-granule cell circuitry in the model cerebellum a specific goal was to create the structure in a way that it elucidates the potential capabilities of the machinery in the granular layer from the viewpoint of this spatial transfer. Two concepts assist this approach: the concept of 'growing' a connectivity-pattern (see below) and the concept of the assumption that the growth of fibers is stochastic rather than deterministic. With these premises the mossy fibers are generated by the following procedure (see Fig. 4):

(1) An abstraction of the mossy fiber is deduced from qualitative morphological observations (Hillman, Llinás; personal communications) thus clarifying the

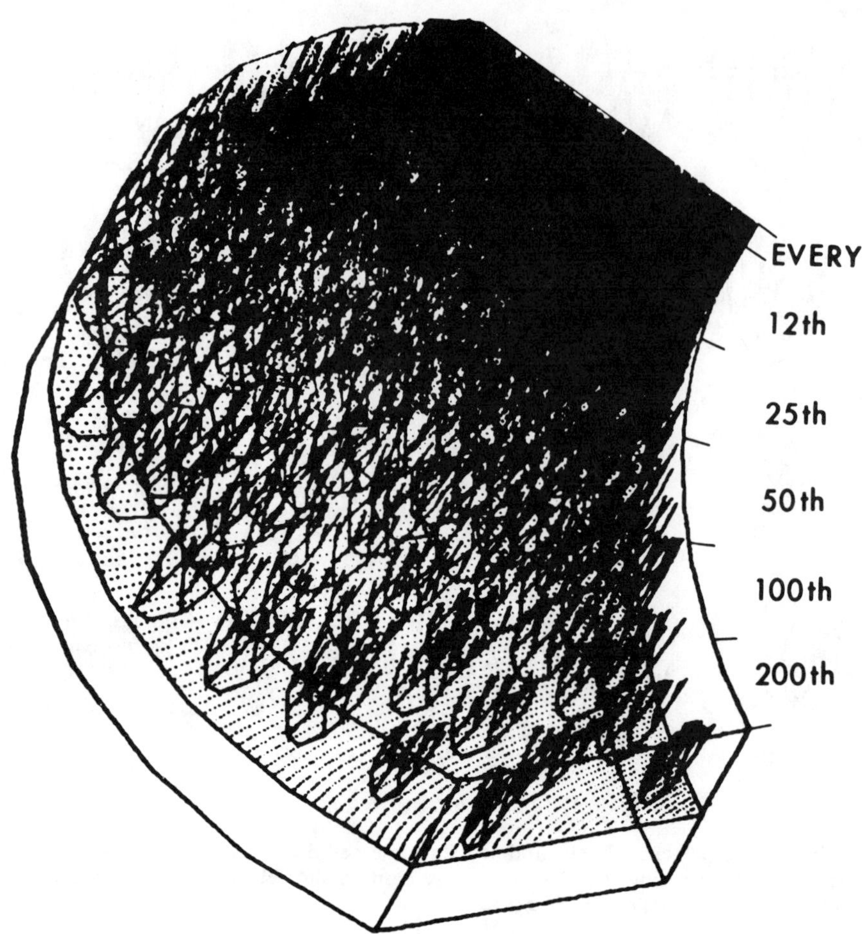

Fig.3 Purkinje cells in the computer model of the frog cerebellum. The staggered arrangement of locations of 8285 Purkinje cell bodies is shown by dots. The number of cells illustrated are progressively reduced in one direction for better visualization.

key morphological parameters: see central inset in Fig. 4: the mossy fiber is simplified to a 'linear trunk' and a triangular isosceles holding a random pattern of endings.

(2) Overall density functions are empirically established for the above parameters (length of trunk, length of fan and means and standard deviations for the pitch, roll and yaw angles of the fan vertex from the direction of the trunk). Since the precise value for the angles could be established only by a grandiose morphological project, for pilot-testing the potential influence of these parameters two sets of the extreme values are estimated: one for a *W*ide, the other for a *N*arrow projection, spanning the estimated range by more than 200% variation in these parameters.

(3) Each particular mossy fiber entering the peduncle at an entry-point of the 145×29 lattice is then generated by drawing a number for each parameter from the density function.

(4) The concept of 'growing' is applied by either considering the generated fiber 'surviving' neuronal element (if it grows within the boundaries of the granular layer) or 'rejecting' it if it extensively protrudes from the body of cerebellum. (Solid diagrams in Fig. 4 show the distribution of parameter-values for those mossy fibers which were kept while the outlining dotted graphs show the distribution for all generated fibers.) Accordingly, the 9820 or 9856 kept mossy fibers (in one peduncle) for the cases of *W* or *N* projections respectively show an elliptical density distribution in the cross-section of the peduncle (see X in Fig. 1).

Morphogenesis of granule cells and their parallel fibers

Granule cells may be located in the model at the orthogonal lattice points of 10 microns spacing

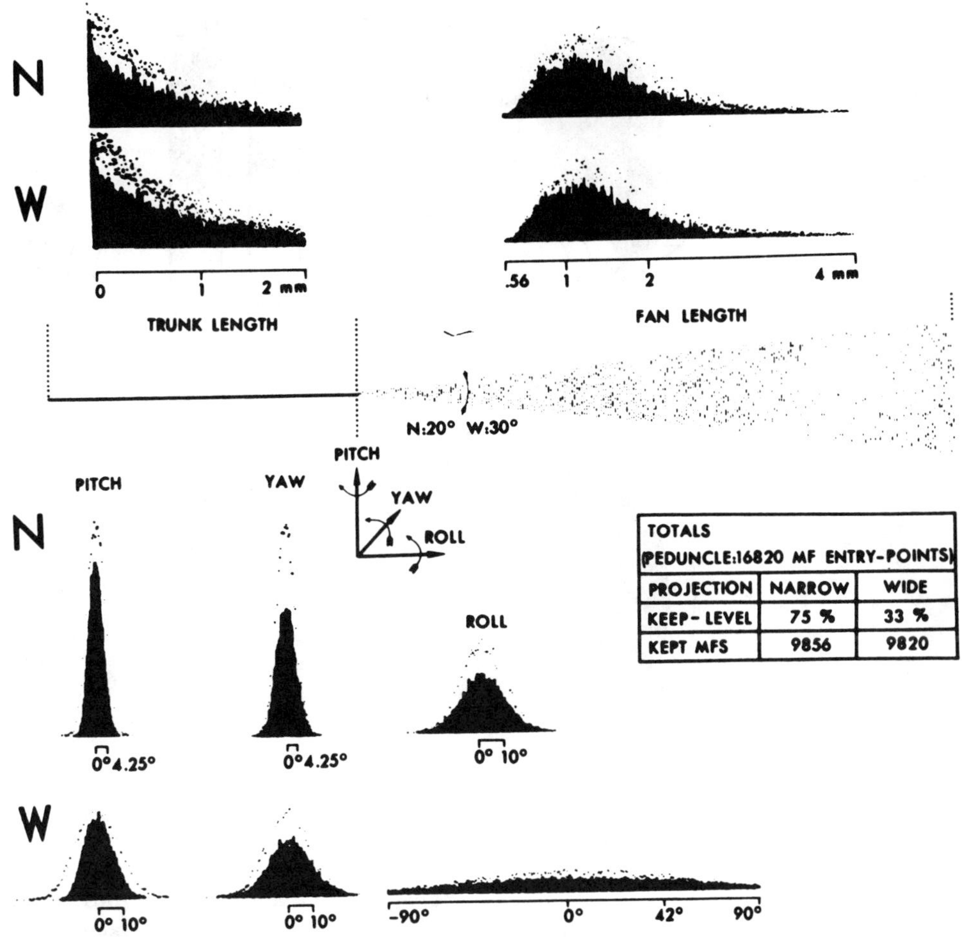

TOTALS (PEDUNCLE:16820 MF ENTRY-POINTS)		
PROJECTION	NARROW	WIDE
KEEP-LEVEL	75 %	33 %
KEPT MFS	9856	9820

Fig.4 Probability distributions for the key parameters for generating mossy fibers. A simplified mossy fiber is shown at the center of the figure. For the trunk length, fan length, pitch, roll, yaw of the angle of vertex of fan from trunk two sets of distributions are given; N is for a narrow, W is for a wide projection. Further description in text.

throughout the granular layer. Therefore, any set of mossy fiber endings (activated by a group of mossy fibers, when thrown on to the potential 1,682,000 granule cell positions, would create a different number of spatially coincident 'active' mossy fiber endings at any possible granule cell locations. According to the varying number of spatially coinciding mossy fiber endings at one lattice point, a 'firing' granule cell could be recognized in the model by simply setting a minimim required number of incoming excitation. The set of activated granule cells at any mossy fiber input combination would then transform into parallel fiber activity.

Parallel fibers (see G in Fig. 1) rise to the molecular layer with the precise translation of the position of the granule cell. There any parallel fiber (along the length of cerebellum) runs symmetrically in both directions. The contact of parallel fibers with Purkinje cells are governed in the model both by the parallel fiber and the morphological structure of the Purkinje cells. For the parallel fibers, separate options are available in the model for determining the length of the fiber (determined by the average length and its standard deviation) and similarly the distribution of synaptic varicosities has an average distance with standard deviation. (In the cases shown an average length of 2 mm was used with standard deviation of 1 mm).

Spatial distribution of the activation generated by clusters of mossy fibers: pathway analysis

Figure 5 shows an illustrative use of the model de-

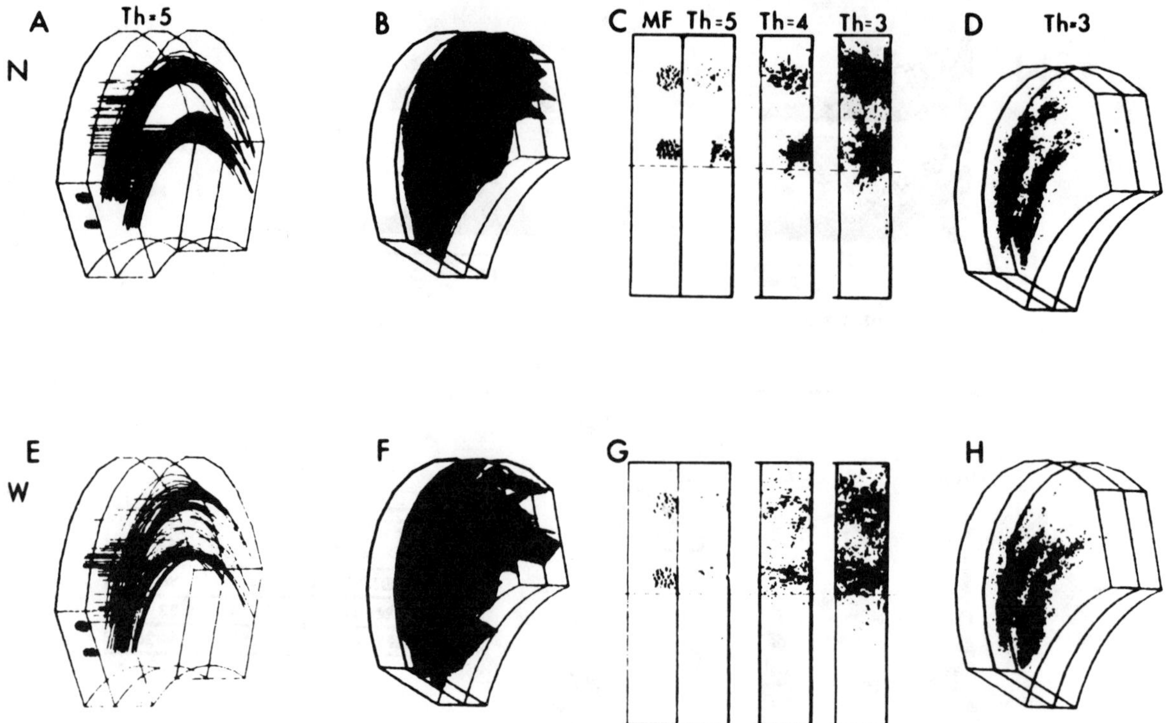

Fig.5 Comparison of the interaction of two clusters of mossy fiber input in the cases of narrow projection (N, upper half) versus wide projection (W, lower half). A and E show the overall spatial distribution of the entry points of two clusters of mossy fibers and the activated beams of parallel fibers. B and F show the mossy fibers. C and G are cumulative cross-sections of the peduncle and molecular layer (MF=mossy fibers), showing the parallel fibers at different granule cell thresholds. D and H are patterns of granule cell activation at the threshold of 3. For further details consult text.

veloped. The two cases (one for a *N*arrow and the other for a *W*ide projection of mossy fibers; on the top and bottom half of the picture, respectively) were generated in order to analyze and compare the potential interactions of the activities triggered by two separated clusters of excited mossy fibers. Compared to the case shown in Fig. 1 where a single cluster of about one hundred closely packed mossy fibers were activated (center of Fig. 1) that threw into action a beam of parallel fibers (G in Fig. 1) on Fig. 5 the difference is that two such clusters of mossy fibers were applied here. These 2×100 mossy fibers are marked on *A* and *L* only by their entry-points in the peduncle. [The end-section of the peduncle (MF) and the projection of molecular layer is shown magnified on *C* and *G*.] As shown, the mossy fiber clusters of 25% density are apart from each other at a distance of one third of the length of peduncle. Insets *B* and *F* show the endings belonging to these two mossy fiber clusters. As seen on Fig. 5 while the 'cloud' of endings (*B*,*F*) appear to be totally integrated (and, therefore, one could expect a maximal interactive activation at their spatial overlap) the activated granule cells are clearly distinguished into separate spindles, with small regard to the threshold set for the granule cells (Th = 3 on *D*,*H*, and Th = 5,4,3 on *C* and *G*.) For the sake of clarity, the parallel fibers, belonging to the granule cells activated at Th = 5 are shown on *A* and *E*. It is of special interest, that this totally unexpected degree of 'spindle separation' of parallel fiber activity (evoked by not too distant centers of mossy fiber activation) is highly insensitive to the more than 200% variation of the angle-parameters with the *W*ide or *N*arrow mossy fiber projection.

Another example of studying gross spatial features of activities over the whole cerebellar cortex of frog is shown on Fig. 6. This case relates to the electrophysiological findings (Precht and Llinás [19,20]) that Purkinje cell activities evoked by ipsi- or contralateral horizontal angular acceleration exhibited quite obvious spatial features: ie. some (so-called type I)

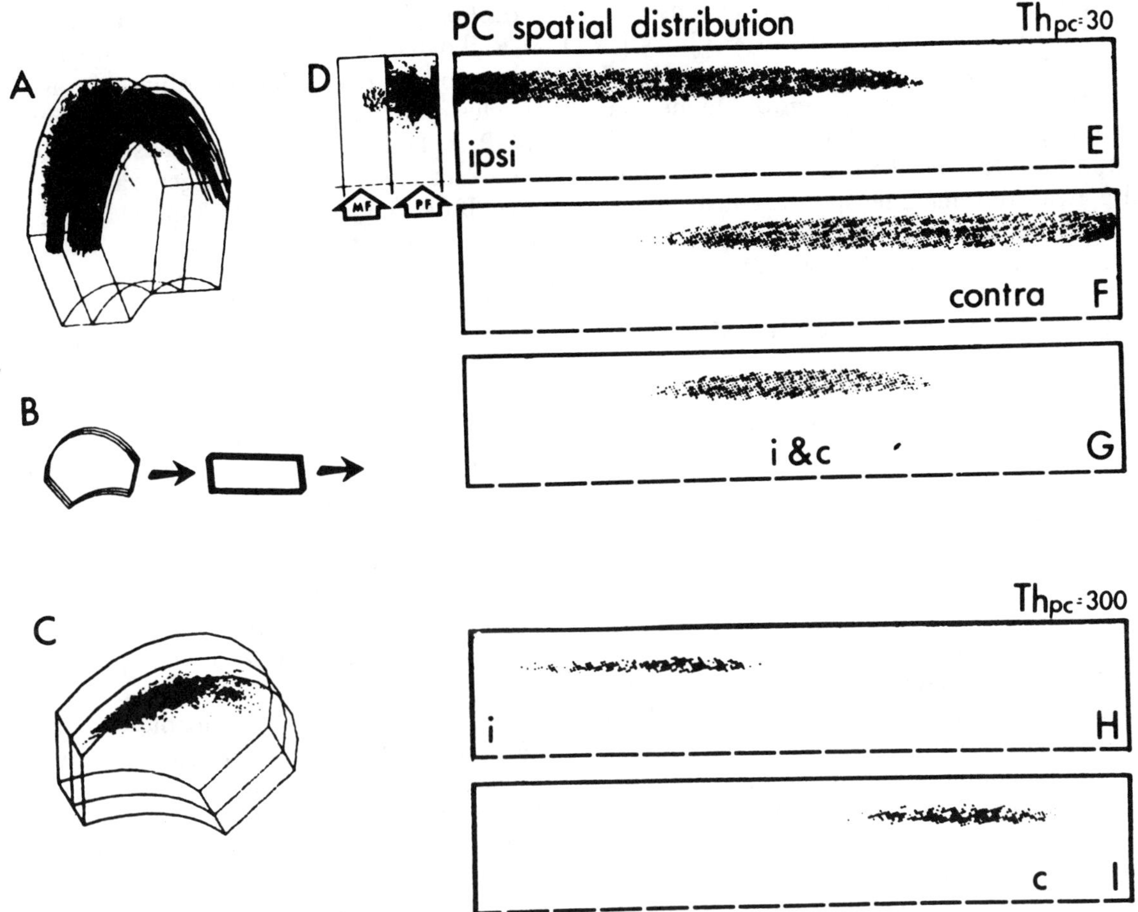

Fig.6 Spatial distribution of the overall parallel fiber activation of Purkinje cells in cases of modeled ipsilateral and contralateral vestibulo-cerebellar activation. A shows a view of monolateral cluster of mossy fibers and the beam of activated parallel fibers. C displays the granule cells, activated if the threshold is 3. The Purkinje cells are shown in Cartesian coordinate system (see B for transformation). Through D–I only the dorsal half of cerebellum is shown. D is a cross-section of the half-peduncle, showing the projections of mossy fibers and activated parallel fibers. On E–F–G dots mark the locations of Purkinje cells that receive more than 30 parallel fibers in the cases of ipsi-, contralateral and both stimulation. H and I show the ipsi- and contralateral cases with Purkinje cell threshold increased to 300. Detailed description in text.

Purkinje cells—mostly grouped at lateral one-third—responded only to ipsilateral rotation, other cells (type II) to the contralateral rotation and the so-called type III cells to both. This finding is accounted for and further analyzed by the model as follows: As seen on A of Fig. 6 a single cluster of mossy fibers [analogous to the peduncular projection of vestibular system (Hillman [25])] generates an active beam of parallel fibers, by throwing into action a spindle of granule cells (see *C* on Fig. 6). Similarly to Fig. 5 an end-section of the peduncle is shown in *D* of Fig. 6, demonstrating the entry-points of input mossy fibers and the projection of their activated parallel fibers. If the parallel fiber action is sampled by the Purkinje cells of the model (cf. Fig. 3) the overall spatial distribution (at the Purkinje cell threshold of 30) is as shown on *E*. (Note, on *B*, that the in-perspective polar-coordinate-display of the body of cerebellum is transformed into a rectangular solid in Cartesian coordinate-system, and from *D* through *I* only the dorsal half of this 'unbent' cerebellum is shown.) The spatial distribution of Purkinje cells, activated by more than 30 parallel fibers, can be compared at ipsilateral input (*E*), contralateral input (*F*), or both (*G*). The overall mapping of the locations of these different types of cells is not at all unlike that the one derived from observations. (cf. Fig. 6 with Fig. 17 in Precht & Llinás [20]).

Moreover, this spatial mapping reveals (see *H* and *I*) that the strongest activation (a tenfold higher parallel fiber input of more than 300 per Purkinje cell) is centered around the distal third of the cerebellum. This kind of feature (emerging spontaneously from the model and confirmed by the physiologists) provides reassurance that the model is predictive for further experiments.

At this point in the modeling of cerebellar cortex, having the ability to handle the morphological feature of 'throwing the mossy fiber input' to the population of Purkinje cells, the urgency of studying the actual *processing* of this parallel fiber input by the Purkinje cell is apparent. Therefore, these studies so far on circuitry level must be combined with a cellular level of detail in modeling the Purkinje cell integrative activities.

2. Multicompartmental Hodgkin-Huxley cable model for the study of simple and complex spike electrogenesis of Purkinje cells

Beyond the fundamental intention that the model of the electrical activity of the single Purkinje cell should be an integer component of the computer model of cerebellum this 'module' has particular significance in many respects: (1) The Purkinje cell exhibits by far the most complex morphological features in the nervous system. (2) While it is the only kind of cerebellar neuron on which the two input systems (mossy and climbing fiber system) converge, these two kinds of afferentations are of extreme character. Namely, one Purkinje cell may potentially establish synaptic contacts with as many as 80,000 parallel fibers (in cat; Palkovits *et al.* [2]), while it receives only a single representative of the other cerebellar afferent the climbing fiber but establishes several hundreds of synaptic contacts with it (Hillman [22]). (3) Accordingly, the Purkinje cell produces a markedly different action potential in response to parallel fiber activation (so-called simple spike; Eccles *et al.* [26]

Both in respect of the simple- and complex-spike activities of Purkinje cells there are crucial problems which make them quite difficult to interpret. For the climbing fiber response (complex spike, see Fig. 11*A*) the highly stereotyped, all-or-none spike, followed by several, normally 3-5 oscillatory wavelets has been observed frequently in the cerebellum (Llinás *et al.* [6], Eccles *et al.* [26]). It is intriguing that while the climbing fiber activation is not only morphologically very well investigated (Szentágothai & Rajkovits [27]) and that an abundance of both intra- and extra-cellular recordings are available (Eccles *et al.* [26], Llinás [1]), moreover, also the EPSP evoked by climbing fiber is well studied, still the exact mechanism which generates this characteristic burst named 'complex spike' has remained hitherto unclear and only highly controversial speculations are available (Martinez *et al.* [28] contra Eccles *et al.* [26]). For the simple spikes, the wide variety of dynamic character of Purkinje cell activities (and its consistency in ipsilateral or contralateral angular acceleration stimulation; Precht & Llinás [19,20]) raises the question if the variance of dynamism is a result of spatiotemporal distribution of afferent activation or, rather, it is an inherent feature of the spike producing mechanism of the cell.

On a general level, the integrative properties of an elaborate dendritic tree are brought into the center of attention especially by the discovery of dendritic spikes (Nicholson & Llinás, [29], 1969). This by necessity, introduces the need of adequate modeling technique which is capable of handling a partially or totally active dendritic tree, contrary to the now classical analysis of passive tree structures (Rall [30]). The developing of such model for Purkinje cells offers the particular advantage that an acceptable model must satisfactorily produce the three distinctly different action potential waveforms, in response to (i) antidromic, (ii) orthodromic and (iii) climbing fiber activations. The requirement of modeling of inhomogenious dendritic tree accords well with the basic policy of modeling of this chapter that the integrity (including that of the morphology) of the neuronal elements be observed. Therefore the central problem-reduction concept of Rall, not dealing with neurons with particular geometry but only with those which can be equated to a uniform cylinder can not be accepted for the present work. Our model follows the trend set recently by Dodge & Cooley [31] when they combined the Rall concept of applying cable equations to idealized (equivalent cylinder) dendritic tree analysis with the application of Hodgkin-Huxley equations (Hodgkin & Huxley [32]) to each segment of the non-uniform cable which represents in their model the abstraction of cellular morphology. Carrying this concept one step further, dealing with particular dendritic tree, introduces significant computational difficulties but it provides the additional advantage that the model can be applied to trees not following the 3/2 constraint. (apparently, according to preliminary measurements by J. McLaren—personal communication—

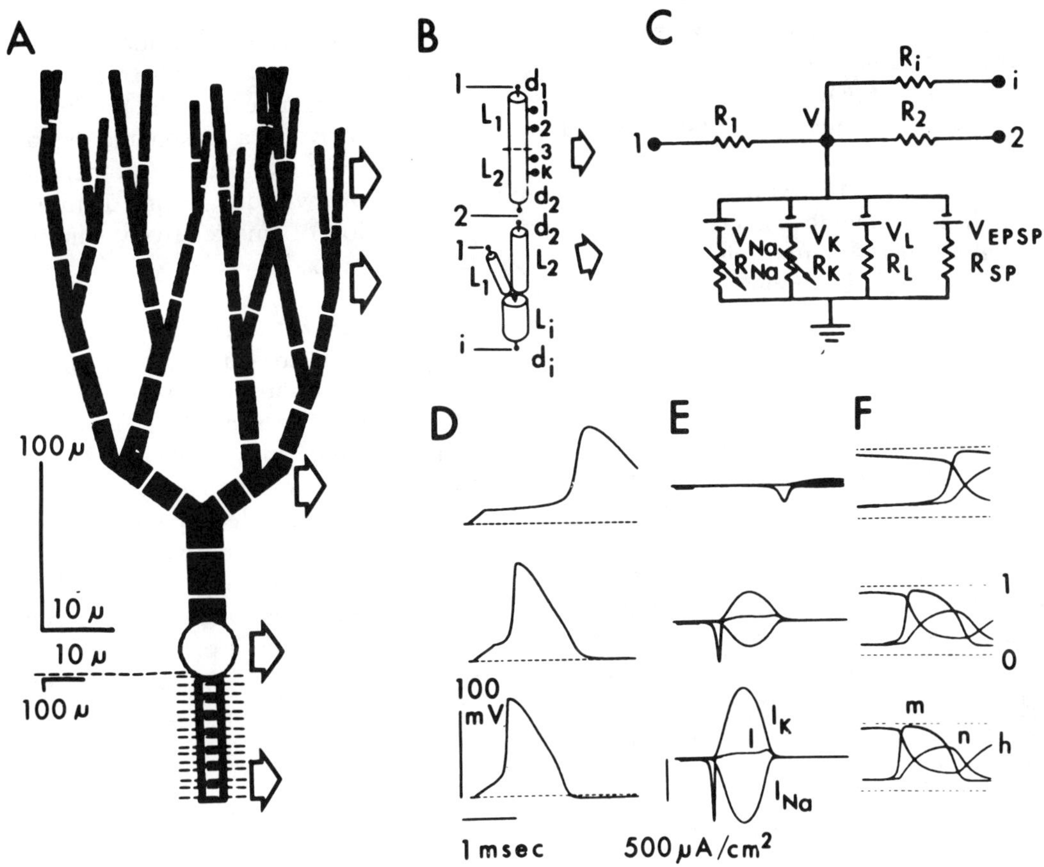

Fig.7 62 compartment model of the Purkinje cell. A show the spatial compartmentalization. B shows how the compartments are simplified to one or three cylindrical segments. C is the equivalent electrical circuit for a compartment, as described by Hodgkin & Huxley. D-E-F show simulated recordings of the membrane potential (D), membrane currents (E), and the m, n, h variables in the H-H equations, as a function of time, for three representative compartments of different excitability: upper row is for a poorly excitable dendritic compartment, middle row is the recordings for the excitable soma, and the lowermost row shows the simulated recordings for the highly excitable node of Ranvier. For further details see text, and Pellionisz and Llinàs, 1976 [17].

this value in frog's Purkinje cell is 1·75 on average).

The basic structure of the model of Purkinje cell (for details see Pellionisz & Llinas [17]) is shown on Fig. 7. As seen on Fig. 7*A* a given structure of neuron (with realistic quantitative morphological parameters) was broken into a multitude of spatial compartments lumped together by the cable equations and the electrical phenomena on each compartment governed by either the Hodgkin-Huxley equations (on compartments representing supposedly excitable sections of the neuron membrane) or by passive *RC* properties. Fig. 7*A* shows 31 mid-branch-compartments, 15 bifurcation points, a soma, an initial segment and 7 nodes and myelinated segments. As seen on *B*, each of these 62 compartments are represented by a cylindrical spatial segment of uniform diameter, or, for the branching compartments, a combination of three cylinders. The equivalent circuit is shown in *C*, where R_i represent the longitudinal resistances of the cylindrical compartments and the rest is the electrical equivalent of the membrane as described by Hodgkin & Huxley [32]. (For detailed description of the equations see Pellionisz & Llinás, [17].) Each of the parameters may be set individually for any compartment according to the different dynamic properties of different regions; any compartment may be assumed as totally or partially excitable. The parameters and constants were selected from Frankenhaeuser & Huxley [18], with only minor modifications. *D,E,F* in Fig. 7 show the assumed electrical properties of three representative compartments: a moderately excitable branching point (upper row in *D,E,F*), an excitable soma (second row) and a very hot, highly excitable axonal node of Ranvier (lowermost row of simulated recordings in *D,E,F*). *D* shows the waveform of membrane action potential, *E* the membrane currents and *F* the *m,n,h* variables in H-H equations, all as a function of time, following a brief current impulse injection (see on *E*).

Antidromic activation of Purkinje cell

Figure 8 shows a spatiotemporal display of the membrane potential waveforms throughout the 62 compartment model in case of antidromic activation. This case (as *all* cases shown in this chapter) represent a dendritic tree of inexcitable character: the main trunk and branches of the tree are passive *RC* cables. A 0.1 msec, 10 nA antidromic impulse, applied to the sixth Ranvier node produces the antidromic invasion. The recordings over the total dendritic tree show a rapidly decreasing degree of depolarization of upper portion of the tree as well as a characteristic after-depolarization of the soma in good agreement with similar electrophysiological findings in frog cerebellum (Llinás, *et al.* [33]).

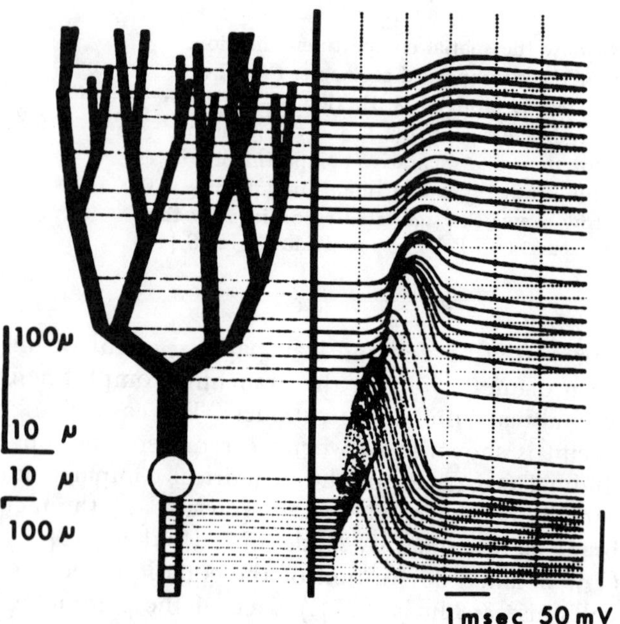

Fig.8 Parallel recordings of the membrane potentials throughout the 62 compartments in the case of antidromic activation of the model. All dendritic compartments are considered inexcitable. Note the long after-depolarization of the soma and the only partial invasion of the upper branches of the dendritic tree.

The reconstruction of antidromic activation plays an important role in the adjustment of geometrical and electrical parameters of the model, because it is unlikely to arrive at a reasonable explanation of electrogenesis of a complex phenomenon, like a climbing fiber answer without getting a suitably adequate representation of this relatively simple, well-known response.

Orthodromic activation by parallel fibers

Simple spikes are generated by the orthodromic stimulation of the Purkinje cell by the parallel fibers. This has long been carefully investigated in experimental preparations (Eccles *et al.* [24], Llinás *et al.* [34]) and also some theoretical considerations were given to the possible impact of differences in integration of input carried by different spatial patterns of parallel fiber input (Llinás [23]). To study the integrative properties of Purkinje cells as a response to spatially different parallel fiber input patterns, on Fig. 9 a horizontal, on Fig. 10 a vertical 'window' of parallel fiber input was applied. The activation of

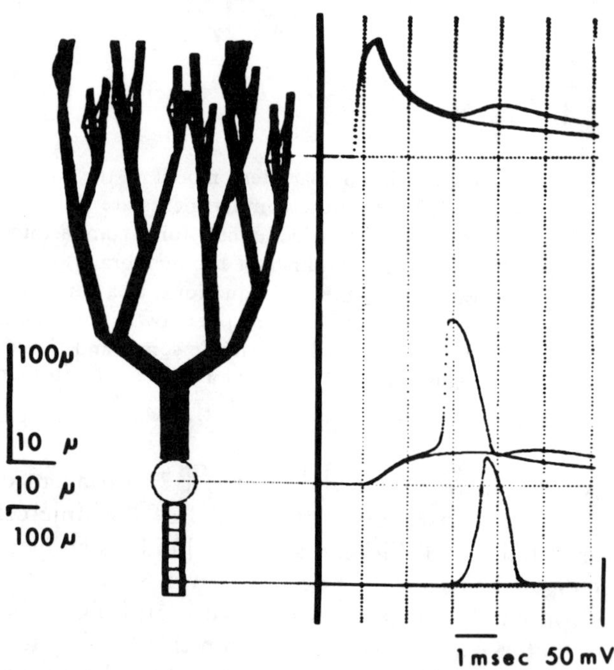

Fig.9 Orthodromic activation of the model Purkinje cell by a horizontal 'strip' of parallel fibers activating the uppermost dendritic branches of the tree. Two recordings are superimposed in the figure: one is subthreshold stimulation, therefore it produced only a slow and prolonged EPSP at somatic level without propagating spike. The input of the same strength, but applied 0.15 msec longer, proved to be suprathreshold, producing propagating spike. For further details see text.

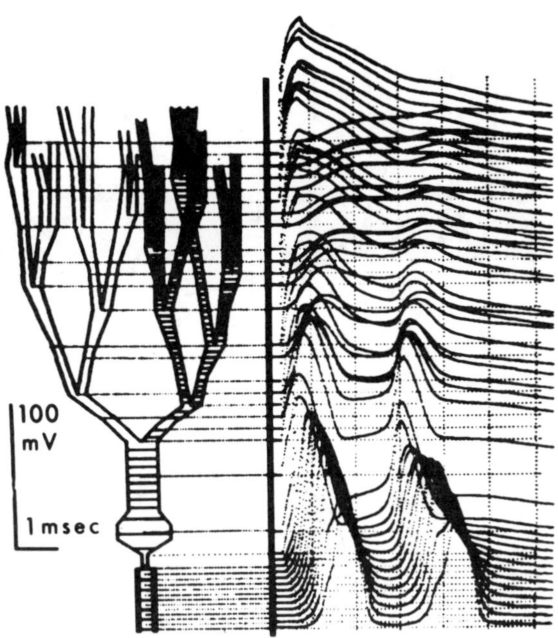

Fig.10 Vertical integration of an orthodromic (parallel fiber) input applied to the right side of the dendritic tree. As seen, this input of the same strength as on Fig.9 but with different spatial pattern produces a highly suprathreshold excitation, with a short latency, two-spike burst. The still photo of depolarization pattern on the branches (left side of the picture) was taken at the time point of 2.5 msec.

dendritic branches by parallel fiber synaptic input was modeled by 'opening' a 100 Mohm synaptic resistance towards the EMF of equilibrium EPSP potential, for a given (0.5 msec) period of time. On Fig. 9 two such parallel fiber input patterns were applied and the recordings were superimposed. The 16 uppermost dendritic branches received synaptic input (by 5 'synapses' each),once during 0.4 msec then for the second time for 0.5 msec interval. As seen, since this parallel fiber input appears to be marginally suprathreshold the shorter impulse evoked only a massive depolarization at the upper branches and a delayed, prolonged EPSP on somatic level which, of course, did not propagate towards the axon. However, the same impulse, held only about 100 μsec longer evoked a full, all-or-none spike at the soma after about 1·5 msec delay and this spike duly propagated along the axon.

For a comparison, an orthodromic activation by the same strength and duration was applied (Fig. 10) in vertical spatial distribution too. As seen, while the horizontal input was only marginally suprathreshold, the same vertical input (applied to the 16 branches at the right side of the neuron) evoked a very short latency, (almost immediate) short burst of two spikes. The off-line playback of generated data provided versatile opportunity for displaying the reults not only in the conventional electrophysiological manner, i.e. in the form of two-dimensional graphs of time functions of membrane potential at one spatial location, but also in a'parallel' manner, by visualization of the degree of depolarization throughout the dendritic tree, by shadowing the dendritic compartments accordingly. This possibility allowed producing computer-made, quantitatively accurate cinemathographic animation (demonstrated in the talk) for visual analysis of spatiotemporal events in different situations.

Climbing fiber response

Figure 11 shows the membrane potential waveforms in the cases of climbing fiber stimulation. Figure 11*A* is a genuine intracellular recording of so-called 'complex spike' from the Purkinje cell of frog (Llinás *et al.* [34], Fig.4). This characteristic response is highly stereotyped, the initial, taller than normal spike is followed with about 2 msec delay by double twins of oscillatory spikes, in each twin the second spike having a smaller (about one msec) delay and smaller amplitude. It is intriguing that the climbing fiber response is surprisingly similar not only from cell to cell but also in different species (c.f. Fig. 11*A* with recording in cat, Llinás [1], although the morphological structure of these cells must obviously exhibit more than trivial differences.

The climbing fiber activation of Purkinje cell was modeled by 'opening' 5 climbing fiber–Purkinje cell synapses on every branch for a duration of 0.15 msec, with a linear shift of onset, zero at somatic level and 0.5 msec at the uppermost branches. The overall membrane potential waveform at somatic level is shown on Fig. 11*B*, which is considered as an adequate analog of the genuine recordings, showing the large initial spike which, before a complete repolarization, produces a second spike of about the 50-60% of amplitude, shortly followed by a small peak then the twin-structure of wavelets repeated once again and the response closed by a final decay of the depolarization.

Figure 11*C* shows the membrane potential waveforms throughout the cell, therefore revealing the mechanism reponsible for the resulting complex spike wavelets. As it is clear by careful observation of the recordings over the spatial compartments the series of peaks of the complex spike are generated by the repetitive firing of the *initial segment* (shortly followed by the somatic firing) in contrast to speculations that it originates in the dendrites or that it is the electrotonically registrated waveform of

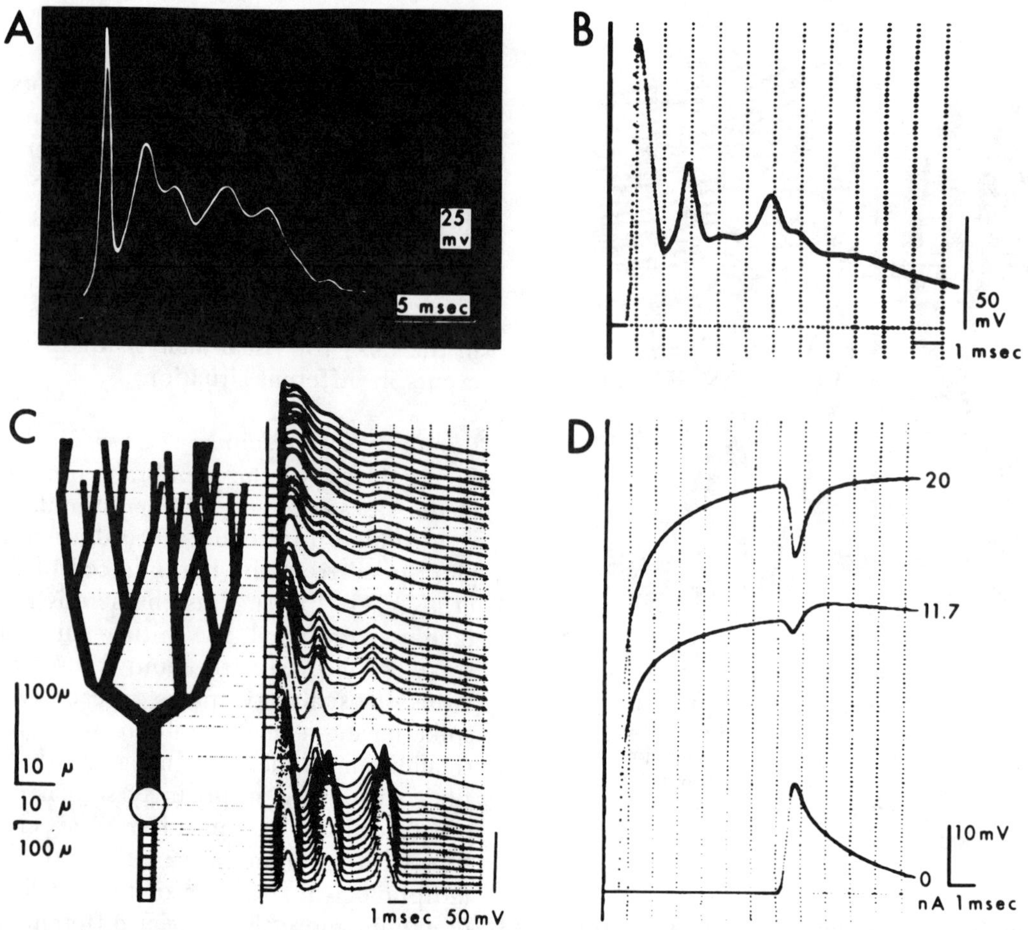

Fig.11 Climbing fiber activation of Purkinje cell with totally passive dendritic tree. A shows a genuine intracellular recording of the climbing fiber evoked, so-called complex spike from frog. B is the comparable recording produced by the model. C is the identical activation, recorded throughout the cell at all of its 62 compartments. Note that the wavelets of the somatic recording are produced by the repetitive firing of the initial segment of axon, and the secondary smaller peaks are produced by the 'backfiring' of the propagating axonal spike into the SD region. D shows the EPSP evoked by climbing fiber (if the model corresponds to a deteriorated, non-excitable cell). By applying different depolarizing current to the soma, the EPSP is totally, or only partially reversed in the intriguing bi-phasic manner known from experimental findings.

dendritic spikes (Martinez *et al.* [28] or Eccles *et al.* [26]). It is clear that in the model this repetitive firing is caused by the delayed repolarization of the dendritic branches, which drives the quickly repolarizing somatic and initial segment compartments into firing again. As a reassurance for the cogency of the model could be interpreted the little detail of the appearance of quite variable twin-peaks which are produced by the model in agreement with experimental findings (Fig. 11*A*). As seen on Fig. 11*C* this phenomenon may be produced by the 'backfire' invasion to the SD region of the propagating axonal action potential, which follows the initial segment firing slightly delayed.

Analysis of EPSP evoked by climbing fiber on Purkinje cell

Also as a reassuring nuance can be interpreted the ability of the model to reproduce (also in quantitative terms) the sophisticated phenomenon of total-phasic or bi-phasic, partial reversal of EPSP evoked by climbing fiber on deteriorated, non-excitable Purkinje cell (Fig. 11*D*). Three modeled recordings are shown in Fig. 11*D*). The EPSP is modified by a current injection to the soma, started at 0 msec with amplitudes of 0·0, 11·7, or 20 nA. As seen, the unmodified EPSP exhibits the well known potential waveform while it could be totally or partially reversed (in the intriguing bi-phasic manner) by 20 or

11·7 nA depolarizing current, respectively.

IV. DISCUSSION

While in going into details of the results one could get an impression about the complexity that the research of this part of the brain entails. In this chapter at least both ends of the spectrum of levels of approaches (within which the particular details would fall) are pinned down by the holistic circuitry model of cerebellar cortex on one hand, and the single unit model of Purkinje cell on the other. It is easy to see that problems such as the overall spatial features of the distribution of Purkinje cells with different dynamism (Precht & Llinás [19,20]) can be directly addressed to the holistic circuitry model while other, such as the interpretation of cellular phenomena, like the climbing fiber response of Purkinje cells fall naturally into the scope of single unit model. Therefore the model can directly help, strictly on these separate levels:

1. By making the complexity inherent in the neuronal *circuitry* manageable, therefore rendering it possible to test rigorously (also in quantitative terms) such questions as gross spatial distribution of activities over the cerebellar cortex of frog. This, exemplified in Fig. 4 suggested such unexpected features that separate mossy fiber inputs, rather than producing an integrative parallel fiber and Purkinje cell output, tend to break into integer spindles of parallel fiber activity. This, beyond that it provides additional support to the assumption that the 'unitary' activation of cerebellar cortex may, indeed, be a laterally inhibited column of fired Purkinje cells, (Szentágothai [7,8]) suggests also an explanation as to why the frog cerebellar circuit lacks the basket cell inhibition (because apparently there is no obvious need for it in the frog's simple, one-folium cerebellum: the parallel fiber spindles seem to separate spontaneously).

2. By making the complexity inherent in one *single neuron* manageable. As it was implied in the foregoing, although all the knowledge necessary to explain the electrogenesis of a spatially distributed structure of Purkinje cell has been available (in the form of H-H equations, Hodgkin & Huxley [32]) for almost a quarter of a century by now, the sheer magnitude of the required computation prevented hitherto giving a better answer than some highly disputable speculations for the location of the electrogenesis of the wavelets in the complex spike. The authors are aware that a period of all-out efforts is being opened by which many curious electrophysiological recordings of the spatially highly complex and specialized neurons will receive adequate interpretations and that will bring into a common platform the morphologists and physiologists who are at present too isolated by the highly different techniques they use in pushing towards their identical goal of understanding the basic features of single neurons.

While the quoted problems are well suited to either the circuitry model or the model of single cell, it is easy to perceive several key questions which must be asked at an intermediary level of approaches. This is, indeed, one of the fundamental merits of the policy adopted in the present modeling: ie. to provide an overall framework for the data *(a)* to be absorbed into a self-consistent system of models, *(b)* to be handled by a method capable of dealing with the resulting complexity and therefore *(c)* to set the theorist free of the separate, narrow levels of approaches that are invariably given by singular experimental methods.

For example, the intriguing problem of highly individual character of Purkinje cells (even nearby) over the cerebellar cortex (Precht & Llinás [19]. Eccles *et al.* [35]) requires first a tool for handling the overall spatial distribution of mossy fiber-granule cell-parallel fiber activities. With this kind of help it can be revealed to what extent can the variety of activities be attributed to the spatial variance of afferentation. However, if (as the case seems to be, Pellionisz & Llinás, in preparation) it is unlikely that the afferentation *alone* can explain the found abundance of dynamism (cf. Fig. 7 in Llinás [36]) then it is necessary to explore the potential factors which can be responsible for an eventual variation in the dynamism of individual Purkinje cells.

The author believes that it may show the heuristic value of this kind of modeling that through the tedious labor of synthetizing the data into a framework of models a hypothesis emerged centered on the very assumption (yet unproven) that the dynamics of the firing of any Purkinje cell is 'shaped' by the deep and prolonged depolarization produced by the climbing fiber activation (Pellionisz [37]). It might be just by the so far developed model as a tool in hand that a workable blueprint of the cerebellum may be produced.

Acknowledgement

This work was supported by the USPHS research grant NS-09916 from the NINDS.

References

1. LLINÁS, R, 'The cortex of the cerebellum', *Scientific American* **232**, 56-71 (1975).
2. PALKOVITS, M., MAGYAR, P., and SZENTÁGOTHAI, J., 'Quantitative histological analysis of cerebellum in cat III', *Brain Res.* **34**, 1-18 (1971).
3. PELLIONISZ, A., 'Computer simulation of the pattern transfer of large cerebellar neuronal fields', *Acta Biochem, Biophys. Acad. Sci. Hung.* **5**, 71-79 (1970).
4. ECCLES, J.C., ITO, M., and SZENTÁGOTHAI, J., *The Cerebellum as a Neuronal Machine*, Springer-Verlag, Berlin, Heidelberg, New York (1967).
5. PALAY, S.L. and CHAN-PALAY, Victoria, *Cerebellar Cortex,* Springer (1974).
6. LLINÁS, R. (ed.), *Neurobiology of Cerebellar Evolution and Development,* Am. Med. Association, Washington, DC (1969).
7. SZENTÁGOTHAI, J., 'Ujabb adatok a synapsis functionalis anatómiájához', *Magy. Tud. Akad., Biol. Orv. Tud. Oszt. Közl.* 6, 217-227 (1963).
8. SZENTÁGOTHAI, J., 'The use of degeneration methods in the investigation of short neuronal connexions', In: *Progress in Brain Research*, Vol. 14: *Degeneration Patterns in the Nervous System* (M. Singer and J.P. Schade eds.), Elsevier, Amsterdam, 1-32 (1965).
9. MORTIMER, J.A., *A Cellular Model for Mammalian Cerebellar Cortex*, Technical Report, Univ. of Michigan (1970).
10. PALKOVITS, M., MAGYAR, P., and SZENTÁGOTHAI, 'Quantitative Histological analysis of cerebellum in cat I', *Brain Res.* **32**, 1-13 (1971).
11. PALKOVITS, M., MAGYAR, P., and SZENTÁGOTHAI, J., 'Quantitative histological analysis of cerebellum in cat II', *Brain Res.* **32**, 15-30 (1971).
12. PALKOVITS, M., MAGYAR, P., and SZENTÁGOTHAI, J., 'Quantitative histological analysis of cerebellum in cat IV', *Brain Res.* **45**, 15-29 (1972).
13. PELLIONISZ, A. and SZENTÁGOTHAI, J., 'Dynamic single unit simulation of a realistic cerebellar network model', *Brain Res.* **49**, 83-99 (1973).
14. PELLIONISZ, A. and SZENTÁGOTHAI, J., 'Dynamic single unit simulation of a realistic cerebellar network model II', *Brain Res.* **68**, 19-40 (1974).
15. PELLIONISZ, A., PALKOVITS, M., HÁMORI, J., PINTÉR, E., and SZENTÁGOTHAI, J., 'Quantitative comparison of the cerebellar synaptic organization of immobilized and normal kittens', Interntl. Union for Pure and Applied Biophys., *IV. Internatl. Biophysics Congress, Symp. Papers, 3,* Acad. Sci. USSR, Puschino 1973, 381-384.
16. PELLIONISZ, A., PERKEL, D., and LLINÁS, R., 'Computer model of the cerebellum of Frog. I. Morphogenesis of cerebellar cortex', *J. Neurobiology* 1976 (in press).
17. PELLIONISZ, A., and LLINÁS, R., 'Computer model of the cerebellum of Frog. II. Integrative properties of Purkinje cells', *J. Neurobiology* 1976 (in press).
18. FRANKENHAEUSER, B. and HUXLEY, A.F., 'The action potential in the myelinated nerve fibre of *Xenopus Laevis* as computed on the basis of voltage clamp data', *J. Physiol.* **171**, 302-315 (1964).
19. PRECHT, W. and LLINÁS, R., 'Functional organization of the vestibular afferents to the cerebellar cortex of Frog and Cat', *Exptl. Brain Res.* **9**, 30-52 (1969).
20. PRECHT, W. and LLINÁS, R., 'Comparative aspects of the vestibular input to the cerebellum', in: *Neurobiology of Cerebellar Evolution and Development* (R. Llinás, ed.), Amer. Med. Assoc., Chicago, 677-702 (1969).
21. LLINÁS, R., 'Functional aspects of interneuronal evolution in the cerebellar cortex', in: *The Interneuron* (M.A.B. Brazier, ed.), Univ. of Calif. Press, Berkeley and Los Angeles, 329-348 (1969).
22. HILLMAN, D.E., 'Morphological organization of Frog cerebellar cortex, A light and electron microscopic study', *J. Neurophys.* **6**, 818-846 (1969).
23. LLINÁS, R., 'Neuronal operations in cerebellar transactions', in: *The Neuro-Sciences: Second Study Program* (F.O. Schmitt, ed.-in-chief), Rockerfeller Univ. Press, New York, 409-426 (1969).
24. ECCLES, J.C., LLINÁS, R., and SASAKI, K., 'Parallel fiber stimulation and responses induced thereby in the Purkinje cells of the cerebellum', *Exptl. Brain Res.* **1**, 17-39 (1966).
25. HILLMAN, D.E., 'Vestibulocerebellar input in the Frog: Anatomy', in: *Progress in Brain Research* Vol.37 (A. Brodal and O. Pompeiano, eds.): *Basic Aspects of Central Vestibular Mechanism,* Elsevier, Amsterdam, 329-339 (1972).
26. ECCLES, J.C., LLINÁS, R., and SASAKI, K., 'The excitatory synaptic action of climbing fibers on the Purkinje cells of the cerebellum', *J. Physiol.* (London) **182**, 268-296 (1966).
27. SZENTÁGOTHAI, J. und RAJKOVITS, K., 'Uber den Ursprung der Kletterfasern des Kleinhirns', *Z. Anat. Entwickl.-Gesch.* **121**, 130-141 (1959).
28. MARTINEZ, F.E., CRILL, W.E., and KENNEDY, T.T., 'Electrogenesis of cerebellar Purkinje cell responses in Cats', *J. Neurophys.* **34**, 348-356 (1971).
29. NICHOLSON, C. and LLINÁS, R., 'Field poten-

tials in the Alligator cerebellum and theory of their relationship to Purkinje cell dendritic spikes', *J. Neurophys.* 34, 509-531 (1971).
30. RALL, W., 'Electrophysiology of a dendritic neuron model', *Biophys. J.* **2**, 145-167 (1962).
31. DODGE, F.A. and COOLEY, J.W., 'Action potential of the motoneuron', *IBM J. Res. Develop.*, 219-229 (May 1973).
32. HODGKIN, A.L. and HUXLEY, A.F., 'A quantitative description of membrane current and its application to conduction and excitation in nerve', *J. Physiol.* (London) **117**, 500-544 (1952).
33. LLINÁS, R., BLOEDEL, J.R., and ROBERTS, W., 'Antidromic invasion of Purkinje cells in Frog cerebellum', *J. Neurophys.* 32, 881-891 (1969).
34. LLINÁS, R., BLOEDEL, J.R., and HILLMAN, D.E., 'Functional characterization of the neuronal circuitry of the Frog cerebellar cortex', *J. Neurophys.* **32**, 847-870 (1969).
35. ECCLES, J.C., FABER, D.S., MURPHY, J.T., SABAH, N.H., and TABORIKOVA, Helena, 'Investigations on integration of mossy fiber inputs to Purkinje cells in the anterior lobe', *Exptl. Brain Res.* **13**, 54-77 (1971).
36. LLINÁS, R., 'Frog cerebellum: biological basis for a computer model', *Math. Biosci.* **11**, 137-151 (1971).
37. PELLIONISZ, A., 'Cerebellar control theory', *Math. Biosci.* (submitted October 1974).

Resonance phenomena in the electrical activity of the brain

E. BAŞAR, A. GÖNDER, M. RONA and P. UNGAN
Institute of Biophysics,
Hacettepe University, Ankara, Turkey

I. Introduction

The method of averaged evoked potentials is one of the most frequently used methods in the study of signal transport in the brain (Rémond [1]; Regan [2]). In previous studies on systems theory of potentials in the cat brain we have studied the time and frequency characteristics of some of the auditory pathway nuclei, reticular formation and hippocampus. Our recent systems theoretical studies on the brain potentials consisted of the following points:

1. Determination of amplitude characteristics of different nuclei of the brain with the help of selectively averaged evoked potentials. (The selective averaging method was introduced by Başar *et al.* [3] and Ungan and Başar [4]).
2. Determination of the dynamics of potentials simultaneously obtained from various brain structures, in order to evaluate the common features of system characteristics of these various brain structures.
3. Determination of the relationship (or interaction) between spontaneous activity and evoked potentials of the brain.
4. Using the results from the studies described above, a working hypothesis for the dynamics of potentials of the brain during waking and sleep stages was outlined (Başar *et al.* [5,6]).

One of the most fundamental features of the results from the systems theory experiments was the demonstration of the existence of an important relationship between the spontaneous and evoked activities in the brain. In the mentioned studies we have searched for the existence of similar activity bands in the power spectra of the spontaneous activity and amplitude characteristics obtained from transient evoked potentials. According to the comparative results of power spectra and frequency characteristics we defined three different kinds of resonance phenomena obtained upon stimulation of the studied brain structures: (1) Weak resonances (occurring between 1 and 8 Hz: delta and theta frequency range); (2) Alpha resonances (frequency range around 10-12 Hz); and (3) Strong resonances (occurring in the frequency range higher than 20 Hz). A direct comparison of both functions (power spectra and amplitude characteristics) is not possible, since the power spectrum reflects the intrinsic activity of a system and the amplitude characteristic represents the response of the system to an excitation. In our recent studies we were able to compare both functions by considering only the relative magnitudes of various peaks revealed by these functions, rather than their absolute magnitudes.

In the present chapter we introduce a methodology consisting of a series of methods which will be applied alternatively to the spontaneous activity and evoked potentials in order to demonstrate the existence of resonance phenomena, by allowing the comparison of absolute magnitudes of EEG and evoked potential (EP) components. In our recent studies we worked with selectively averaged evoked potentials (SAEPs). In the present study, we also apply the selective averaging method; however, we confine our attention mostly to single evoked potentials and to the spontaneous activity.

The results presented in this study show the occurence of the strong resonance phenomena in a more exact manner. The demonstration of the strong resonance phenomena brings a new point of view into the analysis of resonance phenomena in the brain: the strong resonances have a probabilistic nature. First steps towards the understanding of probabilistic occurrence of the strong resonances in the brain are begun in this report.

II. Methods

Surgery: Our investigations were carried out using nine cats with chronically implanted stainless steel electrodes of 0·2 mm diameter in the studied brain center. The derivations were against a common reference which consisted of three stainless steel screws in different regions of the skull. During the experiments the cats were freely moving in an echofree and sound-proof room. The experimental setup is shown schematically in Fig. 1.

Stimulation: The auditory step functions consisted of tone bursts of 2000 Hz and 80 dB (SPL). The auditory step functions lasted 3 sec and were applied at intervals of 16 sec.

The computational methodology

Our new methodology for comparison of the brain's spontaneous activity and the evoked potentials can be briefly described as follows:

1. A sample of the spontaneous activity of the studied brain structure is recorded and stored in the disc-memory of the computer.

2. A stimulation signal is applied to the experimental animal. (In this study an auditory step function in the form of a tone burst of 2000 Hz.)

3. A single evoked response is recorded and stored in the disc-memory of the computer.

4. The operations explained above (steps 1, 2, and 3) are repeated about 100 times.

5. The evoked potential epochs stored in the disc-memory of the computer are averaged using the selective averaging method previously described (Başar *et al.* [3]; Ungan and Başar [4]).

6. The selectively averaged evoked potential (SAEP) is transformed to the frequency domain with the Fourier transform in order to obtain the amplitude frequency characteristic, $|G(j\omega)|$, of the studied brain structure. (For details, see Başar [7], Başar [8], Başar *et al.* [3]).

7. The frequency band limits and band width of the amplitude maxima in $|G(j\omega)|$ are determined, and theoretical pass-band filters according to these band limits are determined using the theoretical filtering method previously described (Başar and Ungan [9] Başar *et al.* [3]).

8. The selected single evoked potentials (single evoked potentials with the help of which the SAEPs are computed; see step 5) stored in the disc-memory are then filtered with theoretical pass-band filters determined in step 7.

9. The epoch of spontaneous activity, which was recorded immediately before the stimulation giving rise to the single evoked response mentioned in step 8, is also filtered with the same theoretical filter applied to the evoked potential.

10. The voltages of maximal amplitudes existing in the filtered spontaneous activity are compared with those of the filtered evoked potentials and the amplification in the evoked potentials is computed.

11. The procedure explained in steps 8-10 is repeated for all the stored epochs.

III. Results

In the previous section a procedure was given for an absolute comparison between the spontaneous activity components and the evoked potentials of the brain. We have applied this procedure to the spontaneous and evoked potentials which we recorded from various structures of the brain; the acoustic cortex, inferior colliculus, reticular formation, hippocampus and the cerebellum. In this chapter we will describe the results briefly, giving only a few examples. For this purpose, let us study the following case: Fig. 2 illustrates a typical analysis of on-going and evoked activities in the mesencephalic reticular formation of the cat brain. At the top of the illustration a selectively averaged evoked potential (SAEP) of the cat reticular formation is presented. This SAEP is the average of five single evoked potentials selected from about thirty single evoked potentials (EPs). The same figure illustrates a single evoked potential, one of the epochs from which the SAEP shown at the top was computed. Figure 2 also illustrates the un-

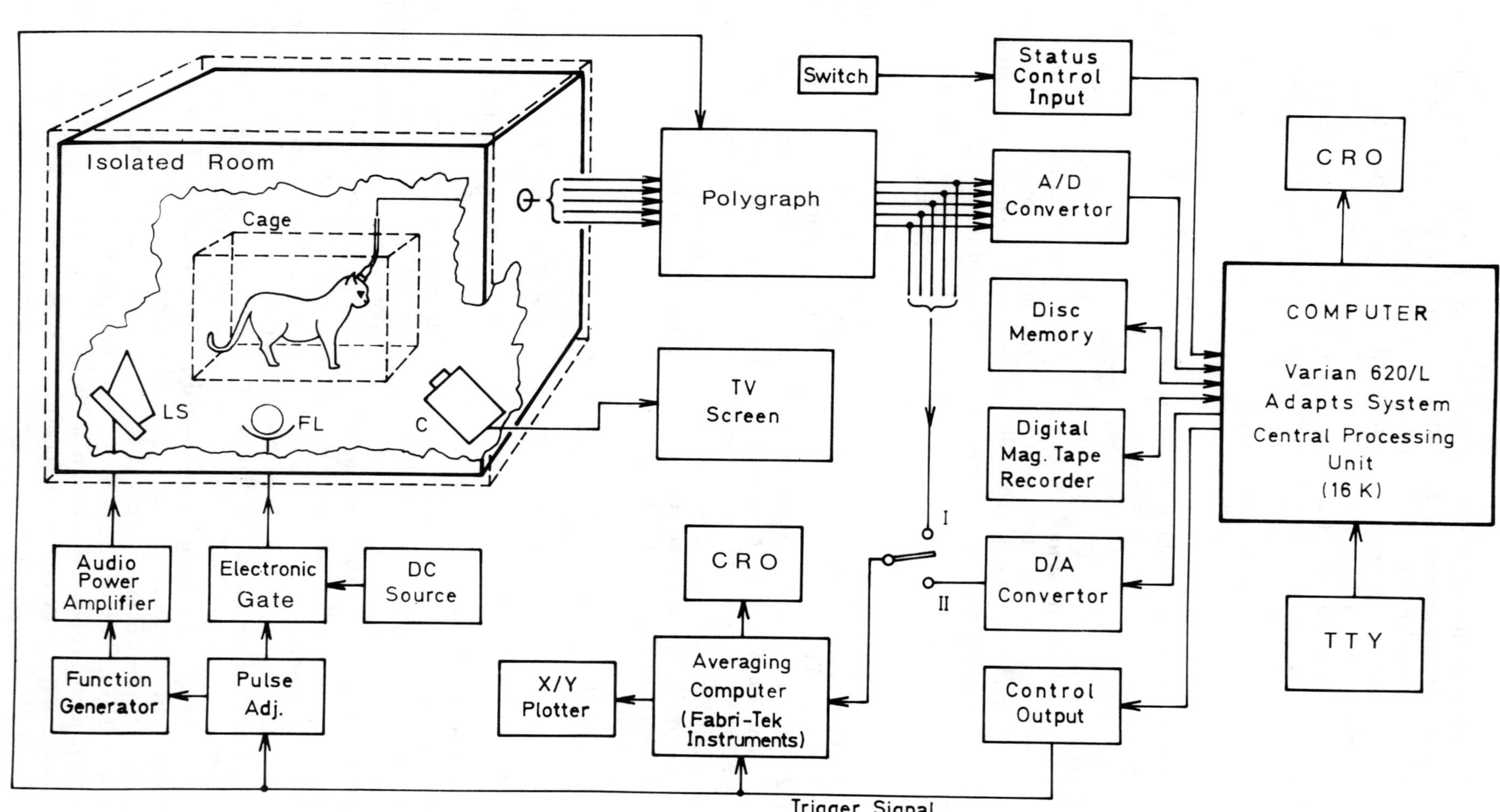

Fig.1 The experimental set-up used for the recording of spontaneous and evoked potentials from the cat brain and for storage of all measured and computed data. The application of various computational methods and plotting of the computed curves are also carried out by means of this set-up.

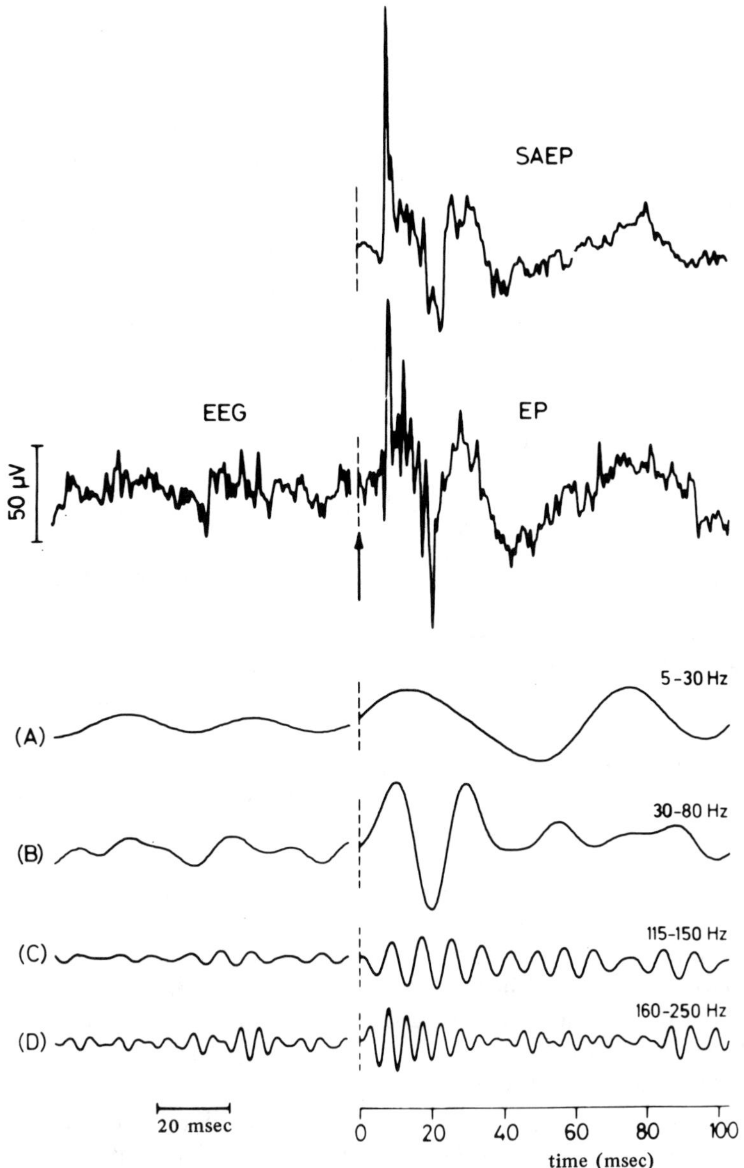

Fig.2 Strong resonance phenomena in the brain as demonstrated by comparison of the evoked potential components with the spontaneous activity components in the time domain. These activities are recorded from the reticular formation of the cat brain. EEG is the spontaneous activity recorded just prior to the stimulus application (indicated by an upward arrow). EP is one of the five single evoked potentials which are used in order to obtain the selectively averaged evoked potential (SAEP) shown at the top. The curves below are obtained from the spontaneous activity (EEG) and from the evoked potential (EP) by means of theoretical filters whose band limits (given at the right side of each curve) are determined from the amplitude characteristic given in Fig.3. The filtered curves at the left column are the EEG components within the specified frequency band limits whereas those at the right are the EP components of the same frequencies.

filtered spontaneous activity before stimulation. The selectively averaged evoked potential (SAEP) of Fig. 2 was used to determine the band limits of the theoretical filters to be applied to the single evoked potentials and to the spontaneous activity. Figure 3 illustrates the amplitude frequency characteristic obtained using the transient evoked response (SAEP) of Fig. 2. In Fig. 3, only the behavior between 1-300 Hz is illustrated. The band limits of theoretical pass-band filters were determined according to the

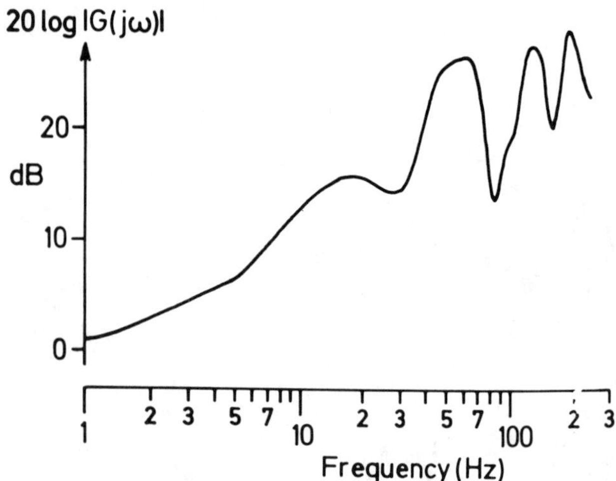

Fig.3 The amplitude frequency characteristic which is computed from the SAEP in Fig.2 by means of the TRFC-Method. Along the abcissa is the frequency in logarithmic scale, along the ordinate is the relative potential amplitude in decibels. The curve is normalized in such a way that the amplitude at 0 Hz is equal to 1 (20 log 1 = 0).

minima of the reticular formation amplitude characteristic of Fig. 3, and then pass-band filters with these band limits were applied to the reticular spontaneous activity and single evoked potentials: (5-30) Hz filter, (30-80) Hz filter, (115-150) Hz filter and (160-250) Hz filter. In this study, we confine our attention only to this frequency range, although higher frequency components are revealed in the reticular spontaneous and evoked activities, as one can immediately check using the procedure given in the section of 'Methods'. Figure 2 illustrates further the filtered spontaneous activity and evoked potentials obtained with the use of the described theoretical filters. The filtered EP epochs elicited with a small delay of 2.5 msec are shown in sections A, B, C, and D, of Fig. 2. The band limits of the pass-band filters are shown at the right side of the corresponding potentials. In comparing the maximal potential amplitudes of the filtered EEG with those of the filtered EPs, an important amplification is seen in the EPs against EEG: in the frequency band of (5-30) Hz the amplification is about 2·5; in the frequency band of (30-80) Hz the amplification is about 5; in the frequency band of (115-150) Hz the amplification is about 3·5 and in the frequency band of (160-250) Hz the amplification is about 2.

Figure 2 is a typical example showing that the frequency components existing in the on-going activity are amplified upon stimulation, or that the spontaneous activity components strongly resonate upon stimulation, as we have previously assumed (Başar *et al.* [5,6]). However, the amplification factors in the given frequency channels are not constant over all the single evoked potential epochs. Although, relatively compared, the amplification seems to be steady in defined frequency channels (also in different cats), during a recording session to obtain a selectively averaged evoked potential, important fluctuations of the amplification factor were observed. Table 1 explains this observation in various channels from an experimental session on a single cat. What can be the source of these fluctuations? This question will be discussed in the next section.

The typical example of the reticular formation, shown in Fig. 2, is chosen from similar observations during experiments with 9 cats. Experiments on strong resonance, which gave similar results, were also performed with brain structures such as the cortex, inferior colliculus and cerebellum. Usually, the strong resonance phenomena are observed in frequencies higher than 20 Hz, as we have previously reported (Başar *et al.* [5]).

We will also give an example from the lower frequency range, where only weak resonance phenomena should be observed, as we recently assumed (Başar *et al.* [5,6]. Figure 4 illustrates the filtered spontaneous activity and the filtered single evoked potential of the cat hippocampus. The band limits of the pass-band filters used were (2·5-6) Hz. We show here only the theta component of the spontaneous and evoked potentials; for higher frequency components of the hippocampal potentials the reader is referred to the references (Başar *et al.* [3], Başar and Özesmi [10], Boudreau [11]). From Fig. 4 one will recognize immediately that there is no increase in the amplitude of the filtered single evoked potential against the spontaneous activity recorded before the stimulation. The amplification factor is about 1.

We also performed experiments during the spindle sleep stage and the slow wave sleep stage, and applied the same filtering procedure. We do not explain here in detail the results obtained. However, we want to mention that the amplification factor was also around 1 for the 1-3 Hz delta waves and for the 10-12 Hz cortex spindles.

IV. Discussion

Demonstration of strong and weak resonance phenomena

Our rationale for these experiments was the exact determination of the existence of strong resonance

Table 1

Maximal amplitudes (μV)			
The pass-band filter applied	EEG components (S)	EP components (E)	Amplification factor (E/S)
5 25 Hz	15.5 13.8 8.3 13.2 10.9 — 12.3 (mean value)	29.6 19.7 13.7 16.3 38.8 — 23.6 (mean value)	1.91 1.43 1.65 1.23 3.55 — 1.91 (mean value)
25 80 Hz	12.1 13.9 9.9 15.3 11.0 — 12.4 (mean value)	42.2 55.0 67.3 62.5 50.0 — 55.5 (mean value)	3.51 3.96 6.80 4.09 4.55 — 4.45 (mean value)
80 150 Hz	17.8 17.1 13.0 14.4 20.8 — 16.6 (mean value)	28.0 24.8 19.3 29.4 60.6 — 32.4 (mean value)	1.57 1.45 1.49 2.04 2.92 — 1.95 (mean value)
150 250 Hz	8.1 8.2 12.2 10.8 11.5 — 10.2 (mean value)	42.1 34.7 24.9 24.8 16.7 — 33.2 (mean value)	5.17 4.22 2.05 2.30 1.46 — 3.02 (mean value)

phenomena in the brain. Therefore, we confined our attention to the time domain and compared filtered single evoked responses with filtered spontaneous activity in the time domain. In fact, the results presented in the previous section show that, in the frequency range lower than 20 Hz, resonances with amplification factors up to 5 do not occur in the brain (Fig. 2 and Table 1). The demonstration is performed using the absolute magnitudes of EEG and EP components, thus showing in an exact manner the strong

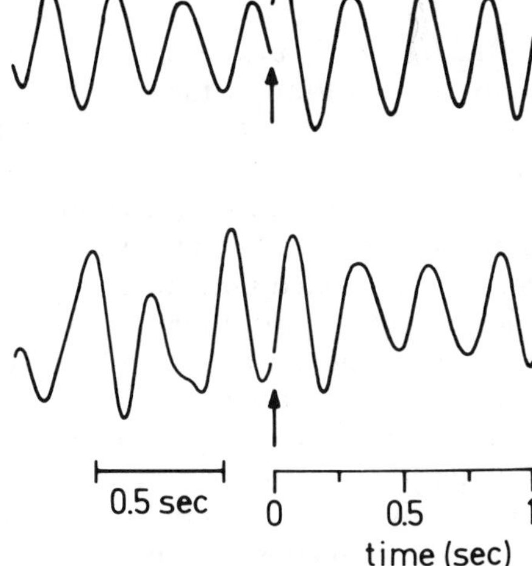

Fig.4 Weak resonance phenomena in the brain as demonstrated by comparison of the evoked potential components (at the right side) with the spontaneous activity components (at the left side) in the time domain. These components are obtained by the application of adequate theoretical filters (pass-band: 2.5-6 Hz) to the activities recorded from the hippocampus of the cat brain. The application of stimulation signal is indicated by an upward arrow.

resonance phenomena predicted in our previous studies, which were based on relative comparisons (Başar *et al.* [5,6]). Also, the weak resonance phenomena in the lower frequency range [delta (1-3 Hz) and theta (3-8 Hz)] are demonstrated using the same methodology.

Although the accuracy and the reliability of the filtering method were discussed in detail in our previous studies (Başar & Ungan [9], Başar *et al.* [3,5] Başar [12], Başar[8]) we include two remarks here directed to the EEG-analyst who doesn't use computer methodology:

a. When one examines the unfiltered EEG-data and compares them with the filtered data (EEG components), one can immediately recognize that filtered components do exist in the unfiltered EEG-data in the form of instantaneous frequencies (Fig. 2). We have already shown the power density function of these high frequency activities (Başar *et al.* [5]). We make the same observation on examination of the single EPs in Fig. 2. The filtered components do exist in the unfiltered EPs as instantaneous frequency components.

b. Frequency components analyzed in the single EPs do exist in the averaged evoked potential (SAEP shown in Fig. 2, see also Başar *et al.* [5]). In other words, high frequency components existing in single EPs are not resulting from individual events as the SAEP of Fig. 2 and also the amplitude characteristic of Fig. 3 show. They certainly result from consistent phenomena.

Probabilistic nature of the strong resonances

An important feature of the observations on strong resonances is their qualitatively repeatable structure along with enormous fluctuations in the amplification factor. The results presented in the previous section showed that the maximal amplitudes of the filtered single evoked potentials displayed a fluctuation leading to amplification factors between 1 and 6 (Table 1), although the experimental conditions were kept as invariable as possible.

with respect to parameters, in addition to its average amplitude, might result in a more causal relationship between a given evoked activity and the immediately preceding spontaneous activity. The choice of the set of such parameters is open for conjecture at this time. With the discovery of each of such 'hidden' parameters, what looks like a fluctuation now might be included in the systematics of evoked potentials. This set, if it exists, will probably be infinite and there will always be a part of the signal that will be classified as a fluctuation.

In order to search the cause of the probabilistic nature of the strong resonance phenomena, we first consider a simple harmonic oscillator upon which an external time-dependent force is applied at time t_0. Now, for a set of given initial conditions and a forcing function of time, the solution giving displacement from equilibrium as a function of time is unique. One can, of course, show this with an experiment.

At this point, we are in a position to decide on the form of generalization to be made in order to choose some relevant parameters for studying the components of the electrical activity of the brain with the help of a simple harmonic oscillator (eg. a pendulum). A tempting step would be to permit the mass of the oscilating particle and the frequency of its oscillations to be functions of time. We instead choose to break away from the mechanical analogy and write the following equation for simple harmonic oscillation:

$$\frac{d^2\psi(t)}{dt^2} + \alpha(t)\,\psi(t) = F(t-t_0) \tag{1}$$

where $\psi(t)$ is the deviation from equilibrium and $F(t-t_0)$ is the external force. Now, for $t<t_0$ and $\alpha(t) = constant$, we can write the solution as:

$$\psi(t) = A\,\sin(\omega t + \phi) \tag{2}$$

This is the general wave expression where A is the amplitude, ω is the angular frequency and ϕ is the phase angle.

For $t<t_0$ suppose we label points on the time axis such that between any two consecutive ones the function $\alpha(t)$ is independent of time and, therefore, the solution is of the type given above (2). If $\alpha(t)$ goes through an abrupt change as time goes through a labeled point on the time axis, we have to choose a new set of A, ω and ϕ.

To study the pass-band filtered spontaneous activity (EEG components), we choose a constant $\omega=\omega_0$ the central frequency of the band. We keep A constant and choose a phase so that at each sampling point, the observed signal is fitted to the same solution. We refer to time intervals during which the phase angle remains constant as the 'coherence time'. During a coherence time α is constant.

The motivation to analyze the data in this fashion is based on two points. Firstly, we would like to analyze the statistics of coherence time. Secondly, we would like to study the fluctuations in the amplification factors (of the amplitudes) in *strong resonance* channels in the light of the phase information of the spontaneous activity at the instant of stimulus application.

To study the evoked activity we use a step function

$$F(t-t_0) = u(t-t_0),$$

where

$$u(t-t_0) = 1 \qquad t > t_0,$$
$$\qquad = 0 \qquad t < t_0.$$

Within a coherence time of the evoked activity $\alpha(t) = constant$ and we can write a shifted solution adding a factor C:

$$\psi(t) = C + A \sin(\omega t + \phi)$$

The justifications and the motivations for introducing a fluctuating phase are the same as in the case of the spontaneous activity.

Let us now try to describe the filtered spontaneous activities (EEG components) and the filtered evoked potentials (EP components) presented in Fig. 2 with the terminology of the wave expression. We have approximated that ω remains constant within the filter band limits. However, the amplitude A and the phase angle ϕ are not predictable entities as one immediately recognizes from the course of the spontaneous waves in Fig. 2. Therefore, we are led to modify our harmonic oscillator model by describing the wave expression in probabilistic form. Since $\phi(t)$ takes on various values during different coherence times in a manner which is probabilistic, we generalize the wave expression to describe the filtered spontaneous activity as

$$\psi\{\tilde{\phi}, t\} = A \sin(\omega t + \tilde{\phi})$$

and the wave expression to describe the filtered evoked activity as

$$\psi\{\tilde{\phi}, t\} = C + A \sin(\omega t + \tilde{\phi})$$

with ϕ being assigned different values during different coherence times in a seemingly random fashion.

Now, we try to fit the filtered curves, which were obtained from the brain, with pure sine functions having the center frequencies of the defined filters. (We should recall that the theoretical filters we applied are ideal filters without phase shift; see Başar & Ungan [9]). Figure 5A illustrates again an epoch of the filtered spontaneous brain wave (recorded in the cat reticular formation). Below (Fig. 5B) we plot the curve which shows the deviation of the phase

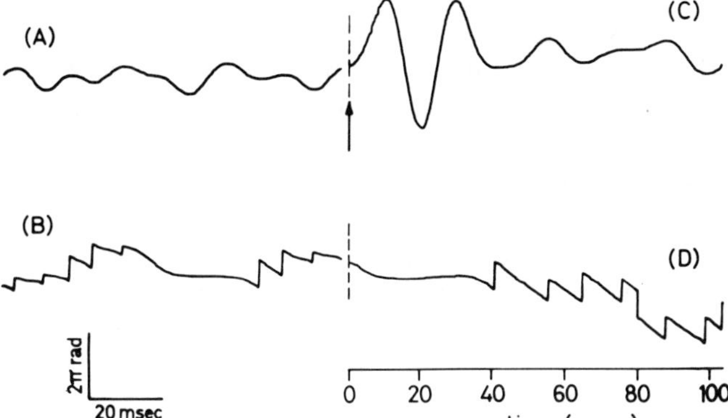

Fig.5 The spontaneous activity component (A) just before the stimulation (indicated by an upward arrow) after the application of which the evoked potential component (C) is recorded. (B) and (D) are the curves which present respectively the phase angle ϕ of the spontaneous activity component and of the evoked potential component relative to the phase angle of a pure sine wave having the same frequency with the components (A) and (C).

angle ϕ relative to the phase angle of a pure sine wave having the same frequency. We immediately see that the relative phase shift indeed has large fluctuations, since irregular jumps and horizontal lines are revealed in this curve. The horizontal lines correspond to a constant relative phase during a coherence time with an almost sinusoidal shape of the spontaneous activity in Fig. 5A. Figure 5C shows the filtered evoked activity (EP component) obtained upon stimulation, immediately after the recording of the wave shown in Fig. 5A. The phase deviation curve in Fig. 5D is almost horizontal for the first 35 msec. After a duration of 40 msec, irregular fluctuations similar to the curve of Fig. 5B are seen. These results may be interpreted as follows:

1. An otherwise pure sine wave with a randomly fluctuating phase due to yet undetermined perturbations can be used to describe the electrical activity components in the brain.

2. These indeterministic phase values play a major role in the response curves (single evoked potentials), since in the response of a harmonic oscillator which is forced by an external signal (here acoustical stimulation), the value of the phase angle ϕ at the moment of stimulus application should be of basic importance for the amplification factor, or the amplitude of the response. (This assumption is contained in the classical wave equation.) As the results of Fig. 5 show, the phase angle ϕ can be described as a problistic entity $\tilde{\phi}$. Accordingly, the maximal amplitudes of the response components (the filtered single evoked potentials) are indeterministic.

3. During a short period after the stimulation, the phase angle of the response fits well together with the pure sine wave function (horizontal line at the beginning of curve D in Fig. 5). It means, on the one hand, that the external stimulus pushes the electrical activity of the brain into the coherent state of a simple harmonic oscillator for a coherence time, after which fluctuations set in again. On the other hand, the phase angle of the oscillations behaves deterministically for a short period, although the amplification factor seems to behave indeterministically.

To the above introduced assumptions, we want to add the following point of view, which we are justified in neglecting in an introductory probabilistic approach to the resonance phenomenon in the brain. The different components of the spontaneous brain activity, which we defined (described) in the form of filtered curves, might represent signals resulting from coupled oscillators. In this case we must modify our model further by considering a number of probabilistic harmonic oscillators with some coupling functions between them. Furthermore, the harmonic oscillator model may also be described by introducing a probabilistic amplitude $\tilde{A}$ instead of $\tilde{\phi}$ or by introducing both as probabilistic functions. The presented study is an introductory step for future considerations in this sense.

References

1. RÉMOND, A., *Handbook of Electroencephalography and Clinical Neurophysiology,* Vol. 5, Part A, Elsevier Publishing Company (1973).
2. REGAN, D., *Evoked Potentials in Psychology, Sensory Physiology and Clinical Medicine,* Chapman and Hall Limited, London (1972).
3. BAŞAR, E., GÖNDER, A., ÖZESMI, Ç., and UNGAN, P., 'Dynamics of brain rhythmic and evoked potentials. I. Some computational methods for the analysis of electrical signals from the brain', *Biol. Cybern.* **20**, 137-143 (1975).
4. UNGAN, P. and BAŞAR, E., 'Comparison of Wiener filtering and selective averaging of evoked potentials', *Electroenceph. clin. Neurophysiol.* **40**, 516-520 (1976).
5. BAŞAR, E., GÖNDER, A., ÖZESMI, Ç. and UNGAN, P., 'Dynamics of brain rhythmic and evoked potentials. II. Studies in the auditory pathway, reticular formation, and hippocampus during the waking stage', *Biol. Cybern.* **20**, 145-160 (1975).
6. BAŞAR, E., GÖNDER, A., ÖZESMI, Ç. and UNGAN, P., 'Dynamics of brain rhythmic and evoked potentials III. Studies in the auditory pathway, reticular formation, and hippocampus during sleep', *Biol. Cybern.* **20**, 161-169 (1975).
7. BAŞAR, E., 'A study of the time and frequency characteristics of the potentials evoked in the acoustical cortex', *Kybernetik* **10**, 61-64 (1972).
8. BAŞAR, E., *Biophysical and Physiological Systems Analysis,* Addison-Wesley Publishing Company, Reading, MA. (1976).
9. BAŞAR, E. and UNGAN, P., 'A component analysis and principles derived for the understanding of evoked potentials of the brain: Studies in the hippocampus', *Kybernetik* **12**, 133-140 (1973).
10. BAŞAR, E. and ÖZESMI, Ç., 'The hippocampal EEG activity and a systems analytical interpretation of averaged evoked potentials of the brain', *Kybernetik* **12**, 45-54 (1972).
11. BOUDREAU, J.C., 'Computer measurements of hippocampal fast activity in cats with chronically implanted electrodes', *Electroenceph. clin. Neurophysiol.* **20**, 165-174 (1966).
12. BAŞAR, E., 'Biological systems analysis and evoked potentials of the brain', *T.I.T. J. Life Sci.* **4**, 37-58 (1974).

The transistor as a model for a living structure

GIACOMO DELLA RICCIA
Laboratorio di Cibernetica del CNR, Naples, Italy, and
The Ben-Gurion University of the Negev, Beer-Sheva, Israel

Introduction

A general approach for the study of 'large systems', which we might call a cybernetic approach, should involve variables like fluxes of information, structure organization, complexity, etc.... There is no general agreement yet on the exact meaning that one should give to these concepts. In this brief communication we shall define some of these concepts in such a way that we can use them as electric variables in a modelling of large systems by electric network analogues.

Under the name of large system we can find biological, socio-economic as well as macroscopic physical systems. Although they belong to seemingly far apart scientific disciplines, one has the feeling that the study of large systems can be conducted under the general framework of thermodynamics, using in particular recent developments in thermodynamics of irreversible processes [1].

We shall apply our electric network modelling technique to a recent work of H. Atlan [2], concerning the amount of information H produced by a biological structure as a function of the dose θ of random excitation to which the structure is submitted. We shall find that the current gain characteristic of a transistor simulates quite well the biological behavior according to Atlan's model.

Further details on the representation of large systems by equivalent circuits will be reported elsewhere [3].

Large systems and thermodynamics of irreversible processes

Let us start by recalling that according to the Carnot-Clausius principle (the second law of thermodynamics), the entropy S of an isolated system in which irreversible processes are taking place is increasing with time, $dS/dt > 0$. Hence for an isolated system, equilibrium is a state of maximum entropy S_0 which, according to Boltzmann statistical interpretation of entropy, corresponds to a state of maximum disorder. As a consequence, if we want to avoid a system reaching a state of equilibrium in which all activity has ceased, we must put it in constant interaction with its environment, allowing thus exchanges of energy and/or matter between system and external world. The thermodynamics of an irreversible process is precisely the science dealing with states of nonequilibrium corresponding to entropies $S < S_0$. This thermodynamics postulates that the temporal derivative of the system entropy can be written as:

$$\frac{dS}{dt} = \frac{d_iS}{dt} + \frac{d_eS}{dt}$$

where d_iS/dt represents the production rate of entropy due to the irreversible processes occurring in the interior of the system, and d_eS/dt is the flux of entropy exchanges between the system and the environment corresponding to various forms of interaction with the exterior. The Carnot-Clausius principle is supposed to apply only to d_iS/dt, thus $d_iS/dt > 0$. The sign of d_eS/dt depends on the nature of the interactions; in fact we can write

$$\frac{d_eS}{dt} = \left(\frac{d_eS}{dt}\right)_1 + \left(\frac{d_eS}{dt}\right)_2$$

where $(d_eS/dt)_1 < 0$ is the part of the flux which tends to reduce the entropy of the system, whereas $(d_eS/dt)_2 > 0$ is the part which on the contrary tends to increase the entropy. The balance equation for the entropy is thus

$$\frac{dS}{dt} = \frac{d_iS}{dt} + \left(\frac{d_eS}{dt}\right)_1 + \left(\frac{d_eS}{dt}\right)_2$$

Of particular interest are the so-called stationary states of nonequilibrium in which the system entropy is maintained at a constant level $S < S_0$; in that case $dS/dt = 0$ and we have

$$\frac{d_iS}{dt} + \left(\frac{d_eS}{dt}\right)_1 + \left(\frac{d_eS}{dt}\right)_2 = 0$$

Let us now, instead of entropy, consider negentropy, $-S$, which as was shown by L. Brillouin [4], is naturally related to information, and talk about fluxes of information $I = -dS/dt$. The condition of stationarity becomes:

$$I_1 = I_i + I_2$$

where $I_1 = -(d_eS/dt)_1$ is an input flow of information, $I_i = d_iS/dt$ and $I_2 = (d_eS/dt)_2$ are output flows of information.

According to our sign conventions an input flow of information tends to decrease the system entropy, whereas output flows tend to increase it. We can thus represent a large system by the general diagram of Fig.1 where, of course, we can imagine that the system is a 'black box' processing electric currents

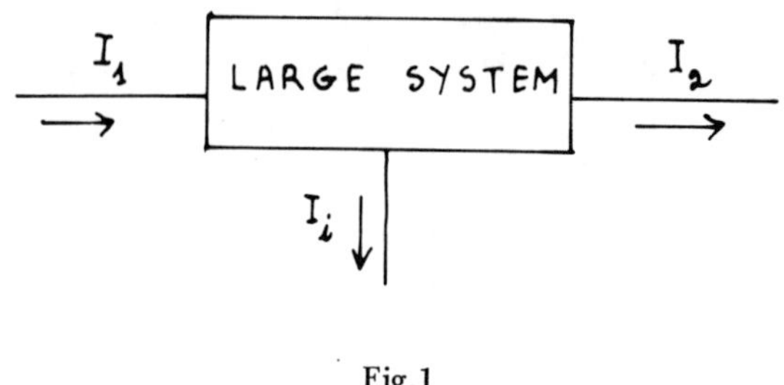

Fig.1

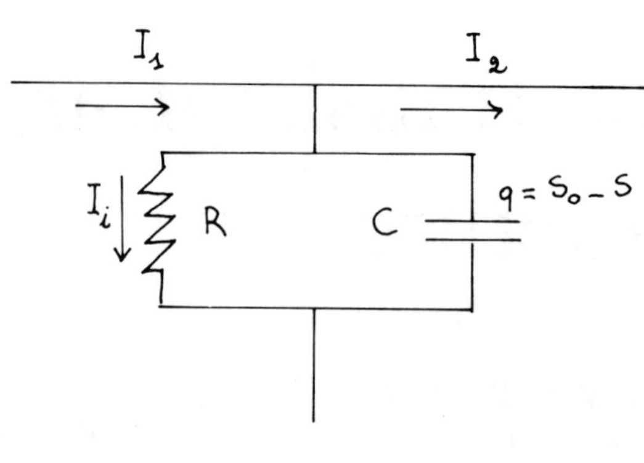

Fig.2

I_1, I_2 and I_i, expressed in bits/second.

The simplest example of an equivalent circuit of a large system is the one shown in Fig.2. In the stationary state indicated in the figure the capacitor C assumes a positive charge $q = CRI_i$. If the system is suddenly isolated, making $I_1 = I_2 = 0$ the circuit will tend towards electrical neutrality according to an exponential decay of the charge: $q = A \exp(-t/RC)$. This suggests that we let the electric variable q represent the quantity $S_0 - S_1$ since according to the Carnot-Clausius principle $S_0 - S$ will tend to zero if we isolate the system. The quantity $S_0 - S$ measures the departure of the system from equilibrium and can also be interpreted as the degree of structure organization corresponding to a particular stationary state. This degree of structure organization in our modelling technique will in general be represented by $q = S_0 - S$ positive charges, expressed in bits, stored in the circuit. We have thus defined and represented by electric variables two fundamental concepts namely: 'functional organization', that is input-output information flows I_1 and I_2 processed by the system, and 'structure organization', that is $S_0 - S$ which depends on the value of $I_i = I_1 - I_2$. The duality between 'function' and 'structure' is a question always present in cybernetics; in biological sciences, for instance, it is the old chicken—and-the-egg problem. It is interesting to notice that our electric analogues emphasize this duality as we can see it on the example of Fig.2 where it is not possible to say whether the positive charge in the capacitor C (structure) is the 'cause' of the information current I_i (function), or the 'effect' of that current through the induced voltage $V = RI_i$.

The modelling of a living biological structure by a transistor

In his work H. Atlan [2], treated a biological structure as a communication channel, and under simplifying assumptions, obtained the dependence of the quantity of information H produced by the structure on the dose θ of random excitations acting on the structure. A typical shape of $H(\theta)$ is shown in Fig.3(*a*). The interesting feature is the presence of a phase during which H increases with θ: the structure adjusts itself to a noisy environment. After reaching an optimal performance H_{max} the second phase starts during which we observe a decay of H: the structure is aging. There exists a remarkable similarity between the shape of $H(\theta)$ and the shape of the curve $\beta = f(\theta)$ where $\beta = I_c/I_b$ is the static current gain of a transistor, that is the ratio of collector current and base current, and θ is a parameter proportional to the density of minority carriers injected by the emitter into the base. Figure 3(*b*) indicates this resemblance; a review of the basic equations of the transistor and a mathematical treatment of the curve $\beta = f(\theta)$ can be found in a previous paper of the author [4]. We believe that the analogy between the two curves is more than a coincidence and should be further investigated [3]. Generally speaking, the transistor shown in Fig.3(*b*) suggests that the biological structure we are modelling is similar to the base B of the transistor which is in interaction with the environment, in that case the emitter E and the collector C, through semiconductor junctions; one can imagine that the current-voltage characteristic of a semiconductor junction could model the relation between entropy flows (information currents) and corresponding thermodynamic 'affinities' playing the role of potentials (temperature gradients, concentration gradients, chemical potential differences, etc.) in various physico-chemical processes occurring between the structure and its environment.

Conclusion

The thermodynamics of irreversible processes describes stationary states of nonequilibrium in terms of fluxes of entropy which we have interpreted as fluxes of information I_1, I_2 and I_i which define the 'functional organization' of the structure. The departure from equilibrium of such a state is measured by $q = S_0 - S$ which we associated with 'structure organization'. A theory of equivalent circuits of large systems where currents I and positive electric charges q are chosen as independent variables could represent a useful tool in cybernetics.

References

1. PRIGOGINE, P., *Introduction to Thermodynamics of Irreversible Processes,* John Wiley, New York (1962).
2. ATLAN, H., *L'Organisation Biologique et la Théorie de l'Information,* Hermann, Paris (1972), see in particular page 271.
3. DELLA RICCIA, G., 'Electric network analogues of large systems', *Annales de la Fondation L. de Broglie,* 2, No. 2, 73-85 (1977).
4. DELLA RICCIA, G., *Annales de Radioélectricité,* tome XIV, 58, 366-374 (October 1959).

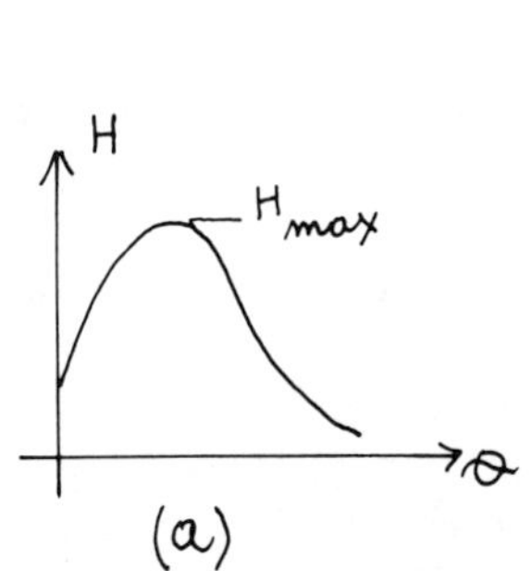

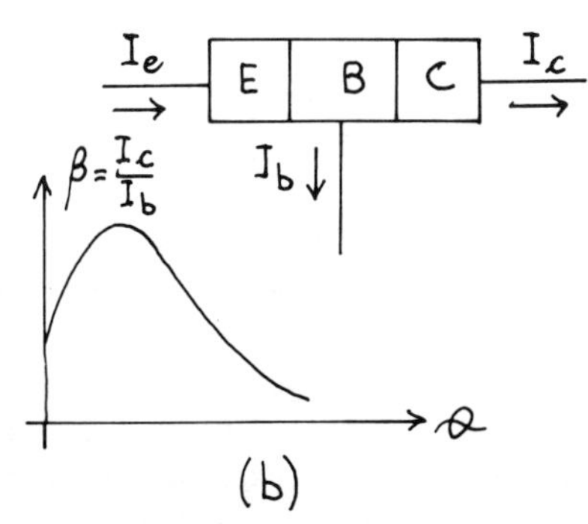

Fig.3

Learning in networks with minimal disruption

A.M. ANDREW
*University of Reading,,
Reading, UK*

Self-organizing systems

The study of self-organizing systems has been a central theme of Cybernetics from the beginning, and has also received much attention under the heading of Systems Research. It can in fact be argued that the emphasis of Systems Research should fall more heavily than it does on self-organization studies. Such a shift of emphasis would clear up some of the uncertainty as to what constitutes a 'system' and therefore what is rightly the concern of Systems Research. A 'system' is not just anything which is large and complex; it must also survive long enough to be noticed. Many systems survive even when all their component parts disappear and are replaced; examples are human bodies, governments, firms, universities.

The systems which are most interesting and important are those which achieve survival by complex processes of adaptation. Most, if not all, of the large viable systems which come to mind are of this kind.

If the changes which occur in a system, constituting its adaptation, are sufficiently fundamental, the system is classed as self-organizing. The importance of flexible adaptive systems, and the relevance of their study to the understanding of the nervous system and other large viable systems, has been emphasised by discussions centered round Ashby's homeostat [1,2]. In a paper given at EMCSR 74 the present writer [3] discussed the importance of studies of self-organization in connection with socio-economic systems, a theme which is treated by Beer, [4,5]. Because of our poor understanding of the adaptive behavior of human organisations, the results of world modelling are of doubtful value. If anything is to be done about the present predicament of the human race it is necessary to know more about complex adaptive systems.

Neural networks

Ashby's homeostat has been valuable as a paradigm of a flexible adaptive system for purposes of discussion, but in fact it makes very poor use of its input data. Another type of adaptive system which has received much attention consists of a network of threshold elements or McCulloch-Pitts neurons [6]. In the studies reported here, networks of elements without delay are consisered, for simplicity. Because of this, the networks cannot be allowed to include circular paths.

There are several reasons for being interested in neural networks. One is that their behavior may constitute a useful model of the behavior of real nervous systems. Another is that McCulloch and Pitts showed that anything which is computable by a finite automaton can be computed by a net of model neurons. Where the model neurons have continuously-variable synaptic weights and thresholds they offer an attractive way of letting the system be modified by a succession of small changes with the possibility that these may eventually produce a drastic change in network behavior.

Training algorithms

It is assumed that adaptation in a net depends on what has been termed, in the context of perceptrons, a 'training algorithm'. That is to say, adaptation is facilitated by mechanisms previously evolved. Such operation is very much in agreement with the idea behind Ashby's [7] 'Design for an Intelligence Amplifier'.

Perceptron training algorithms, and their convergence proofs, are well known, but they are subject to a severe limitation. This point has been discussed in previous publications, and in a paper [8] at EMCSR 72 some possibilities for the introduction of 'significance feedback' into the net were discussed. The aim was to let the adaptive changes occur throughout a complex net, in which some of the variable elements contribute to the output by acting through other variable elements. This is an essential feature of a training algorithm to allow flexible adaptation.

Another feature which is usually expected in a training algorithm is that it should operate locally. That is to say, the adaptation should not depend on an 'adaptation center' having an overview of the entire network; the separate elements should operate autonomously even though the overall effect requires their cooperation. Autonomous behavior of the elements is necessary to conform to the informal, intuitive idea of a self-organizing system. A practical disadvantage of operation depending on an 'adaptation center' is that the necessary computation is likely to be prohibitively complex in a large network.

Another highly desirable property of a training algorithm is that it should find solutions to the problems set to it which are as short and simple as possible. What is meant by a solution in this sense is a state of the network which computes the required function. It is not easy to say what is a good criterion of shortness and simplicity, but solutions which can be realized on a small net are presumably to be rated as more desirable in this respect than solutions requiring a large net.

The reason for preferring solutions which are short and simple is that such solutions are likely to allow inductive inference. The point is discussed in a rather different connection by Rothenberg [9], whose work rests strongly on the approach of Banerji [10].

In the present study it was hoped that the restriction of network size would ensure solutions which can be classed as short and simple. It is of course possible that this use of a small net right from the start is a wrong approach, and that it would be better to find some initial solution on a large net and then to try to derive a concise solution from it.

If the present aim had been merely to find a solution, not necessarily in a concise form, a Simple Perceptron with a large number of randomly-connected A-units could have been used.

Previous experimental work

Experimental studies of the application of the significance-feedback principle in neural nets were reported in the paper given at EMCSR 72. Reference is also made there to the methods due to Stafford [11] which have been investigated more fully in a later study [12]. Yet another possibility is discussed by Widrow [13].

In the experimental studies carried out by the present writer, the attempt has been made to 'train' a small network to compute a logical function. The logical function chosen is one which is not linearly separable. It is the function of three inputs which takes value 'true' if the number of inputs simultaneously active is two and takes value 'false' otherwise.

The studies have shown that none of the training algorithms studied is consistently successful even in this simple task. It appears that something more than 'significance feedback' is needed for effective adaptation.

Correction with minimal disruption

The training algorithms devised by both Stafford and Widrow follow the following general scheme: If the presentation of an input pattern produces the correct response from the network, the network parameters are left unaltered. Otherwise, changes are made in the network which are just sufficient to correct the network output for this particular input.

Comparison of measures of disruption

To allow comparison of the appropriateness of alternative measures of amount of network disturbance, a computer program has been written which simulates a net adjusted by a training algorithm as follows: If the response to an input is correct the network is left unaltered, but otherwise it is adjusted so as to correct the output for this particular input. The set of adjustments is chosen to be that which minimises the particular measure of disruption being considered. The way in which this result is achieved is described in Appendix A.

Strictly speaking, the adjustments are a little larger than is absolutely necessary; a 'safety margin' is specified by the user of the program and the changes in synaptic strengths and thresholds are such that the summations in the neurons fall on the required side of the corresponding threshold levels and separated from them by the specified margin.

The type of training algorithm embodied in this program constitutes a departure from the previously stated rule that a training algorithm should let the network elements behave autonomously. This new method depends very strongly on an 'adaptation center' having an overall view of the network.

It is thought that this approach constitutes an important advance in the study of self-organising systems, as it allows some breakdown of the problem, and hence a more systematic treatment than has been possible up to now.

Methods

The program is written in the programming language ALGOL 68-R [14].

Networks of a specified size are generated pseudo-randomly by the procedure described in Appendix B. A means of testing the generated networks to find whether they are capable of being adjusted to compute the required function is provided. In all of the runs reported here, only network configurations capable of computing the function were employed; the generation procedure was operated until a network satisfying the requirements was found. Approximately one network in every nine satisfies the requirements.

In each test run the attempt was made to train a network by presentation of pseudo-randomly chosen inputs. With three binary input channels the number of possible input patterns is eight, and at any stage of training the network can be tested to find for what number of these possible input patterns it responds correctly.

The tests have been made with rather small networks, containing six elements. In the network shown in Fig. 1 these elements are numbered from 4 to 9, the numbers 1-3 being used to number the inputs. The program allows for networks of arbitrary size, limited only by the total storage capacity of the computer, but it will be clear from the description of the algorithm in Appendix A that computation time must increase steeply with increasing size of network.

Three different measures of amount of disruption in the network are associated with three modes of operation of the program. In Mode 1, the measure is simply the sum of the magnitudes of all the changes which are made in threshold levels and synaptic weights. In Mode 2, the measure is the largest change made in any one weight or threshold, and in Mode 3 it is the sum of squares of all the changes.

Another option provided is that of letting the network operate in 'push-pull' fashion rather than the 'single-sided' fashion described up to now. In single-sided operation a synapse having weight w makes a contribution of w to the summation in the neuron if it is active, and zero contribution if inactive. In push-pull operation the contributions are w and $-w$ instead of w and zero. The adjustment process is modified according to whether the operation is push-pull or single-sided (Appendix A).

The program also allows the option of setting an easier task than the standard one already mentioned. The easier task is to respond whenever two or more inputs are active. The selection of networks is not altered according to whether the standard task or the simple task is used in training.

Results

The main result to be reported is that the training algorithm works very badly. Table 1 summarises its operation in ten runs starting from different randomly-generated network configurations. Each run was continued until the network had been adjusted 40 times.

Table 1 refers to the network operating in Mode 1, without normalization, with single-sided (as distinct from push-pull) operation, and being trained in the standard task (ie. responding if exactly two out of the three inputs is active).

Table 2 suggests that the performance is a little better when the task is the simple alternative (ie. responding if two or more of the three inputs is active). One of the runs actually terminated successfully following the fourth adjustment. However, the network certainly does not show any strong tendency to adapt, and other runs have shown that

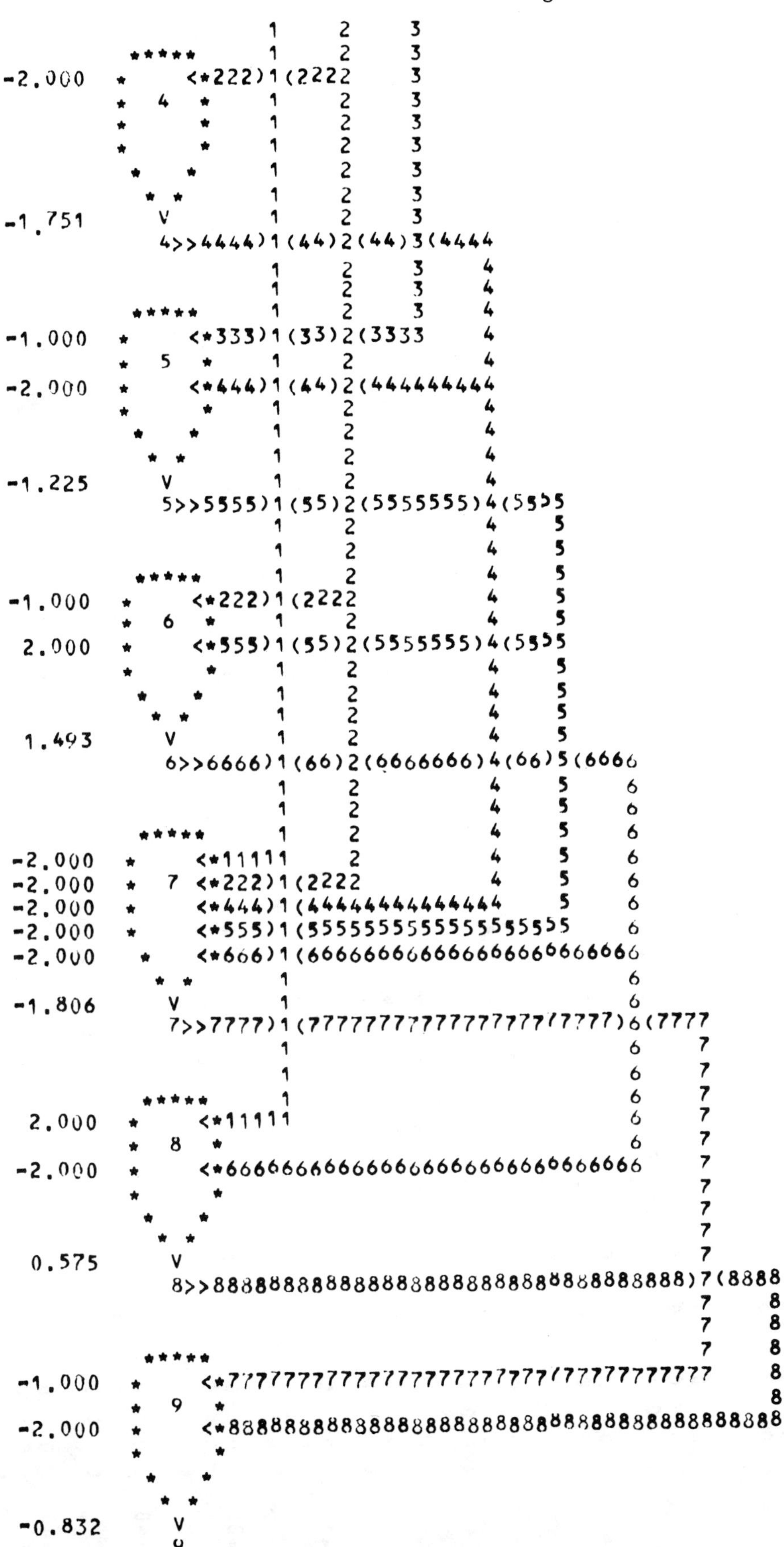

Fig.1 Example of a network generated by the method described in Appendix B. The figures to the left show synaptic weights and threshold values. The three inputs come from the top of the figure.

MODE 1, UNNORMED, SINGLE-SDD, STD TASK, DEPFAC 1.000 1.000 1.000 1.000, MARGIN 0.010, CONFGS 10,

SCORE BEFORE ADJUST

ADJS	0	1	2	3	4	5	6	7	
1	0 0	0 0	1 1	1 0	1 0	1 0	0 0	0 0	4 1
	0 0	0 0	0 0	0 0	0 0	1 2	0 2	0 0	1 4
	(0)	(0)	(2)	(1)	(1)	(4)	(2)	(0)	(10)
2	0 0	0 0	0 0	1 1	2 0	0 0	0 0	0 0	3 1
	0 0	0 0	0 0	0 0	1 0	3 1	1 0	0 0	5 1
	(0)	(0)	(0)	(2)	(3)	(4)	(1)	(0)	(10)
3	0 0	0 0	0 0	0 0	2 0	0 0	0 0	0 0	2 0
	0 0	0 0	0 0	0 0	0 1	5 1	1 0	0 0	6 2
	(0)	(0)	(0)	(0)	(3)	(6)	(1)	(0)	(10)
4-5	0 0	0 0	0 0	0 2	0 0	1 0	0 0	0 0	1 2
	0 0	0 0	0 0	2 0	0 0	10 2	2 1	0 0	14 3
	(0)	(0)	(0)	(4)	(0)	(13)	(3)	(0)	(20)
1-5	0 0	0 0	1 1	2 3	5 0	2 0	0 0	0 0	10 4
	0 0	0 0	0 0	2 0	1 1	19 6	4 3	0 0	26 10
	(0)	(0)	(2)	(7)	(7)	(27)	(7)	(0)	(50)
6-10	0 0	0 0	0 0	1 7	1 1	4 0	0 0	0 0	6 8
	0 0	0 0	0 0	0 0	0 0	23 6	2 5	0 0	25 11
	(0)	(0)	(0)	(8)	(2)	(33)	(7)	(0)	(50)
1-10	0 0	0 0	1 1	3 10	6 1	6 0	0 0	0 0	16 12
	0 0	0 0	0 0	2 0	1 1	42 12	6 8	0 0	51 21
	(0)	(0)	(2)	(15)	(9)	(60)	(14)	(0)	(100)
11-20	0 0	0 0	0 0	2 9	4 6	5 0	5 0	0 0	16 15
	0 0	0 0	0 0	0 0	0 0	42 11	0 12	0 4	42 27
	(0)	(0)	(0)	(11)	(10)	(58)	(17)	(4)	(100)
21-40	0 0	0 0	0 0	4 21	6 9	5 0	10 0	0 0	25 30
	0 0	0 0	0 0	8 0	0 4	82 26	0 15	0 10	90 55
	(0)	(0)	(0)	(33)	(19)	(113)	(25)	(10)	(200)
1-40	0 0	0 0	1 1	9 40	16 16	16 0	15 0	0 0	57 57
	0 0	0 0	0 0	10 0	1 5	166 49	6 35	0 14	183 103
	(0)	(0)	(2)	(59)	(38)	(231)	(56)	(14)	(400)

Table 1 Computer print-out summarizing 10 runs with different network configurations. Each run was continued until 40 adjustments had been made, and groups of three rows refer to different epochs in the runs. For example, the first triple row refers to first adjustments, while the last one summarizes complete runs. The columns correspond to the 'score' of the network before adjustment, i.e. the number of input patterns (out of eight) to which the network makes the correct response. The final column contains the sum of the first eight columns. In each 'cell' of the table there are 5 entries, as follows. The upper left-hand entry is a count of occasions on which adjustment produced an increase of one in the 'score'; the upper right-hand entry refers to occasions when the increase was greater than one. The left-hand entry in the next row refers to occasions when the score remained unaltered, and the right-hand entry to occasions when it became less. The bracketed

entry in the next row is simply the sum of these four entries.

The four DEPFAC ('depth factor') values in the heading refer to a means of letting the 'cost' of a network change be a function of its position in the net. This extra facility is not invoked when, as here, all DEPFAC values are unity.

Table 2 Computer print-out similar to Table 1, but where the task set to the network was the simple linearly-separable task of responding when two or more inputs are active.

MODE 1, UNNORMED, SINGLE-SDD, SIMPLE, DEPFAC 1.000 1.000 1.000 1.000, MARGIN 0.010, CONFGS 10,

SCORE BEFORE ADJUST

ADJS	0	1	2	3	4	5	6	7	
1	0 0	0 1	1 0	1 0	1 1	2 0	0 0	0 0	5 2
	0 0	0 0	0 0	0 0	2 0	1 0	0 0	0 0	3 0
	(0)	(1)	(1)	(1)	(4)	(3)	(0)	(0)	(10)
2	0 0	0 0	0 0	1 0	0 2	0 0	1 0	0 0	2 2
	0 0	0 0	0 0	0 1	1 0	0 2	1 1	0 0	2 4
	(0)	(0)	(0)	(2)	(3)	(2)	(3)	(0)	(10)
3	0 0	0 1	0 0	0 0	1 1	1 0	0 0	0 0	2 2
	0 0	0 0	0 0	0 0	1 1	0 0	2 1	1 0	4 2
	(0)	(1)	(0)	(0)	(4)	(1)	(3)	(1)	(10)
4- 5	0 0	0 1	0 0	1 0	1 3	1 0	0 0	1 0	4 4
	0 0	0 0	0 0	0 1	3 0	0 1	2 4	0 0	5 6
	(0)	(1)	(0)	(2)	(7)	(2)	(6)	(1)	(19)
1- 5	0 0	0 3	1 0	3 0	3 7	4 0	1 0	1 0	13 10
	0 0	0 0	0 0	0 2	7 1	1 3	5 6	1 0	14 12
	(0)	(3)	(1)	(5)	(18)	(8)	(12)	(2)	(49)
6-10	0 0	0 0	0 1	1 0	2 7	2 0	0 0	0 0	5 8
	0 0	0 0	0 0	0 0	11 2	1 3	6 9	0 0	18 14
	(0)	(0)	(1)	(1)	(22)	(6)	(15)	(0)	(45)
1-10	0 0	0 3	1 1	4 0	5 14	6 0	1 0	1 0	18 18
	0 0	0 0	0 0	0 2	18 3	2 6	11 15	1 0	32 26
	(0)	(3)	(2)	(6)	(40)	(14)	(27)	(2)	(94)
11-20	0 0	0 0	0 2	2 0	2 13	5 0	0 0	0 0	9 15
	0 0	0 0	0 0	0 0	27 4	2 0	15 18	0 0	44 22
	(0)	(0)	(2)	(2)	(46)	(7)	(33)	(0)	(90)
21-40	0 0	0 0	2 6	4 0	2 23	7 0	0 0	0 0	15 29
	0 0	0 0	0 0	0 1	49 10	4 2	40 30	0 0	93 43
	(0)	(0)	(8)	(5)	(84)	(13)	(70)	(0)	(180)
1-40	0 0	0 3	3 9	10 0	9 50	18 0	1 0	1 0	42 62
	0 0	0 0	0 0	0 3	94 17	8 8	66 63	1 0	169 91
	(0)	(3)	(12)	(13)	(170)	(34)	(130)	(2)	(364)

there is no marked improvement when Mode 2 or Mode 3 is employed, or when 'normalization' of elements between adjustments and/or push-pull operation are introduced.

Discussion

The poor performance of the training algorithm implemented in each of the three modes of the program is surprising. It suggests there is something basically unsound in the seemingly obvious principle expressed by Stafford [11] as follows: "...it is essential to minimize the effect on previously learned input-output combinations when correcting an error." The same idea is expressed somewhat differently by Widrow [13].

The poor performance may be due to the fact that when a network element is adjusted it is left in a state where, for some pattern of inputs, the summation of input effects is very close to threshold. Hence it is very likely that this same element will be adjusted again a few presentations later, since it allows change at low 'cost'. It is possible that the performance could be improved by increasing the 'margin' employed in making adjustments or by arranging that when a parameter is changed it is then inhibited from further change for some given number of presentations. Variations of the program described here will allow study of these possibilities.

It is important to explore the possibilities thoroughly, to obtain an indication whether a concise solution to a problem can be obtained directly by network adaptation. If it cannot, it will be necessary to look to methods which first produce a non-concise solution and then condense it.

References

1. ASHBY, W.R., 'Design for a brain', *Electronic Engg.* **20**, 379 (1948).
2. ASHBY, W.R., *Design for a Brain*, Chapman and Hall, London, 2nd Ed. (1960).
3. ANDREW, A.M., 'Ecofeedback and significance feedback in neural nets and in society', in *Progress in Cybernetics and Systems Research,* Vol.I, p.228, Hemisphere, Washington (1975).
4. BEER, S., *Brain of the Firm,* Allen Lane, London (1972).
5. BEER, S., *Platform for Change,* Wiley, London (1975).
6. McCULLOCH, W.S. and PITTS, W., 'A logical calculus of the ideas immanent in nervous activity', *Bull. Math. Biophys.* **5**, 115 (1943).
7. ASHBY, W.R., 'Design for an intelligence amplifier', in *Automata Studies*, ed. C.E. Shannon and J. McCarthy, p.215, Princetown University Press (1956).
8. ANDREW, A.M., 'Significance feedback and redundancy reduction in self-organizing networks', in *Advances in Cybernetics and Systems Research,* Vol.I, p.244, Transcripta, London (1973).
9. ROTHENBERG, D., 'Predicate calculus feature generation', in *Formal Aspects of Cognitive Processes,* ed. T. Storer and D. Winter, P.72, Springer, Berlin (1975).
10. BANERJI, R., *Theory of Problem Solving,* Elsevier, New York (1969).
11. STAFFORD, R.A., 'A learning network model', in *Biophysics and Cybernetics Systems,* ed. M. Maxfield *et al.*, p.81, Spartan, Washington (1965).
12. ANDREW, A.M., 'Studies of real and hypothetical nervous systems and their implications for social systems', Conf. on Cybernetic Modelling of Adaptive Organizations, Porto (1973).
13. WIDROW, B., 'Generalization and information storage in networks of adaline "neurons" ', in *Self-Organizing Systems*, ed. M.C. Yovits *et al.*, p.435, Spartan (1962).
14. WOODWARD, P.M. and BOND, S.G., *ALGOL 68-R Users Guide,* H.M.S.O., London (1974).

Appendix A—A training algorithm to minimize disruption

Let a be the number of inputs to the net (afferents) and n the number of elements. Then the numbering of inputs and elements (see Fig. 1) goes up to $a+n$. The process of training the network consists of presenting a series of random patterns to the inputs, and for each letting the network operate. The state of excitation of the element numbered $a+n$ is the output of the network. Where this is incorrect, the network parameters are adjusted as follows.

Each element of the network and each input has associated with it a hypothetical state of excitation. For an input this state corresponds to the actual state. For the element numbered $a+n$ it is the inverse of the actual state. Then all of the 2^{n-1} possible distributions of excitation over the remaining elements are examined. The cost of the parameter-changes to bring about each distribution of excitation is evaluated by considering each element in turn. The internal summation which is performed in the element in normal operation is performed with reference to the hypothetical states of excitation. Where the comparison of summation with threshold for an element gives a result corresponding to its hypothetical state of excitation, no cost of parameter changes is incurred. Where agreement does not result a cost is incurred which can be evaluated according to any of the three modes described. It is assumed, whichever mode is used, that the total parameter-change needed in a neuron is evenly spread over the threshold and the set of synapses contributing to the summation.

The necessary total change in parameters is the magnitude of the discrepancy between summation and threshold, plus the 'margin' which was specified as data. Where the net is operating in push-pull fashion all the synapses of a neuron contribute to each summation, but otherwise it is only those which come from hypothetically active elements.

For each of the 2^{n-1} hypothetical distributions a cost is evaluated according to the chosen criterion. The distribution incurring least cost is selected, and the net parameters are adjusted so that this hypothetical distribution becomes the actual distribution.

Appendix B—Generating the networks

The inputs to the network are numbered from 1 to a, and the network elements from $a+1$ to $a+n$. Since circularity is not permitted, there is no loss of generality in specifying that all synaptic connections go from a lower-numbered input or element to a higher-numbered element.

For the networks used in work reported here, $a=3$ and $n=6$, and the maximum number of inputs to any element, or of connections coming from an element or input is 6.

For each element in turn from that numbered $a+1$ to that numbered $a+n$, a connection is formed from some lower-numbered input or element. The lower-numbered input or element is selected randomly, except that if the selected element is unsuitable because it already has its quota of outputs, a circular search of all the lower-numbered elements is begun at the initially-chosen element. Then for each element or input in turn from that numbered $a+n-1$ down to that numbered 1, if the element or input makes no connection to another element, a connection is formed to a randomly-chosen higher-numbered element. If, at any stage, suitable elements for connections cannot be found the whole process is restarted.

The above steps ensure that every input makes at least one connection to an element, and every element except the last makes a connection to some other. A number is then chosen randomly in the range 1 to $a+n$, with equal probability for all numbers in the range. A procedure which may introduce a new connection is repeated a number of times which equals this number. In this procedure a number i is chosen randomly in the range 1 to $a+n-1$, and a number j in the range $a+1$ to $a+n$. If $j>i$, and a connection is possible without exceeding quotas, a connection is formed from i to j.

For each connection formed a synaptic weight is chosen as -2, -1, 1, or 2, the four values being equally likely. Finally, for each element, the threshold is set at a value which is a selection from a rectangular distribution whose limits are the minimum and maximum summations possible in the element assuming single-sided operation.

Cerebral energy metabolism and memory

H. FLOHR, H. HÖLSCHER, and C. TROCKEL
University of Bonn,
Federal Republic of Germany

Introduction

Information storage in the central nervous system thermodynamically means the formation of time-independent states of low entropy. The physical nature of these states and the physiological processes by which they are generated are not known. In principle, in the central nervous system such states could be achieved in two ways: as metastable states (e.g. by the formation of special macromolecules or new cellular structures) or as a steady-state (e.g. by constituting a spatio-temporal pattern of propagated impulses in some sort of reverberating circulation). In most theories on the engram formation it is assumed that both types of time-independent states play a role. According to these hypotheses the information passes different stores and is maintained in one or more transitory labile forms before it is converted into a permanent and stable one.

Thermodynamically these different states should differ from each other in their energy requirements and hence in their dependence on the cerebral energy metabolism. Experimentally induced disturbances of the CNS energy metabolism, e.g. cerebral hypoxia, could therefore help to identify the nature of the memory traces from a thermodynamical point of view.

In the present study it was investigated whether short periods of (cerebral) hypoxia interfere with the engram formation process. Mice were trained on a passive avoidance task and subsequently subjected to short periods of hypoxia given at different time intervals after the conditioning stimulus.

Methods

340 male albino mice, three months of age, were used. They were kept in groups of 20 and taken to the laboratory 3-4 days before the experiments began. The animals were trained on a one-trial passive avoidance task according to Madsen and McGaugh [1]. The apparatus consisted of a plexiglas box (49 cm × 27.5 cm × 26 cm) in the center of which was an adjustable platform (5·5 cm × 3·5 cm) which could be raised to a height of 17 cm and lowered to a position of 2·5 cm above the floor of the box, which was formed by a metallic shock grid. The grid was connected to a Lehigh Valley solid state shocker with scrambler through which a footshock (1 mA) could be delivered. Each S was placed on the raised platform, which after approximately 2 sec was slowly lowered until it reached a height of 2·5 cm above

the grid. If the animal left the platform the footshock was delivered immediately. Ss not moving off within 10 sec were removed from the apparatus and discarded. Animals stepping off within the criterion time after receiving the footshock were removed and either returned to their home cages or subjected to treatment as described below. A retention test was run after 24 hrs. following training. As a criterion for retention the following parameters were analyzed:

a) The latency of stepping off the platform (in sec).

b) The percentage of animals that avoided stepping off for more than 10 sec.

c) The percentage of animals that avoided stepping off for more than 20 sec.

The animals were divided into 10 groups. The animals of group 1 received a footshock but no subsequent treatment. The animals of groups 2-5 were subjected to periods of 5 minutes of hypoxia (7% 0_2 in N_2) after the footshock. The time interval between the conditioning stimulus and the start of the hypoxic treatment was varied from 0-60 minutes. The animals of groups 6-9 received no footshock after stepping off the platform but were exposed to hypoxia at the above-mentioned time intervals. Ss of group 10 received neither footshock nor treatment.

For statistical comparison of the different experimental groups the Wilcoxon test was applied for the latency times and the chi square test for the 10 and 20 sec criterion.

Results

The results are summarized in Table 1. Naive Ss leave the platform within 4·2 sec ± 0·2 S.E. In the first group the latency increased to 21·1 sec ± 3.1 S.E., the 10 sec criterion from 0 to 68·1 % and the 20 sec ctiterion from 0 to 36·2 % in the retention test. None of the unshocked groups (6, 7, 8, 9, 10) gave conditioned responses. Hypoxia induces a significant retrograde amnesia in a period of 10-15 minutes ($p < 0{\cdot}005$) and 30-35 minutes ($p < 0{\cdot}001$) after the CS. Thereafter no such effect can be proven. Immediate hypoxia has a slight effect which is not significant.

Comments

There exists a considerable controversy concerning the amnesic effects of hypoxia. Hayes [2], Thompson and Pryer [3], Thompson [4], Giurgea *et al.* [5], Sara and Lefevre [6], and Trockel [7] showed in different learning situations that hypoxia interferes with the engram formation. Vacher *et al.* [8] Taber and Banuazizi [9], and Baldwin and Soltysik [10] failed to reproduce these results. The period during which hypoxia disturbs consolidation has not been adequately investigated. For such investigations a one-trial learning situation is particularly suitable since the time of interfering with the consolidation process can be fairly accurately measured. According to Giurgea *et al.* [5] immediate hypoxia given at the end of a multi-trial learned response has significant amnesic effects whereas delayed hypoxia given 6-7 hours after the trial did not interfere with learning. Sara and Lefevre [6] working with a one trial learning model observed a significant retrograde amnesia if hypoxia was given immediately after the CS and if hypoxia was induced gradually by lowering 0_2 concentration from 21% to 3·5% within 15 minutes after the CS. The present results demonstrate that short periods of hypoxia do induce a significant retrograde amnesia. Insofar they are at variance with assumptions of Vacher *et al.* [8] who claimed that it was not possible to produce the necessary cellular hypoxia rapidly enough to interfere with the consolidation process. As the control experiments of groups 6, 7. 8, and 9 show, hypoxia is a non-aversive stimulus and therefore does not interfere with the behavioral expression of retrograde amnesia as has been discussed for the electroconvulsive shock (Coons and Miller [11]). The amnesic effects depend on the time interval between CS and hypoxic treatment. The sensitive period ends after 60 minutes. During that time span the same degree of hypoxia has quantitatively different effects, the immediate exposure is considerably less effective than that during a period of 10-35 minutes after CS. It follows from the experiments that during the engram formation the information is stored in (at least) two thermodynamically different ways: a labile phase in which retention is sustained only if the cerebral energy metabolism is intact and a late form that has the characteristics of a metastable state. It is possible that during the labile phase the information passes through different stores (and codes).

References

1. MADSEN, M.C. and McGAUGH, J.L., 'The effect of ECS on one-trial avoidance learning', *J. comp. physiol. Psychol.* 54, pp.522-523 (1961).
2. HAYES, K.J., 'Anoxic and convulsive amnesia in

Table 1

EXPERIMENTAL GROUPS	n	INTERVAL CS-TREATMENT (min.)	UNTRAINED SS LATENCY-TIME (sec.)	S.E.	RETENTION TEST LT	S.E.	%> 10″	%> 20″
1. Footshock/No Treatment	47	----	4.5	0.3	21.1	3.1	68.1	36.2
2. Footshock/Hypoxia	36	0	4.9	0.4	20.0	4.3	63.9	22.2
3. Footshock/Hypoxia	37	10	4.3	0.3	11.9^{+}	1.9	37.8^{+}	13.5^{+}
4. Footshock/Hypoxia	40	30	4.3	0.3	9.5^{+}	1.3	35.0^{+}	10.0^{+}
5. Footshock/Hypoxia	42	60	4.3	0.4	20.7	4.0	54.8	30.9
6. No Footshock/Hypoxia	34	0	3.7	0.4	5.1^{++}	2.0	8.8^{++}	2.9^{++}
7. No Footshock/Hypoxia	25	10	3.9	0.5	4.1^{++}	0.6	8.0^{++}	0.0^{++}
8. No Footshock/Hypoxia	35	30	4.6	0.4	4.4^{++}	0.7	14.3^{++}	0.0^{++}
9. No Footshock/Hypoxia	23	60	2.8	0.3	3.5^{++}	0.7	8.7^{++}	0.0^{++}
10. No Footshock/No Treatment	21	----	4.0	0.5	4.1^{++}	0.5	0.0^{++}	0.0^{++}

$^{+}$ significantly different at the $p < 0.01$ level from group 1

$^{++}$significantly different at the $p < 0.01$ level from the corresponding FS/Hypoxia group.
Groups 6, 7, 8, 9, and 10 were not significantly different from each other.

rats', *J. comp. physiol. Psychol.* 46, pp.216-217 (1953).

3. THOMPSON, R. and PRYER, R.S., 'The effect of anoxia on the retention of a discrimination habit', *J. comp. physiol. Psychol.* 49, pp.297-300 (1956).
4. THOMPSON, R., 'The comparative effects of ECS and anoxia on memory', *J. comp. physiol. Psychol.* 50, pp.397-400 (1957).
5. GIURGEA, C., LEFEVRE, D., LESCRENIER, C., and DAVID-REMACLE, M., 'Pharmacological protection against hypoxia induced amnesia in rats', *Psychopharmacologia* 20, pp.160-168 (1971).
6. SARA, S.J. and LEFEVRE, D., 'Hypoxia-induced amnesia in one-trial learning and pharmacological protection by piracetam', *Psychopharmacologia* 25, pp.32-40 (1972).
7. TROCKEL, Ch., 'Hypoxieinduzierte retrograde Amnesie', Inaugural-Dissertation, Bonn (1975).
8. VACHER, J.M., KING, R.A., and MILLER, A.T. Jr., 'Failure of hypoxia to produce retrograde amnesia', *J. comp. physiol. Psychol.* 66, pp.179-181 (1968).
9. TABER, R.J. and BANUAZIZI, A., 'CO_2-induced retrograde amnesia in a one-trial learning situation', *Psychopharmacologia* 9, pp.382-391 (1966).
10. BALDWIN, B.A. and SOLTYSIK, S.S., 'The effect of cerebral ischaemia, resulting in loss of EEG, on the acquisition of conditioned reflexes in goats', *Brain Res.* 2, pp.71-84 (1966).
11. COONS, E.E. and MILLER, N.E., 'Conflict versus consolidation of memory traces to explain "retrograde amensia" produced by ECS', *J. comp. physiol. Psychol.* 53, 524-531 (1960).

Complexity and error in cybernetic systems

ROBERT ROSEN
Dalhousie University,
Halifax, Canada,
and Center for Theoretical Biology, Inc.

There is an enormous literature on the complexity of systems, and with attempts at specifying intrinsic measures of complexity. In this chapter, we will take an opposite view; that complexity is not an intrinsic property of a system, but rather manifests our capabilities to interact with the system. Since our capabilities to interact with systems around us are continually changing, so too do their apparent complexities. Some consequences of this viewpoint, bearing on system descriptions, and on the capacity of systems to make errors, will be dealt with below; a fuller discussion of these matters will be presented in another place.

Intuitively, we regard a system as complex if we can interact with it significantly in many distinct ways, and if each of these different ways requires a different mode of description of the system to encompass it. Thus, organisms are complex; we can interact with them at many levels, and describe these interactions in many different ways. For ordinary purposes, a stone is a simple system; we typically interact with a stone in only a few ways, and basically a single mode of description is sufficient to describe these ways. If we adopt this viewpoint, that system complexity refers to our modes of interaction with the system, and to the corresponding number of different descriptions to which these modes of interaction give rise, then a number of questions immediately come to mind:

1. Is there one class of interactions, and consequently one mode of system description, from which all the others can be derived (the problem of reductionism)?
2. Is there any kind of measure of interactive capacity which will allow us to define complexity in other than subjective terms?

In the subsequent discussion, we shall be concerned with some of the aspects of these questions.

As we have repeatedly emphasized elsewhere (Rosen [1] 1972), all system descriptions, however much they may vary in detail, ultimately consist of two basic kinds of notions; those dealing with the specification of an *instantaneous state* of a system, and those dealing with the manner in which this instantaneous state changes in time as a result of *forces* imposed on the system. As a typical example, we may consider the dynamical theory of particle mechanics, from which all other modes of system description have been derived. If we are given a system of N gravitating particles in ordinary 3-dimensional space, then the Newtonian formalism tells us that we can describe this system, at any instant of time, by specifying the displacements of the constituent particles from some convenient origin of coordinates, and by specifying their corresponding velocity or momenta. Thus, a set of *6N* numbers suffices for a specification of the instantaneous state of the system. By this is meant that *any other* quantity pertaining to the system can be expressed as a function of the ones used to describe the instantaneous state, and hence is in a sense redundant. The displacements and momenta satisfy all the properties of a set of *state variables*; they represent a minimal set of observable quantities pertaining to the system, from which all other information pertaining to measurable quantities defined on the instantaneous states can be derived.

Newton's Laws also provide us with a dynamic description of the way in which the instantaneous states change with time, by identifying the rates of change of the momenta of the particles in our system with the forces imposed on the system. Knowing the forces, and knowing an initial state of the system, is thus sufficient to allow us to predict the state of the system at any time (at least in principle), and hence, to predict the value of any observable quantity associated with the system.

In this kind of formalism, we thus identify the abstract states of our system with a given mode of state description; i.e. with a subspace of Euclidean *N*-dimensional space. However, even in this formalism, there is a great deal of room for alternate descriptions. For instance, it is one of the fundamental assertions of physics that every real-valued function defined on the state space represents a potential dynamical variable of the system, and one which could be directly measured by some suitable measuring instrument. Conversely, of course, every measurable quantity pertaining to the system can be regarded as giving rise to such a function. Each such function inherits a dynamical equation, from the dynamical equations which govern the state variables. As we have shown elsewhere, however, (Rosen [2]) it can easily happen that a set of such observables will form a dynamical system in its own right; if we were to interact with the system in such a way as to observe those observables rather than the state variables, we would in effect see a completely different system. There even exist universal dynamical systems, which allow us to find observables inheriting any prescribed set of equations of motion posited in advance. Thus, even in this relatively straightforward situation, there are profound epistemological problems arising when we attempt to give an 'intrinsic' character to the results of measurements or other kinds of interactions with the system.

Another aspect of the crucial concept of interaction between systems is descernible from the fact that the state description, which presumably contains within itself the answer to every question which can be asked about that system, pertains only to a perfectly isolated system. Given two systems perfectly specified in isolation, we cannot, *from those specifications alone*, decide how those systems are going to interact. To deal with the problem of interactions, we must invoke further information, usually in the form of some universal principle (e.g. mass action) which never forms part of the state description of individual systems. This is the primary reason why physics is successful in dealing with isolated systems, but cannot cope with interactions, except in the case in which they are infinitely weak (perturbations).

We have thus seen (a) that even a 'simple' physical system can be interacted with in many ways (i.e. through many different sets of observables of the system); each of these modes of interaction conveys some aspect of reality pertaining to the system, but the system will 'look different' from mode to mode; and (b) the capacity of a system to manifest interaction is never considered a part of physical system descriptions, although it is crucial for such areas as biology. There is no reason why we should expect that two interacting systems will 'see' each other through the same observables through which we find it convenient to interact with them separately, and consequently, there is no reason to expect that our conventional modes of state description will allow us to understand (let alone predict) the results of the interaction.

The above remarks bear heavily on the problem of reductionism, and on the more general problem of approaching complex systems through analytic means. By *analysis* we mean here the resolution of a system into a family of subsystems somehow 'simpler' than the original system from which they were extracted, and attempting to infer the properties of the original system from the properties of the subsystems. The extraction of a sybsystem corresponds formally to a process of *abstraction*, in which a number of degrees of freedom of the original system (i.e. potential interactive capabilities) are excluded, and only a limited number are retained. This process of abstraction can be physically implemented (as when a molecular biologist extracts a fraction of molecular species from a cell, thereby creating an abstract cell) or they can be purely formal (as when an ecologist represents a population of real organisms in terms of predation relations). The basic requirements of such abstractions are the following:

1. The subsystems so obtained must be 'simpler' than the original system from which they were abstracted.
2. The subsystems must be obtained by 'natural' means (i.e. utilizing familiar and justifiable procedures).
3. The properties of the subsystems so obtained must permit the determination of the properties of the original system.

The property (1) is obviously crucial; nothing is gained if we extract systems as intractable as the original system. This has long been recognised implicitly in scientific modes of analysis. Of equal importance is the property (3); any property of isolated subsystems not bearing on the properties of the original system is an *artifact*. The property (2), however, is a purely subjective matter, and refers only to the manner in which we find it convenient to interact with the original system. It thus stands on a different footing from (1) and (3).

Nevertheless, in many empirical modes of system analysis, the greatest weight is placed upon condition (2). It seems to be intuitively hoped that, by relying on procedures which satisfy (2), the conditions (1) and (3) will automatically be satisfied. At the very least, it is hoped that (1) + (2) will imply (3). However, from what we have already said, this is plainly absurd, in general. Indeed, what we learn from the above is that the crucial properties (1) and (3) which must be satisfied by any useful means of analysis of systems, must be allowed to determine what we are to regard as 'natural'. Indeed, 'naturality' must not be allowed to be posited in advance, but only in terms of its bearing on the problems under discussion in a particular context.

A simple example may make this clear. In physics, the three-body problem is complex in a well-defined sense; the dynamical equations governing a system of three gravitating masses in an arbitrary configuration cannot be integrated directly. We could hope to approach this kind of problem by analysis into a family of 'simpler' subsystems, which will allow us to solve the problem. Intuitively, the subsystems available to us are two-body systems and one-body systems. These are indeed 'simpler' than the original system, and are abstracted from that system in 'natural' ways. However, it is clear that we cannot solve a three-body problem in this fashion, for the act of decomposing the original system into isolated simpler subsystems destroys irreversibly the dynamics in which we were originally interested (here again we see the inability of physics to deal with arbitrary interactions). Thus, from the standpoint of solving the three-body problem, our apparently 'natural' decompositions are useless; if analysis is to be successful in this kind of problem at all, the appropriate subsystems [i.e. those which satisfy (3)] must necessarily be of a kind which would appear most 'unnatural' in terms of what we find it convenient to do physically to a system of particles.

An abstraction, in the sense in which we are using the term, owes its efficacy to the removal of degrees of freedom (i.e. interactive capabilities) present in the original system. It represents a simplification in that it faithfully represents what the original system would be like if it could only interact with the world through those degrees of freedom which are retained in the abstraction. Thus, if the abstraction exhibits similar interactive capabilities to those possessed by the original system, we are justified in assigning them to the properties retained in the abstraction. (However, we must note that an abstracted subsystem may exhibit new interactive capacities of its own, which do not pertain to the fact that it is in fact a subsystem of some larger system; always carefully distinguish between those properties which inhere in the original system, and those which arise in the model without reference to the fact that it is a model.)

Let us suppose that we have two different abstractions of the same complex system. How may these be compared or combined to give a fuller picture of the original system than either of them can give alone? We shall touch briefly on this question, and show how it relates to the notion of *structural stability*, which is presently attracting great interest in theoretical biology and elsewhere, largely through the influence of Thom (Thom [3]). Indeed, structural stability is a general framework for comparing one description of a class of objects against another description, usually in terms of measures of 'closeness'. Typically, if two objects 'close' in one description are also 'close' in the other, then the objects are called *generic*; generic objects are thus precisely those for which the two descriptions agree. Those objects which are not generic are called *bifurcation* objects; objects 'close' to a bifurcation object in one description will not be 'close' to that object in the other description. Generic objects thus do not require both descriptions; the descriptions are redundant on these objects. On the bifurcation objects, however, the two descriptions are conveying different information about the objects.

It is usual, in considerations of structural stability, to regard one of the two descriptions being compared as *intrinsic*; the second description is then compared with the intrinsic one. However, as we repeatedly emphasized, (cf., Rosen [4]) the topologies (i.e. the measures of 'closeness') which we typically assign to systems and their states are not intrinsic, but are all contingent on how we choose to interact with a system. Thus, any description is at best conveying partial information about a limited fraction of the interactive capabilities of the system, and all are equally extrinsic.

Nevertheless, if we have an idea that a particular description is intrinsic, and that all other descriptions should be referred to it, we will typically find some bifurcation set where the two descriptions do not agree. Since we tend to prefer a description we regard as intrinsic, it would be reasonable for us to interpret the disagreement between the descriptions as a *mistake* or *error* on the part of the second description. Thus it is at this point in the analysis of complex systems that the concept of error arises in a natural way.

Error has always been troublesome for system theorists. Simple systems do not make errors; it is meaningless to regard a system of mechanical particles as behaving erroneously. Therefore there has always been a relation between complexity and the capability of error; just as with complexity, there has been an enormous literature generated in an attempt to obtain an intrinsic definition of error. We will argue, analogous to our argument regarding complexity itself, that error is not definable intrinsically, but only in terms of the spectrum of interactions available to a complex system. We shall also find that the conventional view, that error depends on complexity, is well-founded.

To put the case simply, we shall assert that error is measured by the deviation observed between the actual behavior of a complex system (with its many distinct interactive capabilities) and the behavior of a simple system or model, which exhibits only a restricted subfraction of the interactive capacities of the original system. The deviation between these behaviors arises precisely because the complex system can be, and in general is, doing many different things at once, and these different interactions interfere with each other in a way unpredictable in principle from the properties of the corresponding simplified system (since of course, the vehicle for these interactions has been abstracted away, and hence permanently lost, in the very process of abstraction which led to the simplified system). Thus, a bridge can collapse because the particles of which it is composed may interact with each other in ways not consistent with the maintenance of cohesive properties; the genes of an organism can mutate because they can interact with elements of their environment in ways not compatible with their coding function, and so on. I would argue that, in fact, all forms of what we interpret as error arise through simultaneous manifestation of modes of interaction in complex systems, which happen to interfere with each other.

These observations have many implications for important areas of control theory and error correction at the technological and social levels, as well as for an understanding of biological and physical systems. For example, we may compare a model of our own regarding a particular biological and social process (i.e. a simple representation of a complex system) with that manifested by natural selection acting on that complex system. A comparison of the two descriptions will reveal bifurcation points, at which the selection behavior manifestly disagrees with our projections for system behavior. These are the points at which great care must be taken in exercising control. For instance, if we decide we want a system to behave optimally, or without error, in a single simplified mode, we can only achieve this at the cost of interfering in unpredictable ways with other modes of system interaction. We have discussed these problems, in a variety of contexts, in other work, to which we refer the reader for fuller discussions (Rosen [5]). See also the papers of Pattee (Pattee [6]). Further, it is important to note that it is in general not sufficient to attempt to replace the missing interactive capabilities which were abstracted away in the initial simplification of the system, by some generalized probability distribution imposed *ad hoc* on the degrees of freedom remaining in the simplified system. This is, of course, the usual procedure adopted in this regard, but from a more general perspective it is frought with pitfalls.

We hope to have indicated, in the above brief survey, the epistemic character of the notions of system complexity, and the related problems of alternate description and of error in complex system behavior. These notions, we feel, are of great importance in attempting to deal with the spectrum of technological, social and biological problems which presently confront us.

Acknowledgement
This chapter was prepared with the support of NIH grants #2RO1HD05136-04 and #1 PO1 HD07328-01, and NASA grant #NGR33015002.

References

1. ROSEN, R., *Mathematics in the Sciences* (edited by (Saaty and Wegl), McGraw-Hill, NY (1969).
2. ROSEN, R., *Bulletin Mathematical Biophysics* **30**, pp.481-492 (1968).
3. THOM, R., 'Stabilité Structurelle et Morphogénèse', W.A. Benjamin, Inc. Reading, MA (1972).
4. ROSEN, R., *International Journal Systems Science* 4, #1, pp.65-75 (1973).
5. ROSEN, R., *International Journal General Systems* **1**, pp.245-252 and **1**, #2, pp.93-103 (1974).
6. PATTEE, H., *Chemical Evolution and the Origin of Life* (edited by R. Buvet and C. Ponnamperuma), North Holland: Amsterdam (1971).
7. ROSEN, R., *Dynamical System Theory in Biology*, Wiley-Interscience (1970).
8. PATTEE, H., *Biogenesis · Evolution and Homeostasis,* Springer-Verlag: Heidelberg and New York, in the press.

Dynamic characteristics of neural network

SHOICHI NOGUCHI, MASATERU HARAO, and TETSUO ARAKI
Tohoku University, Sendai, Japan

Introduction
As a biologically motivated automaton, we consider the finite cellular array with uniform structure as the model of the neural network.

First, in order to discuss the dynamic characteristic of the network we make considerations about the global function induced by the local function.

Secondly, we obtain the number of the stable configuration corresponding to the dynamic memory capacity of a neuron cell for each local function in three neighborhood connection.

1. Definition
In this chapter, the neural network is represented by the finite cellular space in the following way:

The cellular space is $S = (G, \eta, Q, f)$

where $G = \{g_1, g_2, \dots, g_n\}$ is the set of cell label corresponding to the coordinate in the usual case, η is the neighborhood function which defines the interconnection to the cell such that $\eta ; G \rightarrow G^k$, k is the neighborhood index, Q is the finite cell state, and $f ; Q^k \rightarrow Q$ is the local function which decides the next state of each cell in a synchronous way.

The behavior of the network is defined by mapping $c ; G \rightarrow Q^n$ called as the configuration such that

$$c(t) = (c_{g_1}(t), c_{g_2}(t), \dots, c_{g_n}(t))$$

where $c_{gi}(t)$ means the state of the g_i at time t.

The dynamic behavior of the neural network is described by the transition of $c(t)$ and the total set of $c(t)$ is denoted by $C_n = \{c(t)\}$.

In the following, we chiefly discuss the case where each cell is interconnected by a homogeneous directed graph.

Definition 1
Let G be any group and let N be a finite subset of G. A graph with node set G and edge set E_N which is abbreviated to $G_N = (G, E_N)$ is called as a group graph iff, for any g_i, g_j in G,

$$(g_i, g_j) \in E_N \Longleftrightarrow g_i^{-1} \cdot g_j \in N$$

Definition 2
Let G be a finite group with an order n such that $G = (g_1, g_2, \dots, g_n)$, then the neighborhood function $\eta(g)$ corresponding to g is an order set $(g\, g_{i1}, g\, g_{i2}, \dots, g\, g_{ik})$ and $N = \{g_{i1}, g_{i2}, \dots, g_{ik}\}$.

In Section 2, we consider the case of $\eta(g) = (g\, g_1, g\, g_2, \dots, g\, g_n)$ and $N = G$.

Definition 3
The neighborhood configuration corresponding to g cell is defined by

$$c \cdot \eta(g) = \{ c(g\, g_{i1}), c(g\, g_{i2}) \dots c(g\, g_{ik}) \}.$$

Then the transition of each cell g is decided by $f \cdot c \cdot \eta(g)$.

The behavior of S is given by $f \cdot c \cdot \eta(g)$ which uniquely defines the global function F such that

$$F\{c(t)\} = c(t+1)$$

where $F\{c(t)\}_g = f \cdot c \cdot \eta(g)$.

2. Characteristics of the global function and dynamical characteristics of the neural network

In this Section, we consider the properties of the local function which induces injective and surjective map F on the configuration c under the following conditions; $G = N = (g_1, g_2, \dots, g_n)$ and Q is binary. We define an equivalence relation on C_n by $c \sim c'$ if $\exists g \in G,\ c(g\, g_1, g\, g_2, \dots, g\, g_n) = c'(g_1, \dots, g_n)$. This relation induces $C_n/\sim$, and each class on this is denoted by $[c]$.

Next we define the invariant subgroup of c by H^c such that $c(g\, g_1, g\, g_2, \dots, g\, g_n) = c(g_1, g_2, \dots, g_n)$ and obtain the coset leaders by

$$G = H^c g_{i1} + H^c g_{i2} + \dots + H^c g_{ir}$$

where $g_1 = g_{i1}$ is e.

Proposition 1
Let $c' \in [c]$, then $H^c = H^{c'}$ iff H^c is a normal subgroup.

Let $c \cdot \eta = u$, then we define $T[c]$ by the following relation;

$$T[c] = \begin{bmatrix} u(g_1), & u(g_2), \dots\dots, & u(g_n) \\ u(g_{i2} g_1), & u(g_{i2} g), \dots, & u(g_{i2} g_n) \\ \vdots & & \\ u(g_{ir} g_1), & u(g_{ir} g_2), \dots, & u(g_{ir} g_n) \end{bmatrix}$$

By the above relation we define Tc-equivalence between $[c]$ and $[c']$ iff there exists componentwise bijection δ from $[c]$ to $[c']$ such that

$$\begin{bmatrix} \delta u(g_1), & \dots\dots\dots, & \delta u(g_n), \\ & & \\ \delta u(g_{ir} g_1), & \dots\dots\dots, & \delta u(g_{ir} g_n), \end{bmatrix} = \begin{bmatrix} u'(g_1), & \dots\dots\dots, & u'(g_n) \\ & & \\ u'(g_{ir} g_1), & \dots\dots\dots, & u'(g_{ir} g_n) \end{bmatrix} = T[c']$$

where $u' = c' \cdot \eta$. This relation is denoted by $\Delta T[c] = T[c']$.

Proposition 2
If G is abelian, then $[c]$ and $[c']$ is Tc-equivalent iff $H^c = H^{c'}$.

Corollary 2.1
If G is cyclic, $[c]$ and $[c']$ is Tc-equivalent iff $|[c]| = |[c']|$ where $|\ |$ means the number of configuration of $[c]$.

By the above procedure $C_n/\sim$ is classified into the set of Tc-equivalent classes $T_1, \dots, T_p$.
Let $\{c_i\}$ be the representative of the class $[c_i]$ such that

$$c_i = c_i(g_1, g_2, \dots, g_n)$$

then mapping α is defined among the $[c_i]$ in the same Tc-equivalent class and for all T_i and in α, mapping β is also defined as follows,

$$\beta(c_i \cdot g_{i1}, c_i \cdot g_{i2}, \dots, c_i \cdot g_{ir}) = (c_j \cdot g_{i1},\ c_j \cdot g_{i2}, \dots, c_j \cdot g_{ir})$$

where

$$c \cdot g_j = c(g_j \cdot g_1, g_j \cdot g_2, \dots, g_j \cdot g_n)$$

Theorem 3
The global function F is bijective iff α and β is bijective and in β, the invariant subgroups of each element of the domain and range are componentwise equal.

Corollary 3.1
If G is abelian, the global function F is bijective iff α and β is bijective.

Corollary 3.2
If G is cyclic, the global function F is bijective iff α and β is bijective and

in this case, $\alpha[c] = [c']$ means $|[c]| = |[c']|$

Theorem 4
Under the same condition of [theorem 2], let $\alpha[c] = [c']$ and $X \in [c']$, then define the local function f such that $f[c] = X(g_{i1})$,

$$f[c \cdot g_{i2}] = X(g_{i2}), \ldots, f[c \cdot g_{ir}] = X(g_{ir}),$$

then the global function F is bijective and vice versa.

Proposition 5
Let G be a cyclic group of order n, then the cycle index Z_G of G is given as follows:

$$Z_G(x_1, \ldots, x_n) = \frac{1}{n} \sum_{d/n} \varphi(d) x^{n/d}$$

where $\varphi(d)$ is the Euler function.

Proposition 6
Let G be products of cyclic groups of order $m_1, m_2, \ldots, m_k$ then, the cycle index of G is given as follows:

$$Z_G(x_1, \ldots, x_n) = \frac{1}{n} \sum_{d_1/m_1} \cdots \sum_{d_k/m_k} \varphi(d_1) \cdots \varphi(d_k) \times x_{\mathrm{lcm}(d_1, \ldots d_k)}^{\frac{m_1 \ldots m_k}{\mathrm{lcm}(d_1 \ldots d_k)}}$$

where lcm is the least common multiplier.

Proposition 7
The number of the equivalent class of C_n under $\sim$ is given by $Z_G(2,2,\ldots 2)$. As mentioned before $C_n/\sim$ classified into Tc-equivalence class $[Tc_1],[Tc_2],\ldots,[Tc_p]$, and let

$|[Tc_1]| = R(1)$, $|[Tc_2]| = R(2), \ldots, |[Tc_p]| = R(p)$, and $|[c_i]| = r_i$ where $[c_i] \in [Tc_j]$ and $r_i | n$.

Theorem 8
The total number of the bijective global function A_n is given by

$$\prod_{i=1}^{p} R(i)! \times ({}^{r_i}/_{S(i)})^{R(i)}$$

where $S(i)$ is the number of the invariant subgroup corresponding to $[c_i]$.

Corollary 8.1
If G is abelian

$$A_n = \prod_{i=1}^{k} R(i)! \cdot (r_i)^{R(i)}$$

Corollary 8.2
If G is cyclic

$$A_n = \prod_{i=1}^{k} (M_n(r_i)!) \times r_i^{M_n(r_i)}$$

where

$$M_n(r_i) = \frac{1}{r_i} \sum_{r^l | r_i} \mu(r^l)^{r_i/r^l}$$

and μ is möbius function.

Theorem 9
Let G be cyclic group of order n, and its all divisor be $1 = r_1, < r_2, < \ldots < r_k = n$.

Then, the realisable cycle lengths by this network are given as follows:

$$r_i 1, r_i 2, \ldots, r_i M_n(n), r_2 1, \ldots, r_2 M_n(n), \ldots,$$
$$n\ 1, \ldots, n M_n(n)$$

Theorem 10
If G is a cyclic group of the prime order n, then the realisable lengths by this network are

$1,2,3,\ldots, M_n(n), n, 2n, 3n, \ldots, M_n(n) \cdot n = 2^n - 2$.

Corollary 10.1
The maximum cyclic length realizable by the uniform cellular network of n is equal or less than $2^n - 2$.

3. Stable configuration of neural network

Definition 1
Let F be the global function and c be the configuration of the network, then the stable configuration c is defined by $F(c) = c$.

In this section, we consider the binary network with three neighborhood connection in the one-dimensional case.

Definition 2
A transition graph of the cellular network, CN abbreviated TG is a quadruple (V, E, α, β), where (1) $V = Q \times Q$ is a node set, $Q = \{0, 1\}$, (2) $E \subset V \times V$ is an edge set, defined by $v_i = (i_1, i_2)$, $v_j = (j_1, j_2) \in V$, $(v_i, v_j) \in E \Leftrightarrow i_2 = j_1$, and (3) $\alpha : E \rightarrow Q$ is an input label defined by $\alpha(v_i, v_j) = j_2$, (4) $\beta : E \rightarrow Q$ is an output label defined by $\beta(v_i, v_j) = p$ iff $f(i_1, i_2, j_2) = p$.

Definition 3
An edge (v_i, v_j) of TG is stable if $\beta(v_i, v_j) = i_2 = f(i_1, i_2, i_3)$ where $v_i = (i_1, i_2)$ and $v_j = (i_2, i_3)$.

Proposition 1
A CN has stable configurations iff there exits at least one closed path of length n passing only the stable edges.

In the domain C-$\{0^n, 1^n\}$, it is easily shown 100 local functions have non-trivial stable configurations among 256 local functions. Furthermore, 100 local functions are classified into 15 classes.

Definition 4
Two nodes $v_i = (i, s)$, $v_i' = (i', s)$ of any given TG are equivalent if $f(i, s, x) = f(i', s, x)$ for all input x.

The equivalent nodes of TG can be merged and five of the TG's are reduced to the simple forms. By these properties, we have only to discuss the following 10 TG's for calculating the number of stable configurations.

Theorem 2

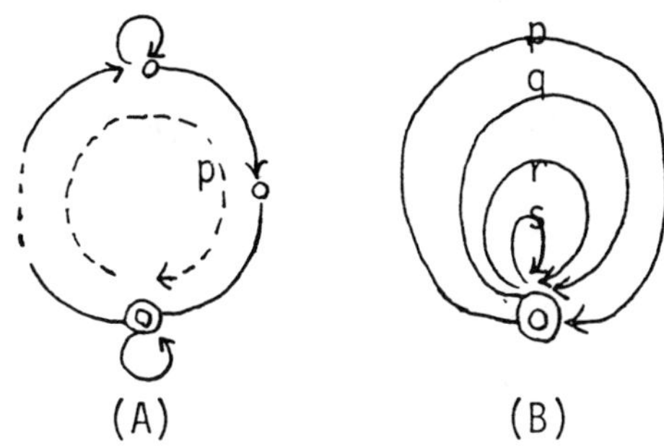

Fig.2 Typical forms of TG

The number of stable configurations of figures (A) and (B) with starting node ◎ are given as follows:

$$\text{(A)} \quad T(n) = \sum_{k=0}^{[n/p]} \binom{n-(p-2)k}{2k}$$

$$\text{(B)} \quad T(n) = \sum_{k=0}^{[(n-pi-qj)/s]} \sum_{j=0}^{[(n-pi)/q]} \sum_{i=0}^{[(n/p)]} \times$$

$$\times \binom{i+j}{i} \binom{i+j+k}{i+j} \binom{(n-pi-pj-rk)/s+i+j+k}{i+j+k}.$$

We may calculate the total number of stable configurations $T(n)$ using these equations and the properties of the transition graph. As another method, the difference equation forms can also be obtained directly by analyzing the transition graphs.

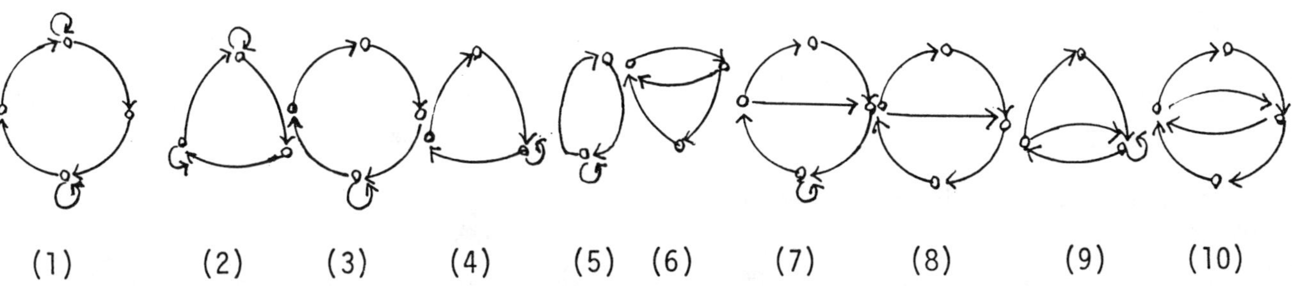

Fig.1 Representatives which generate the non-trivial stable configurations

Theorem 3
The difference equations of any $T(n)$ are given as follows:

(1) $T(n) = T(n-1) + T(n-2) = (5) = (7)$.

(2) $T(n) = T(n-1) + T(n-2) + T(n-4)$.

(3) $T(n) = T(n-1) + T(n-4)$.

(4) $T(n) = T(n-1) + T(n-3)$.

(6) $T(n) = T(n-2) + T(n-3)$.

(8) $T(n) = T(n-3) + T(n-4)$.

(9) $T(n) = T(n-1) + T(n-2) + T(n-3)$.

(10) $T(n) = T(n-2) + 2T(n-3) + 2T(n-4)$.

Information Storing Capacity
Let $r_1,\ldots, r_k$ be the roots of the characteristic equation of the difference equation, then $T(n) = c_1 r_1^n + c_2 r_2^n + \ldots + c_k r_k^n$, where $c_i s$ are constant. Let r_1 be the maximum real number and let $c_1 \neq 0$, then

$$\lim_{n\to\infty} \frac{T(n+1)}{T(n)} = r_1.$$

Hence, $T(n) \approx c_1 r_1^n$.

Under the assumption that the informations are stored in the network as the form of stable configurations the information storing capacity of each cell is given by $(1/n)\log_2 T(n)$.

Hence, for $T(n) = cr^n$, it holds for large n

$$H = \lim_{n\to\infty} \frac{1}{n}\log_2 cr^n = \log_2 r \quad \text{(bits)}$$

Therefore, the value of r is the simple estimation of the information storing capacity.

Theorem 4
The information storing capability of each cell is given by

	1	2	3	4	5	6	7	8	9	10
r	1.618	1.755	1.380	1.466	1.618	1.325	1.618	1.220	1.839	1.696
H	0.674	0.813	0.465	0.551	0.674	0.406	0.674	0.287	0.879	0.762

In the same way, we also obtain the memory capacity of each cell in the ternary case under the two neighborhood condition. The local functions are classified into 14 types and the memory capacities for these cases are given as follows:

H = 1.589, 1.450, 1.389, 1.271, 1.217, 1.168,

1.141, 1.103, 1.000, 0.880, 0.857, 0.811, 0.694,

0.551, respectively.

Conclusion
We have considered the dynamic characteristic of the neural network in the binary case. Most of the results obtained here may easily be extendable to the p-nary case.

Reference

1. ARAKI, T., HARAO, M., and NOGUCHI, S., *Algebraic Properties of the Cellular Automata*, Papers of Technical Group on Automata and Language, IECE, Japan AL 73-1 (1973).
2. HARAO, M. and NOGUCHI, S., *Algebraic Properties of the Iterative Automaton, IECE Trans.* Vol.55-D, No.12, pp.767-774 (1972).
3. KOBUCHI, Y. and NISHIO, H., 'Some regular state sets in the system of one-dimensional iterative automata', *Inf. Science* 5 (1973).

The adaptive neuronlike layer net as a control learning system

RYSZARD GAWROŃSKI[†] and BOHDAN MACUKOW[‡]
[†]*Politechnika Świętokrzyska, Kielce, Poland, and*
[‡]*Politechnika Warszawska, Warszawa, Poland*

1. Introduction

In this report we present the investigation of a learning system, in particular the results of the modelling of some nonclassical learning procedures.

The dynamical properties of an artificial hand or similar executive part of a robot with several degrees of freedom are so complex that the calculation of a set of control signals is very complicated. Many authors used a minicomputer with suitable programs for modelling every situation. The preparation of such programs usually needs some iterative procedures even for open loop control. When the varying situations need multiloop feedback control the on-line optimal control is very difficult to realize with the help of typical minicomputers.

It seems reasonable to use in such situations a specialized adaptive control system. On the basis of some ideas originating from the principles of the motor control in the nervous system we propose a net composed of neuronlike elements arranged in layers with many mutual, controlled connections.

In the present experiment we are interested only in a steady state of the object and the components of the appropriate vectors are real, limited numbers, but when we investigate the dynamic processes the suitable components are appropriate measures of the time functions.

2. Description of the system

The general structure of the system under consideration is presented in Fig.1. Generally unknown and nonlinear object has 'u' inputs (vector Y) and 'l' outputs (vector Z) and satisfies the general assumption of controllability. We shall assume them $y^i \in \langle 0,1\rangle$ and $z^i \in \langle 0,1\rangle$ for every 'i'. The object is described by a set (usually nonlinear) of equations

$$\Phi(Y,Z) = 0 \tag{1}$$

which has one stable solution. Therefore there exists a unique but not known operator F

$$Z = F(Y) \tag{2}$$

as a solution of the equation (1) which is continuous and limited.

Some standard (stereotypic) output signals Z_s are given and for the measure of the error we assume

$$\|Z - Z_s\| = \sum_{j=1}^{l} |z^j - z_s^j| \tag{3}$$

The general task of the learning system is to generate in some number of steps (possibly minimal) such

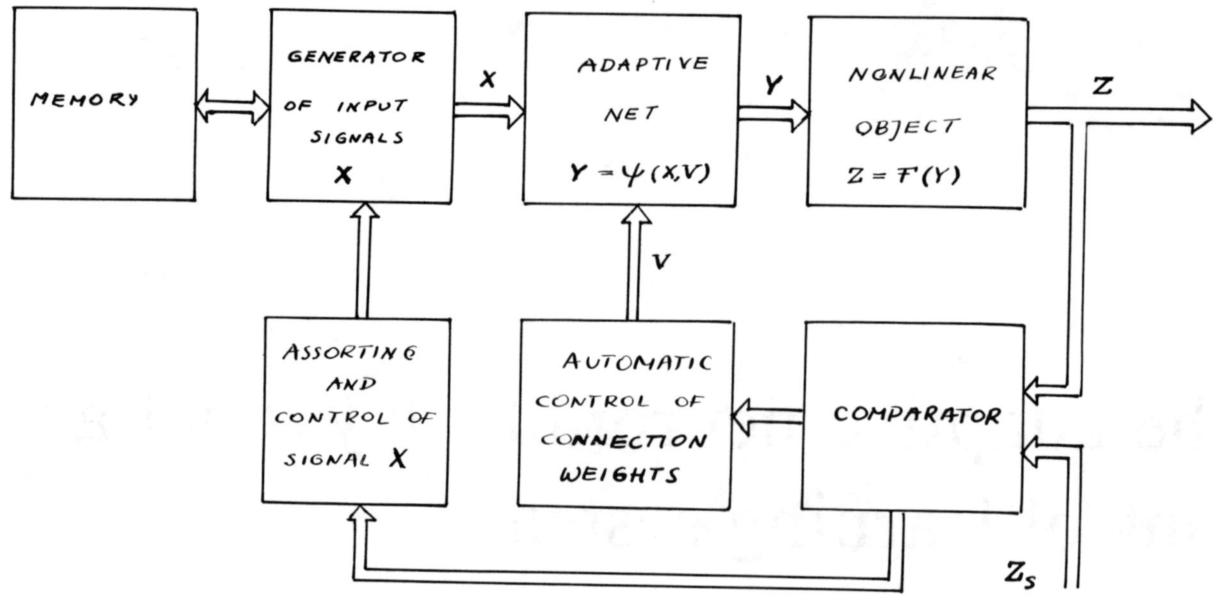

Fig.1 Block diagram of the learning system

values Y_s for every 's' as to get

$$\| Z - Z_z \| < \delta \tag{4}$$

The input signals for the object are generated in the adaptive net which is a set of neuronlike elements arranged in two layers with controlled connection weights W^i_t. For the stationary state we shall assume that the net is composed of nonlinear summators described by relation

$$U = \Phi\left(\sum_{i=1}^{p} W_i \xi_i \right) \tag{5}$$

where

- ξ_i – the signal applied to the input 'i' of the element,
- Φ – nonlinear operator, for example described by a broken line characteristic with threshold and saturation.

The weights parameters depend on the sum of input signals V_s controlling the values of the weight in all 's' steps of the learning procedure

$$W_i = \psi\left(\sum_{s=1}^{n} V_s \right) \tag{6}$$

The adaptive net has two kinds of input:

1. Binary signals $X = \{x_1, x_2, x_3, ..., x_w\}$ where $x_i \in \{0,1\}$, which are generated in a special generator by means of an assorting procedure for choice of optimal input vector [1].
2. Weight control signals which are generated in the block of automatic weight control fitted by a comparator.

In general

$$V^i_t(n) = k^*_t \cdot \varphi(n) \cdot (z^i_n - z^i_s) \tag{7}$$

where

- t – number of the inputs for the appropriate cross connections in the net for the channel 'i'.
- k^* – constant coefficient which characterises the mutual influences in the net.
- $\varphi(n)$ – a non-increasing function of a step number, satisfying the condition: for every $C \leqslant C_0$ there exist such N that for $n > N$, $\varphi(n) < C_0$. (8)

The appropriate weights of the adaptive net are changed according to the relation

$$W^i_t(n) = W^i_{ot} + \sum_{s=1}^{n-1} V^i_t(s) \tag{9}$$

3. Description of the learning algorithm

The learning algorithm for any defined stereotypic

output Z_S is composed in two stages. In the first stage the generator of the input signal X determines an optimal set of input signals X_S. After many investigations of different algorithms it appeared that the appropriate method of assorting the binary inputs X_s^i may shorten ten times (or even more) the learning procedure. In the second stage the error

$$Z_n^i - Z_s^i = \epsilon_n^i \tag{10}$$

signals for every channel 'i' and for every learning step 'n' is used for the generation of signals $V_t^i(n)$ which controls the weights W_t^i according to the rule (9). The procedure of assorting X may also be divided into two substeps and the procedure is based on the idea of simplified identification of the object.

The algorithm of the changing of the weights in the net is divided into four substages. In every substage another group of connections is changed according to the appropriate error ϵ_n^i.

In the paper [1] the learning algorithm is described in detail and the proof of the theorem is shown also.

Theorem

When the learning algorithm satisfies the following assumptions:

1. The object is described by an unknown operator satisfying the condition of controllability.
2. The adaptive net satisfies the relations (5), (7), (8) and (9) and the operator Φ in relation (5) is a monotonic function with threshold and saturation.
3. The signals $V_t^i(n)$ controlling the weights of the net satisfy equation (7).
4. The input vector X is an optimal vector assorting to the special procedure (described in [1])

– then starting from the $n > N$ the learning procedure is convergent and the output signal in every channel satisfies the relation

$$\lim_{n\to\infty} Z_n^i = Z_s^i \qquad \text{for} \quad i = 1,2,3,\ldots,l$$

and for the whole system the relation

$$\| Z_{n+1} - Z_s \| < \alpha \| Z_n - Z_s \| \qquad \text{for } \alpha < 1 \tag{11}$$

is valid.

Proof

Output in every 'i' channel for $(n+1)$ iteration is described by

$$Y_{n+1}^i = Y_n^i + \Phi\left(\sum_{j=-p}^{p} X^{i+j} k_{i+j} V_{n,i+j}^i \right) \tag{12}$$

where

X^t – input signal for element 'i'

k_t – coefficient, connected with magnitude and type of connection.

$V_{n,t}^i$ – the controlling signal connected with 'i'-th channel.

From (12) and (7) obtain

$$Y_{n+1}^i = Y_n^i + \varphi(n)(Z_n^i - Z_s^i)\,\Phi\left(\sum_{j=-p}^{p} X^{i+j} k_{i+j} k_{i+j}^* \right) \tag{13}$$

Then

$$Z_{n+1}^i - Z_s^i = \sum_{r=-m}^{m} \{ f_{i+r}(..)\,.\,\varphi(n)\,.\,k_r\,.\,(Z_n^{i+r} - Z_s^{i+r})\} + (Z_n^i - Z_s^i) \tag{14}$$

where

$$k_r = \Phi\left\{ \sum_{j=-p}^{p} X^{i+j+r} k_{i+j+r} k_{i+j+r}^* \right\}.$$

If

$$B_r(n) = f_{i+r}(..)\,.\,\varphi(n)\,.\,k_r\,.\,\frac{Z_n^{i+r} - Z_s^{i+r}}{Z_n^i - Z_s^i} \tag{15}$$

then

$$Z_{n+1}^i - Z_s^i = (Z_n^i - Z_s^i)\,.\,\left\{ \sum_{r=-m}^{m} B_r(n) + 1 \right\} \tag{16}$$

If the inequality

$$-1 < \sum_{r=-m}^{m} B_r(n) < 0 \tag{17}$$

will be satisfied, then

$$\lim_{n\to\infty} Z_n^i = Z_s^i \qquad \text{for} \quad i = 1,2,\ldots,l \tag{18}$$

From the structure of the object (a negative feedback), a negative sign of k_r and from relation (8),

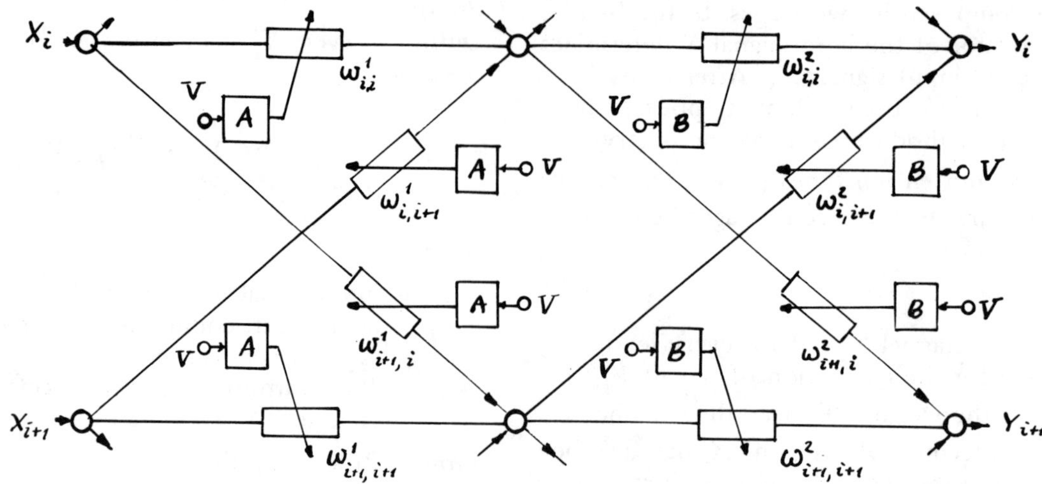

Fig.2 Weights control system
$\omega^1_{j,i}$ – weights in the first layer
$\omega^2_{j,i}$ – weights in the second layer
V – control signal
A, B – local operators

relation (18) with the condition (17) make the procedure convergent for every 'i'.

Denoting

$$\sum_{i=1}^{l} |Z^i_{n+1} - Z^i_s| = \|Z_{n+1} - Z_s\| \tag{19}$$

and

$$\sum_{i=1}^{l} |Z^i_n - Z^i_s| = \|Z_n - Z_s\| \tag{20}$$

we obtain

$$\|Z_{n+1} - Z_s\| < \alpha \|Z_n - Z_s\| \tag{21}$$

where

$$\alpha = \sum_{i=1}^{l} |\sum_{r=-m}^{m} B_r(n) + 1| \tag{22}$$

From the initial condition for procedure, there exist such N that for $n>N$, $\alpha<1$ and from the Banach theorem it follows that the procedure is convergent.

4. Results and discussion

The learning algorithm was programmed on a digital computer CDC 3700 for many different stereotypes Z_s. The modelling was made for ten different threshold characteristics of the net elements and for various objects. Some examples obtained from the line printer are presented in Figs 3 and 4. The asymmetrical object with interrupted internal connections was also investigated. The results obtained show that:

a. The introduction of the assorting procedure for the input signals has the essential influence on the speed of learning. The number of cycles was approximately ten times smaller than in the case of arbitrary assumed *X*.

b. Threshold values have an important influence on the speed of learning. This threshold ought to be approximately equal to the threshold value of the object.

c. To enable the adaptation of one system to the generation of the set different outputs *Z* it is necessary to introduce the special procedure which minimizes the number of changed weights for individual stereotypic signals. This rule gives the possibility of getting more different control signals *Y* corresponding to the needed stereotypes Z_s.

In Figure 3 the results of the modelling of five different stereotypes on the same net are shown. In

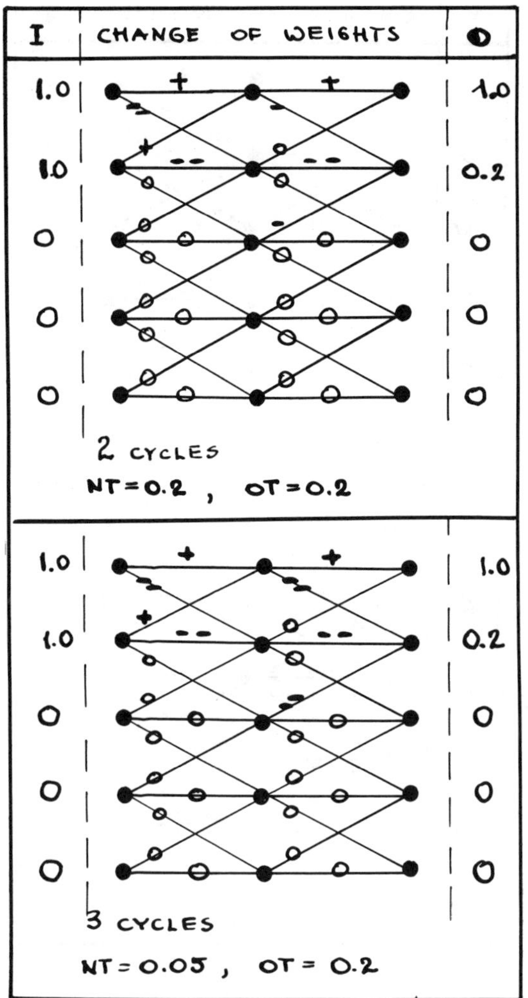

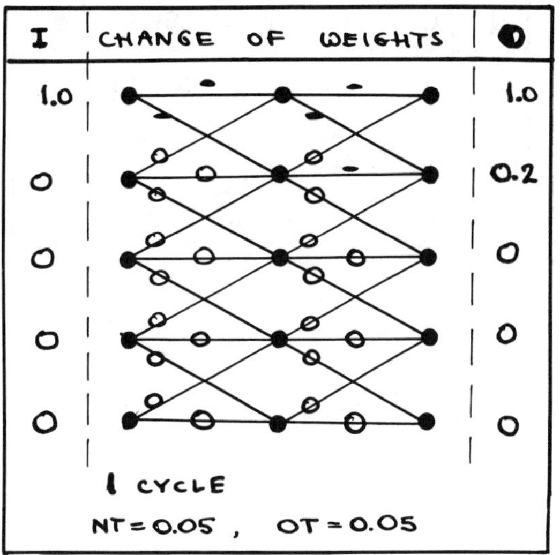

Fig.3 Results and adaptation and

Fig.3 Results of adaptation and learning
I – input, ◐ – output, NT – net threshold, OT – object threshold, ● – nets element.
○ – weight unchanged, +(–) – weight increased (decreased) 0-5%, ++(=) – weight increased (decreased) more than 5%.
The schematic diagram of the net is shown in Fig.2.

that procedure the additional rule was introduced. Every weight which in previous stereotypes changes more than 5, should not change more.

5. References

1. MACUKOW, B. and GAWROŃSKI, R., 'The learning algorithm using an adaptive net to control the unknown object', *Arch. Automat. Telemech.* V.XXI Z.4, pp.503-526 (1976).
2. FU, K.S. and TOU, J.T. (editors), 'Learning systems and intelligent robots', Plenum Press, New York (1974).
3. FUKUSHIMA, K., 'Cognition. A self-organizing multilayered neural network' in *Biological Cybernetics,* 20(3/4) (1975).
4. ANDERSSON, J.A., 'A simple neural network generating an interactive memory', *Math. Biosc.* 14(3/4), pp.197-220 (1972).

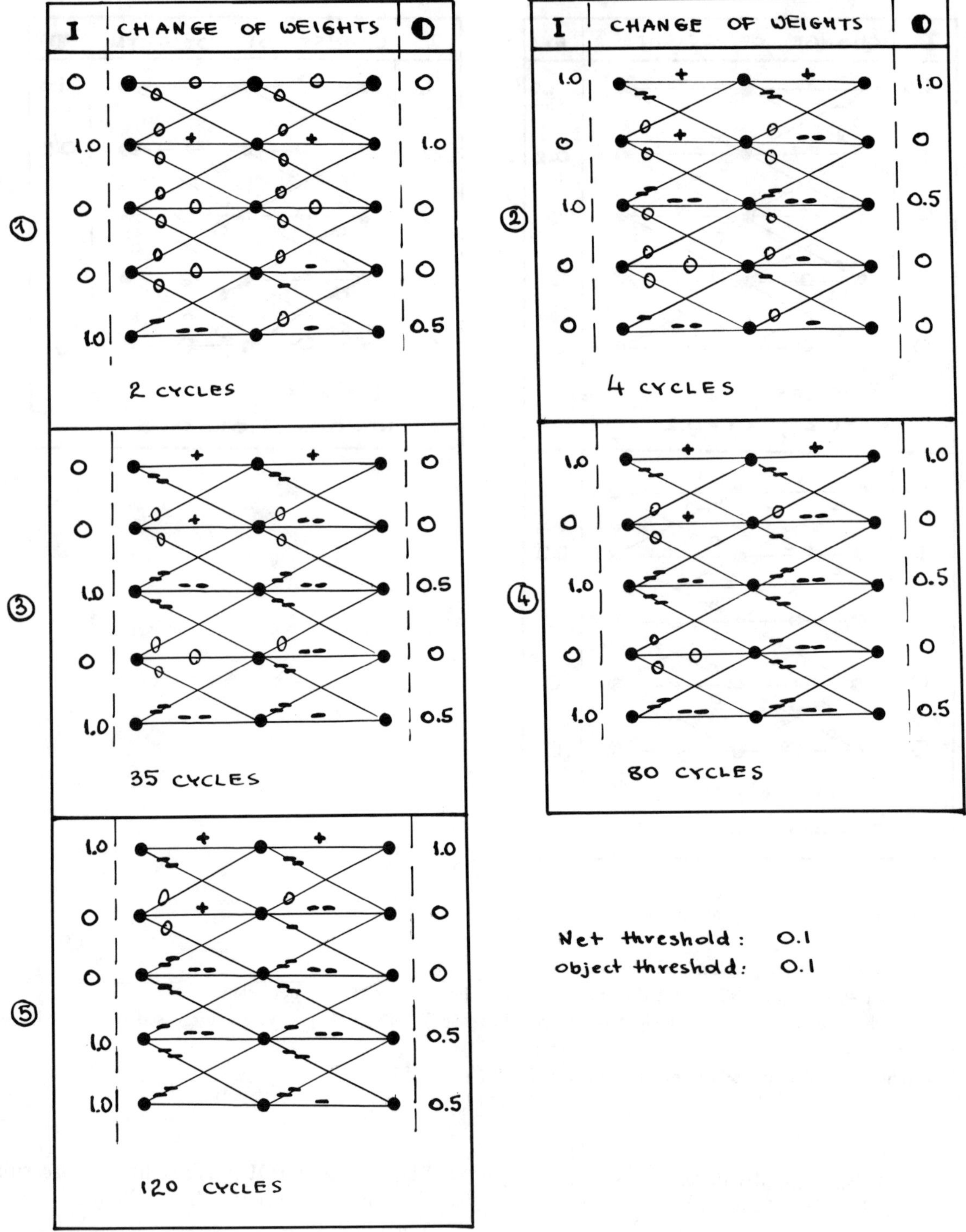

Fig.4 Learning procedure for five stereotypes Z_s on the same net. The schematic diagram of the net is shown in Fig.2.

I – input, ◐ – output, ● – nets element.

○ – weight unchanged, +(-) – weight increased (decreased) 0-5%, ++(=) – weight increased (decreased) more than 5%.

Evolution of self-regulating systems (principles and patterns)

C. BOGDAŃSKI
Paris

The present communication concerns the evolution occurring in the material group of Self-Regulating Events (RES), that is in the RESm group comprising all spontaneous self-regulating systems. It was considered necessary to precede this evolution study by some fundamental data.

1. SELF-REGULATING EVENTS (RES)

1.1 Organization principle of RES

Generally speaking regulation within a spontaneous self-regulating event consists in rhythmic re-establishment of such values as:

1. *Phase* (case of undulatory regulation in the RESo group).
2. *Relative position of structural compounds* [case of models of RESm, especially A and P including delocalized (displaced) structural compounds].
3. *Concentration level of a substance at a definite site of the system* (case frequent in many representatives of B; classical example—glucose level regulation in the blood).

Such regulation is supported by a negative feedback, that is between two channels through which signals pass in opposite directions, namely:

- Lines of opposite fields, e.g.
  - the line of the excitation force and that of environmental resistance (for more details see [4]) *in case* (1);
  - the line of the attractive force and that of the repellent one *in case* (2).
- Trans-substantial channels *in case* (3), e.g. channels at the molecular level: that of adrenalin and that of insulin in the case of glucose level regulation.

1.2 Taxonomy of RES

See Table I and Fig. 1.

1.3 Foundations of the differentiation[1] of RES

The differentiation of RESo and RESm will be successively analyzed.

1.3.1 Differentiation of RESo

As regards *the differentiation of RESo:*

- *Their subdivision into RESoe and RESos spectra* (see Fig. 2) seems to be justified by the difference existing in their affinity to matter. Only the RESos—placed at lower energy levels—require matter for their existence and replication, realized by propagation phenomena.

[1]"Initial" and then evolutive differentiation.

TABLE I Nomenclature of RES and of their fundamental and cardinal properties

NOMENCLATURE OF RES

RES = Self-Regulating Event
RESo = ondulatory RES = a wave
RESo-e = electromagnetic RESo
RESo-s = sound RESo
RESm = material RES
RESm-d = disaggregated-structure RESm
RESm-a = aggregated-structure RESm

FUNDAMENTAL PROPERTIES OF RES

ν = *frequency*
D = *size dimension*, more precisely:
system's average Diameter in the case of each RESm
wavelength (λ) in the case of each RESo
RES *self-organization* and eventually: RES *self-replication:*
= either in contiguous parts of space with a constant velocity of propagation, case of RESo representatives;
= or: supported by a compact (very complex) structural matrix, case of representatives of the 'B' model of RESm.

CARDINAL PROPERTIES OF RESm, each of those properties has antinomic states
= *mechanical properties: 'd', 'c'* and *'s'*
= *informational property: 'i'*

'd' = dynamics$^{(+)}$ of the system's link $\begin{cases} d_{mon} = \text{monodynamical} \\ d_{plu} = \text{pluridynamical} \end{cases}$

'c' = kinetics$^{(++)}$ of motion on the system's internal trajectories $\begin{cases} c_{ster} = \text{stereotypical} \\ c_{aster} = \text{astereotypical} \end{cases}$

's' = statics$^{(+++)}$: system's structure $\begin{cases} s_{dis} = \text{disaggregated} \\ s_{agg} = \text{aggregated} \end{cases}$

'i' = informational$^{(++++)}$ internal channels of system being $\begin{cases} i_E = \text{Extra-substantial} \\ i_T = \text{Trans-substantial} \end{cases}$

COMMENT concerning the antinomic character of the states binarity in each of the cardinal RESm properties where either a such (e.g.: d_{mon}) or a such (e.g.: d_{plu}) antinomic state appears exclusively. It can be formalised, e.g. by the following formula: $d_{mon} \cap d_{plu} = \phi$.

States implicative series being the most typical:	Conjunctures of states	being immanent to:
$d_{mon} \Rightarrow c_{ster} \Rightarrow s_{dis} \Rightarrow i_E$	$d_{mon}+c_{ster}+s_{dis}+i_E$	the dynostatic RESmd models: 'A' and 'P'
$d_{plu} \Rightarrow c_{aster} \Rightarrow s_{agg} \Rightarrow i_T$	$d_{plu}+c_{aster}+s_{agg}+i_T$	the 'B'-model, belonging to RESma
	$c_{ster}+s_{dis}+i_E$	the pluri-'B'='S' model of RESmd

PRINCIPAL MODELS OF RES

RESo models (increasing size order):

RESo-e models:	RESo-s models:
G = Gamma-rays	Ph = Phonons
X = X-rays	Hy = Hypersons
Uv = UV	Us = Ultrasons
V = Visible	So = Sons
Mi = Microways	Is = Infrasons
R = Radio	

(continued)

RESm models (alphabetical order):

Formula — Definition	Antinomic states' code	RES-size implicated position of respective band
'A' – **A**tom, *disaggregated* (nucleo-electronic) system	d_{mon}; c_{ster}; s_{dis}; i_E	10^{-10} - $5 \cdot 10^{-10}$ m
'B' – **B**iotic = each living organism, *an aggregated* system	d_{plu}; c_{aster}; s_{agg}; i_T	10^{-8} - 10^{2} m
'M' – **M**olecular system, *aggregated* (in principle, exception made of molecules having a delocalized electron π)	('M' = pluri-'A')	10^{-10} - 10^{-8} m
'P' – astro-**P**lanetary, *disaggregated* system	d_{mon}; c_{ster}; s_{dis}; i_E	impossible to determine, e.g.: $P_{SOLAR}\phi = 10^{13}$ m
'S' – **S**ocial, *disaggregated* pluri-'B' system	c_{ster}; s_{dis}; i_E	10^{-6} - 10^{7} m

$^{(+)}$System's structural elements—in interaction—are linked either by a single (at least practically) or by many forces corresponding respectively to d_{mon} or d_{plu} state.

$^{(++)}$The c_{ster} state is implicated by d_{mon} state and c_{aster} one by d_{plu} state.

$^{(+++)}$$s_{dis}$ and s_{agg} states validity is delimited by the value of the:

$$\frac{\text{distance between system's constitutive elements}}{\text{constitutive element diameter}}$$

ratio. This ratio being <1 in the case of s_{agg}, but significantly >1 in the case of s_{dis}; s_{dis} should be considered as 'non-aggregated'. Its existence is conditioned by the existence of c_{ster} state in '*c*'.

$^{(++++)}$Channels entering in negative feed-backs (= n.f.b.) which maintain homeostasis of each RESm. The i_T state organization is possible when it exists s_{agg} state and the respective system's structure is sufficiently complex. The i_E state can only exist by s_{dis}. Each i_T channel is placed on a determinate architectural level, e.g.: on molecular, submolecular or supramolecular in 'B' systems, but i_E channels are not necessary on a one determinate level because they are lying on the field-lines (e.g., in force fields by 'A' and 'P' representatives and in electromagnetic or acoustic fields by 'S' representatives).

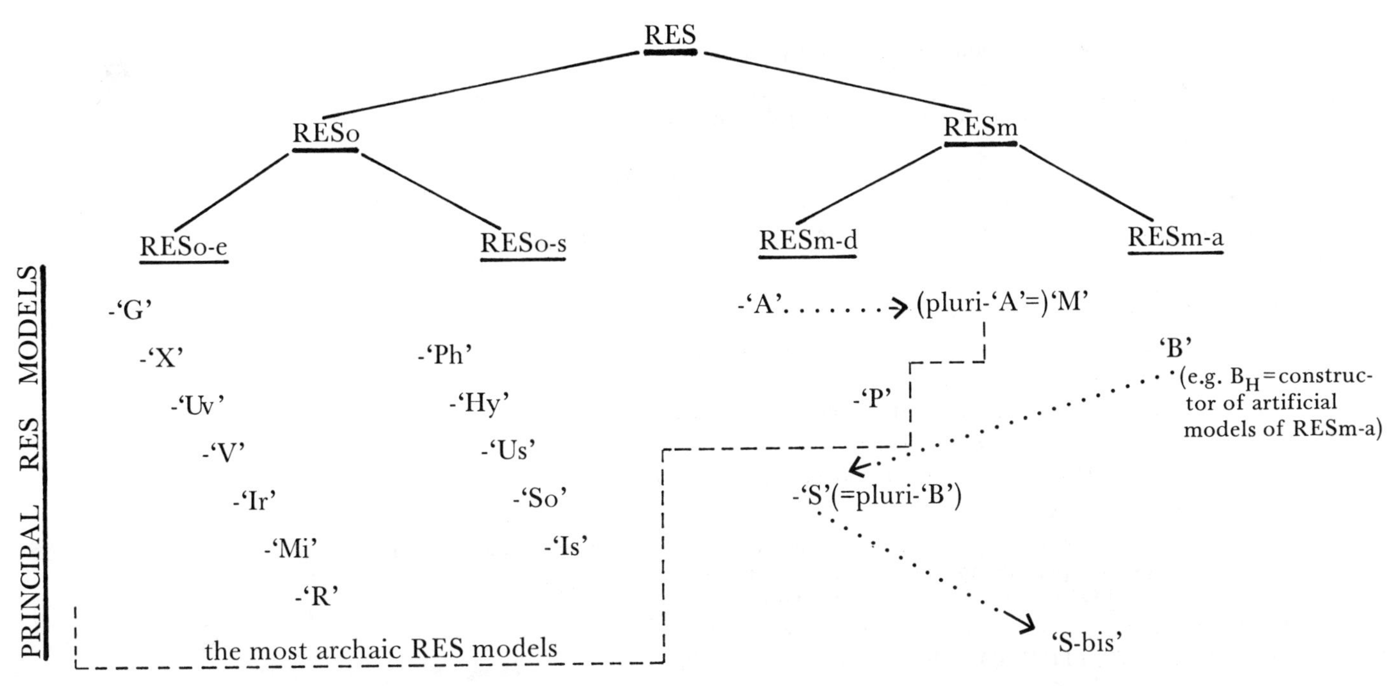

Fig. 1 Multihierarchic taxonomy of RES
(*For the Legend,* see TABLE I)

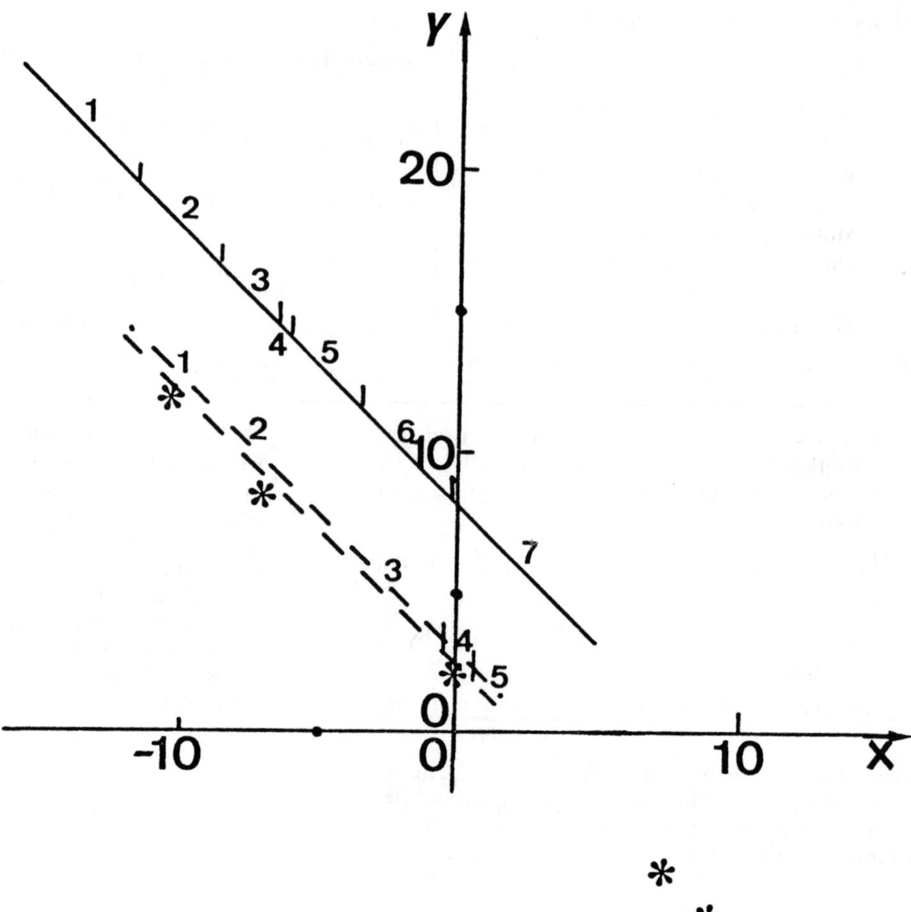

Fig. 2 Relationships between frequency and size by RES

10^x = RES size (meters) wavelength in the case of RESo / average diameter in the case of RESm

10^y = frequency (Hz)

Continuous curve: $y = 8.48 - x$ concerns RESoe group where *material* support for propagation is not necessary; models (for nomenclature see TABLE I): 1–G; 2–X; 3–Uv; 4–V; 5–Ir; 6–Mi; 7–R

Discontinuous curves concern RESos group where *material* support is indispensible for propagation, e.g. in:

air (dry at 0°C) : $y = 2.53 - x$ (lower curve);
lead (at 20°C) : $y = 3.09 - x$ (upper curve);

models (for nomenclature see TABLE I): 1–Ph; 2–Hy; 3–Us; 4–So; 5–Is

Data concerning some RESm (*material* events), from left to right:

* – vibration of the functional group of the molecule: 10^{12} Hz/10^{-10}m
* – enzymatic pulsation in the case of catalase ($1.5 \cdot 10^8$ Hz at 0°C)
* – max. frequency level in the case of anthroporythms: flamenco dancing
* – earth rotation daily rhythm
* – lunar orbital motion rhythm (reported to the size of orbit); and then: orbital motion rhythms of the solar system's planets reported to the respective orbits' sizes:

FIRST (LEFT) PLOTS GROUP (from left to right):

* – Mercury; * – Venus; * – Earth; * – Mars;

SECOND (RIGHT) PLOTS GROUP:

* – Jupiter; * – Saturn; * – Uranus; * – Neptune; * – Pluto.

Plots in both the groups are very finely aligned, but this line is about 20° deviated from the line being drawn discontinuously and valid for the other RESm spots.

- *Their secondary subdivision on models within each of the two spectra* can be justified by the spectral position of each model and is, therefore, the function of the energy level (manifested by frequency, and since $\nu.\lambda$ = constant, closely related with the RES size = wavelength).

1.3.2 The RESm

The RESm are illustrated partially (in Fig. 2 too) by their following representatives:

- *The very classical abiotic examples*—vibration of the functional group of a molecule and some (most typical) examples concerning the planets of the solar system.
- *Some[2] of the biotic cases.*

We find there a plot of points constituting a line, almost straight, discontinuous, situated at about the level where the RESos spectrum lies. Therefore we are inclined to reinforce our affirmation concerning the relationship between the level and the RES affinity to matter by the following statement:

- *The lower level* is occupied by the *material group* of RES (i.e. RESm) and by the group of RESo (namely RESos) which requires material support for its existence and propagative replication.
- *The higher level* is occupied by the RESoe group, that is those RES which have no such need, and constitute, therefore, a purely *energetic group*.

1.3.3 Distance between the two levels

Our search for a *factor which determines the distance existing between the two levels* has led us to the application of Einstein's well known formula: $E = mc^2$, which defines the energy:mass ratio, $E/m = c^2$. Since our presentation—in the Cartesian coordinate system—is one-dimensional (and not three-dimensional) we should write $\sqrt[3]{(E/m)} = c^{2/3}$.

Since $c = 3.10^8$ m.s^{-1} and $\log c \doteq 8.48$; $\log c^{2/3} = (2/3) \times 8.48 \doteq 5.65$. Let us now introduce the latter value into the formula $y_e = \log c - x = 8.48 - x$. After this introduction we obtain: $y_{e'} = y_e - \log c^{2/3} = 8.48 - x - 5.65 = 2.83 - x$ [for x and y see again the legend concerning Fig.2].

As regards the lower level, we should state to be explicit that this level is not univocally defined for the RESos spectrum, and its formula y_s is the function of the kind of matter to be penetrated. Namely:

$y_{s(\text{AIR})} = 2.53 - x$ (in dry air at 0°C with 330 m.s^{-1} propagation speed),

$y_{s(\text{LEAD})} = 3.09 - x$ (in lead at 20°C with 1230 m.s^{-1} propagation speed),

$y_{s(n)} = 2.83 - x$ (in matter n with 676 m.s^{-1} propagation speed).

As compared with the $y_{e'}$ formula the formulae $y_{s(\text{AIR})}$ and $y_{s(\text{LEAD})}$ are similar[3] and the formula $y_{s(n)}$ is identical.

1.3.4 Differentiation of models in RESm quasi-spectrum

Let us now pass to the discussion of the principle which may be responsible for *the differentiation of models within the RESm quasi-spectrum.* This principle cannot be as simple as that responsible for each of the two RESo spectra (RESoe and RESos) because the RESm quasi-spectrum is (see Fig. 3):

- discontinuous,
- heterogeneous, composed of both aggregated (s_{agg}) and nonaggregated (s_{dis}) structure models, namely:
  - s_{agg} are models M and B, belonging to the RESma subgroup;
  - s_{dis} are models A, P and S (pluri-B) belonging to the RESmd subgroup;
- composed of RES the existence of which is not dependent on constant-speed motion (propagation comprising self-reproduction phenomena, case of each RESo); therefore RESm have had possibilities to evolve and to manifest consequently a nonhomogeneous spectrum—the RESm quasi-spectrum;
- composed of such a RES (the material ones) where each event should be integrated by link forces: one force (d_{mon} state) or more forces (d_{plu} state). Since the intensity of each link force, nuclear (F_n), electrostatic (F_e), intermolecular (F_i), or gravitational (F_g) is a function of distance, therefore each of the RESm models is submitted to a different conjuncture of parametral activities of those link forces.

Consequently, the differentiation of models within the RESm quasi-spectrum can be considered as being the function of their scalar positions. This affirmation signifies that there exists here a size-dimensional differentiation factor as is the case (see above) in

[2] No doubt many other points deviate more from the line defined by the function $y_{s(n)} = 2.83 - x$ (because the RESm quasi-spectrum totalizes relatively more evoluted RES representatives than RESo groups: RESoe + RESos).

[3] With very small deviations: minus 0.30 and plus 0.26, respectively.

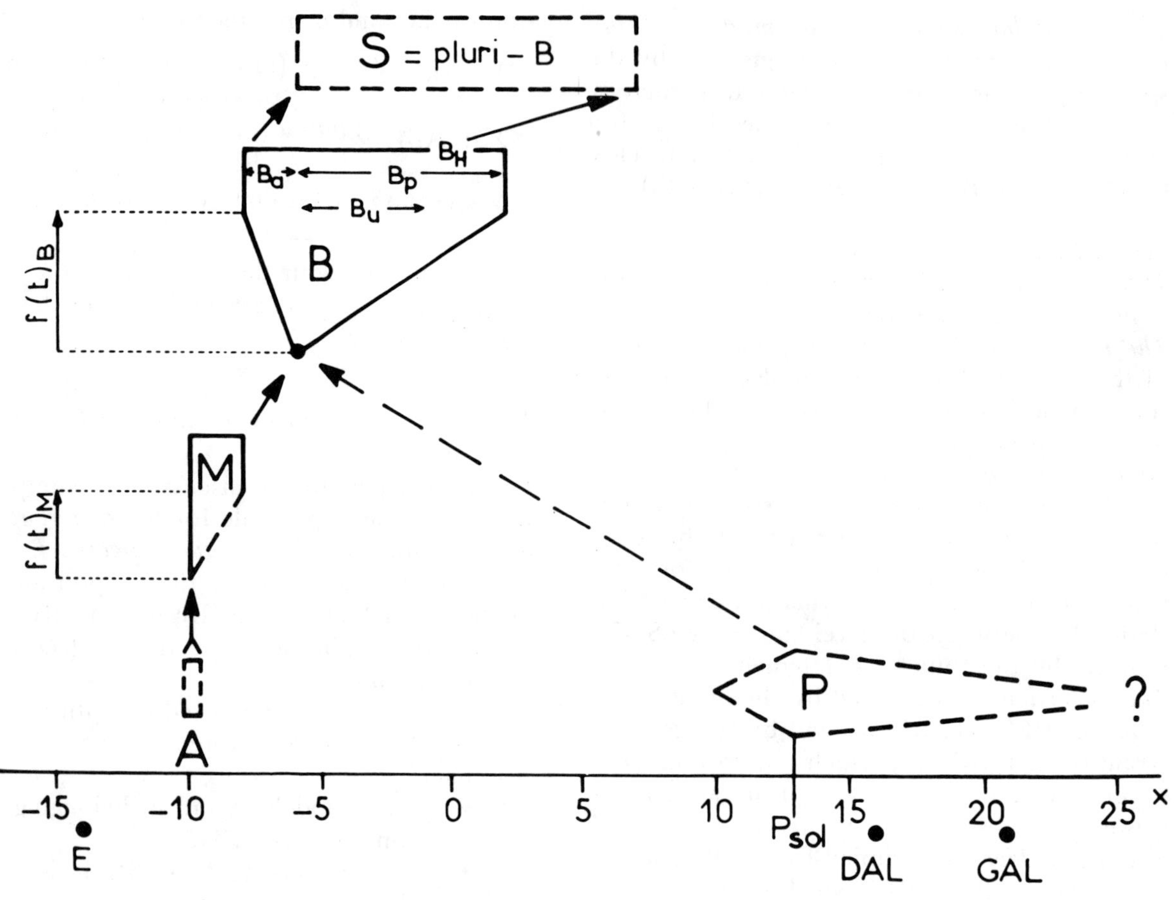

Fig. 3 Bands of RESm quasi-spectrum

Bands corresponding to the representatives of principal RESm-models are circumscribed:

– continuously in the case of RESm-a representatives[(*)]

– discontinuously in the case of RESm-d representatives.

'A', 'B', 'M', 'P' and 'S' models conform to our models code.

B_u = 'B' unicellular; B_a = 'B' acellular; B_p = 'B' pluricellular variants.

B_H = specific 'B' variant: Homo sapiens[(**)]. E = size of Electron.

On the left margin: the arrows mark bands' evolution within time.

The lower point of 'B' signifies the most probable biogenesis site; two arrows are showing there:

– *from the left*: a phenomenon of multimolecular (compact and very complex) agglomeration

– *from the right*: a radiative influence of solar 'P' system

The type 'P' is placed in the position corresponding to the solar 'P' system size (10^{13}m).
The 'P' band cannot be delimited exactly and the '?' sign is placed in the position of the diameter of the telescopically visible cosmos.

DAL = Distance of the light year. GAL = Our GALAXY diameter.

(*)The 'M'-model being an aggregated-structure in principle—but with some exceptions (molecules having π electrons)—therefore the 'M'-band circumference line is partially discontinuous.

(**)The arrow placed near B_H indicates: men socialization phenomena on the planetary scale (=planetization).

each of the RESo spectra, but it is there justified otherwisely.

It should be pointed out that dynostats (A and P, the most archaic models of the RESmd subgroup with equilibrium dynamically supported) are to be found only in those regions of the RESm quasi-spectrum where the d_{mon} state is predominant, namely:

- the A model where d_{mon} is based on a practical exclusivity of F_e,
- the P model where d_{mon} is based on a practical exclusivity of F_g.

COMMENT. The d_{mon} state implies the existence of the c_{ster} antinomic state of c, and the latter state is indispensable for the s_{dis} state (immanent in each representative of the RESmd subgroup) to be maintained.

We can affirm that since d_{mon} is indispensible to the organization of the dynostatic type of RESmd, there should exist a Zi (zone of interdiction of the organization of a RESmd type dynostat) between the $d_{mon\,(F_e)}$ and the $d_{mon\,(F_g)}$ regions.

In Zi the d_{plu} state being predominant, it manifests its apogee in Rpm (Region of maximal dynamic plurality) around the value of 10^{-7}m. Since d_{plu}[4] implies c_{aster}, the maximum of the motion freedom is to be expected in Rpm; where—as far as the structurogenic processes are concerned—the probability that the trajectories will be less stereotypical is very high. Such structurogenic astereotypies lead to the organization of the most compact and complex structures ($s_{agg\text{-}compl}$): the compactness being immanent in each RESma representative and complexity in some of them (especially in the case of the latter being exposed to the Rpm activity—see below).

1.3.5 RESma models

To the RESma subgroup belong models M[5] *and* B; representatives of both models have been known to manifest very spectacular evolutive phenomena which should not astonish us, because being in an s_{agg} state each RESma is rather an acceptor than a donor of information signals, and in the case when $s_{agg\text{-}compl}$ (variant of the s_{agg} state) appears, the considered system has an operational network for treatment of the informational signals (imputed to it).

1.3.6 M *model*

As regards *the* M *model representatives*, their actual scalar positions are comprised within the M band (10^{-10} to 10^{-8} m). It should be noted that the molecular evolution was accompanied by an increase of:

1. *Molecular size* (average diameter of molecule) from 10^{-10} to 10^{-8}m.
2. *Structural complexity* which seems to be not only the function of the evident fact that—when the molecule increases in size—it includes a higher number of atoms, but also of the fact that when the molecule passes from the 10^{-10} to the 10^{-8}m value—*its degree of exposure to the Rpm activity becomes greater* (for more details see [2]).

1.3.7 B *model*

As concerns *the* B *model organization*, we refer here to the statement derived from the field of research on the origins of life, supported by the S. Fox experimental basis, namely, that the first living organism must have been of the size of the contemporary Micrococcus [3], that is 10^{-6}m. The question arises: why not larger or smaller?—to which we reply:

- If it were larger it would need the assembly of too many molecules, having 10^{-6}m diameter, it needs, e.g. 10^6 molecules with a 10^{-8}m diameter or 10^9 molecules with a 10^{-9}m diameter.
- If it were smaller it could not possess a cellular organization. Since the 10^{-8} to 10^{-6}m region of the RESm quasi-spectrum constitutes an Acellular sub-band of the B-band, and the Unicellular sub-band begins with 10^{-6}m extending to higher values, the biogenic position may be considered as
  - the most economic one from the point of view of the number of molecules to have in the assembly,
  - *the most exposed to Rpm activity*,

  within the range of the Unicellular sub-band of the B-band.

Generally speaking, the B model is marked by the following states: d_{plu}, s_{agg}, i_T and c_{aster} (= freedom of motion).

In function of time a biotic evolution occurred and the B scalar position was bilaterally extended from the value of about 10^{-6} to the very large B-band of 10^{-8} to 10^2 m, a 10-magnitude orders span (that is 20 times more as compared with the A band span). Consequently, many of the parameters of determinative factors[6] of the system's characteristics cannot be preserved in their initial biogenic conjuncture and many new conjunctures were organized in function of the occupation of new scalar positions by proliferating B-representatives. No wonder that such scalar changes were accompanied by important changes in properties.

[4] d_{plu} based upon F_e, F_i and F_g and since d_{plu} implies s_{agg}, the basis of such a d_{plu} is supplied by F_c; F_c = force of cohesion.

[5] In principle s_{agg}, but with the exception of molecules having a delocalized electron (π).

[6] The number of molecules assembled, mass quantity, inertia, immanent frequency level, periodism, life span of the system, surface:mass ratio implying both the system's opening degree and metabolic rate.

According to Dean and Hinshelwood [5], a sufficient reason for an organism to change is when, e.g., a unicellular system increases twice its volume: its surface:mass ratio diminishes, and so does consequently, its material requirement. This over-production disturbs the cell metabolism up to the moment of expansion when the given cell is subdivided into two derivate cells with initial parameters restituted. We quote the above mentioned paper because it shows that an increase of diameter from 1 to 1.31, corresponding to a 0.12 unity logarithmic scalar change is sufficient to provoke important biotic changes; in this case the changes become reversible, this not being the case when a living system changes its position definitively.

We can observe actually important changes of the characteristics of living systems from the morphological, anatomical and physiological points of view, to quote the scalar deduction of acellularity of viruses and of the increasing need of both a respiratory organ and a heart with the increase of the animal's size, as suggested in Vienna in 1971 [1]. The s:m ratio of scalar changes constitutes even in these cases an important factor.

Since the surface:mass ratio decreases parallelly to the increase of the system's size, and since this ratio implies directly the metabolic rate (positive relationship), any increase in size risks to diminish excessively the metabolic rate, so it is no wonder that some limits of increase in biotic size are imposed, mainly:

10^1m for animals, and

10^2m for plants.

Such a difference, of the order of one magnitude, can be justified by the fact that the highest plant value concerns trees which are ramified systems (and this implies a certain increase in the s:m ratio).

*1.3.8 Pluri-*B *models:* S *and* S-bis

Transgression of such size limits in the case of B-representatives is not possible otherwise as by means of association phenomena in the form of nonaggregated state systems, that is conforming to the S *(pluri-*B*)* model where the metabolic rate of all structural compounds can remain unchanged and be independent of the S-system size.

The links between the B-compounds of S are ensured by Extra-substantial channels of information. They are not lines of force fields as is the case within the A and P dynostatic models of RESmd, but they are lines of a new type of fields—biotic fields[7] generated by the B-systems. Such a generation becomes possible when the B-systems reach an adequate level of structural complexity.

To sum up, the S model representatives are characterized by the following states: s_{dis}, i_E and c_{ster} (because many kinematic stereotypes are introduced by a social mode of life).

To avoid the inconveniences of some c_{ster} (kinematical sterotypicities) some restricted (s_{agg}) special societies for very privileged persons, conforming more or less to *the* S-*bis model*, have been organized, marked by an increase in the degree of motion freedom (c_{aster}).

* * *

When we analyse the evolution of states in reference to the following RESm model series A + P; B; S and S-bis, we observe that the organization of a new model is accompanied by an inversion of antinomic states in at least two cardinal RESm properties (see Table II) in the previous model.

2. EVOLUTION PRINCIPLES OF SELF-REGULATING SYSTEMS

2.1 Conditions to be fulfilled

I Lack of obstructive factors (see above).

II The presence of favorable factors which promote some RES groups more than other ones (see below the hierarchy of evolutive predispositions).

2.2 Hierarchy of evolutive predispositions

RESo < RESm;

RESmd (dynostats: A and P) < RESma (M and B).

2.3 Principal pattern

As concerns the self-regulating *systems* = RESm, a new model organization is accompanied by an inversion of antinomic states in some cardinal RESm properties of the previous model.

[7] Bioacoustic, bioelectrostatic, bioelectromagnetic, etc.

References

1. BOGDAŃSKI, C., 'Structure des forces auxquelles sont soumis les êtres vivants situés tous dans la bande dimensionnelle de 10^{-8} à 10^2m', First European Biophysics Congress, Vienna, 14-17 September 1971, *Materials*, Vol. VI, pp. 239-244.
2. BOGDAŃSKI, C., 'Etude sur les fondements de la biocinematique', *Comptes Rendus des Séances de l'Academie des Sciences (Paris),* Serie D, Vol. CCLXXIX, No. 11, pp. 943-946.
3. PONNAMPERUMA, C. and GABEL, N.W., 'Current status of chemical studies on the origin of

TABLE II Some states of RES in function of the evolution of RESm-models

Reappearances of identical states are marked by arrows. REMARKS: Reappearances may be considered as atavic manifestations; Models 'prebiotically organized' are the more archaic ones.

Group of RES: / Model: / Property and its antinomic states:			RESo: all models of RESo		RESm: DYNOSTATES 'A' and 'P'	RESm from left to right in chronological appearance order: 'B' (=BIOTIC)	'S'	'S-bis'
			PREBIOTICALLY ORGANIZED				DERIVED FROM 'B'-MODEL	
MECHANICAL:	'd'	d_{mon}			★			
		d_{plu}				★		
	'c'	c_{ster}			★ --->	--->	★	
		c_{aster}				★ --->	--->	★
	's'	s_{dis}			★ --->	--->	★	
		s_{agg}				★ --->	--->	★
INFORMATIONAL: 'i'		i_E			★ --->	--->	★	
		i_T				★		
SELF-REPLICATION SUPPORTED BY THE MATRIX ? :		NOT$^{(x)}$	★					
		YES				★		
VARIANTS of 'B'-MODEL		B_u (=unicellular)				★ ---> ★ (embryo-urgy; ovogenesis)		
		B_p (=pluricellular)				★		

$^{(x)}$but realized—in the case of RESo—by the continuous self-replications in the contiguous parts of space with a constant velocity propagation.

life', *Space Life Sciences,* Vol. I, pp. 64-96 (1968).

4. SCHMIDT, G., *Kompendium der Physik,* Jena, Gustaw Fischer Verlag (1971).

5. DEAN, A. and HINSHELWOOD, C., 'Some basic aspects of cell regulation', *Nature (London),* Vol. CCI, No.4916, pp. 232-239 (1964).

Phosphenes in the light of biohologràphic models

PAL GREGUSS*
Gesellschaft für Strahlen- und Umweltforschung mbH. und Technische Hochschule Darmstadt, FRG

Introduction

The interest of our species to express his relations to nature created two modes of activity, scientific and aesthetic, without expressing any conflict until the time of the Renaissance. Then, the 'man of science' and the 'man of art' gradually separated, and since the beginning of the present century, the two ways of representing experience have diverged so far that a unification of these two modes of activity through understanding each other seems to be impossible. The endeavor of those, however, who have felt dissatisfaction with the separation of science and art, and thought to discover new links between them, created new forms of self-expression, such as kinetic and computer art.

Admitted or not, all these attempts are based on the belief of having discovered one of the relations between a 'state of order' and 'pattern'. However, according to Szent-Györgyi [1], life is based on 'order' and 'pattern', and so if pattern can be considered as the result of 'interference', i.e. an arrangement which is not chaotic but which displays some regularity, perhaps something like gestalts of the psychologists, while 'order' means 'coherence', the processing of these two terms can be described only by applying the theory of holography. The reason for this can be found in the fact that our knowledge of the external world is evaluated from recording and processing scattering. Our eyes record and our vision processes light radiation scattered from objects around us or, when looking at an electronmicrograph, we are confronted with the result of scattering of electrons. And even hearing is a result of evaluating scattered mechanical waves which reach our ears.

The processing of sensory information begins in the sense organs themselves. It is in them that the first step is taken to transform the scattered pattern acting on them into the patterns of nervous action that regulate the activities of the organism in its complex environment. This environment, however, is three-dimensional, and all the receptors are square-law detectors, so that phase-bound information should be lost. Nevertheless, fish are swimming and birds are flying in real three-dimensional space where existence without the ability of simultaneous processing of amplitude-bound and phase-bound information is inconceivable.

I have the feeling that our clumsy understanding of biological information processing as well as the conflict between the man of science and the man of

*Present address: Applied Biophysics Laboratory, Technical University Budapest, Hungary

art is partly a result of the old frustration created by the fact that our world is three-dimensional, in which we acquire information in a three-dimensional format, but record and display it in a two-dimensional format because all recoptors are square-law detectors.

Not until the discovery of holography [2] have we had a means by which amplitude-bound and phase-bound information could be simultaneously recorded, stored, and processed by square law detectors. Therefore it is not at all surprising that, since the revival of holography in the mid-sixties, more and more biologists, physiologists, and even psychologists are wondering whether holography could not help them in answering questions posed by information processing observed in living systems. [3,4,5]

Holographic models are created and questions are raised about whether these models are only functional or structural too. [6] The only way to get the right answer is to look, first of all, for physiological evidence in order to discover whether the model in question is at least functional, and not a fiction, indeed.

2. Bioholographic models

All bioholographic models are based on the assumtion that, similarly to the methods which were used when the results of mathematical analysis of information processing in optics were utilized by analogy in the theory of electromagnetic signal transmission, a living organism can be considered as analogous to an optical system. The incoming information pattern, i.e. its descriptive function, is regarded as a wavefront, and, for example, the dendritic influences are linked to the diffraction of a lens or of wave modulation, when thinking in electromagnetic terms. It is usual to represent the result of wave diffractions as either a Fresnel or a Fourier transform. An obvious difference between these two operations is the appearance of the sum of the parameters in the Fresnel transformation, and of their product in the Fourier transformation. Of greater importance, however, is a similarity of the two operations, as shown in the self-reciprocity of both operations. Since holography is usually delineated in terms of lens (or lensless) diffraction which, however, is represented by the Fourier transform, most bioholographic models use Fourier-transform-based descriptions. This lends itself directly to the assumption that the neural network may act as a Fourier analyzer using holographically matched spatial filters for pattern recognition. This is the point at which we wish to investigate the problem of whether physiological findings could be found which might back up such an assumption.

The principles of holography are rather well known, and since two colleagues of mine, Wess and Röder, discuss today a holographic brain model of associative memory, I think it is enough if we recall that the hologram principle is based on three well-known operations: modulation, frequency dispersion, and square-law detection, which means that phase-bound information is transformed via interference, in a reversible way, into proportional amplitude relations.

It is known from the work of Sokolov [7] that somewhere the sensory input is matched against a comparator before being relayed further in the central nervous system. Bishop, [8] observing cerebral potentials, has found that only those imaging stimuli are recorded that are synchronous with the positive phase of the quiescent period of activity of the brain. This, however, is in close agreement with a hologram type recording and processing of information, since only the component of the complex amplitude is recorded that is in phase with the reference background. [9]

The fact that the sensory input is compared does not mean that the central nervous system needs to share an infinite number of 'sensory models' for recognition to take place. What is required is a set of filters, in action somewhat similar to the filters used in coherent optical data recognition systems. These filters, however, have to be stored in a holographic form since only this ensures the distributedness of the filter, which is neccessary for an information processing model that wishes to be consistent with the well established findings of neurophysiology that incoming information is processed over wide areas of the nervous system. Further, with holographically stored special filters, translation-invariant parallel information processing is feasible, which is necessary if visual perception functions have to be described.

A simple model [10] which consists only of those function groups that are necessary for describing visual pattern recognition processes but does not claim morphological equivalents is shown in Fig. 1. A stimulus pattern is generated by the information pattern acting on the visual reception field, by passing through the function group of neurons A inducing a new stimulus pattern R, which then acts as a reconstructing wavefront for the spatial filters SF stored in a holographic form in another function group of neurons, FG. In the meantime, by a matrix of neurons M, the Fourier spectrum of the information carrying stimulus pattern is formed and projected on one of the reconstructed spatial filters SF of the function group FG, so that direction independent

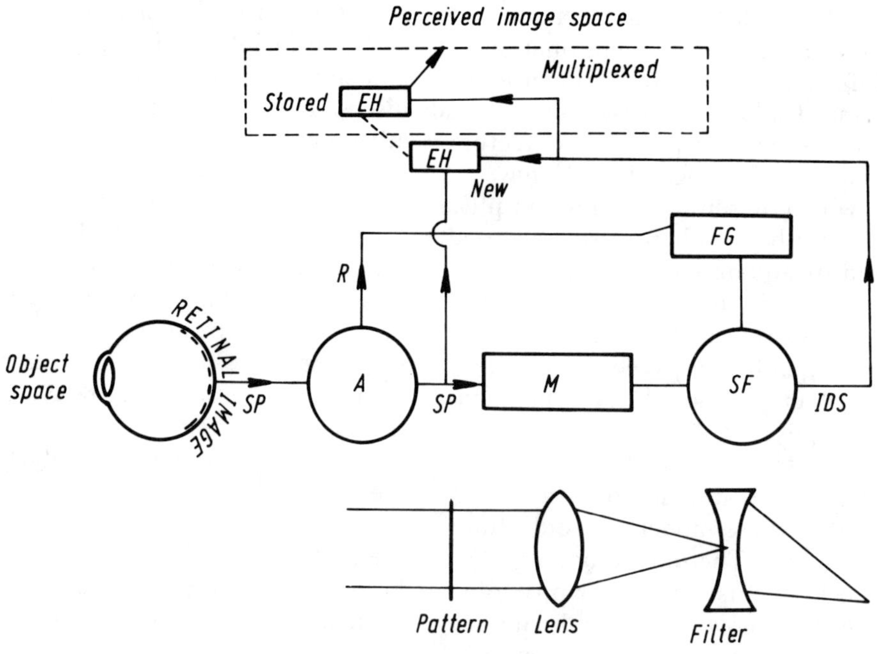

Fig.1 Holographic model of visual information processing

(or direction dependent) differentiation of the information pattern to be recognized is produced. As a result of this operation, an identification stimulus pattern IDS arises, which has two functions.

First, if there is already an engram, a memory trace of this information pattern, stored in a holographic form (EH) in the brain, it reconstructs it, i.e. the information pattern is recognized. The second function of this IDS is to form in a group of neurons a new engram hologram, EH, by interfering directly with the stimulus pattern generated by the information pattern acting on the visual receptors.

It can be assumed that if this model is a functioning one, a non-specific stimulus pattern acting on the function group A may result in a visual sensation corresponding to a pattern issuing from the direction-independent differentiation of the inadequate stimulus pattern.

The perception of patterns not resulting from viewing external objects, but stimulated by non-specific stimuli such as rectangular current pulses or sinusoidal magnetic fields, has been known for a long time. [11,12] Such patterns in the kind of abstract geometric figures, e.g. rings, arcs, waves, etc. are called phosphenes in physiology, and hallucinations in psychiatry. It is generally assumed that the non-specific stimulus acts on the intraretinal optical nerve fibers. This may perhaps be correct but does not explain why phosphenes are produced only when the specific stimulus pattern, i.e. visible light is not present, and that only a restricted number of patterns per person are excited.

3. Characteristics of phosphenes

Phosphenes, sometimes named 'subjective light patterns of the second kind', can be produced with non-specific stimuli other than electric or magnetic stimulation, for example by chemical, acoustical shocks or epileptiform cortical discharge. Experiments have shown that these non-specific stimulations of the central nervous system affect primarily its optical area, and no other areas, e.g. the acoustical one. One of the anatomical reasons for this may be that the neurons of the lateral geniculate body are more readily activated than the occipital cortex, another that the receptive field centers in the cortex are smaller than within the lateral geniculate body and the optical trace. [13,14]

Perhaps the most interesting characteristic of phosphenes is that, independently of the nature of the non-specific stimulus, they can be graded into not more than 15 groups. Figure 2 shows the occurrence frequency of these pattern groups in the case of electrical and magnetic excitation after Seidel [15] Since all these patterns are the result of subjective

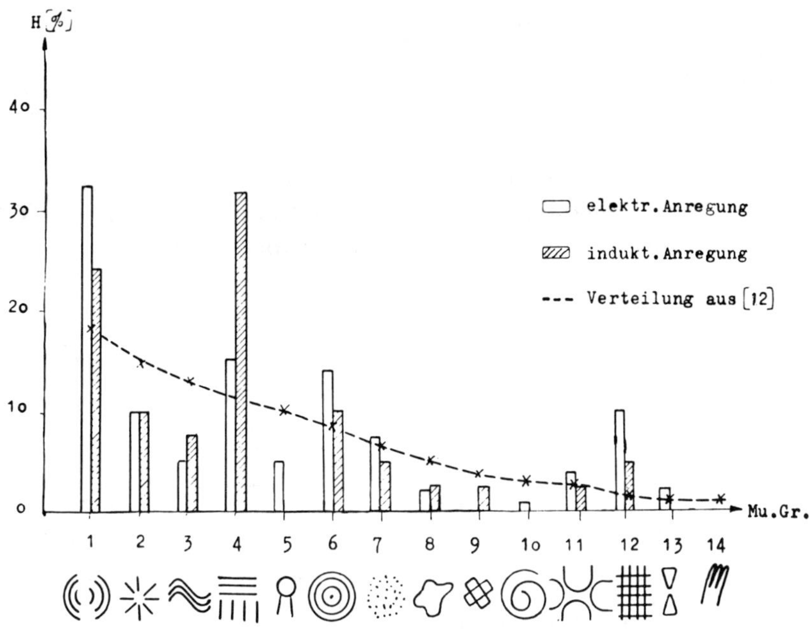

Fig.2 Occurrence frequency of electrically and magnetically excited phosphenes after Seidel

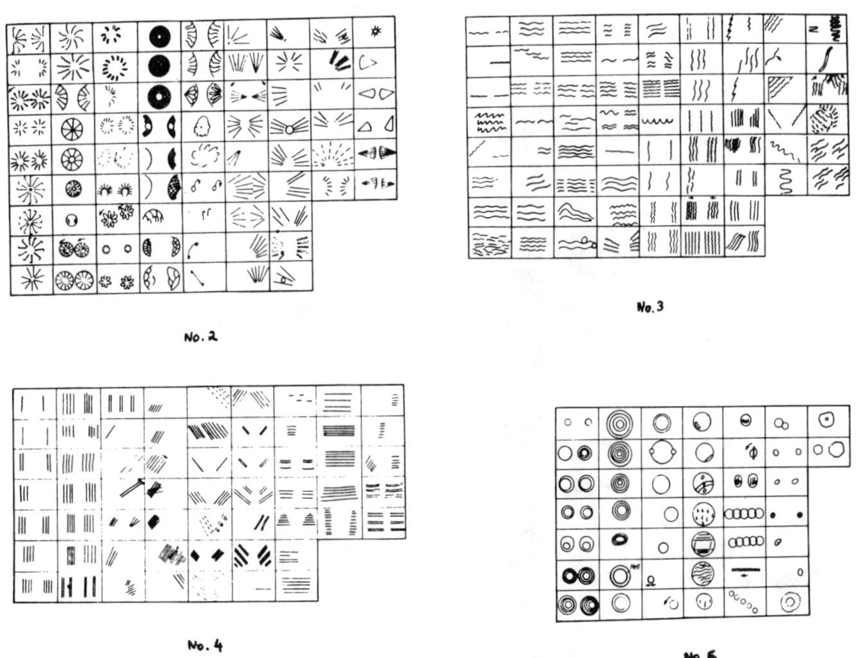

Fig.3 Variations in phosphene patterns belonging to groups Nos. 2-6

(i.e. virtual) perception, they have to be drawn by the subjects themselves or by the experimenter after sketches by the subjects and their corresponding remarks. Since, however, sensations are only interpretable within perceptual context (and not vice versa) the drawing will only be partially identical with the perceived pattern, and this may explain individual differences in the appearance of the drawings. Figure 3 shows variations of the patterns belonging in groups Nos. 2-6.

Another important characteristic of phosphenes is that they can only be perceived with dark adapted eyes or, at least, with closed eyes. This, however, means that the function group A of the bioholographic model shown in Fig. 1 does not receive an adequate information pattern, nevertheless, it is activated at least so far that it induces the stimulus pattern R which, then, reconstructs a spatial filter SF stored in a holographic form. Hence the spatial filter SF is now ready for filtering, and since only the Fourier transform of the non-specific stimulus pattern is available, this is the pattern that will be filtered and then the result perceived as phosphene. Such an explanation can be backed up by the observation of Welpe [16] who reported that the subject, when stimulated periodically by sinusoidal currents and simultaneously viewing *continuously* at a bright area, e.g. an opal glass plate, instead of perceiving one of the 15 phosphene patterns, perceives only more or less continuous elliptic areas as shown in Fig. 4. According to the holographic model presented this can be explained by the fact that the function group A of the visual system is not entirely free of an adequate stimulus pattern and therefore the perceived pattern is not only the differentiation of the non-specific stimulus pattern any more but that of the differentiation of the combination of the adequate and inadequate stimulus patterns.

Fig.4 Patterns of the ellipse phenomenon after Welpe

However, if the adequate stimulus pattern is only of short duration, shorter than the time needed for processing it, it will act only as a non-specific stimulus, although it is a light stimulus and will produce one of the known 15 phosphene patterns, as demonstrated by several investigators. (Fig. 5) They are called optical phosphenes (OP) to distinguish them from the electrically excited (EP) and magnetically excited (EM) ones.

Formen	Häuf. in %	E-Phosphene Höfer-Knoll f=5...60 Hz	O-Phosphene Blum f=5...20 Hz	O-Phosphene Smythies f=5...30 Hz	O-Phosphene Welpe f=20 Hz
1. Kreisbogen	21,8		"Halfcircles" "Arcs"	—	—
2. Radialsymmetr. Figuren	16,9		"Rays"		
3. Wellenlinien	15,7		"Waves"		
4. Strichfiguren	13,3		"Lines"		
5. Kreisfiguren	9,7		"Circles"		
6. Pünktchen-und Vielfachmuster	8,5		"Mottle" "Dots"		
7. Viereckige Figuren	3,6		"Squares"		
8. Spiralen	3,6		"Spirals"		
9. Pol-bzw. Feld-konfigurationen	3,6		—	"Magnetic-field"	
10. Gitterfiguren	1,8		"Mesh"		
11. Dreiecksfiguren	1,5		"Triangles"	"Triangle"	
12. Zackenmuster	1,2		—	—	
13. Kugelstiel-figuren	0,5		—	—	—
14. Strichkonfi-gurationen	0,1		—	"Herring-bone"	
15. Serpentinen und Wendeln	0,1		—	—	—

Fig.5 Patterns of optical phosphenes (OP)

4. Significance of phosphenes

These findings and other (not mentioned here) properties of phosphenes may be regarded as proofs for a possible coherent filtering mechanism in the line of visual perception. However, for understanding how this filtering may work, we have to look a little closer at the relation: retinal image and its perceptual context.

In general, it is assumed that the virtual picture reconstructed by our perceptual mechanism matches the retinal image produced by nature. The question only is which of the retinal images is meant by this since the retinal image, due to the various eye movements, continuously changes. It is therefore more presumable that, in certain respects, the perceived *virtual* image is more like the object itself than it is like any of the retinal images of the object. We believe that this is achieved by abstraction of contours in each of the retinal images. These abstracted contours become then the 'raw material' on which certain space analyzing mechanisms work. [17] However, a certain minimum learning is a necessary condition for the proper appreciation that the perceived virtual image is three-dimensional. To achieve this, the perceptual mechanism in reconstructing the object space has to distinguish between relevant and irrelevant kinds of cues in the various retinal images, and contours may help to do this job. However, as there are no lines in nature, so contours have to be generated by a relatively primitive physiological mechanism. According to our holographic model, the same mechanism that produces the phosphenes is also reponsible for this abstraction.

Before making this statement more probable, we wish to recall that we consider the phosphenes as perceived patterns issuing from direction-independent differentiation of a stimulus pattern other than that resulting from a light pattern. Since the origin of the inadequate stimulus pattern is really not important, at least as long as it acts on the function group A, it could be generated by the central nervous system itself too, just to test how effective one of the stored filters may be in reconstructing picture space and, perhaps, to reinforce this preformed neural network in the visual system.

In general, the result of this self-stimulated, direction-independent differentiation remains subconscious. However, if we realize that all basic geometric form groups produced by scribbling of pre-school children show not only form similarity to phosphenes of adults, but all of them can be classified in one of the 15 phosphene groups, [18] it seems reasonable to assume that the scribbling patterns of pre-school children are nothing more than perceived, self-stimulated phosphenes, used by children to train themselves in space analyzing. From these basic forms the child later composes more complicated line representations of his environment. It is interesting to note that at the end of this period of course of development the number of the basic scribbling patterns that the child further uses reduces to about the same number as many phosphene groups by an adult in general can be excited. Further,

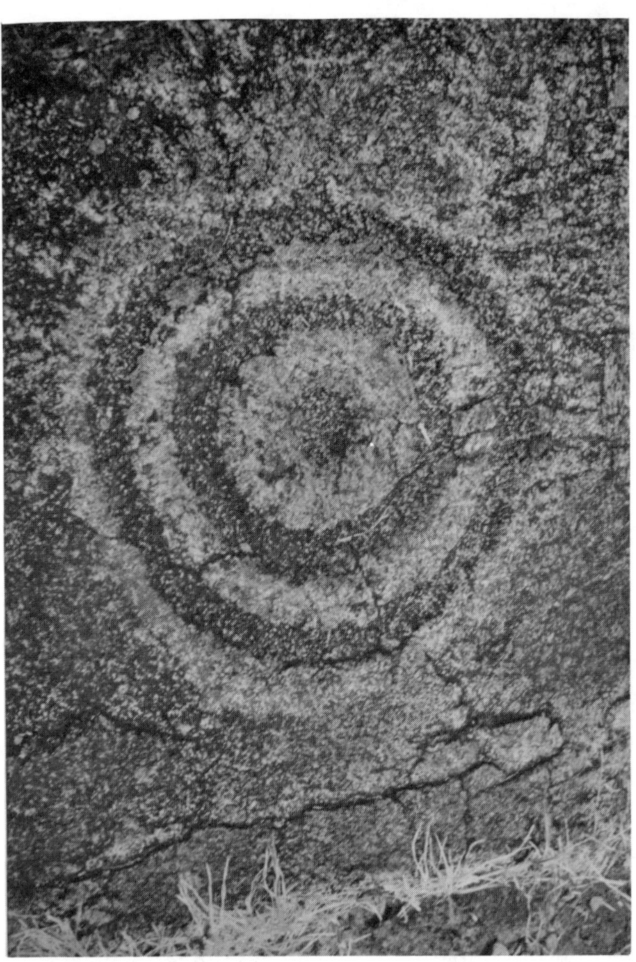

Fig.6 Typical pattern of Hawaiian petroglyphs similar to phosphenes of group No.6

it has been demonstrated [19] that similar relationship seems to exist between the form of phosphenes and forms of certain neolithic rock drawings. Figure 6 shows a typical pattern that can be seen among the petroglyphs of Hawaii. Its similarity to phosphenes of group No. 6 is without question.

From the mentioned examples one can assume that abstraction of contours of retinal images plays an important role in visual processing indeed, and they also indicate that this filter mechanism is involved in phase-bound information processing too. The relative position of the contours resulting from each retinal image contains namely phase information from the object space, without which the perception of a three-dimensional image space is inconceivable. The amount of phase information needed for survival, however, depends on the environment the species is living in, as we have previously demonstrated. [20]

Air space, for example, is an environment where more phase information is needed since the degree of freedom of movement in it is more than on the earth surface. Therefore, the observation of Knoll *et al.* [21] cannot be merely accidental: they have namely found that the number of electrically induced phosphene groups by airplane pilots and pilot candidates without flight experience is more than 100% higher than with population groups uninterested in flight. In terms of the holographic model presented, the frequent phosphene occurrence in pilots may be interpreted as a prevalence of a neural structure—the holographic filter—already present in pilot candidates, which then influences their choice of profession. Further, if we recall that the number of basic scribing patterns of preschool children reduces during the course of their development, and that the average excitable phosphene groups by pilots is always *lower* than by pilot candidates, the space analyzing role of phosphenes, i.e., what they represent—namely the filters—and the way to learn how to use them to perceive image space instead of image plane, is really backed up.

5. Modeling the model

It has already been mentioned that due to various eye movements the retinal image of the object space continuously changes. This can be interpreted as a spatial sampling by a time-compounded lens system, in contrast to the 'spatially-compounded' eye system of insects. Recently we have demonstrated that from a series of two-dimensional images obtained either by a spatially compounded or by a time-compounded lens system a single three-dimensional image space can be assembled when using a holographic multiplexing technique. [22] Several methods exist, but all are based on the same fact that in a single holographic medium many, in their space and/or time co-ordinates different wavefronts can be recorded and stored, and they can all be reconstructed simultaneously. Using this concept, the experimental modeling of the bioholographic model of visual perception seems to be feasible.

According to the holographic model, one of the first steps in processing visual perception is the abstraction of contours in each of the retinal images by direction-independent differentiation of the stimulus pattern resulting from the individual retinal images. In our experimental mode of visual perception, the individual pictures representing the retinal images are either the result of time-compounded spatial sampling by ultrasonic B scans, as shown in Fig. 7, or were taken by a spatially compounded lens system, a so-called fly-eye camera (Fig. 8).

For the abstraction of contours several types of holographic spatial filters can be considered. [23] Our choice was the differentiation filter developed by Wess *et al.* [24] The reason for this will be discussed later. Figure 9 shows the result of the direction independent differentiation of a 2-D picture.

Looking at these contours we cannot escape the feeling that caricatures of cartoonists as well as some pictures of modern artists are the expression of an intermediate product of their visual information process evoked to the consciousness of the artist and then interpreted by him. In general, we are accustomed to artworks where the artist records the average of these filter products, but it is an error to assume that he must be concerned to do so. In some cases, the artwork is a result of abstraction of contours of a *single* retinal image, but to express his feelings the artist may separately record the contour lines of two or more retinal images in the same work, as, for example, Picasso did in his well-known 'Portrait of a Woman', shown in Fig. 10.

If the intermediate product realized by the artist is a result of a filtering which produces broader contour lines, points may be reached at which the two sides of the broad line begin to play distinct roles in the perception of the spectator. The line becomes a pattern rather than a contour, the artwork is no more a drawing in its original sense, but a painting. Nevertheless, in this case too, the artist has the freedom to record several retinal images separately, as seen in Fig. 11, that shows a friend of Picasso painted by the artist himself. This picture does not look naturalistic at all; nevertheless, it has a *visual* resemblance to the person it represents, as demonstrated in Fig. 12, which is a more 'naturalistic' painting of Picasso's. The reason for the visual resemblance is the presence of similar relevant contours emphasized in both pictures.

If we now recall that phosphenes are defined as perceived patterns issuing from direction-independent differentiation of a stimulus pattern other than resulting from a light pattern, and that scribbling patterns may be perceived, self-stimulated phosphenes used in space analyzing, several modern drawings and paintings could be interpreted in terms of rather elementary intermediate patterns experienced by the artist creating these artworks. Further discussion of this question would lead too far outside of the scope of this presentation, and so I wish to make just a short remark. Modern drawings and paintings all involve the spectator in the perception of the artists'

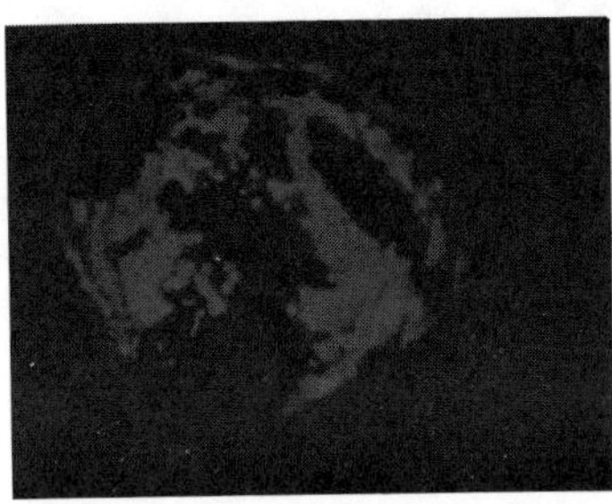
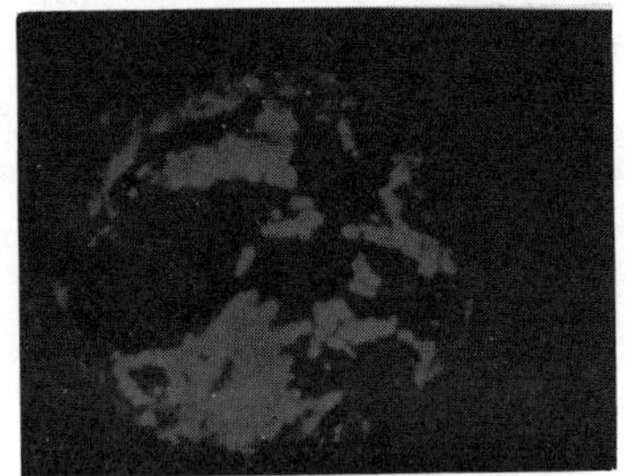
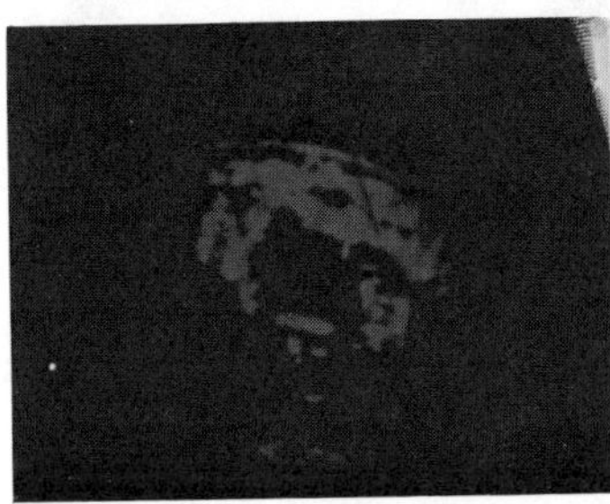

Photographs of reconstructed images obtained from holographically multiplexed B scans of bilateral renal cysts. The pictures differ from each other in a viewing angle of approximately 30°

Fig.7 Time-compounded spatial sampling by ultrasonic B scans

Fig.8 Pictures taken by a fly-eye camera

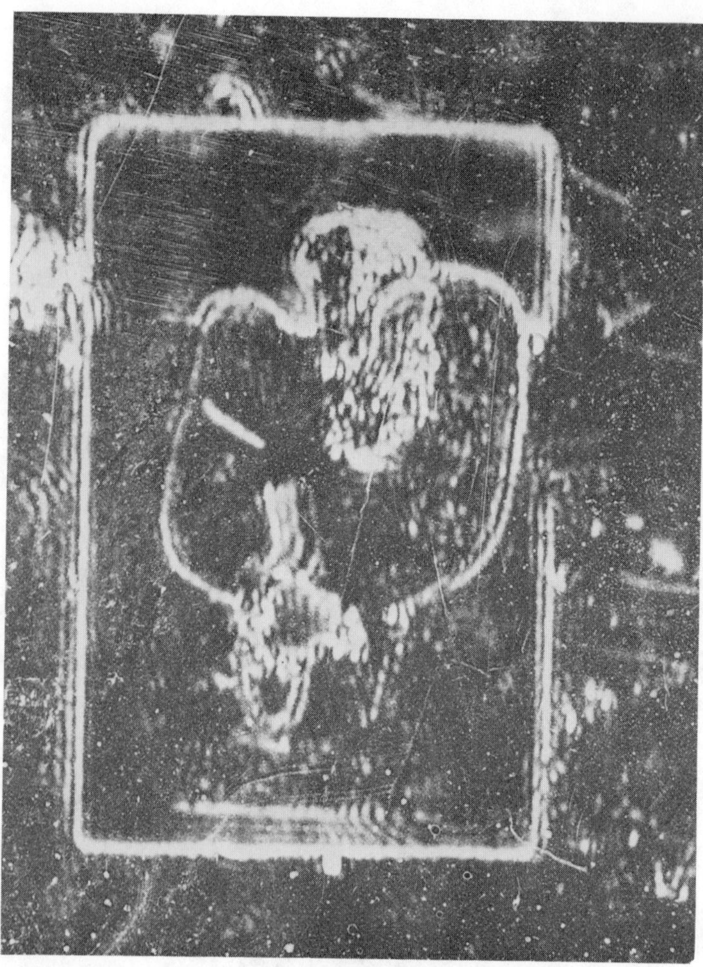

Fig.9 Direction-independent differentiation of a 2-D picture

Portrait of a Woman
1938

Fig.10 Picasso: Portrait of a Woman

Fig.11 Picasso: Portrait of Jaime Sabartès

Fig.12 Picasso: Portrait of Jaime Sabartès

picture planes, (retinal images) which are quite distinct from the artist's perception of the object space, i.e., his virtual image space. A certain minimum learning is therefore a necessary condition for the proper appreciation of these drawings and paintings. This remark, however, does not imply that everyone who looks at these works will be aware of their common qualities, only that everyone who properly understands these works will be aware of them.

The second important step in visual information processing is the creation of perceivable image space that mirrors the object space. Since the latter one is in general three-dimensional, the perceived image space is to be virtual. The simultaneous reconstruction of holographically multiplexed, filtered retinal images of various space and time co-ordinates by an internal stimulus pattern could yield such a virtual 3-D image. According to our model experiments [25] this assembled and reconstructed virtual image space, as in some cases, has the property of giving information not only from those planes–i.e. retinal images–which have been multiplexed, but from other, intermediate, planes too. This, however, is in good agreement with the physiological finding that although a significant part of the retina may be damaged, we still have 3-D perception.

Further, the holographic multiplexing method, [27], allows the reconstruction of one or more multiplexed planes in the virtual image space that mirror their position in the object space. If our visual information processing works in the spirit of the presented holographic model, the possibility of experiencing the reconstruction of single planes of the object space in the virual image space must be feasible. Looking at some modern paintings we believe to have indications that artists do have such experiences indeed. Figures 10 and 11 have demonstrated what we mean by this.

6. Conclusions

The scope of this presentation was to show that if the phosphenes can be interpreted as perceived reconstructions of the direction-independent differentiation of a stimulus pattern other than the adequate light pattern by filters stored in holographic form, they can be used to explain normal space perception as well as drawings and paintings known as 'modern'. The kind of filter we used in our experiments to model this bioholographic concept was a direction-independent differentiation filter of Wess *et al.*, which is produced by the superposition of two off-axis zone plates with slightly different focal lengths. The reason for choosing this type of filter was that, according to Eccles, Ito and Szentagothai, [28] patterns similar to that issuing from the superposition of two off-axis zone plates may be formed in the cerebellar cortex owing to its elaborated neural arrangement, while at most the excitatory synaptic inputs would provide subsidiary organized patterns with the subcortical nuclei. It seems as if the cerebellar afferent signals built up an excitatory background in the subcortical nuclei during the time that they are being integrated in the cerebellar cortex to form a spatio-temporal pattern of Purkinje-cell excitation. By Purkinje-cell discharge this pattern in turn is impressed as an inhibitory pattern upon subcortical nuclei. This idea is diagrammatically shown after Eccles, Ito and Szentagothai in Fig. 13. This means that the cerebellar cortex utilizes the complex inflow of information for the pupose of producing complex interacting patterns of Purkinje-cell excitation and inhibition becasue the output of the cerebellar cortex is mediated solely by the discharges

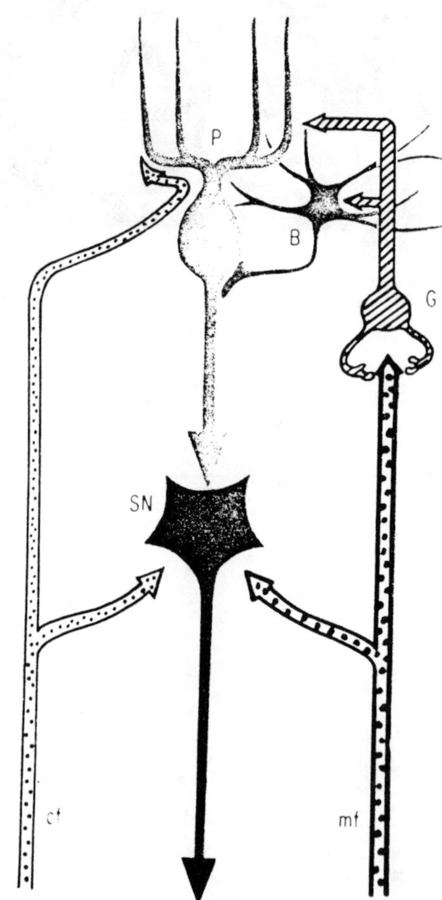

Fig.13 Neural connections between the cortex and subcortical neurons in the cerebellum after Eccles, Ito & Szentagothai

of Purkinje cells. As a consequence, the cerebellar output represents a kind of negative image of the excitatory and inhibitory patterns resulting at any instant from the processing of information in the cerebellar cortex.

By showing Fig. 14 we wish to demonstrate what sort of geometric pattern may result from impulses in a small bundle of mossy fibers (MF) innervating the focus of granule cells (GrC), as shown after Eccles in the transverse section beside the larger diagram where the folium is seen from above. GoC is the Golgi cell distributed to that focus. This focus is shown as a light grey disc that gradually shades off into the dark surround, which indicates zero activity, while white indicates the maximum activity of the granule cells. Because this granule cell activity gives rise to an excited beam of parallel fibers PF, Golgi cells in this beam will be excited and processed to feed back inhibition to the granule cells along

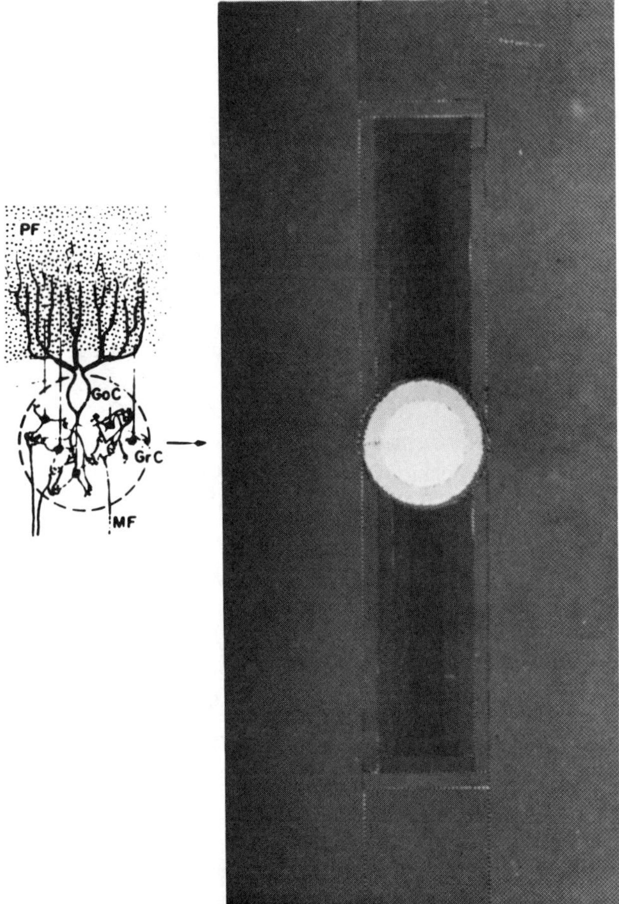

Fig.14 Diagram showing the postulated action of impulses in a small bundle of mossy fibers innervating a focus of granule cells as shown in the transverse section beside the larger diagram where the folium is seen from above, after Eccles, Ito & Szentagothai

the whole length of the beam. There are experimental evidences that excited parallel fibers generate a band of inhibited Purkinje cells directly excited by the parallel fiber volley, so making the pattern of excited regions and inhibited surrounds that is superimposed on the low level background activity of the Purkinje cells. In several places there may be overlaps with the consequence that the excited areas are depressed. This pattern can be regarded as encoding the information fed into this area of cerebellar cortex during a few preceding milliseconds by all the converging mossy fiber inputs.

At this time we are not going into detailed discussion on how to find on this basis a possible morphological equivalent of the presented bioholographic model—this will be done elsewhere—we wish only to point out that attempts have already been made to interpret the phosphenes as the result of the excitation of an ensemble of parallel neural chains [29] and by lateral inhibition [30]. These models, however, are not able to explain the dependency of the perception of the phosphenes on the stimulation frequency or their movement in the visual field. The bioholographic model presented does not lack of this ability.

Acknowledgements

I am grateful to Dr. U. Röder and Mr. O. Wess for their experimental assistance in producing differentiated pictures and for the invaluable discussions we conducted in developing the presented model, further to Mr. B. Gerganow for the fly-eye camera pictures.

References

1. SZENT-GYÖRGYI, A., 'Cell division and cancer', Gordon H. Scott lecture, Detroit, Michigan (8 May 1972).
2. GABOR, D., 'A new microscope principle', *Nature* 161, 777 (1948).
3. GREGUSS, P., 'Bioholography—a new model of information processing', *Nature* 219, 482 (1968).
4. GREGUSS, P., *Bioholography. Developments in Holography,* SPIE Seminar Proceedings, pp.55-83, Boston (April 1971).
5. BARRETT, T.W., 'The cerebral cortex as a diffracting medium', *Math. Biosciences* 4, 311 (1969).
6. GREGUSS, P. (Ed.), *Holography in Medicine,* IPC Science and Technology Press Ltd., Guildford, Surrey (1975).
7. SOKOLOV, E.N., 'Neural models and the orienting reflex', in *The Central Nervous System and*

Behavior (M.A.B. Brazier, ed.), Josiah Macy, Jr. Found., New York (1960).
8. BISHOP, G.H., 'The interpretation of cortical potentials', *Quant. Biol.* 4, 305 (1936).
9. GABOR, D., 'Microscopy by reconstructed wavefronts', *Proc. Roy. Soc.* A197, 454 (1949).
10. LADIK, J. and GREGUSS, P., 'Possible molecular mechanisms of information storage in the long-term memory', *Biology of Memory, Symposia Biologica Hungarica,* Ed. G. Adam Akademiai Kiado, Vol.10, 343-355 (1971).
11. KNOLL, M. and KUGLER, J., 'Subjective light pattern spectroscopy in the electroencephalographic frequency range', *Nature* 184, 1823 (1959).
12. BEER, S., 'Über das Auftreten einer subjektiven Lichtempfindung in magnetischem Feld', *Klin. Wochensch.* 15, 108 (1902).
13. BAUMGARTNER, G., 'Der Informationswert der On-Zentrum und Off-Zentrum Neuronen des visuellen Systems beim Hell-Dunkel-Sehen und die informative Bedeutung von Aktivierung und Hemmung', *Symposium Freiburg 1960,* Springer, Berlin (1961).
14. HUBEL, D.H., 'Single unit activity on lateral geniculate body and optic tract of unrestrained cats', *J. Physiol. Lond.* 150, 91 (1960).
15. SEIDEL, D., 'Der Existenzbereich elektrisch und magnetisch-induktiv angeregter subjektiver Lichterscheinungen (Phosphene) in Abhängigkeit von äußeren Reizparametersn', *Elektromedizin* 13, 194 (1968).
16. WELPE, E., 'Anregung des Ellipsenphänomens durch sinusförmige Reizströme von 162-208 Hz und 287-324 Hz', *Pflügers Arch.* 306, 304 (1969).
17. GREGUSS, P., 'Long wavelength holography or 3-D visualization of information acquired by long wavelength carriers', Proc. Symp. on Engineering Applications of Holography, Los Angeles, Calif., p.271 (February 1972).
18. KELLOG, R., KNOLL, M., and KUGLER, J., 'Form similarity between phosphenes of adults and preschool children's scribblings', *Nature* 208, 1129 (1965).
19. KELLOG, R., 'What children scribble and why?', San Francisco (1955).
20. GREGUSS, P., 'Bioholography—a new model of information processing', *Progress of Cybernetics,* Ed. J. Rose, p.433, Gordon & Breach Science Publishers, London (1970).
21. KNOLL, M., HÖFER, O., and KUGLER, J., 'Fliegerpersönlichkeit und Phosphenhäufigkeit, Szondiana VI, Beiheft zur Schweiz', *Z.f. Psychologie und ihre Anwendungen,* Nr. 50, p.253 (1966).
22. GREGUSS, P., 'State of art of holographic multiplexing', *Technika Kino i Televideniya (Moscow),* No.11, 69 (in Russian), (1975).
23. GÖRLITZ, D. and LANZL, F., 'Ein holographisches Filter zur richtungsunabhängigen Differentiation, *76. Tagung der Deutschen Gesellschaft für Angewandte Optik,* Bad Ischl (1975).
24. WEß, O., MAGER, H.J., and WAIDELICH, W., 'Ein richtungsunabhängiges Differentiationsfilter', *76. Tagung der Deutschen Gesellschaft für Angewandte Optik,* Bad Ischl (1975).
25. GREGUSS, P. and CAULFIELD, H.J., 'Multiplexing ultrasonic wavefronts by holography', *Science* 177, 422 (1972).
26. FALUS, M., CAULFIELD, H.J., and GREGUSS, P., 'Holographic multiplexing of kidney B scans', *Laser and Unconventional Optics Journal* No.51, 1 (1974).
27. GREGUSS, P. and GALIN, M.A., 'The holographic concept in ophthalmology', *Proc. First European Biophysics Congress,* Ed. E. Broda *et al.*, Verlag der Wiener Medizinischer Akademie, Wien, p.97 (1971).
28. ECCLES, J.C., ITO, N., and SZENTAGOTHAI, J., *The Cerebellum as a Neural Neuronal Machine,* Springer Verlag, New York (1967).
29. MEIER, A., 'Die laterale Inhibition retinaler Zapfen als Anregungsmechanismus für subjektive Lichtmuster (Phosphene)', *Z. f. angew. Phys.* 24, 2 (1968).
30. KNOLL, M., KUGLER, J., HÖFER, O., and LAWDER, S.D., 'Effects of chemical stimulation of electrically-induced phosphenes on their bandwidth, shape, number and intensity', *Confin. neurol.* 23, 201 (1963).

A model for the stimulus processing in mechano-receptors (Pacinian corpuscles)

F. GRANDORI and A. PEDOTTI
Center of Systems Theory–CNR,
Institute of Electronics, Polytechnic of Milan, Italy

The Pacinian corpuscle is a mechano-electric transducer with a very high occurrence in many tissues of living organisms. Although there has so far been little information concerning their functional significance in the body, the large diffusion of these receptors suggests they play a relevant role in motor control in general and in the proprioceptive perception. In particular, because of their rapid adaptation, they are sensitive to the quick movements of externally applied rapid pressure and to the dynamics of body in general. Numerous Pacinian corpuscles may be found, for instance, in the mesentheric tissue, in the epithelium tissue and in several articular joints. From many years, due to its easy accessibility and its rather large dimensions, the Pacinian corpuscle has been thoroughly investigated by several authors and many salient features of its behavior have been evidenced [1-4,5-7]. Nevertheless, at present, its functional role has not yet been completely understood, probably because some features of the experimental results about the formation of the receptor potential and the spike generation require further investigations.

An approach based on a detailed and quantified description of the relationships characterizing Pacinian corpuscle behavior is believed to be a constructive step toward a better understanding of the matter and, at the same time, it can give the opportunity of clarifying the role played by these receptors.

On this point mathematical modeling techniques can provide a significant answer. In fact the main variables defining the process can be described by an analytical form and the chemophysical interdependences by suitable mathematical relationships. In this chapter the results obtained by using a mathematical model of the Pacinian corpuscle will be described.

In setting up the model, the classical identification methods, based on the knowledge of the main phenomena involved in the mechano-to-electric transduction and on the analysis of the input-output relationships, have been followed.

In particular the mechanical transformation, related to the lamellated structure, the proper mechano-to-neural transduction producing the receptor potential and the neural processes involved in the spike generation have been recognized as the main steps of the sequence of events causing the transformation of signals. The whole model has been simulated on a digital computer and it has been used to analyze the classical responses of the corpuscle. In the present chapter particular attention has been devoted to the frequency response and the directional sensitivity.

Mathematical model

In order to determine a suitable mathematical model, the main structural features of the Pacinian corpuscle must be taken into account. The nervous part is constituted by a sensitive ending surrounded by an approximately cylindrical core structure consisting in closely packed membranes called lamellae. In turn the core is surrounded by a second set of lamellae; the distances between each other increase from the innermost lamella toward the periphery of the corpuscle. The spaces among all the lamellae are filled with a liquid whose mechanical properties can be properly considered similar to those of water.

On the basis of these general characteristics the simple scheme illustrated in Fig. 1 points out the main processes involved in the mechano-to-electrical transduction in the Pacinian corpuscle and it constitutes a significant block-diagram of the considered mathematical model.

Although this model has been extensively described in previous works [8,9], a brief description of it is reported here for a better understanding of the results obtained.

Referring to Fig. 1, the first block on the left describes the mechanical characteristics of the corpuscle. The displacement of the outermost lamella and the pressure on the core have been taken as input and output of this block, respectively. This part of the system has been described as a linear system [6]. As to the second block, by analyzing the outputs of the intact corpuscle and of the decapsulated nerve terminal, in response to various kinds of mechanical stimuli, a typical non-linear, directional-dependent phenomenon is found [7]. Therefore in the model of the sensory ending a non-linear transfer characteristic has been introduced.

The threshold-like phenomena and the stochastic properties concerned in the generation of the on-off responses were accounted for by means of a time-varying threshold block with a zero mean valued gaussian noise superimposed.

The model has been implemented on a digital computer in FORTRAN language and the usual experimental tests were simulated.

In the Appendix the schematic representation of the whole model and the mathematical relationships are reported.

Results

By using the mathematical model previously described, the most common experiments performed on the Pacinian corpuscle have been simulated. They concern the analysis of the responses to different kinds of stimuli.

The stimuli generally used are compressions of varying rate of increase. They have been applied both to intact corpuscles and to the decapsulated terminal. This condition was easily simulated by driving directly the block of the model (see Appendix) representing the core. The quantities experimentally recorded (which on the other hand constitute the outputs of the model) are the spikes generated at the axon or the receptor potential of the terminal if the spike generation is blocked by tetrodotoxin or procaine.

The results of these simulations, reported in previous works [8,9], permitted us to clarify some aspects of the convection of the mechanical forces and of the transducing properties of the sensory terminal. In particular it clearly appears that the mechanisms of generation of the so called 'off-responses' are the same as those of 'on-responses', and, apart from the influence of an inherent biological noise, it can be stated that the appearance of an impulse is determined by the depolarization of the nervous ending.

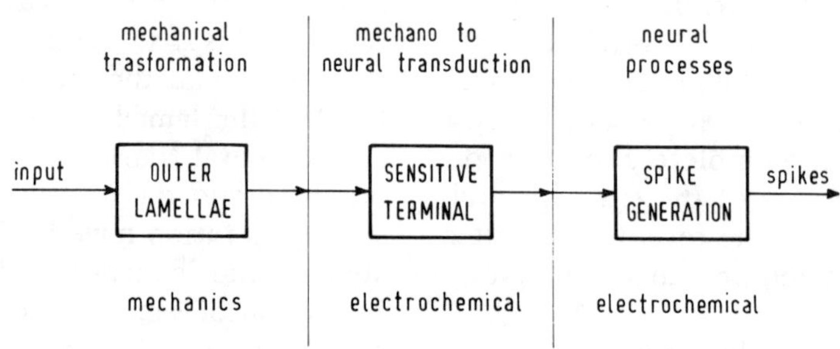

Fig.1 Functional block diagram of a Pacinian Corpuscle

Moreover the remarkable agreement between the experimental findings and the results of the simulation clearly shows the validity of the model, as to the mathematical structure and the parameter values.

In the present work the problems concerning the frequency response and the directional sensitivity are faced, and the results of the model are considered in conjunction with the experimental ones.

Frequency response

Although in the past many experiments dealing with the stimulation of corpuscles by sinusoidal mechanical stimuli have been carried out [10,11], a deeper analysis of this subject seems to be of great interest. First of all it must be noted that the Pacinian corpuscle as a whole is a non-linear system. Therefore, from a rigorous point of view, it would be impossible to define a frequency response curve. On the other hand the knowledge of the outputs (receptor potential) when sinusoidal or periodic stimuli at various frequencies are applied presents a fundamental importance for a better understanding of their functional role. A detailed discussion on this point has to take into account the following considerations.

Referring to the model as to the frequency response, the Pacinian corpuscle can be divided into two parts: the outer lamellae and the sensory ending.

On the basis of an analysis of their mechanical structures [2,6], the mechanical part is described in a linear way and modeled by the present authors, by means of a suitable transfer function. The dynamic behavior of this part is characterized by pass-band filtering properties. In Fig. 2 the diagram of the frequency response is reported in semi-logarithmic scale. As one can see, the maximal gain concerns the range from 600 to 6000 rad/sec. corresponding to 100, 1000 cycles/sec., approximately.

The sensory ending presents a non linear characteristic, $y = f(x)$, given in the Appendix.

Therefore a frequency response curve can be defined only if the sinusoidal stimuli applied have a small amplitude. Then the non linearity can be approximated by a linear non dynamic block with a single gain μ_{nl} given by:

$$\mu_{nl} = \left.\frac{df}{dx}\right|_{x=0}$$

In Fig. 3 the frequency-response curve under these conditions, is illustrated. It presents a maximal gain corresponding to the 250-900 Hz band.

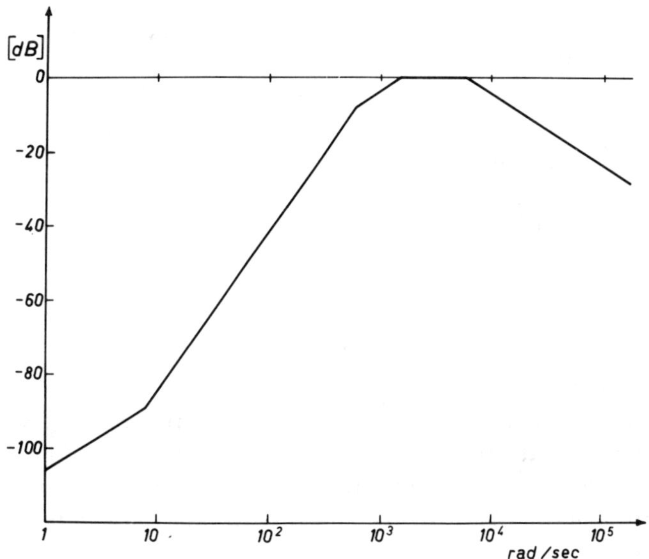

Fig.3 Frequency-response diagram of the whole Pacinian corpuscle for low amplitudes

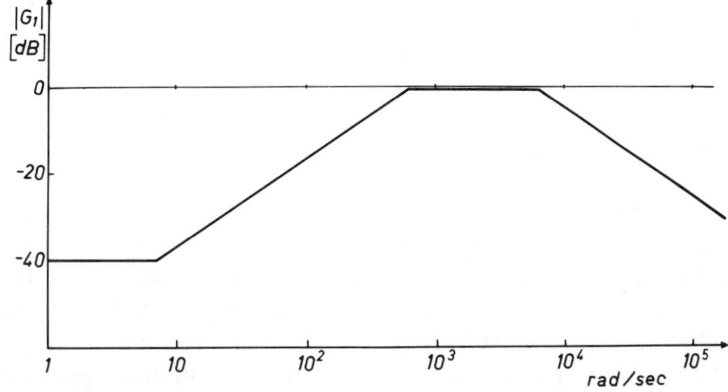

Fig.2 Frequency-response curve of the outer lamellae

This curve can be considered with good approximation the real frequency response of the whole Pacinian corpuscle.

It must be noted that a similar curve was experimentally determined by Sato [11]. In these experiments the recording system contained a capacity-coupled amplifier and a time constant of 1 sec. This fact changes the response obtained. For this reason a significantly appreciable disagreement can be found between those data and our findings. In order to compare significantly the curves, another block simulating the recording system used was introduced in our model.

So the frequency response curve illustrated in Fig. 4 was found and successfully compared with the data of Sato.

As previously noted a frequency response cannot be defined for sinusoidal stimuli with larger amplitude. In this case it is only possible to observe the receptor potential time course in response to the stimuli.

In Fig. 5 the responses to an input at the same frequency (500 Hz) at four different amplitudes are presented. Well noticeable is an increase of the mean value of the response as the amplitude increases.

In Fig. 6 the amplitude of the inputs is the same, while the frequency changes.

As one can see, all the responses present a periodic time course with a mean value which becomes larger with increasing amplitude and frequency.

This *dc*-component of the receptor potential can be considered mainly responsible for the spike generation. It must be noted that the presence of the capacity-coupled amplifier in the recording line did not permit Sato to recognize this important determinant.

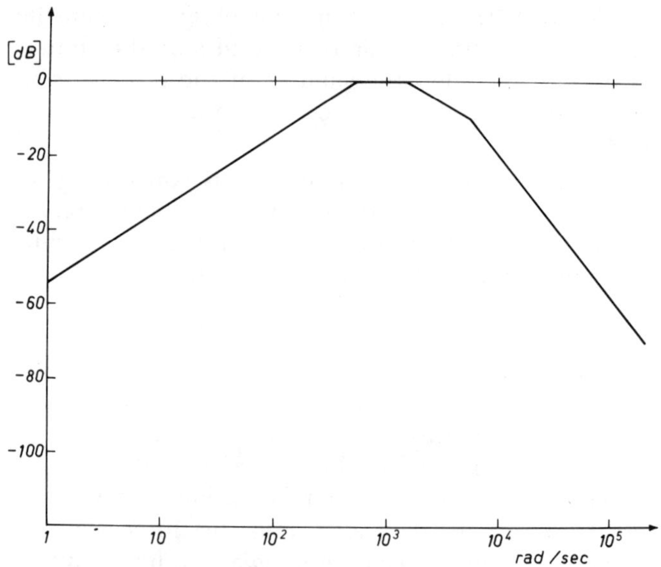

Fig.4 Frequency-response curve including the filtering properties of the recording line as used by Sato

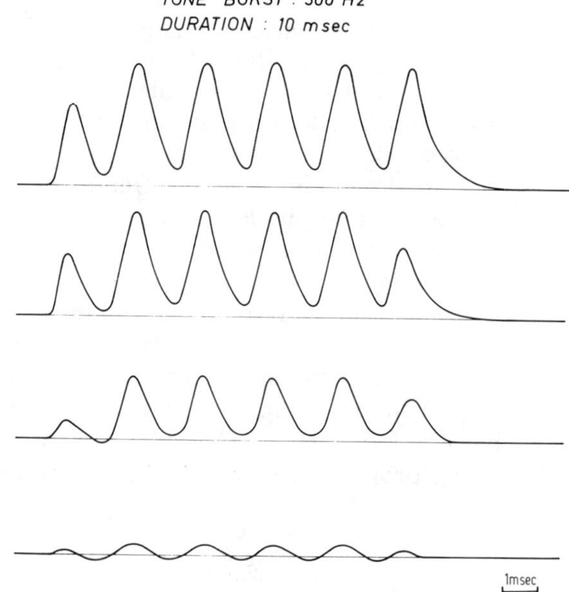

Fig.5 Responses to sinusoidal stimuli at different amplitudes

Directional sensitivity

Ilyinsky [4] showed that Pacinian corpuscles, which had shown the receptor potential (depolarizing response) on compression, generated hyperpolarization in response to a gradually increasing compression after rotating them through 90° along the long axis and vice versa.

These observations were confirmed by other authors [12,13,6]. This phenomenon is probably related to the shape of the non-myelinated terminal which is an elliptic cylinder. Therefore stimuli directed along a different axis produce a different deformation of the surface area of the terminal and then a different permeability of the terminal membrane to sodium ions. In particular stimuli applied along the short axis of the cylinder would produce depolarization, while those falling in the direction of the long axis would evoke hyperpolarization.

In the mathematical model here presented the phenomenon concerning the properties of the membrane are mainly described by the non-linear block. In particular it points out the ultrastructural characteristics of the corpuscle. Therefore a variation of

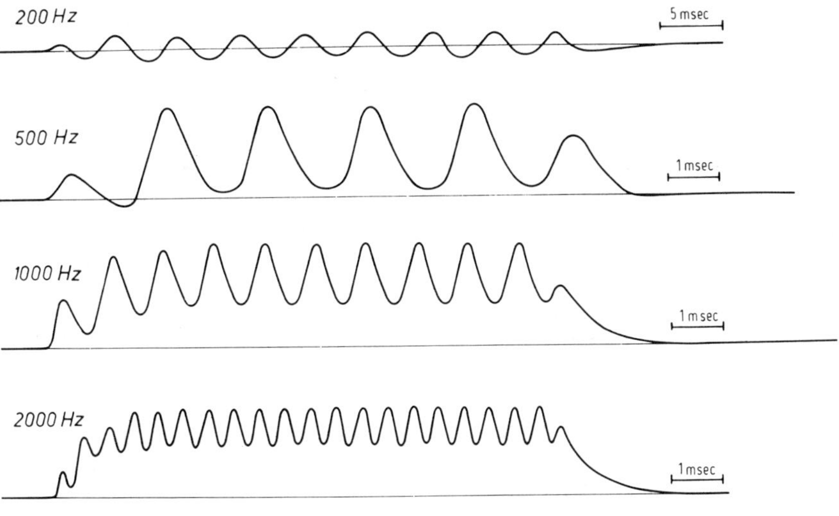

Fig.6 Responses to sinusoidal stimuli at different frequencies

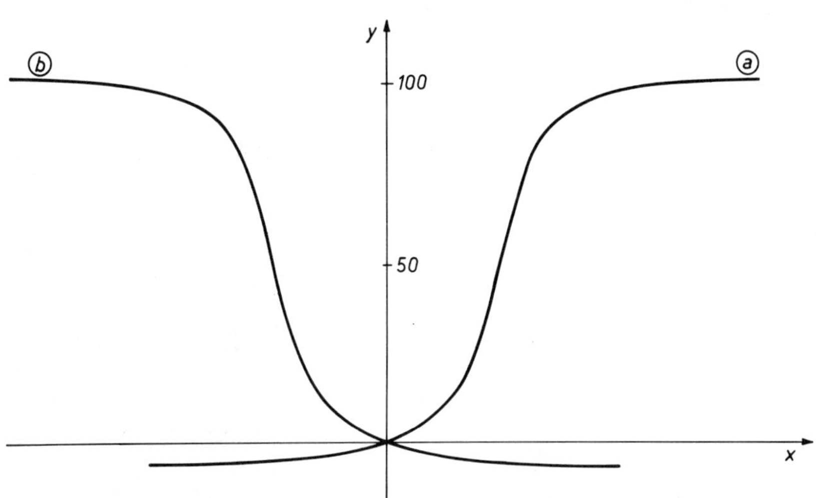

Fig.7 The non-linear characteristic before (a) and after rotation (b)

the stimulus direction is taken into account by a suitable modification of the non-linear curve.

As the only experimental results available on this point concern a 90 degrees rotation of the stimulus direction, this condition has been reproduced by properly changing the non-linear function. In Fig. 7 the two shapes assumed, are illustrated. Curve *a* is valid when the stimulus is applied along the short axis, *b* when it rotated 90 degrees.

The results obtained from the simulation are reported in Fig. 8. They concern three stimuli of different amplitude. On the left, the responses of the corpuscle before rotation are shown. On the right the responses to the same stimuli after rotation are reported.

The good agreement existing between these findings and those reported by Nishi and Sato [6] must be emphasized.

Summary

On- and off-responses of Pacinian corpuscles have been interpreted by a mathematical model. This approach appears to be of primary importance determining the real responses of these mechanoreceptors and therefore in interpreting in a correct way the experimental data, especially those concerning the directional-dependent phenomena and the determinants of the responses to sinusoidal stimuli.

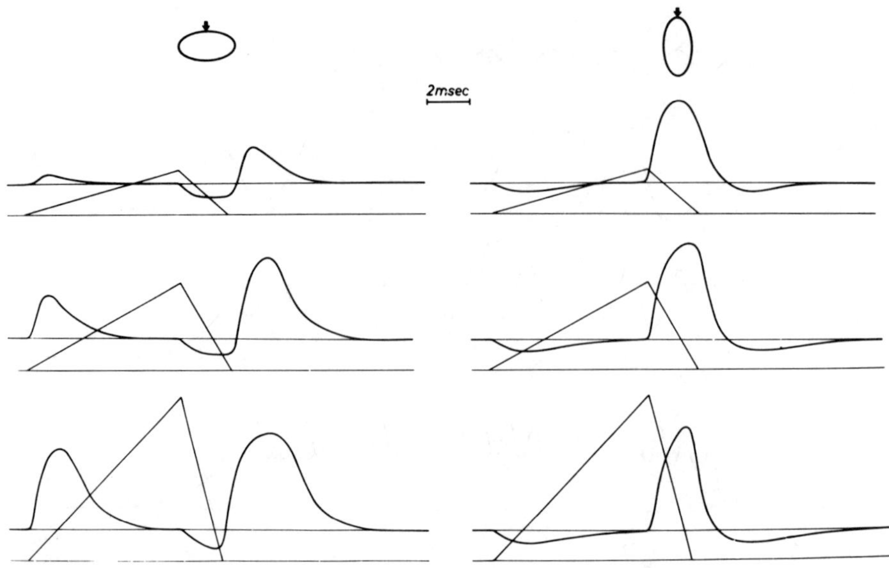

Fig.8 Receptor potential time course before (on the left) and after (on the right) rotation for three different stimuli

Appendix

The complete block diagram of the model is given in Fig. A1, where

$$G_1(s) = \frac{\mu_1(1+sT_1)}{(1+sT_2)(1+sT_3)}$$

and

$$G_2(s) = \frac{\mu_2}{1+sT_c}$$

The analytical expression of the non-linear characteristic is

$$y = (a/\pi)[\tan^{-1}(bx-c)+k\pi]$$

The numerical values of the model parameters are given in [8].

References

1. GRAY, J.A.B. and SATO, M., 'Properties of the receptor potential in Pacinian corpuscles', *J. of Physiol.* 122, p.610 (1953).
2. HUBBARD, S.J., 'A study of rapid mechanical events in a mechanoreceptor', *J. Physiol.* 141, pp.198-218 (1958).
3. HUNT, C.C. and TAKEUCHI, A., 'Responses of the nerve terminal of the Pacinian corpuscle', *J. Physiol.* 160, pp.1-21 (1962).
4. ILYINSKY, O.B., 'Processes of excitation and inhibition in single mechanoreceptors (Pacinian corpuscles)', *Nature* 208, pp.351-353 (1965).
5. LOEWENSTEIN, W.R., ALTAMIRANO Y ORREGO, R., 'The refractory state of the generator and propagated potentials in a Pacinian corpuscle', *J. of Gen. Physiol.* 41, pp.805-824 (1957).
6. LOEWENSTEIN, W.R. and SKALAK, R., 'Mechanical transmission in a Pacinian corpuscle. An

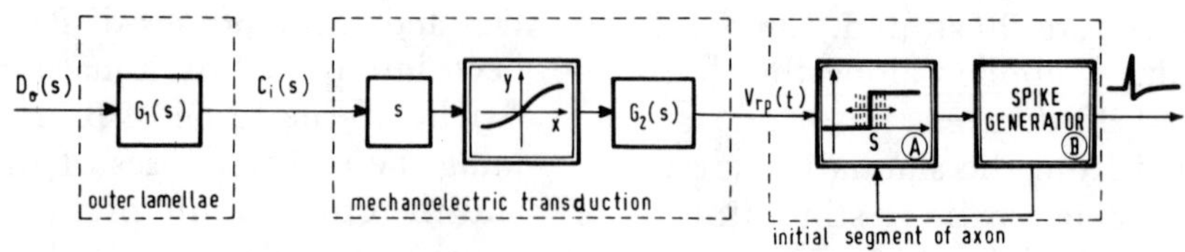

Fig.A1 Schematic representation of the model

analysis and a theory', *J. Physiol.* 182, pp.316-378 (1966).

7. NISHI, K. and SATO, M., 'Depolarizing and hyperpolarizing receptor potential in the non-myelinated nerve terminal in Pacinian corpuscles', *J. Physiol.* 199, pp.383-396 (1968).
8. GRANDORI, F. and PEDOTTI, A., 'A mathematical model of mechano-to-electric transduction in Pacinian corpuscles', Int. Congr. on Cybernetics and Systems, Bucharest (1975).
9. GRANDORI, F. and PEDOTTI, A., 'Mathematical model of neural activity of the Pacinian corpuscle', *Tech. Rept. IEE*–LCA 75-2 (February 1975).
10. LOEWENSTEIN, W.R., 'Generator processes for repetitive activity in a Pacinian corpuscle', *J. of Gen. Physiol.* 41, pp.825-845 (1958).
11. SATO, M., 'Response of Pacinian corpuscles to sinusoidal vibration', *J. Physiol.* 159, pp.391-409 (1961).
12. ILYINSKY, O.B., KRASNIKOVA, T.L., AKOEV, G.N., and ELMAN, S.I., 'Functional organization of mechanoreceptors', in *Somatosensory and Visceral Receptors Mechanisms, Progress in Brain Res.* (Edit. by A. Iggo and O.B. Ilyinsky), Vol.43, pp.195-203, Elsevier, Amsterdam (1976).
13. ILYINSKY, O.B., VOLKOVA, N.K., CHEREPNOV, V.L., and KRYLOV, B.V., 'Morpho-functional properties of Pacinian corpuscles', in *Somatosensory and Visceral Receptors Mechanisms, Progress in Brain Res.* (Edit. by A. Iggo and O.B. Ilyinsky), Vol.43, pp.195-203, Elsevier, Amsterdam (1976).
14. AKOEV, G.N., CHELYSHEV, Yu.A., and ELMAN, S.I., 'Effect of acetylcholine and catecholamines on excitability of Pacinian corpuscle', in *Somatosensory and Visceral Receptors Mechanisms, Progress in Brain Res.* (Edit. by A. Iggo and O.B. Ilyinsky), Vol.43, pp.187-193, Elsevier, Amsterdam (1976).

Reverberations and synthesis of a neural network

MASAKO SATO
University of Osaka Prefecture, Osaka, Japan

I. Introduction

Since Caianiello [1] proposed neuronic equations represented by threshold functions as a model which describes activities of neuron networks, various properties of the equations have been revealed. Of main importance were those of the periodic solutions concerned with thought processes in the brain. A cycle which the periodic solutions make is sometimes called a reverberation using a term of neurophysiology. And the period of the periodic solution, the stability with respect to a change of coefficients appearing in the equations were investigated so far [2]-[6]. It seems, however, that little is known about the relation between parameters of the equations (i.e., coupling coefficients and threshold) and reverberations.

The author has shown in [7] a necessary and sufficient condition that a neuronic equation which describes activity of a single neuron with refractory yields only periodic solutions irrespective of its initial state, and has given also the number of reverberations and the periods of them under the condition.

In the present chapter, we will discuss the conditions on parameters under which every state of the network described by the equations belongs to any one of the reverberations. Such a network is called a reverberating network (abbreviated by an RN simply): First of all, we will show that a coefficient matrix A defined from the equations should have a full rank ($|A| \neq 0$) in order that the network is an RN and that if a network is an RN, then every network represented by the equations whose coefficient matrix is constructed by changing the signs of rows of the original matrix or by exchanging rows is also an RN. Secondly, when a coefficient matrix which gives an RN has a row of a special form, we will discuss a property that the remaining rows have. Finally, we will give methods for synthesis of an RN.

II. Preliminaries and some properties of RN's

Let us consider a neuron network consisting of n elements described as follows:

$$Y = \mathbb{1}[AX - \Theta] \tag{1}$$

where A is an $n \times n$ matrix and is called a coefficient matrix which represents couplings between elements of the network, and Θ, X and Y are n-dimensional column vectors, i.e.

$$A = (a_{ij})_{i,j=1}^{n} = (\mathbf{a}_1', \mathbf{a}_2', \ldots, \mathbf{a}_n')', \quad \mathbf{a}_i = (a_{i1}, a_{i2}, \ldots, a_{in}),$$

$$\Theta = (\theta_1, \theta_2, \ldots, \theta_n)',$$

$$X = (x_1, x_2, \ldots, x_n)', \quad Y = (y_1, y_2, \ldots, y_n)'$$

The $\mathbb{1}[U]$ is called a vector threshold function and is defined as follows:

$$\mathbb{1}[U] = (1[u_1], 1[u_2], \ldots, 1[u_n])'$$

for $U = (u_1, u_2, \ldots, u_n)'$

where $1[u] = 1$ if $u > 0$, and $= 0$ if $u \leqslant 0$. The θ_i, x_i and y_i of Θ, X and Y denote the magnitude of the threshold, the present state and the next state of the ith element of the network, respectively. The X, Y, etc., are called the states of the network.

Let us denote by $\mathbf{X}_n$ a set of all states of a network of n elements. Equation (1) gives a mapping from $\mathbf{X}_n$ into $\mathbf{X}_n$ and we denote it by T, i.e.,

$$T(X) = \mathbb{1}[AX - \Theta]$$

We represent T simply by (A, Θ), i.e., $T = (A, \Theta)$. Thus a neuron network is characterized by a mapping $T = (A, \Theta)$, since it determines dynamical behaviors of the network.

For any neuron network, we can choose (A, Θ) such that every component of $AX - \Theta$ is not zero, since if some of the components, say ith, is zero, we can assign a larger value to θ_i without changing dynamical properties of the network.

Define for any positive integer S,

$$T^S(X) = T^{S-1}(T(X)), \quad T^1(X) = T(X)$$

A state X is said to reverberate if there is an integer S such that $T^S(X) = X$. Let us consider a network every state of which reverberates. Such a network is called a reverberating network (abbreviated by an RN as mentioned in I) if it exists. In the previous paper [8], we showed the existence of such a network and gave a condition for it: If a network is an RN, it is self-dual†, namely, for the mapping T defined by the network, the following relation holds:

† It is reported that the condition is already obtained by Crocchiolo and Drago in 1965, see [2].

$$T(X) = \overline{T(\overline{X})} \quad \text{for any } X \in \mathbf{X}_n, \ \overline{X} = 1 - X.$$

Clearly any self-dual network (A, Θ) can be represented by $(A, \frac{1}{2}A1)$. Thus a self-dual network can be characterized by A only. We mean by a network with A a self-dual network $(A, \frac{1}{2}A1)$ in what follows.

Letting $Z_x = 2X - 1$, we know that $(AX - \frac{1}{2}A1)_i \gtrless 0$ is equivalent to $(AZ_x)_i \gtrless 0$, where $(W)_i$ means ith component of a vector W.

Let us define set $S^{(1)}(\mathbf{a}_i)$ and $S^{(-1)}(\mathbf{a}_i)$ for any row vector $\mathbf{a}_i$ of a matrix A:

$$S^{(1)}(\mathbf{a}_i) = \{X \mid \mathbf{a}_i X - \tfrac{1}{2}\mathbf{a}_i 1 > 0\},$$

$$S^{(-1)}(\mathbf{a}_i) = \{X \mid \mathbf{a}_i X - \tfrac{1}{2}\mathbf{a}_i 1 < 0\}$$

We write $S^{(j)}(\mathbf{a}_i)$, $j = \pm 1$ simply by $S_i^{(j)}$ unless confusion occurs. We write also

$$S_{i_1}^{(j_1)} \cap S_{i_2}^{(j_2)} \cap \ldots \cap S_{i_k}^{(j_k)}, \quad j_s = \pm 1, \ s = 1, \ldots, k$$

simply by

$$S_{i_1}^{(j_1)} S_{i_2}^{(j_2)} \ldots S_{i_k}^{(j_k)}.$$

Let $\#N$ denote the cardinality of a set N.

Lemma 1
A network with A is an RN iff $\# S_1^{(j_1)} \ldots S_n^{(j_n)} = 1$, $j_s = \pm 1$. The proof is trivial.

Corollary 1
If a network with A is an RN,

$$\# S_{i_1}^{(j_1)} S_{i_2}^{(j_2)} \ldots S_{i_k}^{(j_k)} = 2^{n-k}, \tag{2}$$

where $j_s = 1$ or -1 and $i_1, i_2, \ldots, i_k$ $(i_s \neq i_{s'}) \in \{1, 2, \ldots, n\}$ for $s = 1, 2, \ldots, n$ and $k = 1, 2, \ldots, n$.

Proof: Since $S^{(j_1)} S^{(j_2)} \ldots S_{i_{l-1}}^{(j_{l-1})} S_{i_l}^{(j_l)} \cup S_{i_1}^{(j_1)} S_{i_2}^{(j_2)} \ldots$

$$\ldots S_{i_{l-1}}^{(j_{l-1})} S_{i_l}^{(-j_l)} = S_{i_1}^{(j_1)} \ldots S_{i_{l-1}}^{(j_{l-1})},$$

$$S_{i_1}^{(j_1)} \ldots S_{i_k}^{(j_k)} = \bigcup_{\substack{j_s = \pm 1 \\ s = k+1, \ldots, n}} S_{i_1}^{(j_1)} \ldots S_{i_k}^{(j_k)} S_{i_{k+1}}^{(j_{k+1})} \ldots S_{i_n}^{(j_n)}$$

On the other hand

$$S_{i_1}^{(j_1)} \dots S_{i_n}^{(j_n)} \cap S_{i_1}^{(j'_1)} \dots S_{i_n}^{(j'_n)} = \phi$$

for $(j_1, \dots, j_n) \neq (j'_1, \dots, j'_n)$ and $\{i_1, \dots, i_n\} = \{1,2,\dots,n\}$.

Thus

$$\#S_{i_1}^{(j_1)} \dots S_{i_k}^{(j_n)} = \sum_{\substack{j_s=\pm 1 \\ s=k+1,\dots,s}} \#S_{i_1}^{(j_1)} \dots S_{i_k}^{(j_k)} S_{i_{k+1}}^{(j_{k+1})} \dots \dots S_{i_n}^{(j_n)} = 2^{n-k} .$$

Theorem 1
If a network with $A = (\alpha'_1, \alpha'_2, \dots, \alpha'_n)'$ is an RN, then a network with

$$A^* = (\lambda_1 \alpha'_{i1}, \lambda_2 \alpha'_{i2}, \dots, \lambda_n \alpha'_{in})'$$

is also an RN, where $\{i_1, i_2, \dots, i_n\} = \{1,2,\dots,n\}$ and $\lambda_s = 1$ or -1 for $s = 1,2,\dots,n$.

Proof: The proof is trivial since for any α_i

$$S^{(j)}(\alpha_i) = S^{(-j)}(-\alpha_i)$$

Note that networks A and A^* do not necessarily give the same dynamical behaviors.

Example 1
Let us put: $\alpha_1 = (1,-1,1)$, $\alpha_2 = (-1,-1,1)$, $\alpha_3 = (1,1,1)$.
For $A_1 = (\alpha'_1, \alpha'_2, \alpha'_3)'$,

$$\begin{bmatrix}0\\0\\0\end{bmatrix} \rightleftarrows \begin{bmatrix}0\\1\\0\end{bmatrix}, \quad \begin{bmatrix}1\\1\\1\end{bmatrix} \rightleftarrows \begin{bmatrix}1\\0\\1\end{bmatrix}, \quad \begin{bmatrix}1\\0\\0\end{bmatrix} \circlearrowleft, \quad \begin{bmatrix}0\\1\\1\end{bmatrix} \circlearrowleft, \quad \begin{bmatrix}0\\0\\1\end{bmatrix} \rightleftarrows \begin{bmatrix}1\\1\\0\end{bmatrix}.$$

For $A_2 = (\alpha'_3, \alpha'_1, \alpha'_2)'$,

$$\begin{bmatrix}0\\0\\0\end{bmatrix} \to \begin{bmatrix}0\\0\\1\end{bmatrix} \to \begin{bmatrix}0\\1\\1\end{bmatrix} \to \begin{bmatrix}1\\0\\1\end{bmatrix} \to \begin{bmatrix}1\\1\\1\end{bmatrix} \to \begin{bmatrix}1\\1\\0\end{bmatrix} \to \begin{bmatrix}1\\0\\0\end{bmatrix} \to \begin{bmatrix}0\\1\\0\end{bmatrix} \text{ (cycle back to first)}$$

For $A_3 = (\alpha'_1, -\alpha'_2, \alpha'_3)'$,

$$\begin{bmatrix}0\\0\\0\end{bmatrix} \circlearrowleft, \quad \begin{bmatrix}1\\1\\1\end{bmatrix} \circlearrowleft, \quad \begin{bmatrix}1\\0\\0\end{bmatrix} \to \begin{bmatrix}1\\1\\0\end{bmatrix} \to \begin{bmatrix}0\\1\\1\end{bmatrix} \to \begin{bmatrix}0\\0\\1\end{bmatrix} \text{ (cycle)}, \quad \begin{bmatrix}0\\1\\0\end{bmatrix} \circlearrowleft, \quad \begin{bmatrix}1\\0\\1\end{bmatrix} \circlearrowleft.$$

For $A_4 = (-\alpha'_1, \alpha'_2, \alpha'_3)$,

$$\begin{bmatrix}0\\0\\0\end{bmatrix} \to \begin{bmatrix}1\\1\\0\end{bmatrix} \to \begin{bmatrix}1\\0\\1\end{bmatrix} \to \begin{bmatrix}0\\1\\1\end{bmatrix} \to \begin{bmatrix}1\\1\\1\end{bmatrix} \to \begin{bmatrix}0\\0\\1\end{bmatrix} \to \begin{bmatrix}0\\1\\0\end{bmatrix} \to \begin{bmatrix}1\\0\\0\end{bmatrix} \text{ (cycle)}.$$

Caianiello *et al.* [3] have shown that if a network represented by A has a reverberation with the maximum period 2^n, the coefficient matrix A of the network has a full rank, i.e.,

$$\text{rank } A = n \tag{3}$$

The following theorem shows that (3) gives also a necessary condition that a network is an RN.

Theorem 2
If a network with A is an RN, then rank $A = n$.

Proof: Suppose that rank $A = k < n$ and that rank $(\alpha_1, \dots, \alpha_k) = k$. Then there exists $(c_1, c_2, \dots, c_k) \neq (0, 0, \dots, 0)$ such that

$$\alpha_{k+1} = \sum_{i=1}^{k} c_k \alpha_i$$

Let us consider a set S:

$$S = S_1^{(j_1)} S_2^{(j_2)} \dots S_k^{(j_k)}, \quad \begin{cases} j_s = 1, & \text{if } c_s > 0 \\ j_s = -1, & \text{if } c_s < 0 \\ j_s = 1 \text{ or } -1, & \text{if } c_s = 0 \\ \quad \text{for } s = 1,2,\dots,k \end{cases}$$

From Lemma 1, $\#S = 2^{n-k}$.

On the other hand, for any $X \in S$, it holds that

$$\alpha_i Z_x > 0 \qquad \text{for } c_i > 0$$
$$\alpha_i Z_x < 0 \qquad \text{for } c_i < 0 \qquad (4)$$

where $Z_x = 2X - 1$.

Equation (4) implies that

$$\alpha_{k+1} Z_x = \sum_{i=1}^{k} c_i \alpha_i Z_x > 0$$

for all $X \in S$. Thus

$$\# S_1^{(j_1)} S_2^{(j_2)} \dots S_k^{(j_k)} S_{k+1}^{(1)} = 2^{n-k}$$

This contradicts Corollary 1, which follows

rank $A = n$

Let us consider, for an RN with A, a relation between any one row of A and the remaining $n-1$ rows.

According to Theorem 2, rank $(\alpha_1', \dots, \alpha_{n-1}')' = n-1$, since rank $A = n$. Let us put

$$\alpha_i = (\tilde{\alpha}_i, a_{in}), \qquad i = 1,2,\dots,n,$$

where $\tilde{\alpha}_i$ is an $n-1$ dimensional row vector. We can assume that without loss of generality

$$\text{rank } (\tilde{\alpha}_1, \dots, \tilde{\alpha}_{n-1}) = n-1$$

Then there exists $(c_1, c_2, \dots, c_{n-1}) \neq (0, 0, \dots, 0)$ such that

$$\tilde{\alpha}_n = \sum_{i=1}^{n-1} c_i \tilde{\alpha}_i$$

Definition 1
An n dimensional row vector α is said to be dominant with respect to i $(i = 1,2,\dots,n)$, if

$$S^{(j)}(\alpha) = \{(x_1, \dots, x_{i-1}, 1, x_{i+1}, \dots, x_n)' \mid x_k = 1 \text{ or } 0, \ k \neq i\},$$

where according to $a_i > 0$ or $a_i < 0$, $j = 1$ or -1, respectively.

From the definition, if $a_i > 0$,

$$S^{(j)}(\alpha) = \{(x_1, \dots, x_{i-1}, \tfrac{1}{2}(1+j), x_{i+1}, \dots, x_n')\mid x_k = 1 \text{ or } 0, \ k \neq i\},$$

and if $a_i < 0$,

$$S^{(j)}(\alpha) = \{(x_1, \dots, x_{i-1}, \tfrac{1}{2}(1-j), x_{i+1}, \dots, x_n')\mid x_k = 1 \text{ or } 0, \ k \neq i\}$$

for $j = 1$ or -1.

Theorem 3
Suppose that a network with $A = (\alpha_1', \dots, \alpha_n')'$ is an RN. If α_l is dominant with respect to i, then a network with

$$\tilde{A} = (\tilde{\alpha}_1', \dots, \tilde{\alpha}_{l-1}', \tilde{\alpha}_{l+1}', \dots, \tilde{\alpha}_n')'$$

is an RN, where $\tilde{\alpha}_s$, $s = 1,2,\dots,l-1,l+1,\dots,n$ are $n-1$ dimensional row vectors constructed by removing ith components from $\{\alpha_s\}$.

Proof: From Theorem 1, we can assume that $l = n$ and $i = n$ without loss of generality. Suppose that

$$S_n^{(1)} = \{(X1) \mid X \in \mathbf{X}_{n-1}\},$$

where $(X1)$ means a column vector $\binom{X}{1}$.

Since the matrix A gives an RN, putting

$$S_s^{(j)} S_n^{(1)} = U_s^{(j)}(1),$$

$$S_s^{(j)} S_n^{(-1)} = U_s^{(j)}(0), \quad s = 1,2,\dots,n-1,$$

we have

$$\# U_s^{(j)}(1) = \# U_s^{(j)}(0) = 2^{n-2}, \quad \text{for } j = 1 \text{ or } -1 \qquad (5)$$

Suppose that $\alpha_s = (\tilde{\alpha}_s, a_{sn})$. If $a_{sn} > 0$, then $X1 \in S_s^{(-1)}$ gives $X0 \in S_s^{(-1)}$, and if $a_{sn} < 0$, then $X0 \in S_s^{(-1)}$ gives $X1 \in S_s^{(-1)}$. Thus from (5),

$$\{X \mid X0 \in U_s^{(j)}(0)\} = \{X \mid X1 \in U_s^{(j)}(1)\}.$$

Denoting the above set by $\tilde{U}_s^{(j)}$, we have

$$S^{(j)}(\tilde{\boldsymbol{\alpha}}_s) = \tilde{U}_s^{(j)},$$

for the n-1 dimensional network $(\tilde{\boldsymbol{\alpha}}'_1, \ldots, \tilde{\boldsymbol{\alpha}}'_{n-1})'$. And for any $k, i_1, i_2, \ldots, i_k\ (\neq n)$ and $j_1, j_2, \ldots, j_k$ ($j_s = 1$ or -1)

$$S_{i_1}^{(j_1)} \ldots S_{i_k}^{(j_k)} = \{(X1), (X0) \mid X \in S^{(j_1)}(\tilde{\boldsymbol{\alpha}}_{i_1}) \ldots \ldots S^{(j_k)}(\tilde{\boldsymbol{\alpha}}_{i_k})\}$$

That the relation $\# S_{i_1}^{(j_1)} \ldots S_{i_k}^{(j_k)} = 2^{n-k}$ holds implies that the relation

$$\# S^{(j_1)}(\tilde{\boldsymbol{\alpha}}_{i_1}) \ldots S^{(j_k)}(\tilde{\boldsymbol{\alpha}}_{i_k}) = 2^{n-1-k}$$

should hold. Thus a network $(\tilde{\boldsymbol{\alpha}}'_1, \tilde{\boldsymbol{\alpha}}'_2, \ldots, \tilde{\boldsymbol{\alpha}}'_n)'$ is an RN.

III. Methods for synthesis of RN's

In this section, we will give methods of synthesizing reverberating networks.

III.1. Networks characterized by two parameters

To begin with, let us consider a condition that a network with $A = (\boldsymbol{\alpha}'_1, \boldsymbol{\alpha}'_2, \ldots, \boldsymbol{\alpha}'_n)'$ is an RN, where

$$\boldsymbol{\alpha}_i = (b, b, \ldots, b, \overset{i\text{th}}{a}, b, \ldots, b) \tag{6}$$

We assume $a > 0$ because from Theorem 1, if A gives an RN, so does

$$(\lambda_1 \boldsymbol{\alpha}'_{i_1}, \ldots, \lambda_n \boldsymbol{\alpha}'_{i_n}), \quad \lambda_i = 1 \text{ or } -1$$

Theorem 3

A network with $A = (\boldsymbol{\alpha}'_1, \ldots, \boldsymbol{\alpha}'_n)'$ is an RN iff

$$a > |(n-1)b| \tag{7}$$

or

$$-(n-3)b < a < -(n-1)b, \qquad b < 0, \tag{8}$$

where

$$a > 0 \quad \text{and} \quad n \geqslant 3$$

Proof: Let us consider the following four cases:

(i) $a > |(n-1)b|$

Since

$$a > \tfrac{1}{2}\boldsymbol{\alpha}_i 1 = \theta_i = \tfrac{1}{2}(a + (n-1)b), \quad \text{for } b < 0 \text{ or } b > 0,$$

$$a + (n-1)b > \tfrac{1}{2}(a + (n-1)b), \quad \text{for } b < 0,$$

it can be easily shown that $\boldsymbol{\alpha}_i$ is dominant with respect to i, $i = 1,2,\ldots,n$. Since $\boldsymbol{\alpha}_i$ is with $a_{ii} > 0$, $S_i^{(j)}$ can be denoted as follows:

$$S_i^{(j)} = \{(x_1, \ldots, x_n)' \mid x_i = \tfrac{1}{2}(j+1),\ x_k = 1 \text{ or } 0,\ k \neq i\}$$

for $i = 1,2,\ldots,n$, $j = 1$ or -1.

Therefore

$$S_1^{(j_1)} \ldots S_n^{(j_n)} = \{\tfrac{1}{2}(j_1+1), \tfrac{1}{2}(j_2+1), \ldots, \tfrac{1}{2}(j_n+1)\},$$

$$j_s = 1 \text{ or } -1, \quad s = 1,2,\ldots,n.$$

This means that every state $X \in \mathbf{X}_n$ makes the transition to itself. Thus the network is an RN.

(ii) $-(n-3)b < a < -(n-1)b, \quad b < 0$

We have in this case

$$b < \tfrac{1}{2}\boldsymbol{\alpha}_i 1 = \theta_i < 0,$$

$$a + (n-1)b < \tfrac{1}{2}\boldsymbol{\alpha}_i 1,$$

$$a + (n-2)b > \tfrac{1}{2}\boldsymbol{\alpha}_i 1$$

Thus

$$S_i^{(j)} = \{(x_1, \ldots, x_n)' \mid x_i = \tfrac{1}{2}(j+1),\ 1 \leqslant \sum_{s=1}^{n} x_s \leqslant n-1\} \cup (\tfrac{1}{2}(1-j), \ldots, \tfrac{1}{2}(1-j))$$

for $i = 1,2,\ldots,n$; $j = 1$ or -1.

It follows that for $(j_1, \ldots, j_n) \neq (j, \ldots, j)$, $j, j_s = 1$ or -1; $s = 1,\ldots,n$.

$$S_1^{(j_1)} \ldots S_n^{(j_n)} = \{½(j_1+1), \ldots, ½(j_n+1)\}$$

and for $(j_1, \ldots, j_n) = (j, \ldots, j)$,

$$S_1^{(j)} \ldots S_n^{(j)} = \{½(1-j), \ldots, ½(1-j)\}$$

This means that every state $X \neq (0, 0, \ldots, 0)$, $(1, 1, \ldots, 1)$ makes the transition to itself and $(0, 0, \ldots, 0)$ to $(1, 1, \ldots, 1)$, vice versa. Thus the network is an RN.

From (i) and (ii), (7) or (8) is a sufficient condition that A gives an RN.

(iii) $b < 0, \quad a < -(n-3)b.$

We have in this case

$$½\boldsymbol{\alpha}_i 1 = \theta_i < b < 0$$

and for some integer k such that $\frac{n-3}{2} < k < n-2$

$$a + (k+1) < ½\boldsymbol{\alpha}_i 1 < a + kb$$

Thus

$$S_i^{(1)} = \{(x_1, \ldots, x_n)' \mid x_i = 1, \quad 0 \leqslant \sum_{s \neq i} x_s \leqslant k;$$

$$x_i = 0 \,, \quad 0 \leqslant \Sigma x_s \leqslant n-2-k\}$$

for $i = 1,2,\ldots,n$ and $k \geqslant n-2-k$.

It follows that

$$S_{i_1}^{(1)} S_{i_2}^{(1)} = \{(x_1, \ldots, x_n)' \mid$$

$$x_{i_1} = x_{i_2} = 1, \quad 0 \leqslant \sum_{s \neq i_1, i_2} x_s \leqslant k-1;$$

$$x_{i_1} = 1, \; x_{i_2} = 0,$$

$$0 \leqslant \sum_{s \neq i_1} x_s \leqslant n-3-k;$$

$$x_{i_1} = 0, \; x_{i_2} = 1,$$

$$0 \leqslant \sum_{s \neq i_2} x_s \leqslant n-3-k;$$

$$x_{i_1} = x_{i_2} = 0,$$

$$0 \leqslant \sum_s x_s \leqslant n-2-k\}$$

Noting that

$${}_{n-2}C_0 + \ldots + {}_{n-2}C_{k-1} + {}_{n-2}C_0 + \ldots + {}_{n-2}C_{n-2-k} = 2^{n-2} \,,$$

we have

$$\#S_{i_1}^{(1)} S_{i_2}^{(1)} = 2^{n-2} + 2\left(\sum_{s=0}^{n-3-k} {}_{n-2}C_s\right)$$

From $n-3-k \geqslant 0$, $\#S_{i_1}^{(1)} S_{i_2}^{(1)} > 2^{n-2}$, which show that the network is not an RN.

(iv) $b > 0, \quad a < (n-1)b.$

We have in this case

$$0 < a, \; b < ½\boldsymbol{\alpha}_i 1 = \theta_i$$

and for some integer k such that $0 \leqslant k < \frac{n-1}{2}$,

$$a + kb < ½\boldsymbol{\alpha}_i 1 < a + (k+1)b.$$

Thus

$$S_i^{(-1)} = \{(x_1, \ldots, x_n)' \mid x_i = 1, \quad 0 \leqslant \sum_{s \neq i} x_s \leqslant k;$$

$$x_i = 0, \quad 0 \leqslant \Sigma x_s \leqslant n-2-k\}$$

for $i = 1,2,\ldots,n$ and $n-2-k \geqslant k$.

It follows that

$$\#S_{i_1}^{(1)} S_{i_2}^{(1)} = 2^{n-2} + \sum_{s=0}^{k} {}_{n-2}C_s \,.$$

From $k \geqslant 0$, we have

$$\#S_{i_1}^{(-1)} S_{i_2}^{(-1)} > 2^{n-2},$$

which implies that the network is not an RN.

The above (i), (ii), (iii), and (iv) exhaust all cases of a (>0) for any b $(\neq a)$. Consequently, (iii) and (iv) show that (7) or (8) is a necessary and sufficient condition that the matrix gives an RN. This completes the proof.

Although the network with A is rather trivial since each state goes to itself, we can construct networks which may show complex transitions based on A by virtue of Lemma 2.

Example 2

Let us put: $\boldsymbol{\alpha}_1 = (4,1,1,1)$, $\boldsymbol{\alpha}_2 = (1,4,1,1)$,

$\boldsymbol{\alpha}_3 = (1,1,4,1),\ \boldsymbol{\alpha}_4 = (1,1,1,4).$

Suppose that $A = (\boldsymbol{\alpha}_1', \boldsymbol{\alpha}_2', \boldsymbol{\alpha}_3', \boldsymbol{\alpha}_4')'$. Then every state of the network with A makes the transition to itself, i.e.

$$\forall X \in \mathbf{X}_4\,,\quad T(X) = X\,,\quad T = (A, \tfrac{1}{2}A1)$$

While, for instance, $A^* = (-\boldsymbol{\alpha}_2', \boldsymbol{\alpha}_3', \boldsymbol{\alpha}_4', \boldsymbol{\alpha}_1')$ gives the following transitions:

$$\begin{bmatrix}1\\1\\1\\1\end{bmatrix} \to \begin{bmatrix}0\\1\\1\\1\end{bmatrix} \to \begin{bmatrix}0\\1\\1\\0\end{bmatrix} \to \begin{bmatrix}0\\1\\0\\0\end{bmatrix} \to \begin{bmatrix}0\\0\\0\\0\end{bmatrix} \to \begin{bmatrix}1\\0\\0\\0\end{bmatrix} \to \begin{bmatrix}1\\0\\0\\1\end{bmatrix} \to \begin{bmatrix}1\\0\\1\\1\end{bmatrix} \circlearrowleft,$$

$$\begin{bmatrix}1\\1\\0\\1\end{bmatrix} \to \begin{bmatrix}0\\0\\1\\1\end{bmatrix} \to \begin{bmatrix}1\\1\\1\\0\end{bmatrix} \to \begin{bmatrix}0\\1\\0\\1\end{bmatrix} \to \begin{bmatrix}0\\0\\1\\0\end{bmatrix} \to \begin{bmatrix}1\\1\\0\\0\end{bmatrix} \to \begin{bmatrix}0\\0\\0\\1\end{bmatrix} \to \begin{bmatrix}1\\0\\1\\0\end{bmatrix} \circlearrowleft.$$

III.2. Networks characterized by three parameters

Suppose that

$$\{i_1, i_2, \ldots, i_n\} = \{k_1, k_2, \ldots, k_n\} = \{1,2,\ldots,n\},$$

$$i_s \neq k_s\,,\quad s = 1,2,\ldots,n.$$

Let us consider a condition that a network with

$$A = (\boldsymbol{\alpha}(i_1, k_1), \ldots, \boldsymbol{\alpha}(i_n, k_n))$$

is an RN, where

$$\boldsymbol{\alpha}(i_s, k_s) = (b, \ldots, b, \overset{i_s}{a_1}, b, \ldots, b, \overset{k_s}{a_2}, b, \ldots, b) \quad (9)$$

We assume that $a_1 < a_2$, $b > 0$ and $\theta_i = \tfrac{1}{2}(a_1 + a_2 + (n-2)b)$.

Then we have:

Theorem 4

A network with $A = (\boldsymbol{\alpha}(i_1, k_1), \ldots, \boldsymbol{\alpha}(i_n, k_n))$ is an RN iff

$$a_2 > a_1 + (n-2)b\,, \quad (10)$$

$$-(n-4)b < a_1 + a_2 < -(n-6)b\,, \quad (11)$$

where $b > 0$, $a_1 < a_2$ and the network with A is assumed not to be reduced to an RN with two parameters[†].

Proof: Sufficiency. From (10), (11) and $b > 0$, $a_1 < a_2$,

$$a_1 < 0 < b < \theta < a_2\,,$$

$$a_1 + (n-2)b < \theta\,,$$

$$a_1 + a_2 + (n-4)b < \theta\,,$$

$$a_1 + a_2 + (n-3)b > \theta\,.$$

Thus

$$S^{(1)}(\boldsymbol{\alpha}(i_s, k_s)) = \{(x_1, \ldots, x_n)' \mid x_{i_s} = 0,\ x_{k_s} = 1;\ x_{i_s} = x_{k_s} = 1,\ n-3 \leqslant \sum_{t \neq i_s, k_s} x_t \leqslant n-2;\ x_{i_s} = x_{k_s} = 0,\ 2 \leqslant \Sigma x_t \leqslant n-2\},$$

$$S^{(-1)}(\boldsymbol{\alpha}(i_s, k_s)) = \{(x_1, \ldots, x_n)' \mid x_{i_s} = 1,\ x_{k_s} = 0;\ x_{i_s} = x_{k_s} = 1,\ 0 \leqslant \sum_{t \neq i_s, k_s} \leqslant n-4;\ x_{i_s} = x_{k_s} = 0,\ 0 \leqslant \Sigma x_t \leqslant 1\}.$$

It is shown that

$$S^{(j_1)}(\boldsymbol{\alpha}(i_1, k_1)) \ldots S^{(j_n)}(\boldsymbol{\alpha}(i_n, k_n)) = \{(x_1, x_2, \ldots, x_n)'$$

where

(*a*) for $p = n$, $x_t = 1$,

(*b*) for $p = n-1$,

$$x_{k_t} = \begin{cases} 0\,, & \text{if } j_t = -1,\\ 1\,, & \text{if } j_t = 1\,,\end{cases}$$

[†] It is not the case that $b = a_1$ or a_2. Note that even if $b \neq a_1, a_2\ (a_1 < a_2)$, there is a case that the network can be reduced to an RN with two parameters.

(c) for $2 \leqslant p \leqslant n-1$,

$$x_{i_t} = \begin{cases} 0, & \text{if } j_t = 1, \\ 1, & \text{if } j_t = -1, \end{cases}$$

(d) for $p = 1$,

$$x_{k_t} = \begin{cases} 0, & \text{if } j_t = -1, \\ 1, & \text{if } j_t = 1, \end{cases}$$

(e) for $p = 0$, $x_t = 0$,

$$t = 1,2,\ldots,n; \quad p = \#\{j_t \mid j_t = 1, \quad t = 1,\ldots,n\}$$

We omit the verification, since it is too bothersome. Each gives $\# S_1 \ldots S_n = 1$,

where $S_s = S^{(j_s)}(\alpha_{i_s}\ \alpha_{k_s})$, $s = 1,2,\ldots,n$.

This implies the sufficiency of (10), (11).

Necessity: If the matrix A gives an RN, then Corollary 1 holds.
Namely it is necessary that for a case, say $i_1 = 1$, $k_1 = 2$, $i_2 = 2$, $k_2 = 1$,

$$\# S^{(1)}(\alpha(1,2))\, S^{(1)}(\alpha(2,1)) = 2^{n-2} \tag{12}$$

is satisfied. Putting

$$S^{(1)}(\alpha(1,2)) \cap \{x_1 x_2 X \mid X \in \mathbf{X}_{n-2}, x_1, x_2 = 1 \text{ or } 0\} = U_{x_1 x_2},$$

we have

$$\# U_{x_1 x_2} + \# U_{\bar{x}_1 \bar{x}_2} = 2^{n-2}, \quad \text{for} \quad x_1, x_2 = 1 \text{ or } 0 \tag{13}$$

We have also

$$S^{(1)}(\alpha(2,1)) \supset U_{11}, U_{00}.$$

Thus from (13)

$$\# S^{(1)}(\alpha(1,2))\, S^{(1)}(\alpha(2,1)) \geqslant 2^{n-2}.$$

$U_{10} = \phi$ is necessary for (12). Because: If $U_{10} \neq \phi$, there exists an $X \in \mathbf{X}_{n-2}$ such that $10X \in S^{(1)}(\alpha(1,2))$. Since $a_1 < a_2$, $01X \in S^{(1)}(\alpha(1,2))$ and $01X \in S^{(1)}(\alpha(2,1))$.

Thus

$$01X \in S^{(1)}(\alpha(1,2))\, S^{(1)}(\alpha(2,1)),$$

which gives

$$\# S^{(1)}(\alpha(1,2))\, S^{(1)}(\alpha(2,1)) > 2^{n-2}.$$

It follows that $U_{10} = \phi$.

Since from (13), $\# U_{01} = 2^{n-2}$, i.e. for any $X \in \mathbf{X}_{n-2}$, $01X \in S^{(1)}(\alpha(1,2))$ and $10X \in S^{(-1)}(\alpha(1,2))$,

$$a_1 + (n-2)b < \theta = \tfrac{1}{2}(a_1 + a_2 + (n-2)b) \quad \text{for} \quad b > 0$$

Consequently we obtain (10).
This implies that we can write:

$$S^{(-1)}(\alpha(i_s, k_s)) = \{(x_1,\ldots,x_n)' \mid x_{i_s} = 1,\ x_{k_s} = 0;$$

$$x_{i_s} = x_{k_s} = 1,\ 0 \leqslant \sum_{t \neq i_s, k_s} x_t \leqslant k;$$

$$x_{i_s} = x_{k_s} = 0,\ 0 \leqslant \Sigma x_t \leqslant n-3-k\} \tag{14}$$

for some k, $0 \leqslant k \leqslant n-3$. Because if $k < 0$ $(k > n-3)$, $11X(00X) \notin S^{(-1)}(\alpha(i_s, k_s))(X \in \mathbf{X}_{n-2})$, which means that $\alpha(i_s, k_s)$ is dominant with respect to k_s (i_s).

In the case that $i_1 = 1$, $k_1 = 2$, $i_2 = 2$, $k_2 = 1$, $i_3 = 3$, $k_3 = 4$, $i_4 = 4$, $k_4 = 3$,

let us put

$$S(j_1,\ldots,j_4) = S^{(j_1)}(\alpha(1,2))\, S^{(j_2)}(\alpha(2,1))\, S^{(j_3)}(\alpha(3,4))\, S^{(j_4)}(\alpha(4,3)).$$

As A gives an RN, we have

$$\# S(j_1,\ldots,j_4) = 2^{n-4} \tag{15}$$

From (14)

$$S(-1,-1,-1,-1) = \{(x_1,\ldots,x_n)' \mid$$

$$x_1 = x_2 = x_3 = x_4 = 1,\ 0 \leqslant \sum_{s \neq 1,\ldots,4} x_s \leqslant k-2;$$

$$x_1 = x_2 = x_3 = x_4 = 0,\ 0 \leqslant \Sigma x_s \leqslant n-3-k;$$

$$x_1 = x_2 = 1,\ x_3 = x_4 = 0,\ 0 \leqslant \Sigma x_s \leqslant K;$$

$$x_1 = x_2 = 0,\ x_3 = x_4 = 1,\ 0 \leqslant \Sigma x_s \leqslant K;$$

$$K = \min(k, n-5-k)\}.$$

$$\begin{bmatrix}0\\0\\1\\1\end{bmatrix}\circlearrowleft,\quad \begin{bmatrix}1\\1\\0\\0\end{bmatrix}\circlearrowleft,\quad \begin{bmatrix}0\\1\\0\\0\end{bmatrix}\rightleftarrows\begin{bmatrix}1\\0\\1\\1\end{bmatrix}.$$

Thus

$$\#S(-1,\ldots,-1) = \sum_{s=0}^{k-2} {}_{n-4}C_s + \sum_{s=0}^{n-3-k} {}_{n-4}C_s + 2\sum_{s=0}^{K} {}_{n-4}C_s$$

Since

$$\sum_{s=0}^{k-2} {}_{n-4}C_s + \sum_{s=0}^{n-3-k} {}_{n-4}C_s = 2^{n-4},$$

(15) is fulfilled by

$$\sum_{s=0}^{K} {}_{n-4}C_s = 0:$$

It follows that $-2 < n-5-k < 0$ since $0 \leqslant k \leqslant n-3$. If $k = n-3$, the network A under consideration is reduced to an RN characterized by two parameters. Thus it must be that $k = n-4$. And we have from (14)

$$a_1 + a_2 + (n-4)b < \theta = \tfrac{1}{2}(a_1 + a_2 + (n-2)b),$$

$$a_1 + a_2 + (n-3)b > \theta = \tfrac{1}{2}(a_1 + a_2 + (n-2)b),$$

which give (11).
This proves that (10) and (11) are the necessary conditions.

Example 3
The conditions (10) and (11) are satisfied by $a_1 = -3,\ a_2 = 4,\ b = 1,\ n = 4.$

For $A = (\boldsymbol{\alpha}(1,2)', \boldsymbol{\alpha}(2,1)', \boldsymbol{\alpha}(3,4)', \boldsymbol{\alpha}(4,3)')'$,

$$\begin{bmatrix}0\\0\\0\\0\end{bmatrix}\circlearrowleft,\ \begin{bmatrix}1\\1\\1\\1\end{bmatrix}\circlearrowleft,\ \begin{bmatrix}0\\0\\0\\1\end{bmatrix}\rightleftarrows\begin{bmatrix}0\\0\\1\\0\end{bmatrix},\ \begin{bmatrix}1\\1\\1\\0\end{bmatrix}\rightleftarrows\begin{bmatrix}1\\1\\0\\1\end{bmatrix},\ \begin{bmatrix}0\\0\\1\\1\end{bmatrix}\rightleftarrows\begin{bmatrix}1\\1\\0\\0\end{bmatrix},\ \begin{bmatrix}0\\1\\0\\0\end{bmatrix}\rightleftarrows\begin{bmatrix}1\\0\\0\\0\end{bmatrix},\ \begin{bmatrix}1\\0\\1\\1\end{bmatrix}\rightleftarrows\begin{bmatrix}0\\1\\1\\1\end{bmatrix},\ \begin{bmatrix}0\\1\\0\\1\end{bmatrix}\rightleftarrows\begin{bmatrix}1\\0\\1\\0\end{bmatrix},\ \begin{bmatrix}0\\1\\1\\0\end{bmatrix}\rightleftarrows\begin{bmatrix}1\\0\\0\\1\end{bmatrix}$$

For $A = (-\boldsymbol{\alpha}(2,1)', \boldsymbol{\alpha}(3,4)', \boldsymbol{\alpha}(1,2)', -\boldsymbol{\alpha}(4,3)')'$,

$$\begin{bmatrix}0\\0\\0\\0\end{bmatrix}\to\begin{bmatrix}1\\0\\0\\1\end{bmatrix}\to\begin{bmatrix}0\\1\\0\\1\end{bmatrix}\to\begin{bmatrix}1\\1\\1\\1\end{bmatrix}\to\begin{bmatrix}0\\1\\1\\0\end{bmatrix}\to\begin{bmatrix}1\\0\\1\\0\end{bmatrix}\circlearrowleft,\quad \begin{bmatrix}0\\0\\0\\1\end{bmatrix}\to\begin{bmatrix}1\\1\\0\\1\end{bmatrix}\to\begin{bmatrix}0\\1\\1\\1\end{bmatrix}\to\begin{bmatrix}1\\1\\1\\0\end{bmatrix}\to\begin{bmatrix}0\\0\\1\\0\end{bmatrix}\to\begin{bmatrix}1\\0\\0\\0\end{bmatrix}\circlearrowleft,$$

Acknowledgement
The author would like to express her thanks to Professor Ishihara of University of Osaka Prefecture for his valuable discussions and encouragement.

References

1. CAIANIELLO, E.R., 'Outline of thought processes and thinking machines', *J. Theor. Biol.* **2**, 204-235 (1961).
2. CAIANIELLO, E.R., 'Decision equations and reverberations', *Kybernetik* **3**(2), 98-100 (1966).
3. CAIANIELLO, E.R., DE LUCA, A., and RICCIARDI, L.M., 'Reverberation and control of neural networks', *Kybernetik* **4**(1), 10-18 (1967).
4. KITAGAWA, T., 'Dynamical systems and operators associated with a single neuronic equation', *Math. Biosci.* **18**, 191-244 (1973).
5. ARIMOTO, S., 'Periodic sequences of states of an autonomous circuit consisting of threshold elements (in Japanese) Jyoho to Seigyo no Kenkyu', *Trans. IECE,* No.2 (1963).
6. ISHIHARA, T. and SATO, M., 'Variation and stability of reverberations in threshold systems', *Math. Japonica* **19** (1974).
7. YAMAGUCHI, M., 'Characterization of operators and their reverberation cycles associated with a single neural equation', *Res. Rep. No.31, Res. Inst. Fundamental Information Sci.* (June 1972).
8. SATO, M. and ISHIHARA, T., 'Graph theoretical approach to threshold systems', *Math. Japonica,* **19** (1974).

A mathematical analysis of excitable membrane phenomena

GAIL A. CARPENTER
Northeastern University, Boston, Massachusetts, USA

This work defines and analyzes the dynamics of a general excitable membrane. The definition generalizes such special cases as the Hodgkin-Huxley and FitzHugh-Nagumo models of nerve impulse transmission. It describes how any number of microscopic membrane processes are synthesized by a macroscopic potential which, in turn, influences the microscopic processes by feedback. Each microscopic process corresponds to the statistics of switching on and off suitable membrane sites. The macroscopic potential obeys a nonlinear diffusion equation.

The mathematical results classify the types of solutions that can be achieved by particular statistical rules and potential functions. Of critical importance is the number of independent phenomena responsible for the return to rest. The Hodgkin-Huxley model postulates two such processes, the inhibition of Na^+ and K^+ activation; the FitzHugh-Nagumo model combines these two into one. Both models exhibit a single pulse solution (Fig.6A) as well as a continuous family of periodic solutions which converge to the single pulse as the period becomes infinite. The Hodgkin-Huxley model, however, may exhibit plateau and finite wave train solutions (Fig.6A); solutions consisting of periodic bursts; periodic solutions with speed which is greater than that of the single pulse; and even two or more stable pulse solutions, traveling at different speeds. If three or more inhibitory processes occur, there may be periodic wave trains of any length (Fig.7) and yet more complex behavior. This approach illuminates the underlying topological structure of particular models, and helps one to visualize the types of microscopic events that can lead to a given class of observed phenomena. Of particular interest is the qualitative nature of the hypotheses, which allow for the vast array of data one would hope to include in a model which describes the nerve impulse, muscle contraction, and the heartbeat in all species. While the spirit of our analysis is reminiscent of the catastrophe theory methods of Zeeman [1] *et al.*, we associate a global dynamic with its microscopic realization and prove existence theorems rigorously.

We begin with a detailed description of the Hodgkin-Huxley model, as a typical excitable membrane process. We then generalize this system to allow any number of fast or slow subprocesses. Geometric hypotheses are placed on the system as

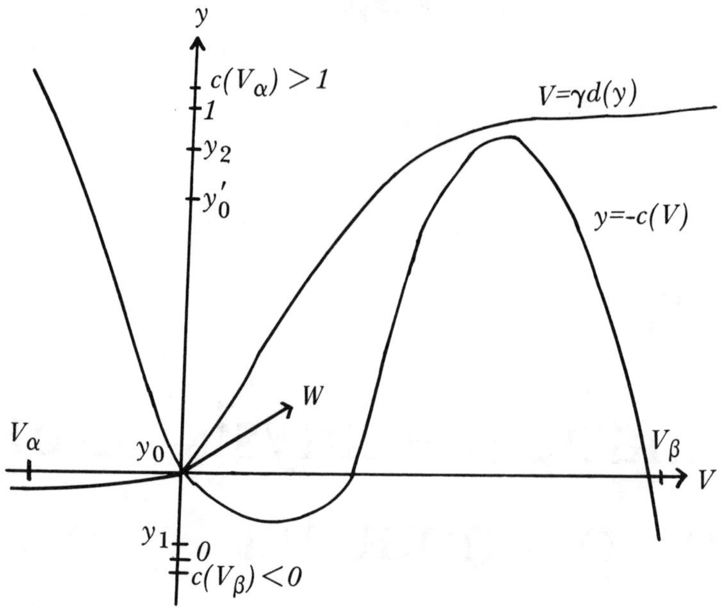

Fig.1 The phase space of the (FN) system. The 'cubic' curve represents the rest points when $\epsilon = 0$

the need for them develops. Our goal in each case is to find a 'singular solution', arising from the different time scales involved. Proofs that the existence of a singular solution implies the existence of a true solution are contained in [2,3]. One proof is sketched at the end of the chapter in order to introduce the methods used.

The Hodgkin-Huxley model

The Hodgkin-Huxley [4,5] equations summarize a model of the nerve impulse in which the axon is taken to be a long cylindrical membrane containing axoplasm (with high K^+ concentration) and bathed in a solution with high Na^+ concentration. When the nerve is stimulated above a threshold sodium ions rush in and dramatically shift the membrane potential, V, from its rest state. Two slower processes, inhibition of sodium entrance and potassium exit, return V to rest. Diffusion of electrons within the axon induces the process to repeat itself in a wavelike manner.

The Hodgkin-Huxley system consists of a nonlinear diffusion equation coupled with equations describing changes in membrane permeability as V varies:

$$\frac{1}{R}\frac{\partial^2 V}{\partial x^2} = C\frac{\partial V}{\partial t} + \bar{g}_{Na} m^3 h(V - V_{Na}) + \bar{g}_K n^4 (V - V_K) + \bar{g}_L (V - V_L)$$

$$\frac{\partial m}{\partial t} = (1-m)\alpha_m(V) - m\beta_m(V)$$

$$\frac{\partial h}{\partial t} = (1-h)\alpha_h(V) - h\beta_h(V) \qquad (1)$$

$$\frac{\partial n}{\partial t} = (1-n)\alpha_n(V) - n\beta_n(V)$$

where x is the distance from the stimulus; t the time since the stimulus; and $\alpha_m, \beta_m \ldots > 0$. Ohm's law, Kirchoff's law, and the Nernst-Planck equations for ionic flux across a membrane are used to derive the first equation. The total membrane current $(\frac{1}{R}\frac{\partial^2 V}{\partial x^2}$, where R is the axoplasmic resistance) is set equal to the capacitance current $(C\frac{\partial V}{\partial t})$ plus the sodium $(\bar{g}_{Na} m^3 h(V - V_{Na}))$, potassium $(\bar{g}_K n^4 (V - V_K))$, and leakage $(\bar{g}_L (V - V_L))$ currents. The equation for m describes the statistics of switching on $(m = 1)$ or off $(m = 0)$ of the active sodium sites. Sodium inactivation (h) and potassium activation (n) are described similarly. The rate $(\alpha_m(V) + \beta_m(V))$ at which m tends to an asymptotic value $\left(\frac{\alpha_m(V)}{\alpha_m(V) + \beta_m(V)}\right)$ for fixed V is large compared to the rates of n and h. Since m represents activation, $\alpha_m' > 0$ and $\beta_m' < 0$.

Similarly, $\alpha_n' > 0$, $\beta_n' < 0$, $\alpha_h' < 0$ and $\beta_h' > 0$.

A traveling wave solution of (2) satisfies a system of five ODE's, where $s = x + \theta t$, $\dot{} = \frac{d}{ds}$, and θ is the (leftward) speed of the wave:

$$
\begin{aligned}
\dot{V} &= W \\
\dot{W} &= \theta W + f(V,m,n,h) \\
\dot{m} &= \delta^{-1}\theta^{-1}((1-m)\alpha_m(V) - m\beta_m(V)) \\
\dot{h} &= \epsilon\theta^{-1}((1-h)\alpha_h(V) - h\beta_h(V)) \\
\dot{n} &= \epsilon\theta^{-1}((1-n)\alpha_n(V) - n\beta_n(V))
\end{aligned}
\tag{2}
$$

In (2) we have set $R = C = 1$; let $f(V,m,n,h)$ represent the ionic current; and introduced δ^{-1}, ϵ to represent the 'fast' and 'slow' rates when δ, ϵ are small.

Generalizing the model

We may now ask: what hypotheses on (2) generate a 'nerve-like' solution, that is, a solution which tends to the (unique) rest state of the nerve as $s \to \pm\infty$? Is it necessary, for example, to assume that f is linear in V or that the Na^+ and K^+ currents are independent, as implied by the form of f in (1)? The answer is no; in fact it is misleading to think of f as linear in V, and the validity of the Nernst-Planck equations (for *flat* membranes) is questionable in the first place. If (2) is to describe the nerve impulse in all species, we would expect a wide range of functions and parameters to yield nerve-like solutions. Moreover, if (2) is generalized to include other excitable membrane phenomena any number of subprocesses should be allowed. The analysis should include not only homoclinic solutions (from a rest point to itself) but also heteroclinic (between two rest points) and periodic solutions.

In order to study this general setting, we replace (2) by (3):

$$
\begin{aligned}
\dot{V} &= W \\
\dot{W} &= \theta W + f(V,y,z) \\
\dot{y} &= \epsilon\theta^{-1}g(V,y,z) \\
\dot{z} &= \delta^{-1}\theta^{-1}h(V,y,z)
\end{aligned}
\tag{3}
$$

where $V \in [V_\alpha, V_\beta] \subseteq \mathbb{R}$, $y \in [0,1]^l$, and $z \in [0,1]^k$; $\delta, \epsilon > 0$ are small; and $f,g,h \in C^2$.

We first consider the 'fast' equation:

$$\delta\dot{z} = \theta^{-1}h(V,y,z) \tag{4}$$

As $\delta \to 0$, (4) makes sense only if $h(V,y,z) = 0$.

Hypothesis (Fast)

There exists $z_\infty(V,y)$ such that $h(V,y,z) = 0$ iff $z = z_\infty(V,y)$. Moreover the eigenvalues of

$$\left.\frac{\partial h_i(V,y,z)}{\partial z_j}\right|_{z = z_\infty(V,y)} \quad \text{are negative.}$$

(FAST) is satisfied, for example, if $h_i(V,y,z) = (1 - z_i)\alpha_i(V,y) - z_i\beta_i(V,y)$, in which case $z_{\infty i} = \frac{\alpha_i}{\alpha_i + \beta_i}$ and the eigenvalues are $\{-(\alpha_i + \beta_i) : i = 1 \ldots k\}$.

If δ is small, then, it is reasonable to study the system (5) in which $z \equiv z_\infty(V,y)$:

$$
\begin{aligned}
\dot{V} &= W \\
\dot{W} &= \theta W + F(V,y) \\
\dot{y} &= \epsilon\theta^{-1}G(V,y)
\end{aligned}
\tag{5}
$$

where $F(V,y) \equiv f(V,y,z_\infty(V,y))$ and $G(V,y) \equiv g(V,y,z_\infty(V,y))$. We shall henceforth consider (5), since it can be shown [2] that all solutions of interest correspond to solutions of (3) for δ small.

Let us now examine the statement: 'A solution of (5) with ϵ small stays close to a solution of (5) with $\epsilon = 0$.' This statement is true *provided* the solution avoids the set where $\dot{V}$ and $\dot{W}$ are near zero. If $\dot{V}$ and $\dot{W}$ become very small, y eventually changes rapidly relative to V, W. Using (1) as a guide, it is reasonable to assume that $F(V_\alpha,y) < 0 < F(V_\beta,y)$ if $y \in [0,1]^l$. If this is the case, interesting solutions of (5) require that F be essentially nonlinear in the sense that, for some fixed y, $F(V,y)$ have at least two (and hence three) zeroes in (V_α, V_β). We shall assume that F has at most three zeroes. This case is often observed experimentally and can be easily extended to include more complicated nonlinearities. [Notice that we are now considering $f(V,m_\infty(v),n,h)$ of (2).] We shall be interested in bounded solutions

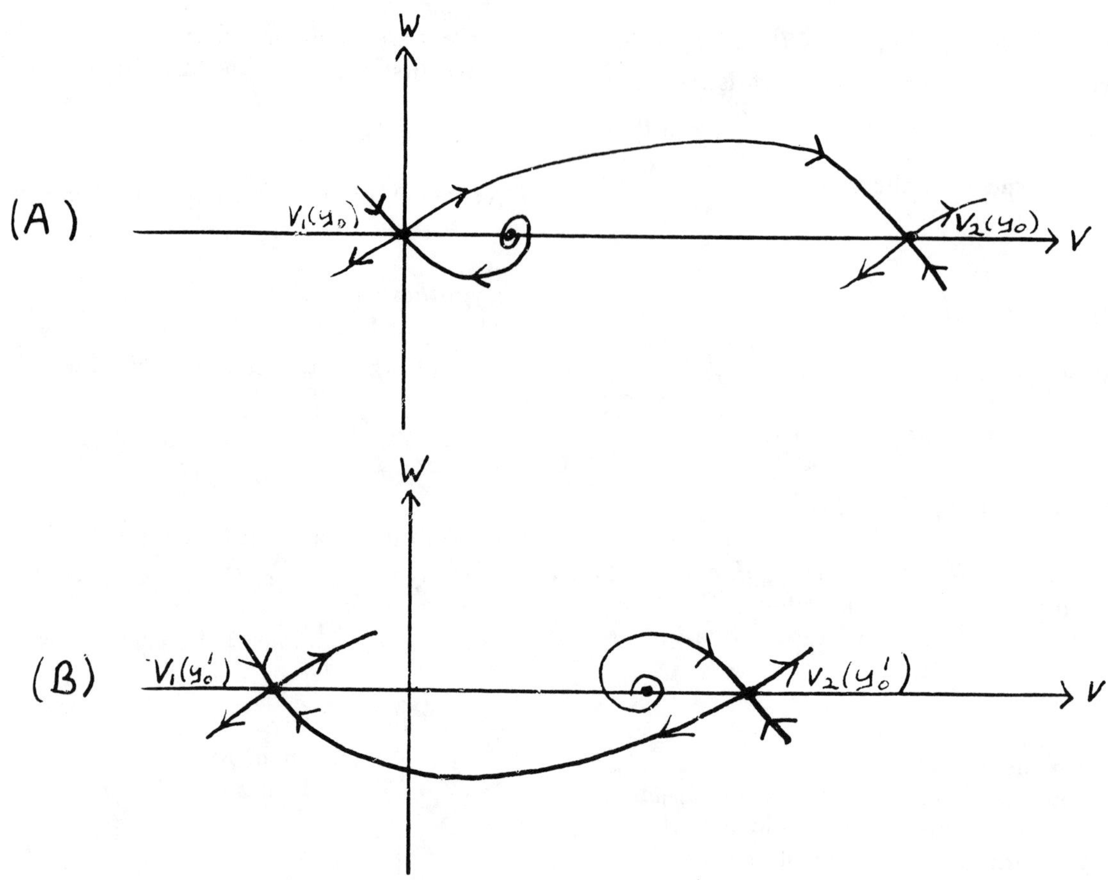

Fig.2 Phase portrait of (5) with $\epsilon = 0$, $\theta = \theta(y_0)$. (A) $y = y_0$; (B) $y = y_0' > y_0$.

of (5) which are alternatively dominated by the $V - W$ and y systems.

Example: The FitzHugh-Nagumo equations.
It is instructive to examine the FitzHugh-Nagumo [6,7] model, in which $l = 1$. The equations represent the tunnel diode system used by Nagumo *et al.* to simulate the nerve.

$$\dot{V} = W$$

$$\dot{W} = \theta W + c(V) + y \qquad (6,\text{FN})$$

$$\dot{y} = \epsilon\theta^{-1}(V - \gamma d(y))$$

where $c(V)$ is the 'cubic' depicted in Fig.1 and $\gamma, d' > 0$.

$\dot{V} = \dot{W} = 0$ precisely along the curve $y = -c(V)$, $W = 0$. If γ is small, $\langle 0, 0, y_0 \rangle$ is the unique rest point of (6). As γ increases (6) acquires a rest point in the right branch of $\{W = c(V) + y = 0\}$. Note that $F(V_\alpha, y) = c(V_\alpha) + y < c(V_\alpha) + 1 < 0 < c(V_\beta) < < c(V_\beta) + y = F(V_\beta, y)$ for $y \in [0,1]$.

Let us examine $\{V = W = 0,\ c'(V) > 0\}$. This set has two components which form the left and right branches of the cubic curve in Fig.1. Restricting attention to the left branch, for each $y \in [y_1, 1]$ there is a unique $V \equiv V_1(y)$ such that $y = -c(V)$. That is, $V_1(y)$ is the lefthand zero of $c(V) + y$ for fixed y. Similarly, $V_2(y)$ may be defined as the righthand zero when $y \in [0, y_2]$. $c(V) + y$ has three zeroes precisely when $y \in (y_1, y_2)$.

Returning to (5), we assume that for fixed $y \in [0,1]^l$ the equation $F(V,y) = 0$ defines a 'cubic' function of V.

Hypothesis (Cubic)

(A) $F(V_\alpha,y) < 0 < F(V_\beta,y)$.

(B) For every y there exist at most *three* V such that $F(V,y) = 0$; for some y there exist exactly three.
Moreover, $\frac{\partial^2 F}{\partial V^2}(V,y) \neq 0$ if $F(V,y) = \frac{\partial F}{\partial V}(V,y) = 0$. (That is, F admits double but not triple zeroes and $\{\dot{V} = \dot{W} = 0, \frac{\partial F}{\partial V} > 0\}$ has two components.)

(C) $\frac{\partial F}{\partial y_j} > 0$ for some j. [(C) may be replaced by a hypothesis on $\langle \frac{\partial F}{\partial y_1} \ldots \frac{\partial F}{\partial y_l} \rangle$.]

(CUBIC) allows us to define $V_1(y)[V_2(y)]$ to be the left [right] zero of $F(V,y)$ for $y \in \Pi_1 [\Pi_2] \subseteq \subseteq [0,1]^l$. It can be shown [2] that for each $y \in \Pi_1 \cap \Pi_2$ there exists $\theta(y) \geqslant 0$ such that $(5, \theta = \theta(y), \epsilon = 0)$ admits a heteroclinic solution from $\langle V_1(y), 0, y\rangle$ to $\langle V_2(y), 0, y\rangle$ if $\int_{V_1(y)}^{V_2(y)} F(V,y)\,dV \leqslant 0$; or from $\langle V_2(y), 0, y\rangle$ to $\langle V_1(y), 0, y\rangle$ if $\int_{V_1(y)}^{V_2(y)} F(V,y)\,dV \geqslant 0$.

Solutions of this system are completely understood; the form of G and choice of parameters θ, ϵ, δ now determine solutions of (3) and (5). We examine a few of the possibilities.

Our attention will focus on the two systems:

$$\dot{y}^1 = G(V_1(y), y), \quad y \in \Pi_1 \qquad (7.1)$$

$$\dot{y}^2 = G(V_2(y), y), \quad y \in \Pi_2 \qquad (7.2)$$

When V and W are small, (7.1) or (7.2) dominates (5).

In the (FN) example, y_0 is a global attractor for the system:

$$\dot{y}^1 = V_1(y) - \gamma d(y), \quad y \in [y_1, 1] \qquad (8.1)$$

If γ is small, the system:

$$\dot{y}^2 = V_2(y) - \gamma d(y), \quad y \in [0, y_2] \qquad (8.2)$$

has no rest point and $\dot{y}^2 > 0$. Thus all solutions leave $\Pi_2 = [0,y_2]$ in $\{y = y_2\}$. As γ increases, (8.2) acquires a global attractor, y_γ.

Let us consider the case $y_0' < y_\gamma < y_2$, where $\theta(y_0') = \theta(y_0)$.

In the nerve, this corresponds to the case where two stable potential states exist. If $\theta = \theta(y_0)$ and $\epsilon = 0$, one solution of (6) runs in $\{y = y_0\}$ from $\langle V_1(y_0), 0, y_0\rangle$ to $\langle V_2(y_0), 0, y_0\rangle$. If ϵ is small, the corresponding branch of the unstable manifold of $\langle 0, 0, y_0\rangle$ will stay near this solution *until* $\dot{V}$ and $\dot{W}$ become small, that is, until the solution approaches $\langle V_2(y_0), 0, y_0\rangle$. If (8.2) then dominates, the solution will move up to the rest point $\langle V_2(y_\gamma), 0, y_\gamma\rangle$. The solution of $(6, \epsilon = 0, \theta = \theta(y_0))$ and the positive half solution of (8.2) from $\langle V_2(y_0), 0, y_0\rangle$ to $\langle V_2(y_\gamma), 0, y_\gamma\rangle$ together form a *heteroclinic singular solution.* There is another possibility, as seen in Fig.3A. If the solution of (6) were to be dominated by (8.2) only up to the point where $y = y_0'$ it could jump back to the left branch, that is, stay close to the solution of $(6, \epsilon = 0, \theta = \theta(y_0))$ running from $\langle V_2(y_0'), 0, y_0'\rangle$ to $\langle V_1(y_0'), 0, y_0'\rangle$. Once the solution approaches $\langle V_1(y_0'), 0, y_0'\rangle$, then, (8.1) dominates and pulls it down to the rest point $\langle 0, 0, y_0\rangle$. This sequence of solutions of $(6, \epsilon = 0, \theta = \theta(y_0))$ and solution segments of (8.1) and (8.2) is a *homoclinic singular solution.* The existence of a singular solution implies the existence true solution of (5) or (6) for small $\epsilon > 0$. Whether the unstable manifold of $\langle 0, 0, y_0\rangle$ forms a homoclinic or heteroclinic solution depends upon the choice of parameters.

In fact, a parameter in a curve (A) (Fig.4) yields a homoclinic solution; and one in (B) yields a heteroclinic solution, where for fixed $\epsilon > 0$ a heteroclinic solution requires a faster wave speed. We note also that there exists a heteroclinic singular solution from $\langle V_2(y_\gamma), 0, y_\gamma\rangle$ to $\langle 0, 0, y_\gamma\rangle$, and a corresponding family of heteroclinic solutions for parameter values in the curve (C).

Periodic singular solutions may be constructed in a similar way (Fig.3B). Fix any $y_p > y_0$ such that $\int_{V_1(y_p)}^{V_2(y_p)} (c(V) + y_p)\,dV < 0$. Then there exist $y_p' > y_p$ such that $\theta(y_p') = \theta(y_p)$. The heteroclinic solutions of $(6, \theta = \theta(y_p), \epsilon = 0)$, linked by finite

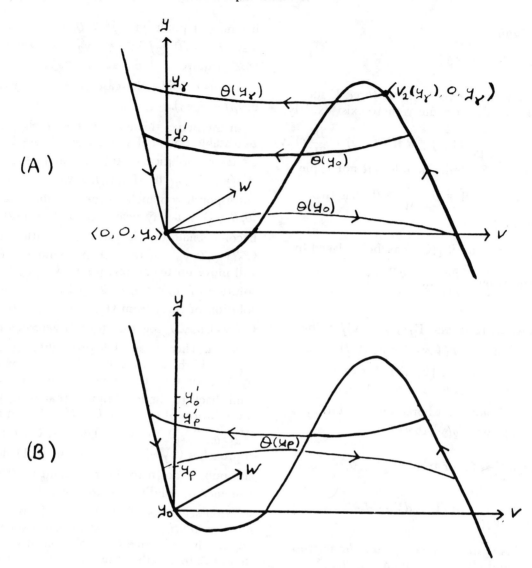

Fig.3 (A) A homoclinic singular solution of (FN) and two heteroclinic singular solutions. (B) A periodic singular solution.

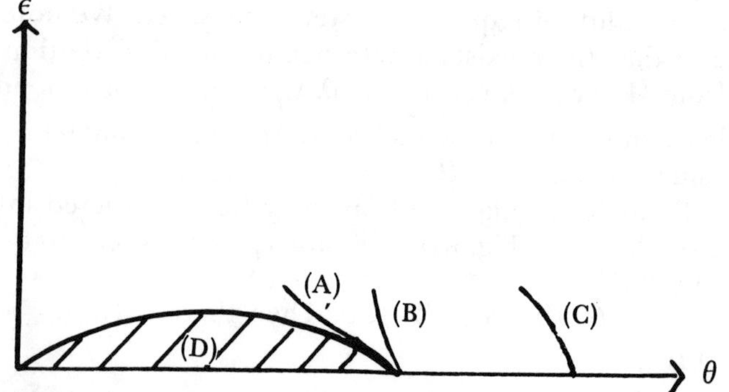

Fig.4 Parameter values for which (FN) admits a homoclinic (A), heteroclinic (B) or (C) or periodic (D) solution.

solution segments of (8.1) and (8.2), form a periodic singular solution. In fact, for any $\langle \theta, \epsilon \rangle$ in the open set (D) of Fig.4 (6) admits a periodic solution. For fixed small $\epsilon > 0$, there exists a continuous family of periodic solutions whose periods range from small (for θ small) to infinite, as θ approaches the value which yields a homoclinic solution. The proof that a singular periodic solution yields a true periodic solution depends critically upon the assumption that $l = 1$, that is, that (5) contain only one slow variable. The correct notion for $l \geqslant 2$ is that of an l-dimensional singular solution, discussed in [3].

The slow manifold

We shall now give examples of hypotheses on the general system (5) which imply the existence of singular solutions. When $l \geqslant 2$, some possible solutions arise which were not seen in the (FN) example.

Hypothesis (Slow)

(A) $G_j(V,y) < 0$ if $y_j = 1$ and

$G_j(V,y) > 0$ if $y_j = 0$.

Hence solutions of *(6, i)* enter Π_i on $\partial[0,1]^l$.

(B) y_0 is globally asymptotically stable in (7.1). That is, all solutions of (7.1) approach y_0 as $s \to \infty$.

(C) $\int_{V_1(y_0)}^{V_2(y_0)} F(V,y_0)\,dV < 0.$

(D) No positive half solution of (7.2) is contained in Π_2.

(B) and (D) are usually verified by means of a Liapunov function.

Hypothesis (SLOW) implies that (5) admits a homoclinic solution for a curve of parameters [Fig.4(A)]. Slightly more restrictive hypotheses, which are satisfied, for example, if (5) is an analytic system, imply the existence of a family of periodic solutions for $\langle\theta,\epsilon\rangle$ in an open set [Fig.4(D)].

Figure 5 illustrates the phase portrait of (7.1) and (7.2) with $l = 2$, e.g. the Hodgkin-Huxley system.

A homoclinic singular solution consists of a jump from $\langle V_1(y_0), 0, y_0\rangle$ to $\langle V_2(y_0), 0, y_0\rangle$; the segment (A) from y_0 to y_0'; the jump back; and the segment (B) from y_0' to y_0. Note, however, that (B), instead of returning to rest, may 'decide' to jump back to Π_2 at y_1, i.e. real parameters may be chosen for (5) to yield a solution with more than one jump. y_1 travels in Π_2 to y_1' (C); jumps back to Π_1; and either returns to y_0 (D) *or* jumps back to Π_2 at y_2. In fact, if the phase portrait is as illustrated, once the extra jumps begin there may be any number of them and hence there exist *finite wave train* solutions of any length $k \geqslant 1$. Similar reasoning shows that periodic solutions may exist for $\theta \geqslant \theta(y_0)$. Neither of these cases is observed if $l = 1$, e.g. in the (FN) example.

Detailed analysis of (7.1) and (7.2) yields further qualitative results. For example, suppose one variable y_j is much slower than the others, as is the case if K^+ exit is inhibited in the nerve [8] or in models of the heart. A long plateau is observed in both the homoclinic and periodic solutions.

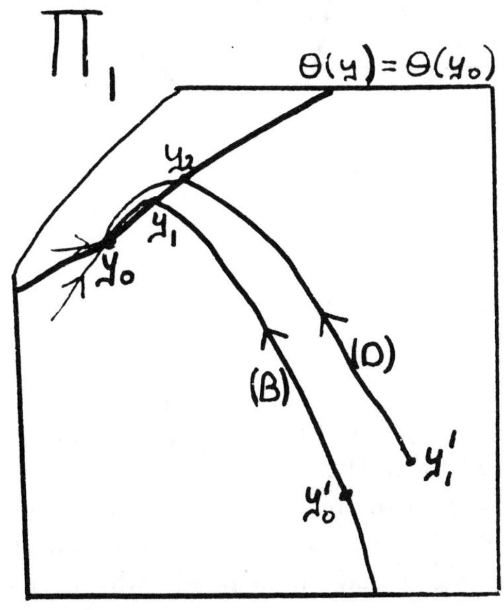

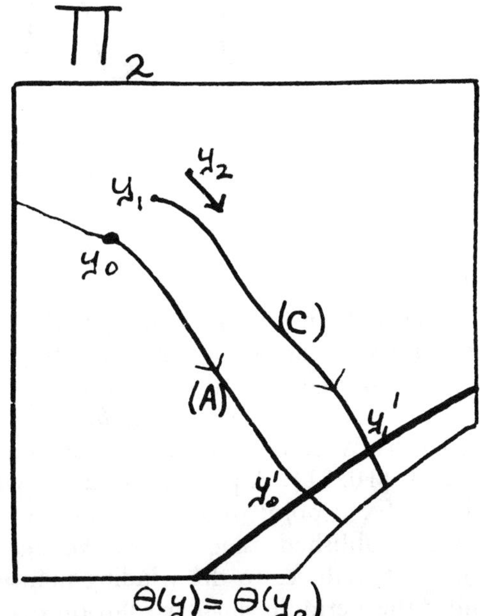

Fig.5 Phase portraits of (7.1) and (7.2) with $l = 2$

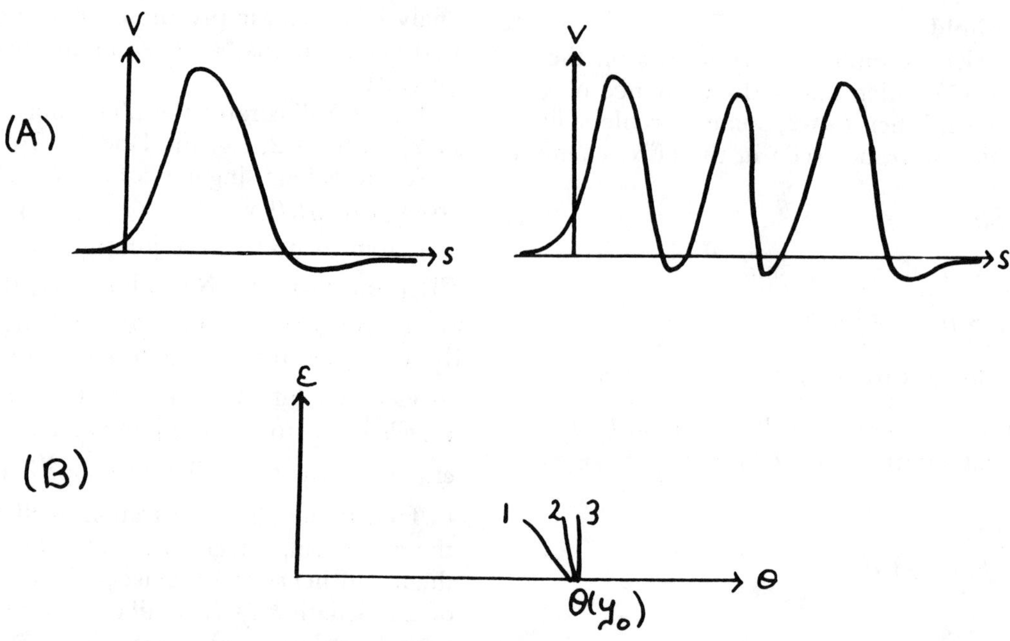

Fig.6 (A) A single pulse solution and a wave train of length three. (B) Parameter values which yield wave train solutions of length 1,2,3...

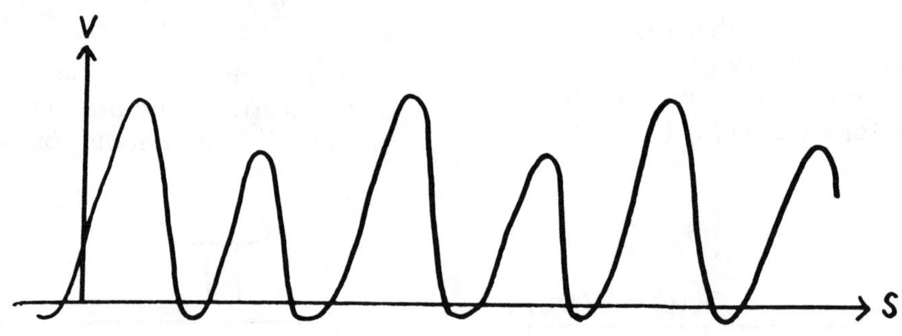

Fig.7 A periodic wave train which may occur if $l \geqslant 3$

A proof

We conclude with an outline of the proof that the existence of a singular periodic solution implies the existence of a periodic solution when $l = 1$. We introduce the notion of an *isolating block,* or *block,* as first used by Wazewski [9] and developed by Conley *et al.* [10, 11]. A periodic solution will correspond to a fixed point of a certain map whose existence is established using *Leray-Schauder degree* which, if $l = 1$, reduces to the winding number [12].

Consider the periodic singular solution depicted in Fig.3(B). We construct two blocks B_1 and B_2 about the left and right legs. For $\theta = \theta(y_p)$ and ϵ small, a periodic solution runs through these blocks and approaches the singular solution as $\epsilon \to 0$.

Let B_1 be diffeomorphic to $[0,1]^3$, as in Fig.8. B_1 may be constructed such that points on the front, bottom, and back of B_1 leave the block in forward time. These points form the exit set, b_1^-.

Points on the rest of ∂B_1 leave in backward time, and form the entrance set, b_1^+. The crucial property of a block is that the map, ϕ_1^-, which sends a point in B to the first point on its forward trajectory in b_1^- is continuous *where defined* [11]. (Observe Fig.9, in which B is a block and A is not.)

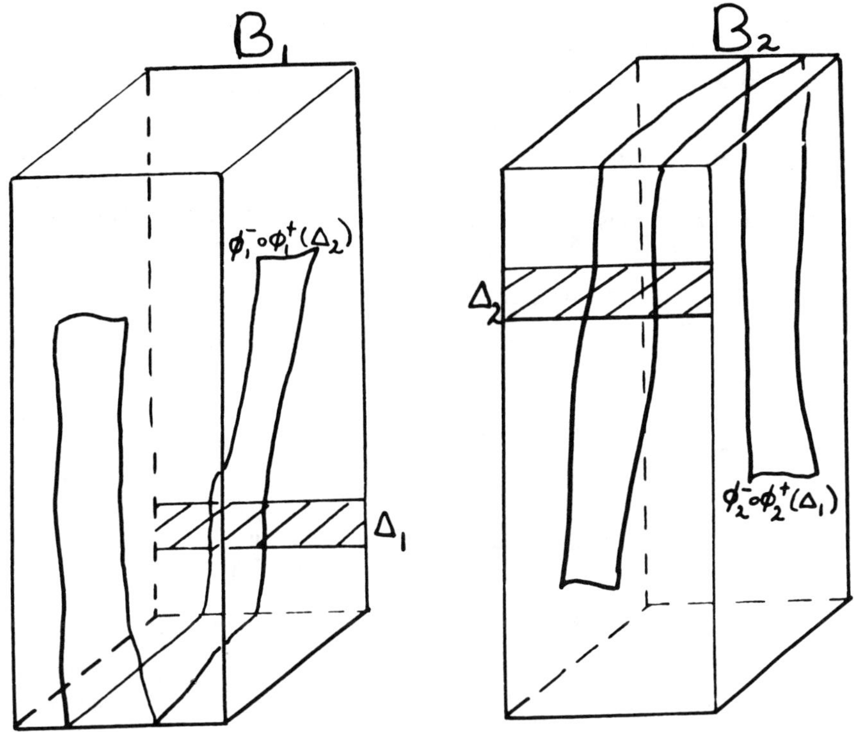

Fig.8 Blocks B_1, B_2 about the left and right legs of the periodic singular solution

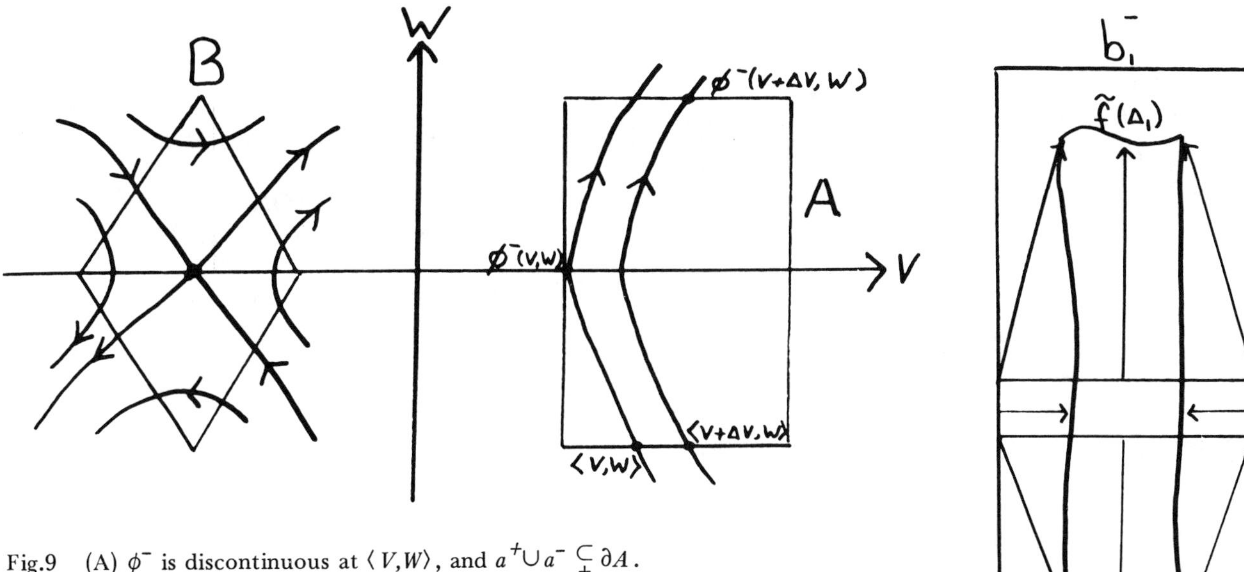

Fig.9 (A) ϕ^- is discontinuous at $\langle V,W \rangle$, and $a^+ \cup a^- \subsetneq \partial A$.
(B) ϕ^- is continuous but defined only off the stable manifold of the rest point

Fig.10 Projection of b_2^- onto $\mathbb{R}^2$

B_1 may be constructed so that no solution stays in it, so ϕ_1^- is defined in all of B_1. ϕ_1^+, similarly, is the map sending a point (in $\mathbb{R}^3$) to the first point if any in b_1^+. ϕ_1^+ is continuous at u if $\phi_1^+(u) \notin b_1^-$. B_2 is constructed symmetrically, with b_2^- the front, top, and back.

The fact that, when $\epsilon = 0$, there exists a solution from $\langle V_1(y_p), 0, y_p\rangle$ to $\langle V_2(y_p), 0, y_p\rangle$ allows us to construct a set Δ_1 spanning b_1^- with the property that ϕ_2^+ maps Δ_1 into $b_2^+ - b_2^-$. Moreover, the *top* of Δ_1 is mapped by $\phi_1^+ \circ \phi_2^-$ into the *back* of b_2^-; and the *bottom* of Δ_1 is mapped into the *front* of b_2^- (and well below $\{y = y_0'\}$). The continuity of $\phi_2^- \circ \phi_2^+$ then implies that $\phi_2^- \circ \phi_2^+(\Delta_1)$ forms a 'ribbon' running over the top of b_1^-; in particular it runs across Δ_2. Again, Δ_2 is defined symmetrically.

We now show that the map $f \equiv \phi_1^- \circ \phi_1^+ \circ \phi_2^- \circ \phi_2^+$ has a fixed point $\bar{u} \in \Delta_1$, i.e. $f(\bar{u}) = \bar{u}$. $\bar{u}$ is then contained in a periodic solution of the system (6)! Let π be a continuous map from b_2^- onto Δ_2 such that π is the identity on Δ_2, and points of $b_2^- - \Delta_2$ are mapped to $\partial\Delta_2$. Then $\tilde{f} \equiv \phi_1^- \circ \phi_1^+ \circ \pi \circ \phi_2^- \circ \phi_2^+$ maps Δ_1 into b_1^-. Project b_1^- onto the plane, as in Fig.10.

Associated with each $u \in \Delta_1$ is a vector $(\tilde{f}(u) - u)$. This vector field is nonzero on $\partial\Delta_1$ and has winding number ± 1. Thus the vector field has a zero in Δ_1, that is $\tilde{f}(\bar{u}) = \bar{u}$ for some $\bar{u} \in \Delta_1$. Since $\phi_1^- \circ \phi_1^+$ maps the top and bottom of Δ_2 into $b_1^- - \Delta_1$, $\bar{u}$ must have been mapped by $\phi_2^- \circ \phi_2^+$ into Δ_2. Thus $f(\bar{u}) = \tilde{f}(\bar{u}) = \bar{u}$.

References

1. ZEEMAN, E.C., 'Differential equations for the heartbeat and nerve impulse', *Towards a theoretical biology 4* (edit. by C.H. Waddington), Edinburgh Univ. Press, pp.8-67 (1972).
2. CARPENTER, G., 'A geometric approach to singular perturbation problems with applications to nerve impulse equations', *J. Diff. Eq.* **23**, pp.335-367 (1977).
3. CARPENTER, G., 'Periodic solutions of nerve impulse equations', *J. Math. Anal. Appl.* **58**, pp.152-173 (1977).
4. HODGKIN, A.L. and HUXLEY, A.F., 'A quantitative description of membrane current and its application to conduction and excitation in nerve', *J. Physiol.* **117**, pp.500-544 (1952).
5. STEVENS, C., *Neurophysiology: a primer*, Wiley, New York (1966).
6. FITZHUGH, R., 'Impulses and physiological states in theoretical models of nerve membrane', *Biophys. J.* **1**, pp.445-466 (1961).
7. NAGUMO, J., ARIMOTO, S. and YOSHIZAWA, S., 'An active pulse transmission line simulating nerve axon', *Proc. IRE* **50**, pp.2061-2070 (1964).
8. TASAKI, I. and HAGIWARA, S., 'Demonstration of two stable potential states in the giant squid axon under tetraethylammonium chloride', *J. Gen. Physiol.* **40**, pp.859-885 (1957).
9. HARTMAN, P., *Ordinary differential equations,* Wiley, New York (1964).
10. CONLEY, C., 'On traveling wave solutions of non-linear diffusion equations', MRC Report #1492 (1975).
11. CONLEY, C. and EASTON, R., 'Isolated invariant sets and isolating blocks', *TAMS* **158**, pp.35-61 (1971).
12. SMART, D., *Fixed point theorems*, Cambridge Univ. Press, Cambridge, U.K. (1974).

Does noise improve learning velocity in a biological net?

E. BIONDI, G.F. DACQUINO and A.E. MÜLLER
Centro per lo studio della Teoria dei Sistemi C.N.R., Politecnico di Milano, Italy

1. Introduction

The present work deals with the problem of clarifying the role of noise in learning a message in the nervous network of a biological system.

As it is well known, the transmission of nervous impulses, which constitute the message, takes place in biological networks by utilizing a relevant number of fibers working in parallel.

Considering the fact that all physiological systems are intrinsically affected by noise, the existence of such a great number of parallel fibers can be related to the necessity for the system to reduce the global effect of noise on the transmitted message, and then to improve the decoding capacity of the central nervous system.

In this work, starting from the hypothesis that the system really makes use of this possibility, the effect of the change in statistical parameters of noise on message learning velocity by the biological system is discussed.

2. Some considerations on the anatomy of sensory system

Sensory systems are nervous networks characterized by a great number of fibers working in parallel (especially in afferent pathways).

The need for such parallel operation among fibers can be explained in different ways. The more simple consideration may be that which leads to the opportunity for the system to reduce the noise level on the transmitted message. In fact it can be demonstrated that the variance of the noise imposed on a constant signal entering the decoder is inversely related to the number n of fibers carrying the signal.

With regard to the acoustic system, for example, by supposing that the codification of nervous impulses takes place at the level of hair cells, the basilar membrane (see Fig. 1) can be subdivided into a number of sections such that the behavior of hair cells referring to one of those sections can be considered almost the same; then the fibers originating from those sections are equally excited (see Fig. 2). Those fibers can be thought of as working in parallel and for each section its number is about one hundred. In such a way for this system the reduction of the influence of noise by reducing its variance is very remarkable.

Another hypothesis on the role of the parallel fiber is the one, which is well known in the medical field, referring to the redundancy phenomenon.

Clearly the existence of parallel elements improves

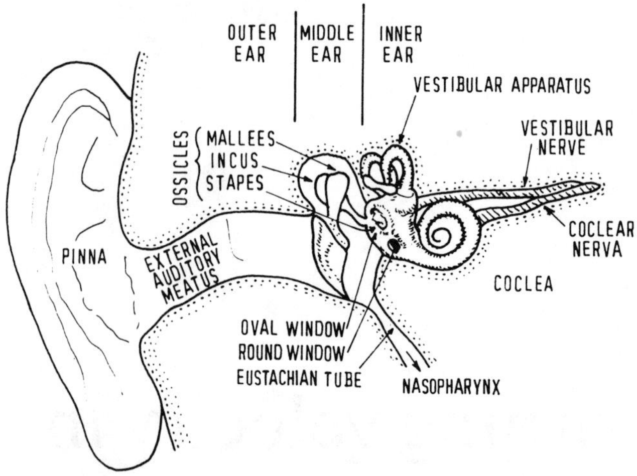

Fig.1 Basilar membrane

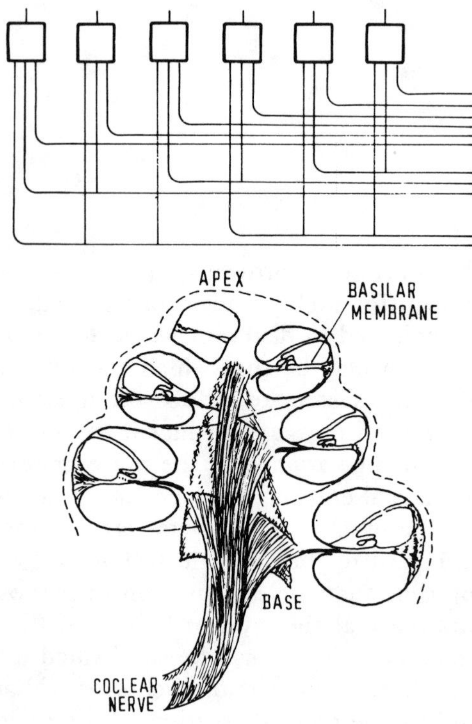

Fig.2 Fibers starting from basilar membrane section

the reliability of the system transmission, but owing to the fact that the fibers are topologically very close to one another this reliability is practically drastically reduced in case of damage. Evidently the hypothesis of reducdancy is not really probable.

On accepting the first hypothesis which has been formulated, it is interesting to note another possibility of working for the system related to this topology. By properly changing the number of fibers actually working in parallel the influence of noise on the signal can in consequence be varied.

Anatomic evidence confirms the existence of nuclei which can perform the indicated operation. Considering for example the acoustic system, such nuclei could be the coclear and olivar nuclei (see Fig. 3).

Such a variation of the influence of noise could be properly utilized by the sensory system during the learning phase in order to reduce the learning period as it will be demonstrated in the sequel.

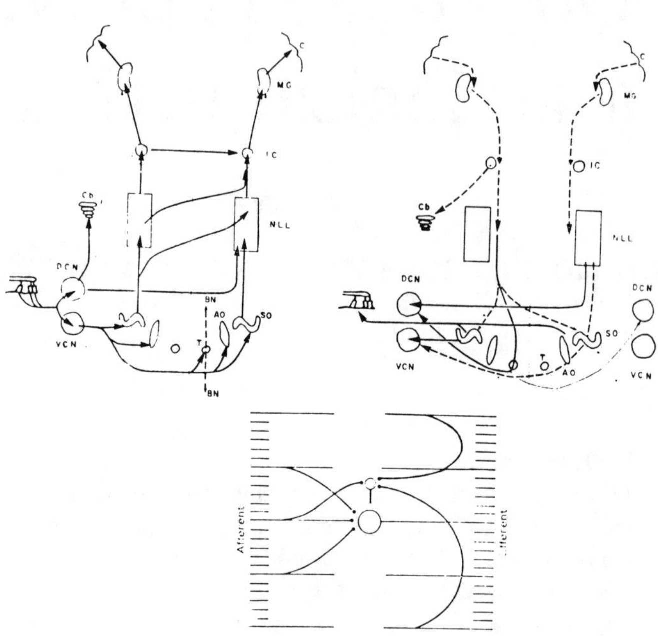

Fig.3 Acoustic system information flow-graph

3. Possible influence of noise on learning velocity

Supposing that the system must discriminate between signals which can be presented at discrete levels, given n signals the problem consists in identifying the incoming signal as coincident with one of the m possible discrete levels in the presence of noise due to the transmission system.

The hypothesis that the input signal to the system is equiprobable among the class of the m assigned levels is assumed in the sequel.

The noise introduced by the transmission system will be assumed Gaussian with zero mean and variance σ^2.

The presence of noise can be schematized as in Fig. 4 where the noise introduced by the tranmission

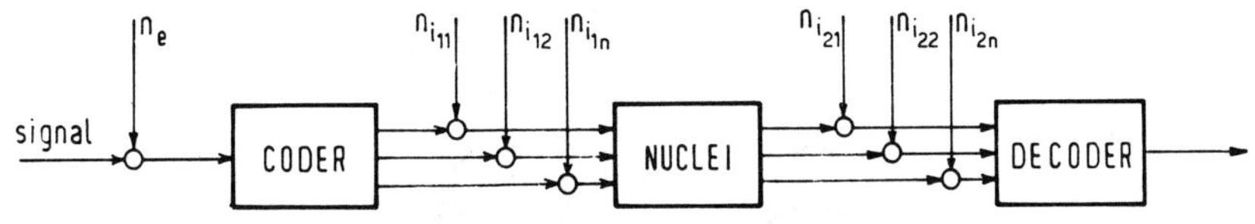

Fig.4 Block diagram of the coding-decoding system

line and the one imposed to the input signal has been considered.

If it is assumed that the system has the possibility to utilize separately the information which comes from the fibers during the learning phase and as a whole (i.e. considering the set of fibers as a single transmission channel) during the recognition phase, it can be assumed, for our purpose, as if the system was affected by a noise with variance σ^2 varying between the extreme limits according to the relation:

$$\frac{\sigma_0^2}{n} \leqslant \sigma^2 \leqslant \sigma_0^2$$

where σ_0^2 represents the variance of the noise related to every fiber and n is the number of parallel fibers.

In the hypothesis that the learning in recognition may consist in the tendency to position optimally separation surfaces in the space of the possible signals as it is schematically shown in Fig. 5, it is possible to demonstrate that the presence of a noise, with variance which is greater than the minimum one obtainable by the biological system, can facilitate the proper setting of separation surfaces.

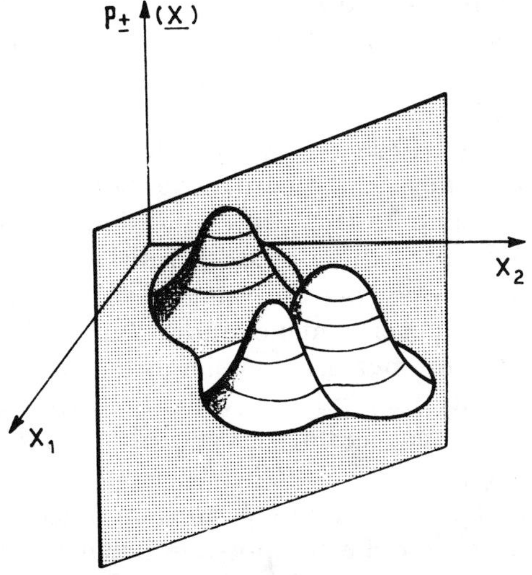

Fig.5 Visual explanation of the recognizing process

In other words by considering $\sigma(t)$ as the control variable of the dynamic process of learning, the condition of minimum errors in minimum time is reached with a control law

$$\sigma(t) = \frac{\sigma_0}{\sqrt{n(t)}}$$

The control law can be determined as a function of a performance index which can take into account the minimum time for learning and other factors as the properly weighted number of faulty recognitions during the transient.

4. Example referring to a particular recognition condition

In the sequel the case of amplitude discrimination between two discrete levels E_1 and E_2, will be considered as an example. Owing to the noise presence in the system, the problem can be schematized as illustrated in Fig. 6.

The levels E_1 and E_2 in the figure have been fixed symmetrically with respect to the origin of axes and their values have been arbitrarily chosen because such an assumption is not relevant for the sequel of discussion. The noise added to the signal has been assumed with Gaussian distribution by supposing applicable the central limit theorem. It has been supposed also that the recognition can be done by fixing a separation line between the possible levels of the signal and by assigning the signal level E_1 or E_2 whenever the considered event lies to the right or to the left of the separation line.

The above hypotheses assumed, it is easy to verify that the optimum position for the separation line (i.e. the one which implies the number of errors to be a minimum) is the one coincident with the origin of the axes.

The hypothesis is made that the learning process is a dynamic process governed by the equation

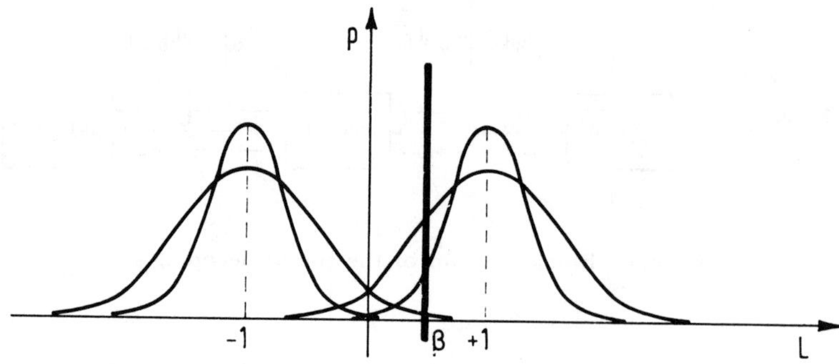

Fig.6 Example of amplitude recognition

$$\dot{z}(t) = -K\frac{\partial}{\partial z}(P_R(z,\sigma))$$

where $z(t)$ represents the abscissa of the separation line defined above, $P_R(z,\sigma)$ is the probability of being correct in the answer and K a constant coefficient.

$$P_R(z,\sigma) = \frac{1}{\sqrt{2\pi\sigma^2}}\left[\int_{-\infty}^{z} e^{-(z+1)^2/2\sigma^2}\,dz - \int_{-\infty}^{z} e^{-(z-1)^2/2\sigma^2}\,dz\right]$$

Referring to the considerations which have been explained in the preceding paragraph, the variance σ^2 can assume values depending on the number of fibers which are connected together in parallel.

If we assumed the square root of the variance as the control variable of the dynamic process of learning, the function $\sigma(t)$ is obtained which is an optimum one with respect to a defined performance index.

In the present work two performance indices have been considered, namely:

a. $I_1 = T$ where $T = \int_0^{T_{FIN}} dt$

b. $I_2 = T+w$ where $T = \int_0^{T_{FIN}} dt$,

$$w = K\int_0^{T_{FIN}} (1-P_R(z,\sigma))\,dt$$

The interval $\overline{O\ T_{FIN}}$ represents the duration of the learning process.

The hypothesis under *(a)* assumes as performance index the only condition of minimum learning time; the hypothesis *(b)* takes also into account a weighting factor related to the faulty recognitions during the whole learning phase. As it is obvious, this second hypothesis shows itself to be more realistic in the case of many neuro-sensory systems.

Figure 7 shows the diagram of σ as a function of time during the learning phase as above defined, starting from two different initial conditions.

It is important to note that with either performance index (I_1 or I_2) the optimum value for the control variable σ is greater than σ_{min} during the learning phase.

As it is clearly shown in Fig.7, by utilizing the performance index I_1, the control variable σ does not reach the value $\sigma = \sigma_{min}$ in equilibrium condition (i.e. when the learning phase is finished) because in equilibrium condition the value of σ is indifferent with regard to performance index I_1. On the contrary by utilizing the performance index I_2 the value of σ during the learning phase is reduced with respect to the previous case and at the end of learning σ tends toward σ_{min}.

The performance index I_2 appears indeed more adequate to describe the possible behavior of the system during the learning phase and besides it utilizes the possibility for the neuro-sensory system fibers to obtain series or parallel connections.

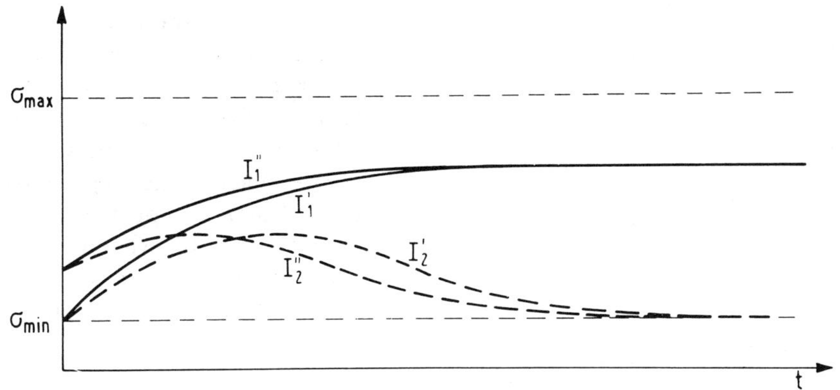

Fig.7 Control variable σ as a function of time in learning process

5. Considerations on the relation between the variance of the noise added on a signal and the number of fibers involved in signal transmission

The considerations in the sequel refer particularly to the acoustic system, but are also applicable to other neurosensory systems.

The action potentials for the single nervous fibers can be thought of as the events of a stochastic process with Poisson distribution, whose parameter λ is a function of the input signal (for example, in the acoustic system the level of sound stimulus).

The amplitude of the input signal E is supposed to be related in a deterministic way with the parameter λ of the Poisson distribution. The hypothesis assumes that the decoding device existing in the neurosensory system utilizes an optimum estimator $\hat{\lambda}$ of λ (and then an optimum estimator of E).

The following must be remembered:

Let $x_1,\ldots,x_n$ be a random sample of size n from a random phenomenon specified by the probability density function $f(x|\theta)$, $\theta \in \Omega$ and $T(x_1,\ldots,x_n)$ be an unbiased estimator of the parametric function $\tau(\theta)$.

By considering the Likelihood function of the sample

$$L(x_1,\ldots,x_n|\theta) = f(x_1|\theta),\ f(x_2|\theta),\ldots,f(x_n|\theta)$$

The Cramer-Rao inequality is written

$$\sigma^2 = (\mathrm{var}\ t) \geqslant -\{\tau'(\theta)\}^2/E\left(\frac{\partial^2 \ln L}{\partial\theta^2}\right)$$

The necessary and sufficient condition that the inequality becomes an equality is that $\{t-\tau(\theta)\}$ is proportional to $\partial \ln L/\partial\theta$ for all sets of observations. This condition may be written

$$\frac{\partial \ln L}{\partial\theta} = A(\theta)\{t-\tau(\theta)\}$$

and thus

$$\sigma^2 = \mathrm{var}\, t = |\tau'(\theta)/A(\theta)|$$

In the hypothesis that the parameter E to be estimated is proportional to the parameter λ of the Poisson distribution, i.e. $\lambda = KE$, it can be obtained

$$\sigma^2 = \frac{E}{nK}$$

and then

$$\frac{\Delta E}{E} = \frac{1}{\sqrt{nk}} = \frac{1}{\sqrt{E}} \qquad (1)$$

where n is the number of fibers involved.

Otherwise the hypothesis that a logarithmic relation holds between the parameters λ and E, i.e. $\lambda = k \ln E/E_0$, it can be obtained

$$\sigma^2 = \frac{E^2}{E_0}\frac{1}{nk}$$

$$\frac{\Delta E}{E} = \frac{1}{\sqrt{nk}\sqrt{E_0}} = \mathrm{const} \qquad (2)$$

where (n) is the number of fibers involved.

Relation (2) justifies the linear part of the Weber law, and relation (1) justifies some psychophysical tests for values of E at threshold level (see Fig. 8).

Clearly justified is the assumption that the system utilizes the parallel pathways for discriminating the signal and also to reduce the noise on the transmission line.

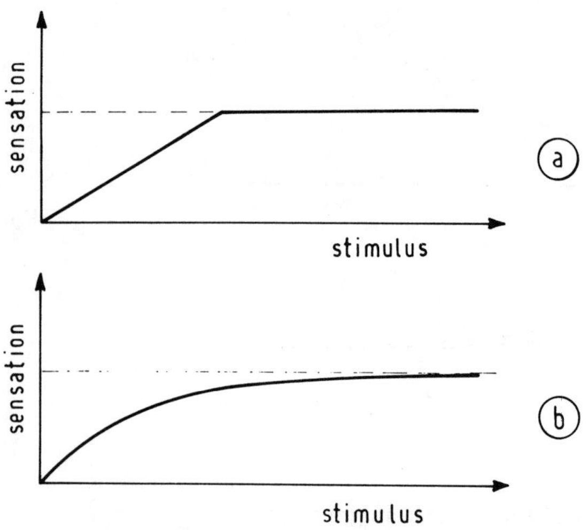

ⓐ theoretical derivation

ⓑ practical findings

Fig.8 Weber law

6. Concluding remarks

From what we have stated above sufficient support is given to the hypothesis which has been made on the utilization by the neurosensory systems of different connections on the same set of fibers. The conclusions reached can be greatly remarkable in the study of acoustic protheses because it is evident that from the knowledge of the way the system operates on the received signals, useful suggestions can be derived to improve their performance.

A dynamic theory of visual perception*

E. HARTH
Syracuse University, Physics Department,
New York,
USA.

That human intelligence could ever attain an understanding of the brain, its own instrument of reasoning, was long considered an impossibility. The stream of information that emanates from our senses is coded and recoded through a network of immense complexity. The physiologist can tap a neuron here and there, bringing its message back to the periphery, our senses of sight or hearing. But the code is inappropriate for that stage, and we are puzzled. What we are attempting may be likened to sampling a spoon of liver bile, and, from its taste, trying to surmise what it does in our intestines.

This brings up the first point I wish to emphasize, namely, that to understand a message we have to know something about the normal recipient of that message. Much lengthy discourse on the relative merits of globalism and atomism, determinism and randomness could be avoided if we kept in mind that the nature of the system receiving the information imposes certain restrictions on the structure of the message and that neural systems exhibit an enormous range of properties in that respect.

The problem I wish to discuss here is a very general one: what are some of the characteristics of sensory perception, and what neural processes can be envisioned as their basis? I shall discuss some of the conceptual pitfalls of the problem and then propose a mechanism by which some of these may be avoided. I shall begin by enumerating certain arbitrary but reasonable assumptions.

Axiomatics of brain function

The following statements will be taken as *axiomatic* in the sense of being propositions assumed without proof. They are meant to be tentative, subject to revision at any time.

First axiom: Neurons carry no labels. The triggering of a neuron can convey meaning only by virtue of the fact that its action potential causes at its synapses the release of transmitter molecules which can affect the firing state of other neurons, or the release of neurohumors having a broader effect on other parts of the nervous system. Thus, *firing* is to be viewed as a link in a chain, never as the end product of information processing.

Second axiom: This may be called the axiom of local interaction. It states that the information arriving at a neuron and expressed in its own firing record can only reflect the information coming through the synapses directly impinging on it. A neuron cannot *see* beyond the most immediate synaptic link. It knows only its immediate milieu. If field effects, e.g. neurohumoral biases, exist, these too can be included in the local conditions.

Feature extraction and feature generation

In all brains we can distinguish between afferent and efferent pathways. On the afferent side relay chains lead from the various sensory systems, where contact is made with the physical world, toward higher and higher cognitive centers. The efferent paths *descend* and converge on the body's muscles and glands which they control.

A functional unit, observed in many sensory systems, is the *feature extractor.* It may be defined, in a general way, as a device which takes global information, i.e. inputs arriving from a large numger of receptors, and signals the presence of specific features. This information is conveyed over a relatively small number of output lines. In the mammalian visual system single cortical neurons with directionally sensitive receptive fields (Hubel and Wiesel [22]) have been interpreted to be feature extractors.

Such devices have interesting counterparts on the efferent side of the brain where global action programs are frequently released by simple commands or triggers. The existence of such *feature generators* is most readily demonstrated in invertebrate systems. In the mollusc *Tritonia* a brief burst from a set of *trigger neurons* will initiate the *swimming escape* reaction which involves activation and precise phasing of different muscle groups, lasting for about a minute or more. This program, once triggered, is played out without further stimulation and without proprioceptive feedback (Willows, Dorsett and Hoyle, [1,2]; Hoyle and Willows, [3]). It is presumed that on the afferent side there exist circuits which can distinguish among inputs from skin receptors, to select those which present a threat, and call for swimming escape. Touch by a starfish, the putative predator of *Tritonia,* is one such stimulus.

The chain of neural control can thus be presented, as in Fig. 1, by a feature extractor (FE) which converts global inputs (the presence of a predator) into a localized command that, in turn, triggers a feature generator (FG), which then releases a stored action program. The dynamics of *Tritonia*'s feature generator has recently been discussed by Harth, Lewis and Csermely [4].

It would be tempting to view the entire nervous system as a network of such coupled feature extractors and feature generators, and to assume that the functioning of the entire system is adequately described by saying that it elicits the motor and endocrine responses appropriate to any set of circumstances perceived by the sensory system. This picture may be correct for invertebrate or lower vertebrate brains, but is probably a very inadequate description of the brains of most mammals. When applied to humans it leaves a large class of phenomena unaccounted for which are not directly concerned with motor or glandular control, and may at times have relatively little to do with sensory input. *Thought* was described by Descartes [5] as "an attribute that belongs to me" and as "inseparable from my nature". It is not always clear what we mean by thought, but it must include an important group of activities that involve our ability and predilection to *simulate* sensory inputs, and to sample possible outcomes of alternative action programs, without necessarily carrying out any of them. The interplay of actions and sensations resulting from actions can be synthesized in our brain spontaneously or at will. "My mind is a vagabond" said Descartes and in Monod's view it is the simulative function which "characterizes the unique properties of man's brain" (Monod, [6]).

The complication introduced by these requirements is enormous. If the nature of central coding of real sensory events is obscure, we know even less about the mechanisms responsible for simulative functions. An economy of assumptions would suggest that the simulation of a sensory experience should be accomplished by producing central activity *resembling* that caused by the real event, the degree of resemblance determining the degree of realism of the simulation.

In this picture, then, the activity at the highest centers determines our sensations, while the more

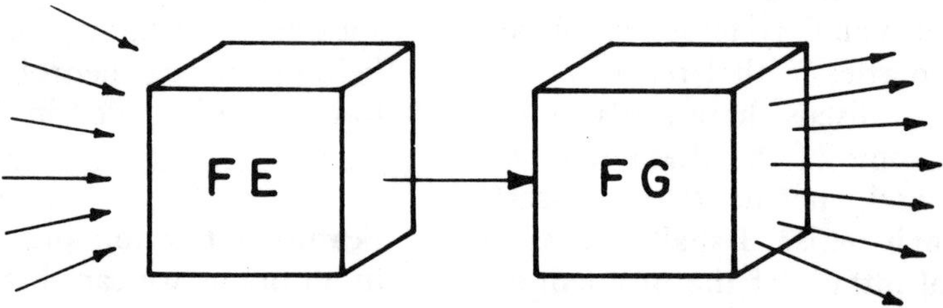

Fig.1 Schematic showing a feature extractor (FE) coupled to a feature generator (FG)

peripheral systems serve mostly to convey information up, and commands down the pipelines of afferent and efferent structures respectively. The central activity that results from the sensory input, after much filtering, cataloguing, and referring to stored information, takes the form of spatially and temporally labeled events involving either single neurons or populations of scattered cells, depending on one's school of thought. The conscious reality of the simulated sensory event, or the simulated action, calls either for a *homunculus* who scans the labeled neural activities in the way a telephone operator scans a switchboard, or suggests that the central neurons themselves are the final sensors, or *sentient atoms.*

The first explanation violates the first of our axioms, since it would require the neuron to convey information by other than established neuronal means. The second ascribes to the individual neuron a knowledge reaching beyond its immediate synaptic environment, thus violating the second axiom.

We are left, then, with this dilemma: if it does not cause an efferent response, a sensory signal moves up through a hierarchy of relay stations, its function obscure, and its final destination unknown. The message appears to exit somewhere, unheard, through an open line. Simulated events begin perhaps more centrally, but must undergo the same open-ended demise. The dilemma is, of course, predicated on the assumption of a unidirectional flow of sensory signals.

Corticofugal control

There is ample empirical evidence that information flow on the afferent side is not always unidirectional. True, the ascending pathways can be followed along a series of hierarchical structures, and a little way into the most central part of the mammalian brain, the cerebral neocortex. There are, however, loops and side branches which complicate the picture. A prominent example of neural feedback has been observed between the cortex and the lateral geniculate bodies (LGB) (Iwama, Sakakura and Kasamatsu [7]; Suzuki and Kato [8]; Angel, Magni and Strata [9]; Hull [10]; Kalil and Chase [11]. Pribram [12] proposes that a "corticofugal efferent control system emanates from the temporal cortex downward to subcortical structures, there to influence by a parallel processing mechanism the visual input". He cites as evidence that stimulation of the suprasylvan gyrus in cats has produced changes in visual receptive fields "as peripheral as the optic nerve", with the most pronounced changes occurring at the LGB. In a series of experiments with monkeys Rothblat and Pribram [13] found that evoked potentials in the temporal lobe and striate cortex show a dependence not just on the stimulus that caused them, but also on the context in which the stimulus occurred. The authors conclude that "attention is truly selective of stimulus dimensions, not just levels of significance". Pribram [12] speaks of a "programmed filter, or program tape" that is able to modify the visual input in response to cortical activity.

Perceptual hysteresis

An interesting special case of preception occurs when a stimulus pattern allows more than one interpretation, as in the Necker cube (Fig. 2), or the figure-ground reversal pattern shown in Fig. 3. In both cases perception switches spontaneously and abruptly back and forth between two stationary states. The perceptual states, moreover, are mutually exclusive. The effect was discussed by Szentagothai and Arib [14] and likened to order-disorder transitions in physics. The random dot stereograms of Julesz [15] are another example of the sudden perceptual change upon recognition of form, and the locking-in on the perceived pattern. The phenomenon of perceptual hysteresis was most clearly shown by Fender and Julesz [16] by observing binocular fusion of single dots of light projected with varying disparity into the two eyes. All of these examples suggest a positive feedback between central and peripheral units.

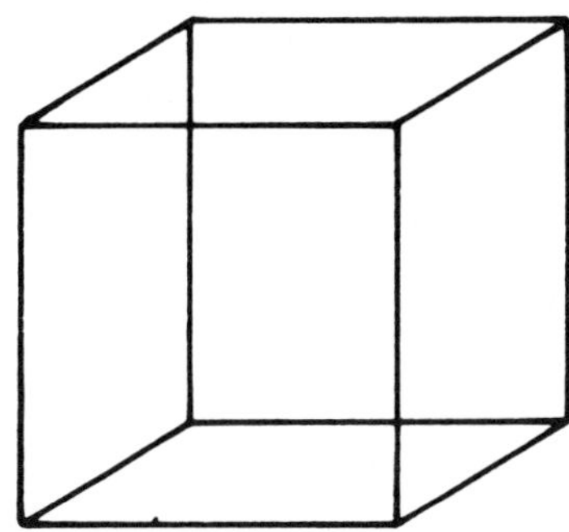

Fig.2 Necker Cube. The orientation of the cube relative to the observer will spontaneously alternate between two possible states

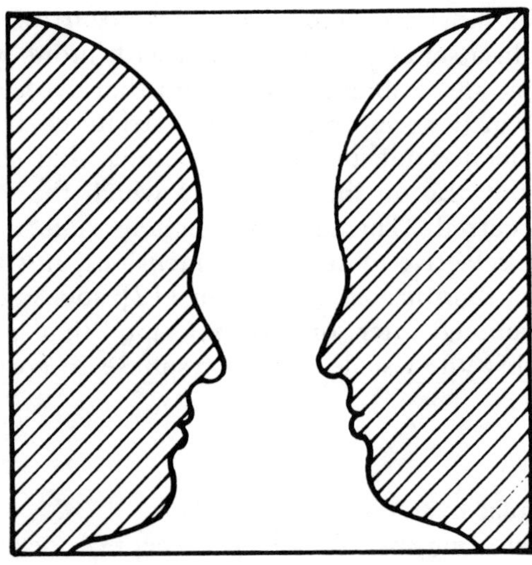

Fig.3 Figure-ground reversal. As in Fig.2 two interpretations are possible.

Neuronal requirements

If we accept all of the foregoing empirical findings, we are led to a visual system that must have the following functional characteristics: if a feature extractor at a given level of sensory processing shows some response, then the neuronal firing pattern at a lower level is to be modified by centrifugal control in such a way as to produce a stronger response in the feature extractor.

The structural requirements of such a device appear at first sight to be formidable. Referring back to the schematic of a feature extractor in Fig. 1, we could, in principle, accomplish the task by coupling the output of a feature extractor whose global output, instead of acting on the motor system, is designed to modify the input to the feature extractor (Fig.4). The two devices would have to be matched, that is to say, the feature generator must produce the kind of input pattern which the feature extractor is designed to detect. It would contain the program tape that Pribram wants to be switched back into the visual system. Thus, for an organism familiar with a large vocabulary of features, many such matched pairs would be required.

The Alopex principle

I digress briefly to recall a principle that was first discussed by Harth and Tzanakou [21] in connection with a proposed new method for determining visual receptive fields. The device described in that paper was called *Alopex.* It uses a computer algorithm to generate a sequence of visual patterns on a

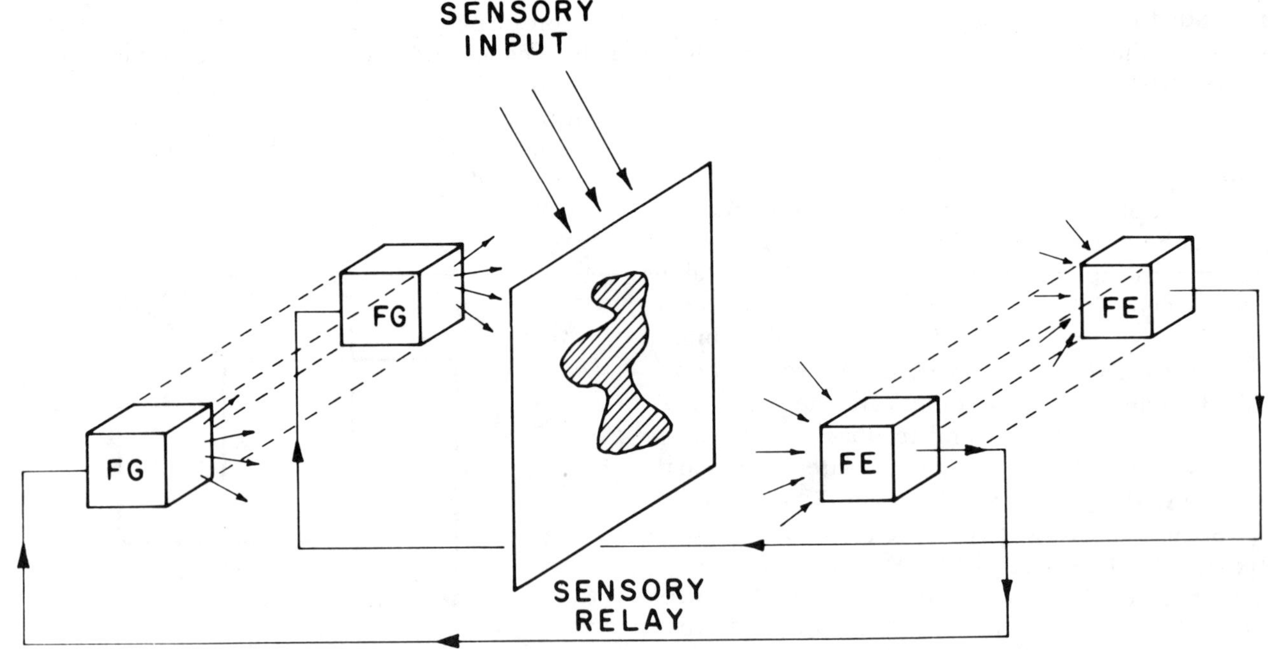

Fig.4 Schematic showing a sensory relay nucleus receiving sensory afferents. The global pattern appearing at the relay nucleus is scanned by a set of feature extractors whose outputs trigger a set of matched feature generators. These, in turn, act *globally* on the pattern appearing at the relay nucleus.

CRT or other display device. The patterns are presented to an animal while a microelectrode records the response of a single cell in the visual pathway. The algorithm which determines the patterns uses the two preceding patterns and responses. It is designed to arrive by an iterative procedure at a pattern resembling the receptive field of the cell. The first two patterns of the sequence are random and there is some randomness in all of the subsequent patterns. A brief description of the algorithm follows.

Consider a visual field divided into small squares forming a square array of $N \times N$ such space elements. Let the nth stimulus pattern be given by the vector $\mathbf{I}(n)$:

$$\mathbf{I}(n) = [I_1(n),\ldots,I_{N^2}(n)] \tag{1}$$

Here $I_j(n)$ is the light intensity at the jth square element during the nth stimulus pattern. These patterns are displayed consecutively on a TV or oscilloscope screen, and presented to the animal. The single cell response to the nth stimulus, as recorded by a microelectrode, is called $R(n)$. The intensities $I_j(n)$ are made up of a *random* contribution $r_j(n)$ and a cumulative bias $b_j(n)$. The value of each $b_j(n)$ is increased or decreased at each iteration, depending on whether, in the preceding two iterations, a change in I_j was accompanied by a change in R in the same or opposite direction. This is expressed by

$$\mathbf{I}(n) = \nu(n)[\mathbf{b}(n) + \mathbf{r}(n)] \tag{2}$$

where

$$\mathbf{b}(n) = [b_1(n),\ldots,b_{N^2}(n)]$$

and

$$\mathbf{r}(n) = [r_1(n),\ldots,r_{N^2}(n)].$$

For the first two iterations all bias values are identical, thus

$$b_j(0) = b_j(1) = b_0$$

for all j. The first two stimulus patterns are thus a mosaic of random light intensities. The factor $\nu(n)$ in Eq. (2) is thus a normalization which satisfies the condition

$$\sum_{j=1}^{N^2} I_j(n) = \text{const.}$$

The vector $\mathbf{I}(n)$ thus represents a constant amount of light flux. The components of $\mathbf{r}(n)$ in Eq. (2) are random numbers whose distribution may be chosen in many ways. The cumulative biases $\mathbf{b}(n)$ are given by

$$\mathbf{b}(n) = \mathbf{b}(n\text{-}1) + c[R(n\text{-}1) - R(n\text{-}2)]\cdot[\mathbf{I}(n\text{-}1) - \mathbf{I}(n\text{-}2)] \tag{3}$$

The constant c in Eq. (3) determines the amplitudes of the bias corrections. Together with the range of random numbers $r_j(n)$ it is one of the critical parameters. Of course, many variants of these basic rules can be constructed.

Figure 5 shows the results of a computer simulation testing this algorithm. A simple linear receptive field is assumed in a 10×10 array in which the 7th column is excitatory and columns 6 and 8 are inhibitory (Fig. 5*a*). The histograms in Figs. 5,6 show the light intensities (summed over vertical columns) for iterations 1, 10, 20, 50, and 100. The pattern, which is random in iteration 1, clearly converges to the assumed receptive field.

Complex receptive fields can be simulated by non-linear superposition of simple fields. In one series of simulation experiments a number of identically oriented *line detectors* were coupled to produce a complex field of the type described by Hubel and Wiesel [17], in which a line of specified orientation but unspecified location is the trigger feature. Applying again the algorithms of Eqs. (2) and (3) caused the illumination to converge on a single line in about the same number of iterations as for simple fields, but this time the location of the line depended on the starting values of the pseudorandom number sequences $\mathbf{r}(n)$. If these numbers were truly random, the location of the pattern would be unpredictable from run to run. However, it was found that, by superimposing some additional bias in one column for a few iterations, the final pattern was shifted to that location, irrespective of the particular sequence $\mathbf{r}(n)$.

Following series of such computer simulations, the necessary electronic hardware was constructed and is now used for receptive field studies in this laboratory. Patterns are generated by a CRT display using 1024 field elements. The method may be extended also to include moving stimulus patterns.

Alopex: The thinking man's filter?

We have touched on a number of problems having to do with visual perception. The ultimate destination and final expression of sensory events, partic-

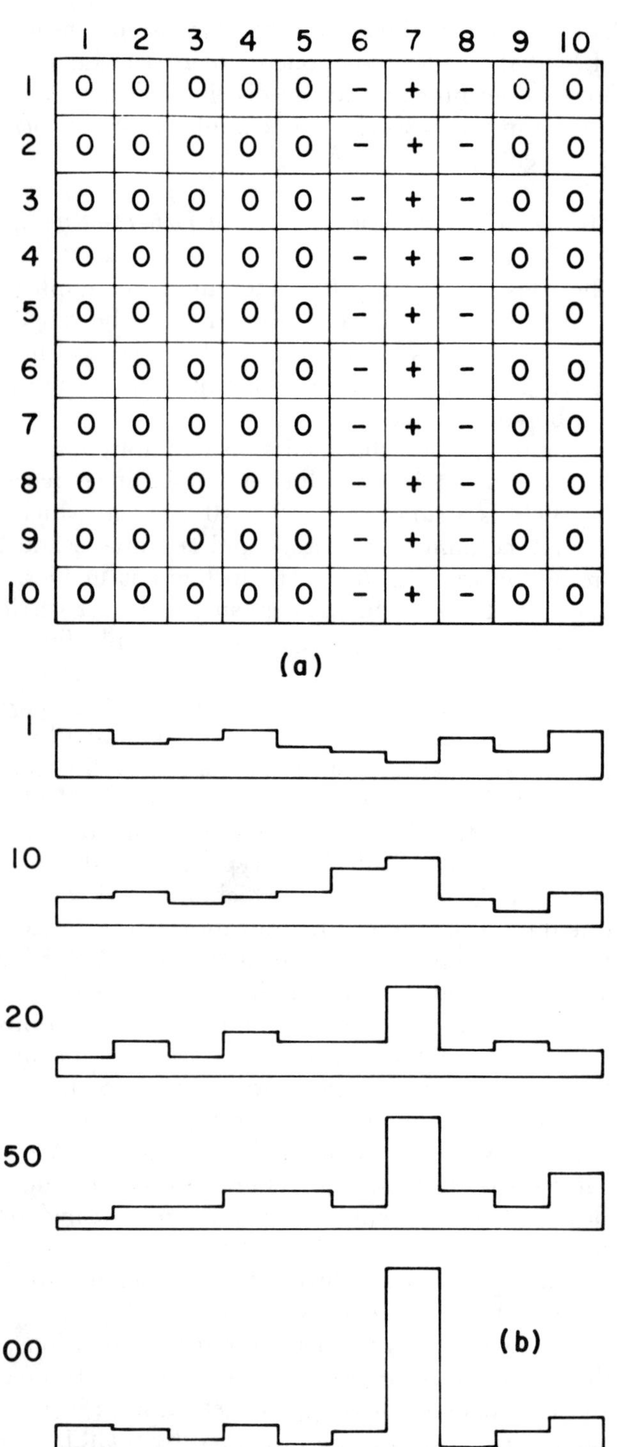

Fig.5 Performance of Alopex in computer simulation. In (a) a simple receptive field is pictured having an excitatory line (column 7) and inhibitory surround. In (b) the stimulus patterns are shown for various iterations (numbers on left). For ease of presentation the ordinates of the histograms show the column sums of stimulus intensities. The first pattern is random. Subsequent iterations show the progressive increase of stimulus intensities along column 7. (After Harth and Tzanakou [21].)

ularly of simulated sensory events, appears mysterious in view of our first axiom. The experiments of Pribram and co-workers suggest that there is not an unequivocal hierarchy in sensory processing, but that cortical control is exerted on the more peripheral parts of the system. According to Pribram's *filter* theory this control is specific to particular features of the input, rather than merely raising overall sensitivity. Finally, the evidence of hysteresis in perception strongly suggests a positive feedback mechanism. We have also pointed out that a straightforward interpretation of Pribram's results would seem to require a complex arrangement in which every central feature extractor is coupled to a matched feature generator which provides the feedback. (Fig. 4)

The Alopex principle provides an interesting alternative solution which would reduce greatly the required complexity of the system. Alopex has the properties of a generalized feature generator whose specificity is determined by the feature extractor that controls it. It follows that a schematic arrangement, similar to the one shown in Fig. 6, can account for most of the phenomena we wish to explain. Here a single Alopex unit (A) receives inputs from any number of feature extractors (FE) which show different responses to a given sensory input pattern. The one with the dominant response will take control over the Alopex unit which then enhances the features appropriate to that feature extractor. Such a feature specific positive feedback might account for the selective filtering ascribed to efferent control, as well as for the abrupt transitions observed in perception. Also, perception would emerge as a dynamic process, involving both central and peripheral sensory areas, instead of being viewed as a unidirectional processing, culminating in a single, final neuronal pronouncement.

The LGB: Not just a relay?

What are the biological requirements for carrying out these processes, and where in the brain might we begin to look for them?

Pribram [12] has singled out the LGB as that part of the visual system of cats showing most strikingly the effects of corticofugal control. It was pointed out by Szentagothai and Arbib [14] that the thalamic relays of the auditory, the visual, and the somatosensory pathways, the *medial geniculate*, the *lateral geniculate*, and the *ventroposterolateral bodies*, possess a remarkable degree of structural similarity. All three consist of two cell types:

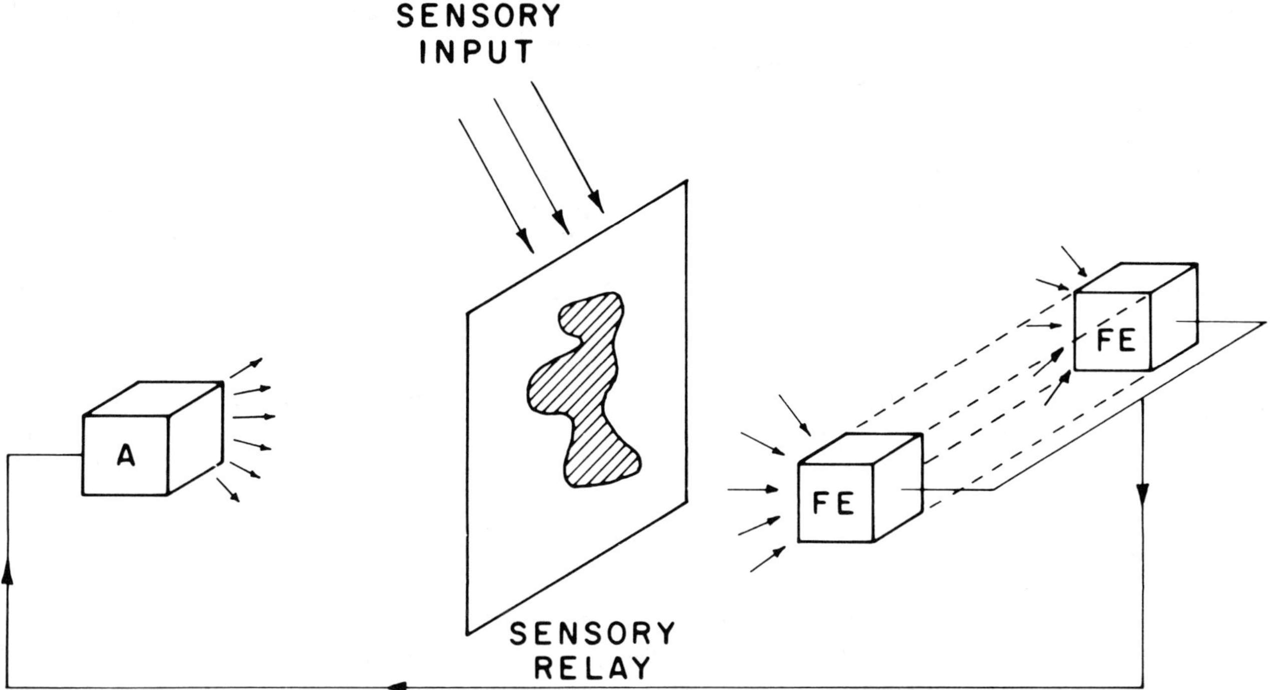

Fig.6 Schematic of a sensory relay nucleus receiving sensory afferents, and being scanned by feature extractors as in Fig.5. The outputs of the feature extractors affect a single Alopex unit (A) which provides the global positive feedback to the relay nucleus.

1) a thalamocortical *relay cell* which receives inputs from the specific sensory afferents, and transmits its output to the appropriate sensory cortex; 2) a Golgi type II interneuron. The synaptic arrangement between these cells is also unique. The sensory afferents synapse with the relay cell as well as with interneuron dendrites which, in turn, form dendro-dendritic synapses with the relay cell. These close *triadic* synapse structures are found in all three of the thalamic nuclei mentioned. In addition, both cell types receive inputs from cortical fibers.

Szentagothai and Arbib [14] express puzzlement over these remarkable features in the relay nuclei. They further conjecture that

> ".....a large number of shaping mechanisms of the incoming sensory pattern by various types of inhibitory interactions could occur in this strucural framework. Unfortunately, both neuron (relay and interneuron) types and the synaptic architecture appear to be rather stereotypic in all of the major subcortical sensory nuclei. Hence, the structure is not conducive to making deductions about specific mechanisms of more sophisticated or global feature-detection strategies."

Alopex makes possible those *strategies* with rather modest structural requirements and *stereotypic architecture.* The retinotopically arranged sheet of neurons in the LGB has many of the properties of the sensory relay pictured in Fig. 6. The activities of neurons on that sheet are determined in part by the sensory afferents, in part by the feedback from the cortex. In addition, these neurons exhibit spontaneous activity which has all the aspects of a random contribution to their firing rate. The algorithm given by Eqs. (2) and (3) requires such a random component, as well as information on the *rates of change* of activity in the sensory relay. Such rates are available in the so-called Y-subsystem of neurons found in the visual system of cats (Enroth-Cugell and Robson, [18]; Stone and Hoffman, [19]). These cells give transient responses from *on-center* and *off-center* neurons.(Brooks and Jung, [20]).

One of the limitations of the process in experiments to determine visual receptive fields, is the time required for convergence. The response R in Eq. (3) is obtained by counting action potentials over preselected time intervals, following stimulus presentation. With firing frequencies of perhaps tens per second, a minimum of several seconds is required to obtain a statistically meaningful reading of R. In practice, because of adaptation, it may be necessary to use repeated presentation of the same stimulus and intergrate responses over a number of short counting gates. Habituation effects may further require the interpolation of waiting intervals or neutral patterns. Thus, the time for a single iteration in the process defined by Eq. (2) may become a minute or more. The number of iterations required for convergence is a quantity that is difficult to predict, since no good stochastic theory of the process exists. With our algorithms, and using parameters which are certainly far from optimal, we generally observe that convergence occurred in fewer than 100 iterations,(Harth and Tzanakou, [21]). Thus, the time for convergence on a single receptive field pattern could be the order of an hour. Several methods to shorten this period are now being studied in this laboratory. In order to make plausible the operation of the same principle in the brain, we must show that this period could be shortened to a matter of seconds.

Cells of the Y-type, with their large diameter axons and correspondingly high conduction velocities, provide a rapid means for signaling rates of change of stimuli and responses. Thus, the measurements and computations, which in the electronic device consume many seconds, could, in principle, be accomplished in the brain in a matter of tens of milliseconds, and the equivalent of 100 iterations, in seconds. It remains to be seen whether the principle is operative in the brain, and by what mechanism. Meanwhile the thalamic sensory relay nuclei are probably the best location where direct empirical evidence may be sought.

Summary and conclusions

I have argued that human sense perception has functions other than linking sensory events to appropriate motor responses. If we accept the principle of hierarchical processing, this raises questions about the ultimate destination of the extracted information. On the other hand, the existence of rigid hierarchies is challenged by anatomical as well as physiological evidence for corticofugal pathways in the sensory system. The existence of positive feedback in sensory processing receives further support by the observation of hysteresis effects in perception.

I have proposed that the feature-specific feedback is accomplished on the afferent side of the brain by the operation of the Alopex principle that was discussed in connection with another problem (Harth and Tzanakou, [21]). It would be premature to speculate on how, in detail, the principle may be realized in the brain. But the necessary neural machinery appears to be available, and many design solutions exist. In the visual system the lateral geniculate body is a possible location for such mechanisms.

In the traditional view the ultimate aim of sensory processing is the turning-on of certain central feature detectors. Instead, if the present theory is correct, perception (real or simulated) becomes a dynamic process, consisting of both central and peripheral neural events with the first providing positive feedback to the second. The unidirectional flow of information toward hypothetical perception centers has thus been replaced by a process in which perception is accomplished by tuning the sensory input, so as to produce a resonance with one or more feature detectors.

References

1. WILLOWS, A.O.D., DORSETT, D.A., and HOYLE, C., 'The neuronal basis of behavior in Tritonia. I. Functional organization of the central nervous system', *J. Neurobiol.* 4, 207-237 (1973).
2. WILLOWS, A.O.D., DORSETT, D.A., and HOYLE, G., 'The neural basis of behavior in Tritonia. III. Neuronal mechanism of a fixed action pattern', *J. Neurobiol.* 4, 255-285 (1973).
3. HOYLE, G. and WILLOWS, A.O.B., 'The neuronal basis of behavior in *Tritonia*. II. Relationship of muscular contraction to nerve impulse pattern', *J. Neurobiol.* 4, 239-254 (1973).
4. HARTH, E., LEWIS, N.S., and CSERMELY, T.J., 'The escape of *Tritonia:* dynamics of a neuromuscular control mechanism', *J. Theor. Biol.* **55**. 201-228 (1975).
5. DESCARTES, R., *Mediationes de Prima Philosophia,* C. Adam and P. Tannery, Paris (1641). English translation: Meditations on First Philosophy, (trans. by L.J. Lafleur), Bobbs-Merrill, Indianapolis-New York (1951).
6. MONOD, J., *Chance and Necessity*, p.154, Alfred A. Knopf (1971).
7. IWAMA, K., SAKAKURA, H., and KASAMATSU, T., 'Presynaptic inhibition in the lateral geniculate body induced by stimulation of the cerebral cortex', *Jap. J. Physiol.* **15**, 310-322 (1965).
8. SUZUKI, H. and KATO, E., 'Cortically induced presynaptic inhibition in the cat's lateral geniculate body', *Tohoku J. Exptl. Med.* **86**, 277-289 (1965).
9. ANGEL, A., MAGNI, F., and STRATA, P., 'The excitability of optic nerve terminals in the lateral

geniculate nucleus after stimulation of visual cortex', *Arch. Ital. Biol.* **105**, 104-117 (1967).
10. HULL, E., 'Corticofugal influence in the macaque lateral geniculate nucleus', *Vision Res.* **8**, 1285-1298 (1968).
11. KALIL, R.E. and CHASE, R., 'Cortical influence on activity of lateral geniculate neurons in the cat', *J. Neurophysiol.* **33**, 459-475 (1970).
12. PRIBRAM, K.H., 'How is it that sensing so much we can do so little?', in *The Neurosciences, Third Study Program*, pp.249-261 (edit. by F.O. Schmitt and F.G. Worden), MIT Press, Cambridge, Mass. (1974).
13. ROTHBLATT, L. and PRIBRAM, K.H., 'Selective attention: input filter of response selection', *Brain Res.* **39**, 427-436 (1972).
14. SZENTAGOTHAI, J. and ARBIB, M.A., 'Conceptual models of neural organization', *Neurosci. Res. Bull.* **12**, 307-510 (1974).
15. JULESZ, B., *Foundations of Cyclopean Perception*, p.406, University of Chicago Press, Chicago (1971).
16. FENDER, D. and JULESZ, B., 'Extension of Panum's fusional area in binocularly stabilized vision', *J. Opt. Soc. Am.* **57**, 819-830 (1967).
17. HUBEL, D.H. and WIESEL, T.N., 'Receptive fields, binocular interaction and functional architecture in the cat's visual cortex', *J. Physiol.* **160**, 106-154 (1962).
18. ENROTH-CUGELL, C. and ROBSON, J.G., 'The contrast sensitivity of retinal ganglion cells of the cat', *J. Physiol.* (London) **187**, 517-552 (1966).
19. STONE, J. and HOFFMANN, K.P., 'Conduction velocity as a parameter in the organization of the afferent relay in the cat's lateral geniculate nucleus', *Brain Res.* **32**, 454-459 (1971).
20. BROOKS, B. and JUNG, F., 'Neuronal physiology and the visual cortex', in *Handbook of Sensory Physiology*, Vol.VII/3 Part B, pp.325-440, Springer, Berlin-Heidelberg-New York (1973).
21. HARTH, E. and TZANAKOU, E., 'Alopex: A stochastic method for determining visual receptive fields', *Vision Res.* **14**, 1475-1482 (1974).
22. HUBEL, D.H. and WIESEL, T.N., 'Receptive fields and functional architecture of monkey striate cortex', *J. Physiol.* **195**, 215-243 (1968).

On a biological system with random parameters

SHUNSUKE SATO and MICHIYOSHI URANISHI
University of Osaka, Osaka, Japan

I. Introduction

When a number of elements each of which changes its state according to a dynamical equation interact with each other, how do they behave as a whole? In biological systems, one may be faced with this problem in various situations. An element without any interaction at each level such as amino acid, protein, organella, cell, organ, or individual spends its own life. However many interacting elements may possess quite different properties as the individual one does. A collection of biological elements (=system) of a certain level yields a new function or has a structure for the next level. Any biological organization seems to have a hierarchical structure. Besides the way of interaction might be different from system to system because the genetic information is not available enough to specify it in detail. In other words, it is more probable that parameters which describe the system are random variables with some distribution function. The mean value and the variance, for instance, of the distribution may only be specified genetically. We may regard any biological system as a dynamical system with random parameters. Now a system must have the same or almost the same properties (e.g. the stability property) as others in order to work as a unit or an element at the next level irrespective of their randomness.

Such a problem was firstly investigated by Gardner and Ashby [1]. They have discussed by computer simulation that the stability of a linear system consisting of n-elements connected at random decreases as the level of connectance increases and have suggested that the stability may suddenly change at a certain level of connectance when the size of the system is large.

May has given a theoretical explanation of their results [2].

In what follows, we will treat an analog of Gardner and Ashby's system somewhat simpler for the convenience of analysis but likely important in some situation, i.e. we will discuss systems where any interaction between elements is symmetric and the connectance is not only constant but is given as a function of the system size.

II. Random dynamical system

To begin with, let us consider a random dynamical system consisting of n elements: Suppose that the system can be represented by

$$\frac{d\dot{x}_i(t)}{dt} = f_i(x_1(t), x_2(t), \ldots, x_n(t); p_i, q_i, \ldots, r_i), \quad i = 1,2,\ldots,n \tag{1}$$

where x_i, $i=1,2,\ldots,n$ denote states of the elements, $p_i,\ldots,r_i$, $i=1,2,\ldots,n$ the parameters of function f_i $(i=1,\ldots,n)$ and each f_i is assumed to satisfy the condition for the solvability of Eq.(1). The system is called a random dynamical system if any one of the parameters of the system is a random variable subject to an appropriate distribution function.

Equation (1) may have equilibrium points depending on the forms of function f_i's. The behavior of the system around one of the equilibrium points $(x_{10}, x_{20}, \ldots, x_{n0})$ can be represented by a set of linear equations

$$\dot{\xi}_i = \sum_j \left.\frac{\partial f_i}{\partial x_j}\right|_{x_j = x_{j0}} \xi_j, \quad i=1,2,\ldots,n, \tag{2}$$

where $\xi_i = x_i - x_{i0}$, $i = 1,2,\ldots,n$.
Since the values of parameters $p_i,\ldots,r_i$; $i=1,\ldots,n$ are subject to some distribution functions, so are those of the coefficients

$$\left.\frac{\partial f_i}{\partial x_j}\right|_{x_j = x_{j0}}, \quad i,j = 1,2,\ldots,n.$$

We assume that for the elements of the $n \times n$ coefficient matrix

$$H = (h_{ij}) = \left.\frac{\partial f_i}{\partial x_j}\right|_{x_j = x_{j0}}$$

(1) $h_{ij} = h_{ji}$, $i,j = 1,2,\ldots,n$,

(2) h_{ij}, $i<j$ are subject to the independent normal distribution function with mean zero and variance σ^2, and h_{ii}, $i = 1,\ldots,n$ are that with mean -1 and variance σ^2.

Thus from (1), the equilibrium point under consideration is a node. From (2), around the point, the system is represented by

$$\dot{\xi} = (-I + B)\xi, \tag{3}$$

where $\xi = (\xi_1(t), \xi_2(t), \ldots, \xi_n(t))$ and I is the identity matrix. Each element of the random symmetric matrix $B = (b_{ij})$ is subject to an independent normal distribution with mean zero and variance σ^2 since $B = H + I$. This implies that each element of the system has a stable property itself. However the behavior of the system as a whole depends on the eigenvalues of B.

Gardner *et al.* considered a lanear system [1]:

$$\dot{\mathbf{x}} = A\mathbf{x}, \tag{4}$$

where A represents connections between elements of the system. Each element is assumed to connect with fraction C of the other elements. (If there exists a connection from ith element to jth, $a_{ij} \neq 0$, otherwise $a_{ij} = 0$. They called the C connectance.) When each non-zero element of the system with connectance C, if it is in non-diagonal, is subject to the uniform distribution function on [-1,1], and if in diagonal, to that on [-1,-0.1], they discussed the probability that a system is stable. Figure 1 illustrates the result by computer simulation. It seems that there exists a critical value of C for the stability property as n becomes large.

Clearly, the necessary and sufficient condition that the equilibrium point (0,0,...,0) of a linear system (3) is stable is that none of the eigenvalues of the matrix B exceed the unity. In the next section, we will discuss, using a computer, how the stability changes depending both on the size n of the random matrix $B = (b_{ij})$ and on variance σ^2 of random variables $\{b_{ij}\}$, and also the case that a system is with connectance C as Gardner and Ashby have discussed, though systems which we will consider have reciprocal interactions for the convenience of a theoretical treatment, while they investigated systems with non-reciprocal interactions in general.

III. Probability of stability

It is likely that a system represented by Eq.(3) is stable if variance σ^2 of the values of the matrix elements is very small compared with the unity, since each element of the system with time constant -1 is stable itself and the ordered eigenvalues of a matrix are continuous functions of values of its elements. Thus the variance σ^2 increases, so does the probability that the largest eigenvalue exceeds unity, which means that the probability of the stability decreases.

Let us mention our computer experiment. An $n \times n$ random symmetric matrix B is constructed by

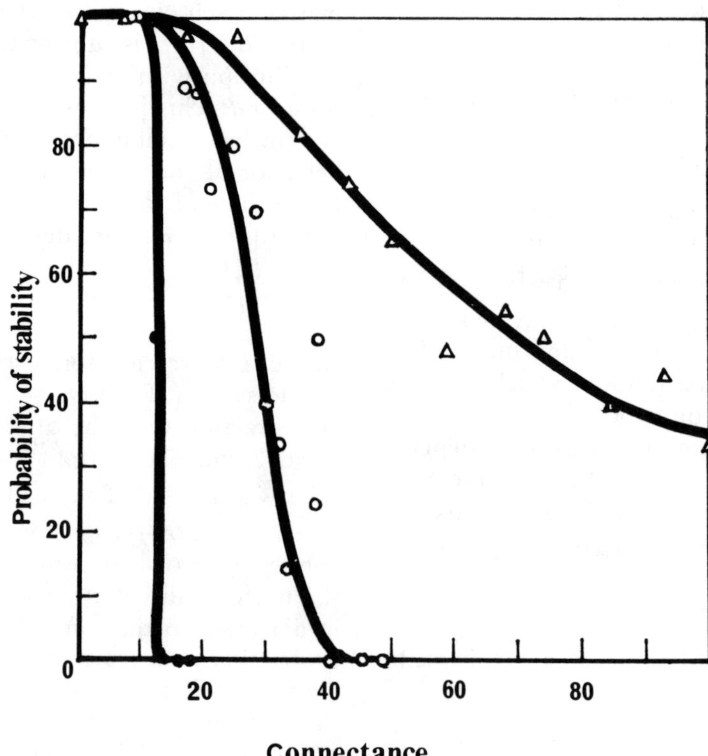

Fig.1 Variations of stability with connectance (from Gardner and Ashby)

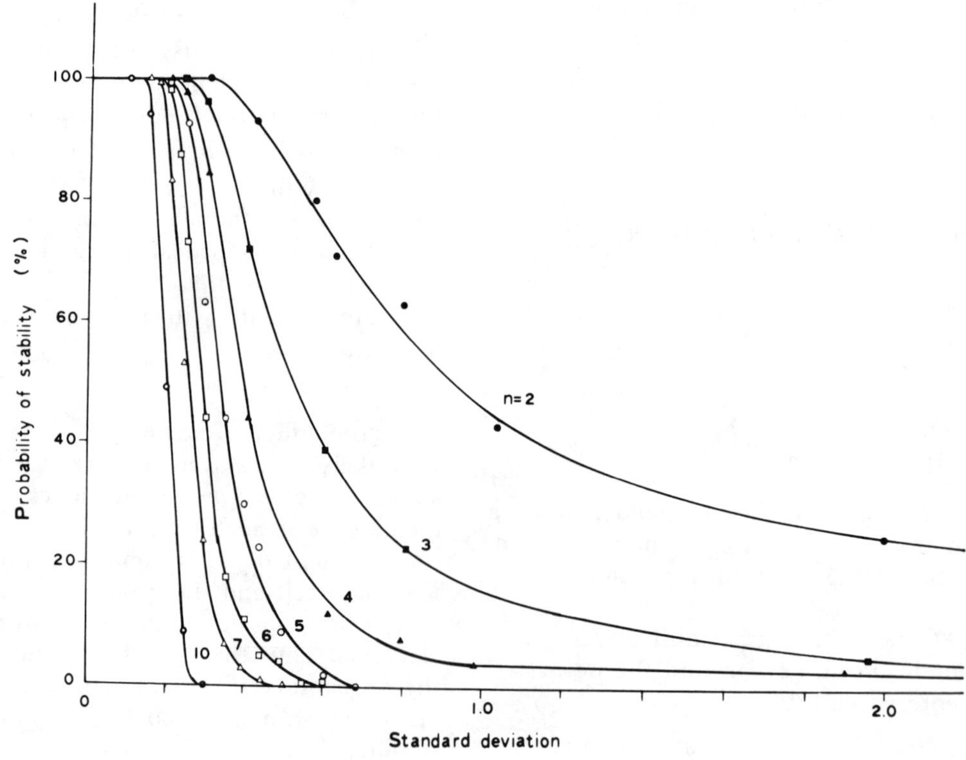

Fig.2 Stability probability vs. variance of elements of random symmetric matrix associated with a system

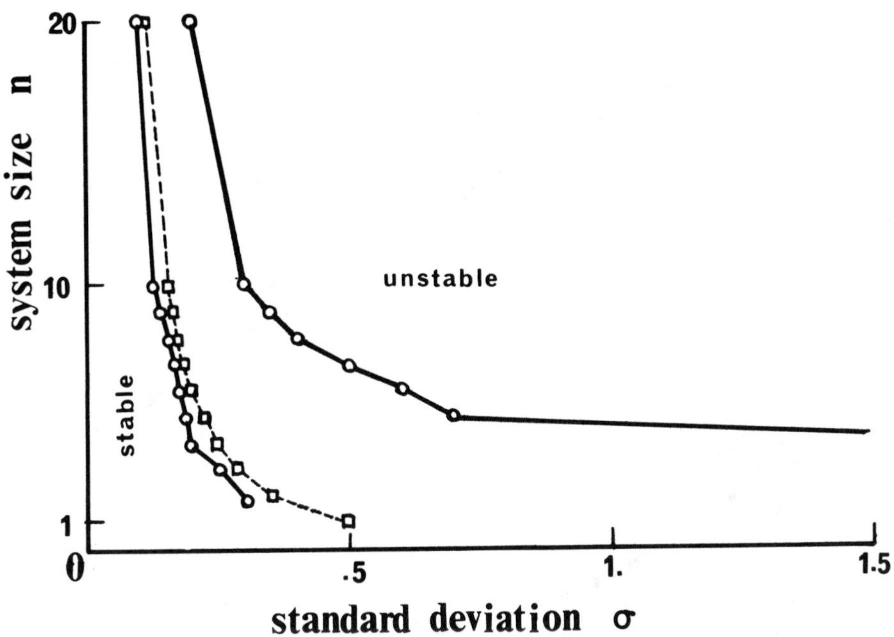

Fig.3 Stable and unstable domains

assigning to the entries $\{b_{11}, b_{12}, \ldots, b_{1n}, b_{22}, \ldots, \ldots, b_{2n}, \ldots, b_{n,n}\}$ $\binom{n}{2}$ random numbers subject to a normal distribution function with mean zero and variance σ^2. The n eigenvalues of the matrix B are calculated by the Jacobi method. If none of the eigenvalues exceeds unity, the system associated with B is determined to be stable, otherwise it is unstable.

A large number of B's are examined and the ratio of the number of B's which make the associated systems stable to the number of B's examined was calculated for various n and σ^2 (see Fig.2). As one may notice, there exist stable systems in a finite ratio for a small n, however, the ratio of the stable systems changes sharply from 1 to zero at a certain value of σ as n increases. We call the upper bound of σ under which the system (3) becomes certainly stable 'a critical value' with respect to σ. Our result shows that there exists such a critical value for the stability as did Gardner *et al.*'s, though their critical value is with respect not to the variance but to the connectance.

Figure 3 shows a domain of (n, σ) plane in which systems are certainly stable and a domain in which systems are certainly unstable.

In Fig.4, we showed a result of the experiment in which the fraction C of all the elements of any symmetric matrix $B = (b_{ij})$ is assumed to be non-zero and any non-zero element is subject to a normal distribution $N(0, \sigma^2)$ with $\sigma^2 = 0.25$. It seems in Fig.4 that for $n = 4$, there exist stable systems in a finite ratio for a fairly large C. While for a large n, the ratio of the number of stable systems decreases sharply as C increases. The critical value seems to be less than 0.5 for $n = 10$, though we did not examine it exactly. This result shows qualitative similarity to that of Gardner *et al.*

IV. Distribution of eigenvalues of random matrices

As mentioned in the previous section, the stability changes very sharply at a critical value of σ for a large n. This implies that the largest eigenvalue of random matrices B's are surely less than some value, say λ_m, and that since the distribution of the matrix elements is symmetric with respect to zero, the distribution density function of eigenvalues is also symmetric and vanishes outside the interval $[-\lambda_m, \lambda_m]$.

Figure 5 illustrates empirical distribution of normalized eigenvalues λ's, i.e.

$$\lambda = (2\sigma\sqrt{n})^{-1}\lambda_B \tag{5}$$

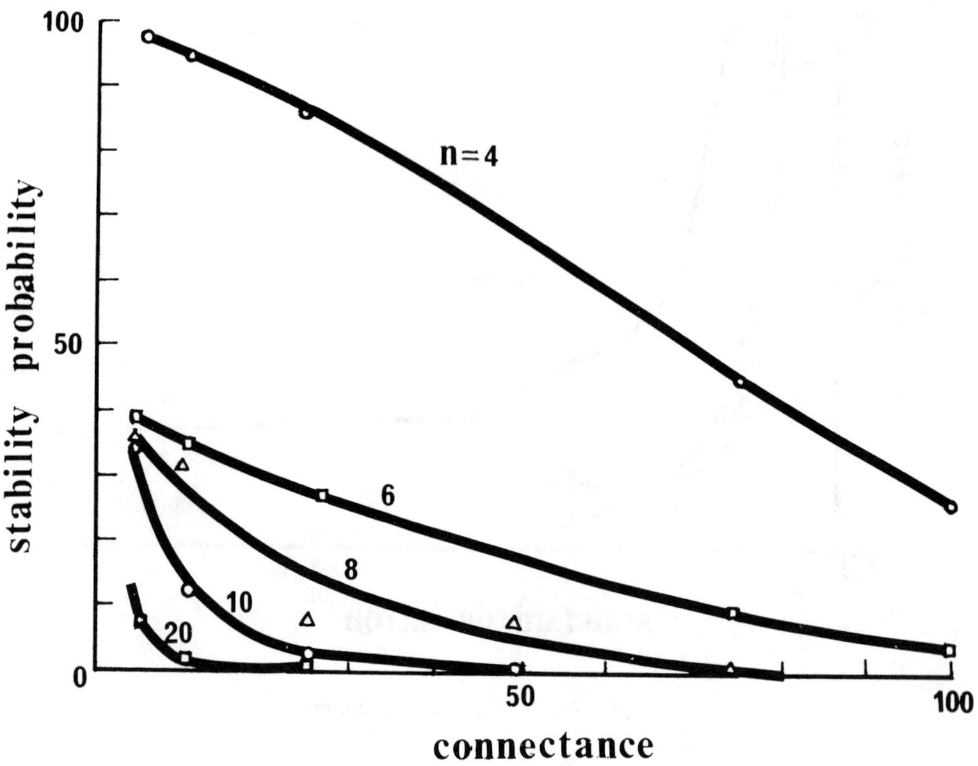

Fig.4 Stability probability vs. connectance (per cent)

for various n and σ^2, where λ_B's denote eigenvalues of the random matrix B. It seems that the empirical distribution of eigenvalues of random symmetric matrices converges to a semi-circle distribution as n tends to infinity. In fact, it is known as Wigner's semi-circle law [3] and the convergency is proved theoretically for several types of distribution functions of matrix elements [4].

According to the law, the asymptotic distribution $w(\lambda)$ of $\lambda = (2\sigma\sqrt{n})^{-1}\lambda_B$ as n tends to infinity is given by

$$\frac{dw}{d\lambda} = \begin{cases} \frac{2}{\pi}\sqrt{1-\lambda^2}\,; & |\lambda| \leqslant 1\,, \\ 0 \,; & |\lambda| > 1. \end{cases} \tag{6}$$

Using this fact, an eigenvalue λ_B of the matrix B is certainly less than $2\sigma\sqrt{n}$ for a large n. Thus if $\sigma < (2\sqrt{n})^{-1}$, then certainly $\lambda_B \leqslant 1$. It is clear that $\sigma = (2\sqrt{n})^{-1}$ yields a boundary of stability in the (n, σ) plane†. A dotted line in Fig.3 shows the relation $\sigma = (2\sqrt{n})^{-1}$, with which our computer experiment shows a good similarity.

While for a finite n, the empirical distribution function differs from a semi-circle. The distribution function of eigenvalues of a random symmetric matrix B for a finite n is known in some cases: The joint distribution function of the eigenvalues $\lambda_1, \lambda_2, \ldots, \lambda_n$ of a random symmetric matrix which has elements subject to a normal distribution $N(0, \sigma^2)$ is given by

$$P(\lambda_1, \lambda_2, \ldots, \lambda_n) = \Omega_n \prod_{i<j} |\lambda_i - \lambda_j| \exp\left(- \sum_i \frac{\lambda_i^2}{2\sigma^2} \right) \tag{7}$$

where Ω_n is determined by the relation;

$$\int_{-\infty}^{\infty} \cdots \int P(\lambda_1, \ldots, \lambda_n)\, d\lambda_1 \ldots d\lambda_n = 1\,. \quad [5]$$

Thus we can know the probability Q that a system

†May showed theoretically an asymptotic upper bound of random matrices necessarily not symmetric. According to him, $\sigma = (\sqrt{n})^{-1}$ gives the boundary for stability for the case.

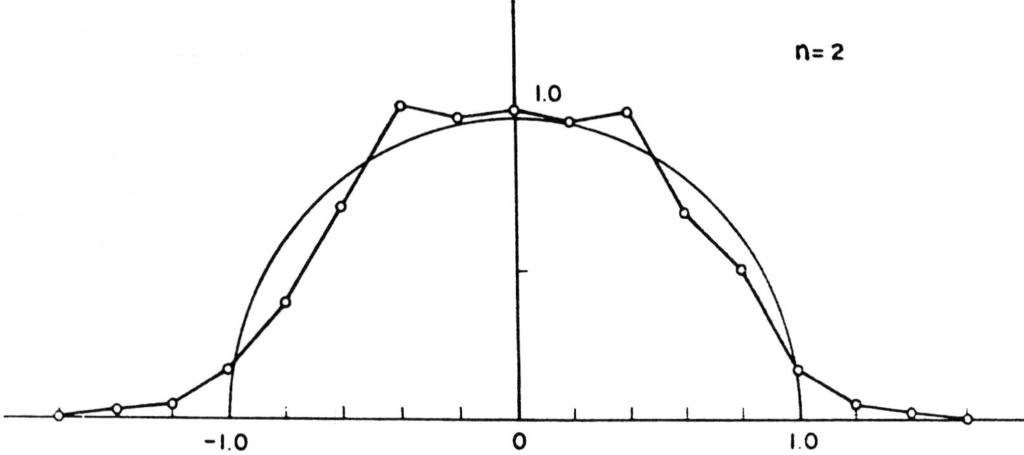

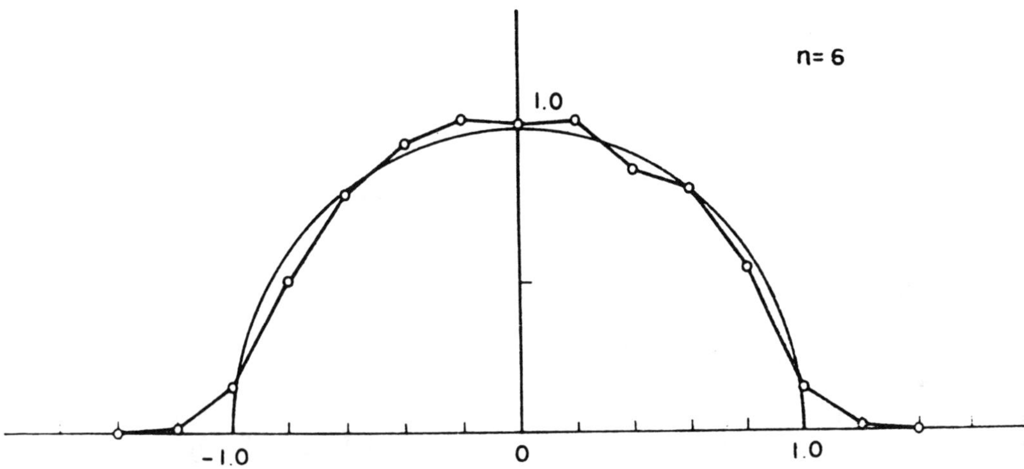

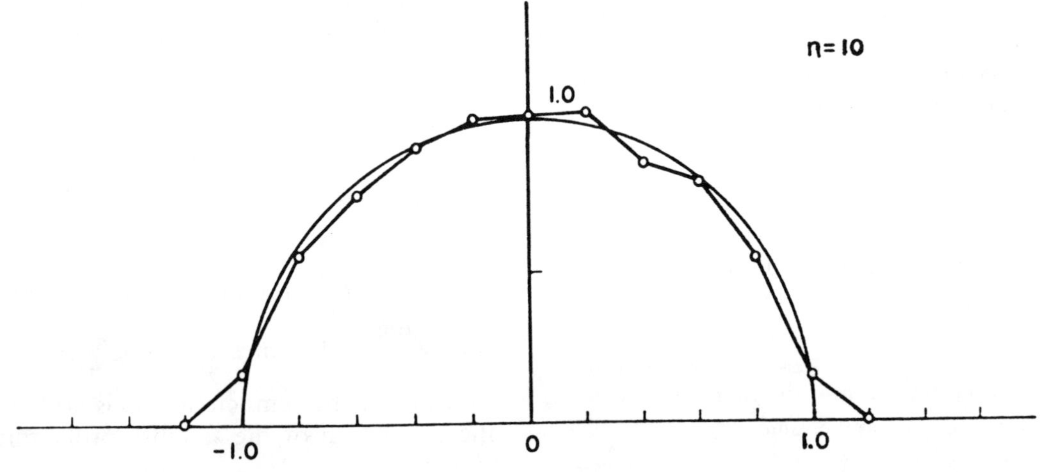

Fig.5 Empirical distributions of normalized eigenvalues of random symmetric matrices

of size n and variance σ^2 is stable by calculating the following integral;

$$\int_{-\infty}^{1} \cdots \int P(\lambda_1, \lambda_2, \ldots, \lambda_n)\, d\lambda_1\, d\lambda_2 \ldots d\lambda_n \qquad (8)$$

It is shown that the integral is reduced to a pfaffian of a matrix. In fact, following M.L. Mehta [6], we know that Q is given by

$$Q = |(q_{ij}(\sigma))|^{1/2}, \qquad (9)$$

where

$$q_{ij}(\sigma) = \int\int_{-\infty < y < x < \frac{1}{\sigma}} \{\varphi_i(y)\varphi_j(x) - \varphi_j(y)\varphi_i(x)\} \times dx\, dy \qquad (10)$$

and

$$\varphi_j(x) = (2^j j!\sqrt{\pi})^{-1/2} e^{-x^2/2} H_j(x)$$

with $H_j(x)$ being jth Hermitian polynomial, $j = 0, 1, \ldots, n-1$.

However our computer experiment shows that the empirical distribution seems to be approximately a semi-circle for $n \sim 10$ more or less, which implies that the semi-circle law can be applicable to examine the stability property of a random dynamical system of size $n \gtrsim 10$.

May has also suggested that the critical value of σ for the stability may be $(\sqrt{nC})^{-1}$ as $nC \to \infty$† for a system with connectance C, since the variance of the matrix associated with it is effectively given by $\sigma^2 C$.

Here we give our result of computer studies on the distribution of eigenvalues of matrices associated with systems with various values of connectance, which shows that for a small C, the empirical distribution of the eigenvalues differs much from a semi-circle. This discrepancy may be caused by the fact that nC for n and C employed in the experiment is not large enough.

In actual situations, it seems that any unit or element of a system can not interact with all the others because of physical or geometrical restrictions on the unit or the system. Thus in general a system is with small C and gives a *sparse* random matrix with Cn^2 non-zero elements.

In what follows, we characterize the associated random matrix B by $2N$, which denotes the number of non-zero elements of it instead of C. Clearly $2N \sim Cn^2$ in Gardner and Ashby's case. Let us give a brief discussion on the discrepancy between the distribution function of eigenvalues of such matrices and a semi-circle distribution if it exists. In some case, it does not depend on a small nC, but it is of intrinsic nature. In this case, May's theory should be applied carefully.

Provided that each of non-zero elements $b_{ij}\ (=b_{ji})$, $i \neq j$ of B is subject to an independent distribution with mean zero and variance σ^2, it can be shown that the distribution of the eigenvalues λ's of a normalized matrix

$$B^* = \left(2\sigma\sqrt{\frac{2N}{n}}\right)^{-1} B$$

possesses the semi-circle property as n tends to infinity if

$$\frac{n}{N} = 0(1)\,. \quad [7]$$

Thus a system is certainly stable if

$$\sigma < \frac{1}{2}\sqrt{\frac{n}{2N}}\,, \qquad (11a)$$

and certainly unstable if

$$\sigma > \frac{1}{2}\sqrt{\frac{n}{2N}}\,. \qquad (11b)$$

Immediately we obtain $\sigma = (2\sqrt{nC})^{-1}$ when $2N \sim Cn^2$, where C is the connectance previously defined. When $2N \sim \alpha n \log n$ ($a > 0$, const., i.e. $C \sim \alpha \frac{\log n}{n}$ effectively), namely each of elements connects the fraction $\alpha \frac{\log n}{n}$ of all the others in average, the condition $\frac{n}{N} = 0(1)$ holds. Thus in spite of the fact that the effective connectance $C \sim \alpha \frac{\log n}{n}$ becomes vanishingly small as n tends to infinity, the semi-circle law is still applicable to the estimation of the stability probability of the

†The critical value given by him differs of course from ours since his system is not necessarily with symmetric interactions. Our case gives $(2\sqrt{nc})^{-1}$ as the value.

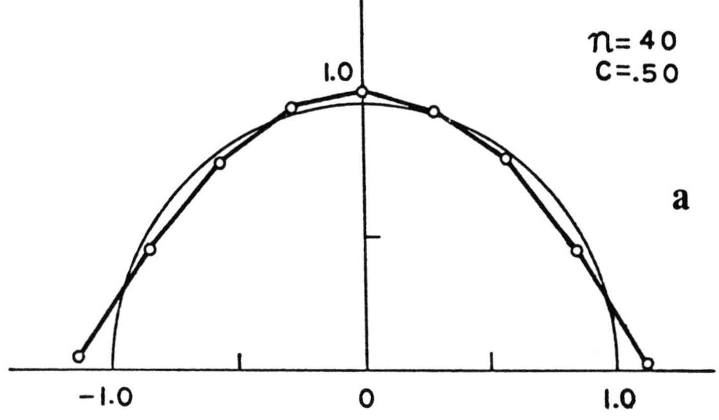

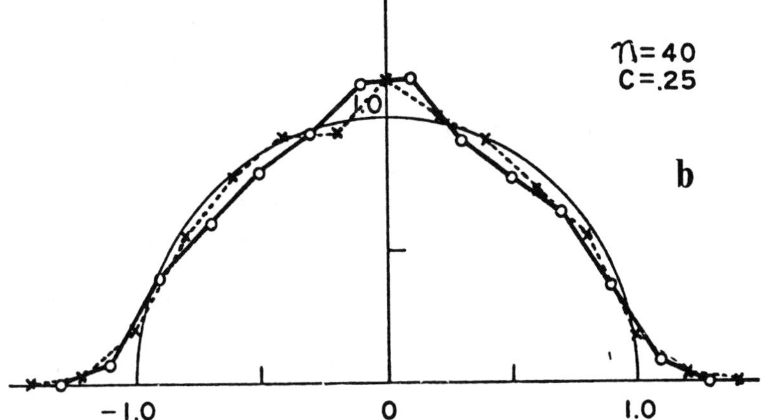

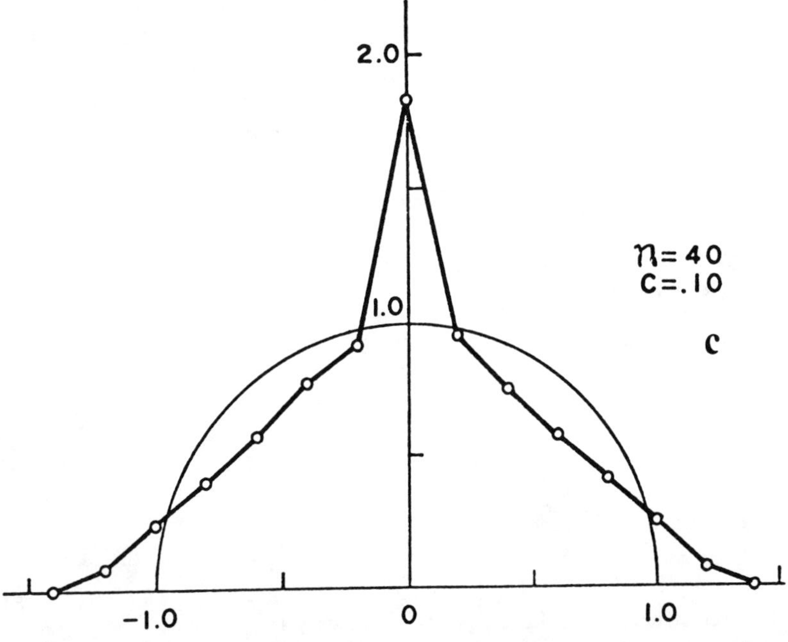

Fig.6 Empirical distributions of normalized eigenvalues of sparse random symmetric matrices ($(2\sigma\sqrt{nc})^{-1}$). a: n=40, c=0.5; b: n=40, c=0.25; c: n=40, c=0.1. A dotted line in b shows that of n=10, c=0.75 for comparison

associated system.

The situation differs completely from those mentioned above when each interacts only with a finite number α of elements in average, i.e. $2N \sim \alpha n$. (Many large dynamical systems seem to be so.) The asymptotic distribution of eigenvalues is not a semi-circle any more. The matrix B degenerates certainly in a finite ratio to n (see Fig.5): Denoting by d_g the degree of degeneration of a random symmetric matrix B, we have, if $2N \sim \alpha n \ (\alpha > 0)$

$$\lim_{n\to\infty} \frac{d_g}{n} > \frac{1}{\alpha} \sum_{k:odd} \frac{k^{k-2}}{k!} (\alpha e^{-\alpha})^k \qquad (12)$$

However we could not obtain so far the exact feature of the distribution function of eigenvalues of such systems theoretically. In consequence, since a system is certainly decomposed into many smaller subsystems (the structures of which can be known from a theory of randomgraphs[†]) when $2N \sim \alpha n$, the stability of the total system for any value of σ^2, briefly speaking, will be improved compared with a system with $2N \sim Cn^2$ or $\alpha n \log n$ for the same variance σ^2 of random variables of non-zero elements.

[†]According to a theory of random graphs, it is shown that, if $2N \sim Cn^2$, (i.e. if a system is with connectance C), then every element of the system is connected to the others directly or indirectly. While if $2N \sim \alpha n \ (\alpha > 0)$, the system is decomposed into many smaller subsystems: For $1 > \alpha > 0$, the structure of any subsystem is a tree if we use a notion of graph theory. For $\alpha > 1$ any system consists of one giant subsystem with n_g elements and N_g connections, and of subsystems with tree structures, where

$$n_g \sim n(1 - \frac{z}{\alpha}), \qquad N_g \sim n(1 - \frac{z^2}{2\alpha}),$$

and z is a solution of the equation $(0 < z < 1)$;

$$z e^{-z} = \alpha e^{-\alpha}.$$

The size of a subsystem with a tree structure can be estimated also [7].

Acknowledgement

The authors would like to express their thanks to Professor R. Suzuki and Dr. K. Kobayashi for their valuable discussions.

References

1. GARDNER, M.R. and ASHBY, W.R., 'Connectance of large dynamic (cybernetic) systems; critical value for stability', *Nature* **228** (21 November 1970).
2. MAY, R.M., 'Will a large complex system be stable?', *Nature* **238** (18 August 1972).
3. WIGNER, E.P., 'Characteristic vectors of bordered matrices with infinite dimensions', *Ann. of Math.* **62**, 548-564 (1955).
4. OLSON, W.H. and UPPULURI, V.R.R., 'Asymptotic distribution of eigenvalues of random matrices', Sixth Berkeley Symposium **3**, 615-644 (1970).
5. GINIBRE, J., 'Statistical ensembles of complex, quaterion and real matrices', *Math. Phys.* **6**(3), 440-449 (1965).
6. MEHTA, M.L., *Random Matrices*, Academic Press, N.Y. (1967).
7. SATO, S. and KOBAYASHI, K., 'Asymptotic distribution of eigenvalues and degeneration of sparse random matrices', *Bulletin of Mathematical Statistics,* Research Association of Statistical Sciences **17**(3), 83-99 (1977).

Optimization concepts in models of physiological systems

RAIMO P. HÄMÄLÄINEN
Helsinki University of Technology, Finland

1. Introduction

Mathematical description of biological systems has become more popular during the past twenty years. The rapid development of modern control theory has taken place at the same time. During that period optimal control and system optimization became central research topics in control engineering. The mathematical apparatus of control theory was soon introduced to the biological field. The terminology and exact concepts used in engineering were adopted for the description of living feedback systems. This resulted in an extensive activity in qualitative as well as quantitative modeling of biological control systems, where analogies were frequently made to engineering problems. Due to the development of high-speed computers dynamic processes in the body became easier to analyze and an increasing interest arose in constructing large-scale simulation models for varying subsystems. The number of books published on this subject attests to the growing research activity, cf., e.g. [1,2,3,4,5,6, 7,8]. No wonder that there soon appeared a trend to make use of optimization techniques in the study of physiological control systems. Engineers and mathematicians, who were attracted by the analysis of biosystems, started to make attempts to apply optimal control theory to these problems. This survey deals with models where the optimization problem is an essential part of the structure of the model. Optimization of models will not be within the scope of this chapter.

The goals of engineering and biological sciences have traditionally been in opposite directions. The former is oriented to applied research and the latter into basic research. The biologist wants to analyze and describe his research objects whereas the engineer aims at a synthesis. In the case of control systems the engineering approach is to design, e.g. an optimal controller with a given criterion. The biologist's aim on the contrary is to find out and explain the performance principles of an operating system. In this sense medical research is not always put into its proper place. Medicine is an applied biological science and its primary aim is to study methods of controlling bioprocesses. For example intake of medical drugs is a way to control physiological systems of the body.

These different goals have given rise to communication difficulties and thus to sceptic thoughts of the usefulness of mathematical description of biosystems. General eagerness of modeling has resulted in a large number of models or mathematical expressions relating different variables to each other, which are formally correct but which miss the basic motivation. Modeling without an objective is useless. Very little is achieved by establishing, for example, linear correlations between variables if these are not put into a suitable framework which either aims at the control or deeper understanding of the system in question. However, there is a general advantage in theoretical modeling in scientific research: even weakly argumented models may become fruitful if they reveal and suggest new experimental problems which have been neglected before and which need more investigation.

Quantitative modeling especially is a field where assumptions of the circumstances where the model is valid determine where the model can be used. These questions become particularly important when we apply optimization concepts to our model. The question is whether we are showing that there exists rationality in the operation or merely using an efficient technique for predicting effects of system parameter changes on certain interesting variables. As will be seen in the sequel most of the published models deal with both of these aspects parallelly, which naturally results in a somewhat confusing ambiguity when interpreting the results.

The basic stimulation for adapting optimization concepts in physiological modeling is the need for quantitative descriptions of decision processes. Living systems usually have a possibility of operating in various ways to carry out a specified task. It is clear that one would like to be able to simulate the related decision process, that is, to predict the system's responses to changing conditions. Biologically a still more fascinating objective would be to understand and explain why decisions are made in the observed way. This, however, leads us to the question how well can we make deductions on our own rational performance criteria. Yet such considerations have received general approval to some extent. Many aspects of these problems are illustrated in an excellent article by Milsum [9] and in a book by Rosen [10].

A major problem in modeling decision processes is how to define the tasks or purposes of the subsystems under consideration and how to take into account their interconnections to other subsystems. For example the cardiovascular and respiratory systems which are usually conceptually separated have very close functional and structural interconnections. It is not at all clear that local control systems can be identified within the real overall system. Yet in practice there does not seem to be any other alternative but to study subsystems and to try to include effects of interconnections into constraint equations through some suitably chosen decoupling and coordination variables.

Typical subsystems that can be studied independently are continuously repeated periodic neuromuscular tasks, which have a relatively short cycle time. These processes can be assumed to operate in an open-loop fashion so that no corrections are applied during a cycle. Each cycle is thus preprogrammed on the basis of earlier information and changes are made only in the future cycles. The use of this hypothesis is supported by the fact that neural information pathways can be too slow to transmit a central feedback control signal to a peripheral periodic process of short duration. The undesired instabilizing effects caused by time-lags are easily confronted. Preprogramming is a type of supervisory control which is used instead of the continuous direct central on-line control. The cardiac, respiratory and walking cycles can be mentioned as examples of such periodic operations.

The definition of the task of a more complex process is intricate. A closely related problem here is how to define the inputs and outputs of a biological system. It is not always clear at all which are the control variables. Actually if we could state the purpose of a subsystem completely we would have the knowledge of both what and how the system should operate. No more models would be needed to understand the decision process. Models could then be built solely for prediction and control purposes on a pure input-output basis.

In general one should be extremely careful when forming the true decision criteria from theoretical optimization models. The cost functionals have very little physiological significance if the system model and constraints used are not correct even if the model predictions match the natural behavior.

The concept of optimization in itself seems to be fascinating and puzzling at the same time in connection with biological systems although it is generally used to describe common phenomena like adaption and decision making. It is quite clear that every decision is optimal with respect to the facts that it is based on. As far as the author knows no such universal biological laws including optimization are found as is, e.g., Fermat's principle of minimal time in geometrical optics.

A model with the structure of an optimization problem has an interesting feature from the point of view of systems analysis as it is an extremely integrated way of expressing information in a mathematical way. This is of course a general feature of analytical models but in these kind of problems, which include decision making, it is not enough to know the output of the physical system as an analytical expression of the input. In addition, a model is needed to tell how the system input is chosen in different circumstances as the input depends on the specific values of system parameters and on external constraints. Optimization problems provide an analytical way of modeling the overall performance of such systems. The model itself becomes very compact although one may sometimes have to employ numerical procedures for the solution. Moreover, no structural hypotheses are needed of the particular neural systems performing the decision process which is being modeled. This is somewhat analogous to a situation where the closed-loop performance of a feedback system can be expressed with the aid of the open-loop response by including the feedback structure already in the mathematical formulation.

Parameter identification is a great difficulty which is always met when physiological processes are studied. Firstly they are distributed in nature. Yet we have generally to be satisfied with approximative lumped parameter representations to avoid the overwhelming computational and experimental burden brought with distributed parameters. Secondly, biological models always become nonlinear. Saturation effects are bound to appear. Once again linearity is the first approximation to be used although sometimes the crucial features which govern the process operation may thus be omitted. Moreover variables and parameters that can be measured directly are not always those of primary interest and many of the indirect estimation methods tend to be inaccurate.

When feedback models are studied the identification of system parameters as time constants is not the most problematic task. Gain coefficients in the feedback loops are very difficult to approach as well as the joint effects (e.g. additive or multiplicative) of several control variables. The corresponding problems with optimization models appear in the selection of weighting parameters for each term in the cost functional. There hardly remain any other methods to find out the feedback coefficients or the weighting parameters than to make a fit by model experimentation. A unique result cannot always be expected. This fact leaves room for scepticism and speculation as to the validity of the entire modeling principle. Fully convincing arguments supporting the use of optimization models are not easily found. The choice of weightings is avoided only in the simplest cases where the objective functionals consist of one single term like in a minimum time problem. The difficulties increase as the criterion becomes more complex. Weightings of an overall decision criterion with a hierarchical structure may easily be quite intractable. The same is true also with complex feedback models.

In the following we first consider some optimization criteria and their possible physiological relevance. Secondly we treat some subsystems to which these concepts have been successfully applied. The interest has been focused on dynamic control phenomena in human subsystems. The most detailed attention is paid to the respiratory system, which is explained by the author's own activity in modeling the control of breathing. Ecological or evolutional processes are omitted. Models of neural information transmission processes are not included. A short note is made on the suggested optimal design principles of organism structures. The survey undoubtedly lacks a number of important papers since this field is presently growing and reports on works are widely scattered in the engineering and medical literature. We hope, however, that the reviewed examples give an idea of the present problems and activities.

2. Optimization criteria

Designing models of physiological systems that have the structure of an optimization problem leads to the difficult task of formulating a criterion that on the one hand is biologically motivated and on the other hand gives good predictions of the observed performance of the system. In addition, a most important feature in practice is the solvability of the resulting problem with reasonable efforts.

Attention is easily paid only to the specific form of the cost functional whereas the static and dynamic constraints on the problem, which are equally important factors when the solution of the problem is considered, are disregarded. It can be said that all physiological variables have saturation constraints due, for example, to the anatomical structure of the organ in question. It is not always clear whether or not and how these become active constraints in the solution. Thus such situations may at least in principle occur where different criteria give similar results if the assumed system dynamics and constraints of variables are somewhat different. From the control theoretic point of view we are dealing with inverse optimal control problems which, as is well known, need not have unique solutions.

A similar problem arises in the selection of the type of variables that are used. So far mostly mechanical entities have been used although, for example, chemical variables might have been more relevant in physiological systems. This is of course due to the methodological difficulties arising in the experimental verification of the models. Description of the system with different variables results naturally in varying cost functionals although these may approximate one common conceptual criterion. The specific task of the subsystem that is being modeled is usually the most important factor that determines the objective functional. Sometimes the task of the system is embedded in the constraint equations which then guarantee the desired performance. In such cases the criterion may reflect the operation principle of some higher level overall system.

In the following some criteria and their characteristic features are described. References at the end of each paragraph refer to works in which corresponding criteria are dealt with.

A. Maximum output

Systems which in one way or another are responsible for the survival of the whole body or for the protection of an organ against unpredictably occurring distrubances may produce efforts of maximal output, e.g. in terms of muscle force. The usual solution of this kind of problem with inequality constraints on the control variables is a bang-bang type of strategy. In practice the length of the time period during which such controls can be applied is limited. It may be that under normal conditions these systems operate according to some other criteria [11,12]

B. Minimum time

Skilled and high-precision neuromuscular processes engaged with safety and rapid information acquisition tasks can operate typically according to the minimum time criterion. The use of this criterion implicitly assumes that there is a well-defined goal that should be reached within minimum time and with bounded controls. Thus also the duration of the transient will be definitely determined unless the goal is moving, that is, unless we have a tracking problem. The solution of minimum time problems is mainly dependent on the available control amplitudes and the boundary conditions [12,13,14 15,16,17].

C. Minimum effort

Minimization of control efforts in continuously repeated periodic tasks may be physiologically advantageous because the local controls become smooth and 'easy to perform'. The systems's insensitivity to instabilizing disturbances is guaranteed because the full control amplitude scale is normally not used. Reserve capacity remains available for compensating disturbances. Minimum effort control may also be motivated because it avoids stress or structural damage caused by 'overload' situations. Computationally simplest criterion is a quadratic form of the control vector [13,18,19,20,21].

D. Maximum efficiency

It is sometimes possible to choose the operation point and environmental conditions of a system so that the performance efficiency is maximized with minor derect control costs such as increases in the energy expenditure. Often in physiological systems such optimum points can be found only experimentally. A typical example is the dependence of the gas exchange efficiency of haemoglobin on blood pH and p CO_2 [22]. These optima can then be used as reference information for other control systems so that their performance criteria give operation points which do not make the basic system deviate from its optimum [22,23,24,25].

E. Minumum mechanical work rate

The performance criterion of many physiological subsystems can be proposed to be the principle of minimum work. In systems where some movements are controlled by muscles it is tempting to consider the work done by the produced net muscular force. The mechanical work done during a specific time period is easily obtained by integration from the total force and velocity, which are often directly measurable. In case of volume displacement or angular motion the mechanical work corresponds to the time integral of the product of pressure and rate of volume change or mementum and angular velocity.

Unfortunately in many cases mechanical work does not have much physiological significance. This can be illustrated with two examples. Consider first that a static muscle force is applied to support a load without moving it. No mechanical work is done although effort is needed to do the task. An extreme example is a situation where movement is opposed; that is, then negative work is done. In fact the latter case appears in simultaneously active antagonistic muscles. Yet odd mechanical work rate criteria have been constructed by summing the absolute values of the negative and positive work [26,27].

The extremal solution of a minimum work problem where the task is to move the state of a linear scalar system from one resting point to another within a given time interval, is a transition with constant velocity. It is clear that such movements are not likely to appear in real systems because they require instantaneous jumps in the velocity [13,20,21,27,28,29,30, 31,32,33,34].

F. Minumum energy expenditure
Oxygen consumption is a more relevant measure of total energy expenditure than the mechanical work discussed above. It takes into account, for example, changes in the efficiency of muscular work. This efficiency depends on muscle length and load. Formulating an optimization problem with oxygen consumption is not straightforward because it has to be related somehow to the other system variables. Some research has been carried out to correlate oxygen consumption with some power-functions of the total momentum, force or pressure, e.g. in processes like locomotion and respiration. Energy expenditure is a criterion which is strongly bound to the considered subsystems and to its assumed tasks and connections to other functions of the body. To put it bluntly: 'To do nothing consumes no energy' [13, 15,16,23,25,35].

G. Optimum stability
It is evident that the performance criterion of each regulating feedback system in the body is to stabilize the process to the desired program of operation. This means minimizing a measure of deviation of the actual variable values from the values that they should follow. With most nonmechanical systems such a measure of deviation would surely be very difficult to find out not least due to the ambiguities in defining the relevant state and control variables. Even though a feedback-controller can be shown to be the solution of an optimal quadratic regulator problem [36], one should be very careful not to overemphasize this optimality unless it can be given a clear physiological meaning [15,16].

H. Combined criteria
Systems that operate according to one single principle are seldom found in biology. Usually a combination of terms describing different aspects have to be included in the criterion. If these terms are opposite in nature the solution can be called an optimal compromise. These kinds of situations occur frequently in connection with systems whose performance is aimed at maximizing efficiency with minimal costs. Which one of these two factors dominates the solution depends on the parameter values of problem constraints describing the higher level controls.

Introduction of penalty terms into the cost functional may either be a computational way to account for some constraints or a way to model real physiological penalties which are thought to exist in the system. Typically, very rapid changes in variables might have negative side-effects and should thus be penalized. Also the effects of muscle fatigue could be modeled with some kind of a trade-off term. A crucial difficulty that is confonted with combined criteria is the identification of the weighting coefficient of each term. A physiologically motivated optimization model may easily turn into a black-box input-output model, if the criterion has more free parameters than is the number of variables whose values are predicted by optimization [26,37,38,39, 40].

I. Multiple criteria
Hierarchical structures have local control systems on different levels which are interconnected to each other via co-ordination and decoupling parameters. Each lower level subsystem has its own performance criterion and constraints which the higher level criteria influence through the co-ordination parameters. On a lower level some specific aspects, which are not included in the higher level criteria may be emphasized. This kind of multilevel structure is typical of living control systems. Attempts to model such an organization lead to the important question of local versus global optimum. How can one distinguish between the various subsystems and overall system with possibly opposite goals? Moreover the weighting coefficients become again a problem. Sometimes if the goals are conflicting the problem can be formulated as a game. Then the concept of an optimal solution is no more clear unless more specific properties required for the solution are defined.

Multiple criteria, however, may occur even in clearly restricted systems if they have different operational modes or phases. The criterion functional may change together with the mode of the system. Examples might be normal versus emergency operation and a shift between two groups of muscles during a control period [23,25,26,38,40].

3. Cardiovascular system

So far there have not been many attempts to model cardiovascular control systems with the aid of optimization concepts. The major interest has been focused on the conventional simulation techniques with dynamic models.

Noldus has quite recently presented a report [39]

of an investigation where the time dependence of the elastance of the left ventricle during the ejection period is predicted with an optimization model. This approach resembles greatly an earlier work [38,40] concerning the modeling of the air-flow cycle pattern in the respiratory system. In both studies the aim is to predict the time course of a cycle during which a given volume change has to be produced with a muscular pump system.

The dynamic model of the ventricle and the arterial system employed during ejection is based on a system of three linear differential equations for ejected blood flow, i, and to ventricular volume, V, and pressure, P. The suggested cost functional

$$J = \int_0^{t_e} (P^2(t) + \alpha P(t)\, i(t))\, dt$$

where α is a weighting parameter, is minimized subject to the system equations with a given value of the ejection time, t_e. The boundary conditions are given so that the desired stroke volume will be achieved.

The performance index has been selected from energy considerations so that the mechanical work done during the cycle is minimized. The role of the first term in the functional is to penalize high amplitude peaks in the ventricular pressure wave. It is interesting to note that there are clear similarities with this criterion and the one used in [38,40].

Solution of the problem yields pressure and elastance $(P(t)/V(t))$ curves which satisfactorily agree with recorded data. The model can also be used to predict effects of varying loading conditions and system parameter changes.

Szücs *et al.* [41] have studied the overall control of blood pressure and analyzed the pulsatile circulatory waves. They postulated that the observed oscillations can be regarded as the result of a searching process which tries to force the operating point to the optimum. The performance criterion remained unspecified but it was thought that energy loss would not be the only significant factor in it.

4. The respiratory system

Human external respiration is one of the first processes where analytical optimization concept studies have been performed. The question of the selecting principle of breathing frequency has long been a challenging problem. There are no structural or other constraints that could explain why a certain cycle pattern is preferred to another. That is why optimization criteria have been developed to get a model of the decision processes in the respiratory control system.

Extensive research has been done on the control principles predicting in the first place only breathing frequencies and secondly flow shape curves, durations of inspiratory and expiratory phases, changes in lung initial volumes and airway dimensions. The first approach taken by Rohrer [42] and later by Otis, Fenn and Rahn [43] to explain observed respiratory frequencies, was based on the principle of minimum respiratory work. They showed that with a sinusoidal respiratory air-flow curve the minimum value of mechanical work done during inspiration is obtained at a frequency that is close to normal breathing frequencies found in human subjects. The system model used by Otis *et al.* was a first order differential equation with a nonlinear airflow resistance. The authors also pointed out that this may not be the only principle involved. In some situations a "principle of maximal comfort" could be more relevant. They also showed that in different conditions there is not a constant correlation between the mechanical work done during inspiration and the total energy turnover obtained from the oxygen consumption of breathing. A criterion of maximum muscular efficiency was also discussed.

Christie [29] has studied experimentally the minimum power requirement criterion, which topic has also been treated by Crosfill and Widdicombe [30]. A strong argument against the hypothesis of minimum work rate criterion is the appearance of an end expiratory pause in normal breathing patterns at rest. Some years later Mead [18] proposed that better predictions for breathing frequencies can be obtained with minimizing the average total driving pressure produced by muscles during the cycle. The author supports his model with the fact that for the measurement of force a less complex sensory mechanism would be needed than for the obtaining of the mechanical work rate. However, this does not seem to be an important matter as the complexity level of the whole principle of decision making by optimization is already so high.

The next step was taken by Widdicombe and Nadel, who derived an expression relating the anatomical dead space volume and flow-resistance to each other. They showed that an optimal dead space volume is found with both criteria presented earlier. This suggested that control of the tonus of bronchial muscles could also be explained by optimization models [21].

No further development could be made with optimisation models that were based on the assumption of sinusoidal flow-curves. The constraints would have to be less restrictive to make possible more detailed predictions of the breathing pattern.

Fincham and Priban [23,25] studied the idea of optimization processes in the overall control of respiration. Their extensive descriptive analysis of this adaptive hierarchical system introduced many new and interesting aspects. Three interacting control loops were considered. The chemical control is suggested to operate so that for any given metabolic exchange requirement of oxygen and carbon dioxide the level of ventilation is kept at a minimum. The pattern of activity of respiratory muscles is selected so that the average expenditure of energy is kept at a minimum. The airways are thought to be controlled so that the energy required to ventilate the dead space is minimized. The open loop nature of the control of muscles during a cycle was established. The proposed assumption of preprograming each breath is essential also in the dynamic and multilevel models developed later.

Tenenbaum [44] presented another descriptive multilevel model where he came to a somewhat conflicting conclusion concerning the performance criteria.

The earlier quantitative models involved only static optimization problems to predict parameter values. The first really dynamic formulations dealing with the air-flow shape of breathing cycle were published not until the early seventies by Yamashiro and Grodins [34], Ruttiman and Yamamoto [20] and by Hämäläinen and Viljanen [38, 40].

The cost functional suggested in [34] for both inspiration and expiration was the integral of the square of volume acceleration

$$J = \int_0^T \ddot{V}(t)^2 dt$$

Only the flow-shape is studied and thus the tidal volume and durations of inspiration and expiration have fixed values determining the boundary conditions for the problem. It is easily seen from variational calculus that the optimal flow solution is an arch of a parabola for both phases and it does not depend at all on the system equation parameters. The authors in [34] have used an approximation technique in the solution procedure and arrived at the incorrect result that the optimal inspiratory flow-shape would be somewhat asymmetrical. Altogether this criterion results in a pattern that is insensitive to system parameters and almost identical to the sinusoidal.

The muscle force criterion with

$$J = \frac{1}{T} \int_0^T P(t)\, dt$$

has been studied in more detail in [20] and shortcomings appearing in the earlier studies [18] concerning the optimal breathing frequency with this criterion were pointed out. Moreover it is shown that there does not exist an optimal inspiratory flow-pattern, either for a linear system model with constant coefficients or with a volume-dependent flow resistance. A ramp function solution is obtained when the quadratic resistance term is included in the model. However, this ramp function does not resemble the observed flow curves at all.

The cycle pattern observed in spontaneous human breathing is very successfully predicted by the performance functionals given in [38,40] for inspiration

$$J_1 = \int_0^{t_1} (\ddot{V}(t)^2 + \alpha_1 P(t)\dot{V}(t))\, dt$$

and for expiration

$$J_2 = \int_{t_1}^{t_1+t_2} (\ddot{V}(t)^2 + \alpha_2 P(t)^2)\, dt$$

again with fixed inspiratory and expiratory durations t_1 and t_2.

A linear first order system model is used

$$P(t) = KV(t) + R\dot{V}(t)$$

where $P(t)$ is the driving pressure produced by muscles, K denoted the total elastance and R the total resistance of the lung rib-cage system with respect to volume changes.

The volume acceleration terms in the above criteria penalize the negative effects of abrupt changes in respiratory air-flow, e.g. on the gas transport and on muscle operation. Mechanical work and the square of muscle pressure account for the oxygen cost of breathing during inspiration and expiration respectively.

It is worth comparing the optimal flow shapes with different resistances to the parabolic curve, see Fig. 1. A significantly better correspondence to characteristics found in human experiments is revealed, e.g. the asymmetry of expiration and the flattening effect caused by increased resistance.

This formulation has also been generalized for a

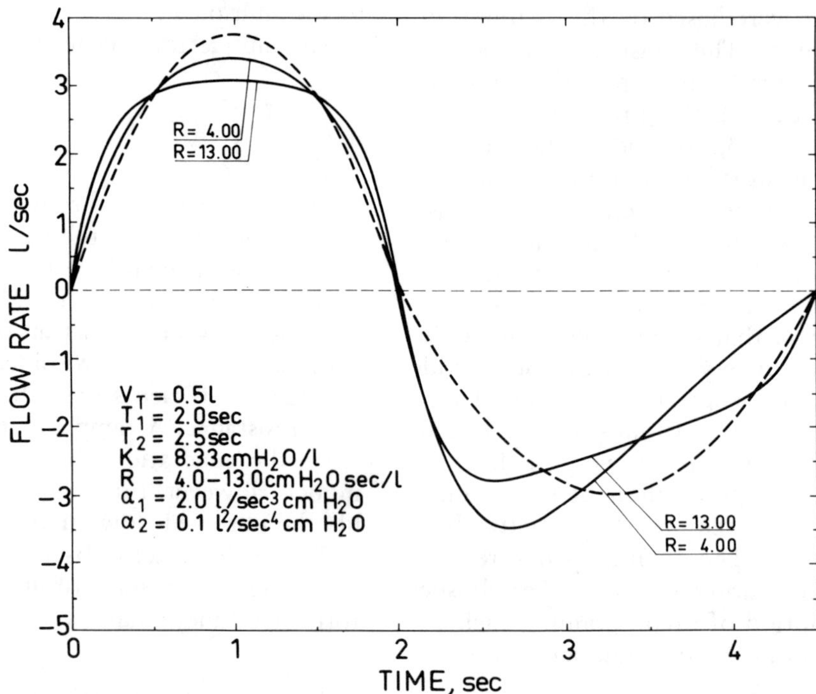

Fig.1 Optimal inspiratory and expiratory air-flow shapes (solid line) compared with parabolic curves (broken line)

model with two degrees of freedom where the relative contributions of rib-cage and abdomen to ventilation are considered [38].

The minimum work rate criterion with a linear system model where

$$J = \int_0^T P(t)\,\dot{V}(t)\,dt$$

yields a rectangular flow shape with constant flow for inspiration and expiration. This solution is again independent of the system parameter values. In spite of the fact that this does not resemble very much the normal breathing patterns, it is proposed to be relevant for higher ventilation rates [20,23].

In the model introduced in [38,40] the cost functionals determining the flow-shapes are only a lower level part of an overall hierarchical model. In this two-level model the solutions of the lower-level criteria become constraints in the higher level problem. On the other hand the higher level criterion determines the boundary conditions for the lower level problem, that is values of t_1, t_2 and tidal volume. The higher level performance index

$$J = \frac{1}{t_1+t_2+t_3}\ \beta_1 \int_0^{t_1+t_2+t_3} P_i^2(t)\,dt + \beta_2 \int_0^{t_1+t_2+t_3} P_e^2(t)\,dt$$

$$+ \beta_3 \int_0^{t_1} dt + \beta_4 \int_0^{t_2} dt + \beta_5 \int_0^{t_1} \ddot{V}^2(t)\,dt$$

$$+ \beta_6 \int_{t_1}^{t_1+t_2} \ddot{V}^2(t)\,dt$$

where P_i and P_e are inspiratory and expiratory pressures, is minimized with respect to the durations of inspiration, t_1, expiration, t_2, pause period, t_3, the dead space volume and the change in the operating level (FRC). Additional constraints are the fulfilment of the given alveolar ventilation demand and the equation relating dead space volume to airway resistance [21]. This hierarchical model is able to predict relatively well the main variable changes in the control of respiratory mechanics both at rest and in

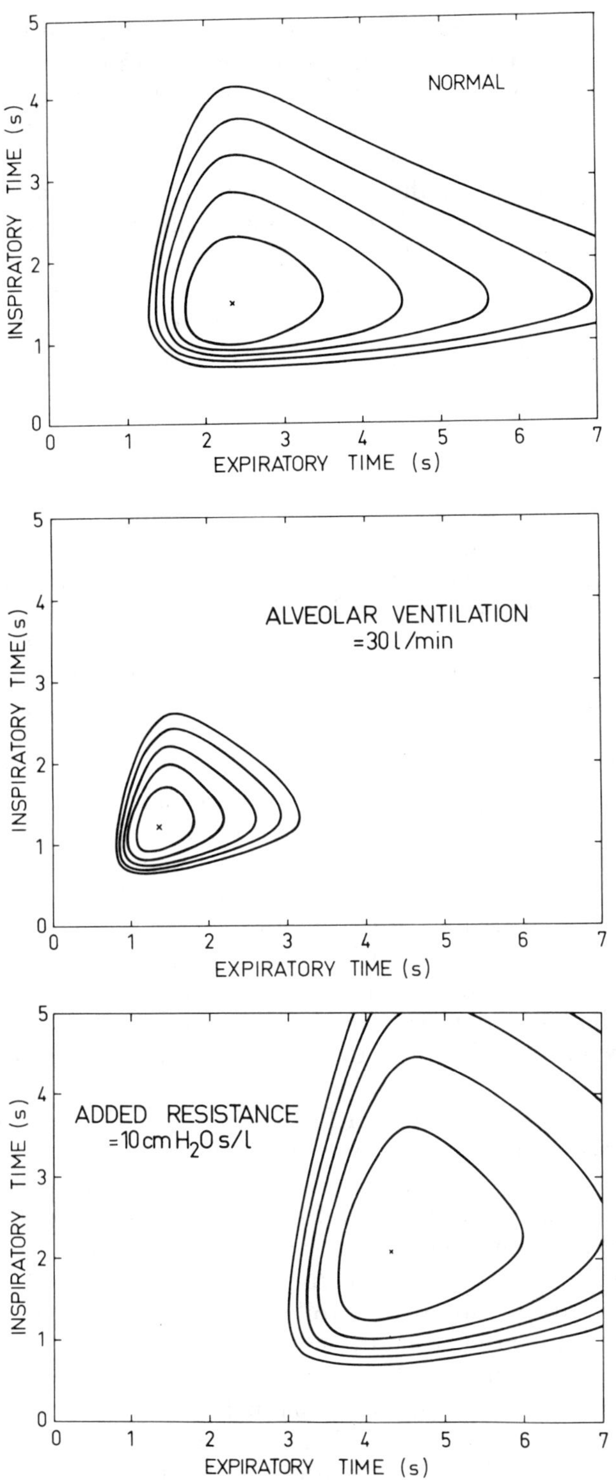

Fig.2 Niveau contours of the two level performance index on the t_1-t_2 plane with other variable values fixed at their optima. The difference between two consecutive contours is ten per cent of the optimal value. (a) normal ventilation, (b) increased ventilation, (c) normal ventilation with added external resistance load.

exercise conditions. Effects of external loads can be studied as well. The cost functionals are proposed to be associated with the efficiency of gas transport and with the parallel minimization of oxygen consumption of breathing. When such a two-level criterion is concerned it may be doubtful whether a minimum can always be found. Figure 2 shows the form of the performance index minimum on the t_1-t_2 plane in three different cases with the same values of the weighting parameters. It is interesting to note that unique optima clearly exist although their steepness vary a lot from case to case.

In spite of the obvious weaknesses of the minimum work rate principle it has still been applied in some recent studies. It is shown in [27] that there is an FRC-level that minimizes the work done against elastance during inspiration if the system elastance is not the same below and above the relaxation volume. A new criterion has been constructed by assuming rectangular flow patterns and by summing the positive and negative mechanical work performed during different phases of the cycle. The predictions thus obtained are more or less unsatisfactory [33].

As has been seen a considerable amount of work has been done in the theoretical modeling of the respiratory control system using optimization concepts. It would hardly be of any use to increase the number of suggested criteria. On the contrary much more experimental work is needed to verify the proposed principles and to point out their physiological interpretation and significance.

5. Visual system

The human eye and its control mechanisms have offered modeling problems where the application of optimal control concepts has turned out to be quite fruitful. A description of the different control loops in the visual system can be found, e.g. in [5].

An interesting system is the visual tracking process and especially the saccadic mode of it. This is an extremely rapid and precise movement responsible for positioning the eye from one target to another; this kind of situation occurs for instance in reading. Control of these saccadic movements has been extensively studied in the literature and different controller structures have been proposed beginning from a simple linear feedback regulator system to varying optimal controllers [11,14,17,45,46]. Examples of different performance criteria that have been used are maximization of muscle output and minimization of the integrated square value of the slip velocity of the target across the retina, which is a measure of the tracking error rate.

Very good results have been obtained with a model with minimum time criterion presented by Clark and Stark [14]. In this model the horizontal eye movements are governed by a sixth-order nonlinear differential system, which also includes an equation connecting the neural and muscular activities to each other, i.e. the nervous signal controlling the muscle and the mechanical force output created by the muscle. This is an interesting feature in the model, because the quantitative relationship between neural and mechanical activities has seldom been modeled dynamically in this kind of models.

The jumping changes from target to target is supposed to happen as rapidly as possible. Thus the control of the eye movement is modeled by a minimum time problem. The control variable is taken to be the nervous signal entering the muscle. In a minimum time problem the control constraints play an important role in the problem solution. In this case the constraints have quite a natural physiological basis as the magnitudes of the muscle force that can be produced are limited.

Results from this model match quite well with the typical saccadic eye movements found experimentally. In some cases the model gives an unnatural oscillatory motion at the end of the saccadic mode. It is probably caused by inaccuracies partly in the process model and partly in the control hypothesis itself. It seems, however, quite obvious that the minimum time criterion plays a dominating role in the true control principle although there may also be other factors included which have minor weightings. Nevertheless this model is a clear example of the promising possibilities that the use of optimization concepts provides us.

Another interesting control system of the eye is the focusing mechanism, or the so-called lens accommodation system. An analysis of its performance principles has been presented by O'Neill, Sanathanan and Brodkey [17].

The plant dynamics, i.e. the dynamics of the lens and ciliary muscle system is described by a nonlinear first-order differential equation. The ciliary nerve firing rate is the controlling variable determining the contraction level of the ciliary muscle and, therefore, the dioptric level of the lens. The authors postulated that the focusing from one level to another is done by minimizing the transition time. And so we have again a minimum time control problem, where the cost functional is

$$J = \int_0^T dt$$

In this case the control bounds are determined by the maximum and minimum frequencies of the nerve signal. The model results in a bang-bang type of control sequence and it is considered to be in good accordance with the experimental findings. Physiologically the minimization of the focusing period corresponds to the minimization of the time interval during which the image is not sharp on the retina. During such 'blind' periods one may get into dangerous situations and thus minimization of the duration of these periods can be essential for survival.

6. Locomotion

In recent years an increasing interest has been taken in the locomotion studies. Works reported in the literature deal both with the analysis and synthesis problems [26,47,48]. The analytical approach tries to simulate and improve the understanding of the performance principles of human walking patterns whereas the synthesis studies aim at designing artificial limbs, assistive devices or other technical biped or multiped locomotion systems.

Many experimental investigations have been performed to find 'optimality' in human locomotion [31,32,35]. Cotes and Meade have shown for example that in level walding with a fixed velocity there is a step frequency that minimizes the energy expenditure. Correspondingly optimal values have been sought for the walking speed and the step-length. Similar studies were performed theoretically by Nubar and Contini [19] and by Beckett and Chang [28].

Chow and Jacobson [13] have presented an analysis of the locomotion system which is of special interest. Their extensive work aims at finding out the basic mechanism and laws governing locomotion. The locomotion control system is considered to have a hierarchical structure and the locomotion activity is supposed to follow a principle of optimality. In formulating the model structure the following problems are dealt with: a mathematical model for simulating the functional behavior of the locomotor system; kinematic and dynamic constraints on the basis of gait information; initial, terminal and trajectory conditions; and optimality criterion determining the patterns of motion.

The system studied consists of a rigid body with two three-jointed legs (ankle, knee, hip). The variables used to describe the motion are the angular displacements of hip and knee and the horizontal and vertical displacements of the joints. The constraints include, for example, initial and terminal conditions for the stance, deploy and swing phases of the normal walking cycle. The control variables are the moments generated by muscle action about the joints.

The performance principle of the control system is supposed to be the minimization of the energy expenditure of walking. The development of the corresponding cost criterion is based on a detailed analysis of muscle characteristics in different situations. They found out that the energy expenditure of the muscle activating system is proportional to the integral of the square of the net moment. Thus the criterion for normal walking was obtained with $p = 2$ from the generalized criterion

$$J = \int_{t_0}^{t_f} \Sigma r_i |u_i|^p \, dt$$

which they presented. Here the index i refers to different joints, u_i is the net moment acting on the ith joint, r_i is a weighting coefficient and p some positive integer. The minimum time criterion with $p = 0$ is thought to be applicable to fast walking.

Due to the three phases of the cycle the formulation leads to a multiarch optimization problem. The solution of this complex model is shown to yield trajectories which agree well with the normal cycle patterns observed in level walking.

Townsend and Seireg have taken a synthesis oriented approach [26]. They studied more complex biped model structures and three different objective criteria. They desired to determine what locomotion characteristics can be expected of given models and what models are most applicable in the study of human and machine biped locomotion control. An aim was also to identify locomotion criterion functions which yield 'desirable' trajectories for system design and recognizable gait characteristics relative to human motions.

The simplest model consists of one rigid body supported by two extensible legs with two joints (ankle and hip). The two other models have two rigid bodies. These bodies are considered cylindrical and in one model also relative rotation between the bodies is allowed. The hypothesized control functions are the joint torques and axial forces of the extensible legs.

The objective functions studied were weighted combinations of the size of the support base (footprint) during each step, an energy expenditure term and the magnitudes of the system's angular motions. The energy expenditure is approximated here with the integral of the absolute value of the mechanical work rate. The angular motion term is included to study the effects of minimizing the fluctuations in the nominal heading of the torso. By changing the weightings in the criterion varying walking patterns can be produced which resemble different types of human gait.

7. Muscle control

Models of the operating principles of muscle systems are often only qualitative descriptions. These aim at explaining how the central nervous system selects the activity of different muscles, e.g. how the coordination of muscles involved in locomotion or in respiration is performed [13,23,25,49]. Usually the control system structure is assumed to be hierarchical so that a higher level decision-making block coordinates the lower levels where the synchronization of muscle groups and the detailed distribution of muscle activity is controlled. Optimization is frequently included in these processes.

Smith [12] has studied the response of the human forearm following step commands. Tests were made which required the hand to move different loads consisting of inertia, friction and spring components. The best responses were observed to approach those of a maximum-effort minimum-time optimal controller.

Muscle tremor is a phenomenon that has been explained to be the result of an optimizing system. Aizerman and Andreeva [49] propose a general control hypothesis which they call the 'simple search mechanism'. This concept has been used in studying the control of respiratory muscles [44] and the muscle tremor [50]. Linn and Vossius [15,16] have suggested that tremor would be a result of a time optimal stabilization of muscular movements by the fast muscle fibers. It is also thought that the slow fibers would operate according to the minimum energy criterion.

8. Anatomical design principles

The above considerations on optimization modeling concentrated only on the functioning principles of control systems with a given anatomical structure of the related organs. There are, however, interesting results obtained for the dual problem suggesting that the design principles of certain anatomical structures related to a given function can also be modeled by optimization criteria. Reference will not be made to this field as a whole but examples from the vascular and respiratory systems are mentioned.

There are a number of works [51,52,53,54], the earliest of which originate from the twenties, where optimality arguments are applied to explain vascular design. Criteria such as minimum blood flow resistance or work rate needed for the flow have been used to predict for example vascular branching and bifurcation angles or the radia of aorta, vascular branches or capillaries. In each case the different assumptions of the physical constraints determine the varying solutions. The discoidal form of the red blood cells is also found to agree with some optimal design formulations [24]. Corresponding analyses have also been made concerning the design of the bronchial tree. Optimal branching patterns and diameters have again been derived [55,56,57]. More detailed discussion on such models can be found, for example, in [10].

9. Practical application of optimization models

The main contribution of optimization models has so far been limited to the improved understanding of the physiological control principles. But when there is enough experimental evidence available to assure the validity of the particular models the scope of applications is likely to widen.

In clinical practice these mathematical models may be used as diagnostic aids in the early detection of abnormal responses and in the following of the recovery process. Models can be used to predict effects of alternative treatment procedures and thus help the physician select the most effective one and avoid experimenting with the patient [58, 59]. The natural performance principles may be taken into account when designing and operating treatment and assistive devices which control some physiological subsystems artificially. This may turn out to be a very important aspect in the future when more and more complex and automatically operating devices as, for example, artificial organs are developed. This kind of approach is needed to minimize interferences in a man-machine system and the side effects of treatment devices.

Optimization models have already been applied in the control of respirators. Wald *et al.* [60] and Jain and Guha [61,62] have made some considerations on the optimization of control patterns of a cycle so that certain harmful side-effects caused by positive pressure breathing are minimized. Mitamura *et al.* [63] and Grevisse *et al.* [64] have designed repirator systems where the breathing frequency is chosen according to a criterion which predicts the normal frequency of spontaneous breathing.

Optimal control of a left ventricular bypass pump assisting the circulation has been studied by Anderson and Clark [65]. The performance criterion used in the design includes terms for myocardial energy expenditure and for mean atrial pressure. These terms can be considered to reflect the natural performance principles of an unassisted system. Such an argument was not stated explicitly in [65], but this interesting note can be made on the basis of the model of systolic elastance described above in section 3.

In the near future the most important practical applications of optimization models will probably be in the design of prostheses. There has been an increasing activity in locomotion studies oriented in the synthesis problems. An objective is to design control systems for artificial limbs which produce normal operating patterns.

The design of technical man-machine systems where high reliability is needed may also require knowledge of the normal performance principles of the visual or some muscular systems. So far most human operator studies with optimization concepts have been limited to the modeling of an airplane pilot [37,66].

10. Conclusion

Optimization models provide a new promising method to study the behavior of biological mechanisms. Good results have been obtained in the modeling of adaption or decision making in a number of physiological subprocesses. Optimization models can be used to predict the performance of a system in various environmental conditions. Some of the successfully applied criteria also have a clear physiological interpretation furthering thus the knowledge concerning the behavior of the organism in question.

This modeling technique has been used already for a long time but only during the past few years has the scope of applications increased significantly. At the same time modeling began to employ advanced mathematical methods of control theory. So far the models have been used mainly to predict average responses for humans. The next step is to make studies on an individual level. One of the main problems will be the identification difficulties of system parameters and weighting coefficients in the performance criteria. Only a few attempts have been made so far to apply these models to practical control or design purposes. However, there is some indication that these aspects will receive increasing attention in the future.

References

1. CLYNES, M. and MILSUM, J.H., *Biomedical Engineering Systems,* McGraw-Hill Book Co., New York (1970).
2. GRODINS, F.S., *Control Theory and Biological Systems,* Columbia University Press (1963).
3. HEINMETS, F. and CADY, L.D. Jr., *Biomathematics,* Vol.I, Michel Decker Inc., New York (1970).
4. MILHORN, H.T., *The Application of Control Theory to Physiological Systems,* W.B. Saunders Co., Philadelphia (1966).
5. MILSUM, J.H., *Biological Control Systems Analysis,* McGraw-Hill Book Co., New York (1966).
6. OPPELT von, W. and VOSSIUS, G. (editors), *Der Mensch als Regler,* VEB Verlag Technik, Berlin (1970).
7. ROSEN, R., *Dynamical Systems Theory in Biology,* Vol.I, John Wiley (1971).
8. STARK, L., *Neurological Control Systems: Studies in Bioengineering,* Plenum Press, New York (1968).
9. MILSUM, J.H., 'Optimization aspects in biological control theory' in *Advances in Biomedical Engineering and Medical Physics,* Vol.I (edit. by S.N. Levine), pp.243-277, John Wiley & Sons, New York (1968).
10. ROSEN, R., *Optimality Principles in Biology,* Butterworths, London (1967).
11. COOK, G. and STARK, L., 'Derivation of a model for the human eye positioning mechanism', *Bull. Math. Biophys.* **29**, pp.153-174 (1967).
12. SMITH, O.J.M., 'Nonlinear computation in the human controller', *IRE Trans. Bio-Med. Electron.,* BME-**9**, pp125-128 (1962).
13. CHOW, C.H. and JACOBSON, D.H., 'Human locomotion studies', *Math. Biosc.,* **10**, pp.239-306 (1971).
14. CLARK, M.R. and STARK, L., 'Time optimal behaviour of human saccadic eye movement', *IEEE Trans. Aut. Control,* AC-**20**(3), pp.345-348 (1975).
15. LINN von, K., 'Der Physiologishe Tremor: Auswirkung eines Abtastregelkreis im Zentralnervensystem?', *Regelungstechnik,* No.9, pp.327-331 (1975).
16. LINN von K. and VOSSIUS, G., 'The relevance of the so-called tremor for the control of voluntary movement', Preprints of IFAC 6th World Congress, Boston, USA, paper 40.4 (August 1975).
17. O'NEILL, W.D., SANATHAN, C.K. and BRODKEY, J.S., 'A minimum variance, time optimal, control system model of human lens accommodation', *IEEE Trans. Syst. Science and Cybernetics,* SSC-**5**(4), pp.290-299 (1969).
18. MEAD, J., 'Control of respiratory frequency', *J. Appl. Physiol.* **15**(3), pp.325-336 (1960).
19. NUBAR, Y. and CONTINI, A., 'A minimal principle in bio-mechanics', *Bull. Math. Biophys.* **23**, 377-390 (1961).
20. RUTTIMAN, U.E. and YAMAMOTO, W.S., 'Respiratory airflow patterns that satisfy power and force criteria of optimality', *Ann. Biomed. Engrn.* **1**, pp.146-159 (1972).
21. WIDDICOMBE, J.G. and NADEL, J.A., 'Airway volume, airway resistance and work and force of breathing theory', *J. Appl. Physiol.* **18**, pp.863-868 (1963).
22. MARGARIA, R., 'A mathematical treatment of the blood dissociation curve for oxygen', *Clin. Chem.* **9**, pp.745-762 (1963).
23. FINCHAM, W.F. and PRIBAN, I.P., 'Autonomic control of respiration', Proc. Second IFAC Symp. The Theory of Self-Adaptive Control Systems, pp.57-63 (1965).
24. LEHMANN, H. and HUNTSMAN, R.G., 'Why are red cells the shape they are?', in *Functions of the Blood* (edit. by R.G. MacFarlane and A.H.T. Robb), Academic Press, New York (1961).
25. PRIBAN, I.P. and FINCHAM, W.F., 'Self-adaptive control and respiratory system', *Nature*, **208**, pp.339-343 (1965).
26. TOWNSEND, M.A. and SEIREG, A.A., 'Effect of model complexity and gait criteria on the synthesis of bipedal locomotion', *IEEE Trans. Biomed. Engineering,* BME-**20**(6), pp.433-444 (1973).
27. YAMASHIRO, S.M. and GRODINS, F.S., 'Respiratory cycle optimization in exercise', *J. Appl. Physiol.* **35**(4), pp.522-525 (1973).
28. BECKETT, R. and CHANG, K., 'An evaluation of kinematics of gait by minimum energy', *J. Biomech.* **1**, pp.147-159 (1968).
29. CHRISTIE, R.V., 'Dyspnoea in relation to the visco-elastic properties of the lung', *Proc. Roy. Soc. Med.* **46**, pp.381-386 (1953).
30. CROSFILL, M.L. and WIDDICOMBE, I.G., 'Physical characteristics of the chest and lungs and the work of breathing in different mammalian species', *J. Physiol.* **158**, pp.1-14 (1961).
31. FENN, W.O., 'Work against gravity and work due to velocity changes in running', *Amer. J. Physiol.* **93**, p.433 (1930).
32. FENN, W.O., 'The mechanics of muscular con-

traction in man', *J. Appl. Physiol.* **9**, p.165 (1938).
33. YAMASHIRO, S.M., DAUBENSPECK, J.A., LAURITSEN, T.N. and GRODINS, F.S., 'Total work rate of breathing optimization in CO_2 inhalation and exercise', *J. Appl. Physiol.* **38**(4), pp. 702-709 (1975).
34. YAMASHIRO, S.M. and GRODINS, F.S., 'Optimal regulation of respiratory airflow', *J. Appl. Physiol.* **30**(5), pp.597-602 (1971).
35. COTES, J.E. and MEADE, F., 'The energy expenditure and mechanical energy demand in walking', *Ergonomics* **3**(2), pp.97-119 (1960).
36. PARK, I.-G. and LEE, K.Y., 'An inverse optimal control problem and its application to the choice of performance index for economic stabilization policy', *IEEE Trans. Syst., Man, Cyb.* SMC-**5**(1), pp.64-75 (1975).
37. BARON, S. and LEWISON, W.H., 'An optimal control methodology for analyzing the effects of display parameters on performance and workload in manual flight control', *IEEE Trans. Syst., Man, Cyb.* SMC-**5**(4), pp.423-430 (1975).
38. HÄMÄLÄINEN, R.P., 'Adaptive control of respiratory mechanics', *Trans. of ASME J. Dynamic Systems, Measurement and Control,* pp.327-331 (September 1973).
39. NOLDUS, E.J., 'An optimization concept of systolic elastance', *Preprints of IFAC 6th World Congress*, Boston, USA, paper 54.4 (August 1975).
40. VILJANEN, A.A., 'Co-ordination of neuromuscular efferent systems in ventilation', in *Ventilatory and Phonatory Control System* (edit. by B. Wyke), Oxford University Press, London (1974).
41. SZÜCS, B., MONOS, E. and CSÁKI, F., 'New aspects of blood pressure control', Preprints of IFAC 6th World Congress, Boston, USA, paper 54.5 (August 1975).
42. ROHRER, F., 'Physiologie der Atembewegung', in *Handbuch der Normalen und Pathologischen Physiologie,* Vol.2 (edit. by A. Bethe *et al.*), Springer, Berlin, pp.70-127 (1925).
43. OTIS, A.B., FENN, W.O. and RAHN, H., 'Mechanics of breathing in man', *J. Appl. Physiol.* **2**, pp.592-607 (1950).
44. TENENBAUM, L.A., 'On control processes of external respiratory parameters', *Automatica* **7**, pp.407-416 (1971).
45. O'NEILL, W.D., 'An interacting control systems analysis of the human lens accommodative controller', *Automatica* **5**(5), pp.645-654 (1969).
46. STARK, L., 'The pupillary control system: its non-linear adaptive and stochastic engineering design characteristics', *Automatica* **5**(5), pp.655-676 (1969).
47. VUKOBRATOVIĆ, M. and JURIČIĆ, D., 'Contribution to the synthesis of biped gait', *IEEE Trans. Biomed. Eng.,* BME-**16**(1), pp.1-6 (1969).
48. VUKOBRATOVIĆ, M.K. and OKHOTSIMSKII, D.E., 'Control of legged locomotion robots', Preprints of IFAC 6th World Congress, Boston, USA, Plenary papers (August 1975).
49. AIZERMAN, M.A. and ANDREEVA, E.A., 'Simple search mechanism for control of skeletal muscles', *Automation and Remote Control,* No.3, pp.452-463 (1968).
50. CHERNOV, V.I., 'Control over single muscle of a pair of muscle antagonists under conditions of precision search', *Automation and Remote Control* **7**, pp.1090-1101 (1968).
51. COHN, D., 'Optimal systems: I. The vascular system', *Bull. Math. Biophys.* **16**, pp.59-74 (1954).
52. COHN, D., 'Optimal systems: II. The vascular system', *Ibid.* **17**, pp.219-227 (1955).
53. MURRAY, C.D., 'The physiological principle of minimum work I', *Proc. Nat. Acad. Sci.* **12**, pp.207-214 (1926).
54. MURRAY, C.D., 'The physiological principle of minimum work II', *Ibid.* **12**, pp.299-307 (1926).
55. HORSFIELD, K. and CUMMING, G., 'Angles of branching and diameters of branches in the human broncial tree', *Bull. Math. Biophys.* **29**, pp.245-259 (1967).
56. RASHEVSKY, N., 'On the function and design of the lung', *Bull. Math. Biophys.* **24**, pp.229-241 (1962).
57. WILSON, T.A. and LIN, K.-H., 'Convection and diffusion in the airways and the design of the bronchial tree' in *Airways Dynamics* (edit. by A. Bouhuys), Charles C. Thomas, Springfield, pp.5-20 (1970).
58. PRIBAN, I.P., 'Forecasting failure of health', *Spectrum*, No.48, pp.9-12 (1968).
59. PRIBAN, I.P., 'Models in medicine', *Science Journal*, pp.61-67 (June 1968).
60. WALD, A.A., MURPHY, T.W. and MAZZIA, V. D.B., 'A theoretical study of controlled ventilation', *IEEE Trans. Bio-Medical Engineering,* BME-**15**(4), pp.237-248 (1968).
61. JAIN, K.V. and GUHA, S.K., 'A study of intermittent positive pressure ventilation', *Med. & Biol. Engng.* **8**, pp.575-583 (1970).
62. JAIN, K.V. and GUHA, S.K., 'Design for positive pressure respirators', *Ibid.* **10**, pp.253-262 (1972).
63. MITAMURA, Y., MIKAMI, T., SUGAWARA, H. and YOSHIMOTO, C., 'An optimally controlled respirator', *IEEE Trans. Bio-Medical Engn.*, BME-

18(5), pp.330-337 (1971).

64. GREVISSE, Ph., DEMEESTER, M. and LECOCQ, H., 'A pulmonary model for the automatic control of a ventilator', Preprints of IFAC 6th World Congress, Boston, USA, Paper 54.2 (August 1975).
65. ANDERSON, C.M. and CLARK, J.W. Jr., 'Analog simulation of left ventricular bypass mode control', *IEEE Trans. Biomed. Eng.* BME-22(5), pp.384-392 (1975).
66. PHATAK, A.V. and BEKEY, G.A., 'Decision processes in the adaptive behavior of human controllers', *IEEE Trans. Syst. Science and Cybernetics* SSC-5(4), pp.339-351 (1969).

Logical probability models and representation theorems on the stable dynamics of the genetic net

ANITA K. BABCOCK*
State University of New York,
Buffalo, New York

Some mathematical theorems were proved in the work reported here (Babcock [1,2]) regarding the stable dynamics of randomly, sparsely connected switching nets simulated numerically by Kauffman [3,4]. Disclosure lengths were derived from a probability model of the genetic net.

The fundamental questions in biology include questions such as: how much do we really know of the mechanism whereby stable dynamics arise in switching networks of arbitrary size? how does metabolic stability arise in nets with interactions between thousands or millions of chemical species? where is the point, as connectance is increased, at which the stable dynamics break down? what properties of the genetic-biochemical network of a cell are responsible for the control of epigenesis in multicellular organisms? Some answers to questions of this sort were obtained as theorems in the work reported here (Babcock [1,2]).

The genetic net is an automaton, approximating the aspects of the stable dynamics of the interactions between regulated genes due to just those properties of the net resulting from the fact that it is a set of positive and negative interactions. The stable dynamics of the genetic net might be responsible for much of the control of epigenetic processes such as cellular differentiation.

The sparsely connected nets observed by Kauffman comprise a class of dynamically stable systems. Since these nets do constitute statistical systems, asymptotic results from both graph theory and combinatorial analysis were among the methods employed in the reported work.

The random net approach to the study of the central nervous system by statistical systems of neural nets, proposed by Rapaport [5], is an example of the earliest studies of such systems. Nerve nets were considered random, with only distance biases on the distribution of connection lengths being specified, because there appeared to be insufficient genetic information to specify fully the neural connections.

Approximating theoretical models are necessary

*Presently at the Laboratory of Theoretical Biology, National Institutes of Health, Bethesda, Maryland.

† The thesis work reported here was completed with Prof. Robert Rosen.

since the data available on the genetic net are insufficient to identify the important network parameters, permitting a detailed understanding of its dynamics. The effector-mediated interactions between regulated genes are hard to analyze mathematically. Statistical mechanics, as a classical ensemble method, can give the average behavior of a system of disorganized complexity. These methods are inapplicable to systems having complexity that is overtly organized, including the genetic net.

The present study gives an analysis of how the properties of a genetic net may be restricted to create control mechanisms. One such property is forcibility. A function is forcible on a particular input if one value of that input completely determines the function value. Such a function will have greater internal homogeneity in its output values if it is forcible on a greater number of its inputs. The presence of many forcible switching functions determining the activity values of regulated genes in a genetic net can impose a reflexive bias on trajectories toward recent antecedent states.

A forcing structure is defined here as a maximal connected subgraph of the forcing digraph, i.e., of the directed graph of the forcible connections in a genetic net. To find forcing structures, should they in fact be present in the genome, DNA-messenger hybridization experiments could be carried out to identify the blocks of genes that are always on (or off) as the forcing structures.

The biological evidence supports the assumption of the restriction of many regulated genes to forcible functions of few controling input variables in bacteria and bacteriophage lambda (Kauffman [6]). In eukaryotes, despite differences from prokaryotes in the molecular mechanisms, the dynamics of functionally similar regulatory circuits might account for much of the cellular metabolic homeostasis. The specificity of macromolecules, due to their geometric complexity, means that only a few inputs are likely to directly regulate a particular gene.

The typical dynamics of circuits of any size or connectance were derived in the study reported here. If the assumptions, based on the present experimental knowledge of the genetic net, approximate the actual biology, then the systems analyzed here reasonably approximate the genetic net in their dynamics. The effect on the dynamics was analyzed of the imposition of a refexive bias on trajectories by the restriction of many regulated genes to forcible functions.

A genetic net is the result of the functional resolution of the genome into a set of interacting elements, such as the repressors or inducers of operons in Monod and Jacob's theory [7]. Genetic nets are generally nonlinear chemical systems having multiple steady states. Their size and complexity makes analysis based on differential equations difficult. For this reason, digital simulations of a switching net as the approximating model were carried out by Kauffman [3], among others. Control circuits implicit in the operon theory were discussed earlier by several authors (Sugita [8], Rosen [9]). Kauffman observed in his sparse nets that a large forcing structure in the net eventually stops changing state. At each transition, only the state of many small trees, which may contain a loop, might change subsequently.

The present mathematical analysis may be compared with Kauffman's numerical experiments on large deterministic nets constructed as though genetic nets were random, since the interactions are too numerous to know at present. A trajectory in a net with N randomly, sparsely connected elements, each computing a randomly chosen function on two inputs, typically contained only about $N^{1/2}$ states. Thus, most trajectories are nonergodic, in that they do not pass through every available state in an arbitrary embedding of the discrete state set into phase space. Since metabolically stable dynamics appear only in nets with low mean connectance, there appears to be a limit to the complexity of stable systems, unless they are arranged in special ways. This is intuitively clear, since additional modes of oscillation become available as complexity is arbitrarily increased.

The density of functions that are forcible on j coordinates $(0<j\leqslant k)$ was determined combinatorially, after which a reflexive bias was imposed on state transitions. An expression was finally derived from the bias for the probability that a trajectory ends in a cycle of a particular length, i, following a transient of some length, j, from an arbitrary starting state. This probability was used in the derivation of the probability that a trajectory ends in a steady state, after which the latter probability was compared with the disclosure lengths observed by Walker and Ashby [10] for sparse nets.

The cycle length densities were found by the technique of counting the number of trajectories with cycles of each length. It was necessary to use this rigorous method to derive the stable dynamics of a system of arbitrary size and connectance because the states of a switching net are without a biologically natural (biochemical or genetic) measure of closeness. The part of the net which is not modeled and the environment both impose the bias on the state transitions of the genetic net.

It was proved that $C_{k,j}2^{2^{j}+1-2^{k}}$ of all mappings on

k input co-ordinates are forcible on exactly j inputs, when the internal homogeneity is $2^k - 2^{k-j}$. Recall that a mapping is forcible on an input if one value taken on by that input determines the mapping value. Newman and Rice [11] proved earlier that 2^{k+1-2^k} of the mappings on k inputs are forcible on all k.

After this expression was obtained, it was found that only mappings forcible on all k inputs correspond *1-1* to the class of mappings with a particular number of identical entries, that number being 2^k*-1*.

A unique, canonical form of a given net was obtained by a series of canonical operations. The net obtained is such that no mapping may be factored through a subproduct of its domain. Each mapping has a unique factorization, $f = \bar{f}Pf'$. The function $f': A^k \rightarrow A^k$ permutes the forcing inputs to the first j of the k input co-ordinates. The projection $P: A^k \rightarrow A^j$ projects the k-dimensional domain to the smaller j-dimensional input set, under the reasonable assumption that at least one forcible input takes on its forcing value. Finally, $\bar{f}: A^j \rightarrow A$ is the nonfactorable function that corresponds to the particular function f that is forcible on j of its input co-ordinates.

The canonical factorizations rely on the reasonable assumption mentioned in the preceding paragraph, that at least one forcible input takes on its forcing value. Under this assumption, a mapping not forcible on the input co-ordinate A^i is factorable through a subproduct of the domain, given by $A^{k-1}(i) = (A_1,...,0,...,A_k)$, where 0 is in the ith position. The representation of a net with nonfactorable mappings was carried out previously by Rosen [12].

The upper bound on the number of forcing connections per element was found. Calculations of a related upper bound were made earlier by Kauffman [4]. The upper bound defined here,

$$k' = k^3 2^{1-2^k}$$

is an expression in terms of k, the mean connectance of the net, i.e. the mean number of inputs per element. Using this expression, the giant component containing most elements in the forcing structure was shown to not be almost surely present. But the probability that a giant component is present was by far the greatest for the lowest reasonable connectance, k, of 2. The latter results followed from the change obtained by Erdos and Renyi [13] in the global structure of asymptotically large random graphs. This change occurs almost surely, i.e. with probability near *1*, when the mean connectance, k, exceeds ½, after which a giant component is present together with many small trees, which can each contain a single loop.

Recall that the reflexive bias on the densities of cycle lengths in sparse nets favors transitions to states that are more recent predecessors. Throughout this discussion, only the autonomous operation of the genetic net is considered. The bias depends upon the distribution of forcible functions mentioned earlier and the probability that a net of a particular mean connectance is strongly connected, a known result in the recent literature of graph theory (Palasti [14]). A strongly connected net is one that has a directed path between every pair of points. A strongly connected net is sufficient to ensure the presence of a giant forcing structure in a net, but it is not a necessary condition. The derivation of the densities of cycle lengths uses these ideas, but the reader is referred to the original work for the description of the entire argument (Babcock [1,2]). Results from the random net literature were used in that argument (Rapaport [5,15]). From the cycle length densities, the probability was derived that a trajectory ends in a steady state, as were several other similar probabilities. The latter results were compared with the results of the numerical experiments by Walker and Ashby [10] on nets with the low connectance, k, of *3*. The comparisons were made in the two cases where functions were forcible on all k inputs, and where functions were randomly selected from all functions on k inputs.

One value of studies of the present type is that several widely different methods, which were applied to very different systems, can be used to give valuable insight into other systems, such as those aspects analyzed here of the dynamics of the genetic net which result solely from its restricted properties as a net of positive and negative interactions.

Until now, only the analysis of such a system was described. A detailed synthesis was also carried out by means of shift registers, using methods from automata theory that are quite different from the random net work on the nervous system that was applied in the present analysis of the genetic net, apparently for the first time.

A unique representation was proposed for the transitions, or state graph, of the net. Shift registers were composed by the canonical method demonstrated by Elspas [16] for linearly realizable machines. The small tree structures were represented, while the giant forcing structure could be neglected, since it became fixed very soon and subsequently never changed state after transitions in the state of the entire net. The small trees with loops are the only part of the forcing graph that are not in large structures, close to coalescence into a giant structure if more connections are added. The large structures were observed in numerical experiments to eventually never change state (Kauffman [4]).

The random net approach was further applied in the work reported here to the problem of determining how several distance biases on diffusion lags might affect the dynamics of the genetic net. An analysis of the same form was applied to biases on the convergence of the transitions in state of nets having external input from the environment.

One advantage in pursuing this type of study is that it organizes many ideas by bringing together methods formerly applied to different structural systems, showing that the parts of the systems modeled have the same dynamics.

The predominance of short cycles derived in the work reported may be interpreted as having an approximate biological identification with the activity patterns of regulated genes that distinguish the cell types of an organism. It is reasonable to make the approximate identification of the short state cycles of the sparsely connected genetic net with the gene activity patterns characteristic of distinct cell types, at the present early stage in the experimental determination of the genetic net.

References

1. BABCOCK, A., 'Logical probability models and representation theorems on the stable dynamics of the genetic net', Unpublished doctoral dissertation, Biophysics Department, S.U.N.Y., Buffalo, N.Y. (1976).
2. BABCOCK, A., 'Stable dynamics of genetic networks', *Progress in Theoretical Biology,* 5, 63-79 (1978).
3. KAUFFMAN, S.A., 'Metabolic stability and epigenesis in randomly constructed genetic nets', *J. Theoret. Biol.* 22, 437-467 (1969).
4. KAUFFMAN, S.A., 'The organization of cellular genetic control systems', *Mathematics in the Life Sciences*, Vol.3, American Mathematical Society, pp.63-116 (1970).
5. RAPOPORT, A., 'Cycle distributions in random nets: II', *Bull. Math. Biophys.* **10**, 145-157 (1948).
6. KAUFFMAN, S.A., 'The large scale structure and dynamics of gene control circuits: an ensemble approach', *J. Theoret. Biol.* **44**, 167-190 (1974).
7. MONOD, J. and JACOB, F., 'Teleonomic mechanisms in cellular metabolism, growth and differentiation', *Cold Spring Harbor Symp. Quant. Biol.* **26**, 389-401 (1961).
8. SUGITA, M., 'Functional analysis of chemical systems *in vivo* using a logical circuit equivalent', *J. Theoret. Biol.* **1**, 415-430 (1961).
9. ROSEN, R., 'Two-factor models, neural nets, and biochemical automata', *J. Theoret. Biol.* **15**, 282-297 (1967).
10. WALKER, C. and ASHBY, W.R., 'On temporal characteristics of behavior in certain complex systems', *Kybernetik* **3**, 100-108 (1966).
11. NEWMAN, S.A. and RICE, S.A., 'Model for constraint and control in biochemical networks', *Proc. Nat. Acad. Sci. USA* **68**, 92-96 (1971).
12. ROSEN, R., 'The representation of biological systems from the standpoint of the theory of categories', *Bull. Math. Biophys.* **20**, 317-341 (1958).
13. ERDOS, P. and RENYI, A., 'On the evolution of random graphs', *Magyar Tud. Akad. Mat. Kutato Int. Kozl.*, Ser. A5, 17-61 (1960).
14. PALASTI, I., 'On the strong connectedness of directed random graphs', *Studia Sci. Math. Hungar.* **1**, 205-214 (1966).
15. RAPOPORT, A., 'Contribution to the theory of random and biased nets', *Bull. Math. Biophys.* **19**, 257-277 (1957).
16. ELSPAS, B., 'The theory of autonomous linear sequential networks', *IRE Trans.* CT-**6**, 45-60 (1959).

Information transfer in a chain of model neurones*

A.V. HOLDEN
University of Leeds, Leeds, UK

1. Introduction

Experimental neurobiologists have provided, and continue to provide, a wealth of detailed information about the electrical activity of nerve cells. One obvious impression to be gained from the experimental literature is of the variety of structures and mechanisms which are to be found in neurones: the typical neurone exists only in textbooks. However, in spite of this diversity, there are two basic properties which appear common to most neurones which have long, thin axons:

(*a*) electrical signalling over long distances is by a series of brief, stereotyped, all-or-none action potentials, which differ only in their times of occurrence. These action potentials can be considered as unitary point events.

(*b*) the action potential trains recorded from neurones are stochastic, in the sense that the intervals between action potentials are random variables.

This suggests that the language of the theory of stochastic point processes is a natural language for discussing problems of information transmission by neurones.

There are a large number of models of the generation of action potential trains by neurones: these have been recently reviewed by Fienberg (1974) and Holden (1976). Here I will consider two simple neural models: the perfect and leaky integrator models. The leaky integrator model was introduced by Lapique (1907) to account for the relation between the strength and duration of a just-threshold electrical stimulus. In the leaky integrator model the effects of inputs sum linearly and decay exponentially with a single time constant τ, for changes up to a constant threshold V_0. When the threshold is reached a pulse is generated and the state variable $V(t)$ is instantaneously reset to its resting value, x_0. Thus, if a pulse was generated at $t = 0$, and no pulse is generated on $(0, t]$,

$$V(t) = \int_0^t x(u) \exp\{-(t-u)/\tau\}\, du \tag{1}$$

for $x_0 < V(t) < V_0$, with $x(t)$, $t > 0$, the input to the integrator, and the interpulse interval t_i is defined by

$$V_0 = \int_0^{t_i} x(u) \exp\{-(t-u)/\tau\}\, du. \tag{2}$$

The perfect integrator model is obtained from Eq.(1) when $\tau \to \infty$:

[†]This work was supported by a grant (B/RG 2195) from the Science Research Council.

$$V(t) = \int_0^t x(u)\,du \qquad \text{for } x_0 < V(t) < V_0 \quad (3)$$

and the interpulse interval of the perfect integrator model is defined by

$$V_0 = \int_0^{t_i} x(u)\,du \qquad (4)$$

These two simple abstract models for pulse train generation have been widely used as neural models (e.g. see Johannesma, 1968), and it is tempting to identify the state variable $V(t)$ with the membrane potential at the action potential initiation site, and the time constant τ with the membrane time constant, given by the product of membrane resistance and capacitance. However, Knight (1973) derives the leaky integrator model as a limiting form of the Hodgkin-Huxley membrane equations, with the time constant τ determined by the sodium activation time constant. These two simple models are best regarded as abstract representations of the pulse generation process, rather than as crude approximations of the neuronal mechanisms, which ignore the effects of neuronal geometry and the non-linear, time-dependent membrane current-voltage relations.

These two simple models are both deterministic: in response to a constant input c a pulse train with a constant rate r is generated, with a linear relationship for the perfect integrator

$$r = c/V_0$$

and an increased sensitivity for small inputs for the leaky integrator

$$r = \frac{1}{-\tau \ln(1 - V_0/c\tau)} \qquad \text{where } c\tau > 1$$

For these models to generate a stochastic pulse train the input $x(t)$ must be a stochastic process.

Wiener (1958) has developed a method of system identification in which the input is a Gaussian white noise $w(t)$, and the response is expressed as an infinite sum of orthogonal functionals. This technique can be applied to the class of nonlinear operators that are Lebesgue square integrable over the sample space of realizations of Gaussian white noise inputs, and some applications of this technique to neuronal systems are discussed in McCann and Marmarelis (1975).

Another kind of stochastic input function to these models is a random train of Dirac delta functions: this might be thought of as representing an input train of action potentials.

In this chapter I will consider some properties of information transmission in a feed-forward chain of these model neurones, and discuss the relevance of these properties to information transmission in a chain of sensory neurones. The method is to characterize the input-output relations of a single model neurone, so that the response of the first unit of a chain to an arbitrary input can be obtained. This response can then be used as the input to the second unit, and its response obtained, and this is then the input to the third unit. Thus the characteristics of the signal can be traced through the chain.

2. Two diffusion processes and their first passage times

A useful input to the model neurones is a white noise process, $w(t)$, which is a delta-correlated process with a Gaussian amplitude distribution and zero-mean. This could represent a sensory neurone subject to a band-limited white noise stimulus (French, Holden and Stein, 1972), or approximate the input to a central neurone which receives a large number of mutually independent excitatory and inhibitory inputs.

In the two models the instantaneous reset of $V(t)$ to its resting value x_0 whenever a pulse is generated destroys all effects of previous inputs. If an output pulse is generated at a time $t = 0$, the next pulse will be generated when $V(t)$ first reaches the threshold V_0. Thus the distribution of intervals between adjacent pulses will be identical to the first passage time distribution $P(t)$ of the process $V(t)$ from x_0 to V_0. When the input to the model neurones is a white noise, $\{V(t)\}$ defined by Eqs (1) and (3) is a continuous, stationary Markov process with a transition probability distribution function $F(y, t \mid x)$ defined by

$$\begin{aligned} F(y, t \mid x) &= \text{Prob}\,\{V(t) < y \mid V(0) = x\} \\ &= \text{Prob}\,\{V(t_2 - t_1) < y \mid V(0) = x\} \\ &= \text{Prob}\,\{V(t_2) < y \mid V(t_1) = x\} \end{aligned}$$

for $t_2 > t_1$, $t_2 - t_1 = t$.

The transition probability density function

$$f(y, t \mid x) = \frac{\partial F(y, t \mid x)}{\partial y}$$

satisfies the backward Chapman-Kolmogorov equation

$$\frac{-\partial f(y,t\,|\,x)}{\partial t} = \tfrac{1}{2}\alpha(x)\,\frac{\partial^2 f(y,t\,|\,x)}{\partial x^2} + \beta(x)\,\frac{\partial f(y,t\,|\,x)}{\partial x} \tag{6}$$

where $\alpha(x)$ and $\beta(x)$ are the infinitessimal variance and mean of $\{V(t)\}$.

The first passage time probability density function $p(V_0, t\,|\,x_0)$ satisfies

$$f(y,t\,|\,x_0) = \int_0^t p(V_0, t-\tau\,|\,x_0)\,f(y,\tau\,|\,V_0)\,d\tau \tag{7}$$

$$\text{for } x_0 < V_0 < y \quad \text{or} \quad y < V_0 < x_0$$

This follows from the continuity of the sample paths of $\{V(t)\}$; if $V(t)$ is to reach a value y at a time t given that it was at x_0 at $t=0$, it must have reached some intermediate value V_0 for the first time at some time $(t-\tau)$, and then changed from V_0 to y in the interval τ. Thus the first passage time probability density function is inside a convolution integral: this can be simplified by taking the Laplace transforms of the densities and using the convolution theorem to give:

$$f^*(y,s\,|\,x_0) = p^*(V_0, s\,|\,x_0)\,f^*(y,s\,|\,V_0) \tag{8}$$

where $f^*(y,s\,|\,x)$ and $p^*(V_0,s\,|\,x_0)$ are the Laplace transforms of $f(y,t\,|\,x_0)$. Taking the Laplace transforms is not just a notational convenience, as $p(V_0,t\,|\,x_0)$ is a solution of a partial differential equation which can be solved by the Laplace transform method, and, for the leaky integrator model, $p(V_0,t\,|\,x_0)$ is not available but $p^*(V_0,s\,|\,x_0)$ can be obtained.

If Eq.(7) is substituted into the backward Chapman-Kolmogorov Eq.(6), and noting that $p(V_0, 0\,|\,x_0) = 0$ for $x_0 \neq V_0$,

$$\frac{\partial p(V_0,t\,|\,x_0)}{\partial t} = \tfrac{1}{2}\alpha(x)\,\frac{\partial^2 p(V_0,t\,|\,x_0)}{\partial x_0^2} + \beta(x)\,\frac{\partial p(V_0,t\,|\,x_0)}{\partial x_0} \tag{9}$$

with initial condition

$$p(V_0, 0\,|\,x_0) = \delta(V_0 - x_0)$$

and boundary conditions

$$p(V_0, t\,|\,V_0) = \delta(t)$$

$$\lim_{x\to -\infty} p(x,t\,|\,x_0) = 0$$

This partial differential equation can be transformed into a second order linear differential equation by taking Laplace transforms:

$$sp^*(y,s\,|\,x_0) = \tfrac{1}{2}\alpha(x)\,\frac{d^2 p^*(y,s\,|\,x_0)}{dx_0^2} + \beta(x)\,\frac{dp^*(y,s\,|\,x_0)}{dx_0} \tag{10}$$

and the problem is to obtain $p(V_0,t\,|\,x_0)$ for the two neural models when the input is a white noise added to a constant level μ.

For the perfect integrator model, $V(t)$ defined by Eq.(3) is a Wiener-Lévy process when $x(t)$ is a Gaussian white noise, and the incremental moments $\alpha(x)$ and $\beta(x)$ are the variance σ^2 and mean μ, and so substituting

$$\alpha(x) = \sigma^2$$

$$\beta(x) = \mu$$

into Eq.(10) and solving using the initial and boundary conditions of Eq.(9) gives

$$p^*(V_0,s\,|\,x_0) = \exp\{(x_0 - V_0) \times [-\mu - \sqrt{(\mu^2 + 2s\sigma^2)}]/\sigma^2\} \tag{11}$$

When $x_0 = 0$, the inverse Laplace transform of Eq.(11) gives the well-known first passage time density of a Wiener-Lévy process as:

$$p(V_0,t\,|\,0) = \frac{V_0}{\sigma\sqrt{(2\pi t^3)}}\,\exp\{-(V_0 - \mu t)^2/2\sigma^2 t\} \tag{12}$$

Note that this first passage time density has no finite moments when $\mu = 0$, when the Wiener-Lévy process has no drift.

For the leaky integrator model, $\{V(t)\}$ defined by Eq.(1) is an Ornstein-Uhlenbeck process when $x(t)$ is a Gaussian white noise

$$V(T+t) = V(T)\exp(-t/\tau) + \exp\{(T+t)/\tau\}\int_T^{T+t}\exp(\theta/\tau)\,dW(\theta) \qquad (13)$$

The first passage time density $p(V_0, t|x_0)$ of an Ornstein-Uhlenbeck process cannot be obtained analytically in closed form, except for the special case of $V_0 = 0$; however, the Laplace transform $p^*(V_0, s|x_0)$ can be obtained (Darling and Siegert, 1953; Capocelli and Ricciardi, 1971; Sugiyama *et al.*, 1970). Rearranging Eq.(8) gives:

$$p^*(V_0, s|x_0) = f^*(y, s|x_0)/f^*(y, s|V_0) \qquad (14)$$

$$\text{for } x_0 < V_0 < y$$

and for these Laplace transforms to exist $\{V(t)\}$ must be continuous. The transform of the transition density, $f^*(y, s|x_0)$, can be expressed as the product of two functions

$$f^*(y, s|x_0) = \begin{cases} u(x_0)\,u_1(y) & x_0 < y \\ v(x_0)\,v_1(y) & x_0 > y \end{cases}$$

and so

$$p^*(V_0, s|x_0) = \begin{cases} u(x_0)/u(V_0) & x_0 < V_0 \\ v(x_0)/v(V_0) & x_0 > V_0 \end{cases} \qquad (15)$$

Setting $x = x_0$ in the backward Chapman-Kolmogorov Eq.(6) and taking Laplace transforms:

$$sf^*(y, s|x_0) = \tfrac{1}{2}\alpha(x_0)\frac{d^2f^*(y, s|x_0)}{dx_0^2} + \beta(x_0)\frac{df^*(y, s|x_0)}{dx_0} \qquad (16)$$

with $f^*(x_0, s|x_0) = 1$ and $\lim_{y\to\infty} f^*(y, s|V_0) = 0$ and so $-f^*(y, s|x_0)$ is the Green's function of (16) over the interval $(-\infty < x_0 < +\infty)$. The Green's function can be expressed as:

$$f^*(y, s|x_0) = \begin{cases} v(x_0)\,u(y) & x_0 > y \\ v(y)\,u(x_0) & x_0 < y \end{cases}$$

where $u(y)$ and $v(y)$ are any two linear, linearly independent solutions of

$$\tfrac{1}{2}\alpha(y)\frac{d^2\phi(y)}{dy^2} + \beta(y)\frac{d\phi(y)}{dy} - s\phi(y) = 0 \qquad (17)$$

which satisfy $u(\infty) = v(-\infty) = 0$.

For the normalized Ornstein-Uhlenbeck process, where $\tau = \tfrac{1}{2}\alpha(y) = 1$, $\alpha(y) = 2$ and $\beta(y) = -y$ in Eq.(17), and Darling and Siegert give $u(y)$ and $v(y)$ as

$$\begin{aligned} u(y) &= \exp(y^2/4)\,D_{-s}(y) \\ v(y) &= \exp(y^2/4)D_{-s}(-y) \end{aligned} \qquad (18)$$

where $D_{-s}(y)$ is the parabolic cylinder function or Weber function defined by:

$$D_{-s}(y) = \frac{\exp(-y^2/4)}{\Gamma(s)}\int_0^\infty \exp\left\{\frac{(-yx - x^2)}{2}\right\} x^{s-1}dx \qquad (19)$$

and $\Gamma(s)$ is the Gamma function.

Thus from (15) and (18) Darling and Siegert obtained the Laplace transform of the first passage time density of the normalized Ornstein-Uhlenbeck process as

$$p^*(V_0, s|x_0) = \begin{cases} \exp\{(x_0^2 - V_0^2)/4\}D_{-s}(x_0)/D_{-s}(V_0)\,; & x_0 < V_0 \\ \exp\{(x_0^2 - V_0^2)/4\}D_{-s}(-x_0)/D_{-s}(-V_0)\,; & x_0 > V_0 \end{cases} \qquad (20)$$

There does not seem to be a known, explicit inverse to this Laplace transform. For the leaky integrator model, with a time constant τ, and the dispersion $\alpha(y) = \mu$, $x_0 < V_0$ and so

$$p^*(V_0, s|x_0) = \exp\left\{\frac{x_0^2 - V_0^2}{2\mu\tau}\right\}\frac{D_{-s\tau}(-x_0\sqrt{\{2/\mu\tau\}})}{D_{-s\tau}(-V_0\sqrt{\{2/\mu\tau\}})} \qquad (21)$$

Some computed solutions for $p(V_0, t|x_0)$ for the perfect integrator model are shown in Fig.1: these are identical to the interpulse interval probability

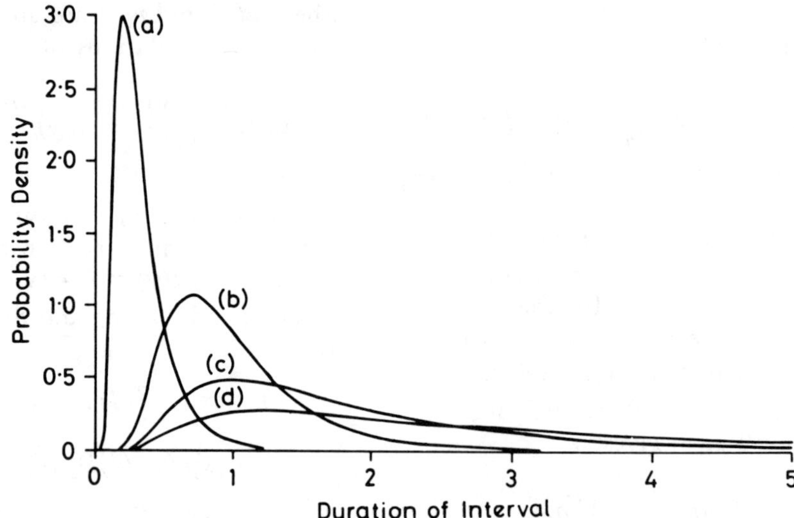

Fig.1 Interspike interval probability density function of perfect integrator model subject to a white noise input. These pdf's were computed using Eq.(12) with means μ and standard deviations σ of the input: (a) $\mu = 3$, $\sigma = 1$ (as in Fig.6b); (b) $\mu = 1$, $\sigma = 0.5$; (c) $\mu = 0.5$, $\sigma = 0.5$; (d) $\mu = 0.25$, $\sigma = 0.5$.

density functions of this model when the input is a white noise. These densities are unimodal and skewed to the right, and are similar in shape to the interspike interval histograms of a variety of sensory neurones.

3. Frequency domain analysis of signal transfer
One method of describing the linear signal transfer characteristics of the two neural models is by their frequency response functions. The frequency response function, $G(\omega)$, describes, as a function of angular frequency ω, the relation between the input signal $x(t)$ and that part of the output pulse train which is linearly dependent on the input. $G(\omega)$ is a complex function, with a magnitude or gain and argument or phase shift at any given frequency. Since the two neural models are grossly nonlinear $G(\omega)$ will not completely describe the input output relation, and the form of $G(\omega)$ will depend on the type of input signal.

If the input to a perfect integrator model with $V_0 = 1$ is a sinusoidal modulation of amplitude a about a constant c, $a \leqslant c$

$$x(t) = a \exp j(\omega t + \phi) + c \tag{22}$$

with an angular frequency ω and phase ϕ at time $t = 0$, the expected density of pulses $h(\phi)$ averaged over a large number of cycles is

$$h(\phi) = a \exp(j\phi) + c \tag{23}$$

Thus the output pulse density is sinusoidally modulated with no change in phase angle and a frequency-independent gain. The output pulse train has a line spectrum, with components at harmonics of the input frequency and many other frequencies. However, if this sinusoidally modulated pulse train is the input to a second perfect integrator model, and k input pulses from the first are required to reach threshold and generate a pulse in the second, its pulse density will be:

$$h(\phi) = \{a \exp(j\phi) + c\}/k$$

Thus the frequency response function of the second perfect integrator model is simply the ratio of the input and output pulse densities, and is a constant:

$$G(\omega) = 1/k \tag{24}$$

This will hold for any periodic signal. However, there will be frequency-dependent changes in gain and phase if the modulation of the instantaneous rate of pulses is considered: Knight (1972) uses linear perturbation theory to derive the frequency response function of the transduction of $x(t)$ to an instantaneous rate $r(t)$ as:

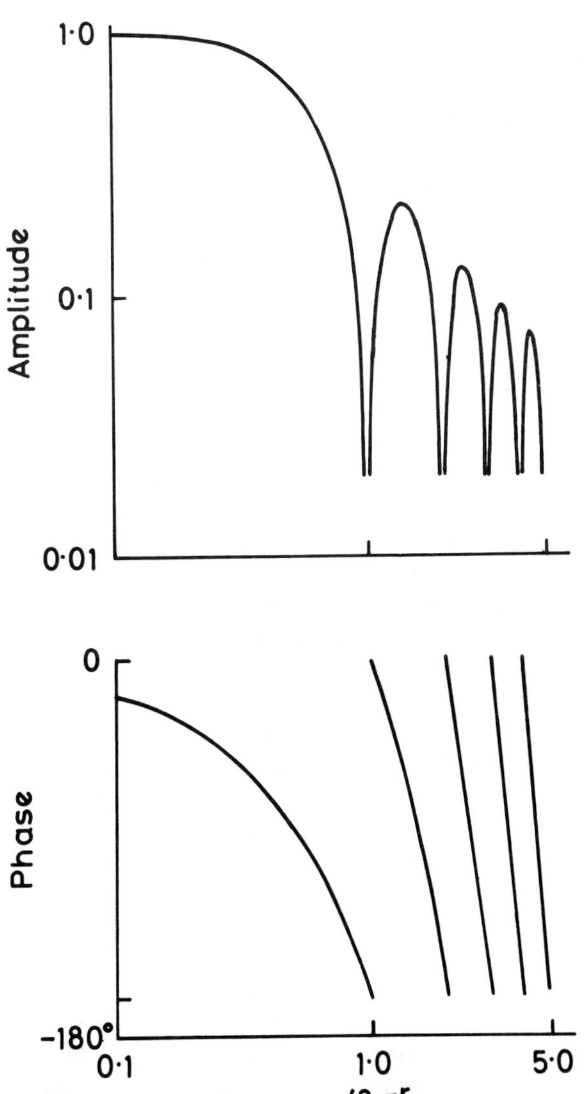

Fig.2 Frequency response function of the transduction of input voltage to instantaneous rate by a perfect integrator model—see Eq.(25).

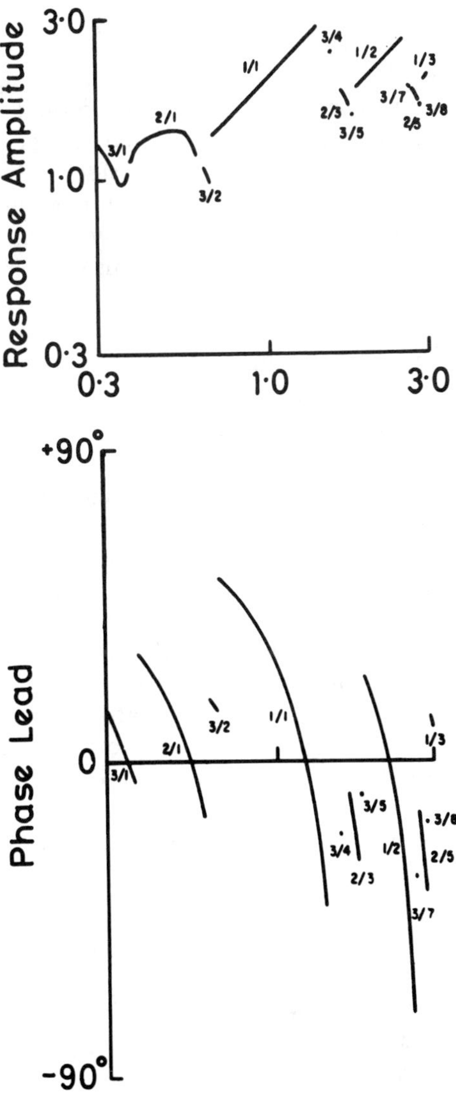

Fig.3 Frequency response function of a leaky integrator model, with a time constant $\tau = 0.3$, obtained from phase-locked responses, with M impulses discharged every N cycles of an applied sinusoid. The ratio a/c (see Eq.22) was set to 0.2 and the mean discharge interval normalized to 1 sec.

$$G(\omega) = \frac{r_0}{c} \frac{1 - \exp(-j\omega/r_0)}{j\omega/r_0} \tag{25}$$

where r_0 is the carrier discharge rate produced by a constant input c. This frequency response function is plotted in Fig.2, and shows no distortion at very low frequencies, a high frequency cutoff as the modulation frequency approaches $2\pi r_0$, and nulls at integer multiples of $2\pi r_0$.

The response of the leaky integrator model to a sinusoidally modulated input is treated in detail in Rescigno *et al.* (1970): the response of the model is a series of pulses which occur at fixed phases with respect to the input cycle, generating a periodic, phase-locked pattern with M pulses occurring at fixed phase angles in every N cycles, where M and N are positive integers. Similar phase-locked responses have been observed in a variety of sensory neurones, and have been obtained numerically as solutions of the Hodgkin-Huxley equations (Holden, 1975*a*). For a given input modulation amplitude, the frequency response function can be readily evaluated at any frequency from the pattern of phase-locked pulses produced at that frequency. The frequency response function of a leaky integrator model, computed from the phase-locked responses to a sinusoidally modulate(input, is illustrated in Fig.3. In this example, simple phase-locking (where M and N are small) extends over most of the frequency range. Such simple phase-locking is most apparent when the product of the time constant and the unmodulated carrier rate r_0 is small. Phase-locking may be considered to produce distortions in the transmitted signal, and the tendency for phase-locking may be reduced by adding a non-periodic auxiliary input or an intrinsic noise source. The resultant stochastic carrier rate $r_0(t)$ will permit a less distorted replica of the periodic input signal to be extracted, by a linear filter, from the pulse train output. This is equivalent to the use of an auxiliary signal to linearize a static nonlinearity (van der Tweel and Spekreijse, 1969), and Stein (1970) suggests that neuronal variability might optimize signal transmission over a bandwidth of interest: for a leaky integrator model with a given time constant, the higher the bandwidth of interest, the greater the noise auxiliary signal needed to reduce these distortions. Knight (1972) has extended this idea to populations of leaky integrator models.

If the input to a perfect integrator model is a Poisson distributed train of excitatory pulses, with a mean rate r, and k input pulses are required to reach threshold, the probability density function $p(t)$ of the output interpulse intervals will be a gamma density function of order k:

$$p(t) = r^k t^{k-1} \exp(-rt)/(k-1)! \tag{26}$$

The spectral density of the input, $S_i(\omega)$, is a constant

$$S_i(\omega) = r/2\pi \tag{27}$$

and the spectral density of the output pulse train, $S_0(\omega)$, is

$$S_0(\omega) = \frac{r}{2\pi k} \frac{(1 + (\omega/r)^2)^k - 1}{\{(1 + j\omega/r)^k - 1\}\{(1 - j\omega/r)^k - 1\}} \tag{28}$$

Since the frequency response function $G(\omega) = 1/k$ is a constant, the cross-spectral density $S_{io}(\omega)$ is simply

$$S_{io}(\omega) = G(\omega)\, S_x(\omega) = r/(2\pi k) \tag{29}$$

The coherence function, $\gamma^2(\omega)$, defined by

$$\gamma^2(\omega) = \frac{|S_{io}(\omega)|^2}{S_i(\omega)\, S_o(\omega)} \tag{30}$$

which is a normalized measure of the linearity of the input-output relation, and represents that proportion of $S_0(\omega)$ which can be accounted for by linear regression on to $S_i(\omega)$, can be computed from Eqs (27)-(30) as

$$\gamma^2(\omega) = \frac{\{(1 + j\omega/r)^k - 1\}\{(1 - j\omega/r)^k - 1\}}{k\{(1 + (\omega/r)^2)^k - 1\}} \tag{31}$$

which approaches 1 at $\omega \ll r$ and $1/k$ for $\omega \gg r$. These spectra and the coherence function are illustrated in Fig.4 for two values of k. At low frequencies, the input and output are almost completely coherent, and the coherence drops as the frequency approaches the carrier rate. The more input pulses needed to reach threshold, the lower the coherence at all frequencies, especially near the carrier rate.

A similar derivation of the frequency response and coherence functions of the leaky integrator model using a Poisson distributed excitatory input does not appear to be possible. The problem is that to obtain $S_0(\omega)$ the interpulse interval density $p(V_0, t|x_0)$ is required. Its Laplace transform cannot be obtained by the method used to obtain Eq.(21) as the diffusion approximation only holds when the input rate is high and the effect of an excitatory pulse is small (k is large): these conditions are necessary to ensure the continuity of $\{V(t)\}$. Stein (1965) considered the limiting case of $r\tau \gg k$, which results in negligible decay and an output pulse train with a gamma interval density given by Eq.(26), and obtained estimates of $p(V_0, t|x_0)$ for lower input rates, when the decay was not negligible, by numerical simulation. These simulated interval histograms could be fitted by a gamma interval density. Estimates of $p(V_0, t|x_0)$ could be obtained by extensive numerical computation, involving solution of:

$$p(V_0, t\,|x_0) = -\frac{\partial}{\partial t} \int_{-\infty}^{V_0} f(x, t|x_0)\, dx \tag{32}$$

where $f(x, t|x_0)$ is the transition density of $\{V(t)\}$; which is a solution of the partial differential-difference equation

$$\frac{\partial f(x, t|x_0)}{\partial t} = \frac{\partial}{\partial x}\left\{\frac{(x - x_0) f(x, t|x_0)}{\tau}\right\} + r\{f(x + V_0/k, t\,|x_0\} - f(x, t\,|x_0)\} \tag{33}$$

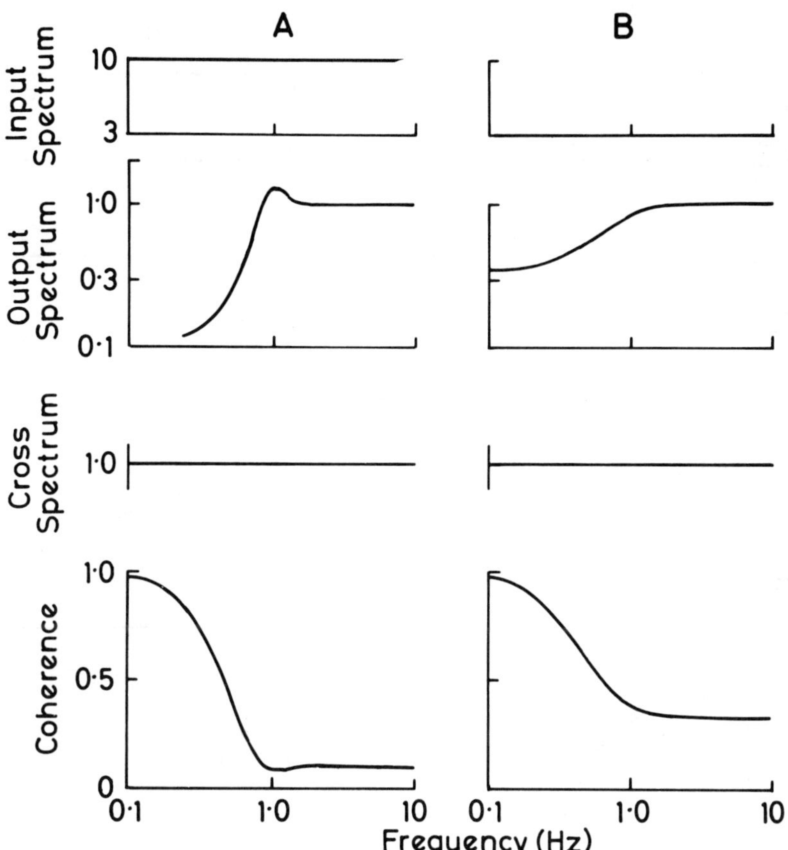

Fig.4 The input and output spectra, magnitude of cross-spectrum (which is proportional to the gain of the frequency response function) and coherence function when a perfect integrator model is subject to an excitatory, Poisson distributed pulse input. These have been computed from Eqs (27)-(31) with (a) r=k=10 (b) r=k=3.

If $p(V_0, t|x_0)$ is estimated numerically by this method, $S_0(\omega)$ can be readily computed. Knox (1974) gives a method for obtaining the cross-correlation function between the output pulse train and the input pulse train if $p(V_0, t|x_0)$ is known, and so the cross-correlation function and hence its Fourier transform, $S_{i0}(\omega)$, could be computed. Thus the frequency response and coherence functions could be estimated numerically.

The case of a white noise input to the perfect integrator model is treated in detail in Stein, French and Holden (1972). The output pulse train has an interpulse interval density given by Eq.(12) and a spectral density

$$S_0(\omega) = \frac{m}{2\pi}\left\{\frac{\exp(2\alpha) - 1}{\exp(2\alpha) - 2\exp(\alpha)\cos(\beta) + 1}\right\}$$

where $\alpha = \dfrac{m}{A^2}(C\cos\theta - 1)$

$$\beta = \frac{m}{A^2} C \sin\theta$$

$$C = \{1 + (2\omega)^2 (A/m)^4\}^{1/4}$$

$$\theta = 0.5 \tan^{-1} 2(A/m)^2$$

and m is the mean discharge rate produced by a white noise input with a spectral density $S_i(\omega) = A^2/2\pi$. For small values of ω, the output spectral density is proportional to the input spectral density, and for large values of ω the output spectral density is proportional to the mean discharge rate. Limits for the cross-spectral density can be obtained by considering the behavior of the cross-correlation function $\rho_{io}(u)$

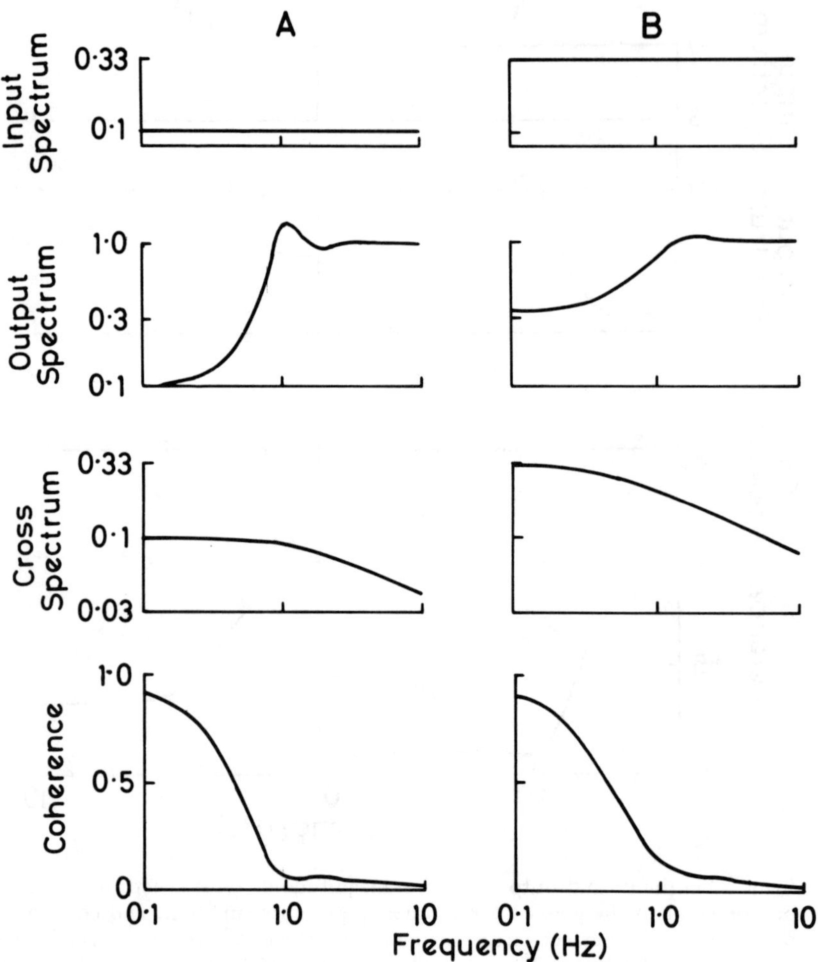

Fig.5 The input and output spectra, magnitude of cross-spectrum (which is proportional to the gain of the frequency response function) and coherence function when a perfect integrator model is subject to a white noise input. The noise levels are (a) $A^2 = 0.1$ (b) $A^2 = 1/3$, and so the standard deviations of the output interspike interval distribution are the same as in Fig.4.

$$\rho_{io}(u) = mE\{x(t-u)\} \tag{35}$$

where $E\{x(t-u)\}$ is the expected value of the input signal at a time u before a pulse. For $t>0$, and for large negative values of t, $E\{x(t)\} = A$. For small negative values of t, the expected value of $x(t)$ can be shown to be

$$E\{x(t)\} = \sqrt{(-\pi A^2/(8t))}$$

which will determine the high frequency part of the cross-spectrum, and so for large ω

$$S_{io}(\omega) = \frac{Am}{4/2j\omega} \tag{36}$$

At low frequencies, the pulse density will be proportional to the input signal, and so the cross-spectral density will be proportional to the input spectral density.

The output spectrum, frequency response and coherence function computed for two white noise levels are shown in Fig.5. The output pulse trains had the same mean rate and variance as those of Fig.4 obtained by a pulse train input, but the spectral density $S_0(\omega)$ is more peaked about the mean discharge rate m. The frequency response function shows a slow decrease in gain with frequency, but the coherence function drops sharply to a negligible value as the frequency ω approaches the mean discharge rate.

The frequency response and coherence functions of the leaky integrator model have not been obtained using a white noise input: however, Stein, French and Holden (1972) investigated the behavior of estimates of the frequency response function of a hardware leaky integrator neural analog using a band-limited white noise input. For a small noise input, there are peaks in the gain at the mean discharge rate and its harmonics; associated with these peaks are large phase changes. These peaks in gain, phase changes and the increase in gain with frequency are less than those obtained using a sinusoidal input. As the input noise level increases, the peaks in gain and phase changes become less prominent. At high noise levels the frequency response function falls at high frequencies with a concomitant phase lag: this is similar to the frequency response function of the perfect integrator model obtained using a white noise input.

Thus the form of the frequency response functions of the two models depends on the type of input signal. In a chain of neural models the input signal is a series of excitatory pulses, and the perfect integrator model acts as a simple *k*-divider. The resultant decrease in number of pulses at every relay in the chain can be compensated by excitatory convergence from correlated units in parallel.

Knight (1973) has shown that all neural models which have deterministic encoders, and which depend only on the input since the last output pulse, fall into two classes: the perfect integrator model and the class of leaky or forgetful models. The activity of a population of perfect integrator models has a population and time average which follows the input signal, for a population of forgetful models neither the time average nor population average of activity follow the input signal. However, variability or noise makes the population and time averages similar to those of the perfect integrator models. Thus the time averaged behavior of noisy, leaky neural models will be similar to that of the perfect integrator models.

4. The small signal impulse response

The small signal step response of the perfect integrator model with a white noise input has been obtained by Rubio and Holden (1975). If $w(\cdot)$ is the white noise input, with a power spectral density $A^2/2\pi$, and the output pulse train is $y(\cdot)$, the small signal impulse response $h(t)$ is, for $t \geqslant 0$:

$$h(t) = \frac{1}{A^2} \lim_{T\to\infty} \int_{-T}^{T} y(\theta)\, w(\theta - t)\, d\theta \qquad (37)$$

This small-signal impulse response is shown to be given by:

$$h(t) = \frac{-m}{A^2} \frac{d\phi(t)}{dt} \qquad (38)$$

when the output pulses have a mean rate m and are brief, with a unit area, and

$$\phi(t) = E\{V(t_i - t)\} \qquad (39)$$

is the expected value of $V(t)$ at a time t *before* an output pulse is generated. The function $\phi(t)$ is evaluated as:

$$\phi(t) = \frac{\int_{-\infty}^{V_0} x\, n(t,x)\{\int_0^\infty f(x,\theta\,|\,x_0)[1-P(\theta)]\,d\theta\}\,dx}{\int_{-\infty}^{V_0} n(t,x)\{\int_0^\infty f(x,\theta\,|\,x_0)[1-P(\theta)]\,d\theta\}\,dx} \qquad (40)$$

where:

$n(t,x)$ is the value of a function which is a density for the event of a pulse occurring at a time t after the occurrence of a value x of $V(\cdot)$. Since any number of pulses can have been generated in the interval $(0,t)$, $n(t,x)$ can be computed by means of an infinite series of convolutions: see Johannesma, 1969.

$P(\theta)$ is the probability distribution function of the first passage times, i.e. the integral of Eq.(12).

$f(x,\theta\,|\,x_0)$ is the transition probability density function of $\{V(t)\}$, and so is a solution of Eq.(6).

Numerical evaluation of $-\frac{m\phi(t)}{A^2}$ gives the small-signal step response: this is illustrated in Fig.6 for two values of the input noise. The noise input which generates a higher discharge rate gives a faster, more oscillatory small signal step response. This dependence of the shape of the step response, and hence the shape of the impulse response, on the amplitude of the input noise, reflects the sharp threshold discontinuity of the model, and implies that if the input-output relations of the model are characterized by an infinite series of Wiener functionals, the form of the kernels will depend on the amplitude of the input signal.

5. Information preservation by stable interval distributions

Information will be transmitted through a chain of neurones as a sequence or train of action potentials

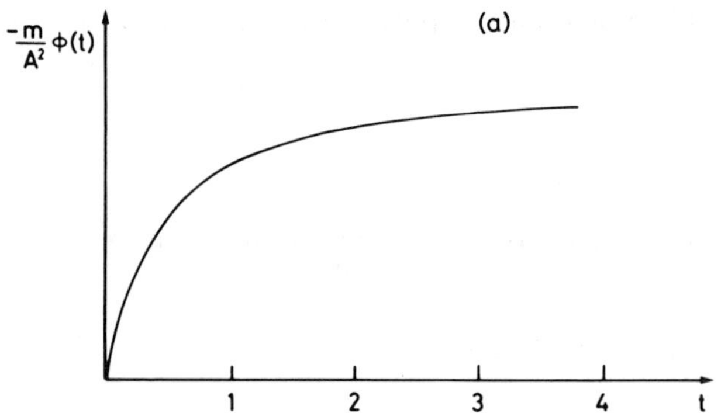

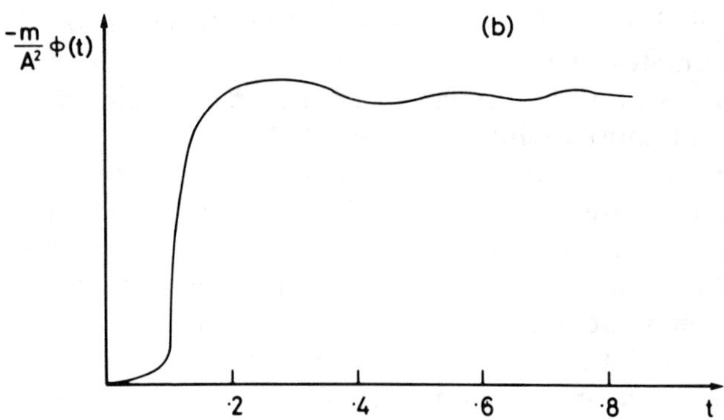

Fig.6 Computed small-signal step response of perfect integrator model $-\frac{m}{A^2}\phi(t)$, for a white noise input with mean μ and standard deviation σ.
(a) $\mu = 1.5$, $\sigma = 2$; (b) $\mu = 3$, $\sigma = 1$.

in each of the component neurones. In any one such component neurone, all the action potentials are identical, except in their times of occurrence, and so the information content of the neural signal depends only on the patterns of times of action potentials. The action potential train can be treated as a stochastic point process, and if the intervals between action potentials are independent, as a renewal process. Such a renewal process is completely specified by its interval probability density function, and so the information content of such an action potential train is equivalent to the information content of its interval density function.

For a chain of perfect integrator model neurones, ..., $j-1$, j, $j+1$, ..., the jth unit will generate an output pulse at every kth input pulse: thus the inteval density of the output from the jth unit will be the k-fold convolution of the interval density of its input, which is the output from the $(j-1)$th unit. Thus information will only be preserved as the signal passes up a chain of perfect integrator models if the interval densities $p_j(t)$ have a form which is invariant under self-convolution. A probability density function which has a form which is invariant (apart from a linear scale change) under self-convolution is said to have a stable distribution function (Levy, 1940; Gnedenko and Kolmogorov, 1968). Gerstein and Mandlebrot (1964) pointed out that the first passage time distribution of the perfect integrator model is a stable distribution, and that the interval distributions obtained experimentally from cochlear neurones appeared to be stable.

This analysis of a chain of perfect integrator models suggests the hypothesis (Holden, 1975*b*):

if action potential trains in a cascade of sensory neurones have stable interval distributions, the sensory cascade is preserving information, and its converse

if action potential trains in a sensory cascade do not have stable interval distributions, information is not being preserved but some information processing is occurring.

This simple idea assumes that the stationary, time-averaged behavior of a single unit in a sensory pathway is similar to the behavior of a perfect integrator model in a neural chain. A sensory pathway consists of a large number of neurones in parallel, with divergence and convergence at each synaptic level. Excitatory convergence will permit correlated activity to pass through a synaptic relay, and the intrinsic variability of the sensory neurones will tend to make the time-averaged behavior of the 'forgetful' neurones similar to the time-averaged behavior of a perfect integrator model.

The properties of stable distributions have been extensively investigated (e.g. see the monograph by Gnedenko and Kolmogorov, 1968), and Khintchine and Levy (1936) have given a necessary and sufficient condition for a distribution function to be stable. Only three non-degenerate distributions which satisfy this condition are available in closed form: the Gaussian distribution, the Cauchy distribution and the stable distribution of order one half. The stable distribution of order one half contains the gamma distribution, and occurs as the first passage time distribution of the perfect integrator model excited by white noise. Thus, for the interval density of an action potential train to be stable and available in closed form, it must be unimodal and either symmetric and locally Gaussian or asymmetric with an approximately gamma density. It is interesting to note that the interval histograms of a large number of first, second and third order sensory neurones may be fitted by a Gaussian or a gamma density function. This would suggest that information is being preserved in these sensory pathways. Further, the interval histograms of a large number of central neurones, believed to be involved in information processing, are multimodal and are obviously not stable.

References

CAPOCELLI, R.M. and RICCIARDI, L.M., 'Diffusion approximation and first passage time problem for a model neuron', *Kybernetik* **8**, 214 (1971).

DARLING, D.A. and SIEGERT, A.J.F., 'The first passage time problem for a continuous Markov process', *Ann. Math. Statist.* **24**, 624 (1953).

FIENBERG, S.E., 'Stochastic models for single neuron firing trains: a survey', *Biometrics* **30**, 399 (1974).

FRENCH, A.S., HOLDEN, A.V. and STEIN, R.B., 'The estimation of the frequency response function of a mechanoreceptor', *Kybernetik* **11**, 15 (1972).

GERSTEIN, G.L. and MANDLEBROT, B., 'Random walk models for the spike activity of a single neuron', *Biophysical J.* **4**, 41 (1964).

GNEDENKO, B.V. and KOLMOGOROV, A.N., *Limit distributions for sums of independent random variables,* Trans. K.L. Chung, Addison-Wesley Publ. Co. Reading, Mass (1968).

HOLDEN, A.V., 'The response of excitable membrane models to a cyclic input', *Biol. Cybernetics* **21**, 1 (1975a).

HOLDEN, A.V., 'A note on convolution and stable distributions in the nervous system', *Biological Cybernetics* **20**, 171 (1975*b*).

HOLDEN, A.V., Models of the stochastic activity of neurones, Springer-Verlag, Berlin (1976).

J OHANNESMA, P.I.M., 'Diffusion models for the stochastic activity of neurons', in *Neural Networks,* ed. E.R. Caianiello, Springer-Verlag, New York, p.116 (1968).

KHINCHINE, A.Ya. and LEVY, P., 'Sur les lois stable', *C.R. Acad. Sci. (Paris)*, **202**, p.374 (1936).

KNIGHT, B.W., 'Dynamics of encoding in a population of neurones', *J. Gen. Physiol.* **59**, 734 (1972).

KNIGHT, B.W., 'Some questions concerning the encoding dynamics of neuron populations', Proc. 4th International Biophysics Congress, Puschino, USSR 422 (1973).

KNOX, C.K., 'Cross-correlation functions for a neuronal model', *Biophysical J.* **14**, 567 (1974).

LAPIQUE, L., 'Recherches quantitatives sur l'excitation electrique des nerfs traitee comme une polarisation', *J. Physiol. (Paris)* **9**, 622 (1907).

LEVY, P., 'Sur certain processus stochastiques homogenes', *Comp. Math.* **7**, 283 (1940).

McCANN, G.D. and MARMARELIS, P.Z., 'Proceedings of the First symposium on testing and identification of nonlinear systems', California Institute of Technology, Pasadena, (1975).

RESCIGNO, A., STEIN, R.B., PURPLE, R.L. and POPPELE, R.E., 'A neuronal model for the discharge patterns produced by cyclic inputs', *Bull. Math., Biophysics* **32**, 337 (1970).

RUBIO, J.E. and HOLDEN, A.V., 'The response of a model neurone to a white noise input', *Biological Cybernetics* **19**, 191 (1975).

STEIN, R.B., 'A theoretical analysis of neuronal variability', *Biophysical J.* **5**, 173 (1965).

STEIN, R.B., 'The role of spike trains in transmitting and distorting sensory signals', in *The Neurosciences—Second Study Program,* ed. F.O. Schmitt, Rockefeller Univ. Press, New York (1970).

STEIN, R.B., FRENCH, A.S. and HOLDEN, A.V., 'The frequency response, coherence, and information capacity of two neuronal models', *Biophysical J.* **12**, 295 (1972).

SUGIYAMA, H., MOORE, G.P. and PERKEL, D.H., 'Solutions for a stochastic model of neuronal spike production', *Math. Biosciences* **8**, 323, (1970).

van der TWEEL, L.H. and SPEKREIJSE, H., 'Signal transport and rectification in the human evoked response system', *Annals N.Y. Acad. Sci.* **156**, 678 (1969).

WIENER, N., *Nonlinear problems in random theory,* MIT Press, Cambridge, Mass. (1958).

Diffusion approximations to neuronal activity and to the time dependent threshold problem*

LUIGI M. RICCIARDI
Istituto di Scienze dell'Informazione, University of Salerno, Italy

1. Introduction

The large degree of variability and randomness exhibited by the responses of most neurons in the absence of specific external stimuli suggests that probabilistic models may fruitfully be used to describe the observed neuron's output. The parameters present in any such model (i.e. the neuron's firing threshold, the decay constant of the membrane potential, the value of the resting potential, the initial value of the neuron's input, etc.) may subsequently be estimated by standard statistical techniques. In this context, one has to compare the first passage time probability density function (pdf) of the neuron's membrane potential through the threshold value with the histograms of the interspike time intervals. The literature on this subject is too vast and too well known to be referenced here. We only wish to emphasize that this methodology involves making inferences about two separate phenomena: the genesis and the nature of the neuron's synaptic input on the one hand; the conversion of synaptic inputs into spike trains on the other hand. Variability in the observed interspike intervals could thus be due to fluctuations in the synaptic input, in the mechanism transforming the chemicals released by the synaptic knobs into postsynaptic potentials or else in the neuron's threshold.

The lack of arguments to select specific types of synaptic inputs when dealing with neurons that are part of complex neuronal networks induces one to resort to asymptotic theorems on the superposition

*Part of the material contained in this chapter was the subject of some lectures delivered by the author at the International School of Biophysics 'Ettore Majorana' held at Erice in May 1974.

of (weakly correlated) random variables. This idea, first proposed by Braitenberg (1965), is motivated by the circumstance that neurons like Purkinje cells receive a large number of inputs from separated sources which in most cases is of the order of 10^5. Under certain constraints pointed out by de Luca and Ricciardi (1967, 1968) one is then allowed to make use of the Central Limit Theorem to describe the neuron's input as a normally distributed random variable and to estimate the deviation of the actual distribution from the normal one.

The drawback of this viewpoint is that the firing mechanism is pictured as a stationary process, which is often contradicted by the observed time dependence of the neuron's firing probability. This is why, after Gerstein and Mandelbrot (1964), models involving stochastic processes have been increasingly used to describe the neuron's output (cf. for instance, Ricciardi and Ventriglia, 1970; Capocelli and Ricciardi, 1971a, 1971b, 1972, 1973; Ricciardi, 1971, 1972, 1975). The feature shared by most of these models is that the synaptic input is a point process of the Poisson type. This is reasonable if one assumes that the activity of each synapsis can be described by a point process and that there is a sufficiently high degree of independence among the individual synaptic activities to be allowed to use Kintchine theorem on the superposition of point processes.

In the sequel we shall briefly discuss two diffusion models for neuronal activity. We shall then take up the more complicated problem of arising to the neuron's input from the knowledge of the interspike distances histograms. The problem of time dependent threshold will be finally considered. This task will be accomplished after a bird's eye look at some of the ideas underlying continuous Markov processes, i.e. diffusion processes. This will be done in Section 2, where the definitions and the notations to be used will be specified.

2. Diffusion processes

Let $X(t)$ be a one dimensional stochastic process of a single real argument t (the time) which varies over the interval $[0,\infty)$. If we take n fixed instants, $t_1, t_2, \ldots, t_n$, the values $X(t_i)$ $(i=1,2,\ldots,n)$ are a family of random variables whose p.d.f. $f_n(x_1, x_2, \ldots, x_n; t_1, t_2, \ldots, t_n)$ is known if $X(t)$ is completely specified.

Inversely, if f_n is known for all n and t_i, $X(t)$ is specified. By choosing the instants t_i sufficiently close to one another, we can replace $X(t)$ by a sequence of random variables $X(t_i)$ with an accuracy which is satisfactory for all practical purposes. Therefore, $X(t)$ can be characterized by an infinite sequence of p.d.f.'s:

$$f_1(x_1; t_1),\ f_2(x_1, t_1; x_2, t_2), \ldots .$$

Of course, the knowledge of f_j allows us to determine f_i $(i<j)$. This situation improves a great deal if one refers to Markov processes, also known as processes without aftereffect. These are characterized by the property that for any n and $t_1 < t_2 < \ldots < t_n$ the conditional p.d.f. of the value $X(t_n)$ of $X(t)$ at the most recent time t_n depends only on the last value $X(t_{n-1})$ and not on the preceding ones:

$$f_n(x_n, t_n \mid x_{n-1}, t_{n-1}; \ldots; x_1, t_1) = f_x(x_n, t_n \mid x_{n-1}, t_{n-1}). \qquad (2.1)$$

For obvious reasons, this function is called the transition p.d.f. of $X(t)$. For any preassigned univariate p.d.f. $f(x_1, t_1)$, if the transition p.d.f. is known one can determine the multidimensional p.d.f.'s of $X(t)$ as

$$f_n(x_1, t_1; x_2, t_2; \ldots; x_n, t_n) = f(x_1, t_1) f(x_2, t_2 \mid x_1, t_1) \ldots f(x_n, t_n \mid x_{n-1}, t_{n-1}).$$

Hence the importance of determining the process' transition p.d.f.

A first step in this direction consists of writing an integral equation for the transition p.d.f. This is readily done because of (2.1). Indeed, considering two instants t_0 and t $(t > t_0)$ one can write:

$$f(x, t \mid x_0, t_0) = \int_{-\infty}^{\infty} dy\, f(x, t \mid y, \tau) f(y, \tau \mid x_0, t_0), \qquad (2.2)$$

where τ is an arbitrary instant between t_0 and t and $x_0 = X(t_0)$. Equation (2.2), found by Smolukowski, can be rewritten in the following, more convenient, differential form:

$$\frac{\partial f}{\partial t} = \sum_{n=1}^{\infty} \frac{(-1)^n}{n!} \frac{\partial^n}{\partial x^n} [A_n(x, t) f], \qquad (2.3)$$

where A_n's denote the moments of the increment $y-x$ occurring during an infinitesimal time interval:

$$A_n(x, t) = \lim_{\Delta t \to 0} \frac{1}{\Delta t} \int_{-\infty}^{\infty} dy\,(y-x)^n f(y, t+\Delta t \,|\, x, t). \tag{2.4}$$

For temporally homogeneous processes these infinitesimal moments are time independent.

A further simplification is attained if one refers to diffusion processes, namely do those processes whose infinitesimal moments vanish from the third order on. In this case Eq.(2.3) becomes the *forward* or *Fokker-Planck* equation

$$\frac{\partial f}{\partial t} = -\frac{\partial}{\partial x}(A_1 \cdot f) + \frac{1}{2}\frac{\partial^2}{\partial x^2}(A_2 \cdot f)\,. \tag{2.5}$$

Solving this equation with the initial condition

$$\lim_{t \to t_0} f(x, t \,|\, x_0, t_0) = \delta(x - x_0) \tag{2.6}$$

should yield the transition p.d.f. of $X(t)$. Since this is also a function of x_0 and t_0, we expect it to satisfy an equation in which the variables are initial state and initial time. This is actually the case, as f can be proven to satisfy the *Kolmogorov* or *backward equation:*

$$\frac{\partial f}{\partial t_0} + A_1(x_0, t_0)\frac{\partial f}{\partial x_0} + A_2(x_0, t_0)\frac{\partial^2 f}{\partial x_0^2} = 0\,. \tag{2.7}$$

The situation, however, becomes complicated if A_1 and A_2 vanish or become infinitely large for some values of the argument, because (2.6) does not suffice any longer to determine the transition p.d.f. Suitable boundary conditions have then to be associated to Eqs (2.5) and (2.7), as shown by Feller (1952, 1954). The results obtained by this author can be summarized as follows. Under the assumptions that $A_2(x) > 0$ and that dA_2/dx and A_1 are defined and continuous in an interval (a,b), the end points of this interval are boundaries for the diffusion process $X(t)$ in the sense that they can be reached in a finite time from an initial point $a < x_0 < b$ (*accessible boundaries*) or the time required for $X(t)$ to reach the end points may be infinite (*inaccessible boundaries*). The end points, of course, do not need to be boundaries of the same nature, and the latter is determined by the integrability properties of functions depending on A_1 and A_2. Without further classifying the boundaries, we only wish to mention that if both boundaries are inaccessible, the transition p.d.f. is the common solution of Eqs (2.5) and (2.7) satisfying the initial condition (2.6). Therefore, under these circumstances it does not matter whether one solves the Fokker-Planck equation or the Kolmogorov equation, with condition (2.6), to determine the transition p.d.f. An example where this circumstance occurs is provided by the unrestricted one dimensional Brownian motion in the presence of an elastically restoring force. In this case A_2 is constant and $A_1 = -\beta v$, where $\beta > 0$ is a constant and v denotes the velocity of the particle subject to Brownian motion. Here the boundaries are the points $\pm\infty$, and they are both inaccessible. In the absence of restoring force ($\beta = 0$) both the moments are constant and Eqs (2.5) and (2.7) become identical to the well-known heat equation. The underlying diffusion process is the Wiener process defined on the whole interval $(-\infty, +\infty)$.

The last step to be sketched before coming to the neurobiological applications of the above considerations consists of showing how the infinitesimal moments appearing in (2.3) can be evaluated in the cases when the infinitesimal transition p.d.f. is not known. This would lead us to discuss the connection between fluctuation equations and diffusion equations, which is a too wide and involved problem to be considered here in its entirety. We shall limit ourselves to pointing out only a few simple instances, in the light of the forthcoming considerations.

Stochastic processes often encountered in biology arise as a refinement of a macroscopic description of the system under study. To be specific, let us assume that the system is described by a unique function of time, $x(t)$, obeying a first order differential equation of the form:

$$\frac{dx}{dt} = f(x)\,, \tag{2.8}$$

where f is a known function. For instance, if $f(x) = \alpha x - \beta x^2$ ($\alpha > 0$, $\beta > 0$) Eq.(2.8) is suitable for representing a logistic population growth in a continuous approximation; if $f(x) = \beta x$ ($\beta < 0$) it can be looked on as the spontaneous exponential decay of a neuron's membrane potential; *etc.* If we assign $X(0) = x_0$, $x(t)$ is uniquely specified—via (2.8)—at all times. It might, however, occur that (2.8) only represents a rough phenomenological description of the actual process, whose time course also depends upon a variety of other causes that, as an overall effect, introduce fluctuations in the measurable quantity $x(t)$. For instance, the actual time course of the neuron's membrane potential is far from an exponential function (even in the neighborhood of its resting value) because of the presence of PSP's. A sensible way of proceeding whenever the input is not known (which seems to be the rule in biology) consists of modifying Eq.(2.8) by adding to the r.h.s. a random function that, in the simplest case, can be thought of as the product of a deterministic function $g(x)$ times a noise term $\Lambda(t)$:

$$\frac{dx}{dt} = f(x) + g(x)\Lambda(t) \, . \tag{2.9}$$

Thus doing, one switches from a single system to an *ensemble* of identical systems described by a stochastic process $X(t)$. Naturally, the choice of $\Lambda(t)$ is critical for determining whether $X(t)$ is a diffusion process. However, in several cases it is reasonable to identify $\Lambda(t)$ with a stationary normal process with zero mean and finite correlation time, so that $X(t)$ (Stratonovich, 1963) is—at least after long times—a diffusion process. This point is sufficiently important for our future purposes to deserve a somewhat detailed analysis. For simplicity, we shall now assume that $\Lambda(t)$ is delta-correlated:

$$E[\Lambda(t_1)\,\Lambda(t_2)] = 2\beta\delta(t_2 - t_1) \, . \tag{2.10}$$

Setting (cf. Ricciardi, 1973):

$$z = \int^{x} d\tau[g(\tau)]^{-1} \equiv \psi(x) \tag{2.11}$$

equation (2.9) becomes:

$$\frac{dz}{dt} = \frac{f[\psi^{-1}(z)]}{g[\psi^{-1}(z)]} + \Lambda(t)$$

so that, denoting by $B_n(z)$ the infinitesimal moments of the new process $Z(t)$, there results (Wang and Uhlenbeck, 1945):

$$B_1(z) = \frac{f[\psi^{-1}(z)]}{g[\psi^{-1}(z)]} \tag{2.12}$$

$$B_2(z) = 2\beta$$

$$B_n(z) = 0 \quad (n = 3,4,\ldots) \, .$$

Therefore, if $g(x)$ is such that (2.11) is a 1:1 transformation, one obtains:

$$A_n(x) \equiv \lim_{\Delta t \to 0} \frac{1}{\Delta t} \int [\psi^{-1}(u) - x]^n f_z[u, \Delta t \,|\, \psi(x)]\, du \, .$$

$$(n = 1,2,\ldots) \, . \tag{2.13}$$

Expanding then $\psi^{-1}(u)$ as a Taylor series of initial value $u_0 = \psi(x)$ and noting that

$$\left.\frac{d\psi^{-1}(u)}{du}\right|_{u_0} = g(x)$$

$$\left.\frac{d^2\psi^{-1}(u)}{du^2}\right|_{u_0} = \frac{1}{2}\frac{dg^2(x)}{dx} \, ,$$

from (2.13) one obtains:

$$A_1(x) = f(x) + \frac{1}{4}\frac{dA_2}{dx}$$

$$A_2(x) = 2\beta g^2(x)$$

$$A_n(x) = 0 \quad (n = 3,4,\ldots) \, . \tag{2.14}$$

In other words, Eq.(2.9) leads to a diffusion equation, for the transition p.d.f. of $X(t)$, of the type (2.5), with moments given by (2.14). This statement, that we have proved under rather restrictive conditions, holds true in much more general instances.

3. Neural models with constant threshold

In this Section we shall be concerned with a problem first studied when Gerstein and Madelbrot (1964) assumed a 'brownian motion-like' mechanism as responsible for the spike generation in a very simplified model neuron. Although several articles aiming at a rigorous formulation of the diffusion approximation for the probabilistic description of the activity of simplified model neurons have appeared in the literature, it is convenient to refer mainly to the paper by Capocelli and Ricciardi (1971). Indeed, some crucial points implicit in the nature of the diffusion approximation for the description of the neuron's activity had not been pointed out before, with the result that some of the statements appearing in the literature are rather obscure.

The main underlying idea is that the state of the neuron can be described by a single variable, x, representing the variation of the potential difference existing across the membrane (*membrane potential,* for short) of the neuron. The state $x=0$ is the resting potential. In the absence of inputs to the neuron, x spontaneously decays with an exponential law to the resting value:

$$x(t) = x(t_0) \exp\left(-\frac{t-t_0}{\theta}\right) ,$$

where θ is the time constant typical of the neuron's membrane.

As is customary, we assume that the neuron's input consists of a sequence of two types of zero-width impulses, Poisson distributed in time with rates α_e and α_i respectively, where the suffixes e and i stand for *excitatory* and *inhibitory*. The effect of an input pulse is supposed to be *instantaneous:* If $x(t)$ is the state of the neuron at time t, the arrival of an excitatory input in the time interval $(t, t+dt)$ induces the transition

$$x(t) \rightarrow x(t) + e \qquad e > 0 ,$$

whereas an inhibitory pulse produces the jump

$$x(t) \rightarrow x(t) + i \qquad i < 0 .$$

Assuming that the neuron releases a spike when and only when the variation x of the membrane potential reaches or exceeds a constant threshold value S, and that the membrane potential is then instantaneously reset to its initial value $x(0) = x_0 < S$, the neuron's output is described by the so-called 'first passage time' p.d.f., to be soon defined. Before coming to some quantitative treatment we wish to remark that attaining the first passage time p.d.f. is to be considered only the starting point toward the description of the neuron's output.

The assumptions made above imply that the membrane potential $x(t)$ is a stationary Markov process in one dimension, so that Eq.(2.3) holds. It is a matter of straightforward calculations to prove that, for this model, the infinitesimal moments are given by:

$$A_1(x) = -x/\theta + m , \quad m = \alpha_e e + \alpha_i i$$

$$A_n(x) = \alpha_e e^n + \alpha_i i^n \quad (n=2,3,...) . \tag{3.1}$$

Unless we add to our model some further suitable assumption, the non-vanishing of the A_n's for all n prevents us from simplifying Eq.(2.3) into a differential equation. This is a crucial point because, as indicated by Capocelli and Ricciardi (1971), approximating Eq.(2.3) by a differential equation may lead to meaningless results. Sensible results are instead found for limiting values of both the magnitude of the jumps e, i and the rates α_e, α_i at which input pulses hit the neuron. Indeed, if we take:

$$\alpha_e = \lim_{x\to 0} C_e/x^2 \qquad \alpha_i = \lim_{x\to 0} C_i/x^2$$

$$i = \lim_{x\to 0} k_i x \qquad e = \lim_{x\to 0} k_e x$$

with C_e, C_i, $-k_i$ arbitrary positive constants and

$$k_e = |k_i| C_i/C_e$$

it is easily seen that (3.1) read:

$$A_1(x) = -x/\theta \neq 0$$

$$A_2 = C_i k_i^2 (1 + C_i/C_e) \equiv \mu > 0$$

$$A_n = 0 \quad (n=3,4,...) .$$

This choice also secures the continuity of the random function $X(t)$. In the limit just discussed, Eq.(2.3) thus becomes the Ornstein-Uhlenbeck equation

$$\frac{\partial f}{\partial t} = \frac{\partial}{\partial x}(f\, x/\theta) + \frac{\mu}{2}\frac{\partial^2 f}{\partial x^2} . \tag{3.2}$$

Hence, the transition p.d.f. describing the 'free' time course of the neuron's membrane potential is given by

$$f(x,t|x_0) = \frac{1}{\sqrt{2\pi\sigma^2}} \exp\left\{ -\frac{[x - x_0 e^{-t/\theta}]^2}{2\sigma^2} \right\} ,$$

having set

$$\sigma^2 = \frac{\mu\theta}{2}[1 - \exp(-2t/\theta)] .$$

Before proceeding further, it is necessary to define the first passage time p.d.f. $g(S,t|x_0)$ through the threshold value S. This is a function that, when multiplied by the infinitesimal time interval dt, provides the probability for $X(t)$ to reach for the first time S at a time $\tau \in (t, t+dt)$ with the condition that $X(0) = x_0 < S$. In the quoted articles by Feller the problem of determining g is fully analyzed. Here we confine ourselves to sketch, in a very loose language, some simple techniques that will allow us to determine this function for our model neurons. We start noting that for a given $x > S$ there results:

$$f(x, t|x_0) = \int_0^t d\tau\, g(S, \tau|x_0)\, f(x, t-\tau|S)\,, \qquad (3.3)$$

having made use of the 'continuity' of the process $X(t)$ and of its time homogeneity. Denoting then by $g_\lambda(S|x_0)$ the Laplace transform of $g\,(S, t/x_0)$ with respect to t, from (3.3) it follows:

$$g_\lambda(S|x_0) = f_\lambda(x|x_0)/f_\lambda(x, S) \qquad (3.4)$$

where f_λ is the Laplace transform of f with respect to t. Taking now the Laplace transform of Eq.(2.7) and making use of (3.4) we find that $g_\lambda(S|x_0)$ satisfies the ordinary differential equation

$$A_2(x_0)\frac{d^2 g_\lambda}{dx_0^2} + A_1(x_0)\frac{dg_\lambda}{dx_0} - \lambda g_\lambda = 0\,. \qquad (3.5)$$

Therefore, $g_\lambda(S|x_0)$ can be obtained as the bounded solution of (3.5) satisfying the initial condition (implied by 3.4):

$$g_\lambda(S|S) = 1\,. \qquad (3.6)$$

Let us now denote by $f_\beta(x, t|x_0)$ the transition p.d.f. of the process obtained from $X(t)$ by setting an absorbing boundary at $\beta < x_0$. From the intuitive relation:

$$\int_t^\infty d\tau\, g(\beta, \tau|x_0) = \int_\beta^\infty dx\, f_\beta(x, t|x_0)$$

we obtain:

$$g(\beta, t|x_0) = -\frac{\partial}{\partial t}\int_\beta^\infty dx\, f_\beta(x, t|x_0)$$

$$= \left[\frac{1}{2}\left(\frac{\partial A_2}{\partial x}\right)_{x=\beta} - A_1(\beta)\right] f_\beta(\beta, t|x_0) +$$

$$+ \frac{1}{2}A_2(\beta)\left(\frac{\partial f_\beta}{\partial x}\right)_{x=\beta}\,. \qquad (3.7)$$

The last equality follows from the fact that (Feller, 1954) $f_\beta(x, t|x_0)$ is the solution of Eq.(2.5) satisfying the generalized absorption condition:

$$\lim_{x\to\beta}\left\{\exp\left[-\int_{x_0}^{x} dz\, A_1(z)/A_2(z)\right] A_2(x) f(x, t|x_0)\right\}$$

$$= 0 \qquad (3.8)$$

and from the assumption that f_β and $\partial f_\beta/\partial x$ vanish sufficiently rapidly at $-\infty$.

We now possess all the necessary elements to work out our neuron models. Indeed, we note that in our case Eq.(3.5) reads:

$$\frac{\mu}{2}\frac{d^2 g_\lambda}{dx_0^2} - \frac{x_0}{\theta}\frac{dg_\lambda}{dx_0} - \lambda g_\lambda = 0\,. \qquad (3.9)$$

As shown in Capocelli and Ricciardi (1971) by a change of both function and variable, this equation can be transformed into the Weber equation

$$\frac{d^2 u}{dz^2} + \left(\frac{1}{2} - \lambda\theta - \frac{1}{4}z^2\right)u = 0 \qquad (3.10)$$

that admits of the linearly independent solutions $D_{-\lambda\theta}(z)$ and $D_{-\lambda\theta}(-z)$, where $D_v(z)$ is the Parabolic Cylinder function. One can thus write down the general solution of (3.9). The function $g_\lambda(S|x_0)$ is finally determined by using condition (3.6) and by imposing that $g_\lambda(S|x_0)$ be bounded as $x_0 \to -\infty$. The result is:

$$g_\lambda(S|x_0) = \exp\left(\frac{x_0^2 - S^2}{2\mu\theta}\right)\frac{D_{-\lambda\theta}\left(-x_0\sqrt{\frac{2}{\mu\theta}}\right)}{D_{-\lambda\theta}\left(-S\sqrt{\frac{2}{\mu\theta}}\right)}\,. \qquad (3.11)$$

Although we did not succeed to calculate the inverse Laplace transform of (3.11), some information on the neuron's responses can be attained directly from (3.11). Indeed, using the identity

$$t_n(S|x_0) = (-1)^n \left.\frac{d^n g_\lambda}{d\lambda^n}\right|_{\lambda=0}\,,$$

where

$$t_n(S|x_0) = \int_0^\infty dt\, t^n\, g(S, t|x_0)\,,$$

it is possible to evaluate the moments of the firing

times. For instance, the average firing time can be proven to be given by:

$$t_1(S|x_0) = \theta\left\{[\sqrt{\pi}\,[\frac{S}{\sqrt{\mu\theta}}\,\Phi(\frac{1}{2},\frac{3}{2};\frac{S^2}{\mu\theta}) - \frac{x_0}{\sqrt{\mu\theta}}\,\Phi(\frac{1}{2},\frac{3}{2};\frac{x_0^2}{\mu\theta}) + \sum_{n=0}^{\infty}\frac{2^n}{(n+1)(2n+1)!!} \times [(\frac{S}{\sqrt{\mu\theta}})^{2n+2} - (\frac{x_0}{\sqrt{\mu\theta}})^{2n+2}]\right\},$$

where $\Phi(z,c;x)$ denotes the Kummer function. It can be seen that it is possible to obtain values of t_1 in good agreement with experimental data for reasonable choices of the parameters S, x_0, θ, μ.

Before coming to a sketch of the second diffusion model, we wish to remark that Eq.(3.2) can be looked at as generated by a Langevin equation of type (2.9) with $f(x) = -x/\theta$ and $g(x) = 1$, and by further interpreting μ as the intensity of the noise $\Lambda(t)$. A question that naturally arises is then whether it is meaningful to derive alternative models for the neuron's activity by directly assuming that specified Langevin equations describe the membrane potential fluctuations. Out of the various possible fluctuation equations we shall here briefly discuss the diffusion model generated by the equation

$$\frac{dx}{dt} + bx - \gamma = \sqrt{S-x}\ \Lambda(t) \tag{3.12}$$

where $S \geqslant x$ is the neuron's constant firing threshold and $\Lambda(t)$ is the Gaussian process earlier introduced. Let us denote by $2a$ its intensity. We then find, in agreement with the procedure outlined in Section 2, the following Fokker-Planck equation for the membrane potential p.d.f.:

$$\frac{\partial f}{\partial t} = -\frac{\partial}{\partial x}[(-bx + \gamma - \frac{a}{2})f] + a\frac{\partial^2}{\partial x^2}[(S-x)f]. \tag{3.13}$$

Here the noise effect 'slowly' decreases as the actual membrane potential approaches the firing threshold. As soon as the latter is reached the noise vanishes and the deterministic action potential generation takes place. As shown in Capocelli and Ricciardi (1973), Eq.(3.13) can be solved and the first passage time p.d.f. can be determined *in closed form* by using formula (3.7). The result is:

$$g(S,t|x_0) = \frac{b[(S-x_0)b/a]^{1/2-(bS-\gamma)/a}}{\Gamma[1/2-(bS-\gamma)/a]} \times e^{bt}[(e^{bt}-1)^{3/2-(bS-\gamma)/a}]^{-1} \times \exp[\frac{b(S-x_0)}{a(1-e^{bt})}].$$

The average firing time and its variance can also be calculated. For brevity, we refer to Capocelli and Ricciardi (1973) for their rather cumbersome expressions. We only want to remark here that this model seems to be appropriate for fitting histograms exhibiting an initial rapid increase followed by a slow decrease to zero after the unique maximum has been attained. The expected firing time tends to be relevantly greater than the mode of the first passage time p.d.f.

We conclude this Section by sketching a totally different method, certainly of greater cybernetical interest, for the description of neuron's activity. This method, that we call the *inverse* method, consists of approximating the interspike histogram by some function; one then imposes that the latter is the interspike distances p.d.f. generated by the actual neuron's input. One thus faces the preliminary problem concerning the existence and the uniqueness of such an input; successively, one has to determine the input itself.

It is easily understood that this second way of approaching the analysis of neuron's activity becomes a sensible one only if the search of the neuron's input is exploited within a preassigned input class. A convenient and fruitful hypothesis is that again the neuron's activity is generated by a continuous Markov process. To provide an idea of the grounds on which this method rests, we approximate the observed histogram by a function $g(S,t|x_0)$, where S is the neuron's threshold and x_0 is its resting potential. Of course, g cannot be an arbitrary function, as the following requirements have to be fulfilled:

$$g(S,t|x_0) \geqslant 0, \quad \text{for all } S,\ t,\ x_0$$

$$0 < \int_0^\infty dt\, g(S,t|x_0) \leqslant 1\,, \quad \text{for all } S,\, t,\, x_0$$
$$\lim_{x_0 \to S} g(S,t|x_0) = \delta(t)\,. \tag{3.13}$$

The question is then to decide whether g can be looked upon as the first passage time p.d.f. for some continuous one-dimensional Markov process $X(t)$ such that $X(0) = x_0$. In other words, one wants to know whether there does exist a continuous Markov process $X(t)$ such that $g(S,t|x_0)\,dt$ is the probability that $X(t)$ passes the value S for the first time in $(t, t{+}dt)$ if $X(0) = x_0$. By imposing a further constraint on $X(t)$ (namely stationarity) one can provide an answer to this question. Moreover, if the answer is yes one can prove the uniqueness *and* actually determine $X(t)$. In other words, one can explicitly calculate, if existing, the coefficients of Eq.(2.7). To get an idea of how one can proceed along the outlined way, we note that condition (3.13) secure the existence of the Laplace transform $g_\lambda(S|x_0)$ of the function $g(S,t|x_0)$. Condition $(3.13)_2$ can then be equivalently expressed as

$$0 < g_\lambda(S|x_0) \leqslant 1 \tag{3.14}$$

and our problem can be set in the following terms: Given a complex function $g_\lambda(S|x_0)$ satisfying (3.14) and such that

$$\lim_{x_0 \to S} g_\lambda(S|x_0) = 1\,, \tag{3.15}$$

can we look at it as the Laplace transform of the first passage time p.d.f. of a temporally homogeneous Markov process obeying the Kolmogorov equation†

$$\frac{\partial f}{\partial t} = A_1(x_0)\frac{\partial f}{\partial x_0} + A_2(x_0)\frac{\partial^2 f}{\partial x_0^2} = 0\ ? \tag{3.16}$$

To answer this question we refer to Eq.(3.5). Setting then

$$\phi \equiv \mathrm{Re}\{\lambda g_\lambda / g''_\lambda\} \qquad \psi \equiv \mathrm{Im}\,\{\lambda g_\lambda / g''_\lambda\}$$
$$\chi \equiv \mathrm{Re}\{g'_\lambda / g''_\lambda\} \qquad \omega \equiv \mathrm{Im}\{g'_\lambda / g''_\lambda\}\,,$$

where g'_λ, g''_λ stand for dg_λ/dx_0, d^2g_λ/dx_0^2, Eq.(3.5) yields

$$A_2 = \phi + i\psi - A_1(\chi + i\omega)\,.$$

The requirement that A_2 and A_1 be real functions finally implies:

$$A_1 = \psi/\omega$$
$$A_2 = \phi - \psi\,\chi/\omega \tag{3.17}$$

where the functions on the r.h.s.'s are known. Finally (Capocelli and Ricciardi, 1972) the uniqueness of the diffusion process defined by (3.17) can be established. Once the coefficients of the Kolmogorov equation have been determined as indicated, the final problem can be easily solved by writing down—via (2.14)'s—the fluctuation Eq.(2.9) and, therefore, the neuron's input $f(x) + x/\theta + g(x)\,\Lambda(t)$, where θ is the time constant of the neuron's membrane potential.

It should be noted that the validity of this methodology rests on the assumption that the neuron's input—whatever it be—does not possess a strong serial correlation. More explicitly, the noise term $\Lambda(t)$ in Eq.(2.9) cannot be an arbitrary one if $X(t)$ has to be a diffusion process. However, the hypothesis that $\Lambda(t)$ is a stationary delta-correlated normal process could be attenuated without drastically changing the nature of our previous considerations.

To conclude this Section, we should mention that Eqs (2.14) would still approximately hold if we assumed $\Lambda(t)$ to be a stationary delta-correlated random process, provided its intensity coefficients were sufficiently small from the third order on. Should one have interpreted $\Lambda(t)$ as an arbitrary noise with small correlation time and small intensity coefficients of order higher than two, Eqs (2.14) would again be approximately describing the actual process $X(t)$, but only after time intervals much longer than the noise correlation time. This explains in what sense it is appropriate to denote by 'approximations' the diffusion models so far considered.

4. Time dependent thresholds

So far we have assumed that the neuron's threshold is constant. Useless to say, this is probably the most unsatisfactory feature of the models earlier described. Nevertheless, the inclusion of a time varying threshold in a neural model still appears to be an over-

†Equation (3.16) is clearly the time homogeneous version of Eq.(2.7).

whelming obstacle to the model's solvability. In this Section we shall briefly discuss this problem and we shall offer certain hints on the methodology that could be proven successful to build neuronal models characterized by non-constant thresholds.

Let us denote by $S(t)$ the function describing the time course of the neuron's threshold, $t = 0$ being the instant at which the last action potential has been released. Moreover, let us assume that the neuron's membrane potential is described by a homogeneous continuous Markov process. It is then easily understood that, for any fixed $x > S(t)$, the following relation holds:

$$f(x,t|x_0) = \int_0^t d\tau\, g[S(\tau),\tau|x_0]\, f[x,t-\tau|S(\tau)]\,, \tag{4.1}$$

where f is the transition p.d.f. of the process and $g[S(t),t|x_0]$ is the first passage time p.d.f. through the threshold $S(t)$. Relation (4.1) is an integral equation in the unknown function g. It clearly reduces to Eq.(3.3) when $S(t)$ = Constant. To our knowledge, with the only exception of the Wiener process and of a *linear* threshold, Eq.(4.1) is of no use for determining the function g, as far as one is concerned with closed form solutions. Nevertheless, we shall point out the usefulness of Eq.(4.1) in our context. However, it is first convenient to reconsider the Ornstein-Uhlenbeck process $X(t)$ that was earlier proven to provide a diffusion model for neuronal activity. We refer to the forward diffusion Eq.(3.2) that, for convenience, we re-write in the following form:

$$\frac{\partial f}{\partial t} = \beta\frac{\partial}{\partial x}(xf) + \gamma^2\frac{\partial^2 f}{\partial x^2}\,. \tag{4.2}$$

The problem is now to determine the function $g[S(t),t|x_0]$. To this purpose, we note that the transformation (Ricciardi, 1976)

$$\begin{aligned} x' &= \frac{1}{\gamma}(xe^{\beta t} - x_0) \qquad f(x,t|x_0,0) = \\ &\qquad\qquad = \frac{1}{\gamma}e^{\beta t} f'(x',t'|0,0) \\ t' &= \frac{1}{2\beta}(e^{2\beta t}-1) \end{aligned} \tag{4.3}$$

changes Eq.(4.2) into the heat equation

$$\frac{\partial f'}{\partial t'} = \frac{\partial^2 f'}{\partial x'^2}$$

and the initial condition (2.6), with $t_0 = 0$, into

$$\lim_{t'\to 0} f'(x',t'|0) = \delta(x')\,.$$

In other words, the transformation (4.3) changes the considered process $X(t)$ into the Wiener process (with zero drift and infinitesimal variance equal to 2) $W(t')$ with $W(0) = 0$. Therefore, there results

$$g[S(t),t|x_0] = g_W[S'(t'),t'|0]\,\frac{dt'}{dt} \tag{4.4}$$

where the r.h.s. contains the first passage time p.d.f. of the Wiener process through the transformed threshold. Making use of (4.3) we thus get (cf. also Ricciardi, 1975):

$$g[S(t),t|x_0] = e^{2\beta t} \times \\ \times\, g_W\left\{\frac{S[\frac{1}{2\beta}\ln(1+2\beta t')]\sqrt{1+2\beta t'} - x_0}{\gamma}, t'|0\right\}. \tag{4.5}$$

Let us now assume

$$S(t) = S_\infty + S_0\, e^{-\beta t}\,. \tag{4.6}$$

From (4.5) we obtain:

$$g[S_\infty + S_0 e^{-\beta t}, t|x_0] = e^{2\beta t} \times \\ \times\, g_W\left\{\frac{1}{\gamma}[S_\infty\sqrt{1+2\beta t'} + S_0 - x_0], t'|0\right\} \tag{4.7}$$

From this formula it follows that if $S_\infty = 0$ a closed form expression is available for g. Indeed, if $S_\infty = 0$, from (4.7) we obtain:

$$g(S_0 e^{-\beta t}, t|x_0) = e^{2\beta t} g_W\left(\frac{S_0 - x_0}{\gamma}, \frac{e^{2\beta t}-1}{2\beta}\,\middle|\, 0\right) \tag{4.8}$$

where the r.h.s. contains the well known first passage time p.d.f. through a constant threshold for the Wiener process $W(t')$. A relation equivalent to (4.8) was first discovered by Siebert (1969).

We now return to our main problem, i.e., the determination of $g[S(t), t|x_0]$ through an arbitrary threshold. According to (4.5), this requires the knowledge of the first passage time p.d.f., through $S'(t')$, for the Wiener process $W(t')$. As shown by Durbin (1971), a numerical solution to this problem can be achieved to a very satisfactory degree of accuracy. In view of the potential great importance of Durbin's algorithm in the neurobiological context we shall provide a sketch of the underlying idea. To this purpose, let us write down explicitly Eq.(4.1) for the Wiener process $W(t')$ and set $x = S'(t')$. Using the well known expression for the transition p.d.f. of such process, we get:

$$\frac{1}{\sqrt{2\pi t'}} \exp\left\{-\frac{[S'(t')]^2}{2t'}\right\} = \int_0^{t'} d\tau\, g_W[S'(\tau), \tau|0] \times \times \frac{1}{\sqrt{2\pi(t'-\tau)}} \exp\left\{-\frac{[S(t')-S(\tau)]^2}{2(t'-\tau)}\right\} . \tag{4.9}$$

This is a Volterra integral equation of the first kind with a kernel which is singular in the sense that it diverges as $\tau \to t'$. It should be pointed out that no standard techniques seem to be available in the numerical analysis literature for solving Eq.(4.9). This task is accomplished by Durbin by approximating $S'(t')$ by straightline segments in subintervals $(\tau, \tau+d\tau)$ and by using known results for the crossing probabilities for straight lines. A detailed exposition of the method is contained in Durbin's quoted paper. Here it will suffice to remark that once a satisfactory numerical evaluation of $g_W[S'(t'), t'|0]$ has been achieved, the function $g[S(t), t|x_0]$ becomes readily available via relation (4.5), which solves our problem.

References

BRAITENBERG, V., 'What can be learned from spike intervals histograms about synaptic mechanisms?', *J. theor. Biol.* **8**, 419-425 (1965).

CAPOCELLI, R.M. and RICCIARDI, L.M., 'Diffusion approximation and first passage time problem for a model neuron', *Kybernetik* **8**, 214-223 (1971*a*).

CAPOCELLI, R.M. and RICCIARDI, L.M., 'Diffusion models for formalized neurons', Proc. 1st Cybernetics Cong., Casciana, 818-836 (1971b).

CAPOCELLI, R.M. and RICCIARDI, L.M., 'On the inverse of the first passage time probability problem', *J. Appl. Prob.* **9**, 270-287 (1972).

CAPOCELLI, R.M. and RICCIARDI, L.M., 'A continuous Markovian model for neuronal activity', *J. theor. Biol.* **40**, 369-387 (1973).

DE LUCA, A. and RICCIARDI, L.M., 'Formalized neurons: probabilistic description and asymptotic theorems', *J. theor. Biol.* **14**, 206-217 (1967).

DE LUCA, A. and RICCIARDI, L.M., 'Probabilistic description of neurons, in *Neural Networks,* E.R. Caianiello (ed.), Springer-Verlag, 100-109 (1968).

DURBIN, J., 'Boundary crossing probabilities for the Brownian motion and Poisson processes and techniques for computing the power of the Kolmogorov-Smirnov test', *J. Appl. Prob.* **8**, 431-453 (1971).

FELLER, W., 'The parabolic differential equations and the associated semi-groups of transformations', *Ann. of Math.* **55**, 468-519 (1952).

FELLER, W., 'Diffusion processes in one dimension', *Trans. Amer. Math. Soc.* 77, 1-31 (1954).

RICCIARDI, L.M., 'Formalized neurons and neural networks', Proc. 1st European Biophys. Cong., Baden, 297-315 (1971).

RICCIARDI, L.M., Diffusion approximations in neuronal modeling and population growth, Proc. 2nd Cybernetics Cong., Casciana, 43-60 (1972).

RICCIARDI, L.M., 'Stochastic equations in ecology and neurobiology', in *Progress in Cybernetics and Systems Research* (R. Trappl and R. Pichler, eds.), Vol.I, 183-188, Hemisphere, Washington (1975).

RICCIARDI, L.M., 'On the transformation of diffusion processes into the Wiener process', *J. Math. Analysis Appl.* **54**, 185-199 (1976).

RICCIARDI, L.M. and VENTRIGLIA, F., 'Probabilistic models for determining the input-output relationship in formalized neurons', I. A theoretical approach', *Kybernetik* **7**, 175-183 (1970).

SIEBERT, W.M., 'On stochastic neural models of the diffusion type', Quart. Progr. Rept. MIT Res. Lab. Electron **94**, 281-287 (1969).

STRATONOVICH, R.L., *Topics in the Theory of Random Noise,* Vol. I, Gordon and Breach, N.Y. (1963).

WANG, M.C. and UHLENBECK, G.E., 'On the theory of the Brownian motion II.', *Rev. Mod. Phys.* **17**, 323-342 (1945).

A holographic brain model for associative memory

O. WESS and U. RÖDER
Technische Hochschule, Darmstadt,
and Gesellschaft für Strahlen- und Umweltforschung, * *Neuherberg, FRG*

1. Introduction

The principle of holography was developed by Gabor in 1948 [1]. It initiated fruitful activities mainly in the field of coherent optics when the laser was invented in 1960. The essential advance of holography is the possibility to store and reconstruct amplitude and phase of a 3-dimensional wave by interference and diffraction. The holographic principle itself is not limited to visible optics only, but can be applied to all those domains where coherent waves are available so that an interference pattern can be formed. Interference patterns which contain the information about amplitude and phase of the waves can be generated in principle by any arbitrary part of the electromagnetic spectrum, by acoustic waves and as we suppose as well as many authors [2-6] by action potentials traveling through the nervous system of highly evolved living beings. At least a holographic brain model can describe some brain functions of which we want to restrict ourselves to associative storage and the formation of sequences of associatively stored information.

What now is the principle of holography? An object wavefield is superimposed with a coherent reference wavefield (in the optical case laser light) and the resulting interference field is stored in a light-sensitive medium (photographic plate, photochromic crystal, etc.). For reconstruction, only the former reference beam hits the stored interference pattern and by diffraction the original object wavefield is restored. (Fig. 1)

Why now does holography allow us to describe functions of neural systems? Figure 2 shows some essential analogies between holographic and neural systems.

First: The spatial distribution of holographically stored information is based on diffraction, scattering and interference effects and corresponds to the results of Lashley's lesion experiments. Each small area of a hologram contains information of all parts of the object. In this way a hologram is extremely insensitive to destruction. Even from small parts of a hologram the total object wave can be reconstructed however with loss of resolution.

Lashley [7] couldn't find localized contents of memory in his experiments and had to assume that the incoming information is spread and stored over wide areas of the cortex, so that destruction of parts of the cortex does not completely erase the information, but only deteriorates the recall of the total information.

Second: A holographic recording medium is capable of storing different holograms simultaneously or subsequently but they can be reconstructed separately. This corresponds to multiple storage which has to be postulated in neural systems when taking

* Present address

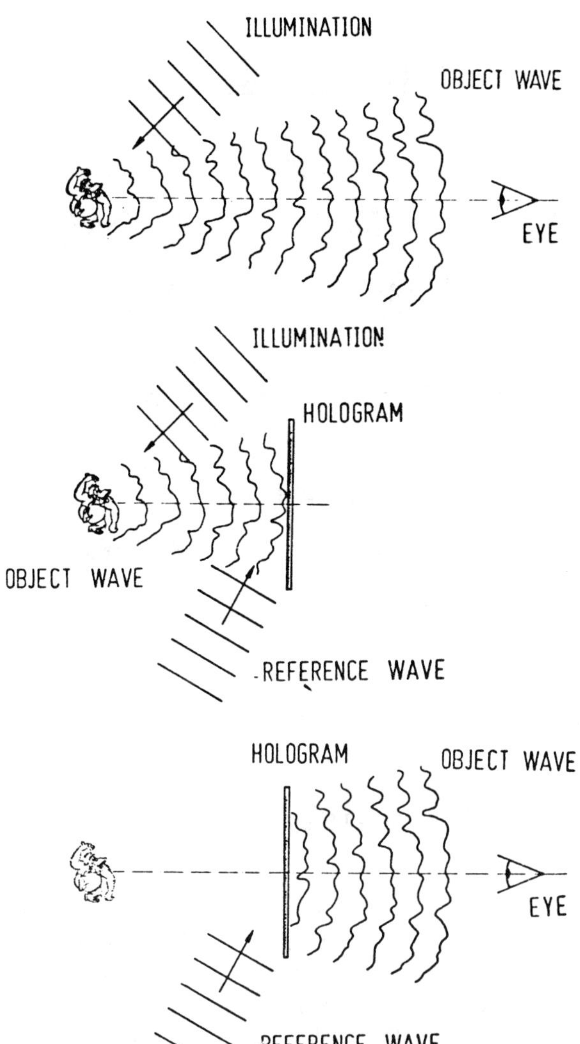

Fig.1 Holographic storage and reconstruction of a tridimensional wave field

Lashley's experiments as basis. This probably means that single parts in the central nervous system are involved in the storage of different contents of memory.

Third: Translation invariant parallel information processing is realized in a special holographic set-up and has similarities with translation, magnitude and direction invariant pattern recognition in neural systems.

Fourth: The storage capacities of neural and holographic systems are of the same orders of magnitude.

Fifth: Last but not least, the point which we want to discuss in this chapter is the possibility of an associative recall of connected events. To develop the concept for an associative holographic memory, we will discuss the basic holographic formalism for associative storage, subsequently we will point out three different mechanisms to link several items to associative sequences and finally we will sketch the possibility of neural holographic processing.

2. Associative storage

As a common view of holography the phase of a complicated object wave is stored by the interference with a simple plane or spherical reference wave. Those waves are easily generated and one can usually reconstruct the stored information out of a two-dimensional hologram with any of those waves at any place, whereas for 3-dimensional storage mediums Bragg's condition has to be fulfilled.

NEURAL SYSTEM	HOLOGRAPHY
NON LOCALISATION OF CONTENTS OF MEMORY (LASHLEY)	SPATIAL DISTRIBUTION OF STORED INFORMATION
THE SAME GROUP OF NEURONS STORES DIFFERENT INFORMATION	MULTIPLE STORAGE TECHNIQUE
PATTERN RECOGNITION INDEPENDENT OF TRANSLATION,MAGNITUDE AND DIRECTION	PARALLEL TRANSLATION INVARIANT INFORMATION PROCESSING
CONTENTS OF LONG-TERM MEMORY 10^{10}-10^{14} BIT	STORAGE CAPACITY 10^{11}-10^{13} BIT
RECALL OF CONNECTED EVENTS ASSOCIATIVE LEARNING	ASSOCIATIVE STORAGE AND RECONSTRUCTION

Fig.2 Neural- and holographic analogies

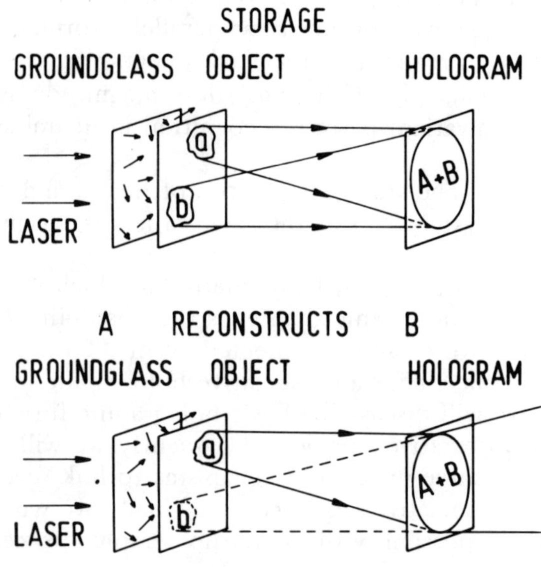

Fig.3 Associative memory (schematic)

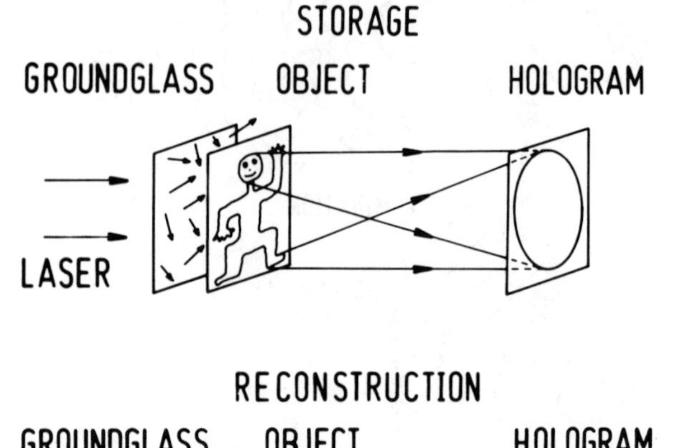

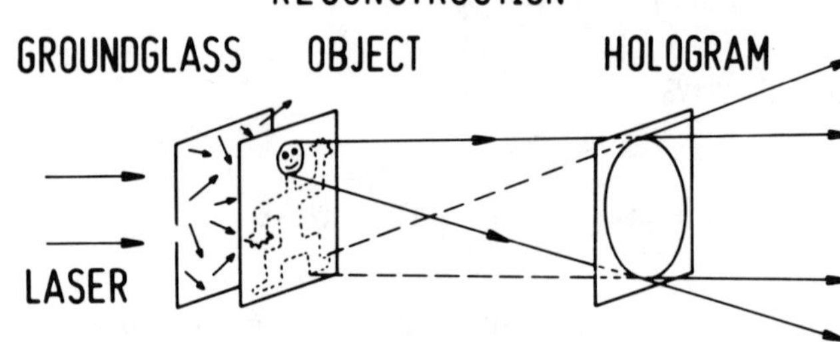

Fig.4 Ghost image (schematic)

To explain the mechanism of associative holographic storage the roles of object- and reference wave can be changed in that way that amplitude and phase of the reference wave will be recorded by interference with the object wave (Gabor [4]). The correctness of this consideration is proved by experiments where illumination of the hologram with only the object wave will lead to a reconstruction of the reference wave which was previously used. If we drop the demand for a plane or spherical reference wave and superpose two complicated object waves to form an interference pattern, amplitude and phase of both waves will be recorded even in this case. Nevertheless wave A can be reconstructed only by wave B and vice versa, whereas before, usually any plane wave was able to reconstruct the information. This is exactly the case we are interested in, because now we can link an associative connection between two different object waves containing information so that one wave is reconstructed *only* by the occurrence of the other and not otherwise. In Fig. 3 this actual situation is sketched.

Object a is associatively linked with object b by their corresponding waves A and B. Essential for this kind of connection is again to record the interference pattern, but this is generated only if a and b exist at the same time and their wavefields A and B are mutually coherent.

An interesting feature of associative holographic storage is the formation of a so-called 'ghost image' predicted by van Heerden in 1963 [2], which is outlined in Fig. 4. Any part of the information (the head of the man for example) reconstructs the whole information (the whole man). To compare it with brain functions it corresponds to the recall of complex clusters of memory by a single associatively stored detail.

The appropriate mathematical descriptions of associative memory are correlation functions which the visual as well as the central nervous system make use of (Reichhardt [8]). An example is the comparison of the input information with previously stored information in the case of pattern recognition or in the case of conditioned response. For a specific type of hologram, the so-called Fraunhofer or Fourier-transform hologram the mathematical formalism is particularly simple. In addition we receive as a present translation invariant processing if two-dimensional recording material is used.

Beyond a simple mutual associative coupling several different objects can be connected to an associative chain as shown in Fig. 5. The upper sequence could denote the sensory region, for example the time-dependent sequence of intensity patterns on the retina, whereas the lower row could be our hologram or the memory itself. Always two inputs must exist

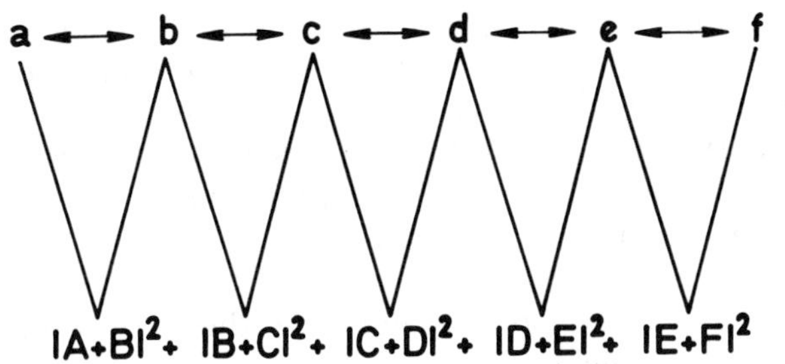

Fig.5 Coupling in pairs to form an associative chain. The upper sequence may denote the sensory region, whereas the lower row could be the long-term memory

simultaneously such as *a* and *b*, then *b* and *c*, then *c* and *d*, etc., to get an interference pattern to be recorded. Subsequently one after another hologram can be stored by incoherent superposition on the same holographic plate. Now it is possible to reconstruct with an arbitrary one of the previously stored inputs the preceding one as well as the subsequent one. That is because *b* for instance is coupled in one exposure to *a*, and in another to *c*, so when illuminating only with *b*, beside *b* also *a* and *c* will appear in the image plane.

In order to reconstruct subsequently more than the nearest neighbors of the chain we used three different experimental methods. Those are optical feedback, non-linear recording and volume coupling. For each of these methods we will show experimental results.

3. Optical feedback

By means of feeding back the reconstructed output to the input plane of the system more elements of the chain can be reconstructed, because the feedback of the just reconstructed element causes the reconstruction of the following, and so on.

An optical feedback arrangement (Mager *et al.* [9]) is shown in Fig.6. The input plane is equivalent with the image plane, which again is imaged to the output plane by means of an auxiliary lens and a beamsplitter. In the case of a passive optical feedback system as we did use in our experiments the succeeding output will have much less intensity than the input because of the low diffraction efficiency of

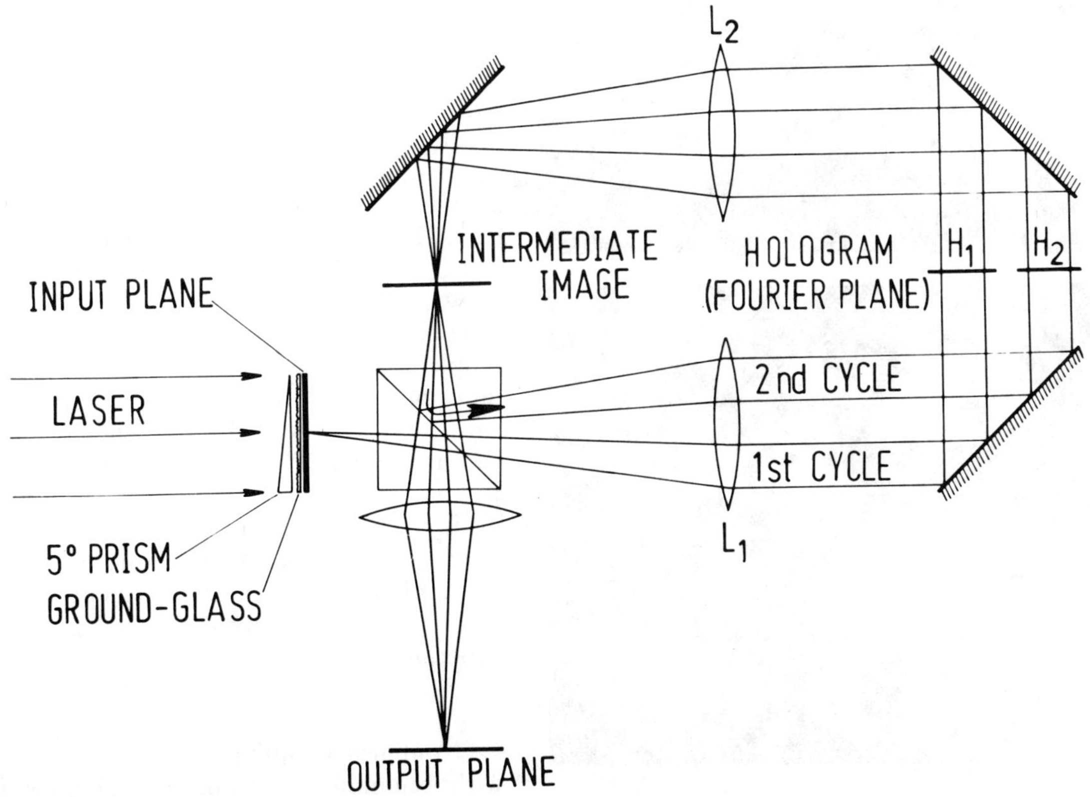

Fig.6 Optical feedback system for two-dimensional complex functions

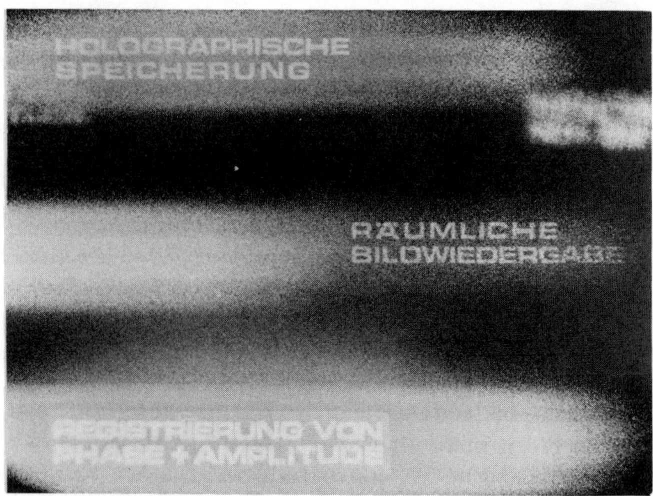

Fig.7 Input, first and second associative reconstruction by aid of optical feedback

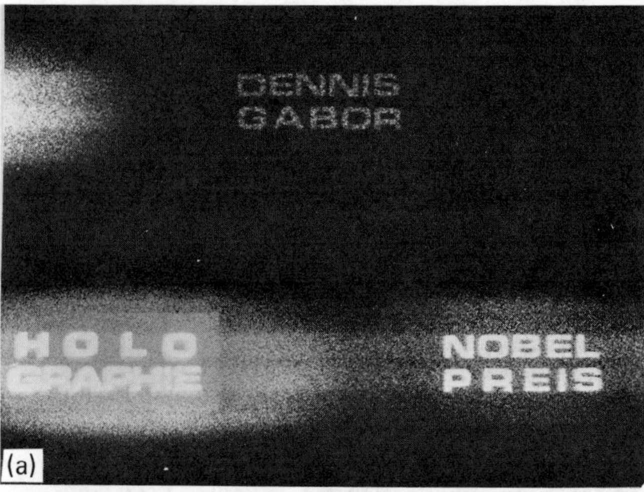

Fig.8 a) Associative reconstruction
b) With a grey tone object

the hologram and dissipation of energy.

Figure 7 shows the input 'REGISTRIERUNG VON AMPLITUDE UND PHASE' as the first element of a chain of three objects and the subsequent reconstruction of the second element 'HOLOGRAPHISCHE SPEICHERUNG' and finally using the second element as input again the reconstruction of the third element 'RÄUMLICHE BILDWIEDERGABE'.

Figure 8 shows another example, the associative connection which is common to all those working on the field of holography: 'HOLOGRAPHIE' as input 'NOBELPREIS' and 'DENNIS GABOR' as first and second output, or instead of the name the portrait of the Nobel prize winner of 1971.

From one reconstruction to the following the loss of intensity is clearly to be seen. Unless no active amplification of the reconstructed waves is applied the number of reconstructed elements will be strictly limited by the loss of energy. Moreover signal-to-noise ratio will become lower by each following correlation unless noise is suppressed by means of a threshold.

Such a system is able to process branched chains of information as it may occur, if different people have the same name. For example starting with the name 'Paul' one may associate on the one hand 'McCartney' and 'Beatles' and on the other hand 'Paul', 'Pope', and 'Pills'. More seriously it is demonstrated in Fig. 9 starting with A followed by B_1 and C_1 on the one hand and B_2 and C_2 on the other.

To emphasize it again this way of forming associative chains is based on the principle of optical feedback. The question of whether we can transfer this principle to the central nervous system is not easily answered.

The present experimental set-up is perhaps too simple, especially as an active amplification and a threshold, which definitely exist in the neural system, could not be realized in the analog optical system at present. This limits a close analogy to neural associative memory and further examination should focus its attention on the development of optical active elements. Another feature of neural systems, the nonlinear characteristic of neurons can be used to couple associatively in an analogous holographic experiment.

4. Nonlinear coupling

With a nonlinear working recording medium—for instance a photographic emulsion or perhaps layers of neurons—it is possible to link information associ-

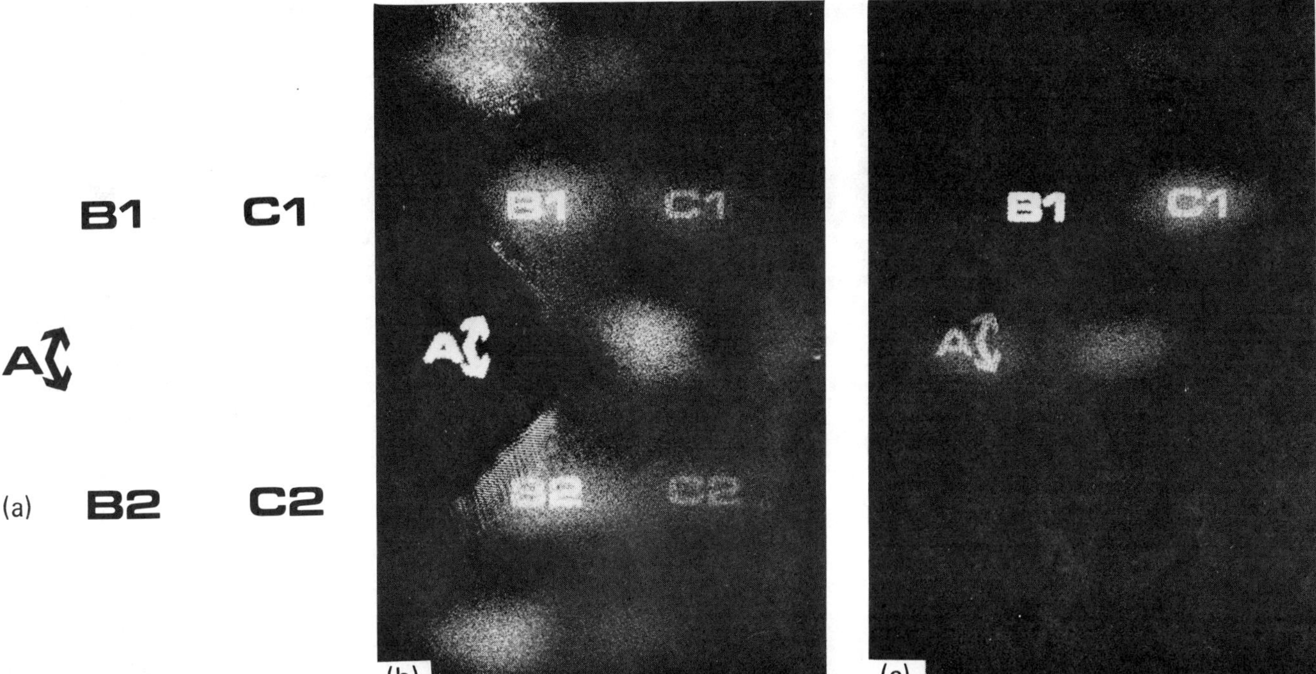

Fig.9 Associative reconstruction of branched sequences of information
a) negative of the branched associative chains
b) reconstruction by illumination with A
c) reconstruction with one element (B1) of a branch results in the reconstruction of this branch only

atively which doesn't exist simultaneously as it is done by optical feedback. Those elements cannot perform an interference pattern by mutual interference, nevertheless they can be reconstructed by each other. In the following part an experiment will be described which makes use of linear and nonlinear coupling properties of the recording medium. In this case it is important again that always two neighbored elements exist simultaneously to form an interference pattern.

Experimental examples are shown in Figs. 10-15. Figure 10 illustrates the whole object viewed through the storage hologram. All those six single elements are stored in pairs and linked to form an associative chain either by linear or nonlinear coupling. An arbitrary element of the sequence reconstructs by linear coupling its two nearest neighbors as usual whereas the following next-but-one neighbors in both directions are reconstructed without optical feedback by nonlinear coupling. To couple more than the next-but-one neighbors still higher order terms in the recording characteristic (which definitely are present in the neuronal case) must exist. [10] The recording materials we used in our experiments were Agfa Scientia 10E75 plates where we couldn't find an influence of these higher order terms.

To compare it with neural systems one might speculate that there is a much stronger influence of nonlinearities so that more extended coupling should be possible.

In general the characteristic of neurons shows strong nonlinear behavior in that way that an incoming action potential generates a subsequent action potential only if a certain threshold is overcome.

Above this value the size of all secondary action potentials is independent of the input size. So the characteristics of neurons might be much more adequate to perform nonlinear coupling than those of photographic plates. The signal-to-noise ratio is reduced by multiple correlation the more elements of the chain are reconstructed.

5. Volume coupling

The final method for associative coupling to be discussed is storage of information in volume holograms.

Fig.10 The 6 used elements viewed through the storage hologram

Fig.12 Example for a nonlinear and linear coupling: Reconstruction with LL from a double exposure hologram (H1 : LL and UL, H2 : LL and LM)

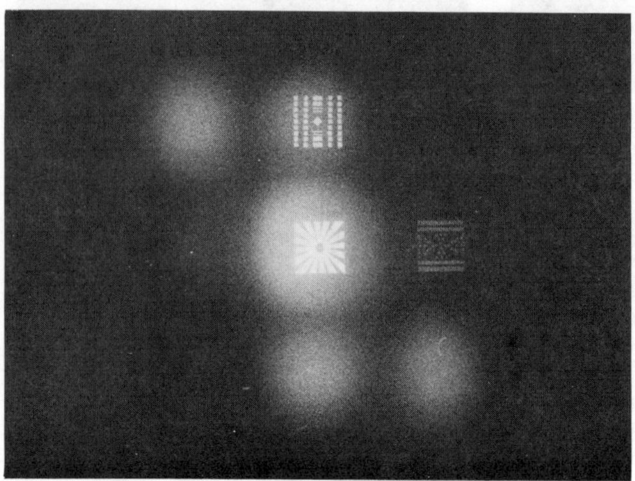

Fig.11 Image plane viewed through a double exposure hologram as an example for *nonlinear* coupling. First hologram: upper row left (UL) and lower row left (LL). Second hologram: UL and lower row middle (LM). The reconstruction is started by LL. The element LM is coupled indirectly by nonlinear terms. Besides the reconstructed waves some crosscorrelation terms appear.

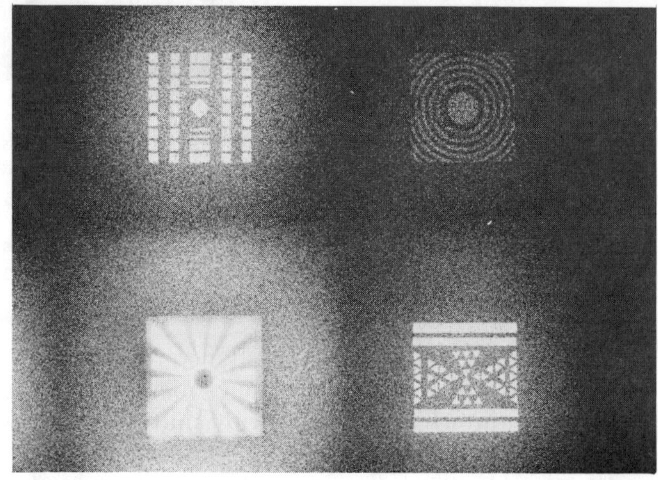

Fig.13 Reconstruction from a triple exposure hologram starting with LL: linear and nonlinear coupling. (H1 : UL and LL, H2 : LL and LM, H3 : LM and UM)

Compared with two-dimensional recording mediums volume holograms have some advantages, like high storage capacity and storage without any chemical developing process (immediate readout).

Associative coupling can be done by volume effects in that way, that the first reconstruction occurs in the first layer of a volume which, itself, while traveling through the volume, will reconstruct the second wave in a second layer of the volume, etc.

In the following experiments we used an Fe-doped $LiNbO_3$ crystal for storing holograms. The principle of creating a phase hologram in this crystal is shown in Fig.16.

In an optical set-up shown in Fig. 17 always two waves are stored in pairs, one exposure with the waves $A+B$, a second with the waves $B+C$, etc. As in the previous experiments element B occurs in both exposures in order to perform the associative coupling of A and C although A and C are temporarily incoherent. The laser beam enters the set-up coming from above. It is split into two beams which interfere at the place of the crystal. After the first

Fig.14 Reconstruction from a fivefold exposure hologram, starting with LM
(H1 : UL and LL, H2 : LL and LM, H3 : LM and UM, H4 : UM and UR, H5 : UR and LR)

Fig.15 Reconstruction from the same hologram as in Fig.14, but starting with UM

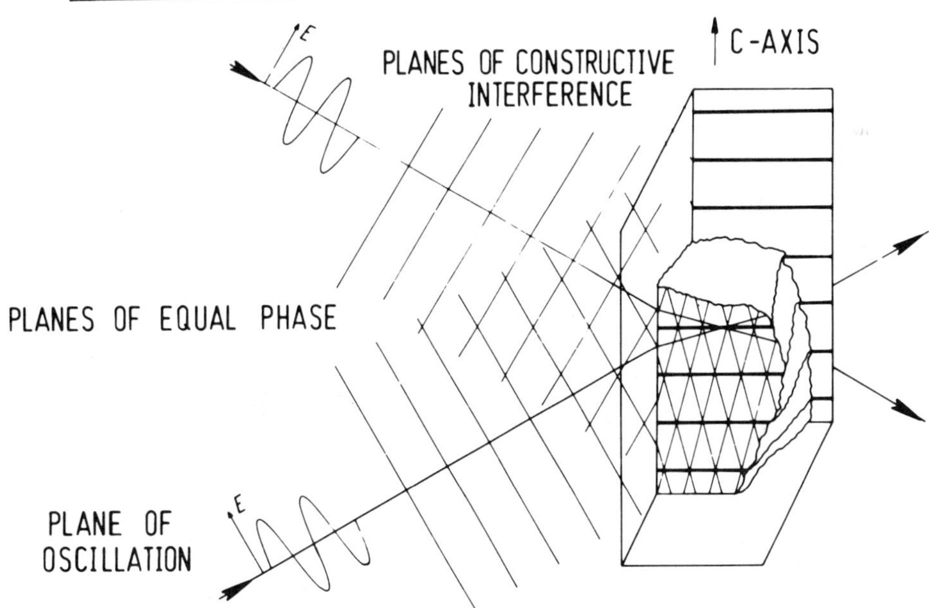

Fig.16 Volume storage in $LiNbO_3$ crystal

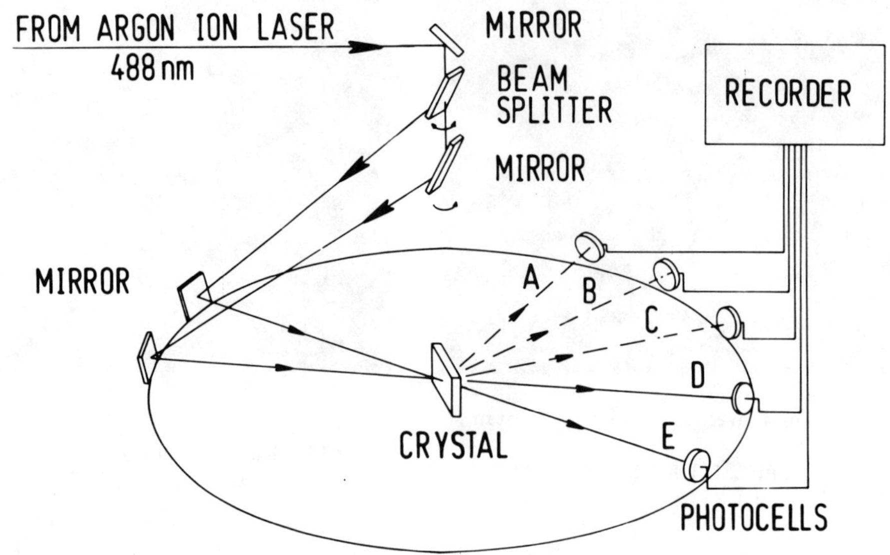

Fig.17 Storage arrangement for associative coupling of plane waves

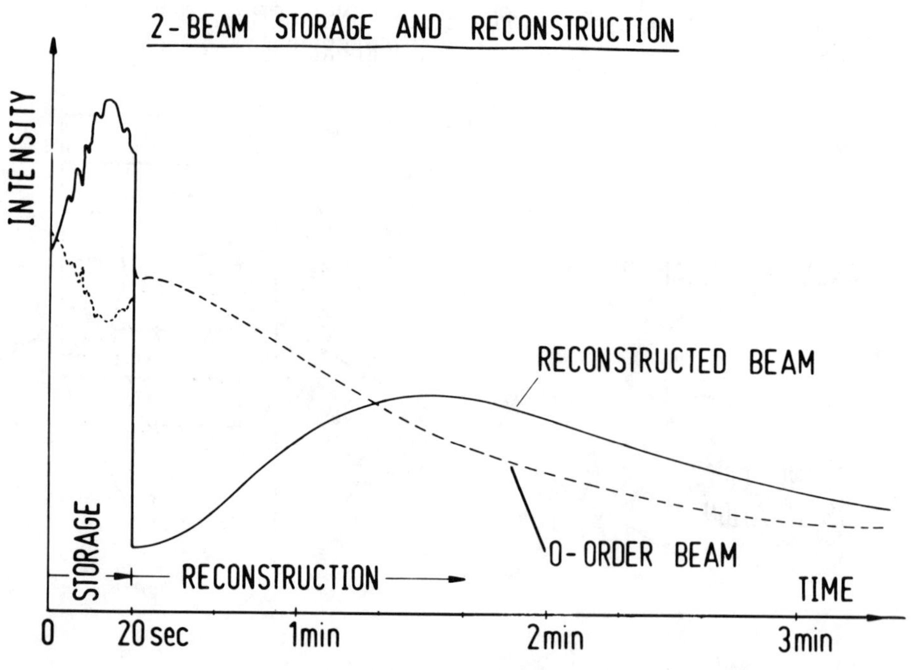

Fig.18 Intensities of two beams during storage and reconstruction in $LiNbO_3$

exposure the angle of incidence can be changed by rotation so that beam A becomes beam C for the second exposure whereas beam B remains unchanged. After the second exposure beam C remains unchanged and beam B becomes beam D, etc.

Because there is no need for chemical processing the stored information can be read out immediately after recording. The time-dependence of storage and reconstruction is shown in Fig. 18 for two beams. The intensities of both emerging beams are measured with photocells behind the crystal. Starting with nearly the same intensities, the energy is partly shifted from one beam to the other during exposure as can be derived from the diverging curves. In the immediately succeeding readout process the intensity of the reconstructed beams increases by selfenhancement from relative small values to a maximum value which can be higher than that of the read-out beam behind the crystal. The diffraction efficiency is growing during read-out and can reach theoretical values of 100%. [11,12] After some time which depends on read-out intensities, energy is dissipated by scattering and the signals decrease.

Selfenhancement means the additional writing of the same hologram during read-out.

An analogy in neural systems can be found in the fact that, during memorizing, a complex cluster of information is getting stronger and more distinguishable.

Figure 19 shows a chain of four waves. The energy of the incident wave is distributed among all reconstructed waves with different rates. As to be seen beam 4 can be more intensive than the one before, which is coupled closer to the incident beam.

It should be possible, if using plane waves, to send the energy of the incoming beam preferably into an arbitrarily chosen beam of an associative chain. In our experiments we were not successful in switching the whole input energy to the last element of the chain, but the trend can be seen, that there is no continuous decrease of signal strength of the reconstructed elements in the associative row. This can be a model for subconscious associations in which one element suddenly comes to consciousness, seeming to be independent of the particular input.

8. Conclusions

We discussed a holographic system capable of performing three different kinds of associative coupling, by optical feedback, nonlinearities and volume effects. We pointed out analogies to some functions of the central nervous system. Conclusions beyond this point are speculated but not proved. That should be stressed because we don't know what actually happens in the brain.

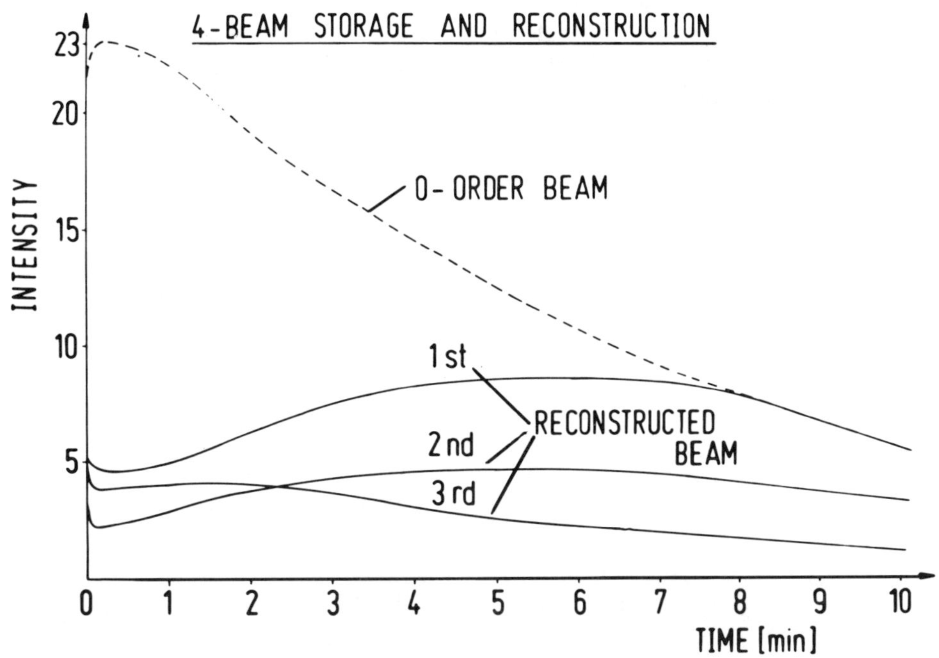

Fig.19 Intensities of 4 beams during reconstruction. Note, that the 3rd reconstructed beam is stronger than the 2nd one during the first two minutes of reconstruction

Nevertheless we want to sketch what might happen in the brain. As far as we know the basic mechanisms for the three types of associative coupling mentioned exist in the central nervous system.

There are feedback mechanisms as reafferent fibers, nonlinear characteristics of neurons and a 3-dimensional architecture of the neural network.

The problem is the formation and storage of a neural interference pattern. Concerning the formation of an interference pattern a phase relation between signals has to be assumed. In the case of storing visual patterns the phase of the incoming light is lost on the receptor level. As outlined by T.W. Barret [13] groups of action potentials travel through a highly interconnected neural network and a relative phase can be introduced by delay times depending on the different lengths of paths. Contrary to other authors, we think that there is no need to have an additional reference source, because one spatial electrical impulse pattern may act as a reference source for the other and vice versa. Depending on the phase difference, signals may interfere constructively or destructively. Exciting as well as inhibiting synapses using different types of neurotransmitters may serve for it. The job can also be done by only one type of synaptic junction if a threshold is introduced.

Concerning the storage mechanism to record such an interference pattern one might think of synapses changing their characteristics by frequent use. The threshold may be lowered and a spatial neural transmittance distribution may be generated. Analogous to optical holographic reconstruction, the spatial electrical input distribution may be multiplied by the neural transmittance pattern to reconstruct associatively stored patterns.

Any way by which neural mechanism correlation of incoming and stored patterns may be performed, correlation and convolution functions seem to be as important for neural as for holographic systems. Moreover, neural systems are much more flexible to perform various tasks because they are based on active elements instead of the passive optical arrangements we used.

References

1. GABOR, D., *Nature* **161**, 777 (1948).
2. VAN HEERDEN, P.J., *Appl. Optics* **2**, 393 (1963).
3. WESTLAKE, P.R., *Kybernetik* **7**, 129 (1970).
4. GABOR, D., *IBM J. Res. Dev.* **13**, 2 (1969).
5. PRIBAM, K.H., *Languages of the Brain,* Prentice-Hall, New Jersey (1971).
6. GREGUSS, P., *Nature* **219**, 482 (1968).
7. LASHLEY, K.S., *Science* **73**, 245 (1931).
8. REICHARDT, W., *Z. Naturforschung* **12b**, 448 (1957).
9. MAGER, H.J., WESS, O. and WAIDELICH, W., *Opt. Com.* **9**, 156 (1973).
10. WESS, O. and RÖDER, U., *Biol. Cybernetics* **27**, 89 (1977).
11. CASE, S.K., *J. Opt. Soc. Amer.* **65**, 724 (1975).
12. WESS, O., Dissertation TH Darmstadt (1976).
13. BARRET, T.W., *Neuropsychologia* **7**, 135 (1969).

Formal description of an analysis system developed for neuronal data

MICHEL DUFOSSE, PATRICK DECANTE and ALBERT HANEN
Laboratoire de Physiologie, Paris, France

We shall consider here stochastic processes. The analysis system computes data derived from stochastic processes by the sampling corresponding to data acquisition. Depending on the type of problem, it will be of more interest to sample the signal in a specific manner. By extent, we will consider the name process to be the set of discrete values associated after the data acquisition.

An analysis system is the systematized set of necessary algorithms which give an image of original data where the non pertinent information will be reduced. We show that an analysis system may be reduced to a quadruplet:

$$S = (T, K, F, G)$$

where:

- T is the set of process types,
- K is the set of numbers of processes of which user may have a synoptic view,
- F is the subset, included in the system, of the process transforms (also called proprocessors),
- G is the same way the subset of the data processors.

T : Process types
Elements of T may be (Fig.1):

a) P_p the set of point processes. A point process is defined by the times of events. The shape of the signal is lost during data acquisition. It will consist of a set of values associated with a time unit.

b) P_e the set of sampled processes. A sampled process is defined by the amplitude of the signal at the sampling time and a sampling function giving discrete values. It will consist of a set of values associated with an amplitude unit.

c) P_r the set of ranked processes. A ranked process is defined by the amplitude of the signal at unknown times. The rank of the data keeps relevance in the time domain.

Other process types might be considered as needed and definable.

K : Numbers of processes
K is the set of numbers of processes of which a user may have a synoptic view. We limited it to $K = \{1, 2\}$ due to the necessary help of visual interpretation of results.

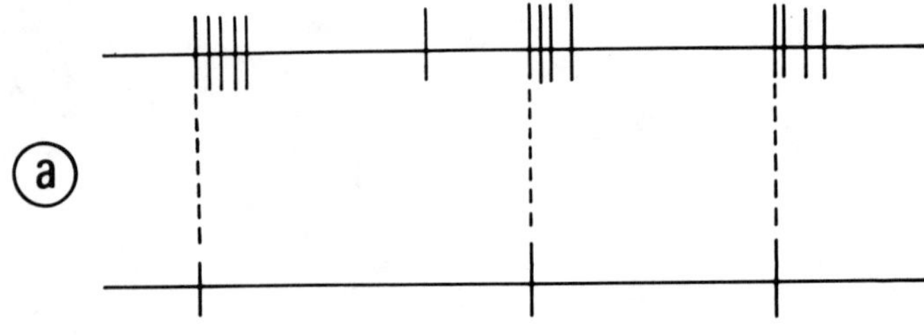

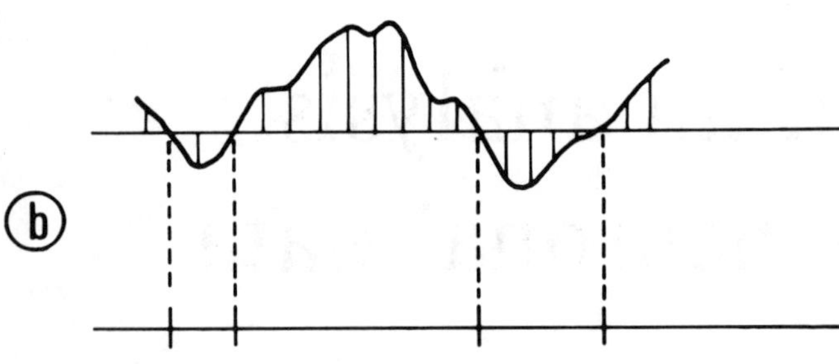

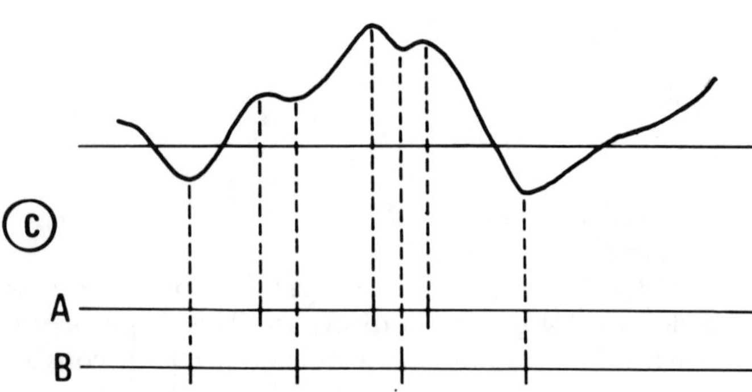

Fig.1 Different process types a) Point process, b) Sampled, c) Ranked

F : Process transforms

A process transform is an application f from P^k into Q^l where:

$$P, Q \in T \text{ and } k,l \in K \qquad \text{(Fig.2)}$$

Some transforms are conservative of the process type $(P=Q)$, such as the lowpass filtering of a sampled process $(P=Q=P_p)$ or the elaboration into a point process consisting in bursts of events extracted from a primary point process (Fig.2*a*).

Other transforms, such as pattern recognition are able to translate a sampled process into point process (Fig.2*b*). Considering rate of events will translate a point process $(P=P_p)$ into a sampled process $(Q=P_s)$.

A transform can modify the number of simultaneously analyzed processes $(k \neq 1)$. For instance taking separately maxima and minima in a sampled process $(P=P_s, k=1)$ will give a double point process $(Q=P_p, l=2)$.

G : Processors

A processor in an application g from P^k into $\mathbb{R}^m$ where: $P \in T$ and $k \in K$. Processing associates to one or more processes a set of values which may be single $(m=1)$ elements or arrays $(m>1)$, which are real numbers supposed to be the aim of the analysis: pictures, statistical coefficients. One can consider that the goal of the analysis is to find f and g such that m be a minimum and the information useful for the problem be still present.

It is of particular interest to group in the same system the three process types, emphasizing the following points:

- Some processors work and are programmed in the same way for the three process types such as histogram.
- A point process and a sampled process can be regarded as ranked processes: the series of correlation coefficients in a ranked process will be the autocorrelogram and the serial correlogram is a point process.
- As the analysis of a problem goes on, it may be useful to go from one type to another.

Elements linkage

An automata is used to associate this quadruplet with a program structure, every processing being a point (P, k, f, g) where f is the equivalent of successive transforms.

The answers to four main questions will select P, k, f, g and will constitute hierarchical executive levels. The automatism will take charge of the transitions between levels by asking questions, giving optimal answers or following a protocol defined by a previous step by step operator controlled analysis.

A matrix gives for the couple (automatism level of execution, question) the 'behavior' of the system. The lowest level asks any question anytime and memorizes the answer in a special array. The highest level using this array, computes the optimal answer. The intermediate levels use the array, ask a question, correct bad answers according to the question.

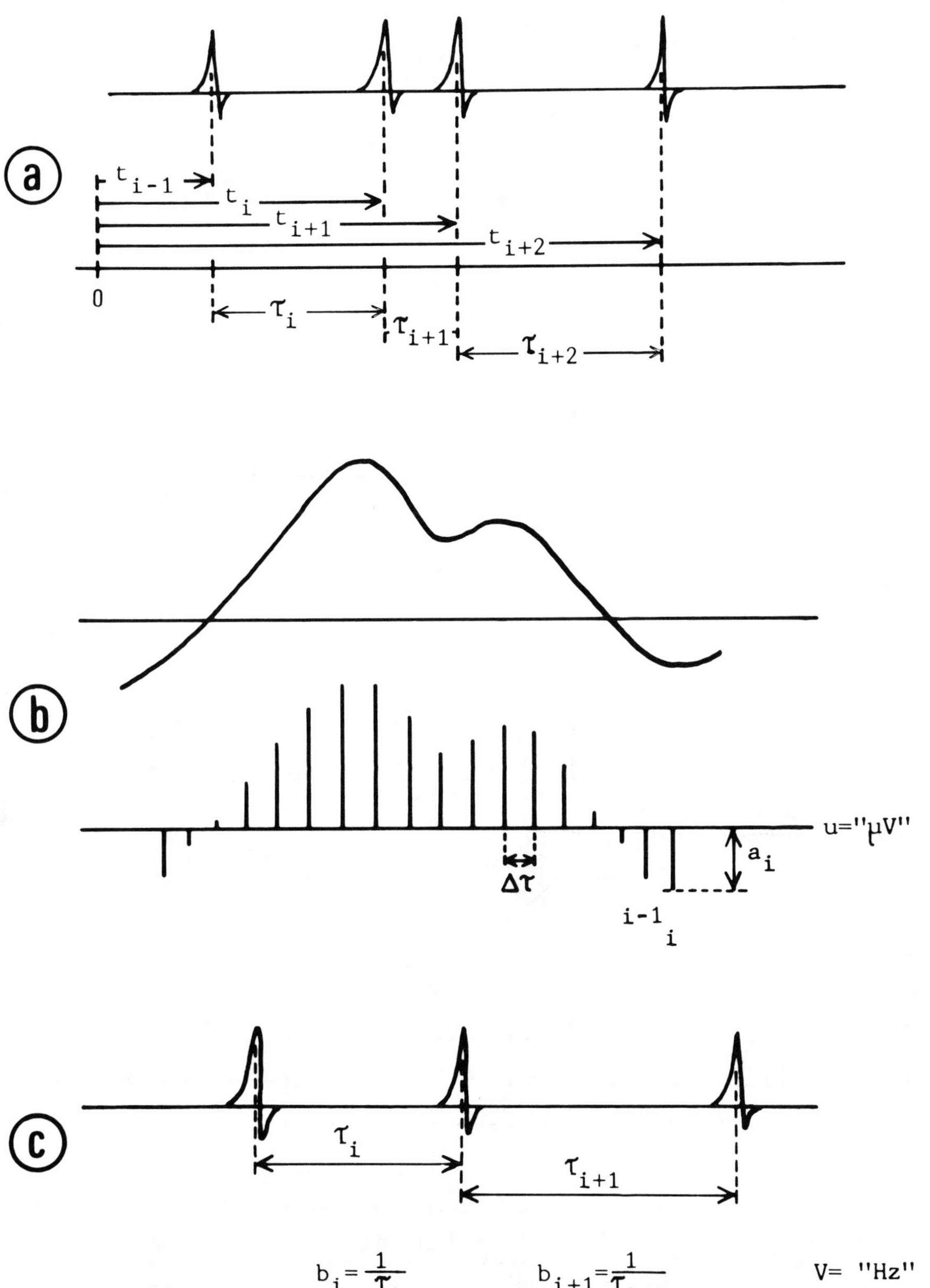

Fig.2 Different process transforms: a) point process into point process, b) sampled process into point process, c) sampled into double point process

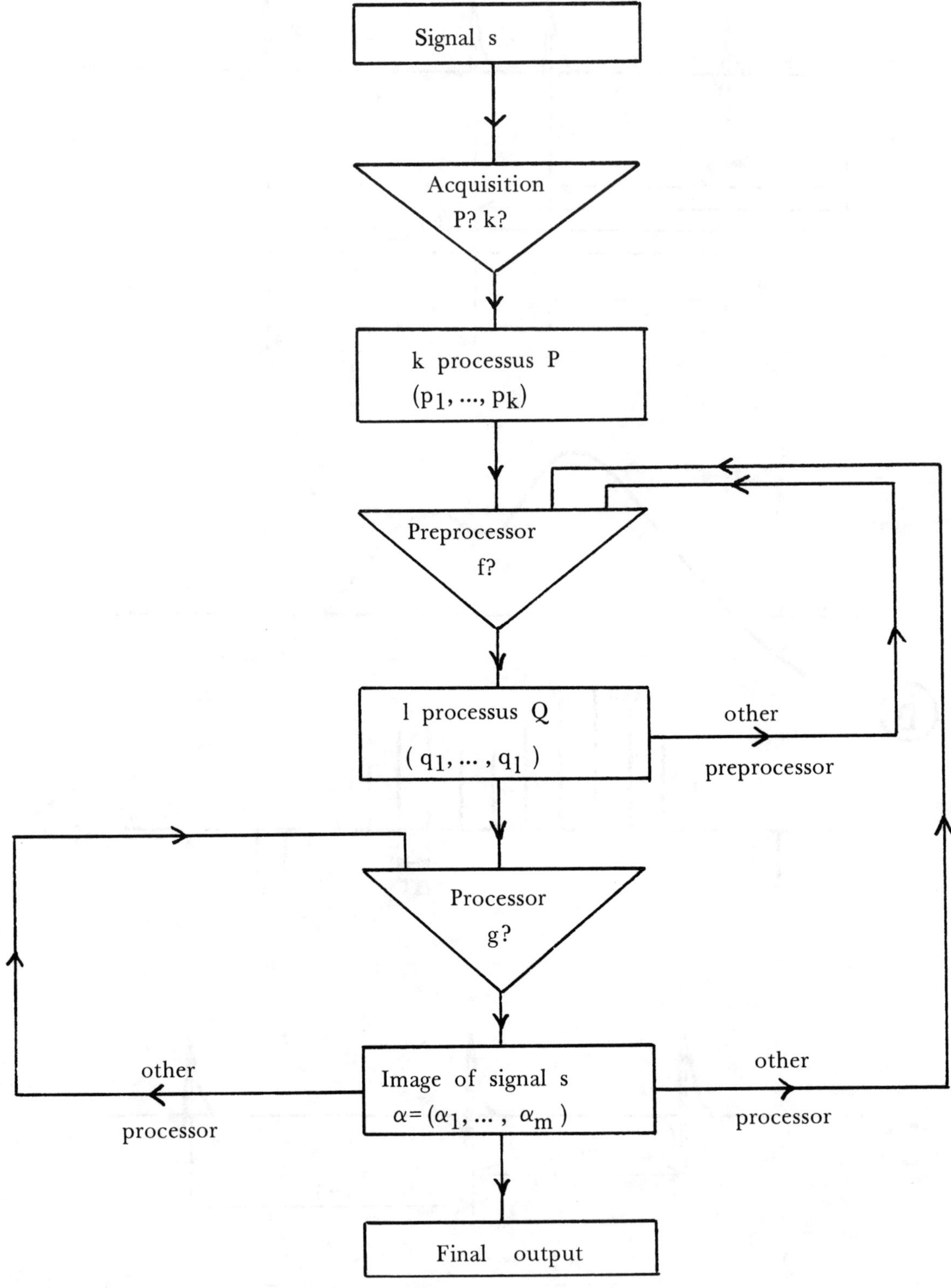

Fig.3 Scheme of an analysis system

Conclusion

This formal approach to the problem is the result of experience in the development of a software package. At a time in the development, when the complexity of the links between subroutines was getting huge, we found the necessity and the usefulness of categorizing algorithms and establishing linking rules between the previous categories. From the practical programming point of view, it is easy to see that we define this way the formats of the links between subroutines and that we give to this type of program a shape adequate for the overlay techniques.

Computer models of the cancer problem

WERNER DUECHTING
University of Siegen,
Siegen, FRG

INTRODUCTION

At a recent medical congress which dealt with leukemia the final summary was as follows: "If we try to classify and interpret the various forms of leukemia discussed here, the result is confusing."

This statement isn't at all surprising when we consider that researchers in the biological and medical fields are facing complex problems in the diagnosis and interpretation of diseases.

However, we know that the development of models is a considerable aid in the description of structure and function of existing biological systems and also in the initiation of new hypotheses.

Medical models have the following aim: to extract a practicable hypothesis or theory from frequently contradictory experimental data where it is impossible to know what is correct or incorrect.

There have been numerous interesting papers written [1-12] dealing with biomathematical descriptions of cell renewal systems, their formulation by differential equations, their presentation by means of block diagrams, and state space techniques and finally the stochastic theory of cell proliferation. The medical world, too, has been carrying out intensive experimental and theoretical tests in the field of cell kinetics of various benign and malignant cell renewal processes [13-40,42,45-47].

Unfortunately, up to now neither method has met with success. Therefore this contribution is an attempt to attack the cancer problem from a completely different side: that is to say with the control systems theory. This poses the provocative question: are cancer diseases really cell renewal systems which have become unstable?

A SIMPLIFIED MODEL OF A CELL RENEWAL CONTROL-LOOP

The very first models which I developed for this purpose were based on a simple block diagram [25] of a single-loop control circuit (Fig. 1).

Starting from this control process with rather simplified assumptions the criterion for the stability of the regulator genes is developed when disturbances have caused irreversible changes in the control-mechanism of the DNA-matrix. This defect leads to a monotonic increase of the cells in an organ (see Fig. 1). Medical colleagues who studied the results of my computer simulations [25] encouraged me to specify this model of the cell renewal process and apply it to the blood-forming cell systems for which the compartment-hypothesis and probably also the chalone-hypothesis are applicable. Therefore the following considerations have been reduced to the special case of the erythropoiesis, that is the production of red blood cells.

Three reasons support such a decision:

1. With haematopoietic cell renewal systems there is a dynamic balance between cell renewal and cell loss which in itself calls for an interpretation by the automatic control systems theory.

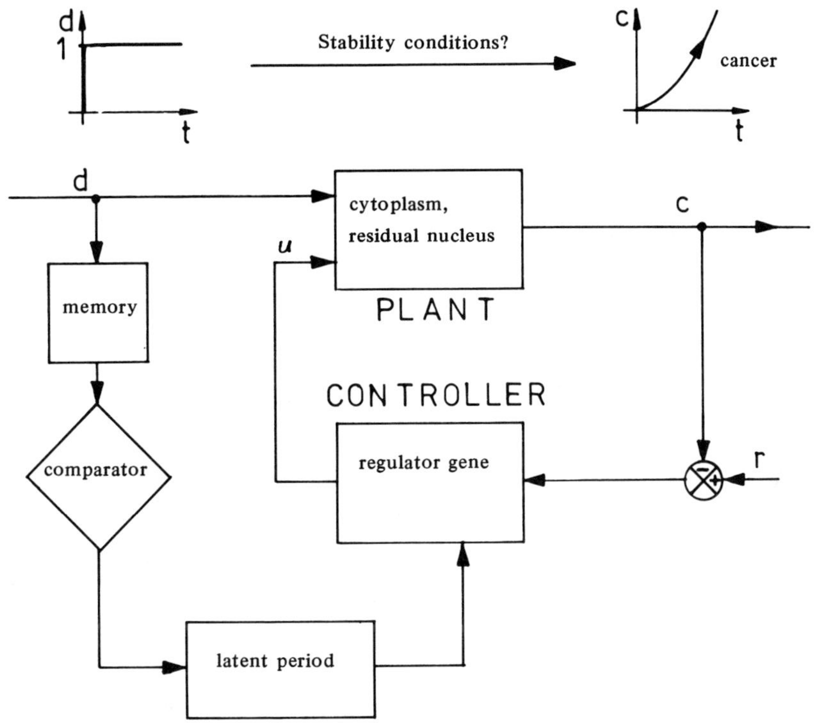

Controlled variable c:	deviation of the number of cells from the steady state
reference input r:	e.g. hormones
disturbances d:	carcinogens, e.g. dumps of tar, mechanical stimuli, radiological rays, viruses, chronic inflammations
control signal u:	production of specific regulation substances, e.g. enzymes

Fig.1 Block diagram of a control process of cell proliferation

2. At present this field offers the best and most data gained in experiments [32,40,42,45,46, 47].
3. Various papers (especially [6] and [46]) have discussed oscillations of the time course of the blood cells (Fig. 2) which are again closely connected with the mentioned problem, and pose the further question: if, in cases of erythroleukemia, i.e. the highly increased erythropoiesis, there is a basic relationship between malignancy and an unstable control-loop in the terms of automatic feedback language.

A MODEL OF THE ERYTHROPOIESIS CONTROL SYSTEM

Development of the block diagram

The complexity of erythropoiesis has been reduced to a simplified model (see Fig. 3) to be developed in the following paragraphs. In the closed loop the reference input R is the required tissue oxygen symbolized by an equivalent number of erythrocytes. If a deviation $E = R - C$ arises between the reference input R and the controlled variable C (momentary number of red cells) the hormone erythropoietin, produced mainly in the kidney affects the determined stem-cell compartment in the bone marrow to feed the proliferation pool with a certain number (Y3) of the determined stem cells. The proliferation pool is likewise to be found in the marrow. It is the place where, step by step, cell divisions and simultaneous cell developments into orthochromatic erythroblasts (Y45) take place. The erythroblasts pass the maturation and function pool in which no more cell division is possible. As reticulocytes and erythrocytes they enter the peripheral blood and are finally removed after a mean life span of about 120 days.

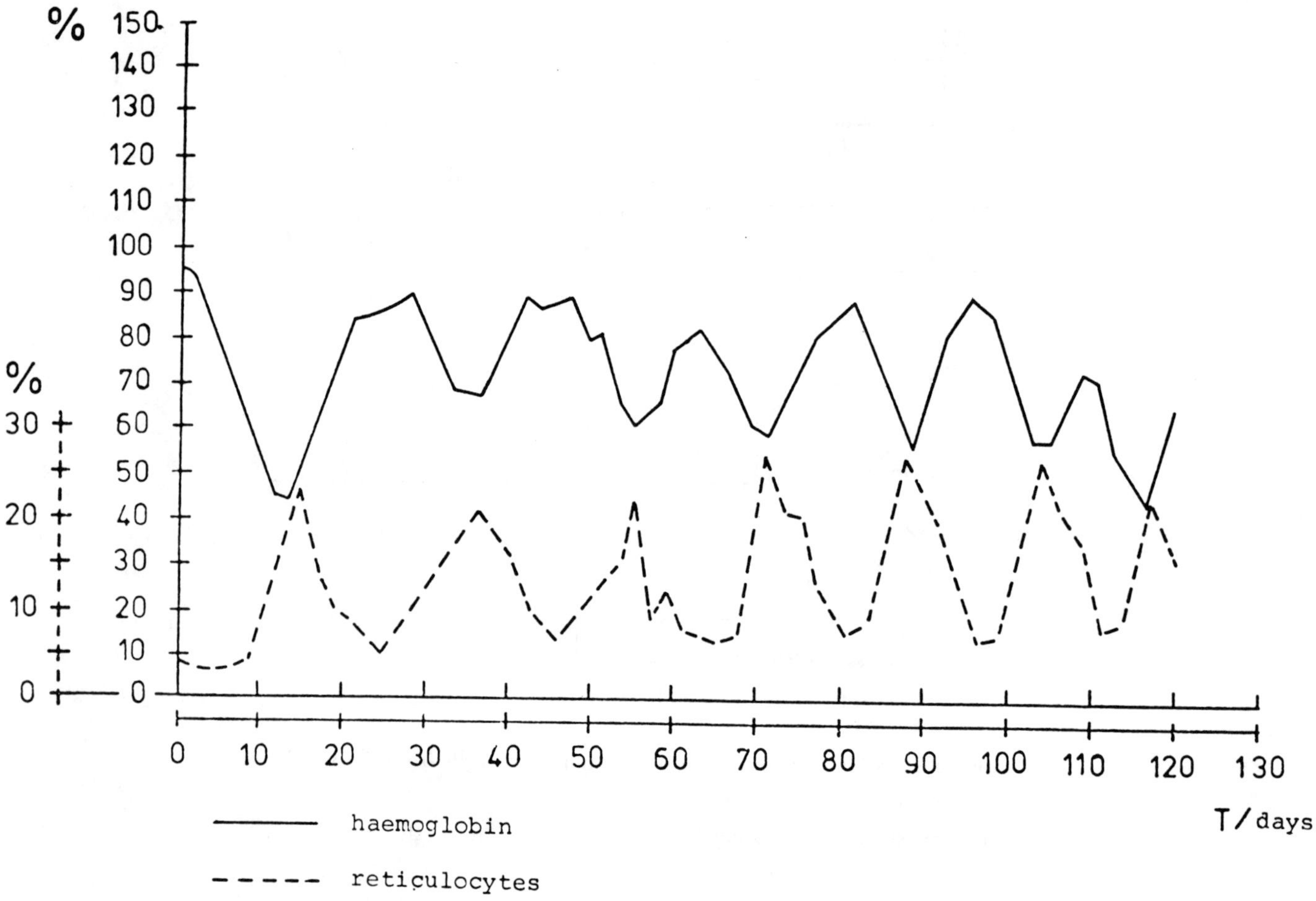

Fig.2 Haemoglobin and reticulocyte responses in a rabbit to a course of iso-antibody (see Orr, Kirk, Gray, Anderson [6])

A specification of the blood-forming process leads to the feedback control blocks which are adapted to a simulation with the block-oriented language ASIM (*A*naloge *SIM*ulation) of the firm AEG-Telefunken [43,44]. The time behavior of the compartment components, stimulated by a step input signal, is described in these feedback control blocks. For further details the reader is referred to special technical literature. [41].

In addition to the symbolic first-order transfer function elements (VZ1) in the proliferation pool there are delay elements (LZ2) in the maturation pool and—for the removal of cells—the delayed derivative element (VD). The output response of the stem-cell pool is symbolized by an integrator INT, and by the switching-function operator, SE2 (comparator). The switching mechanism represents the activity of genes which initiate the dividing process from resting stem cells to daughter cells. The proportional standard-block symbols K1 and K3 indicate the gain of the corresponding phases. In the interest of clarity the points at which the external direct disturbances act on the control-loops are not shown in Fig. 3.

The simulation of the erythropoiesis process was performed on a digital computer (AEG-Telefunken TR 440). A section of the corresponding computer program (ASIM) is shown in Fig. 4. It contains the structure and output notations and the data statements of the parameters. The latter were chosen from the experimental data on cell production, cell division, their transportation and life spans [40,42, 45,46,47] available at this time. For the unknown parameter K1 a feasible value was taken. The time constant (K2) and the gain (K3) alter with the produced quantity of erythropoietin. That is why the parameters are not constant in reality, but are interdependent and timevariant.

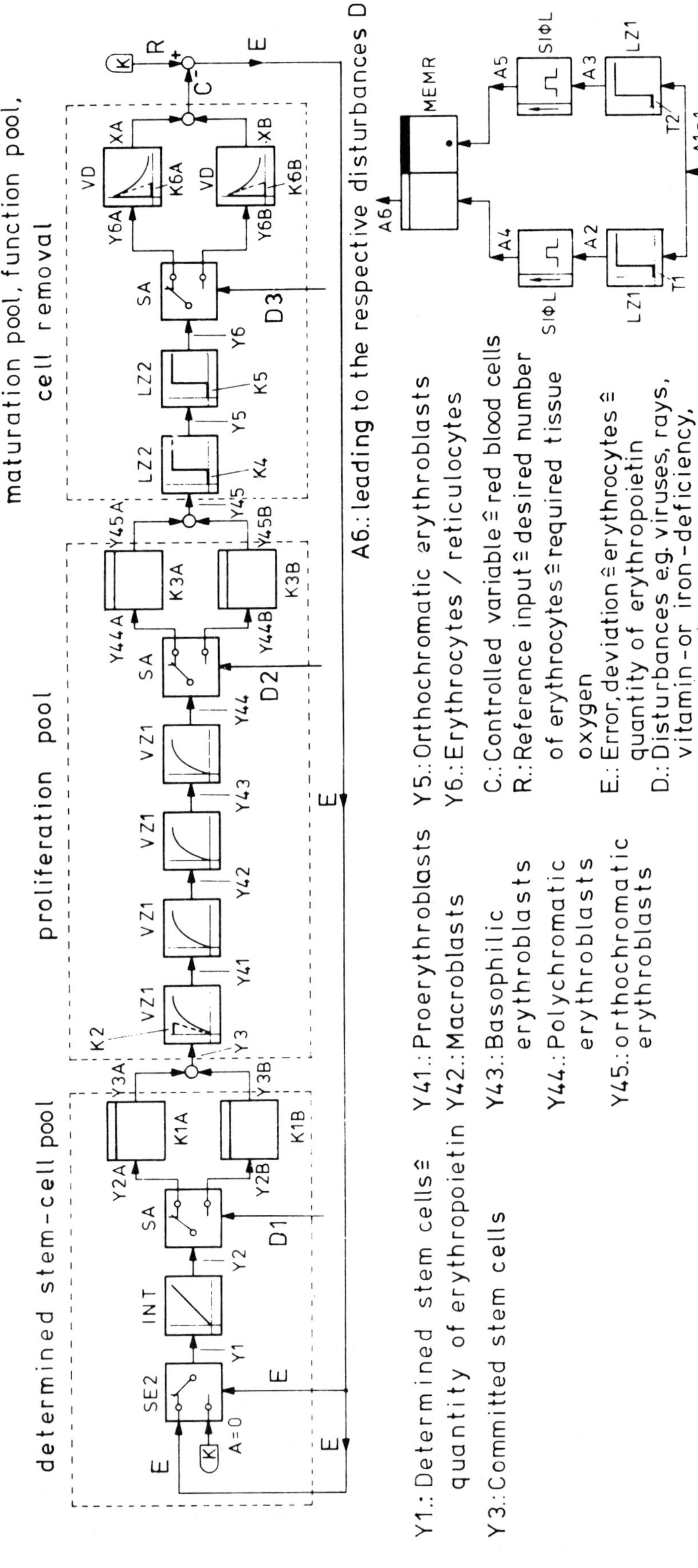

Fig.3 An erythropoiesis control-system configuration representing parameter variations

```
*---------REGELKREIS MIT TOTZEIT UND NICHTSTETIGEM ELEMENT
*---------STRUKTUR
Y1=SE2(XD,A,XD)
Y2 = INT(0,Y1)
Y2A,Y2B = SA(Z1,Y2)
Y3A = K1A * Y2A
Y3B = K1B * Y2B
Y3 = Y3A + Y3B
Y41=VZ1(0,K2,Y31)
Y31=Y3
Y62=K8*Y61
Y61=VZ1(0,K7,Y45)
Y42 = VZ1(0,K2,Y41)
Y43 = VZ1(0,K2,Y42)
Y44 = VZ1(0,K2,Y43)
Y44A,Y44B = SA(Z2,Y44)
Y45A = K3A * Y44A
Y45B = K3B * Y44B
Y45 = Y45A + Y45B
Y5 = LZ2(K4,Y45)
Y6 = LZ2(K5,Y5)
Y6A,Y6B = SA(Z3,Y6)
XA = VD(0,K6A,Y6A)
XB = VD(0,K6B,Y6B)
X = XA + XB
XD = W - X
A1=1
A2=LZ1(TZ1,A1)
A3=LZ1(TZ2,A1)
A4=SIML(0,A2)
A5=SIML(0,A3)
A6=MEMR(0,A4,A5)
*---------PARAMETER
A = 0
W = 100
H1 = ABS(XD) - 1.E10
K1A = 0.02
K1B = 0.025
K2 = 0.25
K3A = 12
K4 = 1
K5 = 2
K6A = 30
K6B=10
K7=0.1
K8=0.2
Z1=0
Z2=A6
Z3=0
TZ1=50
TZ2=0
K3B=10
*---------BEARBEITUNG
SKIP H1
RZEIT (0.,0.1,160.)
PLOTTER (A4Q,T/D,15,X/ERYS,7)0.,160.,T,-200.,200.,X
END
```

Fig.4 Listing of erythropoiesis control-system simulation (computer program ASIM)

Results of computer simulations

For various selected cases the system is stimulated by disturbances and subsequently the time response of the erythrocytes *($C = f(T)$)* is determined by computer simulation. In this simulation the following possibilities of influencing the dynamics of the control-loop have been taken into consideration.

Firstly cells can be removed or destroyed in the compartment by *outside* influences, e.g. by bleeding or irradiation.

Secondly parameter variations can lead to an important change of the time behavior of the variables of the control-loop and thus cause various diseases, e.g. different forms of anaemia.

Thirdly structural changes of the model components are possible, e.g. a defect of the comparator (SE2) of the stem-cell pool (Fig. 3), which can cause the control loop to become unstable. The same effect is given when parameters vary in certain cases.

1. Normal case, all disturbances $D = 0$

If all disturbances are zero and the building-up process is complete there is a resulting constant output curve of the erythrocyte response $C = f(T)$ (broken curve in Fig. 5*a*).

2. Disturbance D3 acting on the function-pool

If when $T = 50$, the disturbance D3 = 25%—a constant loss of blood over 50 days—influences the function—pool, the abrupt decrease in the erythrocytes is restored to a steady state by the control (Fig.5*a*). This disturbance represents a chronic loss of blood, e.g. a stomach haemorhage or an immunological haemolytic anemia. Immunological haemolytic anemias caused by experiments have been described by Orr, Kirk, *et al.* [6]. If at the time $T = 100$, the influence of the disturbance D3 has vanished, a control reaction takes place.

3. Anemia, caused by parameter variations

It is possible that the outside disturbances do not only act directly on the control loop, but also indirectly and decisively alter the dynamic behavior of the control circuit by means of a parameter change.

The assumed switch mechanism (SA) of Fig. 3 enables precise parameter variations in the stem-cell pool (K1A→ K1B), in the proliferation pool (K3A → K3B) and in the cell removal (K6A→ K6B) depending on the disturbances D1, D2, or D3 stimulating the system.

The following example describes the case when perturbation arises in the form of parameter variations between $T = 50$ and $T = 100$, after which it disappears.

General anemia forms with reduced erythrocyte numbers are caused under certain conditions by parameter changes of the factor K1 which is responsible for the production rate of the erythrocytic stem cells. The ineffective erythropoiesis of pernicious anemia can be simulated by reducing the gain K3 in the proliferation pool (Fig.3). For reasons of space limitation the results of the above cases are not shown here but the third possibility of a haemolytic anaemia is depicted in Fig. 5*b*. In the situation of a haemolytic anaemia which can either be inherent or induced by toxines, the life span of the red cells is considerably shortened [6,45]. Therefore, the time constant of the cell removal is diminished by approximately 33% from K6A = 30 days to K6B = 20 days in the control model of Fig. 3. The response curve in Fig. 5*b* shows that, if the parameter deviations are in a small band, the control mechanism is able to restore the previous level of erythrocytes after a certain period. A reduction of the erythrocyte life span by the factor of 20, would, however, have as a result a notably decreased number of red cells because of the increased haemolysis.

4. Anaemia, provoked by structural changes

At present there is no distinction possible (in the experimental field) as to the question of whether anaemias are caused by parameter changes or structural alterations. Nevertheless, the following attempt is made in theory to determine the course of erythrocytes when a structural disturbance arises, for example, a defect in the switch SE2 of the stem-cell pool (see Fig. 3). By this manipulation a disturbed activity of genes, particularly in the recognition mechanism, is simulated. First of all let us consider a parameter change in the control-system configuration of Fig. 3 at time $T = 50$ of K1A = 0.02→ K1B = 0.028 and of K3A = 12→K3B = 16, i.e. of an enormous increase in the stem-cell production rate as well as in the amplification factor in the proliferation pool. If the switch mechanism SE2 is intact, this perturbation of the system is eliminated as shown in the broken curve in Fig. 5*c*.

If, at the same time, there is a structural defect in the switching mechanism of the stem-cell compartment, that is to say if the switch SE2 remains permanently closed, then the simulation run of Fig. 5*c* shows a rising oscillation course of the red cell response with an otherwise unchanged parameter

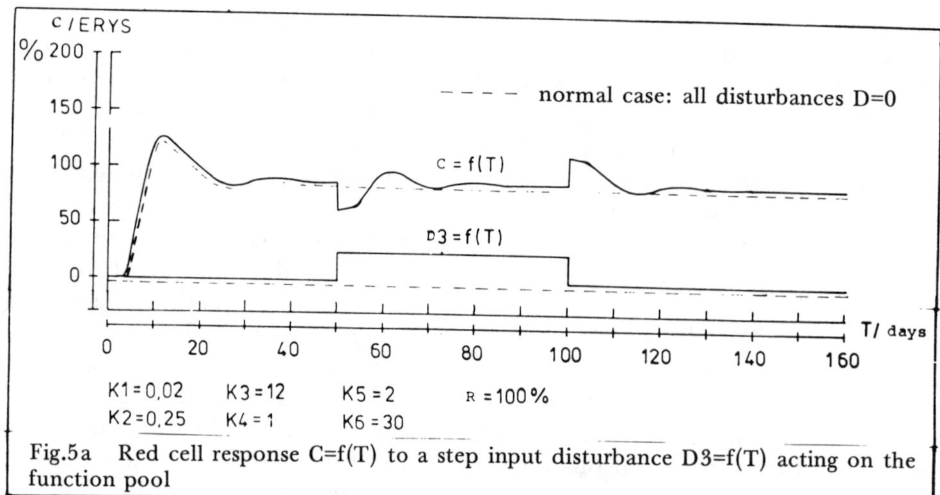

Fig.5a Red cell response C=f(T) to a step input disturbance D3=f(T) acting on the function pool

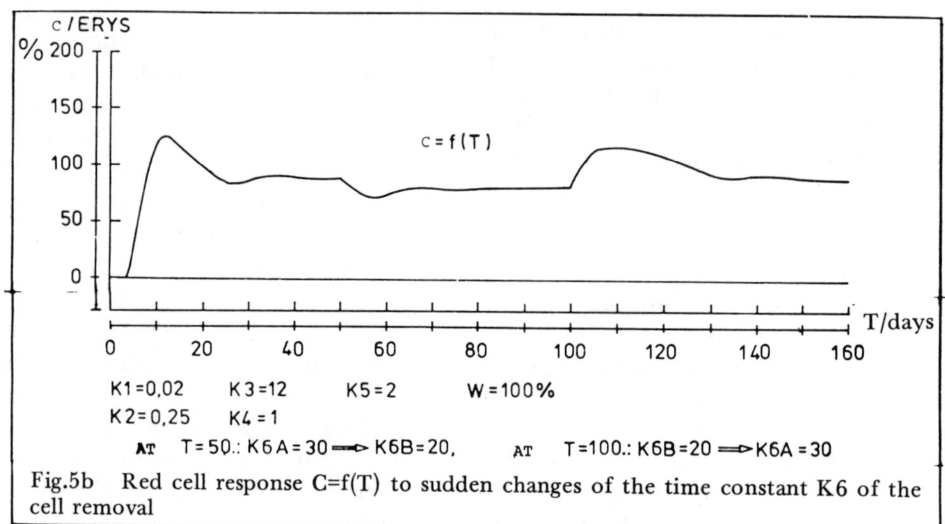

Fig.5b Red cell response C=f(T) to sudden changes of the time constant K6 of the cell removal

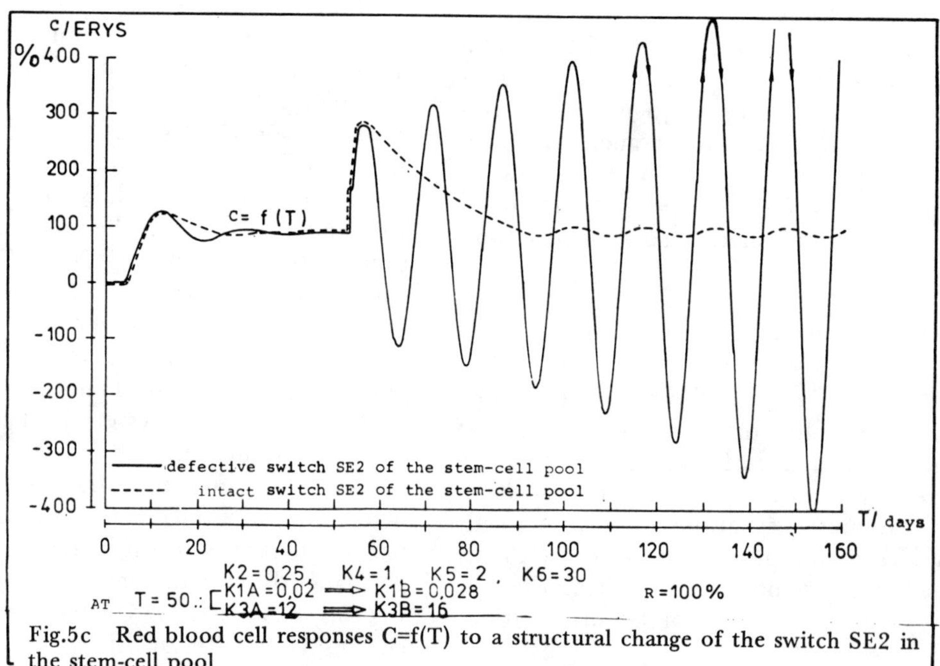

Fig.5c Red blood cell responses C=f(T) to a structural change of the switch SE2 in the stem-cell pool

constellation.

Thus it is discovered that the control-loop system has become unstable in its structure. This result is confirmed by a classical stability analysis in control theory [41]. The rising oscillations outlined in Fig. 5*c* symbolizing certain forms of erythroleukemia have not up until now been clinically observed since experiments on living organisms can only be carried out in their first half- or complete-cycles, as a living being cannot survive erythrocyte changes of that extent.

MULTI-LOOP STRUCTURE OF THE ERYTHO-POIESIS CONTROL CIRCUIT

Block diagram

The previous sections have been devoted to a single feedback loop only (compare Fig. 3). Here a humoral agent—e.g. the hormone 'erythropoietin' produced in the kidney—effected the external feedback action E to the determined stem-cell pool. Several indications [1,2,29,34,40,42] lead to the assumption that in reality there are additional interacting subsystems with erythropoiesis (Fig. 6*a*). This would allow an interpretation of diseases like polycythaemia, i.e. the increased production of red cells [45] or aplastic anaemia, i.e. reduced erythrocyte production. Keeping the parameters unchanged the influence of two different control-loop configurations was observed as in the following cases.

Results of computer simulations

1. Inner chalone control-loop

According to the chalone-hypothesis [6] each type of cell coming into existence produces mitotic inhibitors (regulating substances, called 'chalones') which are believed to be proteins preventing the erythrocyte precursor cells from entering the generative cell cycle by influencing the activity of the respective genes on a molecular level. The erythrocytic chalone is assumed to be present in mature red cells and in normal serum. The chalone mechanism is simulated by the right inner loop in Fig. 6*a*. The results of the respective computer simulations are shown in the response curves drawn in Fig. 6*b*. It is clearly shown that for a weak negative feedback when K8A = 0.02 there are almost no deviations from the normal case of Fig. 5*a* (broken curve) in the control dynamics. If there is, however, a sudden increase in the amplification factor of the inner loop from K8A = 0.02 to K8B = 0.09 at $T = 50$, a rising oscillation in the course of the red cell response $[C = f(T)]$ is to be observed. The course can be stabilized in Fig. 6*b* by removing the disturbance when $T = 100$. It can be seen from Fig. 6*b* that a rising erythrocyte oscillation is not only achieved by a structural change of the control-loop model as was described in the previous section, but also by an alteration of parameters. Thus, the assumption is that the cause of the increased production of red cells, i.e. of the polycythaemia, can be a parameter change in the feedback mechanism of the chalone-loop.

2. Inner stem-cell control loop

It can be further assumed that the committed stem cells have a direct negative feedback on their own compartment. The left inner control loop in Fig. 6*a* represents this mechanism hitherto almost unknown.

The simulation results of Fig. 6*c* (*with* the left inner control loop but *without* chalone loop) prove that a small feedback factor (K10A = 0.1) has practically no effect. An increased feedback factor, however, when $T = 50$ from K10A = 0.1 to K10B = 2 leads to a conspicuously decreased response of red blood cells (broken response curve of Fig. 6*c*), assuming, as this example shows, the distrubance has not been removed when $T = 100$. Thus, it would seem that the general forms of anaemia, i.e. the reduction of erythrocytes, to be found for example in the case of an aplastic anaemia [45], are caused by parameter variations in the stem-cell control loop.

A GENERALIZED MODEL OF THE CELL RENEWAL PROCESS

Whereas the models mentioned so far belong to the group of macro-models, there is another group of micro-models set up by scientists studying the behavior of a single cell with the subsequent phases of its division cycle. Since both approaches represent borderline cases I have made an attempt to set up a generalized cell renewal control loop model. The system describes the dynamics of the cell renewal in the various compartments, but at the same time allows the observation of the division of a single cell in a specific compartment. This model makes it possible to adjust the life span of a single cell. The irradication of a single cell or a whole cell group at any desired moment can also be simulated in this model. Further studies can be made on the

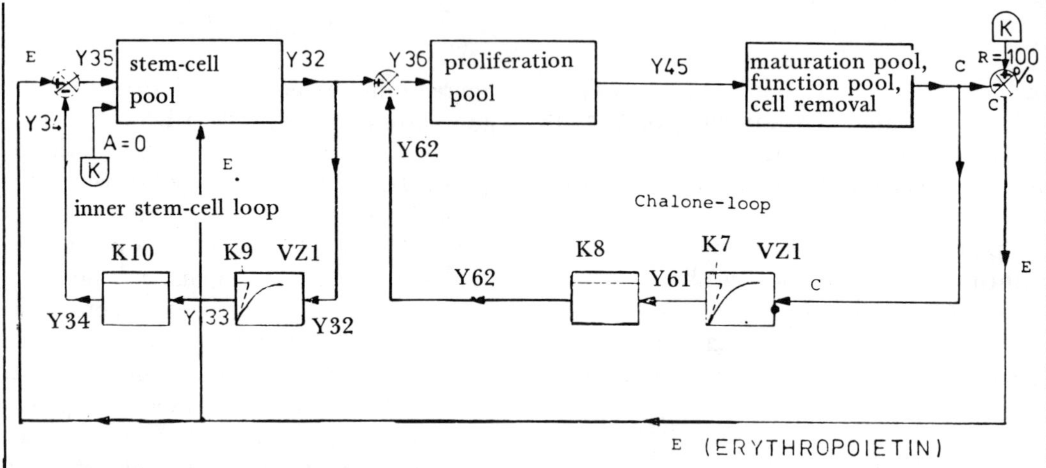

C: Controlled variable ≙ red blood cells
R: Reference input ≙ desired number of erythrocytes ≙ required tissue oxygen
Disturbances: parameter variations of the amplifier gains K8 and K10 of the feedback loops

Fig.6a Multi-loop erythropoietic feedback control block diagram

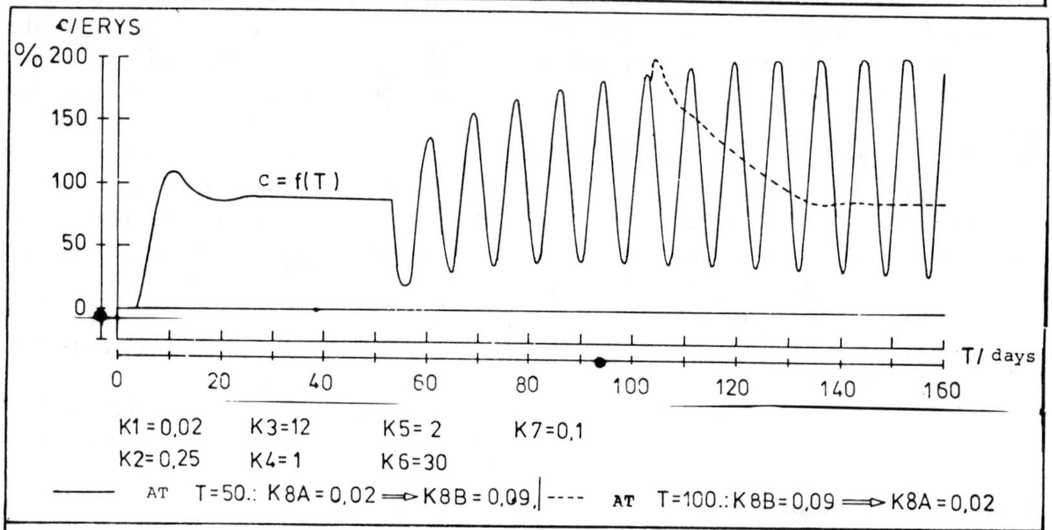

Fig.6b Red blood cell responses C=f(T) to changes of the amplifier gain K8 of the inner feedback chalone-loop

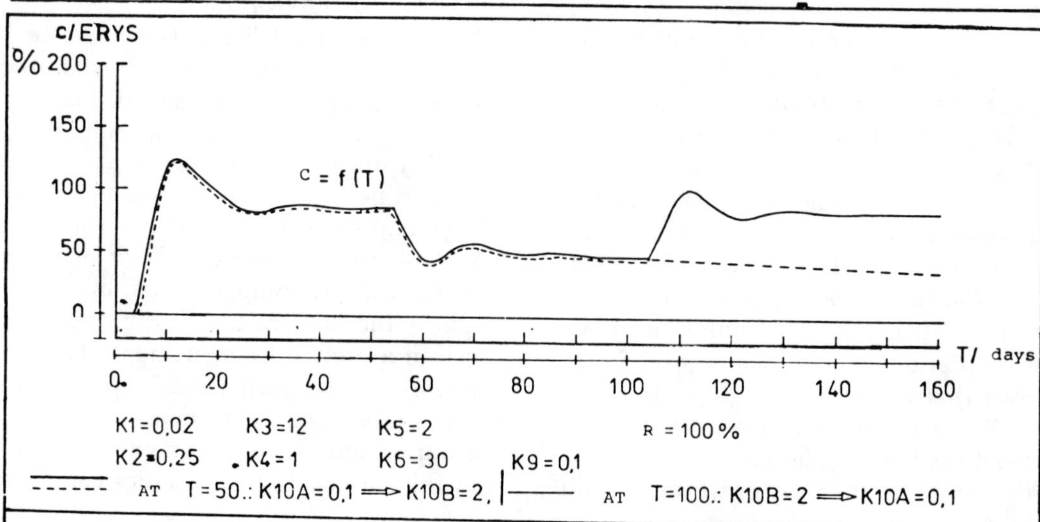

Fig.6c Red cell responses C=f(T) to changes of the amplifier gain K10 of the inner stem-cell loop

transient response of the cell renewal process when additional humoral impulses for cell division have been given to the feedback loop or when single cells have reached the resting phase G_O.

Block diagram of the generalized model

According to the compartment hypothesis, the process of cell division and differentiation takes its course from an undetermined (omnipotent) stem cell in the bone marrow (symbolized by signal A03 of Fig. 7) through various stages of development till they finally become mature cells. An example of these mature cells are the erythrocytes S3 which are sent into the peripheral blood.

Each of the single compartments in the model has an identical structure and they are all connected in series but they contain a limited number of viable cells—the maximal number varying from compartment to compartment. For example in compartment 2 there is a maximum of 12 cells (Z 2.1 to Z 2.12). The strict arrangement of the cells in the outlined block has been made for reasons of clearness. It is not a statement of the real local position of cells in an organ in which the cells, e.g. Z 1.5, Z 2.12 and Z 3.4, could be arranged next to each other. The complicated structure of a single cell, for example of the sub-system of cell Z 1.4, will be discussed under the next heading. Compartment 2 gets the division impulses for forming new cells from the previous compartment 1. In our model special subroutines produce the input signals C 21, when a cell is being divided and is advancing into the adjacent compartment. A digital logic device ascertains and registers the presence of each cell in the specific compartment. The total number of cells present in compartment 2 is symbolized by S2. S2 is compared in the control loop II with the desired number of cells S 22 which represents the required tissue oxygen. If the controlled variable S2 sinks below the reference signal S 22 by a certain fixed value the pulser SIØL sends an output signal (S23)—symbolizing a humoral agent—to the stem cell pool. This feedback signal is called S00. The fictive switch SE1 in the stem cell pool, which symbolizes the gene activities, is temporarily closed and delivers a stem cell in the case of the erythopoiesis for the erythrocytic series.

Apart from the feedback loops I, II and III of the single compartments leading to the stem cell pool there are other inner control loops in the model which are, however. not included in Fig. 7. In the case of a disturbance in a compartment these additional loops directly affect the cells of the previous pool as is shown in the following examples.

The notation of the block diagram has been made in accordance with the digital simulation of the block-oriented simulation language ASIM (*A*naloge *SIM*ulation) [41,43,44]. Besides analogue transfer elements as for instance integrators INT, elements (LZ) with variable delay-times, switching components SE, digital transfer elements as for example logical components MEMR as "memory elements", are also used.

On the input side of a compartment are loops which enable the impulse created by a dividing cell to pass into the next compartment in the search for "vacant spaces" for the newly formed cells. This model has an undetermined stem cell pool in the bone marrow with an unlimited supply of undetermined stem cells symbolized by a permanent signal A03. Figure 7 shows the "initial provision" of the cells in the respective compartments by means of start impulses. Modifications of the general control-loop structure are carried out and discussed later in connection with the special simulations.

In the terminology of the "theory of automata" the developed model would be called an " asynchronous sequential circuit" that is to say the state of the automaton (= the model) at any one point in time is dependent upon its immediate past.

Block diagram of a single pair of cells

In Fig. 7 a survey of the structure of the generalized model is given. Now, the sub-system "cell" as the basic element of a compartment will be developed. For reasons of programming, the block diagram of a *pair* of cells is set up and discussed in Fig. 8. The cells of this pair can be distinguished by the second letters A and B. The existence of cells is proved by the signals AA or AB, symbolized here by the variables logic "1" at the outlet of the memories MEMR. These signals have the effect that the consequent dividing impulses C which arrive in the compartment via the fictive switch SA are at once connected as output signals D with the inlet of the following pair of cells in the compartment. That is how the searching process for vacant spaces of a cell pair is automatically carried out.

After two vacant places have been found for the cell pair and the two cells have been formed, the signals AA and AB affect further control-loop components (IR-blocks) which cause the cells to be divided after the desired mean life span TA or TB, which can be varied. As the mother cell no longer

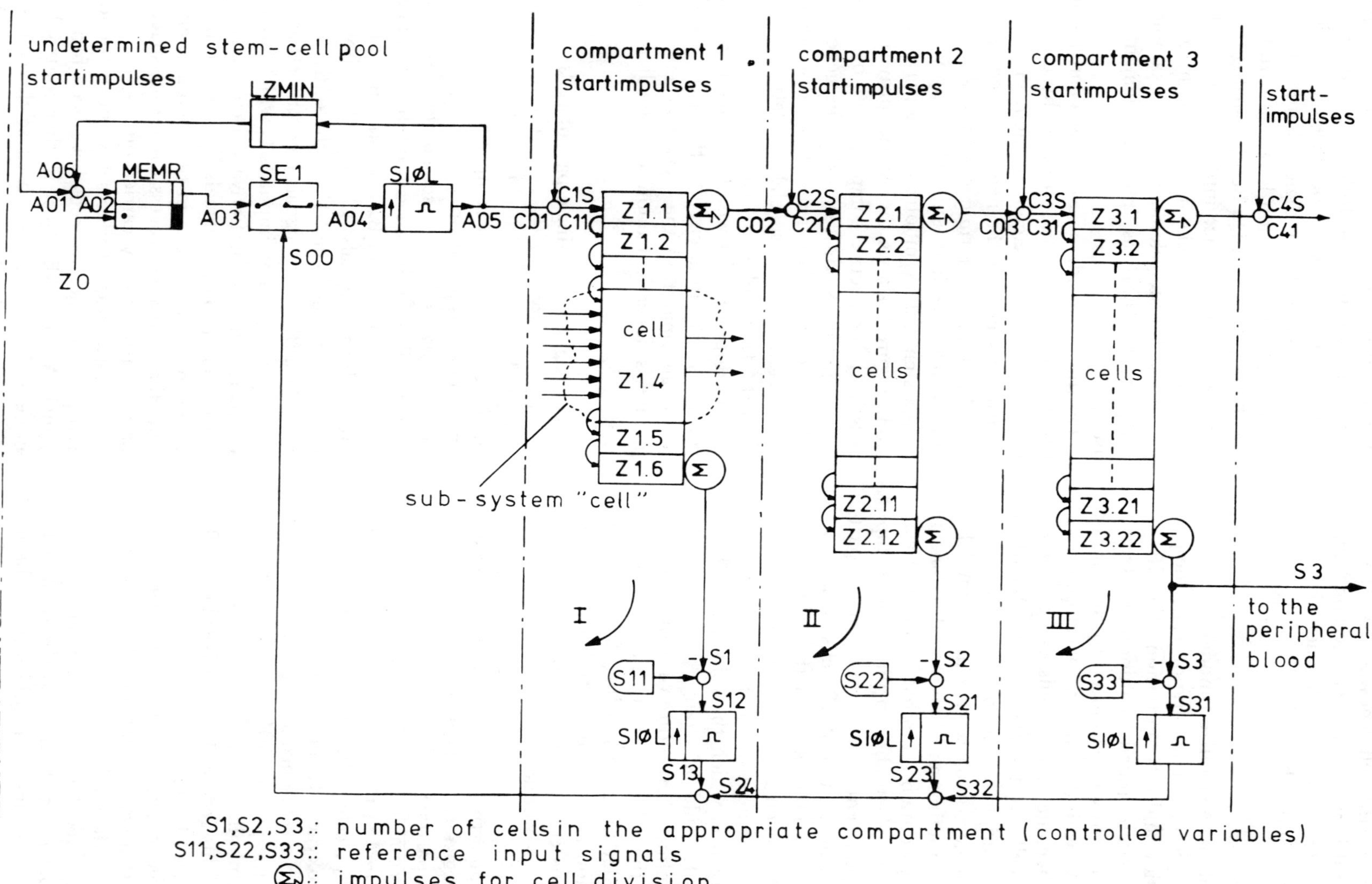

Fig.7 Simplified block diagram of the cell renewal feedback-control system

exists when the two daughter cells are formed, this cell is extinguished in the model by the extinctive impulse (LAA or LBB) to the reset channel of MEMR, after the division has symbolically taken place.

Simultaneously, the newly formed daughter cells enter the following compartments via the outlets RA and RB. If in addition it is desired to alter the life span of the single cells externally, the input channels UA, VA, NA or UB, VB, NB of the block diagram of Fig. 8 can be used for this purpose. The mean life span of a cell in this model comprises the single phases (G_1, S, G_2 and M) of a cell cycle.

The disturbance signals ZA or ZB are very important input variables. They make it possible to destroy any single cell at any desired moment in order to simulate for example an irradiation process; that is to say the output signals AA or AB are changed from value '1' to value '0'. An additional device which it would be too lengthy to describe in detail here prevents another cell from being divided by such a disturbance signal.

Results of computer simulations

In conclusion, some simulations selected from a great variety of possibilities should be considered assuming the following fictive parameters:

Compartment	1	2	3
Number of cells	6	12	22
Mean value $\bar{x}$ of the life span of cells	0.967	1.042	1.050
Standard deviation σ	0.333	0.244	0.197
Range $R = x_{max} - x_{min}$	1.0	0.9	0.9

Before running the simulation the start impulses (C1S, C2S, C3S) for the initial populating of the compartments with cells must be programmed. The average number of cells was fixed to be

two ($\hat{=}$ S 1) in compartment 1,
four ($\hat{=}$ S 2) in compartment 2,
eight ($\hat{=}$ S 3) in compartment 3.

All start impulses act at the same moment. The references input signals S11, S22, S33 in the control loops I, II and III have been adjusted to maintain the number of the cells (S1, S2, S3) in the various compartments. If thus the controlled variables S1, S2, S3 fall below the pertinent reference input signals S11, S22, S33, humoral feedback impulses are sent to the stem cell pool. The result is an emission of undetermined stem cells into the erythrocytic series. Taking into consideration these prerequisites, the results of simulations which were carried out in accordance with the block diagram of the generalized model of Fig. 7 are as follows:

1. Cell renewal process without disturbances

Assuming that the total cell renewal process takes place without disturbances, the curves S1, S2, S3 $= f(T)$ of Fig. 9*a* are obtained. They represent the number of cells in the single compartments and serve as comparison curves for the following simulations.

If the time behavior of the feedback impulses IMP $= f(T)$ of Fig. 9*a* is considered, it can be noted that only the inner loop I produces the feedback impulses S00 = S13, while the two outer loops II and III are inactive. They are always present and they serve as redundant back-up circuits in the case of possible disturbances. This mechanism bears comparison with processes which take place in biological reality, for example immunity processes.

2. Cell renewal process with disturbances to different cells in the compartments

The most frequent case in reality is when single cells of different compartments, being in the DNA-phase, are destroyed directly or indirectly by external disturbances. The arising deviations from the norm are compensated by several interior control loops. Figure 9*b* shows the results of such a case. The graph IMP $= f(T)$ illustrates that only loop I is an active control loop (Fig. 7), while there is no feedback impulse to the stem cell pool generated in the loops II and III.

3. Cell renewal process with unstable response dynamics comparable with tumour-like time behavior.

Figure 9*c* shows results gained when in a disturbed case there are *parameter* changes—for instance the division by half of the life span of cells in compartment 1—as well as *structural* alterations—for example the generation of steady additional humoral impulses. From these curves it can be seen that in the case of a malignancy there is an uncontrollable increase in compartment population. An interpretation is, however, not possible, unless a detailed knowledge of the various kinds of feasible disturbances and reactions of the multi-loop system is available. Some tentative dispositions in this field have been put forward in this contribution.

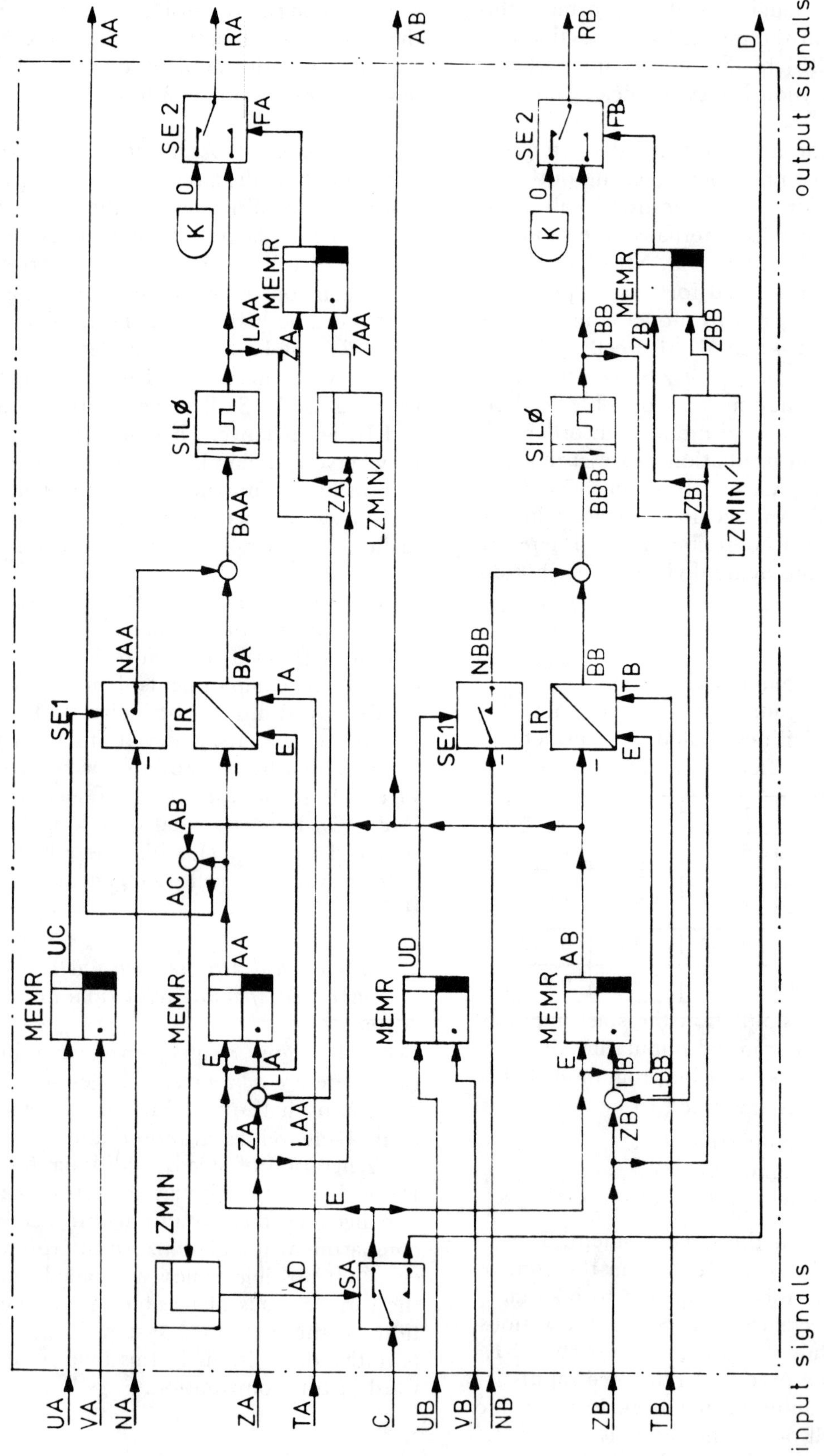

Fig.8 Block diagram of a cell pair (sub-system)

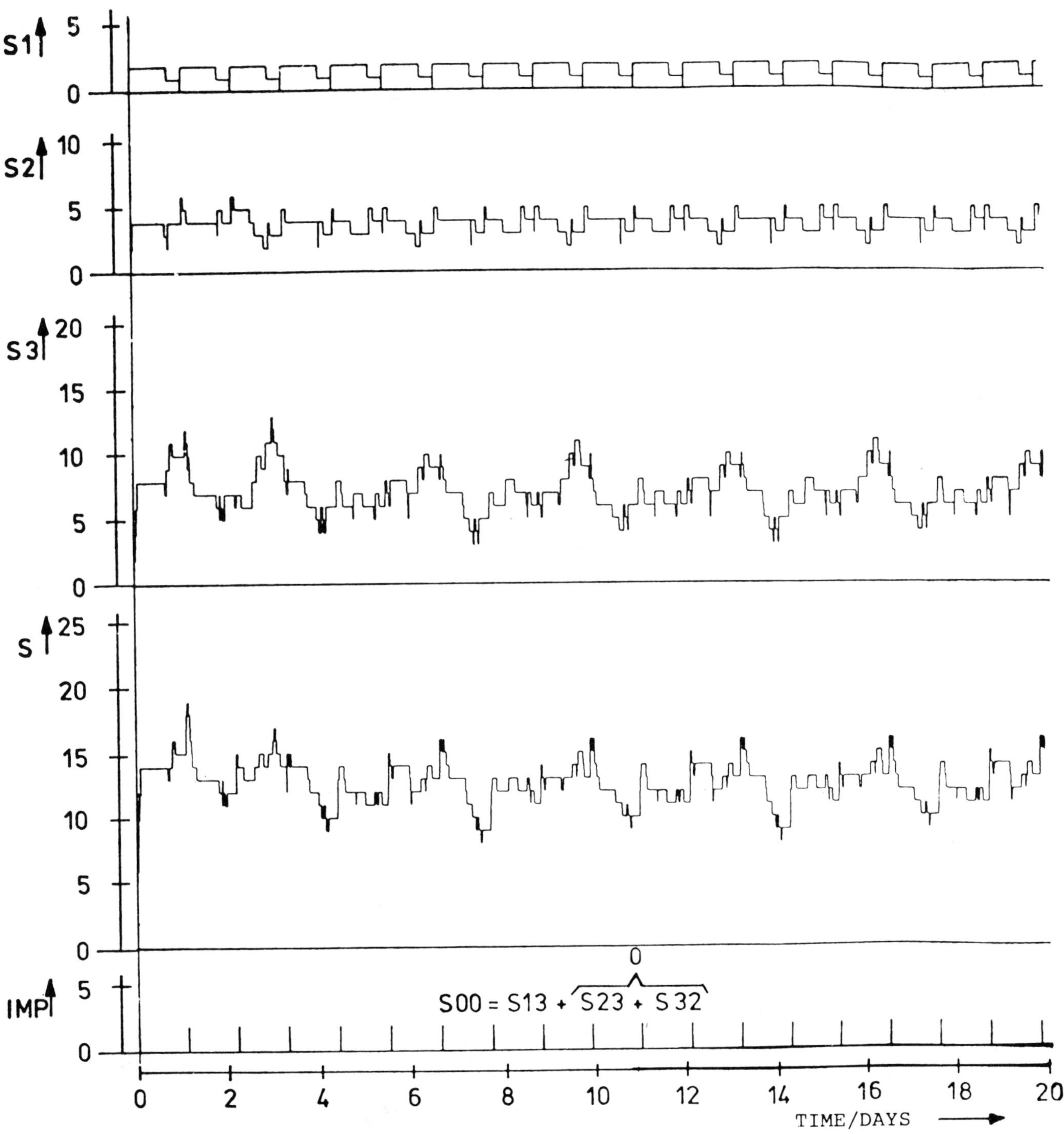

Fig.9a Time behavior of cell renewal process without disturbances

S1, S2, S3: number of cells in the appropriate compartment

S : number of total cells (S = S1 + S2 + S3)

IMP : feedback impulses (S13, S23, S32)

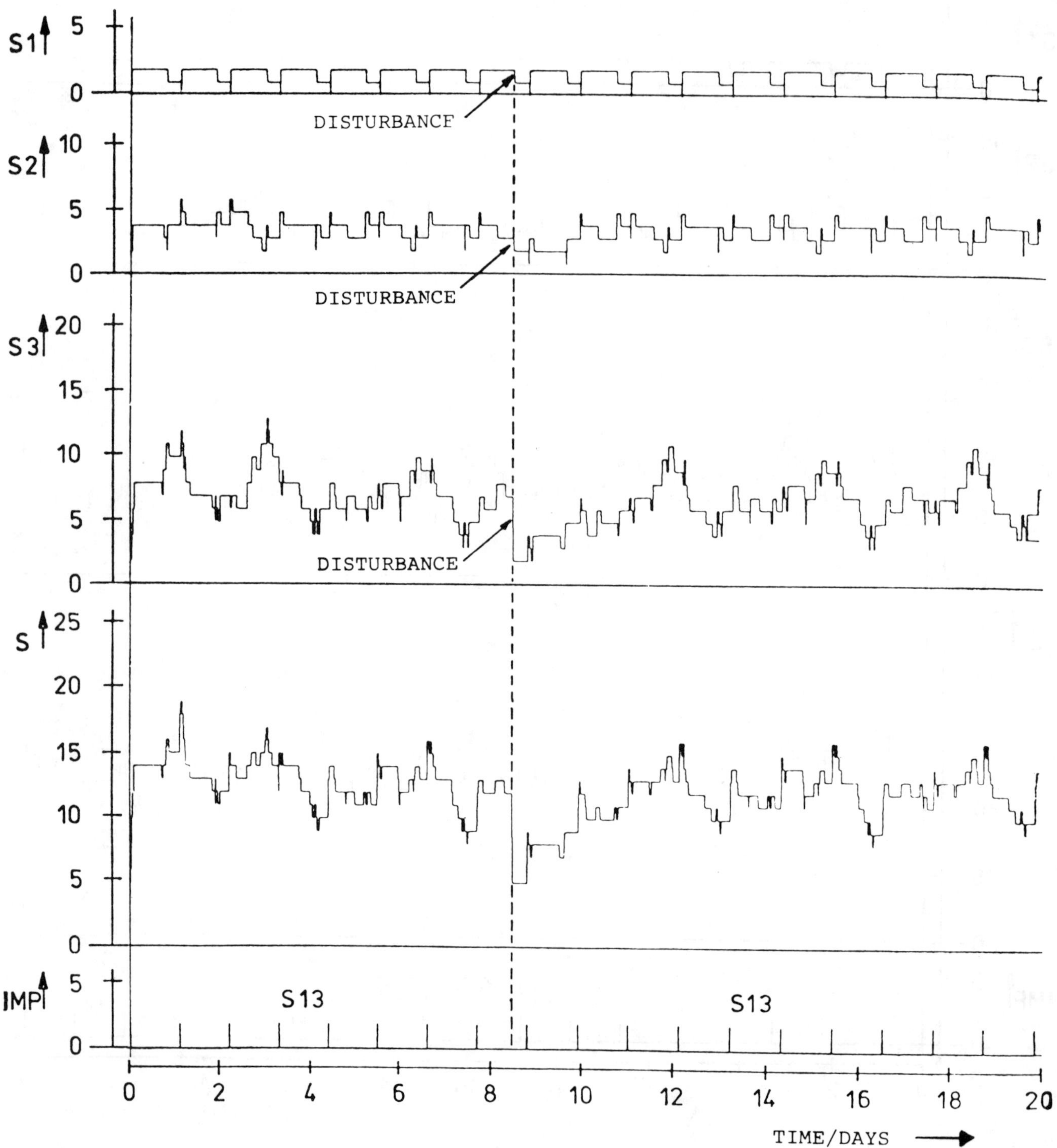

Fig.9b Time behavior of cell renewal process with disturbances to the cells Z1.1, Z2.3, Z3.1 till Z3.5 when T=8.5
S1, S2, S3: number of cells in the appropriate compartment
S: number of total cells (S = S1 + S2 + S3)
IMP: feedback impulses (S13, S23, S32)

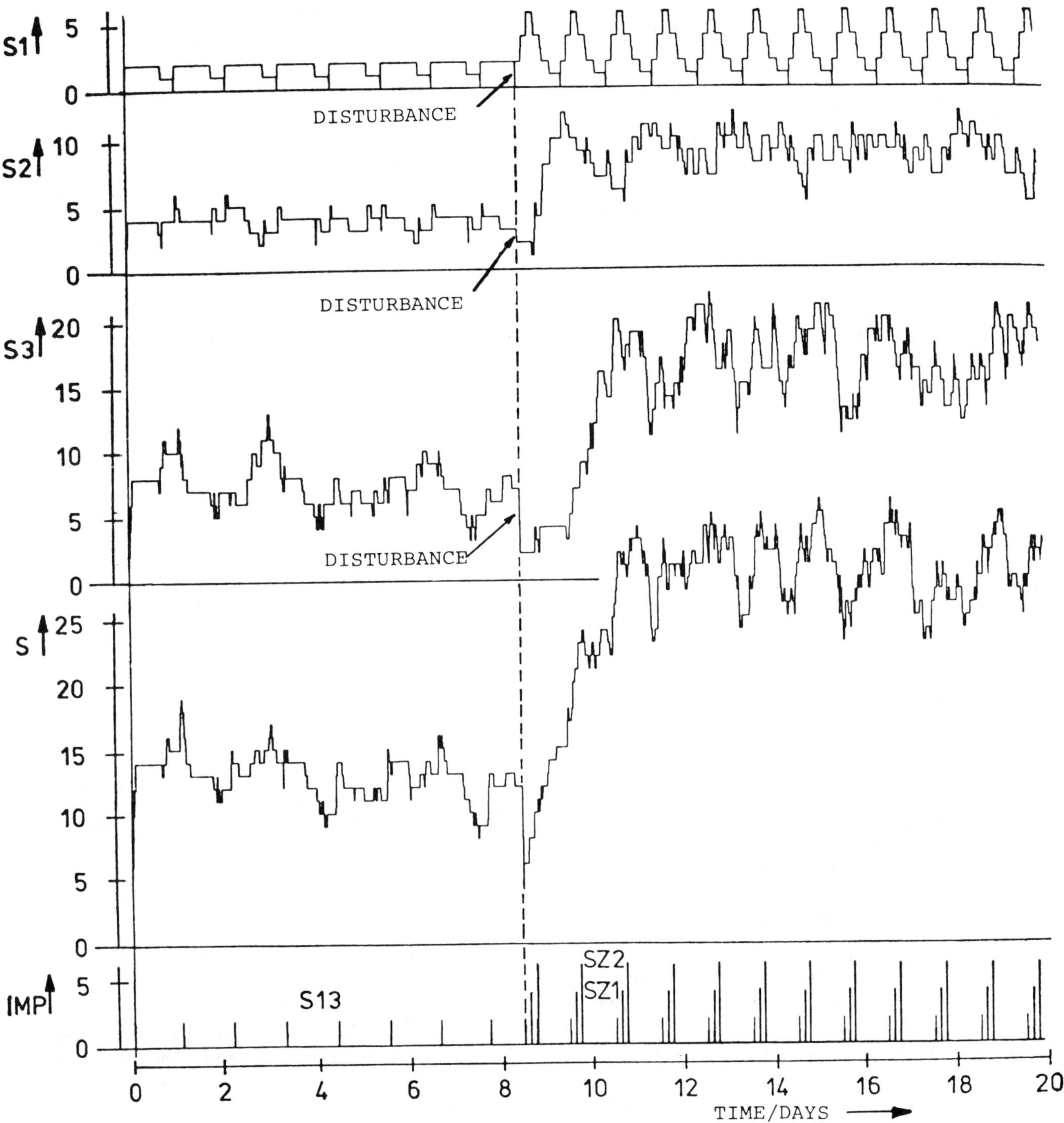

Fig.9c Time behavior of cell renewal process – with disturbances to the cells Z1.1, Z2.3, Z3.1 till Z3.5
– with halved mean life span of the cells in compartment 1
– with two additional regular impulses in the control loop I when T = 8.5

S1, S2, S3: number of cells in the appropriate compartment
S: number of total cells (S = S1 + S2 + S3)
IMP: feedback impulses (S13, S23, S32)

CONCLUSIONS

This chapter deals with various control-loop models and structures set up for the computer simulation of cell renewal processes. By means of simulation the time response of the compartment population was established for normal control-loop systems, but also for those with changes of structure or parameters.

The results show that the cell kinetics of malignant disorders can be interpreted as unstable feedback control loops.

The hypotheses and model presentations developed in this chapter could be tested experimentally by interrupting or blocking the dynamics of the rest of the system by deliberate disturbances of the biological environment. Therefore co-operation between physicians and control engineers should be intensified to answer the questions raised by the computer simulations and to interpret the results in the best manner possible.

References

1. BLUMENSON, L.E., 'A comprehensive modeling procedure for the human granulopoietic system: over-all view and summary of data', *Blood*, Vol. 42(2), 303-313 (August 1973).
2. KIEFER, J., 'Zur mathematischen Beschreibung der Zellproliferation', *Biophysik* **10**, 115-124 (1973).
3. DE VITA, V.C., 'Cell kinetics and the chemotherapy of cancer', *Cancer Chemotherapy Reports*, Part 3, Vol.2, No.1, 23-33 (October 1971).
4. WHELDON, T.E., KIRK, J. and GRAY, W.M., 'Mitotic autoregulation, growth control and neoplasia', *J. Theoret. Biol.* **38**, 627-639 (1973).
5. KIRK, J., ORR, J.S., and HOPE, C.S., 'A mathematical analysis of red blood cell and bone marrow stem cell control mechanisms', *Brit. J. Haemat.*, 15, p.35 (1968).
6. ORR, J.S., KIRK, J., GRAY, K.G., and ANDERSON, J.R., 'A study of the interdependence of red cell and bone marrow stem cell populations', *Brit. J. Haemat.*, 15, p.23 (1968).
7. TARBUTT, R.G. and BLACKETT, N.M., 'Cell population kinetics of the recognizable erythroid cells in the rat', *Cell Tissue Kinet.*, 1, pp.65-80 (1968).
8. WHELDON, T.E., KIRK, J., and ORR, J.S., 'Nonsteady-state analysis of cellular development', *Cell Tissue Kinet.*, 7, pp.173-179 (1974).
9. HAHN, G.M., 'State vector description of the proliferation of mammalian cells in tissue culture', *Biophysical Journal* **6**, 275-290 (1966).
10. VON BRONK, B., DIENES, G.J., and PASKIN, A., 'The stochastic theory of cell proliferation', *Biophysical Journal* **8**, 1353-1398 (1968).
11. HARRIS, T.E., 'A mathematical model for multiplication by binary fission', in *The Kinetics of Cellular Proliferation* (edit. by F. Stohlman, Jr.), Grune & Stratton, New York (1959).
12. TAUTU, P. and JOSIFESCU, M., *Stochastic processes and applications in biology and medicine*, Springer Verlag, Berlin (1973).
13. BASERGA, R., 'The relationship of the cell cycle to tumor growth and control of cell division: a review', *Cancer Research* **25**(5), 581-595 (June 1965).
14. LIPKIN, M., 'The proliferative cycle of mammalian cells', in *The Cycle and Cancer* (ed. by R. Baserga), Vol.1, pp.6-26, Dekker, New York (1971).
15. TAKAHASHI, M., 'Theoretical basis for cell cycle analysis', *J. Theoret. Biol.*, 13, pp.202-211 (1966).
16. BARRETT, J.C., 'A mathematical model of the mitotic cycle and its application to the interpretation of percentage labelled mitoses data', *J. nat. Cancer Institute* **37**(4), 443-450 (October 1966).
17. VALLERON, A.J. and FRINDEL, E., 'Computer simulation of growing cell populations', *Cell Tissue Kinet.*, 6, pp.69-79 (1973).
18. KIM, M., BAHRAMI, K., and WOO, K.B., 'Mathematical description and analysis of cell cycle kinetics and the application to Ehrlich ascites tumors', *Journal of Theoret. Biology*, 50, pp.437-459 (1975).
19. BURNS, F.J. and TANNOCK, I.F., 'On the existence of a G_o-phase in the cell cycle', *Cell Tissue Kinet.*, 3, pp.321-334 (1970).
20. MAERTELAER, v. De and GALAND, P., 'Some properties of a "G_o" model of the cell cycle', *Cell Tissue Kinet.*, 8, pp.11-22 (1975).
21. RAJEWSKY, M.F., 'Proliferative parameters of mammalian cell system and their role in tumor growth and carcinogenesis', *Z. Krebsforschung*, 78, pp.12-30 (1972).
22. STEEL, G.G., 'The cell cycle in tumors: an examination of data gained by the technique of labelled mitoses', *Cell Tissue Kinet.*, 5, pp.87-100 (1972).
23. FRINDEL, E., MALAISE, E.P., and TUBIANA, M., 'Cell proliferation kinetics in five human solid tumors', *Cancer* **22**, 611-620 (September 1968).
24. KIEFER, J., 'A model of feedback-controlled cell populations', *J. Theoret. Biol.*, 18, pp.263-279 (1968).

25. DÜCHTING, W., 'Krebs, ein instabiler Regelkreis' –"Versuch einer Systemanalyse", *Kybernetik* 5(2), pp.70-77 (1968).
26. FRASER, A. and TIWARI, J., 'Genetical feedback-repression', *J. Theoret. Biol.*, 47, pp.397-412 (1974).
27. IVERSEN, O.H., 'Wachstumsregulierung und Krebsentwicklung in der Epidermis von Mäusen', *Archiv für Geschwulstforschung* 32(4), pp.322-338 (1968).
28. RUBINOW, S.I. and LEBOWITZ, J.L., 'A mathematical model of neutrophil production and control in normal man', *J. Math. Biol.,* 1, pp.187-225 (1975).
29. MYLREA, K.C. and ABBRECHT, P.H., 'Mathematical analysis and digital simulation of the control of erythropoiesis', *J. Theoret. Biol.,* 33, pp.279-297 (1971).
30. DÜCHTING, W., 'Entwicklung eines Erythropoese-Regelkreismodells zur Computer-Simulation', *Blut* **XVII**, 342-350 (1973).
31. DÜCHTING, W., 'Computersimulationen von Zellerneuerungssystemen', *Blut* **31**(6), 371-388 (1975).
32. MORLEY, A., KING-SMITH, E.A., and STOHLMAN, F., 'The oscillatory nature of hemopoiesis', in *Hemopoietic cellular proliferation* (edit. by F. Stohlman, Jr.), pp.3-14, Grune & Stratton, New York (1970).
33. STOHLMAN, F., Jr., EBBE, S., MORSE, B., HOWARD, D., and DONOVAN, J., 'Regulation of erythropoiesis. XX. Kinetics of red cell production', *Ann. N.Y. Acad. Sci.,* 149, pp.156-172 (1968).
34. KING-SMITH, E.A. and MORLEY, A., 'Computer simulation of granulopoiesis: normal and impaired granulopoiesis', *Blood,* 36, pp.254-262 (1970).
35. MEURET, G., *Monozytopoese beim Menschen,* J.F. Lehmanns-Verlag, München (1974).
36. MEURET, G., BREMER, C., BAMMERT, J., and EWEN, J., 'Oscillation of blood monocyte counts in healthy individuals', *Cell Tissue Kinet.,* 7, pp.223-230 (1974).
37. KENNEDY, B.J., 'Cyclic leukocyte oscillations in chronic myelogenous leukemia during hydroxyurea therapy', *Blood* **35**(6), pp.751-760 (June 1970).
38. JÄGGI, J., GESSNER, U., and MEURET, G., 'Die Stabilität eines einfachen Modelles der Monozytopoese', *Biomedizinische Technik* **20**, p.39 Ergänzungsband (May 1975).
39. HAMMOND, B.J., 'A compartmental analysis of circulatory lymphocytes in the spleen', *Cell Tissue Kinet.,* 8, pp.153-169 (1975).
40. FLIEDNER, T.M. and HOELZER, D., 'Über die Dynamik leukämischer Zellspeicher', in *Leukämie* (edit. by R. Gross and J. van de Loo), Springer-Verlag, Berlin, p.165 (1972).
41. OPPELT, W., 'Kleines Handbuch technischer Regelvorgänge', 5, Aufl., Verlag Chemie, Weinheim (1972).
42. FLIEDNER, T.M., MESSNER, H., and KUBANEK, B., 'Neue Erkenntnisse der Physiologie und Pathophysiologie der Erythropoese', in *Hämatologie und Bluttransfusion* 8, 1, J.F. Lehmanns-Verlag, München (1969).
43. PROGRAMMSYSTEM-UNTERLAGEN ASIM der Firma AEG-Telefunken (1973).
44. JENTSCH, W., *Digitale Simulation kontinuierlicher Systeme,* Oldenbourg Verlag, München (1969).
45. DÖRMER, P., *Kinetics of erythropoietic cell proliferation in normal and anaemic man. A new approach using quantitative 14 C-autoradiography',* Gustav Fischer Verlag, Stuttgart (1973).
46. MORLEY, A. and STOHLMAN, F., Jr., 'Erythropoiesis in the dog; the periodic nature of the steady state', *Science* **165**, pp.1025-1027 (1969).
47. COVELLI, V., BRIGANTI, G., and SILINI, G., 'An analysis of bone marrow erythropoiesis in the mouse', *Cell Tissue Kinet.,* 5, pp.41-51 (1972).

A computer model of cerebral blood flow dynamics

FRANK MATAKAS, EMANUEL FRITSCHKA, and BERNHARD HOFFERBERTH
Klinikum Steglitz,
Freie Universitat Berlin, FRG

The dynamics of cerebral blood flow are characterized by the fact that a single parameter, viz. CO_2 tension of the arterial blood, regulates a variety of factors the combined action of which determines actual blood flow of the brain. Thus a cybernetic method is necessary to develop a uniform theory of the physiology of cerebral blood flow. There has been no trial so far to do this, except for the model of Himwich and Clark [1], which describes the distribution of blood in the basic arteries of the brain.

Methods

The simulation program was based on the following assumptions:

1. Cerebral blood flow is a function of intravascular perfusion pressure.
2. Intracranial perfusion pressure is the difference between systemic arterial pressure and venous outflow pressure.
3. Because of the thin wall of venous vessels, pressure in these vessels is a function of intracranial pressure. Under normal conditions intracranial pressure is given by CSF pressure.
4. For a given intravascular perfusion pressure cerebral blood flow is a function of CO_2 tension of the blood.
5. The CO_2 tension of the blood is in equilibrium with the CO_2 tension of the surrounding tissue and regulates the diameter of arterial vessels.
6. The actual CO_2 tension in the tissue and the blood is the result of CO_2 production rate in the tissue and the CO_2 transport capacity of the blood.
7. If intracranial vessels are not injured or if intracranial pressure does not exceed a critical value they regulate cerebral blood flow by the autoregulation mechanism. Autoregulation means the ability of the brain to retain its blood flow constant, despite changes in perfusion pressure. The mechanism works by adapting arterial vascular diameter to the perfusion pressure and the desired blood flow.
8. The blood flow which is necessary for the brain depends on the CO_2 tension of the blood and brain tissue. As a consequence, autoregulation works on different levels according to the CO_2 tension.
9. Since CO_2 influences cerebral blood flow by regulating arterial vessel diameter there is a maximal blood flow for a given perfusion pressure. This maximal value is reached if arterial vessels are maximally dilated.

The above-mentioned facts can be expressed by the following formulas:

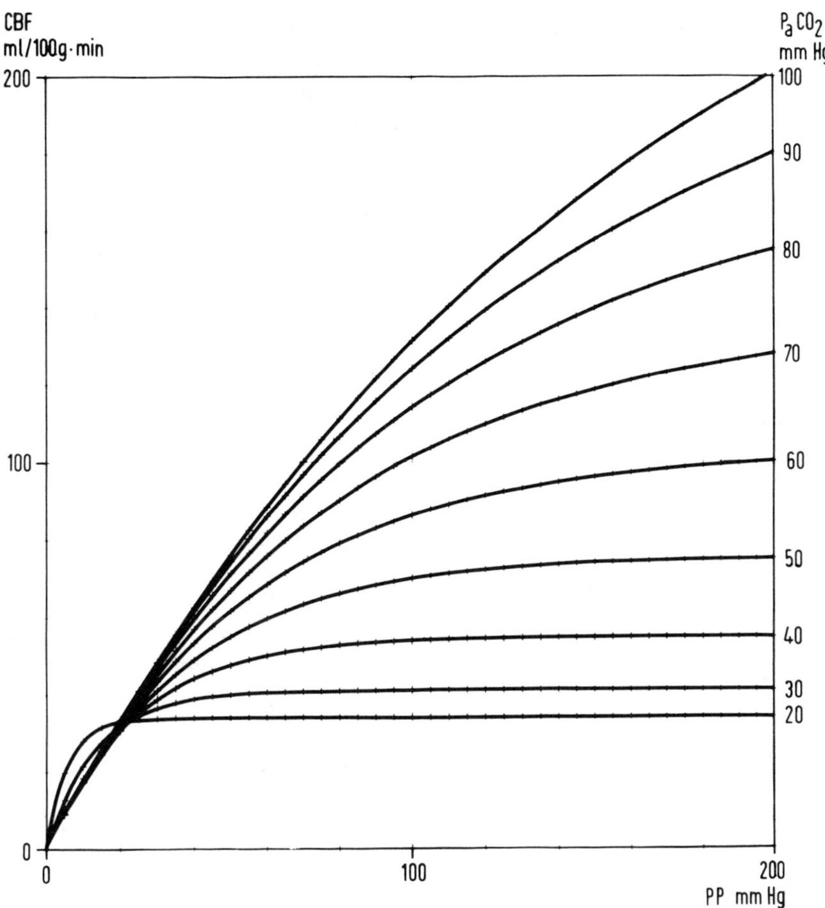

Fig.1 Relationship between CBF and perfusion pressure the CO_2 tension being constant. The curves start at zero. Their first part is nearly linear. At higher values of the abscissa the corresponding value on the ordinate becomes constant

$$CBF = f(PP, pCO_2) \tag{1}$$

$$VP = CSFP + C$$

$$PP = SAP - VP$$

$$pCO_2 = f(CMR, CBF)$$

Where CBF = cerebral blood flow, PP = perfusion pressure, $pCO_2 = CO_2$ tension, VP = venous outflow pressure, CSFP = CSF pressure, c = constant value, SAP = systemic arterial pressure, CMR = cerebral metabolic rate.

According to what we said about the autoregulation mechanism, Eq. (1) must fulfil the following postulation: Eq. (1) must describe curves which begin near zero. The first part of these curves must be linear. This linear part describes the situation where perfusion pressure is so small that in spite of maximal dilatation of vessels, the necessary blood flow cannot be achieved. At a certain level which is given by the CO_2 tension of the blood the curves must approach a parallel to the abscissa, indicating that auto-regulation retains cerebral blood flow constant in correspondence to the CO_2 tension in spite of increasing perfusion pressure (Fig. 1).

A function which fulfils these conditions is:

$$CBF = a(1 - e^{-bx})$$

where a = the cerebral blood flow which is necessary for a given pCO_2, b = the slope of the linear part of the curve, and x = the actual perfusion pressure. Both, a and b, are functions of the actual pCO_2.

For a this follows from the fact that the necessary blood flow value for the brain depends on CO_2. For b this is not evident, but follows from empirical observations. For determining the functions

$a = f(pCO_2)$, $b = f(pCO_2)$

Table 1: Experimental values of a for different values of pCO_2

pCO_2	a	Reference
mm Hg	ml/100g · min	
20	34.0	4, 9, 11
30	42.7	11
40	54	3
90	235	1

we used experimental results. The relationship between a and pCO_2 can be estimated by the results of those authors who investigated the level at which autoregulation begins to work at a given pCO_2 (Table 1).

By polynomial regression analysis of the experimental values the following equation was found:

$$a = 40.400 - pCO_2 + 0.034(pCO_2)^2$$

For determining $b = f(pCO_2)$ the following method was used: to experimental curves which represent the relationship between perfusion pressure and cerebral blood flow, Eq. (1) was applied. Since in this case cerebral blood flow, a, and x were given by the experiment, b could be computed for different values of a and x. By a polynomial regression analysis we found the following equation:

$$b = (30 + pCO_2)(200 - 26\,pCO_2 + 1.500(pCO_2)^2)^{-1}$$

Actual pCO_2 at a given point of an intracranial vessel is the difference between production rate and clearance. Since the cerebral metabolic rate is constant, we assumed that CO_2 production in the tissue is constant, too.

For the simulation program we had the following equations:

$$CBF = a(1 - e^{-b(PP)})$$

$$a = 40.400 - pCO_2 + 0.034(pCO_2)^2$$

$$b = (30 + pCO_2)(200 - 26\,pCO_2 + 1.500(pCO_2)^2)^{-1}$$

$$PP = SAP - VP$$

$$VP = CSFP + C$$

C and CSFP were assumed to be constant so that VP = 15 mmHg. Input values are SAP and pCO_2.

For the simulation we used the Continuous System Modelling Program (CSMP) which is a block-orientated language to simulate analog processes on a digital computer (Fig. 2)

Results

Figure 3 shows the curves representing the relationship between cerebral blood flow and pCO_2, the perfusion pressure being constant. Cerebral blood flow is minimal if pCO_2 = 15 mmHg. The curves are nearly linear in the first part and approach a maximal value with further increasing pCO_2. The corresponding curves which show the relationship between cerebral blood flow and perfusion pressure, pCO_2 being constant, are presented in Fig. 1. If pCO_2 exceeds 60 mmHg maximal blood flow is not reached if the prefusion pressure remains within physiological limits. The curves, which represent the vascular resistance, clearly show that the autoregulation mechanism is essentially the same for all values of pCO_2 or perfusion pressure. If the curves of Figs. 1 and 3 are drawn on a semilogarithmic scale they become linear curves.

From the curves shown in Figs. 1 and 3 it is revealed that the reactivity of the vessel wall to a change in pCO_2 is different at different levels of perfusion pressure. If this is expressed by the quotient ΔCBF/10 mmHg pCO_2, we find that this quotient is greatest if pCO_2 is between 50 and 60 mmHg. Moreover, the greater the perfusion pressure, the greater is the pCO_2 level where a change of pCO_2 has the greatest effect on cerebral blood flow.

Discussion

The simulation program implies some simplifications which want a comment. First, we made the assumption that cerebral blood flow follows the law of Hagen-Poiseuille. This is definitely not correct, but the systematic error is not too great [2]. Second, intracranial perfusion pressure is not exactly equal to our definition. However, there is no definition better than that used, and all experimental observations are based on this definition first proposed by Zwetnow [3]. If, thirdly, intracranial pressure exceeds a certain limit, or if intracranial perfusion pressure becomes so small that cerebral blood flow is reduced below a critical value, the autoregulation mechanism is partially or completely impaired [4].

```
            CONTINUOUS SYSTEM MODELING PROGRAM
     DIGITAL ANALOG SIMULATOR PROGRAM FOR THE IBM 1130/ 1 8 0 0

D A T E
FRI 28. 11. 75      CET   15 * 24 * 28  HOURS

                          CONFIGURATION SPECIFICATION

OUTPUT NAME     BLOCK     TYPE    INPUT 1    INPUT 2    INPUT 3
KONST9            26        K         0          0          0
INT PCO2          27        I        61          0          0
SUMMER3           31        +        33         32          0
CONSTANT3         32        K         0          0          0
SUMMER4           33        +        36         38          0
CONSTANT4         35        K         0          0          0
GAIN3             36        G        37          0          0
MULTIPLIER2       37        X        27         27          0
SUMMER5           38        +        35        -27          0
SUMMER6           40        +        27         41          0
CONSTANT5         41        K         0          0          0
MULTIPLIER3       42        X        27         27          0
GAIN4             43        G        42          0          0
SUMMER7           44        +        43        -45          0
GAIN5             45        G        27          0          0
SUMMER8           46        +        44         47          0
CONSTANT6         47        K         0          0          0
DIVIDER1          48        /        40         46          0
SUM               49        +        48         26          0
CONSTANT7         50        K         0          0          0
MULTIPLIER        57        X        49         50          0
UMK1              58        -        57          0          0
EXPX              59        4        58          0          0
SUMMER9           60        +       -59         61          0
CONSTANT8         61        K         0          0          0
MULT4             62        X        60         31          0
DIVI.             63        /        50         66          0
DIVI              65        /        61         63          0
OFFSET            66        O        62          0          0

                          INITIAL CONDITIONS AND PARAMETERS

IC/PAR NAME     BLOCK      IC/PAR1           PAR2           PAR3
CONSTANT3         32     0.4000E 00     0.0000E 00     0.0000E 00
CONSTANT4         35     0.4000E 02     0.0000E 00     0.0000E 00
CONSTANT5         41     0.3000E 02     0.0000E 00     0.0000E 00
CONSTANT6         47     0.2000E 03     0.0000E 00     0.0000E 00
CONSTANT7         50     0.1000E 03     0.0000E 00     0.0000E 00
CONSTANT8         61     0.1000E 01     0.0000E 00     0.0000E 00
KONST 9           26    -0.4000E-02     0.0000E 00     0.0000E 00
GAIN3             36     0.3400E-01     0.0000E 00     0.0000E 00
GAIN4             43     0.1500E 01     0.0000E 00     0.0000E 00
GAIN5             45     0.2600E 02     0.0000E 00     0.0000E 00
OFFSET            66     0.1000E-03     0.0000E 00     0.0000E 00
```

Fig.2 Print of the program for simulation cerebral blood flow. There are as many computer operations as blocks. Each block has a specification which is given in the third column.
I means integration, x means multiplication, etc.

Thus, the simulation model describes the real situation only if certain pathological facts are excluded. Finally, we must consider the CSF pressure represents the lower limit of venous outflow pressure (VP$\geqslant$ CSFP). The veins are subjected to CSF pressure and may exceed CSF pressure by a varying value.

In spite of these errors the results are in excellent agreement with experimental observations (Table 2). We may thus conclude that the formulas on which the program is based are able to describe the complete autoregulation system of cerebral blood flow. It means that the system can be transferred into a mathematical system. Cerebral blood flow in a given situation can be computed and must not necessarily be determined experimentally.

The program also shows that all important parameters which influence cerebral blood flow can be represented by one three-dimensional function.

Table 2: Comparison between experimental results and computer simulation of CBF

pCO_2 mm Hg	PP **mm Hg**	CBF experimental ml/100g · min	CBF simulation ml/100g · min	Reference
20	80	34.4	33.9	4, 11
25	80	36.7	36.0	9
32	80	42.7	42.9	11
38	87	51	50.1	8
44	80	57.9	58.0	10
48	80	64.1	63.4	10
65	124	85	85.9	6
90	80	121	125	6

Autoregulation and the effect of CO_2 on cerebral blood flow are essentially the same mechanisms.

The CSMP has proven to be of high value for simulating physiological processes. The main advantages are that it is easy to handle, especially for people who are not experts in computer sciences, and that there is no need for normalizing.

References

1. HIMWICH, W.A. and CLARK, M.E. 'Simulation of flow and pressure distributions in the circle of Willis', in *Pathology of Cerebral Microcirculation* (edit. J. Cervós-Navarro, F. Matakas, N. Grcevic, A.G. Waltz), De Gruyter, Berlin, New York, pp.140-152 (1974).
2. .NOELL, W. and SCHNEIDER, M., Zur Hämodynamik der Gehirndurchblutung bei Liquordrucksteigerungen, *Arch. f. Psych. u. Zeitschr. Neur.* **180**, 713-730 (1944).
3. ZWETNOW, N.N., Central blood flow autoregulation to blood pressure and intracranial pressure variations , *Scand. J. Lab. Clin. Invest. Suppl.* **102**, 5:A (1968).
4. MILLER, J.C., STANEK, A.E., and LANGFITT, T.W., 'Concepts of cerebral perfusion pressure and vascular compression during intracranial hypertension', *Prog. Brain. Res.* **35**, 411-432 (1971).
5. HÄGGENDAHL, E., 'Effect of arterial carbon dioxide tension and oxygen saturation on cerebral blood flow autoregulation in dogs', *Acta physiol. scand.* **66**, suppl. 258, 27-53 (1965).
6. LASSEN, N.A., 'Cerebral blood flow and oxygen consumption in man', *Physiol. Rev.* **39**, 183-238 (1959).

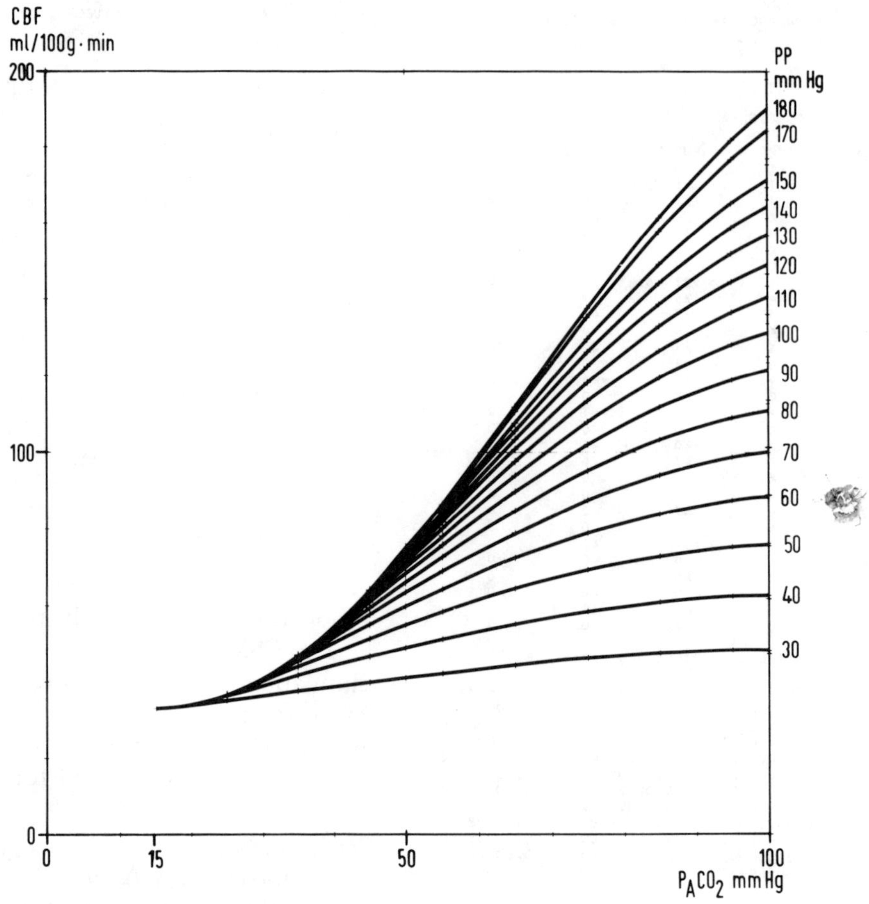

Fig.3 Relationship between CBF and arterial pCO_2, the perfusion pressure being constant

7. MARTINS, A.N., KOBRINE, A.J. and DOYLE, T.F., 'Total cerebral blood flow in the monkey measured by hydrogen clearance', *Stroke* **5**, 512-517 (1974).
8. NEMOTO, E.M., SNYDER, J.V., CARROLL, R.G., and MORIDA, H., 'Global ischemia in dogs: cerebrovascular CO_2 reactivity and autoregulation', *Stroke* **69**, 425-432 (1975).
9. PATTERSON Jr., J.L., HEYMAN, A., BATTEY, L.L., and FERGUSON, R.W., 'Threshold of response of cerebral vessels of man to increase in blood carbon dioxide', *J. Clin. Invest.* **34**, 1857 (1955).
10. SEVERINGHAUS, J.W. and LASSEN, N., 'Step hypocapnia to separate arterial from tissue pCO_2 in the regulation of cerebral blood flow', *Circ. Res.* **20**, 272-278 (1967).
11. SHAPIRO, W., WASSERMANN, A.J., and PATTERSON, J.L., 'Mechanism and pattern of human cerebrovascular regulation after rapid changes in blood CO_2 tension', *J. Clin. Invest.* **45**, 913-922 (1966).
12. WASSERMANN, A.J. and PATTERSON, J.C., 'The cerebral vascular response to reduction in arterial carbon dioxide tension', *J. Clin. Invest.* **40**, 1297-1303 (1961).

Cell interactions during development: array bound simulation of cell clone growth

ROBERT RANSOM*
School of Biological Sciences,
University of Sussex, UK

The majority of pattern forming systems within embryos are studied as if there were distinct control elements governing their development. This is seen, for example, in the search for the controlling factors by which gradients in developmental activity are set up. Chemical or electrophysiological agents have been hypothesized as being responsible for gradients, but none have yet been found.

In this chapter, I will consider the importance of local cell interactions in controlling aspects of development, and ask whether control from *within* an embryological field can determine its own behavior. Parallels will be drawn between control of this kind in developing organisms and the work of Ulam and Maruyama, both of whom have studied the patterns produced by the interactions of abstract units obeying the same or similar rules.

As a particular example of this sort of control process, I will use my own work on the analysis of the behavior of cells in the growing imaginal discs of the fruitfly *Drosophila,* using computer modelling techniques. By making a model with assumed properties of the real system, we can see how closely the final simulation fits the *in vivo* situation. If the two are similar, we may be able to suggest that properties like those in the model are acting in the real system.

In this way, I hope to show that cell division patterns in the imaginal disc may be produced by simple rules of interaction, and not by means of externally mediated control systems. Finally, we will look at a contemporary description of developmental processes, and discuss how interactional control mechanisms are often omitted or have their effects minimized in such descriptions. This may be biasing the way we try to analyze developmental systems.

Developmental control by local interactions

The unfolding of new shapes and patterns by tissues during the development of an organism is thought to occur by means of a variety of different processes. Among these are *differentiation,* the acquisition of specialized differences amongst cells of an initially homogeneous cell line; *communication,* the reception and emission of 'messages' by cells both physically separated and juxtaposed, and *cell division and growth,* giving rise to size and

*Present address: Institut für Biologie III, Universität Freiburg, Schänzlestr. 1, 78 Freiburg-I-Br., BRD.

shape changes both due to instructions generated locally by the cells themselves or by more distant control mechanisms.

There are many situations in development where spatial differences are set up in what appears to be an initially homogeneous cell mass. For example, cells along the developing vertebrate limb bud appear very similar in the early stages, but experiments show that determination of the various regions has already occurred. Cells in tissue culture called fibroblasts may become oriented into parallel bundles with no obvious control mechanism; the seemingly identical cells of insect legs can be made to regenerate a surgically removed section of leg. All these examples indicate that cells somehow 'talk' to each other and arrange themselves into patterns: furthermore, they can recognize an aberrant situation, as in the instance of leg regeneration, and repair the damage.

There are two distict categories of difference acquisition in tissues; those which require physical movements of cells, for example assembly processes like the alignment of fibroblasts, or cell divisions, and the acquisition of differences within a *field,* a developmental subsystem. An example of this might be the spacing pattern of hairs on skin, or vertebrate limb bud development. Whatever the category, we have the problem of explaining how these differences arise.

In the present chapter I will suggest that such phenomena may come about partly because of the interaction of cells within fields. This idea is not novel, but it is in contradiction to the more generally held view that shapes and patterns are usually generated by external control systems; chemical or electrophysiological signals produced by specialized regions of the particular developing area. We will consider the instances in which properties of the cells themselves may act in concert to produce a pattern. One of the few illustrations of this sort of mechanism known at present has been investigated by Elsdale [1]. Elsdale showed that the alignment of fibroblasts in culture exhibits a similarity to the class of 'inherently precise' machines which produce an ordered result from random events, for example the grinding of a spherical lens by randomly rubbing convex and concave glass surfaces over one another. It was demonstrated that 'seesawing' movements of the fibroblasts could bring about aggregation into parallel bundles of cells.

The idea that similarities in unit behavior, like the random movement of fibroblasts, may produce changes in a system was first proposed by

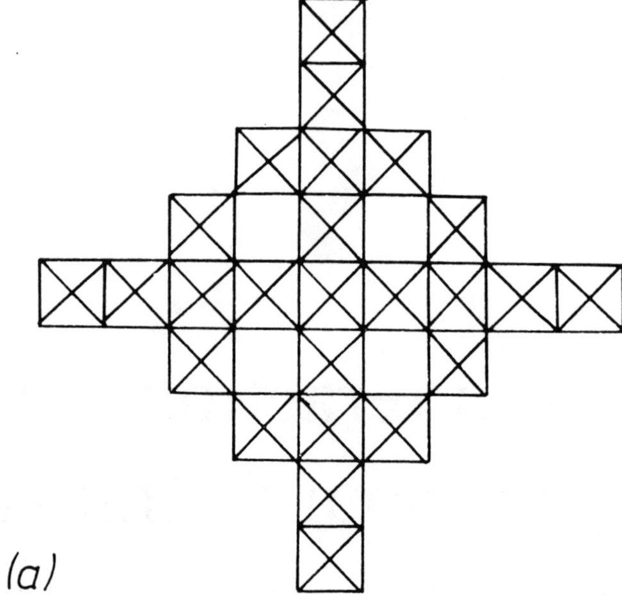

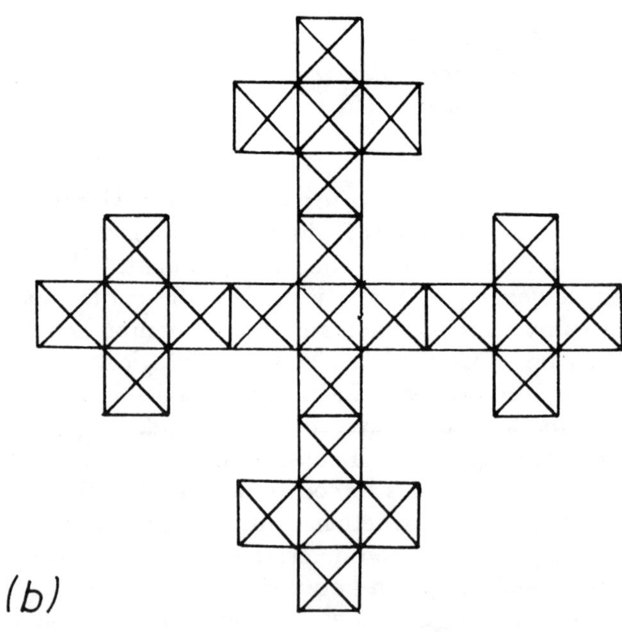

Fig.1 Examples of patterns generated by Ulam using a four neighbor space. We start in the first generation with one occupied square (center) and define the growth rule; given a number of occupied squares in the nth generation, the squares of the nth + 1 generation will be all those adjacent to only one square of the nth generation. (a) shows growth after five generations. If the rule is extended to include inhibition by cells which touch diagonally, (b) is produced after five generations

Ulam [2]. Ulam described how the application of simple rules of growth to regular figures in two or three dimensional space could generate complex patterns, and that after the pattern had been generated it was often extremely difficult to work backwards and discover the rules of interaction (see Fig. 1). Maruyama [3] has called such rules 'deviation amplifying mutual causal processes', to separate them from the more normal type of control system where processes might be studied using the mathematics of deviation counteracting feedback networks. The difference between the two types of control is that deviation amplifying mutual causal processes amplify initial inhomogeneities in a system and build up differences in the pattern of the system in the former case, whilst the latter depend on some separate control element.

If the actual units themselves can generate a pattern, we can ask why an externally mediated control system is necessary. External control has dominated the analysis of developing systems in the past, thus we find the embryological literature full of references to hypothesized 'gradients', and their chemical specifiers called 'morphogens'. The first is merely an observed description of the phenomenon's effect; the latter an explanative attempt. However, numerous examples exist of patterns being set up in single celled systems, either in the egg, or in organisms like the alga *Acetabularia,* which have no units like cells which suggest an interactive function. In such cases, some distinct gradient forming property may well exist (see, for example, Goodwin [4], who discusses membrane gradients in *Acetabularia*).

The application of rules of local interaction to real systems

The first step in applying cell interaction rules to the analysis of developmental problems must be to describe how they might be acting *in vivo.* We have already discussed how difficult it is to extrapolate back from final result to the rules that produced the result even with a 'simple' set of rules like those used by Ulam. This situation has led to the proliferation of computer models currently in the literature describing simulations of developmental phenomena. The rationale is simply that by devising a computer simulation of a system, the rules of the simulation can be amended until they give a reasonable facsimile of the real situation. We can then hopefully get some idea of how the real system works by analyzing our computer rules. Amongst the findings of work of this kind is that different cell types aggregating into cell type specific groups may do so by application of simple recognition rules (Goel *et al.,* [5]; Antonelli *et al.,* [6]; Vasiliev *et al.,* [7].) Similarly, Honda [8] has described how the colony forming green alga *Pediastrum biwae* may attain its mature shape by the random interaction of two types of attraction sites on the individual cells, and not by any externally directed control mechanism.

The 'unit' of multicellular development is the cell, and the central problems of embryogenesis are centred around how the various cell masses called tissues acquire differences in shape and structure. It therefore seems important to look at the application of 'control by interaction' to growing cell populations. This was first done in a computer model by Eden [9] who showed that for simple cell division on a two dimensional lattice using a four neighbour space, growth occurring equally in all four directions gave rise to a circular shaped colony. A later model by Ede and Law [10] extended this sort of model to the growth of the chick limb, demonstrating that a change in a particular cell property, the ability of individual cells to move freely, produced a mutant called *talpid,* with short malformed limbs.

A major failing of these models was that 'cell division' could only be simulated by allowing cells on the perimeter of the two dimensional cell mass to divide. The applicability to the simulation of real cell division systems was therefore reduced. Ede's model in fact possessed a facility by which internal cells mimicked division by placing their daughter cells on the perimeter of the growing mass, but the separation of the two progeny of a division was not very realistic.

Much of the information that we have on cell/cell interactions during division processes comes from examination of marked clones of cells (a clone consists of all the progeny of a parent cell), either in mammalian chimaeras (McLaren, [11]), or in insects like the fruitfly *Drosophila* or the wasp *Habrobracon* (Postlethwait and Schneiderman, [12]); typical clone shapes can be seen in Fig. 2. By studying clone orientations, we can obtain information about the forces that acted on the clone during development, and by observing clone size and knowing the time of clone induction, information about growth dynamics can be obtained. If we wanted to simulate the development of clones to throw light on the forces present in the embryo, a new way of representing cell division on the computer would be needed. Clearly, all cells must be able to divide into their own immediate neighborhood to faith-

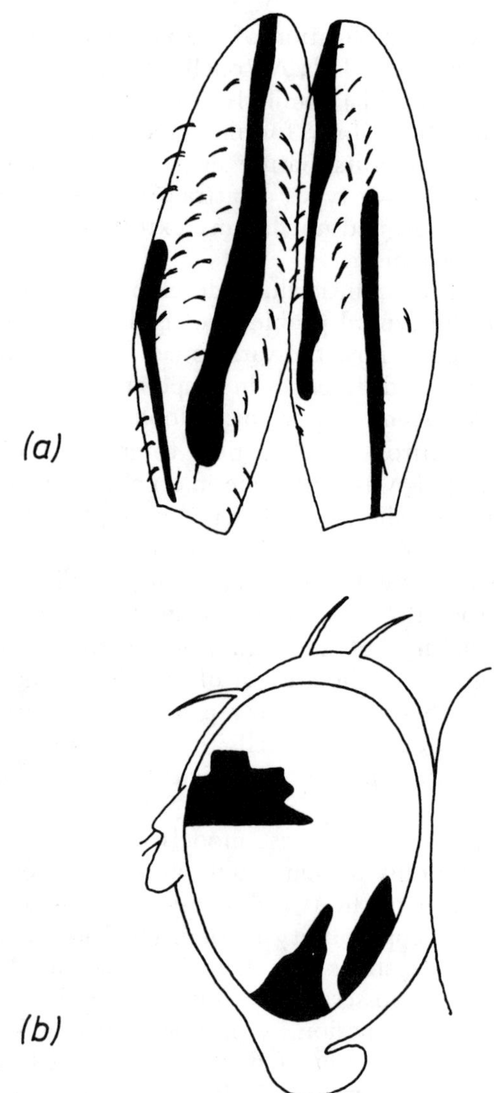

Fig.2 (a) Surface of the femur of the adult leg of *Drosophila* showing clones (black) induced during early larval stages. Both anterior (left) and posterior (right) surfaces are shown. (b) shows the side view of the adult head of *Drosophila* showing the compound eye (inner perimeter) with induced clones.

fully represent clone growth.

A computer model of clone growth

The computer array is a rigid lattice, the use of which gives abstracted cells some kind of spatial co-ordinates: unfortunately, the nature of the array makes it difficult to carry out division as seen in living organisms. Consider a mass of typically amorphous shaped embryonic cells. If one of the central cells in this mass divides, there will be an incremental change in shape of a large portion of the cell mass, as a space is allowed for the increased volume of the two daughter cells. If a computer cell divides into a lattice location that is already occupied by another cell, and this latter cell is moved, it has to travel a whole array position in distance, which means the same thing has to happen to one of its own neighbors in turn, and so on out to the edge of the clone. There is then a marked difference in real and simulated situations, and the modelling of cell division within a clone has to either allow for this 'single cell pushing single cell' phenomenon, or has to model cells as taking up more than one array position so that incremental movements in more than one cell at a time can be introduced. The former is the easiest to program, and this is the technique to be considered here. (See also Ransom [13].)

Division may be programmed to occur as follows. The cell to divide computes the nearest free edge to itself, and all cells in the line between the free space and the cell to divide are moved outwards one array position to create a new free space. The 'daughter' cell can then be placed in this space (see Fig. 3*a*). In this way, cell population growth can be modelled, so that labelling of an individual cell (by a numeric marker) allows the subsequent clone growth to be followed.

A second feature of a clone growth simulation should be a facility for directional growth. If growth according to the rules outlined above occurs randomly, radial growth is seen, and clones inside the mass have a sectorial appearance (Fig. 3*b*). Non-random growth is seen in many of the clone analysis systems mentioned in the previous section, so some method of controlling growth direction must be included. There are two possibilities. Firstly, cells may always orient their division directions by way of some directional signal which individual 'compasses' in the cells can home in on. There is little evidence that this sort of control system works in embryos, although slime moulds do possess a chemical signalling mechanism to attract individual free moving cells to aggregate into a 'slug', the next stage of their life cycle (Bonner [14]). In this case the signal does not normally produce a concentration gradient that the slime mould amoebae swim up; instead, spaced pulses of signal are fired. Alternatively, the cell division orientations may be produced by means of interaction between the cells themselves and surrounding constraints. If plausible rules can be put forward to produce simulated clones like those seen in the living situation, it may be possible to describe how cell interactions can give rise to oriented cell divisions.

Constraints were therefore modelled by introduc-

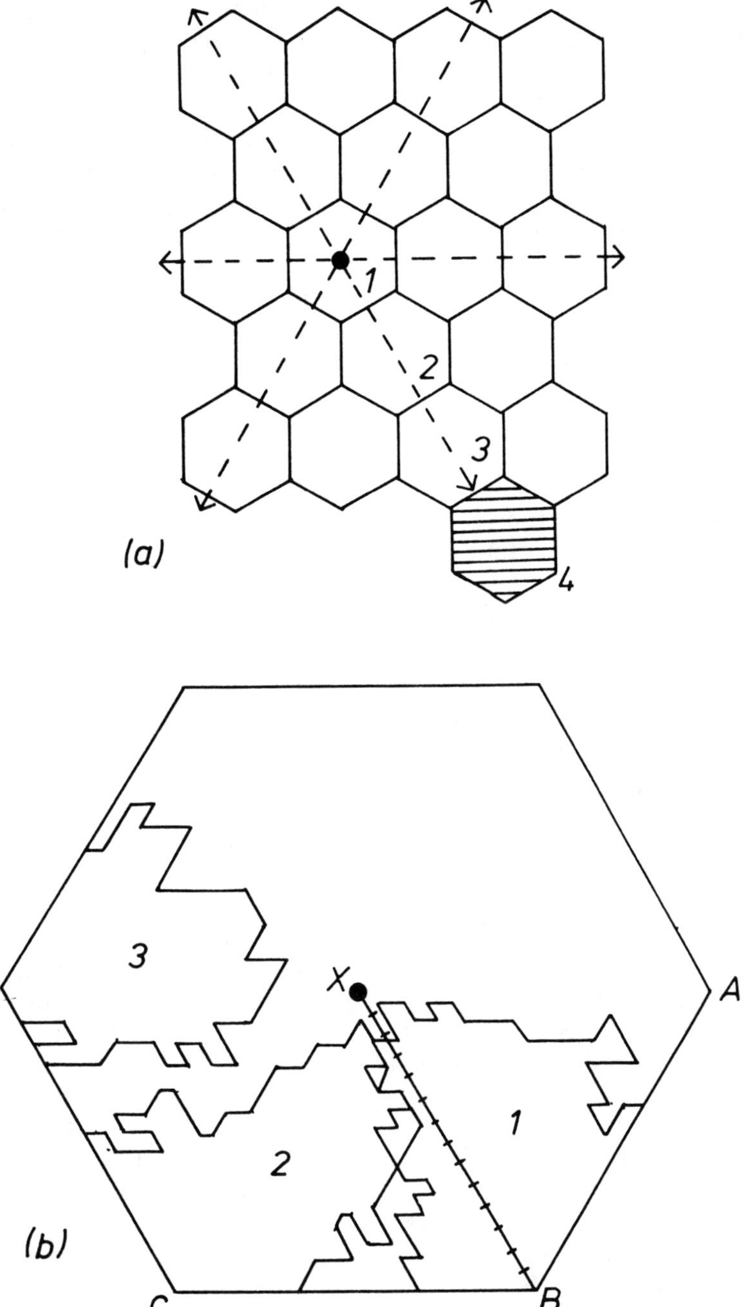

Fig.3 (a) Demonstration of the 'pushing' algorithm using hexagonal cells. Cell 1 is to divide. It decides which is the nearest edge, in this case marked by cell 3. Cell 3 moves out one position, to the previously empty array position 4, and cell 2 moves out in the same way. The space occupied by cell 2 is hence cleared for the new daughter cell. (b) A mass of about 600 computer 'cells' showing the final configurations of 3 clones induced at the 7 cell stage. X = position of initial 7 cell primordium. The graduated scale shows position of successive cells along the line XB.

ing a new class of array location onto the two dimensional array. 'Constraint' array elements were marked with a special numeric marker, and cell division was not able to occur into array spaces so marked; a fresh division direction had to be chosen. For example, a line of such elements would act as a barrier constraining growth in a particular direction. The justification for rules of this kind was simply that adjacent organs, or membranes around the tissue itself, may provide a similar force in nature.

Cell interactions in Drosophila imaginal discs

It was decided that the first test of the model system should be on the ability to simulate clones seen in the adult structures of the fruitfly *Drosophila melanogaster,* this organism being chosen because of the wealth of data on clone shapes in the various organs. The aspects of fly development that concern us here are as follows.

In brief, during the larval period of the fruitfly's life, flat organs consisting initially of a single layer of cells are set aside. These organs, called 'imaginal discs', are undifferentiated and consist of similar cells surrounded by a membrane. Separate imaginal discs correspond to adult structures like legs, wings, head and so on, and during the metamorphosis of the pupa into the adult fly, each disc develops into the adult organ it is predestined to form. The discs begin growth usually with less than 40 cells, and in the mature larva may possess up to 50,000 cells. If clones are induced at an early stage by 'marking' one of the cells genetically, all the cell's progeny will also be marked. We can therefore see that observation of the final clone shapes in the adult will tell us about the forces that acted on the clone during its development.

Taking two particular discs, those representing the compound eye, and a leg, we may examine typical clones (see Fig. 2). In the eye, the clones have a posterior-anterior orientation, whilst in the leg, parallel clones are oriented down the leg. By rough geometrical transformations we can work out the orientations of corresponding clones in head and leg discs (Schubiger [15], Ransom [16]) that gave rise to the adult clone shapes. Examination of a number of clones shows that leg clone growth occurs radially in the leg disc, whilst eye disc growth is in a posterior-anterior direction, as in the eye itself.

The next step is to see if the computer model outlined in the previous section can model clone growth of the two different types. In the case of the leg, no constraint is required: simple cell division in the model gives similar clones to those seen *in vivo* (see Fig. 4*a*). The eye clones cannot be produced in the same way, and so we must introduce a constraint into the model. This was done by embedding a routine calculating the area and perimeter of an ellipse of given radii. This ellipse was superimposed onto the two dimensional array around the growing simulated cell mass, and was grown in steps to represent the membrane surrounding the growing head disc. Comparison of the simulated clones of Fig. 4*b* with real eye clones (Fig. 2) shows definite similarities: the growth of the constraining ellipse which gave these clone shapes is shown in Fig. 4*c*.

Both leg and eye discs possess enveloping membranes. The leg disc's circular growth indicates that no directional constraint is provided by the membrane. The directional growth seen in the eye clones could be explained in two ways: firstly, the membrane itself could bave structural properties biasing its own growth in a forward direction; a second answer is given by looking at the position of the eye disc in the larva. The disc is placed immediately anterior to the larval brain, an organ of large bulk, and it may be that this mass is sufficient to constrain growth of the disc in a posteriorwards direction. Because of this, cell divisions would only be able to occur in the forward direction.

Conclusions

Simple mechanical interactions are therefore seen to be important in particular developmental situations. Using interactions as central elements, we could construct a general model of how development occurs, for example a 'flow chart' of development could be constructed like that in Fig. 5. An essential feature of this diagram is that constant feedback between the various developmental activities occurs, providing stimuli for the next, or parallel stages of embryogenesis. Such a scheme is not universally accepted. Wolpert [17] has outlined a much more linear plan, one which seems to presuppose external control mechanisms and omits interactions. "The spatial pattern of differentiation may be regarded as the process whereby a cell has its spatial position specified—positional information—and it is this which can determine its molecular differentiation. Finally with morphogenesis—moulding of form—the emphasis is on the forces bringing about changes in shape". This suggests that development occurs in discrete steps. First, some control system endows cells in a system with differences, and the cells then undergo differentiation according

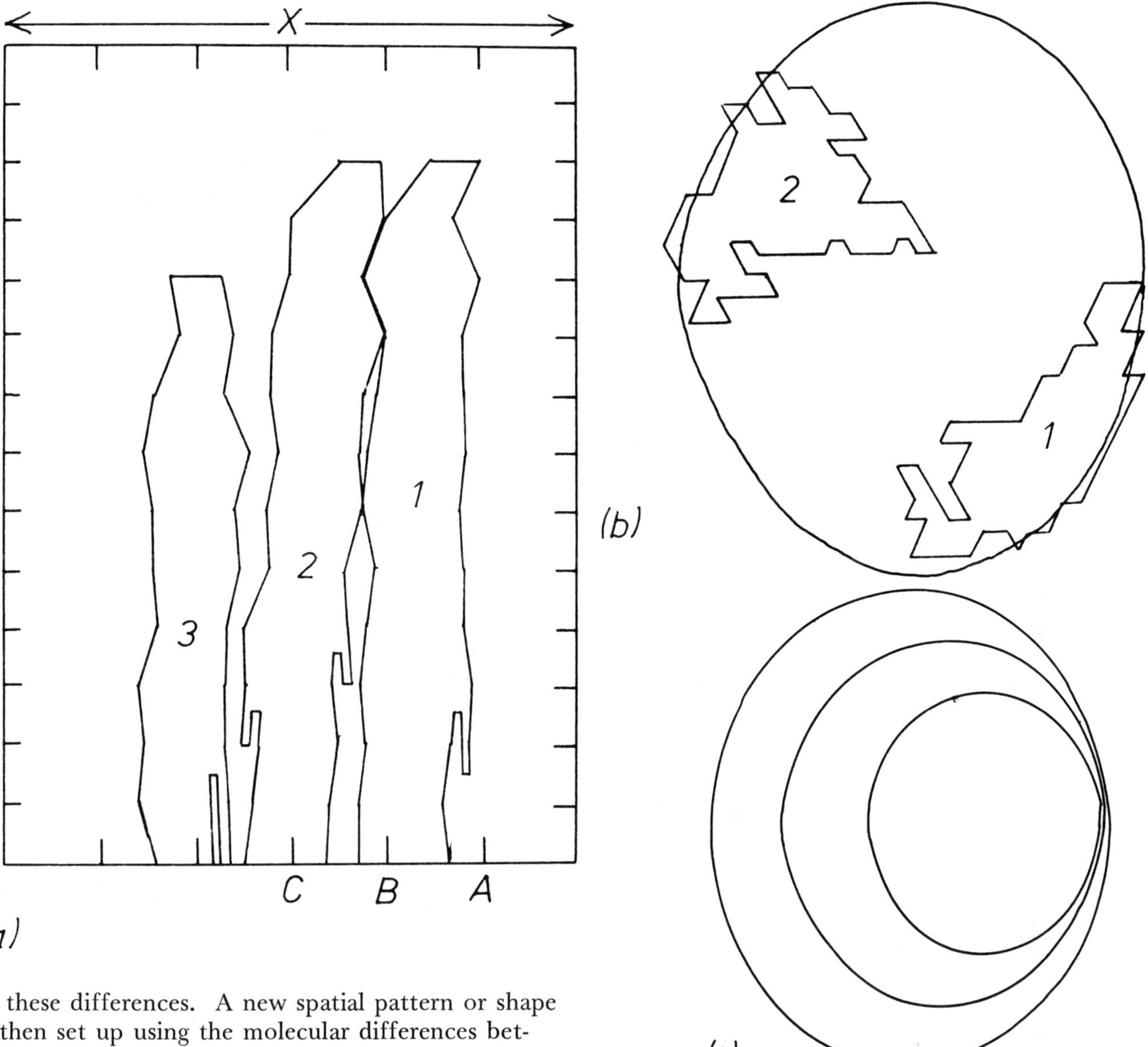

to these differences. A new spatial pattern or shape is then set up using the molecular differences between the cells as its controlling force. In the case of the imaginal disc model described above, this would mean that (1) a mechanism for orienting cell divisions is activated (2) cells orient themselves by the reception of the information provided by the control system (3) the cells divide in the correct direction.

Although these statements do basically describe what happens, the stressing of the control function leads one to the view that cells are merely passive units, 'sheep' to be 'shepherded' by an external control system, and this may not be the most suitable way of describing many developmental phenomena. We have seen that similarities in cell behavior can direct simple activities seen in development. Cells can become aligned in characteristic ways, cells may aggregate into groups showing character-

Fig.4 (a) The leg disc is a flat two dimensional structure that forms an elongated tube, the leg, by a telescoping process during metamorphosis. The centre point X of Fig.3b therefore becomes the distal extremity of the leg. By assuming that the model 'leg' is of 'constant' diameter over its length, the shape of the computer generated disc clone can be mapped to any level on the leg by working out the *proportion* of successive circumferences out from the point X that are occupied by the clone, and plotting the values for a set of equally spaced circumferences (for example, the values of the graduated scale) onto an opened out map of the leg as shown. X is now expanded into the whole top row, as it will have the same diameter as the outermost perimeter of the final disc.

(b) Two clones in a background of 700 computer 'cells' representing the adult eye of *Drosophila*. Growth of the initial 8 cell primordium has occurred in a forward direction because of the constraint pattern (c), of which 3 growth positions are shown.

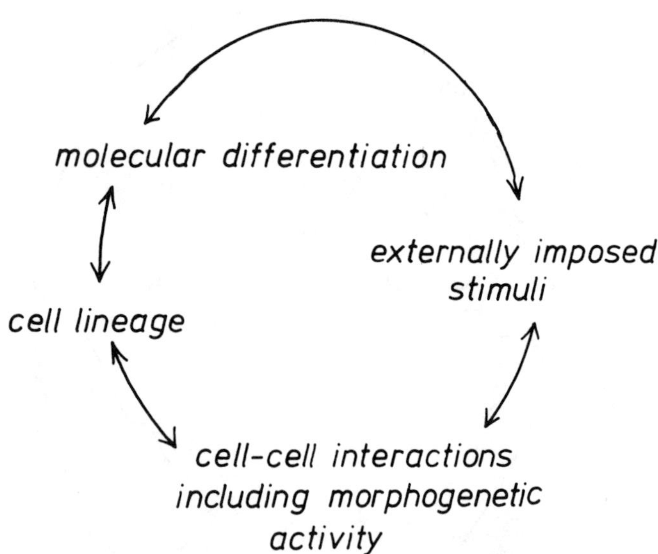

Fig.5 The various processes involved in development. Changes in milieu (cell interactions, external control mechanisms) and cell-internal states (cell lineage and differentiation) produce constant fluxes in developmental activity

istic shapes, and the direction in which cells divide can be explained simply by their interactions during the division process. We have concentrated on a particular category of cell interactions, those involving movements of cells and cell groups, and the next problem is to decide whether more passive processes can be directed in the same way. How this might work is highly speculative and so will not be elaborated on here.

Acknowledgements
I thank Dr. Brian Goodwin for discussing the essential points of this chapter with me. The work was conceived during a years visit to the Hubrecht Laboratory, Utrecht, The Netherlands. During this time I was generously supported by a Royal Society European Programme Fellowship.

References

1. ELSDALE, T., In: *Ciba Foundation Symposium on Homeostatic Regulators* (edit. by Wolstenholme and Knight), pp.291-230, Churchill, London (1969).
2. ULAM, S., *Proc. Symp. Appl. Math.* **14**, 215-224 (1962).
3. MARUYAMA, M., *Amer. Scient.* **51**, 164-179 (1963).
4. GOODWIN, B.C., *Adv. Chem. Phys.* **29**, 269-280 (1975).
5. GOEL, N.S. *et al., J. Theor. Biol.* **28**, 423-468 (1970).
6. ANTONELLI, P.L., ROGERS, T.D., and WILLARD, M.A., *J. theor. Biol.* **41**, 1-21 (1973).
7. VASILIEV, A.V., PYATETSKIY-SHAPIRO, I.I., and RADVOGIN, Yu.B., Preprint, Inst. Appl. Math., Order of Lenin, Acad. Sci., Moscow (1971).
8. HONDA, H., *J. theor. Biol.* **42**, 461-481 (1973).
9. EDEN, M., Proc. Berkeley Symp. Math. Stat. Prob. **4**, 223 (1960).
10. EDE, D.A. and LAW, J.T., *Nature,* Lond. **221**, 244-248 (1969).
11. McLAREN, A., *Nature,* Lond. **239**, 274-276 (1972).
12. POSTLETHWAITE, J.H. and SCHNEIDERMAN, H.A., *Devl. Biol.* **24**, 477-519 (1971).
13. RANSOM, R., *J. theor. Biol.* **53**, 445-462 (1975).
14. BONNER, J., *The Cellular Slime Moulds,* Englewood Cliffs, N.J., Princeton University Press (1967).
15. SCHUBIGER, G., *Wilhelm Roux' Archiv.* **160**, 9-40 (1968).
16. RANSOM, R., *J. theor. Biol.* **66**, 361-377 (1977).
17. WOLPERT, L., *J. theor. Biol.* **25**, 1-47 (1969).

A critical analysis of saturation phenomena in metabolic processes*

M. MILANESE and A. VILLA
Centro Elaborazione Numerale dei Segnali (CNR)
Istituto di Elettrotecnica, Politecnico di Torino, Italy
Istituto Elettrotecnico Nazionale Galileo Ferraris, Torino, Italy

Introduction
The investigation of the saturation phenomena in metabolic processes such as the hepatic kinetics of some dyes (bromsulphthalein, indocyanine green, galactose, etc.), has received particular attention in the past years [1-9].

In fact, the identification of the metabolic functions which could be rate limited, is of great importance for a better understanding of the physiological mechanisms involved in the whole process. Moreover the determinations of maximal removal rates gives a fair quantitative measure of dysfunctions that may be not equalled by other functions tests [9].

For example a great deal of experimental work has been performed, and many values related to maximal removal rates (V_{max}, L_m, T_m) [1-3, 7, 8] have been evaluated by experimental data and correlated to various hepatic dysfunctions; however there is yet some discussions about the interpretation of experimental data when a constant amount of dye is infused (see for example [5, 8]).

Doubts about the meaning and the validity of the evaluation of the transport maximum (T_m) and of the relative storage capacity (S) first proposed by Wheeler [2] and the widely used in the literature, are recently suggested by a critical analysis of experimental data [8].

It can be remarked that, a complete analysis of the most important mechanisms involved in the considered metabolic process, is not explicitly carried out in all these works (except partially in [5]).

In the present chapter a careful modelling of the saturation phenomena is developed, in order to give a satisfactory interpretation of available experimental data, and to clarify many of the problems raised in the previous literature.

1. Problem definition
We consider a metabolic process of a given dye described by the two compartment models of Fig. 1:

*This work has been partially supported by C.N.R.

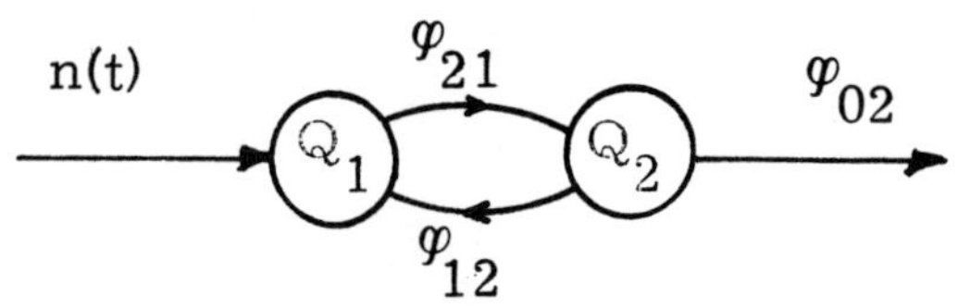

Fig.1 Compartment model of the considered metabolic process

where

$Q_1(t)$ is the amount of the dye in compartment 1,

$Q_2(t)$ is the amount of the dye in compartment 2,

$n(t)$ is the rate of infusion of the dye in compartment 1,

φ_{ij} is the amount of substance transferred from compartment j to compartment i in the unit of time. Generally φ_{ij} is a monotone increasing function of Q_j, such that the following conditions are satisfied:

$$\varphi_{ij}(0) = 0$$
$$\lim_{Q_j \to \infty} \varphi_{ij}(Q_j) = \phi_{ij} \tag{1}$$

Remark:
A model of this type is more or less explicitly used in dealing with the hepatic kinetics of drugs (see for example [10]). In this case compartment 1 is the blood and compartment 2 is the liver; the functions φ_{ij} are given by the Michaelis-Menten law [5, 9, 10], or by more complex laws [11], which however satisfy conditions (1).

It must be remarked that the dye more widely used in the analysis of the saturation phenomena in hepatobiliary mechanisms is bromsulfonphthalein (BSP), which has a very complex metabolic behavior. The model of Fig.1, which is used by several authors for BSP kinetics, is inadequate to represent the very important phenomenon of conjugation of BSP in the liver [12, 13]; on the contrary galactose, indocyanine green, rose bengale, dibromosulphthalein are not conjugated in the liver and the two compartmental models of Fig.1 give a sufficiently good description of the dye's disappearance from blood. Obviously the results of the paper apply only to cases in which the conjugation phenomenon is not relevant.■

It is supposed that only compartment 1 is accessible for measurements, and that the following experimental behaviors [1-9] are obtained when different amounts of dye are infused:

I) If n is less than a certain value n_0, Q_1 reaches a constant value.

II) If $n > n_0$, Q_1 presents a constant rise $dQ_1/dt = D_1$, after an initial transient, and D_1 is an affine function of n.

IIa) If $n_0 < n < n_1$, the removal rate R of the dye from compartment 1 (defined by $R = n - D_1$) is an affine function of n [2].

IIb) If $n > n_1$, R is independent of n [1, 3].

Conditions IIa and IIb indicate that some of the transfer φ_{ij} are working in saturation condition. It has been firstly thought that the limiting process was the uptake (φ_{21}) of dye from compartment 1 [1, 14] [15]. At present, the rising behavior of Q_1 is interpreted as due to a limit in the elimination (φ_{02}) of the dye from compartment 2 [2, 8, 4], while it is usually thought that the saturation of the transfer φ_{21} is not reached in these conditions.

The aim of the present chapter is to find, on the basis of the experimental behaviors IIa and IIb which transfers φ_{ij} are in saturation conditions.

2. The mathematical interpretation

We can prove the following result.

Proposition:
The experimental behaviors IIa and IIb can be *uniquely* interpreted as follows:

a) During condition IIa there is saturation of φ_{02}.

b) During condition IIb both φ_{02} and φ_{21} are saturated.

c) Transfer φ_{12} does not saturate in any condition.

Proof:
The proof is carried on by showing that no other saturation conditions than the one indicated, can reproduce the given experimental behavior.

At first we analyse the saturation conditions defined in *a)*, *b)* and *c)*.

The differential equations related to the compartmental model in Fig.1, under hypothesis *a)*, are:

$$\frac{dQ_1}{dt} = -\varphi_{21}(Q_1) + \varphi_{12}(Q_2) + n$$

$$\frac{dQ_2}{dt} = +\varphi_{21}(Q_1) - \varphi_{12}(Q_2) + \phi_{02} \tag{2}$$

Linearizing Eqs (2) in the neighborhood of certain values $\bar{Q}_1$ and $\bar{Q}_2$ (which may be the mean values of Q_1 and Q_2 during a time interval in which the transient has finished), there results:

$$\frac{dQ_1}{dt} = -\lambda_{21}Q_1 + \lambda_{12}Q_2 + n$$
$$\frac{dQ_2}{dt} = \lambda_{21}Q_1 - \lambda_{12}Q_2 + \phi_{02} \tag{3}$$

where:

$$\lambda_{ij} = \left[\frac{d\varphi_{ij}}{dQ_j}\right]_{Q_j = \bar{Q}_j} \tag{4}$$

The steady state solutions of (3) are:

$$Q_1(t) = D_1^a t + Q_{10}^a$$
$$Q_2(t) = D_2^a t + Q_{20}^a \tag{5}$$

where:

$$\frac{D_1^a}{D_2^a} = \frac{\lambda_{12}}{\lambda_{21}} \tag{6}$$

It also results that:

$$D_1^a = \frac{n}{\alpha} - \frac{\phi_{02}}{\alpha} \tag{7}$$

with

$$\alpha = 1 + \frac{\lambda_{21}}{\lambda_{12}} \tag{8}$$

The removal rate is easily computed as

$$R = n - D_1^a = n\left(\frac{\alpha-1}{\alpha}\right) + \frac{\phi_{02}}{\alpha} \tag{9}$$

Under hypothesis *b)*, the equations are:

$$\frac{dQ_1}{dt} = -\phi_{21} + \lambda_{12}Q_2 + n \tag{10}$$

$$\frac{dQ_2}{dt} = \phi_{21} + \lambda_{12}Q_2 + \phi_{02} \tag{10 contd.}$$

The steady state solutions of (10) are:

$$Q_1(t) = D_1^b t + Q_{10}^b$$
$$Q_2(t) = Q_{20}^b \tag{11}$$

There results that:

$$D_1^b = n - \phi_{02} \tag{12}$$

$$R = \phi_{02} \tag{13}$$

Since, in this condition, φ_{12} is not yet saturated and $Q_2(t)$ does not increase over than Q_{20}^b, this transfer cannot reach the saturation condition. Moreover equations (5, 7, 11) and (12) are consistent with condition II, equation (9) with condition IIa and equation (13) with condition IIb.

We show now that behaviors not consistent with the one indicated by IIa and IIb are obtained under hypothesis different from *a), b)* and *c)*.

Let us suppose that there is only the saturation of φ_{21}. In this case, it can be easily computed that

$$R = \phi_{21}\frac{\lambda_{02}}{\lambda_{02} + \lambda_{12}} \qquad \forall n > n_0 \tag{14}$$

and this relation is not consistent with condition IIa.

Let us suppose that only φ_{12} is saturated. It follows that the steady state solution of $Q_1(t)$ is a constant for any n, so that this hypothesis does not explain the rising behavior of Q_1.

Suppose moreover that both φ_{21} and φ_{12} are saturated. In this case, it results:

$$R = \phi_{21} - \phi_{12} \qquad \forall n > n_0$$

which is not consistent with condition IIa. This means that part *a)* of the Proposition is completely proved.

In conclusion it remains only to show that, in condition IIb, it is not possible to hypothesize the saturation of both φ_{02} and φ_{12}. This results in observing that in such case the steady state solution of $Q_1(t)$ is a constant.

3. Discussion

The results derived in previous sections can be used to discuss some important questions related to the saturation phenomena in the hepatobiliary system.

At first, it can be remarked that the well known method proposed by Wheeler *et al.* [2] for evaluation of the excretory transport maximum (T_m) was derived on the assumption that 'over a limited range of concentration, the quantity of stored BSP (in liver) is directly proportional to the plasma concentration... This factor is designated relative storage capacity (S)'. This assumption is certainly not valid as pointed out recently in [8]. However this does not mean that the T_m, as computed by this method, is incorrect, as it is shown by equation (7). In fact, as suggested by Wheeler, the transport maximum can be actually evaluated from measurements of dye concentration in plasma in the steady state condition, using two infusion values n' and n'':

$$T_m = \phi_{02} = n' - D_1' \cdot \frac{n' - n''}{D_1' - D_1''} \qquad (15)$$

Note that, contrary to the usual method, the evaluation of T_m by means of (15) does not involve the knowledge of the plasmatic volume.

Moreover, Eq.(6) gives the correct interpretation of the relative storage capacity S as the ratio between the rate of increasing of dye quantities in plasma and liver.

It can be emphasized that all these results are consequence of the only hypothesis that φ_{02} is saturated: the validity of such hypothesis is assured by the correspondence between the theoretical results and the experimental behaviors, shown by the Proposition of Section 2.

Another important conclusion, which can be drawn by the above results, is the following: the fact that the removal rate R does not exceed a given value (called L_m [1, 3]) denotes that there is saturation in the hepatic uptake of the dye from blood (φ_{21}) together with saturation in the excretory transport (φ_{02}), excluding the hypothesis of saturation of only φ_{21}. This fact has never been recognized by previous investigators, who tried to interpret the experimental data in terms of the saturation of only one process. On this basis the hypothesis of the saturation of φ_{21}, though suggested by some authors, has been subsequently discarded [3, 5]. From Eq.(13) it results, as suggested by Verschure [1], that the maximal removal rate (L_m) is equal to the excretory transport maximum. However, the L_m values, evaluated from experimental data obtained using BSP, are about twice the T_m values evaluated by the method proposed by Wheeler *et al.* Since the estimation of T_m by direct measurement of biliary excretion of BSP gives in the mean the 90% of the calculated value [2], the discrepancy between the evaluation of L_m and T_m can only be explained by taking into account the conjugation phenomena, which are not considered in the assumed model.

Finally we can show the qualitative behavior of $Q_1(t)$ and $Q_2(t)$, under a constant infusion $n > \phi_{02}$ (for related experimental results see [5]).

Four phases can be considered (see Fig.2):

1) A transient phase, during which the saturation of φ_{02} is reached.
2) A steady state phase, where only φ_{02} is saturated. In this condition, if λ_{21} and λ_{12} are constant, it results:

$$\frac{dQ_1(t)}{dt} = \frac{1}{\alpha}[n - \phi_{02}] = D_1^{(2)}$$

$$\frac{dQ_2(t)}{dt} = \frac{\alpha - 1}{\alpha}[n - \phi_{02}]$$

where $\alpha = 1 + \lambda_{21}/\lambda_{12}$.

The lengths of such phase is depending on the infusion rate n: in particular, for high values of n, this phase may be not present

3) A transient phase, due to the occurring saturation of φ_{21}.
4) A final steady state phase, during which both φ_{02} and φ_{21} are saturated. Now it results:

$$\frac{dQ_1(t)}{dt} = n - \phi_{02} = D_1^{(4)}$$

$$Q_2(t) = Q_{2\,MAX} = \frac{\phi_{21} - \phi_{02}}{\lambda_{12}}$$

Since $Q_2(t)$ increases no more, φ_{12} (as just noted) cannot saturate. It must be remarked that $Q_{2\,MAX}$ is independent of the infusion rate n and it represents the maximum amount of dye that can be stored in compartment 2.

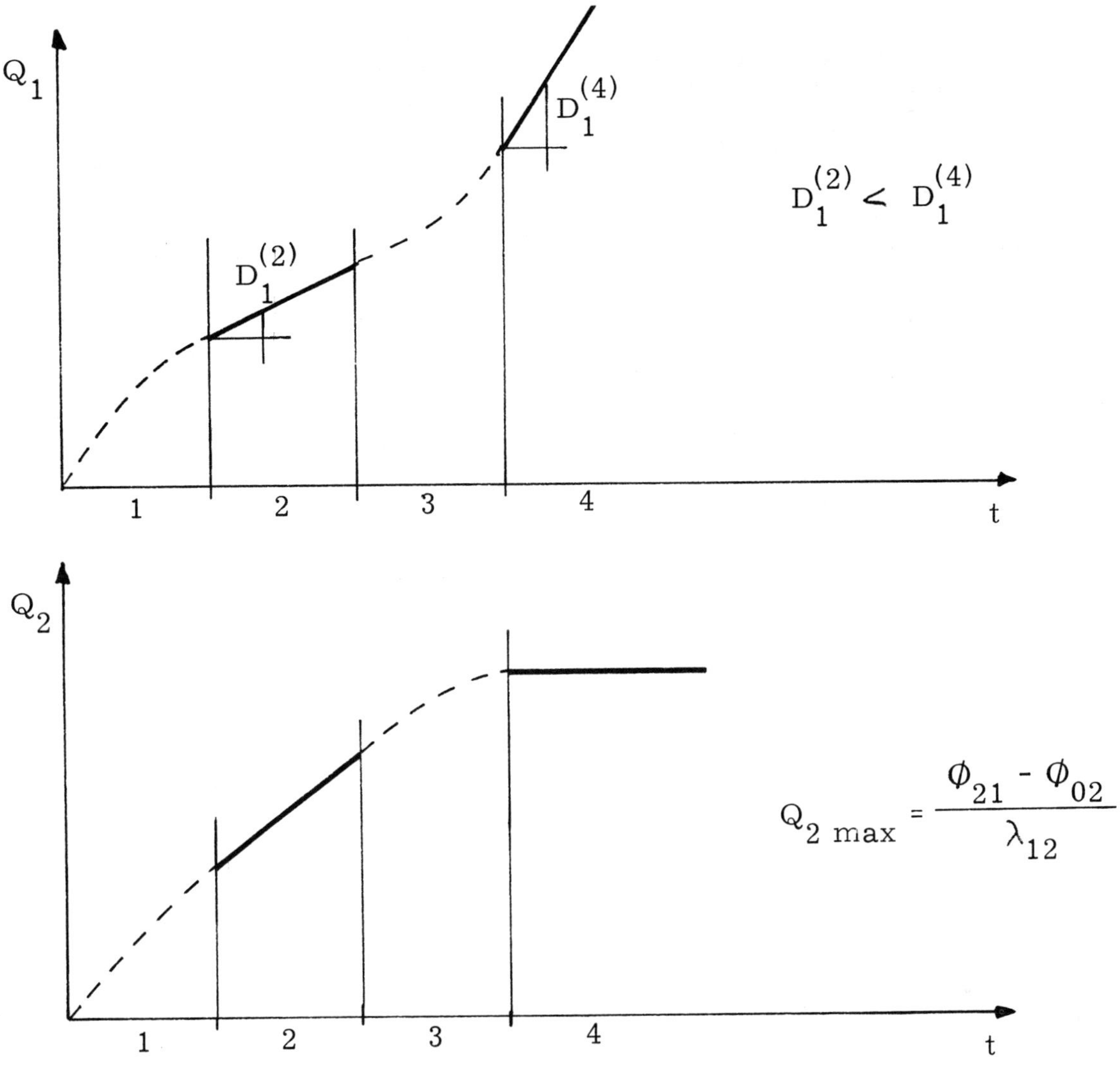

Fig.2 Qualitative behaviors of $Q_1(t)$ and $Q_2(t)$ with constant infusion $n > \phi_{02}$

References

1. VERSCHURE, J.C.M., 'Clinical use of measurements of clearance and maximum capacity of the liver', *Acta Med. Scand.* **142**, 409-419 (1952).
2. WHEELER, H.O., MELTZER, J.J. and BRADLY, S.E., 'Biliary transport and hepatic storage of sulfobromophthalein sodium in the unanesthetized dog, in normal man and in patients with hepatic diseases', *J. Clin. Invest.* **39**, 1131-1143 (1960).
3. PAGLIARDI, E., GIANGRANDI, E. and MOLINO, G., 'Studies on the maximal hepatic sulfobromophthalein removal rate', *Am. J. Dig. Diseases* **8**, 251-266 (1963).
4. SHOENFIELD, L., McGILL, D.B. and FOULK, W.T., 'Studies of sulfobromophthalein sodium (BSP) metabolism in man. III. Demonstration of a transport maximum (T_m) for biliary excretion of BSP', *J. Clin. Invest.* **43**, 1424-1432 (1964).
5. WINKLER, K. and GRAM, C., 'The kinetics of bromsulphthalein elimination during continuous infusion in man', *Acta Med. Scand.* **178**, 439-452 (1965).
6. TYGSTRUP, N. and WINKLER, K., 'Galactose blood clearance as a measure of hepatic blood flow', *Clin. Sci.* **17**, 1-9 (1958).
7. PAUMGARTNER, G., PROBST, P. and KRAINES, R., LEEVY, C.M., 'Kinetics of indocyanine green removal from the blood', *Ann. N.Y. Acad. Sci.* **170**, 134-147 (1970).

8. McINTYRE, N., MULLIGAN, R. and CARSON, E., 'BSP-T_m and S: A critical re-evaluation,' in *The liver Quantitative Aspects of Structure and Functions,* p.417-427, Korger, Basel (1973).
9. WINKLER, K., KEIDING, S. and TYGSTRUP, N., 'Clearance as a quantitative measure of liver function, in *The Liver Quantitative Aspects of Structure and Functions,* 144-155, Korger, Basel (1973).
10. SEGRE, G., 'Kinetics of drugs in the hepato-biliary system, in *Liver and Drugs*, F. Orland and A.M. Jezequel Ed., N.Y. Academic Press (1972).
11. GIANGRANDI, E., GHEMI, F. MOLINO, G.P., CRAVARIO, A., BRUSA, L. and DEFILIPPI, P., 'Il flusso epatico nel soggetto normale stimato con l'uso di Au^{198} e della clearance di piccole dosi di bromosulfonftaleina', *Archivio Italiano delle Malattie dell'Apparato Digerente* **33**, 429-434 (1966).
12. MOLINO, G. and MILANESE, M., 'Structural analysis of compartmental models for the hepatic kinetics of drugs', *J. Lab. Clin. Med.* **85**, 865-878 (1975).
13. MILANESE, M., MOLINO, G. and VILLA, A., 'Pathophysiological information on hepatic functions from standard measurements of the kinetics of BSP-IV IFAC Symp. on System Ident. and Parameter Estim., Tbilisi (1976).
14. MASON, M., HAWLEY, G., and SMITH, A., *Amer. J. Physiol.* 42, p.42 (1948).
15. TALEISNIK, S., *Gastroenterology* **29**, 64 (1955).

Stochastic model of the latency of the conditioned escape response

ANDRZEJ PACUT
Technical University of Warsaw, Poland

Brain and behavioral sciences are among the most rapidly growing branches of biology and medicine. Numerous investigations of the learning processes and their dependence upon pharmacological and surgical treatments are performed. Effects of various parameters of the experimental situations on learning and on performance of the acquired modes of behavior are studied. Mathematical modelling of the learning processes may have important impact on the further theoretical development of brain and behavioral sciences and on their practical applications.

One of the measures of the response is their latency, i.e. the time between the change in experimental situation and the corresponding change in behavior of the organism. Analysis of the systematic changes in response latencies observed in consecutive stages of the experiment is one of the methods of investigation of the learning process.

Models of learning based on the reaction latency were investigated by Estes [1950, 1955], Bush and Mosteller [1955], Bush [1960], Luce [1960], McGill [1963], Sternberg [1963]. However, those authors analysed very simple models of the latency sources; in some of the models the sources of the latencies were not considered. Thus those models are very simple from the mathematical point of view and generally have not biological interpretation. They do not show what is the nature of the latency changes. More fruitful is an approach of Grice [1968, 1971]. Grice treats stimulus as series of 'impulses' which stimulate the threshold mechanism. It is a very simple model but it tries to find sources of latencies.

The aim of this chapter is to propose a model of a conditioned escape reflex learning. The model is based on cybernetic considerations. It has some features of the Grice conception. It describes learning as changes of parameters of response model. The response model describes—in the cybernetic sense—what is happening 'inside' the animal during consecutive stimulations. The model was identified on the basis of experimental data obtained in the Nencki Institute of Experimental Biology, Warsaw.

2. Experimental procedure

The experiment was carried out on 18 cats trained in a cage with a grid floor in order to apply electric shock to the animal's paws and with a bar which terminated the shock when depressed. The experiment consisted of a series of trials (coupled in 10-trial sessions). The trial started with activation of the grid floor (meaning that stimulus was an electric shock) and was terminated by the animal's bar-pressing response. The process of acquisition and consolidation of this response, termed the escape response, was investigated.

The whole experiment consisted of 28 training sessions. During the first 8 sessions sizes of the bar and current intensity were changed according to the cats' behavior. It was done to make the process of incorporation of the bar-pressing response into a set of possible responses of the cat easier. After the 21 session, one half of the cats underwent surgical intervention (prefrontal lobectomy) and the other half rested in their home cages for 10 days' period (control pause).

Latencies of the escape responses, i.e. time durations between switching on the stimulus and the bar-pressing response were measured in the intervals of 0.2 sec. Thus for any animal the experimental data consist of a series of the latencies.

The material and methods were described in detail by Zielinski [1970].

3. General model of the simple learning

The following model of simple learning is proposed. Let us separate two main parts of the process of simple learning. The first one is the response process and the second one is the adaptation process (the term: adaptation process is used in the cybernetic sense and it has different meaning in the biological sciences).

The response process is connected with responding to an external stimulus. The adaptation process is the process of changing parameters of the response process. The response process can be imagined as a connection of the transformation system, the decision system and the executive system (Fig.1).

The transformation system describes transformations of the external stimulus on its way to the decision system. The decision system creates the reaction of the animal and sends the signal to some executive system. The executive system is responsible for producing an external reaction of the subject (for instance: a movement of a limb or salivation). In the experiment described above the executive system elicits the movement of a limb switching off the external stimulus. The output of the transformation system, exciting the decision system will be called the excitatory signal. Any measure of the excitatory signal will be called excitatory strength. Let the output of the decision system, controlling the executive system, be called executive signal.

Then the reaction latency τ is a sum:

$$\tau = \tau_T + \tau_D + \tau_E \qquad (1)$$

where

- τ_T is the transforming time, i.e. the time between the moment of switching on the stimulus and the moment of beginning of work of the decision system,
- τ_D is the decision time, i.e. the time between the moment of exciting the decision system and the moment of taking decision and sending a signal to the executive system,
- τ_E is the execution time, i.e. duration of the execution of that decision by the executive system.

In general all the systems are stochastic systems and the sum (1) is a random variable.

The adaptation process is responsive for changing

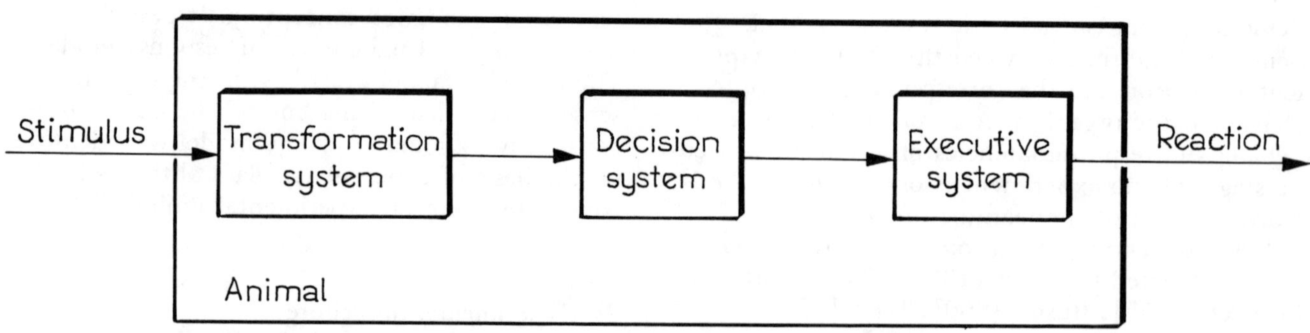

Fig.1 General model of the response process

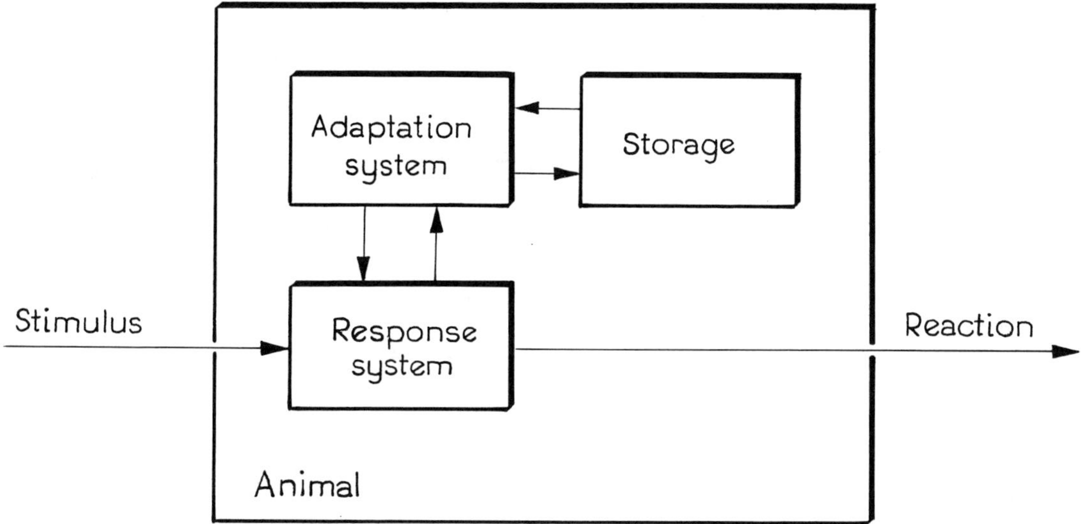

Fig.2 General model of the learning model

parameters of the response process when stimulation is repeated again and again (Fig.2). As a result the parameters of the response process vary during the learning process. Therefore we can write:

$$\tau(n) = \tau_T(n) + \tau_D(n) + \tau_E(n) \tag{2}$$

where the argument n denotes a trial number, i.e. an ordinal number of the repetition of the stimulus.

The foregoing approach to the problem of simple conditioning is very general and is adequate for the very large class of learning processes. The identification of learning process, in the words of this approach, consists of the identification of the transformation system, the decision system, the executive system and the adaptation process. The identification is based on the realization of a stochastic process T:

$$T = \{\tau(n),\ n = 1,2,\ldots,N\} \tag{3}$$

where N is a number of trials in the experiment. The first step towards identification of the whole process is identification of the response process. In the below the model of the response process is built and the parameters of this process are identified for the experiment described in Part 2.

4. The model of the response process

The model of the response process is a part of the model of the learning process. This model allows data transformation from a form of latency times to the form of the model parameters which have a simple biological interpretation. The model of the response process belonging to the class described above is constructed. Three parts of the response process are built as follows:

The transformation system (Fig.3) is represented by the linear first-order differential equation disturbed by the stochastic Wiener process. The state of the system in the moment of switching on the stimulus is a Gaussian random variable. Hence the excitatory strength is given by the stochastic differential Ito equation as follows

$$\begin{aligned} dH(t) &= [k \cdot s(t) - a \cdot H(t)]\,dt + dw(t) \\ H(0) &= H_0 \end{aligned} \tag{4}$$

where

t is a time which is measured from the moment of switching on the stimulus,

$H(t)$ is the excitatory strength,

$w(t)$ is a stochastic Wiener process with $E\,w(t) = 0$ and $D^2 dw(t) = \sigma_w^2\,dt$,

$dw(t)$ is a stochastic differential of the process $w(t)$ (see Gikhman, Skorokhod [1969]),

H_0 is a Gaussian random variable independent of the process $w(t)$ with $E\,H_0 = m_0$, $D^2 H_0 = \sigma_0^2$

$s(t)$ represents the stimulus strength and for given experimental data there is a step function, then

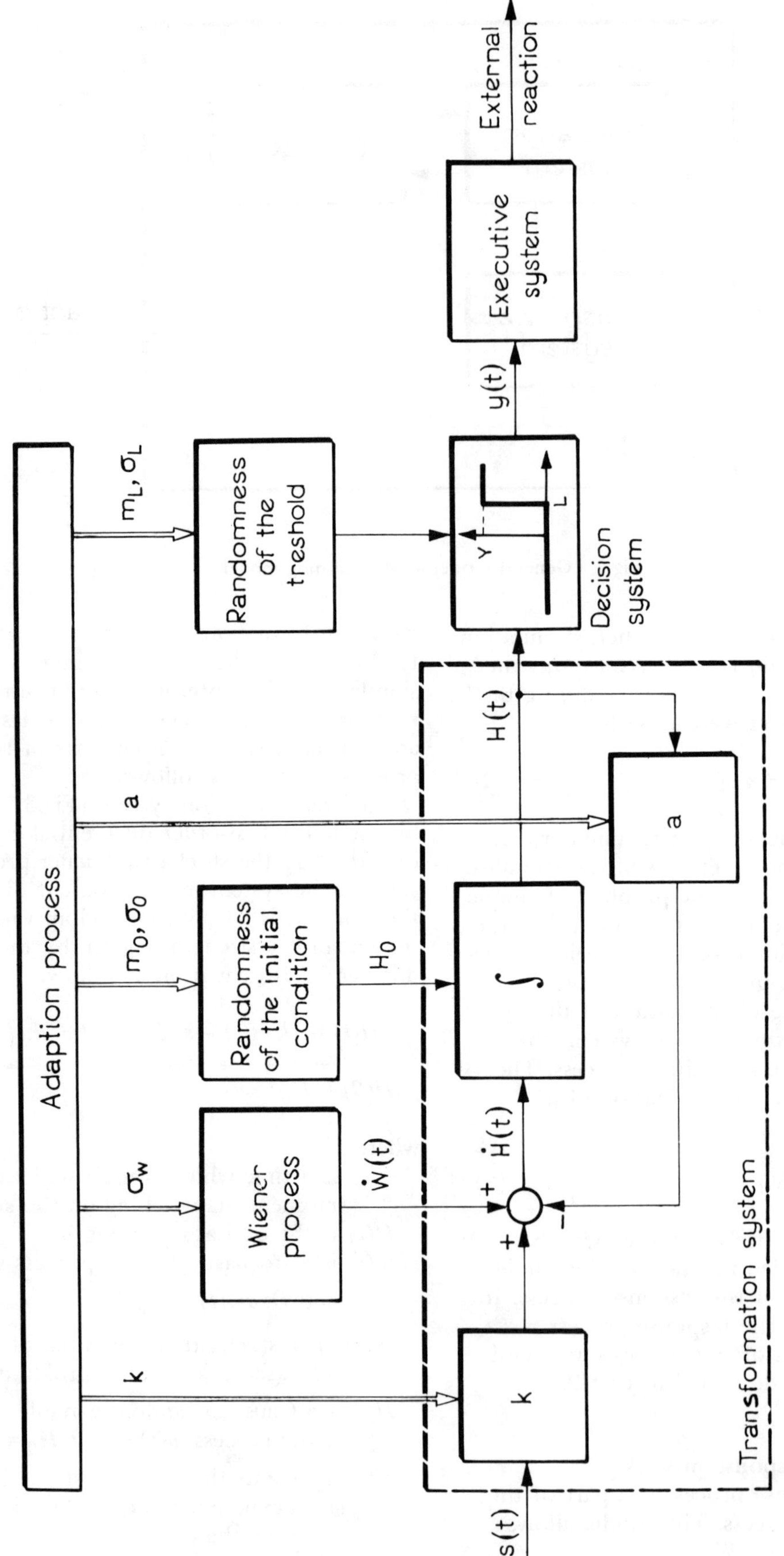

Fig.3 Block-diagram of the response process: ——▶ signals connected with reacting (functions of the time in a trial) ══▷ changes of the model parameters (functions of the ordinal number of the trial)

$$s(t) = \begin{cases} 0 & \text{for } t < 0 \\ s & \text{for } t \geqslant 0 \end{cases} \qquad (5)$$

k is a static gain of the stimulus,

a is a dynamic gain of the stimulus; $a > 0$,

and for the meaning of the symbols E, D^2 see the list of symbols.

Notice that the parameters k, a, σ_w, m_0, σ_0, and thus the function $H(t)$ are dependent on the number of the trial as the effect of the adaptation. Therefore there are functions of n but we drop the argument n for simplifying notation.

Let the decision system be simple random threshold element. Thus (Fig.3):

$$y(t) = \begin{cases} 0 & \text{for } t : H(t) < L \\ Y & \text{for } t : H(t) \geqslant L \end{cases} \qquad (6)$$

where

$y(t)$ is the executive signal,

L is a random Gaussian variable independent of the process $w(t)$ and random variable H_0, with $E\,L = m_L$ and $D^2L = \sigma_L^2$.

Y is constant value of the executive signal.

Notice that the parameters m_L, σ_L and thus the function $y(t)$ are dependent on an ordinal number of the trial n. We drop the argument n for simplifying notation. Assume that if the value of the executive signal is equal to 0 then the executive system does not operate and if the value of the executive signal is equal to Y then the reaction of an animal occurs (i.e. it causes movement of the animal leg pressing the bar in the experiment considered).

It is seen that the decision time τ_D is equal to zero for the random threshold element. The execution time τ_E is not equal to zero in general but we assume that τ_E is short in comparison with the transformation time τ_T. Consequently the latency τ of an animal reaction is approximately equal to τ_T. Therefore from (4) and (6) results:

$$\tau = \inf_{t \epsilon <0,\infty)} \{t : H(t) - L \geqslant 0\} \qquad (7)$$

We shall use the notation $\tau = \infty$ if the set $\{t : H(t) - L \geqslant 0\}$ is empty. Thus the latency of the animal reaction may be calculated through solving a first passage problem for the stochastic process $H(t) - L$. Let us summarize properties of the stochastic process $H(t)$ in two lemmas.

Lemma 1 $\sigma_w = 0$ case

For the case $\sigma_w = 0$ (which means that a Wiener noise does not act) the excitatory strength is a stochastic Gaussian process given by the following equation:

$$H(t) = (H_0 - k \cdot s \cdot a^{-1}) \exp(-a \cdot t) + k \cdot s \cdot a^{-1} \qquad t \geqslant 0 \qquad (8)$$

Mean and a covariance function are given by the equations:

$$E\,H(t) = (m_0 - k \cdot s \cdot a^{-1}) \exp(-a \cdot t) + k \cdot s \cdot a^{-1} \qquad t \geqslant 0 \qquad (9)$$

$$\text{cov}_H(t_1,t_2) = \sigma_0^2 \cdot \exp[-a(t_1 + t_2)] \qquad t_1,t_2 \geqslant 0 \qquad (10)$$

(see the list of symbols for the meaning of cov).

Lemma 2 $\sigma_w \neq 0$ case

For the case $\sigma_w \neq 0$ the excitatory strength is a stochastic diffusion process with drift coefficient:

$$a(t,x) = k \cdot s - a \cdot x \qquad (11)$$

and diffusion coefficient

$$\sigma(t,x) = \sigma_w^2 \qquad (12)$$

Thus the continuous solution of the equation (4) has the form:

$$H(t) = [H_0 - k\,s\,a^{-1} + \int_0^t \exp(a\,r)\,dw(r)] \times \times \exp(-a\,t) + k\,s\,a^{-1} \qquad (13)$$

and the mean and covariance function are given by the following formulas:

$$E\,H(t) = (m_0 - k\,s\,a^{-1}) \exp(-a\,t) + k\,s\,a^{-1} \qquad t \geqslant 0 \qquad (14)$$

$$\text{cov}_H(t_1, t_2) = \sigma_0^2 \exp(-a(t_1 + t_2)) - 0.5\,\sigma_w^2\,a^{-1} [\exp(-a(t_1 + t_2)) - \exp(-a|t_1 - t_2|)] \qquad t_1, t_2 \geqslant 0 \qquad (15)$$

Proofs of the lemmas 1 and 2 are presented in Pacut [1977].

Methods of finding the latency distribution are based on the lemmas 1 and 2 and summarized in the following theorems:

Theorem 1 $\sigma_w = 0$ case
For the stochastic process (8) and the random threshold L given by the equality (6) the latency is random variable given by the following equality:

$$\tau = \begin{cases} 0 & \text{if } H_0 \geqslant L \\ a^{-1} \ln[(k s a^{-1} - H_0)(k s a^{-1} - L)^{-1}] & \text{if } H_0 < L < k s a^{-1} \\ \infty & \text{in other cases} \end{cases} \quad (16)$$

The distribution of the random variable τ is given by the following:

$$P(\tau < 0) = 0$$

$$P(\tau = 0) = \Phi((m_0 - m_L)(\sigma_0^2 + \sigma_L^2)^{-0.5})$$

$$f_T(t) = (4\pi)^{-1}\ a \sin(2\varphi) \exp(-0.5) \cdot$$

$$\cdot\ [1 + (2\pi)^{0.5}\ T\ \phi(T) \exp(0.5\ T^2)]\ t > 0 \quad (17)$$

where

$$\Delta = \Delta_0^2 + \Delta_L^2,\ \Delta > 0$$

$$\Delta_0 = (k\ s\ a^{-1} - m_0)\ \sigma_0^{-1}$$

$$\Delta_L = (k\ s\ a^{-1} - m_L)\ \sigma_L^{-1}$$

$$T = \Delta \cos(\varphi - \psi)$$

$$\varphi = \arctan(\sigma_0 \sigma_L^{-1} \exp(-a\ t))$$

$$\psi = \arctan(\Delta_0^{-1}\ \Delta_L) \quad (18)$$

and for the meaning of P, Φ, f see the list of symbols.

Theorem 2 $\sigma_w \neq 0$ case
For the stochastic process (13) and the random threshold L given by the equality (6) the latency τ is given by the Laplace transform of the conditional distribution function

$$\tilde{F}_\tau(u) = \mathcal{L}P(\tau < t | L = l,\ H_0 = h) \quad (19)$$

as follows:

$$\tilde{F}_\tau(u) = u^{-1} \exp[0.5 \cdot a \cdot \sigma_w^{-2}(h^2 - l^2 - 2 \cdot k \cdot s \cdot a^{-1} \times$$

$$\times\ (h - l))] \cdot$$

$$\cdot\ D_{-u/a}(-(2a)^{0.5} \sigma_w^{-1}(h - k\ s\ a^{-1})) \cdot$$

$$\cdot\ D_{-u/a}(-(2a)^{0.5} \sigma_w^{-1}(l - k\ s\ a^{-1})) \quad (20)$$

where D is a symbol of Weber function (Whittaker, Watson [1963]).

Proofs of the theorems 1 and 2 are presented in Pacut [1977]. The proof of the theorem 2 is based on a theorem of Feller [1954] and results of Capocelli and Ricciardi [1971]. The theorems 1 and 2 give a complete information about the latency distribution. Unfortunately the formulas (17) and (20) are rather complicated and an identification of the parameters based on the experimental latency distribution is a very hard problem. Thus in order to identify the parameters of the model three special cases of randomness of the model are investigated:

a/RT case (random threshold case): $\sigma_0 = 0$, $\sigma_w = 0$, $\sigma_L \neq 0$. This is a case of a negligible effect of randomness of the initial conditions and Wiener noise, and a strong effect of randomness of the threshold.
b/RIC case (random initial condition case): $\sigma_0 \neq 0$, $\sigma_w = 0$, $\sigma_L = 0$. This is a case of a negligible effect of randomness of the threshold and the Wiener noise, and strong effect of randomness of the initial conditions.
c/WN case (Wiener noise case): $\sigma_0 = 0$, $\sigma_w \neq 0$, $\sigma_L = 0$. This is a case of a negligible effect of randomness of the threshold and the initial condition and a strong effect of the Wiener noise.

Three corollaries can be stated and utilized in a numerical analysis of the experimental data. The first one and the second one follow theorem 1 and assumptions *a)* or *b)*, respectively. The third one describes another approach to the first passage problem, which is based on Kolmogorov equations. This approach is simpler than that one utilizing Laplace transformation.

RT corollary
For the *RT* case if $m_0 < k\, s\, a^{-1}$ (which means that increasing excitatory strength when stimulating has been assumed) the distribution function of the latency distribution is given by an equality:

$$F_T(t) = \begin{cases} 0 & \text{for } t \leqslant 0 \\ \phi((k\, s\, a^{-1} - m_L)\sigma_L^{-1} - (k\, s\, a^{-1} - m_0)\sigma_L^{-1} \exp(-a\, t)) & \text{for } t > 0 \end{cases} \tag{21}$$

RIC corollary
For RIC case and if $m_L < k s\, a^{-1}$ (which means that the stimulus strength is sufficiently high to evoke an animal reaction) the distribution function of the latency distribution is given by the equality:

$$F_T(t) = \begin{cases} 0 & \text{for } t \leqslant 0 \\ \phi((k\, s\, a^{-1} - m_0)\sigma_0^{-1} + (k\, s\, a^{-1} - m_L) \times \sigma_0^{-1} \exp(a\, t)) & \text{for } t > 0 \end{cases} \tag{22}$$

WN corollary
For the WN case the distribution function of the latency is given by the limit

$$F_T(t) = 1 - \lim_{n\to\infty} u_n(t, (m_0 - m_L)\sigma_w^{-1}) \tag{23}$$

where $u_n(t, x)$ is generated by the partial parabolic differential equation as follows

$$-\frac{\partial}{\partial t} u(t, x) + a[(k\, s\, a^{-1} - m_L)\sigma_w^{-1} - x]\frac{\partial}{\partial x} u(t, x) + 0.5\frac{\partial^2}{\partial x^2} u(t, x) \tag{24}$$

with conditions:

$$\lim_{t\to 0} u(t, x) = f_n(t, x) \quad \lim_{x\to -\infty} u(t, x) = 0$$

$$\lim_{x\to 0} u(t, x) = 0 \tag{25}$$

where f_n is a twice continuously differentiable function on $(-\infty, 0)$ satisfying conditions:

$$\lim_{x\to 0} f_n(x) = \lim_{x\to 0} \dot{f}_n(x) = \lim_{x\to 0} \ddot{f}_n(x) = 0$$

$$\lim_{x\to -\infty} f_n(x) = \lim_{x\to -\infty} \dot{f}_n(x) = \lim_{x\to -\infty} \ddot{f}_n(x) = 0 \tag{26}$$

$$\lim_{n\to\infty} f_n(x) = \chi_{(-\infty,\, 0\rangle}(x)$$

Proofs of the RT and RIC corollaries follow directly from Theorem 1. Proof of the WN corollary bases on the theorem of Gickman and Skorokhod [1969].

5. Identification of the model parameters

The identification of the parameters of the RT, RIC, and WN model is based on the corollaries RT, RIC, WN. The method utilizes a maximum likelihood technique. The maximum likelihood estimators of the parameters of the model RT and RIC was obtained. A method of searching minimum of x^2 statistics was used for the WN model (see Cramer [1946]). The method of the identification of the parameters may be summarized in the following lemmas:

RT lemma
Define new parameters q_1 and q_2 putting

$$\begin{aligned} q_1 &= (k\, s\, a^{-1} - m_L)\sigma_L^{-1} \\ q_2 &= (k\, s\, a^{-1} - m_0)\sigma_L^{-1} \end{aligned} \tag{27}$$

The estimators of the parameters a, q_1 and q_2 are given by the approximate equations:

$$\begin{aligned} \hat{a}^{-1} &= \bar{t} - \langle t \exp(-\hat{a}\, t), \exp(-\hat{a}\, t)\rangle \cdot \|\exp(-\hat{a}\, t)\|^{-2} \\ \hat{q}_1 &= \overline{\exp(-\hat{a}\, t)}\, \|\exp(-\hat{a}\, t)\|^{-1} \\ \hat{q}_2 &= \|\exp - \hat{a}\, t\|^{-1} \end{aligned} \tag{28}$$

(for the meaning of the symbols see list of symbols).

RIC lemma
Define new parameters r_1 and r_2 putting

$$\begin{aligned} r_1 &= (k\, s\, a^{-1} - m_0)\sigma_0^{-1} \\ r_2 &= (k\, s\, a^{-1} - m_L)\sigma_0^{-1} \end{aligned} \tag{29}$$

The estimators of the parameters a, r_1 and r_2 are given by the approximate equations:

$$\hat{a}^{-1} = -\bar{t} + \langle t \exp(\hat{a}\, t), \exp(\hat{a}\, t)\rangle \, \| \exp(+\hat{a}\, t)\|^{-2}$$

$$\hat{r}_1 = -\overline{\exp(\hat{a}\, t)}\, \| \exp(\hat{a}\, t)\|^{-1} \qquad (30)$$

$$\hat{r}_2 = -\| \exp(\hat{a}\, t)\|^{-1}$$

For the detailed proofs the reader is referred to Pacut [1975].

For the WN case a direct searching of maximum of the χ^2 statistics was carried out. The method follows directly WN corollary and consists in a numerical solving the equation (24) with conditions (25) and (26) and then changing parameters of this equation to minimize x^2 criterion. A detailed description may be found in Pacut [1975].

The numerical analysis of the experimental data has been done on CDC-1370 computer. All the results may be found in Pacut [1975]. The especially interesting results have been obtained for the RT case. We shall outline these results.

There is a freedom in setting zero and unity of the excitatory strength. Consequently the parameters k, a, m_0, m_L, σ_L cannot be determined uniquely for the RT case. In order to define scale values for the RT case let us put

$$m_L = 0 \qquad \sigma_L = 1 \qquad (31)$$

Therefore

$$\hat{k} = \hat{a} \cdot \hat{q}_1 \cdot s^{-1}$$

$$\hat{m}_0 = \hat{q}_1 - \hat{q}_2 \qquad (32)$$

and the estimators of the parameters a, k and m_0 may be found from (28) and (32). These parameters were determined at the few stages of learning. The average results are given on Fig.4. The estimated values for the static gain k, the mean initial value m_0 and the dynamic gain a are plotted vs. ordinal number of the trial for the RT case.

The dynamic gain increases approximately linearly. The mean initial value of the excitatory strength m_0 decreases. The static gain k increases during learning, but it has a negative jump while operation or control pause.

There is a very interesting interpretation of the results. The static gain increases because an animal adapts to the stimulus. After a long training even a weak stimulus having informative meaning is isolated from the other signals. The speed of the increasing of the excitatory strength depends on the dynamic gain a. The effect of growing up the parameter a is faster increasing of excitatory strength and faster reaching of the asymptote by it. As a result a probability density of the latency distribution shifts left during the learning process. An interesting thing is decreasing of the mean initial value of the excitatory strength. The result of this phenomenon is counteracting decreasing of the latency during the learning process. We can say that a sureness of decision making grows up during the learning process which results in more quiet reacting. Thus changes in the two parameters cause decrease of the latencies changes in third one cause increasing of the latencies.

The results seem to be of interest for biologists. Work on modelling of the response mechanism and learning process are continued.

List of main symbols used

- $P(Y)$ probability of the event Y
- f_X probability density function of the random variable X
- F_X distribution function of the random variable X
- E expectancy operator
- D^2 variance operator
- cov_X covariance function of the stochastic process X
- $\mathcal{L}$ symbol of the Laplace transformation
- Φ distribution function of the standardized Gaussian random variable

  i.e. $\phi(x) = (2\pi)^{-0.5} \int_{-\infty}^{x} \exp(-0.5\, x^2)\, dx$

- χ_A characteristic function of A set, i.e.

$$\chi_A(x) = \begin{cases} 0 & \text{if } x \notin A \\ 1 & \text{if } x \in A \end{cases}$$

- χ^2 Pearson's statistics (see Cramer [1946]).
- $\overline{f(t)}$ the mean f based on the experimental results, i.e. $\overline{f(t)} = \frac{1}{n} \Sigma f(t_i)$

  where t_i are latencies and n is a number of the results.
- $\langle f(t), g(t)\rangle$ the covariance of f and g based on the experimental results, i.e.

$$\langle f(t), g(t)\rangle = \overline{(f(t) - \overline{f(t)}).(g(t) - \overline{g(t)})}$$

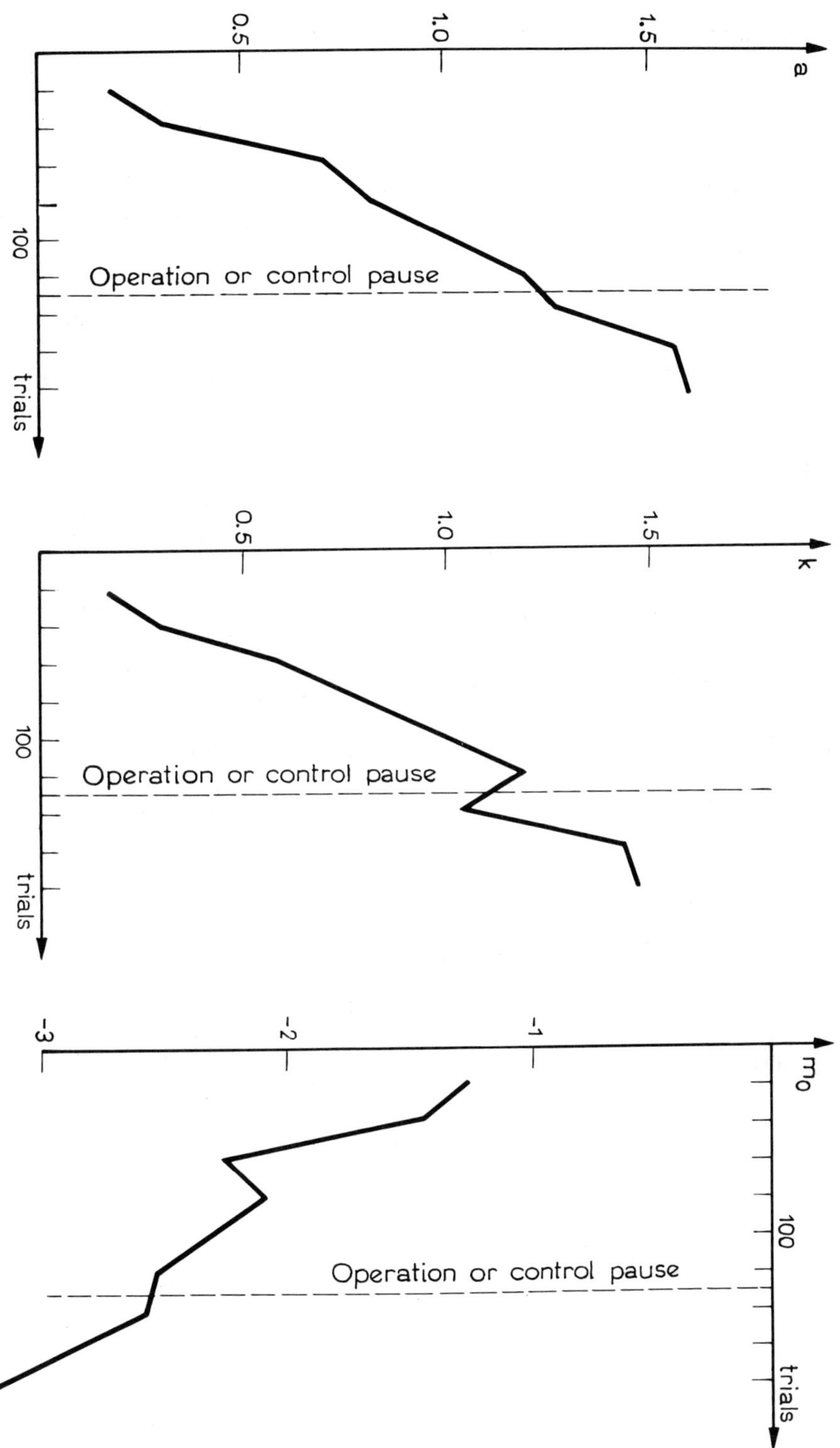

Fig.4 Changes of the model RT parameters during learning

$\|f(t)\|^2$ the variance of f based on the experimental results,

i.e. $\|f(t)\|^2 = \langle f(t), f(t)\rangle$

$\tilde{(\cdot)}$ superscript $\tilde{}$ denotes the Laplace transform of $(\cdot)$

$\hat{(\cdot)}$ superscript ^ denotes an estimate of $(\cdot)$

References

BUSH, R.R. and MOSTELLER, F., *Stochastic model for learning,* New York, Wiley (1955).

BUSH, R.R., 'A survey of mathematical learning theory', in R.D. Luce (ed.) *Developments in mathematical psychology,* Glencoe, Free Press, 122-165 (1960).

CAPOCELLI, R.M. and RICCIARDI, K.M., 'Diffusion approximation and first passage time problem for a model neuron', *Kybernetik* **8**, 214-223 (1971).

CRAMER, H., *Mathematical methods of statistics,* Princeton University Press (1946).

DARLING, D.A. and SIEGERT, A.J.F., 'The first passage problem for a continuous Markov process', *Ann. Math. Statist.* **24**, 625-639 (1953).

ESTES, W.K., 'Toward a statistical theory of learning', *Psychological Review* 57, 94-107 (1950).

ESTES, W.K., 'Statistical theory of distributional phenomena in learning', *Psychological Review* 62, 369-377 (1955).

FELLER, W., 'Diffusion processes in one dimension', *Trans. AMS* 77, 1-31 (1954).

GICKMAN, I.I. and SKOROKHOD, A.W., *Introduction to the theory of random processes,* Philadelphia, Saunders Comp. (1969).

GRICE, G.R., 'Srimulus intensity and response evocation', *Psychological Review* 75, 359-373 (1968).

GRICE, G.R., 'Conditioning and decision theory of response evocation', in G.H. Bower (ed.) *Psychology of learning and motivation, Vol. 5,* Acad. Press (1971).

LUCE, R.D., 'Response latencies and probabilities', in K.J. Arrow, S. Karlin, P. Suppes (eds.), *Mathematical methods in the social sciences. Proc. of the First Stanford Symposium 1959*, Stanford: University Press, 298-311 (1960).

McGILL, W.J., 'Stochastic latency mechanism', in R.D. Luce, R.R. Bush, E. Galanter (eds.), *Handbook of mathematical psychology, Vol.1,* New York: Wiley, 309-360 (1963).

PACUT, A., *Mathematical model of some learning process* (in Polish), unpublished doctoral dissertation, Technical University of Warsaw (1975).

PACUT, A., *Mathematical properties of the linear stochastic model of the latency mechanism* (in preparation) (1977).

STERNBERG, S., 'Stochastic learning theory', in R.D. Luce, R.R. Bush, E. Galanter (eds.), *Handbook of mathematical psychology Vol.II,* New York: Wiley, 1-120 (1963).

WHITTAKER, E.T. and WATSON, G.N., *A course of modern analysis,* Cambridge University Press (1963).

ZIELINSKI, K., 'Retention of the escape reflex after prefrontal lobectomy in cats', *Acta Neurobiologicae Exp.* **30**, 43-57 (1970).

HYDRA simple nervous system and behavior

C. TADDEI-FERRETTI, L. CORDELLA and S. CHILLEMI
Laboratorio di Cibernetica del CNR,
Napoli, Italy

Introduction
The body of the Coelenterate *Hydra* can be schematically described as two layers of cells surrounding the enteron, which communicates with the exterior at the distal end (hypostome or head) through the mouth surrounded by tentacles. The outer layer is composed essentially of epitheliomuscular cells, not yet phylogenetically differentiated into distinct epithelial and muscular cells, and is separated from the inner gastrodermal layer by the acellular mesoglea. The NS consists of bipolar and multipolar ganglion cells, arranged as a net above the base of epitheliomuscular cells or scattered between gastrodermal cells, and of sensory cells, each one of which is apically specialized as a modified cilium and terminates at the opposite side on a ganglion cell. Polarized synapses have been observed [1]. No ganglia are present, but a gradient of the neuron density exists along the animal's long axis, with higher concentration on the head side [2]. The NS controls both the growth and the behavior of the animal: neurohormones were found, which are responsible for the head determination [3.4]; in addition, the hypostomal nervous net has been claimed to trigger the periodic body contractions [5].

Contractions of the longitudinally arranged myofibrils of the epitheliomuscular cells are responsible for the body shortening, which is accomplished in a step-wise way with partial contractions spaced *c.* 1-10 sec one upon each other [6]. To each single body contraction a pulse (here called a CP) of potential variation of some mV is associated, which can easily be recorded [7]. Each CP represents the activity of many elements, is conducted by the ectoderm [8], preferentially along longitudinal pathways [9] and is probably triggered by the nervous net [5,9]. A train of CP (here called CPT) is synchronous with a complete body contraction [6,10]. Similarly, rhythmic potentials (RP) of 0.2 mV amplitude, spaced *c.* 5 sec, are associated with the body elongation [11]. Tentacle pulses (TP) are associated with tentacle contractions [12]. CP's pacemakers, located in the head zone, inhibit RP's pacemakers located in the feet zone and are inhibited by the same RP's pacemakers and excited by TP's pacemakers, which excite one another [13,14,7,11,15]. The CPT's occurrence is "irregular...although regularity of occurrence suggests that rhythmicity is a valid behavior pattern for certain individuals at certain times" [5]. Their repetition period is of the order of minutes. A light pulse affects CPT's occurrence in two different ways, either immediately inhibiting a CPT in progress and advancing in time the next one [5] or provoking a new CPT within some minutes if given during the inter-

CPT interval [16]. A momentaneous CPT mean frequency decrease is caused by an abrupt light-level increase [5,17]. Repetitive stimulation of 2 min light and 2 min darkness locks the CPT's activity to a 4 min repetition period [18,19].

This being the up-to-date state of knowledge, it seemed interesting to investigate the CPT's behavior either in undisturbed or in various stimulation conditions, in order to clarify tentatively the triggering role of the *Hydra* nervous net. Materials and methods are reported elsewhere [20]. CPT's frequency f_H during a time period is here defined as the reciprocal of the average value

$$P_H = \sum_{i=1}^{n-1} \frac{P_{Hi}}{n-1}$$

of the time intervals P_{Hi} between the beginning of two adjacent CPT's in that period. f_s and P_s are the frequency and the inter-stimulus interval for a repetitive stimulation.

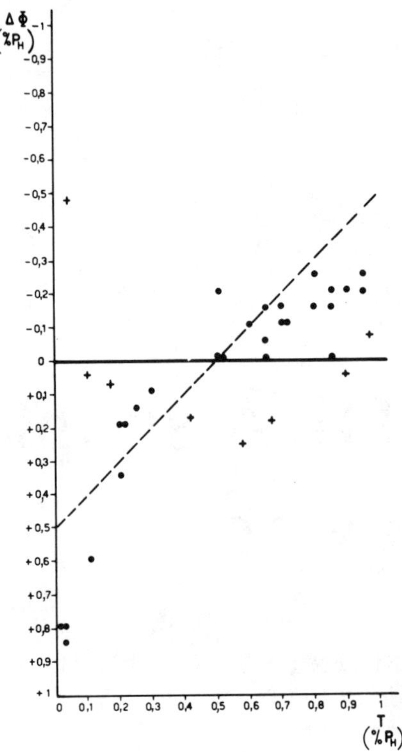

Fig.1 $\Delta\phi$ (advances +, delays –) versus T. ▬▬ subthreshold stimulus. ‒ ‒ ‒ fixed time of response, i.e. $\Delta\phi = -T + P_H/2$. . . . . . 3000 lux, 10s on pulse. ++++ 3000 lux, 10s off pulse. All times are expressed as percentages of P_H.

Results

The irregularity of CPT's occurrence markedly decreases during some P_{Hi} after a light-level variation or a light pulse.

A momentaneous f_H variation is caused by an abrupt light change and its sign depends on the polarity of the light change: f_H increase corresponds to light decrease and *vice-versa*. The higher the relative light-level variation, the higher the f_H variation. For a given light step, the first CPT after it occurs after a time independent of the phase of P_H at which the step occured, if the step was negative; if it was positive, the interval between the step and the next CPT increases with the interval between the last CPT and the step.

As concerns the effect of a single light or darkness pulse on animals which are in the condition of a regular CPT's occurrence, the interval between the time of application of the pulse and the time of occurrence of the first CP of the next CPT is not a constant but is a function both of the polarity of the pulse and of the time $T < P_H$ of application of the pulse after the first CP of the past CPT, i.e. the dependence of the phase shift $\Delta\phi$ (+, advance; –, delay) of the next CPT after the pulse as a function of T is not expressed by $\Delta\phi = -T + A$ (where A = constant $\leqslant P_H$) (Fig. 1).

Note that the T value at which + $\Delta\phi$ is maximum in the case of a darkness pulse is shifted 180° with respect to the corresponding value in the case of a light pulse, and that the first part of the curve for a darkness pulse has a negative slope. For low values of T, the pulse is arriving during a CPT in progress: in such a case, in addition to the fact that the next CPT is shifted, the CPT in progress is immediately stopped by the positive or negative pulse; it is immediately stopped also simply by a positive or negative lightstep.

With stimulation consisting of repetitive light pulses, very rapidly the CPT's are entrained with the stimulation in a ratio 1:1 if $|f_H - f_s|$ is not too high. If it is too high, entrainment is still possible, within certain ranges, but more than 1 CPT alternate with 1 stimulus if $f_s < f_H$, or 1 CPT alternate with more than 1 stimulus if $f_s > f_H$. In all cases a definite constant phase relation between CPT's and stimuli is established, which strictly depends on $(f_H - f_s)$ (Fig. 2), and, for a fixed $(f_H - f_s)$ value, weakly depends on stimulation intensity. After the end of the stimulation, the CPT's activity remains locked to f_s for a time of the order of 5-40P_s, which means also that P_{Hi} variations during such period of time are much smaller than before the stimulation began.

With repetitive darkness pulses, a phase lock between CPT's and stimuli is achieved for a period, then

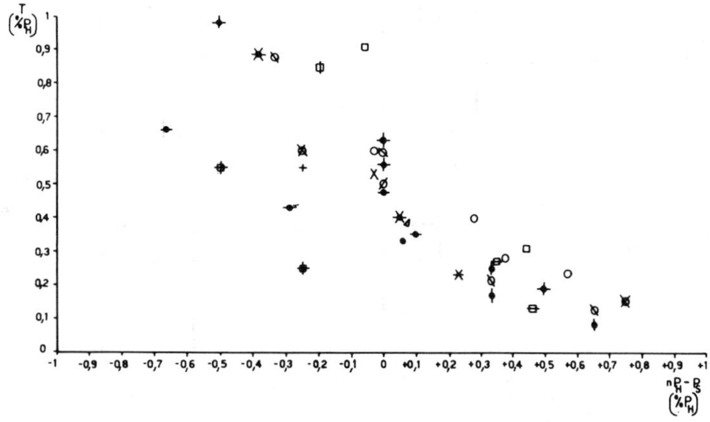

Fig.2 T versus $nP_H - P_s$, for a group of animals. n = number of CPT's occurring during 1 P_s. All times are expressed as percentages of P_H.

the CPT's series slowly shifts along the stimuli series till phase lock is re-obtained, and so on.

In preparations consisting in ½ or $^1/_3$ of an animal obtained with transversal cuts, P_{Hi} deviations from P_H are reduced; in feetless preparations f_H is slightly higher than in the whole animal (conversely, f_H is much lower in headless preparations [5]) and the light-sensitivity threshold, as it concerns the effect over CPT's, is lower. P_H at 7°C is 4 times P_H at 24°C.

Discussion

The described CPT's behavior, especially in the way it concerns the variable sensitivity to light stimuli along the inter-CPT's interval, is typical of an oscillator: the state of the system involved in CPT's triggering undergoes cyclical changes, to which changes of light sensitivity are linked and at a certain phase of which CPT's arise. Thus the CPT's behavior can be analyzed using concepts familiar to the study of biological rhythms, and which have been mainly employed in the analysis of circadian rhythms. The reported results lead one to interpret the triggering system of the CPT's as having the behavior proper to a self-limiting and self-sustaining oscillator [21] that has a repetition frequency affected by stimulation in such a way that it: *a*) varies in its instantaneous value by the application of a single stimulus as a function of the oscillator's phase of sensitivity (see also [22]); *b*) is entrained by a repetitive stimulation the frequency of which falls into definite ranges (see also [23]) that become wider the more intense the stimuli are; the phase relation between the entrained and the stimulating rhythms depend on the difference between stimulating and natural periods.

The different shapes of the two curves of Fig. 1 expressing repectively the shifting effect of a single light or darkness pulse, have been predicted by a feed-back model [24,25] which accounts for many other features of biological rhythms: in such a model, an oscillating variable, which reflects the measurable biological rhythm, oscillates around a reference value and its difference from the reference value constitutes an error signal which, transformed in a non-linear block, delayed and smoothed, affects the oscillating variable itself through an integrator. The behavior of an oscillator similar to that described by our results could arise also from a population of oscillators coupled in an inhibitory way [26]. In this case, the frequency of the oscillator should result lower than those of the single oscillators; such frequency should change after a steady environmental conditions' variation; the effect of a single positive pulse stimulus should be similar to that expressed by the light-pulse phase respose curve of Fig. 1; a random variation of the frequency should be intrinsic to the system and should increase with the difference between the values of the parameters characterizing the single oscillators; the frequency of oscillation of the system should increase in consequence both of the reduction of the number of the oscillating elements or of the reduction of the effect of the coupling coefficients between them.

The hypothesis that the *Hydra* nervous net plays a role in the CPT's behavior [5] mediating the contraction response [27], confirmed by the finding of covert pulses not associated with contractions but probably representing the output of pacemaking neurons [28], was reproposed at least for the CPT's triggering at the neuro-epitheliomuscular level [9].

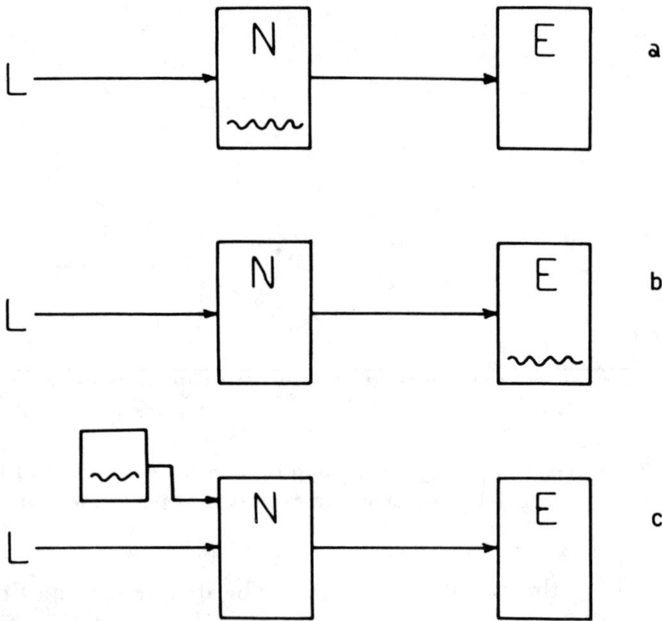

Fig.3 Three possible ways of CPT's generation. L = light, N = nervous net, E = epitheliomuscular cells, ∿∿ = oscillator which generates the CPT's rhythm and which is neuronal in a), chemical in b) and c).

This could lead to a tentative identification of the single oscillators interacting in an inhibitory way [26] with elements of the nervous net (Fig. 3*a*). In order to test such hypothesis, we reconsidered the reported experiments on headless preparations [5] (which contain about 35% of the total neurons [2]) and we performed our experiments with feetless preparations and other preparations of different body zones containing different percentages of neurons, with the aim of varying the number of single oscillators and possibly the percentage of oscillators with specific coupling coefficients and frequencies, thus varying [26] the characteristics of the system of oscillators. As the frequency of CPT's occurrence does not increase simply with the reduction of the dimension of the *Hydra* piece, no matter which zone of the animal is used, but such frequency slightly increases if the feet are removed and highly decreases if the head is removed, contrary to what occurs, for example, with *Aplysia* retinal neurons [29], the hypothesis of neural oscillators interacting in an inhibitory way cannot be directly applied to the *Hydra* nervous net, unless one takes into account more complex interactions between single oscillators, which should reflect the actual situation of excitation of TP's pacemakers among themselves and upon CP's pacemakers, and of mutual inhibition between CP's and RP's pacemakers.

Two alternate hypotheses could be considered at this point. According to the first one, the CPT's rhythmicity is an intrinsic property of the epitheliomuscular cells [30 p.262] based upon biochemical oscillators (Fig. 3*b*): biochemical oscillators have been found which show phase-dependence of the response to phase-shifting stimuli [31,32,33] analogous to that shown by *Hydra*, no temperature compensation as we found in *Hydra* and low frequency (the P_H length is of the order of minutes). Only the inhibition of *Hydra* contractions reported after NS vital staining [34] might be considered to weaken this hypothesis. According to the second hypothesis, the CPT's rhythmicity originates by the secondary activity of the NS, which has the function of being a transducer of primary biochemical oscillations acting as depolarizing input to the NS itself [35 p.46] (Fig. 3*c*). In both hypotheses the NS mediates the response of the oscillating overt system to external stimuli, according to the quoted findings [5,27,9].

Summary

The sensitivity to light stimuli of the system involved in the periodic triggering of the pulse trains associated with *Hydra* body contractions varies during the inter-trains interval. The oscillator nature of the trains' triggering system and the possible role of the NS in the trains' triggering and in modulation of light effects on trains are discussed.

Acknowledgements
We are indebted to Mr. A. Cotugno and Mr. F. Forte for their continuous valuable asistance, and Mrs. M. Izzo Esposito for her skilful typewriting.

References

1. WESTFALL, J.A., YAMATAKA, S., and ENOS, P.D., *J. Cell Biol.* **51**, 318 (1971).
2. BODE, H., BERKING, S., DAVID, C.N., GIERER, A., SCHALLER, H., and TRENKNER, E., *Wilhelm Roux Arch.* **171**, 269 (1973).
3. SCHALLER, C.H., *J. Embryol. exp. Morph.* **29**, 27 (1973).
4. SCHALLER, H. and GIERER, A., *J. Embryol. exp. Morph.* **29**, 39 (1973).
5. PASSANO, L.M. and McCULLOUGH, C.B., *J. exp. Biol.* **41**, 643 (1964).
6. JOSEPHSON, R.K., *J. exp. Biol.* **47**, 179 (1967).
7. PASSANO, L.M. and McCULLOUGH, C.B., *Nature, Lond.* **199**, 1174 (1963).
8. JOSEPHSON, R.K. and MACKLIN, M., *Science, N.Y.* **156**, 1629 (1967).
9. KASS-SIMON, G., *J. comp. Physiol.* **80**, 29 (1972).
10. RUSHFORTH, N.B., *Biol. Bull.* **140**, 255 (1971).
11. PASSANO, L.M. and McCULLOUGH, C.B., *J. exp. Biol.* **42**, 205 (1965).
12. RUSHFORTH, N.B., *Biol. Bull.* **140**, 502 (1971).
13. PASSANO, L.M. and KASS-SIMON, G., *Am. Zool.* **9**, 1113 (1969).
14. PASSANO, L.M. and McCULLOUGH, C.B., *Proc. nat. Acad. Sci. USA* **48**, 1376 (1962).
15. RUSHFORTH, N.B., in *Invertebrate Learning*, vol.1 (edit. by W.C. Corning and J.A. Dyal), Plenum Press, New York, London, p.123 (1973).
16. HAUG, G., *Z. vergl. Physiol.* **19**, 246 (1933).
17. BORNER, M. and TARDENT, P., *Rev. Suisse Zool.* **78**, 697 (1971).
18. RUSHFORTH, N.B., in *Chemistry of Learning* (edit. by W.C. Corning and S.C. Ratner), Plenum Press, New York-London, p.369 (1967).
19. TARDENT, P. and FREI, E., *Experientia* **25**, 265 (1969).
20. TADDEI-FERRETTI, C. and CORDELLA, L., *Arch. ital. Biol.* **113**, 107 (1975).
21. WINFREE, A.T., *J. theor. Biol.* **16**, 15 (1967).
22. WEVER, R., *Kybernetik* **2**, 127 (1964).
23. KLOTTER, K., Cold Spring Harb. Symp. quant. Biol. **25**, 185 (1960).
24. JOHNSSON, A. and KARLSSON, H.G., *J. theor. Biol.* **36**, 153 (1972).
25. KARLSSON, H.G. and JOHNSSON, A., *J. theor. Biol.* **36**, 175 (1972).
26. PAVLIDIS, T., *J. theor. Biol.* **22**, 418 (1969).
27. SINGER, R.H., *Anat. Rec.* **148**, 402 (1964).
28. SELLE, M., KASS, G., and PASSANO, L.M., *Am. Zool.* **8**, 168 (1968).
29. JACKLET, J.W. and GERONIMO, J., *Science, N.Y.*, **174**, 299 (1971).
30. JOSEPHSON, R.K., in *Coelenterate Biology* (edit. by L. Muscatine and H.M. Lehnoff), Academic Press, New York, p.245 (1974).
31. PYE, E.K., *Can. J. Bot.* **47**, 271 (1969).
32. PYE, E.K., in *Biochronometry* (edit. by M. Menaker), National Academy of Sciences, Washington, p.623 (1971).
33. WINFREE, A.T., *Archs. Biochem. Biophys.* **149**, 388 (1972).
34. BURNETT, A.L. and DIEHL, N.A., *J. exp. Zool.* **78**, 697 (1971).
35. PAVLIDIS, T., *Biological oscillators: Their mathematical analysis,* Academic Press, New York (1973).

Phase-locking of the neural discharge to periodic stimuli

C. ASCOLI, M. BARBI, S. CHILLEMI, and D. PETRACCHI
Laboratorio per lo Studio delle Proprietà Fisiche di Biomolecole e Cellule, CNR, Pisa, Italy

Two processes are performed in a single ommatidium of the lateral eye in the Limulus. The first transduces light into generator potential, and the second encodes the generator potential into a neural discharge. The first process occurs mainly in the rabdhomeric membrane of the retinular cells and the second at the level of the second order neuron—the eccentric cell that is electrotonically coupled to the retinular cells. It is the encoder function of this receptor neuron that concerns us in this report.

Early investigation of the dynamic behavior of this system was carried out in linear approximation yielding the transfer functions of the encoding process (Knight *et al.* [1]). More recently a nonlinear feature of the dynamic response has been observed, namely the synchronization of the neural discharge to sinusoidal stimuli, with spikes always fired at fixed positions on the input waveform (Ascoli *et al.* [2]). Sometimes a real phase-locking of the spikes with respect to the stimulus occurs, with entrainment of the firing rate by the stimulation frequency: in particular, when the driving stimulus is modulated at frequencies sufficiently close to the firing rate in the absence of modulation (free run rate f_0) and/or with high enough modulation depth, one spike occurs in each cycle, always in the same position on the input sinusoid. Figure 1 shows the signal recorded

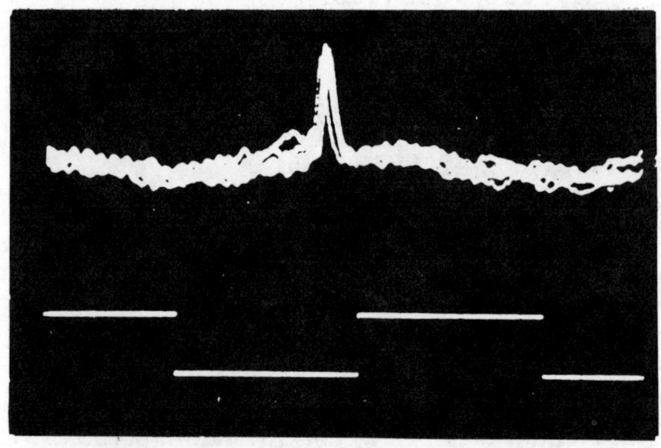

Fig.1 Phase-locked discharge pattern: several sweeps were superposed on the screen of the storage oscilloscope synchronized with the stimulus signal. The stimulus was sinusoidally modulated light reaching its maximum intensity on the on-off transition of the marker signal (lower trace)

from a retinular cell stimulated by light modulated at 5.45 Hz while steadily firing at 5.3 sp/sec. With the help of predictions of the neural encoder model described below, we have performed a systematic investigation of this effect. Using standard techniques (Ascoli *et al.* [2]), recording was done intracellularly from several retinular and eccentric cells in adapted conditions, so that they fired at relatively

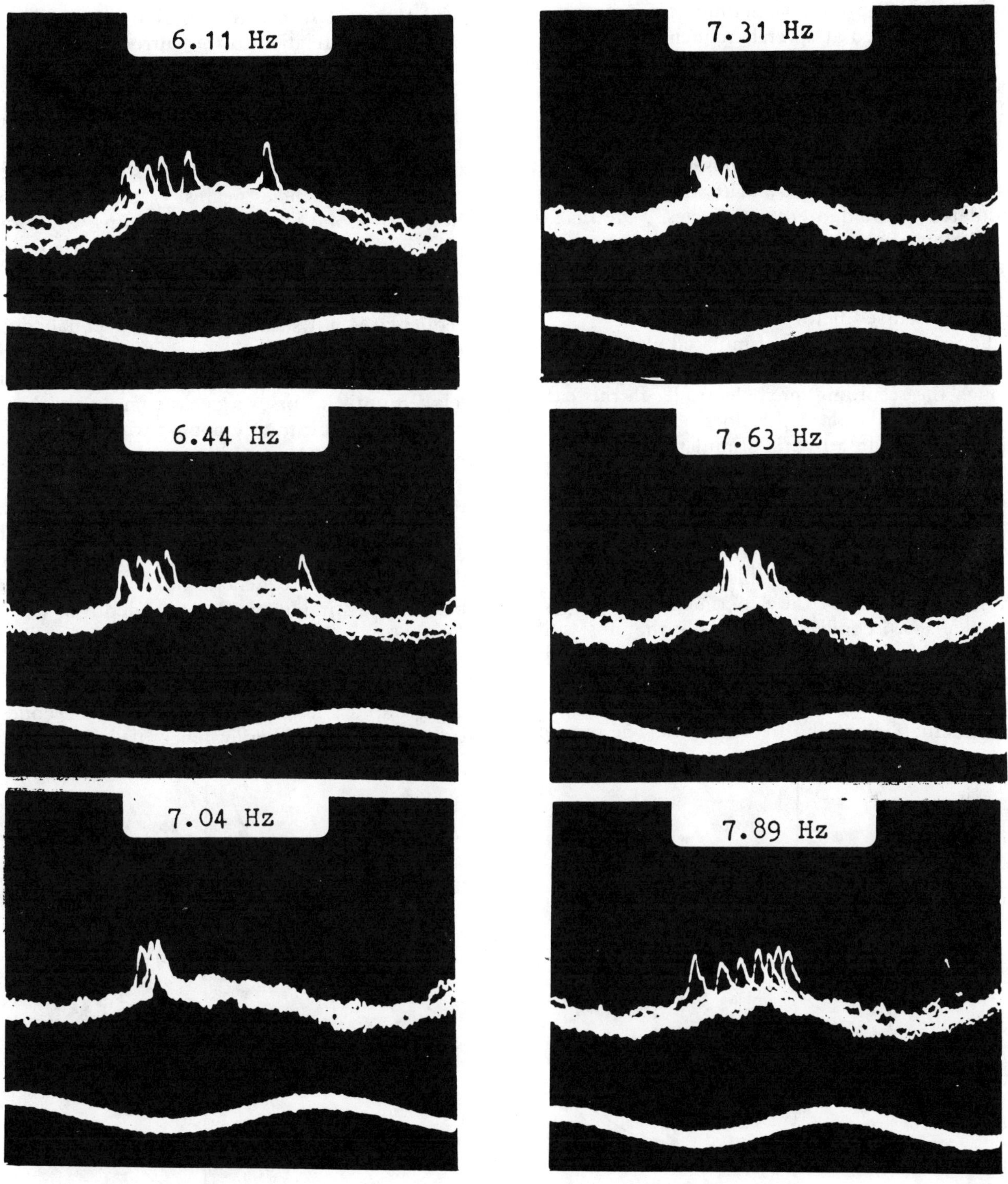

Fig.2 Phase-locked patterns monitored as in Fig.1. The phase φ of the locked spike at each stimulation frequency can be obtained by estimating its average distance from the maximum of the receptor potential. The lower trace represents the light stimulus.

low rates (<10 sp/sec). The stimulus signal was usually modulated at several frequencies ν near f_0 with suitable modulation depths, in order to get the phase-locking condition, and the phase ϕ of the spikes was then measured from the maximum of the receptor potential. Figure 2 shows some patterns obtained by a retinular cell whose free running rate was 7 sp/sec stimulated by light which was 30% modulated at the quoted frequencies. Note that the modulation of the receptor potential is approximately sinusoidal. Figure 3 plots the estimated ϕ values versus ν. The continuous curve is a theoretical fit which is discussed below.

The impaled cells were stimulated also by sinusoidally modulated depolarizing current injected through the recording microelectrode. In this case the input modulation depth does not need to be estimated. Finally a hybrid stimulation was sometimes used: that is, a sinusoidal current superposed on a steady light level. In this case we had to measure the cell impedance and the receptor potential level to estimate the modulation depth of the input signal. On the other hand, this method ensures the long term stability of the experimental preparation in well-adapted conditions. Actually the most suitable cells for these measurements are those which have a fairly regular pattern of steady discharge, that is, with a low variation coefficient and without slow fluctuations of the free run rate. The pure or hybrid current stimulation allows the phase of the locked spikes relative to the current maximum to be directly and accurately measured on the cyclic histograms. Figure 4 shows the cyclic histograms obtained by recording from an eccentric cell current stimulated at several frequencies near the free run rate f_0 of about 5 sp/sec. The histograms corresponding to the frequencies nearest f_0 are very narrow; they correspond to discharge patterns where one spike is fired in each stimulus cycle and always at the same phase. For frequencies farther from f_0 the entrainment of the discharge rate by the stimulus frequency is not so complete. For ν equal to 6.91 Hz, for instance, the cell still tends to fire in each cycle at a given phase but the spike becomes increasingly delayed until it misses a cycle completely. Nevertheless, the spikes can only appear within a given phase interval, and they sweep this cyclically.

An accurate analysis of the experimental recordings allowed us to identify some relevant features: (*a*) for each experiment the phases at $\nu = f_0$, which were usually obtained by interpolation between the ϕ values for the frequencies nearest f_0, are independent of the stimulus modulation depth m (see Fig.5); (*b*) as Table 1 shows, a systematic discrepancy exists between the values of $\phi(f_0)$ obtained by light and current stimulation in retinular cells. The first group are large negative, while the second are near zero. This non-equivalence of light and current

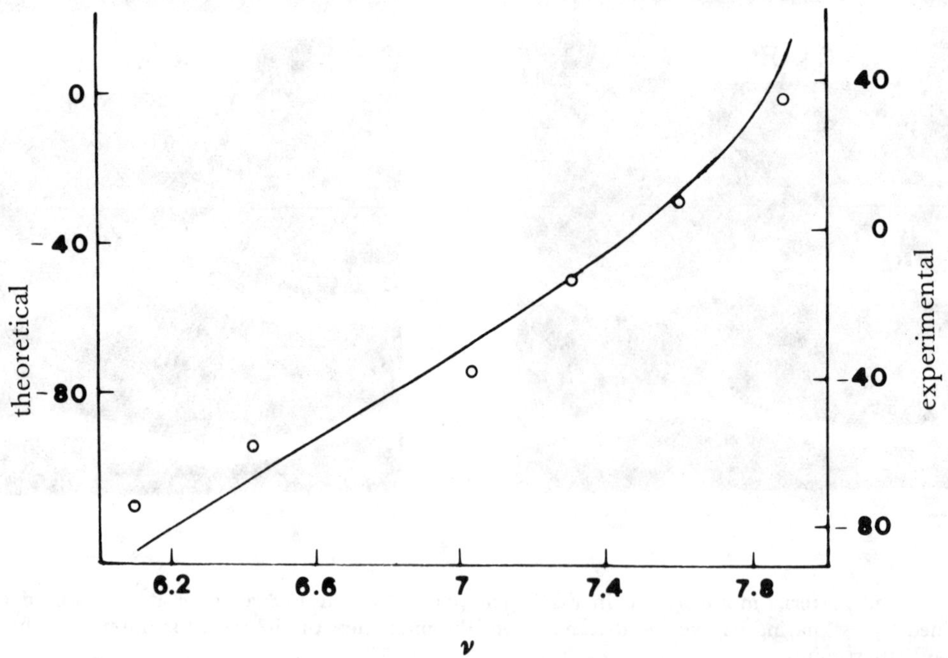

Fig.3 The φ values estimated for the runs of Fig.2 are plotted against ν. The continuous line is a fit by the model with the parameter values K = 2, τ = 0.5 sec, γ = 30 sec^{-1}.

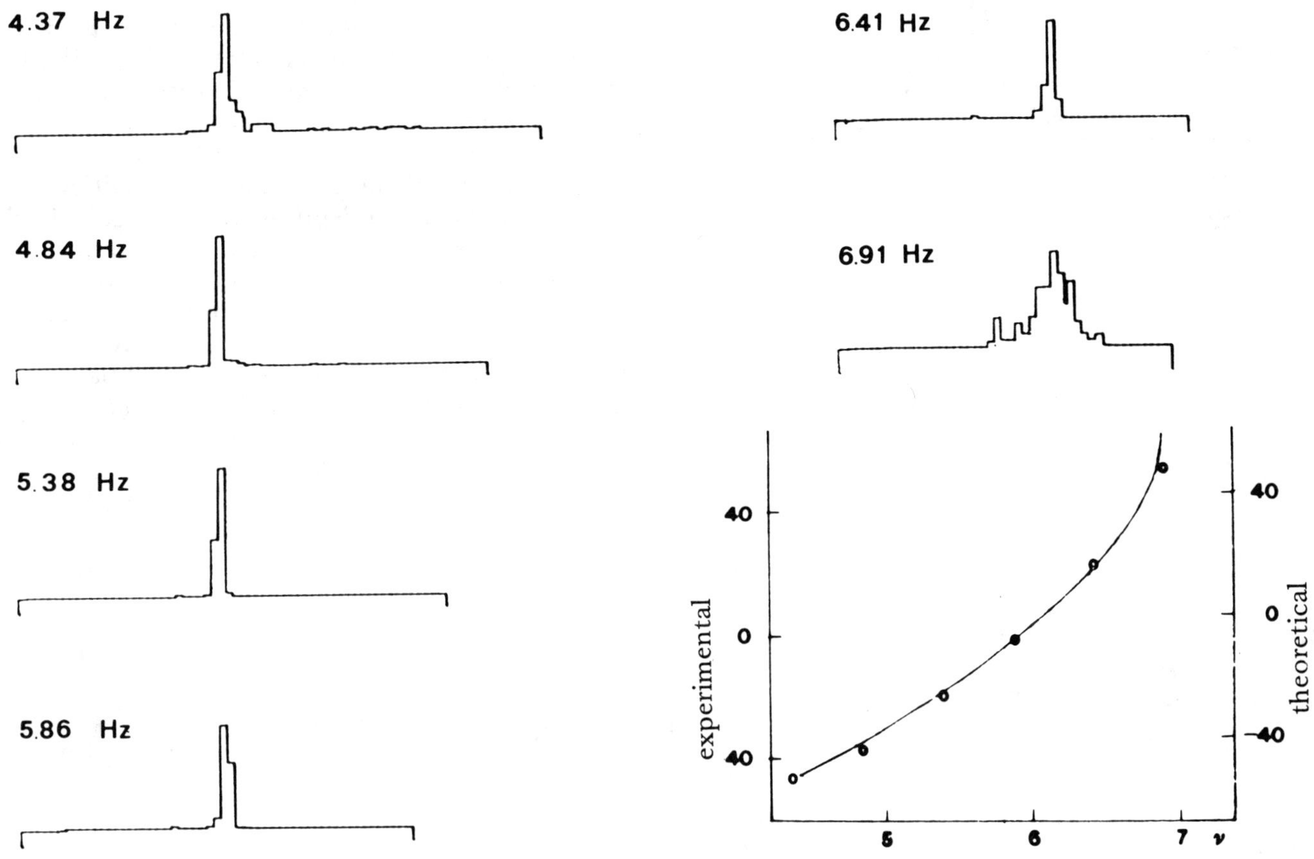

Fig.4 Cyclic histograms of responses of an eccentric cell to current stimulation (m=0.2). Insert: plot of measured φ vs. ν. The delay in detection of the spikes by the processor was taken into account. The fitting curve corresponds to K = 1, τ = 0.5 sec, γ = 23 sec^{-1}.

Table 1

Cell	f_0 (sp/sec)	$\phi(f_0)$ light	$\phi(f_0)$ current
R.73/1	5.3	-50	+6
R.73/4	5		-3
R.74/19	7	-75	0 *
R.75/2	7		-15
	4		-20
R.75/3	5.9	-54	+5 *
R.75/5	7	-65	0 *
R.75/6	2.3	-80	
E.74/16	5		-32
E.74/18	8.8		-34
	6.7		-34

*hybrid stimulation

stimulation in retinular cells could be ascribed to the capacitive couplings between the ommatidial cells. When recording from eccentric cells it becomes difficult to compare the values of ϕ (f_0) obtained with the two stimulation methods; indeed, in the case of light stimulation, the generator potential modulation is completely masked by the after-spike hyperpolarisation (Fig.6) and the phase of the locked spikes cannot be evaluated. By current stimulation in eccentric cells we found values of $\phi(f_0)$ intermediate between those obtained in retinular cells with light and current stimulation; (*c*) in each experiment the ϕ values always increase with increasing ν (see Figs 3 and 4). This holds true independently of the type of stimulation used. Moreover, the slope of the curve interpolating the experimental points depends on *m* and decreases with increasing *m*; (*d*) the frequency range where the phase-locking occurs, even if varying from experiment to experi-

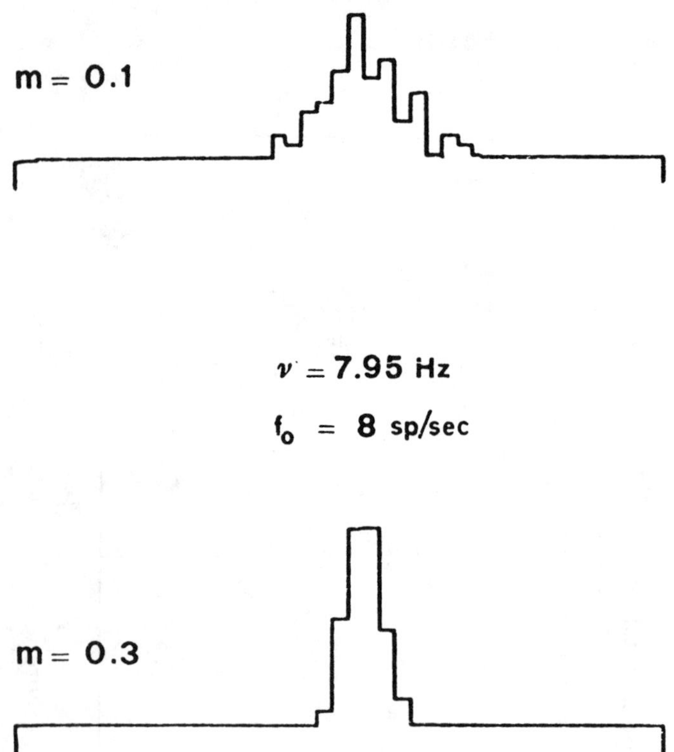

Fig.5 Cyclic histograms of the responses obtained by light stimulation at a frequency very near f_0 with two different modulation depths (quoted)

ment, and, with the modulation depth we used, is of the order of 1 Hz, sometimes reaching 2 Hz. It often turns out to be asymmetric with respect to f_0 in the direction of high frequencies. Moreover, at frequencies below f_0, a peculiar effect sometimes occurs—the loss of the phase-locking condition as the input modulation depth increases (Fig.7). In the end, the range covered by ϕ falls in the performed experiments between 50° and 100° and does not turn out to depend on the type of stimulation.

Now a simple encoder model, the leaky integrator, has been shown to produce phase-locked patterns in response to sinusoidal stimuli (Rescigno *et al.* [3]; Knight [4]). To allow us to draw an effective comparison between the model's behavior and the experimental results, a deeper analysis of the model must be made.

The model is characterized by the first order equation

$$\dot{u} = s(t) - \gamma u \tag{1}$$

where $s(t)$ is the input signal and γ the leakage constant: whenever u reaches a threshold C, a pulse is

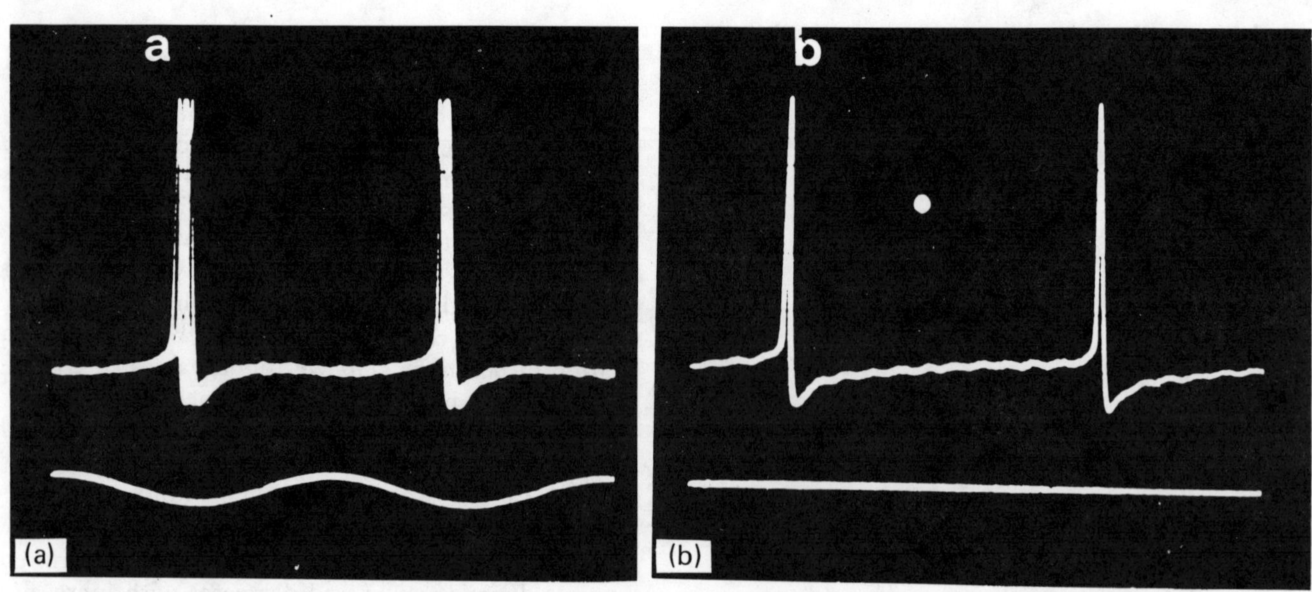

Fig.6 (a) Phase-locked pattern recorded from an eccentric cell stimulated by light with 50% modulation depth; (b) free discharge pattern at the same level of light

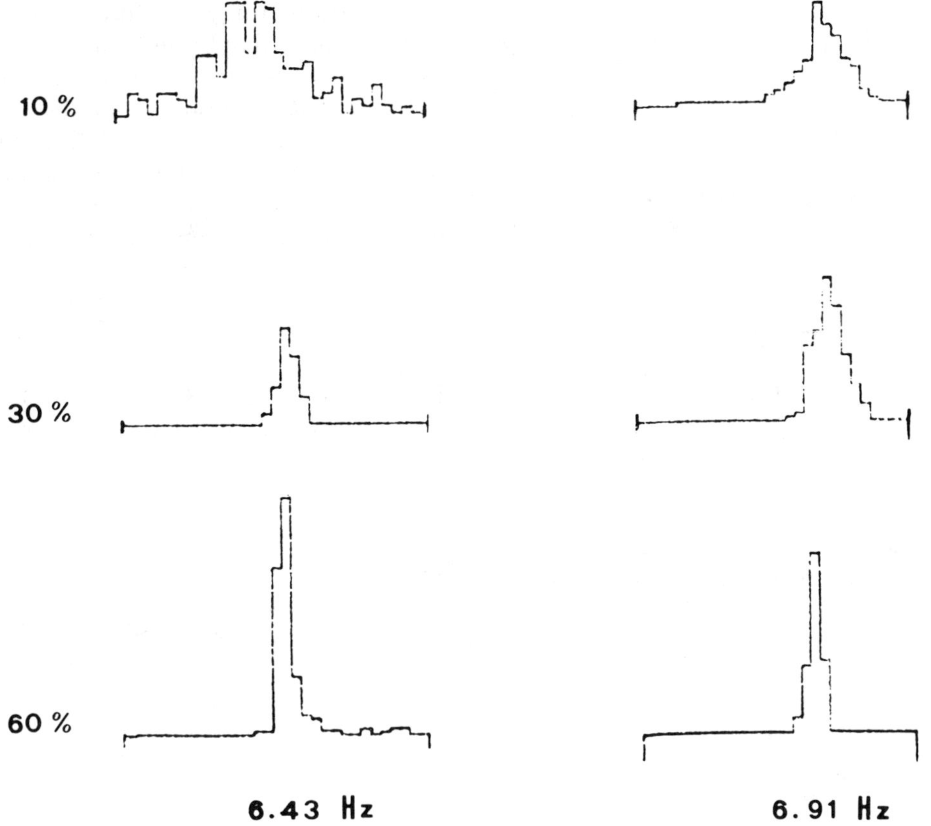

Fig.7 Cyclic histograms of the responses of a retinular cell light stimulated at two different frequencies with the quoted modulation depths. Free run rate f_0 = 6.7 sp/sec.

fired and u is reset to zero. We may ask under what conditions pulse firings will keep in step with the periodic stimulus $s(t)=s_0[1+m\cos(\omega t+\phi)]$, $\omega=2\pi\nu$. Equation (1) is easily integrated, yielding

$$u(t) = \int_0^t s(\tau)\,e^{-\gamma(t-\tau)}\,d\tau$$

or

$$u(t) = \frac{s_0}{\gamma}(1-e^{-\gamma t}) + \frac{s_0}{\gamma}\,m\cos\beta\,[\cos(\omega t+\phi-\beta) - e^{-\gamma t}\cos(\phi-\beta)] \quad (2)$$

$$\beta = \tan^{-1}\frac{\omega}{\gamma} \qquad (0 \leqslant \beta \leqslant \frac{\pi}{2})$$

where, without losing generality, we have assumed that the time origin coincides with the last pulse. Now, by supposing that the time T of the next pulse equals the stimulus period (that is $\nu T=1$), we derive from Eq.(2)

$$\frac{s_0}{\gamma}(1-e^{-\gamma/\nu})[1+m\cos\beta\cos(\phi-\beta)] = C$$

or

$$\cos(\phi-\beta) = \frac{\sqrt{\gamma^2+\omega^2}}{m\gamma}\left[\frac{\gamma C}{s_0(1-e^{-\gamma/\nu})} - 1\right] \quad (3)$$

By means of this equation Rescigno *et al.* [3] and Knight [4] determine the phase ϕ at which the phase-locked spikes occur. However, this phase is acceptable only if the function $u(t)$ firstly crosses the threshold C for $t=T$. This requirement is certainly satisfied for modulation depths low enough to keep $s(t)$ greater than the threshold depolarization for steady firing γC (Fig.8). Indeed, in this case $\dot{u}=s-\gamma u$ is positive at $t=T$, and it necessarily remains positive over the whole interval back to $t=0$. But when this condition doesn't hold, $u(t)$ may be a decreasing function over part of that interval and present a relative maximum greater than C. Figure 9 shows a case where this occurs. In conclus-

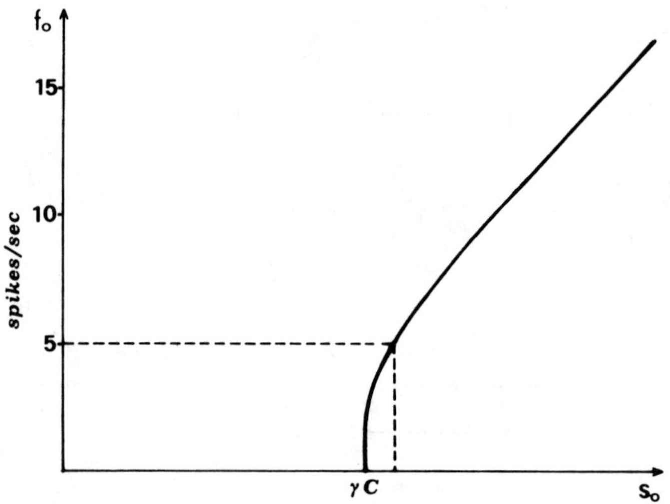

Fig.8 Steady state relation for the leaky integrator. It corresponds to the equation $s_0 = \gamma C/(1 - e^{-\gamma/f_0})$ with $\gamma = 12\ \text{sec}^{-1}$.

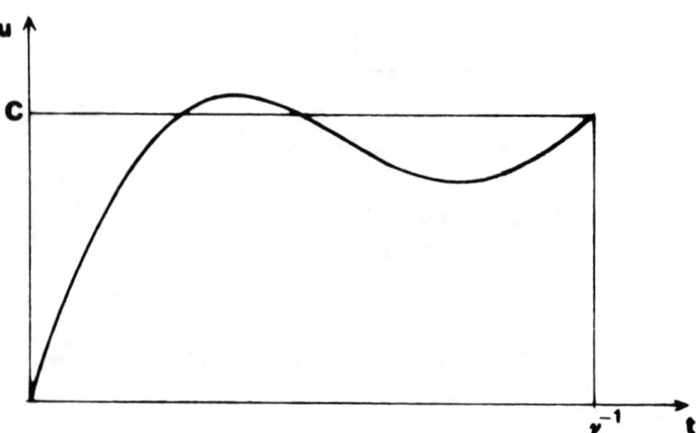

Fig.9 u(t) function corresponding to a solution of Eq.(3) but not to a phase-locked pulse pattern. Parameter values: $f_0 = 5$ sp/sec, $\gamma = 16\ \text{sec}^{-1}$, m = 0.4, $\nu = 3.4$ Hz.

ion if the sufficient condition for the monotonicity of $u(t)$ is not satisfied, the solutions of Eq.(3) must always be checked to see if they correspond to true phase-locking. We did this by numerically computing the function $u(t)$ and by using an electronic analog of the model.

With this proviso in mind, let us now discuss Eq.(3). When the stimulation frequency ν equals the free run rate f_0, the expression in parentheses on the right hand side vanishes. Moreover the right hand side itself is a rising function of ν in a range including this point. Therefore a whole range of frequencies about $\nu = f_0$ exists where the absolute value of the r.h.s. is less than unity and Eq.(3) has two solutions for ϕ. But only one solution will yield a stable fixed point on the stimulus cycle—that corresponding to a positive value of the derivative $\partial u(T)/\partial\phi$. Computing the derivative yields the stability condition $\sin(\phi - \beta) < 0$. Only the solution of Eq.(3) which satisfies this condition can be observed in a biological system, where some intrinsic noise is always present. We conclude that, when m is sufficiently small, real phase-locking occurs (at a given phase of the input signal) for each frequency within a range containing f_0 and delimited by the frequencies ν_{min} and ν_{max}, where the r.h.s. of Eq.(3) equals -1 or +1, respectively. With increasing ν in this range the locking phase increases continuously, passing from the value $\tan^{-1}[(\omega_{min}/\gamma) - \pi]$ to $\tan^{-1}[\omega_{max}/\gamma]$. The phase excursion is therefore

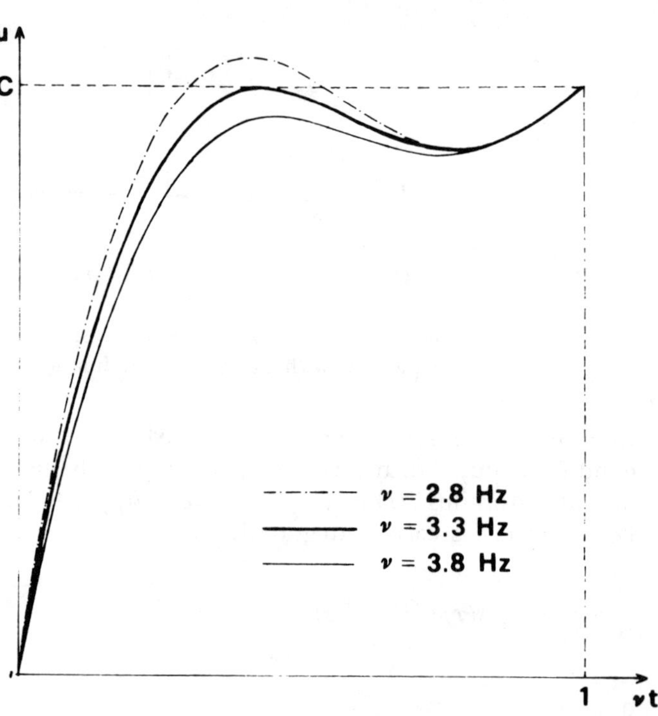

Fig.10 u(t) function corresponding to the solution of Eq.(3) for three different modulation frequencies. Other parameter values: $f_0 = 5$ sp/sec, $\gamma = 16$, m = 0.2.

greater than 180°. For larger modulation depths and/or leakage constants, the stable solution of Eq.(3), must be checked again. It turns out that for frequency $\nu = \nu_{max}$ that solution is really a phase-locking phase, whereas, as frequency is decreased, the function $u(t)$ presents a relative maximum which rises with decreasing ν. The frequency at which this maximum reaches C is actually the lower

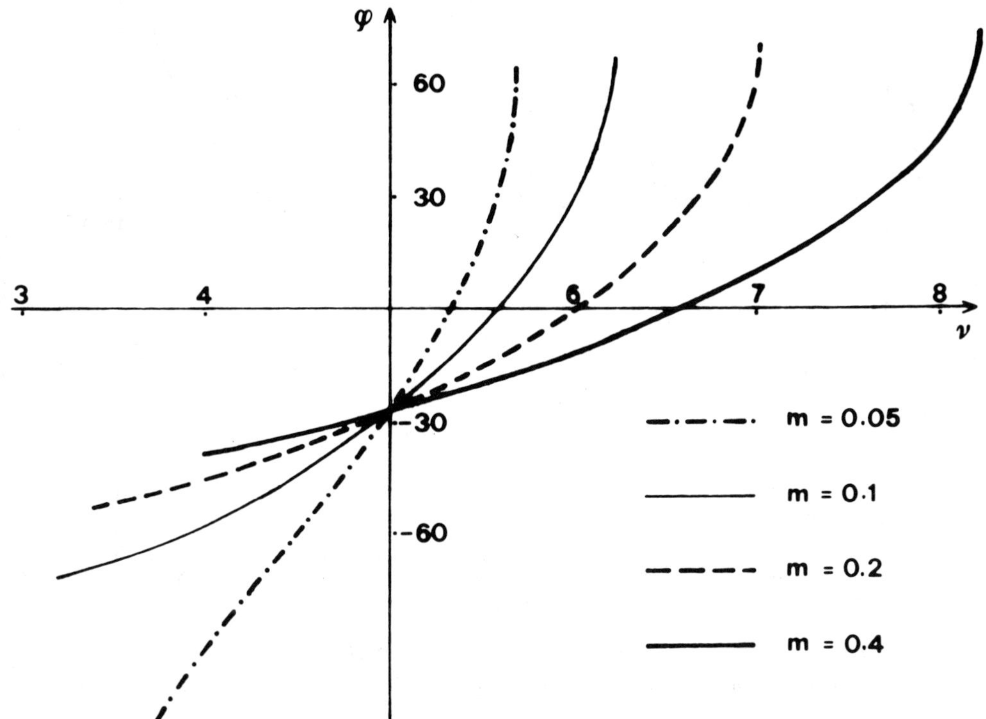

Fig.11 Function $\varphi(\nu)$ predicted by the leaky integrator. Parameter values: $f_0 = 5$ sp/sec, $\gamma = 16$, and modulation depths as quoted.

end of the locking range. This behavior is shown in Fig.10. Finally, in Fig.11 we plotted the phase of real phase-locking vs. ν for the quoted parameter values. As it can be seen from these curves, the model accounts for all the qualitative features of the experimental data. It also appears that for some frequencies ν increases in m destroy phase-locking. Moreover $\phi\ (f_0)$ does not depend on m and it is given, as follows from Eq.(3), by

$$\phi(f_0) = \tan^{-1} \frac{2\pi f_0}{\gamma}$$

This expression could be used to evaluate γ from the experimental data. Unfortunately, the systematic discrepancy between the values of $\phi\,(f_0)$ obtained with different types of stimulation makes them unutilizable for this purpose. However the model could yet reproduce the trend of the experimental phases against the stimulation frequency ν. But now it is worth noting that the leaky integrator cannot account for another relevant feature of the neural discharge in the Limulus ommatidia, namely its rapid adaptation in response to step stimuli. This feature has been ascribed to the self-inhibition process (Stevens [5]). Self-inhibition can easily be included in the model (Barbi *et al.* [6]); this makes it more comprehensive but also more difficult to analyse; however the qualitative behavior of the model is unaltered, as we checked numerically and by an electronic analog of the model. We used this more complete model to fit the shape of the distribution of the experimental points by arranging a vertical shift through a suitable amount in the scale of the theoretical phases with respect to the experimental ones. In this way reasonable fits were obtained (Figs 3, 4 and 12). The γ values which give the best fit turn out to range between 20 and 30 sec^{-1} in all the performed experiments (for the self-inhibition time constant we assumed $\tau = 0.5$ sec and for the self-inhibition strength coefficient K we assumed values between 1 and 3 (Knight [1]). For the cells where the comparison is possible, the values obtained with light and current stimulation seem to coincide; moreover no systematic differences appear between the values obtained with different types of stimulation or cells. In conclusion, we believe that the model works satisfactorily even if experimental difficulties exist in identifying the true driving signal for the neural encoder.

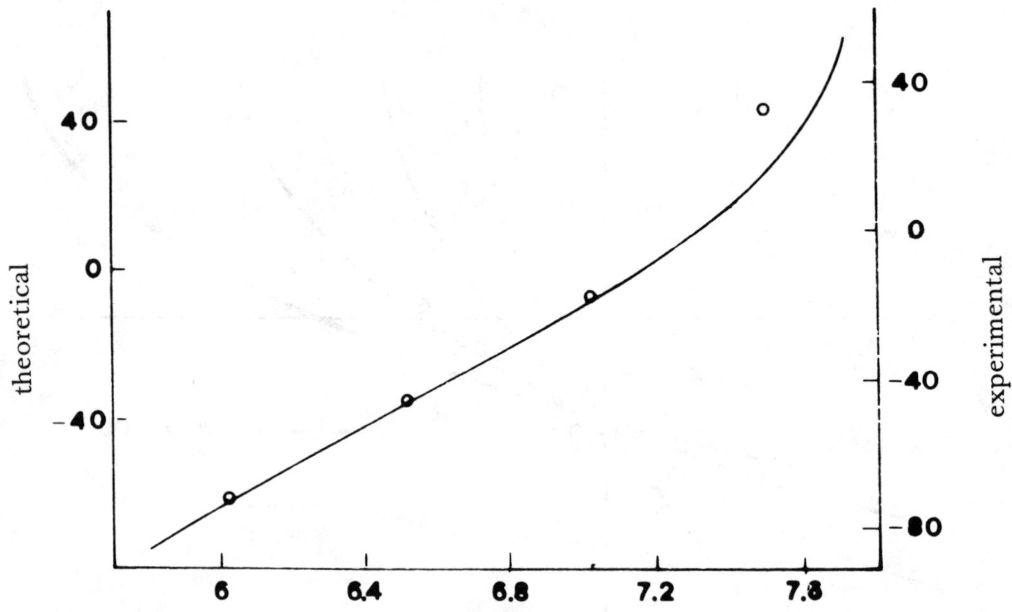

Fig.12 Plot of φ versus ν for an eccentric cell, current stimulated and steadily firing at 6.7 sp/sec. The fit corresponds to the model parameters: K = 2, τ = 0.5 sec, γ = 20 sec^{-1}.

References

1. KNIGHT, B.W., TOYODA, J., and DODGE, F.A., 'A quantitative description of the dynamic of excitation and inhibition in the eye of Limulus', *J. Gen. Physiol.* **56**, 421-437 (1970).
2. ASCOLI, C., BARBI, M., GHELARDINI, G., FREDIANI, C., and PETRACCHI, D., 'Rectification and spike synchronization in the Limulus lateral eye', *Kybernetik* **14**, 155-160 (1974).
3. RESCIGNO, A., STEIN, R.B., PURPLE, R.L., POPPELE, R.E., 'A neural model for the discharge patterns produced by cyclic inputs', *Bull. Math. Biophys.* **32**, 337-353 (1970).
4. KNIGHT, B.W., 'Dynamics of encoding in a population of neurons', *J. Gen. Physiol.* **59**, 734-766 (1972).
5. STEVENS, C.F. 'A quantitative theory of neural interactions: theoretical and experimental investigations', Thesis (1964).
6. BARBI, M., CARELLI, V., FREDIANI, C., and PETRACCHI, D., 'The self-inhibited leaky integrator: transfer functions and steady state relations', *Biol. Cybernetics* **20**, 51-59 (1975).

On certain diffusion approximations to population growth in random environment

R.M. CAPOCELLI and L.M. RICCIARDI
Istituto di Scienze dell'Informazione, University of Salerno, Italy, and Laboratorio di Cibernetica, Nat'l Council for Research, Arco Felice, Naples, Italy

1. Introduction

Stochastic difference equations modeling a variety of dynamical processes have often been proposed and analyzed. Within the population biology context, well-known is the growth model discussed by Lewontin and Cohen [1] to establish how a population with discrete non-overlapping generations would change in size under the assumption that the population is undergoing geometric increase or decrease with a stationary randomly distributed rate independent of the population size. Lewontin and Cohen purposely disregard regulation effects and concentrate on the description of the unrestricted process. This process turns out to exhibit a rather counterintuitive feature, i.e. the possibility for the population to be doomed to extinction whereas the expected population size grows beyond any bound with time. In conclusion, these authors conjecture that this pseudo-paradoxical feature would not characterize the analog continuous-time model. Such conjecture is indeed correct because, *strictu sensu,* in the continuous time model the population size cannot vanish (however, cf. [2]) since the origin is a *natural* boundary for the diffusion process modeling the population dynamics [3]. But if one identifies extinction with the attainment of some small positive 'threshold', as suggested in [3], then precisely analogous counterintuitive effects arise as proved by Capocelli and Ricciardi in the quoted reference.

This brief outline should already indicate that some subtle differences may be involved when postulating that a certain stochastic differential equation is the analog of a given difference equation. In the following we propose to pinpoint such differences by a straightforward argument that should also shed some light on the so-called Ito-Stratonovich controversy on the integration of stochastic differential equations. For the sake of brevity, we shall limit ourselves to considering in some detail two simple population growth models in random environment: the Malthusian model (analyzed in [3] in much greater detail) and the logistic one that has recently originated some controversy in the population biology literature [2, 4-6]. Thus doing we shall also be able to resolve such controversy without much of an effort.

No attempt is made to provide a full bibliography of the numerous papers that, more or less relevantly to our context, deal with stochastic population growth models. We only wish to point out that recently the question of extinction and exponential growth has been nicely discussed by Keiding [6] within the context of branching processes in random environment.

2. Exponential growth in random environment

For a realistic description of the growth of a population, density dependent factors ought to be taken into account [7-9]. Nevertheless, in situations such as, for instance, bacterial growth in an unrestricted environment whereby each individual is equally likely to die or reproduce at any instant, it is reasonable to disregard density dependent components and thus make use of the Malthusian approximation to model the growth phenomenon [3]. In this section we shall make this approximation whilst in Section 3 we shall consider one density-dependent model.

Let $x(t)$ denote the size of the population at time t and $x_0 = x(0)$. Denoting by r the growth rate, under the Malthusian approximation the population dynamics is modeled by the equation

$$\frac{dx}{dt} = rx \qquad x(0) = x_0 \,. \tag{2.1}$$

In a previous paper [3] we constructed a diffusion model for the population growth substituting the growth rate in (2.1) with $\theta + \Lambda(t)$, where θ is a constant and $\Lambda(t)$ is a stationary delta-correlated normal process with zero mean. Thus doing, we obtained a diffusion process, defined in the interval $(0, \infty)$, characterized by drift $(\theta+\beta)x$ and infinitesimal variance $2\beta x^2$ where β is the noise intensity coefficient.

Starting from (2.1), we shall now make use of a different argument to construct the limiting stochastic process that arises when the growth rate is looked upon as a sequence of suitably chosen random variables. To this purpose, let us make a discrete approximation and look at the time variable as a sequence of equally spaced instants by setting $t = n\tau$ $(n=0,1,2,...)$. This can be done in two alternative ways: by considering the solution of (2.1), $x(t) = x_0 e^{rt}$, at the instants $n\tau$:

$$x_{n\tau} = x_0 e^{n\tau r} \qquad (n=0,1,2,...)\,, \tag{2.2}$$

or by approximating equation (2.1) by the difference equation

$$y_{(n+1)\tau} = (1+r\tau)\,y_{n\tau} \qquad (n=0,1,2,...) \qquad y_0 = x_0 \tag{2.3}$$

whose solution is:

$$y_{n\tau} = x_0(1+r\tau)^n \qquad (n=0,1,2,...)\,. \tag{2.4}$$

It is worth stressing that these two procedures are not at all equivalent. Even though they identify in the limit $\tau \to 0$, as far as τ is kept finite the difference $x_{n\tau} - y_{n\tau}$ increases without bounds with n. Clearly, the sequence $x_{n\tau}$ provides a much better approximation of the solution of Eq.(2.1) whilst by approximating Eq.(2.1) one obtains a very poor representation of the continuous process $x(t)$.

Let us now consider the approximation (2.2). Over the time interval τ the change in population size is

$$x_{(n+1)\tau} - x_{n\tau} = (e^{r\tau} - 1)\,x_{n\tau} \qquad (n=0,1,2,...)\,. \tag{2.5}$$

To construct a random process we change $r\tau$ into a sequence of independent identically distributed random variables $\theta_0, \theta_\tau, \theta_{2\tau}$ ... described by the scheme

$$\theta_{n\tau} = \left|\begin{matrix} \sigma\sqrt{\tau} & -\sigma\sqrt{\tau} \\ \frac{1}{2}+\frac{r\sqrt{\tau}}{2\sigma} & \frac{1}{2}-\frac{r\sqrt{\tau}}{2\sigma} \end{matrix}\right| \qquad (n=0,1,2,...) \tag{2.6}$$

where σ is an arbitrary positive constant†. With this choice we have:

$$E(\theta_{n\tau}) = r\tau \qquad E(\theta^2_{n\tau}) = \sigma^2\tau \qquad (n=0,1,2,...) \tag{2.7}$$

†It is worth remarking that the random variable $\theta_{n\tau}$ must necessarily be defined as in (2.6) if one wants to obtain a diffusion process in the limit $\tau \to 0$.

$$E(\theta_{n\tau}{}^{2+p}) = \frac{1+(-1)^p}{2}\,\sigma^{1+p}\tau^{1+p/2}(\sigma+r\tau^{1/2}) = o(\tau)$$

$$(p=1,2,\ldots)\,.$$

In other words, the deterministic quantity $r\tau$ has been changed into a random variable with mean $r\tau$ and variance $\sigma^2\tau-(r\tau)^2$. Equations (2.5) thus become:

$$X_{(n+1)\tau} - X_{n\tau} = (e^{\theta_{n\tau}} - 1)X_{n\tau} \quad (n=0,1,2,\ldots) \tag{2.8}$$

defining a random walk with position-dependent increments. We can now let $n\to\infty$ and $\tau\to 0$ in such a way that $n\tau\to t$ and determine the limiting continuous process. To this purpose, we calculate the moments of the increment over the time interval τ conditioned upon $X_{n\tau}=x$. From (2.7) and (2.8) we obtain

$$\frac{1}{\tau}E\{X_{(n+1)\tau} - X_{n\tau}\,|X_{n\tau}=x\} = [r+\frac{\sigma^2}{2}+o(1)]x$$

$$\frac{1}{\tau}E\{[X_{(n+1)\tau} - X_{n\tau}]^2\,|X_{n\tau}=x\} = [\sigma^2+o(1)]x^2$$

$$\frac{1}{\tau}E\{[X_{(n+1)\tau} - X_{n\tau}]^{2+p}\,|X_{n\tau}=x\} = x^{2+p}\,o(1)$$

$$(p=1,2,\ldots) \tag{2.9}$$

where use of a Taylor series expansion for the exponential has been made. Taking the limit of (2.9) as $\tau\to 0$ we thus conclude [10, 11] that $X_{n\tau}$ converges to the stationary diffusion process $X(t)$ whose infinitesimal mean and variance are given by:

$$\begin{aligned} A_1(x) &= (r+\frac{\sigma^2}{2})x \\ A_2(x) &= \sigma^2x^2\ , \end{aligned} \tag{2.10}$$

respectively. This is precisely the diffusion process derived in [3].

Let us now return to the discrete approximation (2.3) obtained by approximating Eq.(2.1) by a system of difference equations. Changing again $r\tau$ into $\theta_{n\tau}$, from (2.3) we are led to the random walk:

$$Y_{(n+1)\tau} - Y_{n\tau} = \theta_{n\tau}Y_{n\tau} \quad (n=0,1,2,\ldots) \tag{2.11}$$

Recalling (2.7) we thus obtain:

$$\begin{aligned} &\frac{1}{\tau}E\{Y_{(n+1)\tau} - Y_{n\tau}\,|\,Y_{n\tau}=x\} = rx \\ &\frac{1}{\tau}E\{[Y_{(n+1)\tau} - Y_{n\tau}]^2\,|Y_{n\tau}=x\} = \sigma^2x^2 \\ &\frac{1}{\tau}E\{[Y_{(n+1)\tau} - Y_{n\tau}]^{2+p}\,|Y_{n\tau}=x\} = x^{2+p}o(1)\,. \end{aligned} \tag{2.12}$$

Hence, as $\tau\to 0$ $Y_{n\tau}$ converges to the diffusion process $Y(t)$ characterized by the following moments:

$$\begin{aligned} B_1(x) &= rx \\ B_2(x) &= \sigma^2x^2\ . \end{aligned} \tag{2.13}$$

While $X(t)$ and $Y(t)$ have identical infinitesimal variance, their drifts are unequal. For both processes, however, the origin is an unattainable boundary so that extinction, strictly speaking, cannot occur unless, as suggested in [3], we identify extinction with the attainment of some non-zero threshold or 'critical mass'. Nevertheless, there are some major differences in the asymptotic behavior of $X(t)$ and $Y(t)$. Indeed, as it is easily seen by using the results derived in [3], if $-(\sigma^2/2)< r < \sigma^2$ using $X(t)$ as model one sees that the expected value of the population size grows infinitely large in time while the most likely population size decreases to zero. When, instead, the description is in terms of $Y(t)$, the same circumstance occurs for $0 < r < \frac{3}{2}\sigma^2$. In the first case the expected population size grows unbounded in time even though the mean growth rate, r, is negative provided that its absolute value is less than $\sigma^2/2$. Such circumstance does not arise in the second case. Furthermore, the asymptotic behavior of the sample paths is determined by r for $X(t)$ and by $r-\frac{\sigma^2}{2}$ for $Y(t)$. This means that with probability one the population size becomes asymptotically infintely large if $r>0$ for $X(t)$; if $r>\sigma^2/2$ for $Y(t)$. If one uses the description $Y(t)$ one thus concludes that the 'environmental variance' plays an essential role in determining the ultimate fate of the population, a claim made by May [2, 4]. However, as shown in the preceding, the diffusion process related to (2.1) is $X(t)$ and not $Y(t)$. The latter is, instead, appropriate if the starting deterministic model is the one defined by Eqs.(2.3). The relevant question is thus related to the *choice* of the model and *not* to the techniques to be used to integrate the stochastic equation obtained by making the growth rate random.

Let us now return to Eq.(2.1) and set therein $r = r + \Lambda(t)$ where $\Lambda(t)$ is a white noise with intensity coefficient σ^2:

$$\frac{dx}{dt} = rx + x\Lambda(t) \quad . \tag{2.14}$$

It is well known that by using Stratonovich's calculus Eq.(2.14) defines the diffusion process $X(t)$ having the moments (2.10). In Ito's approach an equation such as (2.14) is taboo because it involves the white noise $\Lambda(t)$ instead of the Brownian process $B(t)$. If one wishes to use Ito's calculus, the equation defining the process (2.10) reads [12]:

$$dx = (r + \frac{\sigma^2}{2})\, xdt + xdB \quad . \tag{2.15}$$

whereas the equation for the process (2.13) would be:

$$dy = rydt + y\, dB \tag{2.16}$$

According to our previous considerations, it follows that within Ito's framework the stochastic equation corresponding to (2.1) must be Eq.(2.15). This equation certainly does not admit of a sufficiently intuitive justification to be regarded as the assumption on which the population growth is modeled. If, instead, use of Stratonovich's calculus is made, one can modify Eq.(2.1) to allow for fluctuations of growth rate, thus obtaining Eq.(2.14) in a very natural way. However, there is one more reason emphasizing the appropriateness of Stratonovich's calculus to model situations of biological interest. Namely, one can attenuate the hypothesis that $\Lambda(t)$ is a stationary delta-correlated normal process without drastically changing the nature of our previous conclusions. Indeed, as proved by Stratonovich [13], if in (2.14) $\Lambda(t)$ is simply a stationary delta-correlated random process, one still approximately obtains the diffusion process with moments (2.10) provided the intensity coefficients of $\Lambda(t)$ are sufficiently small from the third order on. Should one interpret $\Lambda(t)$ as an arbitrary noise, the same process $X(t)$ still describes the population dynamics after a time interval much longer than the noise's correlation time. Equations such as (2.15) and (2.16) become instead meaningless if dB is not precisely the differential of the Brownian motion process.

3. Logistic growth in random environment

The discussion of Section 2 will now allow us to provide a definite answer to the question of whether the environmental variance plays any role in determining the asymptotic behavior of a population subject to a regulated growth of the logistic type. More precisely, we refer to the equation

$$\begin{aligned} \frac{dx}{dt} &= \alpha x - \beta x^2 \\ x(0) &= x_0 \end{aligned} \tag{3.1}$$

where α is the intrinsic population growth rate and $\beta > 0$ is a constant. The solution of (3.1) is then:

$$x(t) = \alpha x_0[\alpha e^{-\alpha t} + \beta x_0(1 - e^{-\alpha t})]^{-1} \quad . \tag{3.2}$$

It shows that asymptotically the level $K = \alpha/\beta$ (carrying capacity) is attained. For $\beta = 0$ we clearly recover the situation described in Section 2.

We are now interested in studying the effect on the population dynamics due to random fluctuations of the intrinsic growth rate α. In particular, we want to make suitable assumptions on α as to model the time course of the population size as a diffusion process. To this purpose, again we do not immediately change (3.1) into a stochastic equation but rather consider the discrete approximation to the solution of (3.1). Setting again $t = n\tau$ we thus obtain the difference equations:

$$\begin{aligned} \frac{X_{(n+1)\tau} - X_{n\tau}}{X_{n\tau}} + 1 &= \left[\frac{\beta\tau}{\alpha\tau}(1 - e^{-\alpha\tau})X_{n\tau} + e^{-\alpha\tau}\right]^{-1} \\ & (n = 0,1,2,...) \\ x_0 &= x_0 \, . \end{aligned} \tag{3.3}$$

Following the procedure of Section 2, we next substitute $\alpha\tau$ with the sequence of independent identically distributed random variables $\theta_0, \theta_\tau, \theta_{2\tau}, ...$ described by the scheme (2.6). Equations (3.3) thus become:

$$Z_{n\tau} = [\beta\tau X_{n\tau}(1 - e^{-\theta_{n\tau}})/\theta_{n\tau} + e^{-\theta_{n\tau}}]^{-1} \tag{3.4}$$

having set:

$$Z_{n\tau} \equiv \frac{X_{(n+1)\tau} - X_{n\tau}}{X_{n\tau}} + 1 \tag{3.5}$$

$\{Z_{n\tau} | X_{n\tau} = x\} \equiv \xi_{n\tau}$ is thus a random variable described by the scheme:

$$\xi_{n\tau} = \left[\begin{array}{cc} \left[\beta\tau x \dfrac{1 - e^{-\sigma\sqrt{\tau}}}{\sigma\sqrt{\tau}} + e^{-\sigma\sqrt{\tau}}\right]^{-1} & \left[\beta\tau x \dfrac{1 - e^{\sigma\sqrt{\tau}}}{-\sigma\sqrt{\tau}} + e^{\sigma\sqrt{\tau}}\right]^{-1} \\ \dfrac{1}{2} + \dfrac{\alpha\sqrt{\tau}}{2\sigma} & \dfrac{1}{2} - \dfrac{\alpha\sqrt{\tau}}{2\sigma} \end{array} \right] \tag{3.6}$$

After some tedious calculations, whose details will appear elsewhere [14], one then finds:

$$\frac{E(\xi_{n\tau}) - 1}{\tau} = \frac{-\beta x + \alpha + \sigma^2/2 + o(1)}{1 + o(1)}$$

$$\frac{E[(\xi_{n\tau} - 1)^2]}{\tau} = \frac{\sigma^2 + o(1)}{1 + o(1)} \tag{3.7}$$

$$\frac{E[(\xi_{n\tau} - 1)^{2+p}]}{\tau} = o(1) \quad (p = 1,2,...) .$$

Recalling the previous definitions, Eq.(3.7) imply:

$$\lim_{\tau \to 0} \frac{1}{\tau} \{X_{(n+1)\tau} - X_{n\tau} | X_{n\tau} = x\} = \left(\alpha + \frac{\sigma^2}{2}\right) x - \beta x^2 \tag{3.8}$$

$$\lim_{\tau \to 0} \frac{1}{\tau} \{[X_{(n+1)\tau} - X_{n\tau}]^2 | X_{n\tau} = x\} = \sigma^2 x^2$$

$$\lim_{\tau \to 0} \frac{1}{\tau} \{[X_{(n+1)\tau} - X_{n\tau}]^{2+p} | X_{n\tau} = x\} = 0 \quad (p = 0,1,2,...) .$$

Namely, in the limit as $\tau \to 0$ the random walk defined by (3.4) and (3.5) converges to a stationary diffusion process $X(t)$ characterized by the following mean and infinitesimal variance:

$$A_1(x) = \left(\alpha + \frac{\sigma^2}{2}\right) x - \beta x^2$$
$$A_2(x) = \sigma^2 x^2 . \tag{3.9}$$

This process coincides with the diffusion process defined by the stochastic equation

$$\frac{dx}{dt} = \alpha x - \beta x^2 + x \Lambda(t)$$
$$x(0) = x_0 \tag{3.10}$$

that one is naturally led to associate to the deterministic equation (3.1). We thus find again that if the starting point is Eq.(3.10) then the diffusion process $X(t)$ with moments (3.9) correctly describes the population dynamics. As one can easily prove, this diffusion process always admits of a steady state distribution independently of the magnitude of the environmental variance σ^2, i.e. independently of the magnitude of the noise intensity coefficient. Therefore, the population cannot go to certain extinction.

Let us now return to Eq.(3.1) and approximate it by the difference equations:

$$Y_{(n+1)\tau} - Y_{n\tau} = \alpha\tau Y_{n\tau} - \beta\tau Y_{n\tau}^2$$
$$Y_0 = x_0 \tag{3.11}$$

From (3.11) we then obtain the random walk

$$Y_{(n+1)\tau} - Y_{n\tau} = \theta_{n\tau} Y_{n\tau} - \beta\tau Y_{n\tau}^2$$
$$Y_0 = x_0 \quad . \tag{3.12}$$

and it is straightforward to prove that the following relations hold:

$$\frac{1}{\tau} E\{Y_{(n+1)\tau} - Y_{n\tau} | Y_{n\tau} = x\} = \alpha x - \beta x^2$$

$$\frac{1}{\tau} E\{[Y_{(n+1)\tau} - Y_{n\tau}]^2 | Y_{n\tau} = x\} = \sigma^2 x^2 - \beta^2 x^4 \tau$$

$$\frac{1}{\tau} E\{[Y_{(n+1)\tau} - Y_{n\tau}]^{2+p} | Y_{n\tau} = x\} = o(\tau) \quad (p = 1,2,...) . \tag{3.13}$$

Hence, as $\tau \to 0$ $Y_{n\tau}$ converges to the diffusion process $Y(t)$ characterized by the following moments:

$$B_1(x) = \alpha x - \beta x^2$$
$$B_2(x) = \sigma^2 x^2 \quad . \qquad (3.14)$$

It is now easy to see that this process admits of a steady state distribution only if $\alpha > (\sigma^2/2)$, i.e. only if the carrying capacity K is greater than $\sigma^2/2\beta$. If, instead, the noise intensity is large enough so that $\beta K < \sigma^2/2$, the population is doomed to certain extinction. This is a result essentially due to May [2], [4] and recently criticized by Feldman and Roughgarden [5].

Relying on our procedure we can now conclude the following. If one starts from Eq.(3.1) and changes α into $\alpha + \Lambda(t)$, then the population growth is described by the process $X(t)$ with moments (3.9) and no certain extinction occurs as claimed by Feldman and Roughgarden. If, instead, the starting point is Eq.(3.11), then via (3.12) one sees that in the limit $\tau \to 0$ the correct diffusion process has the moments (3.14). In this case, in agreement with May's statement the magnitude of the environmental variance plays an essential role in determining the ultimate fate of the population. The conclusion is that there is no contradiction between the viewpoints of the quoted authors. They are simply talking about *different* models. The unique source of confusion relies in the so far not enough appreciated difference existing between the discrete approximation of the solution of a differential equation and the approximation of the differential equation itself by a difference equation.

References

1. LEWONTIN, R.C. and COHEN, D., 'On population growth in randomly varying environment', Proc. Nat. Ac. Sciences **62**, 1056 (1969).
2. MAY, R.M., 'Stability in randomly fluctuating versus deterministic environments', *Amer. Natur.* **107**, 621 (1973).
3. CAPOCELLI, R.M. and RICCIARDI, L.M., 'A diffusion model for population growth in random environment', *Theor. Pop. Biol.* **5**, 28 (1974).
4. MAY, R.M., *Stability and complexity in model ecosystems,* Princeton University Press, Princeton, N.J. (1973).
5. FELDMAN, M.W. and ROUGHGARDEN, J., 'A population's stationary distribution and chance of extinction in a stochastic environment with remarks on the theory of species packing', *Theor. Pop. Biol.* **7**, 197 (1975).
6. KEIDING, N., 'Extinction and exponential growth in random environment', *Theor. Pop. Biol.* **8**, 49 (1975).
7. CAPOCELLI, R.M. and RICCIARDI, L.M., 'Growth with regulation in random environment', *Kybernetik* **15**, 147 (1974).
8. RICCIARDI, L.M., 'Stochastic equations in ecology and neurobiology', in *Progress in Cybernetics and Systems Research,* Vol.I, 183, Hemisphere, Washington D.C. (1975), ed. Trappl & Pichler.
9. CAPOCELLI, R.M. and RICCIARDI, L.M., 'A note on growth processes in random environment', *Biol. Cybern.* **18**, 105 (1975).
10. RICCIARDI, L.M., 'Diffusion approximations in neuronal modeling and population growth', in Proc. 2nd Cybern. Cong., Casciana (Italy), 43 (1972).
11. RICCIARDI, L.M., 'Diffusion approximation to neuronal activity and to the time dependent threshold problem', Preprint (1976).
12. STRATONOVICH, R.L., *Conditional Markov Processes and their Applications to the Theory of Optimal Control,* Elsevier, N.Y. (1968).
13. STRATONOVICH, R.L., *Topics in the theory of random noise,* Vol.I, Gordon & Breach, N.Y. (1963).
14. RICCIARDI, L.M., 'On the controversy concerning population growth in random environment', Preprint BTP 035 AD76 9A (1976).

Author Index

Subject Index